Nelson Yuan-sheng Kiang

MUSCLES AND THEIR NEURAL CONTROL

MUSCLES AND THEIR NEURAL CONTROL

Graham Hoyle
University of Oregon, Eugene

A Wiley-Interscience Publication

JOHN WILEY & SONS
New York Chichester Brisbane Toronto Singapore

Library of Congress Cataloging in Publication Data:

Hoyle, Graham
Muscles and their neural control.

"A Wiley-Interscience publication."
Includes index.
1. Muscle contraction. 2. Neuromuscular spindles.
3. Physiology, Comparative. I. Title. [DNLM:
1. Muscles–Physiology. 2. Neuromusclar junction–
Physiology. 3. Muscle contraction. WE 500 H868m]

QP321.H78 1982 599.01'852 82-11045
ISBN 0-471-87709-3

Printed in the United States of America

10 9 8 7 6 5 4 3 2 1

Preface

This book arose directly out of an attempted "revision" of a small book I wrote in 1956 that was published by Cambridge University Press as part of a series of monographs on advancing biological frontiers. Entitled *Comparative Physiology of the Nervous Control of Muscular Contraction,* the book sold out fairly quickly, but the subject was by then moving so rapidly that I asked the publishers not to issue another printing and to wait until I could produce a revision. I started this more than 20 years ago and sincerely apologize both to the Cambridge Press and to all students it might have helped had it appeared. There were many serious gaps in comparative knowledge at that time, especially in ultrastructure, research in which was just beginning to make dramatic strides technically. I decided to try to fill some of them, acquired an electron microscope, and with a lot of skilled help began to collect the needed data. However, evolution had far more tricks up its sleeve than we dreamed of, and it was not long before our library of pictures amounted to several thousand, far more than could ever be published. Experimentally, the subject attracted many young and able investigators, and data accumulated at an ever-increasing pace. From a very modest beginning, it became encyclopedic in barely a decade. What could previously be summarized fully in one short paragraph had now expanded to the length of a large handbook. The data were coming in much faster than I could summarize it, and the goal of a direct revision became lost forever. New, though related, areas of inquiry were opened up by innovations in technique, by far the most important of which was not a muscle technique but the use of intracellular dyes for studying single motor-nerve cells that innervate them. Efferent nerve cells were then discovered – often at the central rather than the peripheral end – some of which turned out to have transmitters, functions, and properties previously unknown.

All of these advances have created new difficulties, particularly for the student, who needs a reasonably comprehensive summary more than ever, but whose needs it is no longer expedient to meet. Textbooks have to stick with a highly selected core of information to meet the needs of the presumed "average" inquirer. In them, controversial matters, much of the rich diversity, and exceptions to rare generalities must be glossed over or avoided to establish a learnable framework.

Another, more basic problem has arisen in the last quarter century. The majority of research in the 1960s and 1970s has very strongly emphasized the reductionistic approach: recognition and esteem were seen to be bestowed on those who discovered a general molecular "explanation" of complex biological events. In the rush to seek out the "heart of the matter" on any subject, it is inevitable that oversimplifications will be made and errors will arise and be perpetuated. Also, pieces of knowledge deemed of secondary interest, together with major exceptions to the popular themes, will eventually become lost completely. It is already well known that muscle research has suffered from this approach. In seeking pithy statements by distinguished muscle scientists about muscles and their control to serve as prologues to the chapters, I found mainly remarks that make their utterers look less than dignified in light of later research. I decided to use them anyway, for they have their own significant story to tell.

Toward the end of the 1970s, the rate of accumulation of new knowledge in comparative aspects of muscle and its neural control at last began to slow down. At the same time, a more realistic perspective on the prospects for further study of the roles of nervous systems in controlling muscles has gradually forced its way into our minds. This is a subject, philosophically new to science, in which it is at last evident that a decade of intensive, inspired work is not going to resolve the principles. The methodology exists to find out in detail how some nervous systems are put together and, at least in a limited way, exactly how they work without recourse to guessing. But it is now certain that it is going to take hundreds of years to gather sufficient data to find out critically what all the general principles might be.

Faced with such a prospect, slowing down was inevitable. One good aspect of this is that we are thereby encouraged to step aside from the flow of everyday laboratory work to take stock of the progress. This book represents one biologist's stocktaking of two decades of knowledge accumulation in his field, which is widely agreed to be one of the more fascinating aspects of animal life: the nature of muscles and how they are controlled by nervous systems at the peripheral level. It is a comparative work: the entire animal kingdom is encompassed without bias as to which are the more "significant" classes.

The book is written primarily for students, and it is my intention to use it as a textbook in courses on comparative physiology, locomotion, and the neural control of behavior. I hope other teachers of the physiological bases of behavior will find it useful in providing basic reading for their students. But most of all I hope it will be read by newcomers to muscle from chemistry and physics.

There was no hope of getting all the relevant unpublished work from my own laboratory prepared for formal publication and submitted to regular journals if I was to complete this book in time for use by my own last generation of students. Accordingly, where I felt there was a gap that could be filled by use of some of our unpublished material, I have taken the liberty of including it here, with the label ORIG. The electron micrographs so labelled were made by collaborators Dr. Pat McNeill, Dr. Myra Elfvin, Dr. Cynthia Jensen, Dr. Ben Walcott, and Melissa Williams, except for a few kindly given to me by Dr. Michael Reedy and Dr. David Smith, to all of whom I am deeply indebted. Previously unpublished physiological data with the label ORIG are my own except where quoted in the legend. Drawings and interpretations are my own except where stated in the legend. Tables are original except where stated.

GRAHAM HOYLE

Eugene, Oregon
October 1982

Contents

MUSCLES AND THEIR NEURAL CONTROL

1

Approaches to Muscle

I have always been ready to defend the proposition that all theories of contraction are wrong—including my own. A. V. Hill†

There is more danger from the rare event of a whole generation of scientists coming to accept an idea which is plausible and attractive but not sufficiently supported by real evidence. A. F. Huxley‡

1.1. Who Studies Muscles?

Muscles occur in all animals, bestowing upon them their single most characteristic feature. Although well over a million different animal species are recognized on the basis of different morphologies, almost from the beginning, experimental biologists have thought that muscles are rather similar. This conclusion was not the end result of intensive individual studies followed by careful comparisons, but rather from the start a form of optimistic wishful thinking. Yet every gourmet knows that no two forms of muscle even taste exactly alike. In Scotland, the better butchers label similar cuts of meat according to the variety of cattle from which they come. A connoisseur will pay more for the rump of Aberdeen Angus than for polled Hereford and will be glad to back up his ability to taste the difference with a bottle of single malt whisky. The differences with which the biologist should be concerned are equally subtle but more functional.

The differences can hardly be attributed to adaptation and selective advantage; like most variation, they are due to happenstance alone – the different but equal survive. Variation is the basis for evolution, but much of it is not associated with the endowment of any advantage: that is, we cannot assume that each variety of muscle has emerged as a form of progress toward perfection. Granted that a million different external forms represent in part genetic "noise," there are still rather widely different movement requirements for an earthworm, a ragworm, a leech, a clam, an octopus, a flea, a fish, or a human. It is one of the many duties of biologists to find out to what extent muscle structure, function, and control have diversified along with the external forms, habitats, and ways of life. Curiously, a majority of those who study muscle feel now, and always have felt, that their calling is to "unravel the secrets of the machinery of motion." They have wanted to be known as "general" physiologists or biophysicists. Only a small handful has been content to be "comparative" and explore muscle diversity. The larger group has paid scant attention to the findings of the latter in the last 50 years, as a glance at the lists of

†A. V. Hill, *Ann. Rev. Physiol.* **21**, 1-18 (1959).
‡A. F. Huxley, *Reflections on Muscle,* Princeton University Press, Princeton, NJ, 1980.

Nobel Laureate A. V. Hill (1886–1977) lecturing at age 79.

participants in muscle symposia attests. This is unfortunate and has, I believe, impeded progress. Comparative and general studies should always proceed together, being continually and reciprocally related. Since evolution has undoubtedly occurred, every living muscle must be the end product of continuous diversification and selection from one or a very few common origins. No one muscle can possible provide all the answers to the core question: What is truly general about muscles? Philosophically, only the totality of knowledge about all the different kinds, with all their variants, constitutes the general knowledge of muscle. Of course, there is another sense in which the term *general* is used, to delineate features found in a *majority* of the known types, based upon the overall state of knowledge at the time. Thus if all known muscles contain actin and myosin but only a few contain paramyosin, the former are general and the latter special. However, when functional aspects are involved, especially when they are not fully understood, as is still the case with regard to the interaction of actin and myosin, it is neither reasonable nor

Nobel Laureate A. F. Huxley, in 1974.

sensible to ignore the special cases. Ideas regarding the general need to be tested in relation to the special to see if they can fit them, as they must if they are true. Conversely, satisfactory explanations of special cases should lead to tests of compatibility with general theories, and if these explanations are incompatible, the general theories should be questioned as well as the special cases.

In the study of muscle we have had a situation, at least during the 30 odd years of my involvement in it, in which a rather large proportion of investigators – certainly the majority – are not familiar with, and act as if unwilling even to consider, available knowledge on more than a very few muscles. The single most significant blindness is that of persons working on muscles of vertebrate animals who exclude all muscles of invertebrate animals from their consideration. It is possible that there are major differences, although there are none of which I am aware. This book will treat all muscles, of all types of animal, warm and cold-blooded, vertebrate or invertebrate, as of interest, and needing to be included in the general understanding of muscle. As a comparative physiologist, I value knowledge of the diverse forms not only for its own sake, but also because it embodies the general truth. Because of this, this book will not be to the taste of some physiologists and biophysicists, although they may be the persons most in need of knowing the information it contains. They would certainly benefit from thinking about its significance. Although I do not deny that comparative physiologists are obliged to "spread themselves thin," I sincerely believe that progress in understanding would have been smoother and faster had those who studied muscle most intensively taken a comparative approach.

1.2. Losing Knowledge About Muscle

Sir Andrew Huxley has pointed out in print (1957, 1974, 1980) that muscle scientists have a long history of getting stuck in dogmatic grooves, firmly believing models that subsequently turn out to be incorrect. This has been true of other disciplines, and perhaps it is excusable on account of human frailty. However, there is a difference: in other disciplines, proponents of different theories may fight bitterly, but they do attend the same meetings and coexist in unresolved strife until new facts reset the battle lines. In the small subdiscipline that is muscle, not only are the dissenters from current orthodoxy forbidden to appear, but also their work disappears from the record by never being quoted. Muscle scientists have a unique record for simply losing the facts. In retrospect it has become known that clear evidence for the occurrence of an internal network of tubules inside many muscle cells, open to the exterior, had accumulated by the turn of this century. This knowledge was neither taught nor in any textbooks about muscle, so it was "lost" by about 1914. Yet it was vital for our understanding of coupling in fast large-diameter cells. But the world's leading authorities on muscle, Nobel laureates and all, were simply not aware of it. Nor did their imaginations substitute an alternative. Sir Andrew also points out that by the turn of the century a substantial, though more controversial, body of knowledge existed about cross striations and changes in I and A bands during contraction. This could have led to the modern sliding-filament theory 50 years earlier, but the early findings were not even discussed.

A possible reason for this was that most muscle research came to be done by persons with physics or chemistry, but not biology, backgrounds. Two generations of muscle investigators were persuaded that the contractile material must be folded or coiled filaments extending from Z to Z (Figure 1.1). Since A/I band differences did not fit this concept, they were deemed irrelevant and consequently ignored. (For explanations of anatomical terms see p. 42.) There are recent additions to this appalling list. Anthony van Leeuwenhoek, the first person ever to see living forms under the microscope, claimed in 1674 that he had seen a spiral in cross-striated muscle cells. Although disputed by some histologists, such spirals were very clearly photographed in a variety of cross-striated muscle cells by Otto Tiegs, and he and his collaborator filmed spiral progression in spider leg muscle cells (Matthaei and Tiegs, 1955). When Peachey and Eisenberg (1975, 1978) examined frog skeletal muscle cells by reconstruction of 0.5–1.0 μm serial thick slices examined with a 1000 kV electron microscope, they found that the invaginated tubular system, and therefore also the contractile myofilaments, were spirally arranged! They wrote they had "always thought that the striations of striated muscle are discrete units, lying in planes roughly transverse to the fiber axis." Their final remark regarding spirals was: "the discovery seems to have been lost as often as made." Lost by whom? Certainly not by the late Otto Tiegs or his colleague, who is still researching. Facts are lost by those

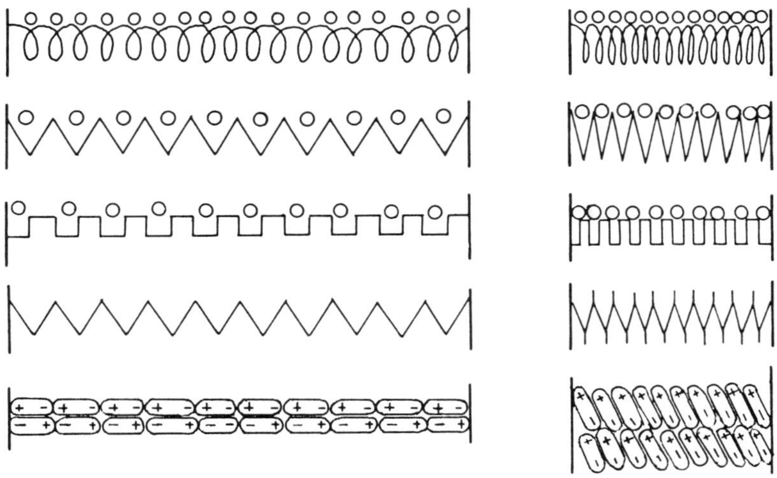

Figure 1.1. Models of contraction mechanisms seriously considered prior to the discovery of sliding in cross-striated muscle in 1954: relaxed or stretched at left, contracted at right (e.g., Astbury, 1950). The coiled or folded chains represent myosin, and the circles actin, which was assumed to be present in globular form. The model fourth from the top was proposed by Nobel laureate A. V. Hill (1950) and represents an assumed conjoint actomyosin. The lower model was proposed by Nobel laureate Albert Szent-Györgyi (1951) on the basis of the appearance of early electron micrographs.

"high priests" of a subject and their armies of followers, who determine its current fashion. They follow the tradition of Lord Rutherford, who is reputed to have said proudly, when challenged about a critic's article: "I don't read the literature, I make it."

1.2.1. Is Anything Being "Lost" in Our Time?

This question may seem absurd. How can one be aware of what has been lost until it is found again? However, as in the case of the meandering nature of invaginated tubules (ITs) and the tendency for some cross-striated muscle cells to have a spiral internal organization, the observations were lost rather willfully, in the making of prejudiced personal choices about what was supposed to be "general". The most glaring example of a very real attempt on the part of a majority of muscle scientists to willfully lose a vital piece of information in the last 25 years has been in regard to ultrathin (T or S) filaments (see p. 90) in parallel with I and A filaments. The evidence for their existence, both theoretical, based on dynamic properties, and visible, based on electron microscopy, has been amply provided (see p. 94). Yet for reasons I have not been able to understand, the majority of muscle scientists have ignored them completely. There have been occasional howls from the frustrated few believers, but these believers have been treated by the majority as nasty pests needing to be destroyed. Why? Without them, T filaments would certainly be getting "lost." With such obstinacy the majority deserve to be proved wrong!

It is also possible that in our supposedly enlightened age, facts are being lost because they are not in vogue. The vogue of our time is cross bridges (X bridges), so to seek the facts that are being lost, one should examine anything that has been discarded because it doesn't fit the X-bridge theory. Since about 1960, X-bridges have dominated center stage. A. F. Huxley (1974) says he heard several of the participants at the large international Cold Spring Harbor symposium held in 1972 say, "The problem of muscle contraction is solved in principle." He went on to write that he was "horrified," offering a list of seven major unknowns regarding contractile mechanism and pointed out that "the whole history of theories of muscular contraction . . . shows that even when a set of ideas seems to be well established, there is a large chance that it will be overthrown by some unexpected discovery."

1.2.2. Why Do Muscle Scientists "Lose" Knowledge?

Sir Andrew suggests that a false lead was responsible for the loss of relevant knowledge. This was a belief that there must be a common molecular mechanism underlying all forms of contractility. It led scientists away from concentration on cross-striated muscle. When they returned to intensive study of vertebrate cross-striated muscle cells in the early 1950s, he believes, light and reason were restored. It is still a widespread belief today that there is a common mechanism, so the "light" from cross-striated muscle (Figure 1.2) is being turned on the darkness in nonmuscle contraction. I consider that it is the passion to second-guess nature, combined with the powerful intellectual appeal of Occam's razor to minds attuned to physics, that is responsible for widespread faith (without evidence) for a common mechanism. The late distinguished Cambridge, England, comparative physiologist Carl Pantin noted in 1956 that "when we consider the extraordinary variety of form and function that has arisen in animals we might reasonably have expected every class to have evolved a distinct kind of contractile machine, but it has long been tacitly assumed *that muscle is muscle* wherever it is found." The American anatomist Arthur Hess, parodying Gertrude Stein, wrote, "A muscle is a muscle is a muscle."

However, it was never more than a tacit assumption: there was not then, nor is there today, any real justification for believing in a common mechanism. There could be more than one force-generating mechanism in a single muscle cell, and many different ones in different muscles, for all we truly know to the contrary. Organisms had many tricks to play with and plenty of competition to encourage their diversification. Granted the rich range of muscles available to investigators, it is at once disappointing that so few have been studied intensively and surprising that the many gratuitous, unsupported statements of uniformity have been made. Confident general summaries that in fact only apply to a very limited type of muscle have been frequent. For example, "When a muscle is stimulated there is an abrupt change from rest to activity," a statement made by J. Murdoch Ritchie (1954a) as the introduction to a paper, should have read: "When some muscles of vertebrates comprising only fast-twitch muscle cells are stimulated, there is an abrupt change from rest to activity" (see p. 257).

Figure 1.2. Early formulation of the ultrastructure of a contractile unit of a representative vertebrate muscle – rabbit psoas – based upon transmission electron microscopy, from The *Contraction of Muscle* by H. E. Huxley. Copyright 1958 by Scientific American, Inc. All rights reserved. (*1*) Actual sections through different regions indicated by planes in the drawing are shown below, and the interpretation between. The left and right sections are photographic duplicates, not actual left and right sections. The interpretation of the image in the I region is incorrect. This is a region of transition to a square array (see Figures 3.17 and 3.27). (*2*) Model of the molecular components of an X-striated muscle cell such as has been adopted in all standard textbooks and many scholarly texts written by muscle authorities. This example is reprinted with permission from L. Goldstein, Ed., *Introduction to Comparative Physiology*, Holt, Rinehart, and Winston, 1977. (*3*) Standard model of X-bridge orientation as presented in 1977 textbook by Arnold G. Kluge. Reprinted with permission of Macmillan Publishing Co. Inc. from *Chordate Structure and Function*, 2nd Edn. Copyright 1977 by Arnold G. Kluge. (*4*) Expected orientation of a X bridge in the contracted state. [Reprinted with permission from F. A. Pepe, *Int. Rev. Cytol.* **24**, 193–231 (1968). Copyright 1968, Academic Press.]

1

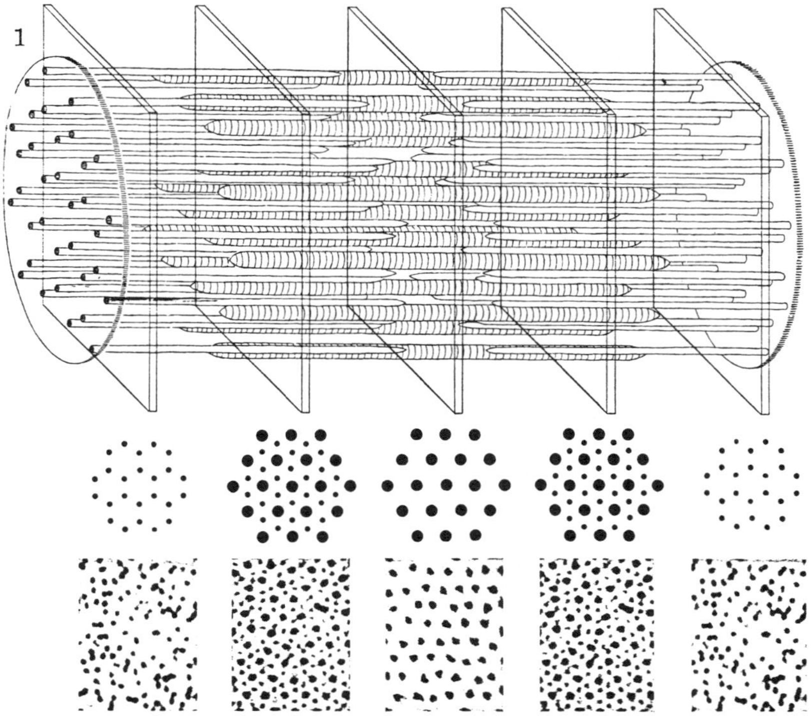

2

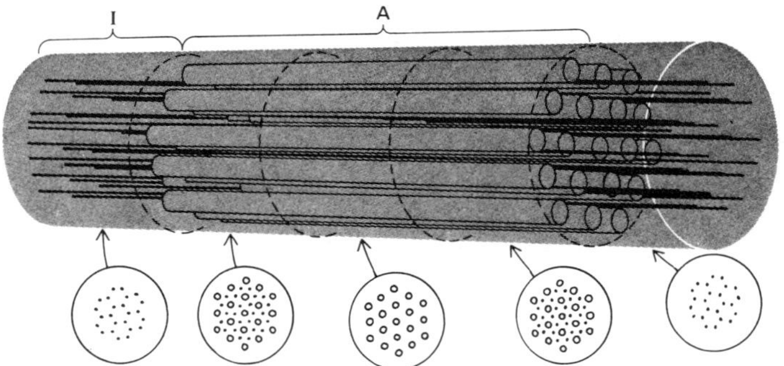

3

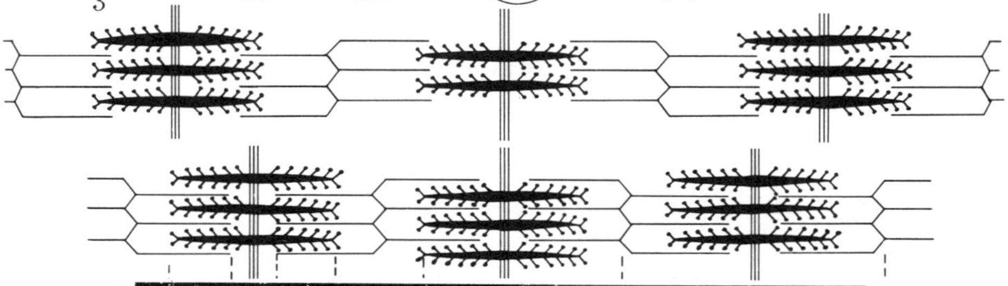

4

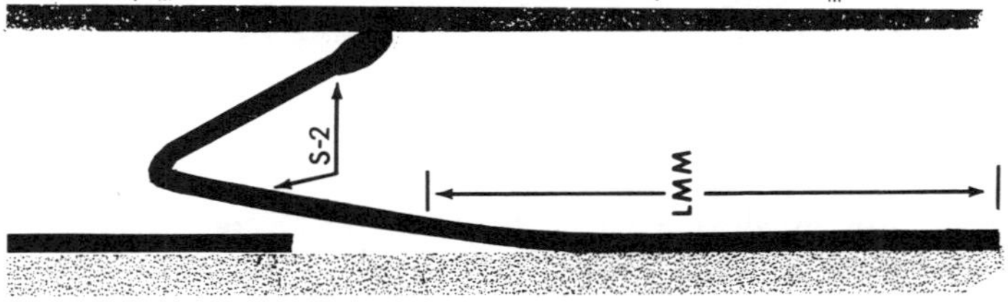

Of greater significance, perhaps, is the competition between scientists for recognition on the basis of their reputations for "excellence." Muscle research, with its limited general goals, has been particularly prone to this unfortunate by-product of the behavior patterns of scientists, and the elegance of the approach has, for many, come to mean more than the biological significance of the findings.

1.3. A Broad Biological Approach

> The forbidding numbers of individual nerve cells that form the central nervous system of all vertebrates . . . have led to the view that single-cell reactions are of little or no value for the analysis of brain function C. A. G. Wiersma†

For the comparative biologist, there is so much more to muscle than the elusive contractile machinery. An enduring body of descriptive data has accumulated on the anatomies and physiologies of several kinds of muscle. Such studies were pioneered especially by the late Cornelis A. G. Wiersma, who loved the invertebrates and was intensely curious about the ways they had resolved their special nerve/muscle problems. Collectively, this forms a broad spectrum of knowledge about a single tissue present throughout the entire animal kingdom that is unmatched for any other tissue or organ. Not only does this provide a fascinating spectrum in its own right, but it is valuable as a double-edged fundamental tool. On one hand, it can serve to test ideas regarding phylogenetic status and relationships. On the other, it permits an insight into how evolutionary pressures have affected the single most vital aspect of animal life. But changing the physiology of an organ has to depend upon genetic happenstance: there is no guarantee that any particular one will have come into line. Probably, many muscles are actually ill-adapted to their present-day tasks. The physiologist's view of this has been stated by Eisenberg et al. (1974): "All animals have muscles of various kinds which have evolved to suit particular functional roles." Maybe, but this statement is only weakly supportable by the existing data. It remains to be determined to what extent muscles of any class have truly undergone functional adaptive radiation. In one instance, recent studies of the fine structure of muscles and innervation of an animal have overturned its apparantly well established phylogenetic status. The onychophoran *Peripatus* has, for a hundred years, been considered to be an interphylum "living fossil, missing link" between annelids and arthropods. However, both the muscle fine structure and the muscle innervation are unique, certainly unlike those of arthropods (Hoyle and Williams, 1980). Comparable studies on a variety of hard-to-classify animals are needed to help resolve their positions.

The study of how muscles and their neural control mechanisms have changed to permit adaptive radiation, response to environmental change, or improved basic functioning is itself an infant subject. In this book I try to develop the evolutionary aspect, though we do not yet know enough for this to have been made the major theme. All types of muscle are considered, although the emphasis is on skeletal muscles. This in part reflects a greater knowledge, but it is also because only skeletal muscles are the organs of overt behavior. My present major interest, and one of the most significant quests for biologists in the future, is to understand how nervous systems generate behavior. As a starting point it is essential to understand muscles, which are the organs of behavior, and how they are controlled.

†From the introduction to Chapter 30 in *The Neurosciences – Third Study Program* (1974).

C. A. G. Wiersma (1905–1979) with *Podophthalmus.*

1.4. Describing Muscles

No prior classifications are adopted but the very general anatomical ones of skeletal, cardiac, and visceral in this book. Muscle is so diverse and there is so much both new and not widely known old material that description must come first. The traditional classifications of muscle cell types were based upon a very small amount of knowledge compared with that available today. Are they adequate? Are they even accurate in light of ultrastructural information? Can we come up with a rational scheme for classification?

Foremost in significance of the various aspects of muscle is, I believe, the overall neural control of each different kind and of the complex mixtures that constitute functional units. This is the basic stuff of behavior, the only part on which evolution acts directly. It matters but little to the student of behavior mechanisms what the molecular events may be. But it is basic to his/her study to understand the principles governing the performance of the muscle in the animal.

Then there is functional morphology: specialization in relation to overall function such as circulation of blood, passage of food, "clamming," jumping, flight, and so forth. These problems have been solved by all or many diverse animals. Are there common themes in their resolution? Muscles with different properties are going to "interpret" a given motor output code in very different ways. In an evolutionary step, does a nervous system adjust its code to match the muscle, or does the muscle change its way of responding?

Finally, there is the molecular aspect that so intensely grips most human minds. Is there truly but a single fundamental molecular event in contraction of muscle cells? Is this a conformational change in a part of the myosin molecule? The known speeds at which muscles operate vary over a 10,000-fold range for contraction and greater than 10^6-fold for relaxation. Given a single basic event, this would put a premium on diversification of regulatory mechanisms for contraction and relaxation. Is relaxation a passive or an active process, a reversal of contraction, or something quite different from it? Are there, perhaps, different types of relaxation mechanisms in different muscles?

The range of forces capable of being developed maximally by different muscles is quite narrow, apparently only about 50-fold (see p. 121), but still great enough to raise questions. There are problems in determining the true force, even in a single cell, because it must be related to the proportion of the cell that is contractile. (I have chosen to substitute the term *cell* for the conventional "fiber" because not all muscle cells are long and slender.) The amount varies from a maximum of about 95% to a minimum of barely 5%, a range that has seldom been taken into account when making comparative statements. In contrast to the force range, the length range is large. There is a muscle in *Amphioxus* that doesn't shorten at all; it only gets stiffer when excited (see p. 33). At the other end of the spectrum is the tentacle muscle of a Portuguese man-of-war (*Physalia*) that can extend down to 70 ft when fully relaxed. When contracted maximally it measures no more than 5 in. long (Parker, 1932; Mackie, 1960), a 175-fold range.

1.4.1 The Range of Contractile Phenomena

It is commonly thought and taught that muscle cells have two basic states: relaxed and contracted, although a special state of stiffness, termed *rigor*, is recognized for vertebrate cross-striated cells that have lost ATP and/or are dead. A different state, termed *catch*, has long been recognized for some molluscan slow-adductor and retractor muscles (Uexküll, 1912). Catch is associated with maintenance of a shortened state and/or stiffness, with high resistance to stretch that requires very little energy for its maintenance (reviewed by Twarog, 1979). A state of extreme shortening, termed the delta state, can be induced in vertebrate cross-striated cells, but this is not followed by spontaneous relaxation. By contrast, some cross-striated cells of both vertebrates and invertebrates can contract reversibly to down to $\frac{1}{6}$ rest length, a capability that is termed *supercontraction* (Figure 1.3.5).

Three universal forms of contraction actually occur in cross-striated muscle cells. The first in rank is the *major contraction* elicited by a membrane depolarization past a threshold, the *excitation-contraction coupling threshold* E_c. Each will be described in detail for a variety of muscle. The second is *basic tonus* (BT), which is still very poorly understood, largely due to neglect that can be traced to the stifling of research, associated with common acceptance of a view based on limited research (see p. 255). BT is strong in chick anterior latissimus dorsi (ALD), where it is responsible for holding the wings folded back against the body (Goldspink et al., 1970). The third is a local contraction that

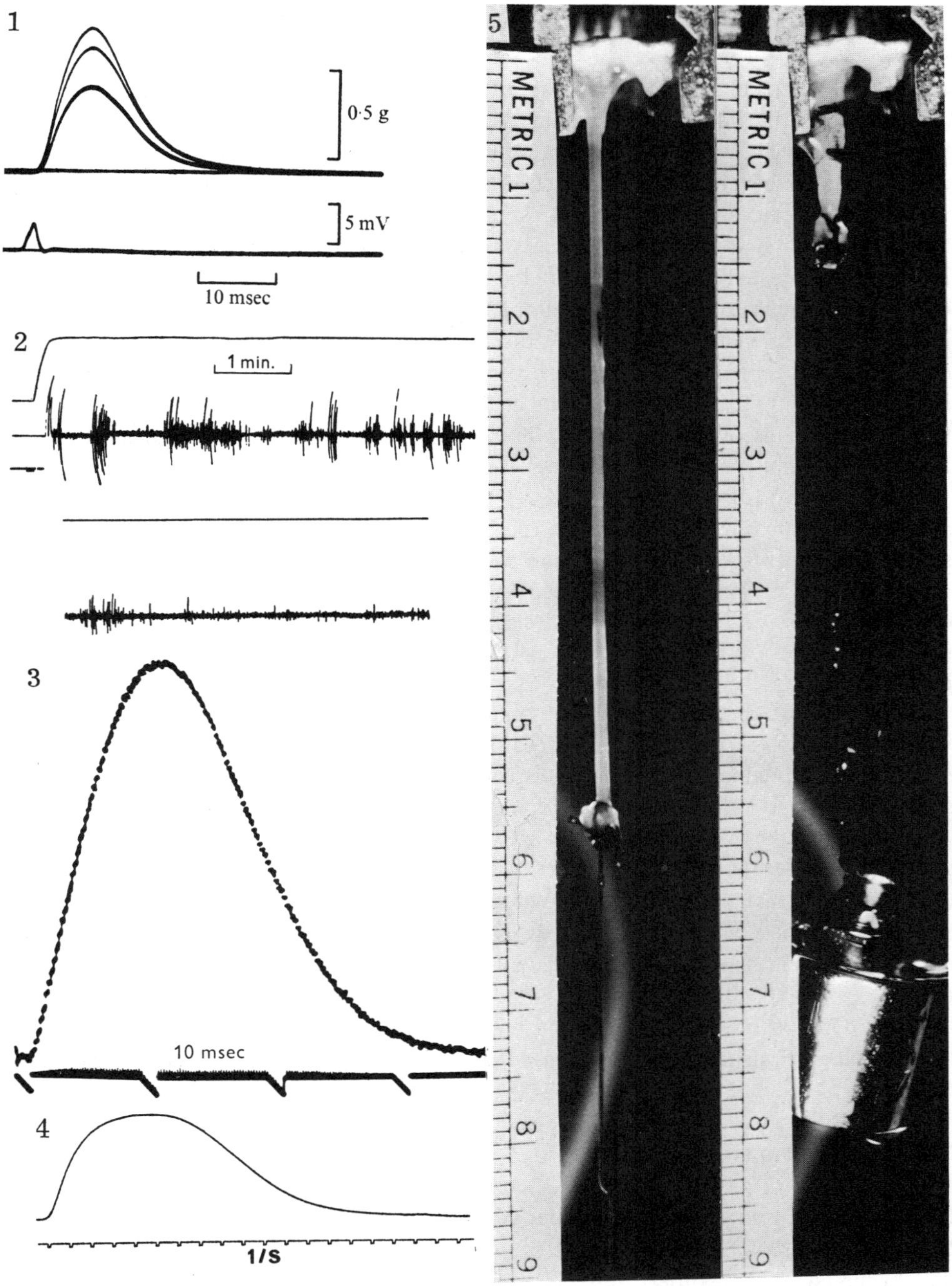

Figure 1.3. Range of contraction capabilities. (*1*) Extreme of speed of an invertebrate skeletal muscle, twitch of singing muscle of a Katydid insect. (From Josephson, 1973.) (*2*) Perfect "catch" state in a lamellibranch molluscan muscle, the *Mytilus* anterior byssal retractor. Even after intrinsic electrical activity has ceased (*below*), there is no relaxation. (From Hoyle and Lowy, 1956.) (*3*) Extreme of speed of a vertebrate muscle, rat extensor digitorum longus, after potentiation. (From Close and Hoh, 1968.) (*4*) Typical vertebrate smooth muscle, rectococcygeus of rabbit. (From Davey et al., 1975.) (*5*) Repeatable supercontractions in a single cross-striated muscle cell of giant barnacle *Balanus nubilus* freely contracting following stimulation under 10-g load: third sequence. Force per contractile unit after maximum shortening under load equal to 38% maximum isometric force capability at rest length. (ORIG.)

develops slowly, being complete in about 5 s, involving from about 5–30 sarcomeres depending on the muscle source. Such a local contraction (Figure 1.4.*3*) is called a *contraction wave* (CW). CWs occur continuously in, and can be seen upon close examination of, many living cross-striated muscle cells. I have watched them in frog, snake, rabbit, various crustacean, and several insect skeletal muscle cells, developing, propagating and fading. Very rarely a double wave occurs, with two CWs moving away from each other. It might be objected that one cannot look at muscle cells in the microscope without either risk of physical damage to them or the use of an unnatural bathing medium and restriction of circulation. But small crustaceans, spiders, and insects have transparent cuticles, and the muscle cells can be observed directly through the cuticle without any risks. If anything, CWs are more abundant in these natural conditions than after dissection. CWs were discovered a very long time ago. They were figured by Bowman (1840) in insect skeletal muscle and by Veratti (1902) in rat, frog, and crab, and were extensively photographed by Hürthle (1909), Tiegs (1955), and Mattaei and Tiegs (1955), but they are virtually "unknown" today. Intracellular recording in the contracted region shows no significant difference in membrane potential from the surrounding region. Quite often the contraction, once developed, propagates. It does so as a whole, moving in either direction capriciously and quite slowly, involving about the same number of sarcomeres. The propagation rate is fairly constant for a given CW, of the order of 10 μm/s.

The extent of the contraction in the middle of the wave is often extraordinary, the sarcomere length (SL) being reduced to as little as $\frac{1}{10}$ normal in insect cells and $\frac{1}{5}$ normal in rat or rabbit, reducing gradually toward the edges. Sarcomeres adjacent to the CW are stretched by up to twice the rest length.

By carefully photographing such waves during propagation along leg muscle cells of a small, thin-legged translucent intact spider (*Pholcus littoralis*), Matthaei and Tiegs (1955) showed that the wave propagated spirally. The CWs have most recently been observed in "skinned" muscle cells, that is, cross-striated cells from which the sarcolemma and surface membrane have been peeled away, from a toad (Natori et al., 1976). We shall return to a further discussion of CWs after considering muscle-cell fine structure (see p. 41). The CWs represent a piece of ancient muscle knowledge that, though frequently filmed and documented, can only be described as "lost" to many muscle researchers today.

1.4.2. Intrinsic Rhythmicity

Intrinsic or myogenic rhythmicity occurs in a wide variety of muscle cells. A majority of the cardiac cells of all animals, most of which cells are cross-striated but of very short length, linked to each other both longitudinally and laterally (see p. 365), possess intrinsic rhythmicity (IR). The rhythms are described in detail in Chapter 10. A few crustaceans, including true lobster (*Homarus*) and rock lobsters (*Palinurus* and *Panulirus*) and the arachnid *Limulus*, have hearts that are synaptically driven from special clusters of a few neurons in pacemaking cardiac ganglia. Some molluscs, notably lamellibranchs and gastropods, do not require such input, but the question of myogenicity cannot be simply resolved because there are nerve cells in the walls of these hearts.

Many of the small nonstriated muscle cells of vertebrate involuntary organs, especially the organs of the alimentary canal, possess intrinsic rhythmicity. The difficulty here is to separate the activity of individual muscle cells from undoubted rhythmicity of whole organs and parts that contain nerve plexuses.

Cultured cells are deprived of the capacity to exercise force development, which alone profoundly affects their characteristics (see p. 402), but they generate periodicities ranging from a few seconds to many minutes, or both superimposed. The slowest ones known are circadian. Some are highly complex yet can be clearly related to life-style. Pieces of the body-wall muscle of worms show various intrinsic periodicities that can be related to life-style. For example, bits of the lugworm *Arenicola marina* show a relatively quick rhythm with a period of about 40 s superimposed on a slower one of 40 min. The latter corresponds to rhythmic defecation by the whole worm, and the former, which shows smooth amplitude rise and fall synchronized with the slower rhythm, represents feeding movements (Wells, 1949). It is not known whether the same or different cells generate the two rhythms. I expect that both simple rhythms and complex polyrhythms that are myogenic will eventually be discovered in a wide variety of muscle cells.

Fairly recently an IR was detected in locust leg muscles (Figure 1.4). No purpose is known for this rhythm, but it probably has functional significance, as emphasized by the fact that it is modulated in both amplitude and frequency by antagonistic nervous actions. One of the modulatory mechanisms, which is inhibitory, consists of a local release of the monoamine octopamine from neurosecretory-type terminals that are situated within the muscle (Hoyle et al., 1980). Further details are given on p. 420. Many species of locusts and grasshoppers have now been examined, and all of their jumping muscles (extensor tibiae) show an IR. It is confined to a small bundle of cross-striated cells situated at the proximal margin of the muscle. Although the IR is usually a synchronous event in participating muscle cells, two to four independent rhythms may occur, showing that there are at least two pacemakers, with intrinsically different rates. Although this muscle has been fairly intensively studied for a century, the rhythm, which can easily be seen even in the intact animal, was not noticed until 1973. One therefore wonders what other subtle contractile phenomena may have been missed. Cross-striated cells from vertebrate skeletal muscle do show spontaneous activity when maintained individually in tissue culture, although the activity is not especially rhythmic. When cross-striated cells are combined with cultured neural cells, neurons innervate the muscle cells; if the innervating cells derive from a cultured ganglion, there is an increase in frequency of contractions that is neurally generated. (Betz, 1976). But these contractions cease following the full development of neuromuscular junctions. The significance of the early spontaneous contractility is not yet understood, but it is possible that it was a very important property of primeval contractile cells and may now be an evolutionary relic. Functioning of primeval single-celled organisms involved not only simple reflex actions but also spontaneous myogenic contractility to generate locomotion to new food sources and to promote respiratory exchange. Spontaneous motor activity is primary not only in protozoa but also in many metazoa. Nerve cells evolve secondarily, to control the contractility, first by modulating it, later by initiating it. Thereafter there was a mixture of myogenically and neurogenically generated muscle activity, the situation persisting in all animals.

1.5. The Question of Size

The functional dynamic problems facing a large animal are vastly different from those facing a small one. Mass increases to the third power of dimension, requiring vastly increased antigravitational forces for postural maintenance and correspondingly greater power for overcoming inertia of heavy skeletal and muscular components. These require special handling at every level: muscle design, muscle-cell composition, innervation and synaptic transmission, control mechanisms, and biochemistry of energy supply.

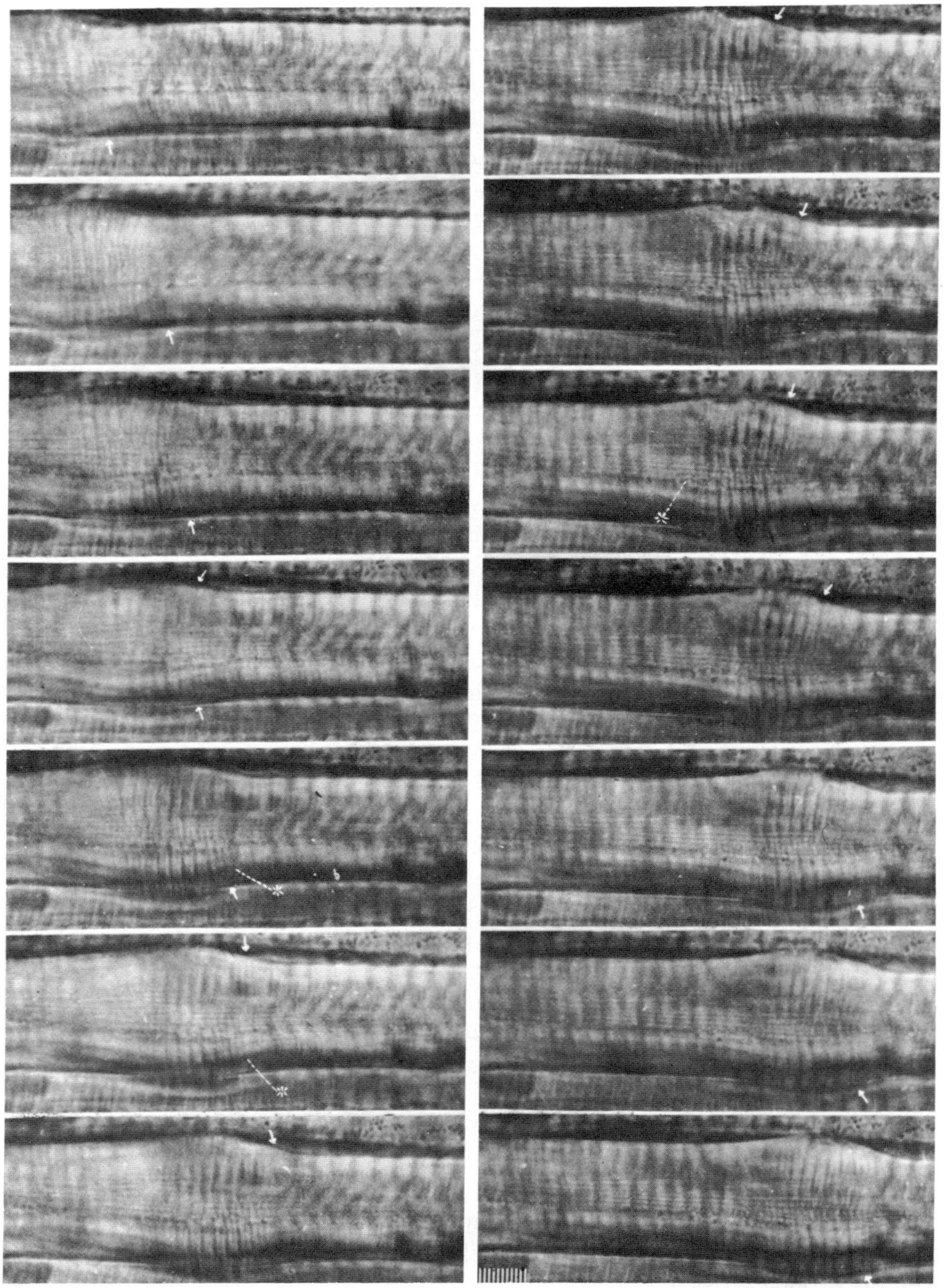

2

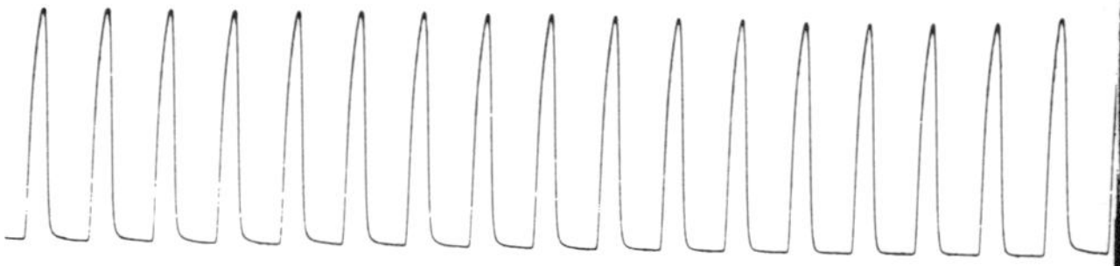

3

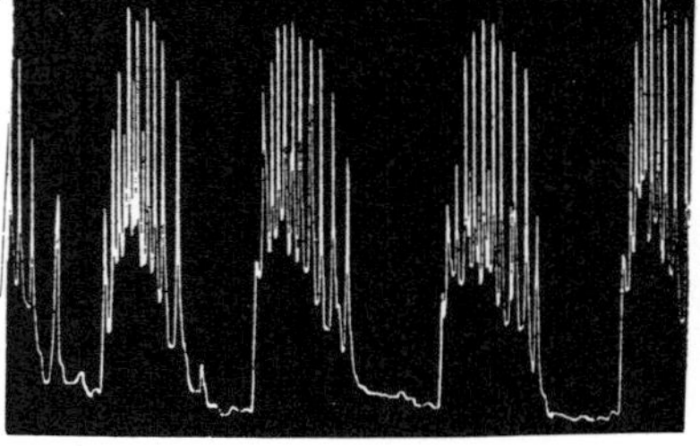

Clearly no phyletic line started out as large animals. Some remained small, and some retained a mixture of intrinsic sizes, whereas others, such as dinosaurs, all got large. There was a tendency of some species, such as tigers and horses, to get progressively larger, but increasing size posed problems, not only for the adult, but also during the animal's development, since the organism must at least start out relatively small. Identical mechanisms of contraction and control could not encompass the whole size range. Accordingly, both progressive change in muscle structure and function during development and even intrinsically different systems of musculature developed.

Size differences pose major differences in requirements for energy utilization, particularly in warm-blooded animals. Here the problem is that heat loss is proportional to surface area, which increases only as the square of linear dimension. A mouse must utilize 100 times more oxygen per gram of its weight than an elephant, either to stay warm or to provide the power required to move an equal distance. On the other hand, it has to develop very much less force to move a limb at a given speed. However, because the limbs of the small animal are smaller, they have to move at a higher frequency. As any given animal moves faster, it requires more oxygen in linear proportion, but the increase is much larger for the small than for the large animal. The mechanical disadvantage increases progressively with limb length, requiring greater muscle power but relatively less shortening. A muscle of a very large animal works almost isometrically. The molecular machinery of muscle was able to cope well with large force development combined with little shortening, and indeed large size was favored because energy utilization is relatively small for isometric-force development. It is shortening that requires a lot of energy. It would seem that all the advantages are with the larger animal, provided sufficient food is available to support the increased biomass. In practice this means that a majority of large animals, both in variety of species and numbers of a particular species, are herbivores.

The problems are a lot different for cold-blooded animals: in their case, small size is no disadvantage energetically. They do not need to generate heat before their muscles can start to function. The problem for them is to evolve nervous systems that are parsimonious in terms of numbers of nerve cells and yet capable of complex control functions, and muscle and neuromuscular systems appropriate for control by few motor neurons. Very small arthropods, especially crustaceans and insects, have done just that. There are no extremely small vertebrates, and especially no small warm-blooded ones, perhaps because their muscles and neuromuscular mechanisms are not appropriate to very small size. Very small cold-blooded animals with efficient nerve-muscle systems have, on the basis of total numbers, biomass, variety of habitats, and numbers of species, already inherited the earth.

1.6. The Question of Life-Style

Although the basic characteristic of animal life is movement, a very wide range of utilization of contractility occurs, and those that move least also tend to move the most slowly.

Figure 1.4. Some rhythmic activities of muscle cells. (*1*) Slow contraction wave (see p. 12). Such waves occur periodically in many cross-striated muscle cells: in small transparent animals they can be observed and photographed in the intact animal. This sequence was filmed in a spider, *Pholcus littoralis,* leg muscle. This particular wave progresses in a spiral, indicating a spiral organization of the contractile material. (From Matthei and Tiegs, 1955.) (*2*) Simple rhythm of isolated proximal muscle cells of locust (*S. gregaria*) jumping muscle. Record duration 3½ min. (From Hoyle, 1978a.) (*3*) Dual rhythm of isolated strip of anterior gut of lugworm *Arenicola*. Record duration 30 min. (From Wells, 1937.)

Chitons rarely move; when going at top speed they manage about 12 in/day. Ghost crabs of similar size are almost constantly active and can run at up to 12 ft/s: 10^6 times faster. An even greater range is seen in repetitive activities. Movements may slow to zero during tonic activity of a muscle, especially in slow adductors of clams, or they can occur repetitively at frequencies up to 1 kHz in flight of small flies (Sotavalta, 1953). Almost every possible frequency in between is achieved by some muscle or other.

The temperature and pressure ranges over which some muscles act are dramatically extensive, whereas other muscles must be within a very narrow range to contract at all. The benthic fish *Coryphaenoides* swims at the bottom of a 2200 m ocean trough near the Galápagos Islands, where the temperature is 2°C and the pressure 3000 psi. (Dreizen and Kim, 1971). Some starfish and sea anemones have recently been found living at more than three times this depth: at ocean depths of 30,000 ft, the atmospheric pressure is 500 times that at the surface. The upper temperature record for sustained metazoan life is held by a species of ostracod crustacean (*Potamocypris*) that lives in 48°C hot springs; for it a cold death occurs at 30°C (Wickstrom and Castenholz, 1973). The other extreme is shown by some spiders living in the Himalayas at nearly 20,000 ft, where the pressure is below one half surface pressure, and the temperature is −20°C. They feed on insects blown up to them. The muscles of warm-blooded animals have a temperature range for functioning of but a few degrees, whereas those of some cold-blooded invertebrates and vertebrates function over a range of as much as 50°C. The speed range is, of course, enormous to match.

One of our tasks is to compare the structure and function of the muscles that are capable of these wide performance ranges and to try to determine what modifications occurred during evolution to permit them.

1.7. The Future of Muscle Studies

The present slowing of the flood of data on muscle is likely soon to be replaced by intense activity. It is amusing to note that more than a decade ago Volkenstein (1969) wrote cynically: "The theory of muscular contraction is partially complicated by excessive amounts of experimental data." The experiments alluded to were designed to test theories cooked up out of the imagination, when muscle was a mysterious black box. Countering Volkenstein's cynicism are questions about the "facts." Paul and Peterson (1977), reviewing muscle energetics, concluded that: "Most of the recent progress has been to undermine what once were considered to be solid facts." It is my hope that the next generation of muscle scientists will be calmer than past ones, willing to start afresh with clear minds, letting nature and natural history guide them toward the unraveling of nature's secrets, *no matter how long it takes!*

2

Contraction Before Muscles

It has long been tacitly assumed that muscle is muscle wherever it is found. C. F. A. Pantin†

The scope of muscle physiology has broadened enormously with the discoveries of actin and myosin in blood platelets, nerve cells, and amoebas. F. D. Carlson and D. R. Wilkie‡

2.1. The Origin of Muscle Cells

We know nothing at all about the possible evolutionary origin of muscle cells, although it has often been hypothesized (e.g., by G. H. Parker, 1919) that dual-purpose cells serving as epithelial or endothelial linings of diploblastic organisms such as *Hydra,* but also having contractile tail regions, served as precursors. In living coelenterates, the ectodermal musculoepithelial cells have their contractile portions arranged longitudinally and serve as retractors that shorten the trunk or tentacles. The corresponding portions of endodermal cells are arranged radially and serve as an antagonistic circular sheet. Synaptically connected primitive bipolar and multipolar nerve cells form a network lying between these muscular sheets. It is not difficult to envisage in the essential coelenterate tubular form the origin of the slow-contracting tubular involuntary muscles that form the gut of all animals. Evolutionists have supposed that the first individual discrete, anatomical muscles developed out of just such a tube. An invaginated portion of the tube became cut off and specialized for digestion, essentially retaining the primitive structure. The outer portion came to serve principally for locomotion. The first true muscles arose in connection with the development of two very different extrusible organs, a pharynx, to aid in food collection, and a penis, to direct male sex cells during mating. These specific organs needed to be withdrawn into the body after use, a function first carried out by parts of the longitudinal muscle sheet. As these regions became clearly differentiated anatomically, so did their bit of body wall, until eventually distinct retractor muscles existed.

Molluscs have evolved additional body-wall projections in the form of tentacles and anus, and some forms have added a simple external skeleton. But overall, molluscs have altered the basic structural form of the tubular antagonistic muscle sheets rather little.

Evolution of muscular systems occurred via other pathways, especially metameric segmentation, that broke up the outer tubular sheets of muscle, permitting the production and utilization of sinusoidal waves for locomotion. Finally, the two great lines of develop-

†C. F. A. Pantin, *British Medical Bulletin 12* (1956) 199–202.

‡F. D. Carlson and D. R. Wilkie, *Muscle Physiology,* Prentice-Hall, 1974.

ment of articulated skeleton, external in the arthropods and internal in the chordates, permitted the completion of segregation into a specific skeletal musculature operating on the joints.

The origin of the metazoa might have been multiple, with parallel lines, but most evolutionary zoologists subscribe to the view that a single form of ciliated/flagellated protozoan is at the base of the evolutionary tree. Whether these ancestral unicellular organisms led to higher forms by cellularization (division within a single organism) or by colonialism (cooperation of clusters of individual cells) is in question, but the former alternative is most commonly favored. In either event, contractility probably put in an early appearance, as a specialization within a multifunctional cellular unit not too far removed from the ancestral cell.

The contractile apparatus could have made its appearance at this time in evolution, but it seems much more likely that it already existed in the ancestor, being modified or emphasized to promote new functions. Thus we should look at contractility and its underlying causes in existing single-celled organisms for clues to the original materials out of which muscles evolved. Cells in which only a small portion of the cytoplasm contains contractile material have been retained not only by primitive coelenterates, but also by a majority of classes including mammals.

2.2. Contraction Before Cells

Movements, both general and of specific regions, occur in precellular forms of life, and a few of these are well understood. For example, the contraction of the sheath of bacteriophage is brought about by the closing in of annuli. But there is no known counterpart to this in any contractile cell, and this is apparently not a profitable area to search for origins. However, withdrawal behavior of slime molds has attracted a lot of attention since 1952, when A. Loewy extracted a protein from the plasmodium of a myxomycete that had the unmistakable properties of those conjugate proteins called actomyosins, which can be extracted from all muscle cells. This discovery initiated a flood of research that is by no means over, searching for the basic muscle proteins actin and myosin in nonmuscle cells.

A few years later, isolation procedures for muscle filamentous actin had been refined, and it was shown that actin filaments can easily be recognized by their appearance in the electron microscope after negative staining (H. E. Huxley, 1963; Hanson and Lowy, 1963). As a corroborative test, when the actin filaments are bathed in a solution containing rabbit myosin, they become "decorated" in the formation of an arrowhead pattern as myosin molecules attach to them at regular intervals (H. E. Huxley, 1963). Actin began to turn up almost everywhere it was sought, even in the growing points of plants. Nachmias and Ingram (1970) obtained *Dictyostelium,* slime mold actomyosin, removed some of the actin, and observed decorated filaments in the remainder that were identical in appearance to ones made from muscle actin and myosin. Their myosin very closely resembles muscle myosin when highly purified (Clarke and Spudich, 1974). Slime mold actin and myosin could be there for only occasional use in retraction, or they might be subserving a more frequent or constant movement, of which the only likely one is cytoplasmic streaming. *Physarum* retraction strands have an ultrastructural appearance closely resembling that of many smooth-muscle cells, being composed of closely spaced uniformly thin filaments (Figure 2.1) with no obvious sign of the sparse myosin. The

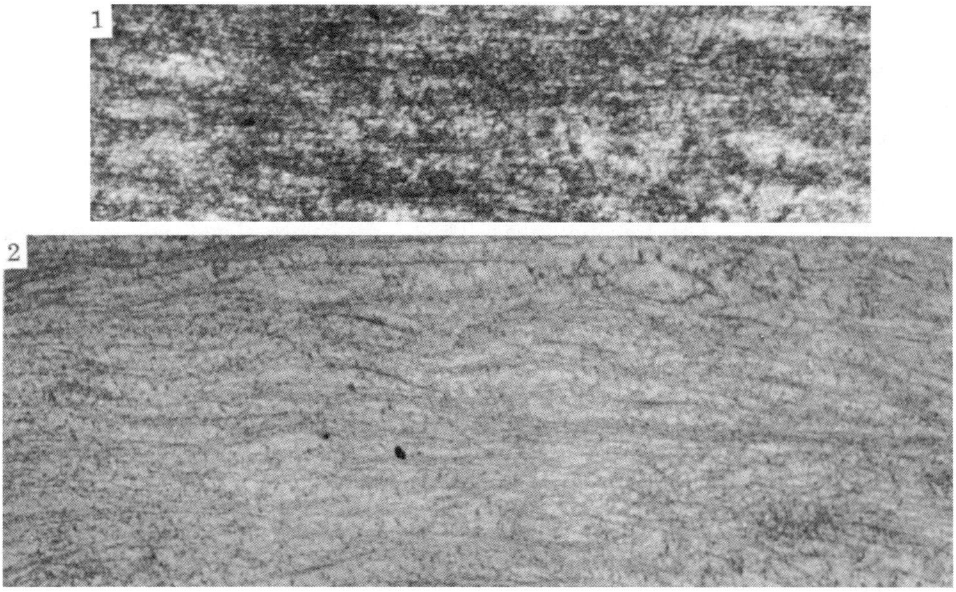

Figure 2.1. (*1*) TEM of retraction strand of a slime mold, *Physarum*, showing parallel, homogeneous thin filaments. Both actin and myosin are present. (From Wohlfarth-Bottermann, 1965.) (*2*) EM of smooth-muscle cell of guinea pig taenia coli: mainly actin, some myosin. (From Needham and Schoenberg 1964.)

myosins of myxamoebae and plasmodia have similar properties, although the modes of motility of the two types of organisms are markedly different (Tariguchi et al., 1978).

2.3. Movements of Unicellular Organisms, Specialized Cells, and Parts of Cells

Movement is a major characteristic of animal cells, but it is not uncommon for single-celled plants as well, and indeed, the distinction between plant and animal life is often blurred at this level. Accordingly, it is desirable to examine all forms of life at this level for the existence of contractile machinery that might have been exploited by the emerging metazoa. It is not just movements of the whole cell that we should consider, but also internal movements, including cytoplasmic streaming, chromosomes during cell division, and acrosomal contraction in sperm.

A conspicuous primitive form of motion is that engendered within cilia and flagella. The forms of their movements are varied, ranging from bending relatively rapidly (compared with recovery) in one direction, through propagated bending to make waves, to rotation. The ultrastructural underpinning of all of these movements except rotation comprises highly organized sets of microtubules formed from spirally wound tubulin molecules. There are two central microtubules surrounded by nine doublet microtubules, of which one subtubule (A) of each is linked radially to the central tubules via a central sheath. The circular subtubule also has outer and inner lateral arms of dynein that can join by a peripheral link with the incompletely circular subtubule (B). Dynein is a protein that exhibits Mg-activated ATPase activity. The form of the monomer is an ovoid, about 14×8 nm, that polymerizes to form strands, 40–500 nm long, clusters of which form the outer arms. Ciliary/flagellar motion is associated with ATP splitting, and in glycerol-extracted "model" systems of cilia, the frequency of beating is proportional to ATP concentration. The ATPase activity is stimulated by passive movement of a cilium. A single small passive displacement of a cilium positioned at the end of its stroke initiates a single complete movement cycle (Thurm, 1968).

A different mechanism, present in some cilia and bacterial flagella, is rotation of the basal body. This source of movement must somehow cooperate with tubule energy and movement. Flagella cut off from both the cell body and their basal body continue to beat for several cycles, stopping only following depletion of ATP. There are thus two sources of power, one associated with the basal body, the other with the microtubules. In some forms, one or the other may be the sole source, or both may cooperate to various extents. Only the tubule mechanism is likely to be of interest to us in relation to muscle. Exactly how the energy released is translated into movement by tubules is not yet known, but it must be due either to molecular shortening or to sliding of one element over another of one side at a time. There is no doubt that the microtubules move past each other, since this can be demonstrated by careful comparison of the tubule terminations, examined electron microscopically after rapid fixation in various positions. Actin occurs in flagellar strands of spermatozoa (Burrini et al., 1980), so is likely to be associated with the mechanism.

Microtubules are often present in situations in which conspicuous movement of cellular constituents occurs, but neither bending nor contraction. They are heavily implicated in the transportation of granules and large vesicles down axons. But how is material formed in cell bodies by the Golgi apparatus transported to the axon? Secretory epithelia have the same problem of getting the particles to the sites of exocytosis. As we have seen, some of these contain significant amounts of actin, and possibly also some myosin, which may therefore provide the movement (Burridge and Phillips, 1975). It is hotly debated whether or not the tubules are responsible for powering transport or are merely rails or guidelines. The maximum rate of fast axonal transport is much the same as that of food particles moving along the outside of axopods of planktonic foraminifera! Microtubules occur in secretory cells, where they are oriented at right angles to the release membrane; here the vesicles are lined up in rows, as if waiting to be propelled to the release sites along the tubules. Tubules are conspicuous during the development of many muscle cells and may be precursors of the thick filaments. Whether they are replaced or merely obscured during elaboration of the thick filaments is not known. In some muscle cells, microtubules are still evident in the final developed state.

2.4. Distribution of Actin and Its Involvement in Movement

Although the two proteins actin and myosin are generally found together, the more ubiquitous is undoubtedly actin, which is almost universal. Actin exists in many cells in the unpolymerized state, for example in the membranes of erythrocytes and in neurons, where it could be involved in some way with active transport. What explains the switch to the polymerized state in other cells is unknown. In the polymerized state, actin filaments are very stable and relatively rigid. If force is indeed transferred by cross bridges (X bridges) to them, they must be almost totally inelastic. Negative staining of actin from a very wide range of sources shows the beaded appearance that first became familiar following the work of Hugh Huxley (1963) and of Jean Hanson and Jack Lowy (1963).

There are, nevertheless, unexplained differences in diameter: actin filaments have been described that are barely 3 nm in diameter, for example, in the acrosomal process of sperm of the sea cucumber *Thyone* (Tilney, 1976). Estimates of the diameter of purified actin from muscle have ranged from 4 to 8 nm, with 7 nm most common. Some actins have been subjected to amino acid sequencing (e.g., Gordon and Korn, 1976), and actins from different organisms have been found to be remarkably similar, but far from identical.

Actin filaments are unequivocally recognized electron microscopically by bathing tissue suspected of containing them in myosin solution, the source of which seem inconsequential. The filaments become "decorated" with myosin molecules, giving a characteristic arrowhead pattern (Figure 2.2). Actin is present in filaments in microvilli from

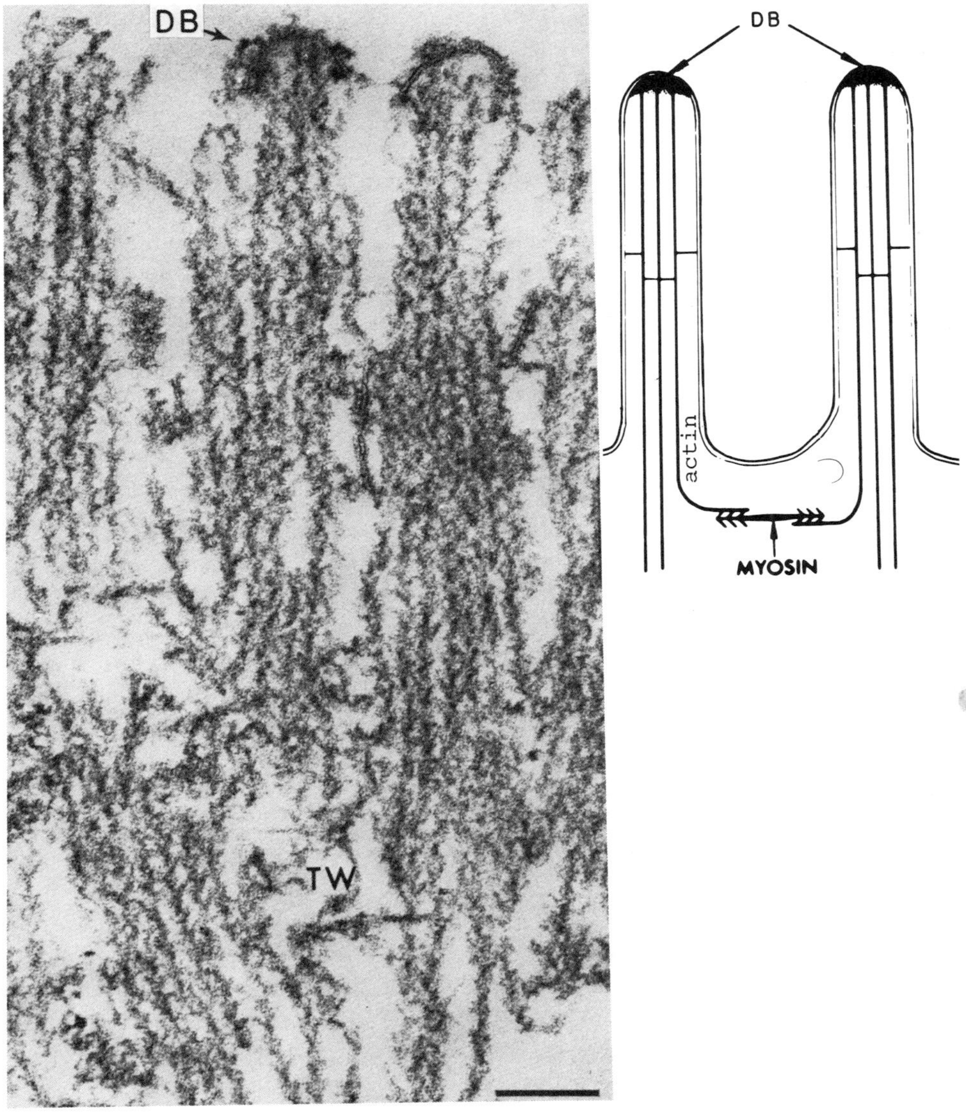

Figure 2.2. (*Left*) TEM of portion of brush border of chicken intestinal cells reacted with myosin. (*Right*) model for the functional organization of actin and myosin in the brush border. (From Mooseker and Tilney, 1975.)

intestinal brush border cells, where they are firmly anchored to the tip. The densely staining points of attachment bear a surprisingly close resemblance to Z bands of vertebrate skeletal muscle, and they have been thought by antibody reaction to contain α-actinin (Mooseker and Tilney, 1975). This is a 2 × 30 nm protein that occurs in vertebrate Z bands (Stromer and Goll, 1972), but cells in tissue culture obtained from human fetal lungs (WI 38) have dense bodies resembling those found in smooth-muscle cells (p. 316). Both contain α-actinin.

In dividing cells, a ring, termed the contractile ring, formed of special cytoplasm underlies the bottom of the cleavage furrow. Tensile forces up to an estimated 10^{-2} dyn occur here. Actin is a prominent component of the ring cytoplasm. In neuroblastoma cells, actin filaments are abundant as well as 10-nm filaments and microtubules, and all also occur in the core of mature neurites (Isenberg and Small, 1978). If microtubules are there to serve active transport and the 10-nm filaments as structural elements, the presence of the actin in addition is mysterious. Perhaps the tubules are only the rails along which material is transported by actin-involved machinery. Retraction occurs in growing tips of neurons when they make pathway errors, and motility of both axon and neurites of a neuron can occur in association with regeneration.

The remarkably stiff, radiating pseudopodia, termed axopods, of the planktonic heliozoan Protozoa are used in propulsion, and they contain actin. Food is moved along the outside of the axopods in some species. In others such as *Actinophrys,* the tips shorten at a contraction speed of about 0.3 mm/s when a particle of food adheres to them (Ockleford and Tucker, 1973). It is possible that myosin is required. Sea urchin eggs contain a lot of actin that is thought to be involved in a very special contraction associated with cleavage. When extracted, it forms a gel that contracts in the presence of ATP. There is myosin in the gel, and when this is removed the extract can still be gelled, but the gel will not contract (Kane, 1980).

Fibroblasts are rich in actin filaments (Ishikawa et al., 1969), and they are very motile; but chondrogenic cells, nerve cells, and epithelial cells are equally endowed yet do not themselves move. Chick fibroblasts have been studied in both stationary and migratory conditions (Badley et al., 1980). When the former were compared with the latter, it was found that both the microtubule and the 10-nm filaments were in similar configurations. But there were marked differences in the configurations in neurons in which it appears the action is involved in bringing synaptic vesicles to their release sites. These observations suggested a skeletal role for tubules, filaments, and actin, but a role in the generation of movement only for actin. The actin itself must have a lot of mobility. Prominent filaments located in fibroblasts parallel to the long axis consist of actin with distributed myosin attached (Figure 2.3). It has been possible to isolate these using microlesions with a laser beam. Irrigation with suitable medium containing ATP causes the isolated bundles to contract (Isenberg et al., 1976).

A seemingly very unlikely source of actin has been found not only to be a rich one but to contain it in a highly organized network; that is the vertebrate eye (Kibbelaar et al., 1980). This finding has led to the suggestion that its function is to alter the shape of the lens to aid in accommodation, but it may be no more than a structural device. An even more regular array of actin has been found in stereocilia of the cochlea of a lizard (DeRosier et al., 1980). Here the actins are cross-connected to each other in a highly regular manner, apparently to make a supporting structure, but one whose stiffness can, in principle, be altered easily.

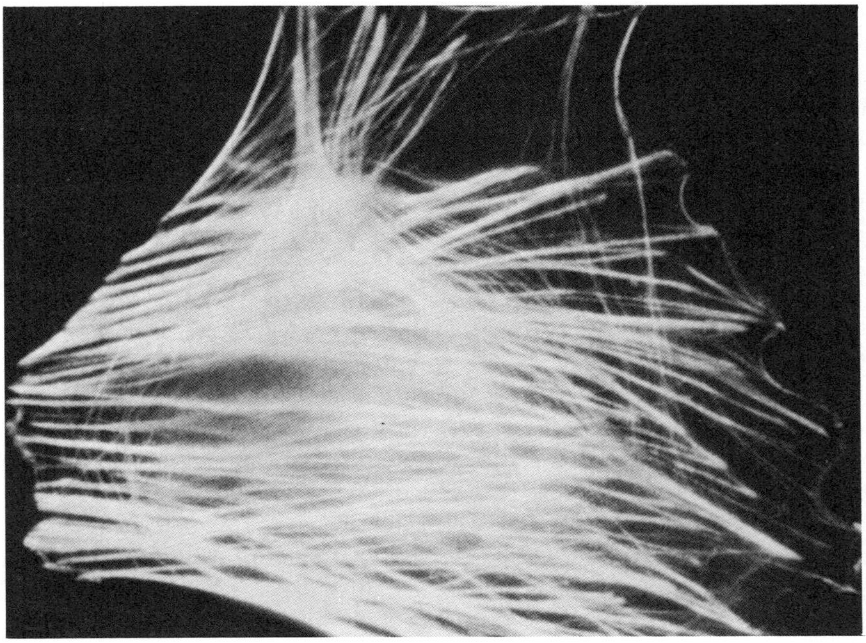

Figure 2.3. Embryonic chick fibroblast stained with heavy meromyosin coupled to fluorescein isothiocyanate to show the oriented array of actin filaments. [Reprinted with permission from Sanger et al. in *Motility in Cell Function.* Copyright 1979 Academic Press.]

A useful tool in the study of actin-related contractile events is the cytochalasins, a family of fungal alkaloids that inhibit cellular processes involving contractile proteins in highly specific interactions without regard to the source of the actin (Howard and Liu, 1979).

It has been claimed tentatively from time to time that more than one form of polymerized actin exists. This is important to know, since all interpretations of the role of actin accept the presumed fixity of the filament length. If a filamentous form of actin possessing elasticity and contractibility were discovered, reinterpretation of many forms of motility would follow.

Actin can exist in monomeric, soluble form (G actin) and may function in this form in association with many membranes, especially those of synapses (Berl et al., 1973). The mushroom *Amanita phalloides* toxin phalloidin combines with G actin and also disrupts some membranes, perhaps because of pulling G-actin molecules toward itself. It has been suggested that the function of actin in synaptic terminals is to produce a local contraction that brings a vesicle toward the presynaptic-membrane release site prior to release (Berl et al., 1973). Actin is present in all brains so far examined and is especially concentrated in the isolated synaptic vesicles and associated membrane material termed synaptosomes.

2.5. Cell Shape and Amoeboid Movement

The earlier view that animal cells in general are membrane-bound bags of liquid (sol) "cytoplasm" is gradually being replaced with the notion that most, including the kaleidoscopic amoeba, have an internal framework (trabecula or cytoskeleton) of filamentous strands. Treatment with osmic acid and a ligand, followed by critical point drying, and use of high-resolution scanning electron microscopy is permitting resolution and identification individually of actin, microtubules, and other filaments, but details are not yet available. The filaments are thought to be elastic, unless they are tubular. The range is from 3 to 10 nm for strands, which are especially prominent in the subplasma membrane "cortical" layer that used to be referred to as "plamagel." Some of the filamentous components may be very dynamic, with rapid reversible depolymerization and repolymerization. Others, especially rather prominent ones averaging 10 nm in diameter, are thought to be stable, stretching from surface point to surface point across the cell contents. These are always referred to as the "10 nm" filaments, and they must be highly elastic. Since they are certainly subject to greater than 10-fold length changes, they should range in actual thickness down to below 3 nm when stretched. One hypothesis proposes that these strands are in parallel with rigid microtubules, which can stretch them as they add molecular units and then use the stored elastic energy to create movement as they subsequently lose units at their ends.

The difficulty of understanding these elementary contractile processes is emphasized when one realizes that there are tens of thousands of published research projects in the field, and hundreds of theories, but nothing definitive that can yet be stated about how amoeboid movements are produced by the networks or other constituents.

It has not been possible to obtain myosin from many of the cell types of which actin is found, and some authors consider that actin is primarily a trabecular structural element. In brain, both are present (Berl et al., 1973), but both are also prominent in the growth cones of developing neurites (Bray, 1973). They are both present in secretory tissues and are especially associated with granules (Burridge and Phillips, 1975), but actin alone is prominent in synaptic terminals. It is possible, therefore, that actin, with possibly various amounts of myosin, is involved in movement generations as diverse as pulling vesicles toward their presynaptic membrane prior to release, active transport, and amoeboid movement.

Amoeba is not as easy to understand as *Paramecium* in regard to the relation between movement and membrane potential, but there is now general agreement that amoeboid movement is brought about by a contraction mechanism. There are many subtle shifts in potential in the giant amoeba *Chaos chaos,* including medium to large spikes (Tasaki and Kamiya, 1964), but they are not clearly coupled to movements, and artificially altered potentials do not lead to clear-cut responses (e.g., van Houten, 1979). However, *Chaos* moves toward the negative pole in a field of 0.9 V/cm, which causes a current of 0.1-0.2 $\mu A/cm^2$ (Nuccitelli et al., 1977). A clear cap develops at the point of entry of the current, followed by initiation of a pseudopod 5-10 s later, in the appropriate direction. This process is dependent on the presence of calcium, as is all muscular contraction.

Two sets of filaments, one about 7 nm in diameter, the other about 16 nm, are present in amoeba cytoplasm (Figure 2.4.*3*), and both are needed together with ATP in cytoplasm for pseudopodial movement (Pollard and Ito, 1970). The thinner ones are likely to be actin; they bind rabbit heavy meromyosin (Pollard and Korn, 1971). The thick ones may be myosin aggregates. A myosinlike enzyme activity is present (Pollard and Korn, 1973). Isolated cytoplasm from the giant amoeba *Chaos* in fluid, elastic (Figure 2.4.*1*), and contractile; its contractility is regulated by calcium ions (Taylor et al., 1973). Contraction is preceded by a marked increase in viscosity, which may be attributed to crosslinks between

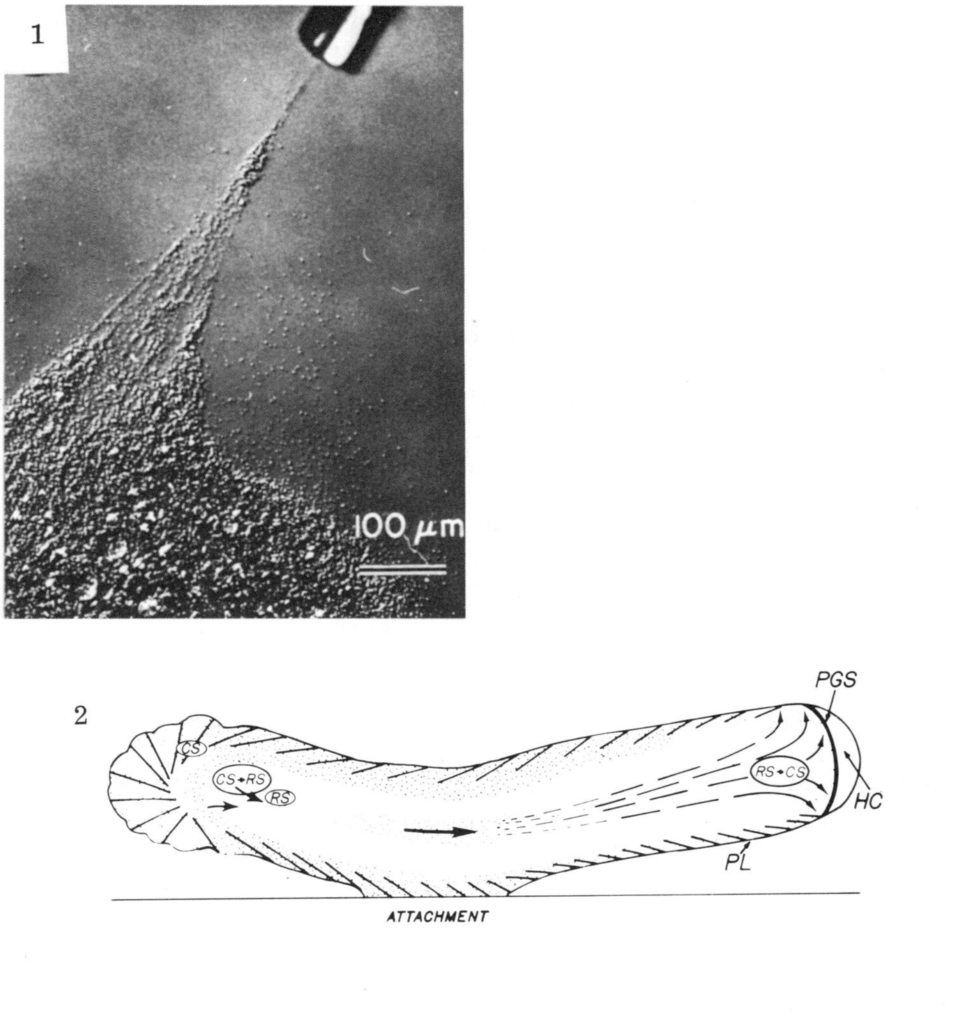

Figure 2.4. (*1*) Demonstration of elastic ultrafine strands in exuded cytoplasm of amoeba *Chaos carolinensis.* (*2*) Diagram illustrating hypothesis of contractility as a cause of amoeboid movement according to the scheme of Taylor et al. (1973). The endoplasm gains in structure as it approaches the tip of an advancing pseudopod. The plasmagel sheet (*PGS*) forms by peeling off the cytoplasmic side of the membrane (*PL*) and passes back over the pseudopod. A transformation takes place from relaxed (*RS*) to contracted (*CS*) state. The ectoplasm is a cytoskeletal as well as a contractile element, decreasing in structure as the endoplasm gains it, possibly due to actin changes. *HC* = hyaline cap. (*3*) EM of parallel 7-nm thin filaments, presumed to be actin, and short thick filament (*arrow*). (From Taylor et al., 1973).

actin filaments (Taylor et al., 1979) and gelation at pH 7.0. At pCa 6, the gel breaks down, contraction occurs, ATP is hydrolyzed, and the pH falls.

Most published researchers since the development of the X-bridge theory for X-striated muscle have regarded actin alone as no more than a combined structural element and myosin ATPase activator. They attribute the power of motion development exclusively to conformational changes in myosin molecules. It is surely premature to accept this as dogma; as Sir Andrew Huxley complained in 1974, it should have only the status of a very promising candidate. Myosin from the variety of cells so far examined at first seemed very similar, but has turned out to exist in more than one form. *Acanthamoeba* has three different forms of myosin, two of which have only a single head, the third having two (Figure 2.5.*1*), like myosin from vertebrate X-striated muscle (Slater and Lowey, 1967). The functional role of conventional myosin in amoebae can only be guessed at, and why there should be three kinds is quite a puzzle. The role of myosin in blood platelets is even more problematical. Within the ranks of two-headed myosin molecules, there are differences in molecular substructure that occur in myosins from different muscles of the same animal having different rates of contraction. A major difference is in the composition of the light chains, but even in a relatively homogeneous vertebrate muscle, the myosin includes several isoenzymic forms (Lowey et al., 1979). Current knowledge of myosin structures is summarized later: it is likely to change rapidly, but so far the differences seem slighter than the similarities and are unlikely to be linked to major qualitatively different functions. Hinssen et al. (1978) compared reconstituted thick filaments obtained from *Amoeba,* the slime mold *Physarum,* and blood platelets of cattle and found them all similar. However, none had a central bare zone or X bridges at their outer tips, as do classically described ones from rabbit psoas (H. E. Huxley, 1963). In this respect they resemble the myosin obtained from vertebrate nonstriated muscle. Amoeba, mold, and some smooth muscle myosin therefore have different ways of assembling molecularly, compared with the cross-striated muscle, in particular making a side-polar short thick filament (Figure 2.5.2).

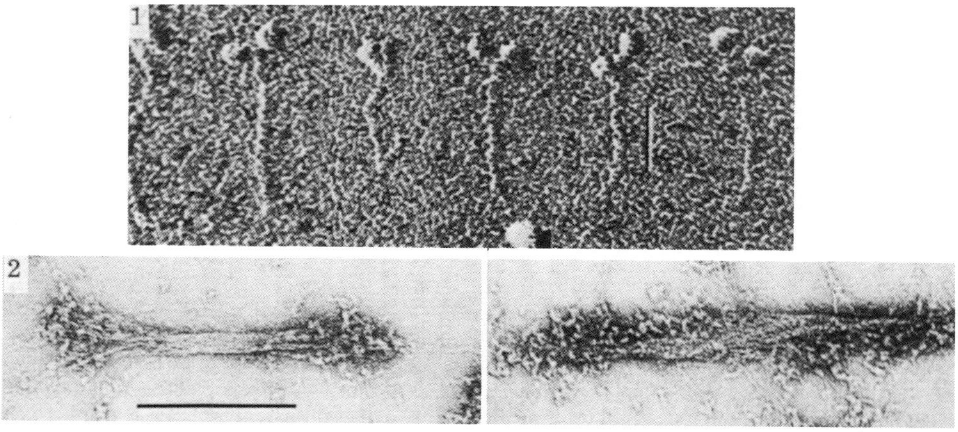

Figure 2.5. (*1*) TEMs of shadowed myosin II molecules from *Acanthamoeba.* Bar = 40 nm. [Reprinted with permission from S. Lowey et al. in *Motility in Cell Function*, pp. 53–67. Copyright 1979 Academic Press.] (*2*) EMS of two different forms of molecular aggregates of calf aorta myosin, negatively stained. Bar = 250 nm. (*Left*) short bipolar. (*Right*) side-polar. [Reprinted with permission from Craig and Megerman, in *Motility in Cell Function,* pp. 91–102. Copyright 1979 Academic Press.]

Nonmuscle myosin may require a cofactor for interaction with actin, but otherwise all myosins share the ability to crosslink actin filaments, even when they only form bipolar filaments with each other (Pollard, 1977). The ratio of myosin to actin in nonmuscle cells can be as little as 1 molecule of myosin to 100 of actin. In such proportions the maximum force that could be developed would be very small, and much of the actin, as it were, wasted. However, for many cellular processes, little force is needed, but there is a real need for there to be a high probability of its being developed when needed. The only way to do this without the risk of too much force being developed is indeed to have a high actin-to-myosin ratio. It is also possible that the actin has other functions, especially structural, or that widely distributed actin is necessary to disperse the motion. A way in which scattered myosin dimers might act in conjunction with scattered actin to promote motion is proposed in the solation hypothesis of Hellewell and Taylor (1979). Scattered myosin but close, long, parallel actin filaments, is the norm in many smooth-muscle cells.

There are other problems regarding regulation of actin–myosin interaction in nonmuscle cells; regulation is achieved in vertebrate striated muscle by "protection" of the actin by tropomyosin. The latter must react with calcium to expose the interaction site. In many invertebrates, as I describe in detail later (p. 224), there is no such protection, but instead, the myosin itself is prevented from interacting until after calcium has interacted directly with it. Some classes of animal have both mechanisms.

2.6. Twitchlike Mechanisms of Specialized Ciliated Protozoa

All students of elementary biology are familiar with the pond-living ciliates *Vorticella* and *Stentor*. When observed quietly, the ciliated nuclear body is extended on a long, thin stalk. A slight disturbance causes a very rapid withdrawal, followed by slow extension. In *Vorticella* the withdrawal is associated with a coiling of the stalk that is not seen in *Stentor*. The *Stentor* stalk contains two distinct systems of about equal size, roughly 0.6 × 2.0 μm running the entire length of the stalk (Bannister and Tatchell, 1968; Huang and Pitelka, 1973). Both systems shorten, thicken, and remain straight, but they behave differently on relaxation. One system, known as the kinetosome (Km) system, consists of a beautifully regular array of rows of interconnected microtubules (Figure 2.6). These start their relaxation after the other, M, or myoneme (also called a spasmoneme), system, causing the latter to be thrown into folds at first as they relax. The M system is composed entirely of a single type of very fine 2-nm diameter ultrathin filaments, closely packed together (Favard and Carasso, 1965). These filaments shorten and thicken during the contraction.

The highly regularly arranged stacks of microtubules constituting the Km system each contain about 21 tubules grouped in a 2 + 19 pattern. X bridges occur between the microtubules of adjacent stacks. In a cross section of an extended *Stentor* there are fewer stacks than in a contracted one. Hence their contraction is associated with the sliding of the stacks over one another. The mechanism in the M system fortunately can be studied in depth because of the existence of a giant of its kind. *Vorticella* has a colonial relative, *Zoothamnium,* in which the cluster of a few hundred zooids, each with its own tiny stalk, is served by one giant extracellular stalk more than 2 mm long (Figure 2.7). This is a remarkable transparent tube containing a translucent thread 1 mm long and 30 μm in diameter, the spasmoneme, which is a fusion of tiny extensions from the myonemes of

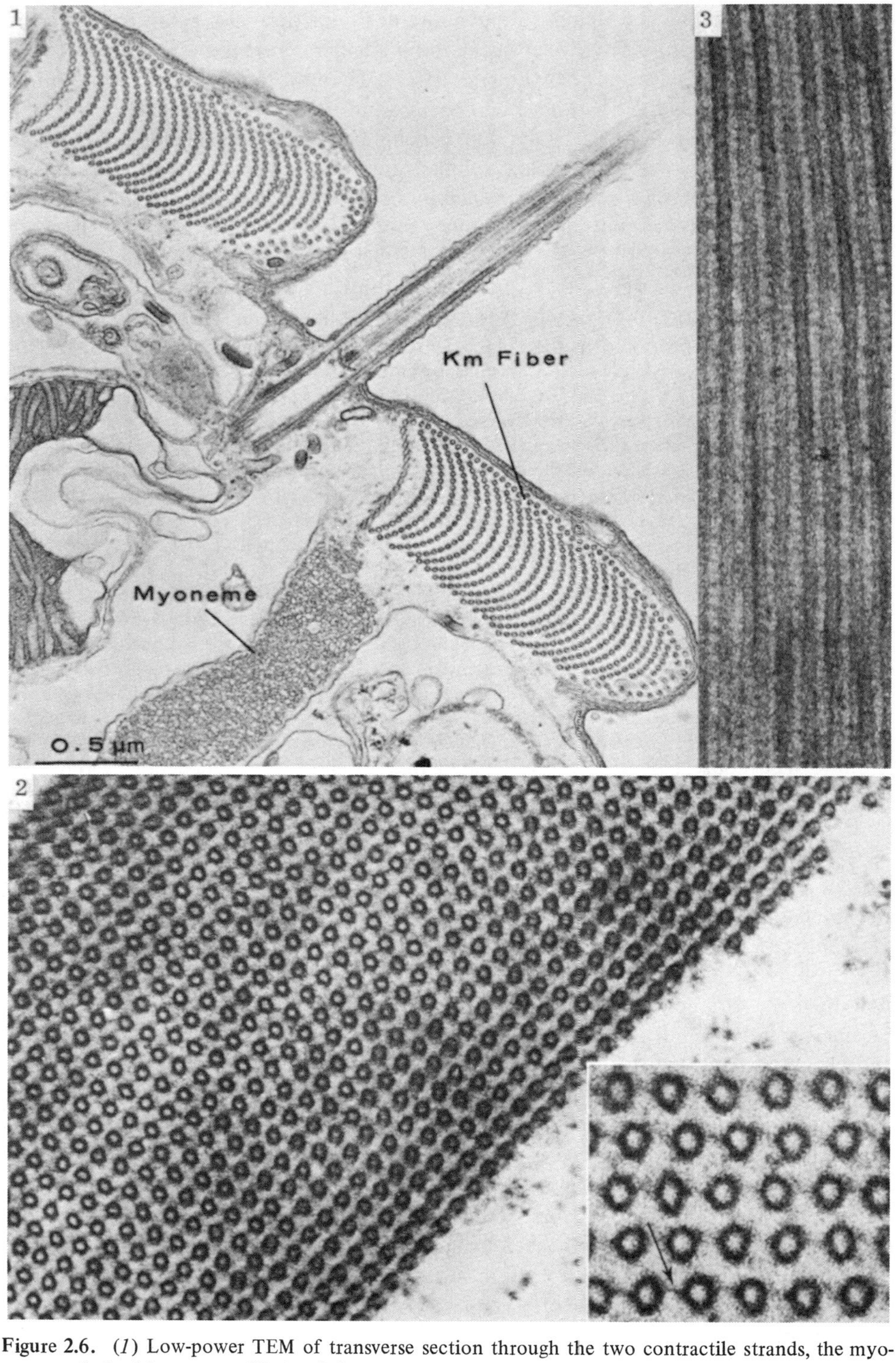

Figure 2.6. (*1*) Low-power TEM of transverse section through the two contractile strands, the myoneme and the kinetosome (Km), of *Stentor coeruleus.* (From Huang and Pitelka, 1973.) (*2*) Transverse (*above*) and longitudinal (*below*) EM sections of contractile axostyle of *Saccinobaculus.* Inset shows bridges (*arrow*) between tubules. (From Grimstone and Cleveland, 1965.)

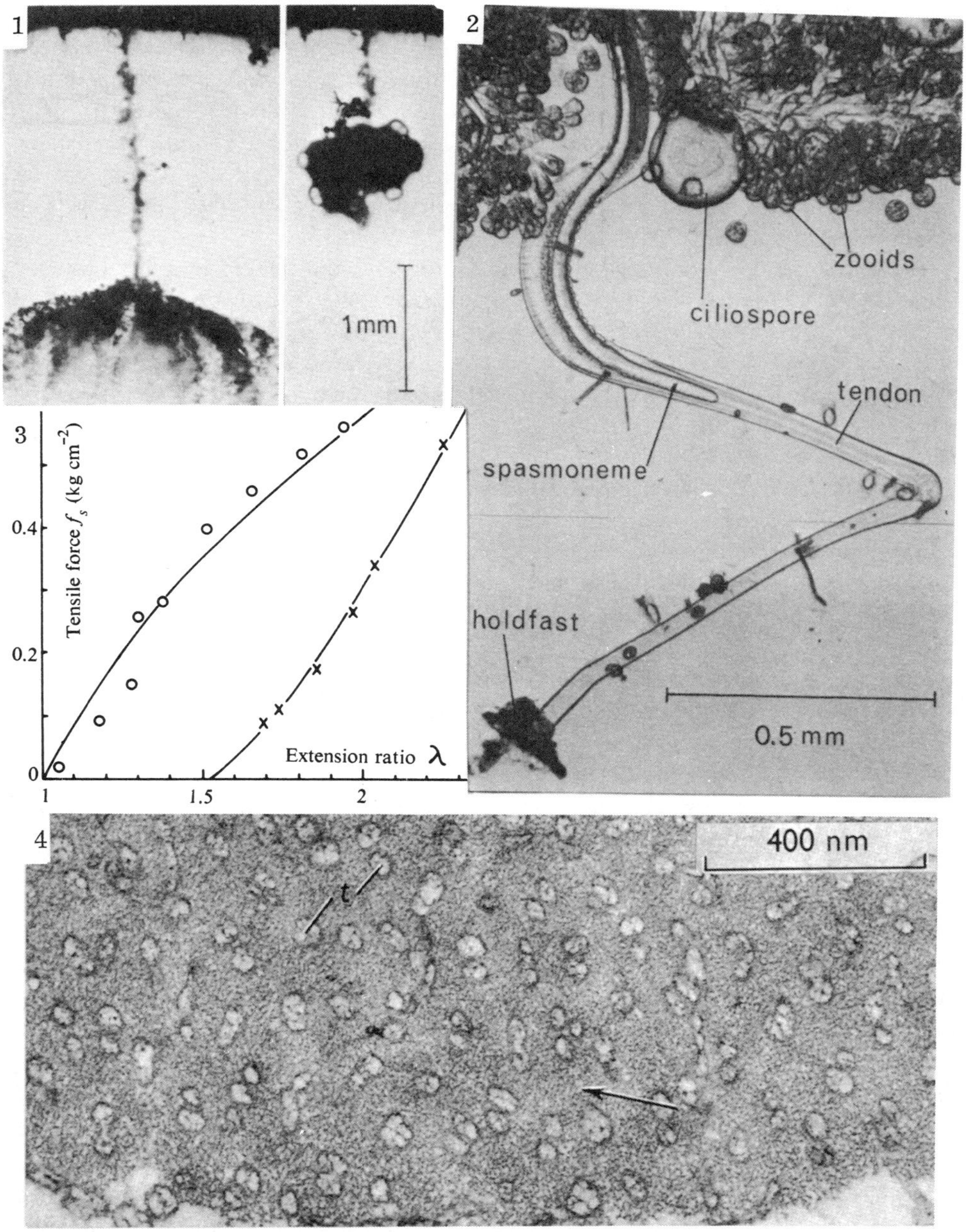

Figure 2.7. (*1*) Living colony of the protozoon *Zoothamnium* in relaxed and contracted states. (*2*) Upside-down enlargement of the contractile apparatus of the stalk. (*3*) Graph of relationship of force to extension (as ratio) at two different calcium concentrations. pCa5 – *open circles;* pCa8 – crosses. [Reprinted with permission from Weis-Fogh and Amos, *Nature* **236**, 5345-5348 (1972). Copyright 1972 Macmillan Journals, Ltd.] (*4*) EM of transverse section through contractile spasmoneme of *Vorticella.* The fine, evenly dispersed filamentar contractile material is permeated with longitudinally running tubules (*t*), which may serve as stores for activator calcium, that are functionally equivalent to sarcoplasmic reticulum. (From Amos, 1972.)

individual zooids and is formed of the same material. At its distal end the spasmoneme is attached to a "tendon" that in turn attaches to the tube near its middle. Close to this point the tube is weak, so that when the spasmoneme contracts, it readily bends there, resembling in action flexion at the tibial joint of a long-legged insect. Torkel Weis-Fogh and W. B. Amos (1972) made an isolated preparation of the spasmoneme and found that it is only necessary to raise the calcium-ion concentration in the medium to $10^{-5}M$ to cause a maximal contraction. Relaxation occurs following return to a solution containing 10^{-8} M calcium. The speed of shortening is remarkable, 110 lengths/s, five times faster than fast vertebrate muscle. The extent of shortening is to about 45% initial length *in situ,* more when isolated.

The spasmoneme shows long-range elasticity in both the relaxed and the shortened state. It could be stretched up to four times its initial length, with complete elastic recovery, even when fully activated, with physical properties resembling those of a slightly vulcanized rubber. It has a more highly organized molecular structure in the shortened state, but reextends spontaneously, without the need to be pulled out. The fine structure is the same as that of the M strand of *Stentor,* comprising clusters of 2-nm diameter filaments. There are, however, longitudinal tubules interspersed among them (Amos, 1972). Although the activation mechanism is not yet known, it seems probable that it will be found to be due to calcium entry and release from these tubules.

Two kinds of action have been observed in *Vorticella:* local, slowly spreading contractions, and the familiar all-or-none rapid ones that always involve the whole stalk (Sugi, 1960). The latency of the latter is 1-3 ms and full shortening takes only 9 ms. By observing the rate of initial straightening of slightly kinked M strands, it was found that the stimulus travels at rates between 5 and 25 m/s (Newman, 1972). Local reactions could be observed, but electrical stimulation provoked only the all-or-none responses.

In fully contracted myonemes of *Stentor,* Huang and Pitelka (1973) found numerous 10-12 nm diameter microtubules that were not there in the extended form. They suggested that in the process of shortening, the ultrafine filaments must be "transformed into tubules." What is surely most likely is that the filaments themselves shorten, and that below a certain range, further shortening occurs by coiling. This could certainly account for their apearance in both transverse and longitudinal sections of contracted specimens. Tubules have not been found in contracted forms of other species (e.g., Lehman and Rebhun, 1971), but that could mean that the contractile filaments have not shortened quite so much. The existence of contractile ultrathin filaments in Protozoa makes such filaments a prime candidate for involvement also in muscular contraction. However, the shortening is the result of a direct action of calcium. It does not involve ATP splitting. The principal energy requirement is therefore in elongation, not contraction.

Huang and Pitelka (1973) attempted to explain the need for the dual systems of microtubules and myoneme by hypothesizing that the tubule system, working relatively slowly and utilizing ATP hydrolysis for energy release, is there to extend the shortened filaments (also see Routledge et al., 1976). Another situation in which microtubule-generated movement has been invoked to explain extension is in elongation of the mitotic spindle during anaphase (McDonald et al., 1979).

Some flagellate Protozoa, such as *Oxymonas,* contain a ribbon-shaped contractile structure called the axostyle. This produces a strong undulatory movement with little actual shortening. It is composed entirely of neatly stacked cross-connected microtubules (Grimstone and Cleveland, 1965).

2.7. Contraction Associated with Ultrathin Filaments

The thinnest filaments ever to have been firmly claimed as responsible for contraction are the 2-nm ones of *Stentor*. Amos (1972) placed the diameter of the *Zoothamnium* ultrathin filaments higher than the 2-nm estimates for *Stentor* myonemes, roughly 4 nm from electron micrographs. The protein of the spasmoneme, spasmin, appears to comprise two units of molecular weight 20,000. Assuming 810 daltons/nm^3 density, the diameter should be about 3.6 nm. It would, of course, be half this for single units.

Ultrathin filaments of the same order of size are responsible for locomotory movements of the axopods of *Sticholonche,* a heliozoan. Each axopod has a minute ball-and-socket articulation at its basal ends that is rotated in its socket to propel the cell through the water. This is done by reciprocally acting bundles of ultrathin filaments (Figure 2.8), 2–3 nm in diameter, attached to the base (Cachon et al., 1977). In the shortened state, the bundles are about $\frac{1}{10}$ their relaxed/stretched length. It was not possible to resolve any details, but there were no tubular forms.

2.8. Other Forms of Movement

2.8.1. Ascidian Tail Resorption

A great deal of work has been done on an interesting contraction that occurs in the tail of ascidian tadpoles. Since ascidians are protochordates, they occupy a lofty place phylogenetically and have received attention to match. The whole of the tail is pulled in toward

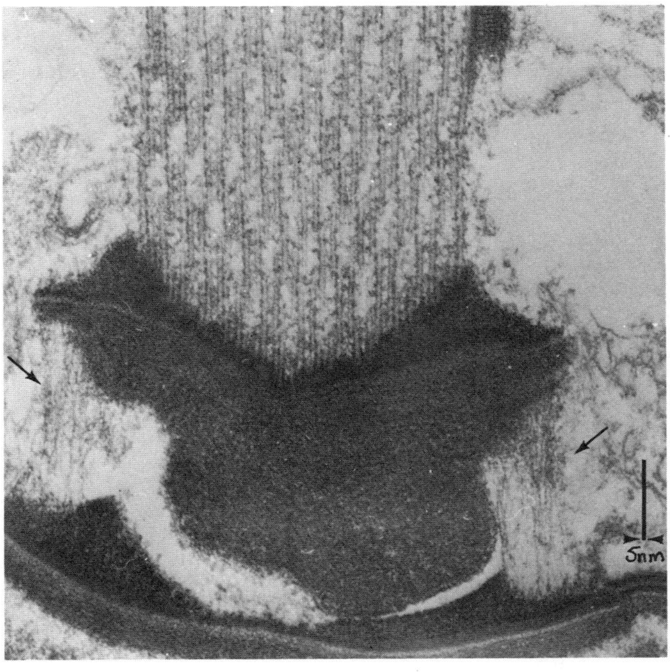

Figure 2.8. Section through the base of an axoneme of the swimming protozoan *Sticholonche.* The dense head of the joint articulates on a capsule. The axoneme is operated by reciprocal quick contractions of antagonistic bundles of ultrathin filaments (*arrows*). (From Cachon et al., 1977.)

the body soon after the free-living tadpole takes up its future sedentary life, the full extent of shortening being to $\frac{1}{20}$ initial length. This is accomplished by a thin sheet of caudal epidermis, 1.1 mm long by only 0.5–5 μm thick. The contractile cells contain numerous coiled or folded fine filaments attached to the plasma membrane at an apical contact cone (Cloney, 1966).

2.8.2. Chromatophore Pigment Dispersion

Widespread among diverse animal groups are chromatophores: cells containing black, red, blue, green, or brown pigment granules that can be dispersed or converged to control the animal's color. The mechanism of granule movement, although slow and steady, at about 10 μm/min, at least matches that of elemental units in some smooth muscles. In all cases the movements are associated with radially arranged microtubules attached to dense bodies at the center. The granules move outward along lines of orientation of the tubules. The lines are, however, not static, but readily break down and equally readily reassemble, each step taking 4–5 s. One theory holds that pigment aggregation is brought about indirectly by tubule disassembly. When the tubules assemble, they are thought to expand the cell, creating strain in elastic filaments of the trabecular network; their breakdown releases this stored elastic energy, drawing the granules toward the center. Others (e.g., Murphy and Tilney, 1974) consider that the granules slide along the microtubules.

2.8.3. Chromosomal Movements

Of greatest general biological interest is the movement of chromosomes during cell division, but in spite of intensive research over the last 30 years by the world's leading cytologists, no mechanism has been agreed upon. Microtubules are definitely involved, but ideas on their mode of operation range from musclelike, with sliding filaments impelled by cross bridges (McDonald et al., 1979), to a complex operation resembling opening a zipper (the "zipper" theory – Bajer, 1973).

2.8.4. Blood-Platelet Plugs

Although the early stages in the formation of blood clots is still not fully understood, blood platelets are certainly responsible. Before the formation of fibrin, blood platelets stick to endothelia of the wound area and after enough are present, suddenly form a plug in which they change shape. Actin and myosin present in the platelets are presumably responsible for this abrupt change, and also for the retraction of the clot that occurs later, after fibrin formation.

2.8.5. Nuclei in Fused Cells

In cultures of hamster kidney cells, the cells tend to fuse, after which the nuclei move, aligning themselves in tightly packed rows in association with 10-nm filaments (Wang et al., 1979a).

2.8.6. Streaming

The most ubiquitous of all cellular movements is cytoplasmic streaming. Before the discovery of actin and myosin in a very wide variety of nonmuscle cells, no plausible theory had been offered for its "generation." With these discoveries, actin/myosin soon became implicated, with Ca serving as activator of Mg actomysin ATPase to provide energy and X-bridge rotation to provide the movement. The actin strands involved would have to be ones that are not attached to the cell membrane or organelles, but able to move around freely. Support for this view came with the demonstration that a movement that closely resembles cytoplasmic streaming, including spontaneously reversing direction of flow, occurs when actin, myosin, calcium, and ATP are mixed and sucked into a fine capillary tube (Oplatka and Tirosh, 1973). It stops only when ATP is fully used up and resumes when more is available. The actomyosin may come from a rabbit or a slime mold with similar results. If the actin is attached in a system that is not otherwise organized, actomyosin formation and activation is only going to lead to increased rigidity or local movement. This is precisely what is thought to happen in macrophage (Stendahl et al., 1980), certain stages in slime molds (Hellewell and Taylor, 1979), blood platelets, and amoebae (Pollard, 1977). Regulation of the interaction in these may be by tropomyosin in *Acanthamoeba,* by phosphorylation in platelets, and by calcium directly in others.

Cytoplasm movement is retained outside a cell (Allen et al., 1960) if ATP is available and the temperature is sufficiently high (Thompson and Wolpert, 1963). The viscosity of such cytoplasm is proportional to the number of actin and thicker, possibly myosin, filaments present (Pollard and Ito, 1970).

2.8.7. Phagocytosis

Immunofluorescence of actin and also of myosin concurrently, have been followed during particle phagocytosis (Painter and McIntosh, 1979). Both proteins move rapidly toward, and accumulate at, the sites of uptake. The migration was not prevented by cytochalasin B, but pseudopod extension around the particle was. Following ingestion, the actin and myosin leave the site and cluster around the nucleus.

2.9. An Important "Muscle" That Does Not Contract

The notochord of the lowly, but evolutionarily very important, acraniate chordate *Amphioxus* is not contractile (Von Ebner, 1895), but if it were classified solely on the basis of its fine structure, it would be called a muscle. It is even in a very unique class, being cross-striated, but with the remarkable sarcomere length of 33 μm (Flood et al., 1969). This length is equaled, though not exceeded, only by muscle in the proboscis of a tiny Syllid worm (see p. 444). The "muscle" cells are innervated by vesicle-containing cholinergic efferent nerves and the subsynaptic region contains acetylcholinesterase.

When the cells are excited by the nerve, however, no contraction occurs, but the stiffness increases drastically. The muscle cells contain large numbers of double-tapered filaments, with diameter from 10 to 150 nm, that possess a giveaway periodicity of 14.5 nm. These are clearly paramyosin filaments (see p. 120). There are also numerous 5-nm actin filaments. The stiffness developed relaxes only slowly, and its rate decreases following longer or repeated stimuli. There is marked summation, increasing the total amount of stiffness, with repeated stimulation at constant strength. I know of no function this mechanism can serve: it is too slow to help in swimming, but it raises many interesting questions.

2.10. Candidates for Supporting Structures – the "Cytoskeleton"

For the attachment of contractile proteins to other cellular elements, we have only one option, but an outstanding one, α-actinin, found in association with actin and its binding to the plasma membrane in even such nonmuscle cells as intestinal microvilli and cells lining lungs.

To provide a trabecular framework, the so-called 10-nm filaments, sometimes called tonofilaments, are the most likely. There are also Keith Porter's (Porter, 1976) range of "dynamic" filaments, from 3 to 6 nm, some of which can merge, with no obvious structural attachment, with subplasmalemmal cytoplasm.

There are so far no molecular candidates for M-band protein or for actin–actin cross-links, which have been seen between actin filaments in nonmuscle cells as well as in a variety of muscle cells.

2.11. The Candidate Molecular Mechanisms for Contraction of Muscles

We may now review the various options that are likely to have been available for incorporation and modification into presumptive muscle cells at the beginning of their evolution. This begs the unresolvable question, of course, whether the known mechanisms of nonmuscle cells available today can be equated with the historically significant ones. But biochemical mechanisms have been highly conservative, so this is a reasonable assumption. From the survey of mechanisms present in nonmuscle, the set of options laid out below (1–4), emerges for the elemental contractile apparatus. Options 5 and 6 have been added because they are physically possible in any living system rather than because there is any convincing evidence for them in examples of nonmuscle contractility. The options are as follows:

1. Movements of bridges attached to an otherwise stiff tubular element formed of polymers of a small elemental unit.
2. Change in length of a tubular element by assembly or disassembly or additional units.
3. Movements generated by a fraction of the long, thin, high-molecular-weight protein myosin, for example by rotation of its "head" section, or shortening of the 140 × 2 nm "tail" section. This molecule must work in conjunction with a molecular rod, especially actin in its filamentous form.
4. Length change by molecular folding or coiling in an ultrathin protein filament of as yet biochemically undefined type, but resembling the elemental unit of spasmin found in certain protozoa.
5. Repulsion and/or attraction in the longitudinal axis alternately caused by electrical-charge change.
6. Repulsion and/or attraction in the radial axis alternately caused by electrical-charge change.

A major objection to option 1 is that in axons, where tubules are believed to be used for transport, there is no sensitivity to calcium ions. Even after 17 h in saline containing zero Ca^{2+}, fast active transport continues normally in frog sciatic nerve (Edström, 1974), and raised calcium depresses it. Also, the rates of movements associated with microtubules are low, though quite fast enough to account for some muscle contractions. There are two basically different types of movement, one of which is the passage of material along the tubules, which occurs at rates of not more than 1.4mm/min. The other is the relative sliding of sets of tubules past each other, which is theoretically about twice as fast. Use of the latter is seen in flagellae of some protozoa, axostyles of others, and stalk elongation of peritrichs like *Vorticella.*

Option 2 is slower still than option 1, but microtubules are quite common in some muscle cells, and no function has yet been assigned to them. They might be involved in overall length regulation or just be for stiffness, but could be involved in contraction, relaxation, or both in slowly contracting muscle cells.

Option 3 is firmly behind current dogma, which regards it as the only mechanism operating not only in all muscle cells but also in a majority of nonmuscle cells for a variety of functions including transport, contraction, streaming, and the movement of intracellular organelles.

Option 4 should, in my view, be given serious attention both as a major component of contraction in some muscles and as a cooperant or accessory in others.

Options 5 and 6 seem unlikely to be primary mechanisms in any known muscle cells, but could be cooperative in either contraction or relaxation, or both if charge reversal occurs, in any or all.

The greatest apparent muscle-length range for the animal kingdom, seen in coelenterate fishing lines, may not represent a true capability of *relaxation* over such a large length span, since the tentacle is hollow and contains circular myoepithelial cells. But there are many invertebrate muscles for which there is no antagonist to promote elongation after supercontraction, yet the muscle naturally and quickly extends from it. For these, an *active elongation* process such as the molecular reversal that occurs in spasmin, or reverse sliding such as has been postulated for some microtubules, is required. Radial elastic forces undoubtedly play a major role in a majority of relaxations, but are not capable of explaining them all. We should be alert to the possibility of coexistence with activable shortening elements of active relaxation (or elongation) processes; i.e., of an independent activable elongating system in some muscle cells.

2.12. Origin of Control Mechanisms

No membrane-potential changes have been detected in association with ciliary (flagellar) beating, but adequate depolarization, at least of *Paramecium,* dramatically affects the cilia, causing reversal of their direction of beat. Suprathreshold polarization produces voltage-related enhancement of the beat. The protozoan has taken advantage of this (Figure 2.9). Touching its anterior end leads to depolarization, proportional to the force, and in turn to graded electrogenesis, which, if suprathreshold, causes ciliary reversal (Eckert and Naitoh, 1972). The same mechanical stimulus applied to the posterior end causes hyperpolarization, proportional to stimulus strength, and graded ciliary-beat enhancement. The strength of reversed beating was found to be proportional to the amount of calcium ions entering in association with the depolarization. Charge transfer per se does not effect the beat, because it is totally ineffective in the absence of external calcium. The calcium channels of *Paramecium* are localized to that part of the surface membrane that surrounds the cilia (Dunlap, 1977). Barium ions convert the quasi-

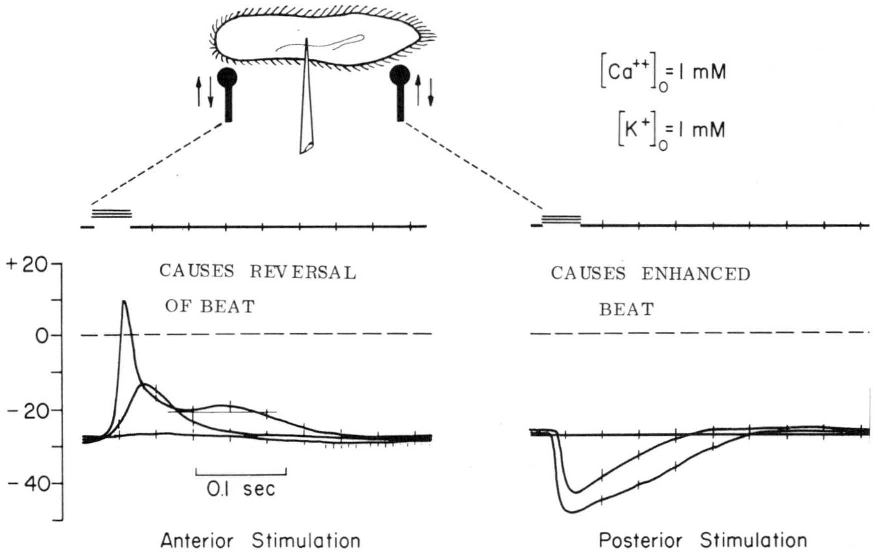

Figure 2.9. Responses to mechanical stimulation of anterior and posterior ends of *Paramecium.* Touch at front end causes graded depolarization (due to increased calcium conductance) and reversal of ciliary beat if above threshold. Touch at rear end causes graded hyperpolarization (due to increased potassium conductance) and enhanced ciliary beat. [Reprinted with permission from Eckert and Naitoh, Bioelectric Control of Locomotion in the Ciliates. *J. Protozool.* **19**, 237–243 (1972).]

regenerative inward calcium current to an all-or-none spike, exactly as in many invertebrate muscle cells. But the membrane itself is not required for movement. Membraneless models made by extraction with Triton X-100 work well; requiring only the addition of Mg^{2+} and ATP, they move forward when the free Ca^{2+} is $< 10^{-6}$ *M* and backward when it is $> 10^{-6}$ *M* (Naitoh and Kaneko, 1972). Pressure injection of calcium into live *Paramecia* also reverses the beat (Saiki and Hiramoto, 1975).

Hyperpolarization in this system does not simply reverse the calcium entry. Instead it promotes an increase in conductance to potassium ions. The inside of the cell is already fully loaded with potassium ions, so increasing the conductance is only going to accelerate their leaving. However, the increased potential somehow promotes the normal beating, presumably enhancing the rate of splitting of ATP (by unknown means) and increasing the frequency, and possibly also the power, of the stroke. It is important to bear in mind that this second set of processes for adjusting intracellular biochemical machinery was available to emergent muscle-type contractile mechanisms.

Graded electrogenic responses to suprathreshold depolarizations, due to activation of voltage-dependent calcium channels (see p. 185), are of special significance for us. This was probably the first form of electric responsiveness to be evolved and is likely to have been the starting point for evolution of the process by which a muscle cell is excited. The direct entry of calcium ions, and especially the ensuing rise in calcium concentration in the cell, provided a direct link with the environment that could be turned to functional use as an initiator of contraction or modulator of intrinsic activity.

The origins of both electrical/mechanical coupling and extrinsic control of contraction are likely to have been in preexisting mechanisms such as those seen today in *Paramecium.*

3

Forms of Muscle Cells

The contractile elements within the muscle fiber are fibrils, which are present in great numbers, in a more or less perfect parallel arrangement. W. F. H. M. Mommaerts†

The muscle fiber contains no preformed fibrils and the contractile matter forms one continuous mass. A. Szent-Györgyi‡

3.1. Origins of Muscle Cell Types

The first contractile cells to evolve that served a specialized role cooperating with other specialized cell types were equivalent to ones present in today's sponges, termed *myocytes*. These cells are not innervated, since sponges do not possess any nerve cells. Each acts independently under local contact, stretch, and chemical control, and cannot be said to be collectively a true muscle. The first primitive muscles were probably evolved independently by diploblastic organisms ancestral to today's coelenterates and other phyla. These muscles comprise(d) extensible connective tissue in which contractile cells, or parts of cells, are (were) aligned at right angles to each other. One set is parallel to the longitudinal axis of the body or organ, forming *longitudinal muscle.* The other is at right angles to the first and therefore runs around the body or organ, forming *circular muscle.* The two act antagonistically to cause shape changes and bending, and, in a coordinated, reciprocal manner, to cause such movements as peristalsis in gut and swimming in some sea anemones (Figure 3.1). This type of musculature has been retained by a wide range of organs, including gut, bladder, spleen, and uterus, as well as the body walls of soft-bodied animals including Coelenterata, Platyhelminthes, Annelida, and Mollusca. The most primitive component cells of all these are bifunctional ones termed *musculo-epithelial* cells, which fortunately are still very much around and can be examined not only in living coelenterates but also in some human organs. The two primitive types are described independently.

3.1.1. Sponge Myocytes

Sponges, with only a single cell layer, are but little advanced above the level of colonial organisms composed of relatively independent protozoalike cellular constituents. It is fairly firmly established that they lack nerve cells, but they do contain contractile cells, termed myocytes, especially around the openings into the body cavity. These cells are

† W. F. H. M. Mommaerts, *Muscle Contraction,* Interscience, New York, 1950.
‡ A. Szent-Györgyi, *Chemistry of Muscular Contraction,* 2nd ed., Academic, New York, 1951.

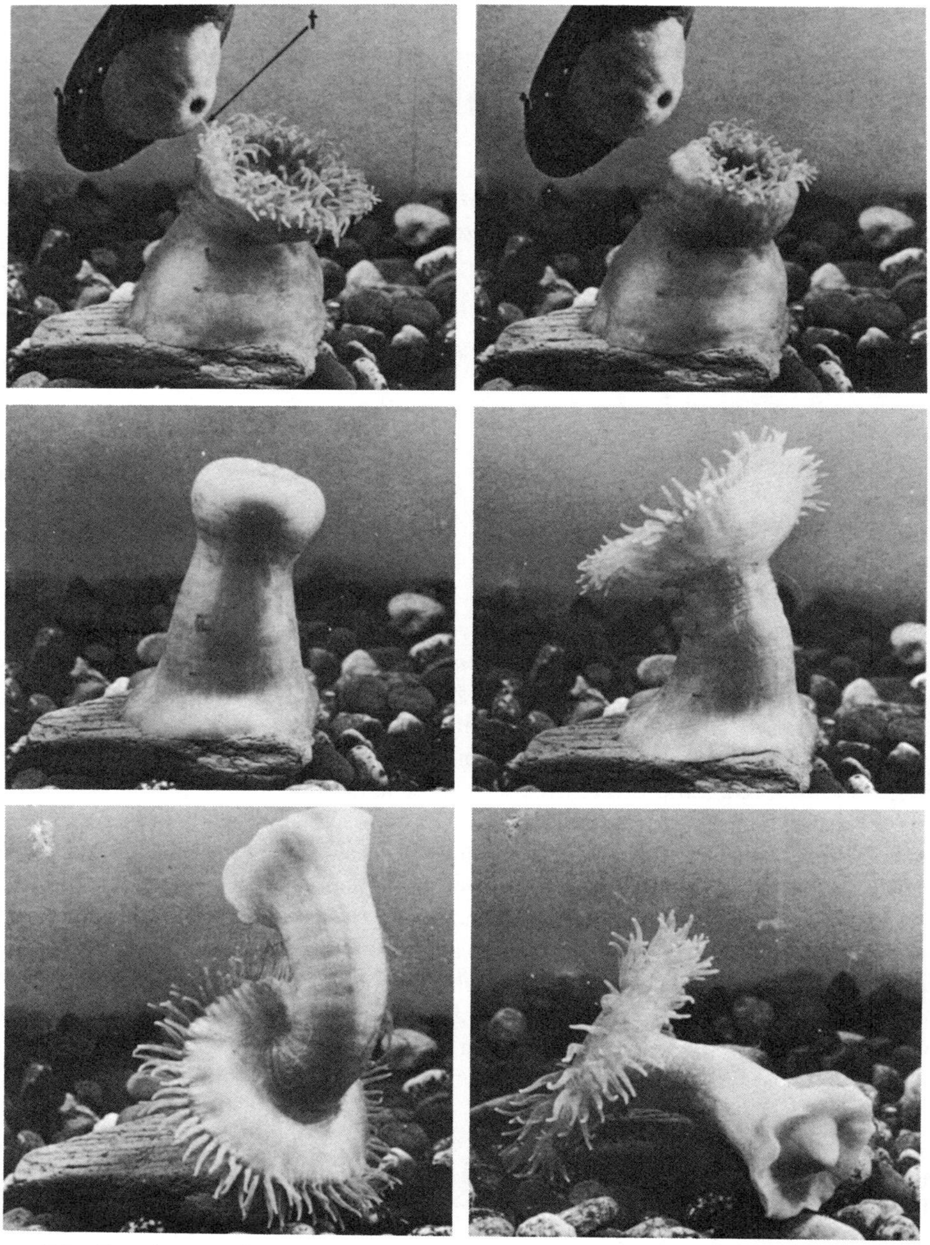

Figure 3.1. Complex coordinated movements of an animal having a diffuse nerve net and lacking brain or ganglia. Swimming escape response of *Actinostola* when but a single tentacle (*t*) was touched by another species of anemone, *Stomphia coccinea*. (From Hoyle, 1975a; original by Ross and Sutton, 1967.)

spindle-shaped, about 2 × 50 μm, and they appear to contain myosin ATPase. Unfortunately, the published micrographs, of two species (Bagby, 1966), are not of high quality, and fixation was poor. This is a pity, in view of the important finding that in one, *Microciona prolifera,* they contain clusters of longitudinally arrayed 25-nm thick filaments in a fairly regular array. Also present are 6-nm thin filaments attached directly to the surface membrane. In a few places they seem to be present in orbits of 6–8. Possible arrays with more than 8 may be myosin bridges. These cells undoubtedly represent the most primitive situation in which true muscle-type contractile material exists. Evolutionary theories that derive coelenterates from spongelike ancestors are strongly supported by this finding.

3.1.2. Musculoepithelial Cells

The cellular elements that provide the contractility of the most primitive tissues recognized as discrete muscles also have other functions, and contraction is not necessarily their primary one. Collectively they are called musculoepithelial cells, and they occur in all polypoid coelenterates, where there are longitudinally oriented tails containing the contractile material in otherwise ectodermal cells and circularly oriented tails associated with endothelial cells that subserve digestion (see p. 423). These cells merge into ones in which the epithelial component is slight. Such cells occur in Coelenterata, Platyhelminthes, Nematoda, Annelida, Mollusca, some Echinodermata, Onychophora, Aschelminthes, and Protochordata. These phyla have few muscles that are as well defined as skeletal muscles; they are principally retractors of extrusible pharynxes or penises necessarily attached to soft tissue and are in part constituents of a larger functional unit. They are also found in contractile soft tissue of mammals, in the sweat glands, the salivary glands, the lung, the lining of small arterioles of the brain (Giacomelli et al., 1970), and especially in the nipple of the female. They have not been well described either functionally or ultrastructurally in vertebrates or most invertebrates.

In no known forms is the contractile material of musculoepithelial cells collected into transverse bands (cross-striations), and the numbers and sizes of myofilament clusters are always small, so that the contractile portion is never more than 1 to a few micrometers in diameter. The myofilament clusters are referred to by some authors, for example West (1978), as myonemes. I think it best to reserve this term for contractile elements of Protozoa. The predominant filaments are 5–7 nm, and presumed to be actin; they are attached to dense bodies structurally resembling hemidesmosomes, all or most of which are associated with the plasma membrane. Thus the contractile elements of these cells closely ally themselves with those of vertebrate unstriated muscle in general.

Musculoendothelial cells from the midgastric region of *Hydra* contain, in addition to actin, microtubules aligned just above myofilament clusters, and many dense-core vesicles (West, 1978). The latter may be associated with gastric secretion or with controlling the motility. Mainly thin, 5–7-nm diameter, filaments occur in the nonstriated cells, and only occasional thicker ones of indeterminate length.

3.1.3. Independent Muscle Cells in Coelenterates

Elongated muscular cells, lacking epithelial or endothelial portions or dual function, also occur in all coelenterates. These are never of large diameter, commonly about 1 μm at rest length. Their length is unknown. Some sea anemones attain lengths greater than 60 cm, and large jellyfish are more than 100 cm in diameter. All such muscle cells in sea anemones are nonstriated, but in jellyfish, in addition to nonstriated ones, cross-striated cells occur in a ring near the margin (Figure 3.2). Contraction of the striated cells is responsible for the quick-twitch contractions used to propel the jellyfish.

3.1.4. Classification of Muscle Cells

No traditional system of classification of muscle cells is truly satisfactory, principally because all such classifications were set up a long time ago without the benefit of the ultrastructural information now available. We can expect that eventually a combination of functional, ultrastructural, and biochemical criteria will be used to formulate a comprehensive system. In the meantime, a rough classification must suffice. The simplest, most commonly used system is that adopted for higher vertebrates, which recognizes only smooth, striated, and cardiac muscle. Vertebrate smooth-muscle cells are indeed vastly different from vertebrate skeletal cells, but when judged in relation to muscle cells of the whole animal kingdom, they must be fitted into a wide spectrum of types. Some of the muscle cells of invertebrates that have been called smooth are also different from all forms of striated muscles: these are ones in which organized thick filaments are hard to find. Only cells of this type are called smooth in the present account.

Many invertebrate cells previously classed as smooth in fact have abundant thick filaments, plus internal attachment sites for thin filaments, and their functional unit is a well-marked thin/thick/thin configuration. In some the filaments have a highly organized arrangement either in a spiral or in oblique planes. These cells exactly resemble those of *transversely cross-striated* muscle when in a heavily contracted condition or when sectioned in an appropriate plane. It is more appropriate to refer to these as *obliquely striated.* In some thin cells, there is but a single oblique layer, but in more cylin-

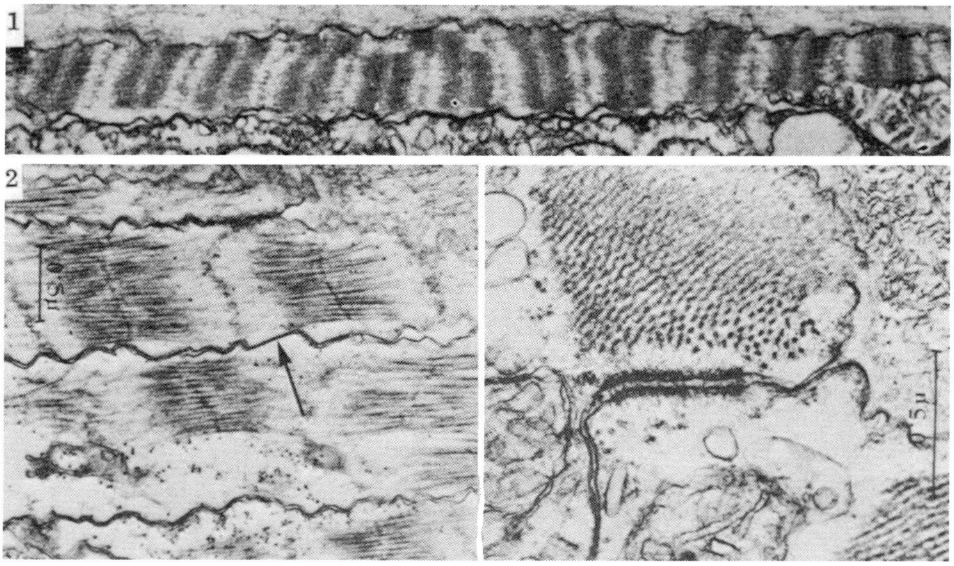

Figure 3.2. Cross-striated muscle cells in coelenterates. (*1*) LS of a very thin striated cell from the margin of the hydrozoan medusa *Amphinema.* (Reprinted with permission from Chapman et al., *Rev. Canad. Biol.* **21**, 267–278 1962.) (*2*) LS and TS of hydrozoan medusa *Stomatòca.* There are tight junctions between adjacent cells (arrow at left); center of field at right. (From Mackie and Singla, 1975.)

drical ones, the obliquity continues around a central core (Figure 3.3). These thin, cylindrical cells have traditionally been called spirally striated or double-oblique striated. There is a variant of this type of muscle in which there is no regular lateral alignment of filaments, but these otherwise closely resemble obliquely striated cells. Some of the variants are both thick (40 μm in diameter) and long (3 mm) compared with vertebrate smooth cells, and they are clearly more closely allied with cross-striated cells when examined ultrastructurally. To keep these types together with the cross-striated ones, I propose to call them *disorderly striated.*

One traditional class of muscle cells still clearly separates itself both structurally and functionally, in vertebrates and invertebrates alike, and that is cardiac striated muscle. This comprises a unique form of short anastomosing cross-striated cells linked end to end by intercalated discs. Here the common requirement for rhythmic pumping over a narrow range of speeds and frequencies has led to a remarkable parallelism.

3.2. General Structural Features of Cross-Striated Muscles

Very thin electron microscope sections through a typical cross-striated muscle cell that has served as the classical modern model, from the backstrap or psoas (= filet mignon)

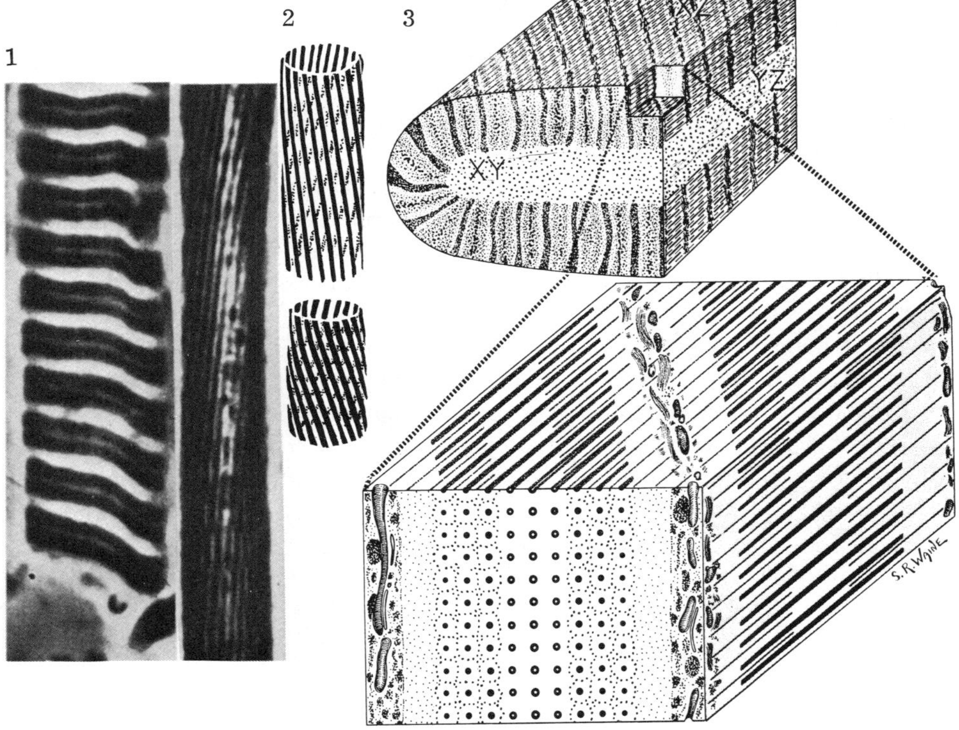

Figure 3.3. Common structural features of transversely and obliquely cross-striated cells. (*1*) The two extremes of common molluscan cells, transversely striated and obliquely striated, occur either in separate bundles or mixed together randomly in clam and oyster adductors. (*2*) Diagram illustrating I band shortening in obliquely striated cell. Reprinted with permission from J. C. Wissocq (1970), *J. de Microsc.* 9, 355–388. (*3*) Drawing of ultrastructure of obliquely striated cell, based on *Ascaris.* Note that in one plane it has a simple cross-striated form. (From Rosenbluth, 1965.)

muscle of the rabbit, are illustrated in Figure 3.4. The principal components of all cross-striated muscle cells are parallel "thick" filaments of equal thickness and shape, very nearly equal lengths, and almost perfect lateral alignment. They are always present in a basically hexagonal array that is faultless in some arthropod muscles. Together the thick filaments form a band that possesses sufficient molecular symmetry to rotate the plane of polarized light (anisotropic); hence it is termed the *A band.* The thick filaments forming this band are referred to as *A filaments.* Interdigitating with the A filaments are "thin" filaments that extend beyond the A band margins as far as the Z band, where they are attached. In a few types only, the Z region is double. The most conspicuous thin filaments are mostly of equal length, and their free ends stop short of the middle in resting cells. Because of the extra density caused by the overlap of thin and thick filaments everywhere except the center, there is, in the middle region, a lighter band, the *Hellerband* or *H region.* Such a region can be seen clearly in some muscles, indistinctly in others, but not at all in a few. At the light microscope level of resolution in a majority, but not all, muscles, the H band is itself transected by a single dense line placed exactly in the middle, hence *Mittelscheibe* or *M band.* In some, to be described below, there are two bands symmetrically placed to either side of center, forming a double M band, and in others there is none. The region, divided by a Z band, from the margin of one A band to that of the next does not rotate the plane of polarized light and hence is termed the isotropic or *I region.* The thin filaments of equal length in this region are termed *I filaments.*

A collection of myofilaments running from one Z region to the next nearest is termed a *sarcomere* (muscle portion, or disc). If all the filaments had equal lengths, and if the muscle cells were of equal diameter along their length, without splitting or branching, then a cross-striated muscle cell would indeed be formed of equal disclike slices as popularly conceived. However, in practice there are many departures. Sarcomere lengths are not equal within a cell. The result is a series of mismatches in lateral alignment. The number of mismatches varies with the muscle: in some there are only a few per cell, in others there is extensive mismatching. A short sequence of mismatching leads to a *vernier* effect.

When Anthony van Leeuwenhoek (1674) made the first illustration of a muscle, it was of the fine structure of a cross-striated cell, and he drew its contents as a series of tiny cylinders of almost equal diameters. Likewise Bowman (1840) in his classical work on frog leg, boa constrictor, and beetle muscle, drew highly regular arrays of internal cylinders. These were termed myofibrils, and to this day all textbooks and even advanced research reviews illustrate striated muscle in general as if it is so constructed. This is quite

---

Figure 3.4. Longintudinal EM sections of rabbit psoas muscle cells. (*1*) Portion of classic micrograph made by H. E. Huxley, *J. Biophys. Biochem. Cytol.* **3**, 631–648 (1957). Cut very thinly in the plane shown in the insert, it provided the basis for universal acceptance of the at-the-time heavily disputed sliding-filament theory. In addition it revealed regularly arranged X bridges between I and A filaments and thereby suggested a possible mechanism for contraction that was completely novel at the time. The image is, however, severely distorted because of longitudinal compression resulting from very slow sectioning of material embedded in a very hard medium, methacrylate. (*2*) An equally thin or thinner section made using Epon and cut with a good diamond knife produced this much less appealing but more realistic undistorted image. The tissue was held at rest length during fixation, which stimulated it. The X-bridges (*XB*) are fixed in a variety of positions, some pointing outward, others inward or laterally. Other items of interest are ultrathin (*T*) filaments and convergence and fusion of two A filaments (*FTk*). (ORIG.) (*3*) Stretched cell, showing wider H zone. Attachments of troponin (*TN*) complexes at regular interval of 38 nm is seen. Other features of interest are T filaments and apparent splitting of some I filaments (*Spl*). (ORIG.)

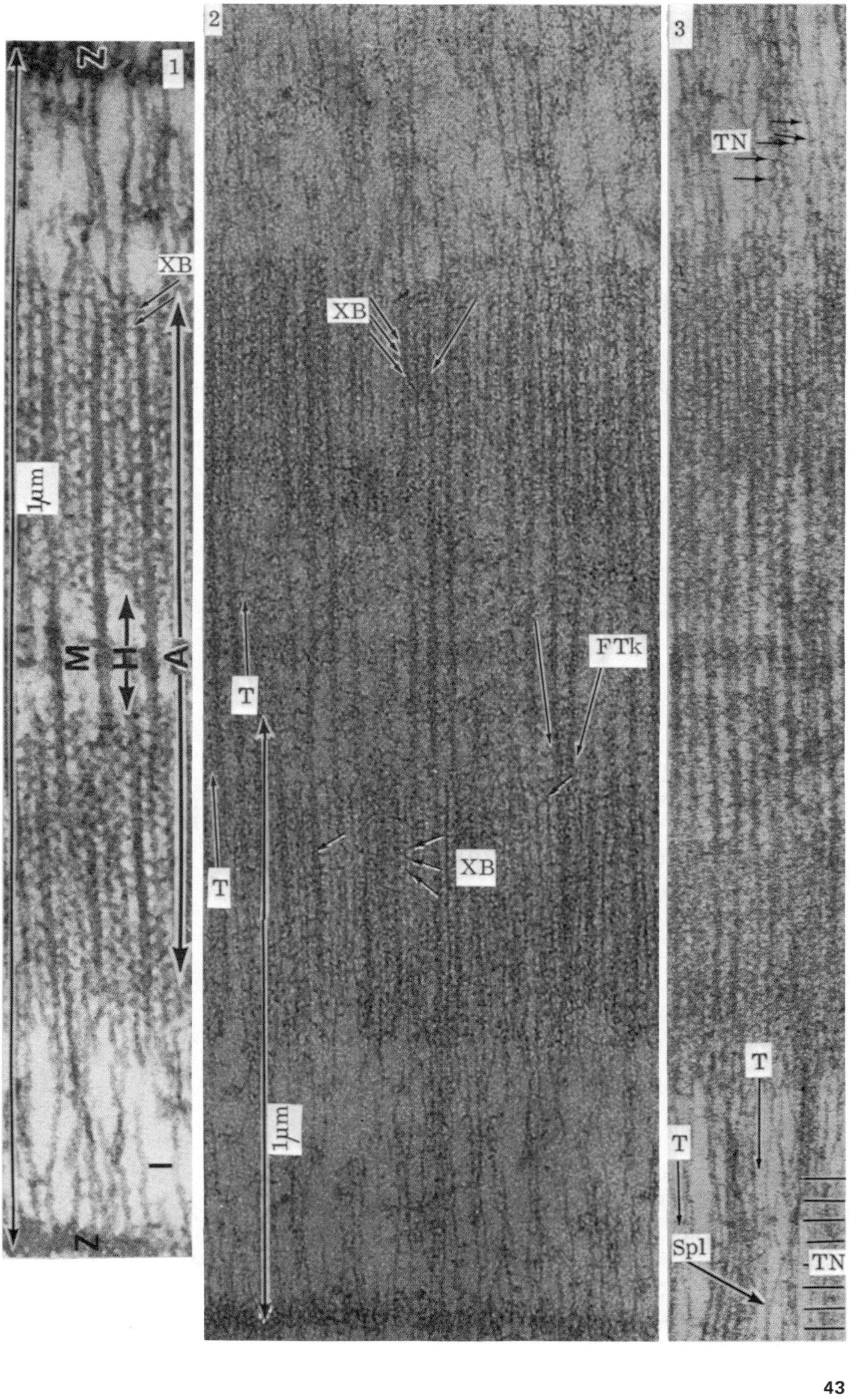
1
Z
XB
1μm
M
H
A
I
Z
2
XB
FTk
T
T
XB
1μm
3
TN
T
T
Spl
TN

wrong. Had Bowman looked at some other muscles of the snake, he would have been hard put to find any sign of internal division of the cells.

It was not until more than a century after Bowman's paper that it was recognized by Krüger (1952) that vertebrate skeletal muscle cells are not all subdivided into independent subunits. Ones that appeared to be so subdivided he called *Fibrillenstruktur* cells; ones that had very little sign of subdivision he termed *Felderstruktur* cells. Examples are shown in Figure 3.5. Modern electron microscopy has shown that the former have an extensive invasion by sarcoplasmic reticulum (SR) and the latter varying amounts, but always much less. *Sarcoplasmic reticulum* is the term assigned to any internal membrane system located within the space defined outwardly by the plasma membrane enclosing the cellular contents of a muscle cell. SR ranges widely in amount and form and provides a major source of muscle diversity, which will be described in Chapter 8. A majority of cells of all kinds contain internal membranes collectively called endoplasmic reticulum. SR is the equivalent anatomically, though not necessarily functionally, in muscle cells. Electron microscope studies on single muscle cells have shown that, even when there is a great deal of SR, it does not completely envelop all the myofilaments in neat clusters. The detailed pattern of SR changes as one moves along the length of a cell. It is not true that there is a subdivision along the length of the cell into component fibrils. Such a division may occur over a small fraction of the length, up to about 20 sarcomere lengths, but that is all. The nearest that any cross-striated muscle comes to being truly fibrillar is

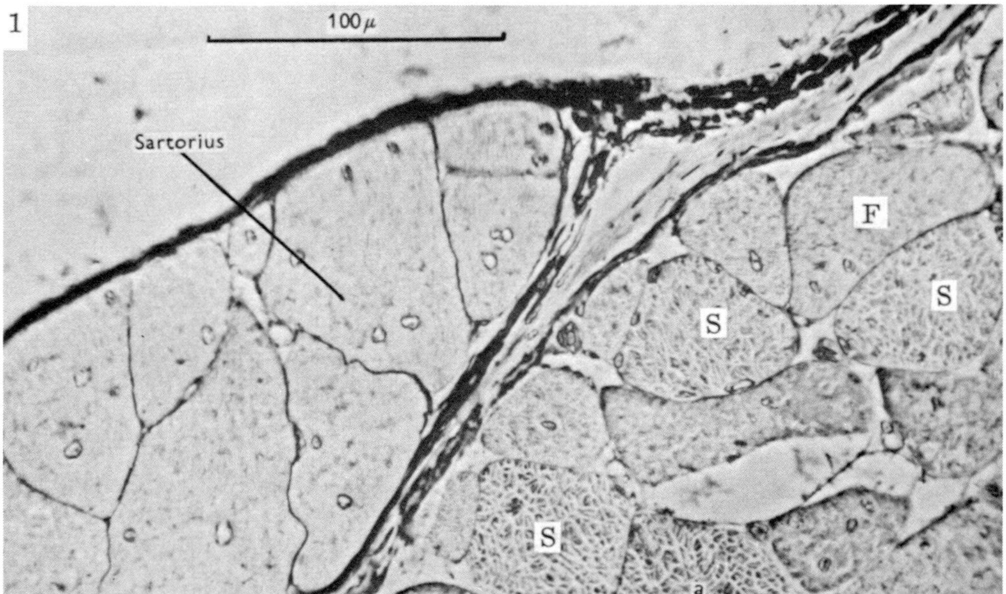

Figure 3.5. *In situ* comparison of frog muscle with homogeneous composition – sartorius with heterogeneous neighbor – adductor magnus. F = Fibrillenstruktur, fast-type; S = Felderstruktur, slow-type. (From Gray, 1958.)

seen in some insect flight muscles, and indeed some of these may meet the requirements of the standard textbook model. When striated-muscle cells are macerated, discrete subunits are obtained that can be reasonably called myofibrils, although they result from the tearing action as much as natural division. In this book the term *myofibril* is confined to these forcefully generated elements, but it should be borne to mind that they are not a natural subunit of striated-muscle cells as has generally been supposed. The natural elements most nearly corresponding to the myofibril concept are termed *myofilament clusters.*

There has developed another general dogma that is not true, a belief that cross-striated muscle cells are discrete and unbranched. "Splitting" was early recognized in developing muscles as a relatively common component of certain diseases and as a possible means of increased size in exercise but was considered to indicate only an abnormal or transitional state. Based upon my own dissections, I consider that a small amount of branching is common and natural in healthy cross-striated muscles of both vertebrates and invertebrates.

3.2.1. Evidence for Spiral or Mismatched Internal Organization

Muscle cells have commonly been referred to as fibers, because they are commonly long and thin. However, since many are neither, and since the term *fiber* connotes nonliving, truly filamentous, structures in other contexts, I have chosen to refer to them only as muscle cells. When observed in the light microscope, living cross-striated muscle cells show localized waves of contraction that spread slowly along their length. These are not an artifact of isolation, since they can also be observed and filmed (first achieved by Hürthle, 1909) in whole small transparent arthropods, and small fish and amphibians that have sufficiently transparent surfaces. The path of such waves can be spiral (Matthei and Tiegs, 1955), showing that either the structural elements or the internal elements responsible for conducting the excitation, or both, must have a spiral form, as found also for frog muscle by Peachey and Eisenberg (1978) using high-voltage electron microscopy. It should not be assumed from these studies that all "cross-striated" muscle cells have a spiral internal arrangement, but it may be concluded not only that some do, but also that the traditionally flat planes of subdivision may be rare, or even nonexistent. Planes have only been assumed, never directly demonstrated in a critical analysis. In a majority of cross-striated muscle cells, the component myofibrils are within ± 10% of each other in length and have a strong tendency toward lateral alignment. They are, however, permeated in an irregular manner in three dimensions by SR. This is a strictly endogenous membranous component, and the amount of it is very different in different types of muscle cell. The range of sarcomere lengths in a single cell can be much larger. I have found a range from 2.0 to 3.0 μm in rabbit psoas, and an astonishing 5.8–22.2-μm range in giant barnacle. The majority of sarcomere lengths within a given cell type are nevertheless closely similar: 2.3 μm in rabbit psoas and 6.0 μm in giant barnacle. The longer ones are rare in both.

Some examples of muscle cells are "fibrils" taken from different divisions of the muscle literature are shown in Figure 3.6: sarcomere heterogeneity is the rule, not the exception. Both sarcomere-length heterogeneity and the effective anastomosis of contractile material bundles dictated by the randomness of SR permeation are well illustrated in the scanning electron microscope image of part of a rabbit psoas cell shown in Figure 3.7.

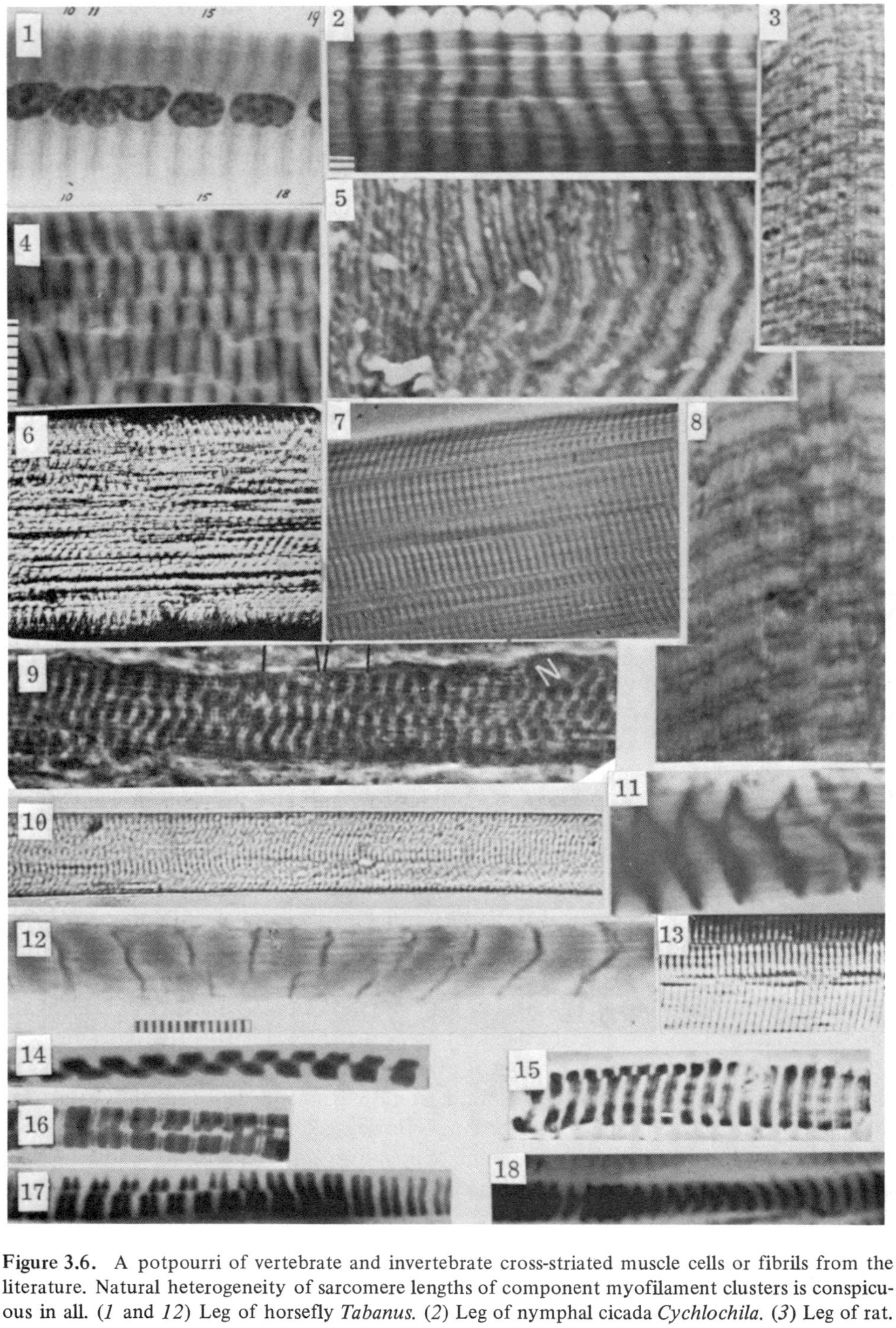

Figure 3.6. A potpourri of vertebrate and invertebrate cross-striated muscle cells or fibrils from the literature. Natural heterogeneity of sarcomere lengths of component myofilament clusters is conspicuous in all. (*1* and *12*) Leg of horsefly *Tabanus.* (*2*) Leg of nymphal cicada *Cychlochila.* (*3*) Leg of rat. (*4*) Wing of homopteran insect *Eurypelma.* (*5*) Frog sartorius. (*6, 7, 10, 13*) Leg of frog. (*9*) Chick thigh. (*12*) Abdominal muscle cell of nymphal cicada. (*11*) Jumping muscle of homopteran *Scolypopa.* (*14-18*) Various molluscan fast adductors. [(*1*), (*2*), (*3*), (*4*), (*8*), (*11*), (*12*) from Tiegs, 1955; (*5*) from D. K. Hill, 1960; (*9*) from Shimada et al., 1967; (*6*), (*7*), (*10*), (*13*) from Washio et al., 1961; (*14*), (*15*), (*16*), (*17*), (*18*) [Reprinted with permission from J. Bowden *Int. Rev. Cytol.* 7, 295–335 (1958). Copyright 1958 Academic Press.] Note in (*1*) the row of large nuclei in central core; in *2*, looping sarcolemma attached to Z regions.

Figure 3.7. (*1*) Scanning electron micrograph (SEM) of partially teased rabbit psoas cell. Note the different sarcomere lengths (compare regions near the bars) and apparent anastomosis of the contractile material. (Courtesy of Dr. J. H. L. Watson, Henry Ford Hospital, Detroit.) (*2*) SEM of portion of bird pectoral muscle cell, showing marked apparent branching of contractile material. This is caused by the irregular intrusion of sarcoplasmic reticulum and glycogen (dark particles). [Reproduced by permission of the National Research Council of Canada from the *Canadian J. Zool.* 47, 771–774 (1969).] (*3*) TEMs of different sarcomere lengths from same rabbit psoas muscle cell at same magnification. (ORIG.)

3.2.2. Sarcoplasmic Reticulum

The SR is truly a discovery of the modern era of electron microscopy, since light microscopical histological procedures generally revealed nothing of it or failed to resolve even its gross patterns. Both the amounts and detailed arrangement or pattern of SR in different skeletal muscle cells are very different. At one end of the continuum are cells in which the SR is so sparse it was at first thought to be truly absent, like snake-rib skin muscle cells (Figure 3.8). It took a strong belief on physiological grounds that some ought to be present, leading to extensive sectioning (Hoyle et al., 1966), before any was found. At the other extreme are a few functionally specialized cells in which – as in the remotor of the lobster second antenna, which is responsible for making a high-frequency sound – the amount of SR is so great it occupies more than half the total volume. The more typical fast-type vertebrate cell is illustrated in Figure 3.9., sartorius of the rabbit. Here there is extensive invasion of the contractile material by SR, which divides it into approximately equal areas, as seen in any transverse section. But the SR does not totally cut off myofilament clusters from their neighbors, as was supposed in the myofibril hypothesis. Only a few sections further down, the detailed appearance is quite different, for the SR is meandering freely through the filaments.

The basic form of SR is two close, parallel sheets of plasma membrane with a large number of membrane-lined perforations. The perforations are never regular in size or shape, and they can be both highly irregular and wide. In the extreme case, the SR is reduced to a set of narrow tubes. Such tubes may even cross over each other, as in SR of toadfish swim-bladder muscle (Fawcett and Revel, 1961). The internal space of SR is narrow, averaging about 20 nm. Although sometimes present as a single sheet, SR is also quite often double or triple; single, double, and triple layers occur within the same cell. In some cells there is no regularity to the double or triple layering, but in others, such as the flight muscle of the dragonfly *Celinthes,* the layer is single except in the H region, where it is double (D. S. Smith, 1968). Also in the H region, there can be fewer or much smaller perforations, as in frog sartorius. The relative volumes occupied by SR in frog sartorius cells has been determined by Mobley and Eisenberg (1975) to be 5% of cell volume, not including the cisterns, which occupy another 4.1%; invaginated tubule (IT) volume is only 0.3% total *cell* volume.

3.2.3. Sarcolemma

All cross-striated muscle cells are surrounded by a layer of extracellular material that is collectively termed the *sarcolemma* (Sl). Earlier the *plasma membrane* (PM) of the cell was included, but more recently the term Sl has been confined to material outside it. There are some muscle cells in which no material adheres to the outside of the PM, others in which there is a very thick layer, up to 20 μm, and cells of all intermediate types are known. The nature of the components of the Sl is not well known for most cells, and since the Sls have a wide range of thickness and inclusions, generalizations cannot be made, except that Sls are all elastic, but with various degrees of stiffness. Some are unwrinkled at the rest length of the cell, and others are highly folded and start to resist only after considerable extension. The elastic quality of Sls of vertebrates is due to a specific elastic substance, *elastin,* together with a contribution from meshes of collagen, which is the major component in invertebrates. There is also a basic matrix, which has been variously referred to as the amorphous layer, cuticular layer, and ground substance

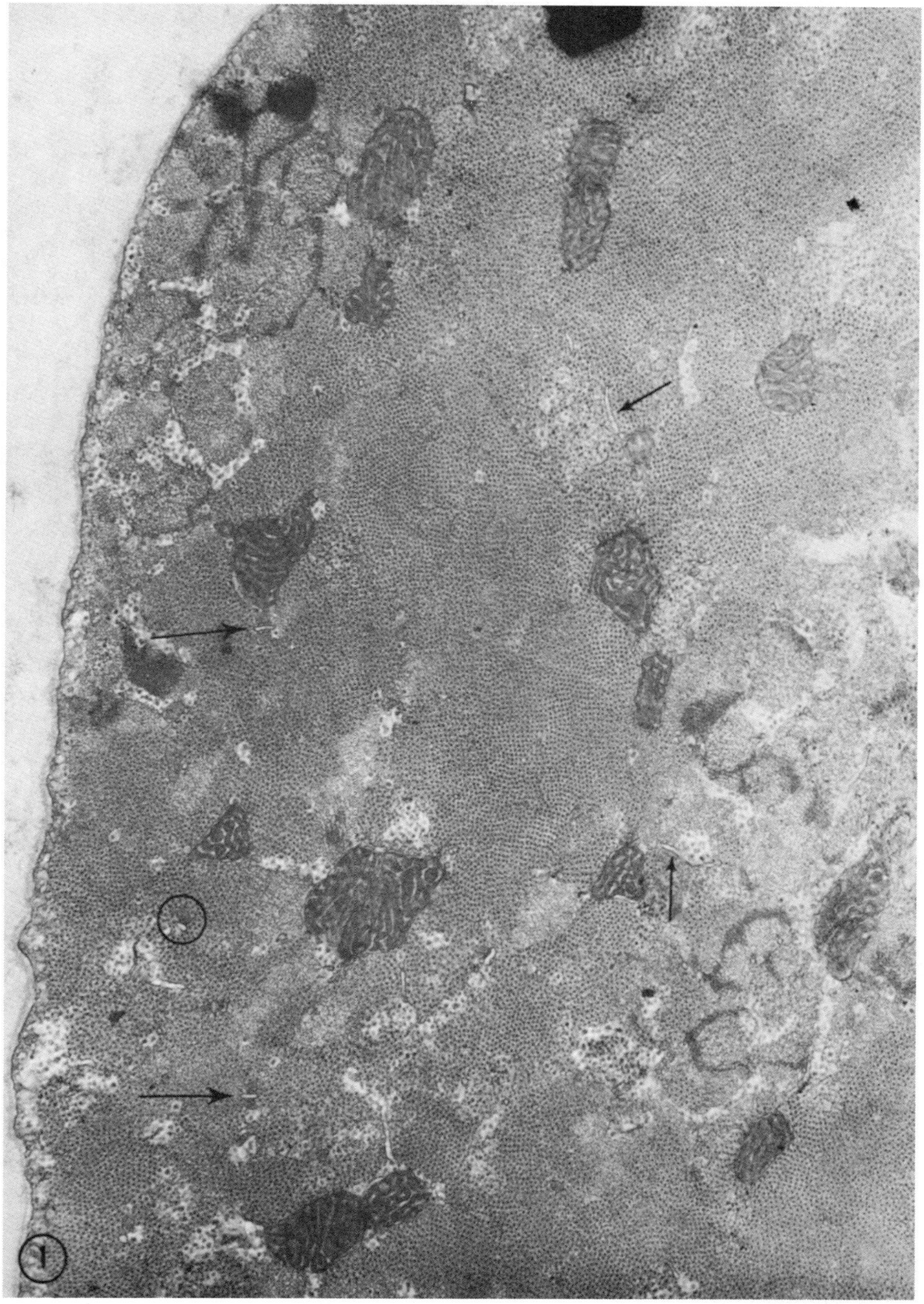

Figure 3.8. Transverse TEM section through slow muscle cell of garter snake *Thamnophis sirtalis*. There are very few invaginated tubules (ITs) (indicated by arrows) and little SR. Coupling elements are longitudinally running dyads and triads. (From Hoyle et al., 1966).

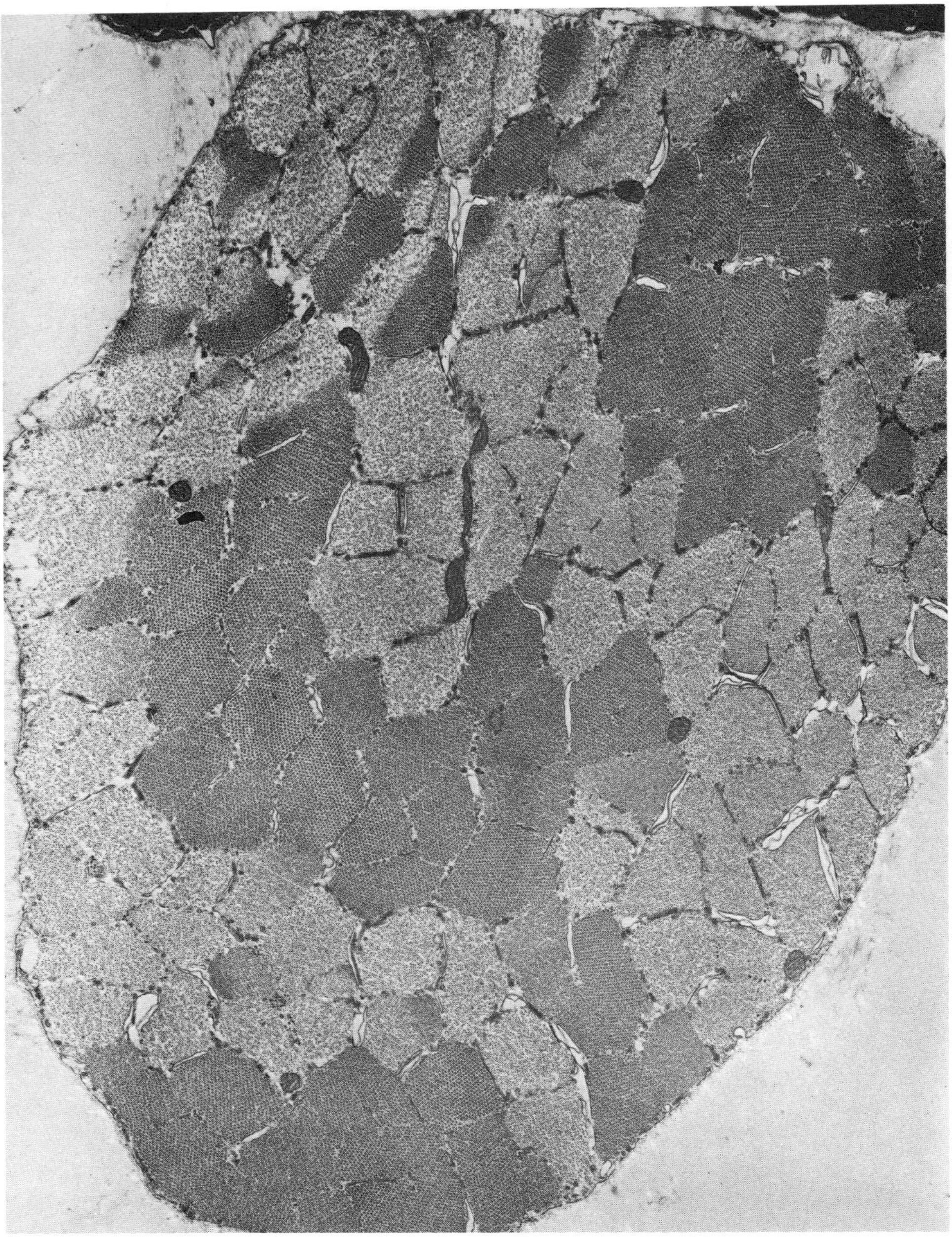

Figure 3.9. Single fast-type muscle cell from sartorius muscle of rabbit. Close scrutiny shows that there is extensive permeation by ramifying SR, but that although this divides the myofilaments into bundles of approximately equal size, they are not isolated from each other. The sarcomeres are not aligned laterally, but slightly staggered, in a clockwise (right-handed) spiral so that different regions abut laterally. (ORIG.)

(Mauro and Adams, 1961); the matrix is a *mucopolysaccharide.* The collagen fibers are numerous in some cells, ranging in size from only a few to over 30 μm in diameter. In insects the Sl thickness is great, from about 20 to over 100 μm, being thickest in muscles that are used frequently and that develop large forces.

Early measurements on the properties of the Sl of frog cells were made by Ramsey and Street (1940), who considered that it alone is responsible for the passive elastic properties of these cells and that force is transmitted by way of it. Buchthal et al. (1951) and Casella (1951) disagreed on both counts. Casella found that handling cells causes a large increase in the stiffness of the Sl, eventually to about double. He concluded that normally the Sl "contributes but little to the tensile force of the resting fibre." Most recent workers on single vertebrate muscle cells consider the Sl to be folded at rest length and not to start contributing significantly to tension until stretch exceeds about 10% mean length in the body. In arthropod skeletal cells, the Sl is regularly attached to Z regions, forming loops at rest length (Figure 3.6.2). It must first flatten upon extension, before contributing to passive tension.

The external material – ground substance, collagen, and elastin – is also present in major and even minor clefts when these are present. In insects, tracheoles enter muscle cells down these clefts, some of which extend to the center of a cell. Also, in some insects and many crustaceans, nerve terminals occur deep in the clefts, so that some of the synapses are located deep within the cell.

The developmental origins of most Sls is unknown. They have often been supposed to be secreted by the cell in some way, but in locust leg muscles there are specialized wandering cells that secrete the mucopolysaccharide (Figure 3.10). These very obviously busy cells are present in adults as well as juveniles and may be replenishing material eroded during contraction, for the Sl must surely serve to protect the PM. Where much mucopolysaccharide is present, it is possible that it can serve as an energy resource.

A most important feature of the PMs of a majority of skeletal muscle cells is a series of tiny perforations. These are regularly arranged in some cells and irregular in others, a feature that is correlated with the general internal structural organization. The holes are the openings to the exterior of functionally important ITs (see below).

In the cells of typical vertebrate skeletal muscles, there are always two wider, non-perforated bands of SR in each sarcomere called *cysternae* (C), which are parallel to each other except in those cells in which they are sparse. Since they occur at the outer margins of discrete lengths of SR, they are sometimes called terminal cysternae. Also in vertebrates, the length of the SR between the C elements is either equal to the average sarcomere length (SL) or to one half this length. In the former, full-sarcomere SR, the C elements are located just to the inside of the Z region, but the Z material itself does not extend into the gap between C elements of adjacent sarcomeral SR. In the latter, half-sarcomere SR, the SR passes across the Z material so that its middle coincides with the Z region. The C elements are located at about one quarter SL away from the Z region, near the center of the overlap region between thick and thin filaments.

The faces of the adjacent pairs of C elements are not only broader compared with SR space, they are also flattened. The space between them is exactly and neatly occupied by a flattened IT (see below). This is a biologically important contact but about whose functioning in detail there is currently almost no knowledge.

Several aberrant forms of SR are known from specialized invertebrate muscle cells. One takes the form of longitudinal tubules penetrating the myofilament clusters (Figure 3.11). There is also a massive penetration of SR among the myofilaments of moth flight muscle (Auber, 1967a, b). These are contacted by ITs running longitudinally to make long, narrow double tubes or even complex rosettes (Figure 3.11.3). All forms of close contact between one IT and one SR element are termed *dyads*. In a majority of both vertebrate and invertebrate muscle cells, the SR simply surrounds myofilament clusters without penetrating them, most completely in some insect flight muscles (Figures 3.11.*2* and 3.11.*5*). In invertebrates, each IT always comes to run between pairs of C elements

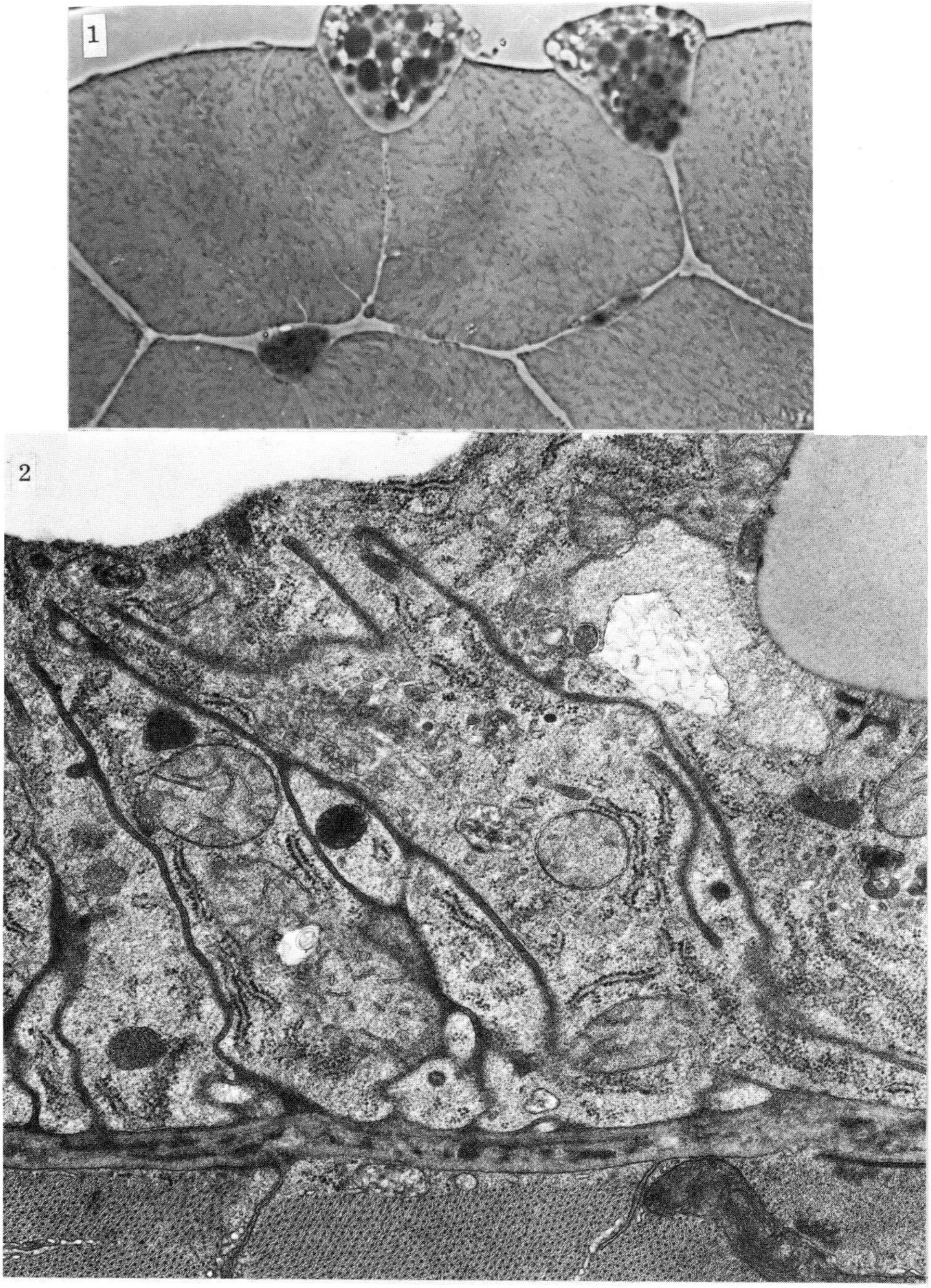

Figure 3.10. (*1*) Small and large wandering cells, heavily laden with lipid globules, secreting the mucopolysaccharide basement layer around muscle cells of locust extensor tibiae. (*2*) EM section of one of the cells above, showing details. Note numerous ribosomes and channels into which the material is secreted. (ORIG.)

and simultaneously contacts them both to form a triple tube termed a *triad.* An exception occurs in some slow-type cells, in which a nearly longitudinal path is made by ITs: then the contacts closely resemble those of invertebrates, and are dyadic or pentadic as well as triadic (Figure 3.11.*1*).

3.2.4. Invaginated Tubules

In all but the smallest-diameter muscle cells, that is, those below about 10 μm in both striated muscles of all types and nonstriated muscles, there may be invaginations of the surface membrane that serve the purpose of inward transmission of membrane-potential changes. There are several kinds of invaginations: major and subsidiary *clefts;* longitudinal and radial *wide tubules,* generally about 150 nm in diameter; and radial, oblique, longitudinal, or spiral *narrow tubules,* which are either flattened or round. The most common and significant, since they occur in all vertebrate and most arthropod cross-striated cells are narrow (20-nm space) tubules open to the exterior. The abbreviation T was applied to ones found in certain muscle cells of vertebrates that have a primarily radial orientation, the T referring to their strong tendency to be present in transverse planes (Andersson-Cedergren, 1951). H. E. Huxley (1964), proved that these tubules are open to the exterior, and A. F. Huxley and Taylor (1958 – see p. 213) proved that their function is to transmit excitation from the surface membrane into the interior. By then it was being realized that this same system enters some cells obliquely and that it also runs mainly longitudinally, not transversely, in many kinds of cells, such as garter snake slow cells (Hoyle et al., 1966). In some cells, it does have a radial orientation but these tubules are connected to each other by longitudinal tubules (Jensen and Mykleburst, 1975), leading to the published absurdity of "longitudinal transverse tubules." Thus I suggested substituting for the abbreviation T the alternative E, for excitatory, or even externally connected (Hoyle, 1965), but the letter T has stuck. Fortunately T is also an abbreviation for tubule, perhaps the abbreviation IT (invaginated tubule) will be preferred.

An intense literature search revealed that these tubules had been discovered by several workers toward the end of last century, and that they had been most beautifully illustrated by Veratti in 1902, but no distinction had been noted between ITs and SR. For a while after rediscovery of ITs, they were confused with the SR, which is actually an independent system that makes intimate contact with ITs at discrete locations, but which is nowhere directly open to the exterior. The SR was called the longitudinal SR, with the intention of distinguishing it from the supposedly transverse SRs, which were the first IT elements to be investigated by electron microscopy; these were at first not known to be open to the exterior. ITs are, however, definitely not part of the SR. The SRs had been carefully and accurately described by Veratti (1902) in skeletal muscle cells of selected crustaceans, insects, fish, reptiles, and mammals. It is apparent from recent electron microscopy that Veratti's staining methods, principally silver, revealed the ITs without affecting the SR. But since SR contacts the ITs at the numerous dyadic or triadic contact sites, the general form of the SR had indeed been revealed. Evidentally the silver actually entered and permeated the ITs throughout each muscle cell. In details revealed by recent electron microscopy there is remarkable diversity in ITs, but there are also distinctive classes with common general features. If this had been fully appreciated, it would certainly have revolutionized muscle physiology. As a student, I heard A. V. Hill, half a century after Veratti, describe a skeletal muscle cell as equivalent

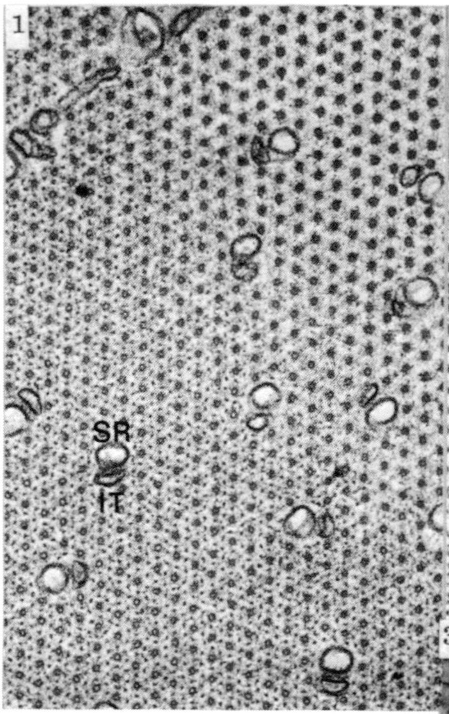

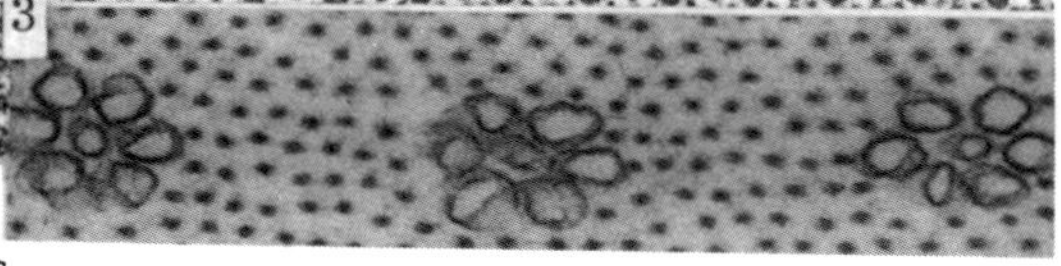

≃1000Å

4

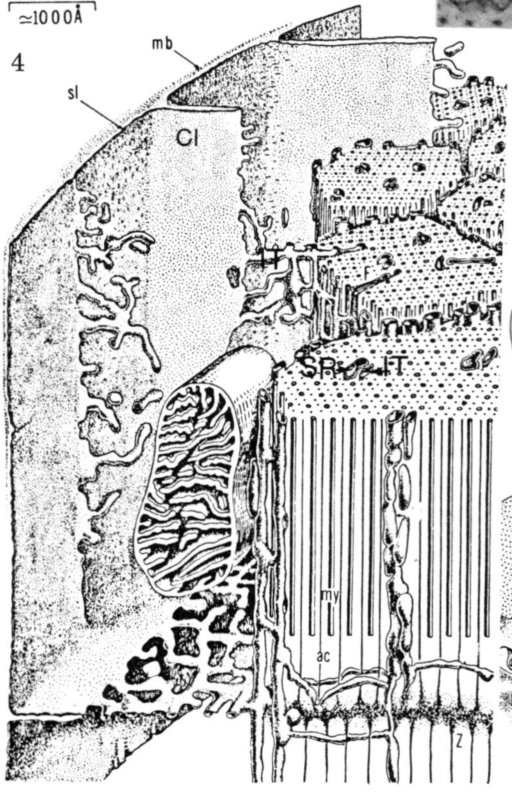

5

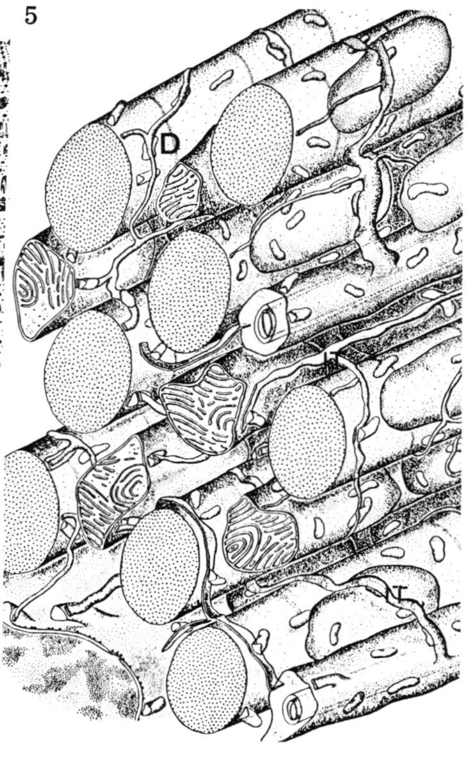

to a cylindrical 100 μm in diameter sausage filled with homogeneous contractile material but with a smooth skin (PM and Sl). He then asked how excitation could possibly travel inward in less than a millisecond when there was nothing along which it could travel inward.

Whenever ITs occur, that is, in all vertebrate skeletal muscle cells and a majority of those of invertebrates, they have a ribbonlike structure, about 20 × 100 nm. Wider, but not thicker, regions occur in the longitudinal direction, and some narrow, cylindrical sections occur. The wider regions of ITs occur at points of contact with SR. In most invertebrate skeletal muscle cells, the IT lumen is quite obviously open to the exterior, but for those of vertebrates, openings are not clear. Several investigators proved them to be open by soaking intact cells in a solution of ferritin, an electron-dense iron-containing protein molecule, and showing this to be present in the IT lumens following electron microscopy (Figure 3.12.*3*).

The same principal types of arrangement of ITs relative to SR are encountered in both vertebrates and invertebrates, although one is predominant in the latter. In one type, the ITs enter strictly radially at the level of each Z region, where they branch to almost surround a myofilament cluster. This type (Figure 3.13) is present in most amphibia, some insect flight muscles, and some arthropod heart muscles. However, in all muscles except those of vertebrates, the flat portion of the tubule runs longitudinally rather than radially, and there is but a single contact with SR. This simple type of junction is a dyad; wheras a triad tends to form a collar around a myofilament cluster, most dyads are little more than patches, roughly ovoid as seen in surface view.

A commoner vertebrate type (Figure 3.13.*2*) has double the amphibian number of ITs, which enter the sarcomere near the center of each A/I overlap region to make two sets of triads per sarcomere. An invertebrate equivalent is seen in dragonfly flight muscles, though of course the contacts are dyads.

The commonest invertebrate form is for ITs to enter obliquely and to run longitudinally in very irregular pathways. They make dyadic contacts principally in the A/I overlap regions, but there may be several sets per sarcomere. In a few insect flight muscles, such as those of the housefly, dyads are made only at the M band. The larger arthropods, crabs, lobsters, and giant barnacles, have extremely thick skeletal muscle cells. The total number remains constant as the animal grows bigger, so the cells get progressively larger. In a large European spider crab (*Maia squinado* – Caldwell, 1958) or a giant barnacle (*Balanus nubilus* – Hoyle and Smyth, 1963), they average 2 mm in diameter, and in a 23-lb king crab (*Paralithoides kamchatica*), they average almost 3 mm (personal observa-

Figure 3.11. Perfect arrays of I and A filaments seen in transverse sections, and some of the different forms of SR associated with them. (*1*) Antennal muscle cell of copepod *Doropygus*, showing numerous longitudinally running invaginated tubules (*IT*) making close contacts (dyads *D*) with longitudinal SR (*SR*) both around and within the myofilament clusters. (ORIG.) (*2*) Small portion of myofilament cluster of insect (*Lethocerus*) flight muscle, with perfectly arrayed filaments. I/A is 3 : 1, as in (*1*). (ORIG.) (*3*) Portion of muscle of an ostracod crustacean, *Cypridopsis*, in which the SR takes the form of rosettes, with six SR cisterns surrounding a single central IT. (From Fahrenbach, 1964.) (*4*) Three-dimensional reconstruction of copepod muscle (Reprinted with permission from: Y. Bouligand, (1963), *J. de Microscopie* **2**, 197–212. The drawing is slightly incorrect, since the IT and longitudinal SR are in fact separate systems. The ITs arise from invaginated clefts (*Cl*). Bm = basement material; Sl = sarcolemma. (*5*) Three-dimensional reconstruction of ITs and SR of a dragonfly flight muscle. The tubes inside the larger longitudinal IT elements are trachae (*Tr*). SR/IT contacts are dyads (*D*). (From D. S. Smith, 1966c).

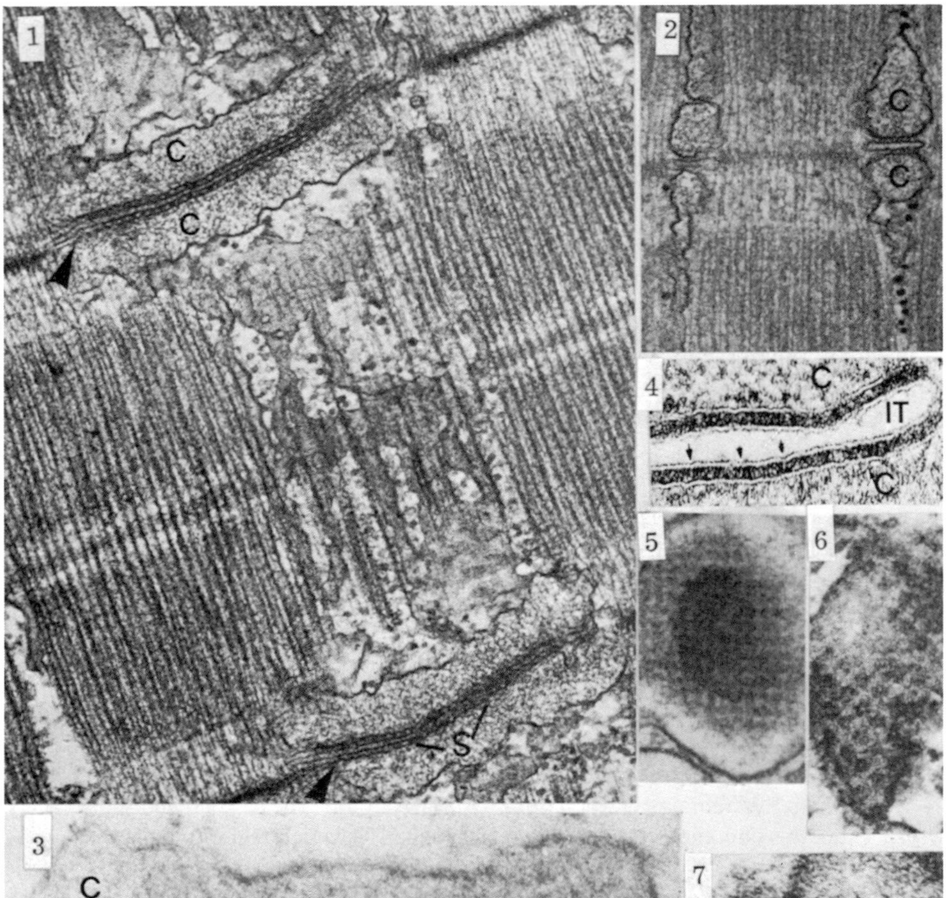
1
C
C
S
2
C
C
4
C
IT
C
5
6
3
C
IT
C
7
8
Tr
P
9

tion). Such cells have a primitive method of carrying excitation inward, part of the way, in the form of longitudinal infoldings of the surface, or *clefts.* In a very large cell, these can be subdivided into *major* and *minor* clefts, depending on their length and depth. These occur in addition to ITs, many of which originate from them. The literature on this and the following aspect is extensive but widely scattered; earlier material has been reviewed by D. S. Smith (1966b, 1972) and Franzini-Armstrong (1973), but no recent comprehensive account is available.

In some insect and crustacean cross-striated cells, there are extensive sets of ITs at each Z region that are wider and more densely staining than ITs entering at other regions (Figure 3.14). They are sufficiently distinctive from ordinary ITs that coexist with them that they deserve a special nomenclature, *Z tubules.* There are special attachments, resembling desmosomes, of Z material to Z tubules. These attachments, which must serve as structural anchoring points, run into the Z regions. Similar Z tubules have been described in some guinea pig muscle cells (Dorn, 1965), but are not widespread in vertebrates.

A few invertebrate cells have wide ITs either instead of, or in addition to, ordinary ITs. Both occur in the onychophoran *Peripatus,* which has muscle cells that are never larger than 40 μm in diameter, with most being 7–20 μm. Each of its larger cells has a dual system of wide tubules, one longitudinal, the other radial. They are completely independent, mingling with, but not touching, each other. Their average diameter is 0.13 μm, almost 10 times that of the nearly tubular parts of their sparse ITs.

3.2.5. Junction of ITs with SR

In invertebrate animals, the contacts between ITs and SR run along the long axis of the cell, never at right angles to it as most invertebrates do. The contacts in invertebrates are almost entirely in the form of an expansion of the IT; the expansion will be termed the E region, coming into close juxtaposition with a corresponding expansion of a single layer of SR, called the C region because it is functionally equivalent to the vertebrate terminal cysternum. Such a contact is called a dyad. When there are two layers of SR, the IT may

---

Figure 3.12. IT/SR contacts. (*1*) LS of a frog sartorius muscle cell showing triads situated at the Z-line level. The gap between the central lumen of the IT and the cisternal elements (*C*) appears scalloped (*S*) or bridged by regular rows of small "feet." Note the fine granular material in the C elements. (From Franzini-Armstrong, 1970a.) (*2*) Same as (*1*) but cut across the triad – revealing the flattened nature of the IT. [Reprinted with permission from C. Franzini-Armstrong, in *The Structure and Function of Muscle* Vol. II, pp. 531–619. G. H. Bourne, Ed. Copyright 1973 Academic Press.] (*3*) The IT lumen of a frog fast cell filled with ferritin after soaking the whole cell in it. [Reprinted with permission from H. E. Huxley, *Nature,* **202**, 1067–1071. Copyright © 1964, Macmillan Journals Limited. (*4*) Details of the IT/SR junction of frog semitendinosus revealed by treatment with tannic acid. [Reprinted with permission from Somlyo et al., (1979) in *Motility in Cell Function,* pp. 195–211. Copyright Academic Press] (*5*) The IT/SR contact region of a spider dyad seen in surface view. [Reproduced by permission of the National Research Council of Canada from the *Can. J. Zool.* **49**, 1549–1556 (1971). (*6*) Same as (*5*) but from newt tail cell. [Reprinted with permission from D. E. Kelly and A. M. Kuda *J. Ultrastruct. Res.* **68**, 220–233 (1979). Copyright 1979 Academic Press.] (*7*) Higher-power view of part of (*6*) showing apparent perforation (*arrows*) and double-ringed fine structure (*circle*). [Reprinted with permission from D. E. Kelly and A. M. Kuda, *J. Ultrastruct. Res.* **68**, 220–233 (1979). Copyright 1979 Academic Press.] Figs. 6, 7, courtesy M. A. Cahill, D. E. Kelly and A. M. Kuda. (*8*) TS of a snake muscle cell showing longitudinally running triad (*Tr*) and pentad (*P*). (From Hoyle et al., 1966). (*9*) LS of a snake muscle cell showing oblique triad and longitudinally running IT. (From Hoyle et al., 1966).

1

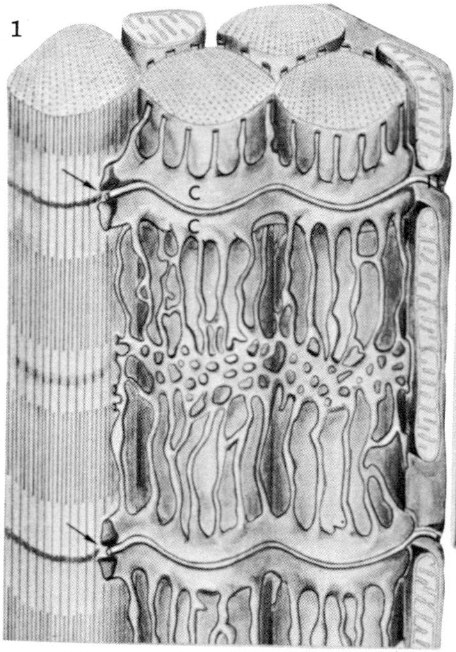
Z
I
A
H
A
I
Z
C
C

2

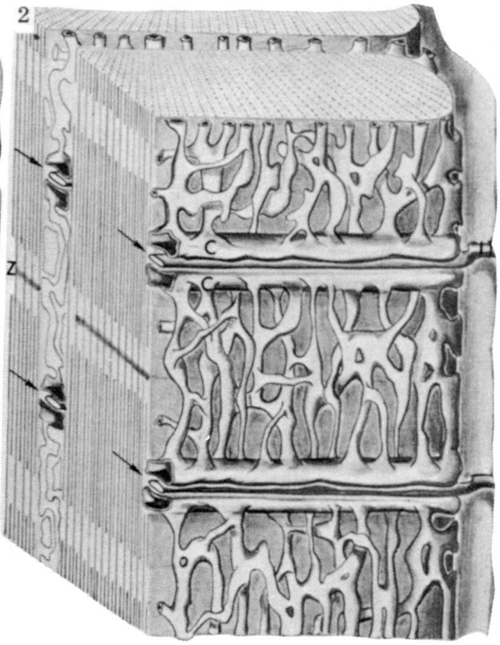
Z
C
C

3

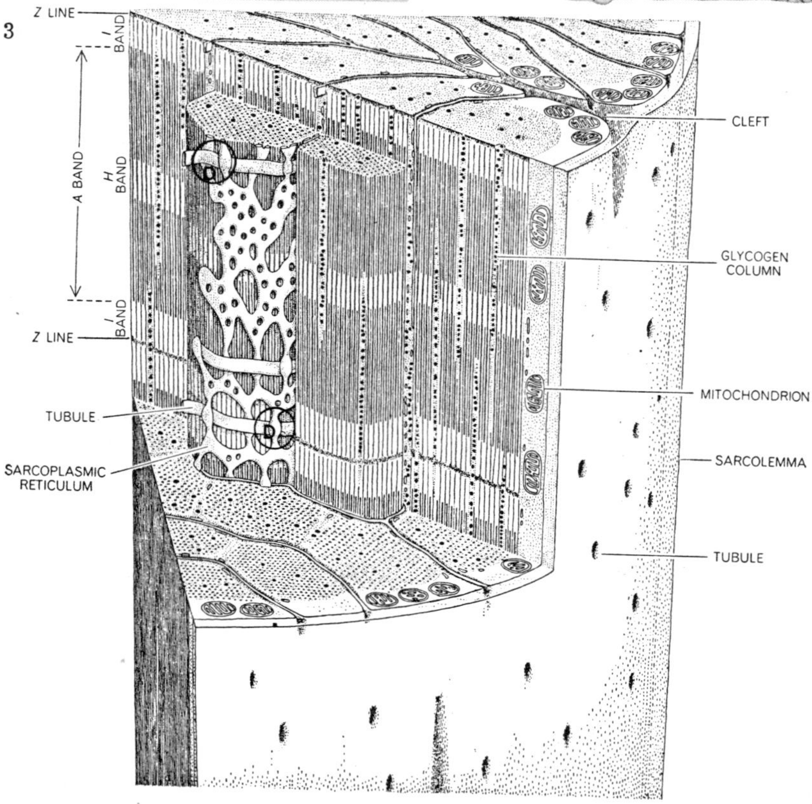
Z LINE
I BAND
A BAND
H BAND
I BAND
Z LINE
TUBULE
SARCOPLASMIC RETICULUM
CLEFT
GLYCOGEN COLUMN
MITOCHONDRION
SARCOLEMMA
TUBULE

run between them and form special contacts with both simultaneously to form a functional unit that is at least triadic in elemental structure, though it does not correspond functionally to the vertebrate triad. A few pentads occur, in which two ITs cross between a triple-layered SR at the same level. More commonly, when an IT passes between double- or triple-layered SR, it alternates between them in making presumed functional contacts.

Basic details of the points of approximation of ITs and SR were first described by Hoyle (1965). They are remarkably similar in fine detail in vertebrates and invertebrates in spite of the overall differences. There is a uniform gap of about 10 nm, which is bridged at regular intervals, as seen in sections, with a tendency to be scalloped at the same periodicity (Figure 3.12). Face views occasionally occur in transmission electron microscope (TEM) studies, revealing a regular tetragonal structure in both an invertebrate (spider – Sherman and Luff, 1971) and vertebrates (newt tail – Kelly and Kuda, 1979; frog – Franzini-Armstrong, 1970a). This is in turn made up of either ridges or dimples, which have also been called feet, formed of concentric circles. Lucent centers to these dimples or ridges and also to the bridges seen transversely suggest the possibility of minute channels between the C and SR cavities. Treatment with tannic acid fills the entire space with dense material except for periodic lucent lines corresponding to the possible channels (Figure 3.12.*4*).

The C elements always contain a dense granular material that is not present in the rest of the SR. The lumens of the elements of the dyad or triad that are derived from the ITs are often clear, but can also contain both small granules and possibly very small diameter (about 5 nm) vesicles. These elements of invertebrates and also the corresponding elements of the vertebrate IT system swell readily, particularly when fixed during or just following excitation (Selverston, 1967; Hoyle, 1980). Under other conditions, the entire IT may swell, but swelling may be precisely confined to regions, strongly suggesting that they have different permeabilities from that of the ajacent membrane.

3.2.6. Nuclei

Muscle nuclei have a wide range of sizes. But regardless of size, they are varied in location, though they most commonly occur only at the periphery, just below the sarcolemma. In some they occur deep in the myoplasm, and in some leg and flight muscles of insects and arachnid legs, they form a neat row down the central core (Figure 3.6.*1*). When the cell has a small diameter, an internal nucleus can really get in the way of contractile material: large nuclei in small-diameter cells totally fill the cross section, requiring tension to be transmitted around them, via the sarcolemma.

Figure 3.13. Typical forms of SR, IT and IT/SR contacts in vertebrates and invertebrates. (*1*) Vertebrate – amphibian frog sartorius. ITs are located at Z-line regions and make contacts with two cysternal elements (*C*) of SR forming triads (*arrows*). [Reprinted with permission from D. S. Smith, in *Muscle.* Copyright 1972 Academic Press.] (*2*) Toadfish, *Opsanus.* ITs are located at each A/I overlap, forming twice the number per sarcomere seen in (*1*). This pattern is common in mammals. [Reprinted with permission from D. S. Smith, in *Muscle.* Copyright 1972 Academic Press.] (*3*) Arthropod – giant barnacle. ITs are located at several places including A/I overlap and near Z bands. They originate both from the surface and from invaginated clefts. [Reprinted with permission from G. Hoyle, *How Is Muscle Turned On and Off?* Copyright 1970 by Scientific American, Inc. All rights reserved.]

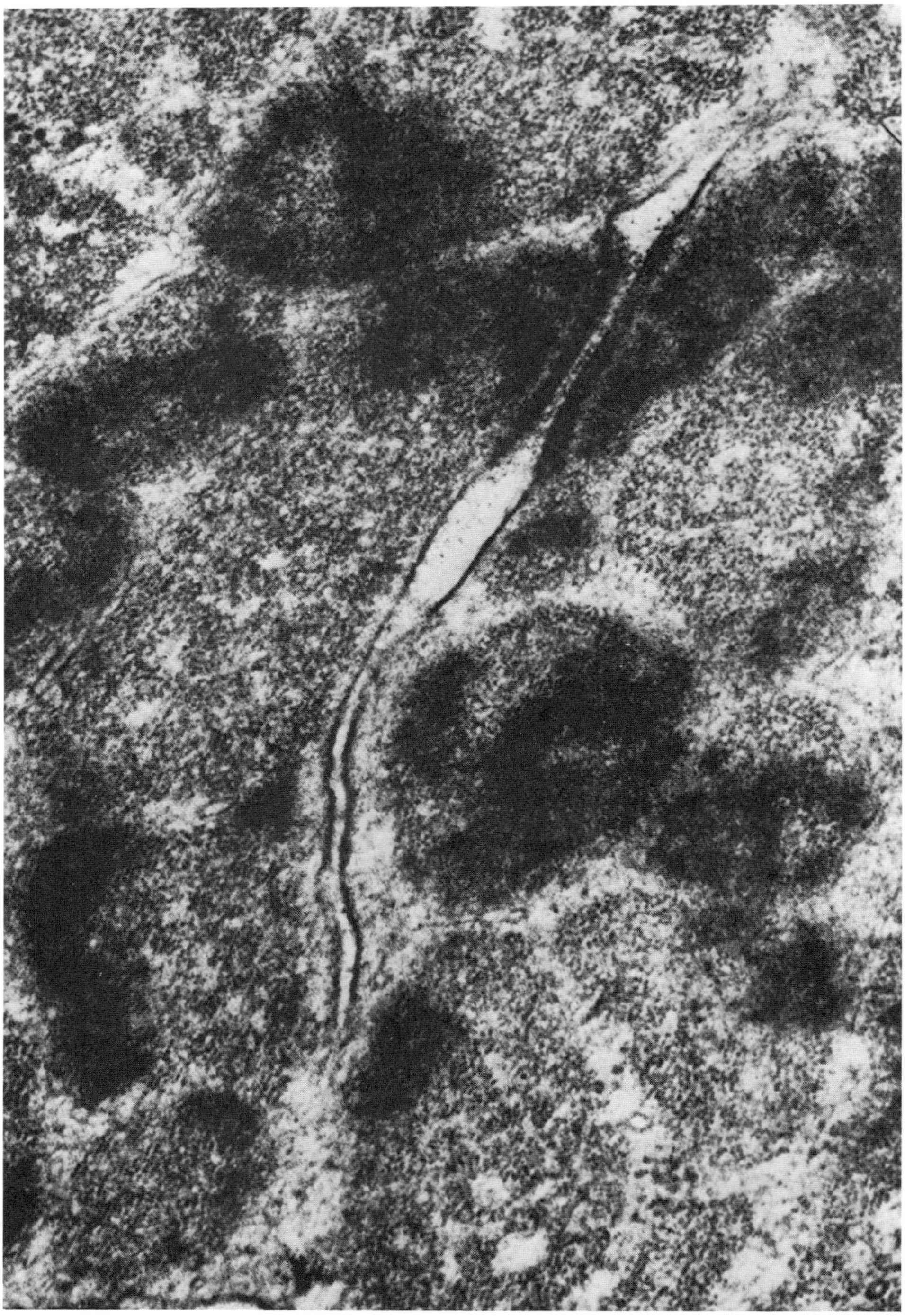

Figure 3.14. TEM of TS through Z region of fast cell of locust extensor tibiae. The denser regions are the centers of the Z regions. Note the approximately square arrangement of I filaments as they approach the Z regions, and also note the diagonal bridges. An invaginated Z tubule passes through the region making desmosome-like contact. (ORIG.)

3.2.7. Mitochondria

Mitochondria in muscle cells have a wide variety of shapes, sizes, and internal patterns of cristae. Some types of muscle cells have been classified on the basis of the detailed appearance of the mitochondria they contain, but there is not yet sufficient knowledge to make a systematic survey worthwhile. The numbers of mitochondria in a muscle cell vary. In some cells, they are both small and sparse, and in others large and/or abundant. For example, in humming bird flight muscle and in squid mantle muscle cells, more than 50% of the entire volume of the cell is occupied by mitochondria. Squid mantle muscle is cross-striated, with cells of small diameter, but the entire core of each is occupied by a single chain of large mitochondria (Figure 3.15). In some arthropod muscle cells, mitochondria are extremely numerous and impart to the cell a distinct pink to red color (Hoyle, 1973). In these cells they are located peripherally, the contractile material occupying the core. The color is not due to spectral emission but to strong absorption of green light by cytochrome *b* within the mitochondria that biases the reflected spectrum to the red. In some insect flight muscle cells, mitochondria are of giant size, up to 30 × 10 × 10 μm, or they may be as small as 0.5 × 0.5 × 0.6 μm.

In insect flight muscles, mitochondria takes forms ranging from almost spherical bodies lying in neat rows to thin flattened plates lying radially like the partitions in a stick of orange rock candy. In other muscles, mitochondria may be simple ovoids or sparsely or multiply branched (Figure 3.15.*3*). The commonest location for nongiant mitochondria in muscle cells is between all myofilament clusters near the Z regions, one in each half sarcomere and touching each other exactly in line with the middle of the Z material laterally. In some insect flight muscles, mitochondria are exactly as long as the sarcomeres (Figure 3.16.*3*). The second commonest location is just below the Sl. Often there is a single layer of mitochondria there, but in muscle cells that are slow and do a lot of work, layer upon layer occur, up to 20 deep. In very large muscle cells of big crabs and barnacles, in which extensive, longitudinal invaginations of the surface membrane, termed clefts, occur, mitochondria line up along the walls of the clefts.

3.2.8. Energy-Reserve Inclusions

Many muscle cells contain substantial amounts of material that represents stores of energy-providing substances, notably glycogen and lipid. Glycogen is in the form of nearly spherical granules, 20–60 nm in diameter, or rosettes, which are often present in neat rows (Figure 3.16.*1*) and are most commonly, and in some muscles exclusively, situated in the I region. Like the presence of mitochondria, which also prefer this region, the presence of much glycogen near Z regions causes marked pinching in of I filaments.

Lipid is present in the form of droplets, which may be small and scattered throughout the myoplasm, or relatively large and regularly arrayed between myofilament clusters, usually near the A/I overlap region.

3.2.9. Z Bodies and Z Bands

Bowman (1840) illustrated Z bands very clearly but did not name them. They were at first called the Quernlinie (Q band) by Krause (1868) and later often named after him Krause's membrane. The letter Q was later used to define a wide dye-absorbing region,

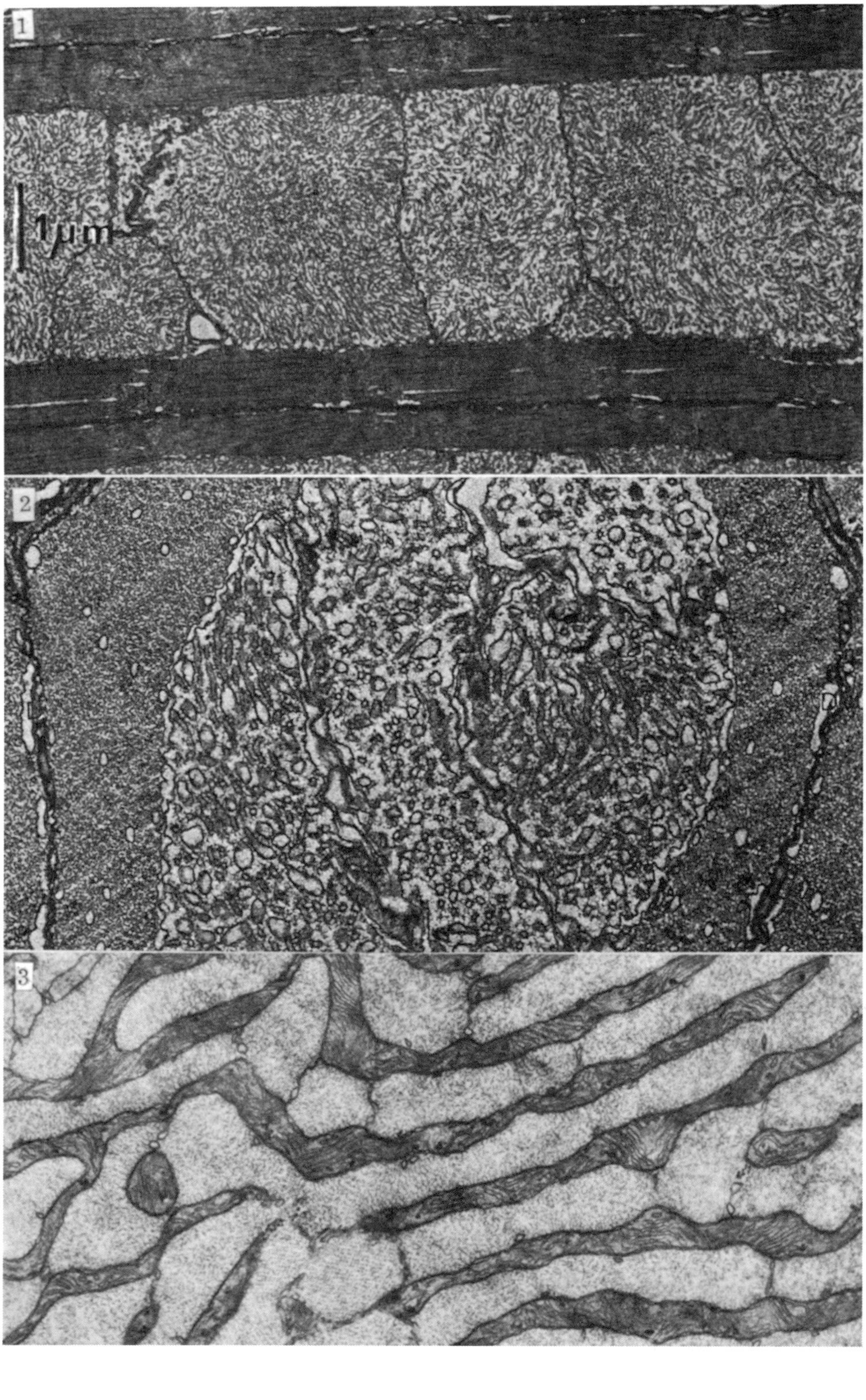
1
1μm
2
3

that at least in part contains the isotropic material of the A band, leading to much confusion. The Z region appeared to widen during contraction to from a "contraction" band. It has also been called Dobie's disc, telophragma, and end disc, but the letter Z (for *zwischen* – intervening, dividing) became the most popular and stuck. I refer to Z material, because in many muscle cells there are dense bodies to which the thin filaments are attached that are clearly serving the same role, that is, attachment sites for polarized thin filaments, and these bodies comprise material not present elsewhere in the cell. These bodies have been variously called just dense bodies, J bodies, and Z bodies. At one end of the spectrum, Z bodies are scattered about in the muscle cell with no discernible pattern. In very small cells, they resemble hemidesmosomes and are located only on the plasma membrane. In larger cells, they are scattered within the body of the cell in addition to being at the surface. Either of these two types would traditionally have been classified as nonstriated, or smooth. Next in order of progression come cells in which the Z bodies are relatively regularly arranged. Commonly these have an oblique or spiral form at rest (see p. 88). Next are cells with a close approximation to lateral alignment, leading to a ragged appearance; these cells are common in arthropods. Cross-striated invertebrate skeletal muscles often have Z regions that appear to comprise only a network of loosely connected Z bodies. In some cells, Z bodies are laterally aligned, even connected together, so that collectively they can be considered to be a form of perforated disclike band. Running through some Z bodies are fine lines that are oriented in the longitudinal axis. These represent either thin filaments running through the Z bodies, or thin Z filaments to which I filaments are attached.

A regular fine structure can be seen in the thinner Z bands of some vertebrate skeletal muscle cells, but the thicknesses of Z regions range over a greater than 10-fold range. The nature of the thicker ones, as well as that of Z bodies and of most invertebrate Z bands, remains to be worked out.

3.2.10. Fine Structure of Vertebrate Z Bands

The first Z structure to be proposed was for frog semitendinosus; it was proposed by Knappeis and Carlsen (1962) based upon EM studies. This supposed that very thin Z filaments (thinner than I filaments) are attached to the ends of I filaments in sets of four. The I filaments of adjacent half sarcomeres terminate just short of the Z region, which is formed by the Z filaments' crosslinking the two sets. They enter this region not in the familiar hexagonal array present in the region of overlap with thick filaments (see p. 82), but in a regular square lattice with 20-nm lateral separation. The set from one half sarcomere is displaced exactly one half of the lattice spacing distance diagonally. A somewhat similar basic plan was soon thereafter proposed by Franzini-Armstrong and Porter

---

Figure 3.15. Unusual forms of mitochondria. (*1*) TEM of LS through a cell from the mantle of the fast-swimming cephalopod mollusc *Sepiola,* in which the entire core, more than 50% of the total volume, is occupied by large mitochondria. (ORIG.) (*2*) TEM of TS of (*1*) Of interest are the longitudinal tubular SR and excitation–contraction coupling dyads (*D*) made by this with the surface membrane. (ORIG.) (*3*) TEM of TS of a portion of an atypical vertebrate cell showing numerous lamellate, branching mitochondria permeating the myofilaments. There is very little SR. Such an organization might represent an adaptation for prolonged strong force generation.

1
Z
T
TR
P
A
SR
P
Z
I
B

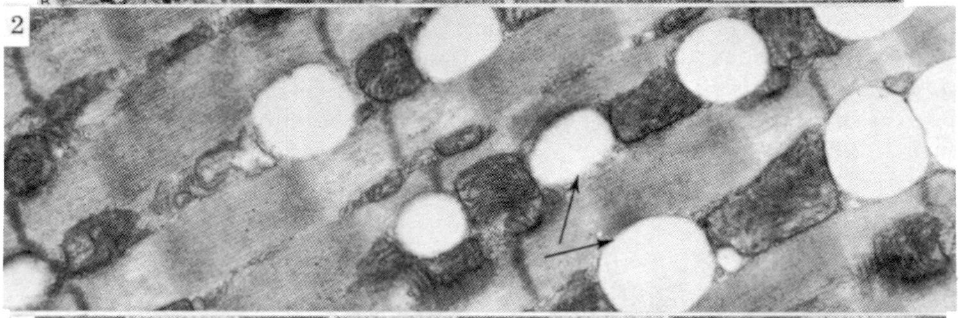
2

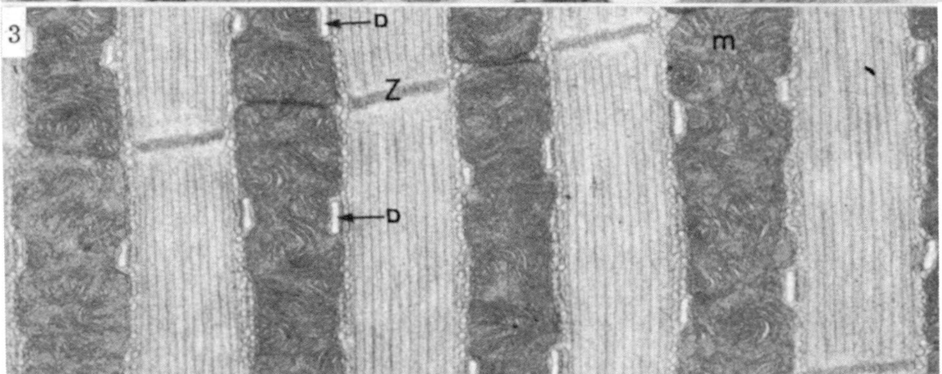
3
D
m
Z
D
D

(1964) as a general model, because it was based upon studies of newt, frog, rat, pigeon, turtle, toadfish, and fish. Their model also had the I filaments stopping short of the Z region, which they proposed to be formed by a perforated membrane folded into square, tent-shaped elements centered on each I filament, with ridges matching the locations of Knappeis/Carlsen's Z filaments. This model clearly owed a debt to Krause, but it was not an accurate representation of any specific Z structure. That same year, Reedy (1964) showed clearly that there really are ultrathin Z filaments, with no evidence of a membrane, in rat diaphragm. He showed them to form a subtle "weave" pattern as seen in face view, in which each set of four Z filaments forms a swastika. To explain this, Reedy produced a model in which the Z filaments unwind from quadruple, helically wound strands, which he referred to as "the terminal portions of I filaments." He had them diverge equally and obliquely to form Knappeis/Carlsen-type Z filaments that continue across into the adjacent sarcomere, where each strand meets three new partners to form I filaments of that sarcomere. This model has not been widely accepted.

Next to tackle the problem was Kelly (1967), who proposed that Z filaments are singlet actin strands, the double strand starting to unwind as it goes into the Z region, then each singlet passing round a partner from the adjacent half sarcomere before returning after making connected loops to become one half of the neighboring doublet. The direction of the twist was made in accordance with Depue and Rice's (1965) determination that actin has a right-handed screw sense.

Meanwhile, Patricia McNeil and I had begun a comparative study that had incorporated muscles of over 40 different species. I pointed out in a review (Hoyle, 1967a) that the variety of Z regions in vertebrate muscle cells alone was so great that it would require drastic flexibility for any single model to be able to explain them all. Except for some frog and fish muscles, the universal feature of all the models available at that time involved a stopping short of I filaments before the Z region. But mammalian and reptilian muscle cells have a marked apparent overlap in the Z region. We found the width of the Z band to be approximately inversely proportional to the intrinsic speed, being narrowest for very fast cells and broadest for slow, tonic ones, the extent of apparent overlap of I filaments across the Z region being greater for the slower ones. An alternative general model was developed by myself in 1967–1968. This was based upon our finding of two successive sets of Z filaments in making extremely thin serial sections across a single Z band longitudinally. They were obvious enough in longitudinal sections (Figure 3.17), but a critical test of the model was to cut two succeeding transverse sections, each of no more than 20-nm thickness, and superimpose them exactly: the window-pattern swastikas should occur in both sections. Pat McNeill and later Myra Elfvin, working in our laboratory, both succeeded in finding two sets in successive sections, which also resolved the screw sense of the two sets of filaments. This turned out to be identical in successive

---

Figure 3.16. Examples of different locations and forms of inclusions – energy-supply reserves and mitochondria. (*1*) Pigeon iris muscle cell has mitochondria only at the periphery but has abundant glycogen reserves (dark particles among the myofilaments). The IT/SR contacts occur at each A/I border in the form of triads (*TR*) or pentads (*P*), Pentads have two IT branches. (From Pilar and Vaughan, 1971.) (*2*) Diaphragm of shrew cell has large lipid reserves located between small mitochondria, principally at the I regions. (From Gauthier and Padyulka, 1966.) (*3*) Dragonfly flight muscle cell has no discernible reserves, giant mitochondria, and very regularly arranged dyads (*D*) and IT/SR contacts. (From D. S. Smith, 1966c.)

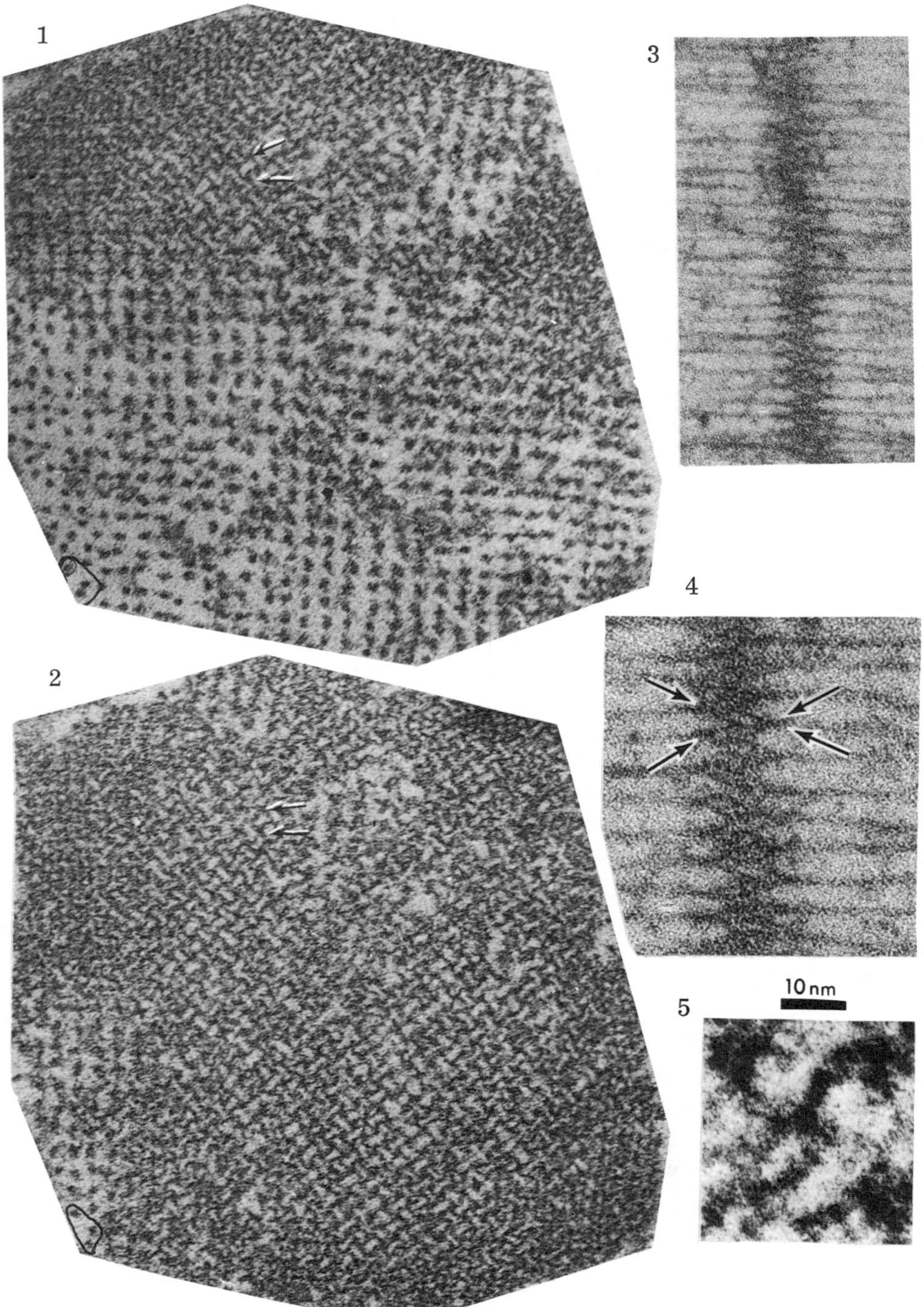

Figure 3.17. Fine structure of Z band. (*1*)–(*4*) Rabbit psoas; (*5*) garter snake. (*1*), (*2*) Successive very thin transverse EM sections through one Z region. In two regions the same "weave" or swastika-field pattern is seen. Arrows point to identical regions. (Courtesy of Dr. Michael Reedy.) (*3*), (*4*) An LS, showing double set of chevrons formed by A filaments (*arrows*). (ORIG.)

sections. Reedy came to a similar conclusion, also with serial sectioning. Reedy's clear electron micrographs are shown in Figures 3.17.*1*, and 3.17.*2*. The screw sense of the Z filaments is *left-handed,* and since that of actin is right-handed, the Z filaments have to be atttached to, or wound round, their ends with the opposite pitch.

Rowe (1971) also found a variety of Z-band thicknesses and proposed a model to explain them using a loop structure as Kelly had done, but without locking of Z filaments from contiguous half sarcomeres. He supposed there to be two sets of loops, with the extent of overlap of the loops determining the basic pattern. He explained the Knappeis/Carlsen image as obtained when the two sets of Z filaments lie very close together with minimal overlap. He later (Rowe, 1973) compared Z bands of rat plantaris cells with different speeds: slow (red), intermediate (pink), and fast (white), finding the same relationships we had. He added one additional loop to his model for intermediate cells and two for the slow ones, making triple and quadruple sets of Z filaments respectively to explain the broadening associated with slow speeds. Goldstein et al. (1979) found broad Z lines, with four sets of Z filaments, in dog cardiac muscle. They proposed a model in which successive swastika formations have opposite handedness. The critical test of their model is clearly to compare the sets in successive ultrathin sections. In the sections made in our laboratory, as in the ones made by Reedy, the screw sense is the same (as in my model) for both sets of filaments. Two opposite screw senses in consecutive Z filaments were used by Goldstein et al. (1979) to explain a commonly encountered small-square appearance. Small squares (Reedy calls this the window pattern) can occur side by side with the basket weave pattern. But as these authors realized, this image could be formed merely by kinking the Z filaments, which will happen if the lateral spacing is reduced. When not clearly associated with I–I bridges, small squares are associated with a reduced lattice size. There is an alternative explanation of small squares. Simple right-angled lateral cross connections occur between I filaments close to the Z region. These must produce a small-square pattern regardless of the Z-filament arrangement. According to all the basic Z-filament models, a simple clear image ought to be common, but never is, possibly because there are the two sets of filaments. Their combined trapping actions for stray molecules could account for the muddy appearance commonly encountered. I now consider it probable that Z filaments are directly attached to the ends of the I filaments (Figure 3.18). The thickness of Z filaments is somewhere between 2 and 3 nm. They are much more electron dense after staining than the components of I filaments. For several years it was thought that Z bands are formed of tropomyosin filaments, but this idea was abandoned when it was shown that labeled antitroponin stains do not touch the Z bands (Pepe, 1966). Tropomyosin was instead shown to be a component of I filaments. A new muscle protein, α-actinin, that was discovered by Maruyama (1965, see p. 125) is now considered to be the principal component of Z bands and most probably is the Z-filament protein.

There is a short, stiff, densely staining region, about 9 nm thick by 40 nm long, between each I filament and the Z filaments. This region represents eight Z filaments twisted together. At the outer end, an I filament is attached. The twisted Z filaments unwind in the Z region to form the swastikas of the weave pattern. Two of the Z filaments from adjacent I filaments combine to make the dense four-stranded core that attaches to the end of the I filament from the next sarcomere.

In this model, the I filaments in rabbit psoas do not overlap, they merely appear to do so on account of the continuity of the multiple-filament strand with the I filament. These

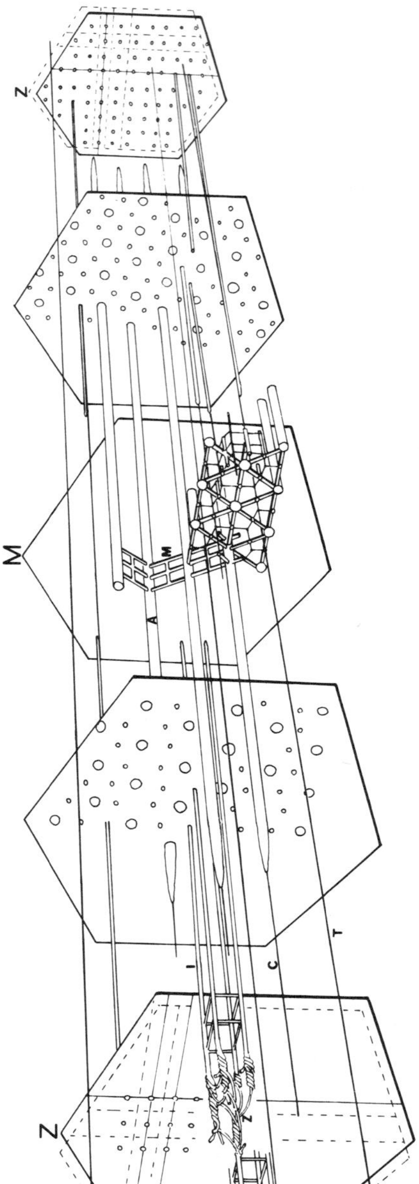
Z
M
M
A
I
C
T
Z

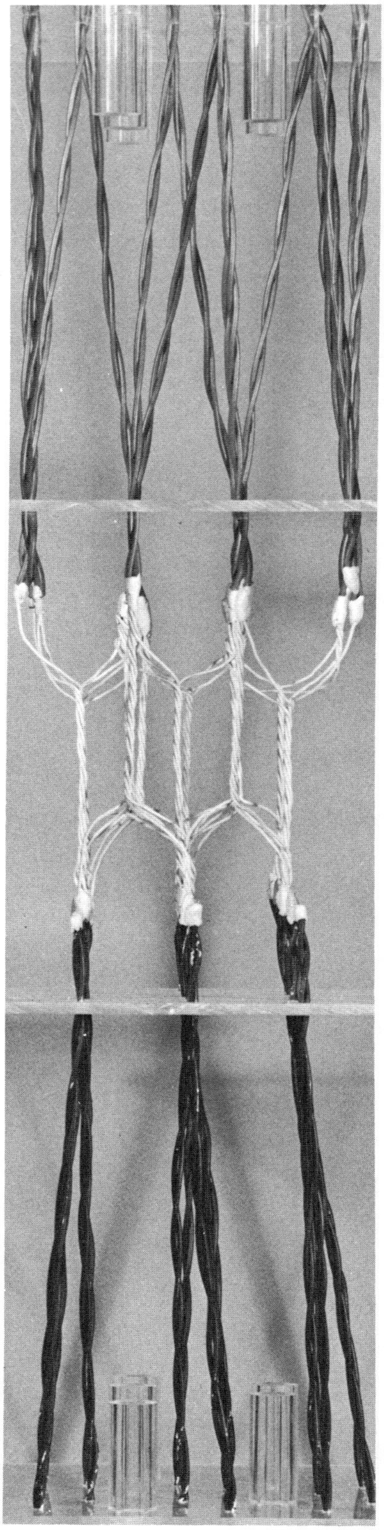

quadruple structures overlap in varying degrees in a majority of vertebrate cross-striated cells. Four Z filaments derived from each of four different sets of quadruple Z's from the neighboring sarcomere converge upon the quadruple strands near their attachments to I filaments. An extra set of Z filaments can be added on one side only, or both sides can add an extra set to make four. It is easy to imagine the addition going on indefinitely. When this happens in humans, it causes the crippling, life-destroying disease termed nemaline myopathy (see p. 416). Each of the dense, transverse bands seen in the thick Z regions of nemaline myopathy cells seems to be formed from laterally aligned sets of Z filaments. An exactly similar appearing Z structure to that occurring in nemaline myopathy has been found by Fawcett (1968) in cat ventricular muscle. In this case, I filaments of normal length are attached to the edges of the more than 1-μm-wide Z region. Spacing apart of the conspicuous longitudinal Z striations is about 20 nm. It is possible to fit most Z bands of vertebrates into the general model proposed if slow-type cells have three, four, or even more sequential sets of Z filaments. Fish fast myotome cells either have only a single set, or double set with very little separation (Figure 3.19.*3*).

3.2.11. Removal of Z Material

The point of attachment of I filaments to Z-filament bundles might be expected to be a chemically vulnerable one, and this is indeed the case. After treatment with 0.1% lactic acid or even just a strong salt solution, the attachments break, leaving only a series of whole "discs" of Z material (Garamvölgyi et at., 1962), alike in rabbit psoas and bee flight muscle. With very low ionic strengths (Stromer et al., 1969), or with 1 *M* urea (Rash et al., 1968), the Z-line density reduces markedly (Figure 3.20.). In urea, I filaments are removed also, leading to A bands connected by very fine filaments (see p. 71). When α-actinin was added to rabbit psoas cells previously treated to remove Z material, most of it got taken up by I filaments, not Z region, in the form of I–I X bridges. Although the Z region drastically loses density, it doesn't disappear, nor do the sarcomeres fall apart. Assuming the Z filaments have dissolved, some ultrathin filament structure must still exist and hold everything together (see p. 90).

A number of procedures have been found that remove Z material without also disrupting sarcomeres. These range from soaking with acetone (Walcott and Ridgway, 1967) to brief treatment with trypsin (Busch et al., 1972) to bathing in a mixture of 1 m*M* deoxycholate with 5 m*M* Mg^{2+} at pH 7.1 with very low calcium (Busch et al., 1972). These treatments, although removing all dense material and Z filaments, leave ultrathin parallel filaments that cross the Z region and hold the two half sarcomeres together. Z material can be returned and attaches to make normal-looking bands (Figure 3.20 – Stromer et al., 1969). The Z regions of different kinds of cells can be markedly different in detail (Fig. 3.19); partial extraction reveals some of the different kinds (Hikida, 1978).

Figure 3.18. (*Top*) three-dimensional drawing of a single sarcomeral myofilament cluster of a typical vertebrate cross-striated cell, based on research in the author's laboratory. (*Bottom*) photograph of a model of the Z region of rabbit psoas cells and the transition zone from the A/I overlap regions of adjacent sarcomeres, but excluding T filaments, I/I bridges, materials attached to Z filaments, and the N band. The thick filaments are represented by thick plastic rods. Z filaments are attached in sets of four to the ends of I filaments and have a left-handed screw sense. The actin components of the I filaments have a right-handed screw sense. Also see Figures 3.17, 3.19, 3.20, 3.21, and 3.27.

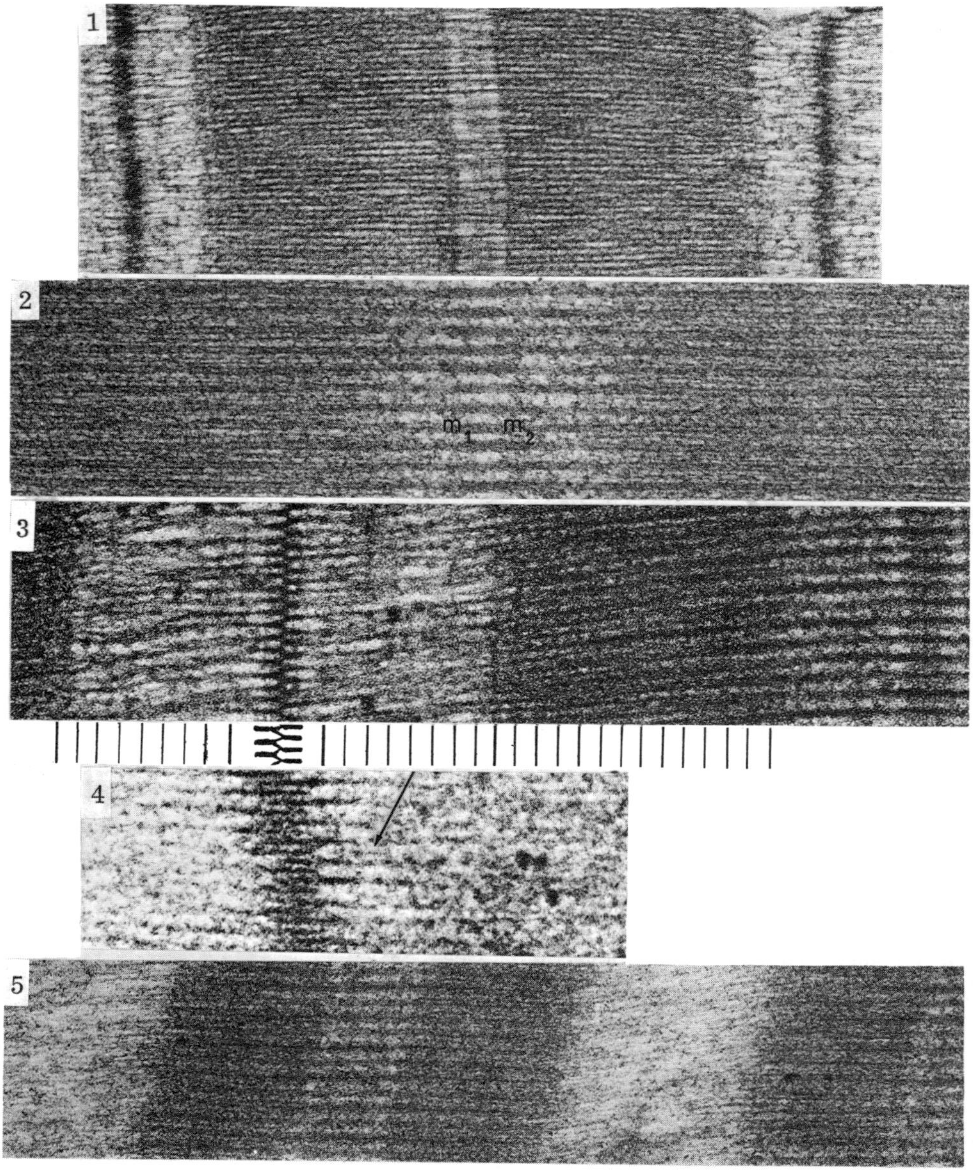

Figure 3.19. Examples of variety of details in M and Z regions. (*1*) Thin section through rabbit psoas in an I-I-I plane, including fragments of A filaments. Note the ultrathin filaments crossing the H zone. (ORIG.) (*2*) The H zone of a slow crustacean cell, eyestalk-raising muscle of *Podophthalmus,* showing weak double set of M bridges. (From Hoyle, 1967.) (*3*) Typical fish myotome muscle Z band. Either there is but a single set of Z filaments or there are two, but with very slight separation. Note the regular short dense regions. (ORIG.) (*4*) Typical Z region of a slow-type vertebrate cell (this one from *Gecko* toe). Note the wide overlap of horizontal portions of Z filaments. Diagonal Z filaments and an ultrathin filament (*arrow*) can be seen. (ORIG.) (*5*) Copepod (*Doropygus*) antennal muscle with barely discernible Z regions. They are bridged by numerous very thin filaments. (ORIG.)

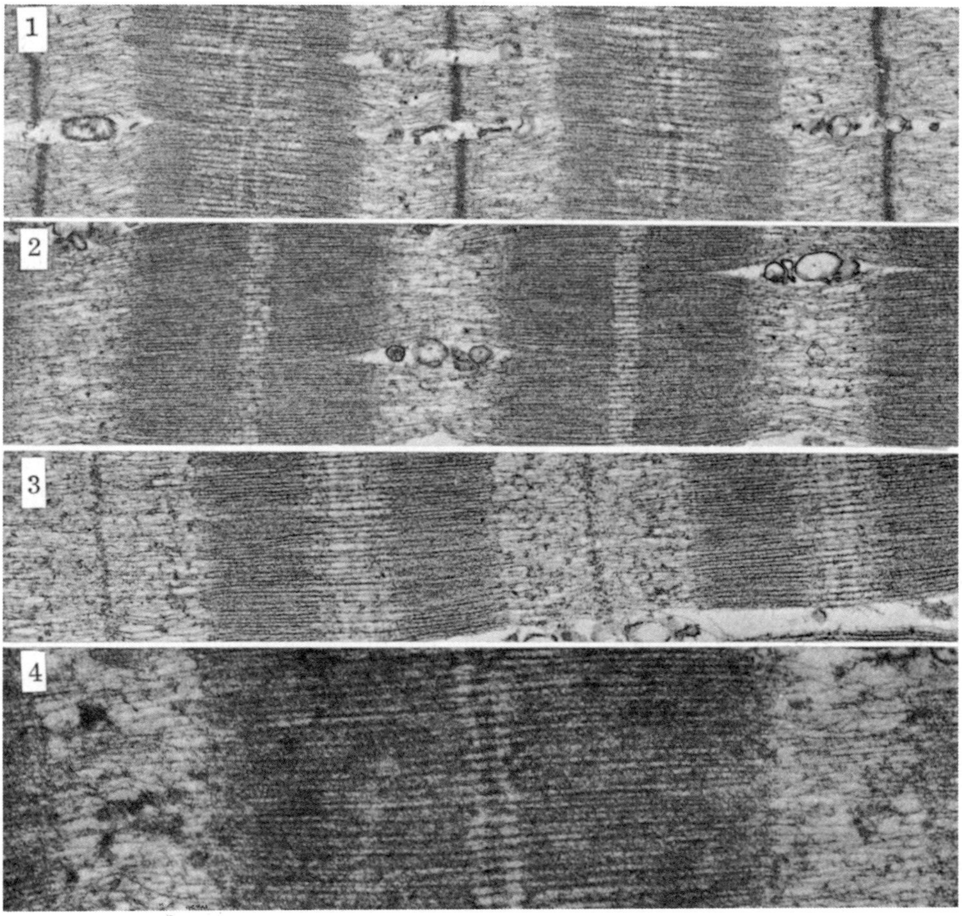

Figure 3.20. Removal of Z material. (*1*) Normal sarcomere of rabbit psoas. (From Stromer et al., 1967.) (*2*) Same as (*1*) after removal of Z material with solution of very low ionic strength. (From Stromer et al., 1967.) (*3*) Sarcomere from which the Z material had been removed by treatment with dithiothreitol, followed by incubation with α-actinin. (From Stromer et al., 1967.) (*4*) Rabbit psoas cell extracted in deoxycholate/Mg solution for 24 h. [Reprinted with permission from Busch et al., *J. Cell Biol.* **52**, 367–381 (1972). Copyright 1972, The Rockefeller University Press.]

The Z regions of invertebrate body and leg muscles are immensely dense. In part this can be attributed to a generally high density of thin filaments (the thin and thick filament ratio is about 5 : 1), but even in the few invertebrate skeletal muscles in which the ratio is only 2 : 1, the Z region details cannot be resolved. An exception is the copepod antennal remotor (Hoyle, 1967a), where the Z region is so faint it can hardly be discerned (Figure 3.19. *5*). However, a few Z filaments very similar to those of vertebrates are present, forming a structure similar to that of fish Z bands. Most invertebrate Z regions are much thicker. A square array of I filaments is often discernible near to the Z region, and a small-window lattice is just discernible. The same is true of the Z bodies of barnacle giant muscle cell, in which the Z region is perforated, but each Z body contains eight repeating substructures, like an early stage of nemaline myopathy.

The I filaments in insect flight muscle do not change from the hexagonal array present in the A/I overlap region as they do in vertebrate muscle cells (see p. 78). There are attachments of the thick filaments to the Z region via so-called C (= connecting) filaments (Figure 3.21) first clearly demonstrated by Garamvölgyi (1963), then by Auber (1967a) and by Trombitâs and Tigyi-Sebes (1974). The chemical nature of the C filaments is unknown. Most insect flight muscle cells cannot be stretched much without breaking them, but Garamvölgyi was able to stretch bee flight muscles well beyond thin–thick filament overlap, and he attributed this to elastic properties of the C filaments. Under stretch, the Z regions become birefringent (Garamvölgyi, 1966b). He suggested connections of A filaments across Z regions by way of C filaments, but Ashurst (1971) disclaimed this, writing, "The filament lattice of the Z-line of insect flight muscle excludes the possibility that filaments are continuous from one sarcomere to the next." The lattice she referred to was one she herself had described. Very recently Trombitâs and Tigyi-Sebes (1979) contradicted her argument by showing that isolated pairs of thick filaments from adjacent sarcomeres of bee flight muscle are linked by C filaments (Figure 3.21.*1*). This means that her model cannot be correct: in her model, based upon studies of *Lethocerus* (Hemiptera), the Z region is formed by overlap of I filaments from adjacent half sarcomeres, together with simple cross links in the transverse plane. Saide and Ullrick (1973) found that three thin filaments from each half sarcomere each enter a triangular "Z tube" that completely traverses the Z band. The orientation of the tubes changes, by a 60° rotation parallel to the axis, in passing from one side to the other. This results in displacement of the filament lattices so that each thick filament of one sarcomere is aligned with the trigonal position of three thick filaments of its neighbor, a conclusion requiring that the C filament shift along the longitudinal axis. However, Z-type filaments with a diagonal aspect occur in *Lethocerus* Z bands (Figure 3.21.*2*).

3.2.13. Functions of Z Bands

It was thought for more than half a century, until as late as 1955 (Tiegs, 1955), that Z bands may provide the route for inward spread of excitation. This view was abandoned following the rediscovery of ITs. In isolated myofilament clusters of giant barnacle retractor muscles, the Z bands expand when irrigated with ATPtaining solution (Hoyle et al. 1965). This finding led to a bizarre hypothesis of the contraction mechanism (Ulrick, 1967) based on active broadening of Z regions. Contraction waves in cells that appear to have a spiral organization follow the same spiral path as the SR structure, as if travelling along a continuous spiral Z line (Matthaei and Tiegs, 1955). Z tubules a subset of ITs,

Figure 3.21. Ultrastructure of Z regions. (*1*) Two isolated thick filaments from contiguous sarcomeres of honeybee flight muscle, negatively stained, are seen to be joined together by a shared C filament. [Reprinted by permission from *Nature* **281**, 319–320 (1979). Copyright 1979 Macmillan Journals Ltd.] (*2*) Ultrathin LS Z band from *Lethocerus*. Diagonal Z filaments resembling those of rabbit psoas are apparent. (ORIG., by B. J. Walcott.) (*3*) Early diagram of C filament of blowfly flight muscle attachment to Z band. [Reprinted with permission from: J. Auber, and R. Couteaux, *J. de Microscopie* **2**, 309–324 (1963).] (*4*) Diagram of face view of vertebrate Z region showing ultrathin Z filaments unwinding from extensions of I filaments of two contiguous sarcomeres, based upon Figures 3.17 and 3.19. (*5*) Side view, same as (*4*). (*6*) Three-dimensional view of vertebrate region. Interpretation and drawing by the author.

1

2

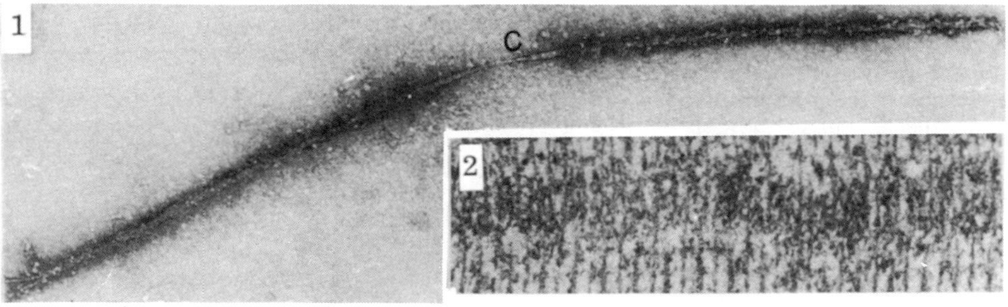

3

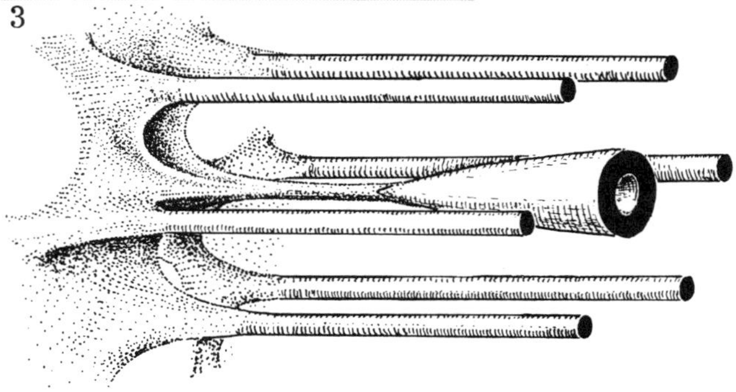

4

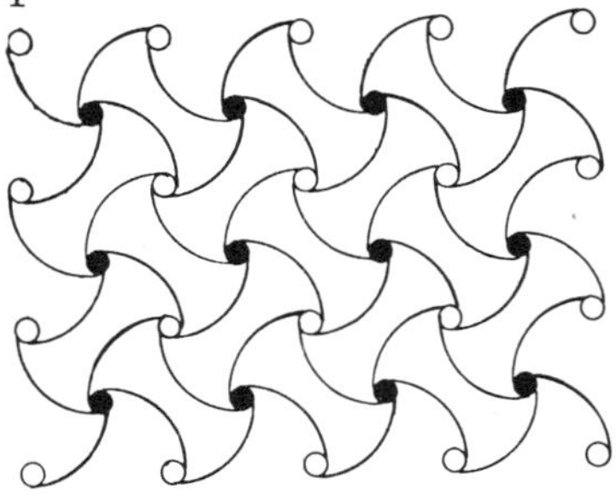

5

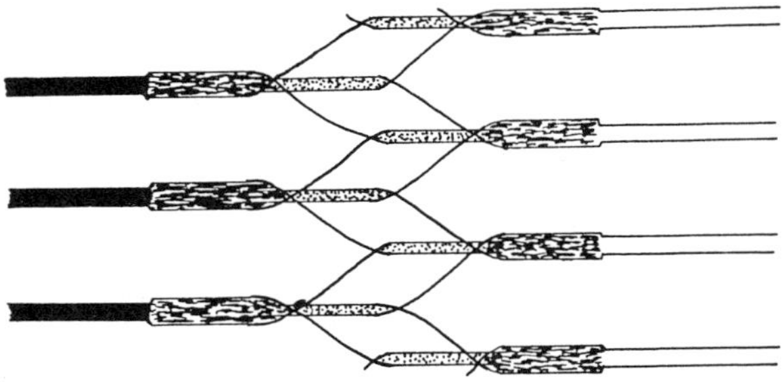

6

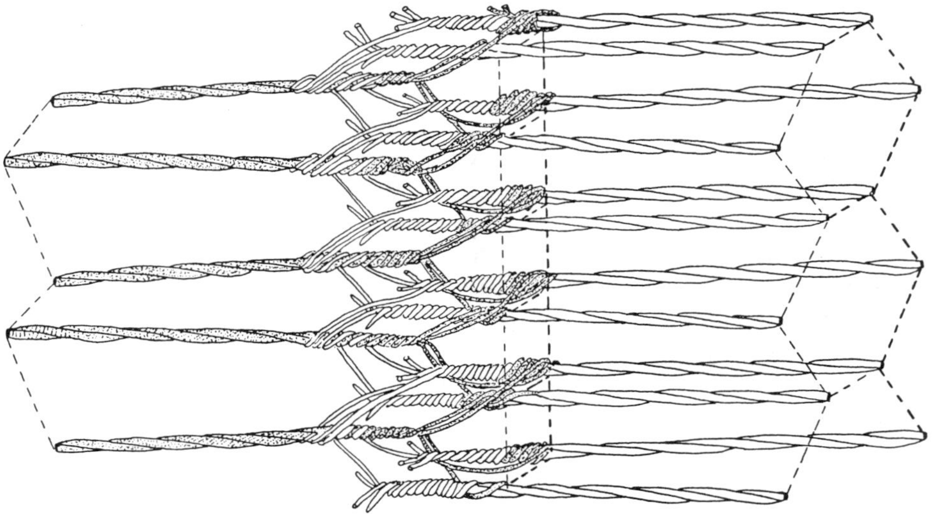

offer a means of inward transmission for some cross-striated cells that is equatable with Z material (Figure 3.14). In spite of these findings, it is generally agreed that the basic function of Z bands is only structural, to anchor the I filaments. The Z-region structure must be sufficiently elastic to enable lateral expansion during extensive contraction, and also to be responsive to strains in the longitudinal axis. In rabbit psoas, strong stretch about doubles the Z-band width.

3.2.14. M Bands

It was long disputed whether there is truly a distinct structure in the middle of a sarcomere: those who believed in it called it an M (for *mittel,* intermediate, middle) disc. It is now realized that in a majority of vertebrate cells there is indeed a distinct complex, M structure, but in a few slow ones this is weak and the details have not been resolved. In some vertebrate muscles, such as chicken striated muscles, in which M bands are not discernible in ordinary TEM sections, fluorescent-labeled antibody to M protein adheres firmly to the center of the H zone anyway. In invertebrates there is a range from well defined to undoubtedly absent. Also in many crustacean cells, there is a clear zone in the exact middle of the sarcomere, but a pair of bands on either side of it forming a double M band (Figure 3.19*.2*). Giant barnacle cells have no M band, the equivalent bridges being distributed widely along the very long A filaments instead of being concentrated in the middle (Hoyle et al., 1973).

The vertebrate form of M band comprises three parallel sets of equally spaced bars. Knappeis and Carlsen (1968) noticed small densities on the M bridges, matched by short bars (M bars) as seen in longitudinal section. Recent micrographs have shown an additional, complex accessory structure (Figure 3.22 – Luther and Squire, 1978). The presence of longitudinal fine filaments between the M bridge was noted by McNeill and Hoyle (1967) and attributed to T filaments crossing the H zone. Extremely fine filaments attach these to the thick ones. Fine bridges from the A filaments onto the T filaments account for the basic structure of the M region of insect flight muscles and many other invertebrate cross-striated muscles (Figure 3.23). These, however, lack the thick M bridges seen in fast-type vertebrate cells.

There are marked differences between cells in regard to details of the M band even within the same muscle. In human muscle, Sjöstrom et al., (1979) recognized four types: 1, 2A, 2B, and 2C. Type 1 had five parallel rows of well-marked M bridges and also had broad Z bands. Type 2A had three strong middle M bridges but weak outer ones, and an intermediate Z thickness. Type 2B had middle bridges only, and these were well defined with rather narrow Z bands. Type 2C had M regions similar to those of Type 2B, but extremely narrow Z regions.

In M regions of insect flight muscle, three rows of M bridges are clear, but five or more can sometimes be seen: the appearance changes from one myofilament to the next, most appearing weak. It is possible to explain these if there is a spiral sequence of some kind (e.g., Pepe, 1971). These thick filaments have a solid, oval profile in the M region, about 30% larger than that in the A region, with different degrees of tilt relative to each other (Figure 3.23*.1*). This appearance is due to attachments of two T filaments to each A filament in the H region, combined with attachments of the fine M bridges (Figure 3.24).

Many crustacean muscle cells have no M bridges in the exact center of the sarcomere, but instead have two sets, one on either side. Muscles with very long sarcomeres may have no M bands, but either sporadically occurring cross bridges between thick filaments along

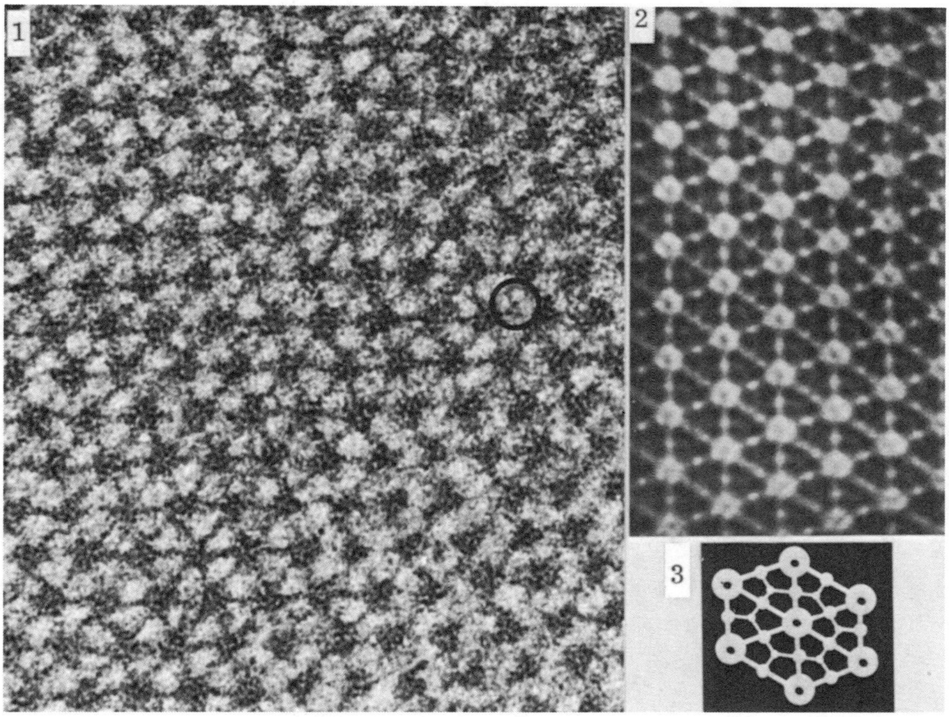

Figure 3.22. Details of the vertebrate M band. (*1*) Rabbit psoas – fresh fixed; untreated negative. (ORIG.) (*2*) Filtered image from frog sartorius. [By P. Luther and J. Squire. Reprinted with permission from *J. Mol. Biol.* **125**, 313–324 (1978). Copyright 1978 Academic Press, London.] (*3*) Interpretation of image shown in (*2*). [Reprinted with permission from *J. Mol. Biol.* **125**, 313–324 (1978). Copyright 1978 Academic Press, London.]

most of their length or localized "M" bands, but with more than one distributed along the length of the thick filaments except at their outer regions.

Antibodies to known muscle components have recently been used to locate them, and these anitbodies have shown that creatine kinase is intimately associated with M bands (Wallimann et al., 1978; 1979).

A means of accounting for some of the diversity in M-band fine structures has been proposed by Pepe (1971). He suggests that as the thick filaments start to self assemble by abutment of opposed myosin molecules (see p. 115), before growing outward from there by progressive addition in sterotyped packing configurations that are specific for the cell type, they may either abut each other end to end or overlap. The M bridges then attach to their ends. Different packing configurations would necessarily be associated with different forms of M bands.

3.2.15. Function(s) of M Bands

The enzyme creatine kinase has been found to be attached to the M region in some vertebrate muscle cells, so it possibly functions in biochemical events. The obvious function is structural, as a device for holding the thick filaments together and keeping them in register. But this role is difficult to uphold in view of the fact that an M band is not

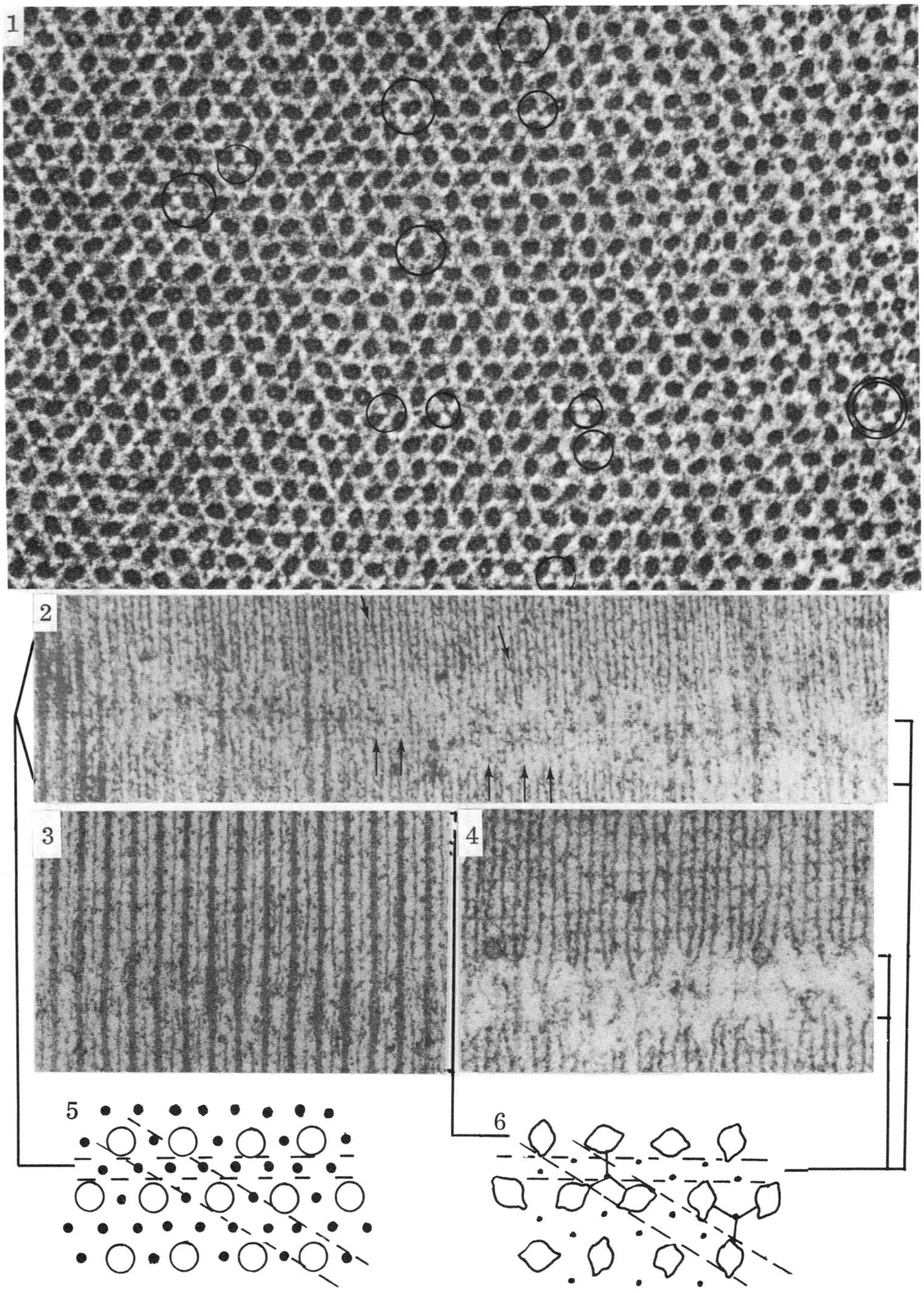
1
2
3
4
5
6

visible at all in many A bands – which nevertheless hold together – or the fact that removal of M bands using low-ionic-strength solution 2 m*M* Tris at pH 7.6, 1 m*M* dithiothreitol (Stromer et al., 1969), or alkaline solutions such as 5 m*M* Tris at pH 8.0, is not associated with disruption. Some bridges between thick filaments occur well away from the M region even in vertebrate cross-striated cells, so bridges specifically associated with the M region may not be the only structural elements holding A filaments together.

3.2.16. N Bands

Light microscopists variously claimed one, two, or no dense bands in the I regions of cross-striated muscle cells; they termed these N bands (N_1, N_2, or N_α, N_β). Electron microscopy has confirmed the reality of these. A single N line was described by Page (1968) in tortoises; it had a variable location at different degrees of stretch, 0.3 μm away from the Z band at rest length or less and linearly farther away in proportion to stretch, being 0.8 μm in cells stretched 60% to SL = 3.8 μm. The distance between N bands of the same sarcomere measured across the M band increased only 44%, compared with an increase of about 300% for the increase in distance between neighboring N bands measured across the Z band. The percentage increase between the Z band and the outer margin of the A band was also about 300. Thus the position of the movable N band of the tortoise muscle is compatible with its representing material attached to a point on a longitudinal elastic element situated in parallel with I filaments. It would be interesting to know where the tortoise N line appears following stretch beyond overlap of I and A filaments. The line becomes broad and poorly defined in extensively stretched cells. Both the progressive shifts and the broadening with stretch strongly suggest association with an elastic element extended during stretch that is independent of both I and A filaments. In very heavily stretched beef muscle, some breaks occur in the attachment of I filaments close to the Z lines. In these broken elements, the N_2 line of the adjacent intact I band undergoes a drastic shift, being displaced toward the Z line (Locker and Leet, 1976b). This strongly suggests that N_2 material is attached to an ultrathin filament that passes directly through the Z line and is not broken even in extreme stretch.

N_1 or N_α is close to the Z line; N_2 or N_β is sometimes seen only under stretch, since it is located in the overlap region. Electron micrographs show dense regions in the two positions (Figure 3.25), but fail to show details of the fine structure. No specific chemical material has yet been associated with N bands. They have generally been ignored, not

---

Figure 3.23. M band of insect flight muscle. (*1*) TS of the M region of *Lethocerus.* At right a few of the inner ends of I filaments are seen (double circle), mainly occupying the in-line position, but with one or two moving toward the trigonal positions. In the H zone, fine M bridges contact thin longitudinal filaments (M filaments) believed to be T filaments linking I filaments. The A filaments are oval in shape in the M region. (Courtesy of Dr. B. J. Walcott.) (*2*) Very thin LS of the M region of (*1*) made through an I-filament layer but grazing A filaments. M bridges onto T filaments are clearly seen. Note also the bending of the inner ends of the I filaments. (*3*) Very thin LS of the M region of (*1*) made through an I-A-I filament layer. (*4*) Very thin LS through the I-filament layer showing bending of the inner ends of the I filaments and the T filaments connecting them. [(*2*), (*3*), and (*4*) Courtesy of M. K. Reedy – who is not responsible for the interpretation] (*5*) Diagram of transverse section through A/I overlap region to show planes for sections (*2*) and (*3*). (*6*) Diagram of transverse section through M region to show planes for sections (*2*) and (*4*).

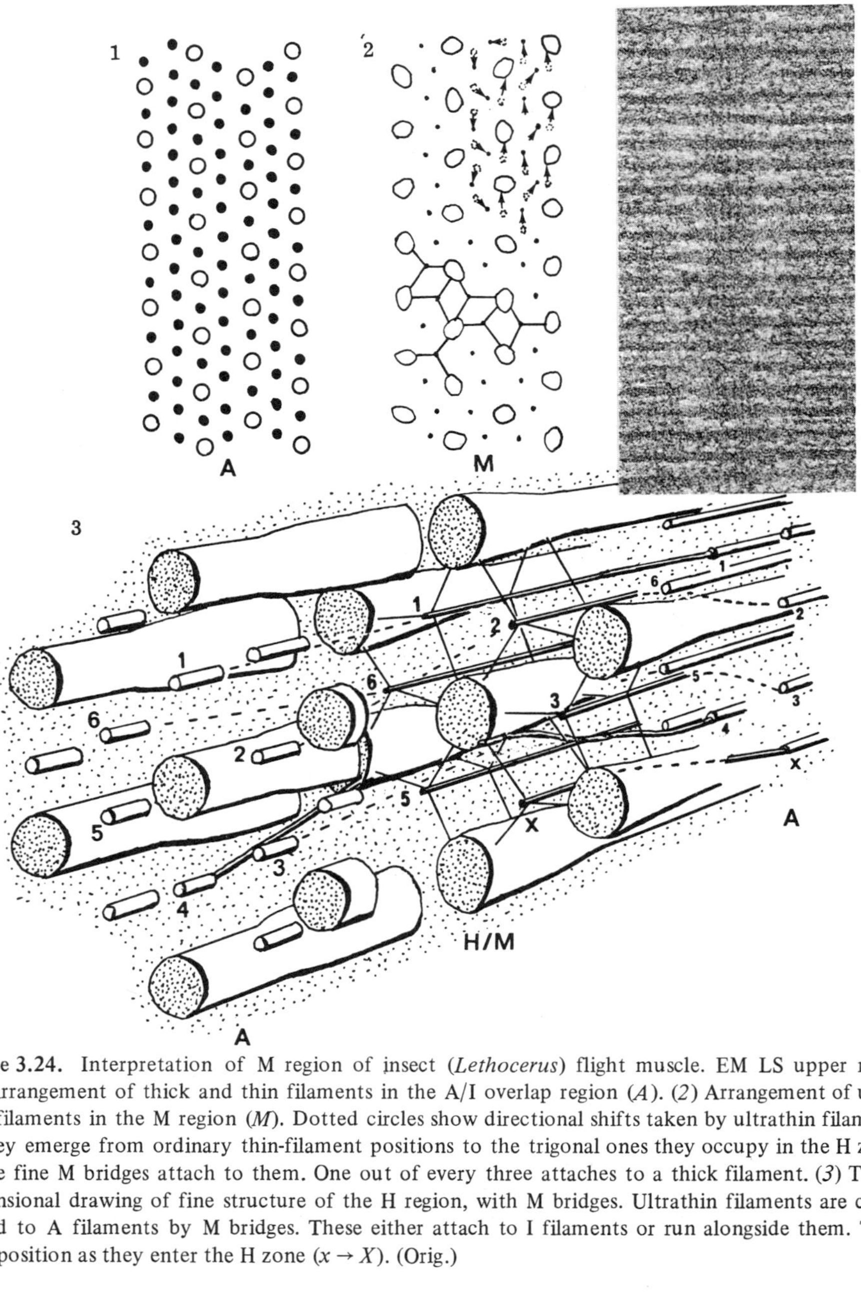

Figure 3.24. Interpretation of M region of insect (*Lethocerus*) flight muscle. EM LS upper right. (*1*) Arrangement of thick and thin filaments in the A/I overlap region (*A*). (*2*) Arrangement of ultrathin filaments in the M region (*M*). Dotted circles show directional shifts taken by ultrathin filaments as they emerge from ordinary thin-filament positions to the trigonal ones they occupy in the H zone, where fine M bridges attach to them. One out of every three attaches to a thick filament. (*3*) Three-dimensional drawing of fine structure of the H region, with M bridges. Ultrathin filaments are cross-linked to A filaments by M bridges. These either attach to I filaments or run alongside them. They shift position as they enter the H zone ($x \rightarrow X$). (Orig.)

only by the textbooks, but also in specialist works on muscle (they are not mentioned in Needham's encyclopedic *Machina carnis,* 1971), and there is even a dearth of speculations regarding both their possible natures and functions. N_1, when present, occurs close to the Z region, 0.18 μm distant in rabbit psoas, and its position is not greatly affected by stretch. In rabbit psoas cells from which myosin has been extracted at rest length, a second N_2 band is seen 0.69 μm from the Z region (Walcott and Ridgway, 1967); its position is stretch sensitive.

The N bands of diverse vertebrate striated cells were compared by Franzini-Armstrong (1970b) using injected ferritin as a location aid. Ferritin tends to become trapped at N_1, and to lesser extent at N_2. She agreed with the early light microscopists that N_1 is fixed, and affirmed and extended their view that the location of N_2 depends on sarcomere length; therefore, the band studied in tortoise muscle by Page was probably N_2. Although N_1 is generally not shifted by stretching, the contrary has been observed in frog heart cells (Staley and Benson, 1969). Electron microscopy of invertebrate cross-striated muscle cells has been of no help in revealing the nature of N bands, although they were first recognized in insect muscles (Engelmann, 1878; Rollet, 1890). A slow crustacean muscle cell that is cross-striated, the esophageal dilator of *Macrocyclops,* has been found to have a very wide, dense N_2 band (Fahrenbach, 1967).

In transverse sections through the region proximal to the Z band, fine X bridges occur between I filaments, together with amorphous material. In longitudinal sections, this region marks the end of well-defined, straight I filaments emerging from the Z region. Next comes a region in which some I filaments take a distinct turn and others seems to split into two, making a Y branch. The N line more distant from the Z region is also a region where some I filaments make prominent bends and where some apparent splitting appears. Strongly staining amorphous material is attached here, but as seen in transverse sections, this is a very indistinct region. It contains fewer well-marked I filaments than occur in either the region close to the Z band, or in the A/I overlap region.

3.2.17. Functions of N Bands

The size of the dense regions present in N bands and their amorphous nature suggest that they might be enzymes. Rabbit muscle is known to contain a protease, cathepsin L, which degrades actin, so one possible N-band function would be that they are sites for such an enzyme, with the function of depolymerizing actin as needed, to prevent jamming of I and A filaments. The functional conversion of actin polymers to simpler forms might be an unsuspected but functionally important operation in some muscles.

3.3. Common Ultrastructural Malformations

Putting together a muscle cell during development must involve many factors. There is also a continuing set of trophic interactions that leads to changes in the protein and enzyme composition in relation to previous activities, so one must not expect to find perfection in structural details. The closest to a fine structure lacking obvious abnormalities is seen in some insect flight muscles.

Four common faults occurring in both vertebrate and invertebrate cross-striated cells are illustrated from rabbit psoas examples in Figure 3.26. Figure 3.26.*1* shows fusion of A filaments, with correspondingly large, raggedly arranged orbits of I filaments. Figure 3.26.*2* shows some of the consequences of either unequal lengths of neighboring myofilaments or their lack of lateral alignment. Some I filaments at the margins form part of the orbit of the neighbor in its H zone. Figure 3.26.*3* shows the effects of contact between thick filaments within the same cluster but with different lattice patterns. Figure 3.26.*4* shows an extremely common error in regions very close to the Z region, where there is a basically square array of the I filaments.

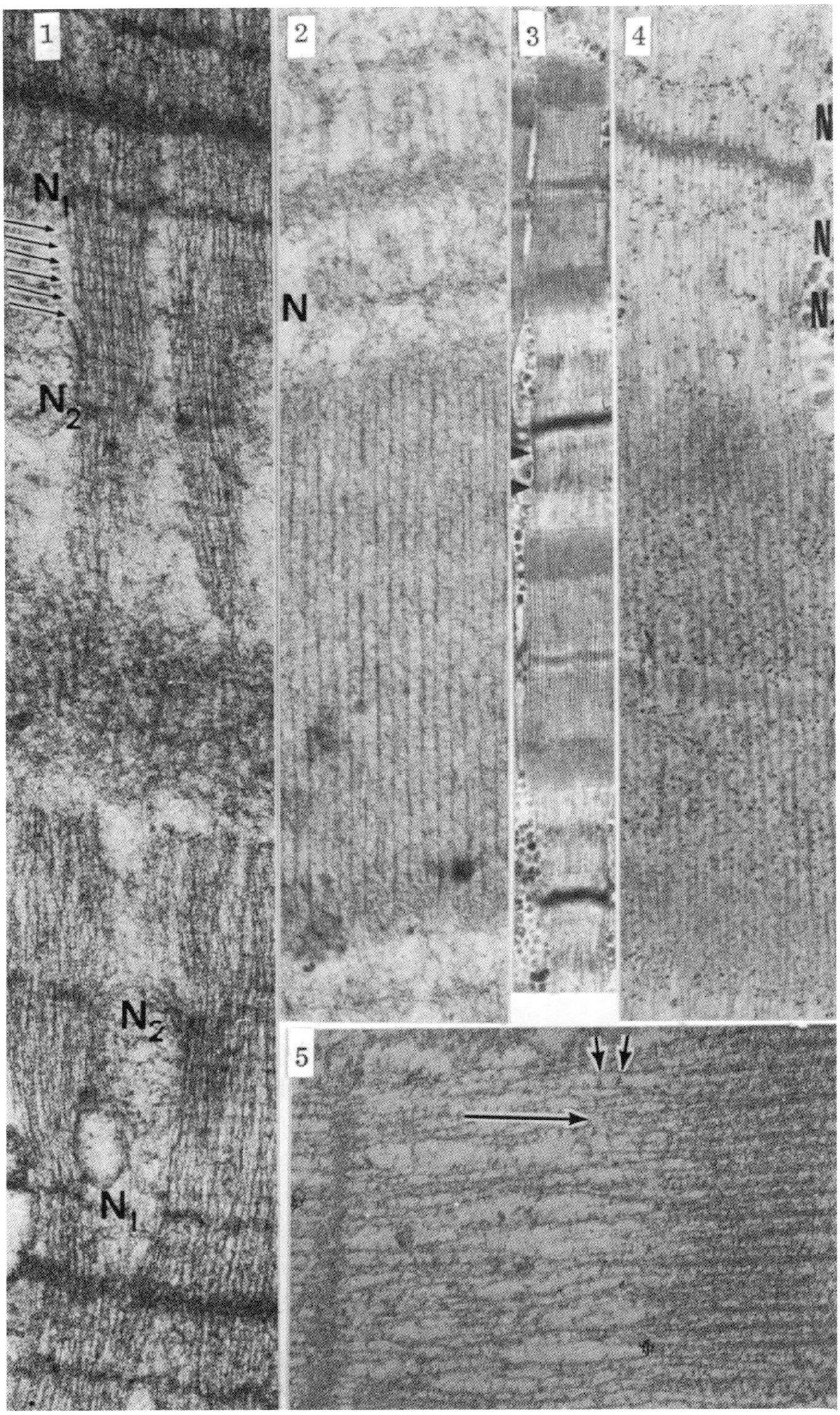
1
N1
N2
N2
N1
2
N
3
4
5

3.4. Structural Architecture of Complete Cross-Striated Sarcomeres

It has been a surprise to me that I could find no published account of the totality of actual fine structure of even a single sarcomere, for any muscle cell. In no account has a complete set of serial sections been referred to. In all accounts, unsupported assumptions have been made, especially about the transition from I/A overlap to Z region.

I decided many years ago to try to fill this gap, at least for rabbit psoas cells, since muscle physiology has leaned heavily on the model of this presented by H. E. Huxley in 1957. I wanted to start with a pair of adjacent Z regions, both of which had a regular square array, but although clusters of myofilaments of all the vertebrate cells that have been examined enter the Z region primarily in such an array, whole Z regions with a perfectly square arrangement have turned out to be quite rare. Furthermore, the shapes of the myofilament clusters in the overlap regions are seldom the neat hexagons commonly depicted, and they are complicated by irregular invasions by SR. In a Z region of a small myofilament cluster, there were up to nine subclusters with lattice planes intersecting at various angles. Where clusters meet, there are necessarily many structural anomalies in which the Z filaments must somehow link together three or five I filaments instead of four. Many filaments on the outer margins are also not able to be part of a square array.

The major problem is shifting the basically hexagonal array of the A/I region to a square one. No simple rearrangement can occur without a major change in shape, but in fact no change of shape accompanies the transformation: the Z-region always has exactly the same outline as that of its neighboring A/I overlap regions. There is a reduction in area, of 10–38%; the space saved accommodates mitochondria, glycogen, extra SR, and the T system, all of which crowd the I/Z region. There are fewer mismatchings of lattice planes in the A/I region than in the Z region, rarely more than three, so that even when the Z region comprises several subclusters, there can be but a single array there. The actual transformation from Z to A can be determined only by the tedious following of individual filaments in serial ultrathin sections. I have found by this means that I filaments of rabbit psoas, garter snake, and frog do not occupy hexagonal orbits in any part of the I region as in H. E. Huxley's diagrams: indeed Huxley's own photograph of this region did not show such an array.

An analysis of the dimensions of myofilament clusters (= fibrils in old terminology) in rabbit psoas cells is presented in Table 3.1. All retained the same shape in passing from A to Z except where splitting occurred, but there was an overall decrease in area averaging

---

Figure 3.25. Locations of N bands. (*1*) Rabbit psoas muscle cell from which the myosin had been extracted, showing one strong band, N_1, fairly close to the Z band, and one weak one, N_2, about midway between Z and M. The positions correspond to the regular banding period of I filaments (*arrows*). [Reprinted with permission from Walcott and Ridgway, *Am. Zool.* 7, 499–504 (1967) Copyright 1967 the American Society of Zoologists.] (*2*) Rat thigh muscle cell sectioned following freezing, without fixation. A single N band is conspicuous. [Reprinted with permission from D. C. Pease, *J. Ultrastruct. Res.* 23, 280–303 (1968). Copyright 1968 Academic Press.] (*3*) Partly stretched frog sartorius muscle cell showing the relatively narrow N_1 and the broad N_2 band. (From Franzini-Armstrong, 1970b.) (*4*) Same as (*3*) skinned, with ferritin, which is aggregated at both N regions, but especially N_1. (From Franzini-Armstrong, 1970b.) (*5*) LS of the I region of stretched rabbit psoas. Large particles (*arrows*) are attached at two of the regular repeat regions, in the vicinity of N_2. The longer arrow shows a T filament. (ORIG.)

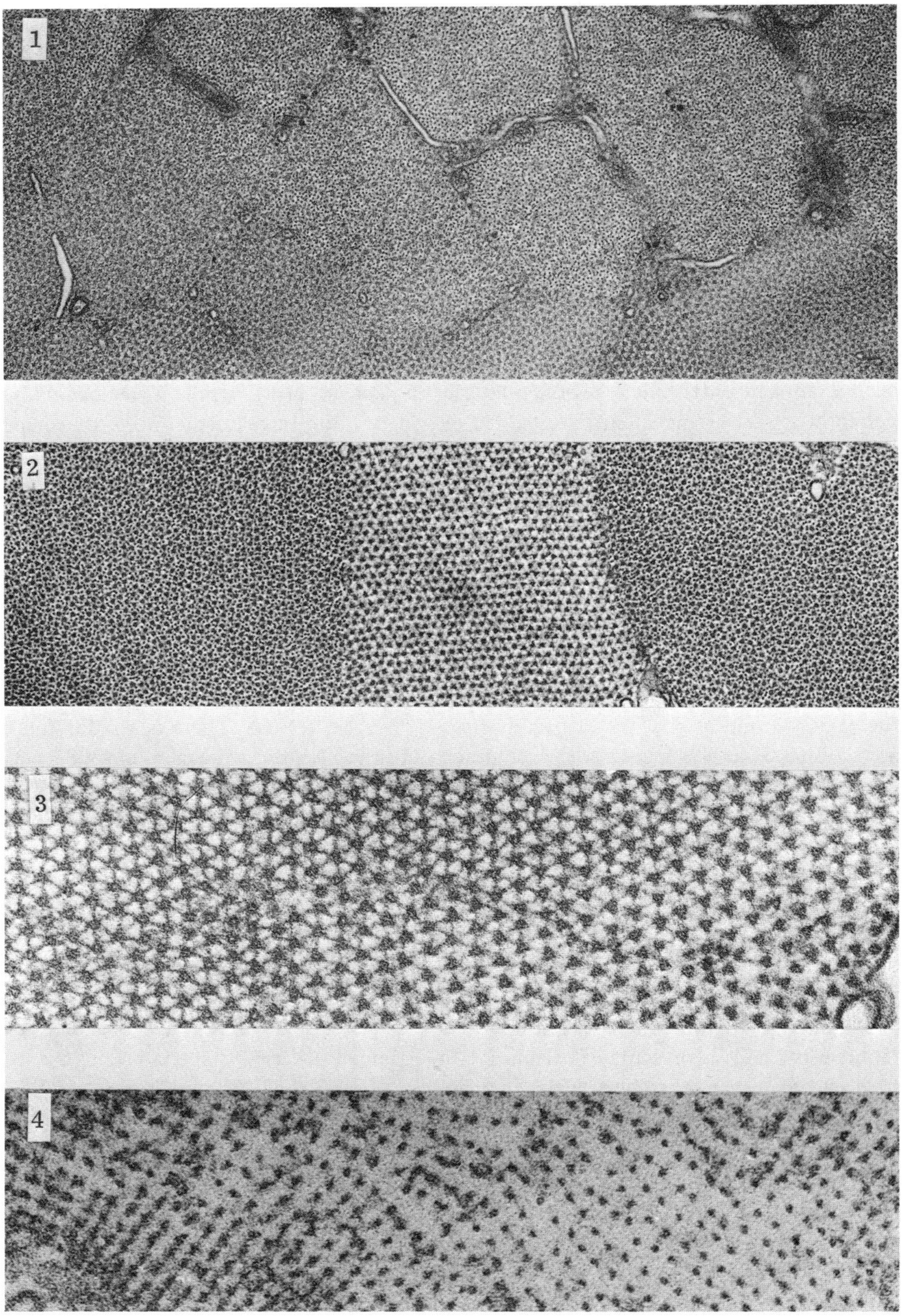

Figure 3.26. Some common structural aberrations seen in most types of muscle cell. All rabbit psoas. (*1*) Clumping together of A filaments, commonly in pairs, but with a few trios and quartets, principally or exclusively in the overlap region. (ORIG.) (*2*) Lack of lateral alignment, leading to some I filaments in an H zone. (ORIG.) (*3*) Imperfect arrays of thick filaments, leading to crowding or gaps, in places. (ORIG.) (*4*) Imperfect arrays of thin filaments, seen especially often in pre-Z regions. (Courtesy of Dr. Michael Reedy.)

Table 3.1 Analysis of Transition From A Band to Z Region for Myofilament Clusters of Rabbit Psoas Muscle Serially Sectioned Electron-Microscopically[a]

Myofilament Cluster Number	Area at Z (μm^2)	Area at A (μm^2)	Percentage Change Z → A	Greatest Length at Z (μm)	Greatest Length at A (μm)	Percentage Change in Long Axis	Greatest Width at Z (μm)	Greatest width at A (μm)	Percentage Change in Width	Number of Rows of Filaments in AZ Plane in A/I	Number of Rows in AZ Plane in Z	Number of Rows in AIZ Plane in A/I	Number of Rows in AIZ Plane in Z	Ratio of Numbers of Rows of Filaments in (A/I)/Z: For AZ Plane	Ratio of Numbers of Rows of Filaments in (A/I)/Z: For AIZ Plane
1	3.13	4.02	+28	2.62	3.01	+15	1.40	1.75	+25	90	68	67	47	1.31	1.42
2	1.12	1.43	+28	1.45	1.71	+18	1.05	1.15	+9					Irregular	
3	2.70	3.41	+26	2.25	2.39	+6	1.58	1.78	+13	52	45	72	51	1.16	1.41
4	1.47	2.19	+48	2.08	2.25	+8	1.33	1.48	+10					Irregular	
5	1.25	1.78	+42	1.50	1.80	+20	1.28	1.38	+8	52	37	35	26	1.40	1.35
6	1.77	2.76	+56	1.75	2.16	+19	1.43	1.63	+14					Irregular	
7	0.93	1.26	+35	1.55	1.80	+16	1.00	1.18	+8	51	36	20	15	1.42	1.33
8	3.19	4.09	+28	2.68	3.10	+16	1.88	2.00	+6	69	50	96	70	1.38	1.37
Summary data															
					Average increase in area: 36.4		Average increase in length: 15%		Average increase in width: 14%			Average increase in number of rows from Z to A:		1.34	1.38

[a]Unpublished analyses by the author of sections prepared by Dr. Myra Elfvin.

36%. The decreases were only slightly asymmetrical, one plane being expanded more than the other. In a simple reorientation from hexagonal to square, there must be a change in shape, one axis increasing, the other decreasing. A square shape becomes a rectangle, with length to width ratio of 1.85:1. Since no such shape change actually occurs, there must be a reorganization of the filaments. All the examples studied in detail were found to follow the same plane, which is illustrated in Figure 3.27. One of the square Z lattice planes also corresponds to one of the three filament planes of the A region that pass through A filaments only; this will be referred to as the AZ plane. The other, corresponding to a plane in which there are only thick filaments, will be termed the AIZ plane. These two planes lie at right angles to each other, and they also correspond to the two axes of the Z region. The total number of rows of filaments was counted for the AIZ and AZ planes. These were then divided by the corresponding numbers in the Z array. The average ratio was 1.34 for the AZ plane, and 1.38 for the AIZ plane. Half the values were 1.4, the others only slightly less. The value 1.4 corresponds to a 7:5 ratio. Effectively, two out of every seven rows of I filaments present in the A/I region are lost during rearrangement, in both planes (Figure 3.27.*1*). The simplest possible pattern for this rearrangement is illustrated by the broken lines in Figure 3.27.*3*. Such a rearrangement is accomplished with negligible change in shape, provided the hexagons of A/I and the squares of Z are geometrically perfect. In the examples studied, this was more true for Z regions than for A/I, in which the hexagons were slightly longer in the AIZ axis than the AZ. This region was not clearly decipherable in any of nine sets of rabbit psoas myofilament clusters I examined. Few single I filaments could be followed all the way from A/I overlap to Z. Several were found to bend sharply just before or immediately after leaving the A region, as seen in longitudinal sections. Others appeared to terminate in Y junctions, the forks of which are thinner and less dense than the stems. In no transverse sections could the majority of the dense dots representing I filaments in the A/I region be traced to corresponding filaments close to the Z region. Most become fuzzy and undecipherable at some point. Because of the Z/A plane difference, a majority of I filaments must cross the plane of section. If there is indeed branching and exchange of components as it appears, this will further complicate the transition. There are two principal regions where the fuzziness occurs, one close to the edge of the A region, the other fairly close to the Z region around N_1. These are regions where abrupt changes of I-filament direction occur as well as the places where some I filaments appear to fork. As the Z region is approached, the transverse image is always imprecise just before the square array

---

Figure 3.27. Transition from A to Z regions in rabbit psoas. (*1*) Plot of all the I filaments in a single well-defined myofilament cluster in the center of the overlap region, indicating two planes intersecting at right angles, one plane along the line of A filament (AZ), the other in which A's alternate with two I's (AIZ). (*2*) Plot of all the I filaments of the cluster shown in (*1*), traced by serial ultrathin sectioning. This region is just outside the Z band, at the same magnification. Note that the orientations of the planes match AZ and AIZ. Arrowheads indicate imperfections in the square lattice (there are usually many more than four). (*3*) Diagram of a group of I filaments in the overlap region to show the basis for transformation. The AZ and AIZ planes are indicated. Each set of 7 × 7 rows of I filaments becomes 5 × 5 rows of pre-Z filaments. (*4*) The same filaments in the pre-Z region. The numbers correspond to the sets of four I filaments indicated in (*3*). (*5*) EM section across the outer margin of the A band. Some of the filaments have already moved into their pre-Z positions. The Z-filament lattice is partly disturbed. At right, note the apparent splitting up of some of the I filaments. (All ORIG.)

1

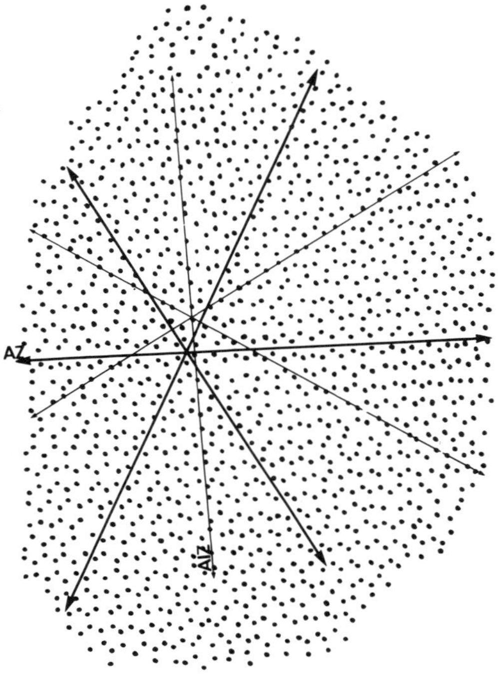
AZ
AIZ

2

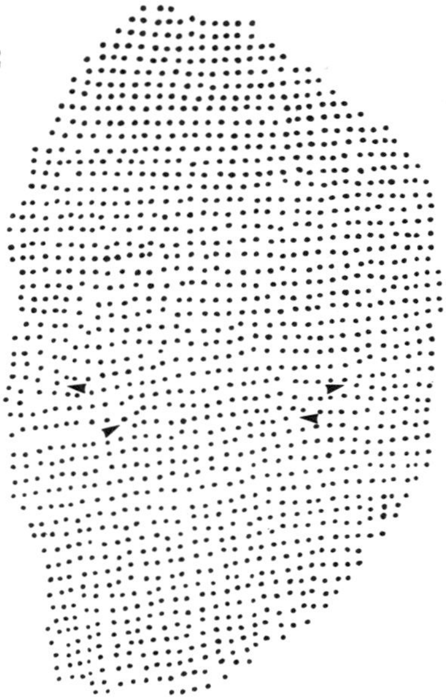

3

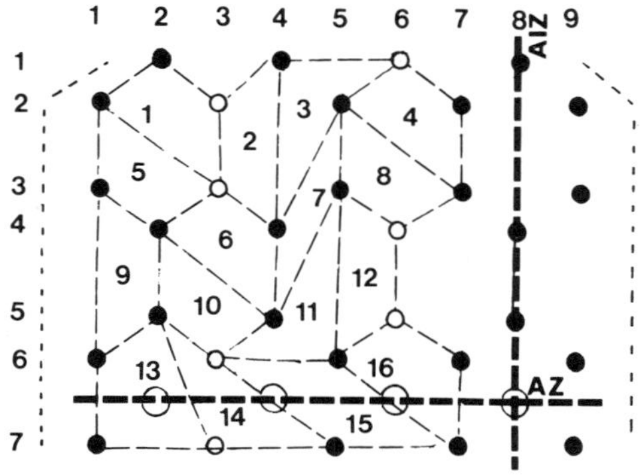
1 2 3 4 5 6 7 8 9
1
2
3
4
5
6
7
1 2 3 4 5 6 7 8 9 10 11 12 13 14 15 16
AIZ
AZ

4

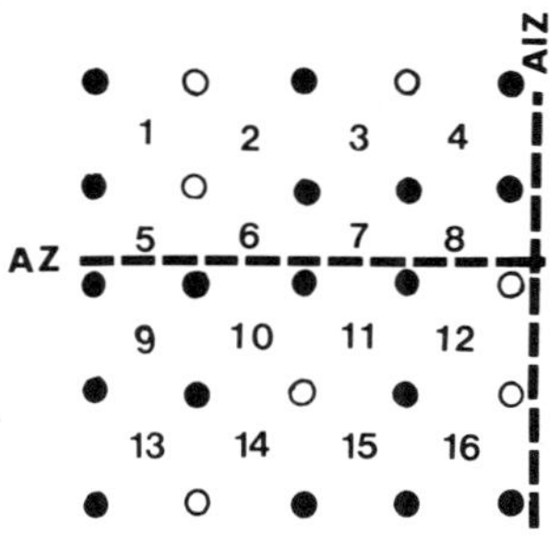
AIZ
1 2 3 4
AZ
5 6 7 8
9 10 11 12
13 14 15 16

5

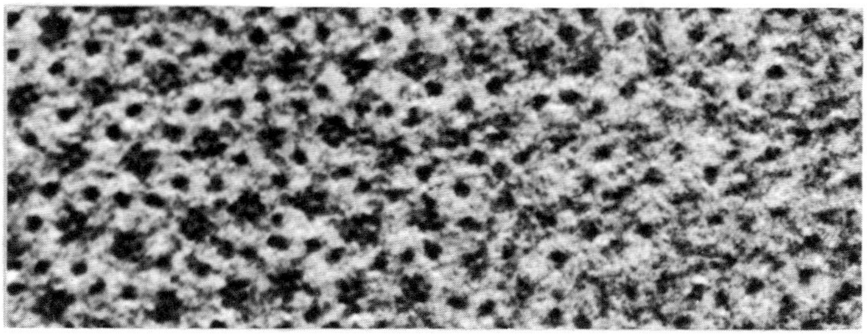

is reached. In the region with square array, ultrathin crosslinks occur between the filaments laterally. Filaments of the square array parallel to the longitudinal axis are both thicker and stiffer than I filaments seen in the remainder of the I or the A/I region. These regions may be formed either of twisted Z filaments only, with I filaments attached at their ends, or Z filaments wrapped around an I-filament core. The number is, however, identical to that of I filaments in the overlap region. At the A/I boundary, custom has led us to expect neat hexagonal arrays, but many anomalies were found. I filaments have already moved out of the hexagonal arrays present in the mid-A/I region before reaching the ends of the A bands (Figure 3.27.*5*), and many A filaments are also displaced from hexagonal array.

I filaments show a tendency to twist as they pass outward from the M region, occupying different orbits in the outer part of the A. In tracing them, many anomalies were noted, including orbits of only 5 I filaments, orbits of 7 or 8, and places where A filaments came close together or even fused into giant thick filaments surrounded by 12 I filaments (see Figure 3.26.*1*). There is almost always a shift in the positions of the Z axes of successive Z regions (Figure 3.28).

There are two regions where it is always impossible to discern individual I filaments clearly: just outside the A band and just before the pre-Z filament region where the square lattice occurs, possibly because the rearrangement of I filaments means that most must run across the transverse plane. However, although this accounts for some of the difficulty in following I filaments, there seems to be another factor involved, which would be of major significance if substantiated: there appears to be some splitting into thinner strands that diverge and then reform. The traditional view of I-filament structure does not permit splitting.

The Z region described had an ideal array; all the I filaments of one side approached it in square array with common planes, and those on the other side likewise, the two sets of planes being in line with each other, and the component filaments offset by exactly half the unit space diagonally. Few Z regions achieve this degree of perfection, even when the two sides are matched. The two component myofilament clusters rarely match, and their planes are seldom at right angles to each other, Z filaments being left to link the mismatched I filaments as well as possible. When the two sets of filaments of adjoining sarcomeres are seriously mismatched, there must be aberrations in fine structure of the Z line. There is usually a blending rather than a compromise, one part following the dominant pattern of one side and the other that of the neighboring one. Although a regular A/I pattern may be formed in an outer region, toward the center of the sarcomere the lattice pattern can shift to one in which the planes rotate part clockwise, part anticlockwise. These planes match the two Z-line planes. This state of affairs persisted across the H zone, but in the other A/I region, the lattice rearranged to a unified one. There was a progressive shift in lattice planes from one sarcomere to the next of from 5 to 30° in all the examples studied closely, but no systematic rotation was discerned. Neighboring myofilament clusters rarely had similar orientations of filaments. It was concluded that each is independent of the others and that the overall orientation within a cell is random.

In addition to all of the above, there must be fitted into the sarcomere the longitudinal ultrathin T filaments (see p. 90), which are in parallel with the I filaments, either separate from or closely adhering to them. These appear to run through the Z regions. When a myofilament cluster is stretched and incompletely activated at the same time, using intact cells under graded excitation or isolated clusters in the presence of ATP, wide

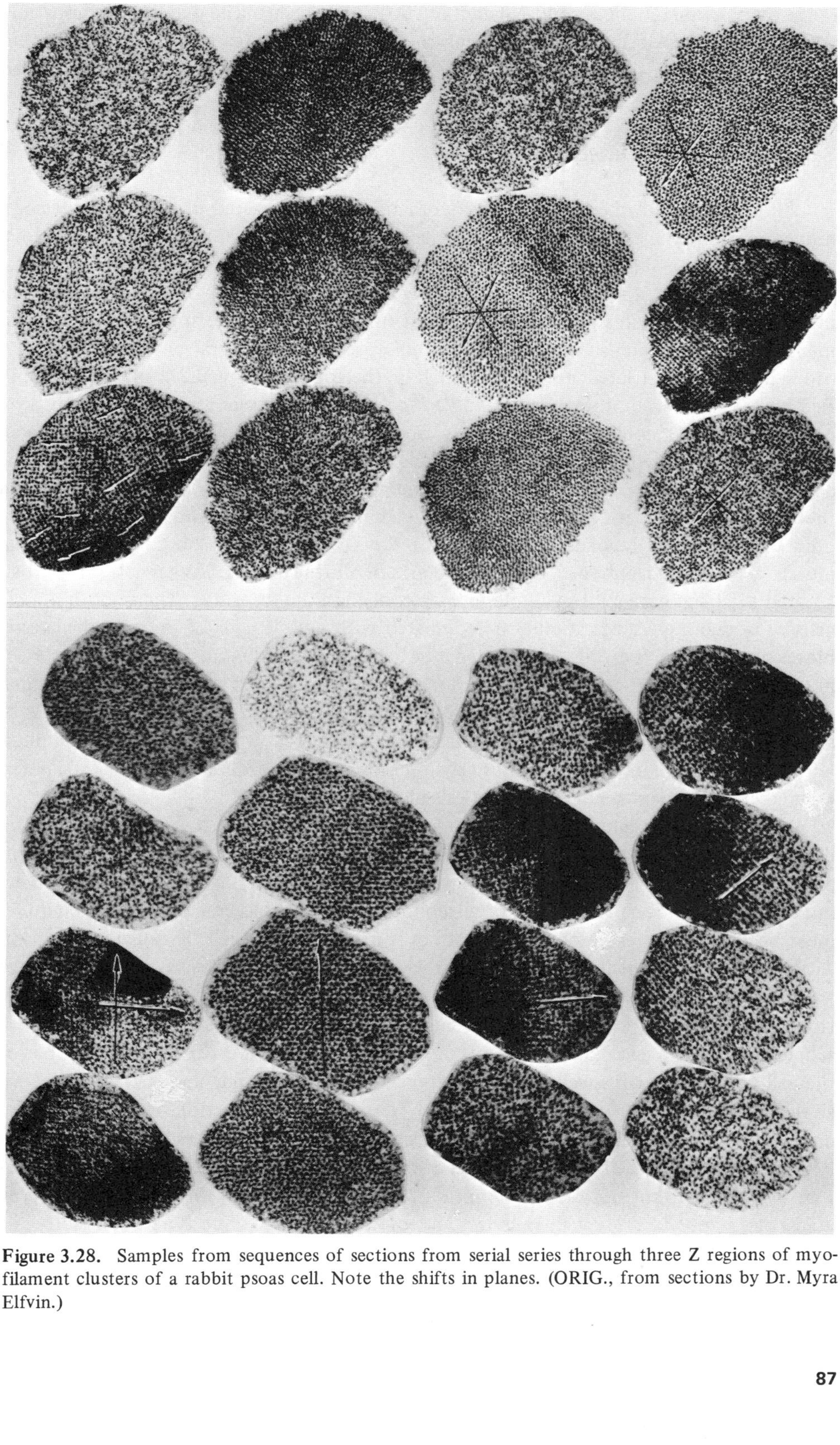

Figure 3.28. Samples from sequences of sections from serial series through three Z regions of myofilament clusters of a rabbit psoas cell. Note the shifts in planes. (ORIG., from sections by Dr. Myra Elfvin.)

ranges of sarcomere lengths and N-band and A/I shifts occur. The details are compatible with continuation of T filaments through the Z. Effects of stretch on N_2-band position (see p. 97) suggest that if N materials are attached to T filaments, these are attached to A filaments but not to I filaments.

3.5. Obliquely Striated Cells

In obliquely striated muscle cells, the myofilaments are staggered in various ways, commonly forming planes that make oblique/acute angles with respect to the longitudinal axis, as we have seen (Figure 3.3.*3*). But there are also spirals, zigzags, and irregular forms. Some of the cells with the latter pattern merge into muscle cells in which the component myofilaments are of nearly equal lengths, but in which there is no apparent lateral alignment. Before the extensive use of electron microscopy, the full range, from the highly-ordered obliquely striated muscle of annelids, the totally disorganized muscle exemplified by the body wall of onychophoran *Peripatus,* were all classified as "smooth" muscle, causing enormous confusion, particularly since one or another was often selected as an experimental "model" substitute for hard-to-study vertebrate smooth muscles. It is now apparent that there is no hard-and-fast line demarcating smooth from striated muscle in the body/skeletal musculature of invertebrate animals. The distinction is clear and simple only for vertebrate animals, though not for the whole phylum Chordata. The skeletal muscle cells of all backboned animals, and also of all true arthropods, are all of the cross-striated type and have a lot in common, as described in the foregoing. The smooth-muscle cells of vertebrates can profitably be compared with muscle cells of the guts of invertebrates (even when the cells are striated), but that is all. Molluscan retractor muscles or body-wall muscle of annelids should be compared with each other and with skeletal musculature in general but not, except for purely intellectual pursuits, with any forms of smooth muscle. The muscle cells of molluscs and annelids show some features in common with those of vertebrate smooth muscle. Common features are relatively small diameter, tapered ends and nexuses or tight junctions between cells.

As viewed in the light microscope, obliquely striated cells were divided into single and double. The latter was a result of obliquity continuing around the whole cell in a spiral, and of the cell being sufficiently thin and transparent for the far side to be seen at the same time as the near. However, not all obliquely striated cells could satisfy this requirement because in some the obliquity does not spiral around the cell. The degree of organization of the component myofilaments provides a major source of variation with a range from cells in which there is so little order that it is difficult to discern any, even in electron microscopic sections, to ones that show strong regular bands even in the light microscope. In all, the angle made between the plane of striation and the longitudinal axis of the cell becomes progressively more acute with increasing stretch, and decreases progressively with contraction (Figure 3.3.*2*). With very strong contraction, the angle can become 90°, when the cell appears transversely cross-striated in all planes.

Almost all of the cells that have been adequately described have some unique features that are best left to individual descriptions. General features of the single-oblique type are seen in annelid body-wall muscles, and of the double-oblique type in adductors of lamellibranch molluscs. What they have in common, aside from obliquity of filament alignment, are a small cell thickness (2–10 μm); thick filaments of great thickness, 23–110 nm in the center, that themselves taper at their ends; and thin filaments attached in clumps to ovoid, dense Z bodies. The thick filaments contain a fine characteristic cross-banding that matches that of crystalline paramyosin. The amount of paramyosin in some of these filaments is such that they are almost entirely composed of this protein and hence may be termed paramyosin filaments. There is very little ordinary myosin in these filaments, mixed in an unknown way with the paramyosin. Their length is different in cells from different sources, ranging from about 1 μm in earthworm body wall (by 2.3-nm-center thickness – Heumann and Zebe, 1967) to over 5 μm in the translucent part of the adductor of the oyster (by 60-nm-center thickness – Hanson and Lowy, 1961).

These thin filaments are attached in clusters of a hundred or so equally to each side of a dense Z body. These bodies are all ovoid, and about 0.7 μm long, but range in diameter from about 0.06 to 0.15 μm, with placements reflecting the overall obliquity. There is a high degree of regularity to the separation of Z bodies longitudinally, less in diagonal placement. The separation is a major variable, averaging around 0.2 μm in relaxed earthworm body wall and 2.0 μm in relaxed oyster translucent adductor, each with a fivefold range. Some ground plan, perhaps of ultrathin filaments, must be present, at least to link the Z bodies, but none has yet been detected. Since the thin filaments are approximately of equal length, which is less than half the Z–Z spacing at rest, there is an H zone.

The thick filaments are staggered precisely so that each is situated equidistant from the nearest Z bodies in the longitudinal axis. On account of the staggering and their gradually tapered shape, in a transverse section, thick-filament clusters grade from relatively thin to thickest and back, in alternating sequences. Thin filaments make orbits around them, the more perfect of which contain 12 filaments. Some contain fewer, and others more. In contracted muscle there are always more, indicating that after sliding over the thick filaments, their inner ends pass around to the other half. I regions containing only thin filaments also occur in the relaxed muscle, but no matter what the condition of the cell at the time of fixation or the region sectioned, many thin filaments are found to occur in clusters between, but not around, thick filaments.

The detailed substructure of dense bodies has yet to be worked out for any muscle. Preliminary examination has shown that all have a longitudinal filamentous structure in which many filaments are indistinguishable from actin itself. There may, therefore, be regions where actin filaments of opposite polarity meet, possibly overlap, come very close together, and are connected to each other end to end and laterally.

3.5.1 Disorderly "Striated" Muscle

At the outer limits of the spectrum of obliquity are muscle cells with similar components but such disorderly filament arrays that they warrant a separate classification. The functional unit comprises dense Z bodies with attached thin filaments that overlap and partially orbit around thick filaments, but the distribution of the bodies is random. They are attached to invaginated tubules, both transverse and longitudinal, to invaginated clefts, and to the surface plasma membrane. In the onychophoran *Peripatus*, the muscle cells (which have disorderly characteristics) though often small range up to 40 μm in diameter and over 1 cm in length without tapering, so they are quite different from the typical "smooth" cell. Their existence clearly draws attention not only to the wide range of fine structural organization present in diverse skeletal muscle cells, but also to the arbitrariness of any system of classification. The extremes of the spectrum may be very different, but when the intermediates are considered, the continuous rather than abrupt nature of the spread becomes paramount.

3.5.2. Functional Advantage of Oblique and Disorderly Striation

Being oblique in fine structure conveys the single advantage of permitting an extremely wide length range. Not only can the thick and thin filaments slide past one another to the same extent possible in cross-striated cells, in addition, realignment of the contractile units from obliquity to cross-banding permits additional shortening by up to the cosine of the angle of obliquity.

3.6. Elastic Elements

All muscles have elastic properties with such an extremely wide range of stiffness and extensibility that many different elements and/or combinations thereof must be responsible in different cells. There are several sources of elasticity, as follows.

1. *Extracellular.* Muscle cells are surrounded by complex meshes of collagen fibers, some of which are wavy and tapering (e.g., Borg and Caulfield, 1980; Hoyle and Williams, 1980). These predominate, but elastin fibers are probably also present around many vertebrate cells. Fine collagen strands extend into the interior of some invertebrate skeletal muscle cells, but in those of vertebrates and arthropods, they terminate within basement material of the Sl or are attached to other ultrafine filaments.
2. *Intracellular-supramyofilamentous.* Ultrafine filaments mesh around myofilament clusters and Z regions, to which they are attached.
3. *Intracellular-intramyofilamentous.* These are longitudinal ultrathin filaments that exist throughout the contractile material in parallel with myofilaments. The ITs and all internal membranes – of ITs and of SR – to which myofilaments are attached are also elastic. The whole Z band is elastic, and so are the fine strands that link I filaments together, M bridges and their equivalents that link thick filaments together passively, and finally, myosin X bridges. Added to all these direct sources of elasticity are electrostatic and molecular attractive forces, which will also contribute stretch resistance.

It has been generally agreed that the widespread vertebrate elastic filament elastin is not involved in elasticity, because elastin is insensitive to trypsin, which readily disrupts myofilament clusters. The nature of intramyofilamentar elasticity has presented the greatest difficulty for resolution and is still not understood. The supramyofilamentar elasticity is attributed to the protein connectin. Direct tests have shown that there is no elastin in a variety of invertebrate skeletal muscles, but there is a spirit-blue-stainable elastic material in all those tested (Elder, 1973).

3.7. Ultrathin (T, S) Filaments

When the interdigitating sliding filament model was first described, it was recognized that a primary elastic element was needed to hold it together. No primary element was visible in early electron micrographs, so it had to be thin and possibly of low electron density or poor affinity for heavy metal "stains." Hanson and H. E. Huxley (1956) stated the following Conclusion in their review of the model:

> The H zone is bridged by fine extensible filaments of unknown composition (the S filaments) which link up the *ends of the actin (=I) filaments* on either side. The S filaments are probably permanently cross-linked to the myosin (=A) filaments at some point in the H zone.

Another of their conclusions started with:

> When resting or plasticized muscle is stretched, the actin filaments are partly withdrawn from the A band, stretching the S filaments in the process.

But one year later, without explanation, S filaments were dropped from the model (H. E. Huxley, 1957). The model is incomplete without some form of fine, but extremely strong, elastic primary filament, yet the incomplete model proposed by Huxley in 1957 became universally adopted. The model could be tested very easily by stretching it beyond the point of overlap of I and A filaments. The 1957 Huxley model falls apart. No cross-striated muscle cell yet stretched so far had done so. The three interacting parts of a sarcomere might perhaps have been held together by external elastic elements, which would be of trivial significance and leave the basic model as shown by H. E. Huxley. But even extracted material reduced to the level of split myofilament bundles that high-resolution electron microscopy showed lacked peripheral elements can be stretched to well beyond overlap in the presence of ATP and returns to normal when released. The techniques are simple. In one, bits of macerated muscle are placed on a microscope slide with a tiny drop of saline so that one end dries out on the slide and the other stays moist. The drop is then further reduced, with a supported cover slip placed over it, until the other end dries out on the cover slip. The remaining moist material can then be stretched by moving the cover slip, while observing it in the microscope (Figure 3.29). It can later be prepared for electron microscopic examination. Another method is to clamp the ends of a large bundle, dissect away all but a single myofilament cluster, pass a hook around it laterally, and pull on the hook (Ernst et al., 1969a). Locker and Leet (1976a) have achieved the record in stretch without breakage, getting an ox sternomandibularis sarcomere to five times its rest length. TEM micrographs of overstretched material from several vertebrate and arthropod muscles all showed that the stretched material is held together by ultrathin filaments (Figure 3.30). They were termed gap filaments by Sjöstrand (1962) because they bridge the expected gap between A and I filaments after stretch. I later called them T filaments (Hoyle, 1967), partly to emphasize their thinness and also to bring them into line with the use of the letters A, I, Z, and M. Sjöstrand (1962) described them as linking the ends of I filaments to the ends of A filaments. This idea lost favor when it was found that in extensive contractions, I filaments overlap each other with no sign of the filaments attached. Several possible alternative locations should be considered. One is that T filaments run in parallel with A and I, either free or attached at one or more places to either I or A filaments, or some might form a central core for A filaments, an idea that has been suggested several times (e.g., Garamvölgyi, 1966b; Squire, 1973; Locker and Leet, 1976a, b), at least for some cells. I have long been convinced that in rabbit psoas they run in parallel with I and A filaments, and that they are attached at the M region but run right through the Z regions.

Another way of uncovering T filaments is to remove major components, notably I material, or A material, or both, and study the remains. Following removal of A material from bee flight muscle, the entire middle region, including the M band, goes. But the I regions are clearly held together (Figure 3.31.*1*). The M band remains when vertebrate

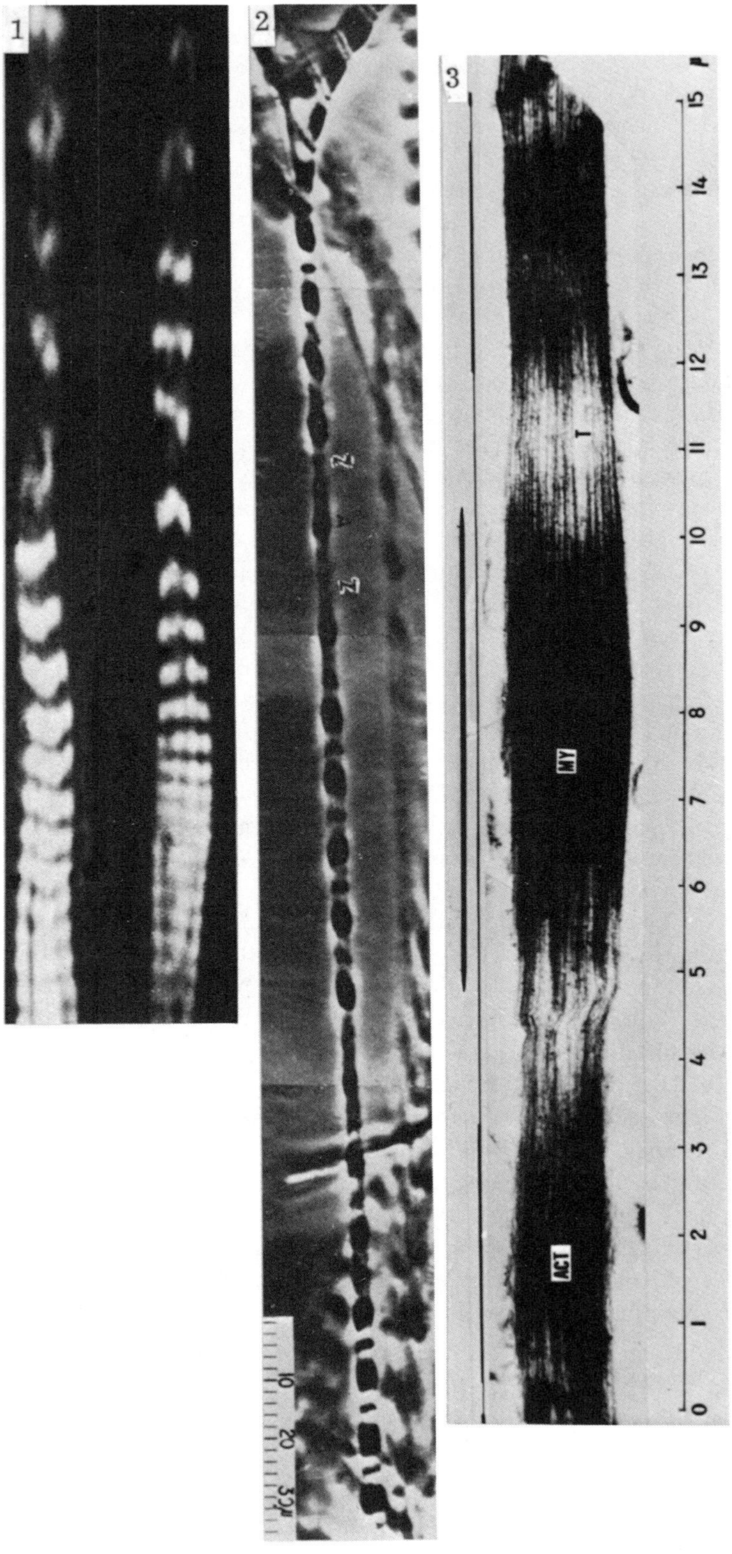
1
2
3

cells are treated similarly, but T filaments are then clearly visible bridging the region between the tips of the I filaments and the residual M region (Figure 3.31.*3*). Even removal of both the Z band and the I material by prolonged treatment with urea does not cause disruption. Instead it reveals the T filaments linking the A bands together across the region where the Z material has been (Figure 3.31.*2*). Locker and Leet (1976b) proposed that beef ultrathin filaments are attached to the M bands and not Z bands which, according to their evidence, the filaments simply pass through. Their reasoning was based on a remarkable occasional occurrence in extensively stretched cells. When an I band breaks away from near the Z line on one side only, not only is the sarcomere held together, but there is also a drastic shift of the N_2 line back toward the Z (Figure 3.32). As pointed out earlier, Page's observations on the shift of N_2 with stretch (Page, 1968) suggest attachment of N_2 material to an elastic element in parallel with I filaments. Complete removal of either Z or M lines does not result in collapse, and in both cases, ultrathin filaments can be seen crossing the gaps. Thus continuity of T filaments from end to end of the entire cell is suggested, with attachments to A filaments near their ends, and to the M region. No single scheme need be exclusive: some T filaments may run down A-filament cores, others outside them; some may be attached to M regions, others to Z, as variants in different kinds of cell.

Each of the above models requires that T filaments at least equal in length to a whole sarcomere be present in macerated muscle, with fragments being about as numerous as are I filaments. Early studies on the components of muscle brei reported the presence only of A and I filaments and Z regions with I filaments attached. However, several authors have noted very thin longer filaments, and close scrutiny of published micrographs of negatively stained material often reveals filaments that were not mentioned. An intensive search in 40 muscles, both vertebrate and invertebrate, but especially in rabbit psoas, made in our laboratory, yielded several ultrathin filaments (Figures 3.33 and 3.34), some of which were more than 2 μm long. In rabbit psoas some were at at least 2.3 μm long (Figure 3.34.1), with a thickness of the component molecules of no more than 2–3 nm. It has long been obvious to me that these filaments might be of great functional importance. Although they could turn out to be purely elastic elements and to not interact with the contractile material, until this has been firmly established, the possibility remains open that they are contractile.

Combining all of our studies on rabbit psoas yields the complete model of a single sarcomeric myofilament structure shown in Figure 3.18.

3.8. 10-nm and "Cytoskeletal" Filaments

These have been little studied in skeletal muscle and indeed may be either absent or present in small numbers in view of the presence of connectin and ultrathin filaments. They are characterized, in contrast to other muscle filaments, by being insoluble in KCl (Cooke and Chase, 1971). In nonmuscle cells they are widespread (see p. 19), and so it

---

Figure 3.29. Myofilament clusters extensively stretched in the presence of ATP, showing the presence of the very strong ultrathin (T) filaments. (*1*) From muscle of stick insect (From Aronson, 1963.) (*2*) From giant barnacle depressor. (From Hoyle, 1967.) (*3*) Low-power EM image of single extensively stretched sarcomere from (*2*) (From Hoyle, 1967.)

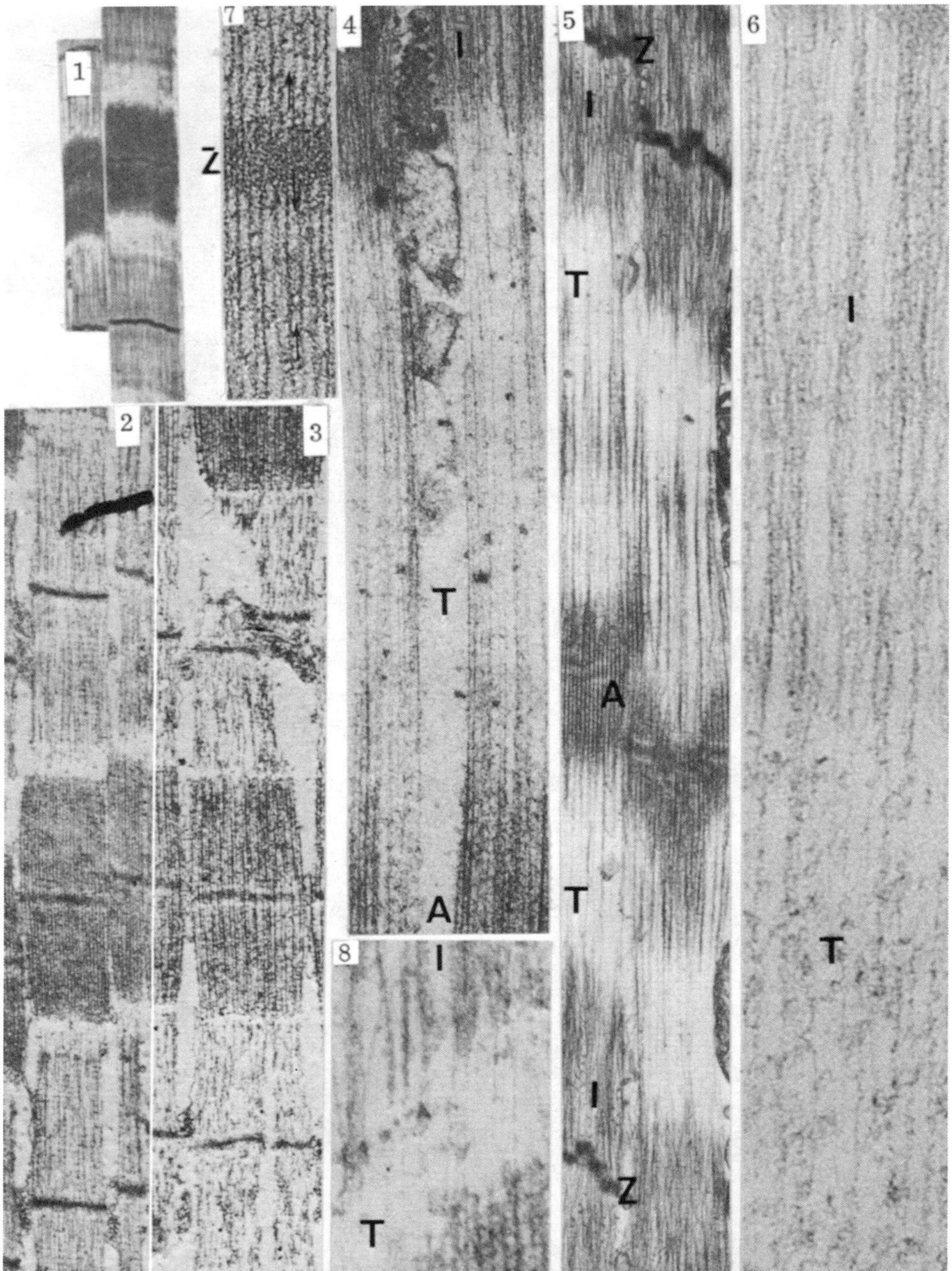
1
Z
7
2
3
4
I
T
A
8
I
T
A
5
Z
I
T
A
T
I
Z
6
I
T
A

would be surprising if they are totally absent in skeletal cells. Certainly they are conspicuous and intensively studied in smooth-muscle cells of vertebrates (see p. 314), where they are attached in clusters to fusiform dense bodies located mainly on the surface membrane; the dense bodies are linked in such a way as to make a helicoidal skeleton. Whether or not they play a functional role is unknown.

3.9. Attachments to Skeletal Structures

Attachment to the body wall or skeletal element is by a wide range of structures, depending on the species and the nature of the skeletal apparatus. Although commonly attached only at the extreme ends of the cells, in soft-bodied animals the obliquely striated or disorderly striated cells are contacted along the entire length of the muscle cell by connective-tissue elements. Fine strands of collagen permeate the Sl and enter the myofilaments. In addition, there are links directly from the muscle cell, in the form of fine arms arising laterally. At first muscular, these become filled with fine filaments of unknown composition that merge with collagen in the body wall. At their ends, muscle cells are attached to the soft body wall in *Peripatus* by collagen strands that pass directly into the cell, running both longitudinally and transversely inside in membrane-bound tubules (Figure 3.35 – Hoyle and Williams, 1980). Except for the fact that the wide ITs of *Peripatus* run a long way into the cell, this is exactly the way cross-striated muscle cells terminate in vertebrates. The collagen in turn melds with bone. Some of the collagen links to terminal Z material, some to thin filaments and possibly also thick filaments. The collagen is mixed with elastin, which merges first with cartilage, then with bone. At the muscle-cell end, the collagen runs into and becomes a major component of Sl, where it connects with the protein connectin. This forms a network around myofibrillar bundles and in some cells forms links between neighboring Z bands.

In arthropods the junction is a neat specialized structure in which the terminal half sarcomere is involved. It lacks a Z region, and its thin filaments mingle with fine filaments of unknown composition, but possibly closely related to collagen. Stout filaments, termed tonofilaments, of unknown composition are attached to the fine-filament region and pass directly into the bodies of the epidermal cells. Some pass through to the outer margin to contact hardened cuticular material. A close mass of microtubules permeates this region in some arachnids, crustacea, and insects. The attachment is often quite inelastic, but in some insect muscles this is far from being the case. The pleurosubalar apodemes (= tendons) of the dragonfly *Aeschna* are highly elastic because of the presence of unique elastic protein used as an elastic element in many insect joints and attachments,

---

Figure 3.30. Exposure of ultrathin (T) filaments in gaps between I and A filaments of extensively stretched cells. (*1*) Two extents of stretch of semitendinosus of frog. (From Ward et al., 1965.) (*2*), (*3*) Frog semitendinosus. That in (*2*) was stretched after fixation; that in (*3*) was stretched before, but not after, fixation. [Reprinted with permission from Galey, *J. Ultrastruct. Res.* **26**, 424–441 (1969). Copyright 1969 Academic Press.] (*4*), (*5*) Beef sternomandibularis extensively stretched. [Reprinted with permission from Locker and Leet, *J. Ultrastruct. Res.* **52**, 64–75 (1975) and **56**, 31–38 (1976) Copyright Academic Press.] (*6*) Rabbit psoas. [Reprinted with permission from McNeill and Hoyle, *Am. Zool.* 7, 483–498. (1967). Copyright 1967 the American Society of Zoologists.] (*7*) Z region of stretched rabbit psoas. The width is a little over twice rest width. T filaments can be seen between I filaments (*arrows*). (ORIG.) (*8*) High-power view of gap from (*3*). [Reprinted with permission from Galey *J. Ultrastruct. Res.* **26**, 424–441 (1969). Copyright 1969 Academic Press.]

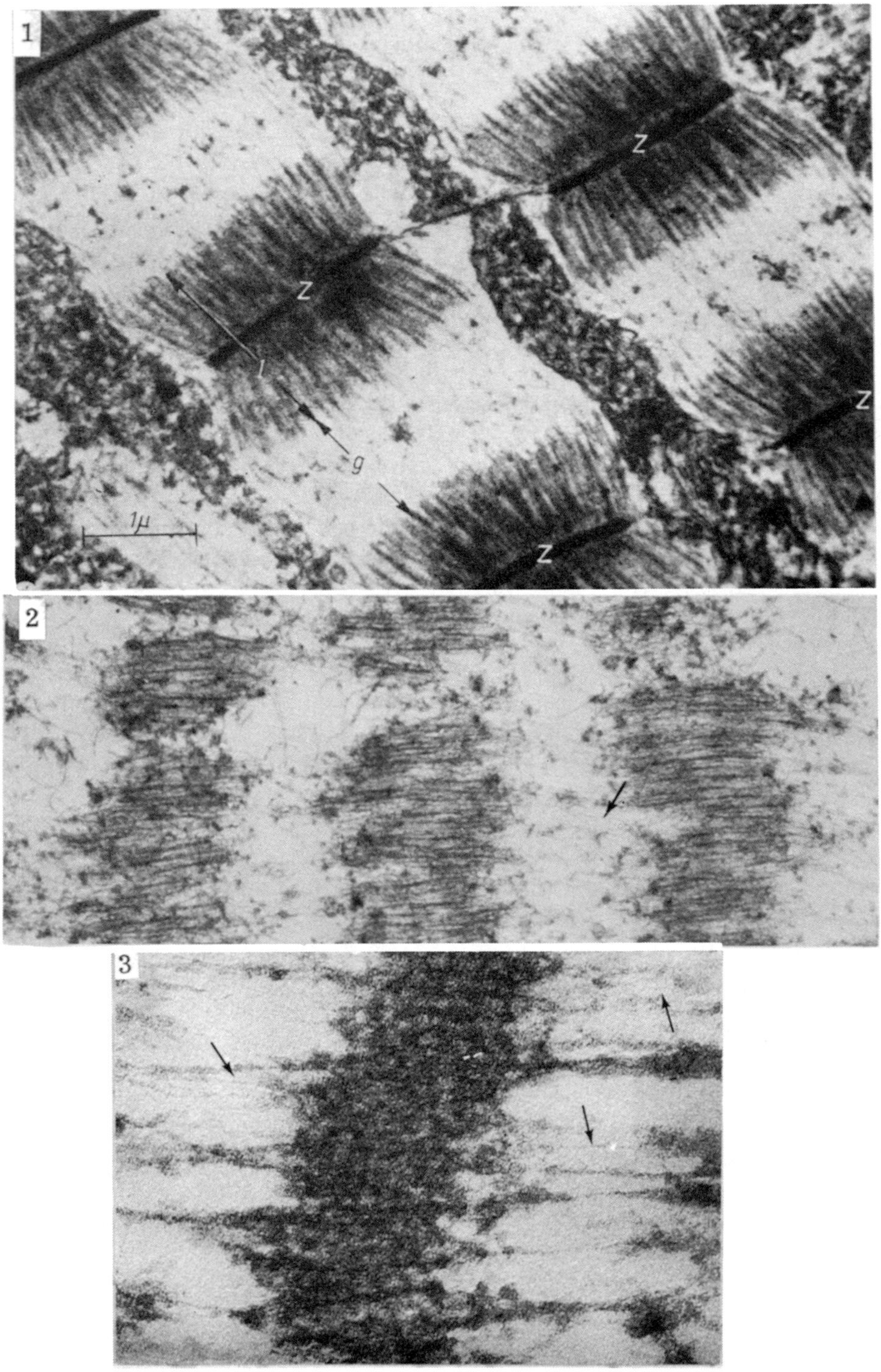

Figure 3.31. Integrity of muscle cells following extraction of major proteins. All reveal the presence of longitudinal ultrathin (T) filaments. (*1*) Flight muscle of a bee following extraction of A band with Weber-Edsal solution. [Reprinted with permission, from Garamvölgyi, N. The ultrastructure of the insect flight muscle fibril ghost. *Biochim. Biophys. Acta* 1 82–88. (1966a), Elsevier Biomedical Press.] (*2*) Chick heart muscle from which all Z material and most I-filament material has been removed by 3-h urea extraction. [Reprinted with permission from Rash et al., *J. Ultrastruct. Res.* **24**, 181–189 (1968). Copyright 1968 Academic Press.] (*3*) H region of rabbit psoas following myosin extraction. The M region remains and is attached to the tips of I filaments by T filaments of only 2 nm diameter. [Reprinted with permission from Wallcott and Ridgway, *Am. Zool.* 7, 499–504 (1967). Copyright 1967 the American Society of Zoologists.]

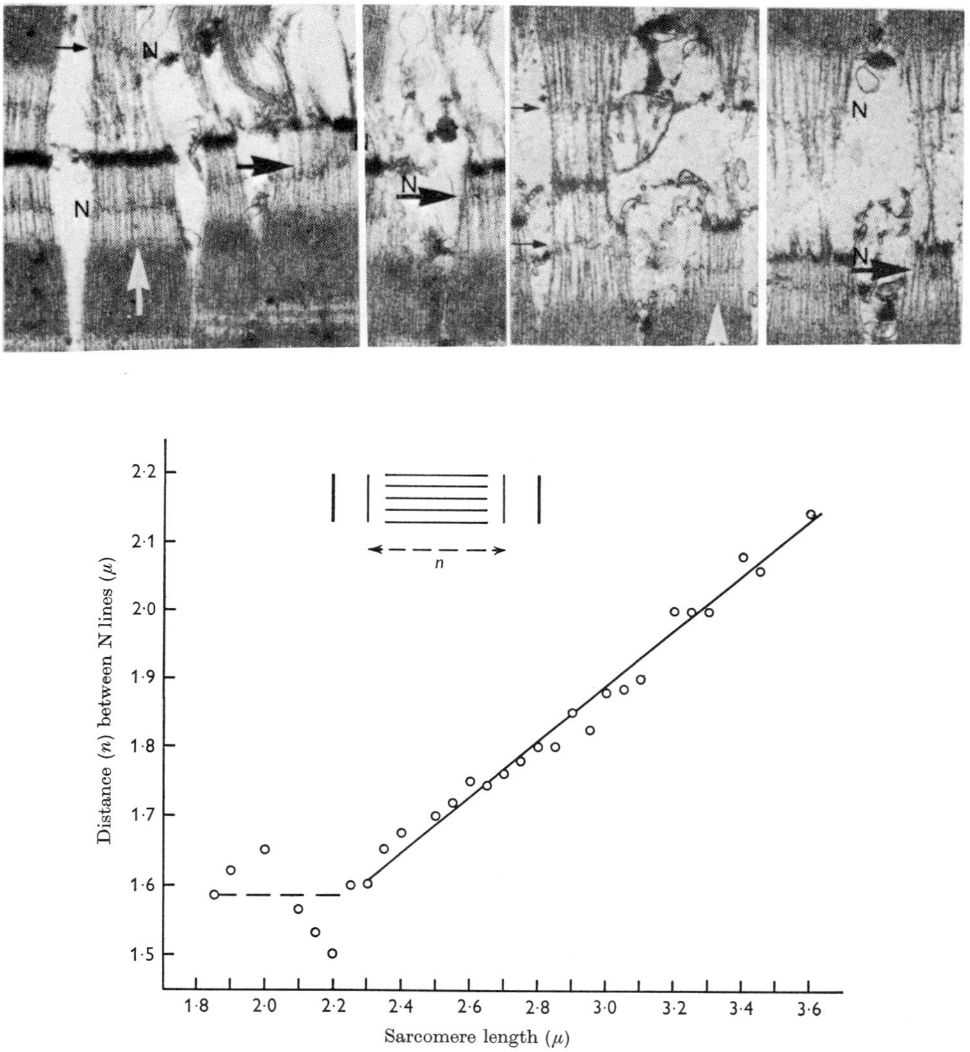

Figure 3.32. Shifts in position of N line associated with stretch. (*1*) Strongly stretched beef muscle cell. When I filaments break on one side of the Z, but not the other, the N_2 line on the intact side shifts its position drastically, suggesting attachment to an ultrathin elastic element that passes through the Z band. [Reprinted with permission from Locker and Leet, *J. Ultrastruct. Res.* **56**, 31–38 (1976). Copyright 1976 American Press.] (*2*) Plot of shift of position of N_2 with stretch in tortoise muscle cell. (From Page, 1968.)

termed resilin (Weis-Fogh, 1960). Between the fine filaments and the tonofilaments there is often a zigzag thickened double-membrane structure that closely resembles the intercalated discs of heart muscle; it is formed by the muscle-cell membrane on one side and epithelial-cell membrane on the other (Figure 3.36). The cuticle may send fine fingerlike processes inward at the region of attachment.

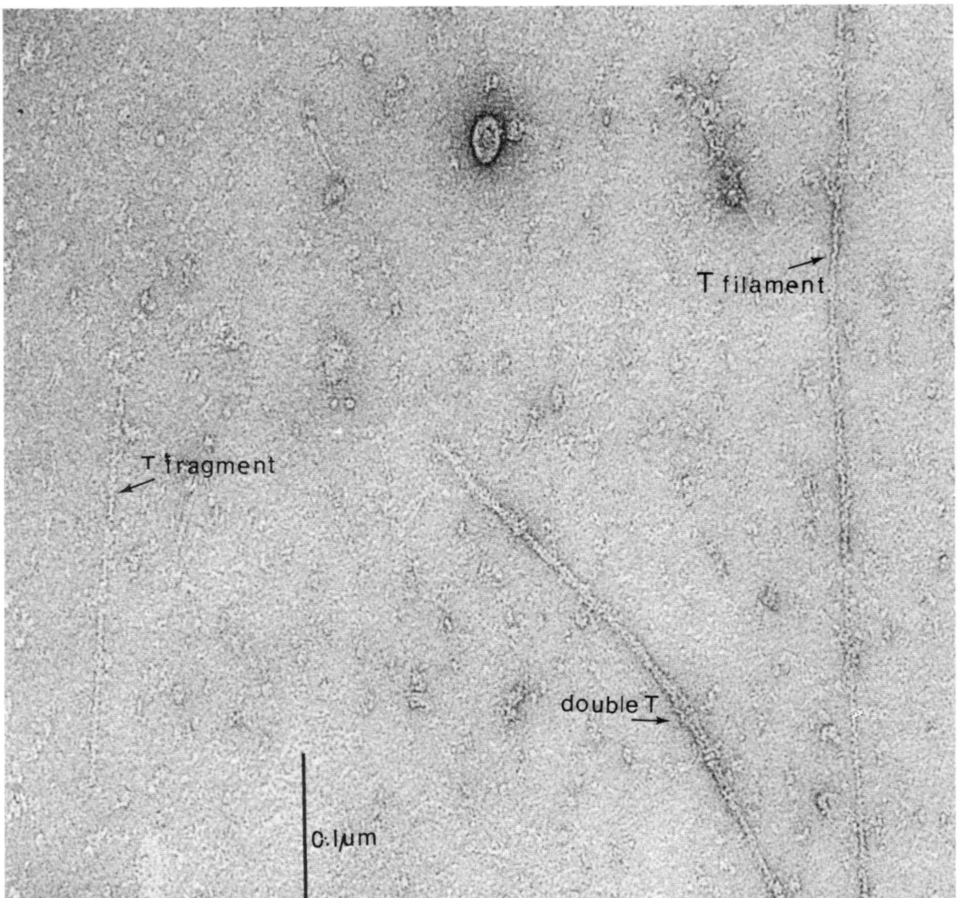

Figure 3.33. Negatively stained images of filaments obtained from dissociated brei of relaxed rabbit psoas. Visible are one intact T filament at right, one fragment of T filament, a closely apposed pair of T-filament fragments, part of an I filament, and an I-filament fragment. T-filament repeat periods in this micrograph range from 35 to 53 nm. The I-filament repeat is 38 nm. × 228,000. (ORIG.)

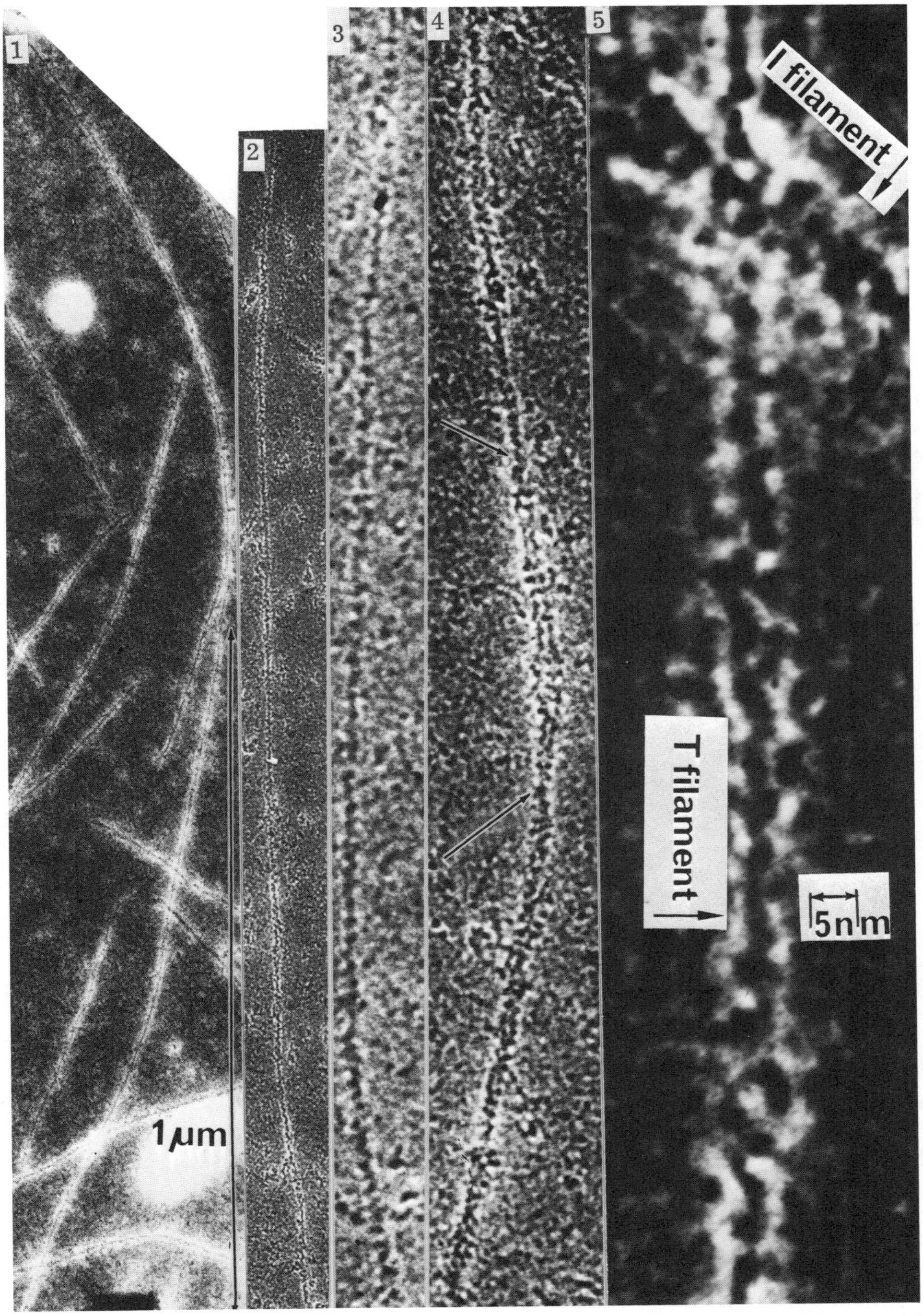

Figure 3.34. Ultrathin filaments from rabbit psoas, negatively stained. (*1*) A T filament at least 2.3 μm long, with several I filaments. (*2*), (*3*) Portions of T filaments showing repeating spiral zigzag structure. (*2*) ×135,000. (3) ×420,000. (*4*) A filament that closely resembles an I filament but is actually a pair of T filaments closely associated over most of their length: fully separated between arrows. × 247,500. (*5*) Portion of a T filament at very high magnification. × 1,400,000. [All ORIG. With Dr. Cynthia Jensen]

Figure 3.35. Typical terminal attachment of a body muscle cell of a soft-bodied animal: dorsal longitudinal muscle cell of *Peripatus.* I filaments are attached to the elongated ovoid dense bodies. These are attached by very fine filaments to a meshwork of collagen strands of different thicknesses. Collagen also forms a major component of the Sl and it enters the cell itself via wide invaginated longitudinal and transverse tubules. (From Hoyle and Williams, 1980.)

3.10. Significance of Striation

Although the striated cells of medusae have not been intensively studied by modern methods, it is evident that they possess all the familiar basic attributes of mammalian skeletal muscles. However, they differ in one significant feature: their sarcomere length is among the shortest known, being 1.3 μm, A = 0.7 μm, I = 0.6 μm (Figure 3.2 – D. M.

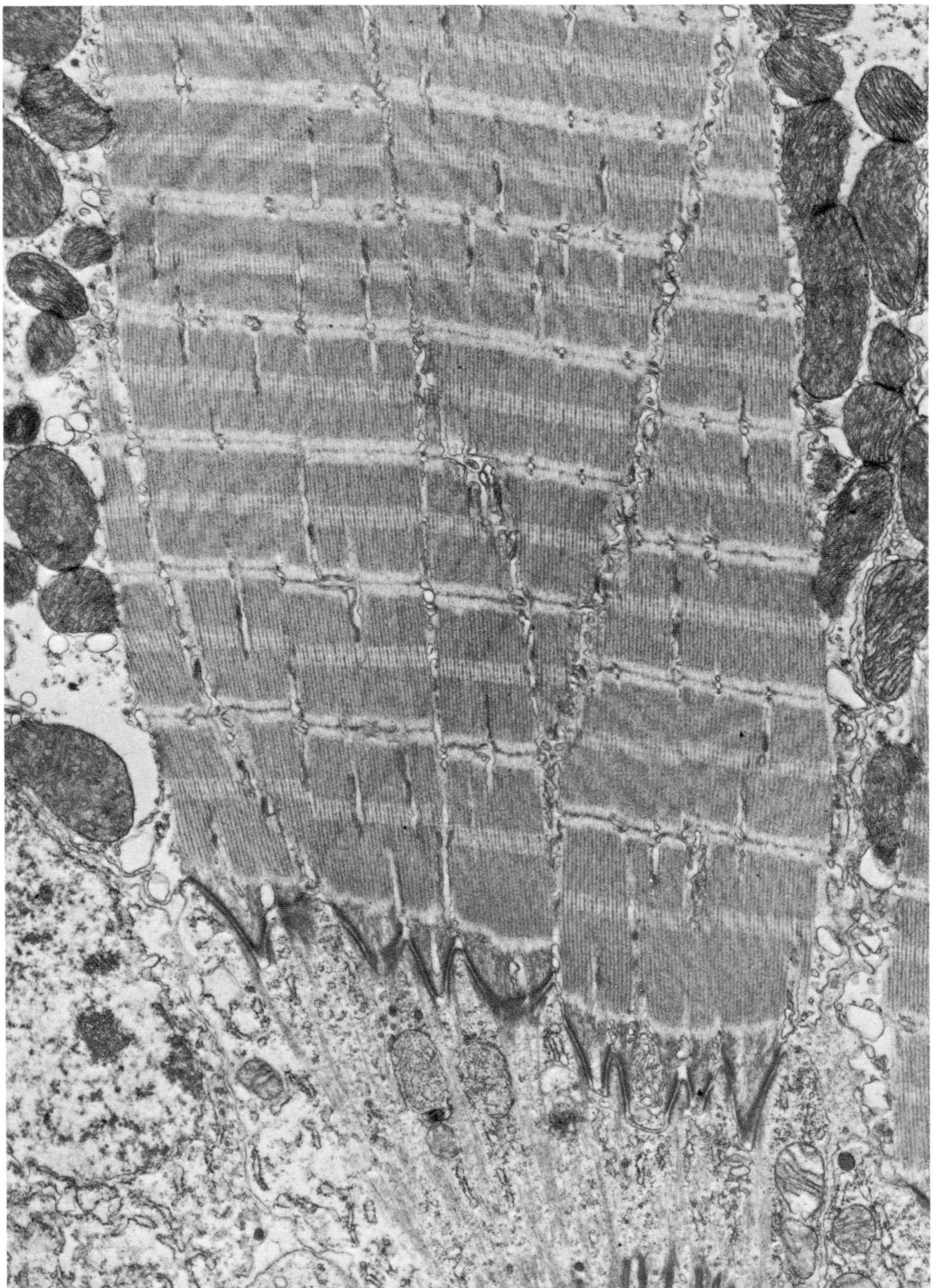

Figure 3.36. Terminal attachment of a single antennular muscle cell of copepod *Doropygus.* Fine filaments of various lengths replaces the terminal Z band and are in turn attached via a region resembling an intercalated disc to fingerlike inward projections from the cuticle. (ORIG.)

Chapman et al., 1962; Mackie and Singla, 1975). Contraction times of these muscles are not well known and are greatly affected by the jellylike matrix in which the cells must operate, but they are quite short. Bullock (1943) found rise times of about 0.8 s and relaxation times of 2 s in *Rhopilema.* The fastest recorded unit response of the relatively fast longitudinal retractor of the sea anemone *Metridium,* which is unstriated, had a rise time of 80 ms and a relaxation time of 300 ms at 15°C (Robson and Josephson, 1969). More commonly in anemones, the fast-twitch rise times are about 2 s. Some of their nonstriated muscles, such as the parietals, which are much slower, have twitch rise times of several seconds and relaxation times of many minutes.

The briefest twitch known from coelenterate muscle is therefore brought about by nonstriated contractile material. However, no twitches caused by striated muscle of coelenterates are more than about 5 s long, whereas the slower ones brought about by nonstriated cells range up to several minutes. Thus we may deduce that cross-striation is not associated with a dramatic increase in capability of either rapid shortening or rapid relaxation compared with nonstriation. However, there was evidently a strong tendency for an orderly alignment of myofilaments to occur in association with the behavioral need for *repetitive quick contractions,* such as occur in jellyfish, in which relaxation between contractions is complete. This was the feature that was achieved by the evolution of regular cross-striation. Muscle cells lacking orderliness may have evolved from the orderly ones rather than conversely, as is commonly assumed. Cross-striation, though it occurred early in the evolution of Metazoa, apparently did not convey any especially dramatic advantage. The medusa form is directly interchangeable with the polyp form in many coelenterates, where the medusa serves as a reproductive stage aiding dispersal. In turn, a young medusa attaches and becomes polypoid. These transitions are accompanied by transformation of the appropriate muscle cells from non-cross-striated to cross-striated and back again as form and function change. The medusa was evolved as a reproductive stage and evidently helped in carrying out the dispersal aspect of reproduction by the switch to striated muscle. For the sedentary forms, smooth cells are better. In the polypoid form the cells must (1) constantly serve a tonic function, (2) rarely contract rapidly, (3) often contract slowly, and (4) sometimes shorten drastically for defense or to expel waste. In the medusa the first, third, and fourth of these functions are not required; instead constant quick contractions are demanded to remain afloat and keep moving. Clearly, cross-striation favors the second function and lack of it the other three.

Cross-striation is unknown in platyhelminths, echinoderms, nematodes, and all molluscs except for a few bivalves. It has, however, become dominant in a few minor phyla, notably in the Chaetognatha and Brachiopoda. It puts in an appearance either in cells mixed with, or occurring in addition to, a portion of non-cross-striated muscle cells in the adductors of a few clams, notably scallops like *Pecten*, that have evolved the ability to swim by repeated fast opening and closing. Once again, repetitiveness in quick action is the distinguishing function with which cross-striation is commonly associated, but there are major exceptions. In arthropods the entire musculature, including that of the gut, is cross-striated, and most or all of the movements are slow. Thus organs that are always somewhat tonically contracted, make larger contractions only slowly, and relax very slowly can also have cross-striation. This phenomenon is not confined to invertebrates: the greater part of the esophagus of the rat is made up entirely of cross-striated cells (Gruber, 1978). There is, therefore, nothing about this organizational feature that limits it to being used for situations calling for repeated quick contractions. But secondary modifications were needed to permit the extreme slowness seen in gut muscles.

3.11. Distinguishing Features of Cardiac Muscle Cells

The principal characteristic of myogenic hearts, which include all vertebrate and most invertebrate hearts, is that they function as an integrated whole. There is a nerve supply, but except for a few neurogenic hearts, the effect of nerve input is only modulatory, to accelerate or decelerate the frequency of beating – which is of myogenic origin – accompanied by small changes in amplitude. Evidently it was found during the evolution of the hearts of most animals to be a more efficient direction to take than that of neurogenicity. To permit this independence, a heart muscle must include a pacemaker and an efficient conducting system within itself. The route followed, apparently to a considerable extent independently, was for component contractile cells to do a major part of the conduction themselves. This they achieved by a combination of branching, even anastomosing, with intimate cell-to-cell contacts having special properties, notably low electrical resistance (Figure 3.37). The most conspicuous, though not universal, special feature of a majority of both vertebrate and invertebrate heart muscle cells is complex structures termed intercalated discs linking the cells, principally at their ends. These run mainly transversely, at the ends of cells, though sometimes obliquely, and contain both desmosomes and less-dense regions of close contact called intermediate junctions. I filaments run directly into these discs. Though it is not certain that all intercalated discs represent low-resistance pathways, it has been directly demonstrated in some instances in which no other pathways were available that a propagated action potential passed across them. Besides these, nexuses, or gap junctions, often wavy, occur between adjacent cells laterally, plus other less-specific narrow clefts that may also function as low-resistance pathways.

3.11.1. Vertebrate

The internal fine structure of vertebrate cardiac muscle cells is basically similar to that of vertebrate skeletal cells, but with specific variations as follows. The sarcomeres are often remarkably short, and the cells are always thin. The shortest vertebrate cardiac sarcomere known occurs in the ventricle of a bird (finch – Bossen et al., 1978) and is only 1.3 μm long; the longest vertebrate cardiac sarcomere recorded is from guinea pig papillary muscle: 2.3 μm. Mouse ventricle sarcomere falls between, at 1.5 μm. The diameters of the cells range from about 7 μm for birds to 15 μm for mammals. The small diameter has enabled the cells of the bird to dispense with ITs; the SR makes extensive contacts with the PM to permit membrane control of contraction. In both some vertebrate and some invertebrate hearts (Sommer and Johnson, 1979) tubules occur around the Z regions and so are termed Z tubules. In the water flea *Daphnia,* the Z tubules have been shown to connect to the exterior via wide ITs (Stein et al., 1966). However, these Z tubules do not make contacts with longitudinal SR, so they are not equivalent to E tubules. Fine radially oriented filaments link Z regions of the opossum heart to each other, also to the PM, and possibly to Z tubules and SR extensions near the Z region. The function of these, and of Z invaginations, may be to serve as a general structural framework. The salamander heart has 30-nm-wide invaginated clefts that are not directly related to the Z lines.

Although Z regions of hearts are relatively wide, they very clearly have a fine structure similar to that described above for rabbit psoas. There are pre-Z filaments that are stiff,

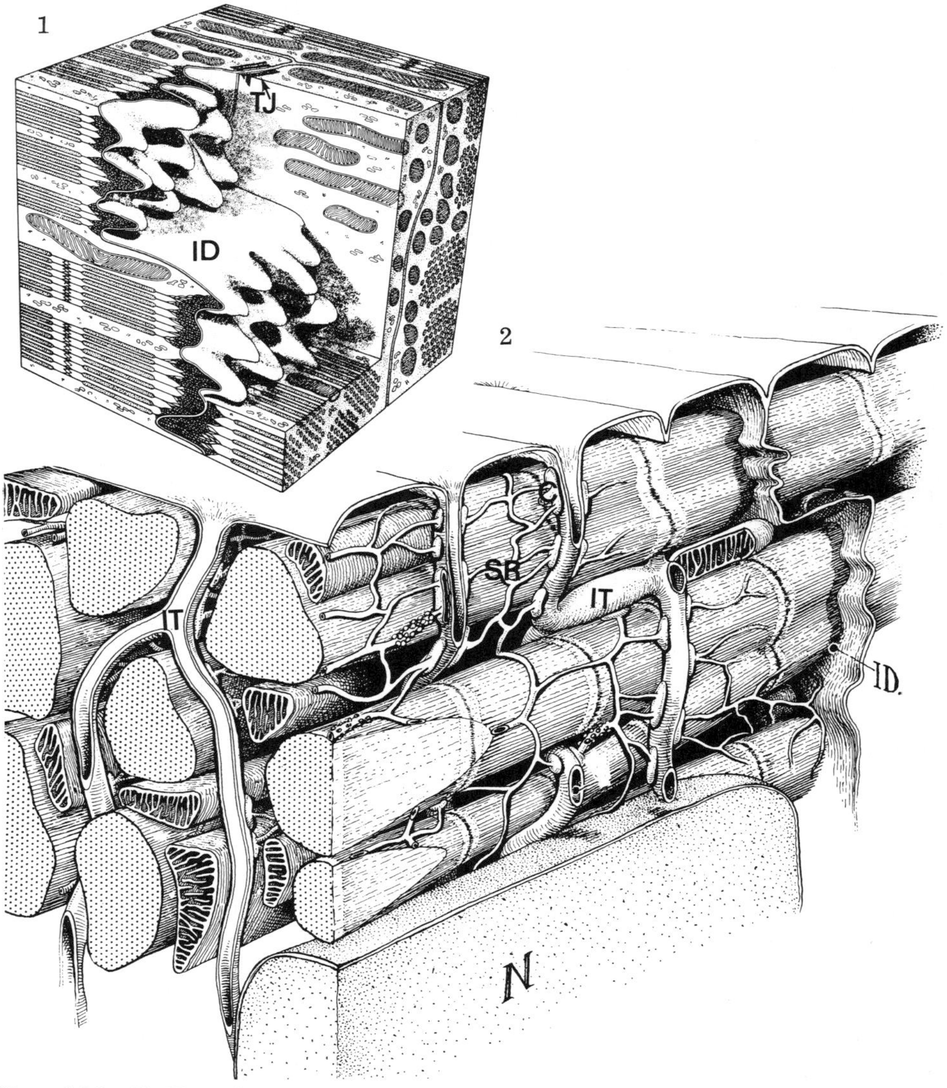

Figure 3.37. (*1*) Three-dimensional view of a generalized intercalated disc (ID) of vertebrate heart. The cells are structurally independent but linked to each other by the tight junctions. Drawing by F. Sjöstrand. [Reprinted with permission from *Physiology and Biophysics,* 19th Ed. T. C. Ruch and H. D. Patton, Eds., Copyright 1965 W. B. Saunders, Philadelphia.] (*2*) Early generalized drawing of vertebrate-heart-cell fine structure. (From Nelson and Benson, 1963.)

densely staining, and even in cells with very short sarcomeres get to a true square array. Double sets of Z filaments can be seen in longitudinal sections.

M bands are conspicuous in some heart cells, weak in others. The former probably are formed by three parallel arrays of M filaments; the latter show more than three components, possibly indicating a staggered array of M-bridge attachments. Thin and thick filaments are uniform and identical in thickness to those of vertebrate skeletal cells.

The amounts of SR and its specific forms vary between species, but all are fairly extensive and invade the myofilamentar clusters in an extremely irregular manner. In mammals there are ITs in ventricular, but not atrial or conducting, cells that are clearly the functional homologues of those of skeletal muscle, but it should be borne in mind that there is a second IT system, the Z system. The ITs are always a lot wider in mammalian cardiac muscle cells than in their skeletal counterparts. Some of the contacts are triadic, at Z-line level only, as in amphibian skeletal muscle, but the SR elements at the contact lack the broad expanses forming the cisternal elements of skeletal muscle. A majority of the contacts are, however, dyadic, resembling those of invertebrate skeletal muscles.

Mitochondria are abundant in vertebrate hearts, as is glycogen.

3.11.2. Invertebrate

In spite of having invented the same special tricks as vertebrate hearts – intercalated discs, low-electrical-resistance contacts, branching, and intrinsic pacemaking – heart muscle cells of invertebrate animals have a myofilamentar architecture that characteristically resembles that of invertebrate cross-striated skeletal muscles in general and is markedly different from that of vertebrate hearts. Invertebrate cardiac cell lengths are generally much greater than those of vertebrates, and the sarcomere length range is also much greater, though not as large as that found in some invertebrate skeletal cells. The longest cardiac cells are the 6.6-μm cells of the horseshoe crab *Limulus polyphemus* (Sperelakis, 1970) and the shortest, the 1.5-μm cells in *Daphnia* (Stein et al., 1966).

There is one unique general characteristic of invertebrate heart cells that distinguishes them from their vertebrate counterparts and that is their Z region. This is not only conspicuously thicker than that of vertebrates, it is clearly composed of only partially fused Z bodies lacking both the obvious zigzag Z filaments of the vertebrate counterpart and a square array. This condition is, however, commonly found in Z regions of all skeletal muscle cells of the more primitive invertebrates, whether cross-striated or not. Like their vertebrate counterparts, all invertebrate heart cells have smaller diameters than those of their skeletal muscles, often being only 5–10 μm thick, and maximally 40 μm in arthropods (where skeletal cells range up to 3 mm thick!).

Intercalated discs are widespread in all phyla and classes, absent only in very small species where the cells are themselves unusually small. In all the larger cells, both regular ITs and Z tubules are present. The ITs make dyadic contacts at various sites on the SR, commonly in the A/I overlap region and never specifically near the Z. Z tubules make attachment plaques to SR that are likely to be serving a structural function.

A filament lengths are constant in the shorter cells, but in the longer ones of *Limulus* heart, they range from 1.7 to 3.9 μm even without the same single cell (Sperelakis, 1970). Some of the sarcomere lengths of neighboring cells are markedly different from each other. The A filaments come in a wide range of thicknesses, as do their skeletal muscle equivalents. In *Daphnia* they are only 12 nm wide, in *Limulus* up to 40 nm, and in molluscan hearts between 20 and 35 nm. In all that have such thick filaments, the thin-thick filament ratio is high. The thin filaments are only partially arrayed in orbits around the thick, so the ratio may be significantly greater or smaller than would be indicated by orbit size only. Orbits commonly contain 10–12 thin filaments, which indicate ratios of 4 : 1 to 5 : 1, but the ratio is usually larger, 6 or 7.5, when all thin filaments are counted.

No strong M lines have yet been described for any invertebrate hearts, and there are counterindications in some. Probably connections between thick filaments occur but are widely distributed.

3.12. Smooth-Muscle Cells

There is left to consider only the general structural features of classical smooth muscle (SM) as present in the alimentary canal, bladder, iris, spleen, and so forth of vertebrates. Smooth muscle is so-called simply because of the lack of striations. These cells are all small, spindle-shaped, and thin, although there is a wide range of sizes, both between averages for particular muscles and within each. The most satisfactory data on sizes are for tissues that have been dissociated into component cells. The length range is from about 100 μm to about 3 mm, and the diameter from 0.1 μm to about 10 μm. The size range thus extends downward about 100-fold below that of average striated cells.

The small size, together with the nature of the muscles in which smooth cells occur, almost all of which lack hard skeletal parts and are often sacs or tubes, necessitates there being junctions between them at their ends and also laterally. There is considerable variety in the nature of these junctions, structural and physiological aspects of the junctions having been major sites for variation, permitting adaptive radiation of SM. Having distinguished a number of invertebrate skeletal cells as obliquely or disorderly striated, the remainder can be lumped with the vertebrate ones as true SM cells.

The single universal feature of SM-cell fine structures is a paucity of thick filaments. All SM cells are tightly packed with actin thin filaments that are essentially identical to those of cross-striated muscle, but lacking tropomyosin and troponin (see p. 118) found associated with thin filaments of cross-striated cells. Unless special treatments are used, thick filaments are not seen in SM cells, leading to a heated controversy about whether or not they exist in the living cell. Myosin is present in all SM cells, though in much lower concentrations than in striated cells. As a rough general average, the molar amount is only one seventh that of striated cells. It is clearly not present in the kinds of thick-filament array present in striated cells. There are two principal alternatives concerning the presence of thick filaments in SM cells. One is that SM cells function without them; the other is that in SM, thick filaments form only following excitation but are essential for contraction. Myosin isolated from SM in which no thick filaments occur after normal fixation forms aggregates indistinguishable from similar ones from cross-striated cells, with a central bare shaft and projections (Figure 3.38.*6*). A small number of thick filaments occur with normal fixation treatment in guinea pig taenia coli (e.g., Cooke, 1976). Ribbons, in addition to a "variety" of thick filaments, were found by Schoenberg and Haselgrove (1974). Treatment of toad small intestine with trypsin leads to the appearance of numerous thick filaments, but curiously these are bundled together and segregated from the thin ones (Rosenbluth, 1971). It may seem anomalous that a proteinase could cause aggregation, but very weak trypsin treatment may merely cause a myosin conformational change that happens to favor molecular interlocking, or may remove a group that inhibits it. Orientation of the trypsin-induced thick filaments is random. Long ribbon filaments of myosin, 15 × 100 nm × up to several μm long, can be made *in vitro,* and they occur in cells, especially ones shrunken in hypertonic solution (Lowy and Small, 1970; Somlyo et al., 1971). A repeat period of 14.3 nm is indicated in living guinea pig taenia coli (Lowy et al., 1973), and this is regarded by some as evidence for the presence of organized thick filaments. In EM sections, Lowy and Small found the ribbons to be well dispersed, and actin filaments to be linked to them by regularly repeating X bridges. However, there was no question of orbiting, the tightly packed thin filaments greatly outnumbering the rare thick ones. The ratio in vascular SM is about 15:1, and in guinea pig taenia coli, a value of 13:1 is claimed (Small, 1977).

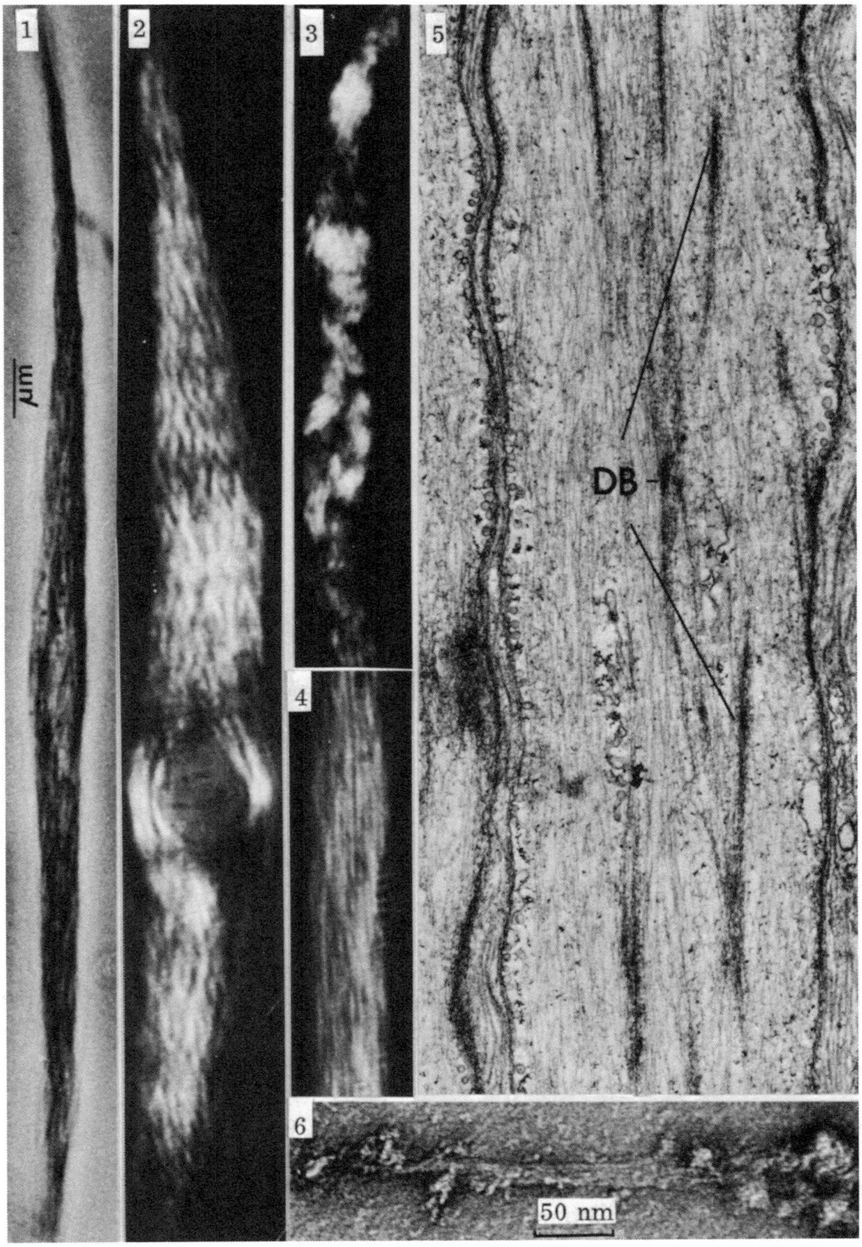

Figure 3.38. Characteristics of SM cells. They are small and spindle-shaped, with a single nucleus located in the middle, and are embedded in a collagen matrix. A spiral alignment of the contents is visible with suitable techniques. Internally only a mass of 5-nm filaments running longitudinally can be discerned, attached at apparently irregular intervals to highly elongate dense bodies. (*1*) Chicken gizzard cell. [By J. V. Small. Reprinted by permission from *Nature.* **249**, 324–327 (1974). Copyright 1974 Macmillan Journals Ltd.] (*2*) Taenia coli cell with phase-contrast optics. [By J. V. Small. Reprinted by permission from *Nature* **249**, 324–327 (1974). Copyright 1974 Macmillan Journals Ltd.] (*3*) Chicken gizzard cell glycerinated and stained with antigizzard myosin. [Reprinted with permission from Bagby and Abercrombie, in *Motility in Cell Function,* pp. 427–431. Copyright 1979 Academic Press.] Myosin locations follow the spiral alignment. (*4*) Part of a stretched taenia coil cell [as (*2*)]. [Reprinted by permission from *Nature.* **249**, 324–327 (1974). Copyright 1974 Macmillan Journals Ltd.] (*5*) Longitudinal EM section of guinea pig taenia coli cell. Note the numerous 5-nm filaments attached to elongate dense bodies (*DB*). (From Cooke and Fay, 1972.) (*6*) Synthetic myosin filament, negatively stained, from rat uterus. (From Hanson and Lowy, 1964.)

After fluorescent antibody to chicken gizzard myosin has been added to gizzard cells, they show a wavy, streaked appearance that indicates the presence of internal organization (Bagby and Abercrombie, 1979), first clearly recognized by Small (1974) during birefringence studies (Figure 3.38.*3*). Small considers that the surface dense bodies of taenia coli cells are arranged in a spiral. Close to these he observed actin filaments in orbits around the few thick ones. Although thick (10–20 nm) filaments were claimed by Rice et al. (1970) in taenia coli, there were no orbits around them; but some, at least, may have been intermediate filaments (see below) rather than myosin. Natural thick filaments of myosin have been isolated from guinea pig taenia coli SM cells by Small (1977), who noted that unlike those of striated muscle cells they have no obvious polarity.

In *in vitro* studies of the self-assembly of myosin molecules from smooth cells, two different types of myosin are formed. One is a long form, in which there is a simple side-polar alignment, one polarity on one side, the opposite one on the other. The other form consists of short bipolar filaments composed of two myosin molecules: double myosin molecules have long been considered to be a common form *in vitro*.

Another kind of filament, commonly appearing about 10 nm thick, is abundant in many smooth cells. These filaments attach to dense bodies present on the surface membrane in such a manner as to suggest that they form a cellular framework or cytoskeleton (Cooke, 1976). They do not attach to either the actin or myosin filaments. Such a skeleton would have to be highly elastic, and of course it could be contractile. A second set of dense bodies also occurs, some attached to the surface membrane, others suspended in the cytoplasm. To all of these, actin filaments are attached, so they are equivalent to Z bodies. A quasi-regular lattice of round 18-nm filaments, with a 70-nm spacing, has been found in smooth cells of the portal and anterior mesenteric veins of the rabbit (Somlyo et al., 1976a).

When guinea pig taenia coli was examined after stretch the intermediate filaments were found to be consolidated into a mass near the center of the cell (Cooke and Fay, 1972a). By contrast, there was no corresponding change or asymmetry in myosin distribution (Cooke and Fay, 1972b). Treatment with EDTA removed actin and myosin but left the 10-nm filaments and their dense-body attachments unchanged. Cooke and Fay's earlier finding is not compatible with more-recent explanations of intermediate filaments as the cytoskeleton. Perhaps as chemical studies progress, the range of types of filament will become clearer. The protein responsible for these so-called intermediate filaments has, like actin, a molecular weight of 55,000, but no specific name has yet been assigned to it. Another protein, which is phosphorylatable and binds to actin, has been isolated from smooth cells and termed filamin (Shizuta et al., 1976).

How all this will turn out remains to be seen, but it seems firmly established that smooth cells do indeed have a cytoskeletal framework, essentially spiral/oblique in overall form, that is highly elastic. The elastic filaments must vary in thickness with their length, but the question of their dimensions at different cell lengths has not been addressed specifically. Campbell et al. (1971) found the range to be 8–11 nm, contrasted with the elastic framework of striated cells, which comprises ultrathin filaments in the range 2–4 nm. No filaments this thin have yet been claimed for smooth cells. Thin filaments attached to the PM at dense bodies are not only intermediate: a second set occurs that is undoubtedly actin. These, attached to different dense bodies, have been "decorated" with myosin and found to form arrowheads pointing away from the attachment. Some α-actinin occurs in SM and may be located in the bodies to which actin filaments are attached. These bodies, which are therefore Z bodies, vary greatly in size, and some are branched structures.

There is SR is smooth cells, but in very different amounts according to source. Not surprisingly, no ITs occur. However, SR in the larger cells makes some contacts with the PM. The surfaces of a majority of smooth cells have rows of large vesicles, or caveolae, which at some times at least are open to the exterior, since they fill with lanthanum or ferritin. The SR does not contact these vesicles, but mitochondria are close to them, so they may be concerned with ion regulation.

The commonest junctions between smooth cells are nexuses, which are also called gap, or tight, junctions. Their abundance differs among cell types: for example, they are abundant in amnion but rate or even absent in nictitating membrane. These differences are functionally related, the cells lacking nexuses being innervated and controlled individually, as are skeletal cross-striated cells. The cells with many nexuses operate along similar lines to those of heart muscle, which requires spread of electric excitation from cell to cell.

4

The Molecular Machinery of Contraction

The loveliest toy ever provided by nature for the biochemist is the contractile muscle fibril. A. Szent Györgyi†

There should be no conflict between the molecular and the holistic approach. W. F. H. Mommaerts‡

. . . *an unfolding and refolding of the actual molecules – the very fundamental kind of local motion.* W. T. Astbury§

4.1. A (Thick) Filaments

Knowledge of the molecular events of contraction has been entirely derived from skeletal muscles of vertebrates, in which the principal constituents are the thick filaments of the A bands. A filaments are arrayed in parallel and are laterally aligned so that the A-band lengths observed in living or fixed muscle to a good approximation equal average individual-filament lengths. A-band lengths of invertebrates range from 0.5 μm in jellyfish swimming muscle to 32 μm in the foregut of the small annelid worm *Syllis.* Vertebrate skeletal muscles, by contrast, have an A length range that is narrow, from 1.2 to 2.5 μm. A filaments often show a regular periodicity of about 43 nm in each half, with a bare center called the shaft region, which corresponds to a lighter region in the center of the H zone, straddling the M band. Unlike the H zone, the shaft region is not affected by stretch or contraction.

The double refraction by A bands indicates a parallel alignment and orderly packing of the molecular components, of which about 95% are the high-molecular-weight protein myosin. Myosin is selectively removed by irrigating myofilament clusters with Guba-Straub solution (0.3 *M* KCl, 0.15 *M* phosphate buffer at pH 6.5) or Hasselbach-Schneider solutions (0.47 *M* KCl, 0.01 *M* pyrophosphate, 0.1 *M* phosphate buffer at pH 6.4). These solutions extract fairly pure myosin, together with other minor proteins associated with A bands, notably C protein and M protein, and they leave myofilament clusters with very little residual A-band birefringence (Hanson and Huxley, 1957). Such material does not fall apart, but retains its original configuration, and it is elastic, indicating the presence

†A. Szent-Györgyi, *Chemistry of Muscular Contraction,* Academic, New York, 1947.

‡W. F. H. Mommaerts, *Arch. Biol.* **76**, 355–363 (1965).

§W. T. Astbury, On the structure of biological fibers and the problem of muscle. *Proc. R. Soc. Lond.* **134**, 303–328.

of a background of fine elastic filaments (T filaments) in addition to the major components, actin tropomyosin, troponin, and α-actinin, which remain. After myosin extraction, the H zones are conspicuous, marking the space between the inner ends of the I (thin) filaments. Following additional treatment, with 0.6 *M* KI, I filaments, most Z material, and residual M material are also extracted, leaving only a fine filamentar background in which faint Z regions are still evident; this is called a myofilament-cluster "ghost."

A filaments dissolve in strong salt solution and reform upon dilution. They self-assemble in a characteristic manner to form filaments similar to the natural ones (H. E. Huxley, 1963). Since these are of a precise length for a particular type of cell, a major question is how the length is controlled. The regulation of length is thought to be the function of special proteins that are intimately associated with the myosin molecules, the most important of which is termed C protein.

4.1.1. Myosin Molecules

The form of a single myosin molecule is shown in a negatively stained EM image in Figure 4.1 (see also Figure 2.5). In each molecule there are two thiol groups, each associated with a dense fragment separable from the other and a "lighter" fraction by differential ultracentrifugation following tryptic digestion. There is a 144-nm-long by 2-nm-wide portion comprising a 79-nm-long region, the entire light meromyosin (LMM) fraction, plus the initial part of the heavy meromyosin (HMM) fraction, the S_2, and one of the thiol groups, which is 67 nm long. (Details are those given by Harrington et al., 1979). This part of the HMM is itself composed of two subfragments, one 22 nm long, the other 43 nm long. The whole is formed of two equal strands with a high helical content twisted around each other into a coiled coil. At the tip of the HMM S_2 fraction are two ovoid-shaped heads, each containing one of the other thiol groups; this is the HMM S_1 fraction. The molecular weight of the LMM is 150,000, that of S_2, 60,000, and of S_1, 120,000. Attached to the S_1 are two light chains, an alkali light chain that is dissociated at high pH and another that is removed by the sulfhydryl reagent DTNB (Lowey et al., 1979). Altogether, this large molecule is 140 nm long, with a molecular weight of 460,000–500,000. Although whole myosin molecules readily self-assemble, the components, the S_1 and S_2 fragments and LMM, show no tendency to associate (Lowey et al., 1969) (Figure 4.1).

We have observed anastomosis quite frequently in thick filaments of rabbit psoas cells and have found a few in which the majority are of twice normal thickness, with orbits of irregular number of thin filaments, averaging 12 (Figure 3.26.*1*). Thus a great deal of flexibility is possible in packing myosin molecules together to make thick filaments, even in vertebrates. A general model that allows for known variants was proposed by Squire (1971; 1972; 1973; 1974), starting with optical-diffraction analysis of TEM plates of longitudinal sections; the analyses provided information about the X-bridge lattices, but Squire's model has not found favor. On Squire's model, the bridges are always swung out laterally. Other models regard this as a secondary "unlocked" position, the true rest or "locked" position being parallel to the backbone.

By using a combination of labeled antibody stains and partial trypsin digestion, Pepe (1967a; 1967b; 1971) has located trypsin-sensitive antigenic sites that are the probable hinge regions of the myosin molecules. These regions, which are parts of LMM, are differently displayed in the main body of the thick filament than the corresponding

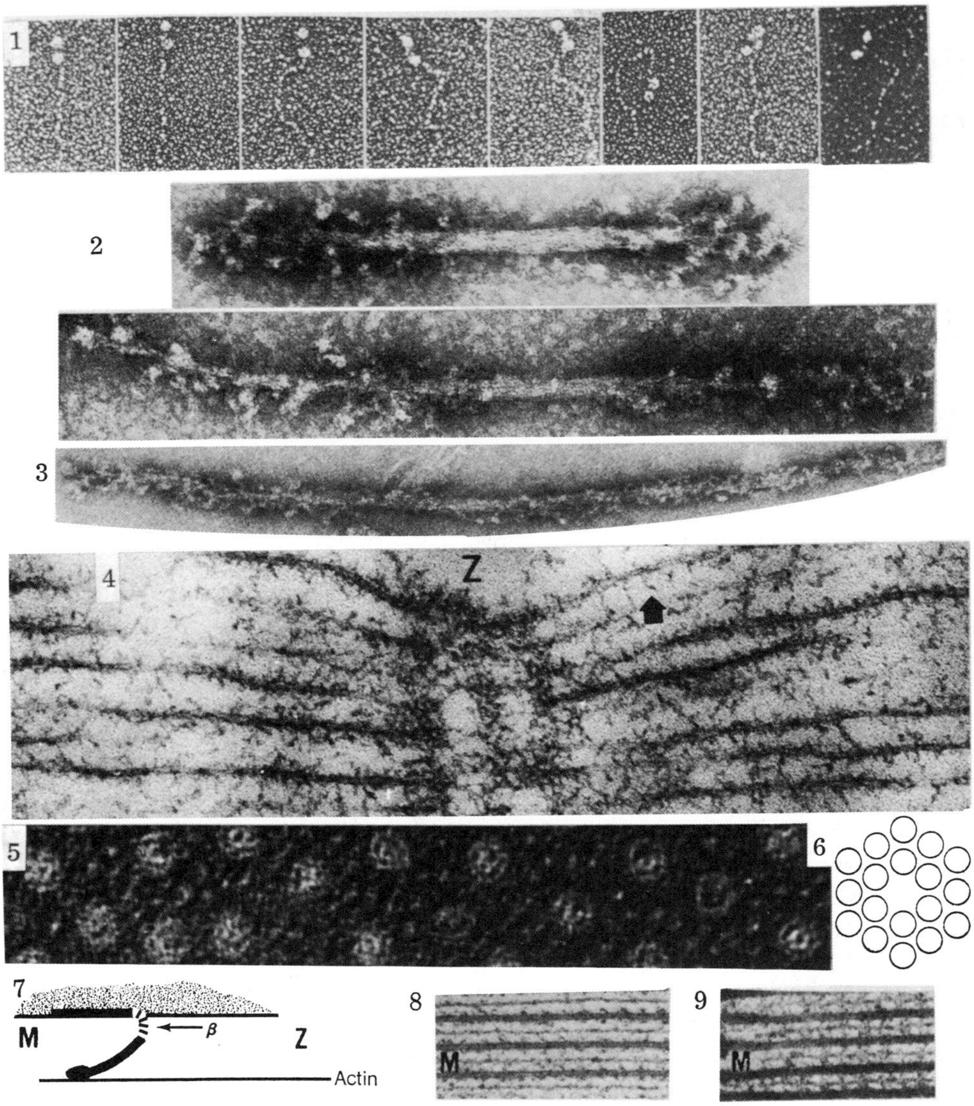

Figure 4.1. Myosin molecules and architecture of thick (A) filaments. (*1*) Negatively stained individual myosin molecules. (From Lowey et al., 1979.) (*2*) Negatively stained self-assembled A filaments from dissolved myosin molecules. (From H. E. Huxley, 1964.) (*3*) Same as (*2*) but assembled in relaxing solution. (From Taylor, 1976.) (*4*) Portion of crab (*Portunus*) muscle cell that lacked I filaments, untreated, fresh fixed with gluteraldehyde. The spiral arrangement of the outer parts of the myosin molecules is readily discerned, and the heavy meromyosin head sections are visible (*arrow*). The outer portions curl around toward the M regions. (ORIG.) (*5*) Reverse-contrast image of a TS of crab A filaments, showing component myosin molecules. (From Gilëv et al., 1968.) (*6*) Diagram of arrangement of myosin molecules in (*5*). (*7*) Location of a portion of the myosin molecule in intact cells, indicated by β, that is sensitive to enzyme action. [Reprinted with permission from F. A. Pepe, *Int. Rev. Cytol.* **24**, 193–231 (1968). Copyright 1968 Academic Press.] (*8*) Portion of relaxed insect (*Lethoceus*) flight muscle cell near the M region (*M*), showing outer portions of myosin molecules with heads visible and turned toward M, but not attached to the I filament. (Courtesy of Dr. Michael Reedy.) (*9*) Same as (*8*) but in contracted (rigor) state. Myosin heads are attached to I filaments. (Courtesy of Dr. Michael Reedy.)

regions in the H zone, on one hand, and the outside edges, where the molecules are terminal, on the other. This makes them both differently affected by the enzyme and differentially available to the antibody. The entire HMM region is free to project laterally or even to be angled inward toward the H zone, which is the position seen in EM sections (Figures 4.1.*3*, 4.1.*8*, and 4.1.*9*). The X-ray data from contracted muscles also indicate inward-turned angles.

4.1.2. C Protein and Its Roles

C protein was discovered during electrophoresis of crude myosin by Starr and Offer (1971). At first thought to be the regulator of A-filament length, it now seems likely to be also an organizer of the myosin molecules (e.g., Miyahara and Noda, 1980). It has a molecular weight of about 140,000, and its presence both slows down thick-filament formation from myosin solution and causes the "synthetic" filaments to be thinner (7–11 nm) and also longer than ones formed in its absence during a very slow dilution. When formed in the presence of C protein, rabbit psoas A filaments have an average length of 1.5 μm – the normal *in vivo* length. C protein is probably normally present on the thick filaments at about 40 molecules per thick filament, binding strongly to myosin, where it is presumed to be important in determining the details of assembly.

Electron microscopy of rabbit skeletal muscle cells labeled with antimyosin shows seven stripes spaced at intervals of 43 nm in the middle one third of each half of the A band (Pepe and Drucker, 1975). This is the same spacing as that shown by myosin molecules, but here reflects the distribution of C protein. It implies a difference in packing of the myosin molecules, affecting their availability for reaction in these regions, which have regular packing. Synthetic rabbit myosin thick filaments self-assembled in the absence of C protein have a 14-nm axial subunit spacing, and a "backbone" with 3–4-nm striation as well as 43-nm repeats with 14.3-nm subunits (Koretz, 1979a). All these are also characteristics of natural thick filaments, but there is a lot of width variation and some ragged, fraying ends and margins in the ones so assembled synthetically. By contrast, when myosin is assembled in the presence of C protein, the width variation and raggedness are eliminated (Koretz, 1979b). If the fraying in the absence of C protein indicates weak myosin-to-myosin binding, this must be promoted by C protein. When assembled in a solution that promotes relaxation, they aggregate differently, without a central bare zone (Taylor, 1976). A rigor-promoting solution leads to the normal appearance. Pepe (1979) has recently indicated where C-protein binding occurs in rabbit muscle and incorporated these data into an earlier structural model he had proposed for vertebrate thick filaments. There are only seven attachment sites in each half sarcomere, at the sites indicated in Figure 4.2. This number indicates that the structure of the myosin "backbone" changes along its length, a fact that earlier models did not incorporate. The diffraction data show a center-to-center spacing of 3.7 nm. Since each myosin rod is only 2 nm thick, this means they are associated in pairs, so each "unit" of thick filaments is two molecules.

In longitudinal array there are two repeats, 43 and 14 nm, the former of which also occurs in paracrystal of LMM and is three times the latter. Therefore, there must be four myosin molecules every 14 nm. A gap occurs in the spacing at the terminal region; it can be explained as resulting from the characteristics of the three equivalent groups of four structural units that make up the model.

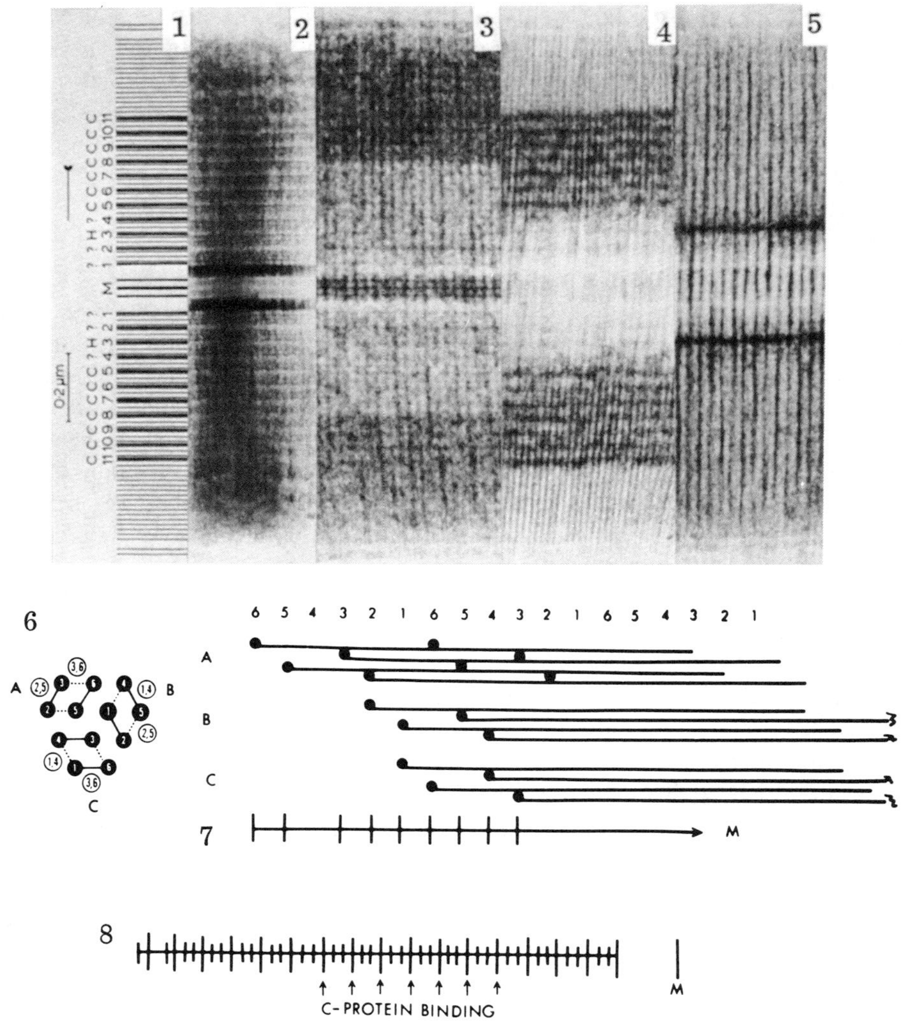

Figure 4.2. Detailed substructure of vertebrate A region according to Dr. Frank Pepe. [Reprinted with permission from F. A. Pepe, in *Motility in Cell Function,* pp. 103–116. Copyright 1979 Academic Press.) (*1*) Model of distribution of components other than myosin. (*2*) Negatively stained A segment showing nonmyosin as white stripes. (*3*) Nonmyosin components shown as stained stripes – frog sartorius. (*4*) C Protein in rabbit psoas. (*5*) H Protein in rabbit psoas. (*6*) (Left) Cross section of the model showing presumed myosin molecule locations. (Right) Longitudinal view of the three clusters (*A, B, C*) corresponding to the section at left. (*7*) Overall model showing position where there is a gap in cross-bridge locations associated with tapering of the filament. (*8*) Locations of the C-protein-binding sites.

A filaments have spindle-shaped ends when fixed in a relaxed state, but in rigor muscle there are fuzzy ends. This suggests that activation may cause component molecules to angle farther away from the LMM "backbone" than they normally do, at least in this region. In solution, C protein also binds to actin, and it has recently been found that myofilaments that already have a full complement of C protein can still take up extra C protein, most of which is added to the I regions (Moos, 1981). This binding is regulated by Ca, suggesting that is may have physiological significance.

4.2. The Natural Vertebrate Thick Filament

The LMM fraction lies largely within the A filament, forming part of a "backbone" for the A filament. It then curls around progressively until the HMM points outward or inward toward the H zone (Figure 4.1). The LMM molecules nearest the H zone either abut their neighbors from the other half of the sarcomere tail to tail or overlap them to some extent to form a staggered array. These portions, in which the myosins are antiparallel, are responsible for a central, clear "shaft" zone, which can usually be distinguished in a myofilament cluster or even a whole sarcomere and is sometimes called the pseudo H zone. Beyond this region, which for rabbit psoas A filaments is 150 nm long, the myosins overlap in parallel, and parts of them project outward. The projecting posts are formed by the HMM portions' turning outward and then curling forward. The point of maximum bending is effectively a hinge and is sensitive to the proteolytic enzyme trypsin.

The heads of the projecting portions of the myosin molecules can attach to I filaments, forming X bridges between the two major filament components. Hereafter, these portions will be referred to as *X bridges* even when not attached, and *attached X bridges* when in the attached state. The ovoid S_1 fractions are enzymatic, splitting ATP. However, when these heads have combined with actin, a sterically different structure is formed that is also an ATPase but one that is very much more powerful, at least 10,000-fold. Indeed, the entire conditions for the enzymatic activation are changed. For example, magnesium suppresses the ATP-splitting ability of myosin heads, but greatly enhances the same activity of the actomyosin combination. Calcium enhances the former, but its presence is critcally essential for the latter. Literature on this subject has been frequently reviewed: readers are advised to consult the book *Machina Carnis* by Dorothy Needham (1971) for details. The functionally critical elements of the molecular components of muscles include the enzymatic properties of the myosin head, conformational changes initiated in as yet unknown sites following actin/myosin interaction, and mechanical properties of the projecting portions of the myosin molecules with their hinge(s), which possess both characteristic stiffness and short-range elasticity.

Myosin from different sources, even different cells of the same muscle, especially when they have different properties, can have markedly different molecular compositions (Lowey and Risby, 1971). Lowey et al. (1979) wrote,

> Even the myosin of a relatively homogeneous fast, white skeletal muscle consists of several isoenzymic forms.

These have widely different rates of splitting of ATP.

4.2.1. Assembly of the Thick Filaments of Vertebrates

A detailed model describing the assembly of myosin filaments in some vertebrate cross-striated cells has been worked out by Pepe (1967a, 1978; Pepe and Drucker, 1979), using extensive combined EM/antibody staining. Especially when enhanced by optical rotation, TEM transverse sections always indicate that there are 12 component strands, each probably a pair of myosin molecules, although they were at first thought of as single molecules. They are commonly packed 3, 4, 3, 2, but may occur in a hollow circular profile in the M region. The projections occur in pairs emerging at the same level, 180°

apart, with a regular helical arrangement between successive pairs so that there is a repeat every three pairs. The subunit repeat, which corresponds to four myosin molecules, is 14.3 nm, and the helical pitch is 42.9 nm. There are altogether about 400 myosin molecules per thick filament. The triple repeating unit must have a counterpart in the M region. Staggered pairs of myosin molecules, with M bridges attached to the points of abutment, would explain the overall array and be compatible with three equally spaced sets of M bands. This array is present in rabbit psoas cells, but it is not universal, so some types probably have different packing arrangements.

Pepe's model utilized straight LMM fragments, abutting in parallel, to make the "backbone." Recent interpretation of the X-ray data suggests that the already-coiled coil structures further coil around each other in a supercoil in the backbone (Wray, 1979).

4.2.2. Myosin Packing in A Filaments of Invertebrates

With the extraordinary 60-fold range of thick-filament lengths found in invertebrates, it is to be expected that a wide range of packing exists, even though the individual myosin molecules appear to be similar in form. Many invertebrate thick filaments include paramyosin; these will be described separately. Of those believed to contain only myosin and minor proteins, the thickness range is from 14 to 28 nm. The thinner thick filaments occur in cells having a low thin-thick filament ratio. Although some arthropod cross-striated cells have been found that have ratios of thin to thick filaments below 2 : 1, this ratio, with orbits of six thins around each thick, appears to be the elemental one. A ratio of 3 : 1 is very common, being present in most copepod muscles, some insect and crustacean leg muscle cells, and all direct insect flight muscles. Ratios of 4 : 1 and 5 : 1 are also common in other insect, crustacean, and arachnid skeletal cells. A few of the slower crustacean cells have ratios of 6 : 1 and 7 : 1. In general, the larger the ratio, the thicker the filament.

The thickness in an insect flight muscle, the dorsal longitudinal, of the hemipteran *Lethocerus* and of the blowfly *Calliphora* is 20 nm. These thick filaments are all very clearly circular and hollow in transverse sections (Figure 4.3) in all regions except the M, which is both solid and ovoid (Figure 3.23). Combined X-ray and TEM techniques (e.g., Reedy et al., 1965; Reedy, 1968) have made this the best known of all muscles structurally, and it is also one of the most thoroughly studied dynamically (see p. 145). The projecting parts of myosin filaments are arranged in symmetrical doubled pairs, four every 14.6 nm. When attached to I filaments, the four form the arms of a "flared X" configuration as seen in transverse sections. There is a 67.5° change in azimuth with every repeat, forming a double helical lattice (Figure 4.3.*6*). There are about 30 repeats per half filament, therefore 640 per whole filament.

Figure 4.3. Arrangement of projections (so-called cross bridges) from thick filaments. (*1*) Arrangement in center of each half of a rabbit psoas A filament. [Reprinted with permission from H. E. Huxley, *Science* **164.** 1356–1366 (1969). Copyright 1960 by the American Association for the Advancement of Science.] (*2*) Arrangement in center of each half of an insect (*Lethocerus*) flight muscle. (*3*) Same as (*2*), represented as a plane projection. (*4*) TEM longitudinal section of (*2*). (*5*) Diagram of sequence of projections viewed end on. (*6*) Flared-X nature of the projections when attached to I filaments, for each of the nine possible forms, corresponding to those shown in (*5*). (4.3.2.–4.3.6. by M. K. Reedy. With permission from *J. Mol. Biol.* **31**, 155–176 (1968). Copyright 1968 Academic Press, London.) (*7*). TEM of transverse section of (*2*), in contracted (rigor) state. (Courtesy of M. K. Reedy.)

1

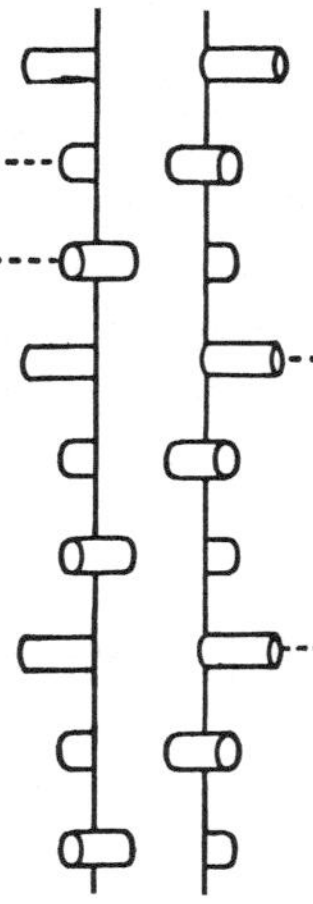

2

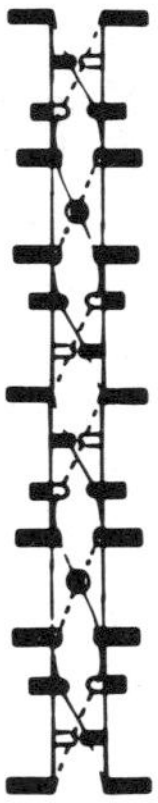

3

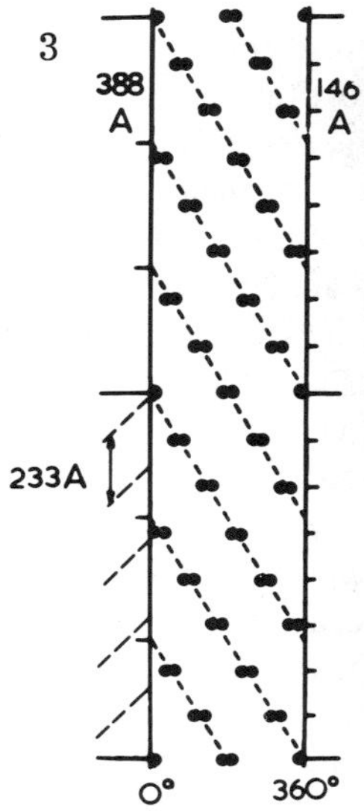

4

5

6

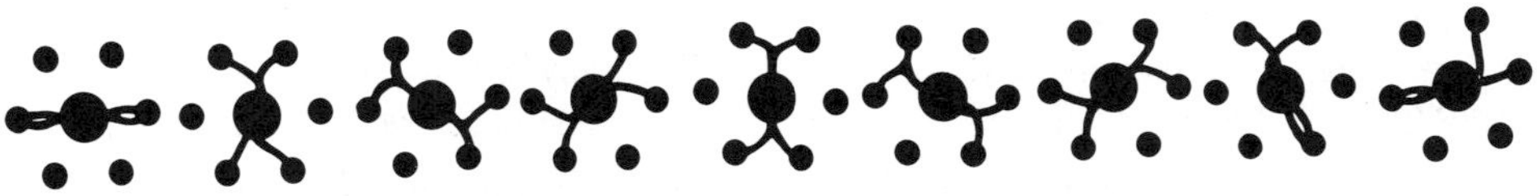

7

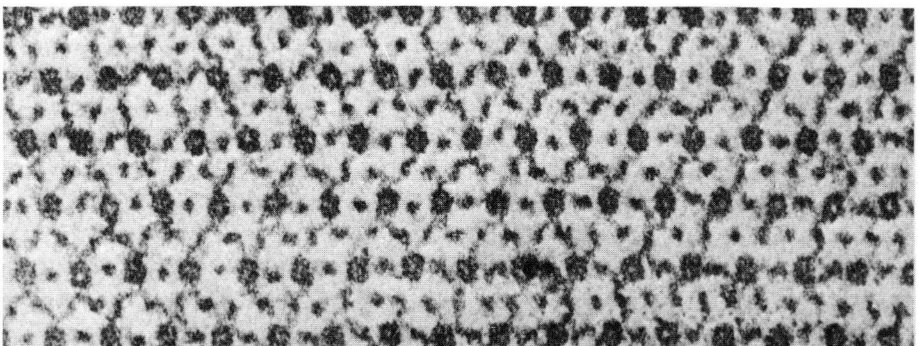

Most of the shorter-SL crustacean thick filaments have a hollow profile except in their M region and outermost tips, and a 20-nm diameter. Some are conspicuously thicker in the center, tapering progressively outward: these have no bridges in the exact middle, but instead have dual sets of M filaments, one on each side of center. These could be associated with overlapping of myosin LMM fractions instead of abutment of tips. In long-SL crustacean cells, there are large variations in thick-filament structure along the length of the cells. In single scutal-depressor cells of the giant barnacle *Balanus nubilus*, A-band lengths range from 2.2 to 20.3 μm, the majority being close to the average of 5.2 μm (Hoyle et al., 1973). The longer individual thick filaments have a series of thin regions, which are hollow in section, alternating with solid ones that are three times as thick: 34 nm compared with only 11 nm. Regions are in register across about 20 thick filaments only, progressively slipping laterally. There are no discrete M bands, but M bridges with central thickenings occur between some of the thicker regions. The molecular assembly of these very long A filaments must contain alternations between basic patterns. There is more than one solid thick region to each thick filament, so for this to be equal to LMM region overlaps, as in the double M types, one half would have their HMM heads pointing in the wrong direction for contraction. That is, unless the heads pointing the wrong way act upon the I filaments of the other half of the sarcomere after shortening has occurred to the point where these neighboring I filaments have reached them. These cells are not only capable of very extensive reversible shortening (see p. 261), but they also have a wide range of length over which maximum shortening velocity under fixed load is constant.

Some X-ray data are available on the thick-filament architecture of selected muscle cells – from which muscles the authors (Wray et al., 1975) forgot to mention, from a crustacean (*Homarus*), a mollusc (*Placopecten* – a scallop), and an arachnid (*Limulus*). They paid particular attention to X bridges, which were all substantially angled to the filament axes. They say their data are consistent with Squire's (1972) model, which permitted prediction of the relation between the rotational symmetry and the diameter of the "backbone." In the *Limulus* muscle, the symmetries are threefold or fourfold for a 19.5-nm diameter filament, whereas in the scallop they are sixfold or sevenfold for a fatter, 22.5-nm filament. The principal conclusion of general interest is that very different X-bridge configurations can be associated with rather similar "backbones. A muscle of the swimming crab *Portunus* gave 76.5-nm layer lines in X-ray analysis (Maeda et al., 1979). The lines had the same positions in rigor muscle as in the relaxed state but were much strong in rigor, in which the X-bridges would be attached.

4.3. Troponin (TN)

Next in functional significance, though not in historical discovery sequence, is the regulatory protein of vertebrate and many invertebrate cells. Prior to 1965, it was already generally believed that the activation of contraction occurs via regulation of the cellular calcium concentration (see p. 196), but the molecular basis for this action was totally unknown. A "relaxing factor" had been isolated from several muscles, and its role appeared to be to sequester calcium released in association with excitation and to keep the level low in muscles at rest. Agents were known that tightly bind calcium, especially EDTA. In low concentration, EDTA inhibits the ATPase activity of "natural actomyosin" of myofilament clusters, but in stark contrast, it fails to inhibit that of "synthetic actomyosin" made by adding purified actin to purified myosin (Perry and Grey, 1956). Something was present on the former that is markedly affected by calcium. The material was isolated by Ebashi and Kodama (1965), who found it to be a protein and termed it troponin. Troponin comprises only about 6% of total muscle protein: that of skeletal muscle of vertebrates has turned out to be a complex of three molecules, T, I, and C, with molecular weights 35,000, 24,000, and 18,000 respectively. TN_T binds the complex to tropomyosin, and therefore to the I filament, TN_I inhibits the ATPase activity of actomyosin, and TN_C is a molecule that itself binds calcium strongly. Crystals of TN_C are readily prepared (Mercola et al., 1975), leading to the hope that a molecular understanding of its role may be possible. The probable functions of these subunits and the whole complex are described below. The complex was localized by the use of ferritin-labeled antibody and was found to occur at regular intervals of 40 nm along the entire length of the I filament (Ohtsuki et al, 1967). This periodicity was well known to electron microscopists (see Figure 3.4.), and now it was seen clearly to correspond to the TN locations. The regular period of 40 nm did not correspond to any periodicity known to be present on actin molecules, except approximately to the crossover points of the two component strands, which are not likely to have any significance. It was clear that some other molecule must be determining the spacing. The riddle was resolved by putting for the first time into its correct place an already well known muscle molecule, tropomyosin.

In heart muscles of vertebrates, there appears to be only a single TN molecule, not a complex. Nothing is yet known about the detailed TN composition in those invertebrate muscles that have it. The complexity of the TN moiety is surprising, and it must have had a sophisticated evolutionary history. The regulation of calcium is, however, a widespread cellular phenomenon, and proteins closely resembling TN_C are found elsewhere, for example, in testis, where the substance is an activator of phosphodiesterase (Dedman et al., 1977).

A protein has recently been extracted from some vertebrate smooth muscles that is considered by its discoverers Mikawa, Toyo-oka, Nonomura, and Ebashi (1977a) to serve a similar function in these muscles to troponin in cross-striated ones. The name leiotonin has been given to this protein (Nonomura, 1980). However, the possible regulatory action is debated at this time (see p. 339).

There is as yet very little data on TN in invertebrate muscle cells. In the swimming crab *Portunus,* the TN repeat period of one muscle has been studied and determined to be 38.3 nm (Maeda et al., 1979).

4.4. Tropomyosin (TM)

The biochemist Bailey (1946) discovered a crystallizable protein in muscle that is a major component of many cross-striated cells; it accounts for about 12% of the total protein in a typical vertebrate muscle. This he termed tropomyosin (TM). He obtained it by soaking the residue from washed, minced muscle in *M* KCl at pH 7.0 and then reducing the pH to 4.5. A precipitate forms at pH 4.5 that dissolves in water at pH 7.0. In the electron microscope, a negatively stained crystal of TM has a regular-square-net appearance that superficially resembles a face view of the Z region (H. E. Huxley, 1963). However, antibody studies showed that TM is associated with the I filaments, not the Z band (Pepe, 1968). TM has a molecular weight of 70,000. Its principal structural feature, as seen in crystalline tactoids, is a 40-nm axial period (Caspar et al., 1969), with alternating smaller repeats of 23 and 17 nm that give the whole a highly characteristic banding that is remarkably similar to that present in annelidan and molluscan paramyosin-containing very thick filaments (see p. 000). The periodicity is due to a combination of bonding end to end, with cross-connection at two sites. The length of individual molecules of TM is 40 nm. This length is after shortening due to coiling of the 90% α-helix basic coil; this is exactly right for TM to serve as the spacing element for troponin if TM is located longitudinally end to end along the I filaments, a suggestion first made definitively by Ohtsuki et al. (1967). It has now been firmly established that two strands of TM molecules occur with each I filament. The TM molecules are located in the two grooves formed by the double strands of actin molecules. The TN molecules are bound to the points of abutment of the TM molecules. Two forms of TM, α and β, are recognized chemically, but not yet related to function.

The presumed locations of tropomyosin and troponin can be understood in relation to the structure of the *actin of the I bands,* with which they are intimately associated.

4.5. Paramyosin (PM)

The discoverer of tropomyosin, K. Bailey, found 10 years later (Bailey, 1956) what he thought, correctly, was a different form of it. Early electron microscopists Hall et al. (1945) had found in clam adductor muscles strange filaments that they called paramyosin. Bailey extracted what he called invertebrate tropomyosin from the adductor of *Pinna,* where he found it to be present in a very high (73%) concentration. It readily forms needle-shaped crystals on lowering the ionic strength to 0.3 from the extraction value of 0.5. For a long time, this protein was called tropomyosin A, and the crystallizable protein found earlier, which is not present in any muscle in more than about 12%, tropomyosin B. They are now known to be markedly different, both in their roles (see p. 124) and in their distributions, as well as significantly different chemically. But paramyosin was at first regarded as being essentially equivalent to what is now regarded as true tropomyosin, the earlier tropomyosin B (Elliott et al., 1957; Hanson et al., 1957). Each is an α-protein, formed of coiled coils of polypeptide chains. Crystals of tropomyosin have a fine cross-banding pattern that closely resembles the patterns found in thick filaments of many annelidan and molluscan skeletal muscle cells (Hanson and Lowy, 1957b), which have a very high content of paramyosin. The actual distribution of paramyosin is still poorly known on account of its close similarity to TM. It has been clearly demonstrated in bee flight muscle (Bullard et al., 1973), where it occurs on each side of the Z band in narrow zones. Nothing is yet known with confidence of its functional roles, though it has long been considered to play a major role in some molluscan muscles which show a "catch" state (p. 477).

4.6. Paramyosin-Containing Thick Filaments of Invertebrates

Paramyosin occurs in substantial amounts in the invertebrate phylum most closely linked with chordates in evolution speculation, Echinodermata, in holothurian (sea cucumber) pharyngeal retractors (Baccetti and Rosati, 1968). It is very widely distributed, being present in small quantities in a wide range of cell types in invertebrates, including body-wall muscle of nematodes (Swanson, 1971; Waterston et al., 1974), sipunculids (De Equileor and Valvassori, 1977), various annelids (Castellani et al., 1978) and specialized muscles of arthropods including lobster claw, barnacle depressor, insect flight, and horseshoe crab telson levator (Levine et al., 1974, 1976; Elfvin et al., 1976).

4.6.1. Very Thick Filaments

The slower skeletal muscles of molluscs, especially muscles capable of sustained shortening under large loads with little energy expenditure, which are known as "catch" muscles, all contain thick filaments that have a cigar shape, with very thick centers, 70-120 nm wide, tapering progressively toward their ends. Similar-shaped thick filaments, with 30-100 nm center diameters, also occur in the ordinary body-wall muscles of most kinds of annelids (Rosenbluth, 1968; Heumann and Zebe, 1967; Wissocq, 1970; del Castillo et al., 1972; Sugi and Suzuki, 1978), nematodes (Rosenbluth, 1965), and echinoderms Baccetti and Rosati, 1968).

These filaments have the fine cross-banding that is characteristic of paramyosin and tropomyosin crystals. Most very thick filaments are considered also to contain myosin. But the proportion of myosin is extremely small and has in no example yet been positively located, though it has frequently been suggested that it forms a sleeve around the outside. This question has been specifically addressed for the *Mytilus* anterior byssal retractor muscle (ABRM) by Hildebrand and Gilloteaux (1979). These authors concluded that the myosin cannot be so located, but must be integrated into cofilaments with paramyosin. We shall return to this question in considering the control of contraction in these muscles. The paramyosin is always in the crystalline array in fixed material. The cells that contain paramyosin thick filaments are mostly spindle-shaped, ranging in overall size from 10-20 × 200 μm long in the opisthobranch *Dolabella* (Sugi and Suzuki, 1978) to 5-7 μm × 1800 mm long in the lamellibranch *Mytilus* anterior byssal retractor (e.g., Twarog et al., 1973). They are even longer in *Ascaris* and the holothurian *Echinoderus*, but they are always very thin.

4.6.2. Filaments of Ordinary Thickness

Paramyosin is not confined to the very thick filaments but occurs also, though in small amounts and in the presence of a normally large amount of myosin, in locust (*Schistocerca*) and bee flight muscle cells, lobster (*Homarus*) and crab (*Cancer*) claw, and tapeworm (Winkelman, 1976). The reason for its presence in these cells is not known, but it may help to stabilize the thick filaments.

Very thick filaments containing paramyosin almost exclusively must have evolved from ones having at least a little in addition to myosin. The presence of paramyosin, even its preponderance, does not prevent normal force development; such muscles can generally generate maximum forces greater than those developed by ordinary cross-striated cells, with 9 kg/cm^2 well authenticated (Abbott and Lowy, 1956b) and claims running to 30 kg/cm^2, compared with 2-5 kg/cm^2 for muscles without any paramyosin. The only consistent accompaniments of the presence of a lot of paramyosin are slow contraction/relaxation rates.

It is possible that most invertebrate thick filaments of whatever intrinsic speed contain some paramyosin.

4.6.3. Is There Paramyosin in Chordate Muscle Cells?

If there is paramyosin in muscle cells of vertebrates, it is certainly only a very small amount. There has been no very serious effort to locate it. Zou and Gong (1979) claimed that they had located paramyosin in chicken breast muscle, and they report finding it in muscle fragments of the primitive chordate *Amphioxus* (=*Branchiostoma*). It seems unlikely that so major a component of some invertebrate muscles would be entirely missing from chordates.

4.7. Actin

Protein was first extracted from muscles by Kühne (1859), using strong salt solution, and was obtained by dilution of the extract. He termed the material myosin. It was soon considered possible that this was actually a mixture of proteins, but it was almost a century later before Straub (1943) succeeded in partially separating the mixture. He first removed a major fraction (mainly myosin) with an alkaline salt solution, then washed the residue with acetone, allowed the solution to dry out, and extracted it with distilled water. This soluble material was termed actin, and it did not react with myosin. It was clearly in a different form than in intact muscle, but Straub found he could get it to polymerize, forming a much more viscous material that did interact, by raising the concentration of the KCl solution in which it was dissolved. The former form of the protein was called G (globular) and the latter F (filamentous) by Szent-Györgyi (1944).

The combination of myosin with F actin is tenacious, and referred to as *actomyosin* (AM). If a solution of AM is squirted from a pipette into distilled water it forms a strand of gel. Such a strand shortens markedly, and thickens, mimicking contraction. The phenomenon diverted an entire generation of muscle biochemists but provided no clues as to the contractile mechanism. The monomeric single unit, G actin, has a molecular weight of 41,700. Modern research has shown that each G actin molecule has a single ATP molecule bound to it (Cooke, 1976). Polymerization is brought about by hydrolysis of the terminal phosphate of this ATP according to the scheme

$$n(\text{G actin} \cdot \text{ATP}) \rightleftharpoons \text{F actin} \cdot \text{ADP} + n\text{P}_\text{i} + \text{energy}$$

Synthetic F actin can be examined by electron microscopy using negative staining and was found to have a structure very closely resembling that of I filaments obtained by maceration of a wide variety of muscles (Hanson and Lowy, 1963; H. E. Huxley, 1963). On the basis of their micrographs, Hanson and Lowy (1963) and Hanson (1967) produced a model of actin that comprises two strings of spherical beads of equal size twisted around each other (Figure 4.4.*3*). Each bead was stated to be 5.5 nm in diameter and to represent a single actin molecule. Uncertainty was expressed only about the number of subunits per pitch, with 15 and 13 being the most likely alternatives, and strong emphasis on the latter. The reported repeat length ranged from 35 nm (Worthington, 1961) to 40.6 nm (Page and Huxley, 1963) and was evidently dependent on the fixative used. Later evidence has placed the repeating functional subunits at one per seven actin monomers (Bremel et al., 1973). An average length of 38.5 nm is now accepted. But the model was shaky and Hanson (1967) herself concluded that "electron microscopy has not yet given reliable information about the form of the actin helix." The substructure of many of the actin filaments in their micrographs is not obviously beaded. Small globules

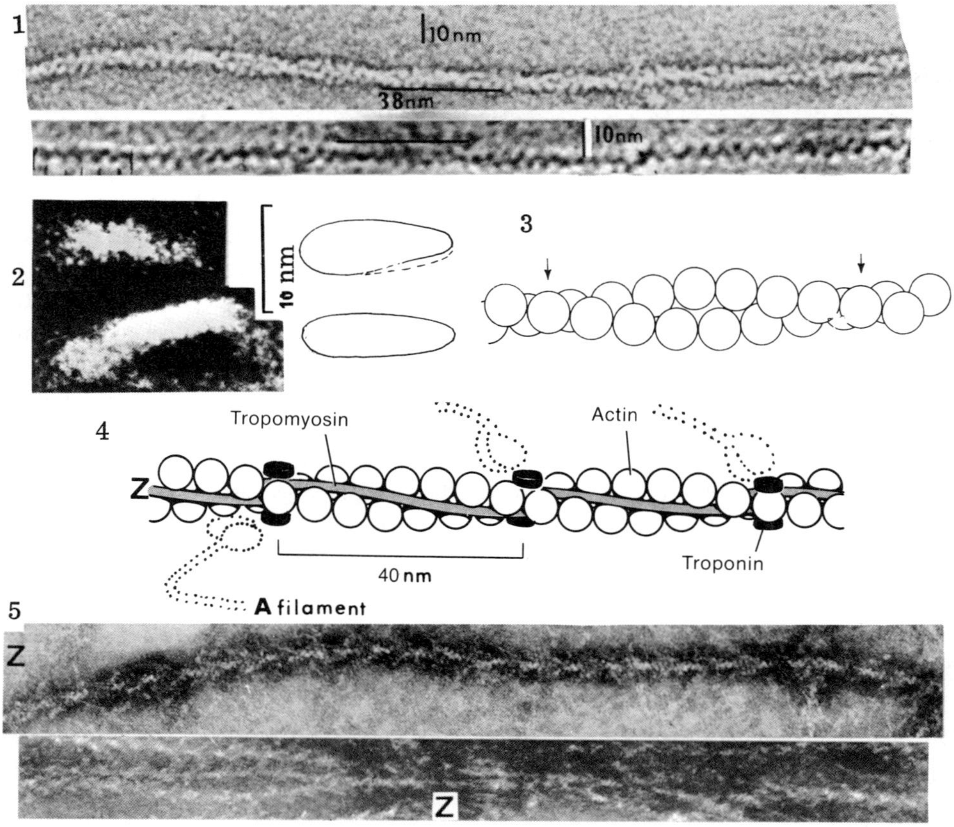

Figure 4.4. The molecular architecture of I filaments. (*1*) Negatively stained images of parts of I filament (*above* – from H. E. Huxley, 1963) and purified F-actin filament (*below* – Reprinted with permission from *J. Mol. Biol.* 6, 46–60 (1963) Copyright 1963 Academic Press, London.] At first these filaments were presumed to be identical, and formed by the twisting around each other of two beads of actin molecules. The individual actin molecules were presumed to be spherical (*3*), though negatively stained images (*2*) indicate an asymmetrical structure. (*2*) Negatively stained images of individual actin molecules. (From Rowe, 1964.) (*3*) Model of actin filament (Reprinted with permission from *J. Mol. Biol.* 6, 46–60 (1963). Copyright 1963 Academic Press, London.] (*4*) Textbook diagram of modern concept of I-filament architecture. Reproduced with permission from *Animal Physiology* by Roger Eckert and David Randall. (Except myosin molecule forms added as dotted lines to illustrate the presumed attachment sites and angles with respect to Z region, *left*, and A filament.) (*5*) Isolated I filaments, as (*1*), after addition of myosin (= decorated I filaments). They are polarized relative to the Z region so that myosin heads can only attach at specific sites, with their tail portions pointing toward the Z region. [By H. E. Huxley, with permission from *J. Mol. Biol.* 7, 281–308 (1963). Copyright 1963 Academic Press, London.]

are present, but even the largest images are not greater than 3 nm wide; none approach the stated value of 5.5 nm. Hanson and H. E. Huxley at first were "confident" about an overall actin thickness of 4.0 nm. Densitometric runs across their micrographs show center-to-center spacing of about 5 nm, but also a mixture of unit widths from as little as 2 nm. There are, of course, great difficulties in making judgments from what amounts to looking at a twisted beady necklace bathed in a shallow mud bath. But if two presumed 5.5-nm spherical units are viewed side by side, the width should be at least 10 nm, and that is without including the two strands of tropomyosin molecules, with attached troponin, that must be added to make natural I filaments. No observed thin filaments register as being this thick in either longitudinal or transverse sections or in negative staining. As has happened so often in muscle studies, the crude actin model was nevertheless adopted forthwith and used ever since as the basis for molecular models of the contraction mechanism.

Several independent studies using different approaches, including UV circular diochroism, have shown that conformational changes occur in actin following the kind of activation that occurs in muscle. Rabbit and some crab cell actins were found to give results so different they indicated basically different actin structures, but both kinds of actin changed markedly when activated (Yanagida et al., 1974). There has long been biochemical evidence that actin shows conformational changes in association with nucleotide binding (Szent-Györgyi, 1965), and there is more recent evidence that there is an actin conformational change during relaxation (Borovikov and Chernogriadskaia, 1979). Muscle actins from different sources have slight differences in their amino acid compositions, but are basically very homogeneous and together are classified as α-actins. Nonmuscle actins are more different and are prescribed by different genes as α-or β-actins (Garrels and Hunter, 1979).

4.8. Natural I Filaments

Whole I filaments have been represented in models of muscle as being rigid rods. Their physical behavior in isolation shows that this is not the case, and that they are pliant, flexible, and likely to curve, although they do seem to have little elastic extensibility. Quite a lot of evidence suggests that conformational changes occur in I filaments during contraction, but this work has all been done on TM/TN-containing muscles, and most or all of the shift has been attributed to movements of the associated TM/TN rather than to actin itself. There is a strong periodicity of TN/TM-containing I filaments, at about 40 nm. This is attributed to the presence of the TN molecules and their spacing by TM molecules. However, in the TEM, this same spacing is often associated with fine filaments attached to I filaments and often bridging adjacent ones (see Figures 3.16*.3*, 3.19). X-ray analysis of TM molecules *in situ* indicates squat structures located very close to the actin molecules, so these bridges must be due to a material other than TN, possibly α-actinin (see p. 125).

Page (1968) noted that there are 24 periods in I bands of an amphibian muscle, which have a single triad per sarcomere (at Z-line level) and a larger number, indicating longer I filaments, in muscle cells having two triads per sarcomere, from fish and mammals. We have referred to the well-known conspicuous banding of I filaments at 37–40 nm in TEM sections. Essentially three different kinds of explanation have been offerred for them. The most commonly accepted is that they represent the troponin, attached at this spacing to ends of TN molecules running in the actin helix grooves. This explanation was strongly favored by Jean Hanson (1973). Another possibility is attached single myosin molecules, which is plausible since this is also the periodicity of "decorated" I filaments (Figure 4.4*.4*). It is also the periodicity of I/I bridges, which are 20 nm long, and which have a longitudinal spacing of about 40 nm.

4.9. Minor Protein Components

It now seems extraordinary that but a few years ago muscle biochemists were confident that muscle consisted of only actin, myosin, and water! The number of known significant minor proteins increases every year, and according to my more-sophisticated colleagues, might eventually reach about 100. The better-known proteins will be considered briefly.

4.9.1. α-Actinin

At about the time of the discovery of troponin in Ebashi's laboratory, he and his wife (Ebashi and Ebashi, 1965) also discovered other new structural proteins. The best-established is α-actinin, which has a molecular weight of 95,000. When a cross-striated cell is treated with trypsin, the Z region is solubilized, and a protein (Z protein) is removed that has properties similar to those of α-actinin (Goll et al., 1969). This protein is exclusively associated with the Z line, and since the only elements in the Z line that are present in sufficient quantity to account for the quantity of α-actinin found are the Z filaments, this protein is a good candidate for them. But one aspect of its functioning has not been explained; the contraction of actomyosin is greatly promoted by the addition of α-actinin (Ebashi and Ebashi, 1965). Perhaps it does so by an organizing action on actin.

The α-actinin molecule is about 2 nm wide by 30 nm long; it attaches strongly to actin, but not to myosin (Briskey, 1966), and according to Podlubnaya et al. (1975), it can attach laterally to actin as well as to the ends of filaments.

4.9.2. β-Actinin

This was discovered by Maruyama (1965) as a protein with 81,000 molecular weight that disperses actin in suspensions and prevents association of F actin. It is associated with I filaments in intact muscle, primarily on their inner, free ends. It promotes polymerization of G actin (Maruyama et al., 1977b) and might be concerned with regulation of the length of I filaments.

4.9.3. M Protein

It is recognized that at least one, possibly as many as three, proteins, with molecular weights 43,000, 90,000, 165,000 (the heavier is the one known as M protein) are associated with the M region. There is now very strong evidence that the functionally important enzyme creatine kinase is attached to the M region (Walliman et al., 1978, 1979). M protein is almost certainly the component making the M bridges, but is probably not the creatine kinase. It is either just structural, or may be involved in regulating ATP production.

4.9.4. Creatine Kinase

Treatment of vertebrate cross-striated cells with anti-creatine kinase Fab results in the removal of the whole M band (Wallimann et al., 1978). Creatine kinase is the enzyme most directly related to the energy supply, apart from Actomyosin (AM) ATPase, because of its role in restoring the depleted high-energy phosphate bond (see p. 132). It is the first molecule in the muscle complex to be studied in place by nuclear magnetic resonance labeling (Brown et al., 1980). It is not known if arginine kinase is correspondingly sited in invertebrates.

4.9.5. Titin

A recently discovered molecule of extraordinary molecular weight, much larger than myosin, has recently been extracted from a number of vertebrate and invertebrate skeletal and also cardiac muscles and has been termed titin (Wang et al., 1979b). It is located in several places, especially in the Z and M bands and at the outer margins of the A band at N_2. There may be three different forms, each in a different location.

4.9.6. Filamin

Filamin is a high-molecular-weight protein, a symmetrical oligomer comprising two identical units of molecular weight 250,000 each. It was discovered in chicken gizzard and is so far known only from some smooth muscles and nonmuscle cells. It interacts with and promotes the aggregation of F actin, and it is a phosphorylatable protein.

4.9.7. F Protein and I Protein

These are minor proteins discovered by Ohashi, with molecular weights of about 110,000 and 50,000 respectively (Ohashi et al., 1977a, b). Both are associated with the A band and probably play roles in details of myosin packing (Miyahara et al., 1980). Maruyama et al. (1978b) suggest that it is I protein that protects ATP from hydrolysis in relaxed muscle.

4.10. The Elastic Proteins

The elastic nature of muscle cells and their extracted "ghosts" shows the presence of highly elastic molecules both radially and longitudinally, and great importance should be attached to the understanding of these materials. In part the elasticity is due to elements intrinsic to the myofilament clusters, and in part to material lying outside them, in addition to the contributions from extracellular material in the sarcolemma. Unfortunately there is no satisfactory knowledge of either the natures or the relative contributions of the various elastic elements for any muscle cell.

The following are the few items for consideration that have so far come to light.

4.10.1. Connectin

A search for candidate elastic proteins led Maruyama et al. (1976; 1977a, b; 1978) to discover a protein they found to represent 5% of total protein in cardiac and skeletal muscles of a wide variety of both vertebrates and invertebrates; they called the protein connectin. They started by extraction in Hasselbach-Schneider solution, followed by prolonged extraction in 0.6 *M* KI, which leaves the "ghosts" referred to on p. 000. Connectin was extracted from the ghosts by 1% sodium dodecyl sulfate, followed by phenol. It lacks any detailed fine structure, but is filamentous, 1-1.5 nm wide. With the aid of fluorescent antibodies, Maruyama and colleagues obtained results that led them to conclude that connectin must be the material that binds together myofilament clusters, both by surrounding them and by crosslinking component Z regions. It does not interact with the contractile proteins.

Connectin is clearly an important constituent of muscle, but it does not appear to be the substance serving the role of primary skeletal element. The ghost myofilament clusters obtained after extraction with *M* KCl followed by 0.6 *M* KI have solubility characteristics suggestive of connectin (dos Remedios and Gilmour, 1978). This runs from Z to Z and would be in parallel with A and I filaments before they were extracted.

4.10.2. Fibrillin

Electron microscopic studies have frequently revealed an ultrathin filament, 2-4 nm in diameter (see p. 94), that might be part of the basic cytoskeleton of muscle cells and provide their resting longitudinal elasticity. A biochemical search for this material, for which a candidate had not turned up in previous protein studies, led Guba (Guba et al., 1968) to isolate a rubbery material he called fibrillin, with a molecular weight of 110,000. Muscle biochemists have neither confirmed, denied, nor extended this work; they have simply ignored it, though we have confirmed that following the recommended extraction procedure with either rabbit or crustacean muscles produces a rubbery material. Electron microscopy shows this material to contain filaments 2 nm in diameter and about 40 nm long. The diameter is the same as that of primary ultrathin filaments, and so fibrillin has seemed potentially a good candidate for these filaments. It is not known whether or not fibrillin interacts with actin or myosin. Isolated ultrathin filaments (Figure 3.34) have a repeat structure that suggests a lower molecular weight than that stated by Guba.

Maruyama et al. (1977b), although not reporting the amino acid composition of fibrillin, reported that its amino acid composition is different from that of connectin, so the two must be regarded as coexisting and sharing elastic roles.

4.10.3. T Filaments

In searching for ultrathin (T) filaments, I came up with several, in negatively stained TEM preparations of various muscle tissues, (Figure 3.34) that although much thinner than most synthetic actin or natural I filaments, clearly have an actinlike periodicity. However, unlike the fixed periodicity of I filaments, this varies over a wide range, from about 20 to over 50 nm, both between filaments and within a single one. This suggests a basis for ultrathin filament elasticity: they are certainly capable of repeat-period length variation. A model of the structure of the T filament, observed at different repeat lengths, is shown in Figure 4.5. The filaments are sometimes found associated together in pairs (Figure 3.34*.5*) that are often hard to distinguish from I filaments. The TEM images are noticeably similar to ones obtained by Hanson and Lowy (Figure 4.4*.1*) from purified actin, but which were interpreted very differently by them. The reality of T filaments has been firmly recognized by the meat industry. After standard forms of cooking, all meats are rendered more tender, though some cuts of some animals are much more so than others. The only common residual structural element in cooked meats is the T filament (Locker et al., 1977). The T filaments "set the tensile strength of the filament", according to these authors.

1
2
3

4.11. Enzymes and Energy Supply

The supply of energy for contraction has an extensive range quantitatively, but the anatomical categories outlined above are not associated with any major qualitatively different energy-support processes. Likewise, although there are some differences in emphasis in enzyme systems between vertebrates and invertebrates, the basic events are common. Consequently the energy systems are considered collectively. Historically, most parts were worked out first on a vertebrate skeletal muscle. As with the ultrastructure and physiology, there was a great deal of jumping to the wrong conclusions, based upon early incomplete findings. But the basic biochemical work went ahead so fast that the true place of earlier findings was soon realized, and a largely correct version of the complex sequence of events established in the mid-1930s. A historical treatment gives a good perspective in this case, so will be adopted for the description. Most early specific references are omitted, but can be found in Needham (1971). Muscle biochemistry is essentially divisible into two parts, the short-term direct supply of energy and its conversion to work in the form of shortening against a load, and long-term recovery following each period of activity. Muscles that work at high rates not surprisingly tend to run out of energy, fatiguing progressively until they cease to function. This is such a striking property, both *in vitro* and *in vivo,* that it was one of the first to be investigated and gave some important early clues to muscle chemistry. Muscles that work less rapidly may either fatigue much more slowly or be able to match their energy supply and its utilization to work perfectly enough to function indefinitely without fatigue. This is true of most smooth muscles, and of many postural muscles, especially those of invertebrates. It was early found that after vigorous activity of muscle, as fatigue occurs, the muscle tissue becomes progressively more acid, due to accumulation specifically of *lactic acid.* The first findings, in 1807 by Berzelius, were on stags exhausted by being hunted down. The accumulation of lactic acid was found by Nasse in 1869 to be at the expense of a matching loss of the glycogen that Claude Bernard had described in 1859 stored in high concentration within the muscle itself. Classical modern work on lactic acid accumulation was published by Fletcher and Hopkins in 1907 and work on the attendant loss of glycogen was published by Parnas and Wagner in 1914. Muscles doing the same amount of work more slowly do not accumulate lactic acid, but they do lose glycogen, so there is a different chemistry depending on the rate of work. There also is a different rate of breathing and oxygen utilization. In very brief bursts of high intensity exercise, not a great deal of oxygen is, or could be, utilized. During rest afterward, the rate of oxygen utilization continues high for some time, in proportion to the amount of work done. Therefore, at high rates of work, a muscle works essentially anaerobically because oxygen cannot be supplied fast enough. There is a chemical process that permits the accumulation of an *oxygen debt,* as it was called by A. V. Hill and Lupton (1922). Since this extent of the debt is proportional to the amount of lactic acid, which disappears during recovery as extra oxygen is consumed, it was reasonable to suppose that recovery was simply burning up of the accumulated acid. Both processes are exothermic, so there

Figure 4.5. Model of possible fine structure of the ultrathin T filament of rabbit psoas, basd on electron micrographs. (*1*) Negatively stained image of a portion. (*2*) Models of sections having different repeat lengths. (*3*) Two forms of association in pairs: the one at right is very difficult to distinguish from either I or synthetic actin filaments. (ORIG.)

began a race to determine if the actual heat produced matched the chemistry. Several factors combined to make this research harder than it may have seemed. Heat changes could not be measured rapidly, because neither the measurement devices nor recorders were fast enough. Eventually partial solutions were obtained, using specially made multiple thermocouples, highly sensitive suspended mirror galvanometers, a low temperature, and large vacuum flasks, by Otto Meyerhof in Germany (1920-1921) and especially by A. V. Hill in England (1913-1928). Meyerhof showed that the heat produced is proportional to the tension developed and work done, and that the conversion of glycogen to lactate gives sufficient energy to be critically capable of supplying the energy of contraction. Hill showed that the heat is produced in two phases, a rapid *initial* heat of contraction, and a slow, later *recovery* heat, the first of which, but not the second, is not affected by oxygen.

Hill and Meyerhof proposed a formal *lactic acid theory* of contraction in which

$$\text{Glycogen} \rightarrow \text{contraction} + \text{lactic acid} + \text{initial heat}$$

Recovery occurs simultaneously with the above when the rate of working is slow enough, or separately after a very vigorous burse, simply by

$$\text{Lactic acid} + O_2 \rightarrow CO_2 + H_2O + \text{recovery heat}$$

The work earned Hill and Meyerhof the 1924 Nobel prize for physiology and medicine, but their theory was quite wrong – another example of sounding off before all the facts were known. The facts that were to destroy the lactic acid theory were already beginning to accumulate. In 1921, Embden and Adler found that a lot of free phosphate accumulates in fluid bathing fatigued muscle, the level rising rapidly during the contraction. Meyerhof himself followed this up, and together with his brilliant student Lohmann published in 1927 research showing that in molecular terms, more phosphate than lactate is released in the early stages of continued contraction.

A source for the released phosphate was soon found by Grace Eggleton and her husband (1927) in the form of a "*phosphagen*", which was quickly identified by Fiske and Subbarow in 1929 as being creatine phosphate in vertebrates and arginine phosphate in some invertebrates. Creatine had been crystallized from beef meat by Chevreul in 1835, but no function had been assigned to it. The phosphagen level in muscle is quite high, and it increases during recovery in the presence of oxygen, and decreases during contraction. These important discoveries required a functional explanation. The hydrolysis of a mole of creatine phosphate by creatine phosphokinase, as Meyerhof and Lohmann together showed in 1928, liberates 12 kcal, an enormous amount of energy and, like glycogen breakdown, sufficient to account for heat production. The mystery was deepened when Lundsgaard (1930-1931) poisoned muscle with iodoacetic acid. Moderate durations of contraction were normal in the poisoned state, *but no lactic acid at all was formed.* Iodoacetic acid inhibits glycolysis by blocking one of the enzymes involved, triose phosphate dehydrogenase..

So now the lactic acid theory was abandoned, and Lundsgaard formally replaced it by a phosphagen to phosphate and creatine/arginine breakdown theory in 1931. However, this too was doomed to a very short existence. Lactate formation could not be denied but could be assigned to anaerobic recovery rather than contraction. Its formation could be hypothesized to be coupled to resynthesis of creatine phosphate. In iodacetic acid-poisoned muscle, it had to be argued that resynthesis was so rapid that work was performed without any lactic acid formation.

These dodges were not required though. Lohmann, in 1929, isolated from muscle a quantity of a markedly different phosphate compound that also easily generated inorganic phosphate. Four years of work were required to show that the base is adenosine: adenosine triphosphate (ATP) had been discovered! When this new compound ATP was added to a muscle extract in gel form, the gel, dramatically, contracted. This it did not do with creatine phosphate. Now both phosphagen and lactic acid were embarrassments! Lohmann further buried the phosphagen notion in 1934 by showing that muscles don't even contain any enzyme that can hydrolyze it. He then correctly deduced that the phosphagen must be contributing by reconstituting the ATP from its breakdown products. Next, while studying crab muscle, he discovered adenosine diphosphate (ADP) and guessed that it may be the first breakdown product. It was his near-namesake Lehmann who soon discovered that there is a reversible equilibrium,

$$\text{ATP} \rightleftharpoons \text{ADP} + \text{P}$$

ATP is broken down rapidly, but it is also reconstituted very rapidly, and it was not until 1962 that Cain and Davies were able to demonstrate that ATP is in fact broken down in a contraction. After developing a sufficiently rapid cooling system, they took advantage of the 1959 discovery that dinitrofluorobenzene poisons the phosphokinase enzyme which promotes the reconstitution of ATP. The breakdown of ATP is a hydrolysis catalyzed by an adenosine triphosphatase, which in muscle is a unique cooperative enzyme that involves both myosin and actin. Although myosin is itself an ATPase, when combined with actin it is about 10,000 times more effective. The overall reaction is often termed the Lohmann reaction in honor of its discoverer.

The breakdown of glycogen, then, is indeed just a part of recovery. In the absence of sufficient oxygen, glycolysis stops short, at pyruvic acid. However, the nicotinamide adenine dinucleotide (NAD) that is present uses the pyruvic acid formed for a little hydrogen transfer, 4 H per glycogen, making one molecule of lactic acid for each.

$$\text{NADH} + \underbrace{CH_3 \cdot CO \cdot COOH}_{\text{pyruvic acid}} + \text{H} \rightarrow \underbrace{CH_3 \cdot C \cdot H \cdot OH \cdot COOH}_{\text{lactic acid}} + \text{NAD}$$

This is a poor trade-off, producing only a little extra energy at the expense of early fatigue. Was it another unfortunate happenstance of evolution? Or did forced rests confer positive adaptive advantage by conserving limited food supplies and encouraging "freezing" into the local environment?

During glycolysis the glycogen molecule is dismembered progressively, with 12 steps in all, each leading to an energy-storing phosphorylation, until only pyruvate is left. By this means its free energy is maximally utilized. When sufficient oxygen is available, the pyruvic acid in effect slips directly into the tricarboxylic acid cycle for dehydrogenation, ending up as CO_2 and, via cytochromes, water and more phosphate bonds. Such lactic acid as has been made waits around until there is sufficient oxygen available before ending similarly. What starts the ball rolling? The initiation of movement is a rise in calcium concentration. The evidence for this will be discussed at length later (see p. 196). Calcium activates interaction between actin and myosin and at the same time promotes the natural enzymatic activity of the complex, which is present on the myosin head or S_1 fraction. This action is to split ATP into ADP and P_i with the release of energy in the amount of 12 kcal/mol:

$$ATP \rightarrow ADP + P_i + E$$

The splitting occurs in a seven-step scheme (Greeves et al., 1979; Tregear and Marston, 1979), with myosin (M) activated in two states by actin (M*, M**), as follows:

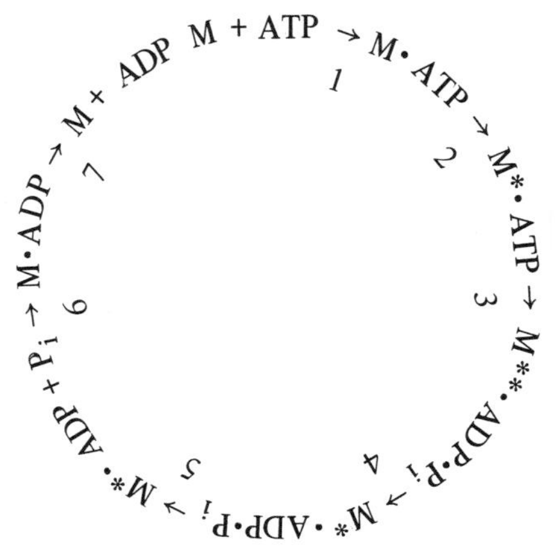

Step 3 produces only 1.3 kcal/mol (Young et al., 1974), so cannot be the principal release step. P is used to denote phosphate whether PO_4, $(PO_4)_2$, or $(PO_4)_3$.

Myosin heads alone split ATP enzymatically, but in combination with actin the process is speeded up at least 10,000-fold. The enzyme action of actomyosin ATPase is promoted by magnesium as well as calcium, but only at concentrations between 10^{-4} and 10^{-5} *M*. The energy released becomes transformed into a combination of work + heat immediately. Although free ATP is always present, its amount is small, no more than sufficient to sustain contraction for a few hundred milliseconds. Nevertheless, the amount remains constant, and as previously noted, it proved to be very difficult to show any ATP loss. Loss there must be, but reconstitution is also very fast. Rapid reconstitution depends upon a rapid direct interaction with creatine (C) or arginine (A) phosphagen:

$$C\,(A)\,P + ADP \rightarrow ATP + C\,(A)$$

This reaction, catalyzed by a phosphokinase, is in fact a coupled equilibrium:

$$C\,(A)\,P + ADP \rightleftharpoons ATP + C\,(A)$$

whose balance is simply upset by the loss of the ATP. There is clearly a problem, and that is how to reconstitute the C (A) P without using up ATP. This is done by an independent enzyme, creatine/arginine phosphotransferase, from C (A) by direct phosphorylation

$$C\,(A) + P \rightarrow C\,(A)\,P$$

Direct studies on the time courses of these reactions became possible with the introduction of methods for following isotopic/nuclear magnetic resonance (NMR) spectra (e.g., Dawson et al, 1977). In frog and toad leg muscles studied at 4°C with 25 s tetani, 20% of the CP was hydrolyzed and then rebuilt exponentially, with a half time of 10 min.

There is now much direct evidence for the involvement of CP in vertebrate muscle; quite large amounts of it are hydrolyzed even in a single twitch contraction, sufficient to account for the entire energy requirement (Infante et al., 1965; Wilkie, 1968; Woledge, 1971). There is a direct relation between external work performed by a muscle and the amount of phosphagen split.

A major new development in understanding the finer details of contraction regulation has come with the realization that myosin itself is phosphorylated during excitation. There are two different light-chain proteins attached to the myosin head. The use of ^{32}P has shown that the only protein that is phosphorylated is an 18,000-daltons light chain that is part of the head (Bárány et al., 1980). Its presumed functional role is to promote actin/myosin interaction after it has been phosphorylated. It possibly increases the rate of combination of the X bridge with actin, putting it into the class of regulators to which calmodulin belongs (see p. 220). The phosphorylation is catalyzed by myosin light-chain kinase, and the reverse dephosphlorylation by phosphatase, as summarized in the following:

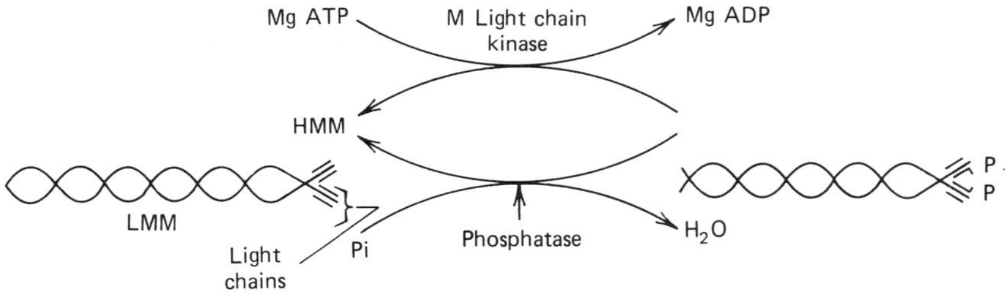

The first step with which a myosin conformational change is associated, as indicated by NMR spin-labeling (Seidel and Gergeley, 1973a, b) or studies made with methylene ATP (Mannherz et al., 1973), occurs when ATP combines with myosin (step 1), that is, prior to hydrolysis.

ATP necessarily enters into other reactions, including the phosphorylation of glucose and the breakdown of amino acids. Also, some is broken down all the way to AMP. AMP and phosphorylase b, which is activated by the raised calcium concentration, together stimulate glycolysis. The latter is not a simple matter and is even now not entirely understood. The principal glycolytic "enzyme" is termed phosphorylase, which is probably a few isoenzymes, of which a and b are the major ones, aided by transglucosylase and glycogen synthetase. Reconstitution of ATP from ADP and phosphagen is promoted by the specific phosphokinase. An important factor entered the scene when Rall and Sutherland (1957) found a factor in muscle that promotes formation of phosphorylase a; this factor turned out to be cyclic AMP (cAMP). It is now known to be such a significant element in membrane-related processes, as a "second messenger", that a role in muscle might also be expected. But no major role(s) for cAMP in muscle has (have) been established. There are functionally important protein kinases present in vertebrate skeletal muscles that are regulated by it, as well as phosphodiesterase, which breaks it down (Walsh et al., 1972). At this time it is correct to regard cAMP as only indirectly involved, in minor ways, in contraction and its regulation.

ATP is in equilibrium with ADP and AMP according to the relation

$$\text{ATP} + \text{AMP} \rightleftharpoons 2\ \text{ADP}$$

The breakdown of ATP is promoted by an adenylate kinase and reconstitution by a myokinase. Removal of the phosphate from the myosin head occurs during relaxation and is promoted by phosphatase.

The enzymes whose actions are described above, actomyosin ATPase, the myokinases, creatine/arginine phosphokinase, phosphorylase b, creatine/arginine phosphatase, phosphotransferases, and adenylate cyclase, are the principal ones specifically associated with muscle. In addition, muscle contains the enzymes associated with the oxidation of pyruvic acid, lactic acid, and fatty acids in general. Some muscles, notably insect flight muscle, contain lipolytic enzymes, especially lipase, and can utilize fat breakdown directly to provide energy. Fat utilization is probably more widespread than it was earlier thought to be in both vertebrate and invertebrate muscles. Prominent among vertebrate muscle enzymes involved in utilizing fats as energy sources are lactic dehydrogenase isozymes, glucose-6-phosphate dehydrogenase, succinic dehydrogenase, γ-hydroxybutyrate dehydrogenase, various esterases, NADH dehydrogenase, NADPH diaphorase, citric dehydrogenase, cytochrome oxidase, and malic-oxidase. To this list should be added coenzymes and decarboxylases. Finally there are the enzymes, especially transaminases, concerned with protein turnover.

4.12. The X-Bridge Hypothesis of Movement

The development of the modern theory of movement generation clearly depended on a landmark event in the history of electron microscopy. By combining very slow cutting with thermal advance of the material, H. E. Huxley (1957) obtained sections and micrographs of unprecedented thinness for their time, probably about 30 nm. This was still a little too thick for clear resolution, but Huxley argued that if, after getting close to a perfect thick-thin-thick-thin arrangement in sections, he rotated the block by 30° and cut parallel to the 1120 planes, he should get a thin-thin-thick-thin-thin filament alternation. Although to be perfect sections should have been even thinner, about 15 nm, the margin of error permitted, which is determined by the separation distance of the two adjacent planes, was in this plane about equal to his thinnest section thickness. He obtained such sections, and when they were published in 1957 (Huxley, 1957), they created a sensational and overnight revolution in thinking about muscle. All opposition to the structural model in which separate A and I filaments interdigitate disappeared, and the *sliding filament theory* was established.

The pictures revealed a highly important feature: *X bridges* between thick and thin filaments situated at regular intervals and clearly emanating from the thick filaments. They had been seen and claimed previously, but Huxley's new pictures made them obvious. They immediately suggested a molecular basis for sliding, and a *X-bridge theory* of contraction. It has often been stated recently that the X-bridge theory replaced, or became the sliding-filament theory: Tregear and Marston (1979) wrote, "The sliding-filament theory was *transposed* into the cross-bridge theory." This was evidently so in the minds of a majority of muscle scientists, but logic dictates otherwise: it is perhaps fair to say that sliding is now more an established fact than a theory, but the involvement of myosin projections to provide the motion remains only a theory. Bridges should be considered separately: sliding could be, but is not necessarily, caused by X-bridge motions.

Two basic consequences of the X-bridge theory are testable: first, that force must be related to the number of X bridges formed; second, that the rate of splitting of ATP must be proportional to the extent of overlap. Several studies on whole muscles and even single cells were inconclusive on the first point. There was a lot of variation, although a roughly linear relationship was found, but tension was still being developed at stretch beyond overlap. Realizing that the results might be confused because of uneven sarcomere lengths, Gordon et al. (1966a, b) devised a servocontrolled spot-follower device automatically controlled by the length of a single centrally placed sarcomere of a frog fast cell; the length is indicated by tiny strips of gold leaf. With this device they obtained the relation shown in Figure 4.6, which met the criteria of accurate following of SL. ATP splitting as a function of sarcomere length had a less steep slope (Ward et al., 1965), but sarcomere length could not be controlled in ATP-splitting tests as it was in the force/length experiments, and allowing for this, the expectations of the overlap-dependent model were realized. These data are also compatible with some theoretically possible non-X-bridge mechanisms, but the success of the experiments has dazzled muscle scientists and blinded them to alternatives. In contracted muscle (Figure 4.7) an extraordinary inward angling of the bridge occurs that is for me a major obstacle to acceptance of the X-bridge theory of force production. But there is now one dominant theory of mechanism: the *X-bridge cycling* theory. Opinions differ only on the "how" of cycling details. All invoke generation of movement by a conformational change in some part of the bridge. The first formal hypothesis concerning how bridges might bring about motion was proposed by A. F. Huxley (1957a, Figure 4.8). His proposal was that each bridge oscillates in the longitudinal axis, alternately attaching and detaching, with a statistically greater probability of attaching to the I filament at the inner limit of its motion and of detaching at the outer. Each movement from inner to outer made while attached would contribute to the contraction.

A later theory (Davies, 1963) supposed that the outer part of the LMM element of the bridge went through cyclical length changes, elongating until attached, then shortening (while attached), followed by breaking and reextending. This nice theory was soon abandoned when it was found that the bridges actually point inward toward the M band in contracted muscle (Reedy et al., 1965). Nevertheless, models with bridges angled outward in rigor continued to emerge (Taylor, 1976). Thereafter it was obvious that the base of the LMM portion must be a rather floppy hinge. Attention focused on a stiff component that might generate tension, the head itself. H. E. Huxley suggested that rotation is brought about by the movement of the abutting faces of the two heads over each other (H. E. Huxley, 1969). A. F. Huxley and collaborator R. M. Simmons (1971) proposed instead that the myosin heads work together, with a rocking motion, as illustrated in Figure 4.8. The myosin and actin were both conceived as having a series of consecutive, individual attachment sites within any particular attachment locations. By combining with each in turn, the elastic X bridge is stretched, transferring force to the A filament. After a brief period in which it shares the load, the head of the X bridge detaches, then "searches" for a new attachment site, which, in the shortening muscle, is

1

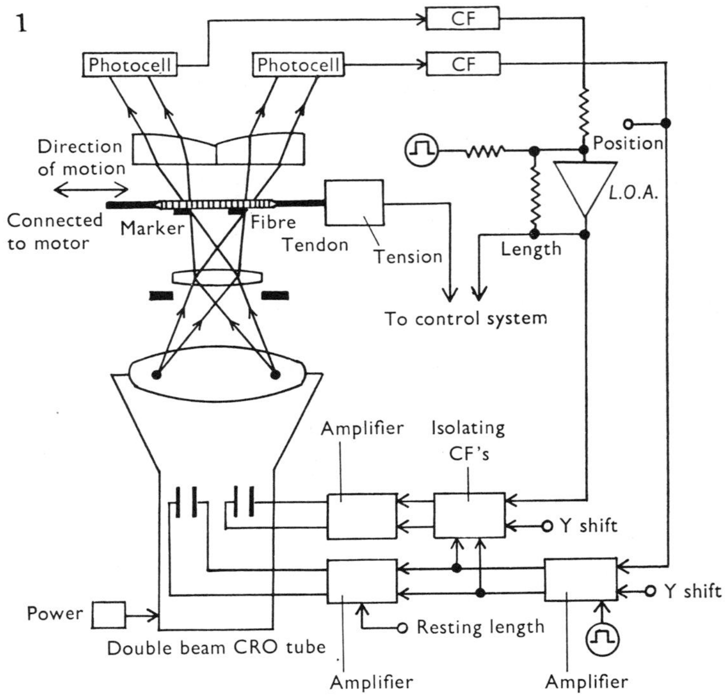
CF
CF
Photocell
Photocell
Direction of motion
Connected to motor
Marker
Fibre
Tendon
Tension
Position
L.O.A.
Length
To control system
Amplifier
Isolating CF's
Y shift
Y shift
Power
Resting length
Double beam CRO tube
Amplifier
Amplifier

2

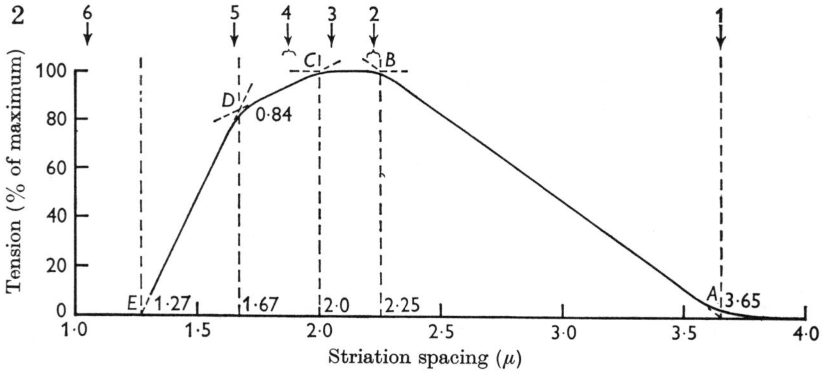
6
5
4
3
2
1
C
B
D
0·84
Tension (% of maximum)
100
80
60
40
20
0
E
1·27
1·67
2·0
2·25
A
3·65
1·0
1·5
2·0
2·5
3·0
3·5
4·0
Striation spacing (μ)

3

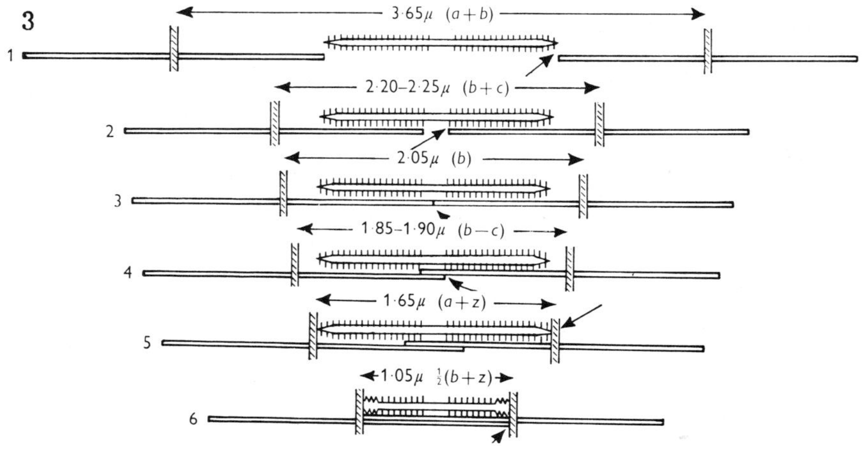
3·65μ (a+b)
1
2·20–2·25μ (b+c)
2
2·05μ (b)
3
1·85–1·90μ (b−c)
4
1·65μ (a+z)
5
1·05μ ½(b+z)
6

nearer the M band. This hypothesis has clearly enjoyed the largest share of attention in the last decade in spite of the awkwardness and inefficiency posed by such a wrist-twisting type of motion. In part the widespread adoption of the theory has been due to experimental underpinning. X-ray analysis of the higher-order equatorial reflections of a vertebrate muscle by Lymn (1975) were compatible with a simple swiveling of the head. The elastic stiffness of stimulated muscle cells *decreases* with increased length proportionally, as also determined by Flitney and Hirst (1974), so it can be related to the number of attached X bridges.

When sudden small decreases in length are made, the lost tension is recovered very rapidly. According to Huxley and Simmons' rocking-head hypothesis, this phenomenon is easily explained by assuming that the heads have been enabled to rotate from the tilted to the flat position, which will be the more stable one, during the quick, short length decrease.

How is the energy of hydrolysis of ATP utilized in this model? As with A. F. Huxley's 1957 model, it is used to break the bond established between the myosin head and the actin. The sequence of steps in the rocking motion are energetically downhill, each a little more stable than the last. The detached myosin head is, as it were, "charged" as a result of the hydrolysis, and ready to resume the rocking cycle if exposed actin binding sites are available. An extensive thermodynamic treatment by T. L. Hill (1975a, b) and Eisenberg and Hill (1978) has shown that this kind of model is energetically feasible, but theoretical objections to the involvement of the presumed elasticity of the S_2 fraction have been made (e.g., Bendel, 1979).

Of course it is desirable that this and other models should be tested on a wide range of different muscle cells before becoming widely adopted, but alas, no other has been tested. The internal SEC for a barnacle giant cross-striated cell, where the tendon contribution is naturally negligible, is no less than 14% rest length (Hoyle and Abbott, 1967). It is not conceivable that X bridges could be responsible for more than a fraction of such a long SEC, and the source of the remainder is unknown. The more elastic muscle cells, of which the barnacle giant cell is an example, can be repeatedly stretched rapidly to several times the length representing that of summed X bridges without affecting tension-generating ability (Hoyle, 1982b). This is an important fact about muscle *in general* that is urgently in need of explanation. The one advantage that results from the bending around back on itself of the long myosin molecule after it has left the A filament is that it facilitates functioning over a wide range of separations of filaments, as occurs between extreme stretch and extreme contraction. The hinge permits the head an operating range from close to the A filament to one in which the bridge has swung out laterally almost at right angles. Flexibility in the hinge presumably promotes the making of "searching" movements by the head and permits a long X-bridge stroke. The maximum stroke length that the tested X bridges of vertebrate cells could make before breaking is 10–11 nm.

Figure 4.6. Test of the dependency of contraction on interaction of I and A filaments. (*1*) Diagram of servocontrolled spot-follower used in conjunction with a single cross-striated cell having relatively uniform sarcomere length. Tiny strips of gold leaf (= marker) are used to monitor the precise length of central sarcomeres. (From Gordon et al., 1966a.) (*2*) Tension/SL curve using the spot-follower. (From Gordon et al., 1966b.) (*3*) Positions of average myofilaments at six positions corresponding to points indicated in (*2*). (From Gordon et al., 1966b)

Figure 4.7. Very thin TEM of an LS of flight muscle cell of *Lethocerus* fixed in a state of rigor contraction. The X bridges, which are conspicuous, are all angled sharply inward, toward the center of the sarcomere (*below*). (Courtesy of Dr. M. Reedy.)

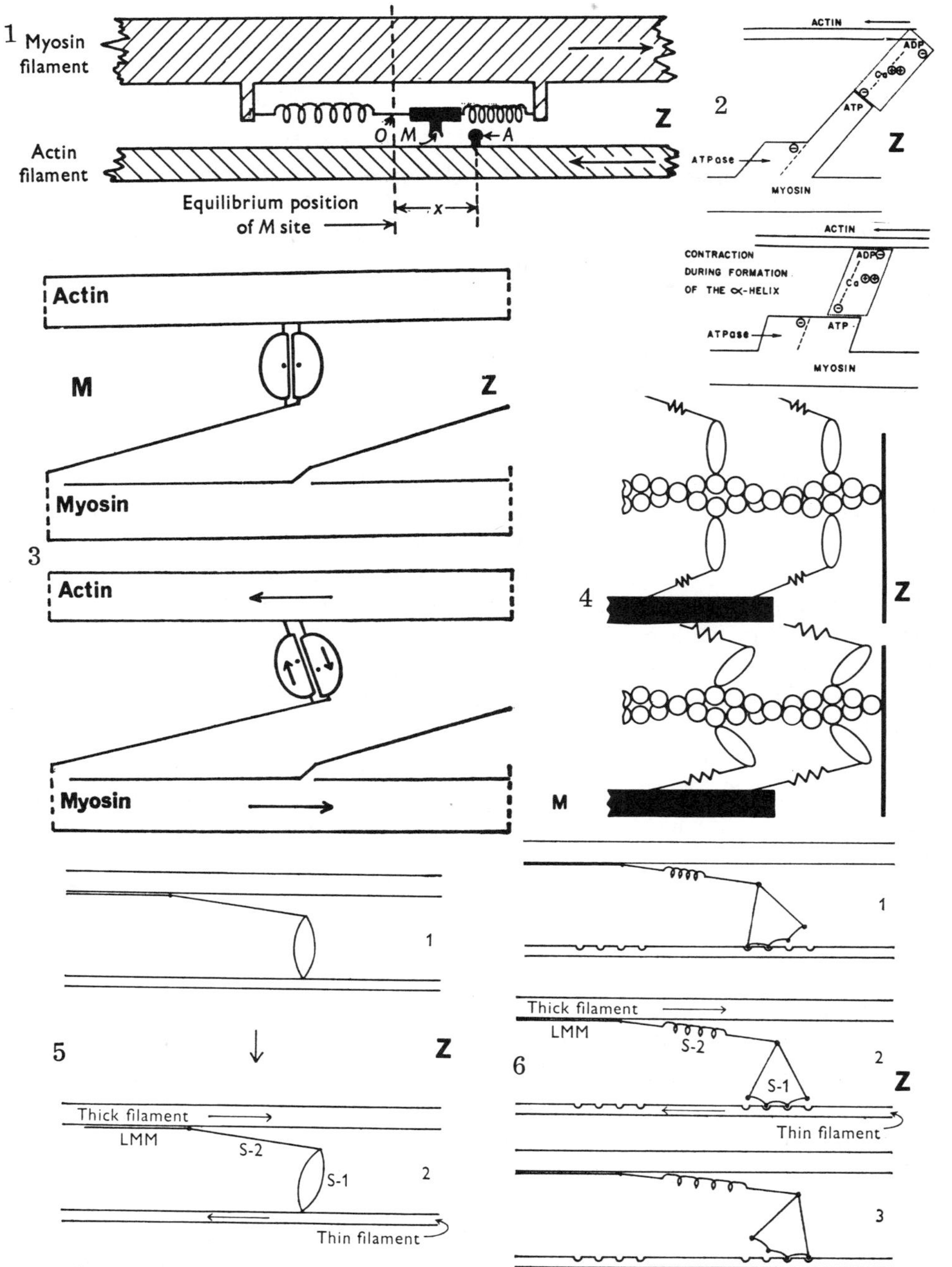

Figure 4.8. Models of contractile mechanism based upon multiple-function myosin heads serving as motile, attachable/detachable X bridges that are not only enzyme locations but also power producers. (*1*) Oscillating attachment site, requiring a greater probability of attachment in the position nearest the Z line. (From A. F. Huxley, 1957a.) (*2*) Bridge-shortening model. [Reprinted with permission from *Nature* **199**, 1068–1074 (1963). Copyright 1963 Macmillan Journals Ltd.] (*3*) Myosin-head rotation, with specific proposal of relative movement of the two heads. (From H. E. Huxley, 1969.) (*4*) Head-rotation-stretching elastic section of myosin molecule. [By permission of the American Heart Association Inc. J. Gergely, *Circulation Research.* **34**, Suppl. III, 74–82 (1974).] (*5*) Head rotation with "hinged" myosin. (From A. F. Huxley and Simmons, 1971.) (*6*) The "rocker" hypothesis, incorporating stretching of elastic section of myosin molecule. Rotation of the head section is brought about by sequential attachments. (From A. F. Huxley, 1974.)

Haselgrove (1975) has suggested, on the basis of axial X-ray diffraction, that the initial change in muscle contraction is not myosin interacting with actin, but rather a change in the myosin backbone. This would represent action of a myosin-activation mechanism, preparing the myosin for interaction with actin at the same time actin interaction sites are being uncovered. The specific change is shown as a shift from a rest state giving strong layer lines and 14.34-nm axial spacing to an active one giving no layer lines but a meridional spacing of 14.48 nm.

A majority of isolated myosin molecules viewed by negative staining are quite straight. How they come to have the pronounced curling that occurs in thick filaments has not been explained. Another problem relates to the siting of active sites on actin for binding with myosin. Actin monomers should have equal capability of binding, yet decoration with myosin is strictly periodic, at 37-nm intervals, close to the actin crossover repeat, which is also the TM/TN repeat. This periodicity persists, even in purified actin or actin from sources lacking in tropomyosin, and this could be responsible for the periodicity in vertebrate I filaments. The rocker hypothesis requires binding sites to be about 6 nm apart, close to the separation of actin molecules. A related problem concerns the two heads of the myosin molecule. Does each link to the same I filament or two different I filaments? Do they work synchronously or alternately? The only data related to this question comes from insect flight muscle fixed in rigor (Reedy, 1968), where the appearance of X bridges is most readily compatible with each head's binding to two different I filaments (Offer and Elliott, 1978). A modified version of the earlier Davies model, in which the X bridge itself is supposed to shorten actively, has been proposed by Harrington (1979). In this hypothesis, although the bridge is angled forward, which is the reverse of that required by Davies's pulling scheme, if the outer part of the bridge were to shorten following attachment – and no other – the Z bands would be pulled inward. Experimentally, testing the scheme requires, as does testing the Huxley/Simmons model, determining the time dependence of the early tension responses following small abrupt changes of length in isometrically contracting single fast-type cells. Harrington carried out such tests and also subjected the muscle to 5° temperature jumps over the range 35-55°C and examined the effects on the relaxation times. The point of the latter was that the results offered insight into the molecular events in the coiled-coil α-helices of the S_2 fraction, which is the most likely to be the part involved if, indeed, the bridge is responsible for force generation and shortening. Harrington found two relaxation times, one in the millisecond range, the other much shorter, corresponding to extending and contracting, respectively, the helices of S_2. This portion could therefore provide not only the elastic aspect of the bridge but also the shortening. The rate constant of the quick recovery following a quick, brief shortening should be determined by the speed of "opening" of the coiled-coil segment, a process termed "melting." With greater steps, a progressively larger fraction of helix is melted until the step length is greater than the length of the region where helix and coil coexist. In the 35-55°C range used, about 85% of S_2 undergoes a helix-coil transition, and the 5°C steps permitted estimation of the rate of helix melting at various fractional contents of helix by using the thermal melting curve of S_2 determined by Tsong et al. (1979). The kinetics of the transitions were found to be compatible with the results of purely mechanical experiments performed by the Harrington group, by A. F. Huxley and Simmons (1971), and by Ford et al. (1977), in association with testing the rocker-head hypothesis. Rotation of the head is included in the Harrington hypothesis, but its only role is as a passive spring, taking up tension during force generation. According to this model, at isometric tension the α-helix and coil regions are in dynamic equilibrium and the head is rotated. Sudden shortening permits the coil region of S_2 to contract, and this is followed by "melting" of the residual α-helical region. Sudden stretching is accommodated by combined head tilting and stretching of the whole S_2 fraction.

The purely dynamic data and S_1-rotation evidence are compatible with both the rocker and Harrington's S_2-contraction hypotheses. Each requires "searching" movements of myosin heads either constantly, as in Brownian motion, which is present continually, or following a rise in calcium concentration, which initiates conformational changes in the myosin heads and therefore promotes their movement. Tests have so far failed to produce evidence for such motions. The extent of movement required is considerable, 5-6 nm between the outer tips of the bridges and the surface of the I filaments, which lie, on the average, on a radius of 18-19 nm from the A-filament center compared with 13 nm for the bridges at rest (Huxley and Brown, 1967). It must occur against the compliance of the S_2 LMM fraction. There is another model, devised by Squire (1972), in which the myosin heads are assumed to reach out laterally, to 18-20 nm in the resting state, so that little or no radial movement is required.

4.12.1. Stereometry of Filaments and X Bridges

No simple general relation can be described for X-bridge positions in relation to I and A filaments because there is such a wide range in A filament structures in different muscle cells. The ratio I to A is 2 : 1 in a majority of lower-vertebrate skeletal cells, but higher values do occur in a number of mammalian muscles, including those of humans. The ratio is uniformly 3 : 1 in insect flight muscle and a few other arthropod skeletal muscles, 4 : 1 or 5 : 1 in a majority of arthropod skeletal muscles, and 7 : 1 in a few very slow ones. The equivalent ratio is extremely varied in smooth muscles, obliquely striated cells, and disorderly striated cells, though always high, from as great as 70 : 1 to 5 : 1. The I filaments occupy triagonal positions, in the centers of A-filament triangles in 2 : 1 cells, but in between A filaments in 3 : 1 cells (Figure 3.11). In 5 : 1 cells, where there are 12 I filaments per orbit, it is as if one set of 2 : 1 and one set of 3 : 1 are combined. With ratios higher than 5 : 1, many, or even most, of the I filaments form self clusters and do not participate in orbits: there are very few I filaments around A filaments in such cells in spite of the high ratio. Alternatively, orbits are made, with as many as 20 thin filaments per orbit.

The A-Filament thickness and attendant myosin molecular packing numbers increase with increasing I : A filament ratio, but no hypotheses have been advanced for the details. The arrangement of the myosin projections that are capable of becoming bridges was worked out from X-ray data by H. E. Huxley and Brown (1967) for A filaments of frog sartorius fast-type cells and by Reedy (1968) for an insect flight muscle. These were shown in Figure 4.3. We can expect that many different ways of packing myosin and other molecules into thick filaments will eventually be found to have been adopted by diverse muscle cells. X-ray data have permitted a reconstruction of the appearance of a myosin head during its combination with actin; the relative positions of the tropomyosin and its three components can be at least estimated. Comparison of X-ray data of relaxed and rigor striated muscle suggest that following the interaction of calcium with the troponin attached to I filaments there is a shift of the tropomyosin. The effect of the TM shift is to uncover sites on the actin that are otherwise masked, so that attachment of the myosin head and formation of the ATP-splitting enzyme complex, which were previously prevented, now become possible.

Other muscle systems in which TM/TN are involved are thought to have similar myosin heads and interaction arrangements. Those that lack the TM/TN complex can presumably make the contact only after some uncovering or promoting effect of calcium on the myosin head.

4.12.2. Locations of ATP In Relation to X Bridges

According to the X-bridge theory, ATP should be freely and equally available for activation along the entire A band. However, both adenine and creatine nucleotides in frog and toad skeletal cells (D. K. Hill, 1960, 1964) are confined to the I band, with a heavy concentration at the A/I boundary. They are in the spaces rather than attached to filamentar components. In insect flight muscle, adenine nucleotide is found attached to the A filaments (see p. 146).

4.12.3. X Bridges and Basic Tonus

Many cross-striated as well as smooth muscle cells possess BT (see p. 255), a state of being continually weakly activated in the unexcited state; this state is recognized because it can be relaxed actively by hyperpolarization. In some cells, such as crustacean leg muscle slow cells, tonus is achieved because the resting membrane potential is just very slightly lower than the threshold for initiation of contraction. In others, including some locust leg muscle cells and chick latissimus dorsi, BT exists even though the coupling threshold is well above resting potential. The X-bridge hypothesis requires that BT be a form of very weak normal contraction, such as would be caused by a suprathreshold level of free calcium in the myoplasm in the resting condition. This would be achieved by reducing the Ca-sequestration rate of the SR, or by continuous weak release of Ca^{2+}. However, there are many alternative possibilities, including ones in which tonus is unrelated to bridges. There is a possibility that a few X bridges are connected to I filaments in most resting muscles, based upon measurements of the short-range elasticity. Whether these residual X bridges are exactly equivalent to the majority of X bridges, which form connections only following excitation, is problematical, and it is possible that there are two populations with different characteristics. This question is very important in relation to the explanation of basic tonus (see p. 256). One notion is that a special set of bridges is present that continually recycles, but very slowly.

4.12.4. Detection of Movements of X Bridges

Following activation, a rise in free-calcium-ion concentration initiates the formation of a large number of regular X-bridge links to I filaments (see p. 197). The cell first passes into a state resembling that of rigor, in which stiffness is drastically increased. But almost immediately, the excited cell starts to shorten, or rather, the sarcomeres do so at the expense of elastic elements in series with them. If there is little load compared with force-generating capability, the load is moved, and the muscle shortens. The rate of change of length (velocity) is a hyperbolic function of the load (see p. 236). But no matter what the nature of the contraction, the theory requires the movement of X bridges, so it has become critical to know whether – and if yes, to what extent – the heads move at activation. This question has been intensively addressed, and there is substantial evidence that significant head movement does occur. In studies of the equatorial X-ray diffraction patterns of vertebrate cross-striated cells, from rest to contraction and rigor in the length range 1.8-2.7 μm, Haselgrove and H. E. Huxley (1973) found that there was indeed a transfer of mass to the I filaments. This movement is not due to calcium shift (Mendelson and Cheung, 1976). It was 15% of the total myosin mass in an isometric contraction, and 45% in rigor. Some movement was stated still to occur in muscle stretched to beyond overlap, where actin-myosin interaction should not occur (Haselgrove, 1975). Later, in a similar experiment, Yagi and Matsubara (1980) claimed there was no movement of the relevant 42.9-nm later line. X-ray data has not always indicated X-bridge movements associated with contraction, and some anomalies have occurred. For example, in chick anterior latissimus dorsi, the changes expected from the attachment of many X bridges were observed in rigor, compared with resting muscle. But for muscle in strong potassium contracture, the X-ray picture was the same as for resting muscle (Matsubara, 1974).

Movements attributable to X bridges can be detected using light diffraction to measure SL in toad cells (Barden and Mason, 1978). Muscles were rapidly released from tetanus, either 1% from 2.5-μm SL or 2% from 3.0-μm SL at 10° C. In the former there was a 12.5 nm/half sarcomere length change, about equal to the working stroke length. In the former case, a simple step shortening occurred, but in the latter case, where the step was longer than a X-bridge stroke, the researchers obtained the same rapid phase, 12 nm in 1 ms, as in the first experiment, then a slow (linear) phase. The velocity in the slow phase was 2.9 μm/s for a half sarcomere, the same as the maximum unloaded velocity (V_{max}); the extent of contraction was 18 nm in 8.5 ms. The initial period of 12-nm shortening presumably corresponded to a single incomplete X-bridge stroke occurring immediately after the elastic response to the length change; a bridge would therefore have to start cycling, to generate normal contraction velocity after unloading, in less than 1 ms. The relative timing of the presumed individual X-bridge movements is likely to be random for a majority of muscles, making it in principle especially difficult to follow them. But, in the case of the insect flight muscle cells whose contractions are not synchronous with action potentials (see p. 547), there is good reason to believe that contractions are brought about by synchronous movements of the bridges. These same cells have elastic C filaments connecting the ends of the A filaments to the Z band (see p. 237), giving them a high elastic modulus even when relaxed. These asynchronous flight muscles contract and relax at a high frequency both in life (Machin and Pringle, 1959) and as models after prolonged glycerination (Jewell and Rüegg, 1964), following addition of ATP. Contraction becomes deactivated when rapid shortening occurs, a condition that is unique to this kind of muscle, and is probably dependent on the synchronous aspect of the X bridges. X-ray diffraction studies of an asynchronous flight muscle of the giant water bug *Lethocerus* were made in the relaxed state and compared with the contracted (rigor condition) by Reedy et al. (1965). A strong 14.6-nm period in the low-angle pattern, together with 7.3-nm meridional reflection, was probably due to X bridges' occupying positions at right angles to the filament axis. In rigor, the meridional reflexions are very weak, and a strong layer line at 38.8 nm is present, strongly suggesting that the X bridges have moved to different positions, but are still in an orderly array. Electron microscopy showed very clearly the X bridges angled inward toward the center of the sarcomere, forming chevrons repeating axially at 38 nm. In exceptionally thin sections showing only I filaments, the same periodicity is clearly present on I filaments because of ends of X bridges attached to them.

Stretch experiments have also provided evidence for the involvement of X bridges. Since arrays of both I and A filaments have a regular periodicity of 38.8 nm, stretch must change the relative axial alignment in a periodic manner so that the tension rise during stretch should show amplitude fluctuations associated with particular lengths. R. H. Abbott and Cage (1978) studied this in insect flight muscle and did find the expected fluctuations, but only when they used calcium concentrations that gave submaximal activation. They interpreted this condition as one in which the sarcomeres are relatively uniform in length compared with the rather heterogeneous mix that occurs at full activation, where slight differences are exaggerated by great shortening of stronger elements. This difference is what, in effect, permitted them to detect the changes. The distances between the peaks corresponded to 38 nm/half sarcomere!

The relatively small percentage of total X bridges actually utilized in a maximal contraction is, perhaps, related to the chance nature of a myosin head's actually contacting an actin binding site. The probability that a head will find one must decrease in proportion to the number of contacts already made. Also, factors may be brought into play that further reduce the availability of sites as a sufficient number of contacts are made. There is no point in producing too much force, because there would be a risk of damaging the insertions. An athlete can rather readily damage a tendon and a grasshopper snap apart its jumping-muscle apodemes, even though they are not using the maximal intrinsic contractile capability.

Very sensitive movement analysis should be capable of detecting tension fluctuations due to X-bridge movements, and several claims of success have been made (e.g., Borejdo and Morales, 1977). The "noise" completely disappears both in the rigor state and during relaxation. Application of pseudorandom white-noise-generated ultramicroscopic displacements to produce less than 1 nm/half sarcomere of stretch has produced evidence that, in "resting" frog semitendinosus, some calcium-activated X bridges are in the linked condition (Moss and Halpern, 1977). Strictly physiological data suggest that myosin heads must remain close to A filaments. The V_{max} is not significantly affected by lattice width (Edman and Hwang, 1977), since it is constant over a wide sarcomere length range, from 1.65 to 2.7 μm in fast frog cells (Edman, 1979). This can only be true if X-bridge kinetics are constant. They should, however, be greatly affected by spacing, if only because of the differences in packing density. The alternative interpretation, of course, is to accept that these observations are not compatible with the X-bridge hypothesis and to seek another.

4.12.5. Velocity in Relation to X-Bridge Theory

The velocity of a muscle is a major characteristic that has an extremely wide range that must be related to X-bridge stroke rate and recycling speed, according to the theory. There is a limit to maximum velocity (V_{max}) that is about 2.3 muscle lengths/per second. In skinned frog cells, the velocity varies hyperbolically with the concentration of ATP available (Ferenczi et al., 1979), indicating that there are at least two steps in the process of detachment. This velocity translates to about 3.0 μm/s for a point on an I filament relative to the nearest A filament, or 250 times the length of a X-bridge stroke per second. Each relatively slowly recycling individual X bridge is imparting its contribution extremely briefly within a cycle during unloaded shortening. The head-rotation or bridge-shortening action providing power is extremely brief compared with cycle time. In an unloaded contraction, according to the X-bridge theory, the myosin heads are imparting what can best be described as a "wristy flick" to the highly mobile I filaments. This is the case to get them both started and once started kept in motion, since it is generally considered that at least in fast vertebrate cross-striated cells, shortening starts immediately at the characteristic velocity for the load. According to my own experiments shortening of arthropod skeletal cells always starts slowly, and later speeds up, but it is also true that they are not activated as rapidly as are vertebrate cells. The X-bridge theory requires that the velocity of unloaded muscle be constant at any given length, regardless of the extent of activation. Although this relation has been shown to hold true for fast-twitch vertebrate muscles and single cells, it has not been extensively tested in other kinds. In particular, it should be tested in systems where there is less extensive release of Ca^{2+} in unitary excitation. In single giant barnacle cells, the relation does not hold, the velocity increasing progressively with increasing activation. It is, however, possible that in barnacle there is progressive longitudinal recruitment of contractile elements. This is true for any striated cell in which the SL is long and the number of ITs small.

4.13. Events in Insect Fibrillar Flight Muscle

Insect fibrillar flight muscle has, in common with some cross-striated cells of small crustaceans, the only truly regularly arranged myofilament structure known. Its regularity, together with oscillatory contraction dynamics (see p. 536), make it favorable material for X-ray analysis on one hand, and correlation of biochemical and mechanical events on the other. The seven-step events of ATP hydrolysis (see p. 132) need to be correlated with actual events in the X-bridge cycle, and the best hope of doing this, in detail, lies with this muscle. General data on the muscle's properties have been provided by Pringle (1967) and White and Thorson (1973). Stretch experiments, done at about the same time as those of A. F. Huxley and Simmons on frog muscle cells, were made by Abbott (1973) on *Lethocerus* muscle. Tension rise simultaneously with a small length change is followed by a very much slower recovery to a value near the original tension, as with the vertebrate muscle. Abbott's interpretation was different from that of A. F. Huxley and Simmons, but at least it was evident that both hypotheses concerned similar molecular events, which they presumed to be X-bridge dynamics.

Almost all the *Lethocerus* work has been done on glycerinated material because it can be stored for up to a year and yet continue to give reproducible results. All the major muscle enzymes are still present and active (Abbott and Leech, 1973) in such material. Major interest was focused on glycerinated *Lethocerus* flight muscle preparations because they produce oscillatory contraction/relaxation cycles when suitably loaded following addition of ATP. Other muscles do not do this. The principal reason for the difference is that when the flight muscle is stretched following activation, there is an increase in the ATPase activity, that is, the length change enhances activation of the contractile machinery – *stretch activation.* A passive aspect is also important, the fact that this muscle is stiff – not easily stretched. The natural dynamic condition in the body, with the muscle attached to the wing, promotes alternate stretch and relaxation. It was for many years believed that oscillatoriness in the flight and cicada drum muscles was due to inactivation upon release of tension. Activation by stretch alone could not do the trick. The reason stretch works this way is that there is a significant delay between the increase in length and the stretch activation (Abbott, 1973). This is sufficient to allow the tension increase to come when the muscle is already shortening because of the nature of the mechanical system to which it is connected.

Investigators of the molecular basis of contraction in insect flight muscle have, to date, uniformly favored a form of X-bridge head rotation. The X-ray diffraction shows clearly the two separate filaments, with a 15-nm-based regularity due to myosin and a 38-nm one due to I filaments. The X bridges are seen clearly in X-ray diffraction images, attached at 55° in groups at each turn of the 38-nm actin helix (Tregear and Marston, 1979). But from X-ray diffraction in ordinary isometric contraction, it can be deduced that the myosin heads are standing away from their regular positions, closer to the I filaments. The radial movement of the heads at activation must be very slight, not more than 10–20% greater than in the relaxed state (Miller and Tregear, 1970).

The *Lethocerus* bridge in its rigor state (r bridge) is the model that has been used for energy-conversion considerations. But in active contraction, the bridged state is so short lived that a different form, which the authors term the p bridge, should be considered. ATP added to the glycerinated flight muscle is quickly and firmly attached, but is immediately converted into ADP (Marston, 1973). The stable bridge is M•ADP• P. Tregear and Marston postulate that an earlier, short-lived p bridge is formed first, and that it is the p → r bridge change that is responsible for the mechanical event. The essential difference is, as in the Huxley/Simmons model, a head rotation, but twisting about a single pivot rather than a rocking motion (Figure 4.8*.4*,5).

The free-energy change in their model is the loss occuring at the moment of p-bridge formation. It is not at its own maximum free energy because it comes via a contact intermediate of brief lifetime. The p-bridge lifetime depends on the mechanical events that follow its formation. Its energy is released as shortening occurs. The p → r transition is itself a function of the ensuing contraction, increasing as shortening occurs. Some energy remains associated with the r bridge and continues to be released in shortening until detachment occurs. This two-state model, for which Tregear and Marston provide experimental evidence, could, the authors themselves argue, be greatly oversimplified, for there may be many intermediates between their two states – enzyme site empty and enzyme site occupied by nucleotides. X-ray diffraction has been carried out on *Lethocerus* flight muscle during superimposed sinusoidal oscillation and activation (Armitage et al., 1975). The change in intensity of the diffraction pattern was compatible with no more than 10–20% of the myosin heads' actually being attached to I filaments. The time course of attachment followed tension generation with a small delay. Not all the myosin molecules attached at any one moment could also be generating tension.

The addition of orthophosphate to glycerinated flight muscle alters the frequency at which the muscle works (White and Thorson, 1972). Pyrophosphate addition relaxes the muscle, producing reversible changes in the X-bridge configurations (Beinbrecht et al., 1972). A theoretical basis for possible translation of the conformational change in the myosin head into mechanical work has been proposed by Marston et al. (1979).

We may ask the question: are stretch activation and possibly deactivation with shortening widespread? Following the unequivocal demonstration by Abbott of stretch activation in *Lethocerus* flight muscle, tests were made for similar effects in vertebrate and other invertebrate muscles, but with generally negative results. With suitable loading oscillatory work was obtained from a crayfish muscle and from frog semitendinosus (Kawai et al., 1977). The frequency must be just right, because the phase delay required in flight muscle is replaced by phase advance, at both lower and higher frequencies. However, it was well known that a slight but rapid shortening step lowers tension to zero, but not whether there was any actual associated deactivation. In giant barnacle depressor cells, a minute (1%) shortening produced a marked fall in calcium release during constant current stimulation (Ridgway and Gordon, 1975). Allen et al. (1977) tested single frog semitendinosus cells, also by injecting the cells with aequorin, and found that a marked drop in the concentration of free calcium occurs during a brief tension drop. There was no increase in calcium concentration during stretching. Ridgway and Gordon pointed out that in the barnacle giant cell, these effects are indirect. They are mediated indirectly, by an effect of the length change on membrane resistance. When they voltage-clamped the cell, the effect of length change was very small. Thus there is no evidence that ordinary skeletal muscles share with water bug flight muscle a direct effect of length on the contractile machinery.

4.14. Matching Energy Utilization to Movement and X-Bridge Dynamics

It is clearly important to be able to relate the presumed molecular events causing motion to known, measurable biochemical ones of energy supply. Early attempts to do this focused entirely on the actomyosin ATPase activities of muscles with different intrinsic speeds. Dramatic differences were found, but the actomyosin ATPases cannot be judged as a set of isoenzymes, because the cooperative involvement of two major and a few minor proteins makes multiple interactions possible. It is not surprising then, that AM ATPase activity rates are very different in different muscles. For a whole muscle, the rate is but an average; if individual cells could be examined the range would be even wider. Bárány (1967) compared rates of glycerinated muscles from cat, sloth, and mouse (extensor digitorum longus – EDL – and gastrocnemius medialis), squid (mantle), *Limulus* (abdomen), sea cucumber (*Thyone* – longitudinal retractor), and dogfish (mesenteric and coracohyoid). The contraction times ranged from 19 ms for the mouse EDL to 20,000 ms for the dogfish mesenteric muscle. The squid time of 68 ms was least for the cold-blooded species. The mouse muscle AM ATPase split 30.0 μmol Pi per milligram per minute, the squid 0.94 μmol Pi per milligram per minute and the dogfish 0.003 μmol Pi per milligram per minute. Many other factors are involved in determining contraction times than this one enzyme (see p. 260). It is unreasonable to assume that it can be equated with a simple enzyme in a purely chemical testing situation. Studying AM ATPase in glycerinated muscle involves the whole complex structural architecture, each part of which adds its contribution. There is no enzyme activity when the X bridge is broken, so the value of the ATP-splitting rate may be as much a measure of X-bridge recycling rate as of actual splitting ability. But the range found in different tissues is dramatic, almost as wide as that in contraction times, making AM ATPase differences a major determinant of intrinsic speed capability.

A related factor is that the AM ATPase rate is also a function of what the cell is doing at the time and what is being done to it in terms of length changes. Stretch changes the structure of the enzyme complex, altering accessibility of substrate as well as the rate of working. This effect is not due to greater calcium binding (Marston and Tregear, 1974). There is an optimum frequency and an optimum amplitude for the tension or work costs, moles ATP hydrolyzed per unit of tension or of work. Direct studies on these relationships have been made only in the experimentally favorable water bug flight muscle (e.g., Pybus and Tregear, 1975), where there is at least a doubling of ATPase activity compared with isometric when a small amplitude (2% of the muscle's rest length) 5 Hz oscillation is applied. The ATPase activity is very sensitive to temperature (Licht, 1964), a factor that must be extremely important functionally for all cold-blooded animals.

The ATP utilization rate is exactly matched by the phosphagen splitting that reconstitutes the ATP, so following this event also enables a direct link between contraction and energy supply to be followed. A major finding of the middle-period research of A. V. Hill was that much extra heat above that in an isometric contraction is produced during shortening (Hill, 1979a) when external work is performed. This is explained on the X-bridge hypothesis on the basis of much more X-bridge recycling when shortening occurs. The amount of creatine phosphate broken down was determined by Curtin et al. (1974) for frog sartorius subjected to similar brief tetani, isometrically and also isotonically; it was found to be much greater when shortening was allowed. The extra amount split was sufficient to account for the work produced. The total creatine content was 22.9 μmol/g and the breakdown rate 0.71 μmol/g during a 1.7 s contraction. Under isometric conditions, the breakdown rate was 0.031 mol/mol total creatine. To make further calculations they made several assumptions: from data of Ebashi et al. (1969) they assumed that myosin splits 0.14 μmol/g, and they also assumed both that there is only one X bridge per myosin and that all X bridges participate. It then follows that each X bridge is reprimed an average of five times during the 1.7 s contraction, so the average cycle duration is

$$\frac{1.7}{5} = 0.34 \text{ s}$$

The X-bridge attachment time during the same contraction but with shortening would result in an average of about half the distance of stroke, and half the chance of attachment. Therefore the average time attached would be one quarter that in an isometric contraction, and the mean stroke time would be 0.12 s.

The equation derived by Curtin et al. (1974) for the amount n of CP broken down per gram is:

$$n = \frac{fgMl}{d}$$

where g = a factor allowing for the effect of bridges' competing for interaction sites,
f = the fraction of possible interactions that actually occur,
M = myosin content,
d = spacing of actin sites, and
l = amount of shortening by one half sarcomere (= 0.38 μm).

If g = 1 (i.e., no competition) and f = 1, then d - 44 nm. The latter value is close to the TN spacing, which equals that of myosin-decorated actin and at least indicates that the calculations were in the right ball park and that the theoretical basis of assumed X-bridge mechanics is not outrageous. The above nice relation was unfortunately confounded by a finding of Dickinson and Woledge (1973). They measured the shortening heat in a series of closely spaced brief tetanic contractions and found that the shortening heat declined progressively. It was reduced to 66% by the third, and only 15% by the tenth contraction, clearly showing that the process producing shortening heat is somehow improved by repetition and that it is essential for shortening.

Also, many questions are raised by the stroke time. There are synchronous muscles, responding on a 1:1 action-potential contraction basis, that can contract at rates over 200

Hz (Josephson and Halverson 1971). Asynchronous tymbal muscle of cicadas and flight muscles of small Diptera reach frequencies in the hundreds of hertz (Pringle, 1954, 1957). There is a wide range either of cycling times or of durations of X-bridge attachments in different muscles. If the latter, there would also have to be a very subtle coordination of X bridges to assure alternation of cycling times of individuals. H. E. Huxley (1972) had concluded from X-ray data that it takes several seconds for all the heads to return, imposing an even longer recycle time. Yagi et al. (1977) explained similar findings on the basis of only 20% of the heads' not returning promptly, the majority returning in less than 0.5 s.

There have already been several attempts to correlate the seven steps in the myosin ATPase hydrolysis cycle (see p. 132) to presumed events in the X-bridge movement/attachment/detachment cycles. Since X bridges are permanently attached in the absence of ATP (rigor state) and since ATP once attached is immediately converted to ADP, it is assumed that bridge detachment corresponds to step 1, the combination of ATP with the myosin head (A) M + ATP → M•ATP + (A) where (A) represents actin. Hydrolysis of the bond at the required high rate requires actin; hence bridge attachment must constitute step 2. The next five steps must all occur during bridge attachment and movement of the head. In an unloaded isotonic contraction in a fast-muscle cell, the attachment time cannot be longer than about 50 ms. Chemical agents and physical events that affect the speed of contraction at this stage, including quick stretch of insect flight muscle, are most likely to do so by affecting step 5, in which phosphate is split off from the activated myosin head:

$$M^{*}\cdot ADP\cdot Pi \rightarrow M^{*}\cdot ADP + Pi$$

or the immediate predecessor, step 4, in which the myosin conformation change occurs:

$$M^{**}\cdot ADP\cdot Pi \rightarrow M^{*}\cdot ADP\cdot Pi$$

The X bridge probably does not contribute to tension development immediately after attaching, but this is the major step in activation. Subsequently, force is produced. These factors, which are attributes of the rocker hypothesis, can be deduced from characteristics of the force/velocity relation (Cecchi et al., 1979).

Independent evidence for the recycling time, in *Lethocerus* flight muscle, has been obtained by examining tension oscillations that follow a relatively long (≈ 6%) quick stretch (Schädler et al., 1971). The argument is that the stretch must detach a lot of X bridges whose cycles were randomly related to each other at the time of stretch, but which are now suddenly ready to start again all at the same moment and so will start off in synchrony. The frequency of this oscillation was different for two species examined: *L. maximus* and *L. annulipes,* respectively, 8 and 14 Hz at 20°C each with a temperature coefficient for 10°C rise (Q_{10}) of about 3.5. The oscillation appears to be heavily damped, lasting for only four to five cycles, as is to be expected in view of intrinsic variability in cycle times. Supporting this interpretation is the finding that the frequency is the same as that at which maximum work is obtained when applying oscillatory driving to the stimulated muscle, which should be correlated with mean X-bridge frequency. The indication of definite and different mean cycle times for different species is in itself very interesting, and the times, of 0.12 and 0.07 s, are in line with other estimates.

4.15. Alternative Theories

In looking back on the history of X-bridge theories, it is clear that they were invoked because of the apparent lack of other material that might bring about contraction more directly, in the traditional assumed way of coiling or folding of long molecular chains. The mainstay of the X-bridge theories, A. F. Huxley's first model, was published in 1957, when H. E. Huxley was abandoning S filaments and five years before Sjöstrand uncovered ultrathin filaments in overstretched cells. By then the element of necessity was lost, but the habit of thinking about bridges was ingrained. When Pat McNeill and I kept finding T filaments and talked about them, we found a very skeptical, unreceptive audience bent solely upon tracking down the nature of the bridge.

But all is not rosy in the garden of bridges. The translation of bridge-head rotation to movement must introduce a compressive, inward force, yet during shortening the lattice expands according to constant volume relations, as it would if operated by a longitudinally arrayed molecule. There is also an outward force, apparent, for example, in a contracting human biceps, that is by no means easy to explain. If for no other reason, alternatives to force development by bridges should be examined.

Since the realization that contraction of striated cells is associated with sliding of filaments, few major alternatives to X bridges have been proposed, and none have been taken very seriously. One, which may be regarded as a category with several alternatives concerning detail, proposes that the structure is driven by electric-charge attraction, as a form of solenoid (Spencer and Worthington, 1960; Ingels and Thompson, 1966; Elliott et al., 1970; Iwazumi, 1970). Either an abrupt or propagated wave of electrostatic charge or charge reversal is postulated to flow over one set of filaments, attracting them inward, toward presumed fixed charges. The major problem with this theory is to link the charge changes to ATP hydrolysis using testable ideas, but ATP hydrolysis might be envisaged as directly leaving charge on one set of molecules (Shear, 1969).

Another theory (Hoyle, 1968) supposes that the ultrathin (T) filaments are not just passive elastic/structural elements, but that they shorten actively by folding as in presliding hypotheses, in the manner illustrated in Figure 4.5. According to this theory, X bridges are formed and are responsible for generating part of the stiffness, but in conjunction with local T-filament shortening. They are also responsible for the availability of energy following the normal schemes. But according to my model, shortening is brought about solely by the T filament in the immediate vicinity of the myosin head, not by head or bridge rotation. It supposes that it is shortening of the T filament, rather than "recycling," that forces breaking of the myosin head/I filament link and prepares the head for reattachment. The transfer of energy to T filaments from the myosin heads presents a problem, but they are so close that even a heat-engine function is possible.

One specific objection to the T-filament theory is that it has been shown by UV disruption of single sarcomeres that as little as one third the A/I overlap region can shorten with ATP (Stephens, 1965). This objection would not apply, though, if the T filament is attached at the A margin and at the M region, as it appears to be (see p. 196). Supporting the hypothesis is the finding of Oplatka et al. (1974a,b), that after all the thick filaments are dissolved, the residual material (= ghosts) will contract in response to a solution containing only MgATP and myosin heads.

The special advantage of the T-filament theory is that there is no difficulty in using it to explain contraction in smooth muscle and nonmuscle in which actin has a similar form. It is proving very difficult to squeeze the rocker mechanism, for example, into explaining contraction of vertebrate smooth-muscle cells (e.g., Figure 2.1.2), some of which generate as much or greater force as frog fast skeletal cells yet have very little myosin which is probably not present in organized filaments. Myosin's role as an *enzyme* is compatible with a low concentration if some other molecule is responsible for contraction. The major problem may then become how smooth-muscle and paramyosin cells stabilize force once it has been developed, and why both are so slow in developing tension, and even slower to relax. For these cells there is the problem in explaining their elasticity: they elongate considerably under full stimulation when placed under heavy load, without breaking. It is possible to understand the elastic properties of many striated cells, such as those of giant barnacles, under stretch following activation if X bridges are not only very easily broken at all times, but are also not the principal load-bearing elements. It is interesting to note that the late Jean Hanson at one time considered dual load-bearing structures a real possibility (Hanson and Lowy, 1965). Tension development by T-filament shortening does not exclude the possibility of a contribution from myosin heads in addition. There could be a spectrum from smooth cells in which ultrathin filaments alone generate tension to insect flight muscle in which only X-bridge heads are involved.

The progression from contractility in nonmuscle to that of advanced cross-striated cells is either via the actin they both contain or via as-yet-unspecified contractile filaments present in both, which for muscle would have to be the T filaments. If the latter were pure actin, in single-stranded form, and if the structure of this is as proposed on p. 128, some of the more obvious unanswered questions are immediately resolvable. The mechanisms discussed above utilize conformational changes in some part of the myosin molecule to generate both force and shortening, yet myosin is not universally available in contractile cells. Some of those cells that have modest amounts of it could be using it in addition to activation of another molecule that is capable of shortening. The possibility that there are combinations of force-generating and shortening mechanisms in the same cell and that there is a spectrum from non-myosin-operated systems to pure-myosin-operated ones, with various mixtures in between in different types, should not be shut off arbitrarily just because there is a sentiment that favors the drastic application of Occam's razor as a matter of principle.

4.16. The Relaxation Process

The spacing between the filaments in a resting muscle cell is determined by the balance between repulsive forces of electrical double layers and attractive ones of London/van der Waals forces, in combination with opposed radial and longitudinal strains caused by the combination of structural and elastic elements. Since these forces are in balance, it can be expected that they will restore the status quo without an energy-requiring active restoring force. But these properties all vary greatly between cells, and so no generalizations can be made. Many smooth and some striated cells have no very clear rest length. They are stretched by a very small fixed load by amounts ranging from 20% mean body length for a barnacle giant striated cell to over 100% for annelid body wall, which is obliquely striated. When the load is removed, the cell may stay at the new length indefinitely. When excited under zero load, they shorten to below mean body length and extend only partially toward mean body length unless loaded. By contrast, an isolated flight muscle cell of locust, or fast cell of a frog sartorius, returns to precisely the same length after 20% stretch, which requires a large force, or following shortening unloaded to 80% mean resting length in the body. The former type of cell can recover rapidly from extents of shortening that are irreversibly crippling to a fast frog cell. The possibility has not been investigated experimentally, but should be considered, that an active extension mechanism is involved in length restoration. Mechanisms that might be invoked include electrostatic-charge repulsion between components, a radial element that shortens actively, and elongation of a folded longitudinal molecule.

4.16.1. Time Course of Relaxation

More-or-less totally ignored compared with contraction, relaxation rate is clearly a very complex process, compounding many factors, the most significant being the rate at which activating calcium is removed. Relaxation after brief stimulation is indeed just that, but muscles also relax as a result of inactivation, fatigue, and both in concert. A decline in concentration of ATP, whose availability is determined by metabolic factors, must also be important, especially when stimulation has been continued for a long time and "fatigue" processes are involved. This can be investigated indirectly by studying the loss of phosphagen, using NMR labeling of the phosphate. The time constant of relaxation of gastrocnemius of frog (*Rana temporaria*) in fatigue was found, using this method, to be determined by metabolic factors and independent of loss of activation or deactivation (Dawson et al., 1980).

All fast-type muscles relax much more slowly when fatigued, partly because fatigue reduces the ATP available for pumping Ca^{2+} back into the SR. The X-bridge cycling may also be slowed (see p. 149).

4.16.2. Energy Relations in Relaxation

When a loaded muscle is relaxing gradually, energy is being consumed. When it elongates as a result of a greater load than it can bear even though excited, which is a common condition for muscles during locomotion, work is done on the muscle. This is termed *negative work*. There is evidence that during the performance of negative work, some of the energy-supply processes work in reverse (Abbott et al., 1951, 1952), to the point of recovery of high-energy phosphate. Although the basic reactions (see p. 132) are indeed reversible, it is difficult to imagine reversal on the X-bridge theory, since it happens while X-bridges are still stroking, even though the I filaments are being forcibly pulled apart.

4.17. Do Some A Filaments Shorten During Contraction?

It is considered firmly established that in all vertebrate and most invertebrate cross-striated cells, the A filaments stay constant in length during contraction under all conditions except extreme shortening. But several instances of possible shortening have been reported. Largely in glycerinated muscle stimulated by ATP, most were shown by electron microscopy to be due to the artifact of some thick filaments in a sarcomere being pulled toward one Z band, others to the other, with a progressive shrinking of the overlap. The overlap was what had been observed in the microscope. However, for the telson muscles of the horseshoe crab *Limulus*, not only was A-filament shortening reported (de Villafranca and Maschaus, 1962), so also was A-band lengthening (Ikemoto and Kawaguti, 1967) following the addition of TM. After a long interlude, de Villafranca's claim has been verified in indisputable fashion (Dewey et al., 1977; Brann et al., 1979). Examples of experimentally shortened *Limulus* thick filaments are shown in Figure 4.9. Maximally, a thick filament decreases in length by 30–40%, increasing in diameter by a corresponding amount. This finding should make it worthwhile examining other claims of change in A-band length carefully, especially the one made by Yarom et al. (1980) that A-band shortening accompanies contraction in humans.

Both calcium and ATP are needed for shortening of *Limulus* thick filaments and alkaline phosphatase promotes lengthening. *Limulus* thick filaments contain paramyosin, which may be involved in their shortening. The probable initiating mechanism is phosphorylation for shortening and dephosphorylation for lengthening, but what happens to the molecules during the actual length changes is unknown. It would be interesting to have experiments similar to these on *Limulus* thick filaments performed on thick filaments from molluscan and annelidan obliquely striated muscle.

The rate of shortening of the A filaments is very slow. The telson muscle supercontracts, and A-filament shortening aids this. Presumably, A-filament shortening although adequately described only in *Limulus*, is not confined to this one species. A recent claim has been made for it in a crayfish muscle cell (Davidheiser et al., 1980). On the X-bridge theory, the generation of tension should decrease when A filaments shorten, but it does not. Contraction of the A filaments provides an alternative to Z-band perforation for supercontraction of *Limulus* muscles. The telson muscles of *Limulus* do indeed supercontract during the most extensive movements of the telson.

4.18. Are Regulatory Mechanisms of Contraction Involved in Acclimation?

As cold-blooded animals adapt to environments with different temperature ranges, they undergo a variety of long-term changes in metabolic activity that are collectively called acclimation. Since the TN/Ca regulatory mechanism governs the muscle enzymatic activity, is it also involved in acclimation? Goldfish were acclimated for five months at either 2 or 31°C, and the Mg/Ca ATPase activity of its AM was determined (Johnston and Lucking, 1978) and found to have acquired markedly different temperature-sensitivity curves. The TNIs of the two were found to differ in molecular weight.

4.19. Problems Caused by Local Heterogeneity of Activation

All forms of testing consequences of the sliding-filament theory and other hypotheses concerning molecular mechanics of motility in living muscle cells must take into account the complexity of fine structure, even within a single cell. A servosystem that regulates sarcomere length by reference to fine markers placed on the surface has greatly aided such studies in fast frog cells (Gordon et al., 1966a,b). But the muscle cell they examined comes as close as possible to the ideal, by having very similar intrinsic myofilament-cluster sarcomere lengths, similar and abundant ITs, with uniform properties, similar IT/SR contacts, and abundant SR. Not many muscle cells parallel frog semitendinosus in these regards, certainly no known mammalian or arthropod cell. Most cross-striated cells are bound to be subject to complication caused by a very heterogeneous set of different local activations even when activated by a large, brief action potential, as I pointed out some years ago (Hoyle, 1966). Many observations have borne out this expectation (e.g., Julian

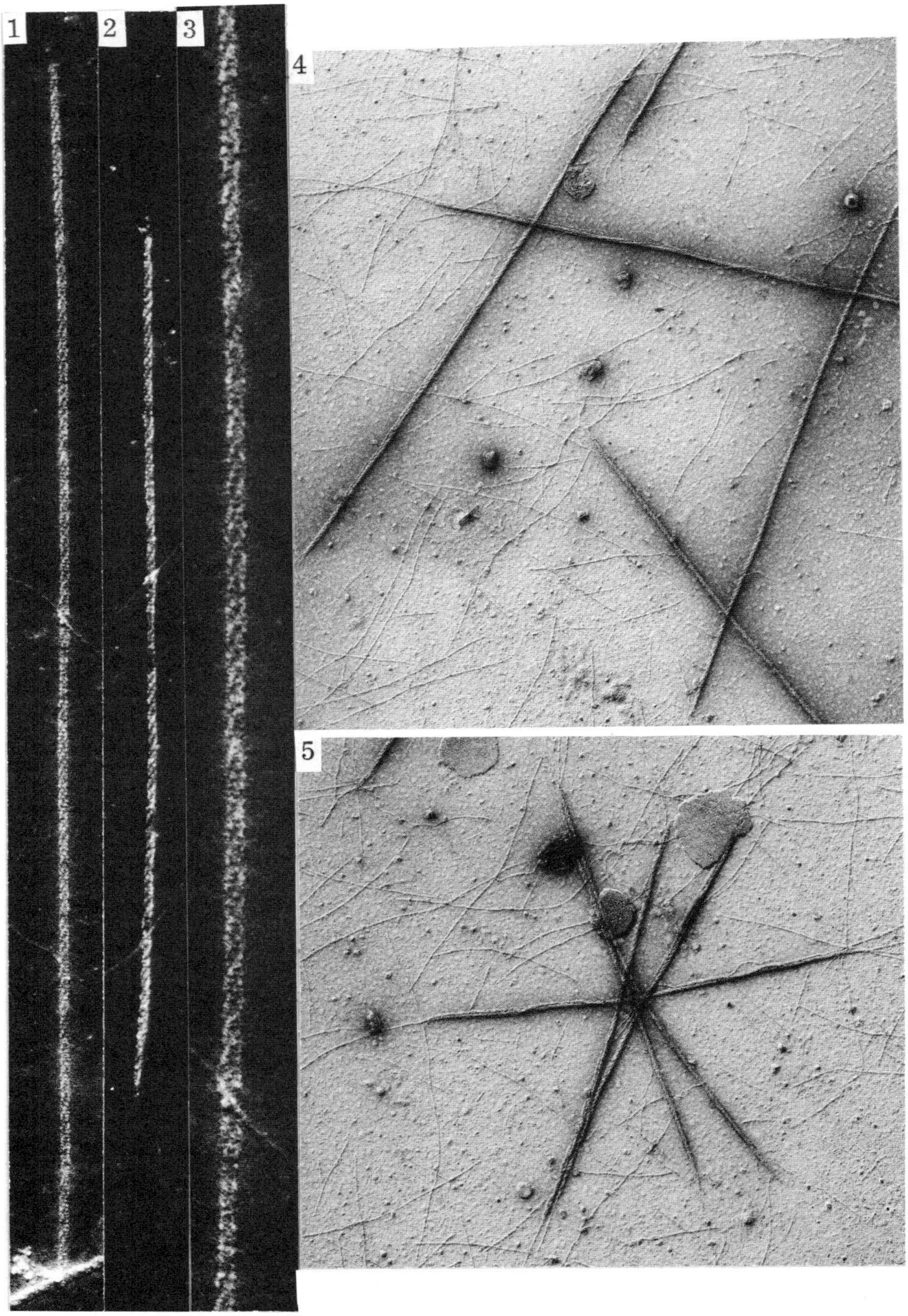

Figure 4.9. Shortening of thick filaments in *Limulus* skeletal (telson) muscle; this phenomenon occurs in the natural condition but has not yet been found in any other species. (*1*) Isolated whole thick filament from relaxed muscle in ATP-containing solution. ×41,000. (*2*) Same as (*1*), but contracted by addition of 2 m*M* calcium. ×41,000. (*3*) Higher-magnification view of one half of (*1*) shows spiral arrangement of myosin molecules in filament. ×90,000. (*4*) Relaxed filaments negatively stained and also shadowed with carbon platinum. ×18,000. (*5*) Same as (*4*), but contracted with 2 m*M* calcium. ×18,000. (All courtesy of Drs. Rhea Levine and Maynard Dewey.)

et al., 1978). Those parts of a myofilament cluster that happen to get the benefit of the earliest rises in Ca^{2+} contract first at the expense of their neighbors, which are stretched. Then a slow period of readjustment follows, early inactivations blending with some late starts. Heterogeneity of local activation is acute in all graded contractions, which are the mainstay of functional control in all arthropod muscles and in vertebrate slow-type cells. The inequality is such that some sarcomeres, or parts of sarcomeres, shorten extensively, beyond the supercontraction stage when this is feasible, at the expense of neighbors that are stretched. Elastic elements within the individual sarcomeres play a major role in distributing the rise in force and in subsequent shortening.

5

Excitation

The state of the skeletal muscle reflects faithfully the state of its motor centre. C. S. Sherrington†

5.1. General Features of Neuromuscular Transmission

There is remarkable diversity in details of neuromuscular synaptic transmission, not only between organisms and muscles but between individual cells in the same muscle. Much of the information available is best covered in the context of the special problems of particular muscles of individual species, as is done in Chapter 12. Here the topic will be introduced and some of the more general aspects presented. The task would have been simpler had muscle cell membranes had similar properties, but this is far from being the case. It is abundantly evident that variations in the properties of ion channels, and in the relative numbers of these channels present in the synapse have played major roles in the adaptive radiation of most phyla, in addition to diversity in innervation and in presynaptic events. Specific excitability features for different muscle cells are also described in Chapter 12, and some are described in considering general aspects of muscle-cell diversification in Chapter 8.

Muscle cell membranes either produce active electrical potential changes spontaneously, in which case they are termed *myogenic* or only following *chemical activation*; the latter includes the majority. Natural substances that initiate muscle membrane activity range from blood-borne *hormones* released at distant sites to *neurotransmitted substances* intimately released a few nanometers away from the muscle membrane by a specific excitatory nerve cell. Myogenicity ranges from extremely slow, even circannual, through circadian and 40–60 min bursting, as appears in some guts, to a second or less in faster hearts. The *primary electrogenesis* in myogenicity is always a regularly recurring spontaneous slow fall in membrane potential that is termed the *pacemaker potential,* because its frequency determines the repetition interval. This may in turn generate *secondary electrogenesis* if the membrane is capable of this and if a threshold voltage is passed. Primary myogenic electrogenesis is especially sensitive to *modulation,* that is, being affected by other naturally produced (and synthetic) substances. Modulators alter the frequency with which either the primary or the secondary events or both occur, increase or decrease, and/or may alter their amplitude, rate of rise, rate of fall, or overall shape. A modulator may be a hormone, a locally released neural transmitter, or any of a number of local factors such as pH, temperature, specific-ion concentration, or product of neighboring-cell action, including damage that releases histamine.

Neurotransmitters act on myogenicity either by causing the primary event to occur early (excitation) or later (inhibition), and by facilitating or, alternatively, suppressing the secondary electrogenesis. The action of a neurotransmitter is simpler in a neurogenic

†C. S. Sherrington, *Q. J. Exp. Physiol.* 6, 251–310 (1913).

system because the transmitter is either itself the excitatory agent for the primary event or an inhibitor of the primary event. There may or may not be a secondary event depending on the intrinsic responsiveness of the membrane and its threshold. Intrinsic responsiveness ranges from nil through all possible shades of *graded responsiveness,* in which response amplitude is proportional to the magnitude of the primary event and/or its rate of change, to triggered *all-or-none responsiveness.* In the latter, a self-regenerative event of uniform amplitude, duration, rates of rise and fall and shape (under constant conditions) is evoked.

Skeletal muscle cells rarely have myogenicity, but even if they do, they are in addition neurally driven, activated via a graded, chemically excited *primary response.* The vertebrate skeletal neuromuscular junction is termed an end plate, so this response is called an *end-plate potential* (EPP). Invertebrate neuromuscular junctions are almost all structurally simpler, lacking the postneural specializations seen in vertebrates, so the response at one of these is termed a *junctional potential* (JP). All known EPPs are excitatory, but the junctional potentials of invertebrates include ones that are either excitatory, excitatory junctional potentials (EJPs) or inhibitory (IJPs). To promote the primary response, hence at a specialized juxtaposition, or *synapse.* All such synapses transmit excitation only from nerve to muscle, never in the reverse direction.

5.1.1. Details of the Synaptic Process

The transmitter substance is released in a single secretory act by the arrival of each single impulse in the presynaptic terminal of the excitatory axon. The released substance has to pass a gap, the synaptic gutter, which is a major variable in fine structure of junctions, ranging from 9 to 100 nm at different types. The gap commonly contains an amorphous, apparently functionless, or perhaps just cementing, material. The postsynaptic membrane forms a simple plain sheet in invertebrates and in fish, but is thrown into folds that are progressively more complex in passing from fish through amphibia to reptiles, birds, and mammals (Figure 5.1). In primitive forms there is merely a groove or gutter in which axonal branchlets fit. In amphibia there are a few long, branched grooves with transverse folds. Reptiles have endings similar to those of amphibia, but with folds running longitudinally as well as transversely. Mammals have more-compact endings in which short branches of nerve take off in all directions. The subsynaptic folds are compound, as in reptiles, but with more transverse folding. The postsynaptic membrane immediately opposite that of the nerve terminal contains receptor molecules that interact with the transmitter. These in turn are attached to ion channels that open when the transmitter attaches. Hence postsynaptic membrane is termed *chemogenic* membrane. The numbers of receptor-activated channels falls off drastically away from the synapse (Figure 5.2), at least in mature animals. The opening action at vertebrate skeletal neuromuscular junctions is brief because the transmitter, acetylcholine (ACh), is very rapidly destroyed by an enzyme, acetylcholinesterase, located postsynaptically.

The synaptic transmitter is released in packets, or quanta, of fairly fixed size, that is, for each an approximately similar number of molecules is released. These combine with receptor molecules situated on the postsynaptic membrane on a random basis before they are destroyed or move away. A packet termed a miniature end-plate potential (MEPP) or miniature junctional potential (MJP) causes a small, 1-2 mV, local depolarization that has exactly the same shape as an EPP, with exponential decay. The MEPP or MJP can be almost exactly mimicked by a very brief iontophoretic release of transmitter

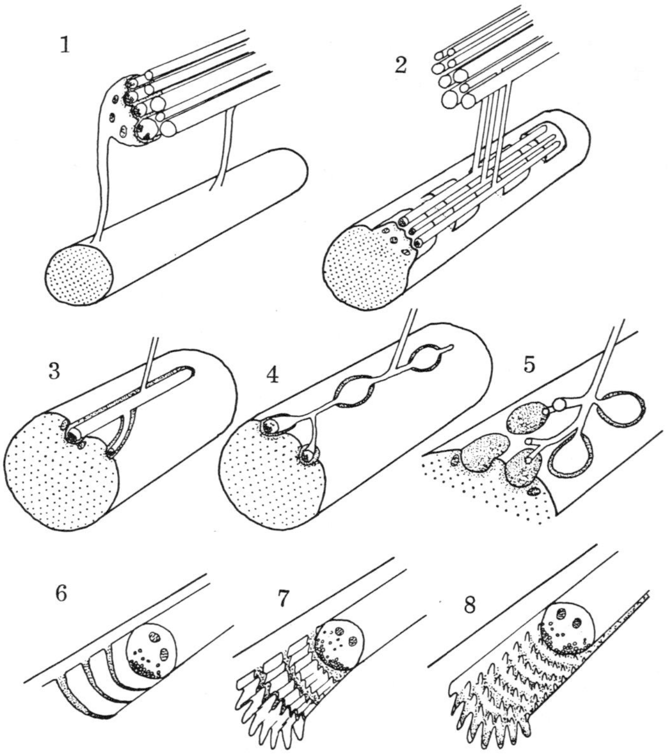

Figure 5.1. Basic types of neuromuscular junction. (*1*) Muscle arms from each muscle cell travel to a motor nerve: each arm expands into a head section that contacts a few motor axons synaptically. Widespread: found in platyhelminths, nematodes, echinoderms, onychophorans, and protochordates. (*2*) Short pillars extend outward from the muscle cell to receive branches from motor axons. Present in some annelids, molluscs, and insects. (*3*) Terminal motor axons and their branches fit into grooves in the surface of the muscle cell. Synaptic juxtapositions occur along the entire length of the axon terminal. The commonest terminal in arthropods, molluscs, annelids, and fish. (*4*) Terminal motor axons and branches make swellings or varicosities that contact the surface of the cell either with or without shallow indentations at each varicosity. Fairly common in crustaceans, insects, some molluscs and fish. (*5*) Termed *en grappe* because it resembles a small bunch of grapes, this is a compact cluster of varicosities present in some kinds of slow vertebrate skeletal muscles, especially those of reptiles, birds, and mammals. (*6*) Groove with transverse infoldings in which a single motor-axon terminal makes functional juxtapositions with each ridge. Characteristic of amphibia. (*7*) Groove with both longitudinal and transverse infoldings. Seen in some reptilian fast muscle cells. (*8*) Groove highly infolded both longitudinally and transversely. Seen in fast mammalian muscles.

substance from a microelectrode tip placed close to a junction, following physical removal of the nerve terminal (Figure 5.3).

A neurally evoked EJP or EPP represents a massive shower of quanta being released all over the terminal nearly simultaneously, the thousands of miniature responses summating. Vesicle membrane fuses with presynaptic membrane, emptying its contents into the synaptic cleft. The vesicle membrane then appears to be incorporated into the presynaptic one near the release site, but a corresponding amount is withdrawn pinocytotically back into the terminal a short distance away, presumably picking up the products of transmitter release at the same time. There should be a large depletion of vesicles in a fatigued

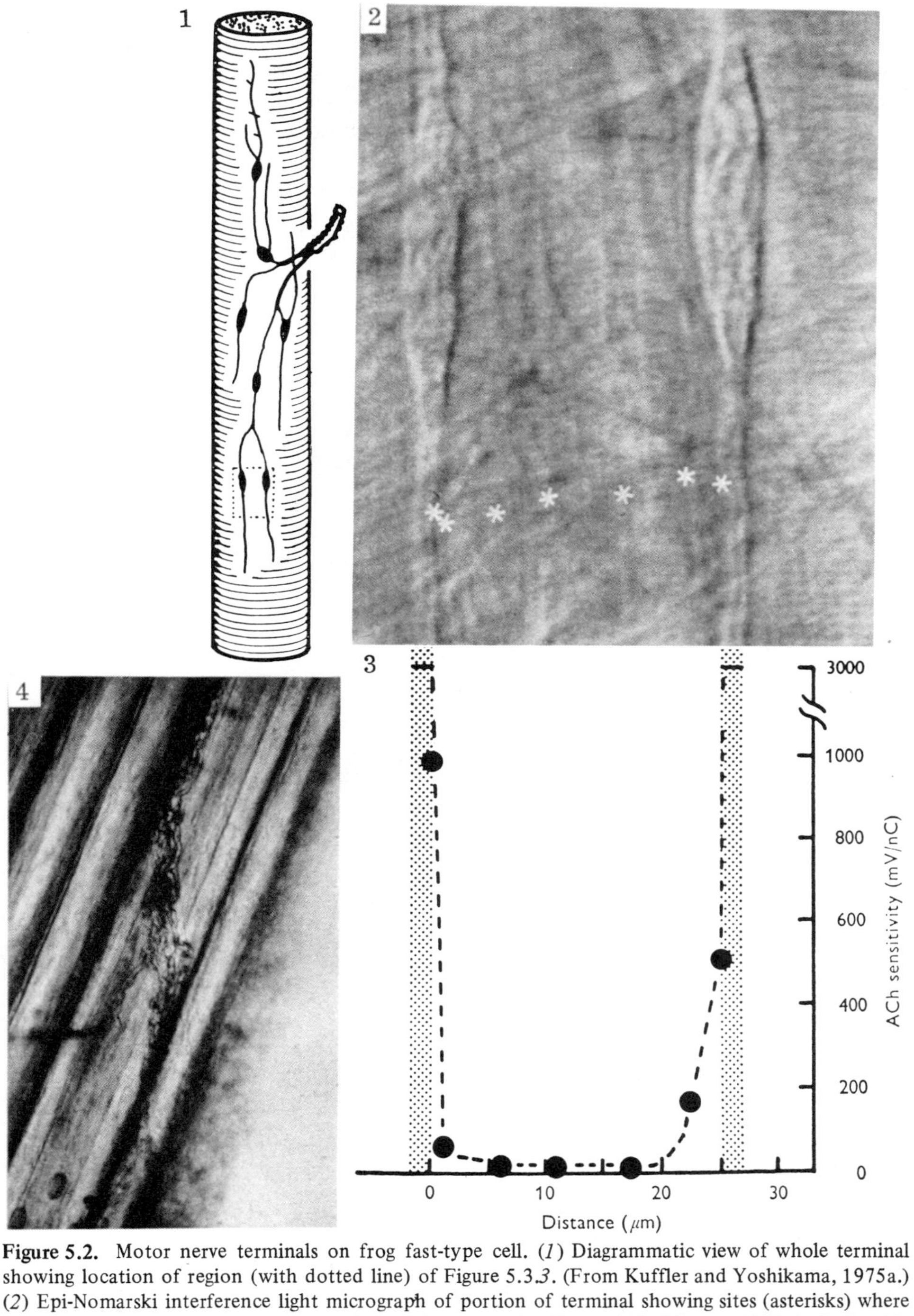

Figure 5.2. Motor nerve terminals on frog fast-type cell. (*1*) Diagrammatic view of whole terminal showing location of region (with dotted line) of Figure 5.3.*3*. (From Kuffler and Yoshikama, 1975a.) (*2*) Epi-Nomarski interference light micrograph of portion of terminal showing sites (asterisks) where acetylcholine (ACh) was applied iontophoretically. (From Kuffler and Yoshikama, 1975a.) (*3*) Graphic representation of sensitivity to ACh at sites indicated in (*2*). The solid circles correspond to the asterisks in (*2*). (From Kuffler and Yoshikami, 1975a.) (*4*) Transmitted light micrograph of whole sartorius junction with intracellular microelectrode inserted. [W. L. Nastuk, The electrical activity of the muscle cell membrane. *J. Cell Comp. Physiol.* **42**, 256 (1953), Fig. 2. Reprinted with permission from Alan R. Liss, Inc. New York.]

synapse, but this has been difficult to detect at most. The only possible explanation is that the reconstitution process closely matches the release, but that the renewed vesicles are empty, or only partially filled. For a garter snake junction, the number of molecules has been estimated (Kuffler and Yoshikami, 1975b) by comparing the local depolarizing action of a single spontaneous quantal event with that evoked by electrically controlled

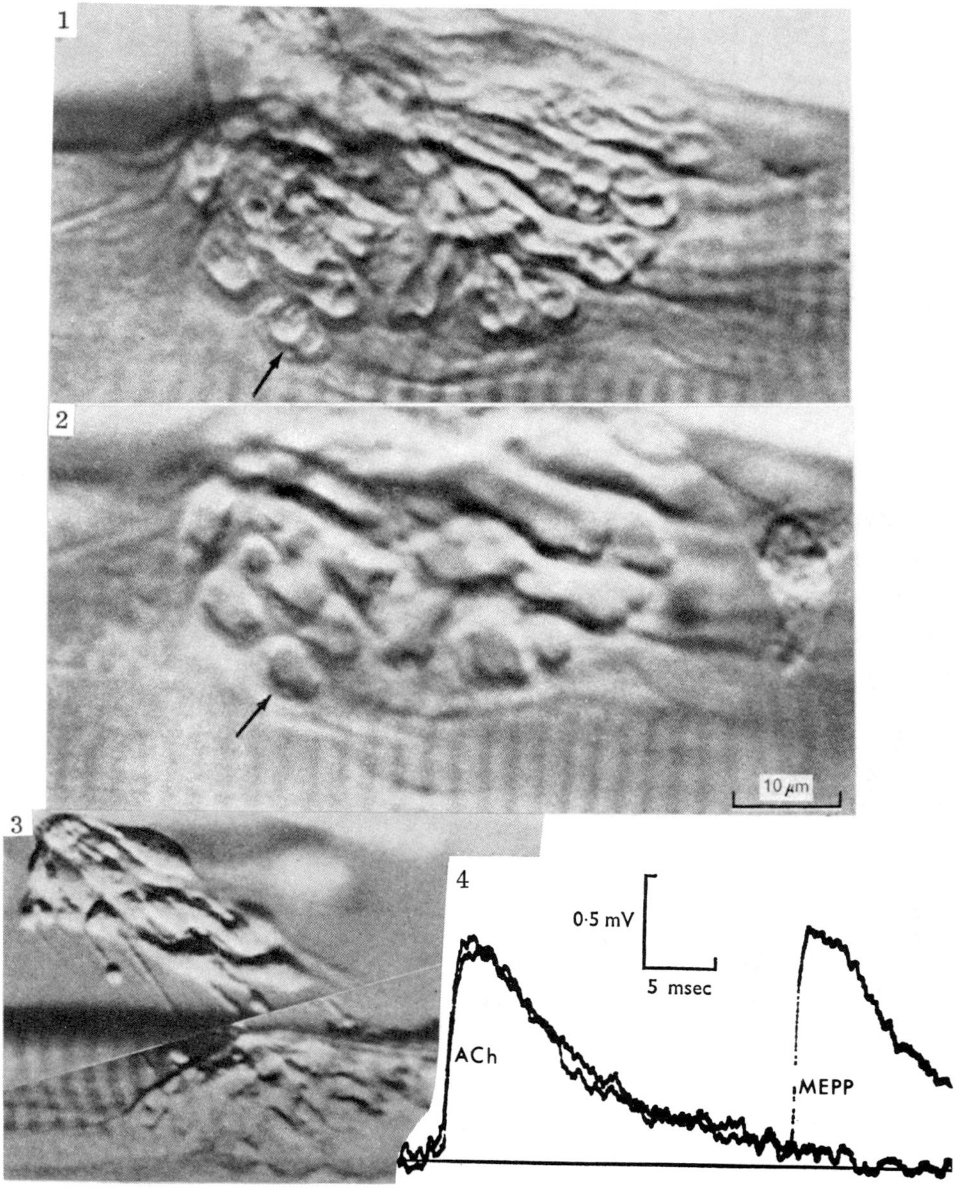

Figure 5.3. Motor nerve terminal on snake fast-type cell. (*1*) Epi-Nomarski interference light micrograph of unfixed terminal. (*2*) Same region as (*1*) after removal of nerve [see (*3*)]. Arrow points to shallow depression in which expanded region of axonal filament was previously lodged [see arrow in (*1*)]. (*3*) Terminal during removal, following treatment with collagenase. (*4*) Responses to iontophoretic application of microdroplets of acetylcholine (ACh) to uncovered postsynaptic region, with same height and time course as natural spontaneous MEPP. (All from Kuffler and Yoshikami, 1975a,b.)

released from a micropipette with its tip placed at a receptor site from which the terminal had been removed (Figure 5.3.4). The output from the pipette was calibrated by collecting the ejected ACh into a tiny droplet of saline suspended in oil and then comparing the depolarizing action of this with that of solutions of known ACh concentration. This method yielded a value of 6–10,000 molecules per quantum. Each molecule that hits a receptor is estimated to result in a flux of at least 3000 univalent ions through the membrane.

The net effect of the ion flux during an EPP or JP is simply to reduce E_m. If there are no electrically excitable ion channels in the nearby membrane, the resulting potential relative to adjacent membrane will simply decay electrotonically, with the characteristic time constant of the membrane τ_m, over a distance determined by the length constant λ. The τ_m is simply $R_m \times C_m$, the effective input resistance times unit length and the capacitance per unit length, respectively.

5.1.2. Initiation of the Secondary, Electrogenic, Response

The vertebrate EPP and larger EJP are large enough to exceed the electrical threshold potential at which contraction is initiated, termed the excitation-contraction coupling threshold (E_C), but only in the relatively minute region close to the ending. It spreads electrotonically from the end plate, with a space or length constant λ. In view of the complex IT systems of most skeletal muscle cells, the determination of λ is complex, but for some muscle cells, a straightforward cable structure can be assumed. As long as cable properties are assumed for the more complex cells for their passive properties only, this simplification does not introduce a very large error. Empirically, λ is easy to determine if two intracellular electrodes at a time can be placed into the same cell, at a few, different, separations. Assuming a single end plate on a cablelike cell, the EPP will have fallen to $1/e$ its peak amplitude after passing along λ nm, since

$$E_x = E_o \, e \frac{-x}{\lambda}$$

where E_x is the magnitude of the potential at a distance x centimeters from that at its source E_o (Figure 000). The value of λ is dependent on the transverse and longitudinal resistivities r_m and r_i as

$$\frac{r_m}{r_i}$$

The specific resistance of 1 cm^2 of membrane R_m is defined by $r_m \times \pi d$, where d is the diameter. The specific resistance of 1 cm^3 of myoplasm R_i is $r_i \, \pi \, (d/2)^2$ so an alternative for λ is $R_m d/R_i$.

The empirical values of λ for muscle cells falls between 0.5 and 2.5 mm, so for a very small organism with minute cells, a single terminal would suffice to elicit contraction. But with the skeletal muscle cells of all vertebrate animals and all except the smallest invertebrates, a single end plate could not evoke useful contractions. In a cell of large diameter – cells up to 3 mm occur in large lobsters and crabs – activation would not even spread around the circumference, so a secondary process is necessary for further transmission of excitation. Two very different methods have been evolved to amplify the local action. One is branching of motor axons so they supply multiple nerve terminals distributed along the entire length and also around the circumference of the muscle cell. The other is a propagated, regenerative electrical response initiated in electrically excitable membrane adjacent to the endings. The muscle membrane of the synaptic region is loaded with receptor/ionophore complexes and has too few voltage-dependent depolarizing channels to be itself the site of origin of an electrogenic response. But current flow into the junctional region from the adjacent membrane, which has few chemically sensitive receptors but many voltage-dependent channels, lowers the potential there past the threshold for opening channels activated by depolarizing current. The basis for excitation transfer is therefore equivalent to local excitation initiated at a small electrode by passing depolarizing current.

The end-plate current (Fig. 5.4.*3*) is the *primary* response in transmission, and sets up an exponentially decaying charge difference locally. The rest of the membrane discharges into the end-plate or junctional region, and as the potential falls, there is initiated an electrogenic secondary response. The flow of ions thereby generated is a mixture of *inward* current (depolarizing) and *outward* current (polarizing), caused by ions' moving down their concentration gradients. Initially the inward current is dominant, but this soon inactivates, and later outward current dominates. Thus the end-plate current, which is equivalent to local-circuit current in nerve, is at first assisted in causing depolarization of the adjacent muscle membrane by the outward current initiated, in a similar manner to the initiation of the nerve impusle.

In vertebrates, multiterminal innervation occurs only on a small number of slow-type cells of fish and amphibia and in mammalian extraocular muscles (see p. 572); peripheral inhibitory axons are unknown. Only minor quantitative differences between species and different muscles of the same species are associated with the physiology of transmission to muscles innervated by a single end plate. A terminal releases sufficient ACh to evoke a large EPP for each single impulse. Although the EPP is principally caused by increased conductance of the postsynaptic membrane to sodium ions, other ion species, calcium and potassium, are also involved. A specific ion channel is called an *ionophore* for that ion. The local membrane potential is driven toward the sum of the contributions of all the activated channels; these contributions are proportional to their equilibrium poten-

Figure 5.4. Basis of neuromuscular junctional transmission. (*1*) Intracellularly recorded response at frog sartorius neuromuscular junction with slightly reduced EPP initiating an impulse after a small delay. (From Fatt and Katz, 1951.) (*2*) Intracellularly recorded response of locust jumping muscle, with graded secondary response initiated by 50 mV EJP. At right is the effect of high Mg, which progressively depresses the EJP, leading to proportionately smaller graded responses. (From Hoyle, 1955a.) (*3*) Frog sartorius junction curarized to reduce EPP size below threshold for impulse: (*above*) potential; (*below*) current, using local voltage clamp. [Reprinted with permission from Takeuchi and Takeuchi, *J. Neurophysiol.* **22**, 395–411 (1959). Copyright 1959 The American Physiological Society.] (*4*) Some MEPPs recorded from frog a sartorius neuromuscular junction cooled to 2.5° C; the MEPPs, are in response to local current through terminal, monitored on lower trace. (From Katz and Miledi, 1967.) (*5*) Statistical nature of MEPPs. Upper histogram shows distribution of sizes occurring spontaneously at frog sartorius junction. Lower histogram shows frequency distribution of the sizes of responses to nervous stimulation at the same junction after block with high Mg^{2+}. (From del Castillo and Katz, 1954.) (*6*) Frequency of MEPPs at rat diaphragm junction as a function of membrane depolarization produced by raised K^+. (From Katz, 1962, based on results of Liley, 1956. (*7*) Naturally small EJP of crayfish opener (*left*) and spontaneous MEJP (*right*). (From Dudel and Kuffler, 1961a.) Ordinate – number of occurrences in particular size range (zero size means no event in unit time); abscissa – size range; I = model unit size, II = twice unit size, III = thrice unit size.

tials and their conductances at each instant in time following activation. The time dependencies of an ionophore are distinctive. For a given ion species, there are different channels with different characteristics of *voltage dependence* (V) and *time dependence* (t). The postsynaptic membrane contains specific transmitter receptors that are coupled to openable ion channels. It lacks the voltage-dependent ion channels that are present elsewhere in the muscle cell membrane. Current flow through the end-plate or junctional region occurring as a result of the end-plate activation is termed end-plate or junctional current.

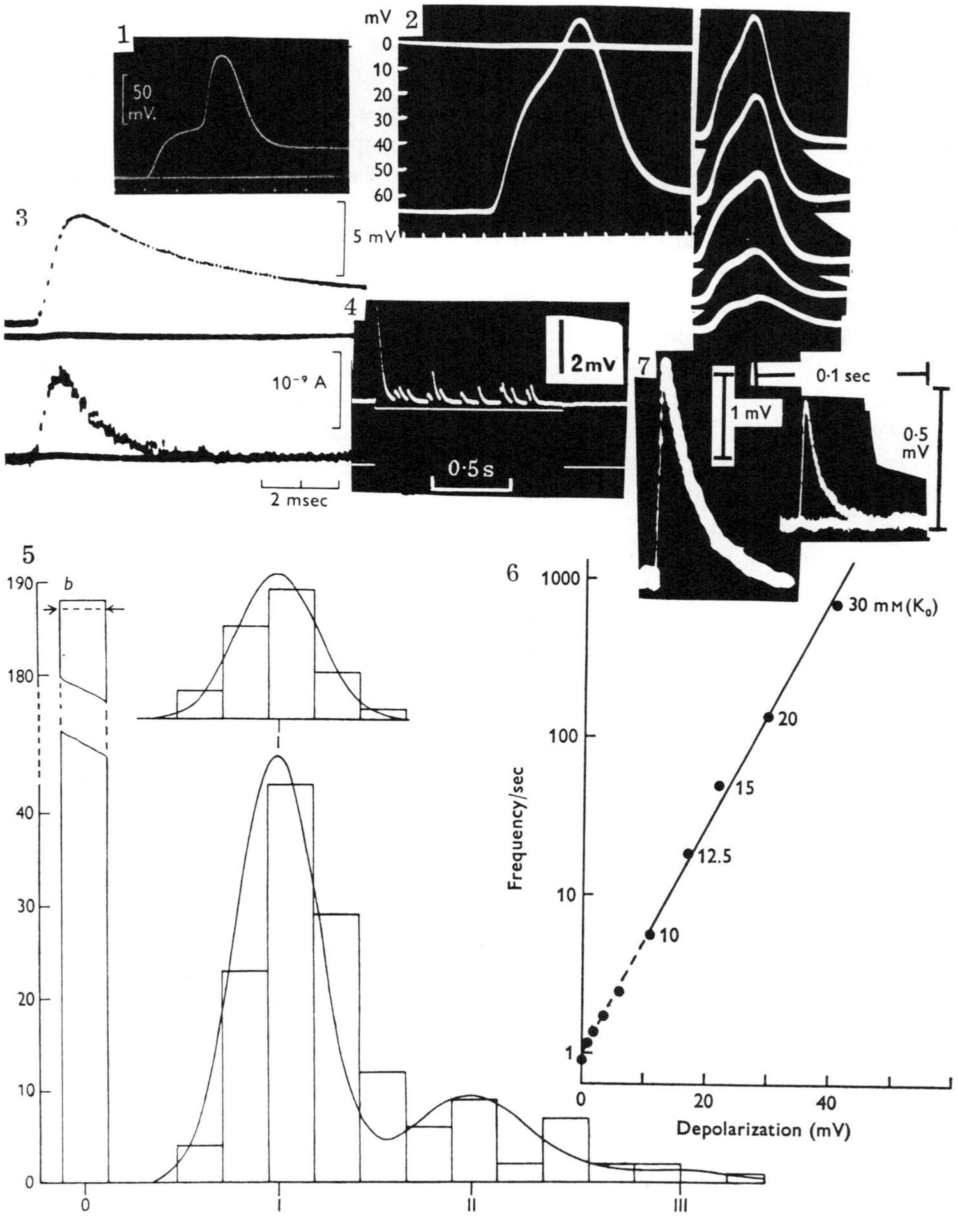

For a typical vertebrate skeletal cell in good condition, with a resting membrane potential of -95 mV, the EPP alone would reach about -20mV. Since the secondary-response threshold is about -56 mV, there is a large safety factor for initiation of an electrogenic response. The only significant variation shown by different vertebrate junctions is in the rate at which the size of the EPP declines during repetition: some junctions decline (fatigue) very quickly, whereas others are relatively persistent. All fatigue very much more rapidly than do most junctions of invertebrates. Most behavioral movement of vertebrates involves only either a low-frequency background or a brief burst of excitation to their skeletal neuromuscular junctions, working within the safety range, so always initiating all-or-none action potentials.

At the majority of vertebrate neuromuscular junctions, the secondary response is an all-or-none action potential because the inward current due to sodium ions, *g*Na, is both large and rapidly rising compared with the slow outward current carried by potassium ions, *g*K. The entire membrane outside the chemogenic membrane of the terminal, which is not also electrogenic, is excited in turn by the action potential. Accordingly, only a single end plate is required, and that is all a majority of vertebrate skeletal cells have.

5.1.3. Peculiarities of Invertebrate and Some Vertebrate Transmission

By contrast to the situation in vertebrates, at invertebrate junctions the secondary response is rarely all or none and is most commonly graded, or even nonexistent (i. e., the cells are totally electrically inexcitable). Such cells need, and indeed all receive, multiple terminals termed *distributed junctions.* This type of motor-nerve supply is termed *multiterminal innervation.* The junctions are often multibranched, with very long, fine, nerve processes that run in all directions along the entire length of the muscle cell and sometimes around the circumference as well, with up to a few hundred contacts per cell when the cell is 50 μm in diameter or more. A few compact junctions also occur, including some with single neural elements parallel to the muscle-cell long axis. Overall, there is a great variety, and individual cell classes within a species can receive markedly different nerve supplies. In none is the innervation comparable to that of vertebrate animals. There is one other general difference between vertebrate and invertebrate innervation: not only are the nerve endings in invertebrates multiple and widely distributed over the surface of each muscle cell, but also more than one motor neuron commonly innervates a single muscle cell, a feature called *polyneuronal innervation.* This feature became functionally feasible secondarily in evolution as a consequence of having first established multiterminal innervation and local, graded activation. On one hand it permitted quantitatively different, but qualitatively similar, actions, thereby independent "fast" and "slow" control of contractions. On the other, it permitted the addition of a qualitatively different antagonistic modulation of excitation, *inhibition.*

A large EJP will exceed the E_C near the terminal, and so will cause a local, graded contraction. If there is a sufficient number of terminals on a single muscle cell, EJPs alone will elicit useful contractions. If not, a secondary electrogenic response is required in addition. The E_C is a significant variable between cells, being near the resting potential for slow-type cells, many millivolts away for the faster ones.

Although IJPs generally cause a polarizing action on the muscle membrane sometimes, particularly in the autumn for crustaceans or after a heavy meal for insects, the equilibrium potential for inhibitory action is less than the membrane resting potential. Then "inhibitory" action alone causes depolarizing JPs. Some cells in the same muscle can give depolarizing IJPs when their neighbors are giving polarizing ones (see Figure 5.9.*2*), and if a sufficient number of inhibitory junctions give large depolarizations the muscle contracts, albeit weakly. But this is rare, and even in this condition, normally caused excitation is opposed by the inhibitory axon except if excitation is very weak, when the inhibitory "excitation" sums with it.

The specific nerve terminals can readily be recognized in TEM sections when a muscle cell receives only one excitatory axon and one inhibitory axon, because the vesicles are rounded in the unstimulated excitatory ending, flattened in the inhibitor, as first recognized by Uchizono (1967 – Figure 5.5). In locusts the vesicles associated with an inhibitory axon are flattened, confirmed by the use of horseradish peroxidase (HRP) marker. A neuron soma is first penetrated in the ganglion with an intracellular electrode and identified by peripheral electrical and mechanical recording. It is then stimulated and the periphery bathed in HRP. The stimulated axon picks up the marker. These studies also showed, however, that vesicles of an excitatory axon become flattened like those of unstimulated inhibitory axons when strongly stimulated (Figure 5.6).

Fast and slow excitatory axons of locusts appear to use the same transmitter substance, L-glutamic acid, and both have rounded vesicles, but those of the fast axon are 69% larger (Titmus, 1981). The inhibitory transmitter substance is very probably γ-amino butyric acid (GABA). For pharmacological details the reader is referred to Usherwood (1978).

The number of excitatory axons supplying a single invertebrate muscle cell ranges from one or two, which is commonest, to nine, and the number of inhibitory axons ranges from none to three. All the axons, no matter how many, tend to branch together, although not all the muscle cells in a particular muscle receive the same supply. On the muscle cell, the innervating axons tend to make their junctional contacts side by side, even very close, although there are exceptions.

The excitatory transmitter substance at neuromuscular junctions of annelids, molluscs, and echinoderms is commonly ACh but is probably L-glutamic acid at all but a very few junctions in arthropods. There is no specific enzyme for destroying the latter: it is attacked by common deaminases; otherwise it simply diffuses away from the junction.

Single EJPs range from ones that are as large as their vertebrate equivalents to ones that are of negligible size (i. e., < 0.5 mV). The former tend to decline quickly in size with repetition, a transmitter-depletion fatigue process that is sometimes referred to as *antifacilitation*. Small EJPs, from neglibible to a few millivolts initial size, grow progressively with repetition, by the functionally highly significant process termed *facilitation*, to a height that may be anywhere from 3 to over 35 mV depending on the junction (Figure 5.7). If the interval between successive stimuli is longer than the duration of individual facilitated EJPs, these *summate* to produce a mean plateau of depolarization. The rates of these dramatic facilitations are the subjects of large variation, being complete in times ranging from only a second, at a frequency of 100 Hz, to an hour or more at 2-5 Hz. The slower the facilitation rate, the greater the resistance to fatigue. We have elicited some insect EJPs at a rate of 30 Hz continually for eight days without any decline occurring. Between the two extremes are all possible combinations of initial size and facilitation rate. Most larger-arthropod skeletal muscles have a complex mixture of EJP sizes and facilitation and fatigue rates in component cells of each muscle, and overall averages for different muscles are markedly different. All these EJPs are caused by the transmitter substance's activating sodium and calcium conductances. The subtle, varied responses are entirely due to differences in presynaptic transmitter-release pro-

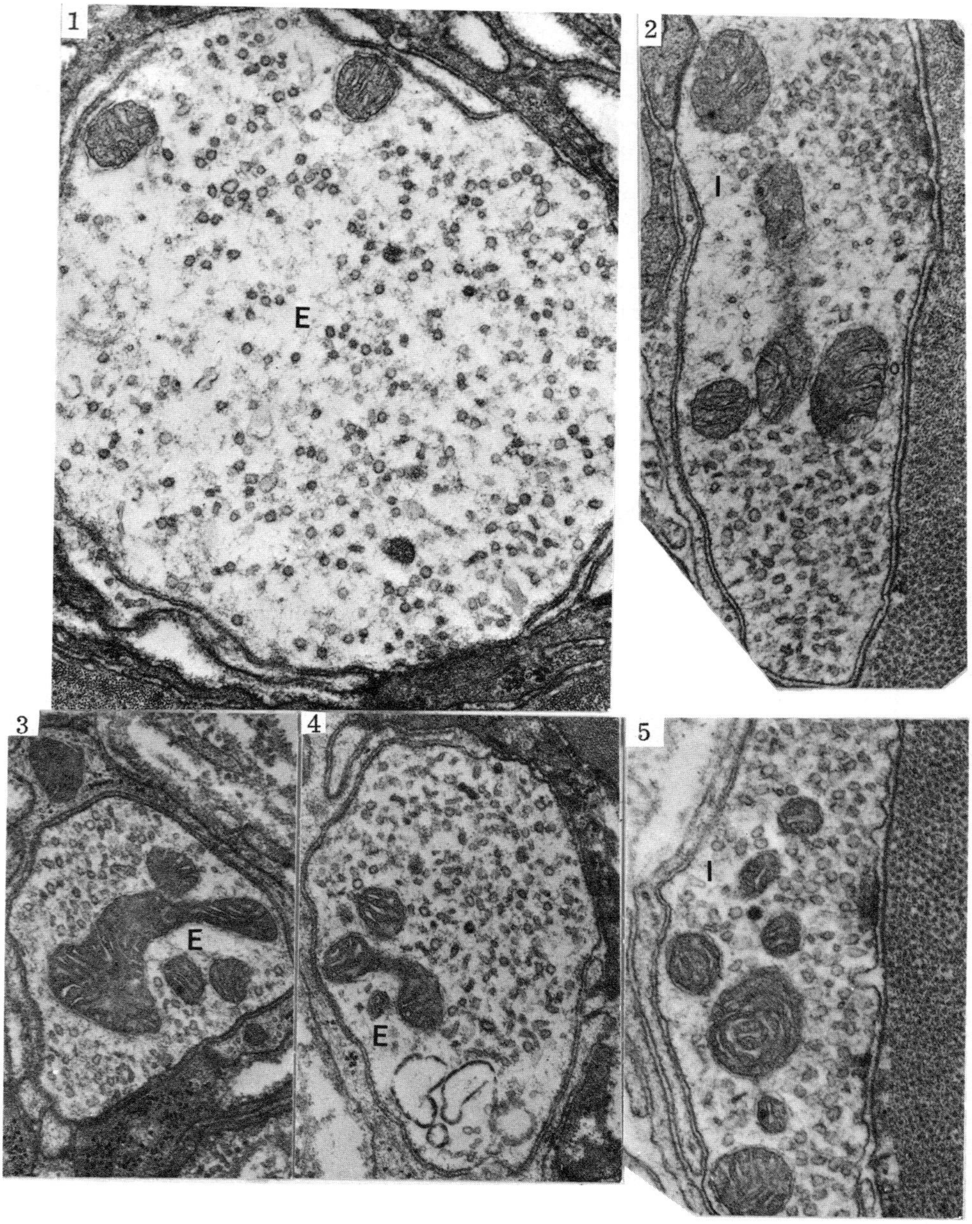

Figure 5.5. Morphologies of excitatory (*E*) and inhibitory (*I*) terminals: Locust jumping muscle. (*1*) Slow excitor to the extensor tibiae (= identified neuron SETi) synapsing at lower right. (*2*) Common inhibitor (CI) ending on same muscle as (*1*). Note lack of postsynaptic granular layer. (*3*) and (*4*) Two profiles of stimulated SETi terminals. Note flattened vesicles, postsynaptic granules. (*5*) Stimulated CI terminal. Note large, decorated omega profiles, where a vesicle is open to the cleft, and complex preterminal bar, which is a probable release site. (All from Titmus, 1977 – see Titmus, 1981.)

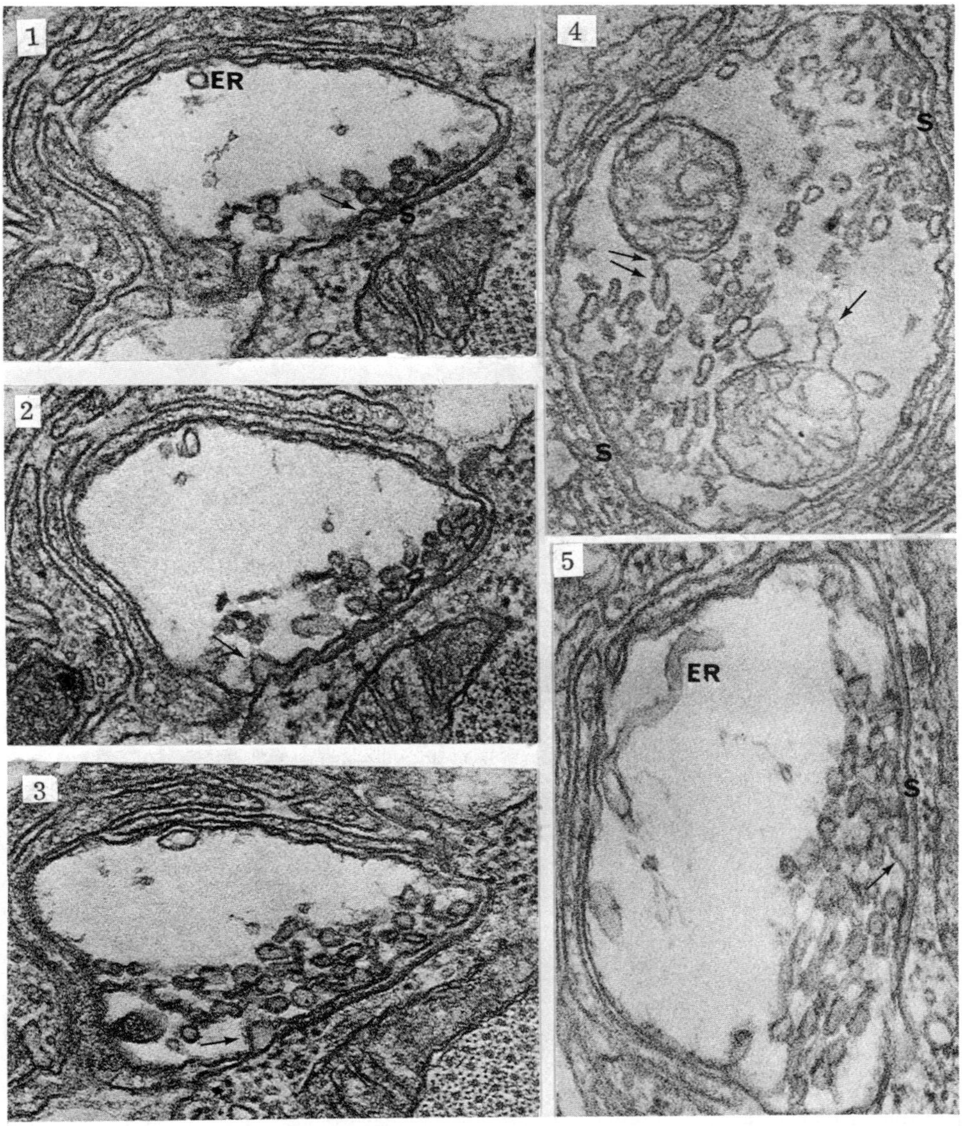

Figure 5.6. Events in strongly stimulated endings. Locust SETi slow excitatory axon to the extensor tibiae (= jumping muscle) stimulated at 100 Hz for 4 h prior to and during fixation. The junction was still transmitting. (*1*)–(*3*) Parts of serial sections with omega shapes in presynaptic membrane. Halos (*arrows*) are believed to represent membrane recovery events. Note the strings of flattened vesicles, longitudinal endoplasmic reticular tube (*ER*). (*4*) and (*5*) Another region showing transversely running ER, strings of flattened vesicles, and apparent production of new vesicles from mitochondria (*arrows*). There are two synaptic sites (*s*) in (*4*) (All ORIG.)

cesses. A major contribution of postsynaptic receptor variations is, however, made in some systems, notably in vertebrate hearts, where cholinergic nerves modulate the heart rate by inhibitory action. The transmitter there is ACh. This may seem paradoxical, since ACh is the excitatory transmitter at skeletal neuromuscular junctions. It has been rendered inhibitory at vertebrate heart sinus venosus cells because of a simple difference in the postsynaptic receptor population: the postsynaptic areas lack ACh-activated sodium

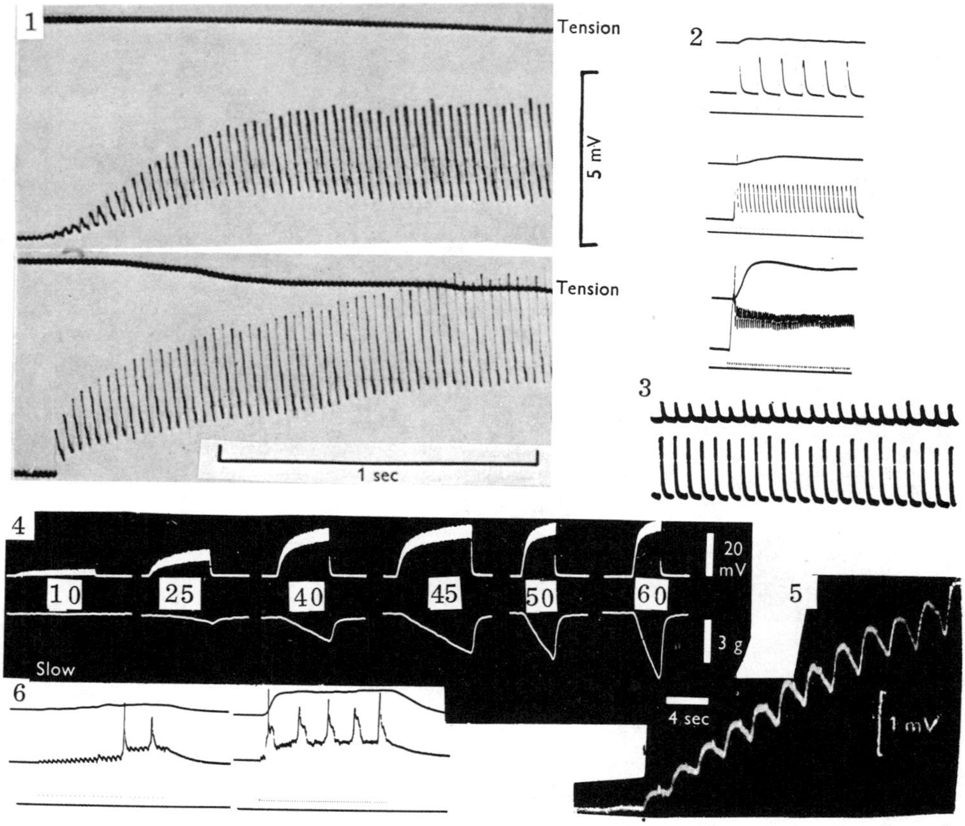

Figure 5.7. Neuromuscular facilitation. (*1*) Intracellular records of EJPs from same muscle cell of crayfish dactylopodite closer in response to the slow axon (*above*) and the fast axon (*below*) stimulated at the same frequency. (From Hoyle and Wiersma, 1958.) (*2*) Slight antifacilitation of EJPs occurring in response to impulses in fast axon to deep-pink portion of power-stroke muscle of swimming crab; tension above, electrical below, at three frequencies. [G. Hoyle and M. Burrows. Correlated physiological and ultrastructural studies on specialized muscles. *J. Exp. Zool.* **185**, Fig. 6a, p. 90, Fig. 4b,c, p. 88 (1973). Reprinted with permission from Alan R. Liss. Inc., New York.] (*3*) Long-term facilitation in stretcher of crab *Grapsus:* top record after 5 min, lower record after 40 min, both at 5 Hz. [Reprinted with permission from R. G. Sherman and H. L. Atwood, *Science* **171**, 1248–1250 (1971). Copyright 1971 by the American Association for the Advancement of Science.] (*4*) Typical crustacean slow-axon responses at increasing frequencies (Hz). Top traces intracellular, bottom traces mechanical from whole muscle. (From Atwood and Hoyle, 1965.) (*5*) Facilitation at paralyzed vertebrate junction, frog sartorius in high-magnesium, low-calcium saline, stimulated at 100 Hz. Reprinted by permission from *Nature* **171**, 1016–1017. Copyright 1953 Macmillan Journals Limited. (*6*) Initiation of graded responses by facilitated EJPs in cell with electrically excitable membrane. The axon to the light-pink portion of the same muscle as (*2*) was stimulated. [G. Hoyle and M. Burrows. Correlated physiological and ultrastructural studies on specialized muscles. *J. Exp. Zool.* **185**, Fig. 6a, p. 90, Fig. 4b,c, p. 88 (1973). Reprinted with permission from Alan R. Liss. Inc., New York.]

channels. Only chloride and potassium are activated, hence the polarizing, inhibitory action, since the equilibrium potentials for each of these ion species is strongly inside negative.

Most IJPs do not grow larger with repetition, but there may be a little summation.

The capability for facilitation is present in vertebrate neuromuscular junctions and is seen quite dramatically after transmission has been drastically suppressed with a combination of high-magnesium, low-calcium saline (Figure 5.7.5). Transmission can be reduced to the point where no more than single quantal units are released occasionally. In normal saline, however, many thousands of quanta are released by each impulse, and the total amount of transmitter released is more than sufficient to saturate the receptors. Even in invertebrates, where neuromuscular-junctional facilitation has been so extensively utilized, the final mechanical response is not necessarily governed solely by the junctional event, since facilitation can also occur in intramuscle-cell excitation-contraction coupling events. It is also further complicated in whole-muscle responses because individual muscle cells or even parts of cells have different initial sizes as well as different rates of facilitation. Furthermore, in gradedly responding cells, large increments of tension occur following each large graded response.

The biophysical basis of facilitation is unknown, but depends upon progressively increasing effectiveness of each impulse in releasing quanta of transmitter substance. There is not thought to be an accompanying progressive enhancement of mobilization of transmitter into the stores (Zucker, 1977).

The fastest type of cell in invertebrate skeletal muscle is also the one with the highest E_C, so a large EJP is required to exceed it. But it is also the cell most likely to be strongly electrogenic, so the EJP is assisted in its depolarizing action. The combined large response always evokes a strong twitch (Figure 5.8). The intermediate cells are always gradedly excitable and give weak twitches that summate at a low frequency to give smooth medium-speed contractions that are highly frequency dependent. The slower cells are generally either electrically inexcitable or very weakly electrogenic. The E_C in these cells is close to E_M and may be greater than it. In this case the resting cell is in a permanent state of weak contraction, sometimes termed a *contracture.* This is responsible for *basic tonus* (see p. 255) in such cells and the muscles in which they occur. The high E_C value explains why these slow cells develop tension with the often very small EJPs that occur in them, and also why tension slowly builds up in the muscles that contain them when they are stimulated at very low frequencies, 4–10 Hz.

Peripheral inhibition has been known in crustaceans since its discovery by Richet (1879) in crayfish and lobster claws. It is essential to normal functioning in these claws because the opener muscle shares a single common excitatory axon with the stretcher. This axon is excited in the familiar startle "defense"/attack reflex response, in which both claws are simultaneously extended, stretched, and opened. In order for the animal to use the claw to pick at food, the stretcher must be inhibited continually and the opener sporadically. This is done by specific inhibitory axons supplying the two muscles. But there are also common inhibitory axons that innervate several pairs of antagonists.

By contrast with the stereotype of vertebrate neuromuscular transmission, the postsynaptic events at invertebrate neuromuscular junctions are as varied as they possibly could be. They have been, and still are, providing a major source for adaptive radiation and evolution of behavior. There is a much larger range in variation for EJPs than for IJPs. The latter, where known, have all turned out to be caused by the release of GABA, which interacts with specific receptors. These are commonly ones linked to chloride ionophores that, following activation, drive the membrane potential toward the chloride

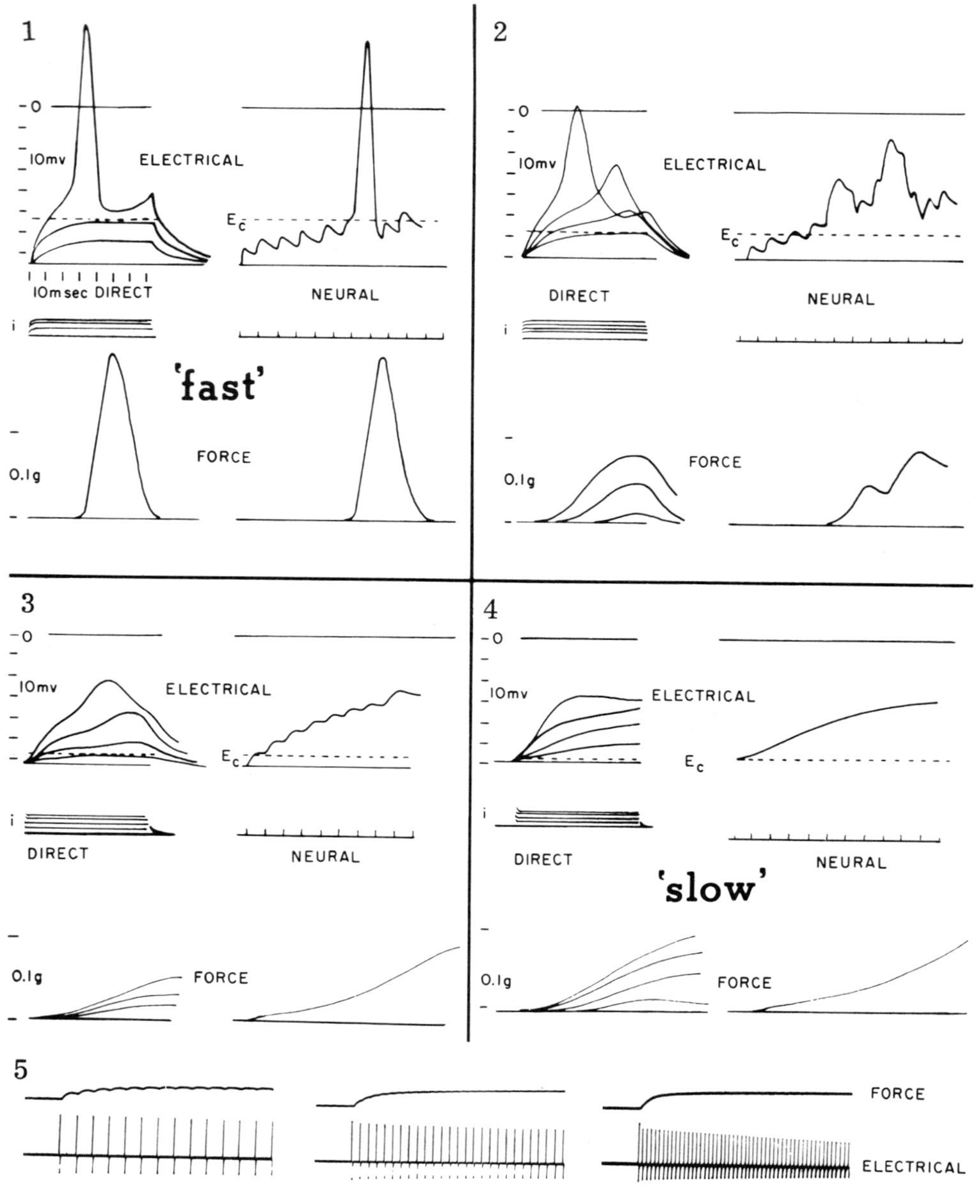

Figure 5.8. Principles of overall neuromuscular transmission events in invertebrate and vertebrates. (*1*)–(*4*) Range of characteristic electrical and mechanical responses seen in invertebrate, especially arthropod, muscles. Individual muscle cells within one muscle, even a muscle comprising but a few cells, may have any of these properties and be either mixed together or segregated into separate bundles. Alternatively, one or another type may predominate in a given muscle, rendering it "fast," "slow," or "intermediate," as a whole. The extreme fast type, with all-or-none action potential is shown in (*1*) and the extreme slow, with no electrogenesis except some delayed rectification, in (*4*). For each the responses to directly applied depolarizing (outward) current are shown at left, and typical EJPs at right, with ensuing mechanical responses below. The E_c is at a lower value for fast than for slow cells. (*2*) and (*3*) Two levels of intermediates, (*2*) being faster than (*3*). G. Hoyle, [Specificity of muscle, pp. 151–167. *Invertebrate Nervous Systems.* University of Chicago Press (1967). Ed. C. A. G. Wiersma.] (*5*) Characteristic vertebrate-muscle-cell responses: all-or-none action potentials, little influence of frequency. Single frog semitendinosus. (From Lüttgau, 1965.)

equilibrium potential. There are also cells in which potassium channels only are opened, and others in which both chloride and potassium channels are activated together. In the fomer, the membrane potential is driven toward the potassium equilibrium potential and in the latter toward the average for both chloride and potassium. The membrane potential toward which inhibitory action drives may be at, or very close to, the resting potential, in which case no potential change results from inhibitory action alone. However, since excitation also causes depolarization when inhibitory action occurs along with excitation, it necessarily opposes the latter's depolarizing tendency. Commonly, the membrane potential is less than the inhibitory equilibrium potential, in which case a (hyper) polarizing current flows, with attendant polarizing IJP. This action is slower and lasts longer than vertebrate excitatory synaptic action because of a lack of a local enzyme that rapidly destroys the transmitter substance. The GABA must diffuse away and be broken down by general deaminases. There is no logical use for such axons except to produce immobility, which might be valuable during molting, or as a defense posture. In addition to excitatory axons insect muscles are also provided with inhibitory axons, termed common because they innervate each of a few muscles, (Figure 5.9) that are known to discharge strongly immediately before each behavioral act and again immediately following one. In insects a major function of inhibition seems to be to promote relaxation of residual tone by way of its polarizing action. However, too little is known of the uses of inhibition during normal behavior of either insects or crustaceans to make worthwhile generalizations. The peripheral inhibitors are capable of producing infinitely subtle reductions in contraction.

5.2. Modulation of Neuromuscular Transmission

Many aspects of synaptic transmission can be altered by external agents, even in miniscule amounts. Sensitivities are difficult to establish with certainty at threshold level, but authentic effects of concentration of a modulator as low as 10^{-12} M have been established for some sites, and even lower values have been claimed. There are two major subdivisions of target areas, which are either *presynaptic* or *postsynaptic.* Not only have animals developed many substances they use within themselves, on their own synapses, *automodulators,* but several animals have also developed ones that affect organisms they attach or need defense from. All of these are harmful, often paralytic or lethal, so are collectively termed *toxins.* Curiously, neither plants nor animals were satisfied with producing but a single toxic compound in their poison exudate once they headed in this direction. Both sea anemone sting material and scorpion venom contain at least four toxic substances each. Materials toxic to vertebrates are generally also toxic to invertebrates and vice versa, often paralyzing synapses, though sometimes in quite different ways in each.

Automodulators are used both to enhance transmission and to reduce it, even to the point of block. Combinations also occur, notably enhancement followed by depression, to achieve a brief burst only of more powerful action. A majority of automodulators act postsynaptically, but some act presynaptically.

The ultimate in modulation of neuromuscular transmission is to block it completely. Curiously, an enormous number of animal species have succeeded in developing agents that attack this most vulnerable site. Even many plants have used it, presumably to prevent being overforaged. Plant alkaloid toxins include curare, strychnine, ryanodine, and erythroidine: all have utilized the postsynaptic membrane, where they lock onto receptor molecules, as do the amaxing α-toxins of snakes, notably bungarotoxin of the krait and cobras. These are all therefore called *competitive* blockers because they complete with natural transmitter for the receptors.

A more subtle action is to block the enzyme that breaks down the transmitter so that natural transmitter hangs around, keeping the junctions depolarized. Many naturally occurring substances, especially ones produced by plants, do this. The best known is eserine, or physostigmine, found in the Calabar bean of West Africa. It was used by natives as an ordeal poison in witchcraft trials. It is known to physiologists because it is an effective antidote after curare poisoning. Curare blocks transmission by reducing the number of receptor sites available to ACh. Eserine attacks the anticholinesterase near

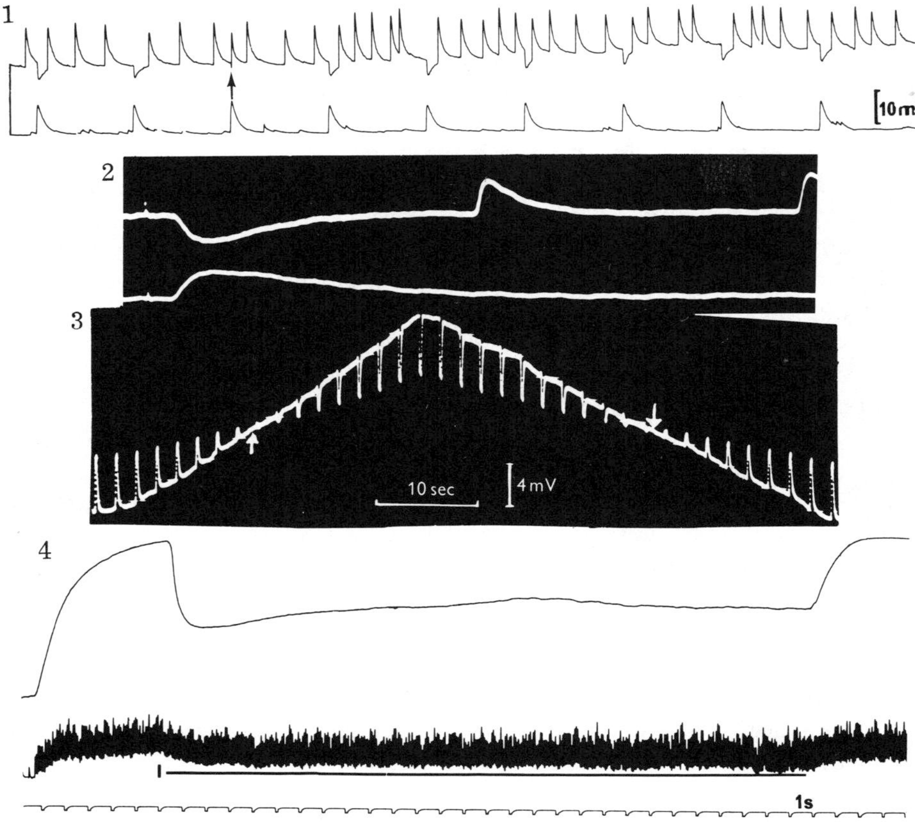

Figure 5.9. Peripheral inhibition. (*1*) EJPs and IJPs from locust coxal adductor muscles, (*above*) anterior, (*below*) posterior. Simultaneous intracellular recordings from two adjacent muscles cells both innervated by the same common inhibitory axon (CI), but different excitors, and bathed in the same saline. One gave a depolarizing, the other a hyperpolarizing, EJP. Coincidence of an EJP and IJP (*arrow*) caused attenuation of the EJP. (ORIG.) (*2*) Simultaneous intracellular recordings from two muscle cells both innervated by a CI, to show longer time course of IJP compared with EJP (*above, right*), regardless of sign. (ORIG.) (*3*) Slow depolarization/polarization sequence superimposed on repeated IJP in a crayfish opener. The IJP reverses at its equilibrium potential (*arrows*). (From Dudel and Kuffler, 1961b). (*4*) Reduced tension (*top trace*) caused in a slow mechanical response to an excitatory axon, the anterior adductor of the coxa (AAdC), and the common inhibitor (CI) when the inhibitor also fired (I-bar). The inhibitor shifted the mean membrane potential away from E_c. EJPs on lower trace. (ORIG.)

these sites, so prolonging the depolarizing action at the few remaining sites available after curare block. The enhancement of the otherwise reduced response can lead to the initiation of impulses again. Eserine has also been used for more than a century in the treatment of glaucoma. The useful action is via cholinergic endings on blood vessels of the eye and ensuing constraint on blood flow.

5.2.1. Actions at Vertebrate End Plates

The neuromuscular junctions of vertebrate skeletal muscles are so keyed into the all-or-none activation of the muscle cells that they are not susceptible to any modulation except block. The one notable natural and functional presynaptic modulator is the magnesium ion, which antagonizes a presynaptic calcium action. Calcium must briefly enter the presynaptic terminal at each impulse in order to initiate as-yet-unknown processes that result in the release of transmitter (Katz and Miledi, 1967a). A magnesium concentration as low as 5 m*M* reduces release to the point where the ensuing EPP may fail to reach threshold for impulse initiation. It is noteworthy that serum magnesium level rises to this level or more in some hibernating animals. It is probably a major means of ensuring autonarcosis to promote hibernation.

5.2.2. Actions at Invertebrate Neuromuscular Junctions

Many invertebrate junctions are enabled to function over a wide range of amplitudes: some have been reduced to a very small initial size to meet subtle functional needs. Those functions rely upon extensive facilitation and summation; therefore, they offer sites at which natural modulation can occur. Individual endings on a single muscle cell are of similar character, but they can be significantly different from each other (Meiss and Govind, 1980): the overall response represents a statistical summation. It is expected that many humoral agents will be found to exert some of their actions via these neuromuscular junctions. One such is octopamine: in some arthropods there are special cells that release octopamine near neuromuscular junctions (Hoyle, et al., 1980). The octopamine diffuses to the junctions, or is carried there by the blood, producing an effect after a minimum delay of 50 ms. It causes a marked increase in the size of the smaller EJPs in the first part of a train, thereby potentiating the contraction initiated by a brief burst (O'Shea and Evans, 1979). This effect lasts for up to about 1 min following a brief release of octopamine. It is only the early EJPs in a long train that are facilitated, later ones are smaller (Figure 5.10).

In invertebrates the neuromuscular junctions are also the targets of many natural postsynaptic blockers. These are present in wasp, spider, centipede, and scorpion venoms, in nematocysts of sea anemones and the notorious siphonophore the Portuguese "man-of-war" *Physalia.* There are also numerous plant-to-invertebrate, especially insect, toxic actions. The alkaloid ryanodine is an insecticide and has been so used commercially. Two principal methods are used on invertebrate targets. The venom of the wasp *Habrobracon reduces* the L-glutaminergic EJP in a manner similar to that of curare at vertebrate junctions (Figure 5.11 – Walther and Rathmeyer 1974). It does not affect the GABAergic inhibitory junctions. By contrast, scorpion and sea anemone toxins produce gigantically *enhanced* junctional responses, including both enlarged EJPs and multiple, prolonged discharges of the presynaptic element. This form of poisoning is equivalent to that caused by eserine in vertebrates.

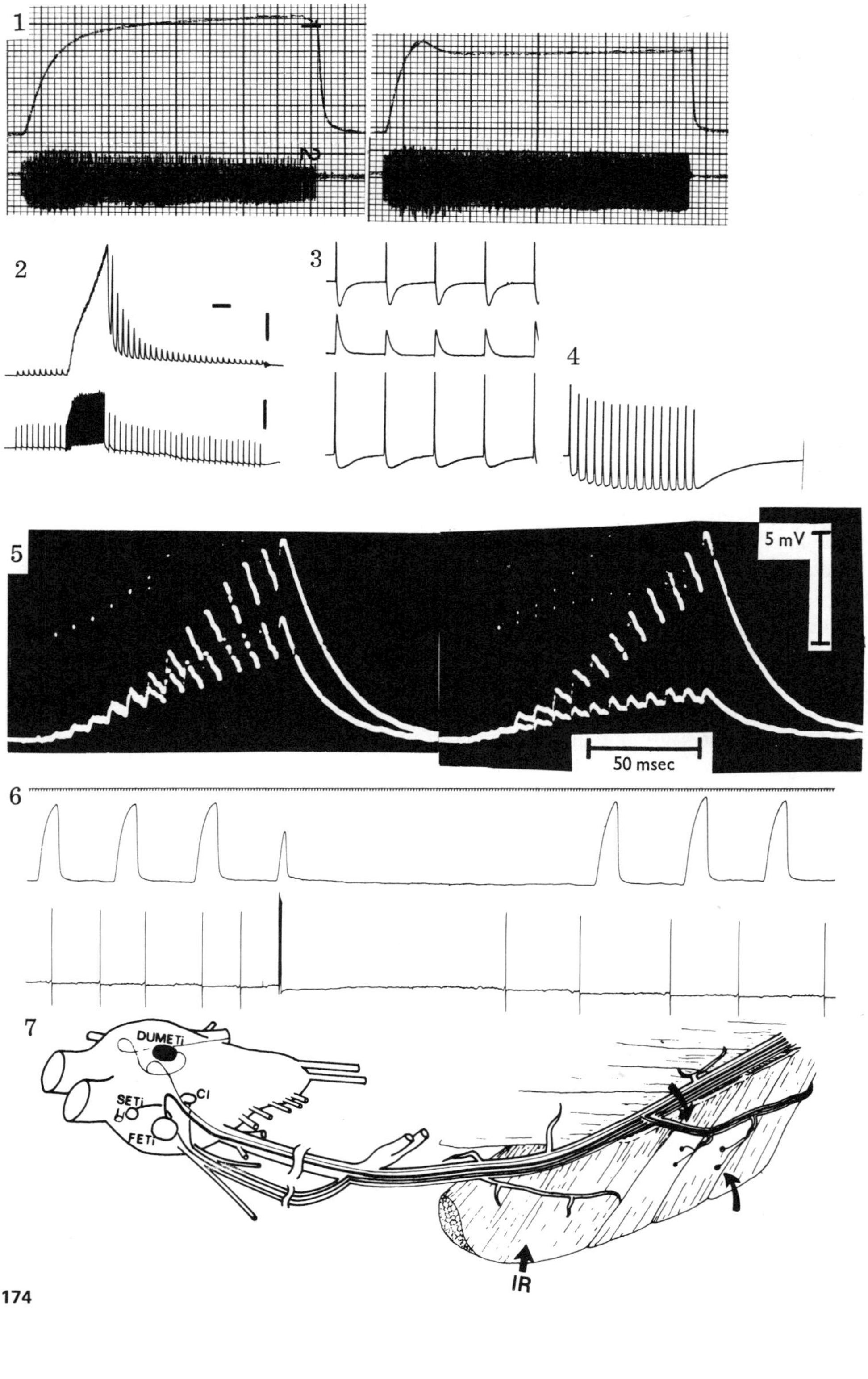
1
2
3
4
5
5 mV
50 msec
6
7
DUMETi
SETi
FETi
CI
IR

5.2.3. Posttetanic Potentiation

Immediately following a brief tetanus, in both vertebrates and invertebrates, there is often a period in which the twitch is enhanced. The enhancement decays with a half-life of a few seconds in vertebrates, to several minutes in some invertebrates. The potentiation is due to a combination of events, one within the muscle-cell, probably a component of the excitation-contraction coupling system, the other at the neuromuscular synapse. In some, such as cat gastrocnemius, the former is predominant; in others, such as soleus, it is the latter (Standaert, 1964): both these potentiations are about 30%. In invertebrates much larger potentiations may occur, and also some that are very long lasting, up to a few hours (Atwood et al., 1975). The largest one observed is in the fast excitor system of the flexor (= closer) of the walking leg dactylopodite of the ghost crab *Ocypode.* Here a gigantic potentiation, of from 500 to 6000% occurs (Figure 5.10.*2*). This potentiation represents a combination of synaptic and contraction enhancements. Even a strictly synaptic transmission event can involve a postsynaptic membrane event (Schulman and Weight, 1976), so many sites may be involved in a potentiation of contraction.

5.2.4. Presynaptic Inhibition and Excitation

An unique form of modulation is known at some arthropod junctions in which there are excitatory (E) and inhibitory (I) axons running very close together. Then the inhibitor may make synaptic contacts with the excitor, and in rare forms the excitor with the inhibitor. In the former, which are always present for example on the opener excitor of the crayfish claw or walking leg, the effect is to reduce the amplitude of the EJP provided the I impulse arrives before the E. If the I arrives slightly late, it increases the rate of repolarization, that is, it takes a bite out of the EJP. These actions produce slightly enhanced mechanical inhibition compared with the action of the IJP on the muscle cell alone. The phenomenon is termed *presynaptic inhibition.* It is interesting physiologically but seems unlikely to be of functional value, nor is *presynaptic excitation* of inhibition (Figure 5.10.*3*), which occurs in both insects and crustaceans. Presynaptic excitation of

Figure 5.10. Modulation by neural action. (*1*) Effects of the neurally produced modulator substance octopamine on EJPs (extracellularly recorded, 30 Hz *lower traces*) and isometric tension (*top traces*) of grasshopper (*Brachystola*) slow extensor tibiae (SETi). (*Left*) normal, (*right*) immediately after adding 1 drop $10^{-6}M$ octopamine in saline. There is a rapid enhancement of EJPs leading to fast-rising tension, but this is followed by a fall in maintained tension. Both effects favor quick, repetitive movements. (ORIG.) (*2*) Dramatic post-tetanic potentiation of dactylopodite fast closer of walking leg of ghost crab *Ocypode*. (*Top trace*) isometric tension. (*Lower trace*) EJPs recorded intracellularly. (ORIG.) (*3*) Simultaneous three-electrode intracellular recording from locust extensor tibiae during SETi stimulation. Two of the cells (*top, bottom*) showed presynaptic excitation of the common inhibitor. (ORIG.) (*4*) Same as (*3*) *lower,* during repetitive stimulation of SETi: note the progressive polarization. (ORIG.) (*5*) Minimum (worst timing), *left,* and maximum (optimal timing), *right,* presynaptic inhibitory effects in crayfish dactylopodite opener EJPs recorded intracellularly. (From Dudel and Kuffler, 1961b.) (*6*) Inhibitory effects of octopamine released from identified octopaminergic neuron, the dorsal, unpaired, median neuron innervating the extensor tibiae muscle (DUMETi) of locust during brief burst. Time trace (*top* is 1/s). Intrinsic rhythm of contraction (IR) is attenuated within 60 ms and suppressed for five cycles. (*Middle trace*) tension, (*lower trace*) intracellular record from DUMETi. [G. Hoyle, A function for neurons (DUM) neurosecretory on skeletal muscle of insects. *J. Exp. Zool.* **189**, Fig. 1. p. 405 (1974). Reprinted with permission from Alan R. Liss. Inc., New York.] (*7*) Location of DUMETi, SETi, CI, and the fast extensor of the tibia and (FETi) neurons in metathoracic ganglion of locusts and grasshoppers, with axonal pathways to proximal part of the jumping muscle. Indicated are site of IR and of octopaminergic endings of DUMETi (*arrow*). (From Hoyle et al, 1980)

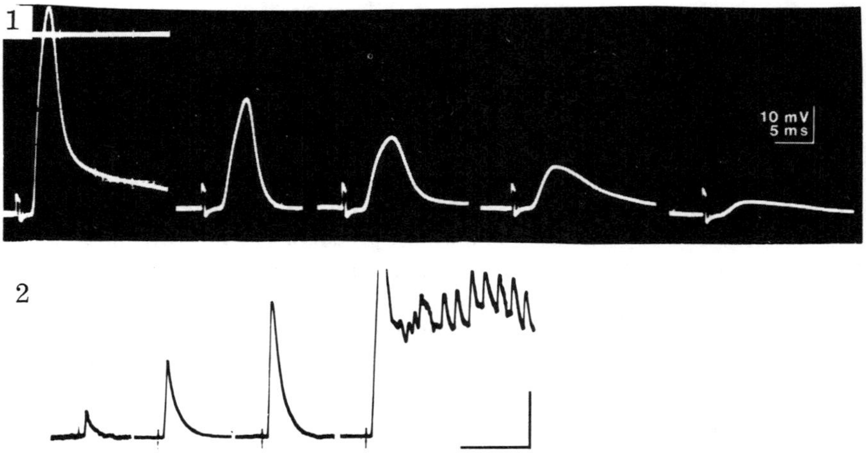

Figure 5.11. Different modes of action of invertebrate toxins on invertebrate junctions. (*1*) Progressive neuromuscular block – total time 6 min – of mealworm *Ephestia* ventral longitudinal muscle by *Habrobracon* (wasp) venom. (Reproduced by permission from Rathmayer and Walther, 1976. Mode of action and specificity of *Habrobracon* venom, in: *Animal, Plant and Microbial Toxins,* Vol. 2, published by Plenum Co., New York.) (*2*) Dramatic progressive neuromuscular enhancement – total time 50 s – of EJP of crayfish opener muscle by a scorpion venom. (Reproduced by permission from Rathmayer et al. 1977. The effect of different toxins from scorpion venom on neuromuscular transmission and nerve action potentials in the crayfish in: *Comp. Biochem. Physiol.* **56C**, 35–39, published by Pergamon Press, Oxford.)

the inhibitor ensures a rapid return of the membrane potential (Figure 5.10.*4*), so it helps promote a series of repeated quick movements. Neither of the two presynaptic actions necessarily affects all the junctions; in a given muscle some junctions can always be found that are not affected.

5.2.5. Modulation at Junctions on Vegetative Muscles

Junctions that are particularly sensitive to a wide range of modulatory actions are found in all vertebrate, and many invertebrate, smooth muscles. Sensitivity to a number of hormones and simpler neuroactive substances, notably adrenaline and noradrenaline and some pentapeptides, has been built into their normal functioning patterns. To a considerable extent, the physiology of smooth muscles can be equated with a study of the modulatory actions associated with them (see Chapter 9). At the beginning of this century, there was a lot of interest in the possibility of the sympathetic nervous system's affecting skeletal muscle tone and performance. The results were unfortunately confused because of indirect actions mediated by altered blood flow, but a direct action of noradrenaline was fairly firmly established. It has been confirmed that at least in high concentration, noradrenaline causes hyperpolarization of frog sartorius fast cells (Bressler et al., 1975) provided there is some external potassium and no ouabain around. Because ouabain blocks sodium transport this action is attributed to the NaK ion pump. A number of drugs may influence membrane potential via this pathway.

5.3. Functional Significance of Diverse Neuromuscular Innervation Patterns

The slower muscle cells of fish, the specific slow-muscle-cell system of amphibia, and some of the extrinsic eye muscle cells of mammals receive distributed terminals on single muscle cells. But these are all exceptions: the majority of vertebrate muscle cells receive but a single nerve ending per cell (Figure 5.12). By contrast, most invertebrate muscle cells receive multiterminal innervation with terminals that are not always well defined but tend to consist of large numbers of very fine branchlets that each touch down on the postsynaptic membrane at many points.

Overall, the difference between the two innervation styles is a fundamental one. This is reflected both in specific transmission events, as described above, and in overall control. Vertebrates rely upon initiating an all-or-none impulse in each muscle cell innervated by a particular motor axon to propagate at a high velocity in each cell, causing a brisk twitch contraction everywhere locally as the impulse spreads. This mechanism is severely restricted compared with that used by invertebrates because the all-or-none quality of the impulse is imposed on a responding unit that is intrinsically capable of infinitely graded responses. The contraction is therefore also all or none, in the form of a brief *twitch.* Twitches sum and fuse smoothly only with rapid repetition of excitation. The summed response is termed a *tetanus.* The smooth tetanic force of a fast-twitch cell is already the greatest the cell can develop: there is no way it can take advantage of the fractional rates it is intrinsically capable of. An intrinsically slower contraction, but with similar excitation characteristics, does permit some tension variation, about threefold maximally, by frequency adjustment above the minimum for tetanus. Therefore, to vary tension, vertebrates have to use different numbers of anatomically based functional units. The total of all the cells innervated by a motor axon, and therefore activated together when the axon fires, is termed the *motor unit* (MU). Variation in force and speed can only be achieved by having large muscles comprising many motor units. This permits varying the *number* of MUs active, that is, by different extent of *recruitment* of MUs. Only to a small extent can variation in frequency of excitation of the units be used in addition to vary force.

Invertebrates, basically by virtue of multiterminal innervation (Figure 5.13), but aided further by polyneuronal innervation and by peripheral inhibition, can utilize the intrinsic capability of infinitely graded force variation that lies within all muscle cells. To do this fully requires graded synaptic events that encompass the full contraction/membrane potential range allowing for summation and facilitation. Then variation in frequency alone provides complete control. The invertebrate method is economical of central connections and is ideal, even essential, for very small animals. The vertebrate method requires complex central nervous system programming to control the many motor neurons per muscle. Their method is suitable, though not essential, for use by large organisms, and vertebrates, in general, are very much larger than invertebrates. There are only a few minor exceptions to these rules. A few skeletal muscles of some vertebrates do possess multiterminal innervation and local activation (see p. 565), and a few muscles of some invertebrates are known that have propagated impulses and/or multiple motor units (see p. 539).

Peripheral inhibition is unknown in vertebrate skeletal muscles, but is widespread in those of invertebrates, in which it has been much touted as a smart device, even leading to the suggestion, because of the integration permitted, that crayfish/lobsters "think in their claws." But evidence concerning how the invertebrates actually use this inhibitory capability is extremely sparse and inconclusive. Certainly the evidence available does not permit the kind of generalization made for frequency-controllled excitation. Central inhibition is also present and definitely used; peripheral inhibition requires subtle timing to be most effective. These matters are discussed in relation to specific muscles in Chapter 12. Peripheral inhibition is, however, used by vertebrates in control of heart and vegetative muscles, which are intrinsically active.

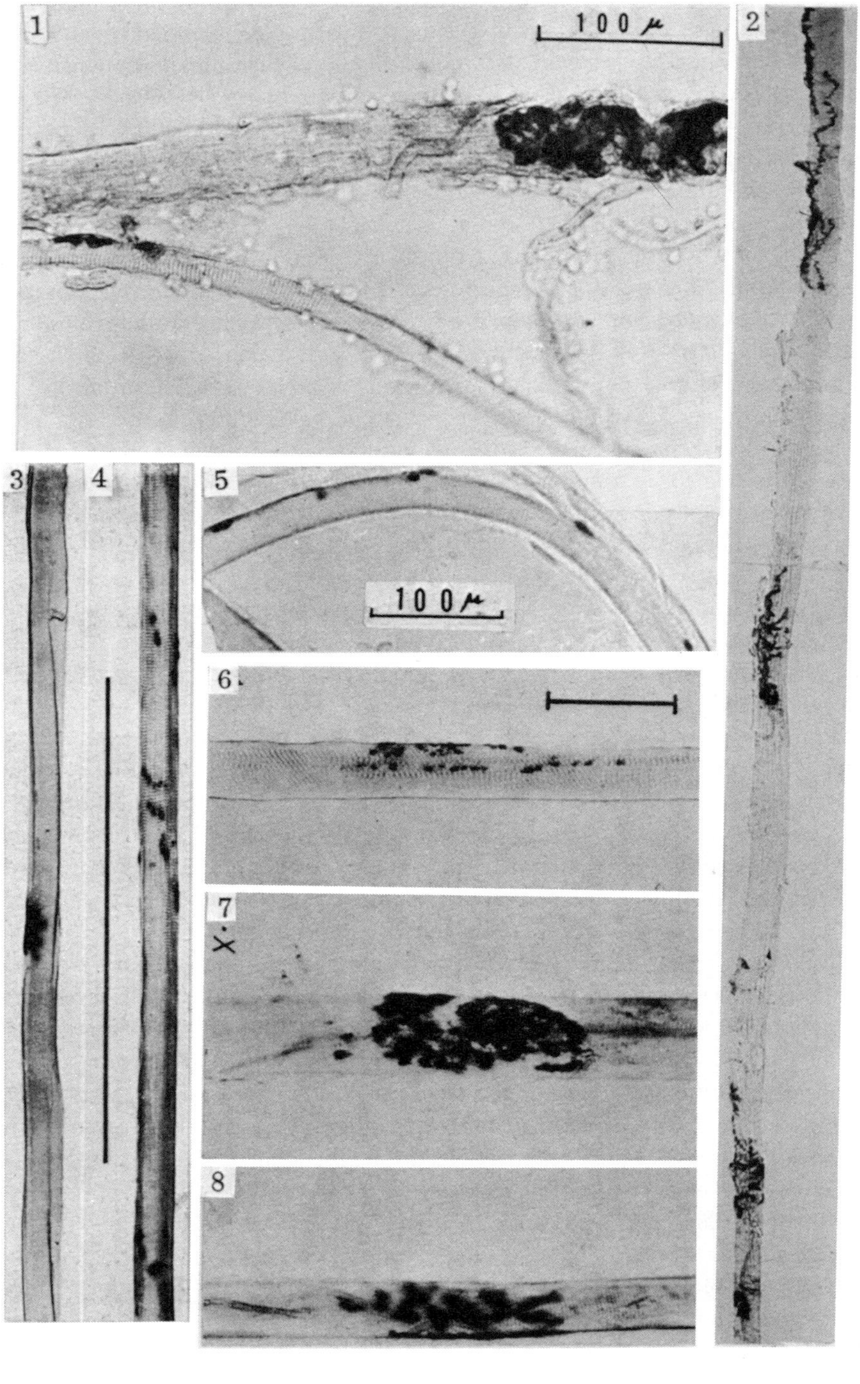
1
100 μ
2
3
4
5
100 μ
6
7
X·
8

Marked differences occur in the details of junctional morphologies of both vertebrates and invertebrates but no functional significance has been detected to be associated with the different forms. In the vertebrates, there is a clear evolutionary sequence in the extent of postsynaptic junctional folding. Yet there is no evidence that the mammalian junctions are more reliable or efficient than those of the reptile, amphibian, or fish! Likewise, in invertebrates, junctions consisting of a simple longitudinal groove may be compared with ones having a mammal-like, complex end plate, or with ones having multiple complex pillarlike muscle-cell extensions. But no advantages of the complex over the simple have been exposed.

All this may merely reflect the state of our ignorance about even so intensively studied a process as neuromuscular transmission. There may be unknown functional features of the structurally simple junctions that have secondarily evolved to give them equal capabilities: there are many different paths to biological success.

5.4. Determination of the Resting Potential

Muscle-cell resting potentials are determined by the summed contributions of liquid-junction potentials of all the ion species to which the membrane is permeable that exist normally on the two sides. These contributions are proportional to the logarithm of the chemical activities, for which concentrations offer a sufficiently close value, times their permeability coefficients. The ions concerned are the cations potassium, which makes by far the largest contribution, sodium, calcium, and magnesium, and the anion chloride. Small additional contributions that are different to evaluate are made by redox activity and by the action of pumps that actively transport some of these ions, notably sodium and chloride, across the membrane. Some actual ion concentrations are given in Table 5.1., together with their equilibrium potentials.

In spite of its complex structure for determination of equilibrium-state liquid-junction electric potentials, a muscle cell behaves as though it is just a bag of electrolyte surrounded by a semipermeable membrane. The principal ions all behave essentially as if in free

Figure 5.12. Some examples of neuromuscular junctions on slow and fast vertebrate muscle cells. (*1*) Extraocular fast and slow cells of Rhesus monkey. [W. Zenker, and H. Anzenbacher. On the two different forms of myo-neural junction in two types of muscle fiber from the external occular muscles of the rhesus monkey. *J. Cell. Comp. Physiol.* **63**, Fig. 2, p. 275, Fig. 9a,d, p. 280 (1964). Reprinted with permission from Alan R. Liss, Inc., New York.] (*2*) Extraocular slow cell of guinea pig. [A. Hess, The structure of slow and fast extrafusal muscle fibers in the extraocular muscles and their nerve endings in guinea pigs. *J. Cell. Comp. Physiol.* **58**, plate 6, facing p. 78 (1961). Reprinted with permission from Alan R. Liss, Inc., New York.] (*3*) Fast chicken (PLD) cell. (From Ginsborg and Mackay, 1961.) (*4*) Slow chicken (ALD) cell. (From Ginsborg and Mackay, 1961.) (*5*) Slow cell of superior rectus of rhesus monkey. [W. Zenker, and H. Anzenbacher. On the two different forms of myo-neural junction in two types of muscle fiber from the external occular muscles of the rhesus monkey. *J. Cell. Comp. Physiol.* **63**, Fig. 2, p. 275, Fig. 9a,d, p. 280 (1964). Reprinted with permission from Alan R. Liss, Inc., New York.] (*6*) Slow cell of ALD of fowl. (*7*) Fast cell of PLD of fowl. (*8*) Slow cell comparable to (*6*) after reinnervation by the nerve to PLD. Figs. 6, 7 and 8 reproduced with permission from Zelená and Jirmanová, slow and fast motor end-plates and their effect upon differentiation of muscle fiber types in young chickens. In: *Structure and Function of Diseased Muscle and Peripheral Nerve.* (Polish Medical Publishers Warsaw Eds. I. Hausmanowa-Petrusewicz, H. Jedrzejowska, pp. 113–117.)

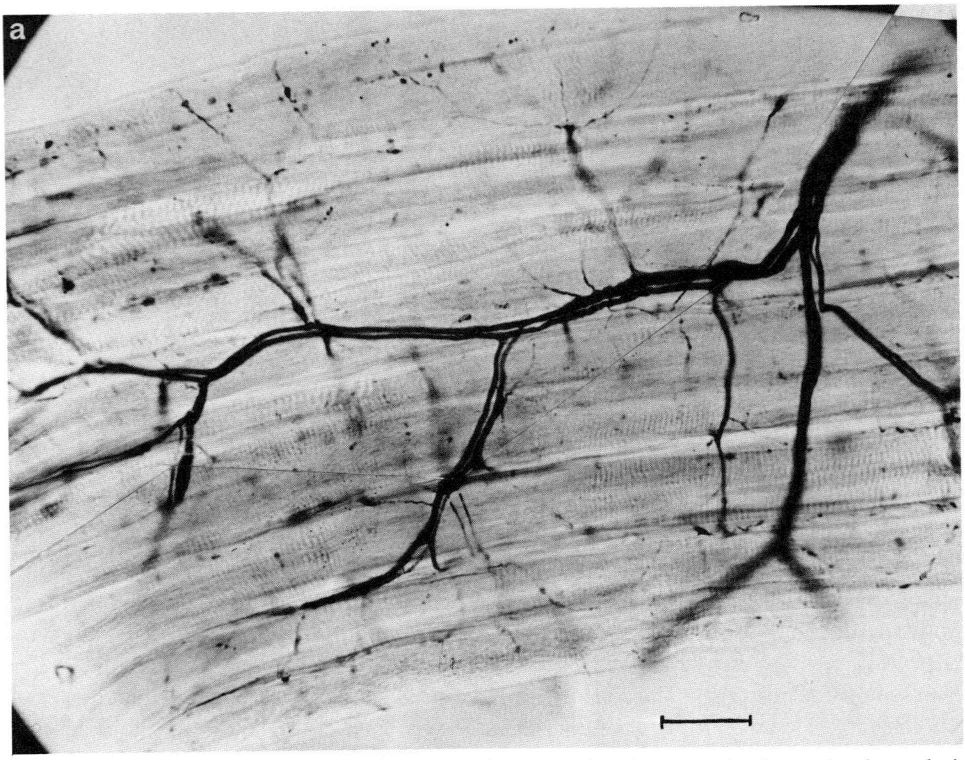

Figure 5.13. Exemplifying innervation of invertebrate muscle cells, especially those of arthropods, is this pattern seen in a coxal muscle of the cockroach *Blattella.* Two axons travel to the muscle, branching together, each supplying many terminals to each muscle cell it innervates (= multiterminal innervation). Some muscle cells receive both axons (= polyaxonal), others only one or the other of the two. One axon is called *fast,* because it causes quick, powerful twitches. The other is called *slow,* because it evokes weak, slow contractions, but only when stimulated repetitively at a frequency of about 10 Hz. Progressively higher frequencies produce progressively faster and stronger contractions. (From Auber, 1960.)

solution (Ashley, 1970). The liquid-junction potential of each is the membrane potential that would occur in the steady state if the membrane were freely permeable to that ion alone, in the concentrations stated. These values are calculated for room temperature, 20°C, by the Nernst equation:

$$E_{\text{ion}} = \frac{RT}{zF} \ln \left(P_{\text{ion}} \cdot \frac{[\text{ion}]_{\text{outside}}}{[\text{ion}]_{\text{inside}}} \right)$$

where R = gas constant

T = absolute temperature

z = valency of ion

F = Faraday's constant (96,500 coulombs)

P = permeability coefficient for the ion being considered

Since each ion makes a contribution, the membrane potential represents the arithmetic sum of their respective contributions. Considering only the monovalent ions this would be

Table 5.1 Examples of Internal and External Concentrations of Common Cations and of Chloride for some Invertebrate and Vertebrate Muscle Cells, Associated Equilibrium Potentials and Experimentally Determined Membrane Resting Potential.

	Internal (mmol/kg H_2O)			External (mmol/kg tissue fluid)			Equilibrium Potentials (mV)			Resting Potential (mV)	
Muscle	K_i	Na_i	Cl_i	K_o	Na_o	Cl_o	E_K	E_{Na}	E_{Cl}	E_M	Reference
Crustacea											
Walking leg											
Carcinus maenas (crab)	120	54	54	12.1	468	524	-58	54	-57	-58	Shaw, 1955
Insect											
Extensor tibia											
Periplaneta americana	110	27	44	27	111	150	-53	36	-57	-58	Wood, 1963, 1965
Locusta migratoria	124	19	40	11–40	103(av)	150	-61	43	-33.4	-60	Wood, 1963, 1965
Sphinx ligustri (moth)	84	21	14.5	50	4	140	-13	44	-57.1	-45	Huddart, 1968
Vertebrate											
Sartorius											
Rana temporaria	125	15	6	2.5	120	115	-99	849	-105	-90	Keynes, 1949
gastrocnemius											Levi and Ussing, 1948
Mus musculus	113	—	—	4.0	—	—	—	—	—	-73	Charlton et al., 1981

$$E_m = \frac{RT}{F} \log_e \frac{P_K [K^+]_o + P_{Na} [Na^+]_o + P_{Cl} [Cl^-]_i}{P_K [K^+]_i + P_{Na} [Na^+]_i + P_{Cl} [Cl^-]_o}$$

For a treatment of the equation including the divalent ions see Mathews and O'Connor (1979) and Sjodin (1980). A simplified equation including calcium is

$$E_m = \frac{RT}{F} \log_e \frac{4 P'_{Ca} [Ca^{++}]_o + P_K [K^+]_o + P_{Na} [Na^+]_o + P_{Cl} [Cl^-]_i}{4 P''_{Ca} [Ca^{++}]_i + P_K [K^+]_i + P_{Na} [Na^+]_i + P_{Cl} [Cl^-]_o}$$

where $P'_{Ca} = P_{Ca} (e^{VF/RT} + 1)^{-1}$
$P''_{Ca} = P'_{Ca}\ e^{VF/RT}$
and V = valency of ion in question

The largest value the membrane potential could have at rest is one due to either potassium or chloride equilibrium potential alone; these are theoretically identical and in practice closely similar. The value for the mean membrane potential is always a little less than the theoretical one because the membrane is never completely freely permeable to an ion, and because K^+ and Cl^- permeabilities are reduced as the voltage rises. The conductances are affected by many factors besides voltage, including each ion's own concentration. Even gK is dependent on K_o (Adelman et al., 1973), creating major effects during activity because of potassium's tendency to get stuck in extracellular spaces. In many kinds of muscle cell, E_m is very much lower than E_K. Many cells of body muscles of soft-bodied invertebrates such as annelids, nematodes, and molluscs and of smooth muscles of vertebrates have resting potential between 30 and 60 mV. This is not a fault, but is gauged either to place the cell in a permanent state of partial contraction or to place its membrane potential so close to the threshold for initiating contraction (see p. 194) that even a very small junctional potential will suffice to initiate a weak contraction.

Finally, cutting off a conductance can also be functionally significant. Detection and specification of ion conductances is complex, depending on a combination of the use of agents that block certain channels specifically (Table 5.2) or that replace an ion without being able to pass through its channels, and a condition in which the membrane potential is held steady at a selected value while the ensuing current is recorded. This requires the use of a feedback amplifier that passes current to oppose any tendency of the membrane potential to depart from the preset value, a method called *voltage clamp.*

In most skeletal muscle cells, active ion-pumping does not directly contribute to the resting potential, but in some invertebrate cells that operate naturally at very low resting potentials, there is evidence that ion-pump actions, which include sodium out, potassium in, and chloride out, make contributions of up to 30% (e.g., Wareham et al., 1974). In cockroach muscle cells, it is estimated that K^+ contributes 42 mV and ion pumps 18 mV to the 60-mV resting potential. Some smooth-muscle cells do not have a stable resting potential; instead the potential oscillates slowly. These fluctuations may be linked to oscillation in the Na^+/K^+ ion-pump activity. All these interpretations assume that the molecular organization of the membrane is static, but this is not necessarily a valid assumption. Slow membrane-potential changes that could be significant in tonic contraction and slow rhythms of contraction may be produced indirectly by cyclic-nucleotide-level changes.

Direct measurements of sodium and potassium fluxes have been possible for giant barnacle muscle cells (Brinley, 1968). As 15°C, 28/pmol/cm^2·s K^+ are entering, compared with 60 leaving, and 49 pmol Na^+ are entering compared with 39 leaving. In these and other crustacean muscle cells, calcium contributes significantly to E_m (Miyamoto, 1964). By contrast, the frog sartorius E_m is very close to E_K, which totally dominates: Na^+ and Cl^- make no contribution (Mullins and Noda, 1963).

The numerous known channels that contribute to the determination of membrane resting and action potentials are summarized in Table 5.2. Only two of the approximately 20 currently known channels appear to be universal: potassium leakage and slow voltage-dependent potassium. An active muscle cell, such as a heart cell, can have as many as seven different channels in one region. Different regions of the same cell may have different channel populations.

5.5. Electrogenesis – Action Potentials

The secondary response of the muscle membrane to current flow and charge transfer associated with the synaptic events is a major variable. It ranges from nil (electrically inexcitable) to all-or-none spiking, with all shades of gradedness in between. All graded responsiveness and a few examples of the all-or-none type that are present in arthropod muscle cells are attributed to activation of a calcium ionophore. A test is provided by barium and strontium, because they cause larger electrogenic responses to occur in cells that have calcium electrogenicity (e.g. Fatt and Ginsborg, 1958; Maruyama and Yamashita, 1962). The lack of dependence of the active responses of arthropod muscle cell on sodium and the involvment of divalent ions was realized in 1957 (Wood, 1957). But it was not until after the introduction into physiology of giant barnacle muscle cells (Hoyle and Smyth, 1963) permitted easy manipulation of the internal contents that it was possible critically to determine the importance of calcium. Hagiwara and Naka (1964) showed that external sodium did not influence the electrogenic response, nor did external calcium, but when they lowered the internal free-calcium concentration progressively with a calcium-chelating agent they got progressively larger spikes. At buffered internal calcium concentration of 8×10^{-8} *M*, the spike finally becomes all or none (Hagiwara and Nakajima, 1966). Activated calcium conductance is a regenerative action, comparable to that caused by sodium activation in vertebrate skeletal cells, but the rate of rise, which is proportional to the initiating voltage, is significantly less than that associated with sodium in nerve cells. Potassium conductance, which rises more slowly still, begins to overtake the rising depolarization generated by calcium current before it reaches its equilibrium value. The rate at which it rises is proportional to the size of the response at any given time following initiation. Hence the graded nature of the calcium response compared with the all-or-none characteristic of the sodium response. The sodium response travels without decrement, whereas a calcium response decrements as it travels except when it is at its largest value, above N 80 mV. When the value is smaller than this, there is a progressive decrement until, if the cell is long enough, it has faded out totally.

A propagated all-or-none action potential occurs in a majority of vertebrate skeletal cells, and it is similar to that which occurs in all axons. It is caused by activation of voltage-dependent sodium channels, which open very rapidly and for a brief time. Even before they close, the membrane-potential reversal has caused voltage-dependent potassium channels to open, adding to electrical and concentration-gradient forces to cause K^+ ions to move out rapidly and thereby recharge the membrane. The characteristics of channel closure are complex, some being automatic after a delay, regardless of conditions,

Table 5.2 Summary Chart of Ion Channels in Membrane

Ion	Current	Dependence	Symbol fast – *f*; slow – *sl*: outward ↑; inward ↓: pump – *P*; chemical – Ch or specific	Occurrence	Function	Blocking Agent	Inert Replacement	Active Replacement
Sodium	Rapid inward	Depolarization Rapid inactivation	↓(V) Na	Axons; vertebrate skeletal, some smooth cardiac muscle	Produces all-or none propagated action potential	Tetrodotoxin Internal zinc slows Saxitoxin		
	Inward	Chemical transmitter is is ACh or L-glutamate activation only	↓(ACh) Na	Neuromuscular synapse	Depolarizing aspect of synaptic and junctional excitation	Vertebrate skeletal – curare Procaine	Choline	
	Outward (*pump* coupled to K^+ in)	ATP driven	↑(P) Na	Axons: vertebrate skeletal, some smooth cardiac muscle	Restores resting state after prolonged activity	Ouabain		
Potassium	Leakage	None	lK	All excitable cells	Determines resting potential, passive current flow			
	Outward	Voltage Slow activation	↑(V, K_o) K	All excitable cells	Recovery after depolarization			
	Outward "C" current	Depolarization Time (delay)	↑(V, t, r_o) K_C	Axon; some arthropod skeletal muscle	Normal rectification reduces response duration	Tetraethylammonium 3-aminopyridine 4-aminopyridine		
	Inward	Polarization	↓(V, r_i) K	Some amphibian skeletal muscle	Anomalous rectification reduces response duration	3,5-diaminopyridine	Cs	Rb
	Outward	Calcium dependent Voltage dependent	↑(V, Ca, K_o) K	Some neuron and muscle cells	Recovery following slow excitation	Cobalt K removal		
	Outward	Chemical transmitter is ACh, L-glutamate, GABA	↑(Ch, K_o) K	Various synapses	Polarizing aspect of synaptic junctional potential			
	Outward "A" current	Voltage	↑(V) K_A	Some molluscan neurons Some arthropod muscle cells	Converts passive to active response when shut off Protects responding to stray excitation	Blocked for a time by prior elicitation		

	Outward oscillatory	Various chemical modulators	∿(Ch) K	Cardiac muscles of vertebrates Some arthropod skeletal cells	Intrinsic rhythmicity cause	Various chemicals		
	Inward pump	Metabolically driven	↓*P*K	Some invertebrate muscle cells and neurons	Contributes to membrane potential	Metabolism inhibitors		
Calcium	Fast inward	Voltage inactivation follows rise in internal calcium	↓*f*(*V*, Ca) K	Most invertebrate skeletal muscle Cardiac muscle Some vertebrate SM Many neurons	Produces graded, decrementally propagated spike	Co, La, Cd, Mn block Ca current		Ba, Sr (enhanced current)
	Slow inward	Voltage, slow rise followed by slow inactivation. May also be chemically activated	↓*sl*(*V*, Ch) K	Some arthropod muscle, some SM, some neurons				Ni, Zn, Mg (reduced current)
Magnesium	Inward	Voltage	↓(*V*) Mg	Some insect muscle cells	Produces graded, decrementally propagated spike			
Chloride	Outward (Cl^- moves in)	Chemical transmitter activation only	↑(Ch) Cl	Synapses and junctions	Polarizing aspect of synaptic and junctional potentials	Some by picrotoxin		
	Fast outward	Voltage dependent	↑*f*(*V*) Cl	Some gut muscle cells Arthorpod ITs	Produces graded, decrementally propagated spike		Sulphate	Br, I, SCN NO_3
	Leakage	None	1 Cl	Some vertebrate skeletal muscle	Contributes to passive current flow			
	Inward (pump) (Cl^- moves out)		↓*P* Cl	Some invertebrate muscle		Metabolism inhibitors		

others closing only after conditions have changed. In either case the process is termed *inactivation.* In heat muscle cells (see p. 377), an almost universal feature that must have been evolved independently in various invertebrates and vertebrates, is a plateau at a depolarized level following the peak of the action potential. Ion channels for sodium, calcium, or a mixture of the two cause a rapid rising phase once their threshold for opening has been exceeded. The process starts in pacemakers, where the membrane potential slowly falls spontaneously during a rest period because of slow sodium- or calcium-ion entry. Two principal factors combine to make the plateau: continued high calcium conductance and a delay in the rise of potassium conductance (compared with that in skeletal cells). The plateau serves two purposes: to ensure a strong contraction and to make this a unitary event, that is, unlike a tetanus, because the heart cannot be reexcited regardless of pacemaker rate until after the recovery of excitability. Functionally, this is necessary to give time to fill arterial blood vessels and to refill the heart. Ion channels may turn on virtually instantaneously when the voltage and/or concentration of critical chemical substance have exceeded threshold, or there may be a discrete, even fairly long, delay. In addition, there are variable aspects of closing. An opened channel may close again almost immediately, even though the conditions for opening it remain, or it may close only after a delay, or it may remain open indefinitely until the conditions causing it to open no longer pertain. Either condition, activation or inactivation, may not be achieved again immediately following restoration. The condition in which it is returning to normal is termed *repriming,* and the time required the *repriming time.*

An alternative to the above interpretations of bioelectric events, which are collectively termed the *membrane theory,* was proposed by Gilbert Ling (1962), a major pioneer of intracellular neurophysiology, in a treatise termed *A Physical Theory of the Living State;* he calls the theory the association-induction hypothesis. This has been largely ignored, but rejected on formal ground by Gulati and Palmer (1977). At the present time the membrane theory is essentially unchallenged. The evidence supporting it is overwhelmingly supportive, and the theory is the accepted basis for interpreting both neuronal and muscle cell bioelectricity.

5.5.1. Controllable Ion Channels

All of the five key ions are able to move across the membrane to some extent, which is to say there are pores or channels wide enough for their passage. Whether or not some of these are permanently open or openness is a statistically determined state is problematical, but the net result is the same. What is firmly established is that there are channels of different types with markedly different characteristics, though the number of different types has not yet been firmly established. Some of the ion channels appear to be quite specific, notably the fast sodium channel. But a hole is a hole, and although its electric charge and shape characteristics may greatly favor one particular ion, others of sufficiently small size are likely also to jostle through it. Channels for calcium permit sodium also to pass, and chloride channels permit some potassium current.

There are two basically different classes of channels, those whose properties are markedly dependent on the voltage across the membrane, and those that are independent. Within each of these classes of channels the response may be either time-dependent or independent of time. In addition, a channel may be dependent on another ion. Dependencies are designated by bracketed abbreviations. A voltage-dependent, time-independent channel becomes $(V)_{\text{ion}}$. The best known and most dramatic is the sodium channel of axons, (*V*) Na^+. This is blocked by the puffer fish poison tetrodotoxin (TTX) which was introduced into electrophysiological research by Narahashi et al. (1960). The size of the (*V*) Na^+ channel has been estimated to be 0.3 × 0.5 nm (Hille, 1972) by the use of TTX. The (*V*) Na^+ channels open abruptly at a characteristic membrane voltage, an action termed *gating.* Because electrically charged molecular components of the ion channel must shift one way when it opens and reverse when it closes, and because any charge transfer must lead to current flow, currents called *gating currents* are associated with the acts of opening and closing. These can be directly, quantitatively studied for some ion channels (Stevens, 1979). The channel may simply switch open at one potential and shut at another with or without time delay, or it may have a secondary dependence on concentration of a local ion. The rapid (*V*) Na^+ automatically shuts off after a variable, but very brief, open period even under constant voltage, an event termed *inactivation.* One of the several potassium channels does not open immediately, but only after the threshold voltage has been on for a time. This channel is therefore both voltage-dependent and time-dependent. Time-dependence is designated by *t,* hence this channel is (V,t) K^+. A channel when open may let an ion pass one way across the membrane but not at all, or to a lesser extent, in the other. This is a *rectifying channel* (*r*) either inward (*i*) or outward (*o*). Since potassium is the major carrier of ions during passive current flow, and since the (V, t) K^+ opens at a particular voltage, it promotes delayed inward rectification and can be given the symbol $(V, t)_{r_i}K^+$. (See Table 5.2.)

Another potassium channel does not turn on until the calcium-ion concentration in its vicinity rises to a sufficiently high value. Calcium does enter many cells following their excitation, raising the local concentration and activating this channel, which is therefore a *calcium-dependent* potassium channel, or (Ca) K^+. Caution is needed in interpreting time-dependence because of possible secondary dependencies. An apparent time dependence could be indicating only the time taken for a critical ion, such as calcium, to rise to a threshold concentration internally. There is not yet sufficient knowledge about the channels in diverse cells to decide if there are any possible natural groupings based upon the population of channels and/or their specific characteristics. There is clearly a very large variety of channels. Furthermore, the ion-channel population is subject to trophic influences, particularly in relation to the differences between subsynaptic and contiguous membrane (see p. 398).

5.5.2. Do Delayed and Anomalous Rectifications Serve a Useful Purpose?

The slower responses of the channels responsible for altering inward and outward current flow with time seem contradictory events that might be merely accidental accompaniments rather than evolutionarily selected, functionally useful ones. This need not be so. Delayed rectification enhances the outward flow of current during the repolarization phase of an action potential, so will aid in recovery and repriming. Anomalous rectification will certainly not aid the individual event, but it must serve to reduce the loss of potassium during a long continuous period of activation.

5.6. Effects of Minor Components of Medium on Properties

The natural environment for a muscle cell is a complex inorganic/organic soup that fluctuates in both irregular and regular manners according to time of day, month, and year, immediate diet and exercise, and long-term diet and exercise. Several hormones affect ionic activities of skeletal, cardiac, and visceral muscle. For example, the sodium efflux is inhibited by natural levels of progesterones, but enhanced by either testosterone or estradiol, sufficiently to affect electrical activity (Misaki, 1964). Calcium fluctuations in the medium exert the most significant effects on account of both direct actions on membrane potential, conductances (Kimizuka and Koketsu, 1963), and physical properties, and indirect ones, by altering internal calcium concentration just inside the membrane. Nonelectrolyte penetration into muscle cells is determined predominantly by a combination of the coefficient of distribution of the substance at a lipid-water interface and by the extent to which it interacts with constituents of the muscle cell (Krolenko and Adamyan, 1967).

Minor divalent ions also can greatly influence membrane properties of some muscles. Vertebrate skeletal cells, although not much affected by strontium and barium, as are those of invertebrates, are strongly affected by zinc. The action potential is significantly prolonged and twitch contraction markedly enhanced by 10^{-6} M Zn^{2+} (see p. 205). Voltage-clamp studies on frog sartorius showed that this is due to a slowing of the sodium-ion-channel gating mechanisms to about $\frac{1}{10}$ normal. Zinc apparently binds directly to the gating molecules themselves.

Ionic strength, regardless of specific ion concentrations or tonicity, affects the properties of some arthropod muscle cells. This was discovered when it was found that adding glucose to saline used for barnacle giant cells depresses inward current that is carried by calcium (Maruyama and Lakshminarayanaiah, 1977). The threshold voltage for initiation of an electrogenic response of a barnacle giant muscle cell is related to the total density of all divalent cations bound to the membrane at the time (Hagiwara and Takahashi, 1967b).

Since membranes have an excess of fixed charges, many of the above effects probably are mediated simply by causing shifts in these charges, thereby altering ion-channel characteristics. Even more-subtle trophic influences (see p. 399) mediated by reciprocal neural/muscular chemical traffic on membrane properties are beginning to arouse attention. The resting potential falls very soon following denervation in those skeletal cells that have been examined. This is attributed to loss of a needed trophic influence on the sodium pump, resulting in sodium entry (Wareham, 1978). The loss is counteracted by catecholamines and by cAMP (Bray et al., 1976). This neurotrophic influence is blocked in the intact system by colchicine (Sellin and McArdle, 1977a), indicating that it is transported by microtubules.

5.7. Some Variants in Membrane Conductance Relations

Potassium-ion conductance dominates the resting potential in a majority of muscle cells, but important exceptions are known. Moth muscle cells operate in a hemolymph that contains barely detectable levels of sodium and three to four times the concentration of magnesium that totally blocks synaptic transmission in vertebrates. The ion-channel characteristics are such that dependence of the resting potential on potassium is very slight, no more than is normally so for chloride (Belton and Grundfest, 1962). Insects generally regulate their hemolymph ion levels to only a small extent, so the ion concentrations reflect diet very quickly. There are large daily and seasonal fluctuations. The nerve axons are protected from these fluctuations by a cellular sheath that does regulate ion levels extremely efficiently, providing the axons with a high-sodium, low-potassium medium that permits functioning in the squid/Hodgkin-Huxley manner (Hoyle, 1953b).

Lepidopterans that as adults will feed only sugar water (nectar), for example, caterpillar larvae eating leaves, and so forth, store sodium in special cells underlying the neural sheath, to release it slowly in the adult.

The muscle cells, lacking such protections, just ride along with whatever the hemolymph contains, with their membrane resting potential fluctuating between wide limits, and graded responses varying likewise. Synaptic events enjoy at least some of the same protection as axonal conduction, since the nerve sheath, or lemnoblast, extends over the synaptic regions.

Many arthropod skeletal cells are totally unaffected by TTX and therefore probably have no (V) Na^+ channels at all. Some sodium does enter during action potentials in these cells, probably through the calcium channels through which the depolarizing current flows. At the other extreme are cells with many (V) Na^+ but no (V) Ca^{2+} channels and others, especially cardiac cells, but also some insect skeletal cells, such as those of the stick insect *Carausius* (Wood, 1957), which have some of each.

The various ion currents across the membrane are differently affected by time, voltage, their own concentrations on each side of the membrane, and those of other ions. The two principal contributors to depolarizing electrogenesis, sodium and calcium, are each very differently affected. Sodium current turns off automatically after a very brief delay regardless of conditions. Calcium current is not inactivated until its own concentration has built up on the inside of the membrane (Ashcroft and Stanfield, 1981). The chloride conductance of muscle cells of the elasmobranch stingray *Taeniura* is 10 times that of potassium (Hagiwara and Takahashi, 1974). The pH can drastically alter permeabilities and in particular, shift the anion/cation permeability relations. In barnacle giant muscle cells, *g*K is very much greater than the almost negligible *g*Cl. But at low pH, the reverse is true (Hagiwara et al., 1971b). Effects attributed to a particular ion, especially chloride, could be due to internal pH change, since the ion may enter the cell in neutral combination and dissociate after entry (Sharp and Thomas, 1979). Tonicity can drastically affect conductances. The *g*Ca is reduced in some crustacean skeletal cells by small increases above normal, and doubling tonicity completely blocks contraction (Suarez-Kurtz and Sorenson, 1977).

The channel characteristics can be markedly different in muscle cells lying side by side in the same muscle. For example, slow cells of frog iliofibularis do not show the delayed rectification property that is bestowed on their fast-type neighbors (Burke and Ginsborg, 1956). A well-defined event as measured by biophysical methods, such as delayed rectification, may be caused by quite different underlying ion channels in different cells. In frog fast cells, it is due to change in IT *g*K; it also occurs in crustacean skeletal cells, where it is due to altered IT *g*Cl (Brûlé et al., 1976).

5.8. Cable Properties of Muscle Cells

The dynamic electrical events associated with excitation of a muscle cell, notably the profiles in time of both synaptic potentials and electrogenic responses, are determined in part by passive electrical properties of the combined surface-membrane and IT system membranes. These can be determined independently only by sophisticated electrical analysis, and this has been done for very few cells.

Some of the many diverse electrical properties of different kinds of muscle cells will be described elsewhere, but one example, from the very careful, detailed work of Falk and Fatt (1964) will suffice to indicate something of the extent of the differences in linear electrical properties. They compared the outer-membrane resistance R_m and capacitance C_m, as well as the IT resistance R_{IT} and capacitance C_{IT}, determined by alternating current (AC) analysis, for frog (*Rana temporaria*) sartorius and carpopodite extensor of crayfish (*Astacus fluviatilis*), obtaining these results:

	Frog	Crayfish
R_m	3100 Ω·cm^2	680 Ω·cm^2
C_m	2.6	3.9 μF/cm^2
R_{IT}	330 Ω·cm^2	35 Ω·cm^2
C_{IT}	4.1 μF/cm^2	17 μF/cm^2

Within the same frog species, C_m for slow-type cells from the iliofibularis is three times that of sartorius, and slow-cell R_m is 10 times as much (Adrian and Peachey, 1965).

5.8.1. Influence of the IT System on Overall Membrane Properties

Overall membrane characteristics of all muscle cells having ITs are to a considerable extent dependent on a variety of properties, passive and dynamic, of the tubules that commonly extend throughout the entire cell (Figure 5.14.*1*). Because the tubules are open to the exterior, there is a continual trend toward chemical equalization of their contents and that of the exterior. But because of the physical constraints, the rate at which equalizations can occur is slow in comparison with the surface events, so the two interact. These considerations greatly affect experimental manipulations such as prolonged hyperpolarization. For example, resting potassium permeability falls progressively during the prolonged hyperpolarization of frog sartorius cells. Under some conditions the loss of K^+ from the tubules dominates the turnoff of K^+ currents, under others it is dominated by depletion of sodium from the tubules (Standen and Stanfield, 1979). A simple, dramatic demonstration of an effect of ITs is provided by comparing the time courses of depolarizations produced by sudden immersion in high-[K^+] saline of single cells with intact tubules and ones in which the tubules have been disrupted (Figure 5.14.*2*, Nakajima et al., 1968). The retention of a lower [K^+] in the tubules greatly delays the reaching of a steady condition.

The contribution of ITs to passive electrical properties is enormous. In frog fast cells the apparent overall capacitance is 8 μF/cm^2, but after tubule disruption it is only 2.2 μF/cm^2 (Eisenberg and Gage, 1967). The same result has been obtained from calculating the capacitance from the drastically altered shape of the foot of the action potential (Nakajima and Hodgkin, 1970). A diagram illustrating the electrical equivalent circuit of a muscle cell with ITs is shown in Figure 5.14.*3*. The above values are, however, not surprising when the relative areas of tubule to surface membrane are considered. Peachey (1965) estimated that for a typical frog fast cell the ratio is 4 : 1.

Because of the extent of IT membrane, properties of the whole cell may be dominated by the IT, rather than surface, membrane. For frog sartorius, Adrian and Freygang (1962a,b) deduced that it is in the IT membrane, rather than at the surface, that most of the gK lies and also the changes underlying anomalous rectification. The effect on ion movement of tubules disrupted by glycerol treatment may be used to determine relative contributions of the two membranes. By studying the effect of disruption on ^{42}K flux,

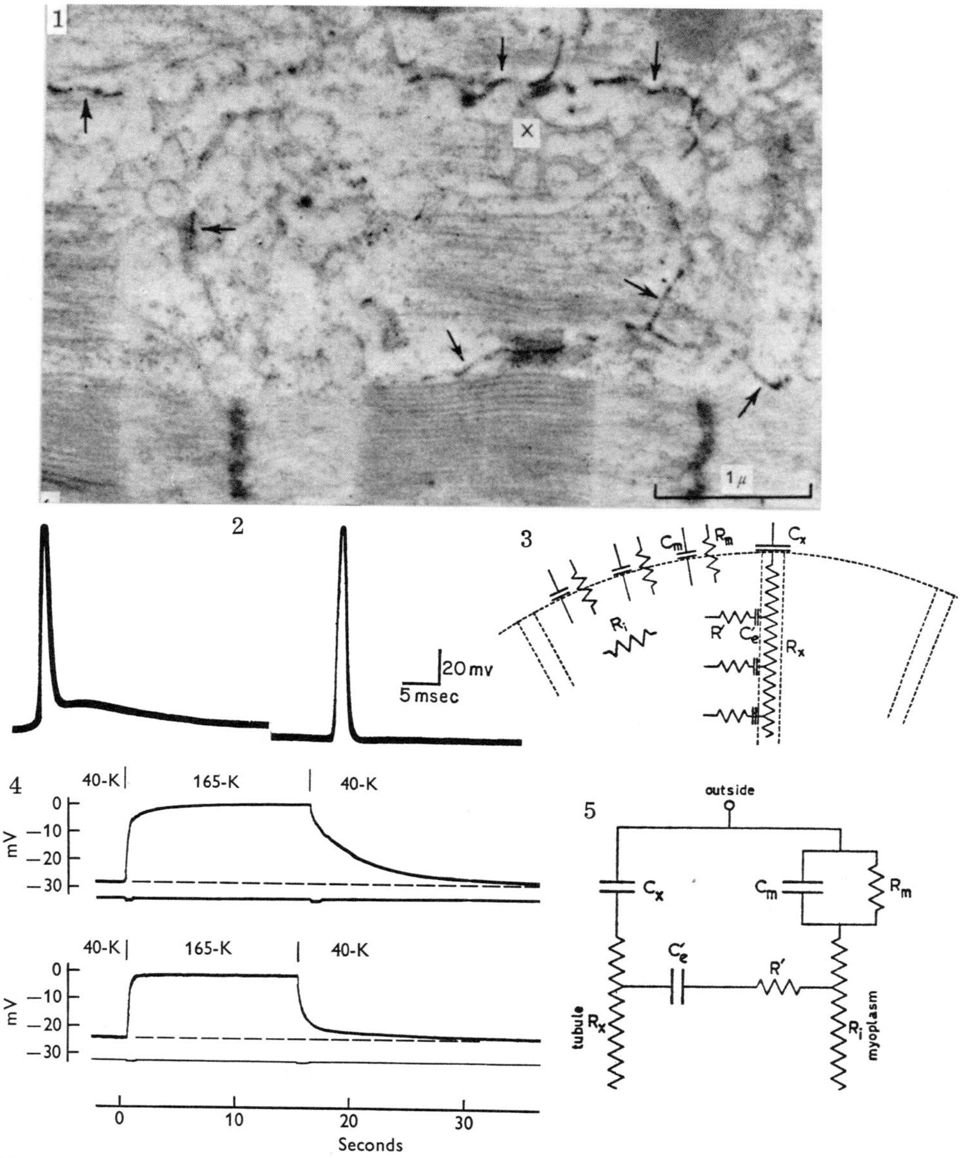

Figure 5.14. Influences of ITs on membrane properties. (*1*) Portions of the IT system of a chicken slow (anterior latissimus dorsi) cell seen in longitudinal ultrathin section EM after prior soaking in saline containing ferritin. Arrows point to portions of the IT network open to the exterior running in all directions within the cell, with major radial and longitudinal branches. Ordinary sarcoplasmic reticulum (*X*) contains no ferritin. (From Page, 1969.) (*2*) Action potential before (*left*) and after (*right*) disruption of ITs of a frog cell by glycerol treatment. (From Nakajima et al., 1968.) (*3*) Locations of electrical properties associated with the ITs (*broken lines*). ITs have an internal resistivity R_x and a capacitance C_e' across their walls. A large capacitance C_x arises where the IT meets the PM. Resistivity of cell contents R_i is coupled to IT via R' in series with capacitance C_e'. (From Falk, 1968.) (*4*) Changes of membrane potential of single frog muscle cells produced by sudden changes in $[K^+]_o$ in Cl^- -free sulphate solution before (*above*) and after (*below*) disruption of the ITs with glycerol. (From Nakajima et al., 1968.) (*5*) Equivalent electrical circuit of membrane and tubular system combination. (From Falk, 1968.)

Henderson (1970) showed directly that there is a large difference between the two membranes in K^+ permeability. The suggestion that inward rectification lies mainly in IT membrane has also been confirmed (Williams, 1975).

A discussion of the relation between accumulating potassium in the tubules, the rectifier properties, and gK in hyperpolarization has been presented by Almers (1972). There are two components to the change of gK with time, a time-dependent rapid one, complete in 200 ms, followed by a slow, diffusion-limited one with a half-time of 0.4 s, which is determined solely by tubular $[K^+]$.

The effects of electrically charging the IT system are sufficiently great that they need to be taken into account in voltage-clamping experiments. A theoretical treatment has been given by Adrian et al. (1969).

It will be shown later that stretch activates the contractile machinery of some muscle cells but not others. Stretch must alter passive electrical properties, and it also drastically alters membrane potential and properties of some cells. Others, including frog sartorius, can be stretched to 170% l_o without changing resting potential (Ling and Gerard, 1949).

5.8.2. Other Membrane Invaginations

The greater part of the surface membrane of a majority of muscle cells (in which the membrane has been satisfactorily examined) is smooth and relatively unwrinkled. But there are some significant exceptions among vertebrates. Frog sartorius and semitendinosus, for example, have large infolded pockets termed caveolae, numbering 36 per μm^2 in sartorius (Dulhunty and Franzini-Armstrong, 1975). The caveolae areas are not considered to have different properties from those of the smooth parts, but they greatly increase the true surface area over a theoretical area based upon assumption of a single cylinder.

In arthropods, all but the smallest muscle cells have longitudinally running invaginated clefts. The ITs arise from the membrane of the cleft lining as well as from strictly surface membranes. Clefts opening directly to the surface are termed major clefts, those branching from those deep in the fiber are termed minor clefts. Clefts can extend to, or beyond, the center of the cell. The main aspect of clefts for membrane function is the great increase in surface area engendered.

6

Excitation-Contraction Coupling

The onset of full activity occurs so soon after a shock that diffusion is far too slow to account for it. A process, not a substance, must carry activation inwards. A. V. Hill†

The events that relate membrane potential of a muscle cell to contraction are collectively termed *excitation-contraction (E-C) coupling.* The functional unit in E-C coupling in a majority of vertebrate skeletal muscle cells is likely to be the anatomical identity termed the triad, or its equivalent in invertebrates, the dyad, for a particular myofilament cluster, but it could be no more than a portion of one. Nothing is known about these matters, since activation of single units is not yet possible. Surface depolarization spreads inward down ITs and necessarily declines progressively with distance from the surface in steady depolarization, though in brief activation of some fast-type cells regenerative graded or even all-or-none IT action potentials may travel inward along ITs (see p. 210) setting up more-uniform excitation conditions.

6.1. The Overall Membrane-Potential–Tension Relation

Starting in the last century, a large number of studies on a wide range of muscle cell types had indicated that agents that lower the membrane resting potential, especially external potassium ions, which largely determine its value following the Nernst relation (see p. 180), promote contraction. Tension takes time to develop, and its peak value in the majority of cells is held only briefly. The rate of development of tension and its maximum value are each functions of both membrane potential and time. The peak tension under prolonged steady depolarization is related to time by an S-shaped curve that was first determined for single cells of frog semitendinosus depolarized by external potassium by Hodgkin and Horowicz (1960 – Figure 6.1.). The threshold membrane potential is termed the E-C coupling potential E_C. It was at an average of -54 mV in the range -48 to -58 mV in the frog cells. Why the range should be so great is not known but indicates either that there is a population of muscle cells with different characteristics or that the value is not constant and depends on the history of the individual cell. Both are true. In arthropods, E_C is a major variable between cells of different functional type. The range is from about -62 mV in some very slow, electrically inexcitable cells to about -40 in some of fast type that give large action potentials.

The E_C has a fixed value for a given cell under constant conditions. It is extremely sensitive to a wide range of influences, including the cell's immediate past history. For-

†A. V. Hill, *Proc. R. Soc. Lond. Ser. B* **136**, 339–420 (1949).

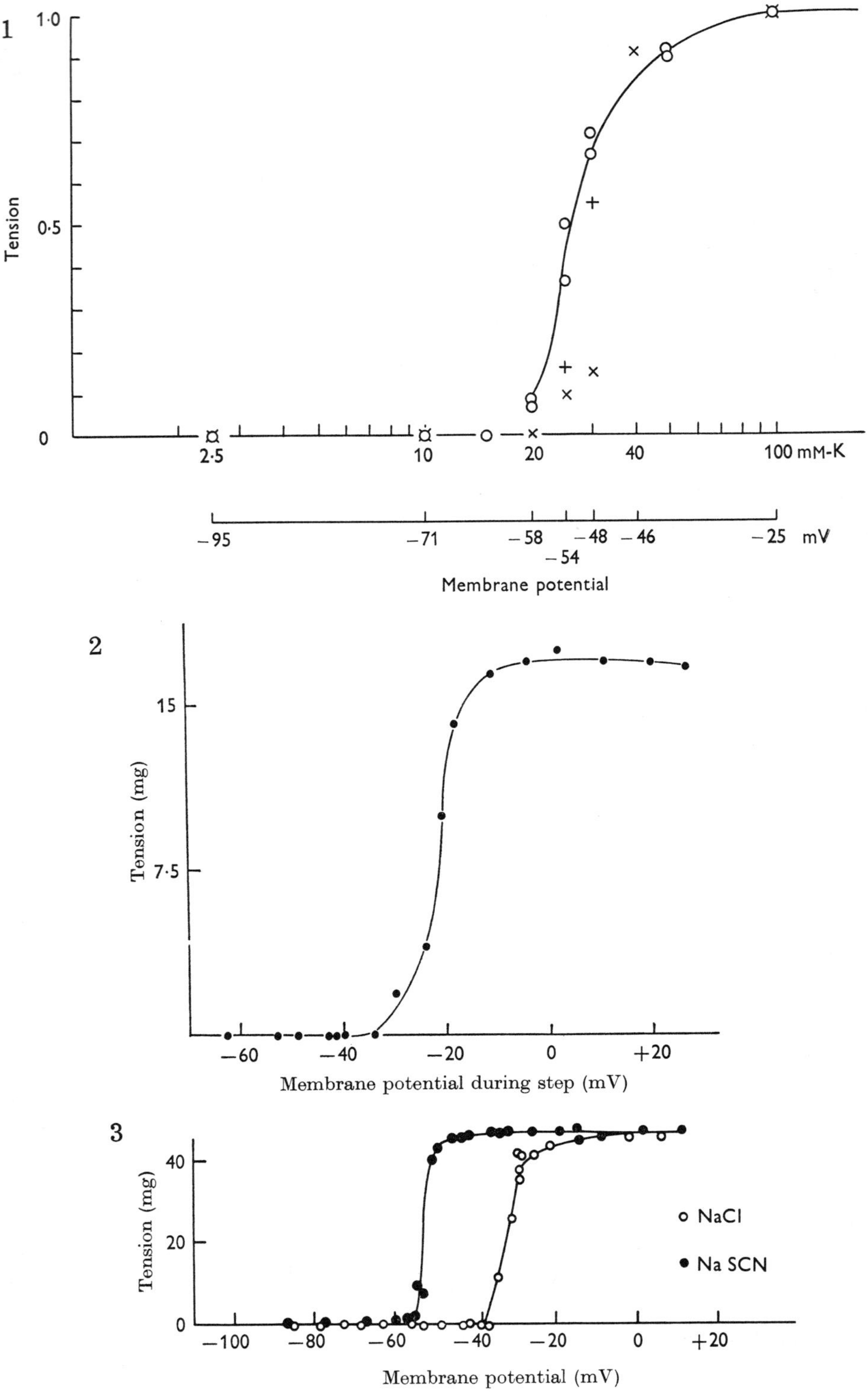

Figure 6.1. Relation between membrane potential and tension. (*1*) Peak tension and membrane potential in single frog cell depolarized with saline in which potassium and choline replaced sodium. Logarithmic scale for $[K^+]_o$ and linear scale for potential. (From Hodgkin and Horowicz, 1960.) (*2*) Same as (*1*), but membrane potential controlled by voltage clamp in fast-type cell of garter snake. (From Heistracher and Hunt, 1969a.) (*3*) Shift in tension–membrane-potential relation when a foreign anion, SCN^-, is substituted for Cl^-; snake fast-type cell type. (From Heistracher and Hunt, 1969a.)

eign ions, especially anions, push it as much as 20 mV away from its basic value. There are major differences between vertebrates and invertebrates in this aspect. In vertebrates, anion substitutions progressively enhance tension in the lyophilic series

$$SCN^- > I^- > Br^- > NO_3^- > Cl^-$$

By contrast in the giant barnacle depressor, for example, tension is optimal in bromide and reduced in the order

$$Br^- > Cl^- > NO_3^- > I^- > SCN^-$$

Neither the basis for these effects nor the reasons for the differences is known.

Beyond threshold, increased tension could represent the progressive recruitment of internally responding units, increased intensity of unit response, or both. Myofilament clusters closest to the surface membrane are excited at lower surface potentials so that a peripheral ring of contraction occurs first, activation spreading inward with time and with increasing surface depolarization. Experimentally, the inner tubules of a single fast-type frog cell can be made to be more depolarized than the outer ones by using a large, 50–60 μm in diameter, pipette applied to the surface. Then the inner myofilament clusters contract first (Sugi, 1974). With a smaller, 20–40 μm pipette, a surface contraction occurs. Sugi noted that sometimes this had an early twitchlike character even when a very long pulse was used that tended to be abolished by adding TTX or changing to sodium-free saline. Inward spread of the slower contraction could sometimes be observed, traveling at a rate of 0.7–2.0 cm/s at 18–26°C. The Q_{10} of the velocity was 2.3. Currents spreading along ITs, especially action currents, should be recordable with suitable intracellular electrodes. This has been achieved by Strickholm (1974), using a single intracellular electrode in combination with a surface-clamp pipette. But Strickholm found that the fast phase of current that he attributed to an action potential could only be obtained when working with spring or summer frogs. In winter frogs it was absent (see p. 213). Thus IT excitability is subject to period influences, hormonal or environmentally related. The tubules of all arthropod skeletal cells, and of many of those of vertebrates, are passive, and surface depolarization may be thought of as passing down the tubules either by electrotonic spread or by current passing down the lumen and across them. Calculations made for me by A. L. Hodgkin show that the voltage vs. time characteristics of depolarization at the abutment with the SR cysternae are identical with either concept. However, the net result is that ionic current flows from the lumen of the ITs across into the cysternae when the surface is depolarized. In crayfish skeletal cells, this current is selectively carried by Cl^- (Reuben et al., 1967), but gCl across the tubules is low in vertebrates. Nevertheless, in frog a low-Cl^- saline or a hypertonic saline cause the ITs to swell (Freygang et al., 1964). In "skinned" vertebrate skeletal cells lacking surface membrane (see p. 202), addition of chloride depolarizes the tubules (Stephenson, 1978), which in turn leads to contraction (Endo and Nakajima, 1973). When a skeletal muscle cell is fixed rapidly, especially during continued excitation, the regions of ITs in contact with SR are greatly swollen both in vertebrates (Rapoport et al., 1969; Birks and Davey, 1972) and invertebrates (Selverston, 1967, Brandt et al., 1968, Hoyle, 1980). It seems unlikely that charge transfer as such or local membrane depolarization causes the next step in coupling. Instead, an associated minor flow occurring at the same time, specifically of calcium ions, is probably what matters. Good evidence for an essential role for the ITs in vertebrate cells was obtained by Gage and Eisenberg (1967), who found that if, after a period of soaking in a hypertonic glycerol solution, a frog fast-type cell is returned to normal saline the ITs have become disrupted. The surface-membrane excitability is unchanged; normal action potentials occur, but the cell shows no trace of contraction. Partial tubule disruption in a crayfish muscle reduces tension partially (Eastwood et al., 1973). Before the tubule disruption experiments, other evidence had pointed to the involvement of tubules. When passing a fine micropette, 2 μm in external diameter, carrying depolarizing current along the surface of single skeletal cells both longitudinally and circumferentially, A. F. Huxley and Taylor (1958) found that contraction occurred only when the pipette was opposite one of a set of sensitive regions. For frog cells these regions were all at the level of a Z line and caused local contractions of the two adjacent half sarcomeres only. For crab cells no such spots occurred at Z lines; all were near the A/I margin, and contraction was confined to that half sarcomere. These local excitations are presumed to have been mediated by depolarization's spreading down single tubules. In the frog the C elements of SR of the two contiguous sarcomeres abutting the tubule terminus would be excited, and in the crab the single C element of the dyad.

6.2. The Calcium Hypothesis

To two generations of muscle scientists by whom knowledge of invaginated tubules had been lost, the process of transfer of excitation from the surface into the center of even an average-sized, 100 μm in diameter, skeletal muscle cell in only a millisecond or two was the greatest of all the many muscle mysteries. One important, though tentative, step was made during this dark period. First Chambers and Hale (1932) made the initial discovery by chance, followed by Keil and Sichel (1936), then Kamada and Kinosita (1943), and finally Heilbrunn and Wierczinski (1947), who alone have been accorded the honor. All found that the only common ion that initiates contraction when injected into muscle cells in a low concentration is calcium. Nevertheless, little attention was paid to this finding until the middle 1960s, once again because of a mistaken belief that electrical current flow associated with the surface action potential somehow triggered contraction. That the flow could not be longitudinal, so must be transverse current, was determined by Sten-Knudsen in 1954. Muscle cells differ greatly in their ability to function in calcium-free media. Some are knocked out immediately, others in several minutes, and a few only after several hours. Because of a presumed need for calcium to maintain normal membrane structure, these findings could not be regarded as critical evidence. Interest in the calcium data began to pick up with two findings. One was the rediscovery of the ITs, independently by Stanley Bennett and by Keith Porter in 1955, which indicated how transverse currents could rapidly spread inward. The other was the development of techniques for making cell fractions, which led to isolation of the SR as a set of fragments called microsomes. These were shown by Hasselbach and Makinose (1961) and Ebashi (1961) to have an extremely strong affinity for calcium ions. They hold the level of ionized Ca^{2+} in a muscle cell to 2×10^{-8} M or less (Weber et al., 1963). The energy supply for the binding is provided by ATP hydrolysis (Hasselbach, 1964; Stössel and Zebe, 1968). Furthermore, when added to muscle cells that are in a contracted state, microsomes caused relaxation, so one previously detected "relaxing factor" (Bendall, 1958) of a cell could be its own SR. Aside from calcium, the only agent found to initiate contraction in all skeletal cells in low concentration was caffeine (Sandow, 1965). When caffeine was added to SR it inhibited its calcium-uptake ability (Herz and Weber, 1965) and caused it to release its bound calcium, so the exciting action of caffeine could be indirect but also fundamentally due to calcium. Caffeine had the effect on muscle cells of lowering the apparent E-C coupling threshold, because as it raised the internal calcium concentration, the additional amounts required to be released by depolarization were lowered. Gradually these findings supplanted an earlier view that calcium was bound to the surface membrane. At about the same time, several persons who played a role in these studies, notably Hasselbach (1964), Annemarie Weber (Weber et al., 1964), and Sandow (1965) proposed a *calcium-release hypothesis* of coupling. An authorative review of the evidence was published by Ebashi and Endo (1969).

According to this hypothesis, calcium is released from binding sites on the SR following depolarization of the ITs past a threshold. Released calcium diffuses to the region of A/I filament overlap, raising the concentration there and causing activation of Mg-ATPase, which in turn causes the onset of contraction. Relaxation is brought about by a combination of cessation of calcium release and/or entry, together with return of the released calcium to the SR. Two major new questions were raised by the hypothesis: how does the depolarization cause calcium release, and how does the calcium activate the actin/myosin? But first there was the need to test the hypothesis and to obtain quantitative data. Direct determination of the calcium concentration in many kinds of muscle cell showed that although there are wide variations between different types of muscle (Hasselbach, 1964), the amount is basically very high and commonly about 10^{-3} mol/kg. Yet when calcium was injected into single skeletal cells or bathed over a "model" system such as glycerinated *Lethocerus* indirect flight muscle + ATP (Chaplain, 1966), a concentration of only 10^{-8} m/l caused detectable contraction. The injection experiments were carried out on very large, 2 mm in diameter, cross-striated cells from legs of a spider crab, *Maia squinado,* and retractors of the giant barnacle *Bàlanus nubilus* (Ashley, 1967). At only 5×10^{-6} mol/kg calcium, maximal contractions were obtained from the crustacean cells, and at only 1×10^{-6} mol/kg in the insect model. Clearly these results lent strong support to the calcium-release hypothesis.

Direct evidence for calcium release was soon forthcoming (Ridgway et al., 1967). A photosensitive protein, aequorin, which is produced by the jellyfish *Aequorea*, was injected into giant barnacle muscle cells, which were then electrically stimulated. A brief flash of light occurred, when onset was earlier than the onset of tension development (Figure 6.2). The light output reached its peak intensity while tension was still rising, then decayed rapidly to a very low level by the time tension had reached its peak. The intensity of the luminescence was proportional to the intensity and duration of stimulating current. This was to be expected in the barnacle cell because it is a gradedly responsive one. Aequorin light output is quantitatively related to calcium concentration. It starts being detectable at a concentration of 10^{-8} mol/kg and reaches saturation at about 5×10^{-6} mol/kg, making it ideally suitable for testing the calcium release in muscle, which is in exactly the same range. Since then, aequorin has been used also in toad and frog skeletal cells (Rüdel and Taylor, 1973) and even in smooth-muscle cells of the toad stomach (Fay et al., 1979). Some examples of results with aequorin injection are illustrated in Figures 6.2.*2*-6.2.*5*). More recently, comparable results have been obtained using a synthetic calcium indicator, Arsenazo III, in frog fast cells (Miledi et al., 1977).

The released calcium has only a brief period when it is free (to interact with aequorin) before it is taken up firmly by the contractile mechanism. During relaxation there is no period when it is suddenly given off, but the later part of the decay of the aequorin light is slower than the initial decay, which is exponential (Figure 6.3. – Ashley and Lignon,

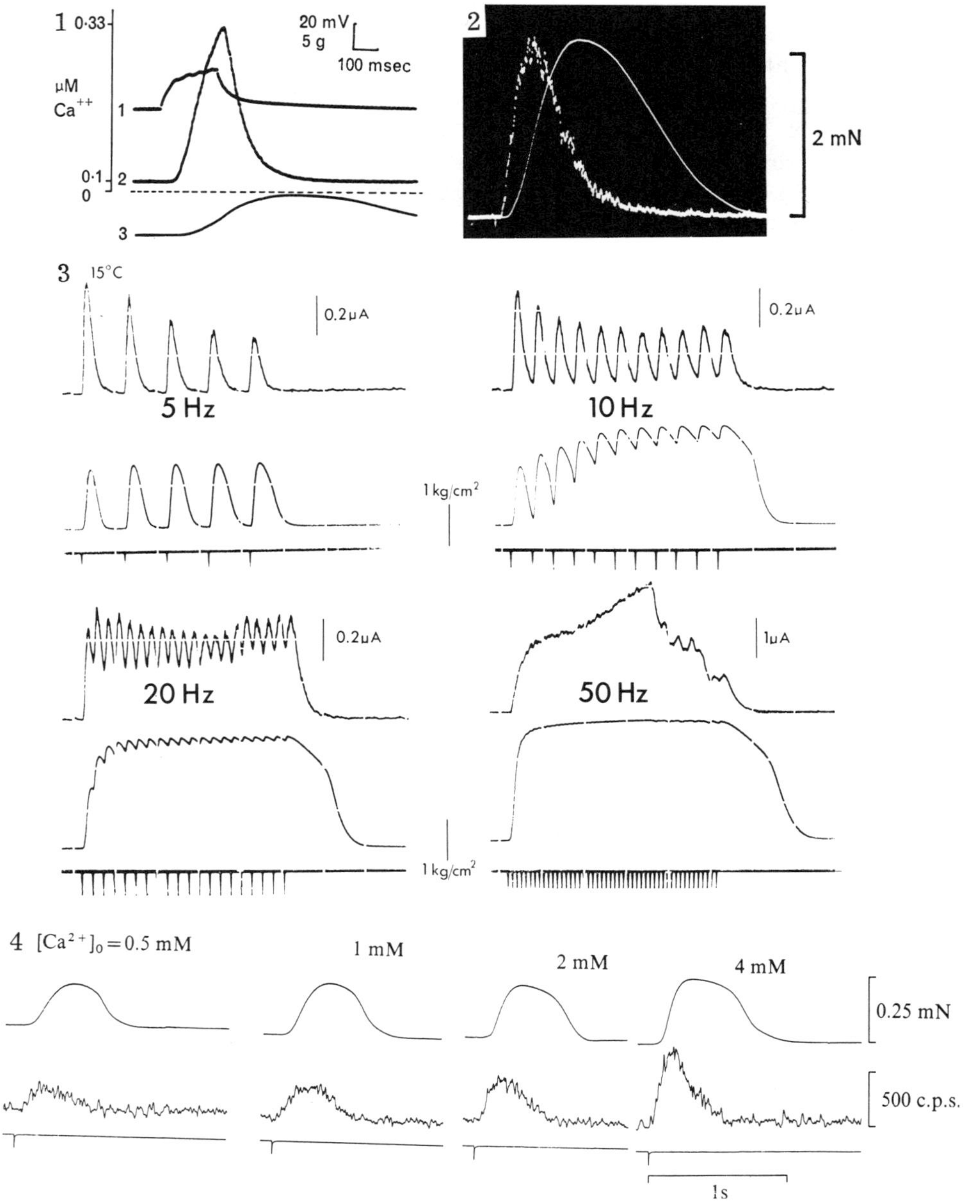

Figure 6.2. Detection of the liberation of calcium following excitation in single muscle cells loaded with the photoprotein aequorin. (*1*) Classical record from giant barnacle cell. (*Upper trace*) membrane potential; (*middle trace*) light output (scale gives equivalent calcium concentration); (*lower trace*) tension. (From Ashley and Ridgway, 1970.) (*2*) Calcium transient (jittery line) in twitch of a frog semitendinosus cell at 10°C, together with tension (From Blinks et al., 1978.) (*3*) Calcium emission (top traces) and tension (lower traces) at increasing stimulation rates in a toad (*Xenopus*) iliofibularis cell. (From Rüdel and Taylor, 1973.) (*4*) Influence of extracellular calcium on calcium transient in single frog cardiac muscle cell: tension *above;* averaged aequorin signal *below.* [Reprinted by permission from *Nature.* **273**, 509–513 (1978). Copyright 1978 Macmillan Journals Limited.]

1981). This period of slower decay is when calcium is relatively slowly leaving the contractile machinery. The calcium may be partly bound to a carrier molecule. Nothing is yet known about the form either in which calcium enters the SR following contraction or in which it is released, that is, whether as free Ca^{2+} ions or as CaX. The pumping of Ca into SR is an ATP-dependent event.

Almost all the calcium is out in the I band at the time of peak tension. During relaxation it moves first into the SR, then returns to the C elements, where most is stored at rest, with a half-time of 9 s at 22°C; Q_{10} for the return is 1.7. This was shown by following the movement of calcium in frog toe muscle cells using ^{45}Ca (Winegard, 1965a, b, 1970). Considering when this work was done, it is surprising that the process of calcium transport within SR is still totally unknown. The principal binding molecule is a protein termed calsequestrin, which has a molecular weight of about 110,000 and is tightly bound to the membrane. It is a calcium-activated ATPase that binds 30-40 Ca^{2+} per mole. Quantitative studies on the calcium-tension activation relation have been made (see p. 203), but all are dependent upon making assumptions about either the amount of free calcium or the contractile-machinery molecular dynamics. The best available indicator of intracellular free calcium is aequorin, and although it is, in effect, a titrator that can be used quantitatively, the calcium-aequorin-light relations are complex and nonlinear (Moisescu and Ashley, 1977). The intensity of the reaction is very sensitive to ions other than calcium, and these ions may also vary during excitation, and some may even cause aequorin light themselves (Allen et al., 1977). This aspect of coupling data is puzzling. In all types of cell that have now been injected with aequorin, the indication is that the free-calcium concentration has fallen much more rapidly than tension rises. The half-time for removal of calcium is only 3 ms (Johnson et al., 1978). The competitive uptake

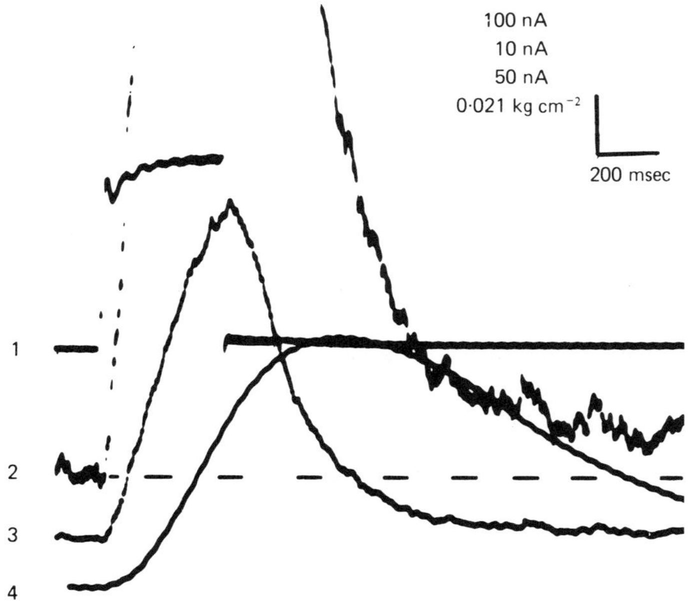

Figure 6.3. Evidence that calcium is given off from the contractile material during relaxation of giant barnacle depressor cell. (*Trace 1*) current; (*trace 2*) light at very high gain; (*trace 3*) light at lower gain; (*trace 4*) tension. The calcium transient decays exponentially by 90% at the onset of relaxation, but the remaining decay is slower. (From Ashley and Lignon, 1981.)

of SR is not nearly sufficient to account for the high rate of calcium removal, raising the possibility that there may be molecules present that have greater calcium affinity than does TN. What, then, are their roles in contraction? Despite these unanswered questions, the data are quantitatively useful and have strongly indicated in the case of the barnacle giant fiber that for each functional unit of contraction, the requirement is for two calcium ions (Ashley and Ridgway, 1970; Ashley and Moisescu, 1977). A comparable relation had been indicated indirectly for fast frog cells (Hellam and Podolsky, 1969). The detailed results favor consecutive actions, one calcium ion first determining an initial event, and the second a following, cooperative one.

6.2.1. Release of Calcium from SR Cysternae

When cell fractions containing SR and SR-cysternae contacts (triads and dyads) are subjected to the passage of current, some calcium is released (Endo and Nakijima, 1973). But release does not occur unless there is also some free calcium present within the tubules. Endo et al. (1970) proposed that calcium passes from the IT lumen across the gap and into the cysternal element. There it triggers the release of the calcium stored in the cysternum. The evidence for the involvement of calcium as a trigger in release of calcium sequestered by SR is now quite strong (reviewed by Endo, 1977) but until the details of the process are better understood, many questions will remain.

It was for a very long time erroneously assumed that Ca^{2+} uptake by SR has a high, fixed rate that is not affected by events of excitation. But it makes no sense, when the goal of excitation is to raise the free-calcium level, to be mopping it up at the same time it is being released. The rate of reaccumulation of ^{45}Ca during stimulation was found by Stephenson (1978) to be much slower than the influx predicted on the basis of data from resting cells. This suggests that there may indeed be a mechanism coupling depolarization to SR transport of Ca^{2+} so that transport is inhibited during excitation. It will be shown later (see p. 206) that a cycle of depolarization and repolarization may be needed to activate calcium uptake, at least in some muscle cells.

Caffeine (Bianchi, 1960; Blum et al., 1957), which causes contracture, and ryanodine (Jenerick and Gerard, 1953), which also causes contracture, but in vertebrate muscle only (Figure 6.4), cause contracture without depolarizing the surface membrane. The caffeine action is not affected by prior treatment that ruptures ITs. These agents work by a combination of inhibition of calcium uptake by SR and direct activation of the C elements to release it. There is a delay in response to external application of caffeine due to the time required for the caffeine to enter the cell and diffuse to calcium-release sites. With intracellular injection, the delay is negligible (Chiarandini et al., 1970). Although the principal caffeine effect is on the SR directly, over a long period it probably also begins to involve external calcium (Ashley et al., 1977). These are not necessarily only insignificant laboratory findings, since human habits have led many to a sufficiently high level of blood caffeine directly to affect these events in muscle.

6.2.2. Direct Entry of Calcium

In many skeletal smooth-muscle and cardiac cells, the outer-surface-membrane depolarization is in part achieved by calcium conductance increase, and significant amounts of calcium enter directly, even in a single excitation. There is some calcium entry in frog fast cells, but it is so slow that it could not be involved in a twitch (Sanchez and Stefani,

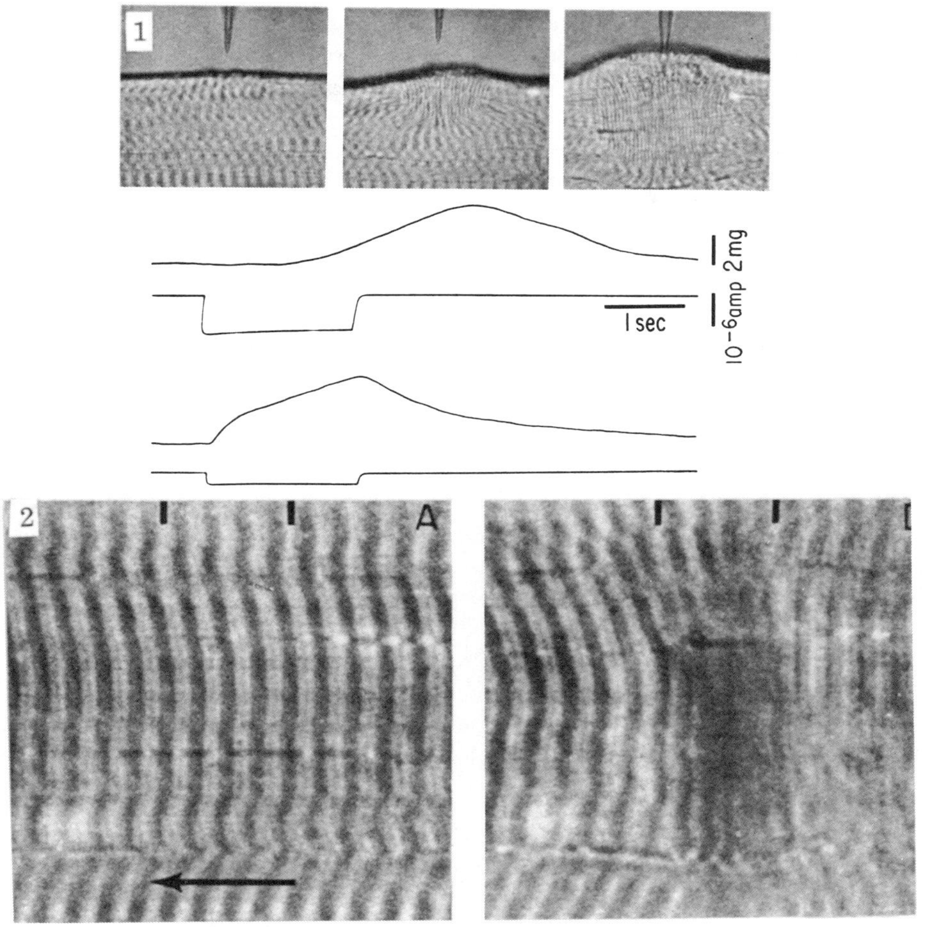

Figure 6.4. Local contractions caused by caffeine release and internal longitudinal current. (*1*) Caffeine released iontophoretically from a micropipette. (*Above, left*) weak current, pipette outside cell. (*Above, center*) strong current, pipette outside cell. (*Above, right*) weak current, pipette inside cell. (*Below, top*) tension record during application of strong current to caffeine pipette outside. (*Below, bottom*) tension record during application of weak current to caffeine pipette inside. (From Chiarandini et al., 1970) (*2*) Contraction in skinned frog semitendinosus cell caused by passing longitudinal current in the direction of the arrow. (From Costantin and Podolsky, 1967.)

1978), but it might be significant in a prolonged contraction or in a contracture. By contrast, in arthropod skeletal and many smooth-muscle cells, the depolarizing phase of the action potential is due to increased calcium conductance, and calcium entry is rapid. For barnacle giant cells, Hagiwara et al. (1968) calculated that about 6×10^{-12} mol of calcium enters for every square centimeter of membrane per excitation. For an average cell, this means 1.5×10^{-13} mol, sufficient to raise the concentration to 10^{-6} mol/kg and for the calcium itself to cause contraction. Direct entry of calcium has also been determined for giant retractor cells of the Chilean barnacle *Megabalanus psittacus.* The actual influx, 474 pmol/60 mV depolarization, closely matched the calculated value of 500 pmol required to raise the internal calcium concentration to 5×10^{-5} *M* (Atwater et al., 1974). However, it would take several seconds for this entering calcium to diffuse throughout the myoplasm, yet a twitch reaches peak value in under 300 ms. The barnacle cell possesses an extensive IT system with cisternal contacts and SR – all the features need for local internal calcium release. Possibly contractures produced by potassium (Gainer, 1968) and prolonged neurally evoked contractions are more dependent on the entering calcium than on released calcium.

For many of the smaller-diameter skeletal cells, such as retractor and columella muscles of molluscs (Huddart et al., 1977), the striated scallop fast adductor, and *Amphioxus* body muscle, there are no ITs. Here, surface-entering calcium may be in part or in whole responsible for contraction. The maximum diffusion distance in the flat muscle cells of *Amphioxus* is only 0.5 μm (Peachey, 1961), a distance that would be traversed by diffusion in only 0.5 ms. Although this cell has a sodium action potential, some calcium current is associated with it also (Hagiwara and Kidokoro, 1971). Calculation showed that 0.7-1.4 × 157 mol Ca/cm^2 enters per action potential, sufficient to raise the internal calcium by from 0.7 to 1.4 × 10^{-5} mol/kg and promote full contraction. *Amphioxus* does have an SR that sequesters calcium and is essential for rapid relaxation; the calcium is released from it by caffeine.

Cardiac muscle cells also have a significant amount of calcium entering during the prolonged action potential, and this must play a role in the elicitation of contraction. It at least adds to internally released calcium, much of which is thought to come from SR cisterns lying directly below the sarcolemma (Bassingthwaite et al., 1973), so the two sources are at the same distance from contractile material. These matters will be discussed in detail in the chapter on cardiac muscle. There is a complex antagonism between sodium and calcium present in the medium bathing cardiac cells that is not seen in any skeletal cells, further suggesting that entering calcium plays a major role in cardiac cells.

6.2.3. Experiments with "Skinned" Muscle Cells

It is important to distinguish between events that are strictly associated with coupling and ones that directly affect the contractile machinery only. Calcium is clearly involved in both. Its subsequent actions on the contractile machinery are interfered with by magnesium, pH, potassium, and many anions, all of which may also affect coupling.

So researchers could bypass the surface-membrane step in E-C and study activation of the contractile machinery more directly, methods were developed, first by Natori (1954), to physically remove the sarcolemma and plasma membrane from single cells by "skinning." The tough collagen in the sarcolemma make this peeling, or skinning, process relatively easy. Alternatively, a mixture of low calcium, salts, and detergent causes disruption of cells chemically: what remains is virtually undamaged contractile material (Eastwood et al., 1979) that is reversibly contractile through many cycles. Skinning reduces the maximum tension obtainable from single frog fast cells from about 2.5 kg/cm^2 to 1.5 kg/cm^2 according to Podolsky and Teichholz (1970). This may be due to damage to surface myofilament clusters and in part to a need for the sarcolemma for local tension transfer. The experiment of Podolsky and Teichholz proved conclusively that most of the tension is transmitted by internal elements, not by the sarcolemma, as some (e.g., Ramsey and Street, 1940) had believed. Skinned cells from frog semitendinosus and iliofibularis contract in response to much smaller amounts of calcium than could by themselves cause contraction (Endo et al., 1970; Ford and Podolsky, 1972). These findings supported the view that calcium initiates regenerative release of more calcium from SR in which it is stored in the resting cell.

Skinned cells nevertheless permit a direct test of the calcium concentrations that cause contraction, without the hazards of injection or use of an indicator. This can be done in spite of the internal release of Ca^{2+} from SR by using a large volume of medium to which the potent calcium buffer ethyleneglycolbis (aminoethylether) tetraceticacid (EGTA) has been added. According to Schoenberg and Podolsky (1972), a concentration of 2.5×10^{-5} *M* calcium is required for full activation of frog fast cells. The tension-*p*Ca relation obtained is shown in Figure 6.5.

Skinned cells can be electrically excited, giving local contractions whose nature suggests that excitation is via the IT network (Figure 6.3.*2* – Costantin and Podolsky, 1967). Most of these responses are graded, but an occasional explosive one was noted that hinted at initiation of an all-or-none response in the tubules. Spontaneous contractions resembling those occurring also in intact cells (see p. 12) occur, as might have been expected in view of the absence of surface membrane-potential changes associated with such spontaneous events in intact cell. The ITs of skinned cells take up ferritin, proving that the ITs remain open (Endo and Nakajima, 1973). These authors first soaked a partially skinned cell in aequorin, then obtained a bright glow after treating it with a very low calcium-EDTA solution, directly demonstrating the calcium release by calcium.

Myofilament clusters of very large cells, such as giant barnacle cells, are as large or larger than whole vertebrate skeletal cells and can be used equivalently to skinned material (Ashley and Moisescu, 1977).

6.2.4. Are Mitochondria Involved in Calcium Regulation and Coupling?

Since the basic excitation process is raising the level of free calcium in myoplasm around the myofilaments, there is nothing unique about the surface-membrane-potential "threshold" in relation to coupling. If the concentration of internal free calcium is high, a small amount of depolarization will cause the release and/or entry of sufficient additional calcium to initiate activation of contraction. There may be a specific threshold for the regenerative release of calcium from SR, but even this is likely to reflect a calcium-concentration effect. The level of calcium will depend upon the external concentration, the activity of the possible surface-membrane calcium pump, the activity of the known SR calcium pump, and calcium sequestration/release by mitochondria. There is no doubt that mitochondria possess this capability, but the extent of their possible involvement is disputed. It is likely to be very different in different types of cell, since the mitochondrial volume ranges from 1 to 70% in different types (Chapter 8), and is also likely to be subject to hormonally mediated seasonal influences.

6.2.5. Overall Regulation of Calcium

A diagram illustrating the various calcium compartments and the flow between them is presented in Figure 6.6. This makes it clear that many factors, including internal ones, can be expected to influence E-C coupling in a mutually interactive manner. Following a caffeine shock that totally depletes calcium from SR stores in crayfish leg cells, there is a 20 min refractoriness to caffeine, but a progressive buildup in response capability to potassium, which therefore causes release by a different mechanism (Chiriandini et al., 1970). The replenishment is from external calcium and possibly also internally from mitochondria. Recovery is accelerated by electrical or chemical depolarization but is not

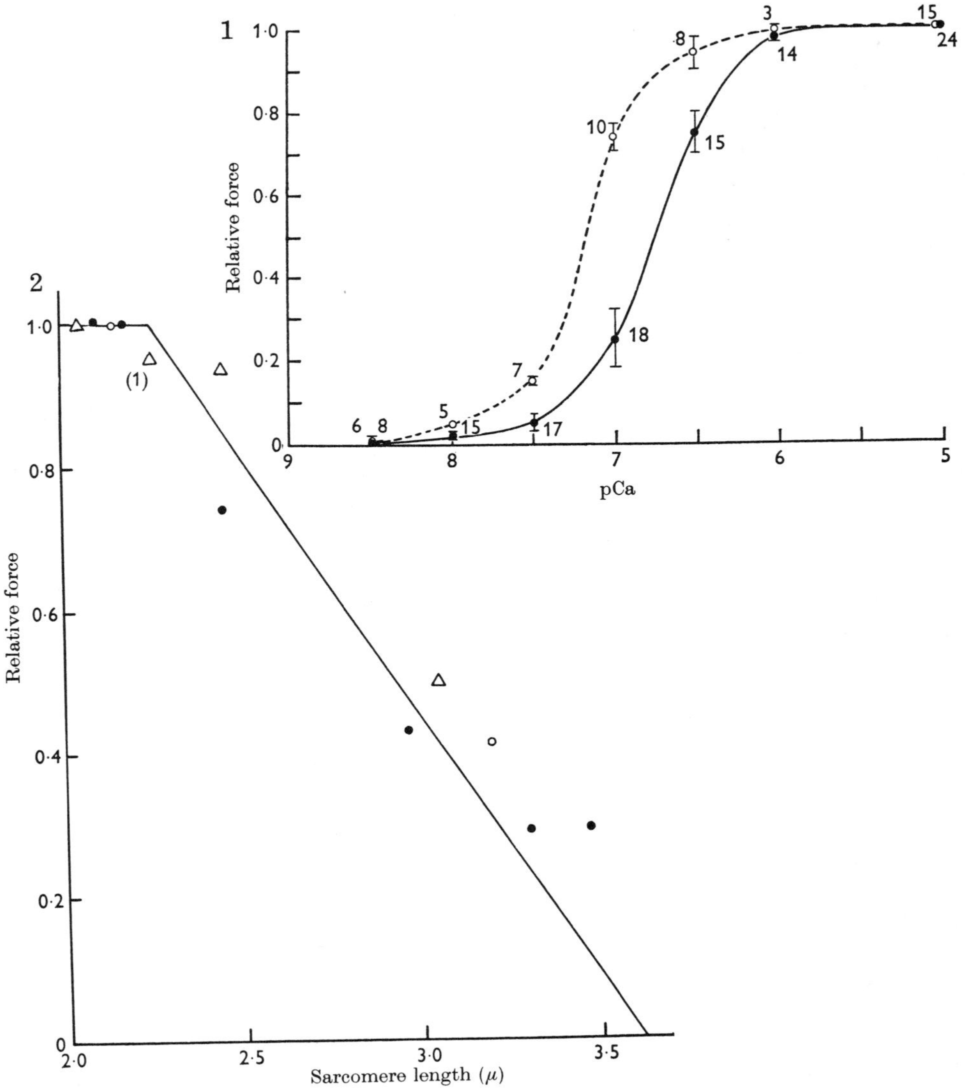

Figure 6.5. Effects of calcium concentration on force at different lengths in skinned cells. (*1*) Force–*p*Ca relation in standard saline (*filled circles*) compared with that in deoxycholate solutions (*open circles*), which inhibit uptake of calcium by SR. (*2*) Force–SL relation. Relaxations produced by *p*Ca 9. (*Open circles*) *p*Ca 6.5; (*filled circles*) *p*Ca 7.5 plus deoxycholate. The line is the relation obtained for electrically activated intact cells by Gordon et al. (1966b). (From Hellam and Podolsky, 1969.)

affected by polarizing pulses. Many other natural factors, including Vitamin D, have been implicated as affecting the SR calcium-uptake mechanism (Curry et al., 1974).

6.3. Control of Relaxation

The flow chart for calcium movements makes it obvious that relaxation as well as contraction can be under membrane, and therefore neural, control. The traditional dogma is that relaxation is "passive" and occurs at rates that are not subject to external influence.

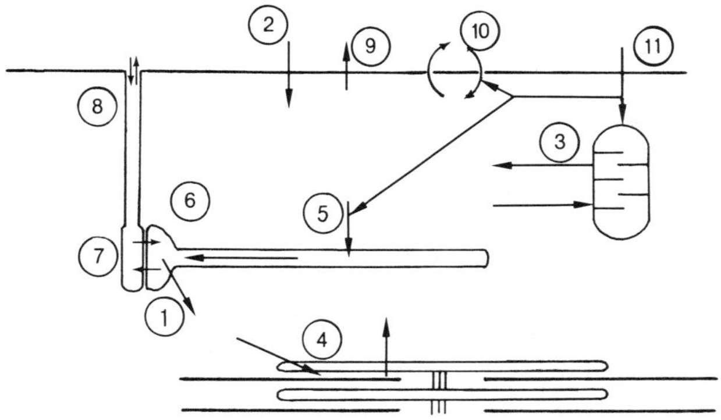

Figure 6.6. Diagram illustrating the calcium compartments of a muscle cell and their involvement in tonus regulation, contraction, and relaxation. (*1*) Cisternal (C) elements of SR, abutting ITs, where a very high concentration of calcium obtains at rest. This is released when the IT is depolarized past a threshold. (*2*) Calcium in bathing medium – the primeval source of activator for contraction. This calcium moves into the cell when the cell is depolarized sufficiently. In many cells there is graded Ca^{2+} electrogenesis. (*3*) Mitochondria sequester or release calcium under humoral controls that are poorly understood. (*4*) The contractile material is capable of taking up large amounts of calcium. (*5*) SR actively takes up calcium. (*6*) SR calcium is transported by unknown means to cisternal elements. (*7*) IT calcium. This calcium passes across to C as part of a not yet understood coupling process. External calcium (*2*) need not be present to initiate contraction in muscle cells with IT/C elements, but IT calcium is required. (*8*) IT calcium is an equilibrium with (*2*). (*9*) Calcium is lost passively to (*2*). (*10*) Calcium is directly and indirectly transported as a consequence of membrane ion pumps. (*11*) Modulators, including dopamine (DA, noradrenaline (NAd, and 5-hydroxytryptamine (5-HT), enter the cell and directly or indirectly promote calcium movement into or out of compartments.

A. V. Hill was once quite abrupt in criticizing the Interdit Singhs (1951) for suggesting otherwise—even for a smooth muscle. Hill's experience was based upon experience with fast vertebrate muscle cells in which the contraction characteristics are dominated by the brief all-or-none action potential and there is a great deal of SR that has a high rate of sequestration of calcium. Even here, slight changes in the duration of the action potential that are caused, for example, by zinc, which gives up to three times potentiation (Isaacson and Sandow, 1963), anion substitutes such as NO_3 and SCN, and adrenalin in mammals, are strongly reflected in markedly different contraction characteristics (Sandow and Preiser, 1964). There are, however, problems with making any simple correlation, because an agent that alters the action potential may also have an internal action on calcium sequestration by SR. Zinc, for example, prolongs the action potential, but it also reduces the uptake rate by SR (Van der Kloot, 1966), both of which effects promote tension enhancement. The relaxation dogma most certainly does not apply to slow-type cells and slow activation. In many arthropod skeletal cells, it is possible to take advantage of the presence of distributed inhibitory (polarizing) junctional potentials to test the effects of membrane potential on relaxation. When IJPs are timed to arrive during the repolarizing phase of the excitatory potential, they accelerate membrane-potential recovery, and the relaxation rate is enhanced. Contractures and natural resting tension are likewise relaxed by IJPs or polarizing electrical current, but it is difficult to separate simple removal of depolarizing activation from a possible activation of calcium sequestration. A "negative" contraction or activated relaxation can be as fast as contraction in these cells (Figure 6.7). They even show a marked facilitation in size of the quick relaxation in response to repeated similar polarizations.

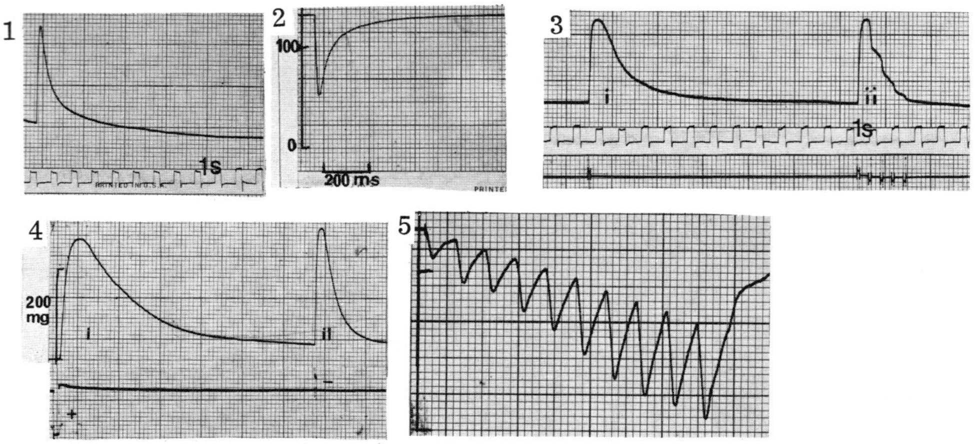

Figure 6.7. Fast activated relaxation. (*1*) Time course of twitch contraction of pink muscle cell of eyestalk-raising muscle of crab *Podophthalmus.* (*2*) Time course of quick relaxation produced by intracellularly applied polarizing pulse. The cell had 125 mg resting tension, most of which was lost in 20 ms following the pulse. Activated relaxation was five times faster than the corresponding tension rise. (*3*) Acceleration by polarizing pulses of relaxation rate of a twitch of a slowly contracting cell from same as (*1*). (*4*) Markedly different time courses of relaxation following contraction initiated by break from hyperpolarization (*left*), compared with direct depolarization (*right*). Tension above, membrane potential below. (*5*) Facilitation of quick relaxations with repetition of equal constant-current polarizing pulses. [All from G. Hoyle, Resting tension, "negative" contraction and "break" contraction in specialized crustacean muscle fibers. *J. Exp. Zool.* **167**, Fig. 1, p. 553. Fig. 2a, p. 554 (1968). Reprinted with permission from Alan R. Liss, Inc., New York.]

Numerous drugs affect the various aspects of contraction. For example, a widely used muscle relaxant, dantrolene, seems to act via the coupling process, though on skeletal muscle only; its clinical use depends upon the fact that it does not have the same effect on cardiac muscle. Dantrolene does not affect membrane potential, but E-C threshold is depressed by as much as 22 mV (Putney and Bianchi, 1974), possibly because the release of internal calcium from SR is inhibited (Van Winkle, 1976; Desmedt and Hainant, 1977). Tetracaine and other local anesthetics also diminish calcium release (e.g., Caputo, 1976) as does heavy water (Cecchi et al., 1979). Nicotine causes contracture that has markedly different characteristics from that caused by potassium (Weiss, 1966), because each has different effects on internal calcium stores. Of the many drugs that affect muscle by way of calcium regulation two, verapamil and D600, both retard the reavailability of calcium and reduce direct entry. There is now a lot of evidence that potassium action is not just to depolarize the surface membrane but also to enter the cell and activate SR and mitochondria. Neither caffeine alone nor potassium in a concentration that completely depolarizes the cell completely depletes SR of calcium, because addition of the other after calcium release has ceased liberates some more (Caputo, 1976). Although internal $[K^+]$ is high and behaves much as it does in free solution, most of it does interact with intracellular anions, so a small additional amount near the surface of the cell, where mitochondria and SR are often located in large amounts, could drive activity.

The extent to which the membrane potential affects subsequent events, including the SR calcium-uptake system, clearly varies greatly with cell type. The range is from that in very fast cells, in which there is little or no significant action, to very slow ones, in which the calcium-uptake mechanism is completely shut off indefinitely by continued surface depolarization.

Another line of evidence for a relationship between surface-membrane potential and calcium uptake is in the great differences in relaxation rate between contractions caused by depolarization and ones caused by release from hyperpolarization, which are several times slower (Figure 6.7.*5* – Edwards et al., 1964; Hoyle, 1968b). Since the latter are not always accompanied by depolarization, they offer a strong suggestion that the combination of depolarization followed by repolarization is needed to activate the calcium-uptake mechanism fully and that it is not constantly working in a resting muscle.

6.4. Effects of Chemical Agents on Coupling

The release of calcium from SR does not cease abruptly following stimulation but decays in a complex manner (Ashley and Caldwell, 1974) reminiscent of decay of graded propagation, in which there is a progressively declining activation. This process may be expected to be especially sensitive to influences of foreign ions.

A wide range of effects on tension is produced by agents that act on both calcium uptake and release, for example, lanthanum, which is a calcium substitute but which also directly alter the kinetics. In frog fast-twitch cells, the twitch is prolonged and its height increased, by lanthanum (Andersson and Edman, 1974a,b). The rate of relaxation is also showed. It is probable that lanthanum acts on several sites and events; on IT/SR, on calcium release from SR, and possibly also on its uptake. The authors suggest also that it either delays the process of inactivation or reduces its rate. The basis of inactivation is unknown.

Acetylcholine's important depolarizing action on the end-plate regions of vertebrate skeletal muscle cells has drawn attention away from its independent ability at the higher, but still "physiological," concentration of 10^{-5} m*M* to cause contraction. It also directly affects contractile properties of heart muscles, in concentrations even below those needed to activate synapses. The contraction is not accompanied by a depolarization, nor is transport of calcium across the membrane stimulated (Coburn, 1979), so a direct action on release from SR is implied.

Anions substituted for chloride *enhance* peak twitch contractions, reduce latencies, and increase the rates of rise of force (Taylor et al., 1969), in such vertebrate skeletal fast cells as have been examined. They do so progressively in the lyophilic series (p. 195). Very different effects are observed in arthropods in general, and there are also major differences within arthropod phyla and cell types. In the neurally excited crab (*Carcinus*) closer, NO_3^- and Br^- depress both slow and fast axon contractions, whereas I^- and SCN^- depress the slow but enhance the fast (Atwood, 1968). The threshold voltages are not changed, but more current is needed to get a particular level of contraction. In barnacle giant depressor cells, direct membrane-stimulated responses are progressively *depressed* in almost the reverse of the lyophilic series (see p. 195). Aequorin tests have shown that progressively less calcium is released in barnacle cells in the lyophilic series, using long-lasting depolarizations to zero E_m. The foreign anions clearly depress coupling. The positive effects in vertebrates are related to a complex combination of effects. The slope of the tension-E_m curve is shifted to lower E_m (Figure 6.1.*3*), the action potentials are longer and calicum uptake by SR is depressed (e.g., Carvalho, 1968). Together these effects could mask a possible fundamental depressing action on later stages of coupling even in vertebrates. One reason why NO_3^-, Br^-, I^-, and SCN^- prolong vertebrate-cell action potentials is that they increase the input resistance (Hutter and Padsha, 1959), probably because they do not support current flow across IT walls as well as does Cl^-.

The relative dynamics of anion substitution are different for vertebrate heart compared with skeletal muscle (Nayler and Chipperfield, 1969) as well as between such vertebrate and invertebrate skeletal muscles as have been examined.

The effects of propionate and isethionate, with variable chloride, have been examined in crayfish leg cells voltage clamped and tension transduced (Dudel and Rüdel, 1969). These ions delay the onset of relaxation and slow it drastically but do not affect tension rise. The effects are here attributed to the need for a flow of chloride ions in the ITs to effect their repolarization. It seems likely that the principal factor governing actions of anions is an indirect one associated with current flow across IT membrane. The differences show that the properties of these membranes are different in different systems and therefore another source of variation in evolution of muscle.

6.5. Influence of Previous Stimulus History on Coupling-Patterning Effects

A number of complex effects of the pattern of stimulation applied to a muscle are known. The most celebrated is that known as the Blaschko, Cattell, Kahn (1931) effect. While stimulating a crab (*Cancer*) claw with a low frequency, they interpolated a single, extra shock. This caused a large additional response, much greater than the background stimulation would achieve. Surprisingly, as long as the background stimulation was continued, the high level was maintained. Experiments with aequorin-injected giant barnacle cells, which also show the phenomenon, show that a large calcium release follows the extra shock (Figure 6.8.*1*). This is only relatively slowly taken up, but the calcium level does return to background level, whereas the tension remains high. There is no explanation yet for this very interesting phenomenon.

The extent of the effect of an interposed pulse is itself a major variable in arthropods. Enhancement is greatest when the extra excitation occurs early in train of excitation. These events explain why, in some arthropods, an exactly similar total number of excitations given in a standard time can produce different tension levels and rates of rise when given in close pairs compared with when evenly spaced.

A related effect is that of immediate history on the magnitude of a directly evoked response. A constant-current depolarizing pulse of fixed duration does not always call forth exactly similar contractions. If the pulse is preceded by depolarizing current that is too weak to exceed E_C, the ensuing Ca^{2+} release and contraction are markedly enhanced (Figure 6.7). For relatively short durations of prepulse depolarization, the enhancement is proportional to both the prepulse depolarization strength and its duration. Exactly the opposite occurs with prepulse polarization (Figure 6.8). The prepulse current presumably affects the subsequent coupling events, perhaps by altering the amount of Ca^{2+} available for initiating release of Ca^{2+} from the C elements.

6.6. IT Properties and Coupling

The IT membrane is directly continuous with surface membrane and was clearly originally a simple inward extension of the latter. Therefore it is reasonable to suppose that it may have similar properties and include electrogenic ion channels. It would then be capable of inward propagation at least gradedly and in the extreme give all-or-none potentials. These would not significantly improve the time for inward spread, but in vertebrate fast-twitch cells might ensure the greatest possible unitary activation at the E-C coupling sites.

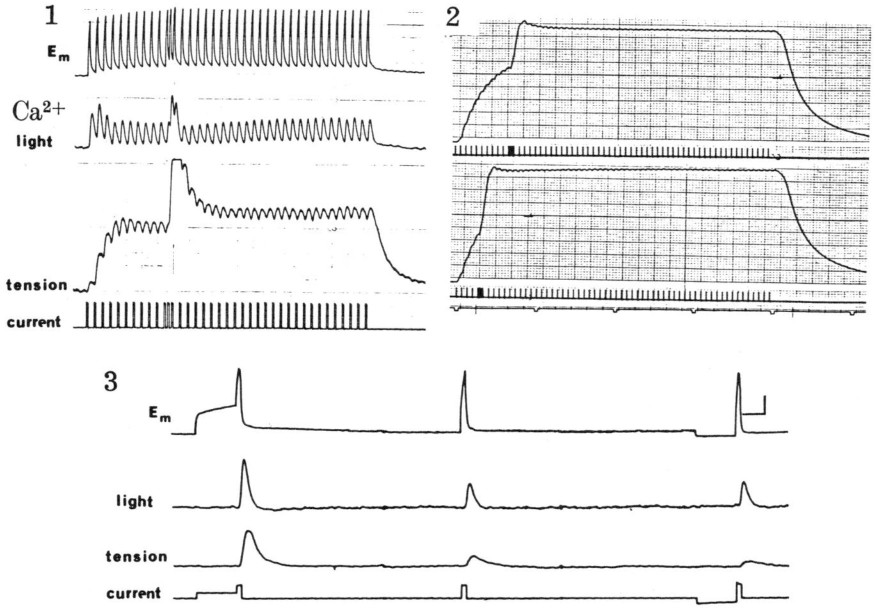

Figure 6.8. Influence of pattern of stimulation on coupling. (*1*) Large tension increment and maintained extra tension during constant background stimulation in single giant barnacle cell. (*Top trace*) intracellular membrane potential; (*second trace*) calcium transient registered as light emitted by injected aequorin; (*third trace*) isometric tension; (*fourth trace*) current monitor. (ORIG.) (*2*) Large tension increments caused by extra impulses applied early (*below*) and later (*above*) to locust coxal adductor muscle stimulated at 15 Hz. (ORIG.) (*3*) Effects of depolarizing prepulse on response to constant current, fixed duration, depolarization applied internally to giant barnacle cell. (ORIG.)

The resting conductance properties of the ITs are different from those of surface membrane, IT *g*K being twice that of the surface, 28 μmhos/cm^2 in frog sartorius (fast) cells, whereas tubular *g*Cl is very low compared with that of the surface. A sudden address of high-chloride solution is commonly used as a stimulus for skinned cells, to depolarize the tubules to initiate contraction (Costantin and Podolsky, 1967; Stephenson and Podolsky, 1977). Slow-IT-conductance changes occur following depolarization, and these are blocked by picrotoxin, which is the well-known agent that blocks inhibitory GABAergic synapses (Robbins and Van der Kloot, 1958). There are two components: a *decrease* in *g*K that is part of the inward, anomalous rectification and is associated with moderate depolarization, and an *increase* in conductance to an ion, possibly also K^+, making the inside of the IT positive and playing the major role in delayed direct rectification. The former channels occur also in the surface membrane, because the same phenomenon exists following tubule disruption (Eisenberg and Gage, 1969). The importance of these events in the present context is that they must play a role in the characteristics of current flow through the coupling region. It is also evident that there is a lot of variation in IT properties among organisms, and that different parts of ITs have different properties. The region opposite the C element is very much more permeable to some ions, and prone to swelling, than the remainder. This is true of both vertebrates and invertebrates, but especially the latter, in which the region opposite the C element appears to be highly permeable to chloride.

In vertebrates possessing a markedly radial array of ITs, inward conduction is fairly directly along radia. The rate of this inward spread may be determined by filming single cells excited at slightly less than rest length. The myofilament clusters are wavy until excitation causes them to shorten, and the wavyness disappears, in normal excitation from the outside in (Figure 6.9). In fast cells of semitendinosus of winter frog, which have passive inward conduction only – no action potential – the velocity was 7 cm/s at 20°C, with a Q_{10} of 2 (Gonzales-Serratos, 1971). This is presumed to be the normal rate of inward spread in the absence of IT action potentials.

The changes that occur in the ITs as current passes through them include charge transfer, ion flow, resistance shift, and changes in ion concentrations in the lumen contents. Some of these occur so rapidly that they influence the late phase of the action potential. A small hump in the early afterpotential was beginning to be associated with active electrogenesis in ITs, but more probably, according to Peachey (1972), represents their delayed repolarization when outward current is flowing through the closing surface potassium channels through the open T-tubule channels. Direct evidence for action potentials in some fast-type vertebrate cells has since been obtained (Bashian and Nakajima, 1974).

The same technique used to study the rate of inward spread, high-speed filming of relaxed cells to determine loss of wavy appearance in the optic midsection, can be used to study effects of agents on inward transmission. There are more wavy regions in the center in low sodium concentrations, supporting a role for this ion, and therefore possible tubule action potentials, in single frog cells (Bezanilla et al., 1972). Indirect evidence using the (V)Na blocker TTX combined with glycerol treatment to disrupt ITs has provided strong evidence for the presence of such channels in fast frog sartorius cells (Jaimovich et al., 1976). Under voltage-clamp, such cells show two phases in decay of sodium inactivation, but following disruption of tubules, there is but a single one, presumably that due to surface sodium only, that due to tubule (V)Na having been lost (Mandrino, 1977). Comparable data are not yet available for arthropods, where IT active channels are unlikely in view of the graded nature of surface excitation, but it has been demonstrated that in locust fast cell ITs there is a high content of sodium matched by low potassium (Njio and Piek, 1977). This observation shows that at least this requirement for a tubule sodium activation, which might occur under intensive fast-axon activation, are met in an arthropod.

Another use of the method is in examining the relation between the space constant of the IT system as a whole λ_{IT} and the effects of changing the diameter by stretch and relaxation and observing the extent of inward spread of wavyness. Radial spread without active membrane boosting cannot fully activate the inner myofilament (Adrian et al., 1969). As the diameter changes, more or less depolarization is required to produce the same extent of inward spread or tension, the latter with a slope of 3 mV/10 μm. The results showed that λ_{IT} cannot be large compared with the cell radius. Passive inward spread of surface potential of an individual cell is significantly affected by its instantaneous diameter and markedly attenuated by inward transmission. The relation between λ_{IT} and function is, however, different in different types of muscle cell. Although that of a fast-type cell may not be able to transmit inward adequately without an electrogenic boost, that of some slow cells, though ITs are sparse, may be efficient at transmitting inward the depolarization initiated by only surface junctional potentials. Direct analysis and examination of model systems persuaded Nicolaysen (1976) to come to this conclusion for slow and fast cells of the hagfish.

Evidence for a state change in some part of the tubule membrane during excitation has been obtained by examining fluorescence changes following stimulation, after staining

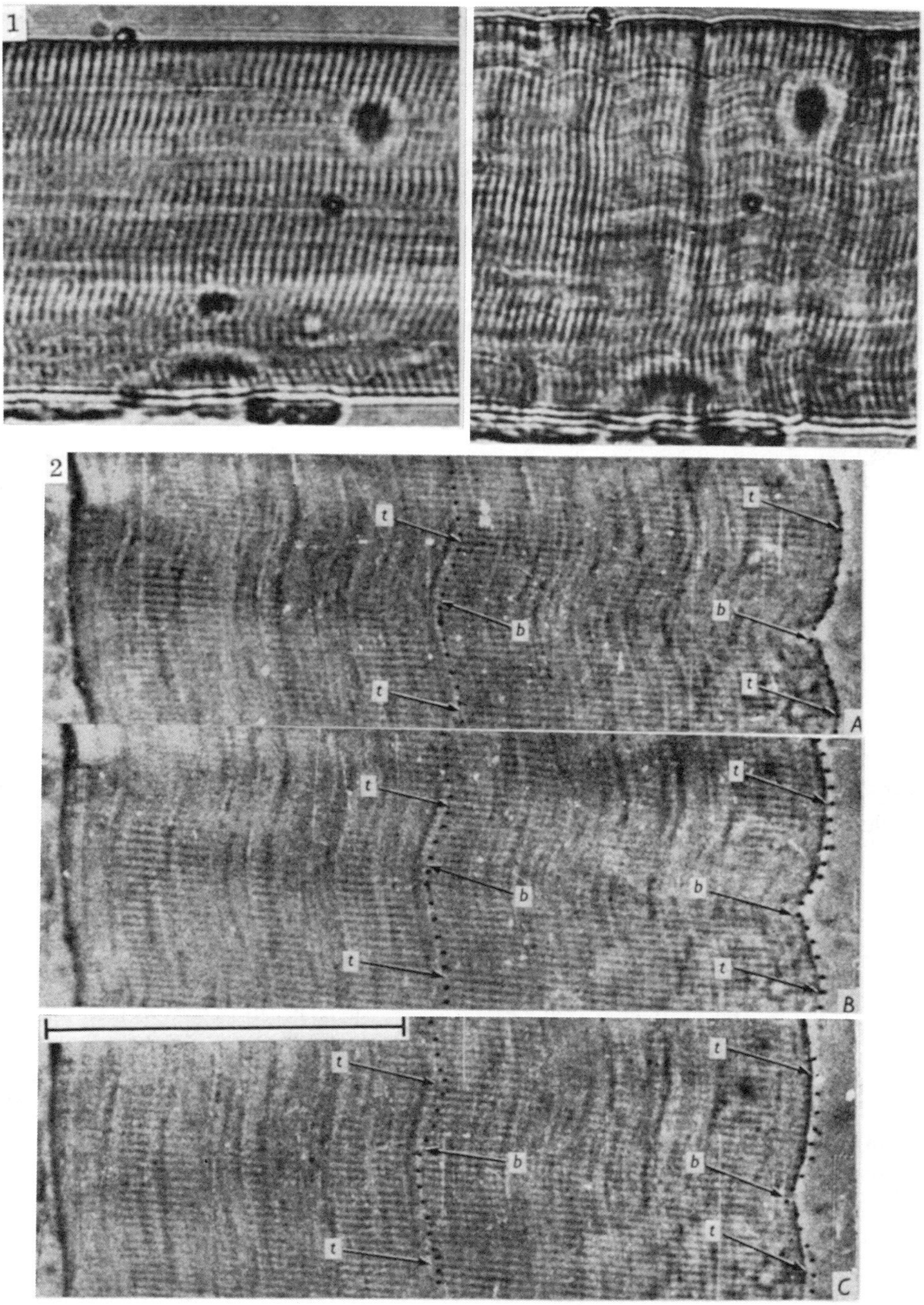

Figure 6.9. Inward spread of activation in cells of winter frog semitendinosus. (*1*) The normal cell (*left*) is placed in gelatin and compressed longitudinally until myofilament clusters are distinctly wavy (*right*). (*2*)–(*4*) The cell is then excited and filmed at 3000 f.p.s. The rate of inward spread can be determined from the relative times of straightening of the wavy myofilament clusters as they contract. Bar = 50 μm. (All from Gonzales-Serratos, 1971.)

with the fluorescent dye merocyanine. Changes of the type associated with some excitable membranes were obtained, but they were eliminated by glycerol treatment that destroyed the ITs without destroying the action potential (Landowne, 1974).

At least two physical changes are detectable before any tension change is detectable in cross-striated cells having ITs. These include a change in optical retardation, associated with membrane changes, and a larger birefringence shift (Baylor and Oetliker, 1977a,b). The latter is associated with a volume change first detected by Abbott and Baskin (1962). Optical transparency also increases (Kovacs and Schneider, 1977). The changes reflect many events of E-C coupling in the ITs, C elements, and SR; the events include a net entry of water into the IT terminals abutting the C elements (Hoyle, 1980). These events may be responsible for the spurious phenomenon, present only in some fast-type cells, of a brief increase in cell length just before the onset of contraction; this increase is termed *latency elongation* (p. 258 = latency "relaxation").

In no invertebrate muscle has there been any indication of all-or-none impulses in ITs: the spread is either passive, or at most aided by decrementing graded responses. Accordingly, even very strong currents applied to the surface activate only myofilament clusters in the vicinity of the stimulus (Figure 6.10.*2*).

6.7. Events at the Coupling Sites

A full understanding of the E-C mechanism demands knowledge of the exact nature of the IT/C contact, but this has not yet been resolved. There has long been speculation of the possibility of channels of communication between IT and C. Many TEM pictures of the contact region have appeared since the first published clear pictures of the gap (Hoyle, 1965) showed that small densities, feet, or projections link the two outer membrane-leaflets. When the gap of the triad of frog semitendinosus cells was stained with tannic acid following glutaraldehyde fixation, the centers of these projections remained clear, though everything else was darkly stained (Somlyo et al., 1979). Thus the center appeared continuous with the cytoplasmic leaflets of the IT membranes and C element (Figure 3.12).

Although nothing is yet known about the events occurring at the IT/SR junction, some experimental testing of major alternatives, such as current flow versus charge transfer, or removal of a plug from a pore, has been made. Glycerol, at concentrations that do not lead to tubule disruption, greatly affects coupling and is known to greatly affect charge transfer. Charge movements are suspected of being involved in coupling (Hui and Gilly, 1979), the most likely site of involvement being the IT/C contact. A model as to how a charged complex might regulate calcium release by unplugging channels has been presented by Chandler et al. (1976).

When frog sartorius cells were treated with neuraminidase, which removes surface-bound sialic acids, there was a shift in E-C threshold that was proportional to enzyme concentration (Dörrscheidt-Käfer, 1979). The pH has long been known to affect the threshold also (April et al., 1968), and part of the action is likely to be by way of charge or tonicity effects.

There is some confusion regarding both the need for calcium in the activation of its release and for the E-C process. Early studies suggested that calcium triggers its own release (Fabiato and Fabiato, 1975), but in studies on isolated frog semitendinosus and iliofibularis cells bathed for long periods in zero-calcium saline, Lüttgau et al. (1977) could still obtain contractions, and Lüttgau and Spiecker (1979) found that magnesium was acting as a substitute in coupling. There was a marked shift in the inactivation without calcium.

Among the complex optical changes occurring during coupling (see p. 212) are some that may be occurring in the SR at the moment of release of calcium. These changes could be due to conformational changes in the Ca/SR binding molecules (Oetlicker, 1980).

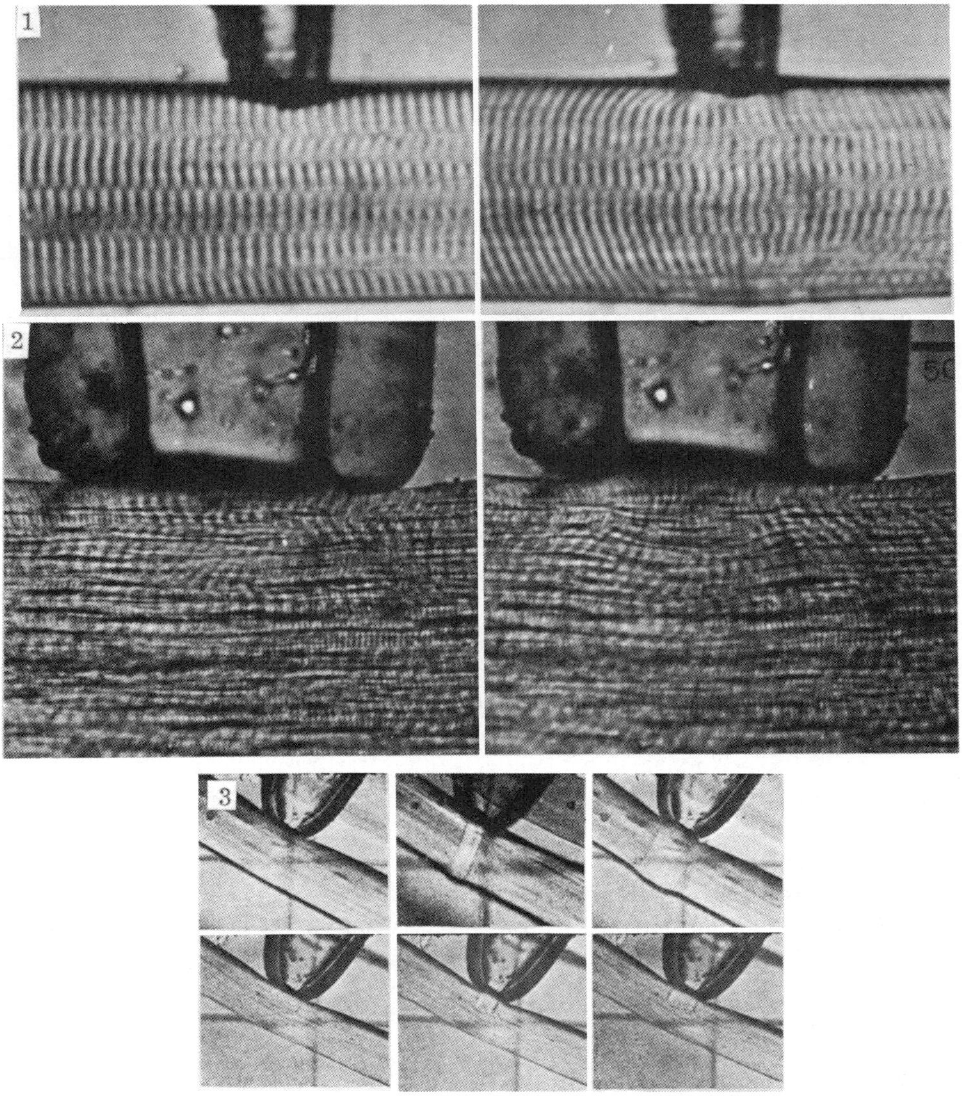

Figure 6.10. Variation in inward conduction. (*1*) Vertebrate fast-twitch cell that probably had an IT inwardly propagated action potential. A small external electrode promotes a ring of contraction in response to a just-suprathreshold local depolarization (*right*). Unstimulated (*left*). (From Sugi and Ochi, 1967b.) (*2*) Arthropod fast-type cell response to even strong depolarizing current passed through a wide electrode with only a local contraction (*right*). (From Sugi and Ochi, 1967a.) (*3*) Lack of propagation in ITs of winter frog (*lower series*) compared with spread around entire cell in summer frog (*upper series*): zero, weak, and strong current, respectively, *left to right*. [Reprinted by permission from *Nature* **212**, 835–836 (1966). Copyright 1966 Macmillan Journals Ltd.]

Calcium itself affects all of the processes related both to its own release and to its sequestration, leading to complex dynamics. One of the effects of a high concentration of external calcium studied by Lüttgau and Spieker (1979) in frog fast cells is its effect on the inactivation process. A high calcium concentration may act not directly, but by impeding the inactivation process. They propose that the plateau of maximum force is proportional to a regenerative Ca^{2+} -dependent shift of the inactivation curve toward more postive potentials.

A calcium ionophore, A23187, is available for experiments but appears to be able to be incorporated into both surface plasma membrane and internal SR membrane, thereby enhancing calcium influx and amount of calcium released from SR at the same time (Desmedt and Hainant, 1976).

A volume change, first detected by Ernst in 1925, can be detected so early in frog sartorius that it must be associated with coupling events. At 20° C it is under way by 2 ms from onset of excitation and has reached its peak at 20ms (Abbott and Baskin, 1962). The magnitude of the volume change is about 10^{-5} ml/g, but is dependent upon initial length (being greater as the length is decreased), extent of shortening permitted, and tension (Baskin and Paolini, 1966). An earlier increase in birefringence that precedes the onset of latency elongation occurs and starts coincidentally with membrane currents. It is thought possible that it is associated with the presumed changes occurring in the IT/C element membrane associated with calcium release (Baylor and Oetliker, 1977a,b).

6.7.1. Seasonal Variations

It has long been suspected that major changes in a variety of muscle-cell properties, but especially E-C threshold, occur on a seasonal basis. In frog fast skeletal cells, all-or-none twitches are readily obtained in the late spring and in summer. Winter frogs are much less excitable, membrane currents are markedly different (Strickholm, 1966), and graded responsiveness is common. The inward spread of excitation is always limited in winter frogs, so probably only in the summer can frogs produce IT action potentials (Figure 6.10*3* – Strickholm, 1974). A 6-mV difference in threshold for initiation of an action potential was noted between winter and summer frog (*Rana pipiens*) sartorius by Jenerick and Gerard (1953). Working with the effects of external calcium concentration on the properties of single frog (*R. pipiens*) cells, Caputo and Gimenez (1967) obtained very different results when using winter (October–April) frogs compared with summer (May–September) ones.

Crayfish muscle cells are also less excitable in winter (Hoyle and Wiersma, 1958), when the chloride equilibrium potential is low, causing depolarizing inhibitory potentials. A seasonal variation in sensitivity of frog hearts to ACh has been noted (Boyd and Palhak, 1965).

6.7.2. Are There Differences in the Rates of Coupling in Different Types of Cell?

Clearly the fastest muscle cells are fully activated within a few milliseconds of suprathreshold excitation, whereas very slow cells require up to several minutes. In part the differences reflect different biophysical properties of the cell membrane and ITs, different anatomical features such as paucity of tubules and SR and long diffusion distances from calcium-release sites to contractile elements, and differences in the contractile machinery. Is the sum of these differences sufficient to account for the differences in activation times, or are there also differences in E-C actions? Comparison of voltage-clamp data on vertebrate fast-type cells with those on a range of cell types obtained from the crayfish *Orconectes* epimeralis muscle (Orkand, 1962a) strongly suggest that speed differences occur in the coupling mechanism also: indeed, this could be a major source of variation.

6.7.3. What Is the Coupling Event in Contractures?

Although contractures in fast-type cells are thought to be due to calcium's directly entering the cell, the E-C coupling processes having been inactivated, in frog toe cells disruption of the ITs abolishes contracture (Kawata and Chen, 1972). Therefore calcium entry in these cells must be by way of the tubules, not the surface membrane. Normal contraction of slow-type cells is comparable to contractures in its slow time course and freedom from inactivation, but the two processes should not be directly compared. Contracture, as classically defined, is a secondary phenomenon of fast-type muscles that may have no functional significance. In slow-type cells calcium not only enters directly across the surface membrane, it may also be caused to be released from SR and mitochondria by the contractural agent.

6.7.4. Is There Facilitation in E-C Coupling?

Facilitation and antifacilitation in neuromuscular systems are often known and described for the overall relation between neural input and contractile output. The relevant events could therefore be either synaptic or intra-muscle cell. Facilitation is an extremely important event functionally at most arthropod neuromuscular synapses, as it is at many vertebrate and invertebrate central-nervous synapses. The mechanism is nowhere fully understood, although calcium is often involved. But many different kinds of molecular events could cause similar-seeming physiological ones. Antifacilitation is less interesting, because it is restricted to fast actions. It could result from depletion, or by accumulation of a by-product that interacts with active sites, progressively blocking them. During recovery the substance would diffuse away and/or be enzymically destroyed.

Pantin (1935) and some other early comparative muscle physiologists studying preparations that showed dramatic overall mechanical facilitation doubted that all of it was due to synaptic events and alluded to "intra-muscle-fiber facilitation" that, in modern thinking, would be in the E-C chain. Sugi and Ochi (1967a), for example, found that two 50-ms-duration direct pulses gave a stronger contraction in a crayfish muscle than did a single 200-ms pulse at the same intensity. This phenomenon has been directly studied in giant barnacle cells injected with aequorin by Hoyle and Ridgway (unpublished – Figure 6.11). The second of a close pair of depolarizing pulses initiates a much stronger calcium response, even allowing for some summation at the shorter intervals, than the first. The effect declines progressively with increasing interval. Likewise, a brief depolarizing pulse superimposed on a background of weak depolarization is progressively more effective the closer it is to the onset of the background (Figure 6.11). This is, perhaps, an equivalent to the increase in transmitter release from the presynaptic terminals at neuromuscular junctions, where the arrival of an impulse not only releases available transmitter but initiates the mobilization of a larger amount that is then released by the next close impulse. Prolonging the depolarization results in continued release but is not as effective

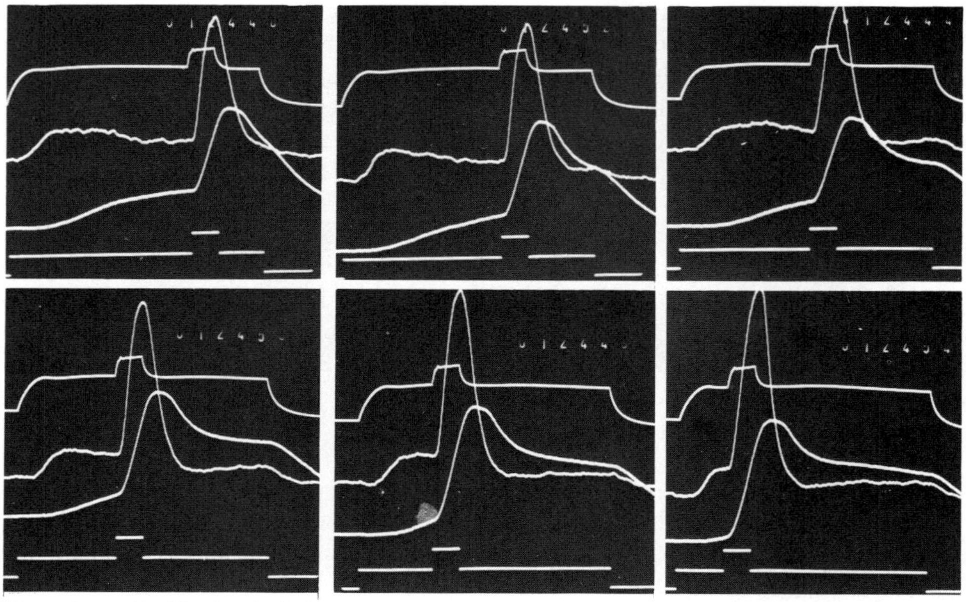

Figure 6.11. Facilitation in E-C coupling. Giant barnacle depressor cell injected with aequorin. The four traces in each record are (*from top*) membrane potential, calcium transient (= aequorin light), isometric tension, and current. A brief constant-current depolarizing pulse was applied at various times during a background long, weak depolarizing pulse. The size of the response to the brief pulse increases progressively as its onset is brought closer to the onset of the background depolarization. (ORIG. – Hoyle and Ridgway, unpublished.)

as switching off and turning on again. At the presynaptic terminal, calcium entry is a key event, and very likely is involved also at the SR cistern.

It is expected that both the links between IT depolarization and calicum release from C elements of SR, and the release will be found to be major sources of variation in diverse muscle cells, as have synaptic transmission and membrane excitability.

6.7.5. Variation in Properties of Individual Coupling Elements and Its Relation to Graded Contractions

It has long been known that thresholds of individual muscle cells of the same type in the same muscle vary rather widely, and in light of what we now know about the complex anatomy and physiology of coupling, this is hardly surprising. During graded activation of a cross-striated cell, parts of half sarcomeres of myofilament clusters range from supercontracted to strongly stretched (Figure 6.12 – Hoyle, 1966), indicating dramatically not only that the functional unit in coupling is the individual IT/SR contact, but also that events at these are activated by the common stimulus in a graded manner, with large individual differences. A similar heterogeneity is also to be seen in vertebrate cells fixed rapidly during graded excitation (Figure 6.12.*1*). The total contraction is the statistical sum of all the graded local events. For a single sartorius cell of a large frog, say 100 μm in diameter by 5 cm long, the number of such units is approximately 6.5×10^6. For a fast-type crab leg cell with the same dimensions, at least twice this number, about 1.5 ×

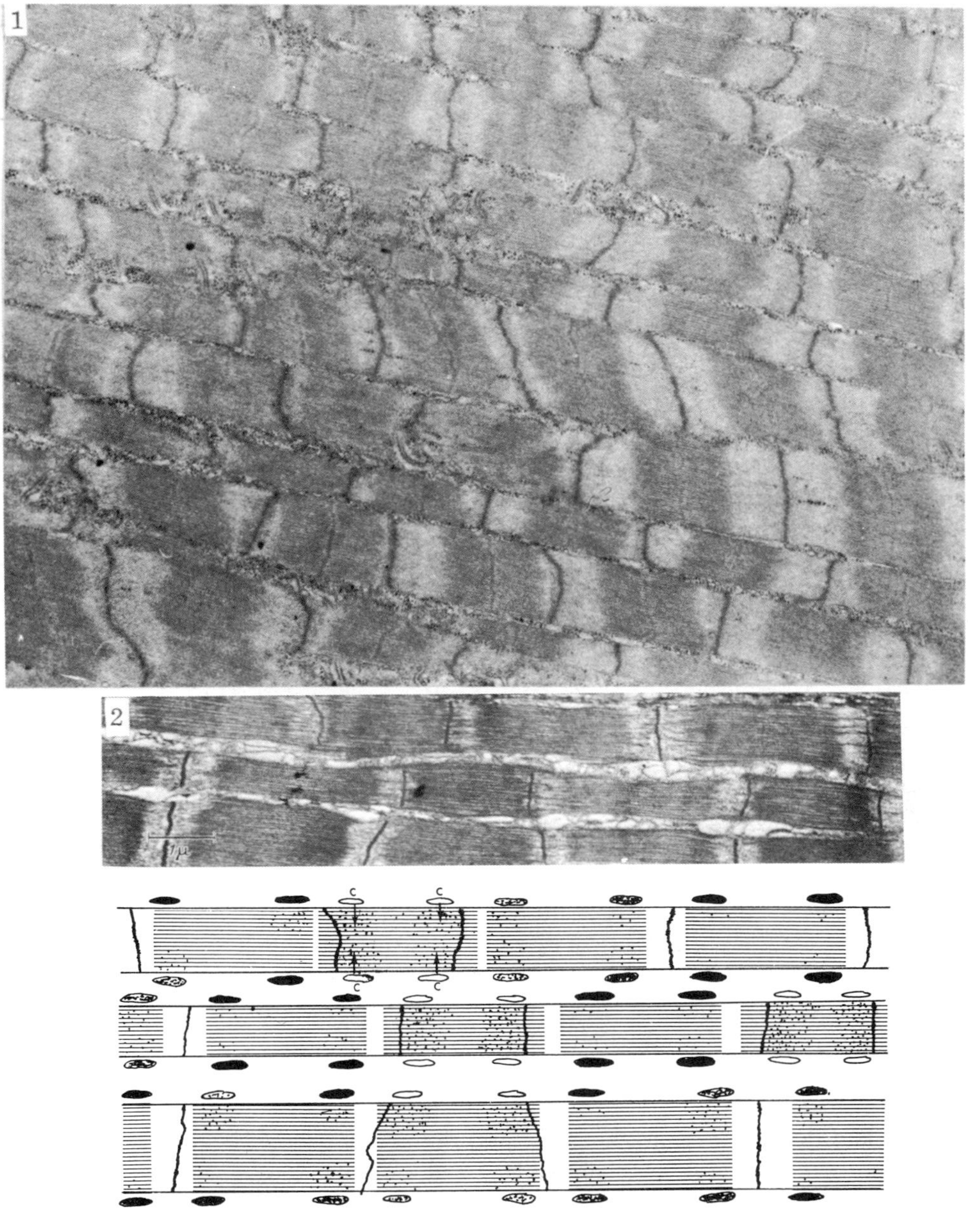

Figure 6.12. Variation in activator release from individual cisternal elements – basis for graded contraction. (*1*) EM of a vertebrate cell, fixed in graded contraction. Individual half sarcomeres, or even parts of them, are contracted at the expense of their neighbors, which are stretched out. (ORIG.) (*2*) Same as (*1*), but of a crab (*Portunus*) cell. The most extensively shortened sarcomeres have gone into supercontraction, with thick filaments passing through the Z regions. An interpretation, based upon different extents of local activation of C elements, with various extents of release of Ca^{2+}, is shown in the accompanying diagram. (From Hoyle, 1966.)

10^7, occur. In slow- and intermediate-type cells, there are fewer, from one quarter to one fiftieth as many, but more than enough to make it very likely that contraction is based upon a heterogeneous mix of local shortenings and elongations.

6.8. Calcium Release in Relation to Design and Location of IT/SR Contacts

There are only two basic designs for IT contacts with C elements of SR, the triad with its occasionally pentadic variant, which is characteristic of vertebrates, and the dyad that is predominent in all invertebrates. An advantage of the former is that two sets of C elements are activated at the same time. This is especially significant when the triad is located at the Z line, as it is in amphibia (Figure 6.13.*1*), since then two contiguous sarcomeres are excited together, aiding synchrony and therefore optimizing smooth tension rise. But this location brings with it the disadvantage that the released calcium has to diffuse over a considerable distance to reach the contractile machinery, located in the A/I overlap region. The total duration of the state of activation in a twitch (for definition and description see p. 232) increases linearly with increasing sarcomere length, as the anatomy predicts, for frog semitendinosus cells (Edman and Kiessling, 1971). The action is, however, further prolonged because of two effects of length that enhance calcium release (see p. 197). This not only introduces a delay, it also creates a need for an exceptionally high calcium concentration at release and so leads to rapid fatigue. Furthermore the diffusion distance is variable, depending on the extent of stretch, being greatest when the least time is functionally desirable, to restrain a contracted antagonist. Reptiles, mammals, and birds have placed the triads at the edges of the A band, trading off the advantage of synchrony for greater activation speed (Figure 6.13.*3*). This shift is correlated with a doubling of the number of release sites per sarcomere.

Invertebrate muscles generally have dyads at several places along the length of the A band, ensuring minimum delays in activation onset (Figure 6.13.*2*). In some slow muscles, however, there are only two sets of dyads, at the extreme edges of the A band, that are attached to the Z regions. There are also dyads associated with Z tubules that may be calcium-release sites also. In such muscle cells, which may have sarcomere lengths over 20 μm, the diffusion distance is long, and as in amphibian muscle, it increases with stretch. Direct tests of the relation, at different lengths, have been made by Gillis (1969) for a 15-μm SL crab (*Cancer*) claw cell. He used a pipette containing calcium to iontophorese calcium onto a Mg-ATP/EGTA-treated glycerinated crab cell to obtain a just-visible contraction at different sites along the sarcomere. The current required, which is an indication of proximity to contractile material, was least at the A/I junction.

6.8.1. Effect of Length on Coupling

The effects of length on different kinds of muscle cells are probably very different. The directly determined effects of stretch on barnacle giant cells are sufficiently impressive to indicate that caution is needed in using length steps as the major experimental variable, as is very often the case in biophysical tests. Constant-current pulses were applied to aequorin-injected giant barnacle cells and the effects of 1% increases and decreases of length noted (Ridgway and Gordon, 1975). A small increase in resistance, associated with about 2 mV depolarization and greater calcium release, occurred with the 1% stretch, and the inverse, decreased resistance with about 1 mV polarization, occurred with 1% shorten-

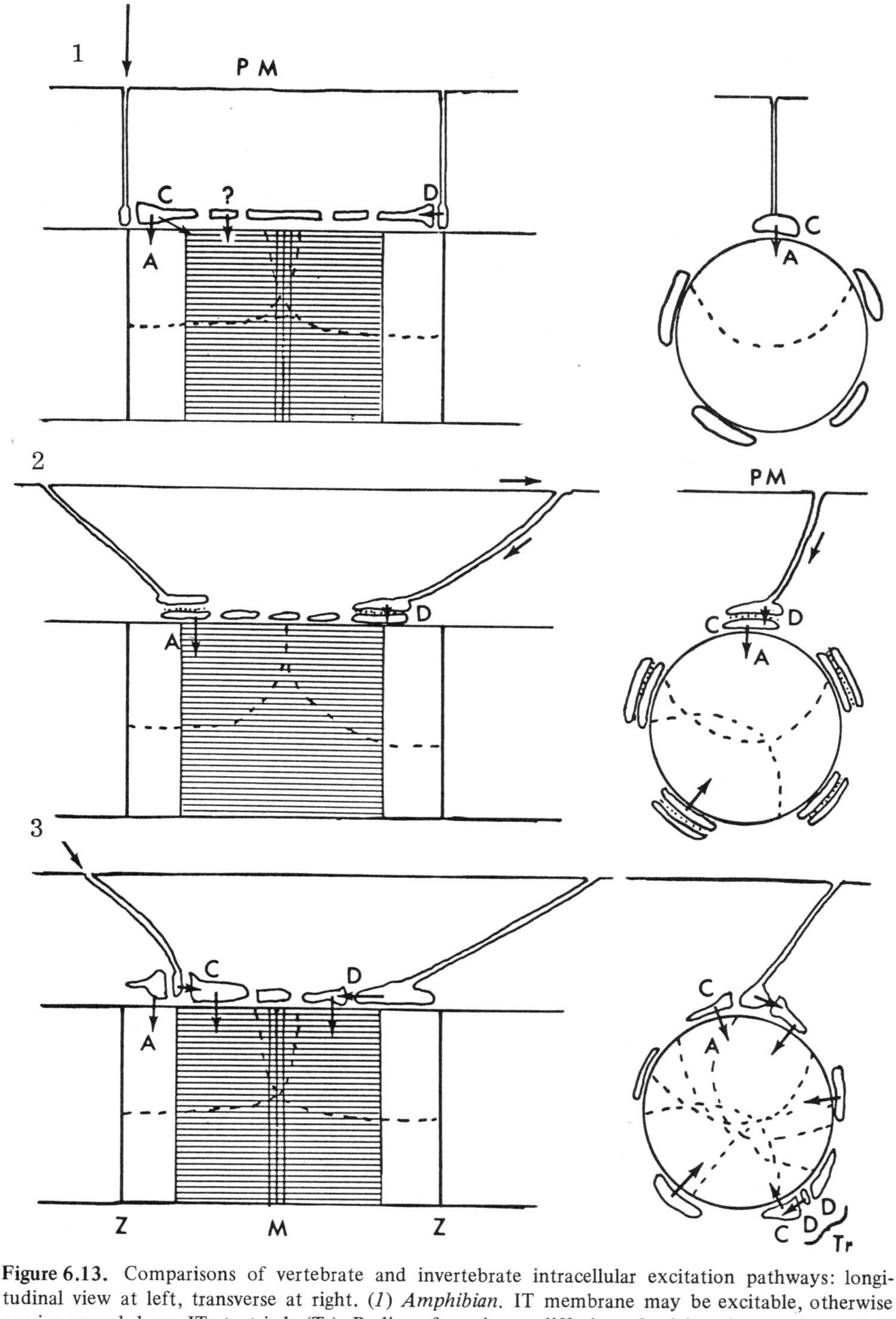

Figure 6.13. Comparisons of vertebrate and invertebrate intracellular excitation pathways: longitudinal view at left, transverse at right. (*1*) *Amphibian*. IT membrane may be excitable, otherwise passive spread down ITs to triads (Tr). Radius of maximum diffusion of calcium from a cisternal (C element) is indicated by broken lines. This is the least efficient because diffusion distance to contractile material is large and increases with stretch. (*2*) *Typical arthropod*. Release is from dyads (*D*) placed very close to contractile material. (*3*) *Mammal/fish/snake*. Triads are located at outer part of A/I overlap region: some radial, some longitudinal. Efficient except for calcium release into I region. (Reprinted from *Invertebrate Nervous Systems*, pp. 151–167. Ed., C. A. G. Wiersma. Specificity of muscle by G. Hoyle. © 1967, by permission of The University of Chicago Press.)

ing. The resistance and membrane-potential changes were probably not the only factors, being superimposed on internal effects. Calcium is released by stretch, even after glycerination, from both rabbit psoas and cat heart muscle (Brenner, 1979).

The internal, and to an extent also external, indirect-stretch-activation processes do have a useful concomitant, since they provide partial compensating mechanism for local inequalities in activation (see p. 153), which are a major factor in determining muscle dynamics (see p. 268).

In calculations based upon the original A. F. Huxley model extended to allow for a percentage of X bridges that change with time, an attempt was made to model the exponential increase in tension to the plateau value in an isometric tetanus (Mittenthal, 1975). Predictions did not agree well with observed dynamics. One reason for this may be that the theory does not allow for the complex effects of intrinsic inequalities in sarcomere length and in local activation.

The amount of Ca^{2+} released per impulse in single frog and toad muscle cells, as determined with injected aequorin, increases with the first slight increase in length above slack length and then decreases progessively with increasing stretch, becoming less than half maximum (Blinks et al., 1978). Why this should be so is a matter for speculation; it may be due to distortion of the IT/SR junctions or to an effect of stretch on SR cisternae. In fast-type vertebrate cells, with a large excess of calcium over the minimum required for saturation of AM binding sites released by each impulse, the changes with length should be insignificant in practise. There might be a slight effect on twitch kinetics. But in a slower-muscle cell such as those of giant barnacles, in which the calcium level is well below saturation in a single excitation, the contraction characteristics are quite strongly dependent on length, assuming a similar effect of length on calcium release. In giant barnacle cells, I found the maximum velocity of contraction to be actually increased, not decreased as expected, by addition of a very small load, sufficient just to stretch the cell by a few percent. Here the improved calcium release more than compensated for the load inertia.

6.9. Calsequestrin, Calmodulin, and Coupling

Relatively little attention has been paid until recently to the material within SR that binds calcium and the regulation of its release, partly because these seemed peculiar to muscle. But there has recently been an explosive interest in a small protein that may have a subsidiary role, because it has turned out to be extremely widespread. The former protein is generally called calsequestrin and the latter calmodulin, names that immediately convey their presumed roles. Calsequestrin is a glycoprotein with a molecular weight of about 110,000, and it is tightly bound to the SR membrane. Various detergents release it. Isolated, it can bind 30–40 Ca^{2+} mol. There are thought to be other forms, with about half the above molecular weight, that are less efficient at binding, and there is also a very efficient nonglycosylated protein that has the same pattern of synthesis in cultures of rat skeletal muscle cells (Marek and MacLennan, 1980).

Calmodulin is a relatively small, 17,000-dalton, protein that is in effect a go-between for calcium and the many proteins that bind it. Itself a strong binder of calcium, calmodulin is caused to change conformationally by the interaction into a form that is in turn bound by any of a number of enzymes. This activates the enzyme to carry out its basic function, so the calmodulin is a regulator of activity. This role is clearly identical to that played by TN in vertebrate skeletal cells (see p. 225). In fact, the two proteins are related, with closely similar amino acid sequences and therefore three-dimensional shape. Calmodulin may play the stimulatory role for calsequestrin, thereby regulating calcium uptake and, what may be of great interest functionally, its release.

In skeletal muscle of vertebrates, an important development has been the finding that the enzyme phosphorylase kinase has as one of its four constituent protein subunits none other than calmodulin (Cohen, 1979).

6.10. Overall Quantitative Coupling Control

It is now abundantly evident that each muscle-cell type, of which the total number is astronomical, has its own individual calcium relations. At one end of the spectrum are very fast twitch vertebrate cells in which every aspect is geared to flooding the contractile machinery with an excess of calcium in the least possible time. The surface membrane gives an overshooting action potential that is as large and fast as any muscle cell can possibly produce. The ITs in spring and summer frogs are also capable of spiking, to rush the excitation inward with the least delay to a very extensive set of primed IT/SR junctions. The C elements are loaded with instantly releasable calcium that has barely a half micron to diffuse to get contraction under way. Dynamic testing using quick stretches (see p. 234) confirms the expectation that the contractile machinery of vertebrate fast-twitch skeletal cells is 100% activated even in a single twitch. For about a century this state of affairs has been regarded as the norm, the standard. It is, in fact, an extreme case far from the median!

At the other end of the spectrum are slow cells, skeletal as well as visceral, striated, and smooth, with electrically inexcitable membranes, no or very few ITs, sparse SR, and few calcium-release sites. Even when activated as rapidly as possible, such cells take up to several minutes to reach the maximum tension they are capable of, and they may never reach a state of full activation. A majority of muscle cells in the spectrum fit in between. They are not even nearly fully activated by a single excitation; it takes several excitations in rapid succession to achieve the highest level.

In all heart muscle, the rate of rise of tension is quite slow compared with that in fast skeletal cells, and tension continues to develop throughout the large and very long depolarization plateau. The calcium entering has time to control the future time courses of the action potential and its own further entry (Bassingthwaite et al., 1976), both of which are activated to the point of saturation in spite of the long time (Edman et al., 1966). This feature has important functional implications. Not only is it well matched to the task, but it leaves something in reserve for activation independently by stretch – hence Starling's law (see p. 372) – and by actions of modulators that are termed *inotropic.*

One of many neat tricks that diverse muscle cells have evolved is displayed by some insect muscle cells that defy simple classification since they have both fast and extremely slow capabilities in spite of receiving only a single excitatory axon that gives quite large EJPs (Hoyle, 1978b). Low frequencies of excitation, 5-10 Hz, evoke a prolonged, but very slowly rising tension. The final tension achieved after very long stimulation at the low frequency is almost equal to the maximum for the cell. The maximum tension is achieved very quickly when a few excitations only are given at a high frequency 120–160 Hz. The feature that bestows this unique ability is in the muscle cell, not in the synaptic physiology. The periphery only is well endowed with short ITs, much SR, and many

IT/SR contacts. This part is saturated with calcium after a single excitation, giving a brisk twitch, but it utilizes only a fraction of the contractile material. The remainder of the cell can be activated only slowly, because calcium must diffuse into it from the peripheral release points across about 50 μm. It can also relax only slowly because it does not contain SR. The core becomes excited only when excitations are repeated, either slowly for a long time or rapidly for a brief one.

The calcium uptake/release mechanisms may themselves be subject to direct, as well as the many indirect, effects of stretch. There is evidence that this is a contributing factor in heart muscle. When aequorin was injected into rat ventricular cells and the light output compared at different lengths over a 27% range, peak output was constant, but the rates of decay declined progressively with increasing length (Allen and Kurihara, 1980). Thus effects on sequestration are implicated, decreased at shorter lengths, and vice versa. The E-C coupling is reversibly blocked by very high concentrations of calcium and calcium buffer, but lower concentrations of the buffer with the same free-calcium concentrations do not block (Barrett and Barrett, 1978). Together with the fact that many muscle cells continue to contract for a very long time in zero-calcium saline, these results suggest that calcium in the ITs is continually replenished from within the muscle cell. This calcium is sufficient to serve the presumed role of that ion in coupling.

Variation in any of these components of coupling can be utilized to help adaptation and acclimation. It is known, for example, that cold-adapted frogs require a smaller extent of depolarization for contraction than do warm-adapted ones: their E_C is higher and E_m lower, than in the warm-adapted (Kovacs and Schneider, 1978).

6.11. Inactivation and Priming

A previously unstimulated muscle cell at its resting potential is "primed" ready for contraction. When suprathreshold depolarization is applied and continued, the response, as we have seen, varies drastically with cell type. Fast-type cells relax spontaneously at various times after onset and at various rates, depending on the type. This relatively abrupt type of relaxation of fast cells under continued stimulation is termed *inactivation.* It is thought to be not associated with a fatigue or accommodation process but rather an active shutting off somewhere in the E-C coupling chain. It is a characteristic of fast-type cells aimed at helping them perform their functions efficiently. A few fast cells recover sporadically under continued stimulation and will contract from time to time, but most require a period of nonstimulation. After repolarization, the ability to contract in response to depolarization returns rapidly, a process termed *repriming.* These cells may secondarily develop another type of contraction, termed a *contracture,* which rises very slowly and is maintained by continued stimulation. This is not the characteristic functional contraction for the cell and is probably caused by slow entry of external calcium. Repriming has been studied intensively in fast-type cells of the garter snake *Thamnophis* under voltage clamp (Heistracher and Hunt, 1969b). Here tension falls after about 3 s, at rates dependent on the extent of depolarization. Polarization to -100 mV restored the ability to give just detectable contraction in 0.38 s, but a repriming time of 4.0 s was needed to just obtain a maximal contraction. The repriming to half maximum tension was 0.68 s (Figure 6.14).

By contrast, in slow-type cells there is no inactivation, so they naturally maintain tension for very long periods, up to a few days or even indefinitely, under maintained depolarization. Futhermore, in some cases the external calcium can be completely re-

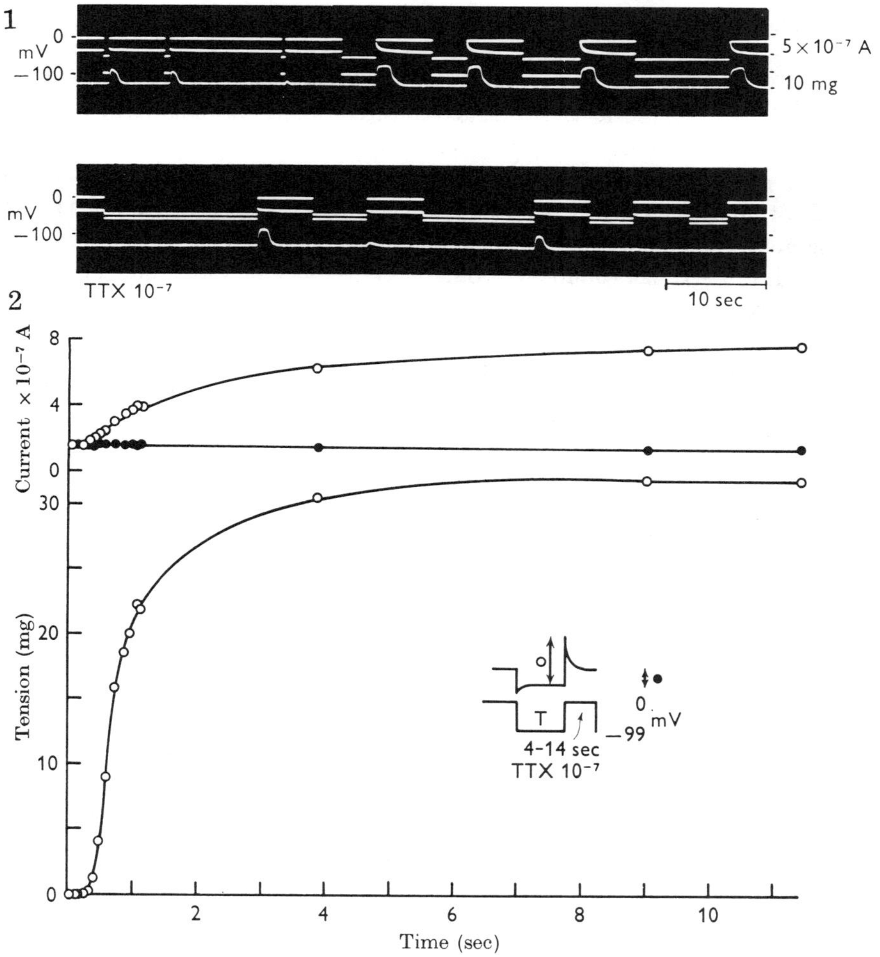

Figure 6.14. Tension repriming in snake fast-type muscle cell. The action potential was blocked with TTX and the cell depolarized by current until the contraction had declined to nil. After passing suprathreshold polarizing current pulses, tension returned. (*1*) The effect of different strengths and durations of polarizing current on repriming: (*upper*) -97 mV, (*lower*) -58 mV. (*2*) Course of repriming at -99 mV. Upper graph shows peak current (*circles*) and steady current (black dots), caused by current pulses. (From Heistracher and Hunt, 1969b.)

moved without impairing the cell's ability to maintain maximal tension (Gilly and Hui, 1980b). This means that in these cells, the internal calcium forms a closed system, at least under depolarization. Calcium is probably not returned to the SR until depolarization ceases, remaining in the myoplasm to continually activate the contractile machinery. Thus it is not to the contractile machinery that we should look for the inactivation process: instead, it must be in the IT/SR link, but where exactly is not known.

At normal operating temperatures, repriming of the contraction ability and repriming of full spiking ability return at similar rates, but as the temperature is lowered progressively, repriming lags behind refractoriness. Eventually, alternate twitches drop out, at 3°C for frog fast cells, even though full action potentials are being produced (Caputo, 1972a,b).

6.12. Regulation of the Contractile Molecules

The next, and probably final, step in the excitation sequence is the activation of the contractile machinery by the released calcium ions. However, it is important to stress that although calcium is certainly the principal determinant of contraction, the way should be left open for the discovery of other factors that supplement or reduce calcium actions. The levels of myosin ATPase activity at the actual concentration of free calcium present in resting muscle are higher than they should be solely on the basis of calcium activation (Perry et al., 1979). The multiple forms of TN present in vertebrate skeletal muscle suggest that several regulatory systems have been – or are still – present to permit the evolution of such a complex arrangement. Quite apart from possible actions on the ATP hydrolysis itself, other agents could affect the uncovering of binding sites, regulating the number of X-bridge contacts made, or their recycling rate.

For striated and cardiac muscle cells, there are two somewhat different ways in which calcium initiates action. The method that has been adopted by vertebrates and higher invertebrates is indirect. Magnesium-activated ATPase, which is the conjugate form of myosin heads with actin, drives the action. The function of calcium is to lead to the physical uncovering of otherwise protected sites that are capable of forming this complex double-molecule enzyme. The sites are on some of the actin molecules, and they are located at intervals on the I filaments. Keeping the actin binding sites covered in a relaxing muscle is the role of troponin; ensuring that TN is in the appropriate place for this blocking role is the task of the tropomyosin to which it is attached.

There are three separate protein components of TN: TN_T, TN_C, and TN_I. The TN_C has a strong affinity for calcium and immediately binds Ca^{2+} after this has been released and diffused toward it. The Ca^{2+} is also attached to TN_T and TN_I, and when Ca^{2+} binds to it, these links are altered, resulting in motion of the complex (Figure 6.15). The motion

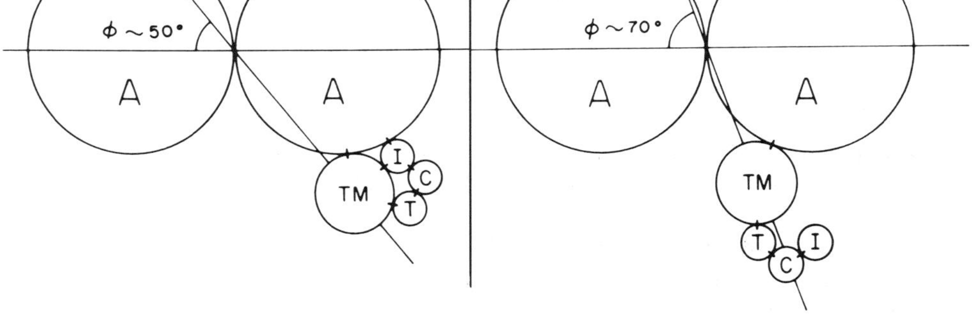

Figure 6.15. Scheme for movement of tropomyosin (*TM*) during activation, which is brought about by calcium ions. The movement (*right*) causes troponin, which comprises three linked molecules *T, C,* and *I*, attached to the TM via the T component, to be swung away from the actin (*A*), uncovering the binding site (*BS*) for myosin. Based on X-ray studies. [By permission of the American Heart Association, Inc. J. Gergely, *Cir. Res.* **34**, Suppl. III, 74–82 (1974).]

uncovers the myosin-binding site on the actin. Rabbit skeletal muscle TN_C binds 4 Ca^{2+}/mol, 2 with high affinity and 2 with low affinity (Potter and Gergely, 1975). In cardiac muscle, a slightly different form of TN_C binds 3 Ca^{2+}, 2 with high, and 1 with low, affinity (Laevis and Kraft, 1978). The two high-affinity bindings may be the ones that are responsible for the evidence that two Ca^{2+} ions are used for each unit contraction site. Proton magnetic resonance of calcium has been used to obtain clues as to how the calcium binding to TN_C induces the conformational change and how this is transmitted through the other components (Levine et al., 1977). Calcium binding to the two high-affinity sites leads to both a change and a stabilization of the structure of TN_C, and that may be the only active change. TN_I is there solely to inhibit the AM ATPase, and TN_T solely for attachment to TM, and myosin in its vicinity quickly attaches to TM_T, permitting the onset of contraction. Changes in myosin have been discussed (see p. 000). There is also evidence that a conformational change occurs in actin (Szent-Györgyi, 1965). Possibly the latter is in association with the binding to myosin. The elucidation of this model was very largely due to Setsuro Ebashi and his associates, in work that has been summarized by him (Ebashi, 1979) and by Carolyn Cohen (1975).

In molluscan muscles, no TM has been detected, so they must use a different mechanism. Their thin filaments cannot regulate rabbit myosin in relation to calcium, so attention focused on possible regulation by the thick ones. It was found that the heads of molluscan myosins can indeed bind calcium. In scallop it proved possible to remove light chains from the myosin heads without also destroying their ATPase activity (Kendrick-Jones et al., 1973). Calcium binding and regulation of ATPase was lost along with these light chains, which must be responsible for control; they have therefore become termed the *regulatory light chains.* Once they have been removed, full tension develops in the absence of calcium in a solution that causes relaxation in the normal muscle. Presumably the light chains cover the myosin-head site, and interaction with calcium exposes it. This mechanism is, in principle, no less efficient, and perhaps more so, than the TM/TN system. It is certainly simpler. The two systems presumably represent two historically different solutions to a common problem, so it is of interest to know which classes of animal have which kind. Lehman et al. (1973) found exclusively myosin-linked regulation in nemertines and sipunculids as well as the molluscs. In polychaetes and heteropteran insects, they found evidence for both myosin and TN regulatory systems. Crustaceans, Chelicerata and other kinds of insects – the majority – have TN. There is no clear story arising from this distribution, except that animals with either an internal or external articulated skeleton have TN, and that myosin regulation is probably the primitive feature, TN the advanced one. This leads one to question why there was TM on the actin in the first place, since this must have preceded the shift to it of the regulatory molecules. One possibility is that TM added stiffness to the actin. This would, perhaps, have conveyed the capability for greater speed at higher loads that is required to make the nearly isometric contractions associated with the marked mechanical disadvantages associated with both forms of skeleton.

Finally, it is necessary to point out that calcium could be having effects on the contractile machinery in addition to those it plays in exposing myosin-actin interaction sites. When studying the action of calcium concentration on the force-velocity relation of very briefly glycerinated frog single fast muscle cells, Julian (1971) found that the relation did not change over a substantial part of the range of SL in which the force generated was increased, in accord with increasing overlap of I and A filaments, when the calcium concentration was relatively low. He considered it possible that calcium may have a direct influence on the X bridges. Another possible target is, of course, the T filaments.

7

Dynamic Aspects of Contraction

The elementary unit of contraction in skeletal muscle is a single twitch. A. V. Hill[†]

Professor Astbury has said that a contracted muscle is like steel, but this is perhaps not literally true; in fact its mechanical properties are more like rubber. M. G. M. Pryor[‡]

7.1. General Dynamic Features of Muscles

Muscles are dual-purpose machines required to provide force on one hand and movement on the other. These are fundamentally conflicting requirements, so there is a mutual trade-off between the two, the maximum force being developed when there is no movement, and maximum movement rate or extent when there is no load. To reduce the relation to measurable terms, muscle physiologists have concentrated their research efforts on working with one of the extremes. Contraction is *isometric* when force only is measured while permitting negligible shortening, and *isotonic* when length change with time, velocity, is determined under constant load, usually all the way from zero to the maximum that can be borne without lengthening. Three different loading conditions are generally employed: *preload,* in which the load is applied before excitation; *afterload,* in which the load is applied only after the muscle has been excited and has shortened by a very small amount; and *auxotonic-load,* which is variable, mediated by an elastic element such as a spring, in which initial resistance to shortening is slight, but load increases progressively with increased extent of shortening. The latter is commonly used for studying heart muscle, where a spring best simulates natural load conditions, but is little used in studies on skeletal muscles.

To obtain a comprehensive understanding of the functioning of a given muscle, the ideal is to have a combined knowledge of whole-muscle properties plus those of representatives of all the major cell types comprising the whole. It is now realized that little analytic value is to be gained from making experiments on whole muscles because of wide differences in the properties of component cells, compounded by effects of connective tissue between them. Even a single cell may be too inhomogeneous internally to permit satisfactory analysis because of different sarcomere lengths of component myofilaments and regional inequalities in the effects of stretch and shortening. To obtain quantitatively reliable relationships of some parameters, experiments must be restricted to small-diameter, optically clear cells with easily visualized striations and in which a region can be found where sarcomere lengths are indeed equal. These are necessary for

[†]A. V. Hill, *Proc. R. Soc. Lond. Ser. B* **137**, 40 (1950).
[‡]M. G. M. Pryor, *Proc. R. Soc. Lond. Ser. B* **137**, 81 (1950).

testing theories of contraction, but for assessing function in the body it is the summed properties of the whole muscle that are of interest, and the effects of the heterogeneity are significant features.

It is relatively easy to measure tension satisfactorily, even at levels of a few tens of millinewtons (mN). By contrast, it is impossible to measure shortening speeds satisfactorily at zero or very low loads because of internal and external viscosity, elasticity, inertia, and friction. Uniform stimulation is achieved by placing the muscle or cell on a grid formed of alternating-polarity platinum wires. Length change is invariably measured by optical methods, commonly by a strip of photographic film gradedly exposed so that it provides a density "wedge," or by crossed polaroid strips. Tension is determined by strain gauges either independently attached or simply part of a light, low-inertia lever system to which the muscle is attached, a silicon rod whose resistance changes with force, or a mechanoelectronic transducer such as the RCA 5734 tube. In the latter, a movable anode projects through the metal case and to it the muscle is attached.

The most characteristic dynamic relationship for a muscle or muscle cell, and the one most commonly examined, is its *load,* or *tension-velocity* relation. Often the load is divided by the mean cross-sectional area because this signifies *force.* However, since there is a wide range of proportions of the cell(s) that constitute contractile material, this value has little significance. The muscle/cell is subjected to equal maximal stimulations, and the maximum velocity is determined at each of a number of fixed loads. From the results the maximum velocity V_{max} versus load curve is drawn: for all muscles it has the shape of a rectangular hyperbola (Figure 7.1). The ordinate, V_{max}, varies for different muscles over a hundred thousand-fold range, whereas the abscissa, tension per unit diameter, has a narrow, not more than 10-fold range, from 1 kg/cm^2 to about 10 kg/cm^2.

In general, the slower muscles also have the larger tensions per unit area of cross section, and the curve is more nearly a straight line. When the muscles are less intensively activated, the hyperbola for these muscles becomes more rounded. It is not reasonable to assume that superstrength electrical stimulation will necessarily cause a maximal contraction, since the extent of contraction depends on electrical properties, internal ultrastructure, available calcium, and so forth, of the cell(s) being tested. For many slow cells, the degree of activation is adjustable over an essentially infinite range, but many cannot be excited to the maximum extent their contractile machinery is capable of, except possibly after special experimental treatment. For such cells, a consistent or meaningful V_{max}-load relation cannot be obtained.

With increasing intensity of activation, there are gains in force but little in velocity. The symbol P was used in early formulations to denote load per unit cross-sectional area, and P_o to denote the maximum load capable of being lifted at optimal length divided by the best obtainable estimate of average area of cross section. Both P and P_o have the dimension of force, and would be true indicators of force if the area of the contractile material were known. As long as it was believed that everything inside the sarcolemma is contractile, an estimate of gross cross-sectional area seemed more than adequate. But ultrastructural studies have shown that this is misleading and for many cells quite erroneous. In some kinds of skeletal cells, more than half the entire volume is occupied by mitochondria that do not share any load; in others SR, whose sole function is to sequester and release calcium, occupies more than 60% of the volume. Even for muscles with few mitochondria and little SR, a cylindrical shape, as always assumed, is in reality relatively rare. Skeletal cells often taper at one or both ends, and they have unequal shapes along their lengths and complex, branching skeletal attachments, that is,

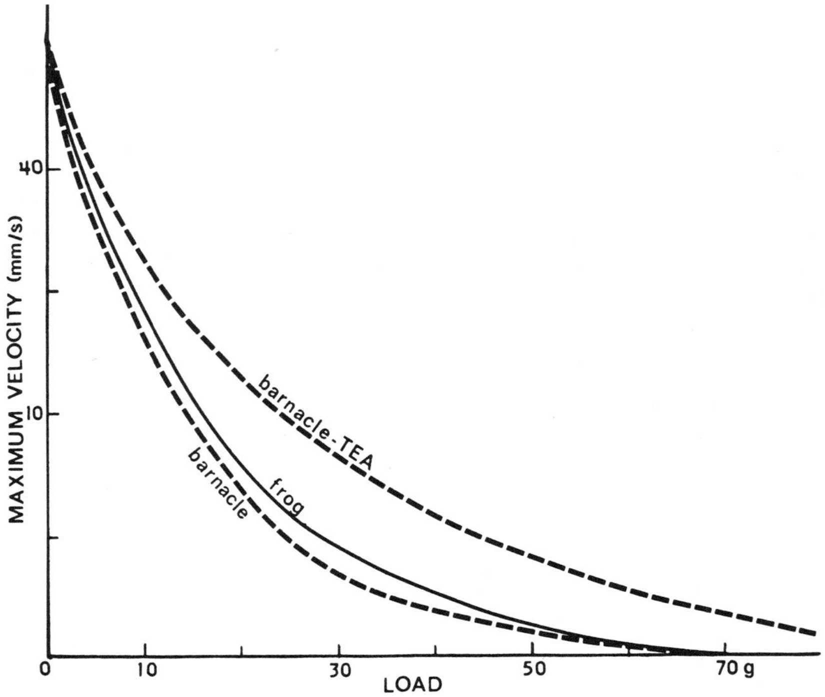

Figure 7.1. Maximum-velocity–load curves for single giant barnacle depressor cell at 15°C in artificial seawater, compared with whole frog sartorius in frog Ringer solution at 0°C. The curves are very similar. The velocity is a little more than doubled for the frog muscle when tested at the same temperature as the barnacle cell. When the barnacle cell is bathed in artificial seawater to which 10 m*M* tetraethylammonium (TEA) had been added, a flatter curve is obtained. In TEA-saline all-or-none action potentials are elicited. The maximum velocity is unchanged, but the load capability is increased! For other muscles, the shape of the curves lies between the two extremes, except that the ordinate values extend either upward up to threefold, or downward to 30,000-fold. The abscissa extends upward no more than 2½-fold for any muscle. Orig.

major departures from the commonly assumed cylindricality. Accordingly, it is preferable to restrict consideration to actual load-bearing capability, until a suitable contractile molecular-based parameter is available to relate it to, for each kind of cell. The symbol L will be used in our account for load, in place of P. The hyperbolic shape of the maximum-velocity-load (V_{max}/L) curve was a contributing factor in the downfall of a theory that dominated muscle physiology for the first quarter of this century, the *viscoelastic theory.* According to this theory, muscle was treated as a body with different states of viscosity and elasticity in the excited, compared with the relaxed, state. The velocity–load relation for a viscous-damped elastic body must be linear. Also, the amount of heat liberated in the change of state should theoretically be constant regardless of the form of the subsequent contraction. This was not compatible with an early finding (Fenn, 1923), that the amount of heat produced is variable over a wide range. Not only is it dependent on the total amount of work done, it is especially dependent on the amount of shortening permitted, being least in isometric conditions (see p. 252). A search for alternative ideas to the viscoelastic theory occupied muscle physiologists for the next half century, first in a formal, or modeling sence, then later, as ultrastructural and biochemical knowledge increased, in terms of presumed molecular events.

7.2. Generating a Dynamic Model

From the V_{max}/L relation, it will be perceived that the velocity at a particular load depends on the difference between the actual force acting on the muscle and the maximum force it is capable of developing. This relation became an early focal point of attention since, by placing clear restraints on the capabilities of the contractile machine, it offered the prospect of understanding it. But first there were some glaring facts about contraction that needed to be understood, in particular, why it is that even in cells that give large twitches, except heart cells, the maximum tension developed in an isometric twitch contraction at mean length in the body, which is given the symbol l_0, is much less than that developed during a tetanus. There is actually a wide range in the tetanus:twitch ratio of tetanus twitch tension depending on the source of the muscle, from about 10 : 1 to 1.5 : 1, being about 2 : 1 for frog fast muscles. A. V. Hill provided the explanation in 1949 by reference to a two-compartment model (Figure 7.2) that can be applied to any fast-muscle cell and is not dependent on any particular contractile model. In this model, the contractile machinery is symbolized by a box *C* in series with which there is a series elastic component (*SEC*). It is not necessary to specify what *SEC* is, and it must include the compliance of the contractile material, but it is at least partially made up of contributions from *passive elements* that are not involved in active shortening, including the attachments to tendons and Z bands. Before development of sliding-filament models, the latter were considered to be the only sources of *SEC*. Now elastic parts of molecules within the contractile system are included also. There are in addition elastic elements in parallel with the contracting elements, T filaments, the supramyofilament-cluster material connectin, and the sarcolemma. It is quite impossible to calculate and allow for the complexities parallel elements introduce, but provided studies are made very close to l_0 and experimentally regulated length changes are kept small, parallel elasticity can be ignored.

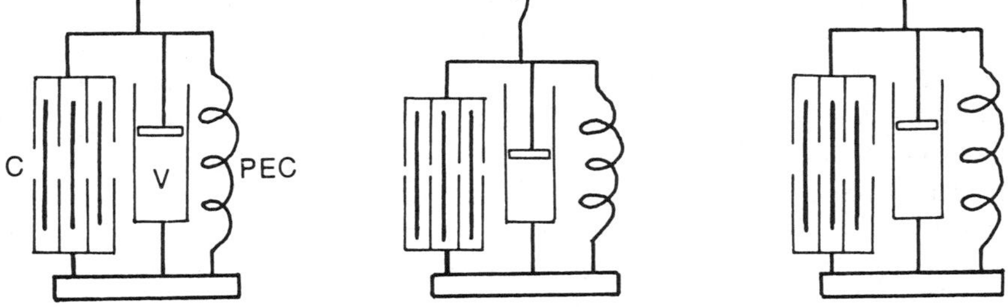

Figure 7.2. Model of muscle element that includes elastic components: series elastic component (*SEC*) and parallel elastic component (*PEC*). Also shown is a simple viscous element (*V*). The contractile component is symbolized by *C*. When the element is stimulated under isometric conditions (*center*) to give a single-twitch response, the tension registered by a gauge is relatively reduced compared with that developed by *C* because *SEC* must be stretched, and peak activation of *C* is very brief, being reduced before *SEC* is fully stretched. If, however, a quick stretch is applied very soon after excitation (*right*), *SEC* is extended, *C* does not shorten, and the full tension in *C* is registered.

Collectively, the sum of contributions to *SEC* may be represented by a spring. In addition, there is the internal fluid in which the structural molecules move, contributing viscosity, which must be taken into account in a full consideration of performance. But there has been no agreement on either the type (Voigt or Maxwell) of viscous element(s) or location(s) for the model that best represents the state of affairs in muscle, and in any case it would be very difficult to allow for a property that must change continually throughout contraction. The Voigt model has been used most often, but the Maxwell model is considered to best fit the modern data. Generally the viscous element is not considered in the model, and again, provided only very small length changes are used, the viscosity can be neglected.

The model is applicable to fast-twitch vertebrate cells and some very fast invertebrate ones such as squid mantle and a few crustacean fast cells. The argument from the model is simple. The state of activation of the contractile material reaches maximum intensity very rapidly following the impulse, within a few milliseconds, but after being maintained for only a few further milliseconds, it declines. Tension developed by *C* is registered externally only through *SEC,* which is stretched, delaying the rise in force. The characteristics of muscle *SEC* are such that relaxation of *C* has begun in a twitch before *SEC* has been fully stretched. By contrast, in a tetanus, *SEC* becomes fully extended after the first 5–10 excitations, so the full tension is seen. The reality of *SEC* and indeed its elastic properties can be demonstrated and determined by suddenly releasing a muscle being subjected to tetanic stimulation isometrically to various loads and following the subsequent length change as it shortens isotonically. With a load so large the muscle can only just bear it, at the moment of release there is a miniscule fast shortening followed by very slow shortening. With a very light load, the initial rapid shortening is much larger and subsequent shortening quite fast. With intermediate loads, the initial rapid shortening is proportionally long, depending on the load, but the speed is identical for each.

The curve of length of this initial rapid shortening versus the load equals the strain-stress curve for *SEC* (Figure 7.3), because during the restrained isometric tetanus, *SEC* will have been fully stretched out by *C,* and when the stop is released and replaced by a load, the length of *SEC* adjusts rapidly, according to its compliance, before any change in length of *C,* which is trying to shorten, can occur. The strain–stress curve was determined by Wilkie (1956) for frog sartorius using six different times of release following the onset of a single stimulus from 100 to 640 ms. The six curves were identical, which Wilkie used as an argument that the contractile machinery itself, which was undergoing activation and decay, could not have been a major contributor to *SEC.* However, this argument was based on the earlier universal view that the contractile machinery is a coiling or folding longitudinally aligned molecule all of which is actively involved in *C.* Had any part of it been elastically stretched by another, it would make a contribution. According to the modern rocker/X-bridge tilting hypothesis, an important part of *SEC* is within the molecular contractile element. The *SEC* for some types of muscle is relatively much longer than that of the frog (Figure 7.3.5): in barnacle giant and other comparable long-sarcomere arthropod cells it is about 7% l_0, almost twice as long as it is in frog sartorius, even though the attachments are less compliant. The apparent strain–stress curve in the barnacle cell is markedly affected by time of release following a single stimulus. This is to be expected in view of extensive internal disorder similar in nature to that shown in Figure 6.12. Internal series/parallel elasticity arising from unequal sarcomere length changes make a large effective contribution to the *SEC* element.

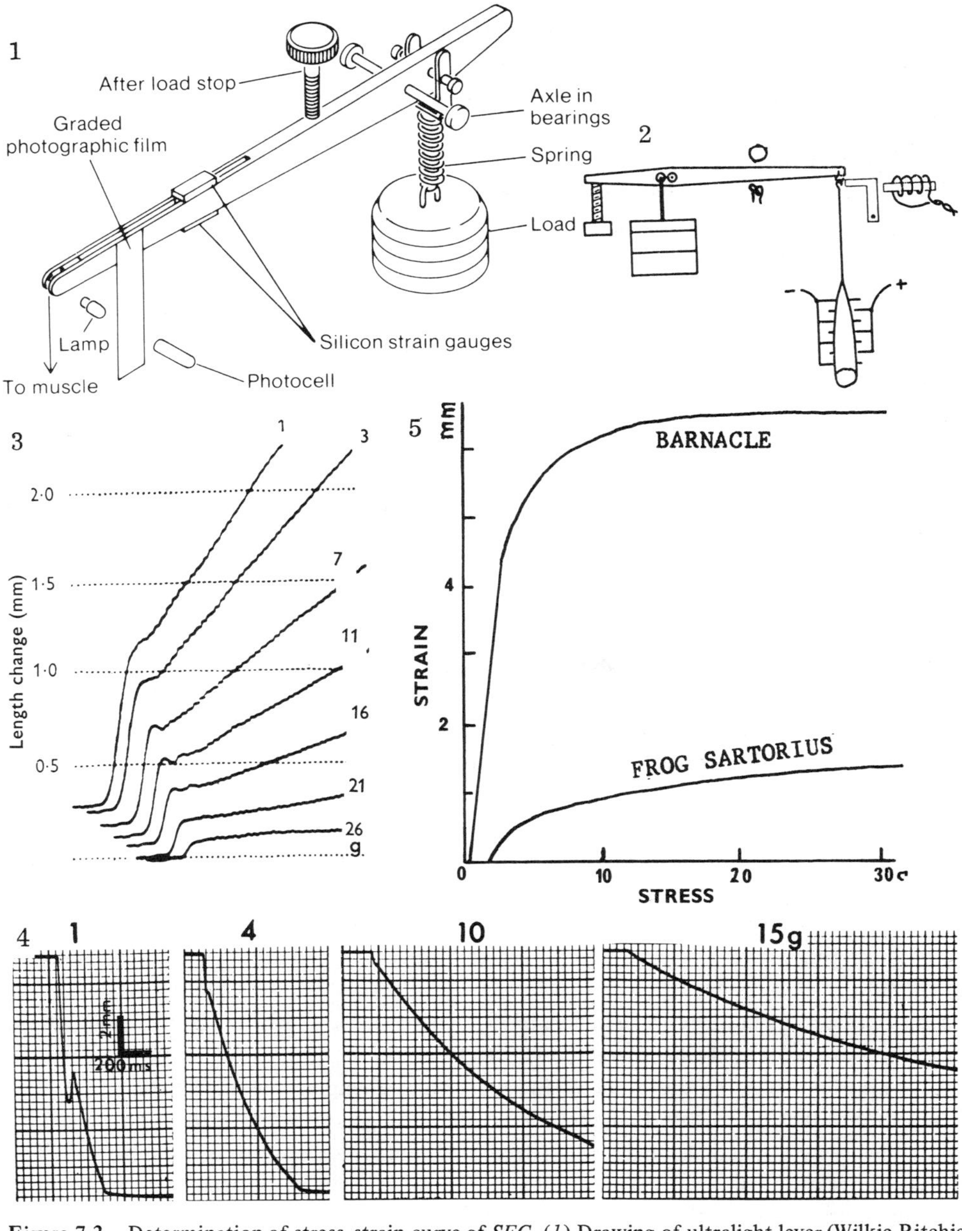

Figure 7.3. Determination of stress–strain curve of *SEC*. (*1*) Drawing of ultralight lever (Wilkie-Ritchie lever) used in determining dynamic properties. (From Carlson and Wilkie, 1974.) (*2*) Diagrammatic representation of lever, with electromagnetic stop and muscle placed on multielectrode grid. (From D. R. Wilkie, 1956, by permission of British Medical Bulletin.) (*3*) Oscillograph records of length change with time following abrupt release a short time following maximal activation of whole frog sartorius. (From Jewell and Wilkie, 1958.) (*4*) Same as (*3*) but for giant barnacle depressor single cell. (From Hoyle and Abbott, 1967.) (*5*) Stress–strain curves of *SEC* for frog sartorius, with stiff *SEC*, and barnacle, with pliant *SEC*. (From Hoyle and Abbott, 1967.)

The actual length of SEC in frog sartorius is disputed, in spite of numerous studies. According to Matsumoto (1967a) it is 5.3% l_0 for *R. pipiens.* His calculations from Wilkie's (1956) data gave 4-5% l_0 for *R. temporaria,* whereas Hill's value was 3-4% l_0. The length is quite likely to be a significant variable between individuals, and for an individual dependent on seasonal variations and even with diet and specific past behaviors. The SEC length has been determined for an obliquely striated muscle containing a large amount of paramyosin – the anterior byssal retractor of *Mytilus* (see p. 472), which behaves in so many respects like vertebrate smooth muscle it was once used as "model" system for smooth muscle in general (Winton, 1937) – to be about 5% l_0 (Sugi and Tsuchiya, 1979), yet this muscle is directly attached to inelastic shell at one end.

A simple test of A. V. Hill's two-component model is to give a muscle or cell a brief, short, quick stretch during the rising phase of tension in a twitch. The stretch will extend SEC, in effect converting it momentarily into a stiff connector that will transmit the time-instantaneous value of *C*. A series of quick stretches applied at different brief times following excitation will permit the true time course of *C* to be determined. The effect of quick stretch is dramatic, especially for a cell that has a high tetanus-to-twitch ratio (Figure 7.4*2*). For a vertebrate fast-type cell, peak tension with a suitably timed quick stretch is exactly equal to tetanus tension. There is a maximum delay to application of stretch following the onset of contraction below which full tension is not reached. Onset of contraction is defined as the moment when there is a sharp rise in tension. For a majority of fast-type cells, this is appreciably delayed after the onset of stimulation, and the first tension event is negative: a brief early elongation of the cell called *latency elongation.* For slow-type cells, there is no latency elongation, and contraction starts early. However, the very early contraction is weak, and onset is usually considered to be the time of onset of the faster-rising portion, which is about 10 ms later.

All of the events outlined above in the section on E-C coupling contribute: spread of electric excitation, activation of calcium stores, release and diffusion of calcium, and AM activation. The total of all these events at all sites at any instant collectively determines the extent or state of excitation of contraction. The total of the local AM events actually contributing to active tension development and capable of contributing to the movement of a load can be termed the *state of activation* (SA). This is approximately equivalent to an old expression used by A. V. Hill (1949), active state, which has come under severe criticism, and has largely been abandoned. It seems to me that the criticisms can be well met simply by reversing the order of the words and changing the expression slightly, as I have done. It is quite true that there is no such thing as an active state as conceived by Hill. What was implied, though, when translated into modern terms, is still valid. The practical definition is useful, and applies to SA, which can be defined quantitatively as the maximum tension that can just be borne by a muscle at a given instant without lengthening overall.

The minimum delay to the maximum tension obtainable with a short quick stretch following onset of contraction is about 3 ms for fast-type frog cells at 20°C and 2 ms for mammalian cells at 37°C. Quick stretches applied at slightly later times than these produce similar peak tensions until, after about 30 ms, less peak tension is produced (Figure 7.5). At this time SA has started to decline, that is, the sum total of all contributing events is less than at peak and declining. According to the X-bridge hypothesis, fewer bridges are now attached and actively cycling than at the peak. The SA rises to the peak about linearly for fast-type cells, being 65% complete in frog single semitendinosus cells at 1-2°C in only 4 ms (Edman, 1970).

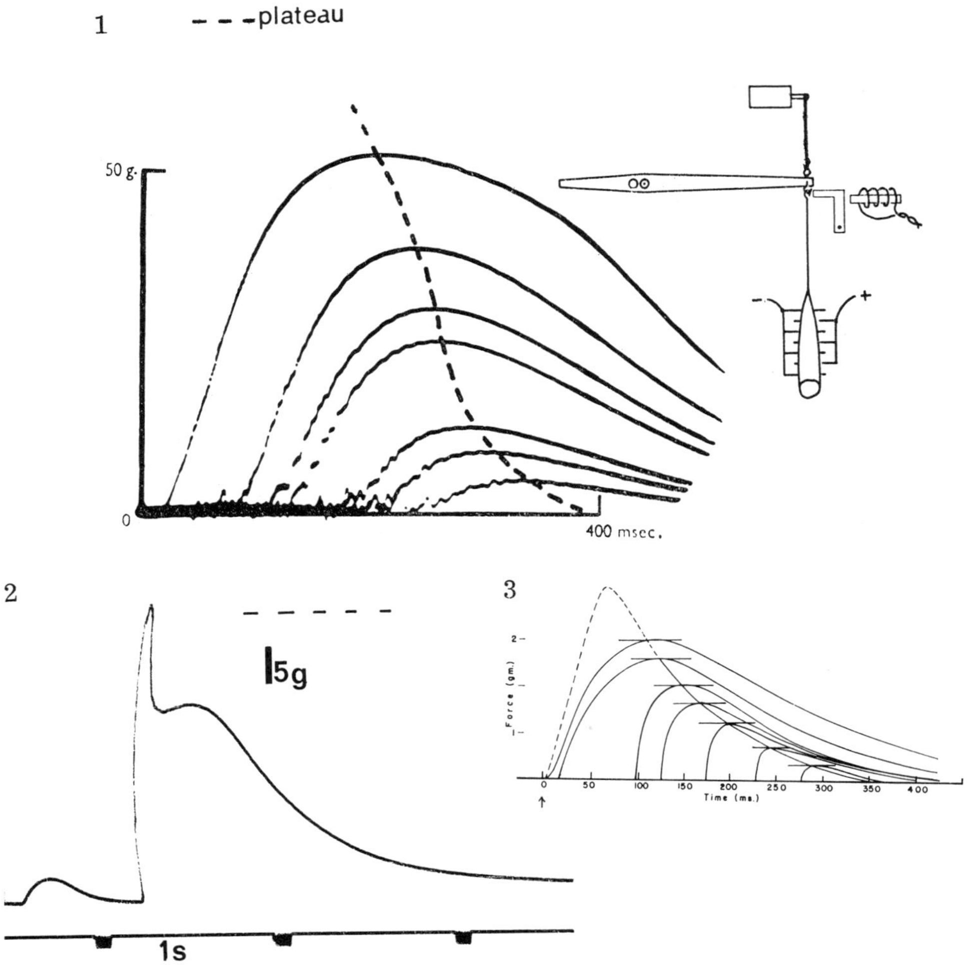

Figure 7.4. (*1*) Method of determining the time course of the later part of the decay of the active state in a single twitch. The muscle or cell is massively electrically stimulated and develops tension isometrically against a stop. The stop is removed at various intervals following the stimulus and allowed to develop tension against a mechanoelectronic force transducer. The peaks of the force-time curves coincide with moments of zero net movement in the contractile elements, hence correspond to points on the active-state-time curve. The broken line is drawn through these points for a frog sartorius muscle. The tetanic level is the short horizontal broken line (plateau), which indicates the duration of 100% activation of the contractile machinery. (Modified, from D. R. Wilkie, 1956, by permission of British Medical Bulletin.) (*2*) Effect of a quick stretch on twitch of a single giant barnacle cell. There is a fivefold potentiation, indicating the presence of a long, compliant SEC. (ORIG.) (*3*) Active-state decay of barnacle twitch determined as in (*1*). (From Hoyle and Abbott, 1967.)

Quick stretches are not useful in determining the declining time course of onset of SA as they are in determining its rise and plateau (when present), but an ingenious method of determining the decline has been devised (Ritchie and Wilkie, 1958). They argued that at the peak of a twitch, the length of *C* in Hill's model, which represents the average value for individual sarcomeres, must be static. It has shortened to its fullest extent against SEC and is about to start lengthening. Therefore, at this precise moment, the tension equals SA tension as defined above. They argued that if they stimulated a muscle cell, restraining

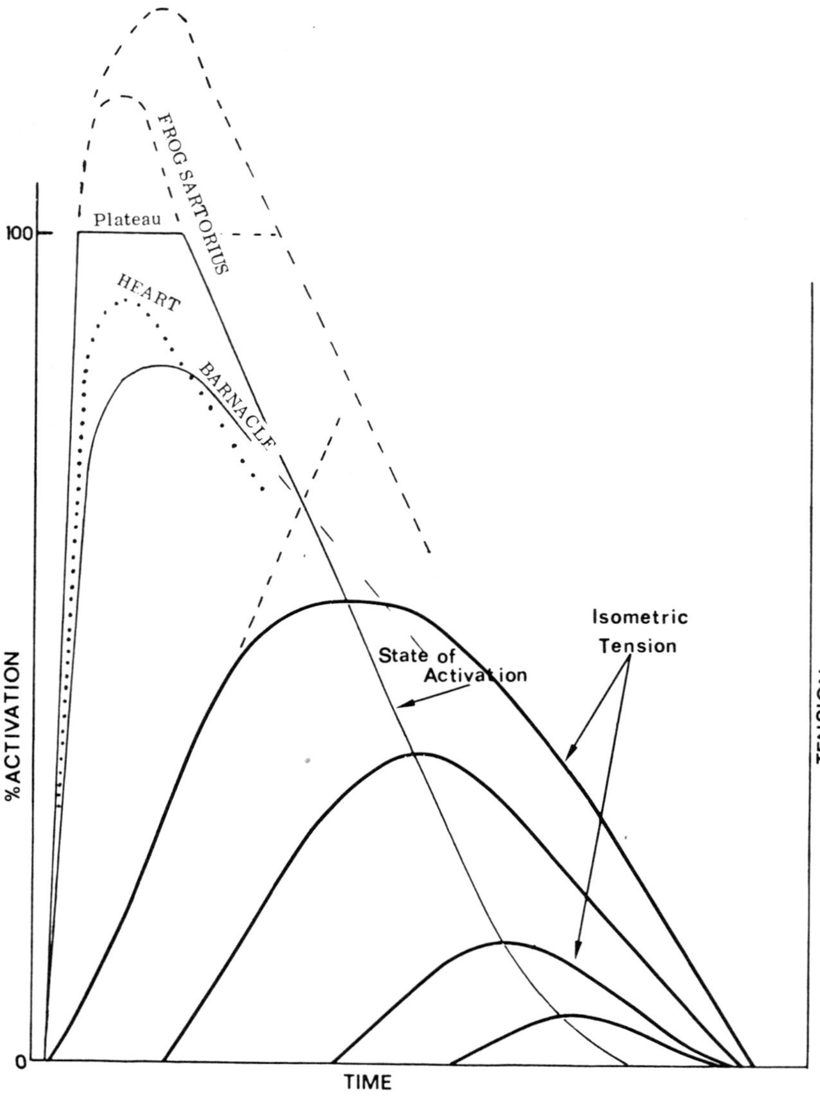

Figure 7.5. Activation–time relations for three muscles or cells: frog sartorius, frog heart, and giant barnacle depressor. The four curves with thicker lines are tension–time relations for frog sartorius isometric twitches, measured after short quick releases at different times following the onset of stimulation. The thinner continuous line drawn through their peaks represents the percent state of activation (SA) at the time. This curve is extrapolated to the time when the maximum state of activation, which is held for about 30 ms, is just beginning to decline. The SA rises rapidly, starting about 4 ms following activation, until it reaches 100%. The E-C coupling events follow exactly the same time course, except that they continue beyond the point at which saturation of the contractile machinery occurs, with the time course indicated by the lower broken line. Saturation lasts for the duration of the *plateau* phase. The *intensity* of activation can be further enhanced by various agents, when it follows the kind of time course indicated by the upper broken line. The effect of time is to increase the duration of the plateau phase and delay and slightly slow down the decline. The rate of rise of SA in heart muscle is slightly slower, and 100% activation is not achieved in a single twitch, leaving plenty of leeway for enhancement (= *positive inotropic action*). The same is true for a giant barnacle depressor cell. (Orig.)

it isometrically for a short time as tension rose, then quickly released it for a very short distance and recorded the rise in tension afresh, there would be a late, weaker twitchlike curve whose peak also would correspond to a point on the SA-time curve. By making a series of releases with different delays, they obtained a plot of the decay of SA (Figure 7.4.*1*). The estimated time courses of SA for some typical muscles are shown in Figure 7.6. The technique of quick stretch during tension rise can only be used for studies on fast-type cells in which there is a large and above all rapid activation.

For many reasons, SA is affected by the events it itself determines. It is, for example, increased with stretch and decreased with shortening, in an approximately linear relation (Figure 7.6).

7.3. Comparative Aspects of Mechanical Activation

Comparable results have regrettably been obtained for very few different types of cell. In giant barnacle cells, the whole time course of events is slower, but there is a major difference in that there is no tension plateau. This is because no matter how strong the stimulation in a single excitation, SA falls short of 100%. The same is true for vertebrate cardiac muscle, probably most smooth-muscle cells, and a majority of invertebrate skeletal cells. In locust flight muscle (Buchthal et al., 1957), SA probably does reach saturation, but this lasts for only a few milliseconds, even at 0°C, so short that very little shortening can occur before it starts to decline. The A. V. Hill model has met with much detailed criticism, for example, from Rack and Westbury (1974) and Julian and Moss (1976). It is,

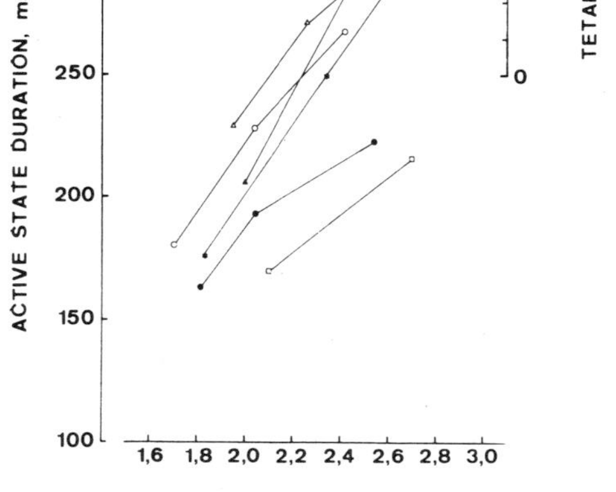

Figure 7.6. Increase in duration of active state of frog muscle with increasing sarcomere length. (From Edman and Kiessling, 1971.)

in effect, superceded by direct molecular models such as the rocker hypothesis, but conceptually it is still valuable. I suggest that the model has value in explaining the tetanus-twitch size difference for a fast-type cell, in conceptualizing the dynamics of contraction activation, and in understanding the effects of agents that affect SA. The concept of an elastic element into which energy is fed, and from which energy can be recovered simultaneously with changes in tension, has become a cornerstone of all thinking about muscle dynamics.

A formal examination of SA, as expressed in the quantitative definition, for a mammalian muscle, rat gracilis anticus, revealed that its duration is proportional to length (Bahler et al., 1967). This would be expected on the basis of different calcium-diffusion times and possibly other effects of length on coupling. Also, the length at which P_{max} was developed in a twitch was different from that at which it was developed in a tetanus, which was probably due to the contributions of sarcomere dynamics (see p. 247) in the tetanus. The V_{max} is a function of muscle length, declining with shortening below l_o in vertebrate fast skeletal cells (Edman, 1979). This possibly only reflects an increased passive resistance to shortening, not any change in capacity to shorten by the contractile mechanism. In barnacle giant cells, there is a broad plateau in the V_{max}-length relation, V_{max} being constant over a relatively wide length range (Figure 7.7) There are other anomalies also. The V_{max} is greater for an extremely light load than for zero load. There is probably some internal folding under zero load so that internal straightening is the initial event. A light load may favor uniform excitation, and it eliminates the waviness. Once contraction is under way V_{max} is attained virtually instantaneously in the unloaded fast vertebrate cell, but it takes at least 40 ms to develop in the barnacle. Several agents affect the duration of SA, including zinc, which in small doses is a harmless minor element. Accordingly, it should be expected to produce improved muscle strength in activities involving brief phasic activations. This possibility has been tested on rats, and it was found that oral intake of extra zinc improved performance and counteracted fatigue (Richardson and Drake, 1979). It would be of interest to know if this applies also to humans.

7.4. Length-Tension Relations

When a muscle is removed from the animal and placed in body fluid or a perfectly matched substitute saline (Ringer solution in the frog) at the same temperature, one of three things happens: (1) it shortens to a length at or below its minimum length in the body; (2) it adopts a position at, or very close, to its mean length in the body; (3) it elongates to, or slightly above, its longest length in the body. The first class comprises two basically very different kinds of muscle, one of which has a natural short length and is passively stretched at mean body length. The other possesses basic tonus, that is, it is in a state of continuous, work excitation. A calcium-chelating agent, or in some cases strong hyperpolarization, is needed to relax a cell with strong basic tonus. The second class includes the majority of those vertebrate skeletal cells that are popular with physiologists: frog sartorius, extensor digitorum longus, and soleus. The third class are slack and loosely wavy in the body until activated. A strong radial compressive force exists in resting cross-striated cells at rest, as directly demonstrated by injecting a small oil droplet into a single muscle cell. Such a droplet immediately takes up an elongated shape, from which the forces can be calculated to be about 1800 dyn/cm^2. During contraction the

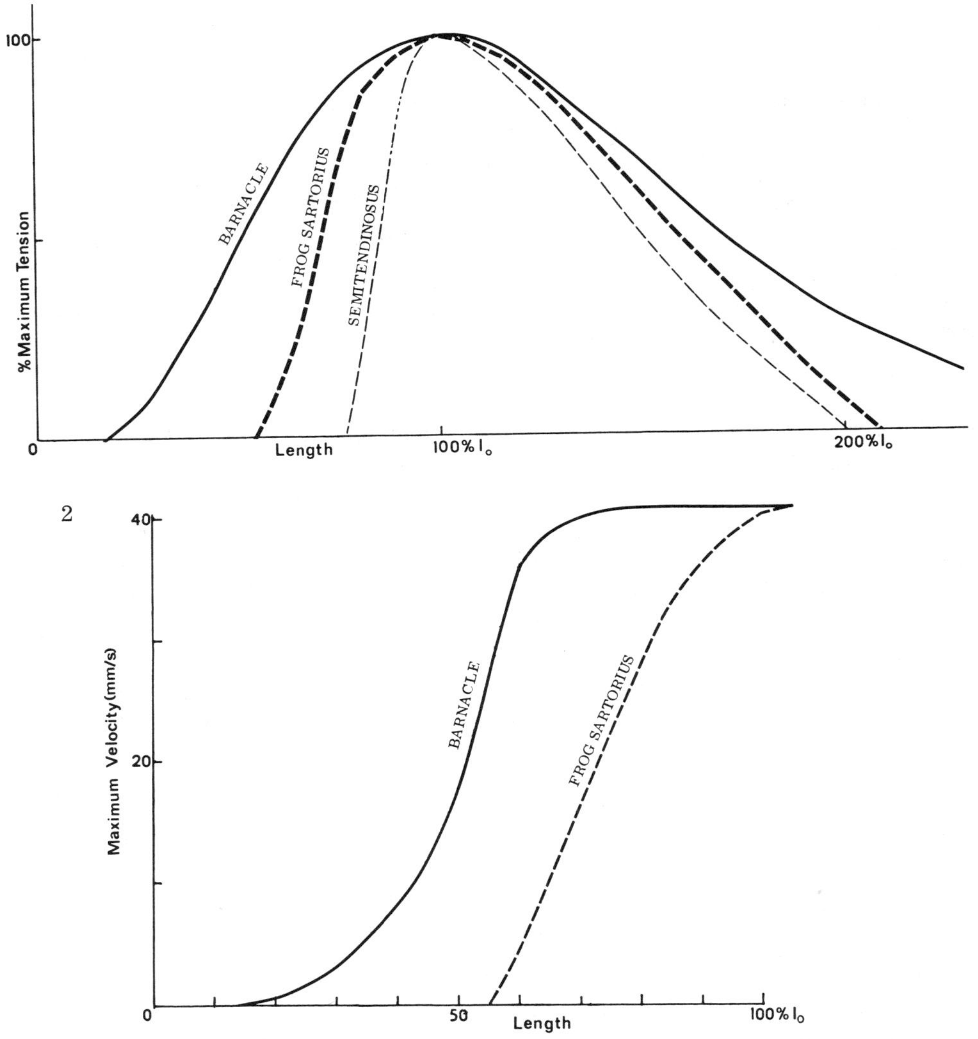

Figure 7.7. (*1*) Tension–length curves for whole frog sartorius and semitendinosus, compared with single giant barnacle depressor cell. (ORIG.) (*2*) Maximum speed of shortening–length curves for single giant barnacle depressor cell compared with whole frog sartorius. (ORIG.)

compressive force doubles (Figure 7.8). There is necessarily a balance between this force, which must play a role in determining rest length, and longitudinal force, which exactly counteracts it.

Stretching by a moderate amount, to 120% l_0, for some skeletal cells requires very little force, but for others it requires a large one, in a few almost as great as one that will tear them apart. In terms of detailed ultrastructure and properties of internal elastic elements, the basis of these differences is not understood. For inextensible flight muscle cells of the giant water bug *Lethocerus*, the stiffness has been attributed to C filaments connecting the A filaments to Z bands (see p. 72). But flight muscles of some bees also have C filaments, yet can be stretched easily. Many factors need to be considered, including such unlikely ones as different properties of different parts of the cell and the extent to which the cell is activated, since basic tonus is likely to be present. But the most

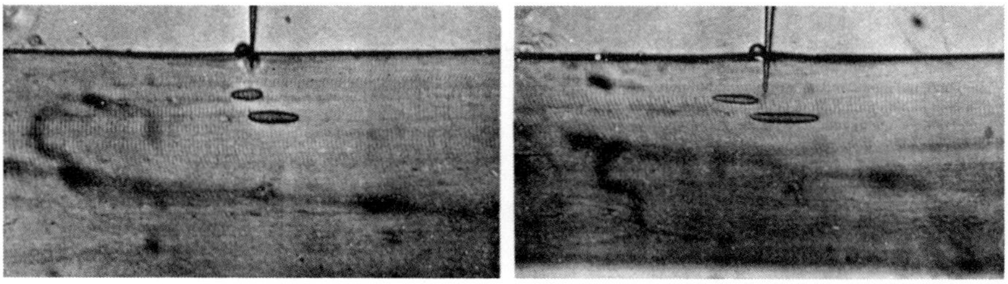

Figure 7.8. Radial compression force in frog biceps femoris cell, increasing during contraction, illustrated by elongation of oil droplets. Resting muscle at left – force is N 1800 dyn/cm^2; isometric contraction at right – force doubles. (From Hiramoto, 1956.)

significant single factor is the concentration of ATP. The ATP is required for supplying the energy of contraction, but in addition it must be present in an amount above about 6 m*M*/l to keep the cell pliable and capable of shortening. The stiffness of a corpse is due to progressive loss of ATP in death. The X-bridge theory contends that the state, termed *rigor*, is due to the making of permanent links between actin and myosin. These links are broken, permitting bridge cycling, shortening, or passive stretch, if [Ca^{2+}] is sufficient, but only if the ATP level is above about 6 m*M*/l. When the ATP concentration of a *Lethocerus* flight muscle cell is low, 5 m*M*/l or less, it becomes highly inextensible because its core goes into rigor (Guth, 1980).

The parallel elasticity of locust flight muscle has an extremely stiff compliance, dominating after only 5% stretch (Buchthal et al., 1957). The basalar muscle of the beetle *Oryctes* breaks at 15% stretch and tension of only 1.5 kg/cm^2 (Machin and Pringle, 1959). At the other extreme, I have found stimulated giant barnacle retractor cells that were still unbroken by a load of 20 kg/cm^2 (measured in the stretched state) when their own length was about 2.5 × l_0. Beef cells have been stretched to 5 × l_0 before breaking and some frog muscles to 3 × l_0. By contrast, the most experimentally exploited muscle, frog sartorius, breaks at a mere 130% l_0.

The starting point for the onset of tension, relative to body length, during stretch, is also a major variable, as are both the tension and relative length at which breakage occurs. Smooth-muscle cells are extraordinarily extensible, as they often need to be for such functions as expansion of the wall of the uterus or the bladder, and filling of the stomach. As discussed above (see p. 106), the distinction between smooth and cross-striated muscle is not a sharp one when invertebrate skeletal cells are considered, especially when the muscles of soft-bodied aquatic animals are included. Both internal and external skeletons restrict natural length change to a small fraction of l_0. This mechanical disadvantage promotes isometric contractions and ballistic movements. Life on land with massive antigravity requirements rendered it necessary for antagonist muscles to provide elastic damping. For all vertebrates since the early ancestral forms, muscles have been exposed to the outer world as an overall covering. They needed to become tough as well as resilient, even when inactive, for physical protection. All of these factors promoted the development of tough connective tissues, even though this interfered with optimal operation.

The tension–length curves of electrically supramaximally tetanically stimulated cross-striated muscles all have a similar basic shape, an arching rise reaching an inflexion, then a very steep rise to breaking point. The inflexion is always at a length a little above l_0, but the breaking lengths vary greatly, from as little as 130% l_0 to 500% l_0. There is also a wide range in the points at which the curves cross the abscissa at short lengths.

Most vertebrate and some invertebrate cells reach zero active force very close to 50% l_0. But for barnacle giant cells and others capable of supercontraction, which includes some human muscles, beef muscle, and several insect skeletal muscles, the curve does not cross zero tension until the length is much lower, the extreme being about 15% l_0. Presumably it is just the physical limitation imposed by a tight Z-band meshwork that restricts some to a small extent of shortening. The other end of the length range is determined by internal elasticity and the extent to which sarcomeres of individual myofilament clusters can slip past each other.

When the passive-tension curve is subtracted from the active one, a bell-shaped curve that is approximately symmetrical is obtained for all vertebrate skeletal muscle cells. This is the active-tension-time relation. For cells that can supercontract, the curve has a long tail toward short lengths. Although many such curves show tension being developed at lengths above twice l_0, when overlap of A and I filaments should stop, the amount is small and can be attributed to a combination of very long I filaments and local length inequalities permitting interaction of noncontiguous I and A filaments.

In the majority of the few muscle cells in which the relative contributions of sarcolemma and internal elements have been assessed, the elasticity at l_0 and up to about 115% l_0 is entirely due to internal longitudinal elasticity. Above 115% l_0 sarcolemmal elasticity is added (Figure 7.9): it is widely different in different cells, as is internal elasticity.

7.5. Elastic Properties

Muscles vary enormously in their dynamic responses to passive stretch, particularly in the rates at which the tension developed decays and the ways both change when the same

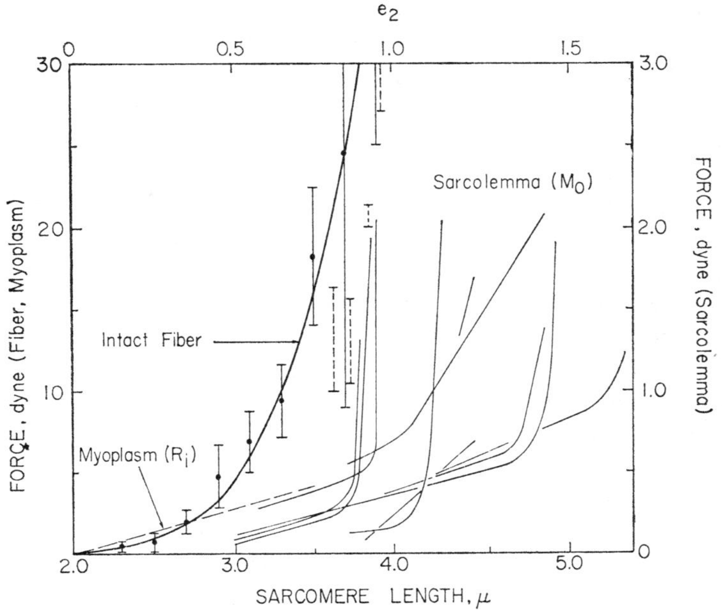

Figure 7.9. Relative contributions of sarcolemma and "myoplasm," i.e., cell contents, to series elasticity of frog semitendinosus cells. The sarcolemmal contribution is highly variable between cells and quite small except at great lengths. (From Rapoport, 1972.)

stretch is repeated. Neither the basic nor the comparative aspects of the complexities have yet been fully explained. The very early fall is due to a combination of X-bridge properties and an assortment of parallel and series elasticities. Most recent studies have assumed that the latter can be ignored for very fast, very short stretches. If a muscle is allowed to come to rest at a length slightly below that at which it develops tension and is then quickly stretched to a length at which negligible residual tension will remain, it develops tension that decays slowly, as does any viscoelastic body. The curves, for a wide range of muscles, are compounded of two exponential decays, fast and slow. The half-times for the fast-decay rates range more than 15-fold for cross-striated cells, from about 0.01 to 0.27 s, and the corresponding slow one from 1.0 to 6.0 s. These times are more than an order of magnitude greater still for cells that are obliquely or disorderly striated and even longer for some smooth muscles. Two different major viscoelastic components must be responsible, but the values are so widely different that different basic elements may be associated with them in different cells. Neither kind of viscoelasticity can be attributed to the sarcolemma, which is folded at the lengths used. Perhaps there are two elasticities, connectin being responsible for one and T filaments for the other. The current popular assumption is that there are two viscosities, one of which is due to friction associated with X bridges in the myoplasm. For frog fast cells Jewell and Wilkie (1960) calculated that there is an area of contact between A and I filaments of 200,000 cm^2/cm^3 or 30 cm^2/cm^2 per half sarcomere, corresponding to a little under 100 gm/cm^2, a value that is similar to that for ordinary limiting friction. Whatever its basis, this component must play the second most important role in determining decay of tension in relaxation, after SA decay. The other viscosity would be that associated with the totality of material: myofilament clusters, mitochondria, glycogen, and so forth, in myoplasm.

When not actively stimulated, striated muscle cells behave in a rubberlike manner when stretched and released. But following stimulation, the behavior is not only unique, it is in several significant respects quite unlike that of rubber. The difference is due to change of the elasticity of the contractile molecules, the formation of cross-linkages, and conformational changes, the passive elastic material presumably having the same properties as before. The relative contributions of the various components are not yet known for any muscle, but are principally associated with tendinous connecting elements, cytoskeleton, and intracontractile elements, respectively. Elasticity of tendons/apodemes as well as surface parallel elasticity is a matter for large individual variation in muscles from the same animal or even the same limb. Tendon properties range in a single animal, the dragonfly *Aeschna,* from ones that are formed of pure resilin, which is as close to being an ideally elastic material of moderate compliance as exists anywhere, to ones that are formed of tanned protein and are inelastic (Weis-Fogh, 1960). For some tendons the compliance is a dynamic variable capable of changing with a slow time course because it is trophically regulated via the motor nerve (Palacek, 1965b). Short-term changes occur in rat soleus, whose tendon becomes stiffer after a period of sustained contraction (Palacek, 1965a).

The diverse range of properties of the cytoskeletons of different muscle cells can be directly determined in so-called muscle cell shells left behind after removing the contractile material (e.g., Westort and Hultin, 1966). The material left behind from a frog sartorius cell is very tough and inextensible, leading to an earlier erroneous belief that the sarcolemma is responsible for parallel elasticity and perhaps even for transmission of tension (Ramsey and Street, 1940). The whole skeletal framework of overall very stiff flight muscle cells of locusts makes but a very small contribution to stiffness (Buchthal et al.,

1957) and it is even less for the highly extensible giant barnacle retractor cells. For these and similar cells, it is important to know the relative contribution of extramyofilamentar elasticity compared with T filaments.

Direct evidence for a contribution of contractile machinery to SEC was provided by showing that as normally determined, SEC properties are greatly affected by the concentration of calcium ions (Wise et al., 1973). Buchthal and Rosenfalck (1960) insisted that SEC must represent elasticity of the contractile element in insect flight muscle, based upon studies in which a small oscillation in length was superimposed on an activated muscle. They discovered a phase displacement, change in length lagging behind force, that is responsible for the characteristic performance of insect flight muscle discussed elsewhere (see p. 535). The specific contribution of the sarcolemma can be studied in many kinds of single cross-striated cells because it is easy to get the contractile material and its associated intrafilamentar and interfilamentar elasticity to break and retract into clots. Detailed studies of the elastic properties of the empty tubes from frog semitendinosus showed that in them the circumferential force is from 1500 to 50,000 dyn/cm, compared with 23,000–225,000 dyn/cm for longitudinal force (Fields, 1970). This stiffness anisotropy can be explained by a helical arrangement of collagen fibers in the sarcolemma.

The compliance of SEC is sensitive to experimental conditions in many ways. For example, it is reduced in hypertonic solutions (Jewell and Wilkie, 1958; Tameyasu and Sugi, 1976). This effect must be mediated by the intermyofilamentar and intramyofilamentar contributions. For single semitendinosus cells, there is no significant contribution of the sarcolemma (Sl) below 3.0 μm, which is about 140% l_0 (Rapoport, 1972). At 3.75-μm SL, 10% of the elasticity is contributed by Sl, whose elastic modulus is 5×10^6 dyn/cm^2, but thereafter the Sl resistance increases drastically, presumably as the collagen coils become straightened. The elastic modulus of myoplasm was 5×10^3 dyn/cm^2. In barnacle giant cells, the Sl remnant also starts to contribute at 140% l_0, but there is no abrupt increase in its resistance to stretch up to $3 \times l_0$.

That there can be radical differences in basic "skeletal" structures of skeletal muscle cells is shown by testing the effect on elasticity of removing myosin. When this was done for *Lethocerus* flight muscle there was a drastic fall in resistance to stretch, whereas in one lacking C filaments, a tergosternal (direct flight) muscle of a locust removing the myosin did not affect stretch resistance (Maruyama et al., 1978). Parallel elasticity, both that due to T filaments and that due to connectin, must be slight in the otherwise stiff *Lethocerus* flight muscle. The relative contribution of series elasticity is more difficult to evaluate: in many muscles there is none except that ascribable to I, A, and T filaments (Figure 7.10). But in others there are large contributions from passive elements closely resembling those postulated theoretically in A. V. Hill's model, that can be equated structurally with tendinous attachments and Z regions.

7.6. Twitch Characteristics

The time course of shortening in a twitch under various loads, the *isotonic* length–time curve, can be determined to a reasonable approximation from the *V*/*L* relation and related to the isometric-tension–time curve as shown in Figure 7.11. Shortening will not start until the tension developed equals the load. Thereafter tension, equal to the load, remains constant, and the load is lifted at a velocity that is determined by the *V*/*L* rela-

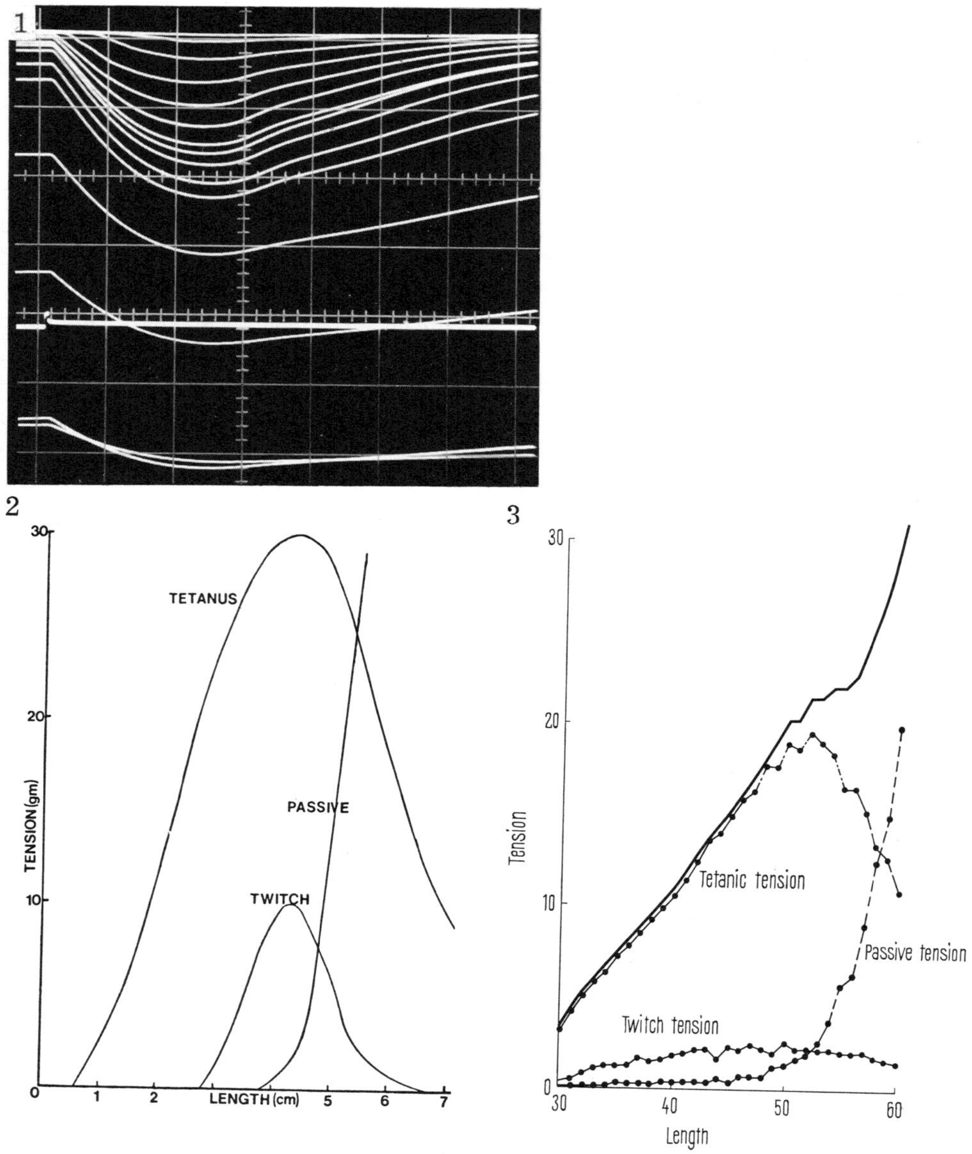

Figure 7.10. Contrasted length–tension relations in different muscles. (*1*) Effect of length on isometric-twitch tension in single giant barnacle depressor cell. The 40-mm long cell was tested in 2-mm length steps from 6 mm below slack length (ORIG.) (*2*) Tension–length relations of giant barnacle cell, same as (*1*). (ORIG.) (*3*) Tension–length relations of the radula protractor of the mollusc *Busycon*. (From R. B. Hill et al., 1968.)

tion. But as the muscle shortens, the relation shifts, because P_{max} declines with declining length (see below). Consequently the velocity declines progressively until lengthening starts or the load is removed. The length reaches zero at exactly the time on the isometric curve that tension has fallen to the value of the load. The further translation of these curves to tetanic ones is simple. The early rise and decay times are identical. Shortening continues for a longer time until a plateau is reached, determined by the steady-state length–tension relation.

There are important differences in the shapes of successive twitch contractions as potentiation develops during repeated stimulation. These are related to activation characteristics. In SA-saturating muscle cells, as exemplified by frog sartorius, the rising phase is extended but not caused to rise more steeply. In giant barnacle and most arthropod leg muscle cells, the twitch has a progressively faster rise time, activation is increased, and the background Ca^{2+} level rises. In the frog fast cell, the SA is prolonged but cannot become more intense, since it is already at its maximum from the first. In the typical arthropod cell, the SA intensity, as well as its duration, increases progressively during

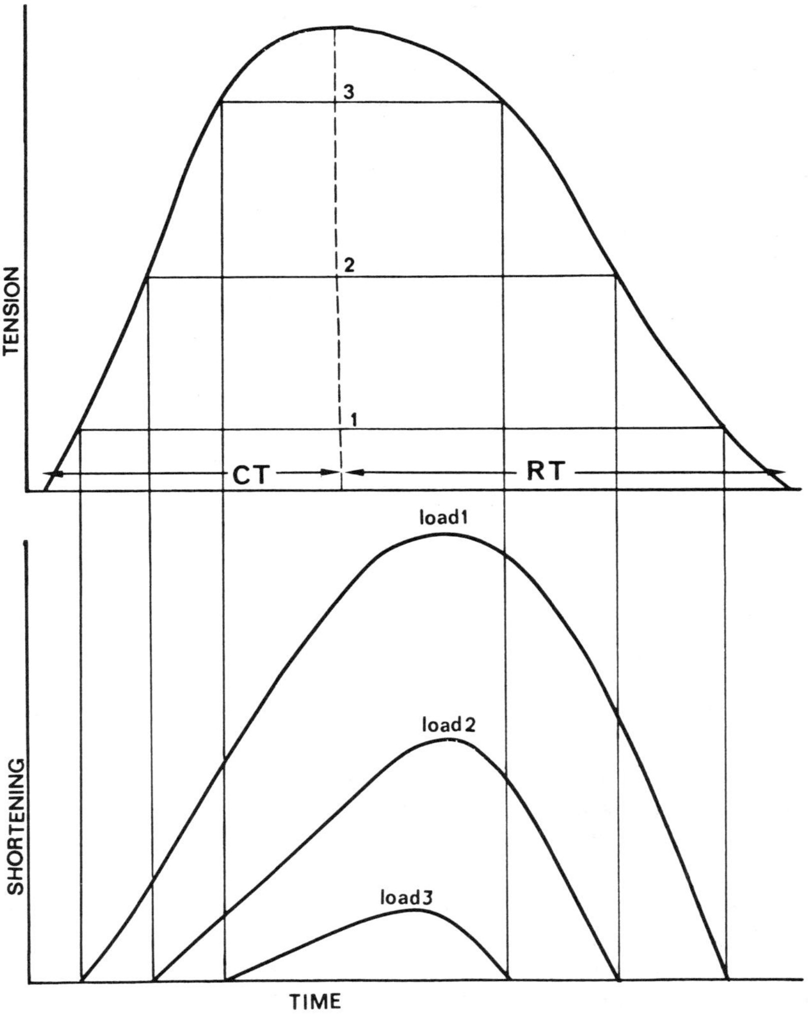

Figure 7.11. General form of the isometric-tension-time curve in response to a single excitation (= *twitch, above*) and of shortening-time curves at different afterloads (*below*). The contraction time (*CT*) is a little shorter than the relaxation time (*RT*). The loads are not lifted until the isometric tension equals the load. Shortening occurs at first at nearly constant velocity, but this then declines with shortening until peak shortening occurs. Return of length is faster than shortening. The load is returned to rest at a time slightly less than that at which declining tension in the isometric twitch reaches the value of the load, in the isometric-tension–time curve. When shortening, *CT* is longer than *RT*.

repetition. Exactly the same is true of heart muscle cells in general, whether of vertebrates or invertebrates.

It is now evident that the original concept of a muscle's having an intrinsic maximal-force capability that would be measured simply by determining plateau tension at l_o during supramaximal electrical stimulation is invalid. The peak of the twitch, like any other part of the tension–time relation, is a complex statistical sum, as Briden and Alpert (1972) and others have shown experimentally. The rough approximation to the average of real events that F_{max} indicates is nevertheless both conceptually valuable and experimentally useful. But earlier assumptions about the lack of direct effects of length manipulations are not. When lengthened forcibly during activation at moderate speeds, all muscles always develop much larger tensions. There was an attempt to explain this on the basis of "locked in" X bridges, but in sum, it is now evident that it is due to a combination of enhanced calcium release and optimal intrasarcomere dynamics associated with stretching the less completely activated sarcomeres. The maximum force obtainable from some arthropod skeletal cells is up to three times isometric tetanus tension, but very close to it in others. The extent of the difference is proportional to sarcomere length, being small in cells with short sarcomeres (1.6–2.6 μm) and large in ones with very long sarcomeres (10–20 μm), especially ones with a wide range of sarcomere lengths in individual cells, such as giant barnacle retractor cells.

The intrinsic capability for greater force explains the potentiating action of potassium contractures on the twitch (Chiarandini and Stefani, 1974) and the fact that potassium can give greater force than can electric stimulation from many skeletal cells. The curves of rate-of-working-tension for fast vertebrate cells determined at different times after stimulation under isometric conditions should theoretically match those calculated from isotonic twitches. However, empirically they do not (Pennycuick, 1964a, b); the disparity shows that the maximum rate of tension-development capability, regardless of SA intensity, is declining as the twitch proceeds.

7.7. Possible Dynamics of X Bridges

If the X-bridge theory is accepted a priori, stretch or very short release can be used to obtain information about X-bridge elastic and dynamic properties. Stretch opposes the attached bridges purely elastically, and their ability to extend and reattach depend on the rate, complicated by effects of stretch on properties of the enzymatic head, and ensuing conformational changes. To determine the extent to which connected X bridges are contributing to the short-range elasticity, it is only necessary to include in the saline sufficient EDTA or another calcium chelator to ensure that the calcium level is well below normal and compare the elasticity with normal. If there is a loss of short-range elasticity, it is reasonable to suppose that X bridges account for it. At least for some vertebrate skeletal and cardiac cells, there is a large difference (Moss et al., 1976). The extent of very rapid shortening required to just bring tension in a X bridge to zero from the isometric value, as estimated from experiments on single frog tibialis anterior cells, is about 3.9 nm (Ford et al., 1981) if filament and Z-line compliances are ignored: such compliance might reduce the value to about 3.0 nm.

7.7.1. Events During Early Tension Rise in Stretch

When fast-type cells are stretched during tetanic excitation within a range over which the resting tension is negligible, details of the tension–time response can be expected to reflect events in X-bridge cycling. This time course is indeed complex and markedly affected by the rate of stretch. At 30 cm/s the fast rise is followed by a steady viscoelastic decay, and there is no late rise in tension (Sugi, 1972). But at greater speed, the decay is delayed on account of the appearance of a delayed rise in tension, which is progressively more marked with increasing speed, up to about 60 cm/s, and which sums with the decay. Then with yet more speed, there is an abrupt fall in tension immediately after the initial fast rise, followed by a slow delayed rise. At speeds of 50–100 cm/s, the muscle bundles sometimes show superimposed tension oscillations at about 10 Hz, at 20°C. At lower speeds, below about 15 cm/s, delayed rise is apparent but it occurs almost immediately following the peak. The late rise is enhanced when the temperature is lowered.

The molecular events responsible for these results are necessarily in question, but the data are expected to afford clues about what these events might be. When the initial elastic-tension rise is included as a first phase and the transition from rapid to slow, which lasts 10–50 ms, as a third phase (Ford et al., 1977), a total of four clear divisions is recognized. The tension rise indicates a compliance that is almost linear and almost undamped, showing that far from the viscosity's being increased after stimulation, in a fast-type vertebrate cell (usually of frog, at 0–3°C), it is less. The early rapid tension recovery, lasting 2–5 ms, is proportional to the step amplitude and to the time course of the length change if this is complete in 2 ms. For a stretch or a release up to about 3 nm/half sarcomere, the early recovery is almost complete. The ratio of the tension reached/step amplitude was nearly constant for very small releases, but for larger ones there was a progressive decline, which reached zero with 14 nm/half sarcomere. To model the behavior, an undamped compliance in series with a damped compliance is preferred to the alternative possibility of an undamped elasticity in parallel with a series combination of viscous and elastic components.

The stiffness of attached X bridges is presumed to be the principal quality measured when small-amplitude high-frequency oscillations are applied to stimulated skeletal cells (Julian and Sollins, 1975). From a study of changes in amplitude as controlled shortening occurs, if the assumption is valid, it is possible to obtain a measure of the number of bridges attached at the time. The stiffness is found to fall progressively with increasing velocity, but not as rapidly as tension, there being considerable residual stiffness at V_{max}. According to the X-bridge theory, this means that at any time during shortening, some attached bridges are impeding shortening, although the majority are aiding it: it is the balance between the two that determine properties.

The quick 3-ms phase is drastically reduced or eliminated if ATP is removed (Guth and Kuhn, 1978): this is the condition in which the majority of X bridges are attached. The authors argue from this that the quick phase cannot be due to detachment of X bridges, since the stiffness is about the same at the onset as at the end of the quick phase. Accordingly, they wish to assign it to conformational change in the head. But alternatively, it could be explained if the bridges are broken, then reattach very quickly during stretch. There is evidence for this in that the stiffness is not constant during either stretch or release.

Most X-bridge theorists, including Guth and Kahn, assume that X-bridge cycling rate is constant, but this is not necessarily so, even for an individual bridge. If X-bridge rates vary with such factors as local calcium and other ion concentrations, instantaneous local mechanical conditions, and so forth, no simple theory could be devised, and treatment would have to be statistical, as in quantum mechanics.

7.7.2. Release

If a tetanically stimulated muscle is quickly shortened over a short distance, tension drops to zero for a brief period before rising at the rate characteristic for a tetanus to the tension appropriate for the new length. The release is made after the contractile machinery has not only become fully activated but also has fully stretched the series elasticity, and sarcomere inequalities have become minimized. Accordingly, the redevelopment of tension has been much used in determining contractile properties. More recently, variants of this experiment are being used in the expectation that they will be testing X-bridge properties.

If an activated fast-type muscle cell is released by various very short distances just before the application of the short quick stretch, there is a critical length of shortening that translates to exactly 14 nm/half sarcomere, which results in the just complete loss of a stiff component of the short-range compliance but leaves unchanged a more compliant one (Edman et al., 1980). The total length is proportional to the sarcomere length, so it corresponds to a constant repeat interval, which is the presumed linear spacing of X bridges.

The validity of the assumption that sudden decreases in length produce effects that may be attributed to X-bridge elasticity has been challenged on the basis of high-speed cinematography of single frog fast-type cells by Sugi and Tameyasu (1979) because they found that the shortening is localized to a small region of the cell near the end that is released.

7.7.3. Relaxation

The cycling events must be reflected also in relaxation rates, which are notoriously affected by the duration of the preceding stimulation, being slowed progressively in proportion to duration. In some invertebrate muscles, such as the pharyngeal retractor of the snail *Helix,* the rate slows more than 10-fold in six or seven twitches, a functionally significant factor (Kerkut and Leake, 1966). The slowing may be interpreted as in part being due to slowing of X-bridge cycling as suggested for human muscles from examination of the heat produced (Edwards et al., 1975).

7.7.4. Work

In glycerol-extracted frog fast cells activated by Ca-Mg-ATP, there is an immediate small force rise with a small (0.1% l_0) quick stretch that falls, giving way to a delayed rise in force that is greatest at about 0.6% l_0. This delayed force decreases with initial mean sarcomere length in the range 2.2–3 μm. Tension lags length when a sinusoidal oscillation of 0.1–1% l_0 is applied, with maximal positive work being done at about 20 Hz (Heinl, 1972).

7.7.5. Comparative Short-Range Elasticities

Almost all studies on muscle-cell dynamics have been carried out on vertebrate fast-type cells, so little is yet known about the comparative properties of short-range elasticity, associated with properties of X bridges. Such data is available for one vertebrate slow-type cell, of the toad *Xenopus laevis* (Lännergren, 1978). Lännergren got an approximation to normal full contraction using immersion in high $[K^+]_o$ for the slow cells. It had the usual two-phase tension–time characteristic, with the rapid phase similar to that of fast cells but with a markedly different slow phase. According to the X-bridge theory, the results could be interpreted on the basis of physically similar bridges in slow cells provided they cycle on the average about 15 times slower.

Slow-type skeletal cells of a toad show variable slopes of the F/V relation. The slope is markedly flattened by raising $[K^+]_o$ when the latter is used to stimulate contraction. For slow-type skeletal cells of the toad *Xenopus*, $V_{max} = 1.1$ lengths/s, the force constant (see p. 250) $a = 0.1\ P_{max}$, and the velocity constant (see p. 251) $b = 0.11$ lengths/s, compared with $V_{max} = 5.2$, $a = 0.38\ P_{max}$, and $b = 1.97$ lengths/s for a fast-type cell, each with SL = 2.3 μm.

7.8. Functional Value of Stretch Effects

We may ask if the effects of stretch are of functional value. The positive effect of stretch on maximum force developed was noted as long ago as 1924 (Fenn, 1924). The plateau value is proportional to both the speed and amplitude of the stretch, and decreases with extent of stretch in proportion to the reduction in myofilament overlap (Edman et al., 1978). One beneficial effect of this action is that it helps to minimize the effects of sarcomere length inequality and local inequalities in activation time (see p. 249). This stabilizing action must be progressively more important as the muscle length increases, that is, in the extended part of the movement range.

An extra contraction interposed during continuous excitation may, like a short stretch, cause local contractions of individual half sarcomeres at the expense of their neighbors because of uneven activation. There would then be a redistribution of load internally that would improve overall tension. Such adjustments have been termed *intersarcomere dynamics* (Julian and Morgan, 1979a,b). They must be important in a general way in all tetanic contractions. There is a progressive slow increase in tension during a tetanus that is termed *creep*, for which the commonly accepted explanation is based upon intersarcomere dynamics. It is assumed to be due to generation of extra tension by a small proportion of sarcomeres and their shortening at the expense of their neighbors, which bear the extra tension as they are stretched. These stretched parts have the advantage of benefiting from the improved energy relations and dynamics associated with negative work referred to below.

An even more significant effect of stretch is seen in whole-body locomotion. In many natural movements, quick small stretches of muscles result from phasic contraction of antagonists, for example when striking an object or on landing following a step. A striking consequence occurs: the stretched muscle, though developing a lot of active tension, consumes little extra energy in doing so. When a stimulated muscle is stretched by a load greater than it can bear, the heat produced is greatly reduced compared with letting it shorten. Instead of the muscle working, work is being done on it, a condition that is called *negative work*. An energy balance sheet shows that the work done on the muscle cannot be accounted for in its heat output (Abbott et al., 1951). All the energy-supply processes are reversible chemical reactions, so it was reasonable to suppose that some of them were actually being driven in reverse during negative work. A dramatic demonstration of the effect is afforded if two persons are set to pedal bicycles connected so that they are in opposition, and the oxygen consumption of each determined. The persons consume similar amounts when no movement occurs, but when there is a net forward pedaling by one, that person consumes more oxygen and the other less, the ratio increasing with pedaling rate, to a maximum of a remarkable 6 : 1 (Abbott et al., 1952; Bigland-Ritchie and Woods, 1976).

The stretch must be applied after full activation, not at the beginning of activation, to be maximally effective. The amount of extra ATP split is detectable in frog sartorius but is too small to be of consequence (Curtin and Woledge, 1979). The extra tension is proportional to sarcomere length, which in these experiments was in the range 2.5–3.1 μm, a range over which the I/A overlap is decreasing, so active tension development is also decreased. The beneficial energetic effect of stretch cannot be attributed to more X-bridge cycles, and indeed at this time no explanation can be offered that directly relates it to X-bridge theory.

7.9. Temperature Relations

The temperature range at which different muscles can operate is extremely large, and most natural ranges can be altered substantially by acclimation. Thus the amino acid compositions of constituent contractile and energy-supply molecules may be able to change to enable the significant conformational changes to occur at different temperatures without affecting the overall characteristics that fit the component cells to their functional roles. No matter what the specific temperature range, the shape of the V_{max}/T curve has an approximately linear rise with increasing temperature, following a Q_{10} of 2, to a peak. At a temperature just above the peak, termed the *critical temperature,* the velocity drops precipitously. The critical temperature changes progressively when an animal is subjected to temperature shift within its tolerance range, as part of the process of acclimation. Any, or all, of the components may play a role, including SR. Tests on SR during acclimation by goldfish to 5°C or 30°C showed that the calcium uptake mechanism was unchanged in its characteristics, but that the area involved had increased greatly at the lower temperature. The average diffusion distance had been reduced and the amount of calcium taken up increased, keeping the relaxation rate about constant.

7.10. Intersarcomere Dynamics

Evidence regarding details of intersarcomere dynamics in single cells is being obtained by high-resolution light diffraction. During shortening under load, continuous smooth changes are not seen in the light-diffraction pattern (Pollack et al., 1977). Instead, small steps occur that are of different patterns in different regions. There are several possible alternative explanations, the most attractive of which in terms of possible X-bridge dynamics is a synchrony of many contiguous half sarcomeres, acting out of phase with other groups. Populations of shortening sarcomeres might be canceling movements of ones being lengthened. Or all the sarcomeres might be shortening continually but the contributions of the shorter ones be decreasing as those of the longer ones increases.

Direct evidence for inequalities and their gradual improvement in a tetanus was obtained by Paolini et al. (1977) when studying single fast-type frog cells. They used computer analysis of light-diffraction spectra to determine changes in a sarcomere length, line intensity, and SL dispersion. In a twitch, line intensity decreased by 60%, with a long recovery time. The dispersion in a tetanus gradually decreased, implying that progressive stabilization was occurring, but the improvement was only about 10%. The first-order diffraction line split into two or more components during initial relaxation, then merged again, indicating that there are different rates of relaxation in individual sarcomeres. The significant adjustments are occurring between sarcomeres of myofilament clusters, but some are visible at the surface of an entire cell. These can be judged by placing fine hair markers at intervals and projecting their images onto a photodiode array (Edman and Höglund, 1981) and scanning them sequentially during a contraction.

The sarcomeres at the ends of a frog fast-type cell are dramatically shortened in an isometric contraction, whereas the center sarcomeres are hardly changed in length. Internal movements continue to occur during raised tension after lengthening (Edman et al., 1978). This argues against one possible X-bridge-theory explanation, which is that the effect is due to some X bridges' becoming "locked on."

The rearrangement of sarcomeres during relaxation and the spectrum of rates in individual sarcomeres must play major roles in determining characteristics of relaxation. These also affect *fatigability,* which can be defined as the coefficient of decrease of tension with time. This value was found in toad sartorius to be proportional to the extent of linear departure from l_0; the extent was different for lengthening than for shortening (Aljure and Borrero, 1968). Laser-light axial-diffraction spots from an actively contracting muscle possess a dynamic component that can only be attributed to continuous variation in thin-thick filament overlap (Bonner and Carlson, 1975).

All of the above considerations should dispel any of the notions that have sometimes been demonstrated in the past, that overall contraction in a real muscle can be treated as a determinant machine. Every muscle cell has internal structural imperfections, inhomogeneities in microanatomy, and associated local diversity of physiological properties, leading to complex intersarcomere dynamics.

7.11. Quantifying Velocity-Tension Relationships

Contrary to earlier opinion, it is now apparent, for reasons outlined above, that there is no unique V_{max}/L relation at a particular length: many factors, including immediate prehistory, affect the relation, and all are subject to wide statistical variations. This problem has been recognized by some biophysicists on the basis of direct experimental evidence from frog skeletal muscle (e.g., Aidley, 1965). Nevertheless averages are the overall determinants, and these are experimentally addressable. Attempts to quantify the relation occupied center stage in muscle studies before the ultrastructural advances of the late 1950s. An equation that is still of direct empirical value was worked out by A.V. Hill in 1938. He started with the observation that the maximum velocity under a particular load is proportional to the difference between the maximum load-lifting capability L_{max} and the actual load L:

$$V = (L_{max} - L)\,k$$

where k is a constant.
When shortening under load, the rate of working

$$W = V \times \text{gm} \cdot \text{cm} \cdot \text{s}^{-1}$$

The amount of heat liberated had been found to be related to the extent of shortening (Fenn, 1923; Hill, 1938) and therefore the rate of energy utilization was proportional to the velocity. A small amount of heat, termed the *activation heat A,* is produced in the coupling and activation processes. Major heat is produced during the maintenance of the state of activation and termed *maintenance heat M.* The M is a combination of an exponential rate of liberation termed *labile heat* and one with a constant rate termed *stable heat.* Heat specifically and only liberated in shortening is termed *shortening heat S;* it includes a component due to thermoelastic effects that is about 10% of the total (Woledge, 1963). The total energy liberated E is the sum of the three components of heat, plus work:

$$E = A + M + S + \text{Work}$$

Since S is proportional to the extent of shortening, it can be expressed as $S = as$, where a is a proportionality constant, the coefficient of shortening heat, and s is the extent of shortening.

The rate of extra heat production $h = aV$, and since this has the dimensions of grams times centimeters, a has the dimensions of force, so may be called the *force constant.* The total extra energy rate = $(L + a)V$. Hill plotted experimentally determined values for E against load and obtained a straight line. He assigned the letter b, which has the dimensions of velocity, hence the *velocity constant*, to the slope of this line. The two energy relations must be equal, hence:

$$(L + a)V = (L_{max} - L)b$$

or

$$V = \frac{(L_{max} - L)b}{L + a}$$

or

$$(L + a)(V + b) = (L_{max} + a)b$$

By making L = zero, V becomes V_{max}, so $V_{max} = L_{max}b/a$. This equation came to be known as the *characteristic equation.* It describes a rectangular hyperbola and, with suitable values for a and b, very closely fits the experimentally determined curve of V_{max}/L (Figure 7.1). The V_{max}/L relation at levels of activation below maximum shifts toward zero, mainly on the load axis. The maximum velocity of unloaded fast vertebrate cells should be constant, allowing for the need to overcome internal viscoelastic resistance to shortening, according to most theories of contraction, providing there are no longitudinal inequalities. This question has been addressed experimentally by Edman (1979) using single fast frog cells. The isometric twitch L_{max} could be reduced by dantrolene to $\frac{1}{10}$ normal without causing a significance reduction in V_{max}.

7.12. Relating Contractile Events to Heat Production

A basis for force comparisons in terms of molecular load-bearing-force-generating units is not yet available for any cell, and anything less cannot be of much critical value, but as with much else in muscle, approximations can be made and many aspects dealt with empirically. The constant *a* can be determined experimentally from measurements of shortening heat on one hand, and by fitting the equation to the V_{max}/L relation. Likewise *b* is the empirical slope of the (work + heat)/*P* curve, and it also can be determined from the V_{max}/L relation. The values of *a* and *b* for fast frog muscle turned out to be similar, as determined by each method, about 7000 g/cm^2 for *a* and 1.7 lengths/s (each adjusted to 20°C) for *b*.

It was later shown that a plot of *E* against the amount of creatine phosphate (CP) split was a straight line, with a slope of 11.1 kcal/mol (Wilkie, 1968). Nevertheless, the amount of CP split is less than the amount required fully to explain the sum of heat plus work (Gilbert et al., 1971; Curtin and Woledge, 1974), so a major factor in the dynamic energetics equation is still missing.

The heat produced in the initial stages of contraction plus relaxation in a short contraction is not matched in a simple manner by that produced during recovery. Instead it is related to a combination of this with the amount of work done. Its time course is closely correlated in time with the consumption of oxygen (D. K. Hill, 1940) and indeed also with the resynthesis of the phosphagen (Dydynska and Wilkie, 1966). For an extended discussion of these matters, the reader should consult Carlson and Wilkie (1974). At least one recent paper has confirmed that the only known net reaction of significance for energetics in vertebrate muscle that is detectable in a brief tetanus is dephosphorylation of CP (Homsher et al., 1979). There have been virtually no published studies that might help relate the above frog sartorius data to different kinds of muscle cells, except for those of another vertebrate, the tortoise.

B. C. Abbott and I carried out the classical types of experiments on single giant retractor cells of *Balanus nubilus,* and some of the data is reported in Figure 7.12. The results are remarkably similar, both qualitatively and quantitatively, to those obtained from whole sartorius of frog. The rate of production of shortening heat was very similar for the tortoise and barnacle, respectively, compared with 320 for frog. The ratio a/F_{max} for frog that of frog sartorius. The shortening heat production was 73.3 and 83.6 g • cm/cm for tortoise and barnacle, respectively, compared with 320 for frog. The ratio a/P_{max} for frog sartorius is 0.25, compared with 0.07 for the tortoise, which produces about the same F_{max} as the frog. For barnacle, a/F_{max} was 0.12, but F_{max} for barnacle is almost twice that for frog, so the value for *a* is the same for both. The rate of working of the contractile machinery is therefore essentially the same in these two tissues, and substantially less in the tortoise. This is reflected in the fact that although V_{max} is greater for frog, the velocities under load are about the same. The greater tension developed by the barnacle cell is clearly related to numbers of X bridges and to sarcomere lengths: the thin-thick filament ratio for the barnacle is 4 : 1; for frog 2 : 1, and the corresponding average sarcomere lengths 6 and 2.3 μm. The V_{max}/L curves of the two cross each other, so at the point of crossing, the performances are identical.

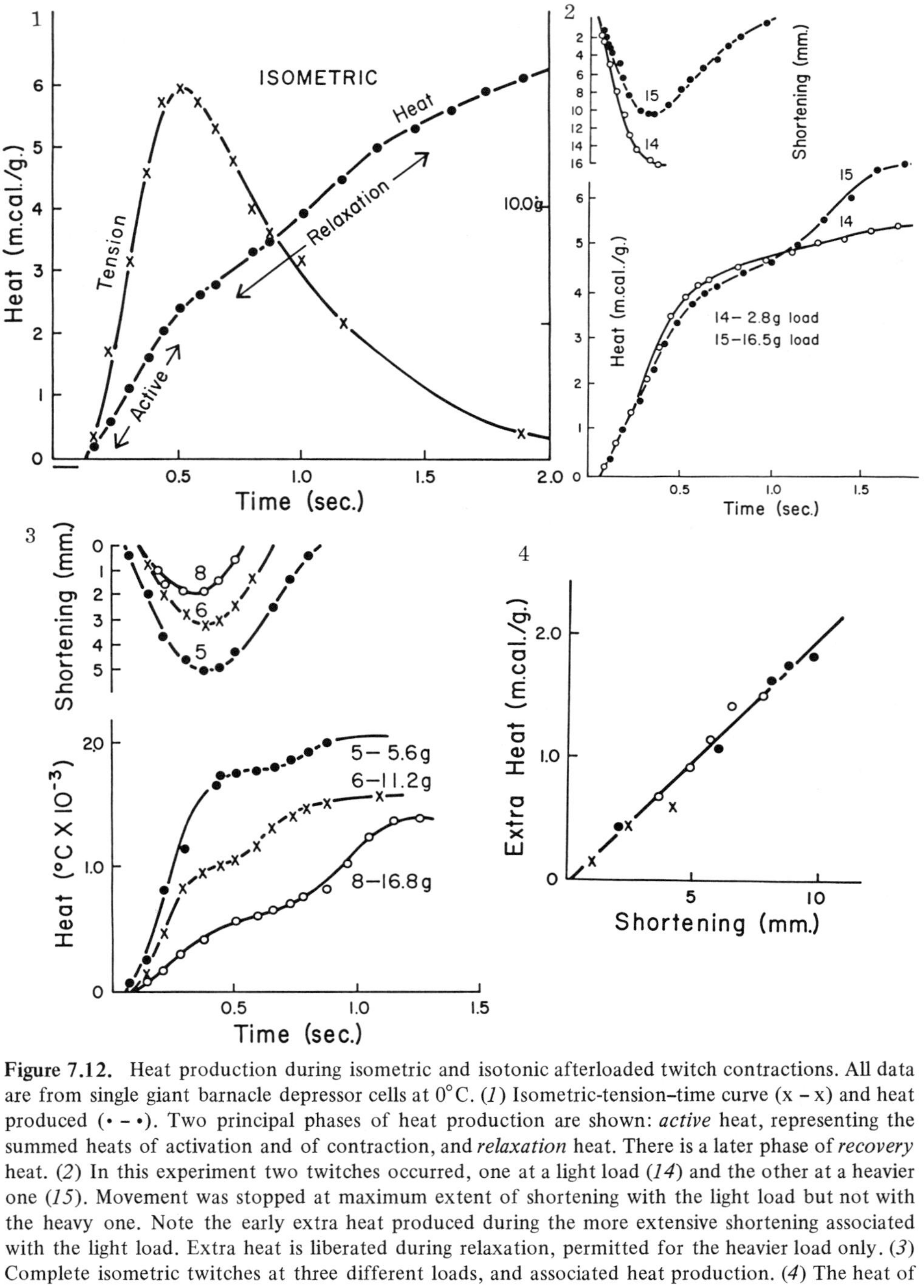

Figure 7.12. Heat production during isometric and isotonic afterloaded twitch contractions. All data are from single giant barnacle depressor cells at 0°C. (*1*) Isometric-tension–time curve (x – x) and heat produced (• – •). Two principal phases of heat production are shown: *active* heat, representing the summed heats of activation and of contraction, and *relaxation* heat. There is a later phase of *recovery* heat. (*2*) In this experiment two twitches occurred, one at a light load (*14*) and the other at a heavier one (*15*). Movement was stopped at maximum extent of shortening with the light load but not with the heavy one. Note the early extra heat produced during the more extensive shortening associated with the light load. Extra heat is liberated during relaxation, permitted for the heavier load only. (*3*) Complete isometric twitches at three different loads, and associated heat production. (*4*) The heat of shortening. (All records from unpublished data by B. C. Abbott and G. Hoyle.)

Constants a and b were determined for locust dorsal longitudinal flight muscle by Buchthal et al. (1957) and also found to be similar to those of the frog muscle. The constants have been determined at lengths below l_o, both being found to be unchanged (Matsumoto, 1967a).

Many problems remain for understanding the relation of heat production to contractile events. Perhaps the most significant is the question of heat given off rapidly during relaxation. If heat production in a twitch in which the muscle is permitted to relax normally is compared with that given off during an identical response, but in which the muscle is prevented from relaxing, 30-35% less heat is given off in the latter (Figure 7.12.2. and see Woledge, 1961). Only a portion of this heat can be due to lost potential energy. If relaxation is permitted, but forced with the aid of an ergometer to occur more slowly than naturally, the heat fails to appear. This paradox is unexplained.

In studies on the energy production by toad sartorius muscle at 6-7°C, used because it is naturally relatively slow, Chapman and Gibbs (1972) could find no strong correlation between isometric heat production and tension-time relations. A. V. Hill (1964) eventually modified his earlier view that the heat of shortening is simply the product of a times distance shortened, and others have concurred (e.g., Brandt and Orentlicher, 1972), suggesting that there is, in effect, a variety of forms of shortening energy. Two propositions are supported for all forms of shortening energy: (1) heat is liberated at a constant rate for a fixed load, and (2) heat varies with load in proportion to the difference $L_{max} - L$.

The constant b is not correlated with muscle length, for muscles of different length, but there is a fair correlation with the total muscle mass. In the premolecular phase, the ratios b/l_o and a/L_{max} had come to be seen as possible universal constants reflecting molecular events. There now seems no justfication for this view.

The only phase of heat production about which a reliable value may be given is the activation heat, but no agreement has been reached on the relative amount that each component contributes. The total activation heat may be determined either by measuring heat output when the muscle is stretched beyond the point at which there is no contraction or by comparing the difference in early heat produced between a single response and a close pair of stimuli. The amount for a frog sartorius is 1.2 ± 3 cal/kg (Gibbs et al., 1966). Assuming 0.1 μmol calcium is bound per gram of muscle, the energy released would be 0.93 mcal/g or 3.9 mJ (millijoules). The enthalpy change of -39 kJ/mol calcium bound is a good approximation to the activation heat (Yamada et al., 1976). The reconstitution of ATP occurs so rapidly that ATP breakdown cannot be detected without specific enzyme inhibition and dramatically rapid cooling. But if a vertebrate muscle is first treated with iodoacetate to prevent glycolysis and then with dintrofluorobenzene (DNFB) to inhibit creatine phosphokinase, the muscle is still able to contract using available ATP, permitting assessment of the heat produced in the blocked Lohmann reaction (p. 131). The major reactions involved in supplying energy are a mixture of exothermic and endothermic processes with a strong net positive heat output. The individual contributions cannot yet be fully assessed, but some parts are separable because of the availability of specific enzyme inhibitors. Since creatine phosphokinase is blocked by DNFB, the contribution of the Lohmann reaction can be eliminated, because there is sufficient ATP for a twitch. The heat produced after poisoning was found by Yamada to have increased because the Lohmann reaction is endothermic to the extent of some 10 kJ/mol CP. So far an energy-chemical-change balance sheet has not been achieved for contraction of any muscle. There are even indications that some significant chemical event or events in the early energy-supply process is (are) still unknown, as concluded by Gower and Kretzschmar (1976) after a study of heat production by rat soleus.

The presumed all-important role of the Lohmann reaction has been ingeniously tested by depleting muscles of CP by dietary adjustment. Inclusion of γ-guanidinopropionic acid in the diet of weaning rats for 10 weeks led to an absence of CP (Petrofsky and Fitch, 1980). The soleus muscle had a 50% slower twitch rise time, reduced peak force, and slower relaxation; however, the plantaris muscle showed unchanged contraction characteristics.

So many temperature-dependent variables seemed likely to be involved in activation, coupling, and contraction that few studies have been attempted on the effects of temperature on the heat produced. It is also technically difficult to study heat production at temperatures above that of melting ice. Effects on activation can be studied relatively independently of other heat production by stretching to beyond overlap of filaments. This was done for frog semitendinosus in the 0-20°C range (Rall, 1979). Surprisingly, the activation heat was found to be independent of muscle temperature in this range. Activation heat presumably corresponds to the coupling-calcium-release–calcium-uptake cycle. In unstretched muscle, except at very low temperatures, which may impair activation, both force and energy liberated decline progressively with increasing temperature, indicating that increased speeds of recycling of force-generating units reduce their efficiency.

7.13. Work

Although a few muscles are concerned only with signaling and are required to develop very little force, the majority are required to move parts or all of the body and therefore to perform work. A muscle does internal, but no external, work during an isometric contraction. The useful work it performs is simply load times distance, and its rate of working is the derivative of this product with respect to time, load times velocity. But we have seen that distance, velocity, and load are mutually antagonistic requirements. It follows that optimal performance of a muscle in terms of work output is a compromise. Optimal work output occurs when the load is about $\frac{1}{2}$ L_{max} for most muscles.

The rate of heat production is about as well known for human muscle as any other. Heat may be determined calorimetrically and temperature by means of a thermistor probe inserted in a hypodermic needle. The initial rate is about 55 W/kg wet weight muscle (Edwards et al., 1975). The rate of heat production falls during continuous excitation to about half its initial rate before there is any significant falling off of work output. This is intriguing, because it suggests that there is a progressive improvement in the efficiency of the chemical events reconstituting ATP.

It is possible to obtain an approximate value for the overall efficiency of muscle in converting chemical energy into work when this is optimal, but a major part of the departure from ideal may be accounted for in the chemical events that provide CP. A closer approximation is to relate work to the amount of phosphagen split. Or the ratio of work to total energy, work + heat, may be evaluated, a term that A. V. Hill chose to call the *mechanical efficiency*. Hill found that whole frog skeletal muscle gave a value of 0.4, whereas the equivalent but slower tortoise muscle gave 0.8, a very high value. On the whole, muscles, in comparison to mechanical heat engines, are highly efficient machines.

7.14. Basic Tonus: What Kind of State Are Skeletal Muscle Cells in at "Rest"?

The length of a particular type of muscle upon isolation is a major variable, as are its elastic and viscous properties. The state of rest is defined as one in which there is no excitatory neural input, included hormonal or other active modulators. Most smooth-muscle cells and all myogenic heart muscles have spontaneously depolarizing surface membranes, and unless their state between actual shortening and lengthening phases is called rest, they cannot be compared to skeletal cells, in which the membrane potential is stable until externally disturbed. Only the latter will be considered here. The classical view of all such cells is that at normal resting potential, they are also completely at rest, with the molecular machinery switched off totally. But this view is giving way to one in which rest is considered to be a state of very low activation, the amount of which varies between cell types and even individual cells of one type, as part of a statistical spread. Had modern knowledge of the molecular basis of coupling (see Chapter 6) preceeded physiological work on dynamic aspects, it would not have been considered possible to make a simple distinction between activity and rest. Almost all interpretations of the effects of stretch on unstimulated muscle aimed at understanding X-bridge movements required the assumption that there is a small amount of recycling, or some residual linked X bridges, perhaps ones different from the others, or both.

When skeletal cells are depolarized by a small increase in $[K^+]_o$, an amount that is much too small to reach E-C threshold, there is nevertheless a substantial increase in oxygen consumption and an accompanying increase in heat production. This phenomenon is called the *Solandt effect* (Solandt, 1936), and it has never been explained. It was earlier assumed that E-C is a unitary event initiated at a discrete threshold voltage. It is more probable there is a steep rise in the curve relating membrane potential to calcium release at the presumed threshold, but that the relationship is in fact a continuous one, and that significant calcium release, and therefore TN/AM ATPase activation, is occurring at all membrane potentials. In most skeletal cells the active tension being produced is just sufficient to balance internal radial elastic forces, so no shortening occurs until the steep part of the relation is exceeded. There is a wide range in behavior of skeletal cells at mean body length. Some do indeed shorten slightly. Supporting this view is the finding that caffeine, which directly causes calcium release from SR (see p. 200), leads to heat production (Hartree and Hill, 1924). Hypertonic solutions also cause heat production and increased tonus (Yamada et al., 1976).

The tension developed at mean body length l_0 has been termed resting tension, which is clearly a contradiction; it is in fact *basic tonus* (BT). Basic tonus was recognized for more than a century, and it played a major role in muscle science prior to 1925 (e.g., Langelaan, 1915; 1922). In 1924 and 1925, Liddell and Sherrington published their classical papers on the stretch reflexes (myotatic reflexes) in mammals, and Cobb (1925), reviewing this and other mammalian research, concluded that the evidence pointed to weak neural activation as the cause of tonus. As has so often happened in muscle research, this view became first the fashion, then the dogma that stifled further inquiry and buried a great deal of work that clearly indicated otherwise. There is no contradiction in any case; *active tonus,* which is due to weak reflex actions and intrinsic tonic discharge of motor neurons, coexists with basic tonus. The two are simply superimposed on each other, and their relative contributions are very different for different skeletal muscles. Very few muscles from any source could be said with confidence to be completely relaxed when deprived of neural input by section or anesthesia. Following anesthesia in humans, a normal level of basic tonus is present that is equal to the basal level occurring in the conscious, relaxed subject (Lakie et al., 1980). Weak basic tonus is possibly a major cause of some forms of the human disease amyotonia congenita, the so-called "floppy" disease; the subjects can have perfectly normal muscle strength in voluntary movement.

Some skeletal muscles have negligible basic tonus; in others it is quite strong. For example, in the eyestalk levator muscle of the Hawaiian crab *Podophthalmus,* where the eye stalk is exceptionally long, in the absence of excitatory neural input, the eyestalk is raised to about 45°. Inhibitory input or hyperpolarizing pulses relax the cells and lower the eyestalk (see Figure 6.7). It can also be quite strong in crayfish claw opener (Hawkins and Bruner, 1979). Basic tonus is altered by many neurohumoral factors: in mammals the principal agent is noradrenaline, which enhances it. In some insects basic tonus is important in determining posture, being generally strong in their antigravity musculature. There it is under natural neurohumoral control, being reciprocally influenced by the biogenic monoamine octopamine (Hoyle, 1978a) and by the pentapeptide proctolin (see Figure 12.48-Piek and Mantel, 1977). Not all muscle cells contribute to basic tonus within a given muscle, indeed, only a small fraction may be participating, even when basic tonus overall is strong. We have seen many examples in which an insect being trained with operant conditioning to alter posture did so using a shift in basic tonus only. No strong action on membrane properties has been detected for either octopamine or proctolin, both of which appear to exert their effect internally. Proctolin induces a contracture in hindgut muscle of cockroach at 5×10^{-9} *M*/l (Cook and Holman, 1979b) that resembles that evoked by caffeine, provided the cells are first depolarized. There is a subtle interplay between the determinants of membrane potential and the modulators. Basic tonus is particularly strong in some vertebrate smooth muscles (p. 338), where its control appears to be mediated by cAMP-dependent phosphorylation (Hirata and Kuriyama, 1980).

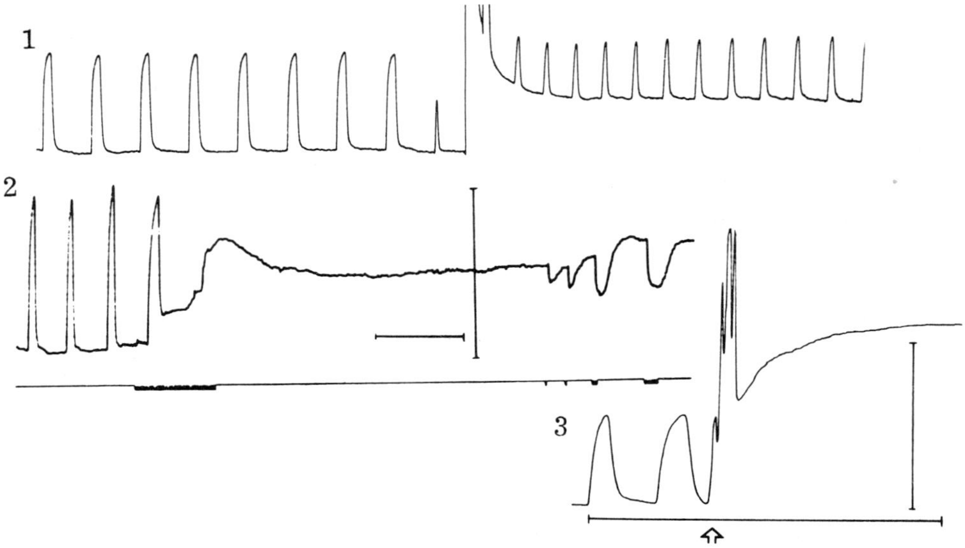

Figure 7.13. Trade-off between intrinsic rhythm of tension, and basic tonus, in locust (*Schistocerca gregaria*) skeletal muscle. Factors leading to enhanced basic tonus, also reduce the amplitude of IR (*1*) or even suppress it completely (*2*), (*3*). In preparations with intact nervous sytems changes such as these occur spontaneously from time to time. Their causes are not known. [(*1*) From Hoyle, 1978. (*2*), (*3*) Reprinted with permission from Piek et al., Peripheral control of basic tonus and myogenic rhythm in the extensor tibiae of *Schistocerca gregaria., Comp. Biochem. Physiol.* **64A**, 565–569, Copyright 1979, Pergamon Press.]

7.14.1. Intrinsic Rhythm of Some Insect Muscles

Known so far only from a small bundle of muscle cells in the proximal region of the extensor tibiae of locusts and grasshoppers is an interesting intrinsic slow rhythm of contraction and relaxation. This intrinsic rhythm (IR) fluctuates in size and can be absent (rarely) or cause quite large movements of the tibia at times (Figure 7.13 – Hoyle, 1978a). These movements are summed with a strong basic tonus that when exceptionally enhanced, completely masks IR. The IR is rapidly and completely suppressed by neurally released octopamine or by infusion in a concentration as low as 10^{-10} M. The existence of IR greatly affects the dynamic properties because results are dependent on the timing of events relative to those of IR. Furthermore, any perturbation in turn affects frequency, duration, and amplitudes of IR events.

7.14.2. Rigor

The well-known state of stiffness termed *rigor* develops in vertebrate skeletal muscle as the ATP level falls. Even extreme starvation causes stiffness to develop consequent upon a loss of available energy for ATP synthesis. However, in some invertebrate muscles, a rigor-like state does not accompany death, but the reasons for this are not known. A rigorlike state can be induced in these invertebrates by chemical excitation, especially a large pH shift. Some agents such as ryanodine that cause rigor in vertebrate muscles produce the opposite result, flaccid paralysis, in invertebrate muscles. This is because although both are depolarized the vertebrate cells have coupling inactivation, whereas those of the invertebrates so far examined, especially insects, do not.

According to the X-bridge theory, rigor occurs in the absence of E-C activation because X bridges form randomly, one by one, and in the absence of ATP they fail to break. There is internal activation due to a rise in calcium as the SR calcium pump, which is driven by ATP, fails. ATP is needed for X bridges to break, so once formed, they do not break naturally. Eventually the muscle goes completely stiff; indeed, this is the reaction responsible for stiffening at death. Bathing a muscle in a state of rigor in a solution containing a high concentration of ATP relaxes it, causing contraction secondarily only if the concentration of calcium is sufficiently high. During the onset of rigor, the low-angle X-ray diffraction pattern associated with the X bridges changes markedly (Huxley and Brown, 1967). The X-ray indications are that the A filaments themselves become less orderly, matching the myosin helices closely to actin, and permitting much easier X-bridge linking than is possible in the living muscle.

7.15. The Onset of Contraction

The first manifestations of a change in state following excitation, excluding birefringence and other changes that may be directly associated with excitation, are not those associated with contraction itself. A very early volume change was detected as long ago as 1925 by Ernst, and was shown in frog fast muscles to consist of an initial small, brief increase followed quickly by a significant decrease that persists. Some water is lost from each cell during contraction. The volume change has been verified and quantified (Abbott and Baskin, 1962; Schäffer and Örkenyi, 1972), but its significance has yet to be explained, and its relation to coupling events is not understood.

Another early finding (Rauh, 1922) was that some vertebrate fast-type muscles, stretched moderately to the point of generating a little tension, showed a brief loss of this tension before onset of contraction. For a long time assumed (without test) to be universal, this phenomenon was eventually shown to be missing both in some very fast vertebrate muscles such as the toadfish swim bladder (Geddes and Mailman, 1973) and both vertebrate and invertebrate slow muscles or cells (Figure 7.14 – Hoyle, 1980). Even the name applied to it, *latency relaxation* (LR), begged a significant question: relaxation of what? When the concept of basic tonus was universally accepted, it was presumably that. It is now increasingly referred to as *latency elongation* (LE). Early onset characteristics for a variety of muscles are presented in tabular form by Hoyle (1980). Elongation, when it occurs, does so during the early phase of volume increase. Explanations had favored changes in the contractile material, and in recent years elongation was often given as evidence for "residual" connected X bridges that had to let go before they could start cycling. However, its lack of universality eliminated the latter theory, and elongation is currently considered likely to be an artifact produced by rapid water movement into terminal elements of ITs, which swell markedly during excitation.

The effects of length on LR are remarkedly different in frog compared with fast insect muscle cells (Figure 7.15). In frog, stretch beyond l_0 progressively and intensively enhances LR, whose peak occurs later, and delays the onset of contraction. By total contrast, in locust the peak is not enhanced and occurs earlier, as does the onset of contraction. These differences can readily be explained on the basis of assumed swelling of ITs (Figure 7.15), because of the different anatomical features of the E-C coupling sites. Mammalian muscle cells have not been tested, but might be expected to follow the pattern seen in the locust, rather than that of the frog.

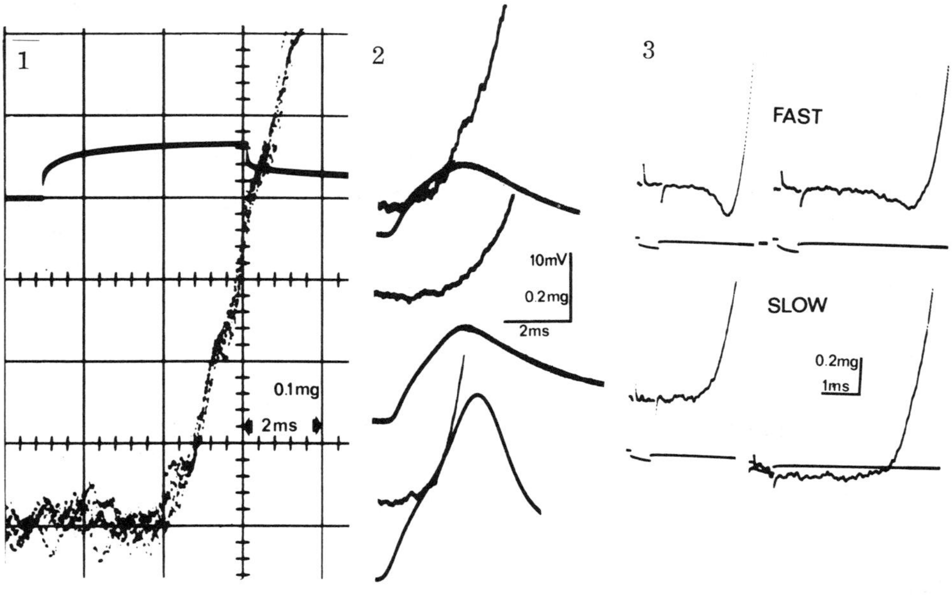

Figure 7.14. Onset of contraction in slow- and fast-type muscle cells compared. (*1*) Barnacle giant depressor cell. (*2*) Locust slow cells, with intracellularly recorded action potentials superimposed. (*3*) Frog muscle cell bundles; iliofibularis at left, rectus abdominis at right. (Reprinted with permission from G. Hoyle, Comparative aspects of the onset of muscular contraction with special reference to latency relaxation, *Comp. Biochem. Physiol.* **66A** 57–68. Copyright 1980, Pergamon Press.)

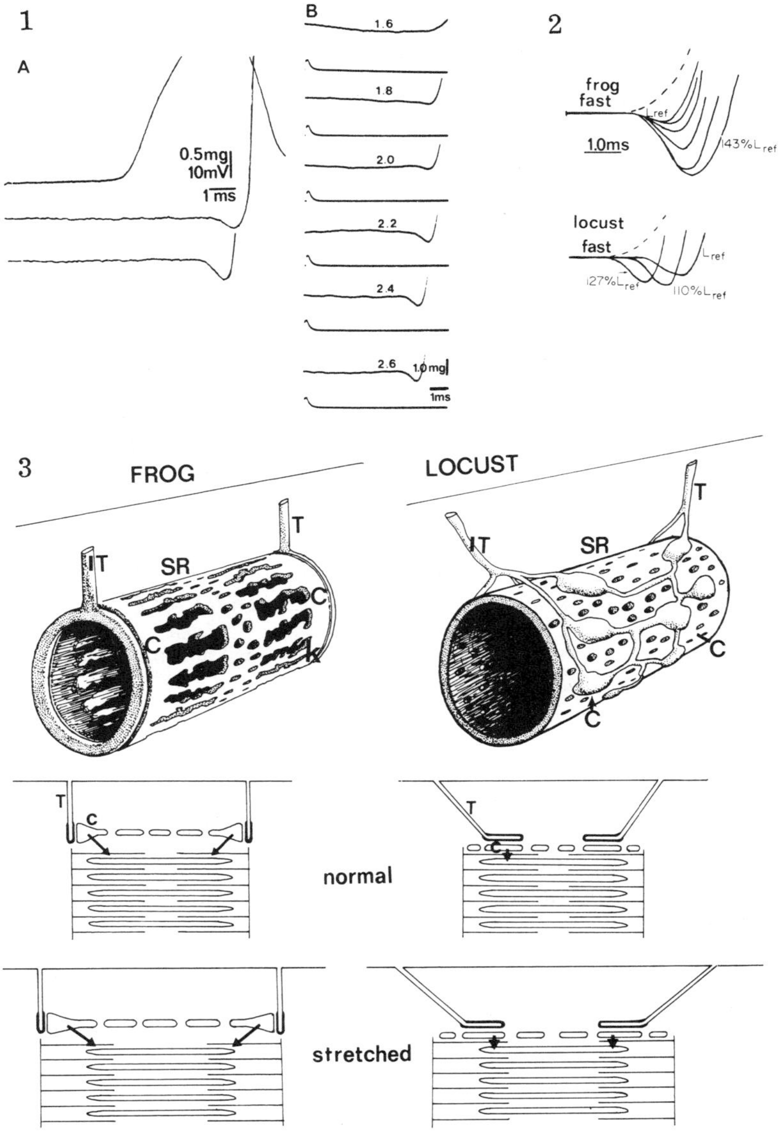

Figure 7.15. Different effects of length on depth of LR in vertebrate and invertebrate fast-type cells. (*1*) Locust extensor tibiae. *A*, LR relative to intracellular action potential (*top trace*), at rest length (*middle trace*), and after stretch by 0.4 mm (*lower trace*). *B*, Tension only, at lengths stated (mm). (*2*) Comparison of LR at progressively greater lengths for frog semitendinosus cell (*above*) with locust fast cells (*below*). (*3*) Different effects of stretch are related to different morphologies of ITs and IT/SR contacts. (All from Hoyle, 1980.)

7.16. Overall Dynamic Relations

The dynamic properties of an excited muscle depend upon the sum of all the various events so far described, which we may enumerate, for natural excitation, as follows:

1. The time course of those changes of membrane potential during excitation that activate E-C coupling.

2. The time course of E-C coupling events initiated by item 1.
3. The mean time course of rise and fall of calcium concentration at TN/AM interaction sites.
4. The sum of force-time and length-time relations for the total of all half sarcomeres.
5. Compliance of series and parallel elastic elements.
6. Internal viscosity.
7. Temperature.

Of the above, items 3-6 are all multiply interactive. Into item 4 are incorporated multiple events including the rate of supply of energy and its translation by the contractile machinery into work and heat.

In considering the control of any muscle in the body, the ways in which movements are made are determined not only by the specific neuromuscular events applying to the muscle in question, but also by the dynamic action of gravitational, inertial, and all external forces (of antagonistic and synergistic muscles acting on it at the time), especially by way of length changes via passive and active viscoelastic elements, that are part of the contractile machinery itself. There is no possibility for worthwhile generalizing to the whole from any one particular muscle, as has so often been the case in the past from experiments on frog sartorius. For example, the satement of Katz (1939):

> Rapid shortening of an active muscle beyond its optimum length is apt to break or weaken permanently parts of the contractile substance. A shortened muscle "gives" and later "reshortens" when tension is suddenly increased.

This has been regarded as the norm for muscle in general and has often been referred to in recent years by X-bridge enthusiasts. However, a quick stretch at several times the speed, and by an amount several times greater, than that which will tear apart even an unstimulated sartorius can be applied to many other fully stimulated muscle cells of both vertebrates and invertebrates without in any way impairing their ability to continue to develop tension. The differences lie in markedly different elastic properties, which must have markedly different fine structural underpinnings that are not yet understood.

Therefore, every dynamic movement by any particular species of animal needs specific examination directly of the properties not only of the whole muscles involved, but of each type of component cell.

7.17. Contributions of Special Cell Capabilities and Secondary Activations

The structural features that permit supercontraction greatly extend the dynamic range, as we have seen. These features are present in some human muscles (Figure 7.16) and beef muscle. It is not likely that the reversible supercontractility as such is functional, because of skeletal limitations, but an accompaniment, a wide plataeu of maximum speed capability over a wide range of length, is. This is especially important in human muscles used in fast skilled actions such as accurate spear-throwing, sprinting, and gymnastic maneuvering, which may have been primevally valuable in arboreal activities.

The Z-band expansion of barnacle giant cell myofilament clusters that precedes shortening following activation (Figure 7.17), although not likely to be a major contributor to force development and movement, must nevertheless make a significant contribution to overall dynamics. This remains to be assessed.

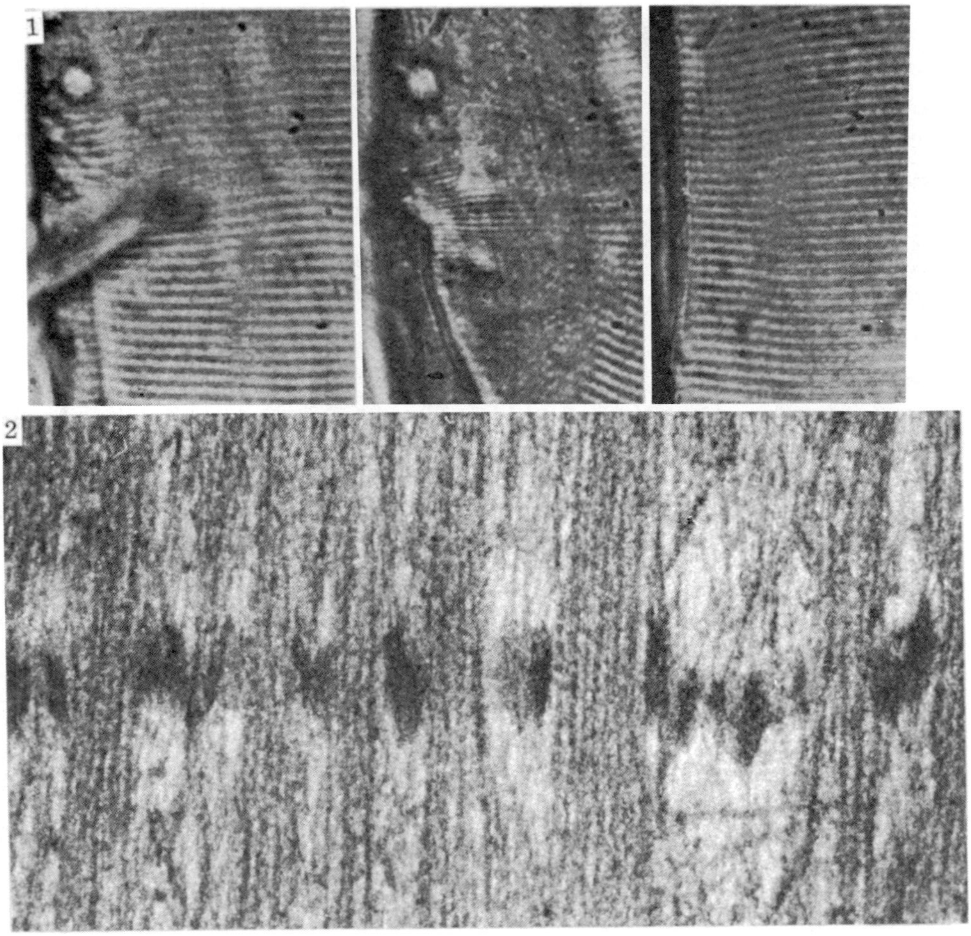

Figure 7.16. Supercontractions of some cross-striated muscle cells. (*1*) Reversible supercontraction in a human muscle cell, filmed before, during, and after the ninth consecutive trial. (Reprinted with permission from R. Gruener, B. McArdle, Brenda E. Ryman, and R. O. Weller, *J. Neurol. Neurosurg. Psychiatry* **31**, 268–283. Copyright 1968, The British Medical Association.) (*2*) Basis of reversible supercontraction is giant barnacle depressor cell. Thick filaments are able to interdigitate across the perforated Z region. (From Hoyle et al., 1965.)

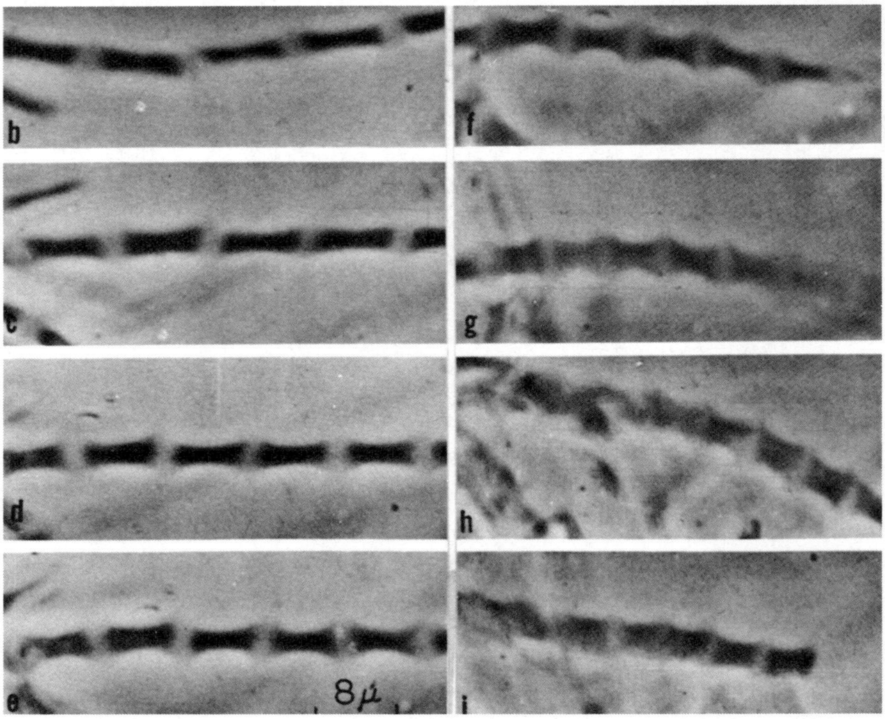

Figure 7.17. Active expansion of Z regions occurs in glycerinated myofilament cluster from barnacle giant cell following irrigation with ATP. The Z expansion precedes shortening. Successive frames of cine film. (From Hoyle et al., 1965.)

8

Muscle Cell Diversity

If the constituent fibres (cells) of a muscle can differ considerably from one another in intrinsic speed that could have a bearing on many of its observed mechanical properties. A. V. Hill†

8.1. Variations on a Structural Theme

Until the introduction of improved fixation methods for electron microscopy in the mid-1960s, little attention had been paid to differences between muscle cells. As long as there was a strong belief in the sausage-with-skin concept and a common contractile machine, there was no point in dwelling upon differences that could only introduce unnecessary complications and retard "understanding." Differences had been recognized, of course, but were not taken very seriously until electron microscopy revealed the astonishing complexity of individual muscle cells. The significance of many of the features is clear, but others are totally unexplained. Many may be the result of happenstance in evolutionary histories and need not be linked to functional explanations. Insect flight muscles, for example, might have been expected to have common features, but there are major differences in overall organization in ones that otherwise function similarly in insects having similar weights, flying speeds, and patterns (Figure 8.1).

8.1.1. Discovery of Intramuscle Cell Diversity

The earliest recognition of different types of muscle in a single animal was, in fact, very old, being attributed to Lorenzini (1678), who noted and compared red and white muscle in the elasmobranch fish *Torpedo*. Such differences are particularly clearly seen in the thin dark "red" muscle of the middle outer region of the body muscle of salmonids and tunas compared with the "white" color of the muscles in the bulk of the body. The red muscle contains red cells packed up to 50% of their volume with mitochondria, as well as much myoglobin. The neighboring white majority contain almost no mitochondria, almost the entire cell volume being taken up with small myofilament clusters and the SR that extensively permeates them. The red muscle is used in all general-purpose cruising (e.g., Bone, 1966; Rayner and Keenan, 1967). The white part, 96% of the total muscle mass, is used only in the brief moments when the fish propels itself upward to "leap" from one level to a higher one and in rare bursts of fast swimming to escape predation. This situation provides a simple example of the diversity that has evolved in originally similar cells to improve behavioral activities. Although fascinating to most zoologists, diversity was only an embarrassment for physiologists searching for "general principles," and little attention was paid to it until quite recently, since electron microscopy revealed

†A. V. Hill, *First and Last Experiments in Muscle Mechanics,* Cambridge University Press, 1970.

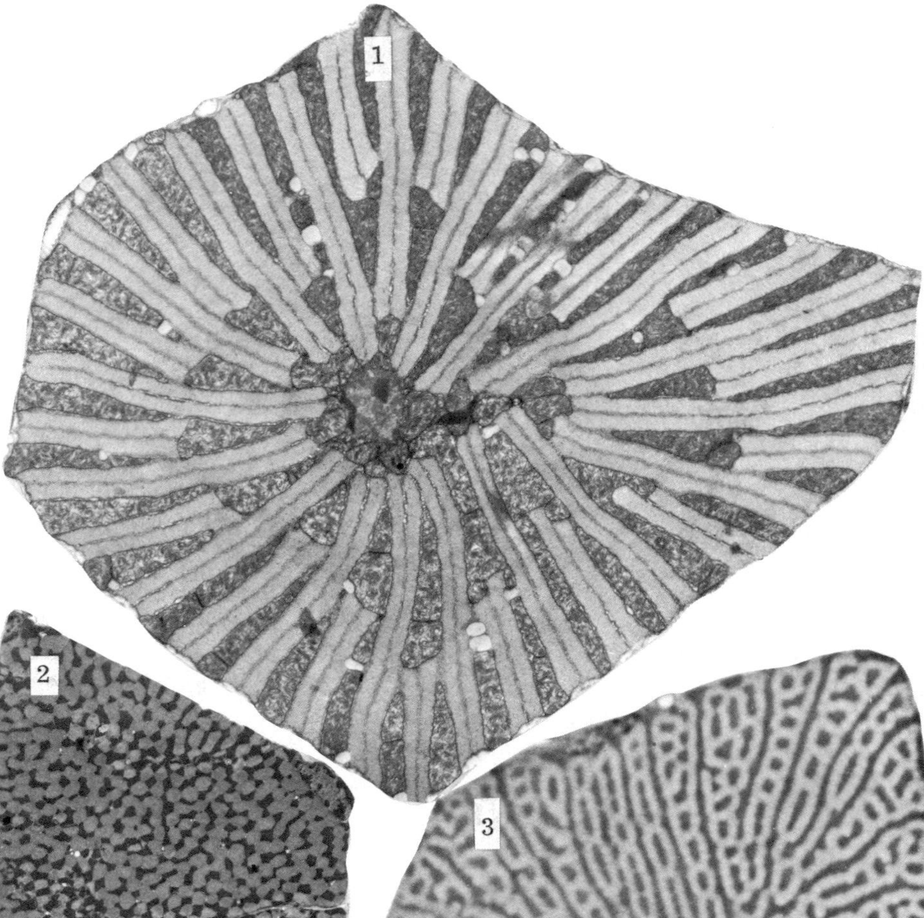
1
2
3

the deep underlying causes of some of the diversity, and since improvements in methodology have permitted biophysical studies to be made on single muscle cells.

Until about the middle 1950s, it was generally assumed not only that differences between muscles within a class were minimal, but that the variety of muscle cells within a muscle was not large. This view should not have survived the early discovery that some frog muscles, notably the iliofibularis, contain a group of smaller-diameter cells that receive an independent motor nerve supply (Sommerkamp, 1928), which were later shown (Kuffler and Vaughn-Williams 1953a, b) to be special slow cells with unique properties. But this finding was regarded as an oddity, and it did not lead to careful examination of other muscles. These slow cells are clustered and therefore separable from the majority, which were supposedly alike, so the discovery did not significantly affect the view of uniform composition. This view was based on a combination of histological ignorance and wishful thinking on the part of physiologists accustomed to working with whole muscles. However, histologists had been aware of different staining qualities of component cells in a number of muscles, and as they began to use methods directly related to the presence of muscle-cell enzymes, more and more examples of cell differences began to be noticed. A landmark of progress was Krüger's (1952) recognition of the *Felderstruktur/Fibrillenstuktur* distinction (see p. 44); thereafter vertebrate muscles were looked at more closely and found to be extraordinarily complex mixtures.

The traditional view might have changed rapidly following the introduction of intracellular electrodes in 1946 because this technique permitted the sampling and comparison of the properties of individual cells relatively easily. All prior electrical studies had necessarily examined the average current flowing in entire muscles and could not have led to the recognition of differences in individual cells. But the expectation of commonality at first caused cells with different properties to be simply discarded from the record by intracellular physiologists. Alas, in invertebrate studies the assumption of uniformity lingered even longer. The late Professor Cornelis Wiersma and I in 1957 got the first hints of a truly remarkable range of diverse cell types in one muscle when intensively studying examples of crustacean walking leg muscles. As it happens, it has turned out that a complex mixture of cell types within one muscle is the rule, not the exception. These matters will be dealt with in some detail in Chapter 13, but it is important to bear in mind the general features in considering the remainder of this chapter.

A majority of the vertebrate and arthropod skeletal muscles that have been intensively examined have been found to contain a mixed population of cross-striated cells that differ in regard to major physiological characteristics including membrane properties, rates of contraction and relaxation, capability for sustained-tension development, and maximum-force capability. These differences are associated with marked differences in coarse structure, fine structure, and biochemistry. Even when a muscle is so small as to contain only three cells, each can be of a markedly different type (Hoyle, 1967a). Only relatively recently has it been realized that cardiac-cell components, in spite of their

Figure 8.1. Transverse sections through representative insect flight muscle cells. (*1*) TEM of radially arranged contractile material, mitochondria, and sarcoplasmic reticulum in dragonfly flight muscle cell, with nuclei in central core. (Courtesy of Dr. David Smith.) (*2*) TEM of locust (*Schistocerca*) dorsal longitudinal flight muscle showing simple mixture of myofilament clusters and mitochondria (darker bodies), with peripheral nuclei. (ORIG.) (*3*) Light micrograph of cicada longitudinal flight muscle cell showing complex radial arrangement with contractile material (*lighter*) surrounding, as well as being interspersed by, mitochondria. (Courtesy of Dr. David Smith.)

anastomoses and intimate connections, are truly a mixture also, but little information is available yet about this aspect.

The first clear demonstration of the extent of diversity was achieved by Dorai-Raj (1964) while examining the electrical properties and histology of cells constituting the tiny, thin, and fan-shaped accessory flexor of the carpopodite of the crab *Cancer magister*. He saw three populations of cross-striated cells, with diameters of about 100, 250, or 500 μm respectively, laid out in a thin, orderly band, thin ones proximal, the thick ones distal, and the others between. But they had a much more conspicuous difference than thickness: the thin ones had sarcomere lengths of about 12 μm, the thick ones 3–4 μm, and the medium ones 4–10 μm. There were correlated electrophysiological differences: thick ones tended to generate large graded electrogenic responses when depolarized, the thin ones did not respond, and the medium ones gave weak graded responses.

A systematic cell-by-cell study of excitability, neuromuscular properties, and contractile characteristics of a larger muscle, the dactylopodite flexor of the same crab species, was made by Atwood, Hoyle, and Smyth (1965) and found to contain a mixture of each of the types found by Dorai-Raj, but instead of being arrayed in bands, the different types were irregularly distributed along the entire length of the muscle. The thick, short-sarcomere cells had brief contraction/relaxation times of about 40 ms to peak with 300 ms decay and so were termed *fast-type cells.* The thinnest, long-sarcomere cells contracted and relaxed much more slowly, taking at least 20 times as long to develop comparable tensions and were therefore termed *slow-type cells.* The cells of intermediate thickness and sarcomere length had intermediate speeds, so were termed *intermediate-type cells.* A summary of the details is presented in Table 8.1.

Since 1965 there have been many studies in which electrical responsiveness, synaptic transmission, and tension development have been made on representative cells, and the physiology correlated with structural studies, including electron microscopy on the same material. These have shown that in crustaceans there is astonishing diversity; no two muscles, even of the same animal, are alike in details. Each muscle has a characteristic set of cell types, located in almost exactly the same regions in different specimens, but each with its own unique variation in detail. The sarcomere length is but one of many features of both structure and biochemistry that distinguish the component cells in a muscle. Major sources of diversity are outlined below.

8.1.2. Grouping

For each of the criteria upon which differences between cells are recognized, the question of their distribution within a muscle is raised. In some muscles, like cells are all located close to each other in a group, and in others they are randomly scattered throughout the muscle. Between these extremes are various intermediates, with scattered partial clustering of like cells. Examples of each extreme, and of many intermediates, are known from both vertebrates and invertebrates. In frog iliofibularis slow cells occur only in the discrete bundle discovered by Sommerkamp; in the large, powerful rotator of the paddle-shaped hind leg of the crab *Portunus* they are confined to a single centrally located bundle (Hoyle, 1973), and in the tail of trout, every myotome has one small bundle of thin, red slow-type cells (Johnston et al., 1975). But in human biceps or in the adductor of the crab *Cancer magister*, slow cells are randomly scattered throughout the muscle. Repeating clusters of cells that are both different within a cluster and different between clusters occur in locust extensor tibiae (Hoyle, 1978) and frog cruralis. In the preceding

Table 8.1 Summary of Muscle Cell Types in a Crab (*Cancer magister*) Walking-Leg Dact[illegible]podite [illegible]

Muscle Cell Type	Histology	Electric Responsiveness	Tension Development[a]	Approximate Percentage	Mechanical Response to Prolonged Depolarization	Resting Potential (mV)	Time Constant (ms)	E_c (mV)	Maximum*[b] Rate of Force Development $(kg/cm^2 \cdot s^{-1})$	Maximum*[b] Rate of Relaxation $(kg/cm^2 \cdot s^{-1})$
Fast	Thick, 3-μm SL *Fibrillenstruktur*	All-or-none spike	*i*	0–20	Series of twitches only when fresh; few twitches plus tonic contraction when deteriorating	70–90	30–40	42 (when fresh) falling to 50	40	24
			ii	None found	–	–	–	–	–	–
	Intermediate thickness, 5–7-μm SL	Large graded response	*i*	30	Few twitches plus tonic contraction which fades rapidly	65–85	30–50	56 (45–62)	16	10
			ii	None found	–	–	–	–	–	–
Intermediate		Small graded response	*i*	15–20	Weak phasic contractions plus prolonged contraction fading slowly	65–80	30–50	56 (45–62)	8	5
			ii	15–20	Tonic contraction only, fading slowly	65–80	50	60 (56–66)	2	1
	Thin 10–12-μm SL	Passive normal	*i*	10	Tonic contraction only, fading slowly	68–80	20–50	60 (54–66)	6	3
			ii	10	Tonic contraction only, maintained indefinitely	60–70	50–100	58 (54–64)	1	1
Slow		Supersensitive	*i*	5	Phasic contraction, plus longer contraction fading rapidly	64–66	15–30	Close to resting potential	8	2
			ii	None found	–	–	–	–	–	–

Total % fast-follower = 70–75%.
slow-follower = 25–30%.

[a]Where *i* = fast follower and *ii* = slow follower.
[b]*These figures are approximate.
[c]From Atwood, Hoyle, and Smyth. *J. Phisiol.*, 1965.

general survey of both structure and function, some significant differences in each were pointed out for a few types of muscle. In the ensuing account we shall systematically examine sources of variation and indicate where they are known to have been utilized functionally. We shall be concerned only secondarily with differences between whole muscles: the intent is to examine some of the currently known details regarding the different kinds of muscle cells, several which are often contained within a single muscle. We shall also look for ways in which the different kinds might usefully be classified.

8.2. Factors Determining Cell Specificity

8.2.1. Sarcomere Length

Sarcomere length (SL) is the easiest and one of the more dramatic sources of variation to recognize. It clearly has played a major role in adaptive radiation in the evolution of invertebrates, where the range is 0.6–40 μm. It has played much less of a role in evolution of vertebrates, where the shortest known average SL is 1.4 μm in bird iris muscle, and the longest 5.0 μm, in cat tensor tympani. Almost as much length variation can occur in a single muscle. The range is from 1.8 to 2.9 μm in chicken breast (Pepe, 1967a), and my collaborators and I have found a range from 1.9 to 2.7 μm in rabbit psoas. The greatest range in SLs in any single class of animals is that present in crustaceans, where the SL ranges from 0.65 to 26 μm (Table 8.2). Although isolated examples of remarkable length are known, such as the 40-μm sarcomeres of the proventriculus of the worm *Syllis*, no other class of animals has exploited SL to nearly the same extent as crustaceans. In insects the range is much less, 2.0–9.3 μm. The relation between SL and tension has not been studied in depth quantitatively, mainly because it is never a unique variable: variations in SL are invariably also associated with different electrical properties, several aspects of ultrastructure, and doubtless also biochemical factors. In the extensor of the carpopodite of the ghost crab *Ocypode ceratophthalma,* a crab muscle in which secondary differences are minimal (except in SR), a logarithmic relationship was found between the total duration of the twitch and SL (Figure 8.2 – Hoyle, 1967b). Rise times and relaxation times both increase with increasing SL, but the contribution of the latter is by far the larger. On the basis of SL alone, other matters being exactly equal, twitch duration should be *linearly* related to SL because of the summative aspect of shortening by elements with similar intrinsic velocities connected in series. The *exponential* relationship must be due to an additional factor, probably the increased diffusion distance from activator release points to contractile material and back to SR in long-sarcomere cells. The quite small range of SLs that exists in chordates is not known to be associated with significant speed differences and has certainly not been exploited by vertebrates as a major variable.

8.2.2. Sarcoplasmic Reticulum

Historically, SR differences could have been fairly intensively studied by light microscopy alone and differences in physiological properties related to them, but the functional significance of SR was not realized by muscle scientists, for the reasons outlined in the introduction, until about the time its fine structure was rerevealed by electron microscopy, in 1955. However, before then, Krüger (1952), comparing mammalian muscles

Table 8.2 Sarcomere Lengths, A-Filament Lengths, and 1-to-A Filament Ratios for Some Cross-Striated Muscle Cells

Phylum, Class Genus, Species	Muscle	Type	Sarcomere Length Mean or Range (μm)	A-Filament Length (μm)	I : A Ratio	Reference
Coelenterate						
Scyphomedusa	Subumbrellar	Fast	0.84–1.0	0.58	?	Chapman et al., 1962 Mackie and Singla, 1975
Annelida						
Polychaeta *Syllis*	Proventricular	Fast	30–40	32	20 : 1	Smith et al., 1973
Mollusca						
Lamellibranchia *Pecten*	Adductor	Fast	2.6	2.1	3 : 1	Millman and Bennett, 1976
Cephalopoda *Sepia*	Longitudinal mantle	Very fast	1.9–2.4	1.3	3 : 1	Hoyle, unpublished
Arthropoda						
Arachnida *Eurypelura* (spider)	Tarsal levator	Intermediate	2.8–3.6	2.5	4 : 1	Sherman and Luff, 1971
	Tarsal depressor	Slow	5.6–6.8	4.0	5 : 1	
Arachnida *Leiurus* (scorpion)	Claw closer	Intermediate	5.0	4.0	5 : 1	
Crustacea						
Ocypode (ghost crab)	Extensor carpopodite	Fast	3.3	2.7	3 : 1	Hoyle, 1967b and unpublished
	Extensor carpopodite	Slow	4.6	3.8	4 : 1	
	Extensor carpopodite	Intermediate	10.0	8.0	5 : 1	
Podophthalmus (white crab)	Eyestalk levator	Intermediate white	7.9	5.9	4 : 1	Hoyle and McNeill, 1968
	Eyestalk levator	Slow pink	13.0	7.5	7 : 1	

Table 8.2 Continued

Phylum, Class Genus, Species	Muscle	Type	Sarcomere Length Mean or Range (μm)	A-Filament Length (μm)	I : A Ratio	Reference
Balanus nubilus (giant barnacle)	Depressor	Intermediate	6.5 (2.6–25.0)	4.0 (2.2–20.3)	3 : 1 to 6 : 1	Hoyle et al., 1973
Hemisquilla (mantis shrimp)	Lateral flexor	Fast	4.3	2.8	3 : 1	McNeill et al., 1972
	Lateral extensor	Slow	7.8	5.1	4 : 1	
Insecta						
Aeschna (dragonfly)	Longitudinal flight	Fast	2.5	2.2	3 : 1	D. S. Smith, 1966a
Schistocerca (locust)	Extensor tibiae	Fast	5.6–6.3	4.8	4 : 1	Hoyle, 1978b
	Extensor tibia	Slow	7.0–7.5	6.0	6 : 1	
Chordata						
Amphibia *Rana* (frog)	Toe	Fast	2.2	1.8	2 : 1	Page, 1965
	Rectus	Slow	2.8	2.3	2 : 1	
	Interhyoideus	Slow	3.3	2.7	2 : 1	Gradwell and Walcott, 1971
Reptilia *Thamnophis* (garter snake)	Ribs – skin	Slow	3.0	2.4	2 : 1	Hess, 1965; Hoyle, unpublished
		Fast	2.0	1.7	2 : 1	
Aves (chick)	Pectoralis	Fast	3.0	1.7	2 : 1	Mendell, 1971
Mammalia (rabbit)	Psoas	Intermediate	2.2–3.0	1.9	2 : 1	Hoyle, unpublished
Mammalia (cat)	Tensor tympani	Slow	5.0	3.0	2 : 1	Fernand and Hess, 1969
	Stapedius	Fast	2.6	1.9	2 : 1	

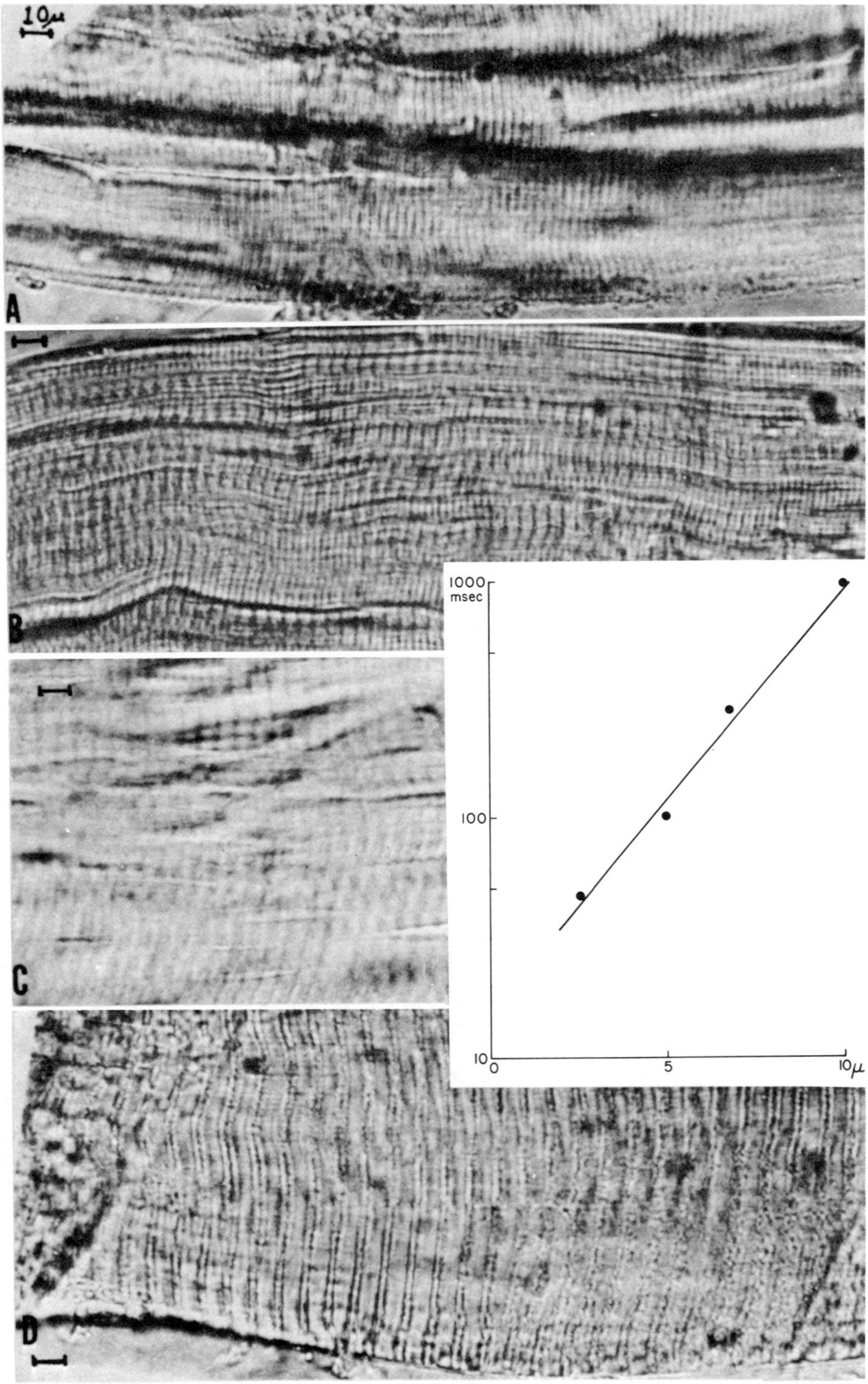

Figure 8.2. Variation in sarcomere length of component muscle cells of the extensor of the carpopodite of the ghost crab *Ocypode:* live single cells all at the same magnification. (*Inset*) relationship between sarcomere length and overall twitch time for each of these cells. Reprinted from *Invertebrate Nervous Systems,* pp. 151–167. Ed., C. A. G. Wiersma. Specificity of muscle by G. Hoyle. © 1967 by permission of The University of Chicago Press.

that are principally postural/*tonic* in function with ones that operate in a ballistic/*phasic* manner, noted that there are marked differences in appearance between these muscles when seen in transverse sections. The tonic ones have a diffuse appearance, which he called *Felderstruktur,* compared with the more familiar punctate appearance of phasic muscle cells, which he termed *Fibrillenstruktur* (Figure 8.3). This difference was subsequently shown by electron microscopy to be due to there being a sparse SR in the former but an extensive one in the latter. The more the SR, the greater the amount of calcium released at coupling, and the less the distance it has to diffuse both to reach contractile material and to be taken up to promote relaxation. Variation in amount of SR has been the principal means of adaptively adjusting overall contractile properties in vertebrates and insects; in crustaceans it has been at least equal in significance to SL variations.

The extreme examples are seen where speed has been accented to the greatest possible extent and the loss of force capability accompanying extensive development of SR was not important. Two vertebrate examples are muscles each concerned with sound production: the swim bladder muscle of the toadfish and the cricothyroid of the bat (Figure 8.4). In the former, which produces a "boat-whistle" note with a fundamental frequency of 250 Hz, 50% of the entire muscle-cell volume is occupied by the mitochondria. A twitch in this muscle is complete in 3–8 ms (Skoglund, 1961). The bat cricothyroid is one of the larger laryngeal muscles whose function is to modulate the sound frequency during echolocation cries of *Eptesicus fuscus* and related bats. The modulations last for 2–3 ms each, at a frequency of up to 200 Hz. Below the Sl is a special form of reticulum

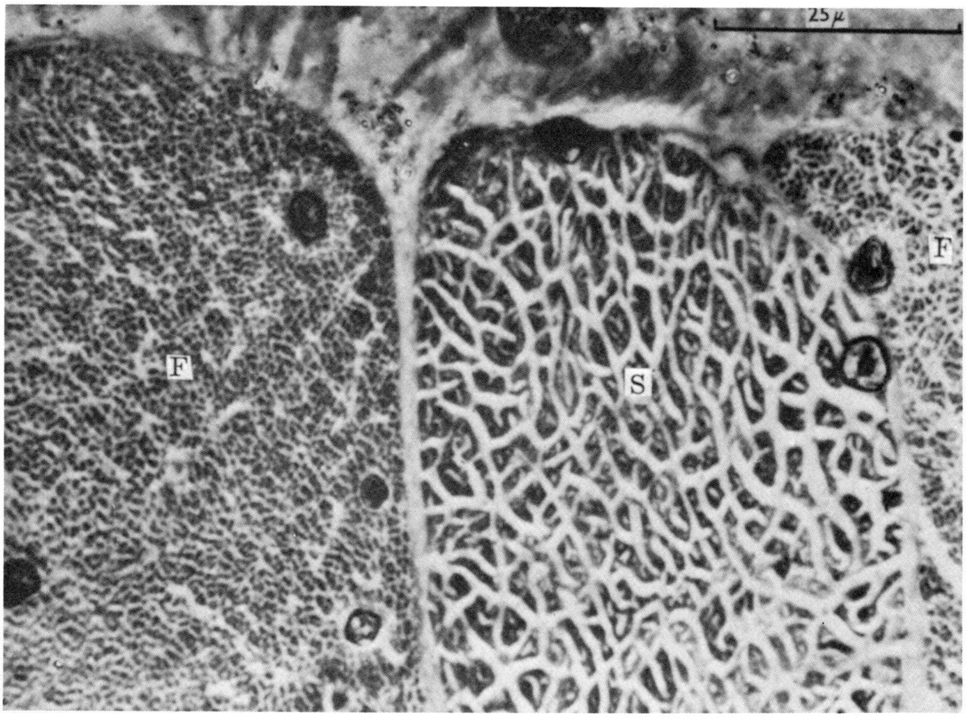

Figure 8.3. Characteristic *Fibrillenstruktur* appearance of fast-type muscle cell (*F*) compared with that of neighboring *Felderstruktur* slow-type of cell (*s*) of frog adductor magnus muscle as seen in light microscope. (From E. G. Gray, 1958.)

Figure 8.4. Ultrastructure of a vertebrate fast sound-producing muscle cell, the cricothyroid of the bat *Eptesicus.* Tiny myofilament clusters are embedded in both abundant SR and large mitochondria. (From J. P. Revel, 1962.)

consisting of bundles of straight hexagonally placed 60–70 nm in diameter tubules. Around them are 12 regularly spaced 7–8 nm filaments. The total aggregate contains up to 1800 tubules and ranges from 1.3 to 3.9 μm in cross section. At their ends, the tubular lumens are connected to each other and to the regular SR, which surrounds myofilaments, and also abut ITs (Cho et al., 1972). These tubules are probably a special form of SR. Together with regular-appearing SR, they constitute 60% of the volume.

An invertebrate example of a fast-contracting, sound-producing muscle is that which operates on the second antenna of the lobster *Homarus americanus,* which is able to generate squeaky sounds by being repeatedly rubbed against the carapace at a frequency basically 100–130 Hz, with occasional peaks of 300 Hz (Mendelson, 1969). The muscle responsible, the remotor, like bat cricothyroid, has an extraordinarily extensively developed SR (Rosenbluth, 1969), fully 70% of the entire muscle volume (Figure 8.5).

By the above criteria, a muscle cell with very little SR should be very slow indeed, ideally suited to the maintenance of postural tone, but of no value for phasic activity. The muscles connecting the backbone of a snake to its skin serve exactly this function. When first examined by electron microscopy, those of the garter snake *Thamnophis sirtalis* appeared to have no SR at all (Hess, 1965), but serial sections showed that they do have a minute amount (Figure 8.6 – Hoyle et al., 1966). Comparable cells have been found in the global region of the superior rectus of the cat (Peachey et al., 1973), where they occur side by side with fast-type cells that are heavily permeated with SR.

If the twitch duration is plotted against the percentage of volume occupied by SR for selected fast-type cells of similar SL and excitability, even though values for widely different kinds of muscle are used, the relation is a straightforward one in which duration decreases approximately exponentially with increasing percentage of volume occupied by SR (Josephson, 1975). Although most cells have a very disorderly SR plan, in some it at least has a spiral form overall, and this must affect its properties somewhat, though the influence appears to be slight (Mathias, 1978).

8.2.3. Mitochondria

The development of extensive SR as an approach to the production of a very high sustained frequency may be contrasted with the approach that has been adopted by the hummingbird pectoralis muscle (Figure 8.6). Here also the proportion of the cell that has remained contractile is small, less than 30%, but the highly developed component is mitochondria rather than SR. The hummingbird is capable of long-sustained flight at a wing-beat frequency of 45 Hz (Hagiwara et al., 1968), and in these cells about half the total volume is occupied by mitochondria. This reflects the enormous energy requirement of one of the fastest repetition rates for complete contraction/relaxation cycles in the animal kingdom. A majority of the mitochondria of vertebrate skeletal muscle cells are quite small and almost spherical, about 0.2 μm in diameter. Branched mitochondria occur in some cells and may be a significant form of variation functionally. Fast-type cells of rat diaphragm have discrete small mitochondria, but the slow-type cells in the same muscle have complex, continuously branching mitochondria. There is no correlation between their numbers in a cell, the need for them, and their size. Some cells contain a small or moderate number of very large ones, others a very large number of quite small ones, each occupying similar total volumes.

In vertebrates it is a general rule that, other matters being equal, the smaller the animal, the larger the number of mitochondria per cell. This is related to effects of the

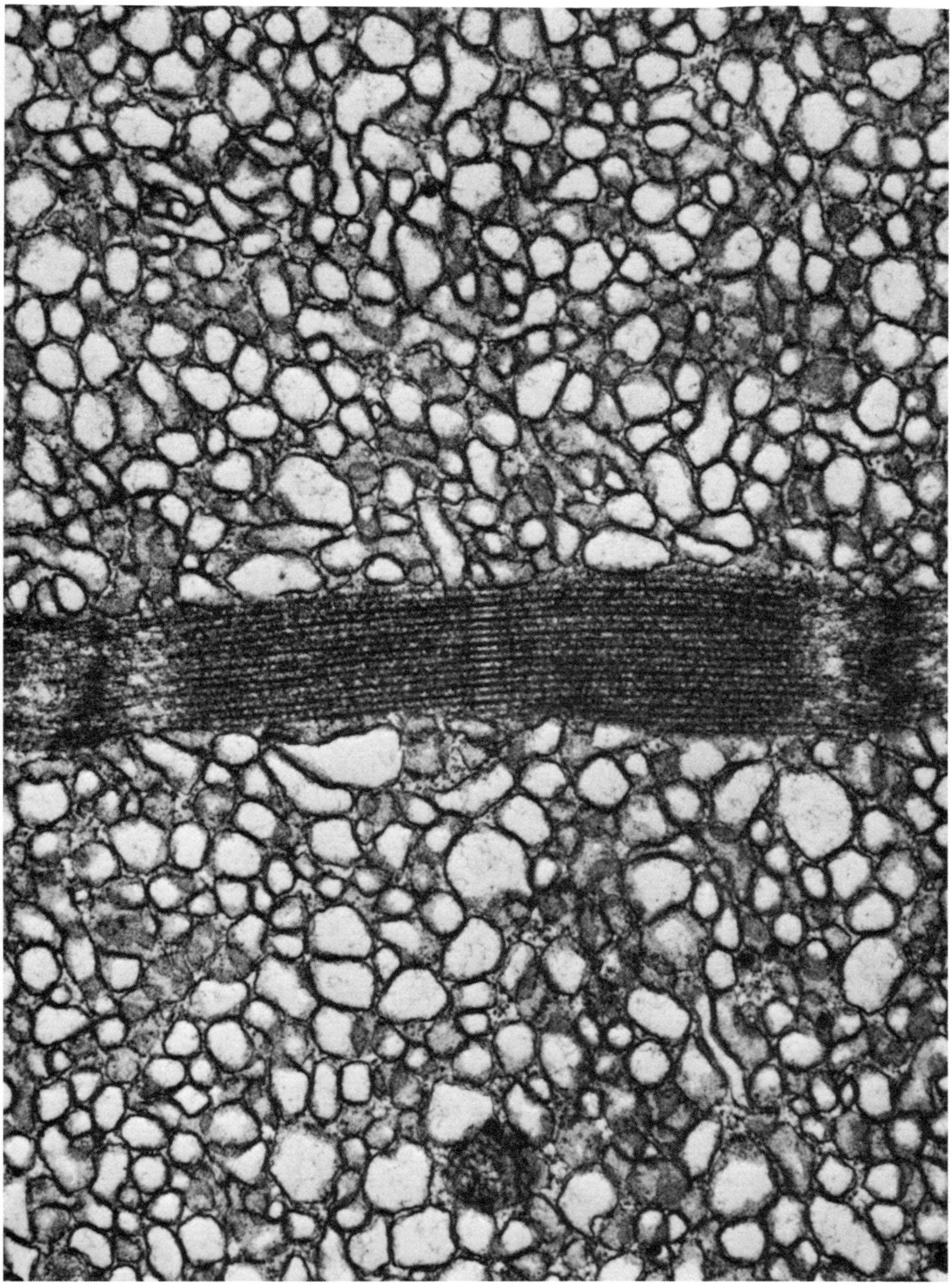

Figure 8.5. Ultrastructure of a fast-contracting sound-producing invertebrate muscle, the remotor of the second antenna of the lobster *Homarus*. More than 70% of the cell volume is occupied by SR. (From J. Rosenbluth, 1969.)

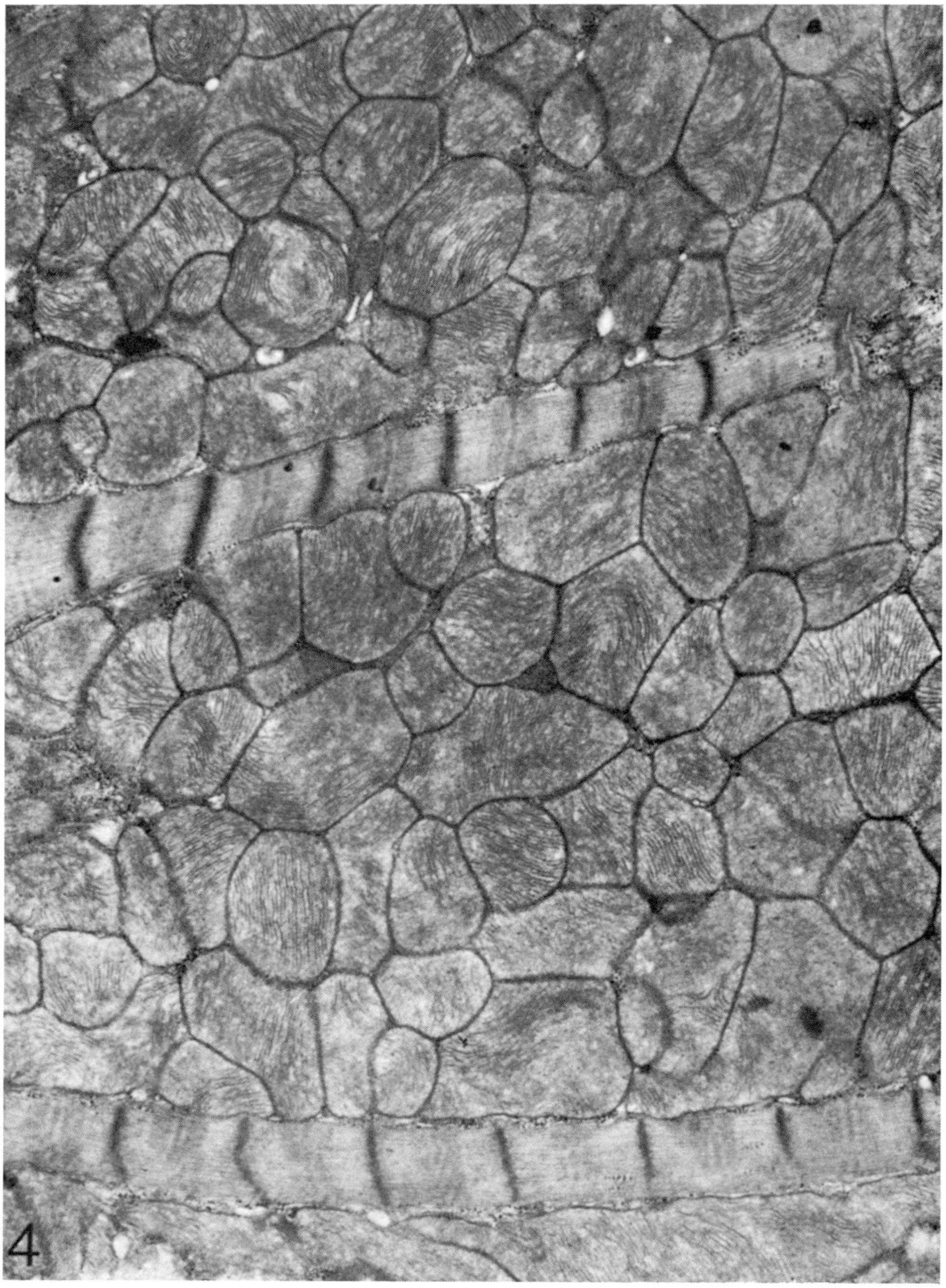

Figure 8.6. Ultrastructure of hummingbird *Archilochus* supracoracoideus flight muscle, showing extraordinarily dense aggregation of mitochondria. (By Grinyer and George, reproduced by permission of the National Research Council of Canada from the *Canadian J. Zool.* **47**, 771–774, 1969.)

well-known inverse relation of the ratio of surface area to volume. But the diameters of cells of like kind functionally and enzymatically increase in proportion to overall size of the animal (Figure 8.7). When cell types are compared within a given muscle, however, the numbers of mitochondria are found to be greatest in the smallest cells and sparse in the largest. The largest are the fastest and the smallest the slowest, but fatigability is high and fast in the former, low and slow in the latter.

In some cells with large numbers of mitochondria, the mitochondria are located almost exclusively at the periphery, the myofilament clusters occupying the core, but in others they are uniformly scattered throughout, as is always the case in insect flight muscles (Figure 8.1), or both peripheral concentrations and central ones can occur.

Mitochondria can provide a simple guide to overall muscle-cell properties in arthropods, since the larger, faster cells, like those of vertebrates, contain few mitochondria, whereas it is certain that a small cell containing many mitochondria will be slow. However, the converse is not always true, since some slow cells have few mitochondria. These are cells in which maximum strength is needed, but which are used infrequently. In crustacean muscles, there are mixtures of slow, fast, and intermediate muscle cells, as we have seen. There is no simple correlation between their contractile properties and the numbers of mitochondria, except that fast cells never have many mitochondria; associated with the lack of mitochondria are antifacilitation and rapid fatigue. Slow-type cells occur in a complete spectrum, from cells with relatively few mitochondria to ones in which more than half the volume is occupied by them. Cells with many mitochondria at the periphery are opaque, often distinctly colored, and easily recognized by comparison with their clear or white fast-type neighbors: the color ranges from gray to red, with generally a pinkish tinge. It is not due to a myoglobin, as in red vertebrate muscle, but to an effect of a lot of cytochrome *b* in the massed mitochondria. The cytochrome has a strong absorption band at green wavelength, and by removing this from reflected light, shifts the impression of the remainder toward red. In subtropical and tropical crustaceans living close to the limits for adequate oxygen in the water, there are relatively more mitochondria, and the color is pronounced. In each muscle a reddish band marks the slow-type cells and a pinkish one the intermediates, when each type is clustered together, as is often the case.

A similar color difference was early noted in some insect muscles, certain ones of the cockroach leg being referred to as "red" and "white" (Smit, 1958; Becht, 1959), though the difference is not as pronounced as in tropical crustaceans. The reddish color is also associated with slowness and tonic activity (Becht et al., 1960); these differences are clearly tilted in the direction of achieving the fastest possible initiation, with highest possible safety factor, in the intrinsically fastest cells. This is not, however, a common trait, and in most insects, slow and fast muscles appear identical to those of the eye. Ultrastructurally, there are marked differences, mainly in amounts of SR, but not in numbers of mitochondria. Mitochondria are fairly abundant in both slow and fast types and are scattered uniformly throughout the cell.

In arthropod and many vertebrate cells that have to produce sustained tension for very long periods, there is a layer of mitochondria below the Sl. This presumably reflects the need to support the continued neural excitation required. Massed location of mitochondria between the contractile elements is associated with sustained high-frequency contraction/relaxation cycles, for example in the singing muscle of the katydid leg (Josephson, 1975). This cricket has a drastically reduced twitch duration at the high temperature (35-40°C) it can develop within itself before singing. The twitch time comes

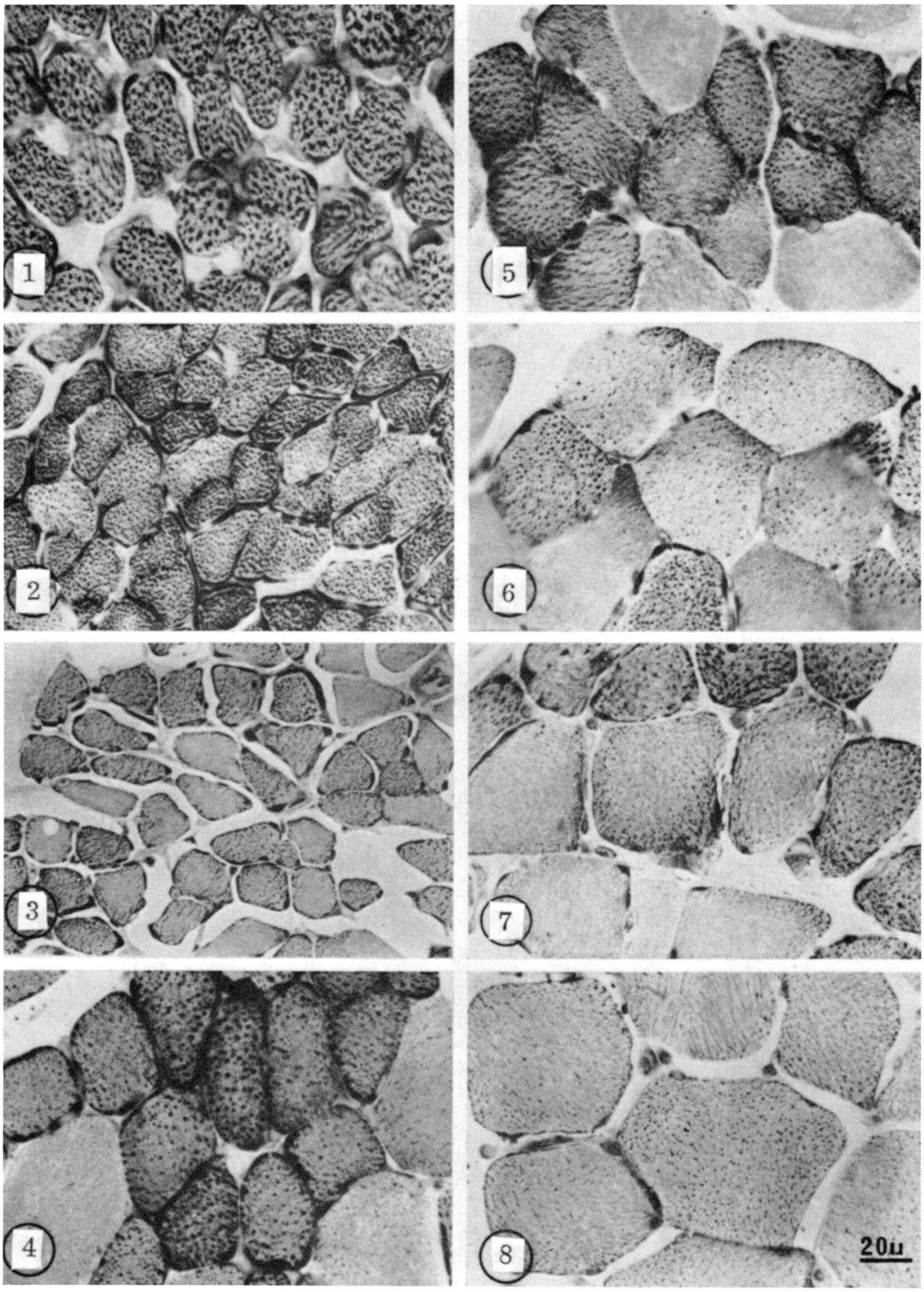

Figure 8.7. Diaphragm muscle from (*1*) Harvest mouse; (*2*) shrew; (*3*) albino mouse; (*4*) albino rat; (*5*) squirrel; (*6*) cat; (*7*) man; (*8*) cow. All are stained with Sudan black for lipid content. In the smaller mammals, cells are uniformly small and rich in both mitochondria and lipid. Cell size increases with body size, and heterogenity of cell type also increases, with accompanying reduction in mitochondria and lipid. All same magnification. (From Gauthier and Padykula, 1966.)

down to about 2.5 ms, permitting singing at a rate of about 190 Hz. Some neurally driven rates higher than this have been reported, 250 Hz for an Australian cicada (Young, 1972) and 280 Hz in the bush cricket *Orocharis* (Walker, 1969), but nothing is known about the ultrastructure in these species. All insect synchronous flight muscles have many massive mitochondria.

One of the more remarkable cellular relationships with mitochondria is that illustrated in Figure 3.15, in muscle cells of the squid mantle. There the contractile myofilaments are entirely peripheral, surrounding a core of giant mitochondria that occupies about 60% of total volume. This is the muscle that is synchronously activated by the giant axons and that produces the extremely rapid jet-propelled backward escape swimming movements. It is not only a very fast muscle, it is unique in the animal kingdom in possessing a tetanus-twitch ratio of unity (Prosser and Young, 1937), which means that a single excitation rapidly saturates the contractile elements with calcium and that a full SA lasts until maximum tension has been achieved. This finding also strongly supports the notion that much of the apparent series elasticity of other muscles is associated with unequal activation of sarcomeres. The short squid sarcomeres are very short and activated evenly.

A closely similar situation occurs in vertebrates, some muscles having a mixed population of cell types recognizable by numbers of mitochondria. For example, in cat soleus, which is a slow muscle overall, slow component cells are homogeneous, and all have similar mitochondrial ATPase activities, but in the fast gastrocnemius, there is a population that can be divided into three types on the basis of mitochondrial numbers: numerous, few, and intermediate (Figure 8.8 – Henneman and Olsen, 1965). Mammalian slow-type cells are all relatively rich in mitochondria compared with their fast-type counterparts (e.g., Schmalbruch, 1971), a trait that is also strongly present in fish (Figure 8.9). Slow cells are associated with the well-known color differences between red and white meat – or cells – from fish to man. Although myoglobin plays a role in determining the differences in colors, the principal cause, as in invertebrates, is numbers of mitochondria, which are distinguishing features of major cell types of mammals (Figure 8.10).

The size, shape, overall location, and internal arrangement of cristae of mitochondria have all been variously and collectively pointed out as being characteristic for some muscle cell or other (e.g., Hagopian, 1967). The internal fine structures are related to oxidative functions (e.g., Mattisson and Birch-Andersen, 1962). But it is not yet known in detail what functional significance can be associated with the differences, although major differences in oxidative metabolism are well known. In muscles where the rate of working is high, for example in flight muscles of insects, the mitochondria are not only very large, up to $1.5 \times 1.5 \times 6.0\ \mu m$, but they completely surround the small myofilament clusters, and it is reasonable to infer a functional connection for both attributes. But in general the detailed ethology of mitochondrial form remains to be worked out.

8.2.4. Enzymes

A major basic source of variation between muscle cells is the actomyosin ATPase (see p. 125). There are comparable differences in all enzymes that have so far been examined directly, or indirectly via substrates. A few examples from the vast list available are given below.

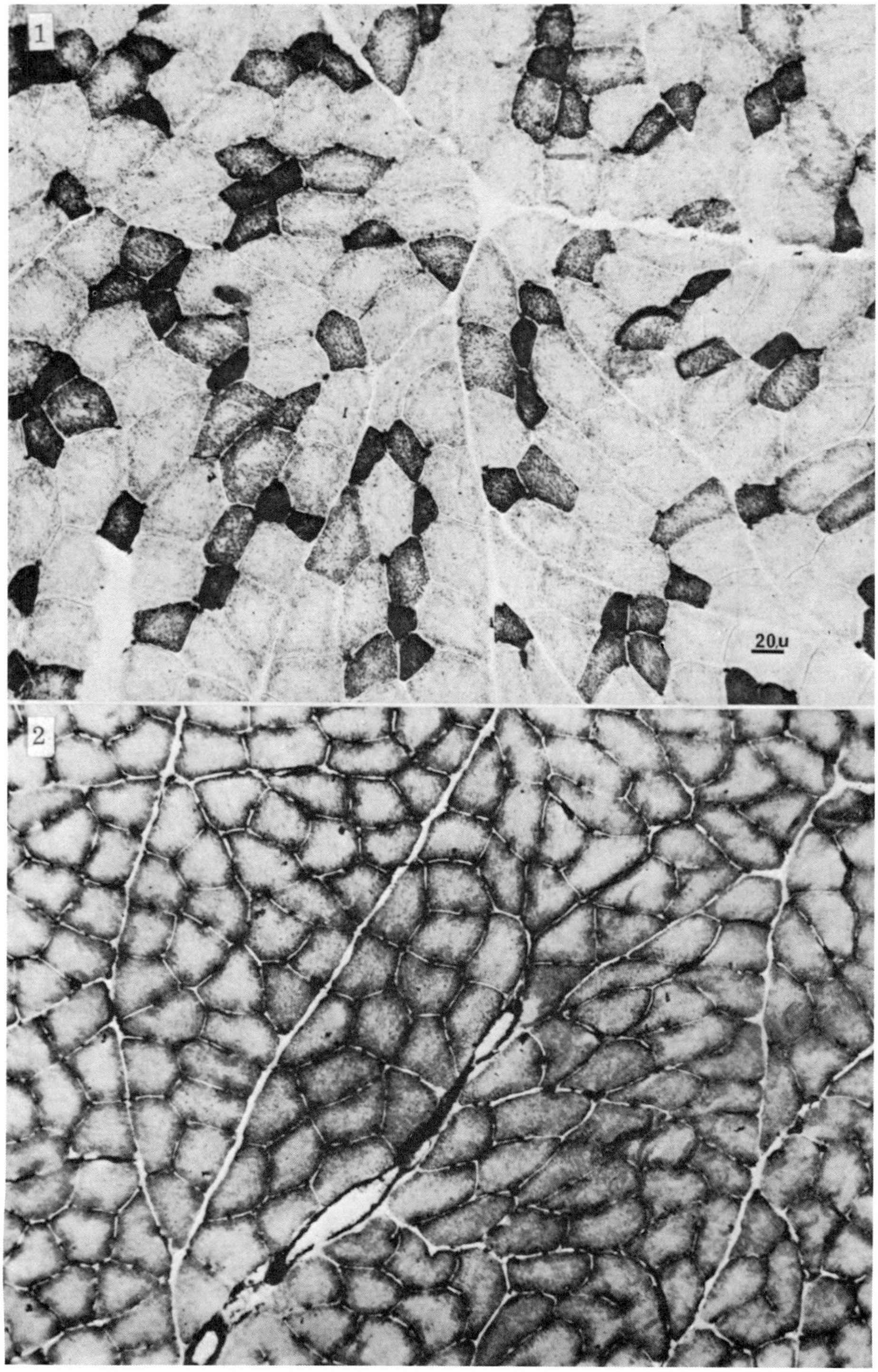

Figure 8.8. (*1*) Distribution of muscle cell types in a heterogeneous mammalian muscle, the medial head of cat gastrocnemius. The different types are indicated by differences in numbers of mitochondria, which have been stained for mitochondrial ATPase activity. The larger cells have few mitochondria. (*2*) Uniform mammalian muscle – soleus of cat – also mitochondrial ATPase. (Both reprinted with permission from Henneman and Olsen, *J. Neurophysiol.* 28, 581–598 (1965) Copyright 1965 The American Physiological Society.

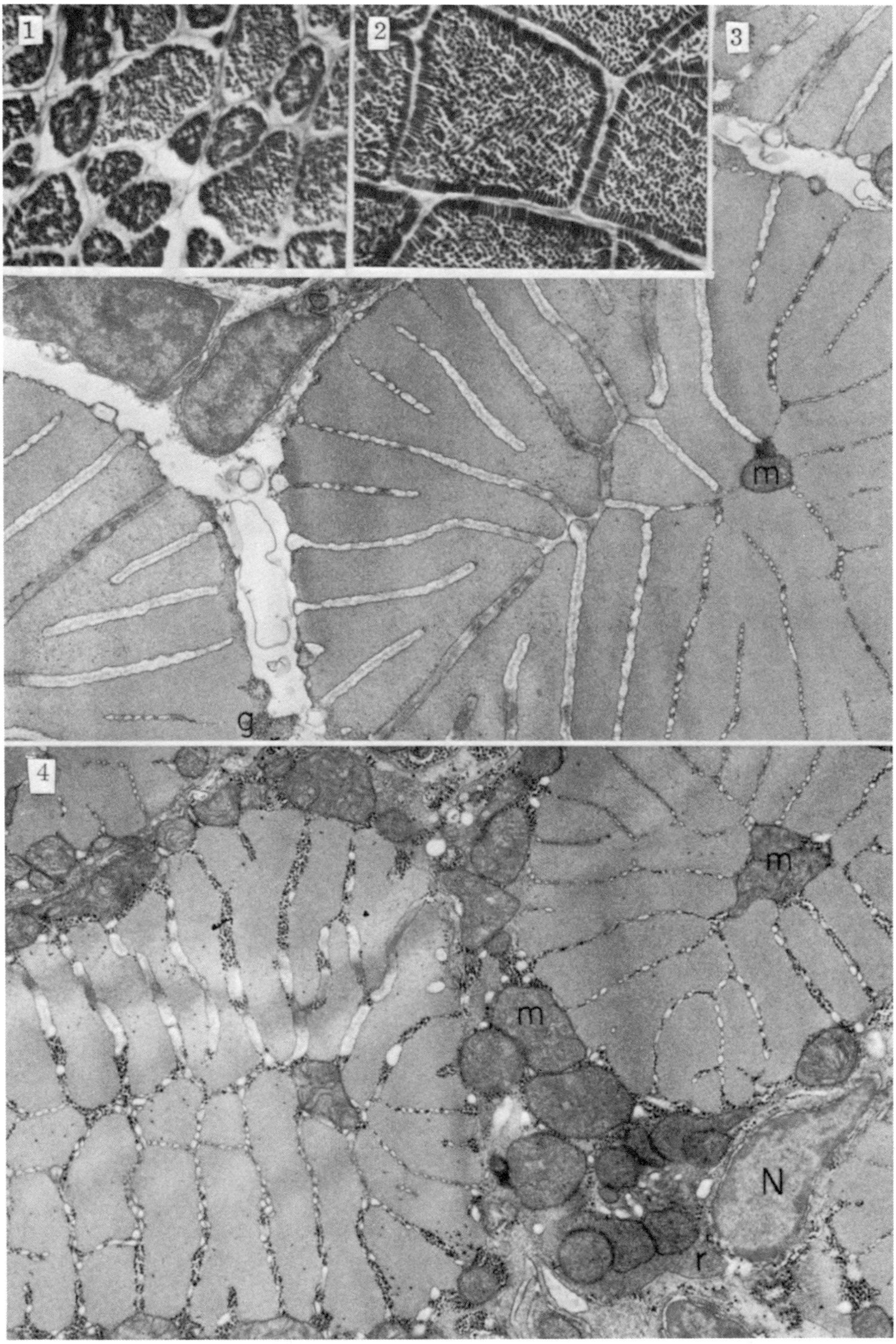

Figure 8.9. Red and white fish muscle cells. (*1*) Levator pinnae pectoralis – red, of snake fish, *Ophiocephalus*. (From Nakajima, 1969.) (*2*) Flexor pinnae pectoralis – white, same fish as (*1*), same magnification. (From Nakajima, 1969.) (*3*) EM of transverse section of white cell from tail myotome of black Mollie. (From Franzini-Armstrong and Porter, 1964.) (*4*) EM of transverse section of red cell, with abundant, especially peripheral, mitochondria. (From Franzini-Armstrong and Porter, 1964.)

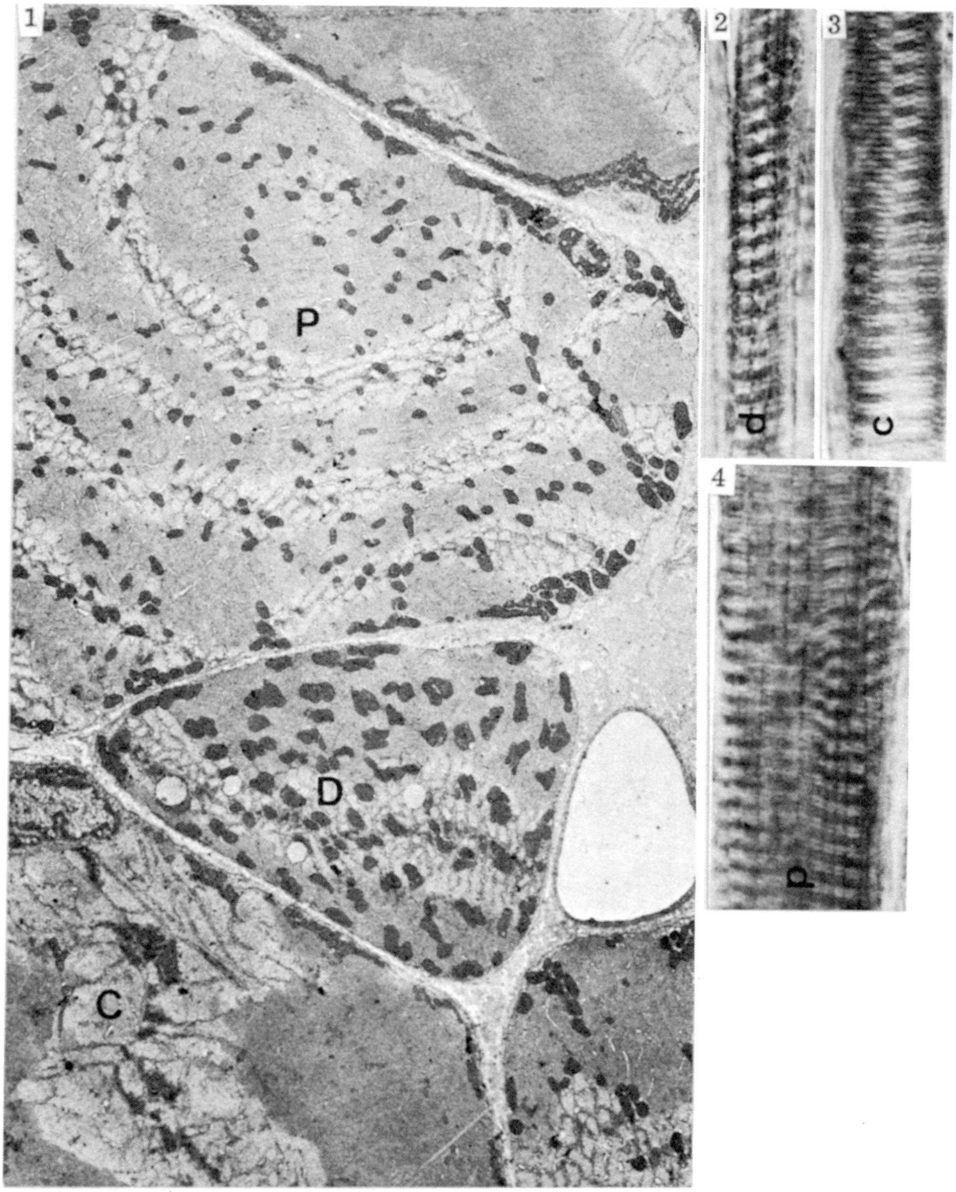

8.2.4.1. Actomyosin ATPase

The activity rates of the actomyosin ATPases of different muscles are markedly different, and that of slow-type cells in general is less than that of fast types, but exceptions are known. No significant difference could be found between ATPase activity of slow and fast cockroach (*Periplaneta*) muscles (Chari et al., 1973). On the basis of histological staining, utilizing, and therefore indicating, intensity of AM ATPase activity, three levels of intensity have generally been recognized in mammalian cells and used as a basis for classification into three types in conjunction with succinic dehydrogenase and other metabolic-enzyme levels, which are correlated. However, a massive analysis of a few

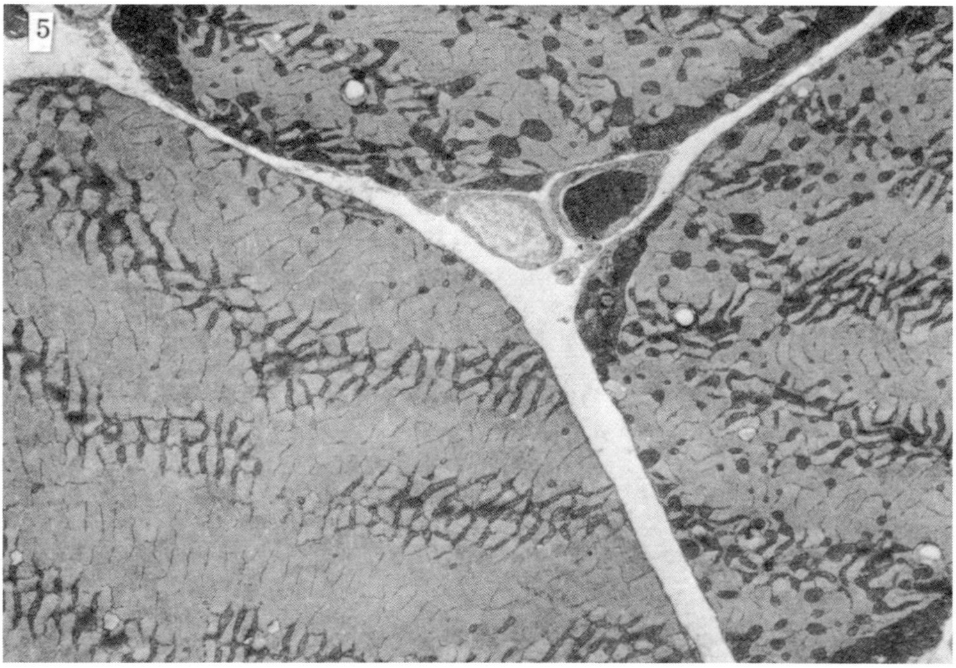

Figure 8.10. Cell types mixed together in external eye muscles of rat. (*1*) TS of a rectus lateralis showing three basic types: clear (*C*), lacking mitochondria, of medium diameter; dark (*D*), small diameter with numerous large mitochondria; pale (*P*), with few small mitochondria and large diameter. (*2*)–(*4*) Examples of whole cells of the three types shown in (*1*), in isolation (From Mayr, 1971.) (*5*) Red (*above and right*) compared with white (*left*) cells of rat diaphragm. Note the peripheral mitochondria in the former. (From Gauthier and Padykula, 1966.)

thousand cells has cast doubt on the validity of a simple tripartite division (Nemeth et al., 1979). A more subtle form of classification is required if all forms encountered are to be firmly grouped.

The amount of work produced per mole, that is, the efficiency, is higher for slow cells. Whole slow soleus of hamster was compared with fast biceps brachialis in this respect and found to be markedly different, the slow producing 191×10^3 g• s/mol ATP per gram split and the fast 30.8×10^3 g• s/mol ATPper gram (Davies et al., 1969).

8.2.4.2. General Metabolic Enzymes

Oxidative metabolism is generally assumed to be similar for the two sexes in all animals, but not on the basis of much actual comparison. In the cockroach, a quite large difference in oxidative metabolism was found between males, in which it is relatively high, and females (Chari et al., 1973). The males have three times as much glycogen in their muscles as females. There is a behavioral correlate in that running and flying ability of male cockroaches is much higher than it is in females.

Training or exercise can alter both total enzyme levels and the balance between oxidative and anaerobic ones. Major changes occur in enzymes – and therefore in effect in cell type – during development of all animals examined and can be drastic in metamorphosing insects. Even in a nonmetamorphosing one, the cockroach, both general and sex-linked changes occur. In all nymphal cockroach muscles of both sexes, succinoxidase content is low, but increases by $6\frac{1}{2}$ times after the molt to adult in thoracic molecules of males only. This is the time they acquire their reddish color, due to mitochondria (Brooks, 1957).

8.2.4.3. Lipase

Lipolytic enzymes such as succinic dehydrogenase and DNP-diaphorase provide a major source of variation. Lipid droplets occur in many muscle cells and are used as an energy source. This is true not only of tonic muscles but also of phasic ones required to work continually for very long periods such as insect flight muscles. All types of cockroach leg muscle contain lipase, but there is more of it in the slow-contracting muscles than in the faster ones (George & Bhakthan, 1961). In pigeon breast muscle, lipase is present in slow-type cells, which are also red in color (George & Scaria, 1958), but not at all in fast white ones. In the quail breast muscle, lipid is present in both types of cell, but there is much more in the thin red ones (Boesiger, 1951). Lipid droplets are abundant in diaphragm muscles of mammals in general, but especially in small mammals such as the mouse and shrew (Gauthier & Padyulka, 1966). With increasing size of the animal, the cells not only become larger and have fewer mitochondria, but they also have less lipid content. In diaphragm cells, lipid droplets alternate with mitochondria between the small myofilament clusters (Figure 8.7).

There is a general reciprocity of amounts of lipid and glycogen between fast and slow cells of both vertebrates and invertebrates, with glycogen highest and fat lowest in the fastest cells and vice versa, but there are exceptions in both. In the clawed toad *Xenopus*, cells are best classified on the basis of their appearance combined with speed, as in Table 8.4, that is, fast-twitch and slow-twitch, with the fast subdivided according to rapid or moderate fatigue rates (Lännergren and Smith, 1966). Slow-twitch cells are clear, and they lack either lipid or the enzymes associated with use of lipid. Fast-fatiguing fast-twitch cells are pale and glycolytic, with only very little lipid. The slower-fatiguing cells are dark, contain a lot of lipid, and high concentrations of lipolytic enzymes. Appearance alone is an unreliable classificatory variable except for the individual muscle.

In crustacean muscles, the cell population as judged by electrophysiological study of junctional transmission forms a continuum with identifiable extremes rather than discrete populations. The differences are matched approximately by amounts of energy substrate stored. Fast-type cells contain small amounts of glycogen, slow-type cells contain a large amount, with appropriate intermediates (Morin and McLaughlin, 1973). The same criterion also applies to many vertebrate fast-twitch–slow-twitch comparisons.

8.2.5. Activation

Number of ITs is a major source of variation between cells, but is usually so closely correlated with amount of SR that is is not used as a factor in classification in the way that amount of SR is. Properties of ITs probably differ between cells, but little is known about this aspect. Fast-twitch cells of summer frogs (*R. pipiens*) sometimes have regenerative responses in their ITs, in which case a local excitation travels equally around and into the nearby region (Figure 6.10.*3*, Strickholm, 1974). Slow-twitch cells from the same muscle never show a uniform spread, so inward transmission in them is either purely passive or at most has a graded electrogenic component.

Major differences occur in the locations of IT/SR contacts in relation to the Z regions both within vertebrate classes and between invertebrates in general and vertebrates in general. But these differences occur across the board, for all types of cell within a particular animal. Vertebrate cells are classified as types I or II regardless of whether the triads are located at Z bands or A/I boundaries. No differences are known in the nature of these contacts in different component cells, though there are major differences in their numbers.

Very little research has yet been done on kinetics of contraction activation of different types of cells, as judged by a major characteristic. It would be desirable to compare ones that are similar in major contraction parameters. A comparison of frog (*Rana temporaria*) slow cells from the cruralis and the pyriformis with fast ones from the sartorius using voltage clamping showed that the average rate of tension development was 10 times higher for the fast cells (Gilly and Hui, 1980a,b,c), but the processes leading to Ca^{2+} release in response to changes of IT membrane potential were similar. The numbers of ITs and of IT/SR contacts were not greatly different, so IT differences are not a major factor in determining speed differences. A major difference was in the complete lack of time-dependent inactivation of the Ca^{2+} release in the slow cells. Gilly and Hui found that the rate of rise of tension in fast cells is constant, as in classical accounts, but not in slow cells, where it is also proportional to duration of the stimulus at constant voltage. The latter is true of barnacle giant cells and of many other invertebrate gradedly excitable cells. These differences are probably related to differences in the numbers of ITs and IT/SR contacts.

At the next step in coupling, actomyosin activation, there may be major differences, but they cannot yet be distinguished from later events. Major differences occur in the tension development brought about by direct address with Ca^{2+} (or Sr^{2+}) after skinning the two major cell types found in rat diaphragm (Hoar and Kerrick, 1979).

There are some properties unique to slow cells that have not yet been explained. Before losing excitability in low $[Ca^{2+}]_o$, they go into prolonged contraction (contracture) (e.g., Irwin and Hein, 1963). When placed in isotonic pure sucrose, they develop a contracture that requires a solution containing both calcium and sodium before it relaxes (Irwin and Hein, 1966). Tension development by fast cells depolarized by a high $[K^+]_o$ are virtually unaffected by $[Ca^{2+}]_o$, but the time course of slow-cell contraction is drastically affected. These properties are also displayed by giant barnacle and other crustacean cells. Evidently external calcium is not necessary for the release of calcium from the abundant internal stores of fast-type cells, on one hand, whereas entering calcium is essential for activation in the slow ones.

8.2.6. X-Bridge Recycling Rate

This factor was mentioned on p. 149, although its consideration requires begging the question as to the exact role of X bridges. On the X-bridge (recycling) theory, the rate of cycling must be a significant variable. Specific tests have been made by Julian and Morgan (1980) to compare the tension transients during a series of short quick stretches and releases in incremental steps. They found the instantaneous stiffnesses to be closely similar, indicating similar elastic properties. But the shortening speeds on unloading were markedly different, showing that the recovery rates – determined by recycling times according to the theory – differ markedly. The X-bridge stiffness does not seem to be a source of variation.

8.2.7. Molecular Components

As more becomes known about the amino acid compositions of major contractile and regulatory proteins, it is becoming obvious that each exists in a number of different three-dimensional conformations (Perry et al., 1979). Two or several of each may occur in a single cell. Eventually, cells may be classified on the basis of molecular compositions. The molecular compositions of myosins from different sources are different in both heavy and light regulatory chains (e.g., Weeds, 1976). A muscle may contain more than one myosin of different molecular weight (Figure 8.11 – Schachat et al., 1977; Lowey et al., 1979). Even a single muscle cell can contain more than one type of myosin, a slow-type and a fast-type (Lutz et al., 1979). Altering the relative proportions of these coexisting molecules, during maturation and perhaps in relation to need, is a major reason why trophic influence of nerve supply is a major factor in natural regulation and can change muscle properties following nerve crossing (see p. 398).

Myosin extracted from a muscle containing a homogeneous population of cells may be purified and used to prepare antibodies. These, in turn, may be used to locate cells in other muscles of the same animal which contain the same type of myosin and to distinguish them from cells that must have another kind of myosin. The antibody specificity is associated with the heavy chains of myosin from the head section (Arndt and Pepe, 1975). Troponin T and C have been compared, using antibodies, for fast, slow, and cardiac muscles of rat and humans (Dhoot et al., 1978). Each was found to be a different protein, so probably provides a major source of functional difference. Either myosin-ATPase-staining (Dhoot and Perry, 1979) or antibody-immunoperoxidase-staining methods may be used to detect the presence of different forms of the TN on different cells within the same muscle (Figure 8.12 – Perry et al., 1979).

8.2.8. Membrane Excitability

Membrane excitabilities of vertebrate muscle cells are highly similar, except in the few instances of true slow cells, which are also differently innervated. The morphological feature of multiple end plates, not excitability, is used to recognize them. Among invertebrates, by contrast, membrane excitabilities of individual cells within a muscle show major differences, commonly over the full available spectrum, from totally electrically inexcitable to large spiking. These provide a major basis for classification (Table 8.1) that is useful, especially when contractile responses are not available. For arthropod slow cells, further subclassification can only be carried out on the basis of electrical characteristics of the membrane.

8.2.9. Passive Electrical Properties

There are major differences between cell types, regardless of the basis used for classification, in both resting potential and passive electrical properties. Differences in the latter are to be expected, since these properties are secondary consequences of structural factors including invaginated clefts and ITs, amounts of SR and mitochondria, and so forth, internally, as well as ion permeabilities. Resting potential is a very significant variable indeed: it is deliberately kept low in some slow cells by the action of ion pumps. By contrast, in some fast cells, it must be near the theoretical maximum for satisfactory functioning.

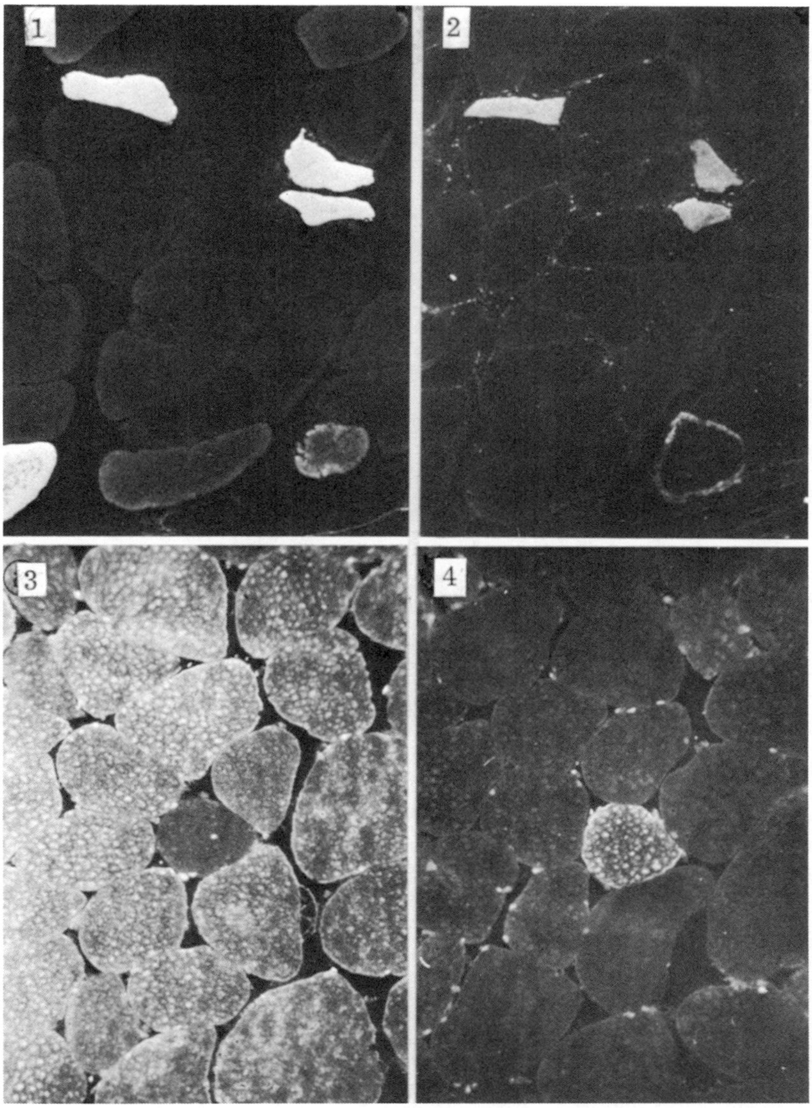

Figure 8.11. Localization of different forms of myosin. (*1*) Chicken anterior latissimus dorsi – a slow red muscle, stained with fluorescent antibody to one molecular form of myosin, A1. (*2*) Same as (*1*), in serial section – stained with a different molecular form of myosin A2. The reactive cells contain more of one than the other. (Reprinted with permission from Lowey et al. (1979) In: *Motility in Cell Function*, pp. 53–67. Copyright Academic Press.) (*3*) Rabbit peroneus longus – a slow red muscle stained with fluorescent-labeled antibodies against the fast and slow myosins. (*Left*) anti-fast. (*Right*) Anti-slow. (Reprinted with permission from Rubinstein et al. (1979) In: *Motility in Cell Function*, pp. 367–370. Copyright Academic Press.)

In the most primitive known living vertebrate, the hagfish *Myxine*, fast-type cells have a mean resting potential of 74.6 mV compared with only 46 mV for slow ones (Andersen et al., 1963). In a primitive elasmobranch, the dogfish, white cells have a mean membrane potential of 85.2 mV compared with 71.1 mV for red ones (Stanfield, 1972). The input resistances are markedly different, 1588 Ωcm^2 compared with 5410 Ωcm^2 respectively, and all-or-none sodium spikes occur in the former (8/27 tested) but not the latter, which

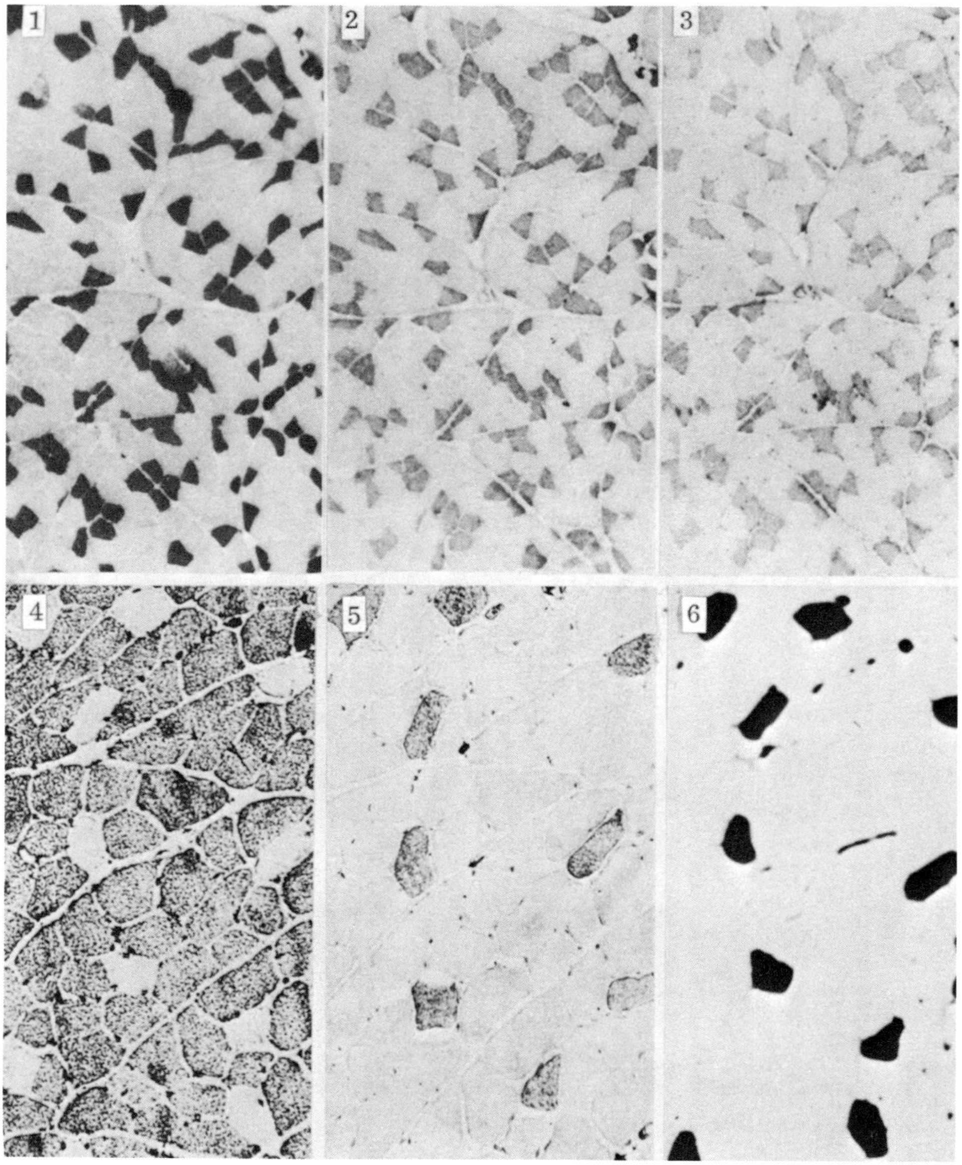

Figure 8.12. Localization of proteins using immunoperoxidase-staining procedures on serial sections of human semispinalis capitis. (*1*) Fast TN_I. (*2*) Fast TN_T. (*3*) α-TM. Localization of slow (*4*) and fast (*5*) forms of TN_I on serial sections of rat gastrocnemius. The fast form is located on cells staining strongly for myosin ATPase (*6*). (Reprinted with permission from Perry et al. (1979) In: *Motility in Cell Function,* pp. 129–145. Copyright Academic Press.)

either have reduced sodium-conductance activation (13/27) or none at all (6/27). An invertebrate example is crayfish slow, compared with fast, abdominal flexors. Fast cells have about 65 mV resting potential, average SL, 2.5 μm; λ, 0.5-1 mm; R_m, 20-30 Ωcm^2; r_i, 2.5 Ωcm, r_o, 30-40 k Ω; and C_m, 3-4 $\mu F/cm^2$. For slow cells, the resting potential averages 59 mV, SL, 8.0 μm; λ, 1.5-12 mm; R_m, 700-1100 Ωcm^2; r_i, 25-150 Ωcm; r_o, 0.2-1.6 m Ω; and C_m, 15-30 $\mu F/cm^2$ (Le Houelleur, 1978). The range was clearly very large for the slow cells, but all slow cells were significantly different in all parameters from fast ones. The fast cells had linear I/V relations, but the slow type all showed delayed inward rectification followed by a prolonged after depolarization, both reduced by TEA. In assessing the above values, it should be borne in mind that slow-type, long-sarcomere cells generally, and in a few crustacean muscles always, have smaller diameters than do neighboring fast ones. This is by no means an invariable rule, and within each class there are wide variations.

A lower mean membrane potential for slow cells compared with that for their fast counterparts is a wholly consistent feature throughout the entire animal kingdom. This might seem paradoxical, since it poises the cells closer to their E-C coupling thresholds. It is probably a primitive feature associated with activation via multiple distributed terminals. By enabling small JPs to initiate weak contractions, it extends the range of tension control possible, especially at low frequencies. The higher resting potentials came later, with evolution of the all-or-none action potential, possibly to aid synchrony of activation by ensuring that it coincided with the steepest portion of the voltage–time relation.

8.2.10. Energy Substrate

Histochemical or direct TEM studies reveal major differences in the amounts of the two major sources of muscle energy, glycogen and lipid, in different types of cells. Cells also differ in the rates at which they will take up the precursors of these, glucose and glycerol. These factors occur in close association with others such as AM ATPase rate and twitch characteristics and although associated with other variables, have not themselves been utilized as a primary basis for classification.

8.2.11. Innervation

It is a significant indication of what is really important for neuromuscular function that innervation patterns, though widely different at both gross and single-cell levels, have not been used as major classificatory categories. Some aspects have, however, been so used for secondary classification in vertebrates. This is because a distinctive type of ending can be seen at the light microscope level, with silver staining, on otherwise-similar-appearing muscle cells. The common ending on mammalian skeletal cells is compact and although quite different in appearance from the clawlike, long-fingered endings of frog muscles (Figure 5.2), could reasonably be grouped with them when being compared to a second distinctive type that turned up first in snake, then frog, and eventually cat. These were anything but compact, consisting of distributed varicosities connected by fine nerve twigs, like a miniature bunch of grapes, so were called *en grappe* terminals, the others being termed *en plaque.* Examples are shown in Figure 8.13. *En grappe* endings are uncommon, but some clusters of cells in frog muscles contain only this kind. The establishment of these terms in common usage has been largely due to the French specialist in junctional histology R. Couteaux. The difference in appearance is certainly dramatic, but so far no functional difference has been associated with it, the characteristics of transmission at the two sites being indistinguishable. One function of the *en grappe* ending may be to ensure excitation of a larger area of muscle cell than in the *en plaque.* Although some *en grappe* endings are compact, many continue over large areas of the cell surface, and it is easy to imagine that in their primeval state they covered the entire surface, as do many motor terminals of invertebrates. Another possible reason for the difference may be some as-yet-unknown relation of structure of the terminal to trophic functions.

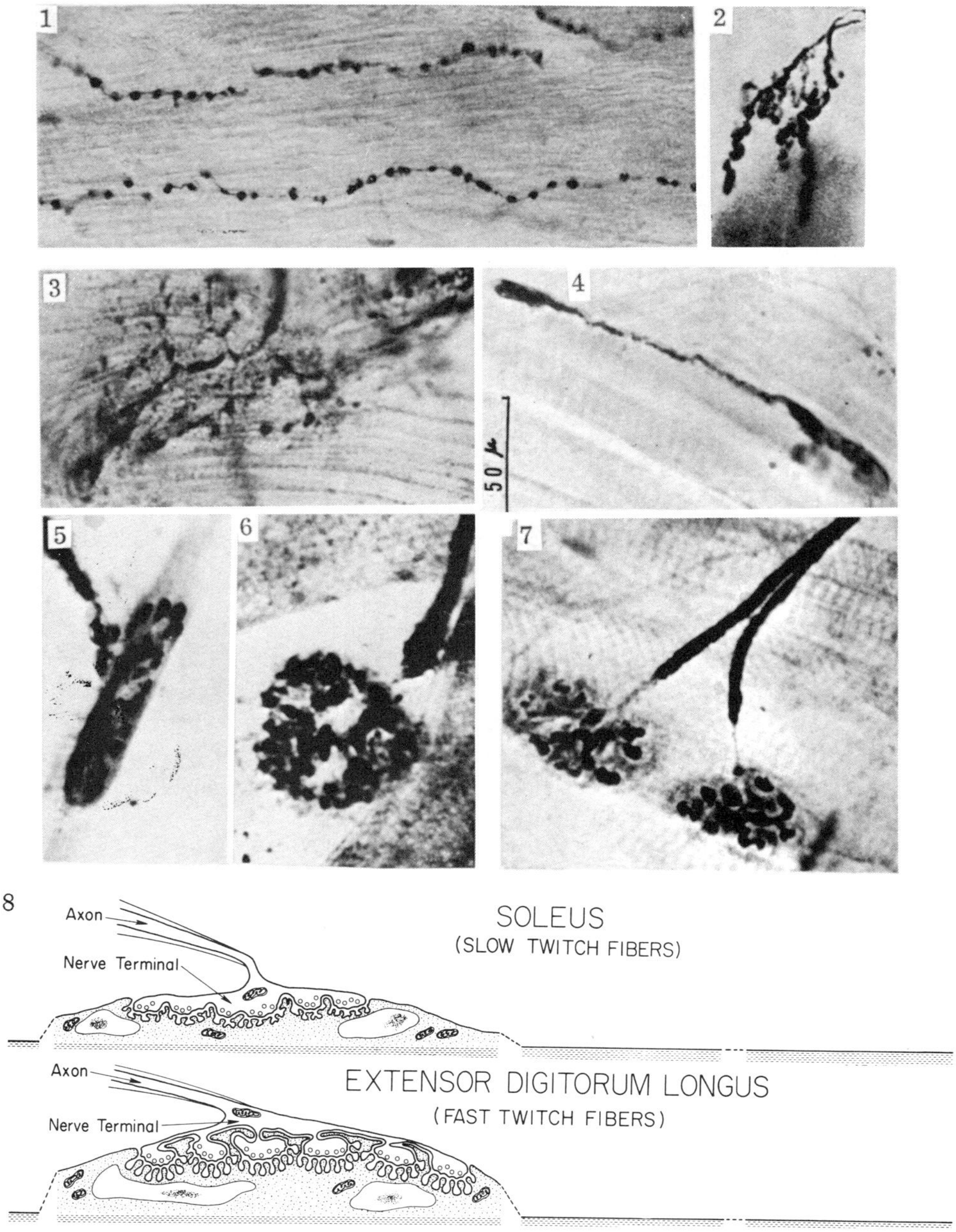

Figure 8.13. Diversity of neuromuscular junctions in vertebrates: (*1*) shows distributed endings; (*2*) and (*3*) are "en grappe" endings; (*5*)–(*7*) are single and double forms of "en plaque" endings. (*1*) Slow cell of hagfish (*Myxine*) myotome. (From Jansen et al., 1962.) (*2*) Fast cell of fish myotome. (From Jansen et al., 1962.) (*3*) Leg muscle of domestic fowl. [W. C. Cole, Motor endings in the striated muscle of vertebrates. *J. Comp. Neurol.* **102**, 708 (1955). Reprinted with permission from Alan R. Liss, Inc., New York.] (*4*) Superior rectus of rhesus monkey. (Zenker, W., and Anzenbacher, H. (1964) On the two different forms of myo-neural junction in two types of muscle fiber from the external ocular muscles of the rhesus monkey. *J. Cell. Comp. Physiol.* **63**, 275. Reprinted with permission from Alan R. Liss, Inc., New York.) (*5*), (*6*), (*7*) Leg muscles of rat. (From Cole, 1955.) (*8*) Diagrammatic comparison of rat soleus with extensor digitorum longus. The broad nerve terminal of soleus is apposed by relatively shallow irregular junctional folds. In contrast, the bulblike nerve terminal expansions of EDL fibers are apposed by more regular, deeper junctional folds. (From Ellisman et al., 1976.)

The endings on frog slow-type cells are entirely *en grappe*. In birds, *en grappe* endings are exclusively located on true slow-type (as opposed to slow-twitch) cells, although slow cells are sparsely mixed among fast ones (Ginsborg and Mackay, 1961). These are especially interesting because they are distributed and multiterminal, as in invertebrates. Some frog iliofibularis cells also have multiple, distributed terminals (Hess, 1963). Multiterminal innervation, with endings scattered over the entire cell length, is found in two places in mammals: on slow-type cells located in the extraocular muscles (e.g., Hess and Pilar, 1963; Zenker and Anzenbacher, 1964), and on some of the intrafusal muscle (spindle) cells (Hess, 1961c). The same has been found in reptiles, especially snakes, on slow-type cells (Hess, 1963).

When close attention has been paid to details of *en plaque* endings on individual cells within a muscle, quite a lot of diversity has sometimes been seen, including multibranches, clublike expansions, and varicosities up to 2 μm in diameter.

Some individual multiterminal endings, such as those found on slow cells of chick anterior and posterior latissimus dorsi, and biventer cervicis, are neither *en plaque* nor *en grappe* but best described simply as diffuse (Hess, 1961a). In some mammals, notably the rat, double endings are common (Cole, 1955). The common appearance of multiple endings on slow-type cells is related to the need to activate these cells during long-continued stimulation that reduces electrogenic responsiveness of the surface membrane (Figure 8.13.*7*). If there are sufficient junctions, the JP alone will cause contraction. There is a progressive increase in the area of the subsynaptic membrane at neuromuscular junctions with evolution of vertebrates, as already mentioned. The subsynaptic surface at fish junctions is a smooth one indented by synaptic swellings. In amphibia there are repeated infoldings along the length of terminal branches of motor axons. Both reptilian and mammalian junctions are infolded along the width of the synaptic gutter as well as along its length.

Variations in the extent and depth of the infoldings have been noted for the junctions of mammalian muscle cells of different types. In rat diaphragm, fast white cells of the type classified as IA (see p. 298) have the greatest total contact area (Padyulka and Gauthier, 1970). Intermediate cells have a medium level of contact area, although the deepest folds, classified as type IB, red slow-twitch cells, have the least contact area.

Rat soleus (slow-twitch) junctions, although having similar microelements (see p. 157) to their fast-twitch counterparts of the extensor digitorium longus, show significant quantitative differences (Ellisman et al., 1976). The junctional folds are deeper, and the final neuronal twigs more deeply embedded, in the fast-twitch terminals (Figure 8.13.*8*). The appearance of the innervation can, therefore, provide an indication of the nature of the underlying cell and has been used as a basis for classification. There are exceptions, as might be expected, in the direction of fast vertebrate cells having multiple distributed terminals. This occurs in some fish, such as the snake fish *Ophiocephalus,* where both fast and slow endings are of *en grappe* type. There is no disadvantage in having multiple *en grappe* endings on fast-type cells, and this is almost certainly the primitive condition, representing the universal state of affairs before reliable propagated potentials and fail-safe junctions evolved.

Of direct classificatory significance in invertebrates are the details of total nerve supply to a muscle cell, for example, slow axon plus fast axon, slow plus fast plus inhibitor, and so forth. Single muscle cells are known that receive as many as nine excitatory axons and two inhibitory axons (in locust flexor tibiae – Phillips, 1980).

A major feature of use in secondary classification, and one that is of direct classificatory value for most arthropod muscle cells, is micro aspects of transmission. The parameters used are initial size of EJP in a train (series) and rate of change, increase or decrease, with long-continued repetition. Miniature end plate potential (MEPP) frequency affords a measure of basic activity in a terminal. When D-glucose enters terminals, it alters MEPP frequency, presumably by directly entering into the metabolic events. The rate of entry, so judged, was markedly higher for fast-type rat cells than for slow-type (Pickett, 1980), suggesting that the slow type have generally slower metabolic activity, larger energy store, and slower transport. These sorts of differences are to be expected for nerve terminals in general. There have been few studies comparing metabolic characteristics of major types, or comparing fast-twitch and slow-twitch vertebrate junctions. Correlated ultrastructural studies on frog junctions having different rates of activity have failed to detect any morphological differences between them (Rose et al., 1978).

A limited amount of the typical invertebrate feature, polyneuronal innervatoin of single cells, has been detected in lower vertebrates such as the clawed toad *Xenopus* (Ridge and Thomson, 1978), raising questions about motor-unit definition. In this case the two units as a whole produced quite different contraction characteristics, so although some muscle cells were shared others that were not were of different cell types.

The excitatory transmitter onto vertebrate skeletal cells is acetylcholine (ACh), and no inhibitor has been found. In invertebrates the commonest excitatory transmitter is the amino acid L-glutamic acid (see p. 491), with ACh a close second. But many invertebrate cells also receive one, two, or three inhibitory axons whose transmitter is γ-amino butyric acid (GABA). However, there is a wide range in the extent of innervation by inhibitory axons. The responsiveness of a cell to GABA appears to be proportional to the extent of innervation and can be measured directly by determining the amount of depolarization produced by applying specific concentrations. A detectably different structure of the postsynaptic membrane for each different transmitter now appears likely (e.g., Rheuben and Reese, 1978).

A new aspect that will repay further study is temperature sensitivity of transmission. This issue was raised by the discovery that the temperature-sensitive *Drosophila* mutant *shibire* (Japanese for "paralyzed") cannot fly above 29°C because at this temperature there is a drastic fall in the height of the EJP to the flight muscles (Siddiqi and Benzer, 1976).

8.2.12. Different Sizes and Shapes of Synaptic Vesicles

Not much is yet known about the details of vesicles in different types of terminal on different kinds of muscle cells, but it is already apparent that there are some major and significant differences in both invertebrates and vertebrates. Quantitative comparisons on the vesicle populations of insect (locust *S. gregaria*) slow and fast excitatory terminals, which are both considered to release L-glutamate (Usherwood, 1977), but with markedly different characteristics (Hoyle, 1974a), have been compared with the GABAergic inhibitory ones (Titmus and Hoyle, 1977). The two sets of excitatory vesicles on the extensor tibiae muscle are both spherical, but those of the fast axon are significantly larger, 52–60 nm, than the slow ones, which are only 43-50 nm. The inhibitory vesicles are aspherical, with maximum-axis–minimum-axis ratio of 1 : 3, but they have similar areas to the fast junctions.

An aspherical shape for inhibitory vesicles may be widespread (Figure 5.5), and was first noticed in cat cortex by Uchizono (1967), who went on to find asphericality in synapses on the crustacean (crayfish) stretch receptor neuron as well. This may be a feature of the response of GABA-containing vesicles to fixation, but slow excitatory vesicles of stimulated terminals change from round to flat: it is a puzzling phenomenon.

In two separate sets of neuromuscular junctions of the phylogenetically important protochordate *Amphioxus*, which sends muscle arms to the nerve cord to receive them, two distinctly different sizes of vesicles occur (Flood, 1966). The small ones occur on muscle arms from lamellae that are full of mitochondria and in many respects resemble slow-type cells, the large ones being on arms that are clearly by comparison to be considered fast-type.

8.2.13. Variation in Motor-Unit Size

The number of cells innervated by a particular neuron is a major variable in the case of identified motor-nerve cells of invertebrates. In vertebrates, the whole is termed the *motor unit,* and its size is a highly significant variable, useful in classification, especially if it takes into account average properties of the component cells. Small units are characteristic of small muscles that require fine control, such as external eye muscles. Large units occur especially in large muscles used for ballistic activities such as jumping. The component cells of a unit are not clustered but rather mixed together, apparently rather randomly, in most mammals in which intensive studies have been made. It is necessary to stimulate single units to exhaustion, then rapidly fix the muscle and examine it histologically, in order to determine the distribution of component cells. Glycogen will have been depleted from muscle cells constituting the unit (Figure 8.14). The use of a variety of tests on serial sections permits determining a variety of characteristics of the depleted cells. These show that a unit comprises cells of similar type (Burke et al., 1973), pointing to a combined developmental/trophic control of characteristics.

For vertebrates, a knowledge of the various motor units making up a given muscle, in terms of both physiological, ultrastructural, and histochemical characteristics, is clearly the most useful. Modern studies are accordingly increasingly trending in this significant direction (Figure 8.15). There are too many choices of parameters to permit all to be displayed on a single graph, so an arbitrary selection based upon available data for a particular muscle must be made. Fortunately there does seem to be a close correspondence between even three-dimensional charts based upon physiological criteria and ones based on enzyme activities. Invertebrates, during this evolution, have been less constrained; the diversity in their motor units is so great that no order can be discerned that would permit acceptable general classifications.

8.3. Classification of Cell Types

Because of the multitude of major factors that vary between muscle cells, any attempt at classification must run into major difficulties. Even for basic descriptive purposes, some words such as *fast* and *slow* have been used that implied acceptance of a semblance of classification. This was reasonable in referring to invertebrate muscles because there has been general agreement to adopt the scheme used by Atwood, Hoyle, and Smyth (1965),

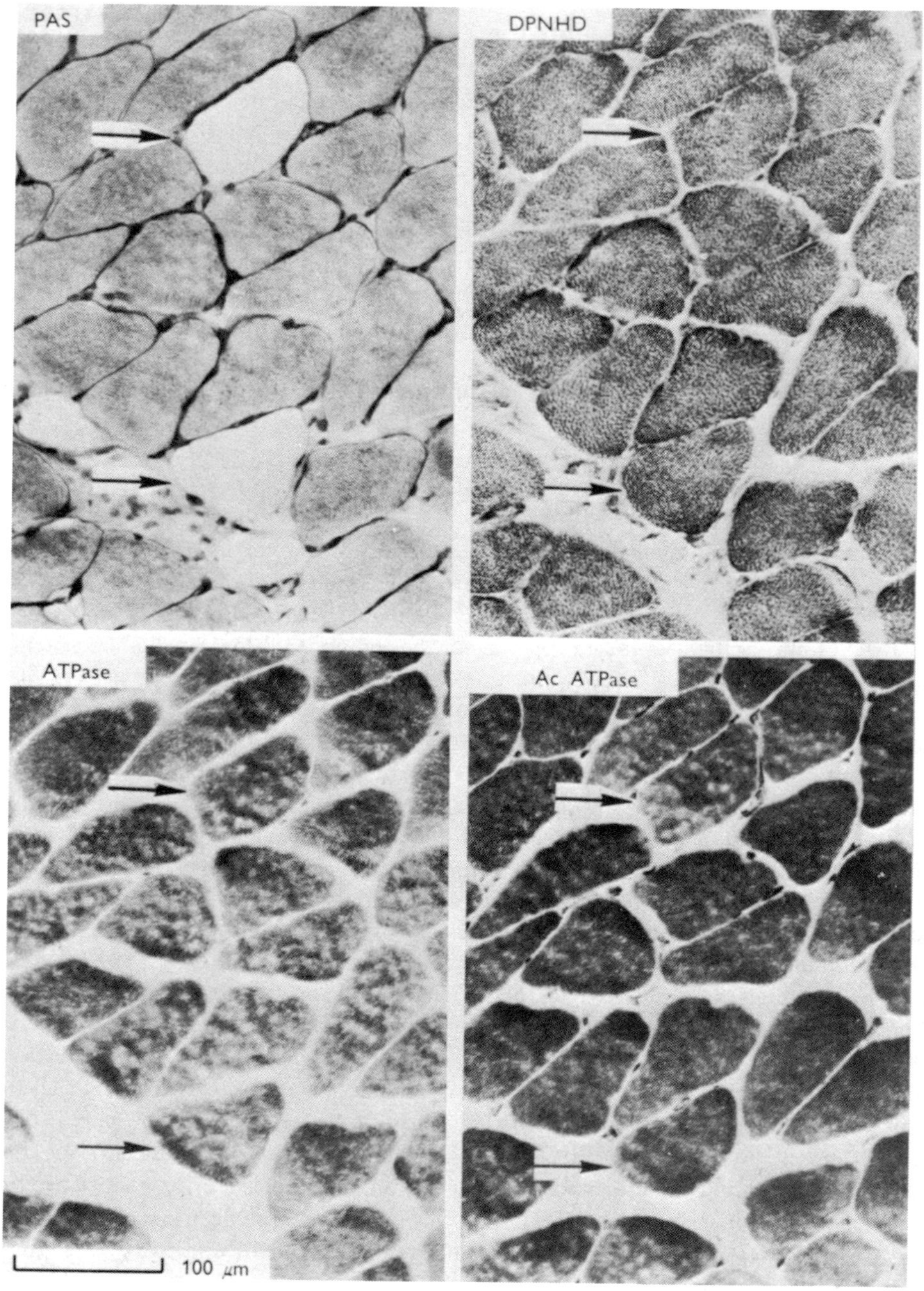

Figure 8.14. Sequential sections of soleus of cat stained as indicated in periodic acid Schiff (*PAS*) for glycogen, for the enzymes reduced diphosphopyridine nucleotide dehydrogenase (*DPNHD*), and myofibrillar ATPase (ATPase), or same with veronal acetate buffer at pH 4.65 (*Ac ATPase*). Glycogen depletion following stimulation of a particular unit (clear in *PAS*) enables histological identification and profile of a particular unit. (From Burke et al., 1974.)

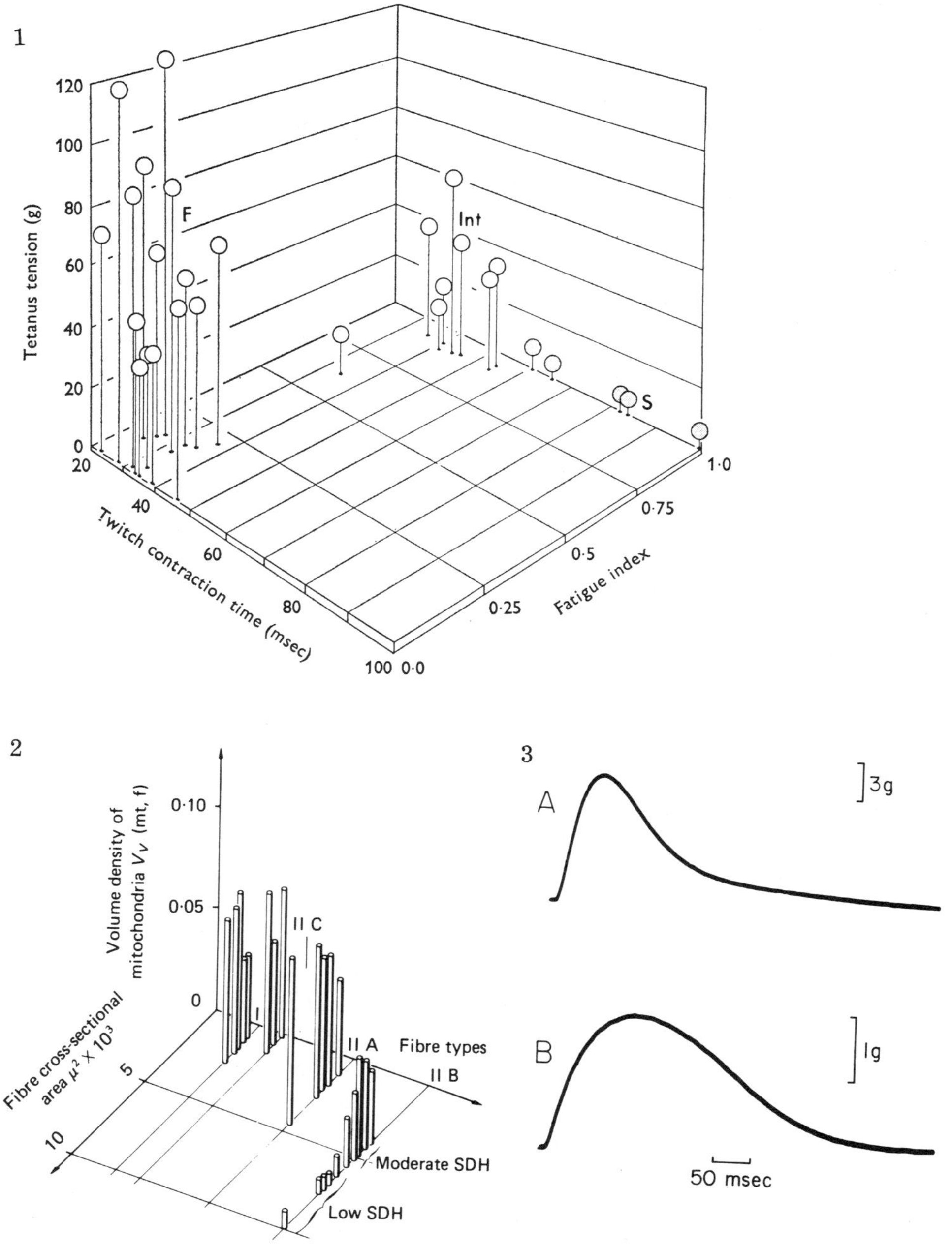

Figure 8.15. Diagrammatic classification of cell types within a muscle with the aid of three-dimensional plot of critical parameters and contractions of different units in human muscle. (*1*) On the basis of twitch time and tetanus tension of single motor units, their fatigability and histochemistry, three types appear in the medial head of the gastrocnemius. Fast (*F*) are strong, fast, and common but readily fatigable. Intermediate (*Int*) are weaker but fast and fatigue resistant. Slow (*S*) are slow and weak but extremely resistant to fatigue. (From Burke et al., 1973.) (*2*) Variations of parameters with cell types, based upon comparison of succinic dehydrogenase (*SDH*), cell size, and mitochondrial contents, in horse semitendinosus. (From Claassen et al., 1981.) (*3*) Comparison of twitch tension–time of a fast-twitch unit, with that of slow-twitch unit, from the same muscle, human multifidus. [Reprinted with permission from Eberstein and Goodgold. *Am. J. Physiol.* **215**, 535–541. (1968). Copyright 1968 The American Physiological Society.]

which distinguished three classes: fast, slow, and intermediate. There are many minor differences, but the principal features the major groups share are easy to recognize and widespread. These are covered in Table 8.3.

However, it should be understood that there are no absolute criteria regarding speed. This is a relative variable: the speed of a particular cell should be compared with that of slower cells within the same muscle. A slow cell of a fast muscle can have a faster speed than the fast cells of a slow muscle. The range of speeds is so great, and there are so many different factors that contribute, that many different routes have been used during the course of adaptive radiation to achieve similar overall results. For example, actomyosin ATPase rates may be identical but amounts of SR different, or amounts of SR may be similar but ATPase rates different. One cell will have achieved slowness by way of a long sarcomere, another by having an inexcitable membrane.

Interpretation for a particular muscle is complicated by its evolutionary history. It should be borne in mind that the evolution is an additive process: reversal has been very rare. A muscle may have been through several cycles of requirement: fast-slow-fast, strong-weak-strong, tonic-phasic-tonic, and so forth. In one set of adaptations, a biochemical event may have been selected, in another an ultrastructural format, and in yet others membrane properties: the possible permutations are astronomic!

Vertebrates have some skeletal muscle cells that can be matched closely with the slow cells of arthropods, but the characteristic cell of vertebrate skeletal muscles, with its single motor end plate and propagated all-or-none action potential is, as we have seen, atypical, a strictly vertebrate feature. The biological intent of turning on the contractile material 100% or nearly so very quickly after stimulation is, however, matched in invertebrates by fast-axon-innervated fast-type cells. The two kinds can reasonably be classified together as fast cells. But vertebrate evolution has proceeded by the progressive differentiation of this type of cell. Within both vertebrate and invertebrate fast cells a mixture of factors has been altered to promote one specific feature or another, such as staying power, fineness of force or length control, extreme speed, or extreme force capability. It therefore seems best to adopt separate systems of classification, one for invertebrates, another for vertebrates. There has not yet been any general agreement about classification of vertebrate skeletal cells, but that adopted in Table 8.4 has been the most common. In this, all-or-none twitch cells are separated from true slow ones that have multiterminal innervation like invertebrate cells. The latter are collectively termed slow regardless of actual contraction/relaxation speeds. The twitch cells are subdivided into relative extremes of fast-twitch and slow-twitch and various intermediates. Because of a major classification adopted independently by histologists, in which there were categories I and II, the latter having subdivisions A and B, these are matched to physiological types in the table.

A fairly frequently used alternative is an alphabetical one in which white type 1A are termed A, reddish IB are termed B, and another subset of type I that are rich in mitochondria and "red" are termed C cells. This was used for example by Schmalbruch (1971) in a comparison of rat, rabbit, and cat soleus (all slow) with rabbit thyreoarytenoid, rat anterior tibialis, and rabbit and cat gastrocnemius, all of which are fast. The cells designated C were very rich in mitochondria and had wide (100–120 nm) Z bands compared with the faster types (50–60 nm), but of special interest was a longer SL than in the other types. Most cells in the soleus of all three animals were types B and C, whereas the faster ones contained many type A, but also some types B and C and several cells, especially in the gastrocnemius, that had mixed characteristics, so could not be classified.

Table 8.3 Classification of Invertebrate (Principally Arthropod) Cross-Striated Skeletal Cells

Feature	Cell Type: Fast	Cell Type: Intermediate	Cell Type: Slow
Color	Clear white	Grey pink	Grey red
Surface-membrane properties	Capability for large graded electrogenic response	Range of electrogenic capabilities: medium to large	Wide range of electrogenic capabilities; many are electrically inexcitable, some gradedly excitable
Neuromuscular transmission	Large EJP; little or no facilitation or antifacilitation; rapid fatigue	Medium EJP; some facilitation and eventual fatigue	Wide range, but generally small EJP; large extent of facilitation, highly resistant to fatigue
IT system	Extensive, many tubules	Intermediate numbers of tubules	Few tubules
Sarcoplasmic reticulum	Extensive ramifications between myofilament clusters	Moderate branching	Sparse, especially near core of cell
Stimulus duration – dependence of speed	Slight	Fairly strong	Strong
Z band	Generally dense and continuous	Diffuse, often composed of ovoid bodies	Diffuse, often composed of ovoid bodies
M band	Pronounced; often dual	Weakly defined; sometimes dual	Generally absent; M bridges present along whole A band
Mitochondria	Few, except in flight muscles	Variable number	Numerous
Latency elongation	Present and strong	Present but weak	Absent
Contraction (maximum)	Rapid rise	Moderately fast rise	Slow rise
Relaxation	Rapid fall	Slow fall	Very slow fall
Response to sustained excitation	Rapid fatigue	Slow fatigue	Contraction maintained
Enzymes	Entirely glycolytic, except flight muscle Weak aerobic activity	Mixed glycolytic/lipolytic Moderate aerobic activity	Ranges from entirely glycolytic to lipolytic with various mixtures Intensive aerobic activity

Orig.

Table 8.4 Classification of Vertebrate Cross-striated Skeletal Cells

	Cell Type			
	Fast-Twitch (Type II)			
Feature	II A Fast Oxidative, Glycolytic White	II B Fast Oxidative,[a] Glycolytic Red	Slow-Twitch (Type I)	Slow
Color	White	Pink	Red	Red
Surface-membrane properties	All or none	All or none	All or none	Large graded to all or none
Neuromuscular transmission	Single end plate, large EPP	Single end plate, large EPP	Single end plate, large EPP	Multiterminal, small EPP, facilitation and summation
IT system	Abundant	Abundant	Range from abundant to moderate	Range from very few to many
*g*Ca inactivation	Present	Present	Weak	Absent
Sarcoplasmic reticulum	Extensive	Extensive	Very wide range, from fairly extensive to almost absent	Range from moderately extensive to sparse
Stimulus duration – dependence of speed	None	None	Some	Strong
Z band	Narrow, sharp	Narrow	Wide, fuzzy	Wide, fuzzy[b]
M band	Distinct	Distinct	Indistinct	Weak or absent[b]
Mitochondria	Very few	Wide range, fairly numerous	Abundant	Abundant
Latency elongation	Strong	Fairly strong	Absent	Absent
Contraction (maximum)	Fast-rise twitch	Fast-rise twitch	Slow-rise twitch	Slow-rise
Relaxation	Fast	Medium speed	Slow	Relatively very slow
Response to sustained excitation	Rapid fatigue	Fairly rapid fatigue	Slow fatigue or no fatigue	Resistant to fatigue
Energy substrate	Glycogen	Glycogen	Fat and glycogen (some have abundant fat droplets)	Glycogen and fat

Myoglobin	Little	Some	Abundant	Abundant
Enzymes				
Actomyosin ATPase rate	High	High	Low	Low
Succinic dehydrogenese	Low	Medium	High	High
Phosphorylase	High	High	Low	Low
Lactic dehydrogenese	High	High	Low	Low

[a]This category is sometimes subdivided into a, the slowest, and b, in which case II A is called c.
[b]Fish muscle cells excepted.
Orig. Compiled from multiple sources.

Extensive data for a large number of histochemical characteristics have been obtained for cat gastrocnemius and related to physiological type on a three-unit scale (Burke et al., 1973). Recognized were fast-fatiguing fast-type (FF), more slowly fatiguing fast-type (FR), and fatigue-resistant slow type (S). The relative enzyme contents are shown in Table 8.3.

There have been many mistakes and confusions in the past and many more are inevitable in the future simply because there are so many independent variables. For example, the anterior latissimus dorsi of the chicken is a slow muscle compared with the posterior latissimus dorsi of the same animal. The half-times to maximum contraction of the whole are 6.0 and 2.7 s respectively, and relexation to half is 11.0 s compared with 2.0 s, but both contain *Fibrillenstruktur* and *Felderstuktur* cells (Bock and Hikada, 1968). These authors examined the hatching muscle of the chick and found it to contain a mixed population. These apparent disparities represent the true facts. A red, slow muscle needs some fast cells for occasional use and vice versa. These "exceptions" had several different routes available for attaining these goals and were not constrained – or even able – to adopt the same ones taken by the majority in their neighboring synergist. The study of one particular structural parameter, no matter how significant functionally, can be very misleading. For example, cells of the cat gastrocnemius have similar volumes of SR, but fast ones contract at five times the speed of their slow counterparts. These examples show clearly the danger inherent in extrapolating from a norm.

The student who perceives general rules to be extractable from the tables will have learned only the average, not a set of laws that can be applied for prediction. All natural variations must be sought by detailed examination of the variant in question.

8.4. Range of Cell Types in Representative Muscles

Very few muscles have been carefully examined in regard to cell-type composition in any of the numerous parameters outlined above, and not even a single one has yet been examined exhaustively for all of them. One or two vertebrate muscles have been intensively studied histochemically and others physiologically, but extensive correlated examination of ultrastructure with electrophysiological and dyanmic contraction characteristics of component cells is not feasible, as it is in some arthropodan invertebrates. But to date only one invertebrate muscle,† the crab leg dactylopodite adductor (see p. 267) has been extensively examined in regard to both electrical and mechanical properties of individual cells, and there are neither ultrastructural nor enzyme data for it. For invertebrate muscles, the crab data and that from the few other intensively studied examples indicate a real need for extensive analysis of individual cells. The need may be much less pressing for understanding vertebrate muscles, because their utilization of propagated action potentials eliminated a major source of variation. There appear to have been other constraints also leading to relative uniformity in vertebrates, notably the relatively fixed SL.

The enormous gap between invertebrates and vertebrates is to some extent bridged by the protochordates, so it is worthwhile taking a close look at the cell contents of their muscles. Although dynamic tesing has not been achieved for cellular elements of *Amphioxus,* fast-, slow-, and intermediate-type cells can clearly be recognized (Flood,

†Combined studies, in which individual muscle cells are numerically addressed, are in progress in the laboratory of D. W. Rathmayer on an *Eriphea* muscle.

1968) as judged by structural criteria. One kind (presumed fast) has wide lamellae with few mitochondria and little glycogen. Another (presumed slow) has narrow lamellae, numerous mitochondria, and much glycogen. There are intermediates in these respects.

In fish, the carryover of a fundamental three-way division of cell types is immediately apparent, as we have seen, in the myotomes. It is also there in their eye muscles (Kumari et al., 1979), where true slow cells occur side by side with fast ones. Single type I red cells of fish have now been isolated, studied individually, and compared with white ones (Flitney and Johnston, 1979). They need frequencies of only 5-10 Hz to develop maximal tension, compared with up to 300 Hz for white ones, which are 12 times faster than the red. As we have noted, several kinds of fish use the 5% or less of lateral deep-red muscle for 99% of all their activities, saving the majority for rare, perhaps once-in-a-lifetime requirements of a leap up a waterfall, or escape from a predator (Rayner and Keenan, 1967; Bone, 1966). Likewise, a grasshopper may spend its entire life without making the jump that it needs in order to escape, but it has set aside more than 70% of its largest muscle solely for this possible occasion.

Of course a fish can afford to devote so much mass to a rarely used muscle because it is streamlined and lives in water, so its body is supported. Land-living vertebrates need all available muscle power for major locomotory exertions. But they evolved from fish and so must have started out with the same control mechanisms as the fish. Recent indications are that there are still major remnants of the primitive system. Apart from the well-known slow-cell bundle of frog iliofibularis, little is known about amphibian muscle cell types. The rectus abdominis undoubtedly contains a mixture of slow and fast cells, but the distributions and proportions are not known.

Vertebrate cells of unique interest were sought, and indeed are present, in the tail of the amphibian tadpole; they are resorbed and not part of the adult musculature, but they are direct evolutionary descendents of the ancestral fish tail. Three characteristic but structurally similar amphibia, the primitive clawed toad *Xenopus,* the toad *Bufo,* and the frog *Rana* have tadpole tail musculature containing the two principal vertebrate cell types. These are fast white cells with large diameter, little SR, and few mitochondria, and slower red ones having smaller diameter, much SR, and large numbers of mitochondria. The slow ones have strong lactic dehydrogenase and succinic dehydrogenase but weak phosphorylase activity, whereas the latter is strong in the white cells (Sasaki, 1974). Both slow and fast cells are also present, mixed together, in larval newt muscles (Watanabe et al., 1979).

Reptiles are clearly endowed with a rich spectrum of muscle cell types, as judged by snake rib skin muscle cells, but there has been no systematic investigation on them. Distinctly different sets of cells occur in the common lizard *Lacerta agilis* (Witaliński, 1974). Small, 20-35 μm in diameter, slow cells are packed with mitochondria and lipid droplets. Large, 50-80 μm, fast white ones contain few mitochondria and no lipid. The latter have single end plates with strong ACh-esterase activity, the small slow cell 2-3 end plates with weak enzymic activity. Of even greater interest, the subsynaptic region of the former contains the numerous infoldings that have become recognized as a typical reptilian form. However, by contrast, the infoldings of the small red cells are simple grooves. In the extraocular muscles of a snake, *Natrix,* true slow-type cells with numerous end plates occur mixed with fast ones (Witaliński and Loesch, 1975).

Birds offer traditional extreme examples of slow/fast muscles as a whole, but also have been little explored at the single-cell level. True slow-type cells were found, mixed with fast ones, in the unique hatching muscle of chicks (Hikada and Bock, 1968), a valid but unlikely muscle for the average muscle scientist to have examined.

The first clear recognition of true slow cells in mammals was in extraocular muscles. The eye muscles must drive very rapid ballistic movements for saccades and also be capable of extremely accurate length change in slow movements used in tracking, so it would be expected that they would have a mixed population, and that is indeed the case. Superior rectus of the rabbit has a mixture of fast-type cells with narrow, straight Z lines, a well-developed IT system, much SR, a distinct M line, and a single compact neuromuscular junction (see p. 158). With them are slow-type cells having zigzag thick Z lines, few ITs, little SR, no apparent M bands, and several motor endings on each (Cheng-Minoda et al., 1968).

A fast muscle of the guinea pig, the vastus, has both red and white cells that are physiologically fast compared with the slow red soleus. Large differences were found between individual muscle cells in each population, but on anatomical features alone there was 90% success in placing representatives with their sources. Soleus cells have a wide Z line, about 120 nm, smaller numbers of ITs, and smaller amounts of SR. Within the vastus, the color range is closely correlated, as it is in invertebrates, with mitochondrial content, ranging from only 0.1% total volume for the whitest, which are also the fastest, to 15% for the reddest and slowest.

For rat diaphragm, Gauthier and Lowey (1977) preferred a classification with four categories, to include a type intermediate between the fastest white and the fastest red cells. In pigs, comparison of cells from two red and two white muscles using only histochemical methods permitted easy resolution of types I and II by ATPase activity, but left a large number of unresolvable intermediates, especially in the red muscles (Swatland, 1979).

An interesting comparison can be made between a rabbit and a hare: although they are closely similar and both very fast runners, the hare has the advantage. The fast extensor digitorum longus of both, which is a "white" muscle comprising type IA cells in most mammals, is actually red in color in the hare, like the slow soleus, with an enzyme pattern that is preponderantly oxidative with high succinic dehydrogenase activity (Bass et al., 1973). The ATPase activities are closely similar, with mixed populations, but strongly glycolytic in the rabbit. The hare has been able not only to equal the rabbit in speed but to greatly improve upon the rabbit in regard to endurance.

Pig red muscle has a very high concentration of zinc, which is low in its white neighbors (Cassens et al., 1967). The reasons for a high zinc concentration are not known but suggest an enzyme complex, though there are few mitochondria in either type of cell. Zinc is, however, a potent potentiator of E-C coupling in frog fast cells, prolonging the activation, and although this effect has been attributed to an indirect action, prolongation of the action potential by way of increased membrane resistance (Sandow and Isaacson, 1966), the possibility of internal effects were not ruled out in the frog experiments.

Perhaps the most unusual vertebrate cells yet discovered are ones found in the guinea pig stapedius, which are concerned with auditory fine-tuning and must be capable of long-maintained force with extreme accuracy in absolute value, with negligible fluctuations, to be maximally effective. There is very little actual contractile material in these cells, but there is much SR and many mitochondria (Wróblewski et al., 1981). The nuclei are placed centrally in the core. Thin filaments are attached to dense bodies, but the sarcomeres are of normal 2.2-μm length. The soleus, always slow overall, but commonly a mixed cell population, in the guinea pig comprises only slow-twitch cells (Gauthier and Padyulka, 1966).

There is unfortunately no known mammalian muscle comprising pure slow cells as defined above, so the only recourse for studying them physiologically is to isolate individual cells. Single slow-twitch cells can be dissected from mouse soleus and fast-twitch cells from extensor digitorum longus (Luff and Atwood, 1972), and chemical dissociation is applicable for others (Bekoff and Betz, 1977). There is little excuse for using whole muscles in physiological studies except where the properties of the muscle in question as a functional whole are of interest.

A majority of all human movements are manipulations, not massive exertions. It has long been assumed, and we certainly have the mental impression, that we are utilizing the whole of each muscle in the gentler activities of everyday life. It has been believed since Sherrington (1906) that there is a rotation of active motor units. But it now seems possible that we humans leave much of the work to a few ("preferred") units only (Monster et al., 1978; Thomas et al., 1978). These units are ones comprising entirely type-I slow-twitch cells with a lot of mitochondria. The percentage of this type of cell in a major muscle can be as low as 3%. Presumably the other 97% of the cells are used fish-fashion, to supplement the tiny slow-twitch cell population when needed for the occasional high-power-requiring activity (see p. 606).

In the senses already referred to, human muscles are considered to be relatively uniform, but there are sufficient differences in maximum speeds to distinguish biceps, deltoid, soleus, and tibialis anticus as slow compared with the faster triceps, quadriceps, and gastrocnemius. All, however, contain a mixed population of cell types. The relatively recent deployment of small-diameter syringe needles as core-sampling devices, combined with electron microscopy, is bringing rapid advance in knowledge of the cell types in human muscle and how they are affected by exercise, disease, and so forth. Each individual has a unique pattern of distribution of cell types, which are, in general, very unequally mixed in different muscles. All human muscles appear to contain a broad spectrum of cell types. This is by no means random, because the same mixtures appear in the same places of specific muscles of identical (monozygous) twins, but not in dizygous twins (e.g., Karlsson and Komi, 1976). The cell-type composition can be altered by exercise (see p. 410), but only to a limited extent: genetic factors determine both the basic pattern and the limits for change (e.g., Karlsson et al., 1978).

Oxidative-enzyme activity as studied by histological procedures permits recognition of three types in vertebrates, and these can be matched to ultrastructure (Payne et al., 1975). They conform to the general vertebrate classification shown in Table 8.4. In *Octopus,* enzyme profiles of cells match their diameters: thick ones are fast and oxidative, thin ones slow and more or less anaerobic (Ladanyi and Leray, 1967).

It has turned out that there are unexpected general muscle differences between human males and females. Comparisons were made in the vastus lateralis between male-female twins (Komi and Karlsson, 1978). Females had 7% fewer slow-twitch cells and also different enzyme activities. Muscle speeds were much slower in the females, with twitch rise times twice as long in the women.

Fresh biopsy tests of 10 muscles, rectus, multifidus, pectoral, interior and exterior oblique, biceps, sternocleidomastoid, semispanilis, deltoid, and abductor hallucis, showed two dominant populations of cell types in each (Eberstein and Goodgold, 1968). Fast-twitch cells had a high membrane potential (88.4 ± 8.7 mV), and slow-twitch cells had a much lower membrane potential (57.8 ± 7.5 mV). The surveys were insufficient to permit estimates of proportions of the two in different muscles, but presumably the fast–slow ratio is high in the overall faster muscles and low in the overall slow ones.

Anatomical investigations using immediate postmortem biopsy show that in most human muscles, the cells of different types are very widely interspersed, with a minimum of grouping (Warmolts and Engel, 1972). Large type IIA cells are the most likely to be grouped. They have a high ATPase activity and are probably used only in brief bursts of phasic activity. When the freeze-fracture technique is used, striking differences between human types I and II cells are seen in respect to rectilinear arrays of particles, which are virtually absent in type-I cells but numerous in type II (Shafiq et al., 1979). Surface caveolae are distributed uniformly in the former, clustered at the I band in the latter.

Patients with myopathies tend to be deficient in one or another type of cell, depending on the particular disease (see p. 414). Recognition of cell types in humans becomes much harder if there has been any one of several motor diseases.

8.5. Proportions of Cell Types in Representative Muscles

A very extensive study of a muscle is needed before a faithful estimate of cell populations can be established, and nothing less than 100% testing is essential for the more complex small ones. Unfortunately, such a survey is usually impractical. It is quite possible that small pockets of cells with special properties are mixed during intracellular electrode and TEM sampling. Yet just a few special cells can be extermely important functionally. Failure in an intracellular survey to find a little deep pocket of slow cells led me at one time to support an old view that in crustacean muscle, some cells can be directly activated by neurally released activator (Hoyle and Wiersma, 1958). The hidden cells were found eventually (Atwood and Hoyle, 1965).

The proportions of unequivocally recognizable types of muscle cells within a muscle can have just about any ratios, in both vertebrates and invertebrates. Some examples, presented in Table 8.5, will give an idea of the mixtures currently known. Physiologists are bound to hope that muscles that are truly homogeneous in cell type will be found in representatives of different orders of animals. Among amphibia, the best contender for a pure fast-type muscle is still the sartorius, which will give all physiologists a sigh of relief. There is no pure slow-cell amphibian muscle known, the tonic bundle of the ilio-fibularis discovered by Sommerkamp (1928) still being the best option.

In mammals, a claim has been made for the omohyoideus of the rat as a convenient source of pure fast-twitch cells. It should be noted that true slow cells, which are rare in mammals, are not to be confused with slow-twitch cells, which are common and the basis for different properties of so-called slow muscles. By far the commonest muscle used as a representative "slow" muscle, that is, one comprising mainly slow-twitch cells, is the soleus.

8.6. Muscular Portions of Stretch Receptors

There have been few more heated subjects for mutual dismemberment of reputations in the muscle field in recent years than in the classification and resolution of the nature of the muscular components of spindles, carriers of stretch-receptor sense organs, present in vertebrate, especially mammalian, muscles. There are clearly two extremes, which may be termed slow and fast *as muscle elements,* based upon innervation and general ultra-structural characteristics. The slow, at least in rat, are the nuclear-bag elements. Both Z and M bands of these cells are weakly defined, and they have little SR and few ITs, all of which suggest slow, sustained contraction. By contrast, the contractile parts of the

Table 8.5 Approximate or Estimated Proportions of Cells of Different Characteristic Type in Different Muscles, Expressed as Percentage Total Number

	Fast-Twitch	Intermediate	Slow-Twitch	Slow	Source
Vertebrates					
Hagfish (*Myxine*) myotomes	70	–	–	30	Andersen et al., 1963
Dogfish (*Scyliorhinus*)	82	0	18	0	Bone, 1966
Frog (*Rana pipiens*) sartorius	100	0	0	0	Bone, 1966
Frog (*R. pipiens*) iliofibularis	65	0	15	20	Hoyle (unpublished)
Frog (*R. pipiens*) rectus abdominis	0	70	30	0	Hoyle (unpublished)
Cat tensor tympani	0	20	60	20	Erulkar et al., 1964 and others (author's estimate)
Rhesus monkey external rectus	70	10	–	20	Zenker and Azenbacher, 1964 and others
Rat diaphragm	40	30	30	–	Padyulka and Gauthier, 1966 and others
Invertebrates					
Accessory flexor (*Cancer magister*)	20		55	25	Dorai-Raj, 1964
Dactylopodite closer walking leg (*C. magister*)	25		55	20	Atwood et al., 1965 and unpublished
Crusher claw raised normally (*Homarus americanus*)	0		0	100	Lang et al., 1978
Cutter claw (*H. americanus*)	70		–	30	Lang et al., 1978
Crusher claw raised on smooth bottom dish (*H. americanus*)	50		–	50	Lang et al., 1978
Cutter claw (*H. americanus*)	50		–	50	Lang et al., 1978
Retractor of scutal plates giant barnacle (*Balanus nubilus*)	0		70	30	Hoyle (unpublished)
Locust extensor tibiae (*Schistocerca gregaria*)	60		25	15	Hoyle, 1978b and unpublished

nuclear-chain elements have well-developed SR, many ITs, many large IT/SR contacts, and well-defined Z and M lines (Ovalle, 1971). But arguing against these aspects are very few, small mitochondria in the former and numerous large ones in the latter. Most likely the requirement is for the fast elements to do a lot of work and to continue to be able to give unchanged responses when repeatedly activated. The paucity of mitochondria in the slow ones may indicate only that they are used infrequently. Once again we see the capability of a cell departing from the average in a single major characteristic.

The stretch receptors of crustaceans and insects show remarkable parallels in certain respects to those of vertebrates. Each has a central element where sensory terminals pick up the strain, with innervated contractile elements on each side. In the well-studied crayfish/lobster tail, there is a pair of such cells on each side of each segment of the abdomen (Alexandrowicz, 1951). One member of each has a single slowly adapting sensory cell, the other has a single rapidly adapting sensory cell. The muscular portions and also the innervation and neuromuscular junctional transmission match those of the appropriate cell. The slow cell is a thin electrically inexcitable cell with long sarcomeres, many mitochondria, and small, facilitating nonfatiguing distributed junctional potentials. The fast cell is a thicker, gradedly excitable cell and has narrow sarcomeres with few mitochondria and large rapidly fatiguing, distributed JPs. Here there is a clear indication of infrequent use in strictly phasic activity – rapid-escape tail flexions for the fast system, sustained posture for the slow. In all respects the two cells of each pair are to be regarded as typical slow and fast muscle cells.

8.7 Use of Differences in Muscle Cells in Taxonomy and in Evolution Studies

Muscle ultrastructure is increasingly being used in taxonomy and in studies on adaptive radiation – the biological consequence of diversity. The process works two ways. Studies on the comparative structures of the cells may suggest natural evolutionary relations, correcting artificial classifications based upon more-superficial external forms that are merely the results of convergence. Where difficulties of classification are encountered using ordinary criteria, a close examination of muscle structure is recommended. One example is the famous "living fossil," *Peripatus,* long presumed to be a "missing link" between annelids and arthropods. Although it is usually classified as an arthropod (e.g., Manton, 1977), a minority of taxonomists has steadfastly refused to accept it as such. An intensive study of its muscle fine structure (Hoyle and Williams, 1980) has now shown quite clearly that it is not an arthropod.

Fine structure of the tail musculature of larval ascidians seems likely to help in taxonomic classification of this phylogenetically important (because chordates) group whose adults have such different form from their larvae (e.g., Cavey and Cloney, 1976). The ultrastructures of muscles of two small nematodes, a parasite form (*Dictyocaulus filaria*) and a free-living saprophyte (*Endorylaimus obtusa*), were compared to see if there is a correlate of life-style (Zmoray and Guttekova, 1970). The parasitic form has multiple invaginations of the membrane, with layers of SR, neither of which was present in the other species. The difference would seem to favor occasional rapid movements in the parasite, which is the opposite of what may have been expected but could lead to a better understanding of its behavior. More-obscure, but presumably also related, differences were seen in the mitochondrial cristae, those of *D. filaria* being spiral and those of *E. obtusa* being parallel.

The receptor molecules for the common transmitter substance of vertebrates are already known to be systematically different. That this might be so was suggested because the venoms produced by many snakes and advanced lizards, cobrotoxin and α-bungarotoxin, block neuromuscular transmission of other vertebrates by combining with receptors at neuromuscular synapses. How is it that they are not poisoned by their own toxin? The answer is that the snakes and phylogenetically advanced lizards have toxin-resistant ACh receptors (Burden et al., 1975). It was because these novel receptors had by chance evolved in higher reptiles that some were enabled to evolve the toxin. The resistant receptors function in an identical manner to that of normal ones. Testing the sensitivity to snake venom of neuromuscular transmission in reptiles provides a novel clue to the reptile's phylogeny.

8.8. Adaptive Radiation and Evolution by Muscle-Cell Change

All the major parameters of variation in structure and physiology must be associated with significant differences in the contractile response of the muscle involved. Each will cause a behavioral variation in response to a fixed motor output. Accordingly, since reflex actions and stereotyped motor patterns can be altered by peripheral changes, the latter offer a means of adaptive radiation and evolution. There are some indications as to how this has occurred in specific instances in arthropods.

The most extensive available study on interspecies diversity is the Ph.D. thesis of Rhodes (1977) on the muscle-cell populations of different muscles of different species and sexes of the fiddler crab *Uca.* Males of different species wave their enormously enlarged signaling claw at a species-typical frequency in order to promote species recognition. In association with this stereotyped behavior, major differences have evolved in muscle-cell characteristics and in the mix of cells. The A-band lengths were found to increase in length progressively with growth of the crab, attaining a maximum of 15 μm in *U. minax* claw levator, making it important to compare only muscles of equivalent age.

A summary of Rhodes's results is presented in Table 8.6. There is a strong correlation between the rate of waving of the claw and the average A-band length of cells in the principal muscle involved, the levator, suggesting that relative SL was a major variable of adaptive radiation.

In the locust, the hind leg is markedly different functionally from the nonspecialized prothoracic and mesothoracic legs, the former being greatly enlarged: this permits jumping. All three pairs of legs have a single extensor tibiae muscle innervated by two excitatory axons, slow (SETi) and fast (FETi), and a single common inhibitor (CI). Each of these neurons has been filled with dye, permitting detailed comparison (Figure 8.16). In the first two thoracic segments, SETi is located posteriorly and laterally, emerging from the fifth nerve (N5), whereas FETi is anterior and medial, and emerges in N3 along with CI (Wilson and Hoyle, 1978). The anatomical forms of the two homologous pairs of neurons are very similar in the two segments. But FETi of the jumping muscle is located posterior and laterally and leaves in N5, whereas SETi emerges in N3 with CI. The anatomy of the metathoracic FETi closely resembles that of the prothoracic and mesothoracic SETi and the metathoracic SETi closely resembles that of the prothoracic and mesothoracic FETi. Therefore, during the evolution of the jump, the metathoracic extensor muscle changed so drastically – as it increased in size – that the neurons switched basic functional roles, the slow becoming fast and vice versa. The principal change as more

Table 8.6 Features of Muscle Cells of Different Species, or Sex, of Fiddler Crabs[a]

Relation of courtship claw-waving rate to sarcomere lengths for muscles waving large claw in fiddler crabs at moderate intensity.

Species	Approximate Mean Wave Duration (s)	Mean A Band Length (μm)			
		Levator	Remotor	Extensor	Stretcher
Uca *speciosa*	0.5	4.7	5.9	5.9	6.2
Uca *pugilator*	1.25	5.5	7.4	6.2	6.1
Uca *pugnax*	2.1	7.5	6.5	7.6	7.2
Uca *minax*	2.5	8.9	6.9	7.6	7.1
Uca *rapax*	3.5	9.2	7.5	7.2	7.1
Uca *rapax* longis	5.6	9.0	6.6	7.8	7.5

Sex differences. Only the male has a large claw.

		Muscle			
		Mean SL, Major Cheliped (μm)		Mean SL, Minor Cheliped (μm)	
Species	Sex	Levator	Remotor	Levator	Remotor
Minax	♂	17.8	14.4	11.9	10.8
	♀	15.0	–	14.3	–
Rapax	♂	17.9	14.6	10.7	8.1
Pugilator	♂	13.6	13.6	10.9	–

[a]All data from Rhodes, 1977.

muscle cells were added was that these were each of the fast type and innervated by the transforming SETi, not by the regressing FETi. The synapses were of the fast, not slow, type and the SETi neuron increased in size to accomodate the increase in number of muscle cells. Since dually innervated muscle cells are located primarily in extreme distal and proximal locations (Figure 8.16) we can deduce that to evolve jumping, new muscle cells of the fast type were added in the middle of the femur joint. The diameters of cells in this muscle range from about 40 to 120 μm, but their sizes are correlated with their positions in the joint, not with the type of cell. The terminals of the enlarged, originally slow, motor neuron became also functionally strong, and their neighbors on cells retaining dual innervation became weaker at the same time. The original fast and slow neurons had, because of purely peripheral changes, switched roles!

The periphery, including muscle-cell numbers, sizes, kinds, and innervation, neuromuscular junctions, and synaptic-transmission properties are all plastic. The central-nervous patterns, pathways, and functioning are conservative by comparison.

The primitive muscle cell in vertebrates was almost certainly similar in properties, and above all in innervation, to an average arthropod muscle cell, with polyneuronal, multiterminal innervation and graded responsiveness. True vertebrate slow-type cells as seen in frog iliofibularis slow bundle or external eye muscles represent the ancestral type. Vertebrate evolution was made possible in part by a combination of peripheral and cen-

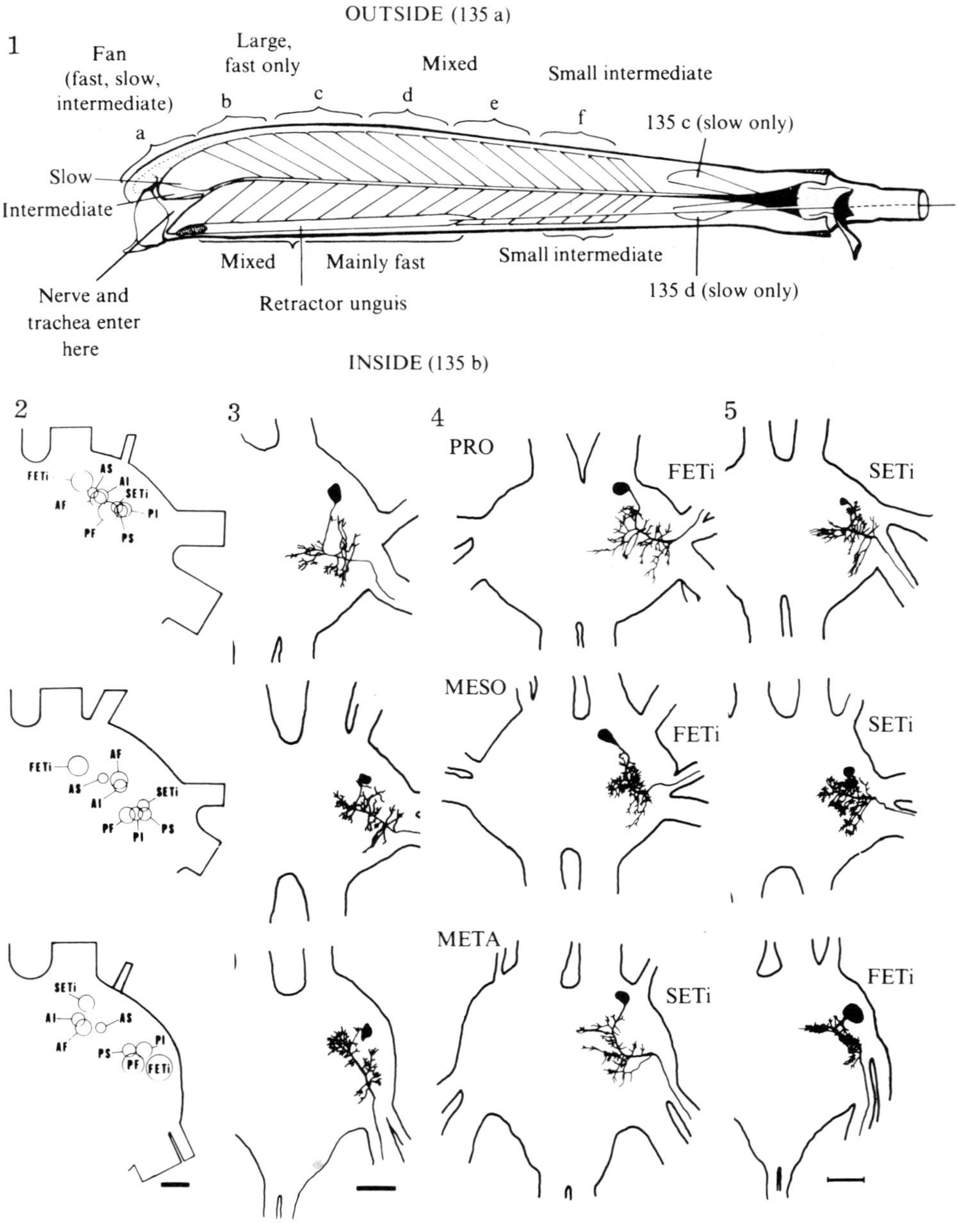

Figure 8.16. Evolution of a muscle, the extensor tibiae (ETi), jumping muscle of grasshopper/locust. The slow and fast axons switched roles as innervation, muscle-cell structure and function, and neuromuscular transmission properties, changed from ones characteristic of slow function to fast function, and vice versa. Numbers 135a–d refer to classical anatomical division. (*1*) View of the muscle from the ventral surface to show distribution of muscle cell types. (From Hoyle, 1978.) (*2*) Serially homologous identified motor-neuron cell bodies located with intracellular dye-filled electrodes: slow extensor (*SETi*), fast extensor (*FETi*), and six flexors. [Reprinted by permission from *Nature* **274**, 377–379 (1978). Copyright 1978 Macmillan Journals Ltd.] (*3*) Posterior fast flexor (*PFFl*) showing similar morphologies. [Reprinted by permission from *Nature* **274**, 377–379 (1978). Copyright 1978 Macmillan Journals Ltd.] (*4*) Metathoracic *SETi* matches prothoracic and mesothoracic. *FETi*s. [Reprinted by permission from *Nature* **274**, 377–379 (1978). Copyright 1978 Macmillan Journals Ltd.] (*5*) Metathoracic *FETi* matches prothoracic and mesothoracic *SETis* Bars = 0.5 mm. [Reprinted by permission from *Nature* **274**, 377–379 (1978). Copyright 1978 Macmillan Journals Ltd.]

tral events. The former included evolution of the sodium channel, with its rapid opening and automatic, rapidly inactivating characteristics, permitting the fast-conducting action potential. This permitted eliminating all but one junction per cell. The latter demanded development of central-nervous system information-processing circuitry that could handle the switch from peripheral integration to central control via numerical and statistical programming.

Differentiation into more than one type of muscle cell per muscle probably had an equally or more-ancient origin than innervation differences. The separation into relatively slow for posture and relatively fast for ballistic action, using similar innervations, would have conferred the distinct behavioral advantages that it does throughout almost the entire animal kingdom. The distinction is probably present (see p. 561, in *Amphioxus*) and it is very strong in primitive elasmobranch fish (Bone, 1966).

8.9. Dynamic Nature of Cell Populations in a Muscle

The balance of cell types in a particular muscle is subject to constant change during adult life, though within genetically prescribed limits. The final adult composition is itself dramatically related in some muscles to developmental experience. An extreme example is seen in the two claws of the lobster *Homarus americanus,* which are unequal in shape and function, one being sharp with fine teeth, the cutter, the other being blunt with rounded teeth, for crushing. The crusher closer muscle contains 100% slow-type cells, but the cutter has 65–75% fast-type cells and only 25–35% slow (Lang et al., 1977a, b). These balances are dependent upon early experience. When the lobster is raised on smooth-bottomed containers, the morphological difference is missing and both claws have at least 50% fast-type cells (Lang et al., 1978).

Transformation of cell types occurs in vertebrates (see p. 399), where it has been shown that both impulse traffic and two-way chemical exchange between the motor-nerve cell and the muscle cells it innervates are involved (e.g., Salmons and Sreter, 1976). Among the muscle parameters already known to be involved are myosin, tropomyosin, troponins, Z band, M band, several enzymes, and the calcium-sequestration capability of SR. By using type-specific antigens, it is becoming clear that each major protein is present in all the types of cells in each of the forms that are characteristic for the particular type, but with the dominant type in by far the highest proportion. Tenotomy, denervation, exercise, impulse-pattern history, nerve-crossing, and altered environment (see p. 400) during development all exert their effects by altering the balance drastically rather than by initiating novel syntheses. Innervation is not the only factor able to influence cell type. When a developing fast-type chicken muscle, the posterior latissimus dorsi, was minced and exchanged with a slow-type, the anterior latissimus dorsi, the two sets of debris acquired features of the muscle from the sites to which they were transferred (Ashurst and Vrbová, 1979).

Fifteen different human muscles were examined by Monster et al., (1978), and the percentage of type-I cells was compared with that of type II in relation to independent studies on the percentage of time each muscle was active during a variety of daily activities. The number of type-I cells was 20% in the least-active muscle, and it increased progressively with increasing percentage of time active to a maximum of 85% in the most active. This distribution suggests a strong functional/developmental role in this particular differentiation. Not all muscles seem equally amenable to modification of cell populations as a result of experience. Forced-exercise programs, promoting almost continuous movement including regular jumping and/or kicking except during feeding, for locusts from hatching to adulthood have so far failed to cause cell-type changes in experiments in my laboratory. The subjects were compared with insects reared in isolation in small cages, with plenty of fresh food always available so that exercise was unnecessary for survival. Although there were small gains in the force capability of the exercised individuals, there was no detectable difference in the proportions of the different kinds of muscle cell.

Of course slow/fast-twitch cells of vertebrates are not strictly comparable to their locust namesakes, and it is to be expected that a full range exists, from muscle-cell populations that are strictly genetically determined and hardly influenced by activity to highly plastic ones whose differentiation is solely determined by use. These questions will be addressed specifically in Chapter 11.

9

Smooth Muscle (SM) and Other Vegetative Muscles

Since, as far as we know, nervous activity is discontinuous, the problem of how plain muscle yields a tonic response to discontinuous stimulation becomes an important one. F. R. Winton†

Muscle cells that lack orderly striations are collectively termed *smooth*. For almost a century, the British Physiological Society insisted on calling them *plain* muscles, but then in 1975 switched to smooth, which had persisted in common usage regardless. Neither term is apt in relation to modern knowledge of ultrastructure, but in the absence of a truly suitable term, I propose to retain the term but restrict its application. Eventually, when the cellular aspects, of nonstriated, or "smooth" muscle cells are better known, it is probable that smooth muscle (SM) will be regarded as comprising a series of distinct, specialized subpopulations with subtle variations as complex as those of cross-striated cells. The principal feature all SM cells have in common is an association with either body walls or soft-bodied organs that are capable of three-dimensional shape changes. SM cells are all small, with a variety of myogenic and cyclic neurogenic contractions, and the time scale of all activities is in the slow category. SMs receive both intrinsic innervation from local neurons of a nerve net and/or ganglia or plexuses, and a direct nerve supply from the central nervous system. Literature on SM has almost all been addressed to some presumed general topic rather than to a particular muscle, but a few specific muscles have been fairly intensively studied. In this book I have endeavored to collect the available data under major topics. Only very recently have techniques been developed that permit the physiological study of component SM cells individually. The procedures for isolation are somewhat drastic and are likely to have affected the properties of isolated cells, but in any event many cooperative factors combine in the natural functioning of SM cells, so the isolated SM cell cannot be operating exactly as it must in the intact animal. The current data suggest that many discrete vertebrate SMs such as the gizzard of chickens and in invertebrates the hindgut of the cockroach are indeed composed of similar cells. Unlike the situation with striated muscle, structural differentiation of the cell population within a muscle does not seem to have played a major role in adaptive radiation, though there are very large differences between intrinsic electrical properties of particular SMs, and therefore different classes of component cells are recognized. There are classes that have a stable, relatively high membrane potential, and others with an oscillatory, basically low one. Some cells are electrically inexcitable, some give all-or-none spikes, and others give graded ones, but so far no classification along these lines has been attempted.

For many organs operated by SM, there is more than one anatomical component, such as outer circular, inner circular, longitudinal, and outer and inner oblique. Seldom can the component layers be studied separately. Differences in average properties of component

†F. R. Winton, *J. Physiol.* 88, 493 (1937)

muscle cells, as well as local differences in nerve supply, can exist and cause markedly different behaviors of regions of a single organ operated by SM. The anterior part of the dog stomach, the fundus, has very different physiological responses than the large antrum. The fundus serves as a reservoir during ingestion, when it must greatly enlarge, but it is important also in emptying the stomach. A very small amount of depolarization causes the fundus to contract, but not the antrum. The latter is spontaneously active, but the former is not. The fundus acts more like a readily stretched voluntary muscle than a typical vegetative one.

The ensuing survey is only partly comparative, mostly separating vertebrate from invertebrate vegetative muscle. Most invertebrate muscles that have traditionally been treated as SMs will be excluded for reasons given on p. 88; the fine structure of the anterior byssal retractor muscle of *Mytilus* (ABRM), for example, allies it with striated rather than smooth muscle. Even the body-wall musculature of *Peripatus* (see p. 100) should be allied with cross-striated muscle rather than vertebrate SM because it has a very clearly marked repeating Z-I-A-I-Z progession of components longitudinally, even though these are not aligned laterally. I shall define an SM cell as one in which Z-I-A-I-Z sequences cannot be discerned ultrastructurally. Such cells are the principal components of involuntary organs of vertebrates. In the present state of our knowledge, this definition restricts SM to vertebrates only, because none are known – except by default resulting from poor electron microscopy – in invertebrates. Their functional equivalents in invertebrates are often cross-striated or obliquely striated, but should be considered with the vertebrate gut rather than with skeletal muscle. Such cells will be termed *vegetative muscle cells.*

9.1. Fine-Structural Features of SM Cells

Some general features of SM cells were described on p. 106. Vetebrate SM is characterized by a lack or extreme paucity of A-filament structures, probably because there is insufficient myosin. There was a long dispute in the late 1960s, about the nature of myosin in living SM. Using certain fixatives or treatments, especially on taenia coli, ribbon-shaped 8 × 35 (range 20–110) nm by up to 3-μm-long filaments of myosin are obtained (Lowy and Small, 1970). The thick filaments and/or ribbons are very sparse, but are fairly evenly, though not regularly, distributed through the myoplasm. A variety of cylindrical profiles, from 10 to 20 nm, has been found; the smaller ones, now often termed intermediate filaments, are possibly structural rather than involved in active contraction. The sparse SM thick filaments are generally similar to thick filaments of cross-striated muscle (Figure 9.1 – Rice et al., 1970; Campbell et al., 1971) and some are tapered. In cells from the lower small intestine of the toad, after treatment with trypsin, 25-30 nm in diameter filaments of 0.5-1.0 μm length, with 14.3-nm axial periodicity occur (Rosenbluth, 1971). A mixture of ribbons and circular filaments was seen in guinea pig taenia coli without special treatment (Schoenberg and Haselgrove, 1974). There are some changes in internal distribution of myofilaments associated with contraction, including the appearance of thick filaments in some SM cells (Figure 9.2), but the exact nature of the changes is disputed.

Highly purified SM myosin can form thick filaments by self-assembly (Figure 3.38.*6*) that resemble those from striated muscle in having lateral projections with a 14-nm repeat and lengths up to 1.5 μm. But they are of variable thickness and never with a central shaft or bare zone (Sobieszek, 1973). Two principal types of assembly occur, a short bipolar

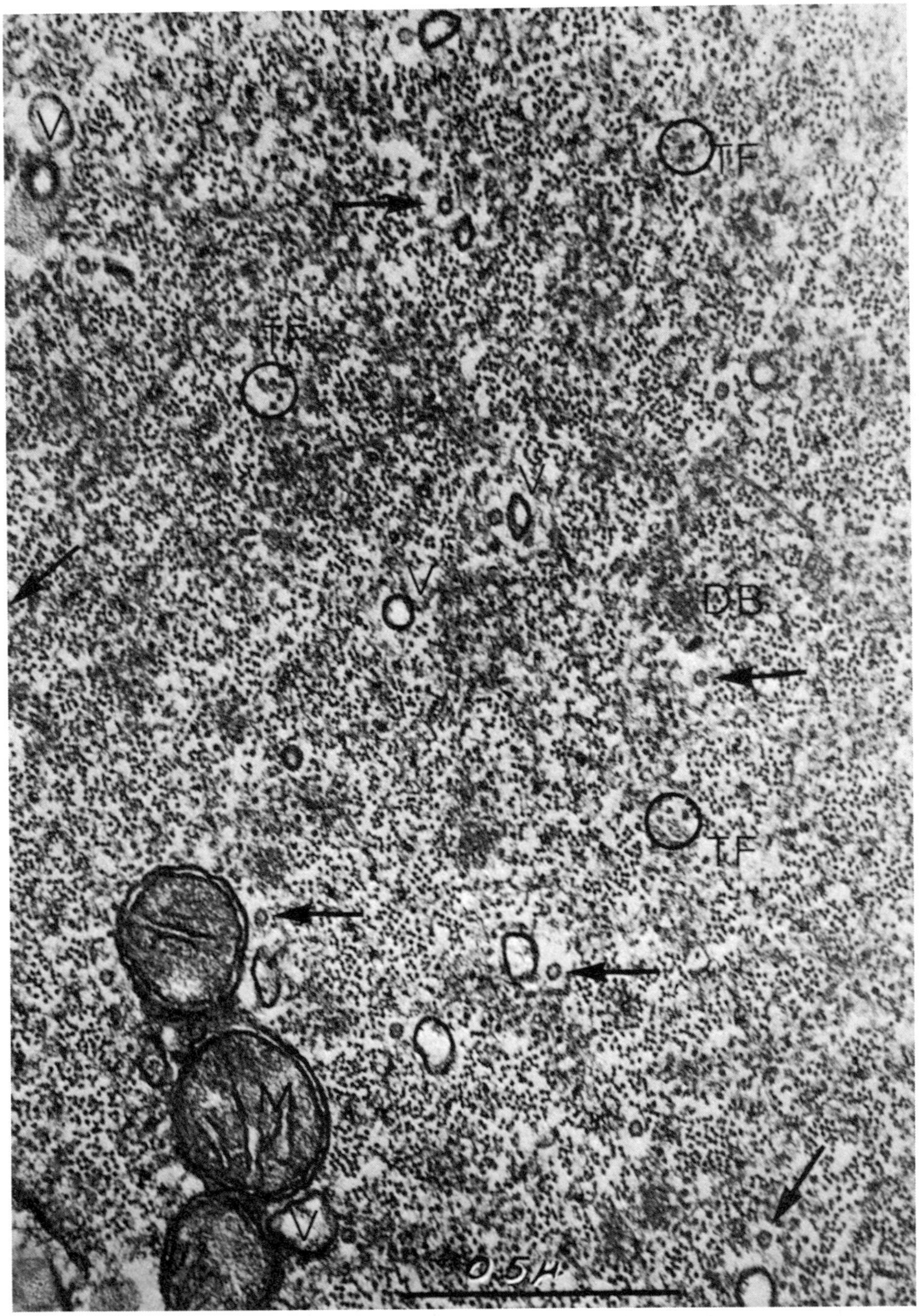

Figure 9.1. Myofilament components of guinea pig taenia coli seen in TEM of transverse section of relaxed (with adrenaline) muscle. There are very large number of thin filaments but few thick (12 nm) ones present in small separate clusters (TF-*circles*). The SR is represented by separated vesicles (*V*), close to each of which is a microtubule (*arrows*). Thin filaments are attached to dense bodies (*DB*). (From Rice et al., 1970.)

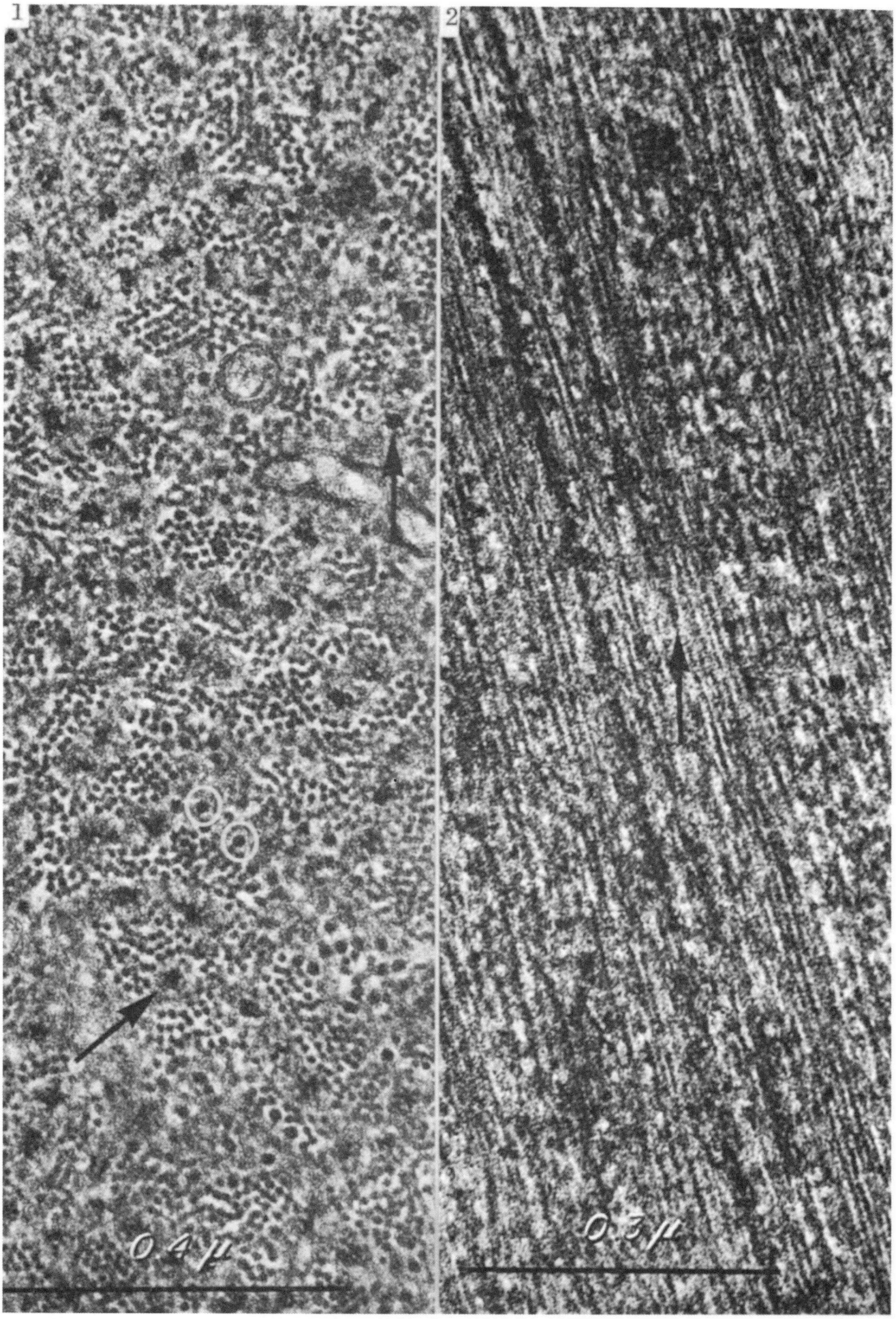

Figure 9.2. Transverse (*1*) and longitudinal (*2*) EM sections of contracted guinea pig taenia coli. Note regular close arrays of thin-filament clusters, occasional 20-nm-thick filaments (*arrows*) and 12-nm intermediate filaments (*circles*). (From Rice et al., 1970.)

form and a long side-polar form in which there is a single polarity of bridges on one side and the opposite on the other (Craig and Megerman, 1979). The forms of these SM thick filaments are highly dependent on a number of factors including the concentrations of Mg^{2+}, Ca^{2+}, ATP, and specific light-chain kinase. In chicken gizzard, where thick filaments are readily observed, the ease of their formation may be due to a rich supply of light-chain kinase (Ebashi, 1979).

Conspicuous in many SM cells are dense bodies (DB) very similar in appearance to those of primitive invertebrate cross-striated muscle (see p. 63); 5-nm thin actin filaments are attached to, or pass through, SM DBs in profusion. In most vertebrate SMs there are two distinctly different sets of DBs, one comparable to the above and called *primary*, to which actin filaments are attached, and one called *secondary*, to which the thicker 10-nm filaments are attached (see p. 93). The secondary type, though it does occur within the cytoplasm, is especially associated with the surface membrane. Primary DBs are randomly scattered within the cytoplasm, and only a few are attached to the surface. Thin filaments of some kind run through them, very tightly packed together (Figure 9.3.*2*).

The number of DBs in a cell is a source of systematic variation, there being relatively few in SM cells of some muscles, such as uterus and bladder, but many in others, such as aorta and taenia coli. A small number of DBs must indicate the equivalent of a relatively long sarcomere length and therefore be associated with extreme extensibility and also slowness, as is certainly true for uterus and bladder. The organs in which DB numbers are high are stiffer, tougher, and much less extensible, though not faster, than those with few.

There are thin filaments other than actin in SMs, in a range from 2 to 10 nm (Figure 9.3). The larger of these filaments are considered to be the major components of a cytoskeleton, partly randomly oriented (Campbell et al., 1971) but with a helical aspect (Figure 3.39 – Cooke, 1976). In support of this or other alternatives to filament sliding. Ohasi and Nonomura (1979), when examining by electron microscopy guinea pig taenia coli that had been in a state of contracture due to high Ca^{2+} for 30 min, found no sign of thick filaments of any kind. The actin filaments had shifted their positions and were now concentrated near the outside instead of being uniformly distributed. The 10-nm filaments of SM are formed by a 55,000-molecular-weight protein. These filaments are conspicuous early in the development of chicken gizzard, guinea pig and zebra finch ureter, vas deferens of rat, and taenia coli of guinea pig (Uehara et al., 1971). A strongly helical aspect for them was first clearly noted by Cooke (1976), supported by Small (1977), and firmly established by Bagby and Abercrombie (1979). The latter authors, however, suggest that the helical aspect need not be due to helically arranged attachment sites, but instead could be due to clusters of myofilaments having curved paths. According to the helical-attachment notion, the clusters themselves are thought of as being straight. A possible alternative is that both attachments and pathways could be helical. Certainly in phase-contrast microscopy of isolated SM cells, a spiral arrangement of some sort is strongly apparent (Figure 3.39).

9.2. SM Contractile Mechanism

On the basis of the sparse, controversial, ultrastructural observations on SM cells, the idea arose that SM works in a like manner to that of cross-striated muscle, but that myosin is normally present in the form of either monomers or dimers at rest. Following excitation, it aggregates to form filaments and ribbons around which thin filaments of actin as-

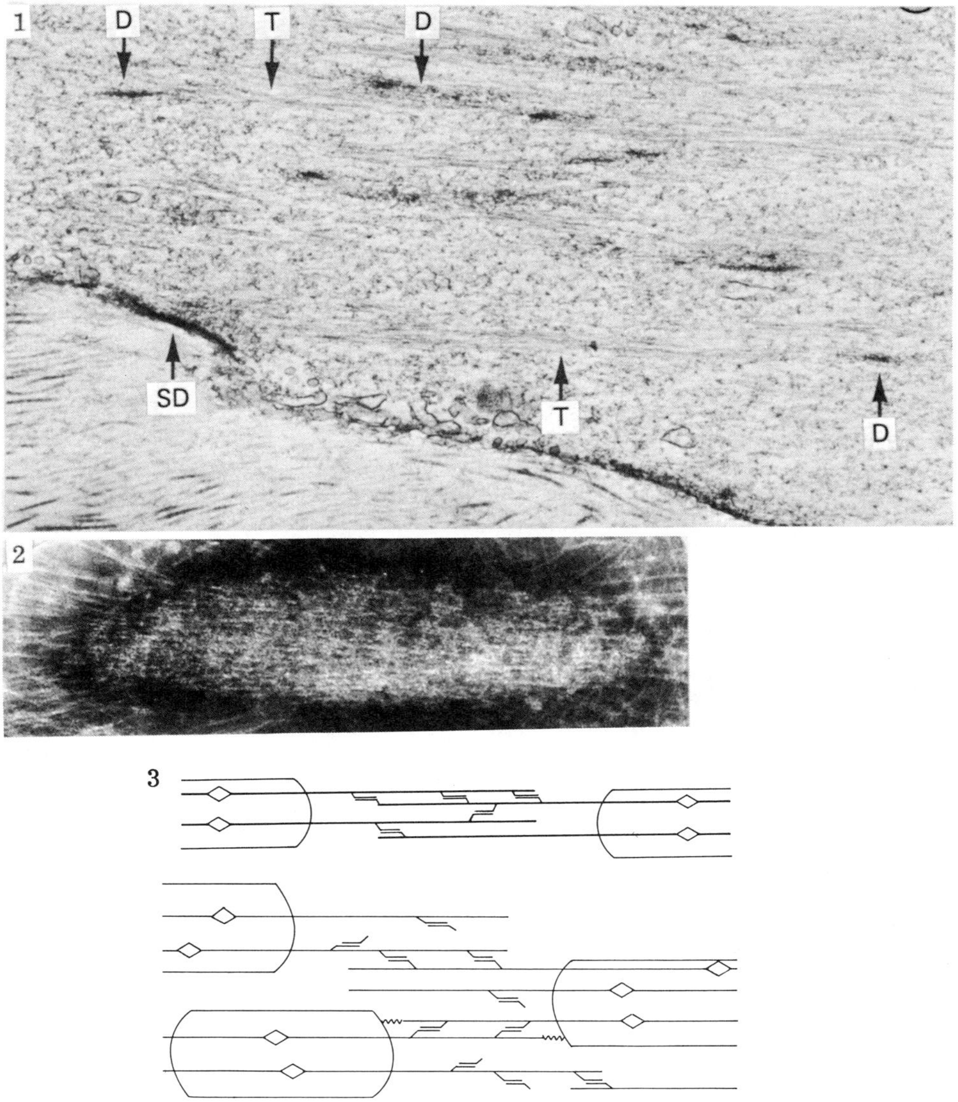

Figure 9.3. A TEM of an LS of guinea pig taenia coli cell showing general organization. (*1*) Thin filaments are attached to dense bodies (DB) in the cytoplasm and to occasional dense areas of the surface membrane (SD). Some ultrathin filaments (T) can be seen. (From P. Cooke, 1976.) (*2*) Negatively stained DB of turkey gizzard is composed largely of longitudinal fine filaments from which thin filaments emerge. (From Panner and Honig, 1967.) (*3*) Model of SM contraction mechanism based upon relative sliding of sets of thin filaments brought about by scattered myosin dimers. The two molecules of myosin constituting each dimer are assumed to fit together in antiparallel order (From Panner and Honig, 1967.)

semble. They then slide past each other by a X-bridge mechanism. However, whether thick filaments are present normally or only after excitation, there are at best 70 thin filaments to one thick one in some taenia coli. The most favorable ratio yet claimed has been for vascular (arterial) SM, where it is 15 : 1. A majority of thin filaments are clustered together in close arrays, and even when near a thick filament, they are not obviously arrayed in orbits around it.

An alternative view regards the thick filaments as being rare or artifacts, but still invokes a X-bridge mechanism. This view places the myosins together in dimers, utilizing the polarity of the molecule to explain shortening (Figure 9.*3*.*3*). The contractile units in this hypothesis are the clusters of 5-nm filaments, attached to DBs, that happen to be close enough together to interact, and of opposite polarity as determined by their attachments. Shortening of ultrathin filaments is another possibility, coupling between actin and myosin molecules being required only in a random manner to provide energy for the transformation by ATP hydrolysis.

9.3. Mitochondria

The size range of mitochondria in SMs [0.1–0.5 × 0.1–1.5 μm] is much less than that seen in striated cells, and they are simple ovoids, never with multibranching. The numbers range from very few to dense aggregates occupying the entire cell volume, but systematic association between numbers and properties has not been attempted. The numbers and sizes vary greatly in relation to many factors and are influenced on a short time scale by activity. There are few mitochondria in uterine SM cells until midpregnancy, when their number increases drastically (Yamamoto, 1961).

There seems to be a tendency for mitochondria to move within SM cells, perhaps in a systematic manner. There are two states, one in which mitochondria are scattered throughout the cell, the other in which they are all packed together in the core (reviewed by Huddart, 1975; Huddart and Hunt, 1976). What this movement may be associated with functionally is not known.

9.4. Cell Assemblies of SMs

Individual cells of true SM are all in the shape of an elongated cylinder with progressively tapering ends. The largest diameter at the center is about 8 μm, and the overall length about 0.5 mm. Any given SM is likely to have a mixed population of cell sizes, most being in the range 4 μm × 0.2 mm. Systematic size differences between cells of particular SMs are not known to be a significant factor functionally. The small lengths of individual SM cells means that except in very small blood vessels, they must join each other longitudinally. Two basically different forms of contact exist, one in which the cells are locked together by a *peg and socket* arrangement, projections from one entering invaginated pockets of the other, and the other, by far the most common, comprising a variety of close juxtapositions only (Figure 9.4). Pegs and sockets are seen especially in vas deferens (Lane and Rhodin, 1964). The other type has been generally referred to as a *nexus*, and two principal subcategories are recognized: side by side, in which the cells just come into very close contact without sending out a special arm, and plaques, in which one cell sends out a lateral branch that abuts the neighbor. However, although some SM cells have

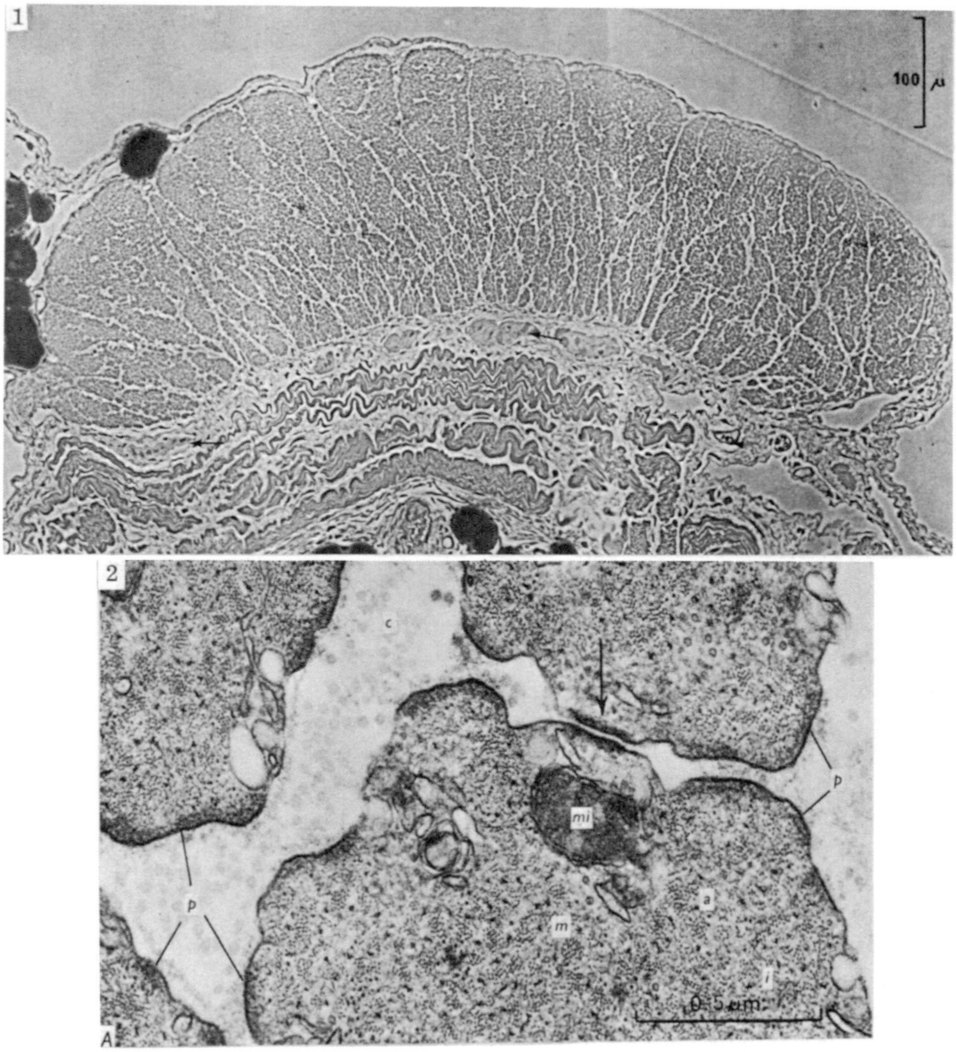

Figure 9.4. Organization within SM. (*1*) Montage TS of taenia coli of guinea pig. Ganglia of the myenteric plexus are visible (*arrows*) (*2*) TEM of TS of (*1*) showing details of contact between two cells (arrow) and also caveolae at surface membrane, c = collagen, mi = mitochondrion, p = PM. (From Gabella, 1976.)

smooth surfaces at their mean length, others show a variety of infoldings, some of which are extremely irregular (Figure 9.4.*2*). The infoldings are intensified with contraction, reduced with stretch, and presumably are strongly associated with the extreme extensibility shown by a majority of SM cells. Surface caveolae, of much smaller dimensions than infoldings, are also common in SM cells.

At the points of membrane juxtaposition, extremely fine filaments link the cells together. At the grosser level, the cells are held together principally by collagen bundles that weave around them in loose coiling bundles. The amounts and qualities of connective tissue associated with an SM-invested organ vary enormously, from small, soft, pliant and extensible in bladders, to abundant, hard-rubbery, and inextensible as in chicken gizzard.

9.5. Electrical Properties

9.5.1. Resting Potential

All SMs have low "resting" membrane potentials, but the term is not very meaningful for most because it is actively kept not only low but often also fluctuating spontaneously in response to internal chemical rhythmicity of one kind or another. Furthermore, it is impossible to define rest length for an SM, and in many the membrane potential is highly sensitive to stretch. For example, at its estimated average *in situ* length, guinea pig taenia coli mean E_m is 60 mV; when the muscle is allowed to shorten maximally, E_m rises to 78 mV, but if it is stretched fully, E_m falls to only 43 mV (Bülbring, 1955). The corresponding limits in the salamander stomach are 38 and 77 mV (Greven, 1953). Basic E_m for the guinea pig bladder is -40 mV, for the vas deferens, 63 mV, and for aorta, trachea, and pulmonary artery, 60 mV (Redfern and Thesleff, 1971). Basic tonus (see p. 255) is high in all SM cells and must be regarded as a major functional property.

The SM membranes are almost freely permeable to potassium and chloride, but the internal chloride concentration is high, about 70 meq/kg, so that E_{Cl} is only some 20 mV. Together with the reverse potential contribution of sodium, this lowers the potential well away from E_K, which is 90 mV, to the basic E_m of 57–60 mV. Also contributing to these low values is a weak Na/K-pump action that is more efficient at removing sodium than in pulling potassium in. Therefore, the pump action contributes to the membrane potential, and agents that affect the pump directly affect E_m.

9.5.2. Electrogenesis

For the moment we shall leave aside the matter of spontaneous membrane-potential fluctuation and consider SM in a similar manner to striated muscle regarding response to applied current. Good intracellular recordings from SM were first obtained by Edith Bülbring (1954) using guinea pig taenia coli and excellent ones in the more favorable sphincter pupillae of the rabbit eye (Bülbring and Hooton, 1954). But with cells only 2-7 μm in diameter, it is not possible to do two-electrode or three-electrode voltage clamps. An indirect method of effective intracellular recording was invented by the Swiss researcher R. Stämpfli, who had worked out the mechanisms of saltatory conduction in myelinated axons with A. F. Huxley. The aim of his method is to create an artificial "node" by coating the cell(s) with a high-electrical-resistance isotonic sucrose solution (Stämpfli, 1954). On one side the tissue is bathed in normal saline and on the other, in depolarizing isotonic KCl. The method is effectively intracellular, via the depolarized end with respect to the one in saline, and yields membrane potentials similar to the better intracellular ones. The method was modified for voltage clamp by Julian et al. (1962) and improved by Moore and Narahashi (1967). Originally used for nonmyelinated axon bundles, it proved to give good results with strips of many kinds of SM.

Smooth-muscle cells have turned out to be as diverse in excitability as arthropod skeletal cells, ranging also from electrically inexcitable to all-or-none spiking, with a range of gradedness between. However, there seem to be relatively homogeneous populations of cells within a particular location. For example, in guinea pig, all cells in the aorta, trachea, and pulmonary artery are inexcitable. The stomach has differently responding cells in different regions, the antrum giving all-or-none spikes and the fundus no responses. Full spikes occur in the taenia coli, bladder, and vas deferens, and graded ones in jejunum, ileum, rectum, and portal vein (Redfern and Thesleff, 1971, and others). Other sources of graded responses include bird gizzard (Bennett, 1970), rabbit carotid artery (Meketa, 1971), cat duodenum (Kobayashi et al., 1967) and small intestine (Burnstock and Prosser, 1960a), rabbit colon (Gillespie, 1962), and a number of invertebrate SMs. So far no obvious relation of the kind discerned in arthropod skeletal muscles has been found between function and membrane electrogenesis in SMs.

The SM cells are short compared with their length constant, which is from about 0.5 to 2.1 mm, so an action potential is not needed for propagation along the individual cell. It may be needed either to continue depolarization past the E_c threshold or to activate the contraction, slow though it is intrinsically, as rapidly as possible. However, another important reason is in relation to an SM phenomenon that is totally absent in all skeletal muscles, that is, to propagate excitation from cell to cell. This serves at least in part to replace motor axons, which are often either not involved at all in SM, innervate only a small fraction of the cells, or serve only to provide modulatory influences. Examples of SM electrical activity are illustrated in Figure 9.5. Each SM cell has its own peculiarities. Neural control is drastically different from that of skeletal muscle of vertebrates but closely resembles that of invertebrates. There are small, facilitating EJPs, and peripheral inhibition is common.

The contacts between SM cells are often of low electrical resistance, permitting current to flow from a depolarized cell to its neighbors. There will be decremental spread of potential in systems with either electrically inexcitable or gradedly excitable cells. For cells with all-or-none electrogenesis, propagation of an all-or-none wave over the whole is possible in principle, but there will also be a tendency for dying out at sites distant from the initiating one if the safety factor at junctions is low. This depends on specific overall membrane resistance and capacitance and resistance at individual contacts, together with the total area of contacts and threshold for spike initiation. Specific resistances are in the range 300–1100 Ω/cm^2 and capacitances effectively very high, 4–20 $\mu F/cm^2$, probably because of infoldings and caveolae. Spike thresholds are at about -40 mV, but for graded responses, they start at about -50 mV and are proportional to voltage up to a maximum of about -25 mV.

For weakly gradedly responding and inexcitable cells, current–voltage relations are linear (e.g., Kobayashi et al., 1967). The more excitable cells have a transient inward current for depolarizations past threshold, plus a delayed outward current. Steady-state current is linearly related to voltage: SM cells do not show rectifications.

Vertebrate membrane physiology was long dominated by the notion, paradoxically derived from an invertebrate, the squid (giant axon), that sodium currents are responsible for active membrane responses. Changes in external sodium did affect SM properties, including spikes, but it was by no means clear how. The SM experiments that indicated that some other mechanism was involved were shelved as "mysterious" until Hagiwara (Hagiwara and Naka, 1964) clearly demonstrated that graded electrogenesis in giant barnacle muscle cells is entirely due to activation of calcium conductance. Tests were thereafter made on a variety of SMs, and it was soon apparent that Ca^{2+} is the principal carrier of current in all. However, the situation was not nearly as simple as in the barnacle cells, calcium having a variety of secondary effects on SM that in turn affected excitability, membrane potential, and other characteristics. Sodium, which drastically alters axonal spikes but doesn't affect the resting potential, does affect the E_m of SMs. Replacing external sodium, even with sucrose, leads to large increases in E_m (Brading et al., 1969). The action potential is markedly altered by $[Na^+]_o$, but TTX doesn't change it.

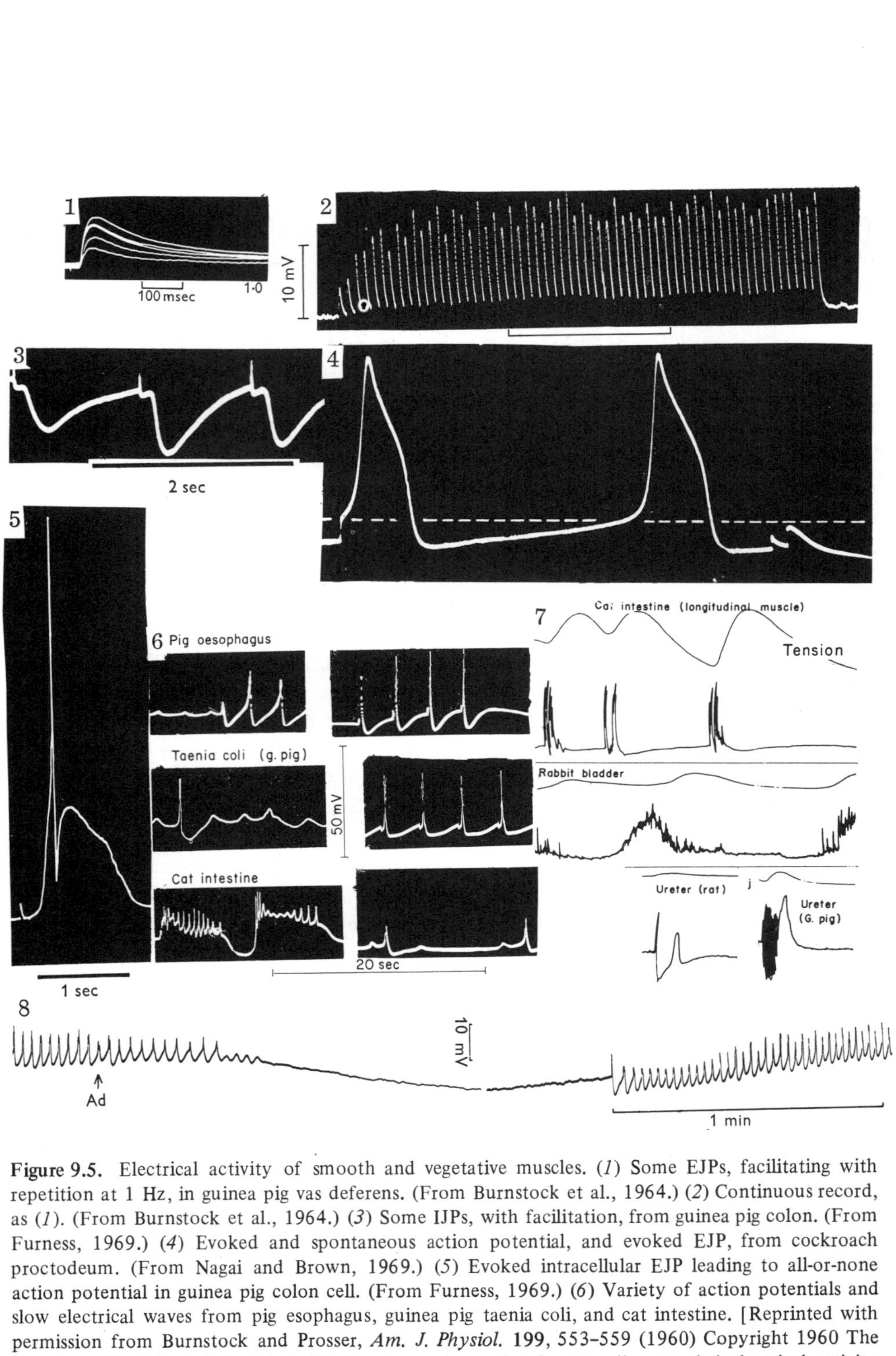

Figure 9.5. Electrical activity of smooth and vegetative muscles. (*1*) Some EJPs, facilitating with repetition at 1 Hz, in guinea pig vas deferens. (From Burnstock et al., 1964.) (*2*) Continuous record, as (*1*). (From Burnstock et al., 1964.) (*3*) Some IJPs, with facilitation, from guinea pig colon. (From Furness, 1969.) (*4*) Evoked and spontaneous action potential, and evoked EJP, from cockroach proctodeum. (From Nagai and Brown, 1969.) (*5*) Evoked intracellular EJP leading to all-or-none action potential in guinea pig colon cell. (From Furness, 1969.) (*6*) Variety of action potentials and slow electrical waves from pig esophagus, guinea pig taenia coli, and cat intestine. [Reprinted with permission from Burnstock and Prosser, *Am. J. Physiol.* **199**, 553–559 (1960) Copyright 1960 The American Physiological Society.] (*7*) Tension and associated externally recorded electrical activity from cat longitudinal intestinal muscle, rabbit bladder, and ureter of rat and guinea pig. [Reprinted with permission from Burnstock and Prosser, 1960.) (*8*) Effect of adrenaline (*Ad*) on electrical activity of guinea pig taenia coli, with recovery. (From Burnstock, 1958.)

External calcium and sodium have antagonistic effects on spike amplitude, but it is completely blocked by manganese, which specifically affects calcium channels. When the Na_0-Ca_0 ratio was held constant and calcium concentration changed up to threefold, spikes of cat intestinal SM cells were unchanged (Connor and Prosser, 1974). Lanthanum, and cobalt, potent calcium-channel blockers, reduce or abolish inward current depending on the particular SM; they are only weakly effective on uterus (Anderson et al., 1971).

Some of the effects of changing either $[Na]_0$ or $[Ca]_0$ are on resting potential, others are on threshold and time dependencies. Allowing for these, it is apparent that the principal carrier of inward current in both graded and all-or-none spikes of SMs is the calcium ion. At one extreme this alone is involved, but some sodium conductance is also not only present in some, but required for full-size spikes in a few, such as those of uteri. Internal sodium concentration is much higher than in axons or muscle cells, averaging about 75 mE/kg (Burnstock et al., 1963). Quite likely the Na^+ and Ca^{2+} share the same single channel.

The size of the spike, from small graded to all-or-none, must therefore depend on the time course of repolarizing electrogenesis, which in all known cells is due to a delayed rise in gK^+. The dependence on gK^+ can be tested using TEA, which blocks the K^+ channels, during voltage clamping for all-or-none cells and under ordinary conditions for graded ones. Tests without clamping are valid for gradedly-responding cells because graded responses become converted, first into all-or-none, then into prolonged spikes (Inomata and Kao, 1976). However, strontium, which slows the rate of repolarization when substituted for calcium, has a similar action in Na^+-free TEA medium (Inomata and Kao, 1979). The effect of strontium is on the rate at which the inward current inactivates. Thus calcium that has entered the cell in turn affects a significant part of the potassium conductance, the calcium-dependent gK (see p. 185). The balance of components in these multiple feedback events determines the exact time course of the response to depolarization.

The time course of repolarization following a spike is itself a major variable, ranging from recoveries that are faster than the rising phase to ones having a long plateau somewhat resembling that of heart-cell action potentials. The former recoveries are as brief as 8 ms total, the latter last as long as 1-5 s.

The ionic basis of the plataeu spike potentials obtained from some SM cells, for example in the stomach of the guinea pig, has not yet been resolved, but appears also to involve activities of ion pumps (Tomita and Sakamoto, 1977). There are three components to the spike: a basic wave, the spike, and the plataeu. The first phase of depolarization is equivalent to the basic wave and due to changes in the sodium/potassium pump and indirectly to calcium ions. Hence the inward movement of calcium may be directly linked to the Na/K system in the form of a three-ion pump that reverses, ejecting calcium and potassium as sodium enters. The principal depolarization of the large spike is due to increase in gNa. The abrupt initial phase of recovery is due to inactivation, but leaves the cell depolarized in the plateau stage. In such a model, recovery represents not potassium activation, but stimulation of the tripartite ion pump to work in the normal direction, pumping sodium out and potassium in.

9.6. Spontaneous Contractile Activity – Slow Waves

Peristalsis of gut musculature is universal and in a majority of guts occurs in the absence of neural input. Such spontaneous activity is common in many kinds of SM, especially those of the stomach and intestine, and those of the blood vessels, bladder, ureter, and vas deferens, but of course it would be disastrous in a uterus at midterm, although very important later. There is a very wide range of intrinsic activity in different SMs that is subject only to modulation by neural inputs. Neuromuscular blocking agents may alter the characteristics of a rhythm, presumably by interfering with ongoing local reflex arcs and modulations thereof, but they do not completely arrest periodic contractions. The residual rhythms are therefore *myogenic*.

All possible parameters of the slow waves of depolarization that arise spontaneously in SM cells – frequency, amplitude, rise and decay profiles, and durations of waves – are the subject of intrinsic variation associated with cell source, and within cells from a particular source by modulation. Some waves start from a flat baseline, others recur in increasing continuum, with shapes ranging from sine waves to linearly rising ramps. The waves are often further complicated by spikes, graded or all-or-none, that are initiated by the depolarizing waves. In some cells these spikes simply summate electrically with the waves, but in others the ionic events producing the waves are themselves affected by the entering sodium/calcium and by activation of potassium currents.

Electrical waves are not confined to rhythmically contracting SMs, but occur also in SMs that have a steady tension. Either these waves are failing to pass E-C threshold, which is unlikely, or they are asynchronous in different regions so that their effects are averaged.

Mean slow-wave recurrence rate is in a range from 1 per few hours for beginning-to-be-active uteri, to 1/s in some taenia coli. Amplitudes range from zero to about 20 mV in the absence of accompanying electrogenesis. A possible classification into three types for contractions has been suggested (Golenhofen and Loh, 1970):

Type 1 comprises waves of 3–20/min, seen in taenia coli.

Type 2 includes medium frequencies of about 1/min, seen in blood vessels.

Type 3 encompasses slow periods of 1 per several minutes, or hours, seen in uteri and ureters.

A fourth category is added for rhythmically occurring spikes not individually associated with a contraction wave.

The action potentials initiated by the slow waves may be single events, one per wave, or occur in bursts of up to 40 spikes per wave, depending on the source. Tension is related to the mean amplitude of the wave and is accentuated by the action-potential bursts.

For detailed accounts of different types the reader is referred to reviews by Huddart (1975) and Prosser (1974).

9.6.1. Ionic Basis of Slow-Wave Activity

During the electrical waves, there is very little change in resistance, unlike the situation with cardiac and nerve-cell pacemaking. The wave in an SM *is* the pacemaker change; it does not lead to a further slow change, even in cells that appear to be oscillating sinusoidally. The situation seen in the slower SM pacemakers, in which the membrane potential is steady between depolarizations, is the norm; it is just that the frequency is so high the waves run together. These cells are distinguished by having a depolarization wave that is not affected by the occurrence of spikes. An exception is rat mesenteric artery (Steedman, 1966), in which the potential slowly falls until spike threshold is reached. The repolarizing phase returns the potential abruptly to baseline or above as in pacemaker neurons. But for the majority of SMs, the membrane is not a relaxation oscillator: it is driven by a primary internal event. Under voltage clamp, spontaneous inward currents occur with the same periodicity as the natural ones. In cardiac and neuronal pacemakers, there are no fluctuations in membrane current under voltage clamp because it is the swings in voltage that cause each following event.

The most likely possible source underlying the current shifts in SMs seems to be the sodium/potassium pump (Connor et al., 1974). If this is so, the currents should be affected by lowering $[K^+]_o$ and by agents that specifically affect the pump, notably ouabain. Both do indeed reduce and eventually abolish the waves, accompanied by a fall in resting potential. But this finding only pushes the matter one stage further, to the question, what causes the pump to fluctuate? The waves are also sensitive to $[Ca^{2+}]_o$, which is slowly leaking into the cell. Connor (1979) has proposed the hypothesis that the rising calcium concentration inactivates the sodium/potassium pump. At the moment of inactivation, the membrane potential falls, producing the wave. But eventually $[Ca^{2+}]_i$ rises to a level at which it activates mechanisms resulting in its own sequestration internally and/or its active extrusion. When the calcium level has fallen sufficiently, the sodium/potassium pump is released from inhibition, restoring the membrane potential. This is a nice idea, and well suited to producing the intermediate and slow ranges of wave frequency, but it is hard to imagine its being responsible also for the fastest ones.

9.6.2. Rebound

Another type of action is inhibition followed by rebound excitation as the tissue recovers from it. This is not a dual action but has the equivalent effect. It is seen, for example, in longitudinal muscle of guinea pig caecum (Campbell, 1966), including biphasic responses such as initial excitation followed by depression, and in stomach of toad and dog (Sato, 1960). The cause of action(s) depend, among all the other things, on the amount of basic tonus occurring at the time. When strong, it is inhibitable but not excitable, and when weak, excitable but not inhibitable. A basic-tonus inhibitor will appear to be inactive in the latter case, or even seem to be an excitor because its direct action, though negligible, is followed by rebound excitation. Another source of similar paradox is in relation to ongoing activity, especially rhythmic activity. When spontaneous electrical activity is absent, an agent may initiate it – in addition to direct action – but when spontaneous activity is ongoing and strong, the direct action may dominate and result in inhibition of the rhythm.

9.7. Innervation

The innervation of most vertebrate SMs is highly complex, and so extremely difficult to unravel that it is doubtful if we yet have an accurate extensive knowledge of any SM. Anatomical studies alone were not successful and have often been misleading. The systems of innervation are not at all like those of any skeletal muscle. Even such eminent pioneers as Cajal mistook fibroblasts for neurons. Only by a combined use of several methods, gross and fine anatomy, ultrastructure, histochemistry, intracellular electrophysiology, and pharmacology, on the same tissue, will it be possible to sort out the pieces of the puzzle and assemble them into a comprehensive picture of that particular muscle.

9.7.1. Evolutionary History of SM Innervation

As discussed earlier (see p. 37), the evolutionary starting point for all muscles was probably a soft-bodied tubular animal not unlike a present-day *Hydra,* with a pair of antagonistic sheets of fine, small nonstriated contractile cells or parts thereof. These were organized into locally controlled clusters with intrinsic activity and superimposed reflex action mediated by local neurons. They could also function cooperatively, because of connections between the local neurons and through-conducting "giant" neurons innervating them collectively. The local links forged the neurons into a nerve net. The through-conducting pathways in neurons with larger, longer axons permitted simultaneous action of the whole on demanding occasions. Skeletal muscles evolved out of such diffuse assemblies as this only following the evolution of hard skeletal elements to which muscle cells could be attached. In turn, concentration of nerve cells became set aside and "centralized" to control the evolving "skeletal" muscles.

But the gut retained the primeval organization, as did many other organs that evolved out of the tube by being pinched off while acquiring other functions such as storage, secretion, and transport of body fluid. The major element is the gut, but it is appropriate to call all these various organs vegetative. Each has its own local nervous system, in the form of a nerve net, a plexus, or ganglia, or combinations thereof. The local nervous system that controls them can then be called the *vegetative nervous system.* An example of such a ganglion is shown in Figure 9.6. The useful, independent, "involuntary," myogenically driven activities were retained and modified for particular purposes, as was local reflex action. The two muscular systems needed, the *skeletal,* controlled entirely by a central nervous system, and the *vegetative,* entirely or largely autonomous could, and probably originally did, function independently of one another.

The evolving central nervous systems of higher invertebrates and vertebrates eventually formed links with the vegetative system, probably via outpushings from the central nervous systems. The primary control outputs of vertebrates appear to be those associated with a chain of ganglia situated close to the spinal cord, the *sympathetic* nervous system. This is distinguished by having one or two cholinergic synaptic relay stations located in ganglia principally present in an interconnected chain. The chain starts anteriorly with the superior cervical ganglion. These are secondary relays also, the coeliac and superior and inferior mesenteries, for the more posterior outlets. The most important distinguishing feature, however, is that the secondary neurons and tertiary neurons release a different substance, *noradrenaline* (NAd) also called *norepinephrine,* than do the primary ones emanating from the central nervous system. NAd is itself released by the terminal neuron within a visceral organ, which includes eyes, salivary glands, bloods vessels, heart, trachea, alimentary canal, pancreas, spleen, bladder, and urogenital tracts. The same organs receive a second innervation from the central nervous system that probably evolved independently and later, the *parasympathetic* nervous system, also with one or two interneuronal relays. There are two major differences, however: the final relay occurs not in a ganglion but in the tissues of the organ itself, and the final substance released, like that at the intervening synapse(s), is ACh.

There are significant additions to be made to the above: blood-borne hormones and related substances, especially the product of a gland situated above the kidney (hence epinephral or adrenal) that secretes the substance adrenaline (Ad), also called epinephrine, and sympathin, which either itself affects the same sites as does NAd, or does so after being converted locally to NAd. There is a voluminous literature showing that nerves

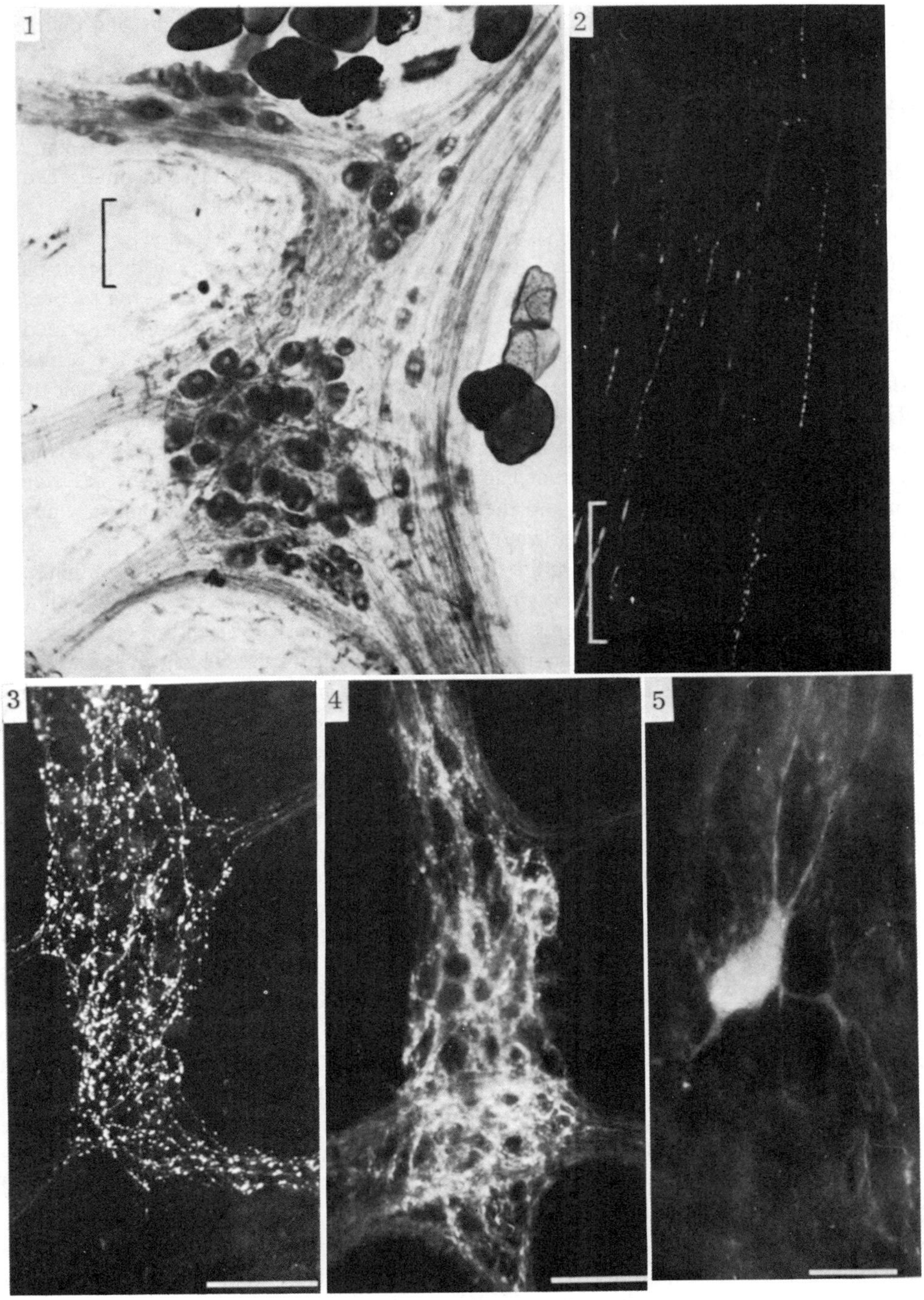

Figure 9.6. (*1*) A vertebrate nerve plexus – rectal wall of rabbit. (From King et al., 1977.) (*2*) Adrenergic innervation of a rabbit hindgut, rectococcygeus, muscle revealed by histofluorescence. (From King et al., 1977.) (*3*) Myenteric plexus ganglion of guinea pig stained by glyoxylic acid method to show noradrenaline-containing and 5 HT-containing neurons. (From Furness et al., 1980.) (*4*) Same as (*3*) stained to show dopamine-containing neurons. (From Furness et al., 1980.) (*5*) Single dopamine-containing neuron from (*4*). (From Furness et al., 1980.)

releasing 5-hydroxytryptamine (5HT) are widespread, at least in stomachs, and there is a lot of histological evidence for dopaminergic (DA) endings also.

Therefore, converging on a muscle cell in a vegetative organ are purely electrical inputs from neighboring cells and various amounts of the chemical transmitters ACh, NAd, Ad, 5HT, and DA, a variety of humoral modulatory substances including histamine, and the unknown synaptic transmitters of local independent nerve cells. The incoming neural activity acts in two ways, directly on SM cells and indirectly on local neurons.

Each tissue has differences in detail regarding the types and locations of the nerve endings, and there are differences between species for a given organ and different parts of an organ. The major nerves to the gut, certainly the posterior ones, the pelvic and pudendal, contain afferent branches that must be associated with sense organs located in the gut itself. Though the details of the innervation are inadequately known, it is probable that many are associated with stretch receptors and that some are pain communicators. They may mediate local spinal-reflex actions and play a role in higher-level overall autonomic control. Contraction of rectal longitudinal muscle is a reflex driven by activation of such stretch receptors. The mammalian hindgut is certainly the most heavily researched organ in which SM is operating and therefore iş the best known. A general scheme for innervation of its muscle cells is shown in Figure 9.7.

There is such a multiplicity of amines released by neurons onto SMs that a combination of techniques is needed to sort out which ones may be present. The tissue is subjected to a modified Falck-Hillarp fluorescence test (Falck and Hillarp, 1962). In this test the tissue is subjected at room temperature to dry formaldehyde vapor, which condenses with amino groups of catecholamines to form a product with an additional ring structure. Each of these has a characteristic fluorescence, principally yellow, emitting at 470–530-nm wavelength when excited by a 420–425-nm wave. Unfortunately, a similar intrinsic fluorescence is given naturally by some tissues, and the fluorescing reaction products of NAd, Ad, and DA all look very similar. It takes longer for Ad to develop the reaction, so this can usually be distinguished from the others. And NAd is the principal substance released by sympathetic nerves based on other criteria, so the fluorescent layers seen in SM are largely due to NAd contained in sympathetic-nerve terminals. In the modified method, the tissue is first treated with phentolamine, which blocks only NAd and octopamine (OA), to determine if a change in activity occurs. If it does and if fluorescence occurs, NAd is present, since OA does not give a fluorescent product. The NAd can be separated from 5HT and DA since it is also blocked by propranolol, the others not. Miaserin is used to separate 5HT from DA because only 5HT is blocked by it. Experts claim they can see a color difference between 5HT fluorescence and that of DA, which is more green. In addition to these amines, an increasing number of small peptides are recognized to be involved in gut control, and also the natural opiods β-endorphin and the enkephalins. The whole history of opiods is more closely linked with their soothing action on the disturbed alimentary canal than with pain-killing or euphoria. The 14-amino-acid peptide somatostatin, which promotes secretion of growth hormone, has been found using radioimmunochemical methods in neurons of mouse intestine (Costa et al., 1977), though it may promote secretion rather than motility. In invertebrates, a pentapeptide, proctolin, has become recognized as a major natural neural controller of the gut (p. 555).

9.7.2. Neuromuscular Junctions

Whatever the overall details for gross innervation, the junctions on SM cells are always small and very simple. Several light microscopists thought they saw nerve terminals

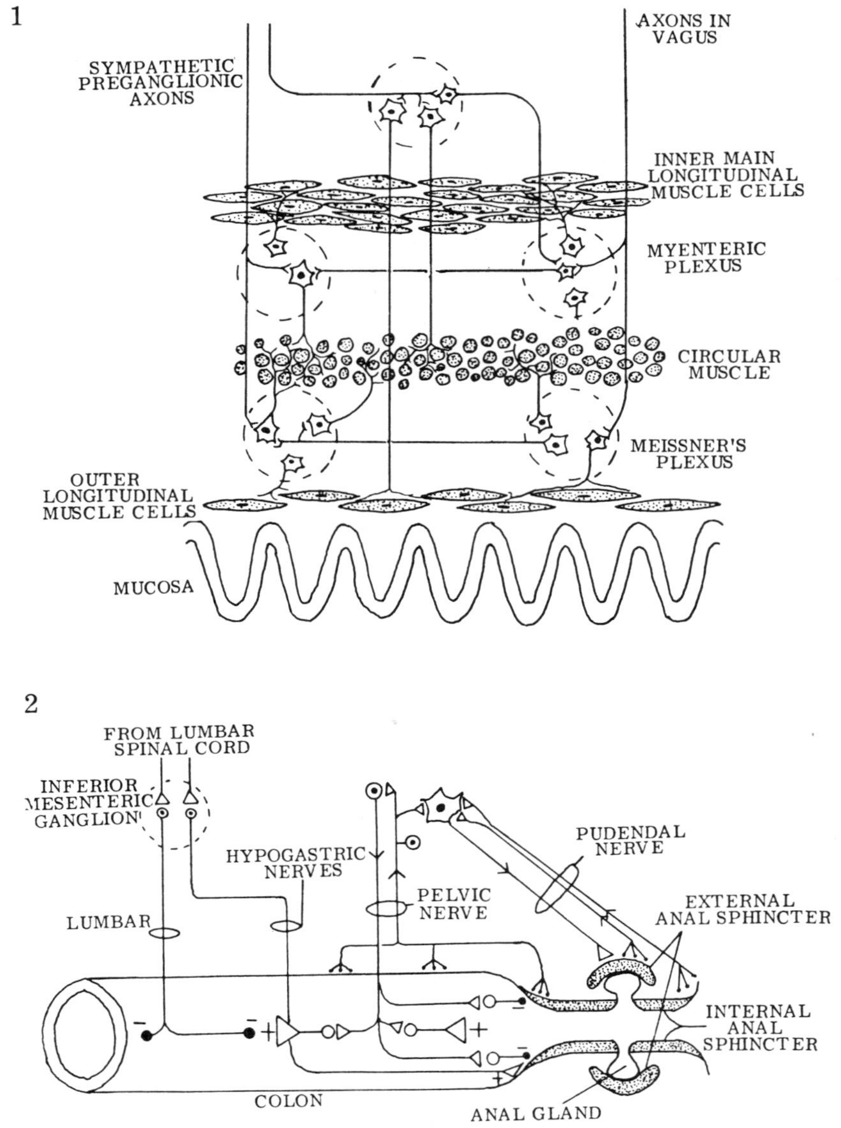

Figure 9.7. Basic innervation patterns of major regions of mammalian gut. (*1*) Intestine. (Based on Schofield, 1968.) (*2*) Lower viscera, rectum, and anus – based on cat. $\overset{-}{\circ}$ inhibition/relaxation; $\overset{\Delta}{+}$ excitation/contraction. (Based on notes by the late R. C. Garry and several sources.)

actually entering individual SM cells, and this has been confirmed by electron microscopy, but it is a rare event, essentially caused by the muscle cell's wrapping itself around the nerve varicosity at the point of contact, and does not represent a characteristic type of junction. Commonly, fine branches emanate from an innervating axon, and these continue, without further branching, over several smooth cells, passing at right angles to them. As each is passed, a small varicosity is formed, 1-2 μm in diameter, containing synaptic vesicles and making appositions without infoldings. These may be single, or with two or three adjacent cells simultaneously. The varicosities are spaced apart as little as 20 nm in vas deferens, to 1000 nm in some blood vessels. The 5HT-containing endings make similar contacts to those presumed to contain other substances.

The presence of numerous dense-core vesicles indicates presumed dopamine-containing endings. At the TEM level, three types of varicosities can be distinguished from one another by their contents. Each contains a mixture of clear vesicles, small granular vesicles, and large dense-cored ones, but the relative proportions of each vary. Small granular 30–50 *nm* vesicles predominate in sympathetic neurons and are presumably NAd-storage sites. They are visualized in light microscopy by fluorescence histochemistry using the Falk and Hillarp method (1962). An example is shown in Figure 9.6.*2*. The images of fluorescence microscopy of SM show very complex tangles of fine lines interrupted by long strings of intensely fluorescent beads that correspond to the varicosities, which are at synaptic sites. Earlier histologists had variously called what they could see of this system a plasmodium (Lawrentjen, 1934), a ground plexus (Boeke, 1949), and a terminal reticulum (Meyling, 1953; Stöhr, 1954). Two schools of thought developed, one believing there to be an asynaptic syncytial net (Meyling, 1956), contradicting the neuron doctrine, and a traditional view led by Nonidez (1944) that claimed the others mistook fibroblasts and connective tissue for neurons. In light of resent research, it is evident that the elaborate networks envisaged by Stöhr and Meyling were indeed fabrications. But the terminals of sympathetic- and parasympathetic-nerve endings are very diffuse indeed, as they need to be to innervate the enormous number of tiny SM cells. Nevertheless, a more orderly innervation than the one that actually exists could have been evolved.

Another earlier controversy concerning the number of SM cells in a muscle or tissue that is directly innervated is still not fully resolved for all SMs. Cannon and Rosenblueth (1937) produced the hypothesis that only a few cells, which they termed *key cells,* are innervated, a view that was contested by Fischer (1944). Although this is virtually impossible to resolve, modern research has supported Fischer: most or all SM cells receive a direct nerve supply of some kind, though from which neurons is impossible to resolve. A chemical tool used in the quest is 6-hydroxydopamine, which is taken up selectively by NAd-producing neurons and then prevents them from producing NAd: eventually they are destroyed and the breakdown products are easily located histologically.

The general concept has arisen of there being functional (motor) units in SM comprising the total of cells innervated by a single efferent axon. Although this remains a possibility for some SMs, it is clear that multiple innervation of single cells, or at least physically associated clusters of cells, confounds the picture. Very rough estimates place the average number of SM cells directly innervated by an average sympathetic neuron at about 100.

A bewildering number of pharmacological agents modulate synaptic transmission in SMs, and several help in the resolution of the nature of the transmitter(s). Noradrenaline and/or Ad interact with *adrenergic receptors* on the postsynaptic membrane. Two basic classes are recognized: α and β, with several subdivisions. The basic distinction is due to the early finding that circulating A causes dilatation of SM of blood vessels in skeletal muscles, but constriction in those of viscera. The behavioral advantage of the difference is abundantly clear.

To relax SM cells, as with other excitable cells, they must be polarized, and for contraction they must be depolarized. A general alerting agent preparing the body for flight/aggression had to be able to do both, in different tissues. This difference in response makes classification tricky and renders broad generalizations suspect, but it is paralleled by the reversed responses of cardiac pacemaker cells, compared with skeletal cells, to the neurotransmitter ACh. It is the specific combination of ion channels that is activated by the substance that determines the current flow, and therefore the membrane-potential effect. The different kinds of receptor are recognized in the first instance by the final actions they evoke, which may be relaxation, contraction, secretion, or lipolysis, and for rhythmic activity, initiation, amplitude/frequency modulation or suppression, altered conduction velocity, and basic tonus.

For the parasympathetic transmitter ACh, it is known that receptors at skeletal muscle synapses have different properties from others located on the same cells but extrajunctionally, and the same is true in insects, where the transmitter is L-glutamate (e.g., Cull-Candy, 1976a, b). The receptor mentioned above for heart pacemakers must be different from either of the ones on skeletal cells. A good system of subclassification has not yet been devised for receptors. The responses of a given SM to ACh are generally exactly opposite to the ones to NAd or Ad when the cells have receptors to both, which includes the great majority. However, there are a few instances in which both cause the same reaction but one more intensively than the other, as in the ureter, where basic tonus and rhythmic contractions are both enhanced by all three agents. Or one may cause tonic contraction and the other(s) rhythmic contractions, as in the male sex organ.

Table 9.1 summarizes some of the more important actions of the sympathetic and parasympathetic systems. By and large the systems are antagonistic, acting reciprocally with the few exceptions noted, but each is about equally concerned to cause contraction or relaxation.

9.7.3. Neuromuscular-Junctional Transmission

Good neuromuscular preparations of SM, with intracellular recording, were first achieved by Burnstock and Holman (1961) using an isolated guinea pig vas deferens complete with its autonomic hypogastric nerve supply. Some sucrose-gap recordings had given results almost as good. Threshold electrical stimulation of the nerve evokes a small, 1 mV, EJP from some of the cells. The EJPs summate during repetition, but they also facilitate dramatically, as do many slow EJPs in arthropods. When sufficiently large depolarization has been achieved, there is a strong tendency to elicit a spike, which in this case is large, 57-90 mV, mean 68 mV, definitely all or none, and overshooting. Increased stimulus strength evokes progressively larger EJPs up to a total magnitude of 20 mV. Therefore, there is polyneuronal innervation of individual cells. As many as 10 steps occur (Burnstock et al., 1964) but it is not possible to ascertain the exact numbers of excitatory axons innervating individual cells because of their electrical connectedness. There are latency variations up to 40 ms, which may in part be attributable to electrical spread. However, the vas deferens does not show spontaneous waves and is in this and some other respects not a typical SM, but can be thought of as transitional between vegetative and skeletal muscle. Miniature EJPs occur in every vas deferens muscle cell, so the innervation is widespread in this SM. The transmitter that activates the postsynaptic membrane must be NAd. IJPs are not seen. The taenia coli of the guinea pig is markedly spontaneously active, with waves recurring at frequencies ranging from 0.1/s to 1/s. Unspecified nerves can be seen in the preparation and electrically stimulated. They each evoke IJPs (Burnstock et al., 1963) with long latencies, 150-300 ms, and are long-lasting, about 2 s each. They summate to give a total peak polarization of about 10 mV, but those of taenia coli do not show facilitation. In the colon of the same animal, very large IJPs occur in some cells; in others there is facilitation (Figure 9.5*.3* – Furness, 1969). During inhibitory action the myogenic waves cease, returning immediately following cessation of inhibition at a higher frequency than the initial one. This frequency increase is enhanced progres-

Table 9.1 Principal Effects of Vertebrate Sympathetic and Parasympathetic Outflow[a]

Organ/Tissue[b]	Sympathetic	Parasympathetic
Tongue	Contraction	Relaxation
Iris	Dilation (relaxation)	Contraction
Ciliary muscle	Relaxation	Contraction
Lacrimal gland	Relaxation	Stimulation
Salivary glands	Stimulation	Stimulation
Sweat glands	–	Stimulation
Esophagus	Relaxation	Constriction
Cardiac sphincter	Contraction	Relaxation
Heart		
pacemaker	Accelerates rate	Slows rate
muscle	Enhances contractile force	Reduced auricular force No effect on ventricle
Pulmonary system	Relaxation	Contraction
Spleen	Contraction	Relaxation
Stomach	Relaxation	Contraction
Intestine	Motility reduced	Motility enhanced
Colon	Relaxation	Contraction
Rectum	Relaxation	Contraction
Anal sphincter	Contraction	Relaxation
Bladder	Contraction	–
Veins	Constriction	–
Major arteries	Constriction	Relaxation
Arterioles, gut	Contraction	Relaxation
Arterioles, peripheral	Some relaxation, some contraction	–
Genitalia	Orgasm	Tonus increase
Uterus, normal	Some contraction	–
Uterus, pregnant	Strong contraction	–
Hair muscles	Contraction	–

[a]From several sources.
[b]The glands contain local muscle cells.

sively with increasing duration of inhibition and also with increasing intensity of inhibition, by higher frequencies of stimulation of the nerve. The presumption is that the nerves are parasympathetic and that they release ACh.

9.8. Conduction

Since nerve nets and plexuses are almost universal in SM, it is possible that they are involved in conduction of excitation in addition to the muscle's capability for conduction from muscle cell to muscle cell. Nerve cells are, however, absent in chick amnion SM in which conduction occurs, and removal of all surface neural elements fails to block propagation in guinea pig intestine or ureter. The speed of propagation with the nerves present, 5-10 cm/s, is significantly less than that which occurs in the isolated nerves. Most critical is that nerve-blocking drugs can be applied in high concentrations to a wide range of vegetative muscles without affecting conduction velocity. The wave of excitation passes a region of mechanical immobilization, and it is not influenced by pharmacological agents that affect contraction. Cholinergic and adrenergic agents neither weaken nor enhance it. The velocity is a function of the number of cells per unit distance, so although an electrical wave propagates outward in all directions, it does so with a much lower velocity laterally than longitudinally for longitudinal muscle, vice versa for circular muscle.

All these considerations make it evident that in vegetative muscle of vertebrates, the nerve elements control contraction locally only and are not concerned with propagating excitation. Propagation is universal and plays a major role in coordination of action, but it is from muscle cell to muscle cell. The propagated action is, however, subject to superimposed local reflex actions, local intrinsic activities, and several forms of modulatory activity, both blood-borne and neurally released, in addition to direct synaptic excitation and inhibition.

9.9. Compound Interactive Effects of Antagonistic Elements

An SM is rarely a discrete contractile system but commonly contains at least two layers, circular and longitudinal, that act both reciprocally and cooperatively according to immediate functional requirement. Surgical separation of the layers is almost impossible in some tissues and has been attempted in very few, and there are large differences for the same organ of different species. In some organs, such as the small intestine, cells of the circular and longitudinal layers are actually connected to each other via junctions of low electrical resistance, so synergism is ensured directly. However, pacemaking in the cat intestine is located primarily in the longitudinal layer (Connor, 1979). The mean amplitude of the wave in the isolated longitudinal layer is only 13 mV and is confined to isolated regions, but when in similar cells in the intestine, it is 27 mV and occurs all over. This is explained by an effective amplification and dispersion by the outer circular cells. These are excited by the pacemaker cells to give a much larger depolarization that not only spreads among all the circular cells, but it also spreads from these back into the longitudinal ones, hence the greater size in the intact tissue. This type of action is not known to be common in SM organs, but it illustrates the kind of complexity that has evolved to meet changing needs. For the experimenter exploring a whole organ with intracellular electrodes, it illustrates the need for extreme caution in interpretation.

9.10. Dynamic-Tension Capability

In view of the low myosin-actin ratio in SMs, the maximum-tension capability is of special interest. Surprisingly, in some SMs maximum tension per unit are of cross section equals or exceeds that of the majority of cross-striated cells. Only slightly lower values occur in SMs at the bottom end of the range, but there is a relatively wide range.

In part this may be due to the great difficulty in estimating force capability satisfactorily and to major differences in activation. But in part it appears related to different relative amounts of actin. The amount of myosin in SM cells is always very small, but there is nevertheless at least a fourfold range in the ratio. Some values for the maximum tension are shown in Table 9.2. They range from 1.9 kg/cm^2 (10^5 N/m^2) for rabbit bladder to 6.1 kg/cm^2 for taenia coli of guinea pig (Murphy et al., 1977). Earlier authors have given substantially lower maximum values in most cases: the table compiled by Gabella (1976) lists values roughly half these. These values, because of the overall structure of SM, are likely to be on the low side by a factor up to half.

For guinea pig taenia coli, it has long been agreed that the peak active tension-unit area is reasonably similar to that for frog sartorius, permitting an interesting comparison (Lowy and Mulvany, 1973). The maximum power, work per unit time, of taenia coli is $\frac{1}{80}$ times that of sartorius because of the corresponding difference in speed. The range of length over which 90% tension is developed is twice as great for the SM, an unexplained finding that cannot be reconciled with a simple X-bridge/sliding-filament mechanism. The length range for active tension development by an SM is very different for different types and in the higher ranges is dramatically greater than for cross-striated cells, even those that can supercontract. The differences may be as much related to differences in connective tissue and junctional properties as to contractile machinery. The maximum range for the supercontractile cross-striated giant barnacle retractor is 15-fold; that for feeding tentacle retractors of the Portuguese man-of-war, an SM, is 178-fold, the largest range known for any muscle. Most vertebrate SMs have a much longer range than the fourfold one shown by vertebrate cross-striated cells, with about 20-fold for uterus being greatest. Segments of arterial wall from rat intestinal wall approximately 0.7 × 0.1 mm subjected to quick stretch or release experiments apparently have little intercellular compliance (Halpern et al., 1978). They develop maximum tension of 350 mN/mm^2 and a shortening velocity maximum of 0.17 lengths/s. Tension increment declines logarithmically after a short quick stretch. The active dynamic stiffness examined under varying $[Ca^+]_o$ was proportional to active tension but not temperature dependent. By contrast, the time taken to recover tension was temperature dependent. These findings are compatible with a dual elastic contribution: one from attached X bridges, the other a compliant passive component in series with it, but they leave a major problem regarding identification of the enormously elongatable, highly compliant SEC.

Table 9.2 Maximum Force Capabilities of Some Smooth Muscles[a]

Tissue	Maximum Force F_o (10^5 N/m^2)
Carotid media (pig)	3.7
Carotid artery (cow)	3.7
Carotid artery (dog)	2.4–3.5
Mesenteric artery (rat)	3.6
Trachea (dog)	1.9
Bladder (rabbit)	1.9
Superior mesenteric vein (calf)	2.0
Taenia coli (guinea pig)	4.0–6.1

[a]From Murphy et al., 1977

9.10.1. Contraction Speed

Unlike a cross-striated vertebrate cell, an SM cell, like arthropod skeletal cells, is capable of being excited to produce any velocity in the range from zero to its maximum capability, under zero load. Values for the speed of unloaded contraction of SM are difficult to obtain because of the morphology, and ideally should be measured on single isolated cells. Common values for the maximum are surprisingly similar, regardless of source; trachea, bladder, uterus, and aorta, all produce about 0.2 muscle lengths/s. This may be compared with the value of about 10 muscle lengths/s for the fastest cross-striated cells. Thus the range of maximum intrinsic speeds of shortening by cross-striated cells, which is at least 100-fold, does not seem to be matched in smooth muscle. These values are not to be confused with rise times or relaxation times, which are very diverse in both kinds of muscles. The V_{max}/load curves for SMs closely resemble in shape those for striated muscles. By contrast, the times taken to reach maximum tension or maximum extent of shortening under a large load are widely different for different kinds of SMs. For example, least contraction time of turtle retractor penis is 0.4 s, but that for its intestine is 0.5 min, and relaxation time is 3 s for the former but more than 4 min for the latter.

9.10.2. Energy Utilization

Energy utilization by SM is notably efficient, very little ATP being consumed in its maintenance. Why this is so is not understood and provides a major obstacle to acceptance of X bridges as the sole contractile element for muscle in general. The period of force development or duration of the state of activation following excitation provides another problem. Guinea pig taenia coli uses about 0.35 μmol/g additional high-energy (creatine) phosphate during the initial period, the same as frog sartorius. But the energy utilization for maintaining the same amount of tension in frog is 250 times greater (Butler et al., 1977).

The X-bridge specialists have looked to stress relaxation characteristics of X bridges that are attached but not cycling to explain the tension, thinking of it as an indefinitely slow relaxation. The early stages of relaxation are normal and proportional to the time course of Ca^{2+} removal. In arterial SM it has been suggested that dephosphorylation of a myosin light chain underlies the making of a noncycling bridge (Dillon et al., 1981). Phosphorylation of the light chain is a necessary step in SM activation (see p. 225).

Tests of SM properties have recently been carried out in the only way that is truly satisfactory on individual cells, using methods of isolating and handling developed by F. S. Fay (1977). The amount of stiffness at peak force is very different for a whole organ or piece of tissue than it is for single cells from the same material, justifying the extra effort required for working with isolated cells. Sinusoidal length changes applied during tension alteration permit assessment of stiffness and tension increase, which turn out to be simultaneous. If X bridges had to attach prior to the onset of tension development, an increase in stiffness should precede the tension increase.

The stress relaxation is an extremely large effect for all SMs, and it is very calcium-dependent. It can be thought of as existing in addition to basic tonus, but it may have a similar molecular basis. According to X-bridge theory, it may be that active X-bridge cycling slows progressively, at a rate that exceeds the rate of loss of tension. The ability to redevelop tension also disappears more quickly than does tension. It must follow that during relaxation, X bridges can maintain tension without using ATP, that is, without cycling. It will be recalled that some of the short-range elastic properties of unstimulated cross-striated cells are attributed to such latched bridges (see p. 142).

Slowly cycling X bridges supported by noncycling attached ones could, in principle, maintain tension without significant extra energy turnover as the muscle slowly relaxes. This process will stop when the cells reach the mean resting tension appropriate to the new length, as determined by passive elastic elements. However, at the new length, stiffness remains that is associated with the contractile apparatus, and this is calcium dependent. This may be due to a dependence of residual connected latched bridges. Alternatively, it would be necessary to look to other explanations, either ones directly involving ultrathin-filament shortening in tension development or ones involving the equivalent of sarcomere dynamics, in which there is slow stretch of cells previously shortened at the expense of stretched ones.

9.10.3. Latency

It has long been known that SM contraction has a late and very slow onset. The tension does not rise abruptly as it does in cross-striated cells, and velocity increases progressively. Both have been attributed to connective tissue and viscoelastic properties rather than to fundamentally different coupling and activation or contractile mechanisms. Now that single isolated SM cells are being used more, exact data, unencumbered by the connective tissue and complex substructure, are being obtained, and surprisingly the very long latency has turned out to be genuine. Even a supramaximal stimulus has a minimum latency for toad stomach cells of 200 ms with a mean of 416 ms, which is reduced by only 34 ms when applied on top of a submaximal contraction that could have been expected to take up any SEC slack (Figure 9.8 – Fay, 1977). The delay is much less in mammalian (dog antrum) muscle, but there is still quite a long time before the very slowly developing tension is detectable – appreciably longer than in any striated one (Neering and Morgan, 1980).

The source of this extraordinary delay is uncertain, but it cannot be due to slow rise in $[Ca^{2+}]$, because diffusion time in these small-diameter cells must be brief, and dispersion should be 90% complete in 10 ms even if only surface entry/release is involved. The long latency is likely to reflect the time required to activate the actin-myosin interaction. Since the latency is so long, it strongly supports the hypothesis that SM activation is regulated via phosphorylation of a light myosin chain; the requirement for phosphorylation introduces a much greater delay than the time needed to roll aside a TM/TN chain as occurs in skeletal cells.

9.10.4. Stretch Sensitivity

A majority, though not all, of SMs are caused to develop additional active tension when stretched (Bozler, 1947). Some are extremely stretch sensitive, and it has been established that this is a significant aspect for them functionally. Some peristaltic waves appear to be propagated by the direct effects of stretch in longitudinally connected chains.

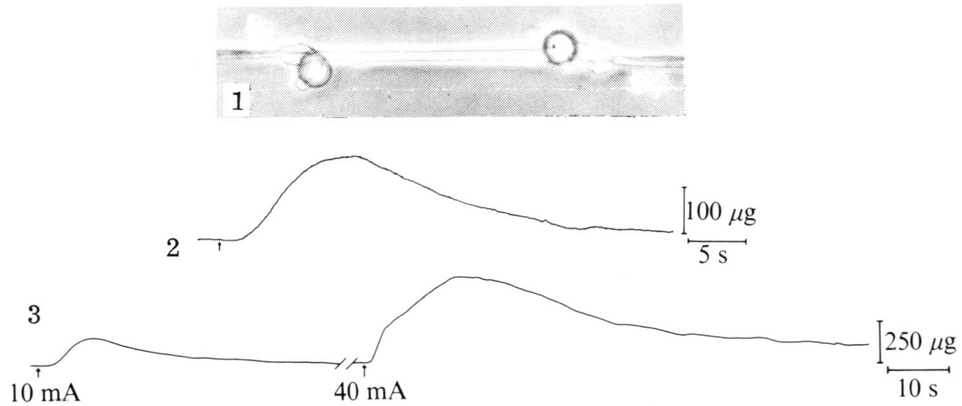

Figure 9.8. Experiments with single isolated SM cell from toad *Bufo* stomach. (*1*) Phase-contrast micrograph of cell knotted at each end to permit force determination. (*2*) and (*3*) Isometric contractions to electric pulses, delivered at arrows, with different strengths. Note the very long latency. [Reprinted by permission from *Nature* **265**, 553–556 (1972). Copyright 1972 Macmillan Journals Ltd.]

Stretch activation follows with a delay, after the fast direct viscoelastic tension increase, reaching its peak 2–4 s after the stretch is completed in portal vein (Basar and Eroğlu, 1976). The mechanical responses of organs containing SM to direct or indirect electric excitation are compounded of the direct response and the stimulating effect of the ensuing stretch on some of the components. The second wave of contraction propagates mechanically and so is slower than the initial electrically propagated one, provided it does not result in depolarization of the membrane past threshold for initiating a propagated spike. The interactions of the slow wave with intrinsic pacemaker activity further complicate the response by altering intrinsic rhythmic contractions.

It is difficult to hold a microelectrode tip in a small SM cell at any time, but especially so during applied stretch. However, it is clear that the stretch effect is always mediated via membrane potential. Stretch depolarizes SM cells, with accompanying fall in resistance, showing that sodium/calcium channels are opened. Release of tension is associated with polarization and increased resistance. SM cells that are discharging rhythmic action potentials fire faster during stretch. The relation between tension and frequency of spikes depends on the overall state of the SM being studied. Sometimes, or in some muscles, spikes propagate evenly through the entire sheet of cells; otherwise they are localized. Only in the former condition can their frequency be related to tension; when it occurs the relation is approximately linear (Bülbring, 1955; Holman, 1958).

In SM that lacks spikes or has erratic or complex spikes, the tension is related principally to the mean membrane potential, following an S-shaped curve similar to that seen in skeletal cells depolarized by high $[K]_o$ (see p. 193). The tension in all SMs is proportional to $[K]_o$, mediated by E_m (Figure 9.9).

However, most SMs have local sensory cells that could be stretch sensitive and mediate the stretch-reflex action indirectly. Until single cells have been examined and shown to possess stretch sensitivity, it will remain possible that such sensitivity is not a property of individual component SM cells but rather a reflex action mediated via stretch-sensitive sensory cells. These may work via a local ganglion, network interneurons, or even by direct local arcs onto the SM cells.

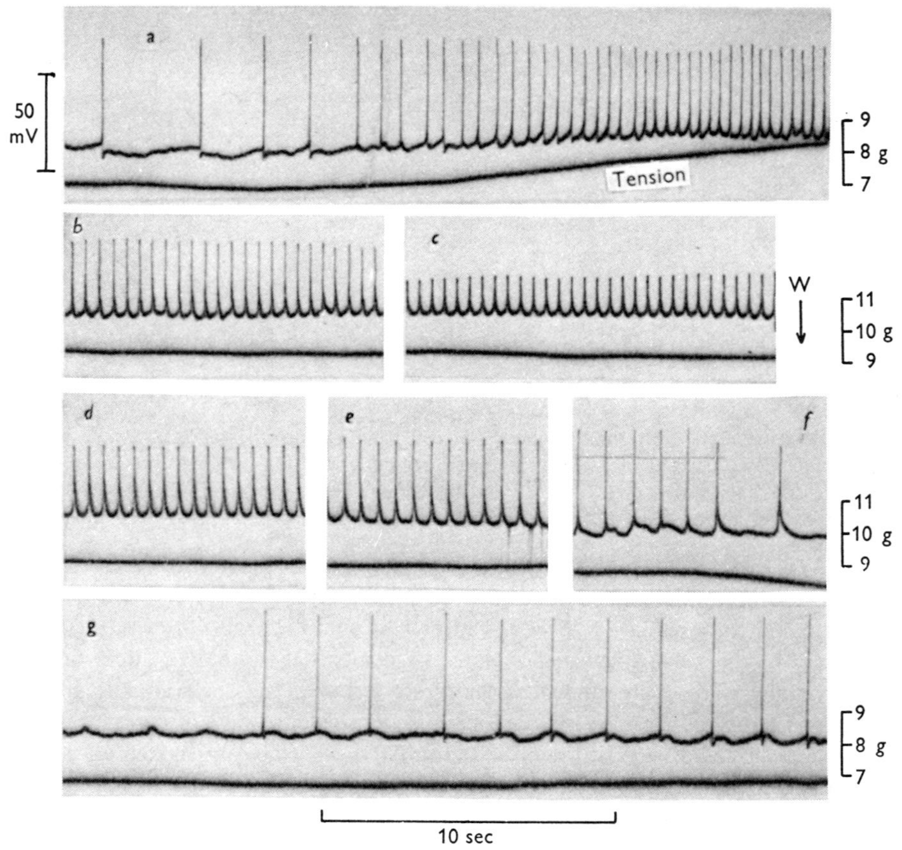

Figure 9.9. Changes in mean membrane potential, spike frequency, and tension during depolarization of guinea pig taenia coli with potassium sulphate, and recovery. (From M. E. Holman, 1958.)

9.11. Basic Tonus (BT)

Basic tonus is the tension that is maintained actively, that is, as opposed to viscoelastically, in the absence of excitatory neural input. Unlike active shortening, it is maintained economically as far as energy utilization is concerned; it is small in amount and the shortening it causes when released is slow. With many whole SMs, it is difficult to assess true BT because a small amount of neurally evoked weak contraction is often also present.

The BT is actively initiated, principally in association with the intrinsically low resting potential and related high calcium conductance, although the amount of free internal calcium in SM is not high. The free calcium is actually very much less than the calcium sequestered in striated cells. But that is because there is much less total calcium in SM. There is a continuous movement of calcium in and out that maintains the free level sufficiently high in the vicinity of the contractile elements to weakly activate the myosin phosphorylation and production of latch bridges continually. The BT falls with decreasing $[Ca^{2+}]_o$, eventually to zero, but SM can still be excited in zero-calcium solutions, indicating both that sequestered calcium is still present and releasable and also that calcium is not essential for the release of this calcium.

9.12. Excitation-Contraction Coupling and Activation

All SMs are affected by $[CA^{2+}]_o$, and only ones with a large number of mitochondria or large amount of SR that stores calcium can function in its absence externally. Smooth-muscle mitochondria have been described as making complex contacts with the surface membrane, so there could be a form of coupling directly via mitochondria, and the amount of SR in some is as high as 0.8 m*M*/kg cell mass. Cleftlike invaginations of surface membrane are rare and often absent, but surface caveolae are common. A flash of light is emitted by an SM cell into which the bioluminescent protein aequorin has been injected (Fay et al., 1979) as in skeletal cells, when electrically excited. This is coincident with the membrane-potential fall below threshold, with less than 1 ms delay, and rise to peak in 4.8 ms in a 4.3 μm in diameter cell obtained from toad stomach.

The rising calcium is due to a combination of entering calcium with SR and mitochondrial-released calcium (Popescu, 1977). The extensive caveolae and microvesicles associated with the surfaces of SM cells not only provide an extensive surface area, but they provide sites at which high concentrations of extracellular calcium occur. The internal "buffering" capacity for calcium is small in all SM, probably in part accounting for the relatively great sensitivity to $[Ca^{2+}]_o$ (Mangel et al., 1979). The free concentration at "resting" potential is above that required to initiate contraction in most SM cells.

There need be no fall in the membrane potential to cause contraction in a SM cell, though. Guinea pig stomach cells are activated by ACh and arteries by NAd without a change in E_m: these agents must enter the cells and activate calcium release directly from internal sites (Prosser, 1979). Thus although coupling is due to a raising of the level of free internal calcium, this calcium is not released in exactly the same manner as in skeletal muscle. Large action potentials are required to produce the larger, faster contractions and these are due to increased *g*Ca.

The next step is activation, and this is clearly different in molecular details in SMs to that occurring in skeletal muscles. At this time the details are still highly controversial. As in molluscan (scallop) cross-striated adductors (see p. 463), there is no skeletal-muscle-type troponin in SM. A controversy arose because there is evidence in some SMs for the presence of troponinlike material resembling TN_C and TN_I. This was discovered by the prominent Japanese muscle biochemistry group led by Setsuro Ebashi – the discoverer of TN – (Mikawa et al., 1977a, b). First categorized as a protein of 80,000 molecular weight, it now appears to be a complex with three components, A, B, and C (Nonomura, 1980) to which the basic name leiotonin has been assigned. Leiotonin differs from TN in having an affinity for actin rather than for TM. Without it, SM actin and myosin cannot interact in the presence of ATP; Ca^{2+} is conceived as a collaborative positive activator, along with the new molecule. In cross-striated cells, Ca^{2+} is an inactivator of otherwise strong masking of A-M binding sites. A group of sideline watchers are patiently believing that a characteristic SM form of TN will indeed be found (e.g., Dabrowska et al., 1979).

The opposition camp's view has developed out of studies on the cross-striated adductor of the scallop *Pecten,* which lacks TN. In this muscle, regulation of myosin interaction with actin is achieved by the uncovering not of the actin-binding site, but of that of myosin. In normal cross-striated muscle, the myosins-binding site is open, but the actin-binding site is covered by TM/TN. In the scallop, the site on the actin is open but that on the myosin head is covered, by one of the light-chain-kinase molecules. This becomes phosphorylated in the presence of calcium, and in the process the binding site is uncovered. Dephosphorylation restores the relaxed condition. An exactly similar mechanism was suggested for SM (e.g., Kendrick-Jones and Jakes, 1977; Mrwa, 1979).

The protagonists agree that there is a lot of calcium-dependent light-chain kinase (LCK) present, but they disagree about its being involved in activation. The Ebashi group believes that contraction takes place only when leiotonin is present. Further characterization of their 80,000-molecular-weight protein is needed before the issue can be resolved. There is one other complication in regard to the extent of phosphorylation required by the regulatory light chain. There are two of these per myosin head, and both seem to be needed in the "open" state to maintain myosin unreactivity.

Biochemical studies on chemically "skinned" SMs support the LCK regulation hypothesis (Kerrick et al., 1980) but the most compelling support comes from physiology. The very long latency (see p. 337) may readily be explained by the time required to phosphorylate the LCK. There is no reason why this delay should exist on the model proposed by Ebashi unless, of course, leiotonin is concerned with LCK phosphorylation and this is the regulatory pathway.

9.13. Modulation of Vertebrate SMs

Modulation is defined as a naturally occurring alteration of the characteristics of a natural activity produced by a substance that has an action other than short-range chemical synaptic transmission. Because SMs have so much intrinsic activity and local autonomy, their control and coordination have to be particularly subject to adjustment by modulations.

An initial distinction will be made between agents that affect SM cells directly and ones that affect the nerve cells that directly or indirectly transmit to them. The former will be called *direct modulators* and the latter *indirect modulators.* Direct modulators of tension developed by SMs may be divided into two major classes: (1) *surface-acting*, which work without entry into the cell, by way of changes in ion conductance that shift the mean membrane potential (the majority), and (2) *intracellular,* which exert their action only after the substance has entered the cell, usually without significantly changing E_m. The natural agents Ad, NAd, 5HT, ACh, and histamine all cause relatively large membrane-potential shifts, at least in a majority of SM cells. Spiking cells follow the potential changes with associated spike-frequency changes. It is speculated, though not yet proved, that the *coupling resistances* between SM cells can be altered by some agents. These would constitute a third major class.

Temperature, pH, and many other factors also cause significant changes in membrane potential. Intestinal cells, for example, are polarized by warmth and depolarized by cold. As pointed out above, the response to a particular substance depends on the receptor. The E_m is most commonly lowered by ACh released from parasympathetic cells, increased by NAd or Ad released from sympathetic nerves. One of the more interesting effects is that of the sex hormones on uterus. These have small effects on nonpregnant uterus, large, even opposite ones, in the condition of midpregnancy, and again different at full term (e.g., Kuriyama and Csapo, 1961). Following ovariectomy, the membrane potential falls, and no action potentials occur. In this condition, however, the uterus is highly sensitive to estrogen, which leads to polarization and resumption of rhythmic spiking. Progesterone leads to even greater polarization, with cessation of discharge and low tonus. The final convulsive contractions at birth are due to neurally released ACh and prostaglandins. In part, ACh action is due to synaptic excitation by EJPs, leading to spikes, but ACh also exerts an overall depolarizing action in pregnant uterus and possibly in addition an internal one. By contrast, taenia coli is not depolarized by ACh (Sperelakis and Prosser, 1959). The ACh action, when direct, is due to a general conductance increase somewhat resembling that due to ACh at vertebrate skeletal muscle end plates, but with calcium as well as sodium and potassium playing a major role (reviewed by Burnstock et al., 1963). There are presumably many extrajunctional ACh receptors in some SMs, at least at certain times or under suitable modulatory/hormonal influences. In the presence of acetylcholinesterase inhibitors, ACh works on such material at a concentration below 10^{-8} *M*.

There is also a very wide range both of sensitivities and responses to adrenaline. Adrenaline is an excitant for initiating contraction of pregnant uterus, anal sphincter, vas deferens, and spleen, but it is a relaxant for intestine. But Ad action, whatever its direction on a particular muscle, is associated with increased metabolism. However, a variety of tests has failed to establish any simple ionic-conductance change: several occur at once, including specific ion-conductance increases, especially to potassium but Na/K/Ca pump(s) are also activated.

Histamine action is similar to that of ACh, depolarizing the cells, but with a slower action, and causing even stronger contraction. The SMs sensitive to 5HT, such as taenia coli, respond to very low concentrations (10^{-9} *M*), associated with slow depolarization. These kinds of slow actions are probably mediated via cAMP released after entry of the agent into the cell. The most primitive living fishes such as the lamprey have neurons in their guts containing 5HT, DA, and NAd (Baumgarten et al., 1973).

9.13.1. Purinergic Nerves

Another major candidate as a transmitter or modulator released by autonomic efferent nerves of vertebrates is ATP itself, perhaps as a special case for the more general one of purine nucleotides. Motor nerves to guinea pig taenia coli, stomach, and bladder are not obviously either adrenergic or cholinergic, but appropriate actions are mimicked by adding ATP. Accordingly, Burnstock (1972) has proposed that some autonomic nerves are purinergic.

9.13.2. Opioids

The relatively recent dramatic rise of interest in receptors to morphine and synthetic opium derivatives arising out of the discovery of the natural peptides, endorphins and enkephalins, whose receptors they attack, has drawn attention to ancient interest in opium actions on vegetative organs invested with SM. Opium has been known for at least 2000 years to relieve digestive-tract motility problems. It does this by increasing BT and enhancing pendular movements while suppressing peristaltic waves: the whole progress of food through the alimentary canal is slowed by opioids. The increase in tonus in the colon of man maybe sufficient to produce a tonic "spasm."

Smooth-muscle preparations, especially from alimentary canals, were from the start of modern experiments used as bioassays for enkephalins and endorphins (Hughes et al., 1975), the body's natural opioids. Independent assay methods for brain opioids using specific radioimmunochemistry showed that the enkaphalins – leucine and methionine as well as β-endorphin – are present in the alimentary tract naturally, in different cells

(Figure 9.10). The tone of bladder, bronchi, and uterus is increased also, but that of peripheral blood vessels and heart is relaxed. The latter phenomena are the cause of another well-known effect of opium, that on body temperature, which necessarily falls. Similar effects are produced in other primates and in rabbits, mice, and birds. But as is so often the case with any agents affecting SM, exactly opposite actions occur in cat, goat, beef, and horse. In guinea pigs and rats the response is first one, then the other.

The site(s) of action of the opioids on SM has(ve) not been elucidated, and they are likely to be multiple. Natural-transmitter receptors on SM cells is a possibility, as are intracellular actions that ultimately affect Ca^{2+} sequestration. But the opiates also affect cholinergic, dopaminergic, and 5HTergic synapses, some or all of which are present in the plexuses, so an indirect neural action exclusively, or in addition, must also be considered.

9.13.3. Substance P and Other Peptides

Another ubiquitous peptide that affects SMs is substance P, now known to be an 11-unit peptide, discovered by U.S. von Euler while working on intestines in the early 1930s, though it is also present in brains and dorsal roots of spinal cords and is released by

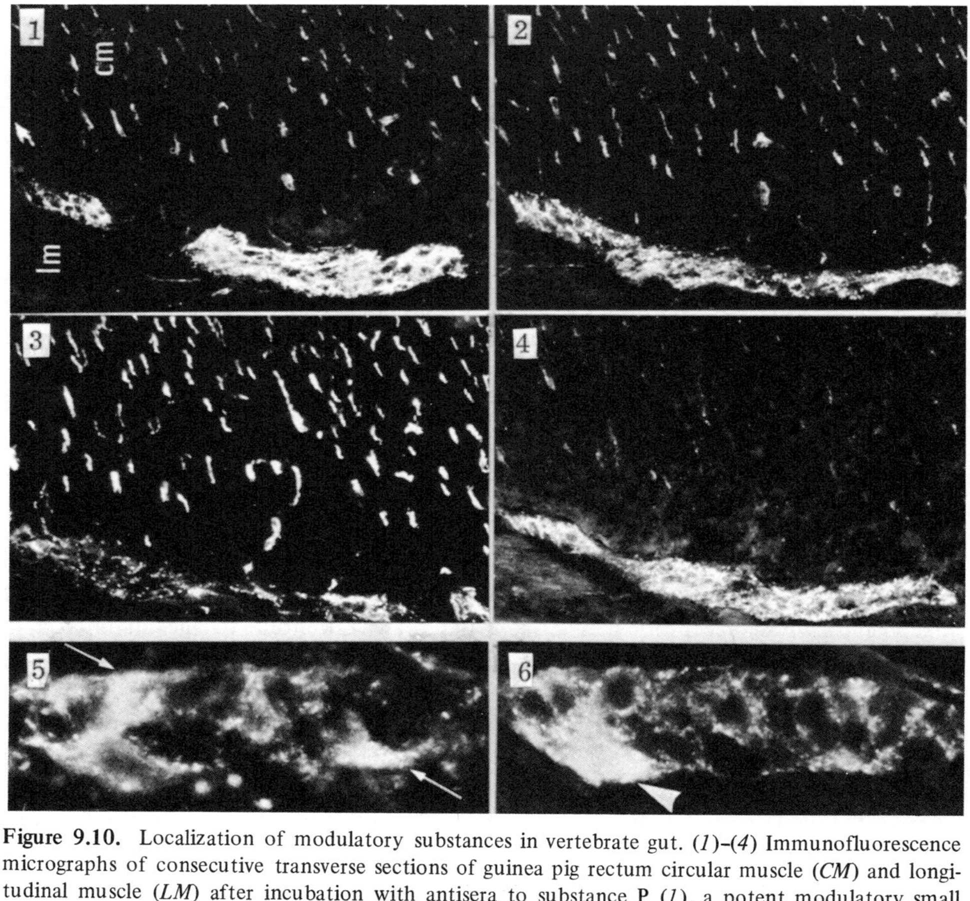

Figure 9.10. Localization of modulatory substances in vertebrate gut. (*1*)–(*4*) Immunofluorescence micrographs of consecutive transverse sections of guinea pig rectum circular muscle (*CM*) and longitudinal muscle (*LM*) after incubation with antisera to substance P (*1*), a potent modulatory small peptide, vasoactive peptide (*2*), enkephalin (*3*), and somatostatin (*4*). (*5*) and (*6*) As above, but proximal colon, enkephalin (*5*) and substance P (*6*). (All from Schultzberg et al., 1980.)

stimulation of the vagus. Some of its many actions may be indirect, because it appears to promote histamine release from most cells. One example is its potent vasodilator action. Substance P–releasing neurons (P-ergic neurons) innervate stomach, intestine, and hind gut (Figure 9.10.*1*) possibly ending around ganglion cells of the myoenteric plexus. The entire spectrum of contractile activity, pendular to peristalsis, is promoted by substance P.

A possible extensive antagonist to the substance P–releasing neurons has been discovered, also releasing an undetermined peptide. This peptide relaxes mammalian intestines (Polak and Bloom, 1979) and guinea pig trachea (Matsuzaki et al., 1980). Neurons containing peptides are widespread in both autonomic nervous systems and accessory ganglia (Schlutzberg et al., 1980).

9.13.4. Oxytocin and Parathyroid Hormone

Oxytocin is a nonapeptide that is synthesized by neurons in the supraoptic and parventricular nuclei and stored in the posterior pituitary, the neurohypophysis. From there it is released directly into the blood. Several SM targets are activated by oxytocin, notably the uterus and mammary-gland muscle cells. Like other hormones, that from the parathyroid gland exerts both direct and indirect actions. One of the former is dilation of the major coronary artery (Crass and Pang, 1980).

9.13.5. Vasoactive Intestinal Polypeptide (VIP) and Bombesin

These are substances that have indirect actions on intestinal motility. The VIP is a 14-amino-acid peptide present in the hindgut at least (Figure 9.10.*2*). There is a secondary innervation of the gut that is collectively termed the "diffuse neuroendocrine system" (Polak and Bloom, 1979). It forms a collection of fine, branching neurons with neuropeptide release sites, for example, in the myenteric plexus. This system is compact in healthy organisms but sprawling in disease. The full implications of this sytem are not yet realized. The substances released either have a *paracrine* action, that is, local diffusion-limited action, or a *neurocrine* action, that is, classical cell-to-cell close transmission. Both VIP and bombesin are products of this system.

9.14. "Behavior" of the Alimentary Canal

This book is concerned with the principles operating in nervous control of muscles, and space does not permit detailed descriptions of each SM organ, each of which has individual peculiarities. The general aspects in which they differ from skeletal muscles are exemplified in the control of the gut, which will be described here.

Historically the oldest, always the largest and most functionally significant, the gut is the most complex of the SM-operated organ systems. Fortunately, for descriptive purposes, vertebrate guts are anatomically very similar to each other in organization, and a few general features occur all the way from fish to mammals. Three major movements occur, two of which are rhythmic: longitudinal slow waving or writhing that is termed the *pendular movement,* and circular waves of constriction that travel slowly along the tubular organ, *peristaltic waves.* The pendular movements are myogenic, the peristaltic ones neurogenic, originating in, and mediated by, various enteric nerve plexuses. Nonrhythmic movements are characteristic of the region itself: gulping and swallowing by the esophagus, long-maintained contraction with occasional relaxation for pyloric and anal sphincters, and strong longitudinal contraction of rectum. All of these are neurally driven in the manner of skeletal muscle, but they are all utilze cell-to-cell electrical fast-wave propagation as well as direct action. Peristaltic waves are strongly dependent on the stimuli of stretch, leading to slow propagation. All active contractions are superimposed upon basic tone, which may be steady or fluctuate very slowly to produce patterns of contraction.

9.14.1. Peristalsis

The most signficant and widespread of all waves is normal *peristalsis,* the analward progression of a sequence, first of relaxation, then of contraction, through almost the entire alimentary canal. It is capable of reversed direction, *antiperistalsis,* in conditions of direct invasion of the canal by foreign bodies, toxic substances, extreme pH change, and conditions of general nervous stress that affect the entire sympathetic nervous system. In the isolated gut, it is generally possible to initiate peristalsis simply by local distension, but this does not work in some, such as cat, salmon, and tortoise. In the latter, however, addition of 5HT before stretch starts peristalsis (Burnstock, 1958). In the guts that are difficult to start by distension the distensibility is low because of a large amount of collagen. These observations together suggest that peristalsis is initiated in local stretch by the combination of a direct action of stretch and a local reflex action.

The stretch-reflex arc promotes a contraction anterior to the stretch and a relaxation posterior to it. The arc is thought to involve the local sense organ, with local cell body, a single interneuron that communicates with the nerve net or plexus, and motor cells of the plexus. The interneurons have uptake mechanisms for catecholamines and indoleamines and so are probably acted on by extrinsic controlling neurons (Furness and Costa, 1979). One of the final motor elements releases 5HT, causing the first phase, which is relaxation. This phase may be termed *preparatory* and the contraction *emptying* (Trendelenburg, 1917). The relaxation is of both longitudinal and circular muscles locally. Contraction starts in the local longitudinal elements first, and is then followed by the dominant circular contraction. The relaxation must be due to local release of an inhibitor, since stretch causes contraction. Peristalsis is therefore determined by local neural activity. It is eliminated by blockers of cholinergic transmission, but ACh, though certainly used as another of the final motor transmitters, is also possibly used at the sensory–interneuron synapses in the local plexus. The vagus nerve acts on these plexuses also. In mammalian gut, either Auerbach's or Meissner's submucous plexus could be the first integrating element, with an edge toward Auerbach's because a submucous plexus is missing in fish and amphibia (Gunn, 1951). Although most of the above applies equally along the entire alimentary canal, regional differences occur and have long been recognized, even in fish (Young, 1936).

9.14.2. Plexuses

The primevel nerve net presumed to innervate the SM cells has done quite a lot of evolving on its own, especially in vertebrates. In all forms there are concentrations of nerve cells in the principal muscle layers forming the myenteric plexus (Auerbach's). In reptiles, birds, and mammals, there is also a major plexus below the mucous layer, the submucous plexus (Meissner's). There are probably corresponding nerve cells in fish and amphibia, but they are scattered in net form, not concentrated into ganglia.

The links to the central nervous system (CNS) have shown continuing evolution. The vagus innervates the stomach and anterior canal in all forms, but stops there in teleost fish. It innervates also the pyloric sphincter and anterior duodenum in amphibia and higher forms and the anterior part of the small intestine in addition in mammals. The vagal efferent axons synapse with neurons of the myenteric plexus, making adrenergic envelopments of ganglion cells there.

Although there are no nerve cells bodies in the submucosal layer of teleosts, fine nerve filaments occur there. Although these may be entirely sensory, it has long seemed possible that there are also efferent terminals in the submucosa, but their function is likely to be to promote secretion rather than movement. Higher forms have unique small multipolar nerve cells in addition to the fine sensory/efferent elements.

The neural organization within the plexuses has in no case been deciphered, but it is clear these plexuses include reflex pathways with at least some single interneuron relays. They are organized so as to promote excitatory action cephalically and inhibitory action caudally. Guts with nets and plexuses are multiply affected by many modulators, since these act not only on the SM cells directly, but also on local neural components. Sorting out the effects is always difficult and often impossible.

9.15. Vegetative Muscles

It will have become obvious that each major organ system, and in some cases the same organ system in different species, has evolved special features, so that very few generalizations across the board can be made for vegetative muscles. The gut is the major single vertebrate organ that utilizes SM, and it is also the major vegetative organ of invertebrates. To permit comparison with the foregoing account of vertebrate guts, a brief survey of some invertebrate guts will be presented.

9.15.1. Invertebrate Guts

The component cells of invertebrate guts other than those of arthropods are nonstriated. They resemble those of vertebrates and are assembled in a similar manner. As mentioned earlier, I consider that most of the invertebrate skeletal muscles that have traditionally been grouped with vertebrate SM because they lack cross-banding should be treated either separately or with skeletal muscle in general. These muscles and their innervation and control are described in Chapter 12, which deals with the skeletal musculature of the different phylogenetic classes. Most are indeed skeletal, and they nearly all have many thick filaments with some form of orderly alignment thereof. It was equally absurd, in the converse direction, to avoid equating vertebrate SMs with the visceral muscles of arthropods simply because the latter happen to contain cross-striated muscles cells. Both the common historical origins and the common functions of gut muscles require that they be compared with each other, not with skeletal muscles selected on the arbitrary basis of their striping. Even a cursory examination of arthropod vegetative muscles emphasizes the conclusion made on p. 582 that physiological features of a muscle are determined by many factors, but cross-striation is not one of the more significant ones, because the musculature of the gut, reproductive tract, and vegetative organs of arthropods is entirely cross-striated, and yet some or all have the same special features associated with SM of vertebrate guts. There are three other major anatomical differences. One is that the component cells have a very wide range in diameter, from 0.5 to 50 μm from different sources, but even within any particular section, there can be a 20-fold range. The second is that the cells do not have obvious ends to them. Toward their extremities they make many branches that terminate in collagen strands that are in turn attached to body wall or to other soft tissue. The third is anastomosis, which is extraordinarily extensive and complex in all, and includes lateral and reverse branching to form a netlike form overall.

The sarcomere lengths also have a remarkably wide range, from 2 to 22 μm even in a single species (e.g., Morison, 1927). Although nothing is known about the functional relations of SL in gut muscle, it obviously formed a major secondary basis for functional, presumably speed, variations. The striations fall where they may in relation to the branches, which include complete sarcomeres only when the branches are sufficiently long. Within a sarcomere the ultrastructure is indistinguishable from that of many arthropod slow-type skeletal cells. Thick filaments are present in an orderly hexagonal array, with neat orbits of generally 12 thin filaments surrounding them. The Z region is thick, when seen in longitudinal sections, and extensively perforated as best seen in transverse sections. There is no well-defined M region. The mitochondria are small, generally unusually so, and although present throughout the sarcoplasm are concentrated, as they often are in striated cells, just below the surface membrane and in the I regions.

The thinner cells and branches, up to about 4.5 μm, do not have any ITs, but all the thicker ones have invaginations running spirally inward. There are, however, very few of these (D. S. Smith, 1966b). Along their lengths, typical cisternal elements abut the ITs to form dyads of characteristic form. There is, however, only a little SR. Several reports have been of larger-diameter cells in a wide range of species, and all were for inner circular muscle cells. The gut of the onychophoran *Peripatus* has not been examined by modern electron microscopy, but the body wall has, and some very thick 40 μm in diameter cells occur in it (Hoyle and Williams, 1980). This may be a widespread feature of inner circular cells.

There is a progressive transition in the alimentary canal from anterior musculature that must include a phasic component for rapid swallowing, through the medium-speed grinding process in the gizzard or in some the "stomach," to the intestine, where only tonus and very slow movements are required. Finally, in the last part, relatively fast movements are needed for voiding.

For phyla other than the Arthropoda, data on vegetative musculature is even scarcer, but some attention has been paid to the cephalopods among molluscs, which provided some very interesting surprises. The buccal mass, esophagus, and stomach of the cuttlefish *Sepia* all comprise obliquely striated muscle cells (Figure 9.11 – Amsellem and Nicaise, 1980). They are spindle-shaped, roughly triangular in cross section, with axes 3.5–8 μm. There are very few mitochondria, and these are located below the surface membrane. As with all molluscan and annelidan skeletal obliquely striated cells, the thick filaments are

Figure 9.11. Ultrastructure and innervation of muscle cells of the gut of the cuttlefish *Sepia.* (*1*) Neuromuscular junction in plexiform layer of stomach, with detail of synaptic membranes in inset. Note extremely wide range of synaptic vesicle (sv) sizes. *t* = Teloglia. (*2*) LS of esophageal circular muscle cell, showing oblique array of short filaments. (*3*) LS of obliquely striated cell of plexiform layer of stomach with long filaments. (*4*) Paramyosin ribbon in thick filament of (*3*). (*5*) Details of dyadic coupling site (above d) between surface membrane, with collagen outside, and SR cisternal element. Note regular arrays of feet between the membranes (*arrowheads*). (All from Amsellem and Nicaise, 1980.)

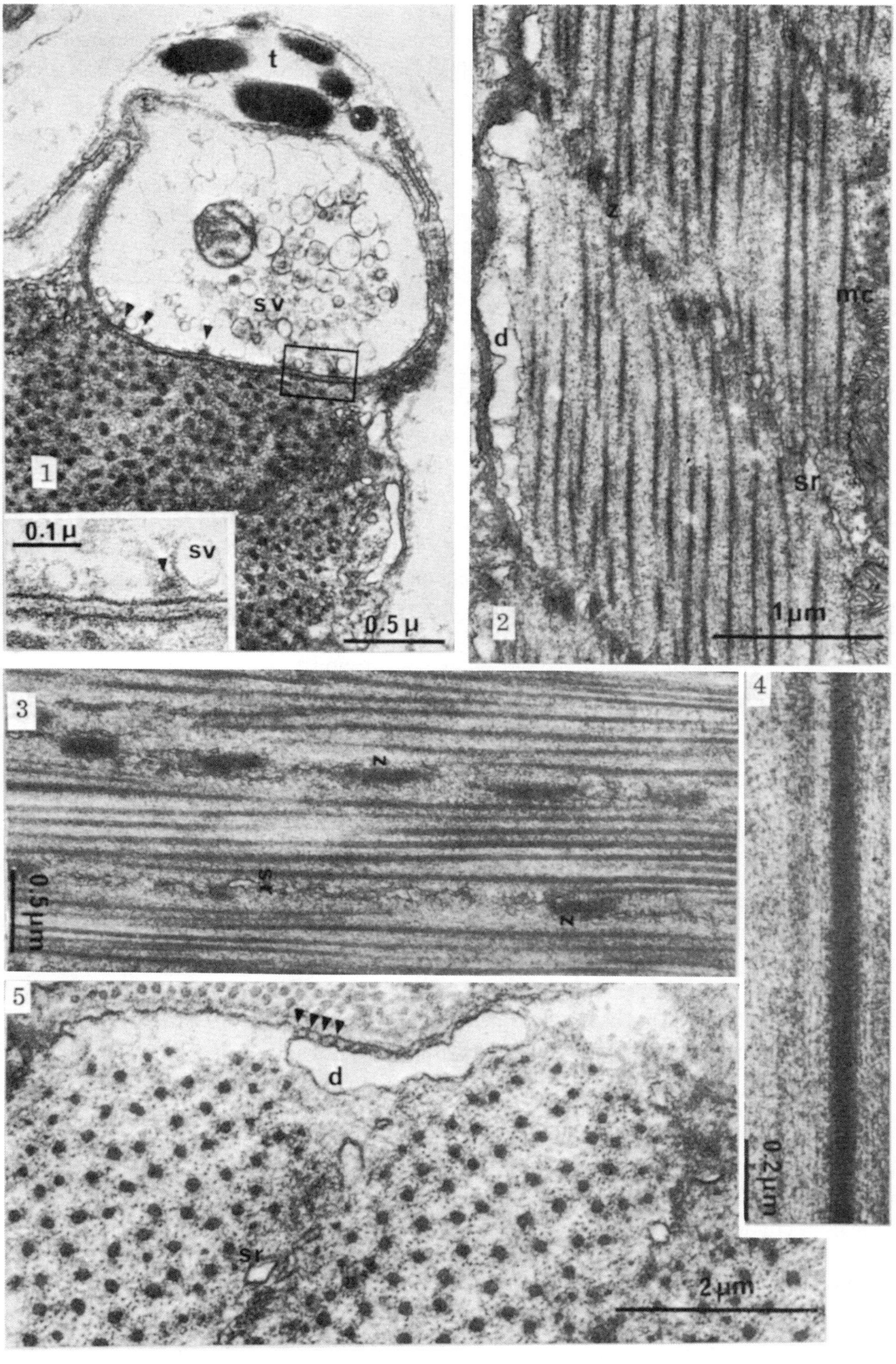
1
t
sv
0.1 µ
sv
0.5 µ
2
z
d
mc
sr
1 µm
3
z
sr
z
0.5 µm
4
0.2 µm
5
d
sr
2 µm

tapered at their ends, with very thick central regions about 58 nm in diameter. They are not in orderly array, but have a wide range of separations. Thin filaments are interspersed irregularly between thick ones, being heavily clustered round some, not at all around others. They are attached to a staggered array of Z bodies. Paramyosin banding is clearly seen in the cores of the thick filaments.

The cells come into close contact at a few sites, forming desmosomelike junctions. Neuromuscular junctions are of a very simple type, lacking both grooves and muscle extensions. They are distinguished by containing an extraordinarily wide range of sizes of vesicles. There is a sparse SR, and although the cells lack ITs, as do most vegetative cells, they have large, clearly defined cisternal elements. These directly underlie the surface membrane, and there are regularly-arranged feet between this and the outer face of the C element, exactly as seen in dyads and triads (Figure 9.10.5). Signficant differences occur in details in the three tissues. In the esophagus there are more mitochondria, more SR, and a more regular overall organization. All these factors make it a little faster, as is appropriate for phasic swallowing.

9.15.2. Nerve Supply

Most early histologists examining arthropod vegetative muscles such as insect ventral diaphragm, and oviduct (Morison, 1927) claimed they saw many local ganglion cells, probably in a network, and some thought a syncytium. There is little modern data on these, and in view of the fate of the Meyling school of thought about such systems in vertebrates, we should be sceptical about them in arthropods.

A direct innervation of gut muscle exists in insects, so if there is a peripheral nerve net as well, a view that was accepted by Cook and Holman (1979a), there is at least a dual, more probably multiple, innervation of individual cells. The transmitter substances of these cells are likely to be different from each other. The anterior parts of the guts, crops, gizzards, and stomachs are innervated from a medial nerve that is connected to a frontal ganglion located at the meeting point of two nerves from the forebrain that form an anterior commissure. In insects this nerve first sends off a branch dorsally that innervates the corpus cardiacum, a neurosecretory body. Along this medial nerve, in crustacea there is a small ganglion, the stomatogastric ganglion, that has received a great deal of attention since 1970 and is probably the best-known of all single neural entitities. It contains only about 30 neurons, and more than half are motor neurons innervating muscles operating the stomach (see p. 525). These motor neurons and the interneurons of the ganglion in turn receive many hundreds of neural inputs from the brain and from paired commissural ganglia, and the neurons in turn send branches to the ganglia in addition to innervating the muscles. However, these muscles are discrete and extrinsic, and although excited rhythmically by pacemaker activity generated in the ganglia, and more closely allied with skeletal muscles than with vegetative ones. The equivalent region in insects is the foregut, and it is operated only by intrinsic musculature. The commissural ganglia are located in esophageal connectives, and four nerves arise from each: superior and inferior esophageal, the posterior stomach nerve, and the mandibular nerve. When electrically stimulated, the first three all have marked actions on the parts of the gut they innervate but each contains hundreds of very fine axons, some of which have cell bodies located in the esophageal ganglia, others in the brain. The whole is termed the stomatogastric system.

The midgut is less densely and less complexly innervated than the foregut, principal innervation coming from the posteriormost end of the nervous system, the last ganglion of the abdomen. In some insects there may be partial innervation in addition from the stomatogastric system (Orlov, 1924). There do not appear to be any ganglia associated with the paired posterior nerves, but the two join each other and form complex multibranched anastomosing plexuses in which motor axons ramify, possibly mixed with sensory axons. The hindgut receives a dense, complex motor innervation from the plexus. The final motor-nerve terminals from all the gut nerves are simple, without infoldings, but with bridge abutments, or sometimes covering enclosures, from the muscle cell at the region of contact. Each muscle cell receives a set of axon terminals located side by side. To what extent there is division into motor units is unknown.

9.15.3. Physiology

There is very little modern data on the physiology of invertebrate guts, but in the few that have been examined intensively all the characteristic vertebrate features are to be found. Basic tonus is invariably present and can be quite strong. Some are arhythmic, displaying only steady tonus. Others are rhythmic, with both peristaltic waves and longitudinal slow waves, all parameters having somewhat lower spreads of frequency and duration for waves than do their vertebrate counterparts. However, this may only reflect the small number of invertebrate visceral muscles for which we have data. Invertebrate guts are extremely varied anatomically and functionally and each needs to be decribed individually. They do have in common engulfing, swallowing anteriors and storing as well as voiding posterior extremities. But some of the former are eversible, others not. The gut in aquatic animals, especially of Holothuria, is also concerned with respiration. Prosser et al. (1965) described the overall activity of several guts in a comparative survey. Action potentials occur spontaneously in the esophagus at a mean frequency of about 1 Hz, but they are highly erratic and commonly asynchronous, so that tension, though fluctuating, is not correlated with even the larger waves or spikes. But occasionally waves, spikes, and tension come together, indicating that overall control superimposes on local activity. After the guts had been treated with tetracaine to block neural activity, there was a tendency to synchronization of both spikes and waves, so it is local neural inhibition that is responsible for the asynchrony.

The intrinsic rhythms can each be complex, especially in foreguts, and they interact to further the complexity. Contractions in the foregut of the earthworm *Lumbricus,* a relatively advanced annelid, are simple periodic longitudinal contractions. The amplitude, duration, and frequency of contractions are all sensitive to various pharmacological agents, temperature, pH, and ionic concentration of medium, as also is BT-level changes. These movements ensure the automatic engulfment of soil as the earthworm moves forward. By contrast, the marine annelid *Arenicola* extreme foregut has a complex, behaviorally sophisticated, that is, functionally relevant, polyrhythm (Figure 9.12.*1*). Periodic, relatively fast, extend-contrast-extend sequences that progressively increase and then gradually decrease in amplitude are superimposed on BT fluctuations in phase with the amplitude swings. This rhythm is associated with proboscis eversion/retraction cycles that serve both to pull the worm forward in the sand and to engulf and swallow sand for feeding. The detailed patterns are different, and characteristic, for different species.

Other material examined by Prosser et al. (1965) included intestines of *Thyone* (Holothuria), *Arbacia* (Echinodea), *Spisula* and *Busycon* (Mollusca), *Libinia, Callinectes,* and *Homarus* (Crustacea). The researches' principal purpose was to determined the extent to which conduction along the organ in response to local electrical excitation is caused by

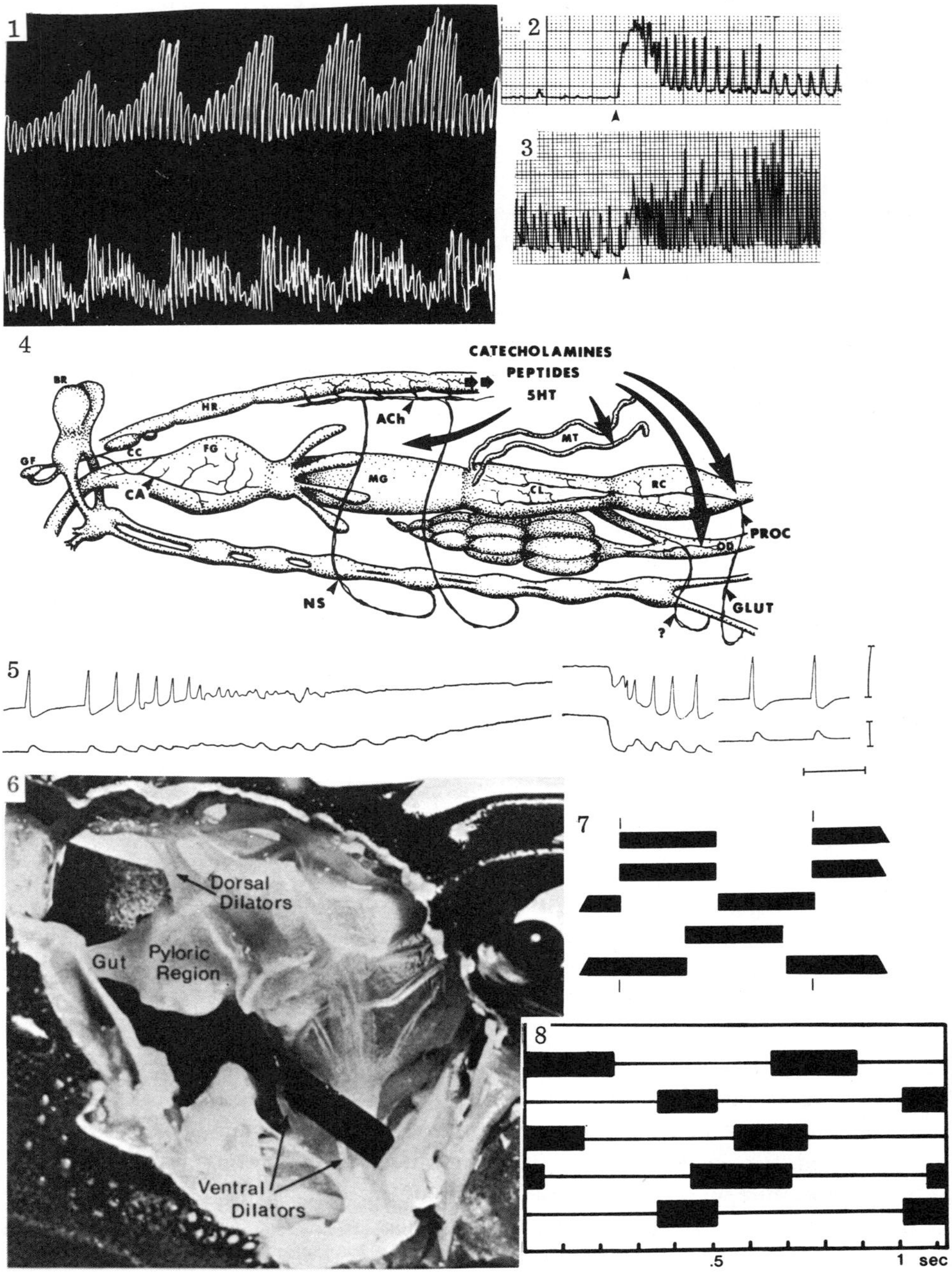

Figure 9.12. Activities of invertebrate guts. (*1*) Simultaneous recordings of movements of body-wall (*above*) and anterior extrovertible portion of gut (*below*) of lugworm, *Arenicola marina.* (From Wells, 1949.) (*2*) Response of cockroach *Leucophaea* to 3×10^{-10} *M* proctolin. (From Holman and Cook, 1979.) (*3*) As (*2*), but 4×10^{-8} *M* 5HT. (From Cook and Holman, 1979a.) (*4*) Diagram of cockroach vegetative organs, with indication of sites of action of modulators. *HR,* heart; *RC,* rectum; *PROC,* proctolin; *GLUT,* L-glutamate. (From Cook and Holman, 1978.) (*5*) Intracellular (*upper trace*) and mechanical (*lower trace*) responses of cockroach *Leucophaea* hindgut longitudinal muscle to high $[K^+]_o$, and recovery. (From Nagai, 1972.) (*6*) Foregut of spiny lobster *Panulirus,* showing externally visible accessory muscles. (From Maynard and Selverston, 1975.) (*7*) Typical timing pattern for bursts of activity in five different motor neurons generating the posterior, gastric cycle [of (*6*)]. (From Mulloney and Selverston, 1974.) (*8*) Typical timing pattern for bursts of activity in five neurons generating anterior, pyloric cycle [of (*6*)]. (From Maynard and Selverston, 1975.)

neural elements rather than the muscle itself. It is also possible that both routes could be utilized, but they would be expected to have markedly different velocities. As we have noted above, in vertebrate viscera the muscle cells are electrically connected to each other and in many types produce all-or-none spikes. Hence electrical conduction occurs without decrement by muscle-cell conduction. The nerve plexuses are concerned mainly with providing strictly local control, rarely with the production of larger patterns.

In all known invertebrates guts, an elicited contraction is conducted only decrementally, being reduced significantly after only 2 cm of conduction. The range of velocities of the propagation at room temperature is narrow, from 4 to 6 cm for *Thyone*, the slowest, to 12 cm for *Loligo*, the fastest. Following treatment with the nerve anesthetic procaine, conduction is blocked in all but the echinoderm *Thyone*, which means that conduction in *Thyone* is along muscle cells as it is in vertebrates. Students of evolution have long recognized from comparisons of early larval forms that chordates are related to echinoderms and to no other invertebrates. Intestinal muscle cell-by-cell conduction is characteristic of the echinoderm-chordate line and phylogenetically very old indeed.

The contraction is single and fast in the squid *Libinia* and in lobster, single and slow in the echinoderm. In the clam *Spisula*, dual slow and fast contractions occur in response to currents of different durations. Stimulation of local ganglia can produce inhibition.

9.15.4. Transmission to Vegetative Cells

Very few SMs of invertebrates have yet been studied with intracellular methods. Among the few isolated examples, the best are preparations of the gills of molluscs, which are invested with longitudinal and circular SM sheets that cause a variety of movements. There is innervation from both local-nerve-net elements and from centrally (ganglion) located neurons, paralleling the situation found in many vertebrate vegetative organs. Three neurons have been identified in the abdominal ganglion of *Aplysia* that innervate its gill and cause movements: L_7, which causes closing movements of the two gill halves, called antiflaring: LDG_1, which causes rostral rotation and pinnule bunching: and LDG_2, which causes flaring (Carew et al., 1974). The muscle cells involved are ones that also surround blood vessels or "veins." Some of these cells have been recorded intracellularly during stimulation of the motor-neuron cell body. Both L_7 and LDG_2 produce EJPs initially of small size (1 mV) that both facilitate and summate with repetition; LDG_1 produces larger (12 mV) nonfacilitating EJPs. The latency, but not the amplitude, is constant in the face of changing concentrations of calcium and magnesium, showing that the innervation is direct, not via the nerve net.

Some of the *Aplysia* vegetative muscle cells receive dual innervation, from both L_7 and LDG_1. And IJPs also occur, but their origin was not identified. Both LDG_1 and LDG_2 are cholinergic, but L_7 is not. The activity of L_7 is modulated by L_9, which appears to be dopaminergic (Lukowiak, 1979).

These observations are unique, not only for invertebrate guts, but also because they are the only organisms in the entire animal kingdom from which the intracellular response of an SM cell has been evoked by intracellular stimulation of an identified central motor neuron. There is a strong similarity to arthropod skeletal neuromuscular transmission in general, as well as to the responses of guinea pig taenia coli to central nervous output (see p. 331).

The biggest gap in our knowledge of SM control relates to the activities of neurons of the local nerve net and local ganglia, and their cooperation. They are responsible for modulation of BT and intrinsic rhythmicity. The central elements can be thought of as direct excitors and inhibitors producing patterned movements of local regions in a manner analogous to control of arthropod limb muscles. Central nervous controls are superimposed upon the local, poorly understood nerve net/ganglia modulations. The central elements do not exert their control via the peripheral elements. Complex "behavior," including habituation and dishabituation, can occur in the molluscan gill independently of the central elements (Peretz et al., 1976). It is probable that this is a fairly general rule applicable to a majority of true vegetative muscles.

Intracellular recordings of excellent quality have been made from longitudinal cells of the cockroach hindgut (proctodeum) by Nagai and Brown (1969). Mean membrane potential was 48 mV in the range 34-64 mV. In a nerve-muscle preparation (Figure 9.5.*4*), spontaneous neurally generated small EJPs were obtained. Several steps in EJP height could be obtained by graded stimulation of the nerve, showing polyneuronal innervation of individual cells. These summated and some also facilitated, others did not. Their amplitude was 2-10 mV, with a rise time of 5-7 ms, decay to half of 20-30 ms, and to baseline about 250 ms. All-or-none action potentials occurred spontaneously (Figure 9.11) or were evoked by neural excitation. These spikes had different forms, even in the same cell, ranging from simple roughly exponential decay to plateaus of various durations. Each action potential was followed by a contraction rising to peak in about 0.6 s and decaying to baseline in 2-3 s. These authors observed small local contractions occurring rhythmically and others occurring arhythmically. All regions contracted together strongly at defecation. They presume that the latter contractions are initiated by synchronous multiple neuronal discharges.

9.16. Invertebrate Gut Control

Annelid gut musculature is, like that of vertebrates and arthropods, innervated by nerves from opposite ends. The anterior organs, buccal cavity, pharynx, and esophagus, are innervated by anterior nerves that arise from the circumesophageal ring. The crop, gizzard, and intestine are innervated together by nerves that have attracted some interest because stimulation of the anterior one produced relaxation, whereas stimulation of the posterior one produced contraction in *Lumbricus* (Wu, 1939a,b). The nerves makes complex plexuses, and anteriorly there is an enteric "ganglion." There is reasonably good older evidence for an intrinsic nerve net (e.g., Millott, 1943a,b) and at least a dual set of superimposed nerves of central origin that probably exert antagonistic actions. These nerves are mixed, so the effects observed by Wu represented different relative dominances of one set over the other when both are stimulated together. In the more primitive specialized leeches, extensive plexuses with ganglia occur along the wall of the gut. Physiologically, the annelid gut, because of its phylogenetic interest, has attracted some lofty attention, for example, from von Euler (1961), who found Ad and NAd in a 1 : 5 ratio in *Lumbricus* gut. Later 5 HT was also found there, in a high concentration (Welsh and Moorhead, 1960).

It is regretable that modern physiological techniques have not yet been used on these systems. The larger muscle cells should be quite easy to record intracellularly. At the moment we can only guess at the control systems, but they seem to be remarkably similar to those of vertebrates, with the possible exception of proctolin and other peptides. This may, however, only reflect inadequate knowledge of the vertebrate system, but I think it unlikely that the pentapeptides are used as direct synaptic activators. Glutamate and GABA may well be the direct transmitters from neurons for excitation and inhibition respectively, with proctolin serving as an enhancing "modulator" at specific excitatory synapses. And 5HT is probably active in the periodic-wave mechanism, enhancing amplitude without greatly affecting frequency. This action is very likely mediated by the ubiquitous second messenger cAMP. Acetylcholine may act as a depressing modulator and NAd or Ad by accelerating rhythmic actions.

9.16.1. Modulation of Invertebrate Vegetative Muscles

Modulation of central-nervous control and the intrinsic myogenic, local nerve net or ganglion of invertebrate vegetative muscles has only relatively recently been the subject of active research, but it is already apparent that it is at least as complex as the equivalent for vertebrate guts. Invertebrates have had very much longer than vertebrates, in terms of numbers of generations, to evolve diverse rhythmicity and control mechanisms. The same basic substances appear to have been utilized, notably ACh, DA, 5HT, and Ad, plus various small peptides. In insects, ACh changes the contractile activity of guts (e.g., Freeman, 1966), but at rather high concentrations. The skeletal excitatory motor neurons release L-glutamate (see p. 492), and this excites the hindgut of the cockroach at 1.8×10^{-5} *M* (Holman and Cook, 1970). So also does L-aspartate, which has been postulated as a transmitter for abdominal muscle slow excitors. Glutamate also potentiates neurally evoked contractions and eventually desensitizes junctions to natural transmitter (Holman and Cook, 1970).

The evidence for involvement of biogenic amines, esepcially 5HT and DA, is much more convincing. Both have been found by direct analysis (Kusch, 1975), at least in cockroach. Nerves containing DA as seen by fluorimetry occur in locust foregut (Klemm, 1972). In cockroach hindgut, DA inhibits both myogenic and neurogenic activity at 8×10^{-5} g/ml (Holman and Cook, 1970). And 5HT has strong actions at very low concentrations, enhancing rhythmic contractions of the cockroach foregut at 10^{-9} *M* and the hindgut at 10^{-7} *M* (B. E. Brown, 1967). In *Locusta,* DA increases the amplitude and frequency of foregut contractions (Freeman, 1966): both Ad and NAd depress them.

While B. E. Brown (1967) was studying the cockroach hindgut, he found that electrical stimulation of their nerves produced a slow contraction of the longitudinal muscle only. And 5HT, ACh, Ad, NAd, GABA, and glutamate did not mimic this natural activity, so he correctly believed that he had discovered the presence of a novel transmitter, which he provisionally termed *gut-factor.* He later made the important identification of the substance as a pentapeptide and named it *proctolin* (B. E. Brown, 1975). It was isolated, identified and synthesized by Starratt and B. E. Brown (1975) and immediately became the standard for the new class of peptide transmitter/modulators. Its amino acid sequence is Arg-Tyr-Leu-Pro-Thr. Cook and Holman (1979b) consider that it is only one of a few peptides released by the gut nerves. Both 5HT and peptides are stored in the corpus cardiacum and act generally on the gut and other visceral organs via the blood. There are also neurons in the thoracic and abdominal ganglia of the cockroach that released proctolin when directly stimulated (O'Shea and Adams, 1981). In some molluscs a tetrapeptidamide called FMRFamide has been detected (Greenberg and Price, 1979). Both proctolin and FMRFamide are effective at very low concentrations and affect not only guts but also act on appropriate hearts, which they stimulate to beat when quiescent and to beat faster and more strongly when active, in addition to affecting BT in some SM cells. FMRFamide is especially potent in promoting prolonged tonic contraction of the radula protractor of the whelk *Busycon*. It is suspected that these peptides do not act directly on ion conductances but indirectly via cAMP.

The involvement of DA is probably widespread but has been authenticated only for the stomatogastric ganglion of the lobster. This is innervated by a DA releasing neuron (Kushner and Ono, 1978). Following isolation of the ganglion, one of the two separate rhythmic outputs generated by it, the pyloric rhythm, is sometimes present but often not. However, it can always be initiated by addition of 10^{-6} *M* DA and once so initiated continues as long as the DA is present (W. W. Anderson, 1980). The stomatogastric ganglia that show the rhythm probably have a naturally high DA level at the time of isolation. Thus DA is a modulator of cells capable of generating the rhythm in collaboration. Not only is it responsible for gating activity, but the detailed pattern is a function of the amount of DA above the threshold. The other rhythm generated by the ganglion, the gastric mill rhythm, is always present in the isolated ganglion and is not sensitive to DA.

An extremely interesting neuron recently located and identified in two lobster species is likely to revolutionize studies on the neural control and modulation of invertebrate guts. This is a large unpaired motor neuron whose cell body lies in the esophageal ganglion, with a T branch whose axons innervate left and right esophageal dilator muscles (Moulins and Nagy, 1981). It is addressable by an intracellular electrode and gives small 30-mV spikes that correlate with EJPs of about 10 mV in dilator muscle cells. The cell is termed the first opener dilator (OD_1). Impulse initiation zones occur not only on the principal neurite but also on each of the two axons. These permit independent or unified left and right dilation control and the coordination of phasic and tonic activities of the anterior gut.

9.16.2. The Roles of Prostaglandins

Prostaglandins (PG) are ubiquitous, being produced by all major tissues, and there are many different ones forming a family of substances. Knowledge of their functional relations is still in its infancy, and well-defined roles have not been ascribed to most. Multiple actions and mutual interactions are apparent. Several have profound, often antagonistic, actions on SMs, especially of the uterus, intestine, and respiratory tract. The PGFs contract, but the PGEs relax, respiratory-tract muscle, but both strongly contract the pregnant uterus and are thought to be involved in a major way in parturition. Both also promote contraction of longitudinal muscle of the gut, but whereas PGFs also contracts circular muscle of the gut, PGEs relax it.

9.17. Patterned Activity Controlling Specific Gut Movements

Parts of the alimentary canals of some invertebrates have evolved an extent of sophistication in function and neural control that is unknown in any vertebrate. Among molluscs there can be a complex opening, gulping, biting sequence that occurs in many nudibranchs such as *Tritonia* and in *Octopus* and squid. In predatory gastropods such as *Navanax*, the prey is literally sucked into a very rapidly expanding pharynx and completely engulfed. Phytophagous browsing molluscs have a complex eversible cyclically rasping multitoothed radula. All of these are operated by paired buccal ganglia that have been the targets of much recent research because they contain identifiable neurons of large size and because they continue to function following complete or partial isolation.

Although they are of SM type ultrastructurally and operate within a cylindrical soft body, within the buccal masses several discrete anatomical muscles are easily discerned. These include dorsal and ventral paired retractors and protractors, dilators, compressors, elevators, depressors, and tensors – a degree of complexity matched only by the entire leg or arm musculature of a vertebrate. The complete description of one such system alone can fill a large Ph.D. thesis and has been very popular for that purpose, so that already a dozen or so have been described. The controlling ganglia in each have clearly evolved with the sole function of controlling the anterior gut in its food-procuring and swallowing activities. They have done this by developing a combination of individual cellular properties and cooperative interactions based upon connectivity, with not only chemical synaptic excitation and inhibition, but also electrical coupling. Among these, rebound excitation from inhibition appears to have been especially extensively used. This is particularly true in the stomatogastric ganglion of the lobster (Selverston et al., 1976).

Selverston has likened the lobster stomatogastric ganglion to a model motor programming element within a central nervous system. He thinks of it as analogous to an element controlling rhythmic limb movements, as in locomotion. If we accept this view, internal coordinated motor activity mincing up food and external locomotory activity as in walking and swimming are both to be regarded as forms of animal behavior presenting similar basic neural problems. The only difference is that the external movements cause interactions with the environment, which must therefore add, by visual, tactile, proprioceptive, and chemical inputs, an element of superimposed control that is absent for functioning of the internal organs.

With this view, control of these complex invertebrate gut regions should be compared with control of skeletal muscle rather than with control of SM, but once again the traditional distinction, based upon a histological feature, is seen to be without direct classificatory value.

9.18. Invertebrate Blood Vessel Control

Acetylcholine may be a modulator for molluscan blood vessels. All early attempts at recording from *Octopus* brains were frustrated, largely due to an inability to keep the brain alive, until it was discovered that surgery causes brain blood vessels to constrict. It was then discovered that the constriction can be prevented with a small amount of ACh, which presumably polarizes the circular SM cells of the blood vessels and counteracts the unknown agent causing constriction. Aortal blood vessel SM cells of *Aplysia* are the only ones of invertebrate to have been penetrated intracellularly, and they are unique in that motor neurons innervating them have been identified and studied directly. The muscle cells are innervated by one or another of three motor neurons whose cell bodies are located in the abdominal ganglion: LB_{VC1}, LB_{VC2} and LB_{VC3}. Each of these is cholinergic and excitatory, causing vasoconstriction. The electrical response is a facilitating EJP, increasing from 2-5 mV to 5-12 mV at about 20 Hz. The rise and decay rates are extremely slow, a single EJP having a duration of about 250 ms. There is no electrogenesis. Each cell receives similar synaptic inputs, and the natural firing patterns they receive are closely similar, ensuring synchronous activation of the muscle cells.

9.19. Invertebrate Oviduct

Among the few vegetative organs other than guts to have been studied is the oviduct of a moth (Cook et al., 1980). The muscle cells, which are cross-striated, and their innervation were examined in a light microscope and scanning EM. The Z regions of the 2–8-μm SL cells are composed of well-separated elongated Z bodies, suggesting supercontractibility. The thin–thick filament ratio is 5 : 1. The organ posses intrinsic rhythmicity, with two different rhythms: an overall periodic shortening and peristaltic waves. Different types of electrical activity, bursts of spikes or slow waves are associated with each.

10

Cardiac Muscle

Movements of the auricles and the ventricles occur successively but so harmoniously and rhythmically that they appear to happen together. William Harvey†

10.1. General Features of Hearts

Soon after the evolution of the first forms of three-layered organisms, cavities within the middle layer become useful as reservoirs of metabolities and the oxygen needed to convert them to energy. When lined with contractile cells, these cavities became better defined, and contraction/relaxation of the muscle cells aided transport of the contents. The muscle-lined cavities eventually differentiated into tubular elements strictly concerned with transporting the fluid, and others, termed hearts, to be used actively for driving fluid into and along the tubes. The former almost universally retained the primitive SM cell structure and control mechanisms described in Chapter 9, becoming blood vessels. Two principal system forms evolved. The first is *open,* in which the blood, propelled by the heart, emerges from open-ended arterial vessels to percolate through the tissues, from which it is collected by veins and transported by them back to the heart. The other is *closed,* the arterial vessels leading into smaller, multibranched tubules that converge to form veins leading back to the heart, which, in almost all forms, developed the unique characteristics that greatly enhanced its power and speed compared with the muscle of the vessels. Ultimately it became distinguished from other forms of muscle as a specific cardiac type. The circulatory systems of all molluscs and arthropods are open, those of vertebrates are closed. Although probably evolved independently by the major classes of invertebrates and by chordates, certain features, both anatomical and physiological, of their hearts are remarkably similar. The most notable of these is a unique form of cross-striation in cells that otherwise retain many of the aspects associated with SM cells. Another is action potentials that, though overshooting, propagated, and all-or-none, have a long-lasting phase of depolarization before a relatively abrupt late repolarization restores the membrane potential.

Beyond these points the similarities fade. Initiation of the heartbeat is myogenic in the majority of hearts, but there is a split within the arthropods: the hearts of arachnids and the higher crustaceans-decapods (crabs and lobsters) are all neurogenic, relying upon specific local ganglia to pacemake. No special advantage seems to be associated with either mode: both are subject to satisfactory pacemaker control from the central nervous system and modulation from neural and blood-borne substances. Likewise, whether the system is open or closed has not bestowed any clear advantages: some disadvantages are attributable to both types.

†W. Harvey, *De Motu Cordis,* 1628.

Most circulatory systems are unidirectional, because of both the mechanics of the heart pump and the presence of valves in the heart and return vessels. But the primitive condition in invertebrates, as exemplified by some oligochaete annelids, insects, and tunicate protochordates, is bidirectional.

The morphologies of hearts are highly varied. Invertebrate hearts are dorsal, those of chordates, except balanoglossids, ventral. In polychaete annelids there are paired hearts, with separate auricle and ventricle, but five pairs of simple undifferentiated tubular "hearts" occur in the earthworm. These are located dorsoventrally, and commonly collect blood from a ventral vessel and propel it into a midline dorsal vessel. The anterior part of this vessel is contractile and more truly deserving the term heart; indeed, the lateral contractile vessles are often called pseudohearts. Blood flows from it anteriorward, thence down through the pseudohearts and back along the two ventral vessels. Hearts of protochordates are themselves very different. That of adult tunicates is a simple tube. It beats several times in a similar manner peristaltically, driving blood in one direction. Then the beat suddenly reverses for about an equal number of contractions, driving blood in the opposite direction.

Balanoglossids and related protochordates have a simple heart at the anterior end of the dorsal vessel, but pulsing of the blood seems to be achieved by contraction of the muscular wall of the pericardium rather than the heart itself.

Molluscs have chambered hearts in which one or two atria pump blood into a single ventricle that in turn drives it into short, open-ended arteries. The veins are contractile, driving blood forcibly back into the atria. Separate gill hearts occur in cephalopods and drive the blood through the gills. The main heart has separate anterior and posterior aortal exits from the ventricle.

Arthropod hearts consist of a long dorsal tube, commonly the entire length of the body, with simple one-way openings from the surrounding pericardial sac. Blood collects in the sac from around the tissues after the contraction of the heart, enters the heart, then exits from all of several arterial vessels at the same time during contraction. Crustacean hearts are generally broad, squat thoracic organs, with a few dorsally placed openings from the pericardium, and anterior, posterior, and large centrally placed ventral arteries. They are very tough compared with the thin, transparent, delicate hearts of insects.

Arachnid hearts occur in the abdomen, tapering posteriorly; they send blood into the thorax through an anterior aorta. A long dorsal cardiac nerve runs along the heart and sends branches into its musculature, where there are local ganglion cells. The *Limulus* heart is innervated by paired lateral nerves, but these are in turn connected by paired branches to a central narrow ganglion complex running the entire length of the heart. The details of this complex are not known, but many ganglion cells are included. By contrast, the heart of the largest lobster is operated by a minute ganglion containing only nine neurons, the smallest discrete neural network known in the entire animal kingdom.

These differences all point to there having been late, independent evolutionary origins of hearts and their control systems. Surprisingly, none seem to have utilized reflex actions, either for producing rhythmicity or for modulation. A nearly complete autonomy for hearts has been a universally desirable functional "goal" in their evolution.

The primitive vertebrate heart, as seen in elasmobranch fish, is an anterior, ventral S-shaped tube that collects blood from dorsal, lateral, and ventral veins into a dorso-posterior sinus. From there it enters a single atrium before passing into the strongly muscular thick-walled auricle. Amphibia have a three-chambered heart, with two atria (auricles) and a single strongly muscular ventricle. A combination of morphofunctional devices ensures separate blood fractions. Reptiles have a partially divided ventricle, whereas birds and mammals have four-chambered hearts. But even with the evolution of a separate pulmonary circulation, an essentially very primitive pacemaking and control system has been retained. All component cells except perhaps some of those set aside for conducting excitation, which were independently evolved by only birds and mammals, are myogenic.

The pumping region of the vertebrate vascular system comprises cells that are all cross-striated. They merge almost imperceptibly into SM-cell tubes at input and output sites, emphasizing the ease of transition between the two. As we saw in the last chapter, for many organs invested with SM, the SM functions as a whole rather than as a complex of contributing units. Yet nevertheless, there are units within the whole that can both function independently and be controlled independently. The largest number of units that can be said to exist in a large vertebrate heart is five, and these are so tightly interlocked functionally that their independence is negligible compared with their dependence. In their evolution, factors that contributed to unified operation were paramount. It is probable that evolving hearts progressively lost innervation they primevally possessed because it impeded subordination of parts to the whole.

Likewise with ultrastructure. Intercalated discs are a structural innovation peculiar to cardiac muscle that act like a form of super-leaky junction, and may have started out as such. Branching and anastomosis were also probably absent primevally except as occasional genetically caused aberrations, but since selective functional advantage was bestowed on hearts possessing them, they evolved to the wide extent currently seen. The neighboring tubular muscle of the vessels retained the advantages offered by SM-cell properties – slowness, economy in maintenance of tone, and cooperative action, with local reflex control – yet superimposed central nervous and humoral control overall.

Finally a word of caution: heart muscle should not be compared with the average skeletal muscle in normal operation. In all except exceptional circumstances, most skeletal muscles are only partially activated. In a heart, all the cells are activated to almost the fullest extent of which they are capable, with every beat.

10.2. Ultrastructure of Cardiac Cells

10.2.1. Myofilaments

Somewhat surprisingly in view of their probable evolutionary origin, the myofilaments of cardiac muscle cells are identical to those of their skeletal counterparts. In vertebrates the SL is uniform overall, the largest reported range in a single tissue being only 1.6–2.0 μm, in cat ventricle (Sperelakis, 1970). Mean SLs range from 1.3 μm for finch ventricle (Bossen et al., 1978), through 1.5 μm for a teleost fish atrium (Santer and Cobb, 1972) to 3.0 μm in the amphibian (newt) ventricle (Hirakow, 1971). The majority average 2.2 μm, as in vertebrate skeletal muscle. A band length range associated with the above is 1.0–2.4 μm, sufficiently great to mean that the range is of functional significance. The greater lengths occur only in a few cold-blooded vertebrates where a very slow heartbeat may offer a desirable economy. Some examples are compared with some values for representative invertebrates in Table 10.1.

The vertebrate-heart-cell ultrastructure closely resembles that of vertebrate skeletal muscle cells, except for SR, which is markedly different (Figure 10.1). A filaments are

Table 10.1 Sarcomere and A-Band Lengths and Thin–Thick Filament Ratios of Cardiac Muscle Cells

Phylum, Class	Genus	Sarcomere Length Mean or Range (μm)	A-Filament Length (μm)	I : A Ratio	Reference
Mollusca					
Lamellibranchia	*Cassostrea* (oyster)	1.3	1.0	12 : 1	Hawkins et al., 1980
Gastropoda	*Busycon* (whelk)				
Cephalopoda	*Rossia* (cuttlefish)	3.5	2.7	11 : 1	Sanger, 1979
		2.0–4.5	1.7–3.9	6 : 1	Jensen and Tjønneland, 1977
Arthropoda					
Arachnida	*Limulus*	2.5–6.6	2.5	7.5 : 1	Sperelakis, 1970
Crustacea	*Daphnia*	1.5–3.0	1.3–2.6		Stein et al., 1966
	Squilla	3.6–4.5		4 : 1	Irisaw and Hama, 1965
Insecta	*Hyalophora* (moth)	2.0	1.8	11 : 1	Sanger and McCann, 1968a
Chordata					
Pisces	trout atrium	1.5	1.2	2 : 1	Santer and Cobb, 1972
Amphibia	newt ventricle	3.0	1.6	2 : 1	Hirakow, 1971
	salamander	1.5	1.2	2 : 1	Sommer and Johnson, 1979
Aves	finch ventricle	1.3	1.0	2 : 1	Bossen et al., 1978
Mammalia	rat ventricle	1.5	1.2	2 : 1	Page and Fozzard, 1973
	guinea pig papillary	2.3	1.5	2 : 1	Simpson et al., 1973

always about 18 nm in diameter, with hexagonal array and 40–45-nm center-to-center spacing. I filaments are, as in all muscles, about 7 nm thick, in exact hexagonal array and 2 : 1 ratio. There is a conspicuous M region in which up to five sets of M bridges are discernible in longitudinal sections. The Z bands are strongly marked and clearly formed, as are those of rabbit psoas (see p. 73), from two sets of Z filaments. The I filaments enter the Z region in a square array and are crosslinked to those of the contiguous sarcomere by an arrangement of the Z filaments exactly like that of skeletal muscle, with conspicuous swastika patterning.

A single N line is commonly present in vertebrate heart cells fairly close to the Z line. Much I-filament branching seems to occur at this region.

There is much more variety in invertebrate heart cell structures than is seen in vertebrates, although there are also some remarkable similarities, notably between those of insects and gastropod molluscs. The known SL range is from 1.7 μm, in several insects, to 5.0 μm in a spider, but is narrow compared with that of skeletal cells. A majority of the SLs are narrow, barely 2.0 μm wide; slow hearts are not especially valuable. The thin-thick filament ratio is generally higher than 2 : 1, as it is in invertebrate skeletal cells, except in lobster heart (D. S. Smith and Anderson, 1972). Moth hearts have a 4 or 5 : 1 ratio (Sanger and McCann, 1968a, b), like most arthropod skeletal cells. Both moth and insect hearts have their I filaments attached to ovoid dense Z bodies, giving the impression of a capability for supercontraction (Figure 10.2).

Individual cell diameters vary greatly because, as in vertebrate hearts, they anastomose. Diameters reach 10 μm in the mollusc *Busycon,* 25 μm in the moth *Hyalophora,* and 28 μm in the xiphosuran *Limulus.* None possess an IT system in the strict sense, but in insects and crustaceans there are deep invaginations at Z-line level. These are functionally equivalent to ITs because they give off longitudinally running branches that make dyadic and even triadic contacts with SR in the lobster (D. S. Smith and Anderson, 1972).

In mollusc hearts, which lack equivalent invaginations, some have an extensive SR that makes close contacts with the surface PM such as are common in SM cells. But even this SR system is missing in the heart of the clam *Venus* (Kelly and Hayes, 1969), which has a DB thin-thick-thin filament organization, but with the Z bodies scattered, as in *Peripatus* skeletal muscle. The authors wanted to call these a special class of "smooth cardiac" cells, but in the classification adopted in this book (see p. 89) they are to be considered as disorderly striated. Lack of alignment of contractile units need not confer any speed or other disadvantage provided E-C coupling calcium can be got to them quickly. However, in the absence of SR and ITs, diffusion from the surface membrane is the only possibility, and this is necessarily slow.

10.2.2. SR and ITs

Invaginated tubules occur in mammalian heart cells but not in those of birds or other lower vertebrates, which rely on SR's contacting the cell membrane for coupling. This is interesting, suggesting that IT systems in general have evolved independently and are not of ancient evolutionary origin. Mammalian cardiac ITs enter radially at the levels of peripherally located Z-bands, but make longitudinal branches that link up with their neighbors. These may be straight or follow tortuous paths, but they soon bend or branch until at about the outer margins of the A regions they make transversely aligned links between dual terminal cisterns of SR. The resulting triads are very similar to those of mammalian skeletal muscle, with regularly spaced small projections between the three

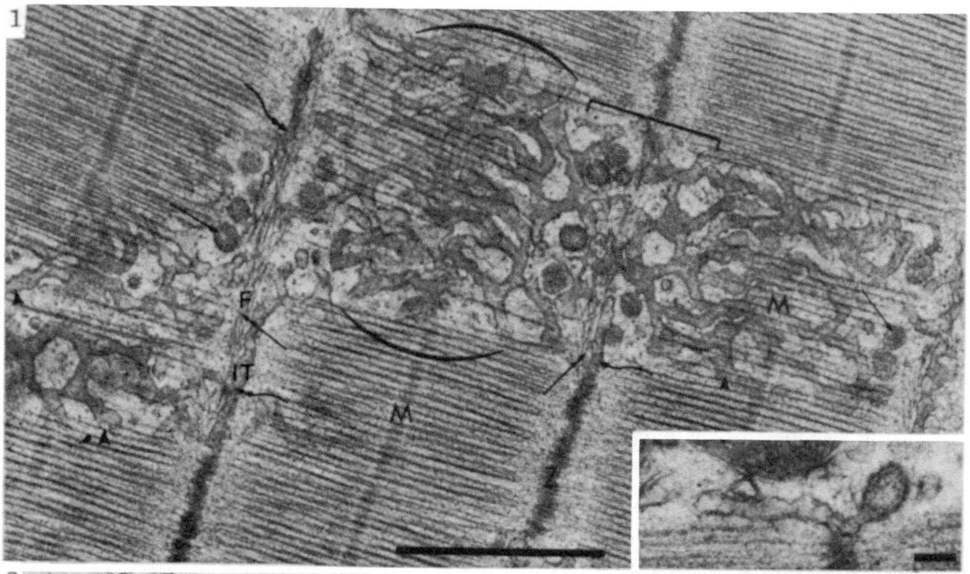
1
F
IT
M
M

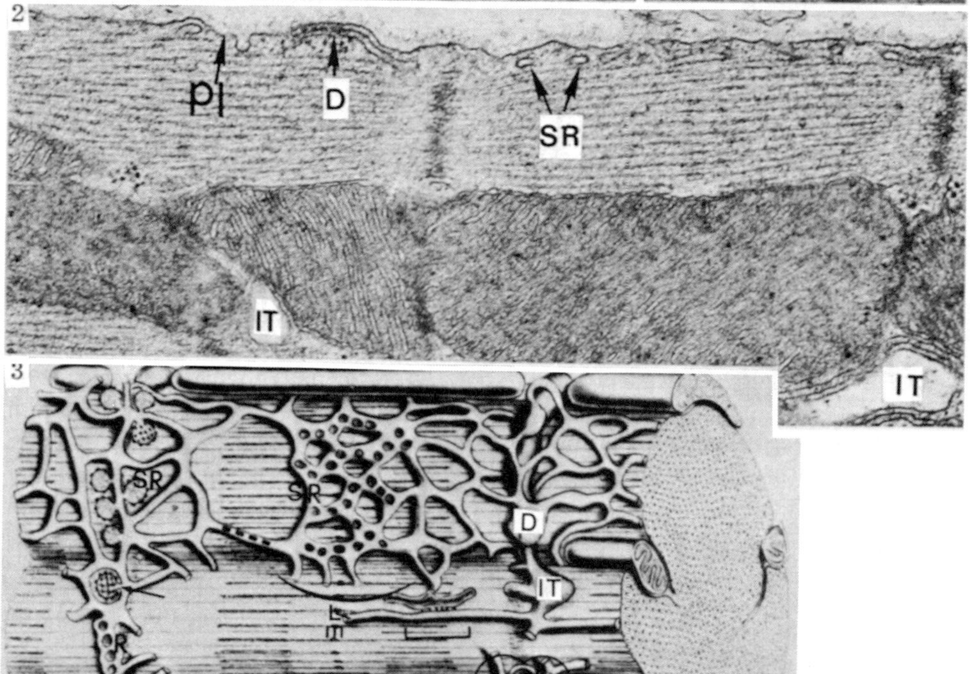
2
pl
D
SR
IT
IT

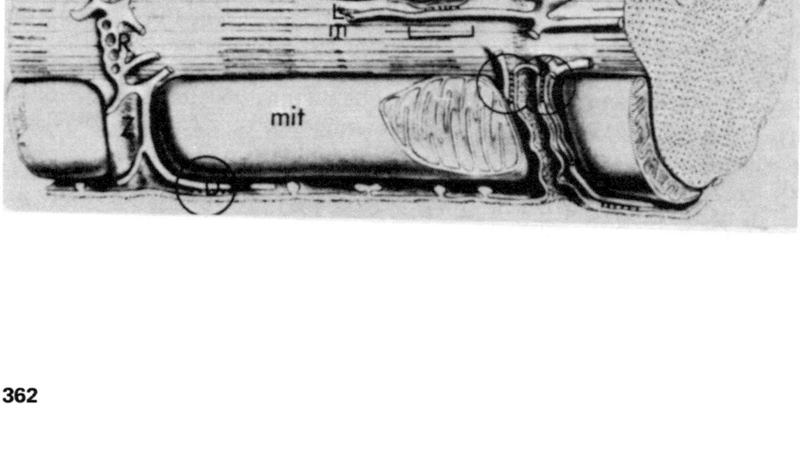
3
SR
SR
D
IT
mit

tubules. In heart cells with ITs, SR/IT contacts can occur almost anywhere in addition, but especially at branch points of the longitudinal IT elements.

At the Z region, a characteristic fine network of SR is formed, with many small coated sacs or globules. To this unique network with sacs, occurring in mammalian cardiac cells, the term *corbular* (beaked – after blackbird) has been applied. Elsewhere SR is heavily perforated and closely similar in appearance to that of skeletal muscle.

In vertebrate and invertebrate cardiac muscle alike, there is commonly a lot of SR, regardless of whether or not there are any ITs. Although there are no really reliable estimates, SR appears to occupy about 30% of the total volume. But some hearts have much less than this, even in closely related species.

10.2.3. Is There an Internal Framework in Cardiac Cells?

In many cardiac cells, structural fine filaments connect Z bands to each other (Figure 10.1.*1*) and peripherally-located ones to the surface membrane (Sommer and Johnston, 1979). Together these filaments and the Z bands must form an internal elastic skeletal framework. Much of the space between cardiac cells is occupied by collagen strands, which form their own independent gross framework.

10.2.4. Mitochondria

Mitochondria might be expected to be rather uniform in an organ having such uniform function, but there are wide variations. These are principally seen in sluggish cold-blooded forms. The newt has few, small mitochondria but the more active frog has numerous, larger ones. In birds and mammals generally, mitochondria are both large, about 2 × 1.5 × 1.0 μm and abundant: the largest reach 6 μm long. They are located between myofilament clusters, commonly in continuous rows, an arrangement otherwise seen only in insect flight muscles. The total volume occupied by mitochondria ranges from about 2% in the amphibia to 35% in most mammals. Branched mitochondria are not known in heart muscle.

10.2.5. Substrate Particles

Glycogen, often in the form of rosettes, is present in large quantities in cardiac cells. The particles lie in rows between and among myofilaments and myofilament clusters. Located near the glycogen particles are numerous small lipid droplets. Large granules of unknown composition and function sometimes occur in mammalian hearts. Some are lipofuscin, protein-breakdown products associated with aging. Golgi bodies are common, with associated vesicles, but what is being synthesized by them is not known.

Figure 10.1 Ultrastructure of vertebrate cardic cell. (*1*) Opossum cardiac cell showing corbular SR, see above (p. 363), characteristic of vertebrate cardiac cells. *Inset* shows single globule of corbular SR continuous with a tubule invaginated at Z-line level, one of which may also be seen at left (*double-headed arrows*). The fine fibers (*F*) near the Z line at left are anchoring fibers running from Z to Z; the outermost Zs attach to the S1. (*2*) Mouse left ventricle cell showing SR near surface membrane (*Pl*) and making dyadic contact (*D*); these cells also have wide ITs. (*3*) Three-dimensional drawing of mammalian cardiac cell. The IT system underlies SR, makes dyadic contacts with it, and connects with other invaginations via longitudinal branches (*L*). Surface dyad is encircled. [Reprinted with permission from Sommer and Johnson in *Handbook of Physiology*, Vol. 1, Chap. 5, 113–186, Copyright 1979 The American Physiological Society.]

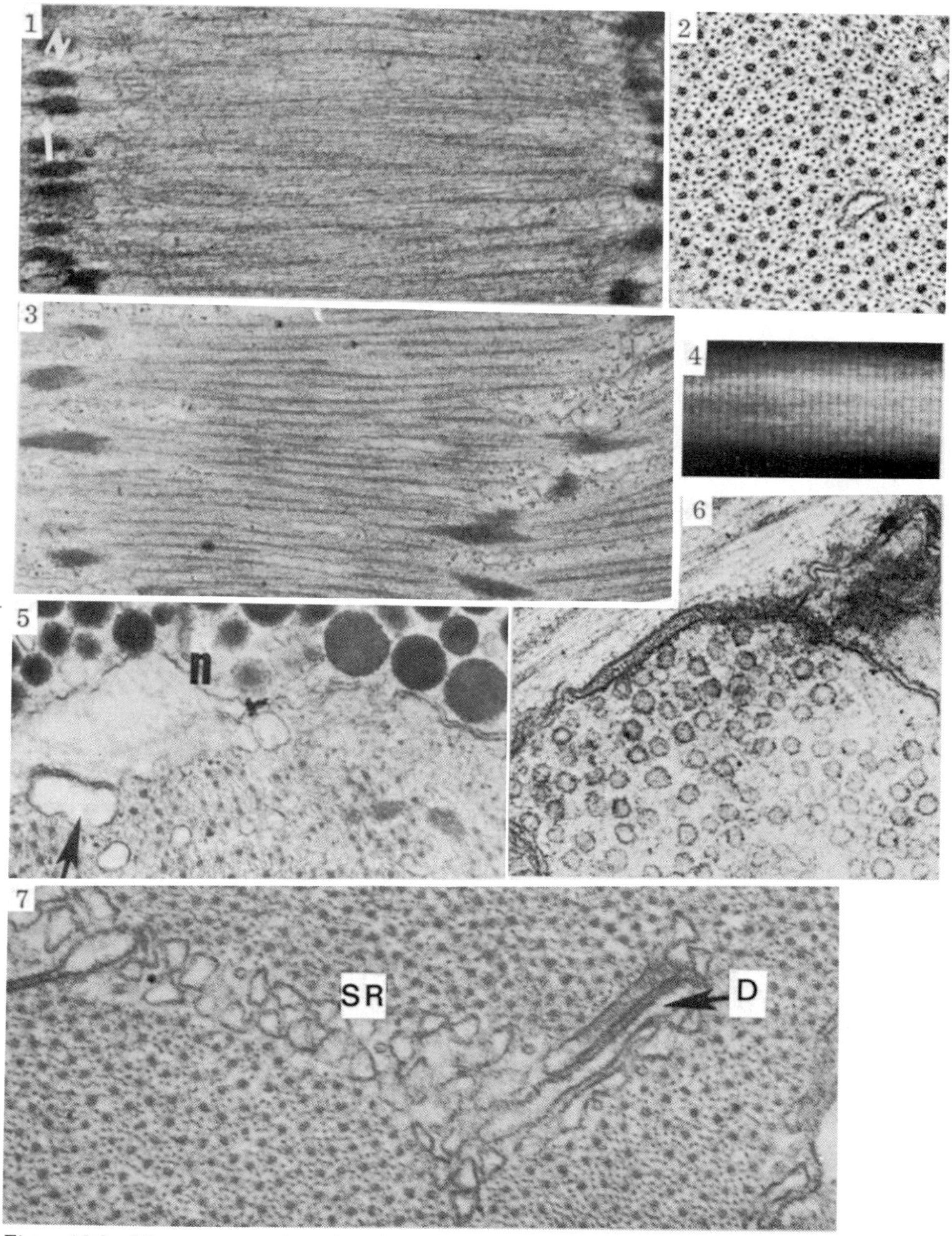

Figure 10.2. Ultrastructure of moth and mollusc hearts. (*1*) EM of LS heart (alary muscle) of moth *Hyalophora cecropia*. Note ovoid Z bodies suggesting capability of supercontraction. (From Sanger and McCann, 1968b.) (*2*) EM of TS same as (*1*). (From Sanger and McCann, 1968b.) (*3*) EM of TS heart of gastropod mollusc *Busycon*. Note remarkable similarity to (*1*). (*4*) Portion of crystal of paramyosin extracted from (*3*). (*5*) Nerve terminal containing large DCVs, from (*3*). (*6*) Nerve terminal containing clear vesicles, from (*3*). (*7*) EM of TS of (*1*) showing SR and internal dyad (*D*). [3–7 reprinted with permission from J. W. Sanger, *Am. Zool.* 19, 9–27 (1979). Copyright 1979 the American Society of Zoologists.]

10.2.6. Organization of Different Regions

Very few detailed investigations of the organization of component muscle cells have been made, and it is usually taken to be similar in different regions. But in the mollusc *Dolabella,* the cells of the ventricle are arranged neatly (Kuwasawa, 1979), whereas in the atrial region, they are interwoven in a complex manner. There is more contractile material in the ventricular cells: a similar distinction may be widespread.

10.2.7. Cell Contacts

A truly remarkable convergence in ultrastructural/functional evolution has occurred in the form of special contacts between component cells of hearts of unrelated animals, as was first realized during an ultrastructural study of the heart of a moth (Sanger and McCann, 1968a). This is true both of complex end-to-end juxtapositions and lateral ones. At the terminal abutments, there are many complex foldings with broad dense regions on the inner sides of the two membranes. The membranes do not come close at these regions as they do at nexuses, but there is fine granular material between them. It is these folded regions that are responsible for the appearance that has long been termed in vertebrate hearts an *intercalated disc* (Figure 10.3). Fine filaments are associated with these areas, which are therefore termed *interfibrillar areas.* There are sometimes small regions in such contacts, where the membranes extend into the space and touch, called *spot nexuses.* End-to-end cell contacts are much commoner than lateral ones.

Intercalated discs tend to come near Z regions, but they also cut across the middle regions of some sarcomeres. Lengths of thin filaments are adjusted to accommodate the extensive infoldings. Although present in fish, characteristic fibrillar areas of intercalated discs can be absent, as is the case in the newt *Necturus.* Very similar ones, shown in Figure 10.3.*2*, occur in lobster hearts (Anderson and Smith, 1971), but no broad or spot nexuses occur within them, the gap reducing maximally to 7.5 nm from an average of 40 nm. Dyes do not pass across these discs, but electric current does so readily. Zigzag intercalated discs are also found in some hearts of molluscs (Figure 10.3.*3* – Kuwasawa, 1979), which also have a variety of junctional complexities. By contrast, hearts of ascidian protochordates have only simple tight junctions. Curiously, similar discs occur in the fast adductor of the scallop (Sanger, 1979), the only cross-striated skeletal muscle cells known to possess them. Although I filaments are attached to widely separated ovoid DBs in most mollusc hearts, they are attached to Z bands that are continuous sheets that are not perforated in hearts of scallops. Thick filaments of most molluscan hearts are not especially thick, but in *Busycon* they reach 60 nm and contain paramyosin (Figure 10.2.*4* – Sanger, 1979). There is a very high thin–thick filament ratio, up to about 30 : 1, most thin filaments clustering together and only a few making orbits around the thick ones.

Lateral junctions are made between heart cells of both vertebrates and invertebrates. They are commonly identical to simple tight junctions found in many tissues, in which PMs of abutting cells are in very close contact. Such a junction is generally called, as in SM, a nexus; it may occur in addition to more-complex ones. One level of greater complexity is a contact very closely similar to the desmosome, as seen in transverse section. These contacts may occur anywhere, but the desmosomes are especially prominent at lateral contact points. The individual contact nexus, like other tight junctions when seen in surface view, forms a regular hexagonal array, each hexagon having a tiny central dot. In some contacts there is a marked patterning of nexuses, as viewed from the surface in freeze-fracture images (Figure 10.4.*6*).

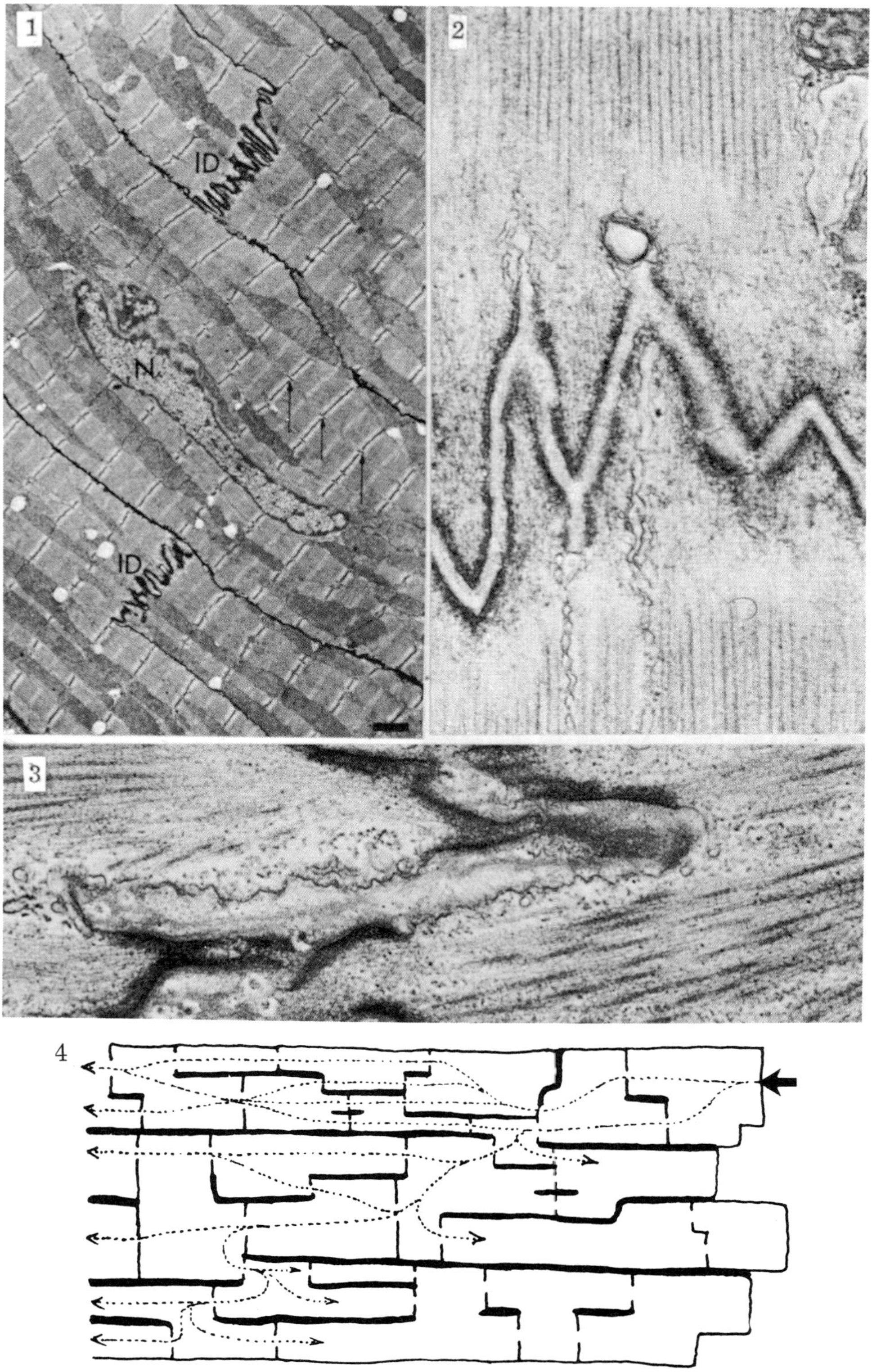
1
ID
N
ID
2
3
4

Blood vessels are not just connective-tissue tubes, but regulatable muscles. The muscle cells are, however, smooth cells quite unlike cardiac cells. There is therefore a region where the two kinds are linked. At this boundary, each kind is heavily interlaced with collagen and it is the common collagen that actually holds them together.

10.3. Functioning of Cardiac Cells

A majority of hearts, including all those of vertebrates, are myogenic, but a few invertebrate hearts of diverse phyla have evolved quite differently and are neurally driven, either from an independent local cardiac ganglion, from a central ganglion, or even from the entire nerve cord, as in the leech.

10.3.1. Neurogenic Hearts

No major fine-structural features distinguish the neurogenic hearts from myogenic ones. The details of innervation and neural control of these hearts are poorly known, the best known being that of *Limulus*. Each myocardial cell is innervated by at least six motor axons, three from the nearest ipsilateral segmental nerve, three from an adjacent nerve, and probably others from more-distant nerves (Abbott et al., 1969). Each of these axons produces an EJP, a single one of which causes the cells it innervates to twitch weakly. The excitatory transmitter is probably L-glutamate, as it is at skeletal junctions. The ganglionic chain, with a few-score motor neurons altogether, produces a train of impulses lasting altogether about 2 s, at a declining frequency, with a mean of about 10 Hz, in each motor neuron. At the level of the contractile units, which themselves appear to contribute no electrogenesis, the many arriving EJPs summate with each other to produce a plateau of depolarization. After a few nerves in the chain are cut, these may be reduced in number to the point where the contribution of a single neuron can be seen (Figure 10.5.*2*). The whole produces in each component cell simultaneously, an overall depolarization in which most individual EJPs cannot be seen. This summed event bears a remarkably close resemblance to the unitary cardiac action potential occurring in an electrically fully excitable vertebrate or invertebrate heart cell. It serves exactly the same purpose, but it is produced very differently indeed.

The ganglion chain serves as a substitute for three functions: pacemaking, conduction, and coordination. Inhibition and modulation are the functions of a totally independent nerve supply. The inhibitory nerve works like a typical arthropod inhibitory axon, by being distributed to all the muscle cells and having distributed junctions close to the excitatory ones, releasing GABA, as do peripheral skeletal muscle inhibitory axons. The GABA reduces the depolarization produced by the EJPs because it activates polarizing

Figure 10.3. Intercalated discs (IDs) (*1*) EM of LS of Vertebrate (finch) heart. Cells lack ITs but branch and join end to end via (*ID*). (Reprinted with permission from Sommer and Waugh (1976) *Am. J. Path* **82**, 192–232. Copyright Lippincott/Harper & Row.) (*2*) TEM of LS of Invertebrate (lobster) heart, showing details of an ID. (From Anderson and Smith, 1971.) (*3*) Tortuous IDs of mollusc *Busycon* heart. (From Sanger, 1979.) (*4*) Diagram of expected mode of current flow through heart muscle cells via low-resistance routes associated with IDs, following injection into cell at upper right. (Reprinted with permission from Sommer and Johnson (1979) In: *Handbook of Physiology*. Vol. 1, pp. 113–186. Copyright The American Physiological Society.)

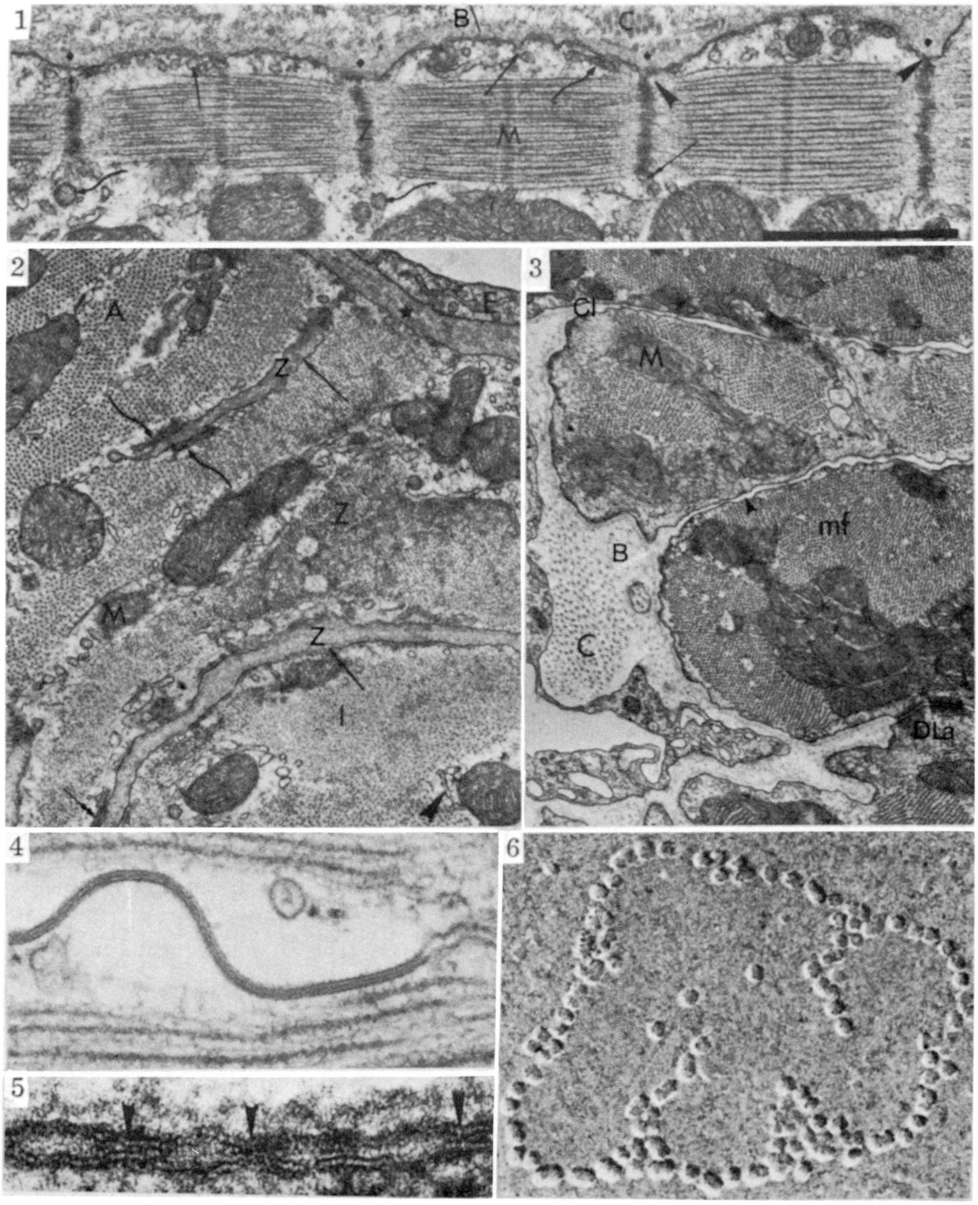

Figure 10.4. Ultrastructural special features of vertebrate hearts. (*1*) Outer membrane of cell of opossum heart showing looped Sl, attached to Z regions (arrowheads), a condition that is common in invertebrate skeletal cells. Immediately outside the membrane is a basement material (*B*), and outside this are numerous collagen fibers (*C*). SR is closely associated with Z material also (small arrows). (*2*) Z tubules (*Z*) with dense contents, a feature also of many insect skeletal cells (Figure 3.14) in cardiac muscle. Specific densities occur on it at intervals. (*3*) Clefts (*Cl*) invading cells of salamander cardiac muscle, and desmosome lateral connections (*D La*). The basement material (*B*) continues over a cleft. [1–3 reprinted with permission from Sommer and Johnson (1979) In: *Handbook of Physiology*. Vol. 1. pp. 113–186. Copyright The American Physiological Society.] (*4*) Nexus of dog cardiac muscle. (From Kensler et al., 1977.) (*5*) High magnification of nexus of frog cardiac muscle with spot contacts (*arrowheads*). (From Kensler et al., 1977.) (*6*) Same as (*5*) but freeze-fracture face view of disposition of contacts in nexus region. (From Kensler et al., 1977.)

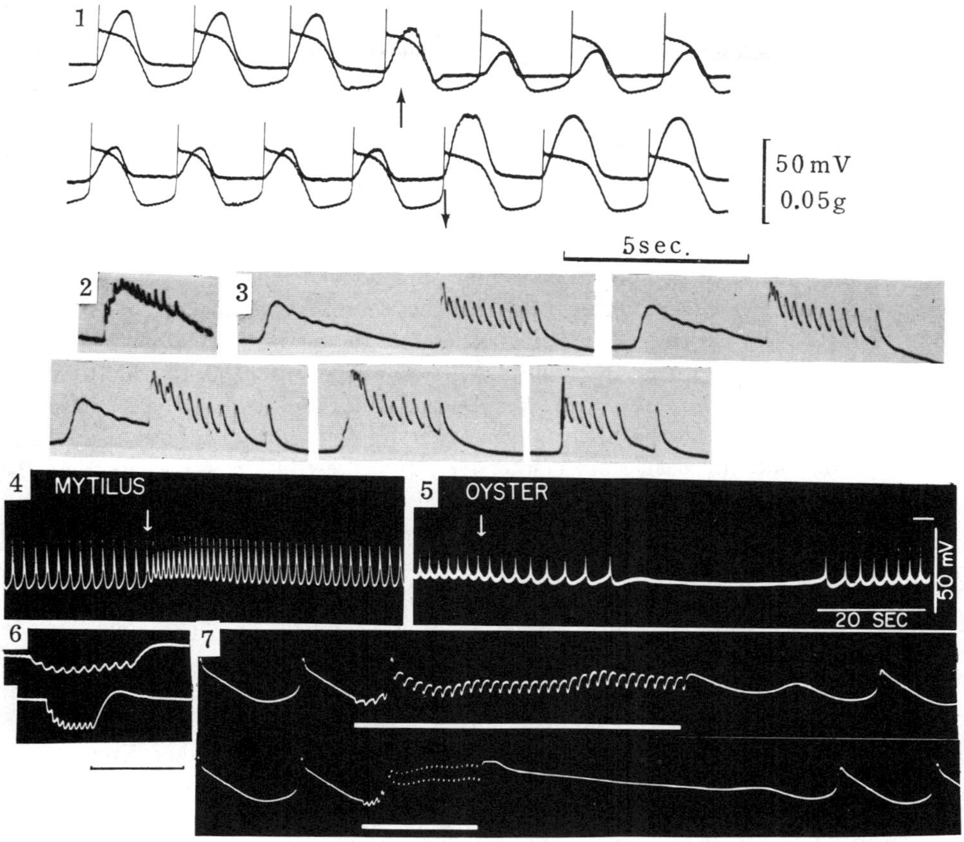

Figure 10.5. Electrical activity in some invertebrate hearts. (*1*) Oyster (myogenic), intracellular record (*lower trace*) with tension (*upper trace*), and the effects of stretch, between arrows. (From Nomura, 1963.) (*2*) *Limulus* (neurogenic), intracellular record. (From Abbott et al., 1969.) (*3*) As (*2*), but after sectioning of ganglion chain causing loss of coordination. Dual inputs to recorded region reveal that the exciting wave is formed by bursts of EJPs that sum initially. (From Abbott et al., 1969.) (*4*) Accelerating action of ACh on myogenic *Mytilus* heart. (From Shigeto, 1970.) (*5*) Depressing action of ACh on myogenic oyster heart. (From Shigeto, 1970.) (*6*) IJPs in oyster heart. (From Shigeto, 1970.) (*7*) Neural inhibition of mollusc *Dolabella* heart, with bursts at different frequencies. The heart cell was not hyperpolarized. [Reprinted with permission from K. Kuwasawa, *Am. Zool.* **19**, 129–143 (1979) Copyright 1979 the American Society of Zoologists.]

chloride conductance. It may also in part suppress the ganglionic mechanism that generates the EJPs, since bath-applied ACh drastically reduces the output from the ganglion.

The cardiac ganglion of lobsters and crayfish has been a presumptive model of an integrating nervous system for the last 20 years since it comprises a total of only nine neurons, the smallest discrete neural cluster known in the entire animal kingdom. However, in spite of much intensive research and modeling, its operation is still not fully understood. This is principally because of difficulties in working with the five smaller cells. In each of at least the four larger motor neurons, the ganglion spontaneously generates bursts of impulses that are relatively synchronous, with a mean frequency of about 100 Hz, for a total duration of about 0.5 s. Exactly how the output excites the muscle cells of the heart is not known, but the mechanism is presumed to be similar to that of *Limulus,* with multiple innervation of each cell by the motor neurons emanating from the ganglion, and summating JPs. The heart cells are electrically inexcitable.

The lobster/crayfish cardiac ganglion receives a complex innervation that includes an inhibitor that can slow the beat progressively down to complete stoppage and also suppress the amplitude. There are also accelerators that can increase the frequency up to about threefold without amplitude enhancement (Wiersma and Novitski, 1942).

Not all crustacean hearts lack electrogenesis of component cells. In the mantis shrimp *Squilla*, there is a graded component that augments the first large JP, but even in *Squilla* the plateau is due solely to continuing summed junctional activity (H. F. Brown, 1964). The wood louse *Porcellio,* a terrestrial isopod, has produced yet another variant in the form of a very slow electrogenesis that is responsible for the maintained plateau. This is initiated by, but not formed of, summed EJPs. The ionic mechanism is due to mixed activation of calcium/sodium conductances (Delaleu et al., 1972).

Some annelids also have neurally driven "hearts"; the control of those of the leech *Hirudo medicinalis* has been studied intensively (Calabrese, 1979 – review). Here the operation is controlled by cells located in a majority of ganglia of the central nervous system rather than by a peripheral ganglion, and many of the neurons involved, both motor neurons and interneurons have been located, identified, and studied extensively. The leech has a closed vascular system, and it does not have the normal single-heart structure. Instead there are paired left and right lateral vessels, in which blood flows in one direction, plus dorsal and ventral vessels, in which it flows in the other. Blood can flow in either direction depending on the pattern of pulsation in the lateral tubes that constitute the "hearts." Each tube's beat is actually a forward-progressing peristalsis, lasting 10–30 s, on one side, but a nearly synchronous one on the other. Every 10–50 beat cycles, the two sides reverse their pattern. The asymmetry is presumed to be a device aimed at preventing blood from one side from being delivered into the heart of the other: instead it must enter the dorsal/ventral vessels and be delivered to the tissues. This bizarre mechanism further illustrates the late and independent evolution of hearts. The two contractions are driven by excitatory motor neurons innervating the whole length of each heart wall that are activated in a sequence on one side but synchronously on the other. The details of neuromuscular transmission are not known.

10.3.2. Myogenic Hearts

The innervation and control mechanisms for myogenic hearts are vastly different from those of neurogenic ones, and they pose many problems independent of those of any other muscle systems. We must presume that myogenic hearts originally had all the complex neuronal paraphernalia associated with vegetative tubular structures: local reflex arcs, nerve nets, and plexuses. Their extensions and inputs, the major blood vessels, still possess these, but it has evidently been to the advantage of myogenic hearts to lose these trappings, either in part or in their entirety. Their innervation in vertebrates is only from the autonomic nervous system, either parasympathetic only or parasympathetic plus sympathetic, and it is largely concentrated onto cells of the pacemaker sinoatrial region. Since all component muscle cells are intrinsically rhythmic, it is remarkable in itself that this innervation is constrained to those cells with the fastest rhythm, to which the majority, in a healthy heart, are subordinate. There must have been a progressive loss of local innervation in evolving myogenic hearts until only this one region of nerve supply remained. There are varying degrees of residual complexity in those few hearts that have been thoroughly studied. However, as this evolutionary trend developed sensitivity, presumably based in part on membrane-located receptors to some neurotransmitters and numerous modulators, it was retained and perhaps also underwent further evolutionary change.

10.3.3. The Nerveless Protochordate Heart

Intensive anatomical study of the thin-walled myogenic heart of the tunicate protochordates has shown that there is no direct innervation of even a local region of this heart (Anderson, 1968; Kriebel, 1968). It pumps blood several times in one direction before reversing and pumping an approximately similar number of times in the other. There is not one but two pacemakers, one located at the junction of the heart with the stomach blood vessels, the other at the junction of the heart with the ventral blood vessel. A tunicate heart completely isolated, but with these junctions intact, continues to show beat reversal (Kriebel, 1968). The reversal apparently occurs during a pause, when a backflow occurs. It is the blood pressure that directly affects the pacemaker, possibly by an action of stretch, the rise at one end promoting acceleration, the fall at the other leading to deceleration; then the other side takes over.

The pacemaking appears to be due to an independent oscillatory mechanism in a few cells that produces a relatively small, 10 mV, slow depolarization recurring at about 0.5 Hz. This spreads electrotonically through the nexuses, and it is likely to exceed threshold for firing an all-or-none overshooting action potential in some. The spread of the larger potential, summed with the smaller one, increases the probability of involvement of the remaining cells, but some failures occur. The action potential has a slow recovery, and although lacking a strong plateau phase, takes 0.75–1.25 s to return to baseline. The approximately 0.4 s of depolarization past contraction threshold ensures a strong contraction for each heartbeat. The frequency of the beat is accelerated by up to twice normal by Ad at 10^{-5} g/ml and slowed by ACh at 10^{-8} g/ml, inhibitory action that is blocked by atropine.

10.3.4. Innervation of Vertebrate Hearts

By contrast to tunicates, the most primitive living vertebrate, the cyclostome hagfish *Myxine,* which also lacks heart nerves (Jensen, 1965), is insensitive to ACh. The closely related but slightly more advanced lamprey however, does have a cholinergic nerve supply, stimulation of which increases heartbeat frequency, but it does not have an adrenergic innervation in addition (Jensen, 1969). In all other vertebrates there are mixed cholinergic/adrenegic axons, and ACh is an inhibitor, NAd an accelerator/enhancer of amplitude. The present-day innervation of the heart evolved relatively recently as novel central outgrowth following a complete loss of the primeval local nerve supply, but not of its receptors to a variety of transmitters. Unfortunately, rather little is known about the details of innervations. Partly this is a matter of lack of interest, although some domestic animals like the cat, dog, and cow have been studied fairly extensively because of the attentions of veterinarians. Hearts of higher vertebrates, like blood vessels, have a rich sensory innervation in addition to motor nerves, but precisely what the role of the sensory cells may be is unknown. The difference between the structural details of the sensory nerve branches compared with some of efferent supplies is by no means obvious. Many presumed sensory endings have been illustrated in the books by Grigor'Eva (1962) and Khabarova (1963).

The existence of possible reflex pathways complicates interpretation of experiments on semiintact cardiac systems, but because isolation eliminates all cardiac innervation, the situation is at least a lot simpler than it is in the gut! Both guts and hearts share the additional complication of variable, unknown extents of influence from muscle actions other than that of the organ itself.

A heartbeat is profoundly affected by cardiac volume and pressure, which are related to the state of the entire vascular system. In an era when any new action was termed a "law" at the drop of a hat, Starling's (1914) discovery that more work is done by an overfilled heart than one less extensively dilated became his famous "law of the heart." Only a year later Bainbridge (1915) injected a slug of saline into the vena cava or the atria of frogs and obtained increased heart rate. The response was eliminated by bilateral section of the vagi. In spite of the fact that nobody has ever been able convincingly to repeat his experiments, the phenomenon has been accepted as a significant part of control of hearts and termed the Bainbridge reflex. An opposite result was obtained by McDowall (1924) in cats depleted of blood and with both vagi sectioned. More recent work has used successive inflation of three balloons placed in the left atrium, left upper, and middle vein in atrial junctions of dogs. Distension of each in turn produced an increase in heart rate (Kappagoda et al., 1975).

There is at least general agreement that the neural pathway for many reflexes involved in heart control is by way of sympathetic ganglia. Discrepancies in interpretation of heart control mechanisms are related in part to interactions between reflexes and nonreflex ongoing activities, as was also the case with control of guts.

The lamprey heart has the simplest innervation functionally, since it is only innervated by cholinergic nerves. Acetylcholine accelerates the heartbeat as it does also in arthropods (see p. 384). But unless this action is indirect, via local ganglion cells that release something other than ACh, which cannot be Ad or NAd to which the heart is insensitive, this is the opposite of the response of all other vertebrate hearts. Here is further evidence for independence and very late evolutionary origins of autonomic innervations.

The nerves innervating hearts of all higher forms are mixed, and the gross distribution is not of value in tracing motor axons. The sympathetic-nerve branches of higher vertebrates are derived from the anterior cervical, thoracic, and cervicothoracic ganglia, via the common stellate ganglion. There is considerable variation in detail, but in general the nerves from left and right sides, although they predominantly innervate the heart ipsilaterally, also cross the midline and innervate the contralateral side. Parasympathetic innervation is largely from the vagi, but some comes from the recurrent laryngeals. The nerves become distinguishable as they approach the heart and can then be called coronary nerves, which ramify as a coronary plexus. There are sympathetic ganglion cells along the course of these nerves as well as in the plexus. The final branches supply two major foci, the sinoatrial node and the atrioventricular node, whereas others supply the surfaces of each of the heart chambers. There is very little information about the proportions of sensory, sympathetic, and parasympathetic (and other) axons.

Motor actions of the nerves may be classified in three divisions: pacemaker rate, contraction intensity and duration, and conduction within the heart. Each influences secondarily the amplitude, speed of shortening, and time course of relaxation. They include not only short-distance neurocrine actions but also medium-distance paracrine and long-range endocrine ones.

Presumed cholinergic- and adrenergic-vessel-containing branchlets occur close together in TEM general sections of atria. But specific synaptic areas do not seem to occur. It may be that there are indeed none, paracrine release serving better than a neurocrine one. This is a very significant area in which the state of knowledge is abysmal.

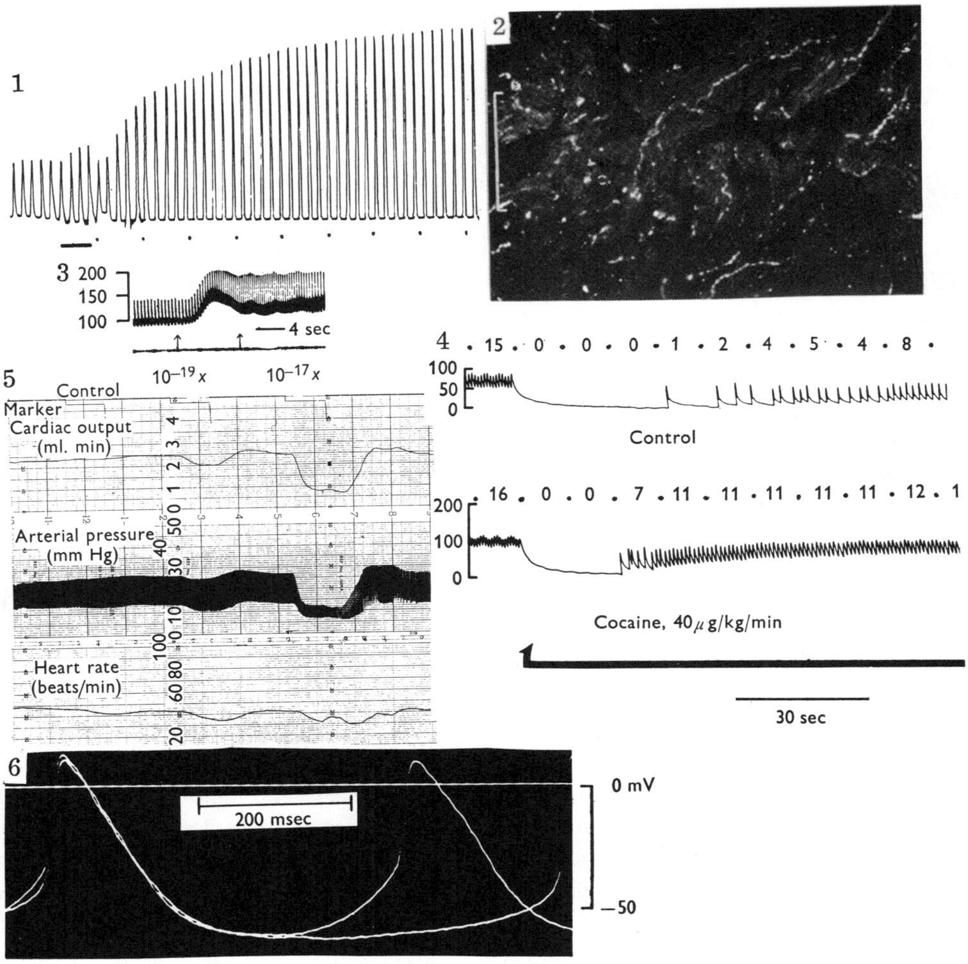

Figure 10.6. Positive and negative inotropic relationships of vertebrate hearts. (*1*) Potentiation of driven twitches of ventricle of frog heart by 10^{-6} g/ml adrenaline. (From Graham and Lamb, 1968.) (*2*) Histofluorescent localization of noradrenergic terminals in trout atrium. (From Gannon and Burnstock, 1969.) (*3*) Effects of vagal stimulation on cardiac output of dog heart. (From A. M. Brown, 1967.) (*4*) Effects of a small quantity of cocaine in counteracting vagal inhibition of dog heart. Numbers above indicate mean beats per minute. (From Campos and Urquilla, 1969.) (*5*) The ultimate in sensitivity: depressant effects of acetylcholine at 10^{-19} *M* and 10^{-17} *M* respectively, on cardiac output. (From Boyd and Palhak, 1965.) (*6*) Slowing action of ACh released by vagus stimulation on pacemaking cell of rabbit sinotrial node. (From N. Toda, 1968.)

The density of autonomic axons even in the sinus venosus of fish (trout) is so great that in a typical section as many can be counted as muscle cells (Yamauchi and Burnstock, 1968). Each muscle cell could receive innervation from many axons, but since neither junctions have been seen nor JPs recorded, it is more likely that there are just local paracrine-release sites. There is very little hope of a precise description's being obtained for the fine details of nerve supply to any vertebrate heart. Fluorescence histochemistry reveals abundant catecholaminergic neurons in sinus venosus of trout hearts (Figure 10.6.*2*), and a fairly rich supply in both atrium and ventricle as well (Gannon and Burnstock, 1969). Atropine treatment uncovered an excitatory response to stimulation of the vagus, which was mimicked by 10^{-7} g/ml NAd or 10^{-8} g/ml Ad.

There is no doubt that there is widespread sympathetic innervation of blood vessels of the heart wall, and in some cases the same nerves may continue into the myocardium, providing both direct and indirect access for the amine.

10.4. Transmission to Heart Cells

In 1956 the first intracellular recordings were made from amphibian sinoauricular nodes (Hutter and Trautwein, 1956; del Castillo and Katz, 1955c) and there has been surprisingly little work since on these functionally important events. A single impulse produces a small IJP in a parasympathetically innervated muscle cell after a long latency, $\sim$ 0.5 s at 16°C. The polarization is very slow, reaching peak in 1.5-2.0 s and returning to baseline even more slowly, in 5-7 s. The slowness suggests a paracrine rather than neurocrine release/response situation.

In mammals discrete inhibitory events have not been described, either because target cells have not been hit or because the action is modulatory rather than direct. The heartbeat is slowed or stopped during vagal stimulation, depending on the frequency (Figure 10.6.*4*). This inhibition is followed, in some species such as rabbit, by a delayed, long-lasting increase in rate, with attendant decrease in maximum tension. The second effect could be due to NAd released at the same time as ACh. The ACh action is clearly to slow the rate of depolarization in pacemaking (Toda, 1968).

The inhibitory polarizing potentials summate, but the total polarizing-potential change is always relatively small and in some cells none occurs; nevertheless, in such cells there is not only an arrest of spontaneous action potentials, but also an attenuation of the first three or four occurring following cessation of vagal inhibitory stimulation. The released ACh must cause an increased conductance to an ion or ions with a negative membrane potential, chloride or potassium or both. Increased conductance to these ions will cause a potential change only if the resting potential is less than their equilibrium potentials. However, the increased conductance will clamp the membrane potential at resting potential and so prevent depolarizing prepotentials from developing. We may wonder why any vertebrate has a strong inhibitory innervation of its heart. The literature is replete with information on stimulatory action on cardiac pacemaking, but not a single instance of neural arrest except in hibernation. Perhaps it is an evolutionary relic.

The cholinergic parasympathetic nerves of the mixed vagus nerve dominate the noradrenergic sympathetic ones when the vagus is driven electrically, and it is not possible effectively to stimulate the latter selectively. However, this can be done pharmacologically since the ACh receptors are competitively blocked by atropine. Atropine sulphate at 10^{-6} M suffices to block them completely. When the nerves are treated with this chemical, there is no inhibition but instead an increase in frequency to a maximum of about doubling, at a vagus stimulation rate of 20 Hz. The action is brought about by a much faster pacemaker potential during NAd release. It is accompanied by a larger amplitude of overshoot of the action potential.

10.5. Conduction of the Heart Action Potential

From the initiation point in the sinoatrial node, the wave spreads over the atrial cells electronically, being reinforced as it initiates local action potentials. The wave then converges on the atrioventricular node. If it has weakened or failed to spread to this point, a wave will be initiated at this secondary pacemaking site. In cyclostomes, and probably also in fish, most cells of the single atrium can pacemake and take over if an earlier-generated impulse has failed to reach them. The cells in these species are smaller than those of higher vertebrates, with fewer nexuses and higher electrical resistance, so spread is less effective, and intracellular records from some cells show no action potentials as the overall wave passes.

The spread of excitation over the fish heart is quite slow, though different in different regions, being especially so in certain cells of the ventricle. These transmit excitation to the tip faster than do other cells to the main body of the ventricle, so that the tip contracts first. This represents an independent beginning of a specialization for conduction within the heart such as has been extensively developed in birds and mammals.

Transmission of the excitation wave in amphibian hearts can easily be followed because after the start in the sinus venosus it is delayed in three places: first before entering the atrium, second before spreading to the ventricle, and finally before entering the conus arteriosus, which also contracts. In fish the conus does not actively contract, though it has strong elastic tonus.

Prepotentials indicative of intrinsic pacemaking are absent even in many sinus venosus cells of amphibia, and it seems probable that some are specialized conducting cells, but they are not obviously collected together into tracts. The delay at the atrioventricular junction is associated with the presence there of cardiac muscle cells having special properties leading to low-amplitude, slow-rising action potentials (Hoffman et al., 1959). The dense autonomic innervation of this region suggests that the delay can be increased or decreased at this point, which is naturally sensitive to modulation.

Little is known about the electrophysiology of the reptile heart. Birds, which have independently evolved a four-chambered heart, as have mammals, have distinct tracts of specialized conducting muscle cells on the inside of the heart. A ring of cells named after their discoverer, Purkinje, occurs at the atrioventricular junction, embedded in connective tissue. These cells are of larger than average size, and run from the node through the ventricular septum, where the tract divides into left and right branches. These in turn fan out into an extensive network under the entire surface of the ventricle. The corresponding system in mammals lies entirely on the inside of the thick heart wall. It starts at the sinoatrial node (bundle of Bachman) branching to supply the left atrium (Bundle of Thorel), which goes on to supply the right ventrical (bundle of Kent). The left branch supplies the left auricle and continues along the septum (bundle of His). Additional, separate bundles link the left auricle to the left ventricle.

Purkinje cells conduct rapidly because of their greater size, with attendant lower resistance and need for fewer terminal nexuses. The arrangement ensures a synchronous contraction of ventricular muscle cells, which is especially important in a large heart. The pacemaking capability is retained by Purkinje cells, but compared with the rates of the others, their rate is quite slow. In the event of both sinoauricular and auriculoventricular nodal block, ordinary muscle cells of the ventricle are responsible for pacemaking, but Purkinje cells still do the conducting. Parasympathetic cells do not terminate on or influence Purkinje cells, but they receive some sympathetic innervation, and Ad and NAd affect them.

10.6. Pacemaking

There is a voluminous literature on cardiac pacemaking, as there is on the ionic bases of cardiac action potentials, with ever-changing opinions about them. The characteristics of pacemaking for cells of different regions are different in detail, and there are significant species differences, accounting for some of the confusion. There is an inequality in knowledge because Purkinje cells, because they do not contract, are much easier to work with than the others, as are certain species of animals. The overall events are significantly different in cold-blooded animals, whose hearts are temperature sensitive but can at least function over a wide temperature range, which hearts of mammals cannot. The evolution of the ionic mechanisms underlying the action potentials in a bird or mammalian heart has clearly been a wondrous sequence, the subtlety of which we are only now beginning to appreciate. Granted a basic set of processes capable of generating a recurring action potential with the characteristic shape, we still have a heart with a lot of local pacemaking and erratic performance, like an insect or oyster heart. Without the permitted slop provided by their open circulatory systems, this would be very poor, as doubtless such hearts were: they did not come fully synchronized and coordinated. There is a great deal of ion-channel-programming strategy involved in having the sinoatrial node cells just a little faster than the rest.

Was this achieved by speeding up the sinoatrial pacemaking, lowering the firing threshold there, or what? It now seems as though it was done by slowing down the other heart cells, but again the question arises – how?

Pacemaking for a particular type of cardiac cell occurs only with a limited membrane potential range that is characteristic for location and species. It is between -60 and -40 mV for frog sinus venosus at 20°C and -65 to -45 mV in frog atrial cells. For Purkinje cells of sheep heart at 36°C, the range is much higher, -90 to -60 mV. Within this range the potential changes spontaneously and when clamped, ionic current flows. This current has a reversal potential, which rules out ion-pumping mechanisms as possible causes, and the reversal potential is both negative and close to the potassium equilibrium potential, so it is probably a potassium current. As we have seen (see p. 184), there are several different potassium currents. An outward potassium current i_{K_1} is responsible for repolarization after an action potential, so changes in a potassium current other than this must be involved. There is more than one candidate, and in atrial cells the current is clearly not the same as that used by Purkinje cells. In the latter it is i_{K_2}, a time-dependent channel, that is gradually inactivated. Not only does this lead to a falling membrane potential, it also unmasks the time-independent background sodium current ($i_{Na(b)}$). This leaves inactivation of the time-independent channel i_K as a likely candidate for voltage drop during pacemaking, at least in some atrial cells. This current is usually given the neutral term i_p, p being for pacemaker. In sinoatrial and Purkinje cells of rabbit and sheep, by using voltage-clamped cells and examining the effects of brief pulses applied during clamping in the pacemaker range, it has been found that there is an inward activating current, which has been named i_f (Di Francesko and Ojeda, 1980). It is activated in the same range in both sinoatrial node and Purkinje cells (Di Francesco, 1980). At first thought to be solely a potassium current, this appears to be a combined sodium/potassium current. In high $[K^+]_o$ the i_f reversal potential is about -20 mV and varies with $[Na^+]_o$, so it is partly carried by Na^+, explaining the sodium dependence of i_{K_2}. Here, i_f is the determinant of pacemaking rather than inactivation of i_{K_2}. By using one or another of these time-dependent currents singly or in combination, the needed variety of pacemaker rates in different regions was obtained. It was a relatively recent evolutionary development, and different species came up with different combinations that achieved the same desirable goal.

Adrenaline affects pacemaker rate by increasing i_f: in Purkinje cells of sheep heart,, ACh has the opposite action.

These various ionic events presumed to occur in some Purkinje cells during pacemaking and action-potential generation are shown in Figure 10.7. Alternatives that may apply to other types of cardiac cells or other species can be imagined without difficulty. An intensive study on rabbit SA node concluded that shutting down of i_K channels could not be involved there (Maylie et al., 1981). These authors used double-barreled K^+-sensitive microelectrodes to directly measure potassium activity under voltage-clamp conditions. They would accept activation of an inward current with a large positive equilibrium potential, sodium or calcium or both, or activation of a sodium/potassium pump as alternatives.

10.7. Cardiac Action Potentials

When the membrane potential has reached a critical value the (V,t) sodium channel i_{Na} opens up, leading to a rapid upstroke overshooting zero. This action in turn opens up the calcium channel – through which some sodium also passes – and both the voltage-sensitive, time-independent barium-blocked potassium channel i_{K_1} and the (V,t) i_{K_2}, which is the pacemaker channel for atrial cells. We may presume that the latter situation is the more primitive one evolutionarily, and that the one operating at the SA node is a later one that improved performance by initiating the beat more systematically. In turn this was followed by evolution of the conducting pathways. The later-evolved i_f current is inactive after depolarization.

There is a brief rapid-repolarization phase, as after a normal nerve or skeletal-muscle-cell action potential, but then repolarization in the cardiac cell slows drastically. In both vertebrate and myogenic invertebrate hearts, this is due to depolarizing current carried by the still-open calcium channel. It is only possible, however, because there is a delay in the rise of potassium currents, one of which is distinguishable from i_{K_1} and i_{K_2} by its time dependencies, and termed i_X, with subvarieties i_{X_1} and i_{X_2}. It is the i_{X_1} and i_{K_2} channels that are responsible. This delay lasts for about 400 ms. Then, as these currents rise in conjunction with the rise in i_{K_1}, the membrane repolarizes, i_{Ca} inactivates, and a condition of undershoot or full repolarization is achieved. The potassium channels then inactivate and the cycle repeats. The secret of the shape of the cardiac action potential is the combination of calcium activation and delayed rise in potassium conductance.

The closing of some channels is associated with a chain of events that follows the rise of internal free calcium, one of which events is a rise in cAMP. Links in the chain provide a target for catecholamines and some other modulators of the heartbeat.

The upstroke in rabbit sinus venosus cells is accomplished even after the fast sodium channel has been blocked by TTX (Yamagishi and Sano, 1966; H. Brown et al., 1979), showing that the calcium channel alone can carry a full depolarizing potential. Under voltage clamp the effect of A on the currents can be determined, and the calcium-channel current is found to have been more than doubled. There are complex interactions between entering calcium and inactivation of i_{K_2} and also activation of i_f channels that lead to more-rapid pacemaker-potential rise, notably by altering their voltage-dependence relations. Converse actions are produced by ACh, resulting in a slower rise of the pacemaker potential, presumably especially activation of one or more of the potassium channels (Noma and Trautwein, 1978), the net effect being to oppose any depolarizing shifts.

The threshold membrane potential at which the all-or-none action potential is generated is also different in different parts of a heart. The direction is reciprocal to that for the rate of pacemaker-potential change, the threshold being at a large membrane potential

1

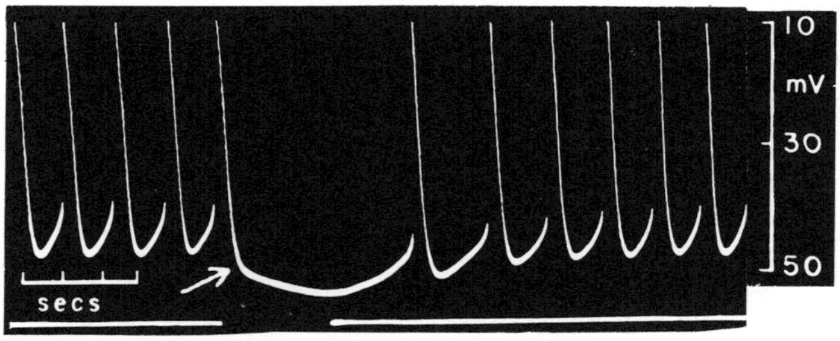

10
mV
30
50
secs

2

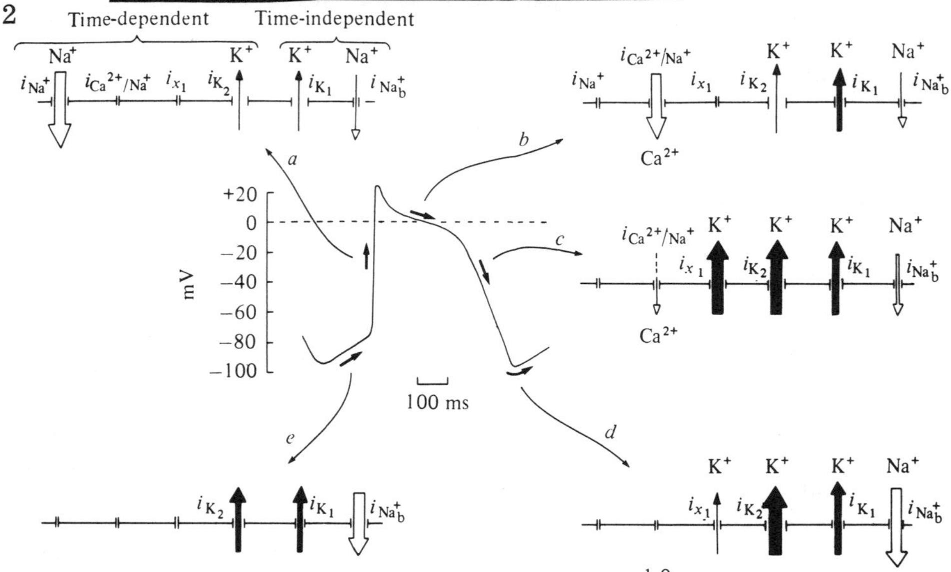

Time-dependent
Time-independent
Na+
K+
K+
Na+
i_{Na^+}
i_{Ca^{2+}/Na^+}
i_{x_1}
i_{K_2}
i_{K_1}
$i_{Na^+_b}$
Ca2+
a
b
c
d
e
+20
0
−20
−40
−60
−80
−100
mV
100 ms

3

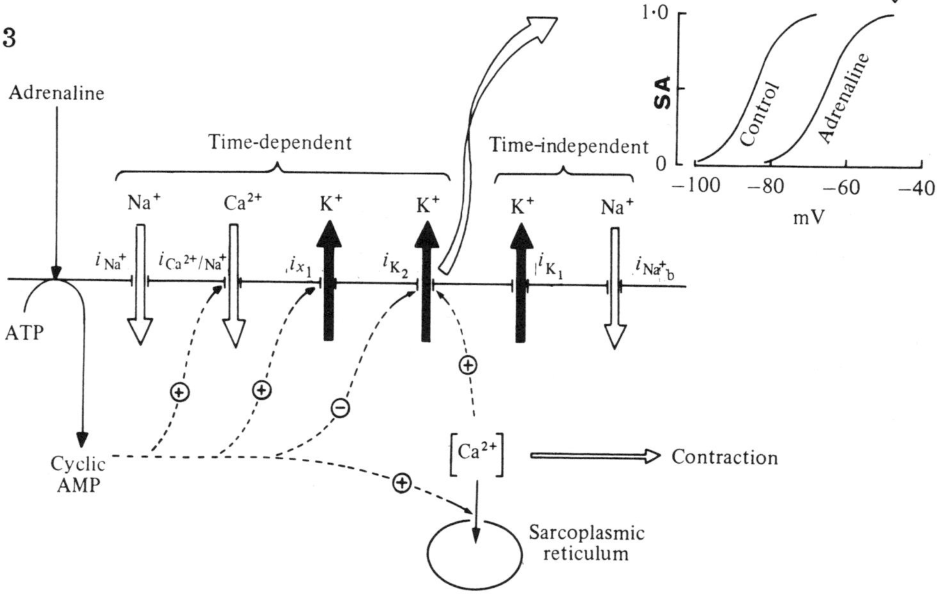

Adrenaline
ATP
Cyclic
AMP
Time-dependent
Time-independent
Na+
Ca2+
K+
K+
K+
Na+
i_{Na^+}
i_{Ca^{2+}/Na^+}
i_{x_1}
i_{K_2}
i_{K_1}
$i_{Na^+_b}$
[Ca2+]
Contraction
Sarcoplasmic
reticulum
1·0
SA
0
Control
Adrenaline
−100
−80
−60
−40
mV

in the sequence sinoauricular-auricular ventricular. These differences are interesting and clearly aimed at ensuring synchrony, but at the risk of breaking up coordination because of independent activity by local pacemaking. It is therefore significant that local pacemaking does occur readily in either a stressed or a diseased heart.

10.8. Arrhythmias

Collectively termed arrhythmias, abnormal heartbeats occur in a very extensive variety, the commonest being *flutter* and *fibrillation.* The former refers to sequences including brief periods of high-rate beating; the latter refers to lack of coordination and multiple initiation of asynchronous contractions initiated by local pacemakers, either atrial, ventricular, or both.

A slower-than-normal overall rate is termed *bradycardia* and a faster one *tachycardia,* and each can be due to either normal SA or A pacemaking. Pacemaker wandering occurs to some extent at the site of initiation of the beat, moving not only from one cell cluster to another within the sinus venosus but also from the sinus to the auricle and back. Additional, occasional contractions of A or V only or both can be initiated by a local pacemaker. Failure to initiate an action potential, or block of conduction of an action potential may occur at any point, resulting in a full or partial missed beat.

10.9. Invertebrate Myogenic Pacemaking and Action Potentials

There are two remarkable features of the heart muscle cells of diverse invertebrates that parallel those of vertebrates. Invertebrates also have evolved action potentials that are rhythmically generated within heart pacemaker muscle cells, and a long plateau of depolarization. Unfortunately, invertebrate hearts are nearly all fragile and have not been intensively studied, especially with intracellular voltage-clamp techniques, so little is known about the ionic currents. They differ, even with a class, in the extent to which they are innervated. The moth heart, for example, is not innervated (Sanger and McCann, 1968a), but that of the cockroach is extensively innervated by a combination of local nerve cells and segmental neural inputs (Miller, 1968). There is no obvious connection with life-style, the life of a moth being generally much more erratic and disturbed than that of a cockroach.

There does not seem to be any regular localization of pacemaking in insect hearts: all the cells are capable of pacemaking at about the same frequency. Whichever cell is leading at the time wins; a beat may begin in the center or at either end, although the postero-

---

Figure 10.7. Pacemaking and its regulation in vertebrate heart cells. (*1*) Intracellularly recorded events at the initiating pacemaker site (sinoauricular node) of frog heart. Spontaneous depolarization initiates an all-or-none action potential with undershoot, after which the cycle repeats. Stimulation of the vagus nerve (*arrow*) releases ACh, which produces summed small IJPs, preventing onset of depolarization. (From Hutter and Trautwein, 1956.) (*2*) Summary of sequence of ion-conductance changes (determined by many investigators) occurring in conducting heart (Purkinje) cells. The all-important plateau phase, which prevents heart cells from fused contraction, is due principally to delayed rise in time-dependent *g*K. (From Berridge and Rapp, 1979.) (*3*) hypothetical changes brought about by Ad. It acts by promoting synthesis of second-messenger cAMP, which in turn affects ion channels, especially K^+ channel i_{K_2}, which is inhibited. This shifts the activation curve, as shown in inset. (From Berridge and Rapp, 1979.)

anteriorward direction is more common. There are probably endocrine modulators circulating in the blood that accelerate or slow down the moth heart. The principal candidates at this time for natural excitation are octopamine, a phenolamine with a structure very similar to that of Ad, which is also excitatory to insect hearts, and widespread in insects, and proctolin, a pentapeptide discovered in the gut (see p. 353) (Collins and Miller, 1977). No natural inhibitors of insects hearts are currently known. The cockroach, with its complex innervation, was expected to be subtler in its responses to neural stimulation, but the nerves supplying it all seem to be excitatory. However, these nerves probably serve to coordinate the contractions, not just to modulate them (Miller, 1968).

All molluscan hearts are believed to be myogenic and to be innervated, some with local nerve-cell bodies as well as axons. Many hearts found in lamellibranchs, gastropods, and cephalopods alike have two atria and a single ventricle. They are extremely sensitive to pressure, which probably promotes direct stretch action: weak beats occur unless the hearts are fully distended. When the hearts are incompletely filled, it can be seen that local contractions can occur at any point, suggesting universal pacemaking capability. The nerve supply includes sensory cells, permitting reflex involvements. Because it is possible to record from identified neurons of many molluscs, some of the cells specifically concerned with heart control have been identified and their transmitters and effects determined. The heart of *Aplysia* is a simple two-chambered tube, but with valves to ensure unidirectional blood flow. Three large efferent blood vessels emerge from it. The heart is myogenic, but innervated by a trunk of the abdominal ganglion, the pericardial nerve, which makes atrial and ventricular branches.

Two neurons that directly excite the atrium have been identified: one labeled LD_{HE} causes a brief increase in heart rate, up to about doubling, from 20/min to 40/min; the other, labeled RB_{HE}, produces a long-lasting 0.5–3-min increase of up to 25–30% (Figure 10.8). A burst of action potentials is needed in either cell to produce the characteristic action. The LD_{HE} produces EJPs of about 20 mV in atrial muscle cells, RB_{HE} smaller ones of about 5 mV. A train of impulses in LD_{HE} occurring immediately after a beat delays the onset of the next beat. A later-timed train initiates a premature weak beat and resets the rhythm (Figure 10.8.*2*). The RB_{HE} fires tonically at about 1.5 Hz, and when it is hyperpolarized to reduce this frequency, there is a corresponding decrease in heart rate down to 2–3/min from 20/min. The transmitter/modulator released by this neuron has been identified as 5HT (Mayeri et al., 1974). The agent released by LD_{HE} that causes a short-lasting increase in rate is not known: it is not ACh, because this causes only inhibition.

Two neurons that innervate the ventricle have been identified: LD_{HI_1} and LD_{HI_2}, both of which are cholinergic (Mayeri et al., 1974). Three neurons that have been found that innervate the major blood vessels, LB_{VC_1}, LB_{VC_2}, and LB_{VC_3} are also cholinergic. The two ventricle innervators are so labeled because each is inhibitory and reduces the heartbeat frequency to the point of complete inhibition when they are discharging for a few seconds. In the pulmonate gastropod *Helix pomatia,* a total of 22 cells are located in the visceral and the right parietal ganglia, which are in this species drawn together in a common ganglion.

10.10. Modulation and Inotropy

The most famous, most used, and most useful of all positive inotropic agents is digitalis. It slows ventricular rate when fibrillation and flutter are present, as in congestive heart

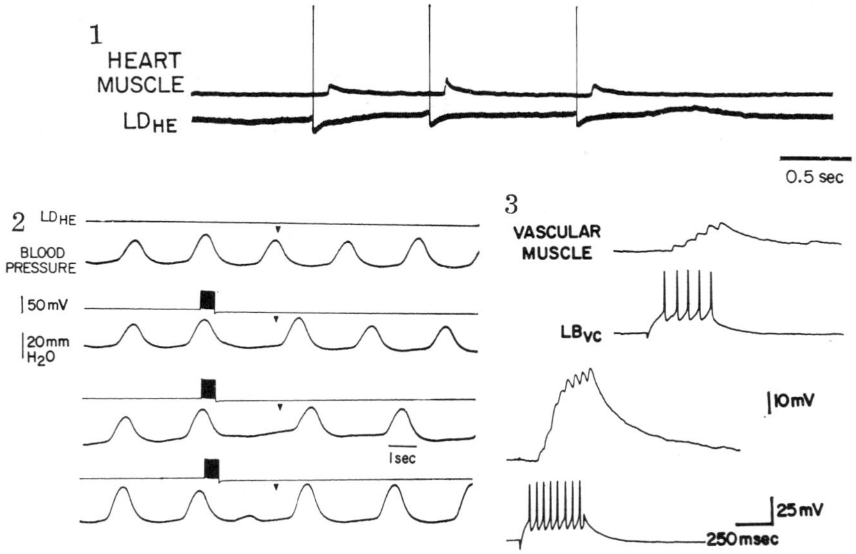

Figure 10.8. Direct neural regulation of *Aplysia* heart by identified neurons of abdominal ganglion. (*1*) EJPs in heart muscle cell caused by action potentials in heart excitor motor neuron LD_{HE}. (*2*) Effect of burst of activity driven in neuron LD_{HE} on heartbeat as indicated by blood pressure, at different times in the cycle. (*3*) Direct EJP responses of a muscle cell to impulses in identified vasoconstrictor cell LB_{VC}. [All reprinted with permission from Mayeri et al. *J. Neurophysiol.* 37, 458–475 (1974). Copyright 1974 The American Physiological Society.]

disease, considerably increasing contractile force of all parts of the heart. Digitalis is but one of a large chemical family termed the cardiac glycosides. They are formed by a combination of an aglycone, the active agent, with one to four sugar molecules. The basic chemical structure is a cyclopentane-phenanthrene nucleus, which relates them structurally to sterols and therefore to sex and adrenocortical hormones. It is not surprising then that all of these affect cardiac performance. The sugars attached do not affect performance because they are hydrolyzed off, but they do affect lipid and water solubility and therefore access to the heart cells.

Cardiac glycosides might have had a single effect aimed in this direction, but in fact several actions, each independently promoting the beat, occur. Even the electrical activity of the heart is affected in a complex manner: the mean membrane potential falls because of a combination of reduced sodium extrusion and reduced potassium intrusion. Calcium extrusion falls and $[Ca^{2+}]_i$ rises. Pacemaker potentials that would previously have failed to reach firing threshold, causing a missed beat, exceed it. The rate of depolarization during the spike increases, and of course the delay to its onset decreases. Hence tension rises more rapidly, and because of the increased background, calcium activation is more intense: the whole load–maximum velocity curve is shifted upward.

The lowered mean membrane potential leads to additional background activation that has been incorrectly termed "resting" tension; it is presumably enhanced basic tonus: BT of cardiac muscle is not steady; because it is determined by weak activation, it must fluctuate in synchrony with the heartbeat. Each large polarizing undershoot following an action potential leads to a fall in BT (Scherlag et al., 1966).

It is possible that glycosides and other inotropic agents may in addition directly affect the molecular machinery of contraction (e.g., Hibberd and Jewell, 1979). If they indeed do so, this may be via phosphorylation of the myosin light chains. Whether or not these must be phosphorylated before actin and myosin can interact, their phosphorylation, and that of TN also, may at least aid in the linkage (Chapman, 1979). Finally, even the calcium-entry mechanism is affected by length, more entering per unit area under stretch (Lakatta and Spurgeon, 1980).

The strong dependence of cardiac tonus and contractility on its ATP-dependent membrane-located ion pumps is very likely to have been evolved in conjunction with the partial, relatively slow activation of contractility. Between them they permit a wide range of modulation of performance by both direct neural means and indirect ones via circulating hormones. These enable precise linkage to functional needs.

Mechanically, cardiac muscles always appear to have something in reserve. Neither extra external calcium nor this in combination with several additive potentiators, even releasers of SR calcium, which together must saturate the contractile mechanism with free calcium, prevent a positive effect of stretch. Therefore the effect of stretch may be due to improved access of calcium to the contractile sites or a direct kinetic action on the contractile molecules. The inotropic actions of different substances are additive to some degree, which is not surprising since there are several possible target sites. For example, the action of NAd, powerful though it is, is further enhanced by cAMP (Tsien and Weingart, 1976). The action is most strongly positive when cAMP is applied to a cut end, permitting it to more readily enter and get passed around: when applied externally it has a weaker or even negative action.

Inotropic agents may also act via mitochondrial effects and SR sequestration influences. The latter are particularly difficult to assess. The range of amounts of SR in cells from different parts of a heart differs, being greatest in the regions from which the greatest tension is demanded. But the range in hearts of different species, even in ventricles only, is quite wide, from only a few percent of total volume to almost 50%. Examples of such hearts have unfortunately not been compared systematically to see if there are associated speed, rate of beating, or maximum-force capabilities. It would also be of interest to determine if there is any relation to life-style.

10.11. Comparative Physiology of Modulatory Actions

It might have been adequate for modulatory purposes for any organism to concentrate on altering ion conductances by the common neural agents ACh, Ad, NAd, 5HT, and DA. But invertebrates, like vertebrates, have in addition taken to using an as-yet-inestimable number of other substances, especially small polypeptides. The pentapeptide proctolin (Arg-Tyr-Leu-Pro-Thr) is definitely used by insects in heart rate and amplitude control. At 10^{-9} *M*, proctolin not only enhances the heartbeat frequency, but it also produces a strong positive inotropic effect. Insect hearts are innervated by nerve cells that release octopamine, and 10^{-9} *M* octopamine greatly increases the frequency of the beat, but does not enhance the amplitude as does proctolin. Molluscs use the tetrapeptidamide Phe-Mer-Arg-Phe-NH_2, which is termed FMRFamide (Greenberg and Price, 1979) and possibly many others. An FMRFamide concentration as low as 10^{-9} *M* induces beating in a quiescent molluscan heart and increases amplitude, frequency, or both of active ones. All these effects are believed to be related to a second-messenger, cAMP-mediated, action affecting calcium, either internally or at the surface membrane or both. FMRFamide was "peak C" of a five-unit chromatogram of excitable substances occurring in several molluscan tissues (Agarwal et al., 1972). The others have not been specified, but "peak C" is the only universal one in the molluscan nervous systems.

The change in pacemaker rate of molluscan hearts that is brought about by the discharge of neurons releasing 5HT appears to be mediated not by direct but by secondary action. The 5HT promotes the synthesis of cAMP, which in turn promotes phosphorylation of several proteins, including ones present in SR, which is thereby activated to give up calcium: it is this action that in the final step accelerates pacemaker depolarization (Higgins and Greenberg, 1974).

Direct studies of modulation of pacemaking and of individual beats of the mollusc *Dolabella* have shown that both are affected by ACh, which is released by cardiac-nerve efferent neurons. The membrane potential is low, apparently due to a high resting sodium conductance, but nevertheless overshooting spikes are produced, with a following plateau potential. Independently of its action on pacemaking, ACh shortens the duration of the plateau phase (R. B. Hill, 1974), which it does not do in any vertebrate, and 5HT has the opposite action (Figure 10.9.): nerves containing 5HT occur in *Dolabella.* The force of the ventricle beat is correspondingly decreased by ACh and increased by 5HT; since tension rises slowly, its peak is proportional to the duration of the plateau. The ionic conductance changes underlying these effects have not been worked out, but they are likely to be indirectly initiated by way of calcium. This action of 5HT in the heart is the opposite of its action elsewhere in molluscs, where it is a potent promoter of relaxation.

The pacemaker effect of ACh in molluscs is complex, because although the inhibitory

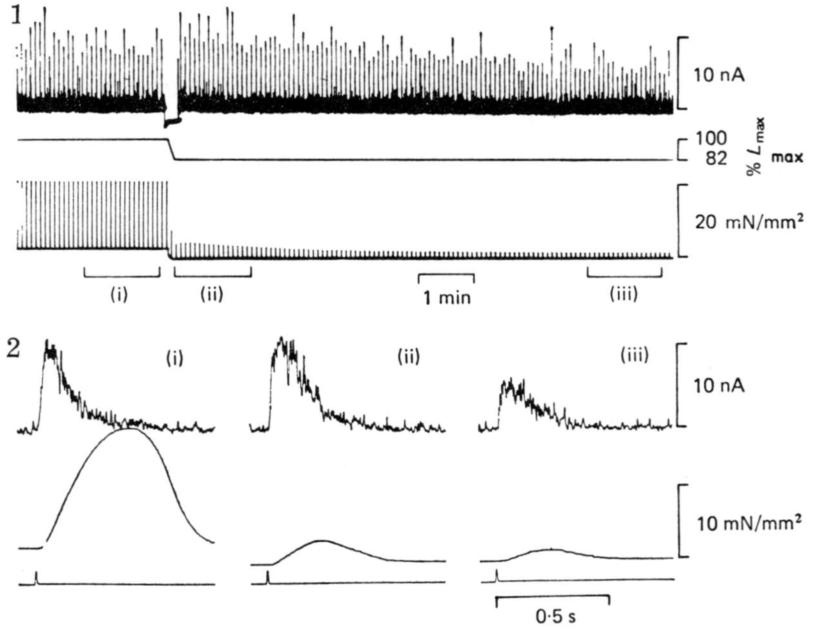

Figure 10.9. Calcium transient in cat papillary muscle detected with injected aequorin, and the effects of length. Tension is markedly reduced at shorter length, but calcium transient is increased. (*1*) *Upper trace,* light; *middle trace,* length; *lower trace,* tension; 12 stimuli/min. (*2*) Summed light and tension from 16 individual responses each, at times indicated in (*1*). (Both from Allen and Kurihara, 1980.)

nerve produces discrete hyperpolarizing IJPs associated with each ACh release, it at the same time produces a general slow conductance increase and depolarization. But the latter effect is confined to low (10^{-6}-10^{-8} *M*) concentrations of ACh when its action is tested by bath application. Higher concentrations hyperpolarize the cells at the same time as the resistance falls. Thus the lower concentrations activate depolarizing currents, probably sodium and calcium; higher ones in addition activate polarizing potassium and/or chloride currents, causing net hyperpolarization. Pacemaker effects of ACh in molluscs are not consistent, for although they are commonly inhibitory, they can also be excitatory (Figures 10.5.*4*, and 10.5.*5* – Shigeto, 1970) as they are, for example, in the lamellibranch *Mytilus.* Small differentials in ion conductance changes, both between different cells in the same organ and between homologous cells in different species, can lead to wide differences in response, even to the point of reciprocal actions. Nowhere is this principle seen more clearly than it is in the hearts of molluscs. These hearts also show that there is no universal association between hyperpolarization and inhibition. A general conductance increase that causes moderate depolarization can also lead to inhibition of a pacemaker mechanism. However, the principle that contraction is linked to the maintained lowering of membrane potential is inviolate.

10.12. E-C Coupling and Activation

It has long been realized that cardiac muscle cells in general, like the various kinds of SM cells, have evolved aspects of contraction coupling and contractile activation that differ from the ones utilized by skeletal muscle cells of the same animal. The one universal feature is that the calcium ion is the fundamental determinant of contraction. However, in no cardiac muscle does the free [Ca^{2+}] rise as rapidly as it does in fast skeletal cells of vertebrates, nor does it saturate the contractile machinery to produce 100% state of activation (see p. 234) in a single twitch.

There is a reasonable teleological explanation in that incomplete activation is compatible with bidirectional modulation, permitting either enhanced or reduced heartbeat tension as needed. This capability is in turn explained by relatively poor calcium-supply/storage membrane elements, especially the absence of ITs and clefts in small (N 7 μm) cardiac cells and paucity of ITs in large ones. Mitochondria are abundant in all heart muscle cells, much more so than in the majority of skeletal cells, and they probably play a major role in sequestering cardiac-muscle calcium. The SR is moderately abundant and serves its usual role, sequestering calcium, but in the absence of cisternal concentrating and releasing sites, it is more of a buffer in cardiac cells.

The surface membrane of cardiac muscle is not only relatively permeable to calcium, but cardiac-action-potential plateaus are supported by increased calcium conductance. A relatively large amount of calcium enters the cell directly during the cardiac action potential. Therefore, playing a significant role in cardiac-cell activities is the removal of calcium, which is necessarily achieved by a calcium ion pump. This is metabolically and ATP driven (Sulakhe and Dhalla, 1971), and, because sodium has also to be removed and potassium recovered, it is possible that a common pump for these ions is present.

The surface membrane of a cell is itself a sequestration site for calcium, and it is extensively infolded with caveolae in most heart cells, as it is in SM cells. Many calculations of the amount of directly entering calcium (e.g., Niedergerke, 1963; summarized by Chapman, 1979) have agreed that this is capable of raising the internal free calcium concentration by only a fraction of the amount required. From where, then, is the extra calcium forthcoming? It is released from the combination of sites that includes surface membrane, all forms of SR, and mitochondria, but undoubtedly by far the largest reservoir is non-cisternal elements of SR. The entering calcium triggers calcium release from C elements of SR as it does in skeletal cells, but probably also over the entire SR surface, not just at the sparse C elements. About 1-2 μmol of calcium per liter of tissue water enters per beat in "average" heart muscle. And 50-60 μmol calcium per kilogram of tissue are needed for activation. The surface-entering calcium can be thought of as "activator calcium." When it ceases to flow into the cell because the action potential has terminated, no more calcium is released and relaxation follows.

Aequorin has been injected into cells of frog atria and the rise in free cellular calcium studied directly (Allen and Blinks, 1978). There is a light flash associated with the calcium transient, as there is in skeletal muscle (Figure 10.9). But the rise time is much slower, which also means it is a more accurate representation of the true time course of calcium change. The $[Ca^{2+}]$ rises to a peak of about 2×10^{-6} *M* according to the light intensity, which is proportional to $[Ca^{2+}]_o$, emphasizing the importance of entering calcium in triggering contraction.

Vertebrate heart muscle contains actin, myosin, tropomyosin, troponins I, C, and T, and associated enzyme systems, all of which are closely similar to those of skeletal muscle. There are some slight differences though, notably in the calcium dependence of interaction between TN_C and TN_I (Ebashi et al., 1975). Also, there are much greater effects of phosphorylating agents in heart muscle, suggesting that phosphorylation of myosin is a regulatory event there. But much of the evidence points to a TN regulation mechanism similar to that of skeletal muscle: possibly both are utilized cooperatively. Alternatively the phosphorylation could be associated with calcium sequestration rather than contraction regulation.

10.13. Dynamic Aspects

The amplitude of a cardiac contraction is remarkably sensitive to initial tension, as embodied in Starling's law. This was derived without reference to heart muscles in general, but fortuitously a strong effect of length has turned out to be widespread, not confined to vertebrate hearts. It has been nicely demonstrated in the oyster heart and in the gastropod *Dolabella* (Nomura, 1963). Here the tension enhancement with stretch and reduction with shortening are in part mediated by a marked effect of length on the duration of the action potential. This is not the case for vertebrate hearts.

Whatever indirect effects stretch may have, such as increasing the amount of calcium entering or increasing the amount released internally, which is unlikely, there is a more direct one, on the molecular activation, comparable to that which occurs in insect asynchronous flight muscle (see p. 146). That is, cardiac muscles in general have a length dependence of actin-myosin-calcium interaction that directly affects ATP splitting and hence tension production. While calcium concentration was directly followed with the aid of injected aequorin, a marked increase in tension development occurred in response to a stretch under conditions when both the extent of filament overlap and the calcium concentration were decreasing (Allen and Blinks, 1978). The relative magnitude of this effect of stretch is therefore large.

Attempts to treat cardiac muscle like skeletal muscle in dynamic experiments lead to bewildering results, though some basic similarities are indicated. For example, in rabbit papillary muscle in afterloaded isotonic contraction, more energy is liberated than in an "equivalent" isometric one (Giggs and Gibson, 1970); there is a large energy-liberating effect of shortening. But no simple linear correlation exists between energy output and work done in cardiac muscles. The discrepancies are likely to be the result of relatively greater effects of length-change history on activation in cardiac muscle.

The complex dynamic properties of heart muscle appear to be in part due to the existence of some degree of excitation in the unstimulated condition. The "resting" tension or BT is very high, starting at a short sarcomere length of about 1.6 μm. At 2.2 μm, this tension is about half that which can be developed maximally by stimulation (Brady, 1967). X-ray studies on diffraction patterns of heart muscle from cat, goat, guinea pig, and rabbit at different lengths all indicated the existence of some actin-myosin interaction leading to weak myosin layer lines (Matsubara and Millman, 1974).

The ultimate attachments of heart muscle cells are necessarily to each other, and they are made via a large number of enmeshing collagen filaments. The length of SEC of rat papillary muscle has been estimated on the basis of light diffraction studies to be about 7% of total length (Krueger and Pollack, 1975). When SL and A-band length were plotted against diameter and SL respectively during stretch, they were found to increase linearly (Winegrad, 1974), but the effects are not likely to be genuine but rather due to imperfectly aligned A filaments, a factor that must introduce additional compliance during stretch of cardiac muscle.

The concept discussed above for skeletal muscle, in which V_{max} and F_{max} are determined in parallel by the statistical average of all the local molecular activations, applies also to cardiac muscle. Quantitative determinations of the average state of activation (see p. 233) can, of course, be made. But as pointed out by Brady (1968), the value obtained is "only as good as the assumption of mechanical homogeneity." Just because of its physical complexity and structural inhomogeneity, it may be deduced that there must be even larger contributions from intersarcomere dynamics in heart muscle than is the case in skeletal. Light-diffraction-pattern changes during contraction emphasize that this is indeed the case (Krueger and Pollack, 1975).

The outstanding feature of cardiac muscle in general is the extent to which the state of activation (SA) is altered by both minute quantities of a host of chemical agents and by mechanical manipulations. The condition of modification of SA is, somewhat inappropriately, termed *inotropism:* agents that enhance it are positively inotropic, those that depress it, negatively inotropic. Adrenaline is a potent natural positively inotropic agent, and it has long been known that it does not act by directly promoting actomyosin interaction (Katz, 1967) but indirectly, probably due to multiple factors which all lead to more calcium being available. The single most significant factor is a rise in cAMP initiated by entering Ad (Stull and Mayer, 1979). By far the largest single factor mediating modulatory influences on cardiac contraction is the free calcium. Even the progressively increasing strengths of successive contractions that occur after a period of rest, the well-known intrinsic *staircase* inotropism, is explained by a progressive rise in available calcium with each succeeding contraction. Membrane potential-calcium-tension relationships are fundamentally similar to those for skeletal cells (Morad and Orkand, 1971), but a heart action potential is so long and complex in shape that the quantity of activation, SA times time, produced by it is much more than that of the brief transient of skeletal cells. Therefore, to investigate membrane potential-tension relations for cardiac cells, voltage-clamping must be utilized. Although this is not possible with single cells, for small bundles of cells satisfactory clamping can be achieved with a double sucrose gap. When applied to frog atrial muscle, a dual response is obtained (Einwächter et al., 1972). An initial transient phasic contraction occurs, but this is superimposed upon a sustained tonic contraction about half as strong. The early peak probably reflects the equilibrium concentration of calcium summed from partly entering and partly internally released calcium. The latter is clearly not inactivated with time, as it is in skeletal fast-type cells. In this respect cardiac muscle resembles both SM in general and the slowest types of skeletal muscle cell. This difference must be associated with different properties of their SR/calcium-releasing mechanisms.

There are two inward currents during a cardiac action potential or a voltage-clamped depolarization: fast and slow. The latter, which is the calcium current, is much more important in determining the final contraction strength (Léoty, 1974). This calcium current, all of which is effective both directly and indirectly in coupling, lasts between 200 and 400 ms per action potential (Bassingthwaite et al., 1973). But there is a significant antagonistic action of sodium on the effectiveness of the stimulus. Cardiac contraction is enhanced by lowering $[Na^+]_o$ without altering $[Ca^{2+}]_o$, and this is not due to an effect on the time course of the action potential (Lüttgau and Niedergerke, 1958). What happens is that with less sodium externally, more calcium enters the cell.

Using injected aequorin, the relation between length and actively developed tension can be studied at the same time as the calcium transient, which might itself be length-dependent. This turned out to be the case (Allen and Kurihara, 1980), but the calcium transient was actually enhanced by a shortening of small extent, which was associated with a very large fall in tension (Figure 10.9).

Many of the more important drugs that influence contraction of hearts, for example antazoline, work like caffeine in skeletal muscle, by liberating calcium from SR. Some others affect calcium entry into the cell, and multiple actions also occur. For example, ouabain has strong effects on hearts that appear to involve not only sodium-pump action but also intercellular communication. Following the inhibition of the sodium pump, there is an increase in internal calcium, whose extrusion may be linked to it, together with a delayed large increase in input resistance of the cells. This suggests that as $[Ca^{2+}]_i$ rises, nexal conductance decreases (de Mello, 1976; Weingart, 1977). This would drastically alter current flow through the heart cells and serve to activate them more uniformly as well as more intensively. The normal size of these low-resistance pathways has been estimated by comparing radio-labeled potassium, TEA, and Procyon yellow diffusion properties to be about 1 n*M* (Weingart, 1977).

In spite of the complex three-dimensional geometry of heart muscle, the intercalated discs and nexuses dictate a functional unity that leads to a pattern and time course of current spread more or less like those expected for a cable with a long space constant. When treated as such a cable, λ for an average vertebrate heart is about 0.5 mm (Bonke, 1973).

11

Long Term Factors Affecting Muscles and Motor Nerves—Plasticity

To arrive at any understanding of neural ontogeny as an integrated whole it is essential to cultivate an awareness of the interdependence of the parts, and of the subtle balance of their relations in space and time. Marcus Jacobson†

11.1. Development, Maintenance, Trophic Influences, Exercise, Disuse, Disease, Aging

It is not within the scope of this book to cover each of these important topics in depth, but in the interest of providing the reader with a reasonably complete overview of muscle, some basic matters of each will be described in this chapter. When I attempted to categorize some of the less specialized literature, it became apparent to me that they are, somewhat surprisingly, quite closely related to each other. The single common theme is the plasticity of muscle.

In summary, there are some characteristics of a muscle cell that, having been set genetically and developmentally, are not subject to alteration. Superimposed on these are ones that are dependent on one or more of at least six dynamic variables: mean length, passive tension, active tension, tension–time relations, impulse pattern, and trophic chemical factors. The characteristic myosin-light-chain regulatory factors related to speed are primarily dependent on impulse pattern. Distribution of receptor molecules on the surface is governed principally by trophic chemical agent(s), though muscle excitation alone has a partial similar action. Total protein produced is dependent upon combined passive and active tension. Sarcomere length and longitudinal growth are partly regulated by mean muscle length. ATPase activity of myosin is determined by a mixture of isoenzymes ranging from fast to slow, and overall activity depends on the relative proportions: the synthesis of fast or slow types is switched on or off according to dynamic activity dictated by neural-input pattern.

11.1.1. Development

Except for the diploblastic coelenterates, where many muscle cells are also functional epithelial or endothelial components, muscle cells of each of the three classes – striated, SM/vegetative, and cardiac – are derived embryologically from mesoderm. Indeed, the fate of a majority of the cells of the middle germ layer is to form muscles and their related

†M. Jacobson, in *Developmental Biology*, Holt, Rinehart and Winston, New York, 1970.

supporting tissues. In segmented animals the mesoderm becomes divided into blocks that will form the somites, and in these somites, muscles form. For a considerable time during early development, presumptive muscle cells cannot be distinguished from other types, and it is quite late before the three fundamental muscle cell types are differentiated from each other. The first sign that they are to become muscle is marked elongation of the presumptive muscle cell. At this stage each is still mononucleate and contains a lot of rough endoplasmic reticulum and numerous ribosomes but no myofilaments. In vertebrates these cells have been called *myotubes,* but this historically based terminology has no ultrastructural support and will be discarded. The cells are simple, elongated, and do not in fact have tubular internal substructure. Cells destined to become skeletal muscle soon start to fuse laterally, becoming thicker, multinucleate cells. Nuclei commonly pass to the periphery, but in some cells they go to the core and in others remain scattered.

Once the presumptive muscle cell is capable of synthesizing myosin, it is termed a *myoblast.* At this time it contains large polyribosomes with 50–60 ribosomes per chain, and also numerous mitochondria (Allen, 1978). Once the lateral-fusion process for presumptive skeletal cells has accumulated about the right number of components, a surface coat – the basement material, basal lamina, or glyocalyx – is formed around the cluster. It may have the function at this early stage of preventing further fusions. It is, however, further developed and maintained throughout life, perhaps serving as a membrane-protecting layer and inter-muscle-cell lubricant.

In arthropods the number of such cells for a skeletal muscle is very narrowly prescribed. The smallest muscles of the smaller animals comprise a single cell; the next larger two cells, three cells, and so forth. As the animal grows, which for some species of barnacles, lobsters, and crabs can be to truly giant size, the number of cells per muscle remains about constant. In consequence, cells of 2–3 mm diameter or more are the rule in their larger muscles. This does not occur in vertebrates, in which skeletal muscle cells do not increase in diameter beyond about 0.1 mm, even in a whale. When the animal grows larger after its muscle cells have reached this diameter, more are produced. The precise origins of all the new cells is still uncertain; some are produced from existing cells by lateral splitting followed by growth of the splinter fractions. It is probable that some increase in numbers of skeletal cells of arthropods and other invertebrates also occurs by splitting but mainly in early growth stages. Individual muscle cells are characterizable and re-identifiable in smaller arthropod muscles.

11.1.2. Myofilaments

Following fusion, the first myofilaments start to appear, first at the periphery (Figure 11.1.*2*), explaining the early apparent tubular appearance. It is abundantly clear that in cross-striated muscle cells from all sources, thick and thin myofilaments appear independently, though at about the same time. The first ones to appear occur in random orientation relative to the presumptive longitudinal axis, but they soon move into positions parallel to this, and filaments appearing later are formed only along this axis.

Unpolymerized molecules, especially of actin, are present before myofilament formation, and some are probably present before any myosin is formed. In chick myotome cells, thin 6-nm filaments appear slightly ahead of any thick ones. It is not yet known whether or not thin filaments have their TM and TN attached at the time of their appearance, but they are present chemically at about the same time (Hirabayashi, 1971). The ratio of thin to thick filaments does not start out at its definitive value. There are com-

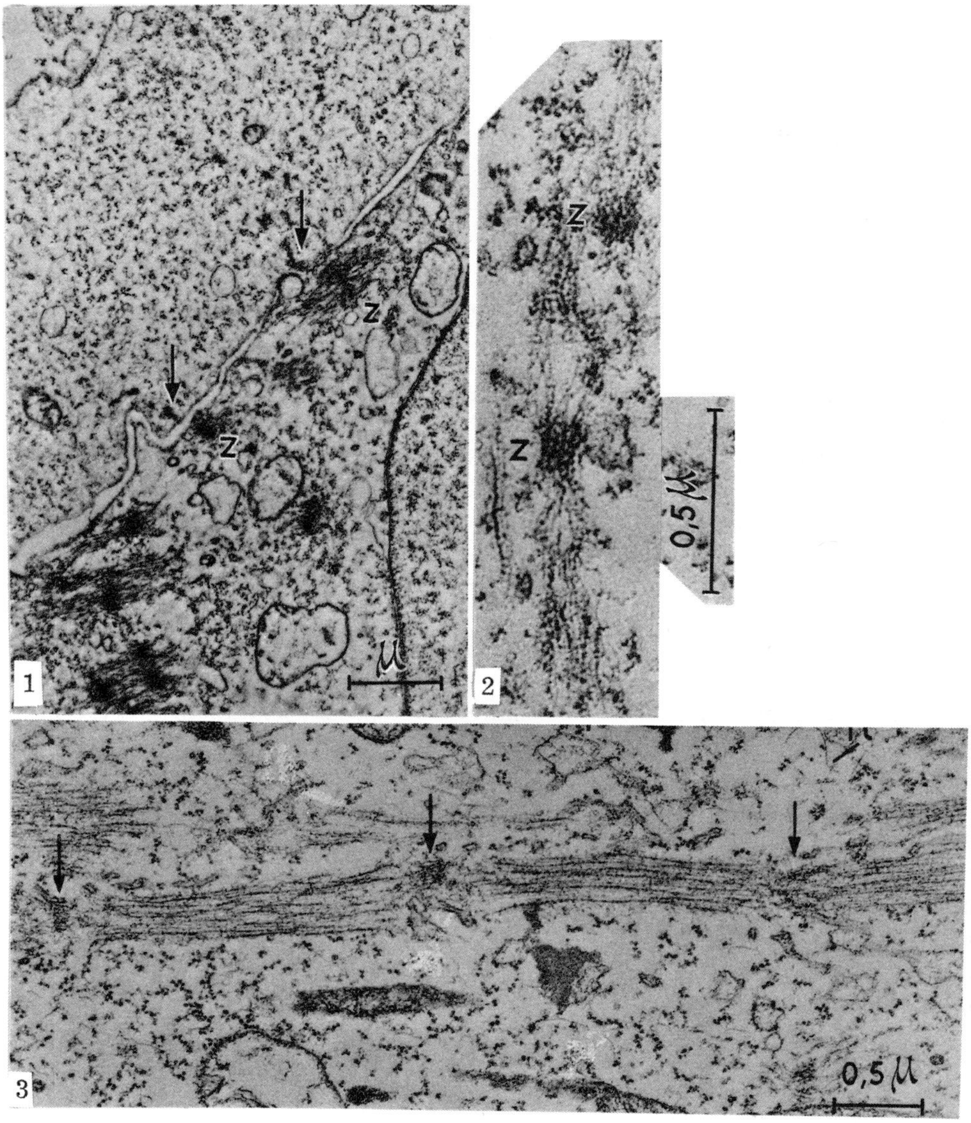

Figure 11.1. Development of a mammalian (rat back muscle) cross-striated cell. The first elements to appear are parts of presumptive Z bands, with very short lengths of thin filaments attached. These appear in two locations, shown in *1* and *2* respectively. (*1*) at the surface membrane, where there are regularly spaced Z-origin sites on the lower cell. Note the beginnings of similar sites on the neighboring cell (*arrows*). (*2*) A second set of sites occurs randomly in the myoplasm. (*3*) Later A filaments appear, parallel to the longitudinal axis, and the small presumptive Z regions (*arrows*) align themselves with these, the short thin filaments interdigitating with them. The inter-Z spacing is already at its definitive length. Thin I filaments are present in association with A filaments before the Z elements attach to them. [Reprinted with permission from J. A. Heuson-Stiennon, *J. de Microscopie* 4, 657–678 (1965).]

monly too many thin filaments to start with, especially in vertebrate muscle. In chick 12-day embryos when orbits first appear, the ratio is 7 : 1 (Fischman, 1967), but it gradually reduces to the standard vertebrate 2 : 1.

11.1.3. Z band

The relative time of appearance of Z material is different in different striated cells and so, therefore, are the whole early patterns of formation of the filamentar architecture. In rabbit, many thin and thick filaments have formed before any Z material does, and until some Z material has appeared, there is no organization and much inequality in filament lengths (Kilarski and Jakubowska, 1979). In chick, the Z material also appears late and not until after there has been considerable lateral alignment of both thin and thick filaments into their presumptive positions for cross-striation (Figure 11.2 – Fischman, 1967). In human embryonic muscle also, many thick and thin filaments have been formed by the time Z material appears, and they also do not orientate until after the Z material has appeared. The Z material appears in patches throughout the cytoplasm and appears to be interconnected into an irregular network even before it associated with the myofilaments (Saetersdal et al., 1980). In strong contrast, in one of the first cross-striated muscles to be described, from the back of a rat, Z material appears first, or at the same time as thick filaments (Figure 11.1 – Heuson-Stiennon, 1965). Small patches of Z material develop at regular intervals along the length of the cell, at the surface membrane. For neighboring cells these regions where Z patches appear are adjacent (Figure 11.1.*1*). These have very short lengths of densely staining thin filaments emerging from them in the longitudinal axis. The small patches then migrate inward (Figure 11.1.*2*), where they eventually attach themselves to myofilament clusters (Figure 11.1.*3*) after these have formed. At this time the definitive Z–Z spacing is already established.

By contrast, in many invertebrate cross-striated muscles, large changes in SL occur before maturity. In a developing crustacean muscle, that operating the second antenule of the copepod *Doropygus,* the myofilaments are perfectly aligned before any dense Z material appears. However, ultrathin Z filaments 2-3 nm in diameter are already present, crosslinking I filaments of neighboring sarcomeres. The characteristic dense material associated with arthropod Z bands is added later. Thus there must be fundamental differences between the natures of arthropod and vertebrate Z bands.

11.1.4. Organization Within Cells

There appears to be no commonality to the modes or times of appearance of molecular components of cross-striated cells in general, perhaps reflecting their independent evolutionary histories.

Several studies have located longitudinally aligned ultrathin filaments appearing before thick A or ordinary thin I myofilaments in a variety of cell sources. Heuson-Stiennon (1965) found ultrathin filaments to be the first in rat cells. In chick myoblasts, 10-nm filaments occur first (Ishikawa et al., 1968). In some invertebrate muscles, small microtubules occur before any myofilaments. When the thick-filament lattices form, some of these thin tubules occupy presumptive thick filament sites (Auber, 1969 – Figure 11.3), suggesting that thick filaments arise out of or around them. Such tubules are also present in the blood-sucking hemipteran *Rhodnius,* where they can be disrupted, as are other microtubules, by colchicine. Treatment of dedifferentiated muscle with colchicine,

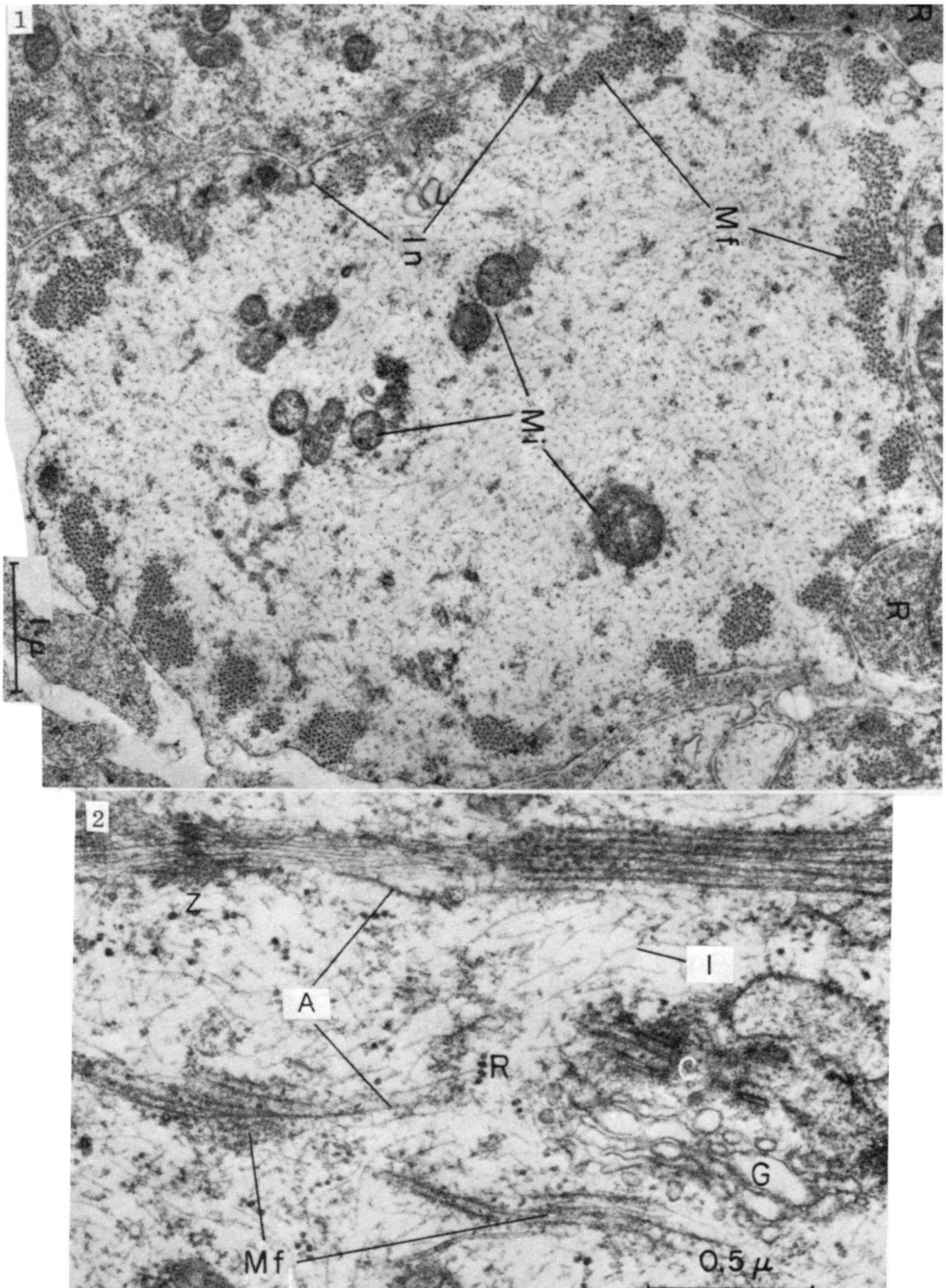

Figure 11.2. Development of a chick leg muscle cell. (*1*) EM of TS of a single cell, showing peripheral origin of myofilament (*Mf*) clusters, central location of mitochondria (*Mi*), and invaginations (*In*) of unknown function. (*2*) EM of LS showing independent clustering of thick (*A*) and thin (*I*) filaments. At this stage, I filaments meander in all directions. Ribosomes (*R*), Golgi apparatus (*G*), Z band (*Z*) and centriole (*C*) are also present. (From D. A. Fischman, 1967.)

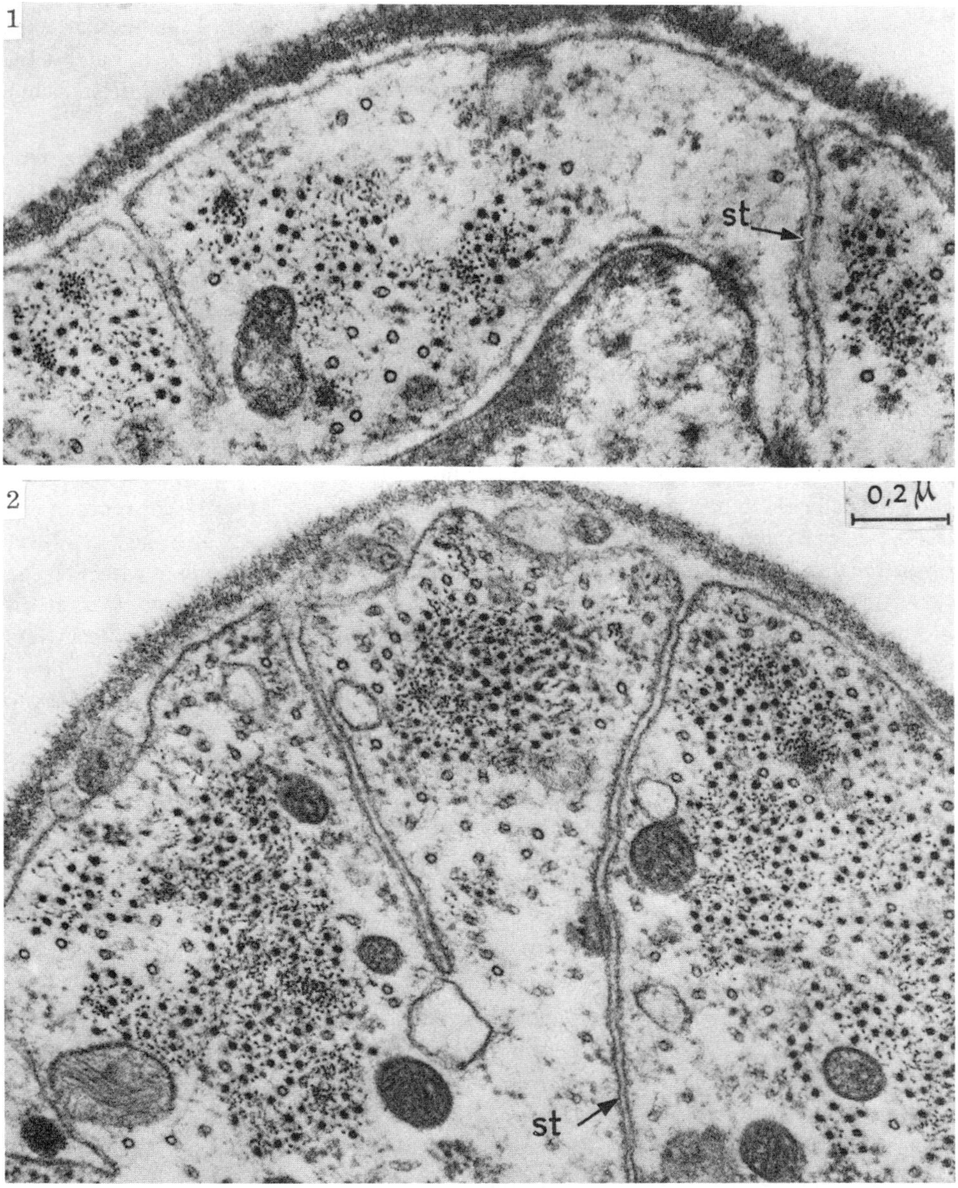

Figure 11.3. Redevelopment of muscle cell of abdominal wall of the blood-sucking hemipteran *Rhodnius.* (*1*) Cell 20 h after the meal that triggers redevelopment; (*2*) 48 h after (*1*). [Reprinted with permission from M. Auber-Thomay, *J. de Microscopie* 6, 627–638 (1967).]

however, disrupts the tubules but without affecting either normal thick-filament or lattice formation (Warren and Porter, 1969). *Rhodnius* provides unique opportunities for studies on muscle-cell formation because many muscles are lost before molting, to permit expansion, then reformed after a blood meal, which promotes the next molt. As in rat back cells, the process starts with the formation of patches of Z material under the surface membrane. As these are formed, thin filaments extend a short way out of them. Thus whether these Z bodies exist independently of bits of I-filament material or whether they result from attachment of these to the membrane cannot be resolved. The Z bodies leave the membrane and move inward: additional similar material with I filaments attached moves beside them to complete the myofilament or array. At first, though, I and A filament clusters do not interdigitate in an orderly manner (Figure 11.3). Final orbits, which will be an average of 12 round each thick, arrange themselves much later.

There is little data on M-band formation, but in all striated cells its appearance, if any is to occur, is relatively late, after Z-band and myofilament formation and alignment are about complete.

Studies on development of SR are complicated by the early presence of rough endoplasmic reticulum (ER). The future SR is probably independent of this and of much later origin. The SR is beginning to appear at the same time as myofilaments, but continues to form well into late stages of embryonic development. It does not reach its final form in any animal until some time after emergence.

Invaginated tubules are first apparent as shallow, narrow grooves in the surface membrane. These break up into regularly spaced invaginations that also penetrate farther inward (Valvassori et al., 1978). The final locations of the openings, the pathways inward, and especially the inner terminations of ITs are very different for different types of cells (see Chapter 6) and sufficiently different for individual cells to provide a fingerprint-level of difference. The IT/SR contacts are the last elements to appear, and as with SR, continue to form for some time after emergence. There are very few IT/SR contacts at first in rat skeletal cells, and these contacts are longitudinal and oblique, not transverse (Schiaffino and Margreth, 1969). The first ITs do not go deeply into the cell and there is little SR. When SR development does occur, it does so almost explosively, and is very closely paralleled by IT development.

The major means of longitudinal growth of cross-striated muscle cells is the formation of new sarcomeres at their ends. The attachments somehow permit functioning while growth is going on. The end sarcomeres are irregular in length because a zig-zag structure resembling an intercalated disc occurs there (Figure 3.37). The half sarcomeres at the attachments are especially long. It has been suggested that occasional very short sarcomeres in the middle of cross-striated muscle cells may indicate that some growth is going on there also (Jahromi and Charlton, 1980). During increased growth of all vertebrate and most invertebrate cross-striated cells, there is a nearly constant sarcomere length as the numbers increase, except for local inequalities. But for a few crustacean muscles that will have long sarcomeres, longitudinal growth is in part also achieved by progressive sarcomere elongation after starting out short (Bittner, 1968).

Cross-striated cells are well aligned laterally at the moment of their initial formation, because regions of formation of Z material by adjacent myoblasts are in line with each other as if a zone of Z origin in one influences those of abutting neighbors. But during later addition laterally, and elongation, mismatching is common. The faster the cell, the more SR it has and the more frequent the mismatching. A slow cell with little SR fulfills the classical conception of a striated cell, but one with much SR may not even look striated as seen in the light microscope.

11.2. Differentiation of Cell Types

The muscle-cell population of a muscle is commonly highly heterogeneous, with details of fine structure, enzyme content, and contractile machinery being markedly different for various cell types (Chapter 8), posing significant questions as to how this diversification is accomplished during development. Fortunately, myoblasts are relatively easy to culture, and details of the differentiation process are beginning to emerge. The current view is that different cell types originate from a single cell lineage, and that each starts out with the propensity for being any type, certainly any subcategory. For example, cultured myoblasts from the origins of the fast-type pectoralis major of the chick can synthesize myosin light chains not only of characteristic fast type, but also of the slow type associated with the anterior latissimus dorsi (ALD) and vice versa (Stockdale and Baden, 1979). Slow and fast isoenzymes of myosin coexist in all cells, even of a muscle destined to become fast-twitch (Gauthier et al., 1978). As development continues, production of the slow-type myosin is progressively "switched off." Mammalian muscle cells are all slow contracting at birth and become differentiated progressively thereafter (Figure 11.4). Large increases in contraction speed occur in those cells destined to be predominant in fast-type muscles.

Comparable flexibility is that exhibited by the structurally and functionally different crusher and cutter claws, either of which may be the left or right claw of the lobster

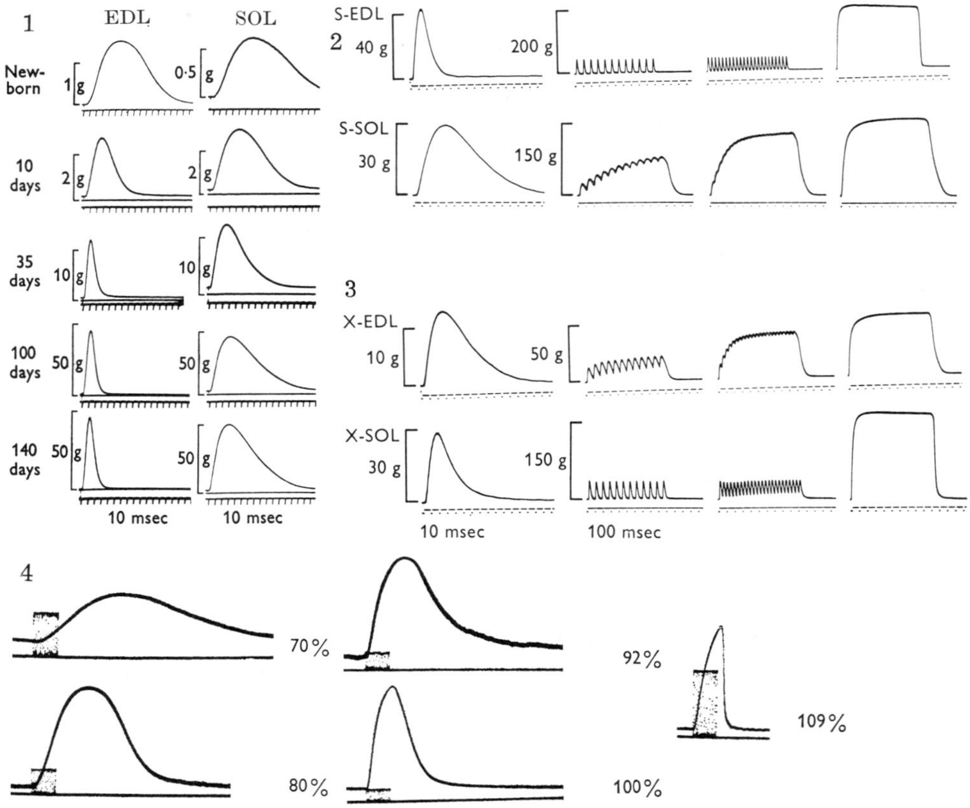

Figure 11.4. Changes in properties of muscles during maturation and crossed reinnervation. (*1*) Comparison of changes in normal mammalian (rat) slow soleus (SOL) and fast extensor digitorium longus (EDL) muscle twitches with time after birth. (From R. Close, 1964.) (*2*) Properties of the same muscles as *1*, self reinnervated, after 16 months. (From R. Close, 1969.) (*3*) Properties of the same muscles as *1* cross-reinnervated by each other's nerves, after 16 months. EDL has become slow and SOL fast. (From R. Close, 1969.) (*4*) Progressive increase in speed during embryonic and early posthatching development of abdominal intersegmental muscle of locust *Schistocerca gregaria*. (From N. M. Tyrer, 1973.)

Homarus. The crusher has no fast-type cells, only slow and intermediate, but the cutter has 60–70% fast cells (Lang et al., 1977a). A lobster raised without hard material on which to practise cutting has equal claws with similar mixtures of cell types. Thus the potential for becoming either extreme type, in which there are differences up to three-fold in sarcomere length, is common and the actual state dependent on an influence related to behavioral experience during development. Some enzyme activities, such as those of ATPase and NADH diaphorase change in relative amounts during the transformation.

Another, quite different way the muscle-cell population is changed is by selective loss from within a spectrum. In dog pectineus muscle, for example, at birth there is a mixture of types I, II, and intermediate. But after 12 months, a total of 32% of cells by numbers is lost. The size of the ones that remain is greatly increased, but the proportions are changed because the loss is principally of intermediate and type II cells (Ihemelandu, 1980).

Some of the changes that occur progressively during development can be explained simply on the basis of the late formation of calcium-releasing and sequestering membrane. This is the principal cause of the initial universal slowness, alike in mammals (e.g., Close, 1964) and insects (Tyrer, 1973). In locusts the process is essentially complete by emergence, so to see the difference embryonic muscles must be studied (Figure 11.4). By contrast, in rats the process is not completed until 14 weeks after birth. As might be expected, a presumptive fast-type muscle such as the extensor digitorum longus of the paw (EDL) shows the greatest change, twitch time reducing from about 200 to 40 ms, with no change in maximum twitch tension (Figure 11.4.*1*). By contract, soleus (SOL) speeds up little, from about 250 to about 160 ms twitch time, also with no change in peak tension. These differences are mediated by neural factors.

For other factors it is evident that a propensity for developing different characteristics is there to permit activity during developmental stages to play a role in determining the final outcome. But in addition, many kinds of flexibility are retained to permit a truly remarkable ability to adjust throughout the major part of the life span to changing functional needs. The balance of slow and fast characteristics of a muscle as determined by the individual cells can only be prescribed genetically to a limited extent. Each is determined by several factors, one anatomical – location – the others functional. These factors include average length, passive load, the types of activities in which the cells participate – sustained high load and repetitive, brief phasic lightly-loaded contractions being the extremes – pattern of motor-nerve impulses received (regardless of contractile performance resulting), and some chemical characteristics of the nerve cell(s) that innervate the muscle, together with consequences of the mutual interaction of each muscle cell with these factors.

11.3. Regeneration

It may be argued that since some of the examples quoted above are derived from regenerating material, they are atypical, and that regeneration is in principle different from generation. The counterargument is that the same gene controls should be occurring. However, they are being expressed in a very different environment from that which is present during development, so there are probably significant differences. This does not detract from a real interest in the problems of regeneration *per se,* but I concur that they should not be treated as if they provide a model for development.

Many invertebrates can regenerate entire limbs, and when the limbs are fully formed, they are difficult to detect from normal ones. But I well remember the scorn with which C. A. G. Wiersma, with whom I shared a research project in 1956 at Caltech, interrupted an experiment with an expletive and the remark: "its abnormal – must be a regenerate!" The capability to regenerate an appendage exists up to the higher taxa, although some insects have completely lost it. There is complete retention of the ability in lower chordates, and partial regeneration, repair, occurs in all damaged vertebrate muscles if the damage was not too extensive. New muscle cells are formed from residual mononuclear precursors indistinguishable from fibroblasts (Sloper and Partridge, 1980) and even from completely undifferentiated cells. In tissue culture many cell types may turn into muscle cells.

11.4. Plasticity in Nerve Terminals

Very little is yet known about morphological changes associated with synapses during development. The introduction of techniques for studying the progress of central development of connections by following identified neurons is likely to change this situation in the near future. At the peripheral level, physiological studies have been made on mammals and insects, both of which show a large increase in speed of contraction as development proceeds.

In the lobster, development of excitatory innervation to the claw closer muscle has been examined ultrastructurally (King and Govind, 1980). The study took advantage of the difference in excitatory synaptic vesicles compared with inhibitory ones. In the cutter claw, the fast and slow axons innervate exactly in parallel. In the first larval stage, only four discrete locations have excitatory innervation. Therefore, this innervation has to spread enormously to reach every muscle cell in the adult. There is also a large increase in synaptic size. This is accomplished by a combination of lateral fusion of small synapses and enlargement of others. The mean size of presynaptic dense bars and their mean number per synapse remains fairly constant.

11.5. Development of Ion Channels

Ion channels cannot be expected either to develop all at the same time or to remain constant in individual or relative numbers throughout life. They undergo changes in relation to maturation and possibly in relation to function as an adaptive measure. The latter could be regarded as a form of primitive learning. However, as yet there have been very few studies addressed to these matters. An exception is in *Drosophila* flight muscle, where it has been shown that early in development there are no (V) K^+ channels. Eventually fast, rapidly inactivating channels appear, that is, those carrying the ion flow termed A current, appear. Later, (V) K_A^+ channels mediating delayed rectification appear. One singular advantage offered by this finding is that it permits study of the A current independently of other K^+ channels. This has permitted the confirmation of the belief that a prior activation of the A-current channels leads to a long-lasting suppression of these channels.

11.6. Trophic Influences

The capability for plasticity in muscle function is often remarkable but was recognized very late in the history of muscle studies.

It had been recognized by the middle of the last century that nerves exert some chemical controlling influences over the properties of the muscles they innervate. But it has only been relatively recently that solid evidence for these actions has been obtained or even that the nature and extents of these influences has been appreciated. Then for a long time it was naively supposed that such long-term influences must be entirely dependent upon chemical communication, probably mediated by long-range orthograde transport within the motor axons. That some or even all of the transport and release might be related to impulse traffic, both to numbers and patterns, did not occur to early investigators. The starting point for modern interest in trophic factors was twofold. Axelsson and Thesleff (1957) during a study of an already well-known phenomenon, denervation supersensitivity (Cannon and Rosenblueth, 1949), discovered that within four days of denervation of mammalian skeletal muscle the ACh sensitivity, which is normally virtually confined to the end-plate region, starts to spread from there, until eventually, long before atrophy, the entire surface is sensitive. This process reverses one that occurs during development. It is now known that before completion of innervation, the developing muscle cell possesses ACh receptors along its entire surface, but very soon after innervation. the area reduces until it is largely confined to the end-plate region (Prives et al., 1980). Only one year after the Axelsson/Thesleff paper, a landmark pioneering study was published by Buller, Eccles, and Eccles (1960), who found that crossing of the motor nerve supplies to two muscles in the same limb that have different muscle-cell populations and contraction speeds results in significant changes in the properties. They had crossed the fast flexor digitorum longus (FDL) and the SOL, which is quite a bit slower, in cats. The experiment was done also in a young animal (kitten). The dramatic result was that not long after successful reinnervation, the FDL was much slower, while SOL was faster. These experiments have since been repeated and confirmed by several investigators (e.g., Close, 1965, 1969; Amphlett, 1975 Figures 11.4.*2* and 11.4.*3*) as well as extended to other species including amphibia and birds (Close, 1968; Zelená and Jirmanová, 1978) and other slow/fast muscle pairs. Biochemical and ultrastructural (Figure 11.5) as well as contraction characteristics have been examined and also found to have changed, fast-type becoming at least partly changed to slow and vice versa.

There is, therefore, a set of dynamic relationships superimposed on genetically determined constraints, within which flexibility of properties is possible. Constraints start to be exercised during development, but they continue, to some extent at least, throughout life for some muscles. Chemical influences of a nerve on the muscle cells it innervates, which exist in addition to ones associated with the elicitation or modulation of contraction, are termed *trophic.* Research on such influences has been confounded by the discovery that it is a two-way natural interaction, the innervated muscle cells influencing the properties of the nerve cells that innervate them, so that they in turn affect the trophic input they receive. To further complicate matters, it has been found that even the pattern of impulses received affects the chemical constitution and contractile properties of the muscle. Finally, what the muscle does physically and also its general chemical environment exert influences in what is clearly a very involved multiple set of parallel interactive feedback processes.

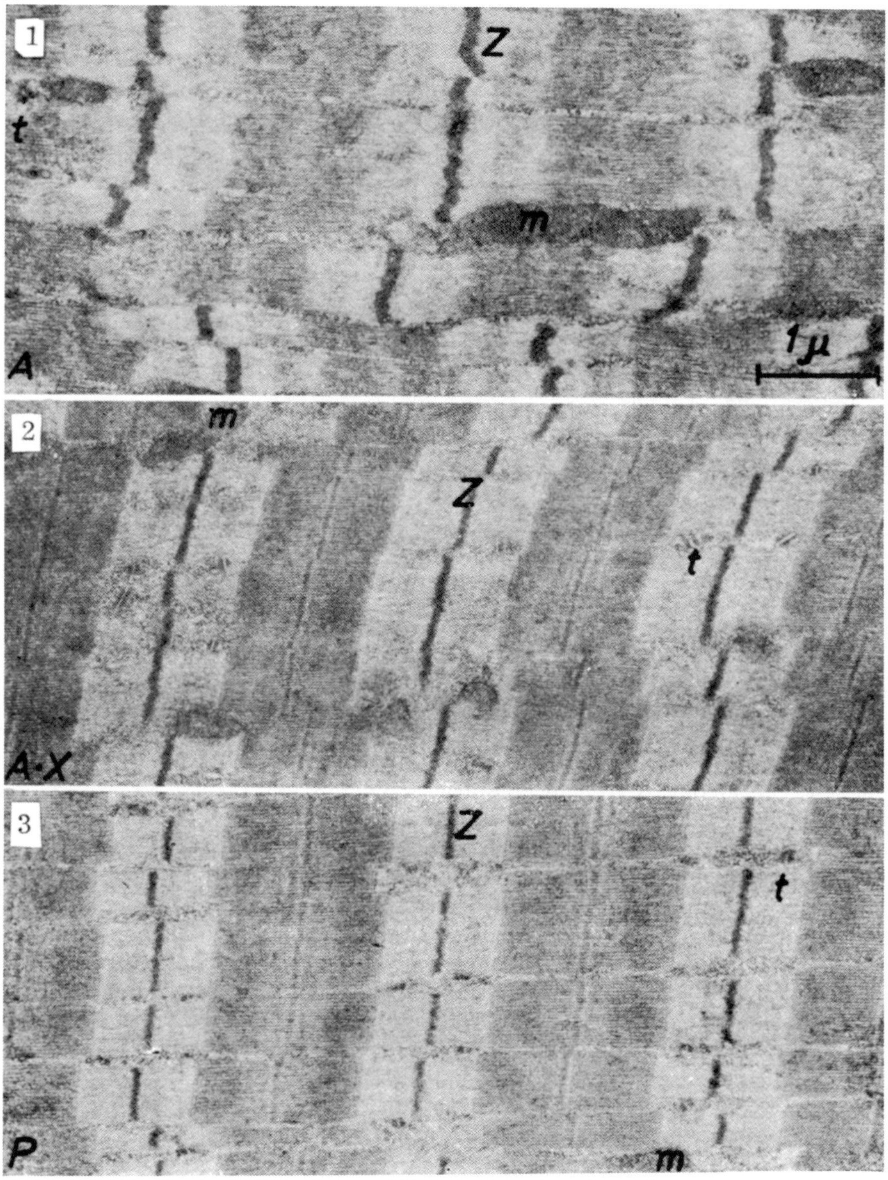

Figure 11.5. Ultrastructural changes associated with successful cross-innervation in chicken: slow anterior latissimus dorsi (ALD) crossed with fast posterior latissimus dorsi (PLD). (*1*) ALD control. (*2*) Cross-innervated ALD. (*3*) PLD control. (From Zelená and Jirmanová, 1974.)

Some previously unsuspected fundamental aspects of muscle-cell physiology have been revealed as a result of experiments aimed at examining trophic influences. For example, although vertebrate skeletal muscle action potentials are sensitive to TTX, showing that they are dominated by sodium-conductance activation, denervated muscle cells produce spikes that are TTX insensitive. These spikes are due to calcium-conductance increase; they are smaller than sodium spikes because a fall in resting potential accompanies denervation, and they have 42% slower rise time because of the slower activation of calcium channels. This capability for a switch in properties of the excitable membranes of vertebrate skeletal cells following denervation was realized only relatively recently. The switch in rat leg muscles is complete after only 4 days (Albuquerque et al., 1972).

11.6.1. An Influence of Motor-Nerve Impulse Pattern

Although Buller et al. (1960) correctly attributed the changes in muscle properties that followed cross-reinnervation to dynamic characteristics dependent on chemical qualities of the respective nerves, they failed to consider it possible that the different natural impulse-patterns in the two nerves might be a significant variable. This variable was not allowed for, nor was the important reciprocal possibility, that the muscle can impose properties on the nerve that innervates it, leading in turn to a closed-loop feedback system.

When a denervated mammalian muscle is simply directly stimulated sufficiently regularly, the spread of sensitivity to ACh is at least partially reversed (Drachman and Wizke, 1972; Lømo and Rosenthal, 1972). A complete physiological transformation of EDL and SOL of rabbit occurs after cross-reinnervation, even to the point of change in light-meromyosin-molecule type, after 11–12 months. But an exactly similar transformation of EDL to SOL type occurs after only 15 weeks of chronic stimulation of EDL, which is naturally phasically activated, with tonic activity (10 Hz) such as is characteristically received by SOL (Sréter et al., 1975). Rat SOL, when subjected for 3–6 weeks to either brief trains at 100 Hz or longer ones at 10 Hz to achieve in each case the same average of 2 Hz, became fast in character (Lømo et al., 1974).

The latest technique is to implant small stimulators chronically, impose patterned stimulation for various periods up to several weeks, and sacrifice the animals (rabbits) at various times for testing results (Eisenberg and Salmons, 1980). Tibialis anticus (TA), which is of fast type, was contrasted with SOL. Subjected to continuous 10-Hz stimulation, there was a rapid increase in mitochondrial volume to above control level of either TA or SOL. The Z-band thickness increased until after two weeks it was equal to normal Z thickness of SOL. The IT number declined to the SOL level in two weeks with a concomitant dramatic 75% reduction in number of triads. The SR proteins must have changed also, because the rate of uptake was slower.

A difference in array of 6-nm particles present in orthogonal clusters in the membrane of fast-type rat EDL has been noted in freeze-fracture electron microscopy (Ellisman et al., 1978). These are absent in slow SOL but the latter acquires them when cross-innervated by the EDL motor nerve.

However, in contrast to these results, application of colchicine, which blocks axonal transport without affecting nerve impulse chronically, in a silastic cuff placed around the nerve is followed within about four days by the appearance of extrajunctional ACh sensitivity – like that following denervation (Albuquerque et al., 1972; Lømo, 1974). These researchers studied the relative contributions of microtubule-mediated orthograde transport in rabbit SOL and EDL (Alburquerque et al., 1972) confirming by electromyography that impulse activity had not changed. Within four days the muscle resting potentials had fallen, the rate of rise of the spikes had fallen 42%, and their total size had diminished 17.5%. But of greater significance, these action potentials were changed in character, being now of the TTX-insensitive variety, that is, calcium spikes. The changes coincided with the spread of ACh sensitivity away from the end plates.

The converse experiment, to retain transport functions but block impulses and not substitute artificial neural programs, has been achieved for rat EDL and SOL over a 21-day period. The nerve was blocked by TTX released from a capillary implant (Bray et al., 1979). The muscle-cell resting potential fell by 40% for EDL and 50% for SOL, the reasons for which are not known. Direct evidence that trophic material moved by axonal transport is responsible for the ACh-receptor control has been obtained in experiments in which organ-cultured rat EDL cells were treated with material accumulated in the swelling above the cut of the sectioned motor nerve (Younkin et al., 1978). Some interesting changes take place in frog slow-type cells following denervation. Acetylcholine sensitivity spreads away from the distributed nerve terminals as it does away from the single endings on fast-type cells. But excitability changes occur that are confined to localized patches, not uniformly distributed. These are patches of spiky membrane with high depolarization-activated sodium conductance (Schalow and Schmidt, 1979). Thus when the trophic influence is removed, the membrane responds by protein synthesis, especially of ACh receptors and sodium channels.

Denervated cat EDL and SOL before onset of atrophy at 28–35 days are both slower than normal (Kean et al., 1974), and maximum tension capabilities are lower. In EDL, but not SOL, the number of sarcomeres per cell has increased, to some extent compensating for intrinsic loss of shortening speed. The changes are even less when the denervation is carried out in kittens, where the muscle properties are still changing drastically. Significant differentiation can therefore occur in the absence of immediate nerve influences.

The ability for sarcomere lengths and numbers per cell to change is remarkable and was not earlier suspected. It can occur quite rapidly. When a denervated muscle is fixed in the shortened position, changes in both structure and function follow very quickly. After only seven days, not only is passive elastic extensibility greatly reduced for cat SOL, but the total number of sarcomeres per cell is down (Hayat et al., 1978).

Denervation affects the calcium-accumulating and ATPase activity of SR in rabbit skeletal cells, which activities are increased by about 1.5 times (Lopina, 1976). Coupled with the spread of surface sensitivity to ACh, this makes denervated vertebrate muscles give large contractures in moderately low concentrations (10^{-6} M). It has recently been shown that such a contracture is about doubled when as little as 10^{-13} M β-endorphin is added. This secondary action is likely to be mediated via the calcium-sequestering/release mechanism also.

Some very long term experiments on cross-reinnervation are under way: for example, nerves to cat SOL were crossed with those to fast flexor hallucis longus and left in place for a minimum of three years (Lewis et al., 1980). The changes in properties seen in shorter-term experiments not only remained but were even further enhanced. Furthermore, the motor units as a whole were changed to the point where it was apparent there had been some reorganization in the direction of supplementing the fast-unit size in the originally slow muscle now receiving innervation from a fast nerve, and vice versa.

The characteristic myosin light chains of EDL and SOL are switched by cross-reinnervation (Weeds et al., 1974), but the relative significance of impulse pattern vs. chemical agent(s) has not been determined yet. In cell culture, myosin synthesis is increased simply by repetitively stimulating the cells electrically (Brevet et al., 1976), but there is no evidence that the type can be changed in association with pattern.

Arthropod muscles, by contrast to those of vertebrates, do not atrophy following denervation, but they do undergo a number of changes. In particular, as in vertebrates, there is a spread of sensitivity to the presumed excitatory junctional transmitter L-glutamic acid (Cull-Candy, 1976a; Usherwood, 1977), and ibotenate, which mimics inhibitory action, produces a much-larger-than-normal hyperpolarizing response (Cull-Candy, 1975). However, there are some partial degenerative changes, considerable weakness, and even eventual loss of contractility, though all of these are quickly reversed following reinnervation. The changes include progressive losses of SR, ITs, and IT/SR contacts, mitochondria and even myofilaments (Rees and Usherwood, 1972).

11.6.2. Effects of Muscle Action

The influences of the static and dynamic tension forces generated by a muscle and acted upon by other muscles and by gravity have been examined in a variety of ways, some of which are drastic. The commonest and simplest is to cut the distal tendon, an operation termed *tenotomy,* so that the muscle receives normal trophic inputs and some normal motor impulses and can change length but cannot develop force. Another is to denervate its synergist(s) so that its load is greatly increased. Also used are immobilization, either at fully extended length, fully shortened length, or at midrange, or tenotomy followed by immobilization of the tenotomized muscle only. All these may be done either in mature adults or in developing young ones.

Tenotomy produces effects ranging from no more than a temporary small fall in membrane potential to a rapid complete atrophy, depending on the muscle and species. Rabbit slow-twitch soleus is an example of the type that atrophies, but before this is under way it becomes much faster in both contraction and relaxation times. This interesting result was, however, confounded by an accompanying weakness, suggesting that it might be merely the result of an earlier loss of the slower cells. However, it appears to be a genuine effect (Bagust, 1979).

Some drastic changes have been reported in slow ALD of pigeons one year following tenotomy, although by this time the muscles had reattached after shortening by 25%. They were smaller and faster, the Z lines had changed to a thinner type charactersitic of fast cells, and there were many more triads (Hikada, 1972). In invertebrates, tenotomy can have much more drastic consequences than does denervation. A tenotomized crayfish opener atrophies within 50 days, although there is virtually no change in the same muscle after denervation for five times as long (Bittner, 1973). However, the nerve terminal (with its sheath) doesn't degenerate either, and may be able to synthesize the "trophic" agents required for survival even though cut off from the cell body.

Tenotomy followed by immobilization at minimum length results in degenerative changes in ultrastructure in some locust skeletal cells starting at nine days (Figure 11.6 – Jahromi and Bloom, 1979). Myofilaments are disrupted, some being lost, both SR and Z bands degenerate, and mitochondrial structure alters. A muscle can be subjected to tests of the effects of force change only by immobilizing it *in situ,* without preceding tenotomy. This is a natural experiment in humans afflicted by paraplegia, just as tenotomy happens naturally in athletes who break their Achilles tendon. Immobilization at mean body length produces less drastic changes than do immobilizations at the corresponding extended or shortened lengths, but they are nonetheless significant. One change is a "disuse atrophy," in which total cell numbers are reduced as well as cell sizes (Booth and Kelso, 1973; Anderson, 1979). The extent to which a muscle changes with use and disuse, if at all, is probably quite different for different muscles, even in the same animal.

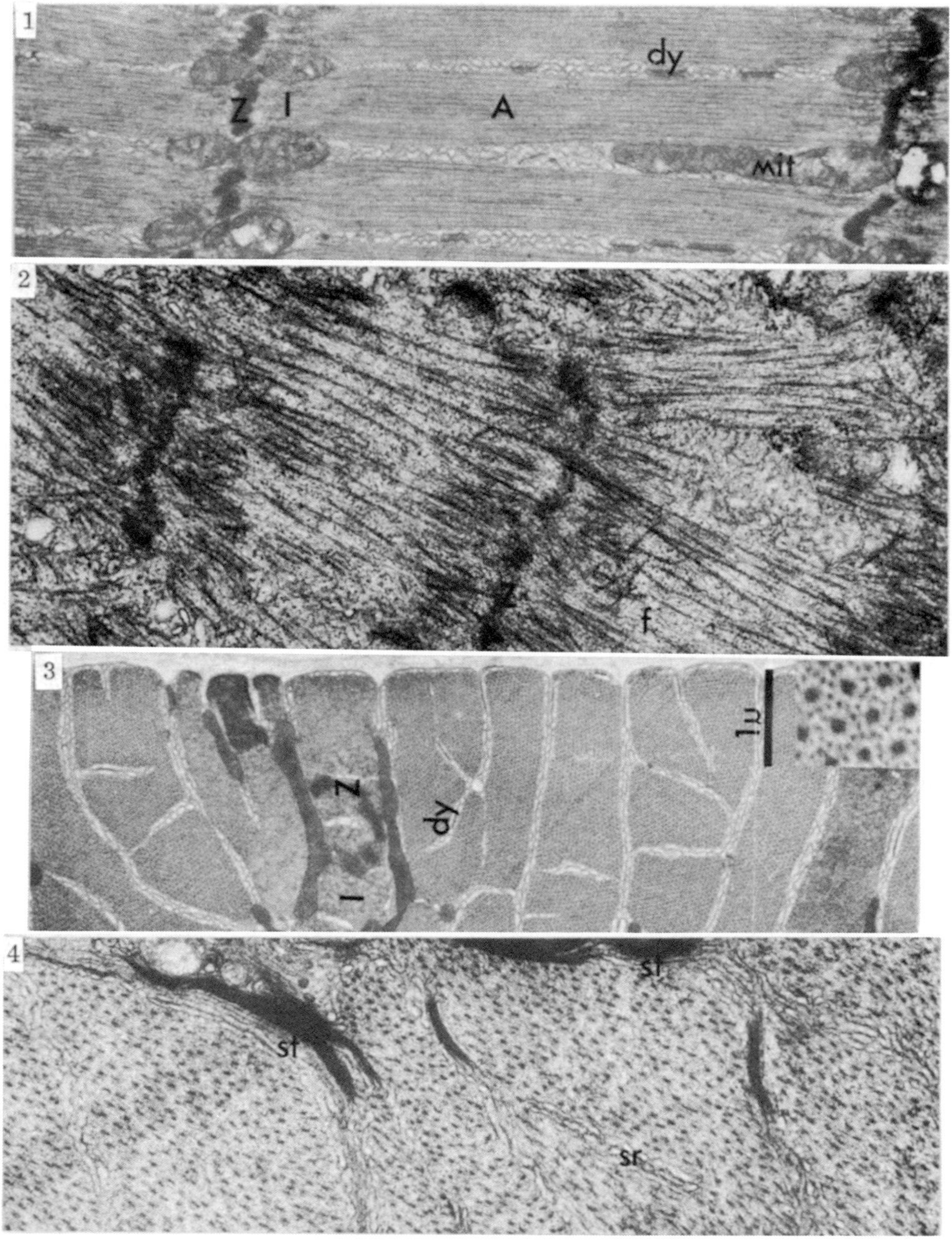

Figure 11.6. Degeneration of muscle following tenotomy – flexor tibiae of locust *S. gregaria*, immobilized 45 days at minimum functional length. (*1*) Control EM of LS. (*2*) Tenotomized – thick filament and SR are drastically disorganized. (*3*) Control EM of TS. From Jahromi and Bloom (1979).

11.6.3. Passive Tension

Effects of passive tension alone have been little studied but are probably important, especially during development: they may change from positive to negative following maturation. Developing chick muscle fixed in a stretched state shows increased accumulation and incorporation of amino acids and proteins including myosin (Vandenburgh and Kaufman, 1979). By contrast, adult muscles show at least partial atrophy when fixed in any state. The initial determination of muscle cell type in chickens occurs at a time soon after innervation, the differences being due to factors introduced by the nerves that successfully innervate them (Gordon et al., 1977).

Tibial muscles of young rabbits have been held immobilized during the major growth period to see how this affected their development. They have greater force capability than partially immobilized ones, but although they grow in length somewhat, they end up 10–20% too short. The difference is accounted for by their having a mean sarcomere length some 20% less.

Both sarcomere length and cell diameter of crawfish skeletal muscle are heavily dependent upon maintenance of a minimum length (Boone and Bittner, 1974). It has become abundantly obvious that there are influences of both trophic agents and impulse pattern on each of these structural parameters. There is no point in researchers' dividing themselves into rival groups: "antitrophists" and "trophists" (c.f., Vrbová et al., 1978), as they seem bent upon doing, unless of course it impels the trophists to do the sophisticated biochemistry needed to isolate the molecules they have come to believe exist. There are some effects that are due to impulse and activity patterns, others that are due to chemical inputs, and some may turn out to be interchangeable.

What is remarkable is the extent to which so many aspects of muscle biochemistry and structure and their functional consequences are plastic, tuned to meet behavioral needs in extraordinarily subtle manners that were absolutely not understood by earlier generations of muscle scientists. If one may venture a prediction of the ultimate outcome, it is that it will transpire that each desirable functional goal can be achieved by diverse routes. There are in effect three outputs from a motor neuron that affect the muscle cells it innervates: substances that are transported and released to some extent independently of impulses; substances that are dependent on impulses, including impulse pattern; and the impulse pattern itself. The latter has numerous indirect effects, ranging from muscle-membrane activation to ensuing contraction dynamics. Each of these modulates the same targets: the various protein-synthetic regions, so each alone will be seen to influence the target when treated singly under open-loop conditions. But it should be borne in mind that not only must all three act in concert in the intact animal, but all are mutually interactive, and all are present in closed-loop systems because there are both chemical and physical feedback information channels. Nothing is known at this time about how the latter function.

11.6.4. Changes in Genetic Expression

Following denervation of any mammalian muscle it is harder to recognize the different types of cells, showing that the factors responsible for differentiation are dependent upon either the pattern of neural input or on chemical agents released by the nerves. There are several general changes, expecially in SR and SR/IT contacts, mitochondria, and ribosomes, in all types of cell, but these changes are always more marked in fast white cells than in other types (Gauthier and Dunn, 1973).

The basic phenotypic expression for both fast- and slow-type mammalian cells in the absence of any external influence appears to be for fast-type cells. When excised and cultured, muscle cells from presumptive slow- and fast-types synthesize both the heavy and light meromyosin chains characteristic of fast-type cells only (Rubinstein and Holtzer, 1979).

When rat diaphragm muscle cells are developed in culture, they have uniform overall sensitivity to ACh, but not if RNA/protein synthesis inhibitors are present. However, there is no effect of the inhibitors on established sensitivity (Fambrough, 1969). It follows that when the nerve causes restriction of ACh sensitivity, it is doing so via a factor that alters gene expression in the muscle cell. Transformations of the different myosins of basic cell types occur during trophic shift of rabbit peroneus longus from slow to fast characteristics occurring as a result of stimulation. The cells were examined using fluorescein-labeled antibodies against the fast and slow myosins. Unstimulated cells treated with anti-fast were nearly all stained, but only a few were stained by anti-slow. When again examined with similar staining methods following chronic stimulation for three weeks, many of the cells failed to take up anti-fast, wheras almost all showed intense staining by anti-slow (Figure 11.7 – Rubinstein et al., 1979).

11.6.5. Transformation

The remarkable switch that occurs in form, function, and muscle-cell content of one or the other claw of a larval lobster, depending only on which one, if any, is the first to encounter sufficiently hard material, represents an extreme example of a widespread plasticity in potential for varied gene expression by muscle cells. Following transplantation of appendages in many crustaceans, the transplants show progressive changes in form from their original one, in the direction of matching the appendage appropriate to the new site. But this capability is bettered by an even more dramatic change, or *transformation,* in an existing nonspecialized claw. This is particularly prominent in the snapping shrimp, in which one claw develops as a rather small pincer, the other as a much larger specialized one that produces audible snapping sounds. The sound is produced by an initial pressing together of opposed disclike regions of the dactylopodite and propodite, followed by a sudden forceful "opening" movement. Thus it requires morphological, behavioral, neural, and muscular modifications.

If the existing snapper claw is damaged, or if its nerve is cut, the pincer claw transforms into a snapper (Figure 11.7.*2* – Mellon and Stephens, 1978). Significant changes are observed with 24 h, but a molt is needed to complete the process, which takes 18–20 days. The motor neurons of the new snapper claw increase in diameter up to 90%, and one of them undergoes a shift in position. The muscle-cell number remains constant, but muscle-cell size increases up to sixfold. Various characteristics of synaptic transmission and muscle-cell properties are changed in the direction of decreased overall speed with relatively greater force production.

It follows from these results that the trophic control mechanisms can be abruptly altered to a new, different, essentially permanent relationship by a single event. Determining how this is achieved at the cellular level presents a number of major challenges to investigators.

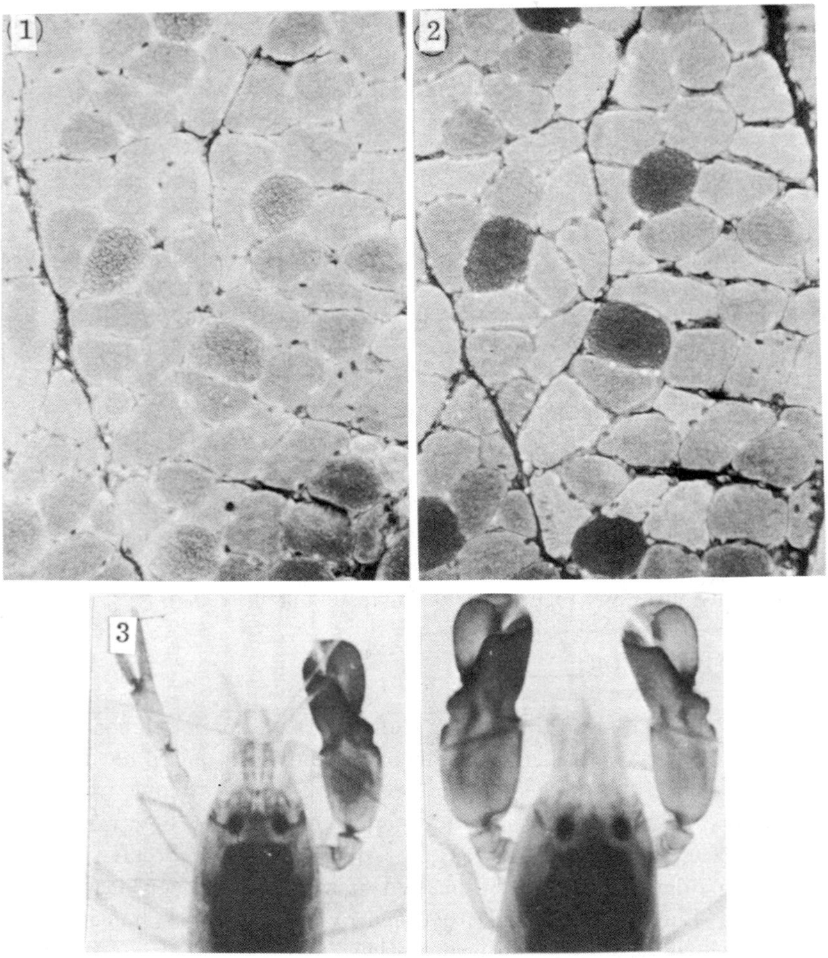

Figure 11.7. Changes in muscle myosins with altered stimulus pattern. Peroneus longus of rabbit chronically stimulated for three weeks, stained with anti-fast and anti-slow antibodies labeled with fluorescein. (*1*) Unstimulated control. (*2*) Stimulated. There is a decrease in anti-fast and increase in anti-slow. (1, 2 reprinted with permission from Rubinstein et al. (1979) In: *Motility in Cell Function,* pp. 367–370. Copyright Academic Press.) (*3*) Transformation of whole crustacean limb occurs when a snapper claw of the snapping shrimp *Alpheus* is damaged. The small pincer (*left*) is transformed into a snapper (*right*). (Reprinted by permission from *Nature* **272**, 246–248. Copyright 1978 Macmillan Journals Ltd.)

11.6.6. Membrane-Excitation Pattern

I reported above that pattern of stimulation alone can result in transformation of overall muscle characteristics. However, the extent to which the results depend upon possible effects of membrane excitation alone, compared with the effects of dynamic performance, has not been tested. Not only are there common effects on cell properties, there are relative effects on the different types of cells in a given muscle. For rabbit fast tibialis anticus and slow SOL, a follow-up of the possible changes in balance of cell types has been undertaken (Pette et al., 1975). With intermittent 10-Hz treatment for three weeks,

8 h daily, SOL was unchanged, but the TA was slowed to half. This was not associated with any myosin ATPase change, but the whole energy-supplying enzyme system was shifted in the direction of the slow-twitch type. Arthropods might seem to provide ideal model systems for experimenting with the effects of neural patterns on development of muscle cells and subsequent dynamic maintenance. However, in this respect, when compared with vertebrates, arthropod systems are fraught with disadvantages in view of the existence of multiterminal innervation. The neural-pattern effect is different at junctions having a large extent of facilitation compared with ones having little or even negative facilitation, as can be the situation in adjacent muscle cells receiving the same axon. It seems that in early development, a special dynamic relationship must exist that balances muscle-cell and synaptic-terminal properties. Thereafter there are still some dynamic capabilities, but others are self-perpetuating in the different branches *of the same motor neuron.*

11.6.7 Summary of Trophic Events

Each of the 13 major events illustrated in Figure 11.8 plays a continuing significant role in determining properties of the adult muscle. First in importance is the genetically determined output from the nucleus. But the gene expression is in part modulated by all the inputs, both to the cell body from its local environment and from its dendrites and synapses, and from the peripheral terminals. There are at least two different orthograde-transport systems in the axons distinguished by different transport rates: fast – 0.3 mm/min – and slow – below 1 mm/h (Ochs, 1974). Also present are one or more retrograde-transport systems that transport substances all the way to the cell body which have been picked up by the terminals. Some singularly potent toxins, such as tetanus toxin, travel into the central nervous system by taking a free ride on the retrograde-transport systems of motor nerves. The retrograde transport makes the whole into a dynamic closed loop with positive and negative feedback. The terminals too, are more than just communicating devices: they also provide a secondary synthetic site for some of the trophic substances. As long as the terminals remain alive, in arthropod muscles following nerve section, muscle function is normal.

That retrograde transport can influence motor neuron properties is not just a conjecture. Cat soleus motor neuron properties have been compared, using intracellular techniques, for cells innervating untreated muscles, denervated muscles, and muscles with TTX blocking of distal propagation of impulse traffic (Czeh et al., 1978). The TTX block does not impede either orthograde or retrograde transport. Within eight days following conduction block, the afterhyperpolarizing undershoot of the cell body decreases. The same thing happens after nerve transection. But if the nerve peripheral to the TTX block is stimulated for a while each day, there is no change. When the same pattern of stimulation is applied proximally to the cut, the same change does occur. Peripheral stimulation also prevents atrophy and even initial weight loss. These experiments prove that, if muscle excitation is maintained something is different in regard to retrogradedly transmitted materials in preparations with severed nerve cell compared with preparations in which the same impulse activity is shared by the cell body/dendrite system and the motor terminals. An abbreviated undershoot would possibly have the effect of increasing the mean motor frequency in response to fixed inputs.

The retrograde action can presumably influence the peripheral junction as well as the terminal. In arthropods there is often a remarkable match between the properties of an

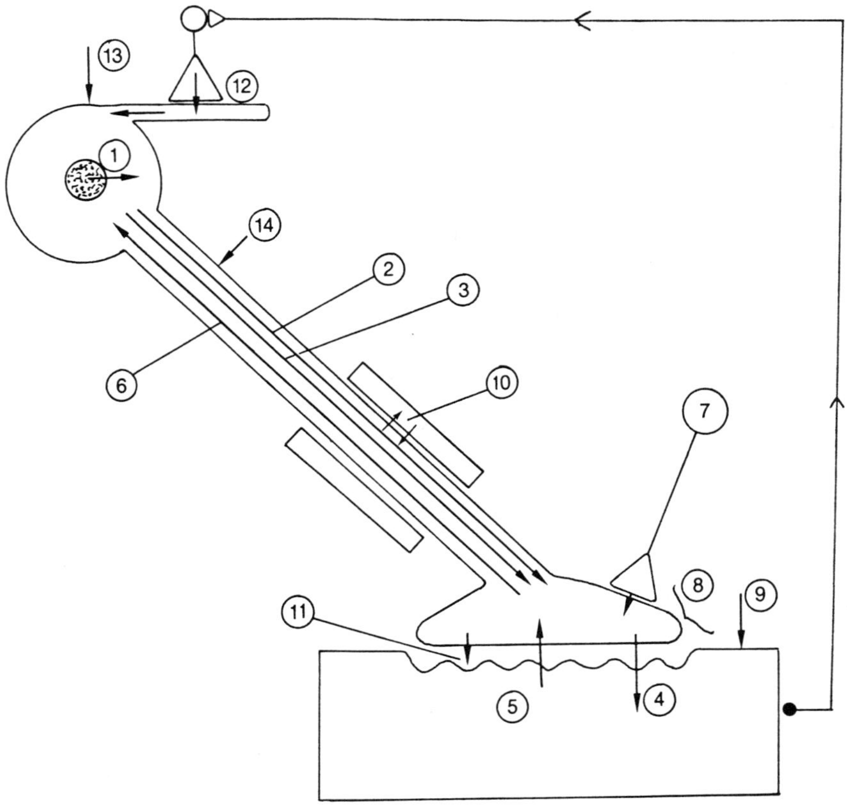

Figure 11.8. Summary diagram illustrating the sources of trophic influence on a muscle cell and reciprocal feedback influences of the muscle on the motor nerve. (*1*) Primary genetic factors. (*2*) Fast orthograde-transport system. (*3*) Slow orthograde-transport system. (*4*) Synaptically released trophic substances. (*5*) Retrograde pathway to synapse. (*6*) Retrograde transport to cell body. (*7*) Presynaptic influences. (*8*) Chemical environment of junction. (*9*) Chemical environment of muscle membrane. (*10*) Axonal sheath. (*11*) Long-term changes in transmission. (*12*) Feedback via interneurons. (*13*) Chemical environment of neuron. (ORIG.)

individual muscle cell and those of its associated nerve terminals. The details of synpatic transmission at single junctions on a muscle cell are generally nearly identical, but can be markedly different from those of a neighbor in either a vertebrate or an invertebrate (Frank, 1973; Hoyle, 1978b). Since each arthropod muscle cell is both polyneuronally innervated and has multiple, distributed terminals, this strongly suggests that the muscle cell influences the properties of the neurons that synapse with it. In summary, "nerve terminals are as different as the muscle cells they innervate." (Pickett, 1980).

The transport systems are dependent upon structural integrity of microtubules and microfilaments in the nerve. Dense-core vesicles, of which even the major motor axons have a few, transported along the tubules are considered most likely to contain the trophic substances. These tubules are disrupted, without interfering with other aspects of nerve function, by colchicine, providing an experimental tool for examining the different effects of excitation of the terminals and excitation of contraction, and the combination of strictly muscle excitation effects with those due to tubule-transported substances. Some of the trophic influences are lost with colchicine, others are retained, provided the impulse-traffic pattern continues unchanged. However, a direct assay of possible protein release from mouse phrenic nerve, which innervates the diaphragm, gave positive results (Musick and Hubbard, 1972), showing protein release accompanying ACh release.

11.7. Training

With so many closed-loop dynamic systems regulating muscle structure and properties, it is not surprising that movement practices affect the long-term nature of the muscle, opening the way for effects of exercise training. The extent to which muscles of an individual animal are affected by either training or diet are remarkably varied: there is no formula that works except statistically! Some humans respond to sustained isometric exercise by enormous gains in strength and hypertrophy of major skeletal muscles. Others, given identical training and the same diet, are unaffected, and a few actually lose weight. A human subject who can be termed "athletically competitive" is not simply psychologically specialized: such a person can be defined physiologically as one whose muscle cells respond strongly in the positive direction to training. Training is not always successful. I recently listened to a lecture in which the impression was gained that the lecturer had developed a foolproof regime for improving cardiac performance in cats. On asking detailed questions, I learned that the training had not been successful on *all* the cats. A large number had died! Increases in skeletal muscle performance range up to more than threefold in force development and about doubling in speed, in the better subjects. Regardless of animal species, genetic factors play a major role. By far the greatest amount of data available is on humans, but there is always a real problem of controls in human research. Rats offer better controls and have been extensively studied, but appear to respond very differently as a whole from humans. An entire species may have evolved a particular attribute such as endurance at the expense of maximum speed, and these attributes may be compared in closely related forms such as rabbit and hare. During all types of training, the muscle both shortens and develops force, each of which may independently and differently affect muscle properties. A wide spectrum of muscle properties is clearly present in humans, where there are drastic differences between individuals, and this is even more obvious in some domesticated species of animals such as dogs and horses.

Much interest lies in comparing effects of training regimes aimed at increasing force production with ones aimed at improving speed or accuracy. Short-term intensive sprint training of rats produces decreases in twitch times of both fast and slow muscles and an increase in both carbohydrate oxidative and glycolytic enzymes (Standte et al., 1973). Endurance training increases principally enzymes associated with fatty acid metabolism.

A naturally sedentary animal such as the domestic pig offers an interesting opportunity to test the effects of exercise, and specimens have been trained independently for distance running on one hand and sprints on the other (Fitts et al., 1973). Specific muscle characteristics, contraction time, relaxation time, isometric tension, and twitch-tetanus ratio, were not changed significantly, but nevertheless, ability to sprint by the sprint-trained and to run long distances by the distance-trained were greatly improved. There is good evidence from some mammalian muscles, notably rat leg muscles (e.g., Ho et al., 1978), that the number of cells increases during training in part by splitting and increase in size of the fragments. But in these, as in many, even most, other forms of training, it is in the energy-supplying mechanisms that changes are most markedly manifest. There is abundant evidence in humans that transformation of subgroup types, notably type II B to the high-oxidative type II A (Anderson and Henriksson, 1977) readily occurs with suitable training. Likewise, in research appropriately supported by the Horserace Betting Levy Board, sprint training in horses has been found to increase greatly the activities of both aerobic and anaerobic metabolic enzymes (Guy and Snow, 1977). Several studies show increases in mitochondrial size following sprint training. The increased work-load "training" of removing the gastrocnemius leads to 40% weight increase in rat SOL in only six days (Goldberg et al., 1975). The hypertrophy in SOL occurs even while the animal is starved and neighboring muscles are wasting away, showing that the effects of work take precedence over hormone-regulated factors. In starvation, passive stretch alone will retard loss of protein by SOL, and electrical stimulation does better. Tension thus seems to be the most critical single factor in determining the promotion of muscle protein production. The same is true for connective tissue, especially collagen (Alzaid et al., 1980), which makes a frequently loaded muscle harder.

In suitable human subjects, exercise that leads to hypertrophy is generally agreed to increase greatly the diameters of existing muscle cells and to increase enormously the amount of collagen and other connective tissue. But it seems unlikely that new cells are added.

In a rare experimental series on a lower vertebrate, exercise training was carried out on brook trout by Johnston and Moon (1980). The fish were obliged to swim continuously for 21 days at a high speed, 3 body lengths/s. This is fast enough to ensure involvement of some white muscle in addition to the small lateral red band. The red cells hypertrophied by an astonishing 70%, though neither aerobic-enzyme activity of the cells nor their mitochondrial volume was altered. The white-cell glycolytic enzyme activity declined, but these cells were otherwise unchanged.

The source of the new or extra protein is increased synthesis within the muscle cells, and it now seems likely that muscle cells are themselves also responsible for the accompanying glycoprotein, collagen, and elastin, by secreting them (Jones et al., 1979). Following cross-reinnervation, changes can be detected not only in various aspects of the contractile machinery, but also in the sarcolemma (Ellisman, 1978).

11.7.1. Can Training Alter Invertebrate Muscles?

There have been few attempts to study the effects of training in invertebrate-muscle structure and function, which is disappointing in view of the ease of extensive, accurate analysis that is possible at both muscle-cell and identified-neuron levels. In my own laboratory we are especially familiar with locusts/grasshoppers, which can produce an extraordinarily powerful, rapid extension of the hind-leg tibia to make a defensive kick or a jump. The excitation is accomplished by a single motor axon, FETi (see p. 529). We have been surprised that a carefully raised "coddled" locust that has not, to our knowledge, used FETi once in its entire life following emergence can still at any moment make a normal-appearing jump when required to. Since about 70% of the muscle cells in the jumping muscle are innervated exclusively by FETi, the majority of these muscle cells could not have been exercised previously. A burst of impulses directly applied to the motor nerves via chronically implanted leads produces extension directly without the complex cocking motion associated with jumping. Such bursts were applied to locusts embedded in wax, with their tibia attached to a spring/strain gauge force transducer or scale pan with weights. The muscles were exercised to exhaustion once each day or each second day. Gains in maximum force capability of 20–30% occurred. Of the insects that had received the training, some put themselves out of the experiment by breaking the tendon. It has long been known that locusts, like highly trained human athletes, can snap their extensor tendons during the onset of jumping. Since a totally unexercised locust jumping muscle is already capable of producing almost as much force as its skeletal attachment can bear, there is little potential biological advantage to be gained by exercise!

The above experiments were followed up by histological and electron microscopical examination. No change had occurred in exercised insects in either muscle cell size or structure, but their mitochondria were significantly fatter than those of unexercised controls, bringing locusts into line with several studies on vertebrate animals.

In arthropods, the effects of training may be manifest in altered neuromuscular transmission rather than in muscle-cell structure. Not all of the synapses are functional, some being dormant (Acosta – Urquidi, 1978). Training can both bring dormant synapses into use and improve efficacy at existing ones.

It is unreasonable to suppose that one can generalize from the small number of experiments so far made on arthropods in regard to all matters concerning the plasticity of their muscles.

11.7.2. Effects of Training on Cell Types

One aspect in which training might play a role is in regard to the balance of cell types within particular muscles. A drastic method of forced training in animals is to surgically remove an agonist from a cluster of synergists, which forces the other muscles to take the extra load in normal movement. After 60 days without the gastrocnemius, the plantaris and SOL of rat, which between them must share the gastrocnemius functions, were each changed, but in different ways. The plantaris was hypertrophied 100% and had about a 30% increase in numbers of cells, made up entirely of type-II cells (Ianuzzo et al., 1976). The SOL was 45% larger and composed virtually 100% of type-II cells, whereas there are only 83% in controls. In mice, when the EDL had to do dual work following cutting of TA, after 116 days it was found that the substantial hypertrophy had largely occurred in only type-II cells, which were much thicker (James, 1979). This type of cell seems to be much more plastic in several respects than other types. In overloaded gastrocnemius and plantaris of mice, the hypertrohy is due to thickness increases unaccompanied by increasing numbers of cells (Vaughan and Goldspink, 1979).

11.7.3. Are There Muscle Changes During Hibernation?

It might have been the case that changes in muscle-cell composition or proportions of cell types would occur in association with hibernation. This possibility has been checked for a bat, which is totally inactive in winter (Armstrong et al., 1977). The pectoralis, which is the principal flight muscle, is homogeneous, comprising only fast-type cells having high aerobic metabolic profile and ATPase activity. Two flight accessory muscles, the acromiotrapezius and the acromiodeltoideus, by contrast each contain a mixture of three types. However no change, either biochemical or histological, was detected in any of the cells of each of these three muscles during hibernation.

11.7.4. How Important Is Longitudinal Cell Splitting in Adaptive Changes?

Longitudinal splitting of cells has long been recognized as a common accompaniment of muscle disorders. This may suggest an active attempt to adapt to adversity by producing some new cells of potentially favorable type. However, investigators of the shifts in balance of cells as a result of selective training have not considered that this was happening to a significant extent. But it may be worthwhile devoting more attention to the possibility, since selective splitting and subsequent growth of the fractions offers a kind of cloning process for cells of a particular type that could quickly alter overall balance in conjunction with some atrophy of other types.

11.8. Hormonal Influences, Aging, and Related Long-Range Effects

Drastic changes in muscle mass occur in response to hormonal influences, especially from the thyroid, and major differences between sexes are probably all hormonally determined. There has recently been an attempt to study the possible influence of thyroid secretion on cell types and properties in rat muscle. Induced hyperthyroidism that was not sufficient to affect overall growth rate nevertheless led to significant changes in cell content (Bruce and Nicol, 1981). In EDL there was a 12% increase in fast oxidative glycolytic cells at the expense of slow ones, but no major contractility change. In SOL there was a dramatic increase in the speed of contraction, to twice normal, as well as 67% larger twitch tension. These changes were occurring in advance of a parallel switch in muscle-cell phenotype to the fast type. The early changes are probably associated with SR binding/release characteristics for calcium.

There have been dramatic recent improvements in electromyography and histological sampling. Minute syringes permit focal recordings from small clusters of muscle cells, and the same ultrafine tubular needle that is first used for electrical recording is then used as a sampling device for picking up cells from the region. These technical advances, termed needle biopsy, are not only valuable as aids in diagnosing muscle diseases, but they are also enabling improved understanding of details of motor units, neural control, training, disuse, diet, and so forth, as well as aspects of gene control. Comparisons of homozygous twins indicate genetic control over some aspects of the localization of cell types within a particular muscle. These aspects are different in different unrelated individuals but similar in the twins (Karlsson et al., 1978), who appear to have similar, and quite marked, preferences in use of motor units. Skilled tasks requiring light force tend to be achieved by the same few motor units (see p. 576). The muscle cells in these units have appropriate sets of contractile characteristics and enzyme systems for the specific task, doubtless a result initially of genetic factors, which are being continually reinforced by the feedback systems outlined on p. 408.

In aging of all animals, maximum muscle speeds slow markedly and maximum-force capability falls. There may be a reduction in total muscle mass: this is subject to large individual variation. The progressive changes are multiple: there is a shift in cell type from fast to slow, but this is because there is a loss of SR, oxidative enzymes, and mitochondria in all cell types.

Although the effect is not too well studied, it is known that insulin can affect muscle properties. In potassium deficiency, insulin lowers the resting potential in rat diaphragm to the point of paralysis (Otsuka and Ohtsuki, 1965). The action is abolished by replacing sodium in the bathing solution with choline, showing that insulin increases $g\text{Na}$. Pituitary hormones affect BT and may also affect enzyme levels. Old rats (29 months) have Z-lines that are sometimes as expanded longitudinally as they are in the disease nemaline myopathy (see p. 69). The normally sharp distinctions between cell types have been lost, with a general leveling of enzyme levels, especially myosin ATPase (Gutmann and Hanzliková, 1973).

The commonest of all accompaniments of aging is the accumulation of lipofuscin granules, and although not confined to muscle cells, they are very prominent in them. These are protein-breakdown products that have not been voided: they take up valuable space and impede shortening.

11.9. Disease

Because muscle cells are so dynamic throughout life, capable of showing dramatic changes in structure and enzymes, it is not surprising that they are also vulnerable to both genetic abnormalities and acute and chronic diseases. Several major general classes of muscle disease are recognized, and it has become possible to perceive and understand the underlying abnormality in terms of departures from the structural and functional norms described above. A few of these diseases will be described because they help illuminate our overall understanding of muscle function and its neural control.

Congenital myopathies are ones in which the ultrastructure is abnormal or asymmetrical, in many instances from birth. They all cause muscle weakness, which is especially noticeable as low basic and active tonus, hence termed *hypotonia. Muscular dystrophy* is a class of genetically determined progressive degenerations of otherwise initially normal muscle. *Myotonia* is a class of disorders in which relaxation following contraction is slowed drastically. In some types repetition of a particular movement is followed by improvements, in others by becoming worse. *Myasthenia* is characterized by extremely rapid onset of fatigue, with gradual slow return of normal contractibility during rest. *Paralysis* represents a wide range of varieties of loss of contractile ability or motor-neuron activity, ranging from permanent loss to daily fluctuations in *periodic paralysis.* Some subsets of each of these will be described briefly. Most are known only from humans, but there are animal forms of some, especially dystrophies and myasthenia. Several "muscle" diseases such as tetany – episodes of violent involuntary contraction spasms – are in fact secondarily produced by motor-neuron malfunction or by circulatory defects and immunocomplexes.

11.9.1. Myopathies

The commonest of all muscle diseases, the myopathies include every conceivable malformation of individual muscle cells, ranging from abnormal cell types to ones with peculiar mitochondria, missing sarcomeres, or lack of longitudinal alignment. An excellent comprehensive review is presented in the book by Dubowitz (1978) on muscle disorders in childhood. There may be abnormal proportions of slow/fast-type cells (Fardeau et al., 1975), either paucity, excess, or abnormal forms of mitochondria, or paucity of myofilaments, especially in the core. Abnormally small cells occur, and are associated, interestingly, with abnormally large nuclei situated in the central core, a condition characterizing the normal condition of several invertebrate cells (e.g., Askanas et al., 1979). Some abnormal cell types are illustrated in Figure 11.9.

A particularly interesting form of aberrant ultrastructure is seen in *nemaline myopathy.* This is a human and animal disease (Figure 11.10) that is progressively debilitating and eventually fatal in which some Z bands progressively thicken. They do so symmetrically about the center of each Z region, which expands at the expense first of the I region, then the A as well, and can continue until the entire sarcomere is filled (Price et al., 1966). The existence of this disease is significant, though neglected, for understanding general aspects of muscle ultrastructure and especially the nature of the Z band. The spacing of the fine bands that dominate the nemaline-disc image as seen in longitudinal section is exactly twice that between I filaments near the Z region, and this may be a significant feature. The components of a nemaline disc are distinctly thinner than I filaments.

In advanced stages of nemaline myopathy, the Z bands of heart muscle cells are often also affected, and abnormal Z bands occur exclusively in heart cells in some specifically cardiac diseases (Izumi et al., 1978).

11.9.2. Muscular Dystrophy (MD)

The dystrophies encompass the most heterogeneous cluster of muscle diseases, and the least well understood. There are several dystrophies for which comparable forms exist in experimental animals, facilitating research on them, but there is also an abundance of human biopsy tissue, since some forms, notably Duchenne, are widespread in human populations, reaching a maximum of 3 per 10,000 (Dubowitz, 1978).

Many different ultrastructural changes are associated with MD, the commonest of which is a trend toward smaller cells with larger and central-core-located nuclei. Other features of some MDs are enlarged mitochondria, scattered loss of myofilament, and loss of SR, ITs, and IT/SR contacts. The sarcolemma is ragged on cells in the commonest form, Duchenne MD (DMD) (Bradley and Fulthorpe, 1978).

In an animal model, dystrophic chicken muscle cells have an enlarged IT system with caveolar evaginations, as if an attempt is being made to compensate for calcium-regulation deficiency by increased area (Crowe and Baskin, 1978). A hypertrophy is common in childhood dystrophies. In dystrophic chickens, there is also a shift in type of TM and a markedly changed TN_C (Wilson et al., 1979). A difference in calcium sequestration has been indicated in comparative studies on pectoralis compared with the posterior and anterior latissimus dorsi (PLD and ALD). The calcium sensitivities of the latter were little different in dystrophic forms, but that of pectoralis was greatly reduced (Kerrick et al., 1979).

Although current research suggests calcium regulation as a primary factor, many other

---

Figure 11.9. Cell-structure correlates of some muscle diseases. (*1*) Normal human neuromuscular junction after incubation in ^{125}I-labeled α-bungarotoxin and staining for acetylcholinesterase: compact receptor area. (*2*) and (*3*) Same as (*1*) but from two different myasthenia patients. Receptor areas are both decreased and dispersed. [1–3 reprinted with permission from T. A. Levitt et al. *Science* **210**, 550–551 (1980). Copyright 1980 by the American Association for the Advancement of Science.] (*4*)–(*7*) Abnormal muscle features of children with excessive muscle weakness. [Reprinted with permission from V. Dubowitz: *Muscle Disorders in Childhood.* Copyright 1978 W. B. Saunders Company.] (*4*) A chain of large nuclei occupies the core of each muscle cell. Lack of structural organization is localized. (*5*) to the core of the muscle cell. (*6*) Abnormal mitochondria, and abnormal internal membranes, indicated by arrows. (*7*) Abnormal whole muscle cell side by side with normal one.

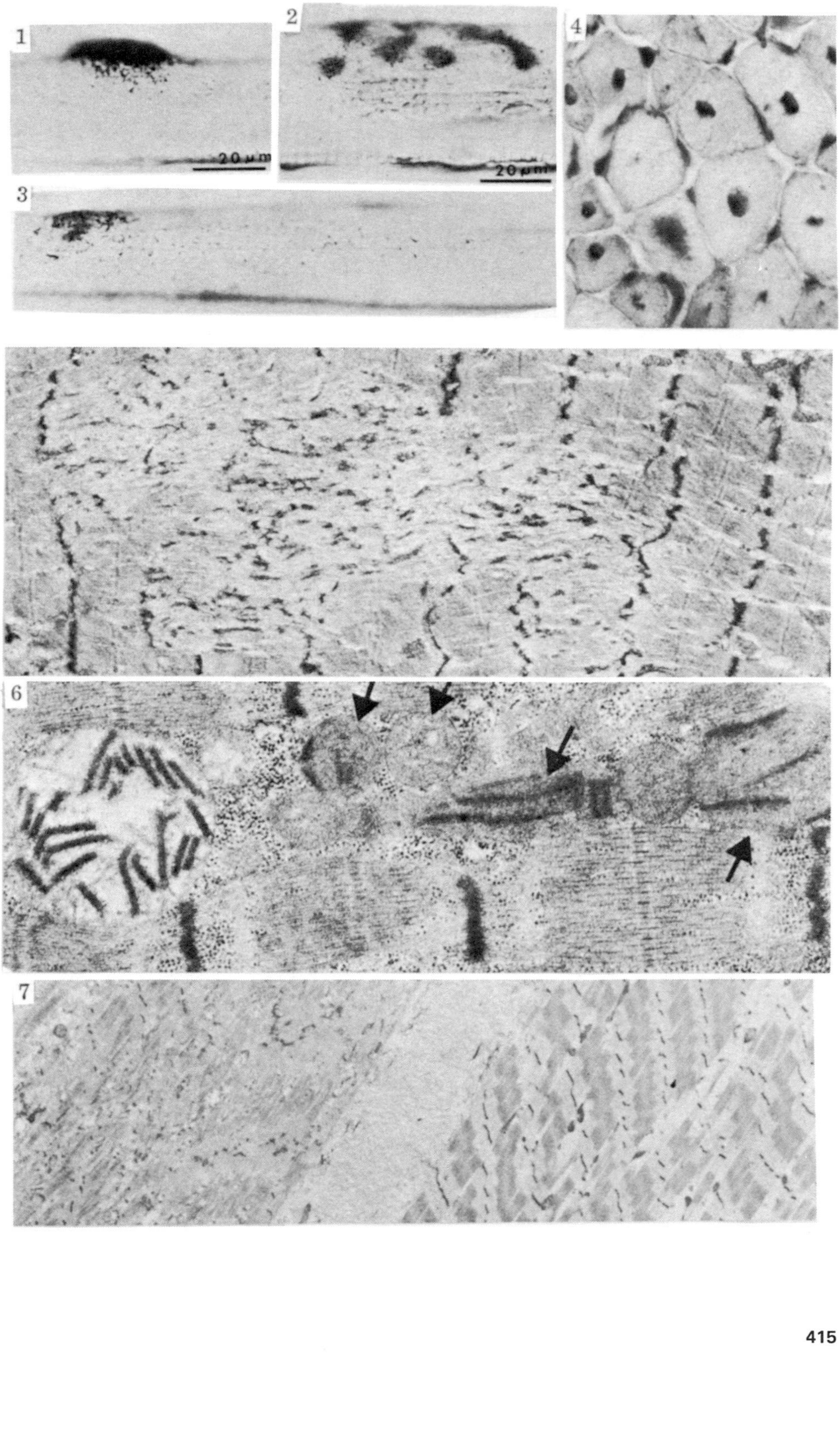
1
2
3
4
20 µm
20 µm
6
7

Figure 11.10. Anomalous Z regions in some cardiac cells of cat. Their appearance is exactly similar to that seen in the relatively common human disease nemaline myopathy. The EM of a LS includes some normal regions for comparison. Inset is a high-power image of the region showing the regular transverse banding. The Z expansion occurs at the expense of I filaments. (From D. W. Fawcett, 1968.)

features are abnormal in MD; for example, there is a large increase in collagen (Duance et al., 1980). A relative importance of an intrinsic muscular basis in MD has been advocated by some (e.g., Peterson, 1974) but contradicted by others such as Douglas and Cosmos (1975) on the basis of cross-innervation studies. A compromise is that there is a strong neural influence, but that it must become active very soon after innervation even though it will take a long time before becoming expressed. It may be noted that the virally determined neural disease Kuru may take 20 years or more to be expressed following infection.

Abnormal biochemical factors have long been implicated in DMD, and it has recently become apparent that mitochondrial membranes are abnormal in the disease from a very early age, before major symptoms have developed (Scholte and Busch, 1980).

11.9.3. Myotonia

This is the most prescribed and easiest to recognize of all muscle diseases. The ultimate myotonic person – or animal – goes rigid upon being startled, then collapses in immobility for a time, before falteringly getting up and continuing to move. A strain of goats exhibits myotonia and has been very useful in understanding one aspect of this disease class, but it has become clear that earlier expectations that there is just a single elemental underlying mechanism are unfounded. In the goats and humans, a prolonged discharge with waning amplitude follows a single motor impulse, even in an isolated nerve-sectioned preparation. Even after denervation and curarization, repetitive discharges occur to stimuli applied to the muscle. The weakness is therefore a combination of abnormal excitability with inability to recover normally following excitation. The surface membrane(s) are implicated, but the ultrastructure is normal, focusing attention on transmission or postsynaptic ion conductances. The most significant basic change is a low chloride conductance (Bryant, 1973), but its consequences are indirect. The trouble is caused by potassium accumulation in the ITs, leading to a long-maintained depolarization that is sufficient to initiate self-maintained spiking.

There is a possibility that deficiency of a zinc-containing structural compound, based upon presence of an abnormal ligand, is an underlying cause (Horrobin and Morgan, 1980). This is in turn associated with control by insulin (Tevaarwerk et al., 1979), which fits with therapeutic values of cortisone and related endocrines in treatment of myotonia. Unfortunately, the basic involvement of zinc in muscle is not at all understood.

11.9.4. Myasthenia

The muscular weakness collectively termed myasthenia is in a sense the converse of myotonia: the subject goes limp upon being startled. And the disease has a remarkably consistent set of symptoms in its clinical varieties. The site of the disease has long been known precisely to be the neuromuscular junction, which fails rapidly under a prolonged neural barrage and which recovers only slowly from a brief one, that is, the disease is a form of posttetanic exhaustion (Desmedt, 1957). It was shown after the introduction of modern intracellular recording methods that the MEPP frequency is below normal in myasthenia humans (Dahlbäck et al, 1961). The basis has been positively identified as a shortage of ACh receptors (Fambrough et al., 1973; Fambrough, 1979). The number is drastically reduced under the nerve ending itself, and others occur some distance away as in an intermediate-stage response to denervation (Figures 11.9.*1*, and 11.9.*2*). The size of the individual transmitter package is probably normal, because no features indicating a possible presynaptic-release abnormality can be detected. It is estimated that at a normal human end plate, a single packet opens about 1500 channels, whereas at an average myasthenia one, this is reduced to about 600. The size of the first few EPPs in a train is still large enough to fire impulses with a good safety factor, but later ones, with fewer ACh molecules per package, fail progressively (Cull-Candy et al., 1979; 1980).

The fall in EPP size is brought about by an autoimmune attack directed against the receptors (reviewed by Drachman, 1979). Why the body should be so silly as to initiate the destruction of its own receptors is unfortunately not known, but thymus-gland malfunction is heavily implicated. Thymectomy is the recommended treatment, with results so spectacular and rapidly successful they have been extensively dramatized by the popular press. Less drastically, but more expensively, plasma exchange (plasmapheresis) is used to remove the antibodies. The antibody does not bind to the ACh receptor site, as might have been expected, but to some other part of the receptor molecule; the effect is to stop the receptor from operating as a gateable channel. Unfortunately for the possibility of a simple cheap treatment, though, the antibody is of the polyclonal type. The autoimmune reaction is reduced in experimental animals by vitamins, notably B_1, C, and E (Peeler et al., 1979).

Humans in which the disease is relatively weak generally learn motor-nerve patterns/behaviors that to a considerable extent compensate for the debilitating effects of the disease.

11.9.5. The Potassium-Related Paralyses (kalemias)

For muscle physiologists a remarkable, diverse class of largely inherited muscle diseases leading to various forms of cyclical or permanent paralysis has turned out to be indirectly caused by faulty regulation of serum potassium. This is too high in one subclass – *hyperkalemia* – and too low in the other – *hypokalemia,* but a human can suffer at different times from both (Adachihara and Takagi, 1974).

11.9.5.1. Hyperkalemic Periodic Paralysis (Also Called Episodic Adynamia)

In this hereditary disease, potassium-level regulation is inadequate, and following any meal that is high in potassium salt, which is to say almost any meal, since both plant and animal cells have a high potassium content, the blood potassium concentration rises. The rise is particularly acute after an evening meal, when regulation is generally slowed down, but takes effect in early morning. The person is partially paralyzed upon wakening in the morning and is unable to move around. Later in the day the symptoms wear off. Potassium had long been known to be probably involved, but this was not firmly established until an unique critical experiment was performed on volunteer patients (Cruezfeldt et al., 1963). The investigators told me that their experiments were in part the result of a suggestion I made during a lecture given at the California Institute of Technology in 1957. This arose out of my finding that a state of rest, to the point of paralysis, occurs in locusts following a heavy grass meal, attributable to muscle-cell depolarization and neuromuscular block. This finding explained some of the behavior of locust swarms, but it was possible that it applied to some extent to humans and all animals and would be expected to cause temporary paralysis in extreme cases. The researchers made an incision in a forearm muscle of the volunteers, allowing a pool of lymph to collect. The arm was then immobilized and an intracellular electrode inserted into representative muscle cells. The incision was sealed, then reopened and the recording repeated at various times including during peak paralysis. The resting potentials of the cells were very low at the peak of an attack (∼ 69 mV compared with 87 mV normally), resulting in small EPPs and reduced excitability. Paralysis occurs because the smaller EPPs, combined with reduced excitability, fail to generate propagated action potentials. The serum potassium was sufficiently elevated (to 5.3 m*M*) to account for the depolarization simply on the basis of the Nernst relation (see p. 180). However, there are cases in which the serum potassium is not obviously responsible (Poskanzer and Kerr, 1961). These must represent a fundamentally different disease in which there is periodic malfunction within the muscle cells, probably due to disturbance of membrane-potential-generating ion pumps and/or conductance. When serum potassium is the immediate cause, a simple control of diet and eating habits, combined with water intake plus a diuretic agent each night is sufficient to prevent the paralysis. Agents that promote activity of the muscle-cell sodium/potassium pump directly or indirectly are also beneficial.

11.9.5.2. Hypokalemic Periodic Paralysis

There is no doubt about the association of this converse of the hyperkalemias with potassium deficiency because relief is brought rapidly by intravenous injection or oral administration of KCl. Normally amounts of up to 10 g are given orally. Paralysis occurs because with too high a membrane potential, background gK is low, reducing end-plate current, and the EPPs just fail to depolarize the membrane past the threshold for impulse initiation.

The cause is a relatively low basic potassium intake, combined with very efficient removal by the kidneys, so that attacks can occur at any time, not just in early morning. The reservoir of potassium in the muscle cells themselves is enormous, and potassium storage and release are major balancing mechanisms for serum potassium. Many hormonal factors are involved in control of muscle-cell sodium/potassium ion pumping, and it is possible that enhanced activity of the pump could bring on a hypokalemia paralysis.

11.10. Intermediate-Term Influences – Modulation

This section is necessarily short, but only because modulatory influences on skeletal musculature are of recent discovery. I expect that eventually there will be a large amount of material to deal with. Modulation of hearts and SM have long been recognized and were described in the appropriate chapters. It is also possible that modulation can have only minor behavioral consequences in vertebrates because of their propagated all-or-none action potentials.

The first clear notion of the presence of neurons having the function of modulating excitatory neuromuscular transmission and tension development was in arthropods and arose out of morphological findings. In some insects, somata of a few large neurons were found in the dorsal mid-line of ganglia that were unusual in not being paired but instead made symmetrical left and right forks, termed T junctions. The two principal neurites were traced into nerves leading from the ganglion, and in two cases were traced into muscles: the locust extensor tibiae – jumping muscle (Hoyle et al., 1974) – and dorsal longitudinal flight muscle (Hoyle et al., 1980). There the axons terminate after first branching and making several varicosities. The varicosities do not synapse with membrane of the muscle cells, but they are located very close to them (Figure 11.11.*4*). The vari-

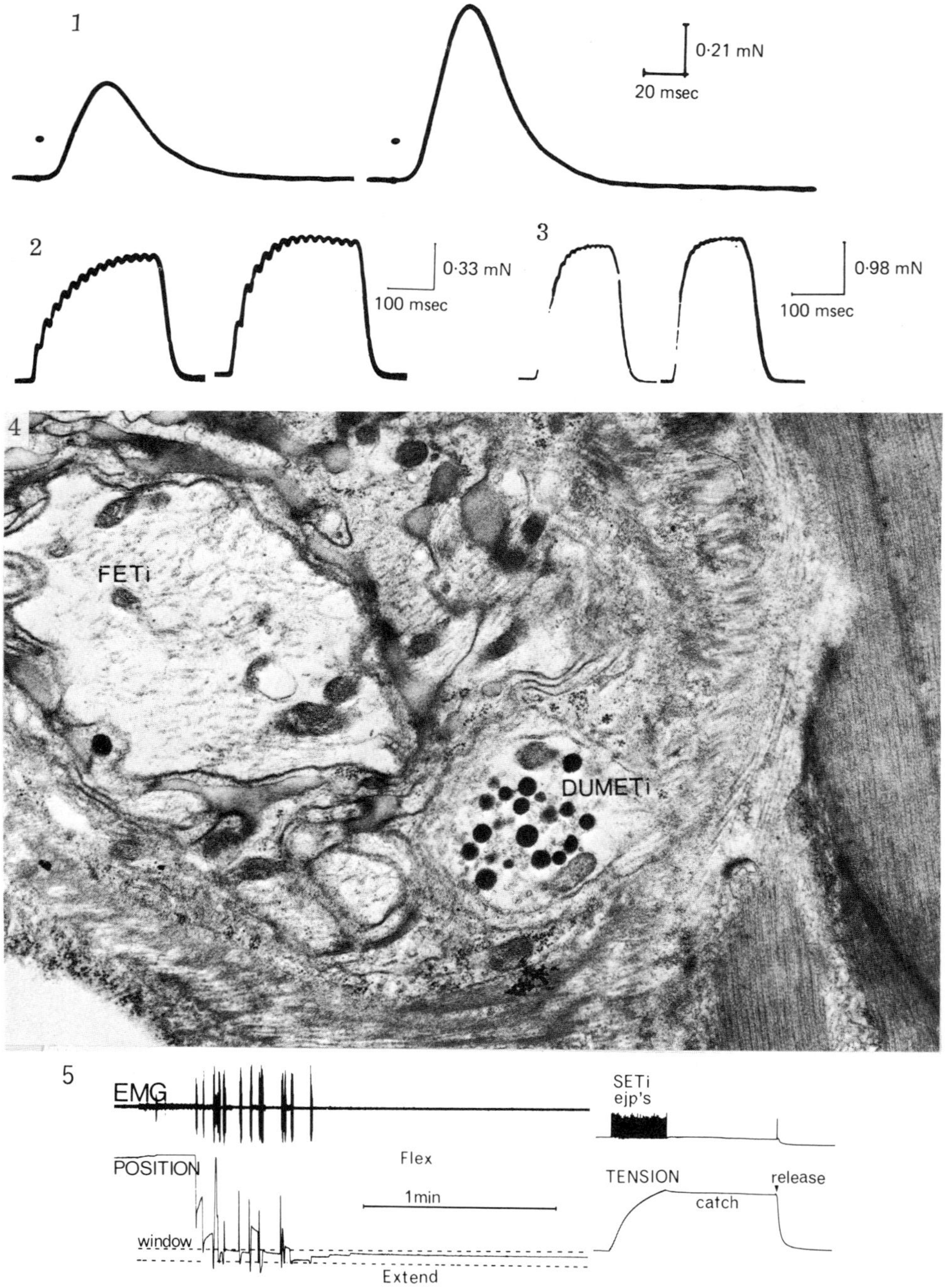

Figure 11.11. Modulatory actions in muscle function. Positive inotropic action of 10^{-6} *M* adrenaline on twitch of fast frog semitendinosus cells in the presence of cAMP. (*1*)–(*3*) Effect on tetanus at 40 and 70 Hz respectively. (From Gonzales-Serratos et al., 1981.) (*4*) Final nerve branch of an identified insect modulator of skeletal muscle DUMETi on a fast muscle cell of the locust *S. gregaria.* This cell releases the insect equivalent of adrenaline, octopamine. It shares a pad of protective collagen with a preterminal twig of the fast excitatory axon (FETi). (From Hoyle et al., 1980.) (*5*) Catch mechanism in leg position learning by primitive stenopelmatid insect, the New Zealand 'WETA' *Hemideina femorata.* The hind-leg tibia adopted a position in a pre-set "window", linked to an aversive loud sound, indicated by the broken line, after operant-conditioning. However, no electrical activity occurred during the occupation, yet about 7 gm tension was required to maintain this position. The behavior is explained by operation of a catch mechanism, shown at right in a nerve-muscle preparation. A brief discharge of DUMETi must occur before SETi for catch setting to occur, or alternatively, a drop of 10^{-8} *M* OCT gives a similar result. (ORIG.)

cosities are filled with large dense-core vesicles. These nerve cells contain and make a biogenic amine, octopamine (see p. 328) first discovered in the brain of *Octopus* (Erspamer and Boretti, 1951). Octopamine is also present in vertebrate brains, including those of humans, but no function has yet been ascribed to it there. In the locust it inhibits intrinsic rhythmicity at only 10^{-9} M, but its major significance in the present context is that it causes, briefly, a marked enhancement of excitatory transmission by the slow excitatory axon SETi (see p. 529).

Octopamine-containing neurons are present also in some crustaceans, notably the lobster (Evans et al., 1975) and in a neuron of the mollusc *Aplysia* (Saavedra et al., 1974), though effects comparable to those occurring in the insects have not yet been demonstrated in these species. However, in *Aplysia* a dramatically strong neuromuscular modulatory action by 5HT has been demonstrated. The 5HT is contained in a large identified neuron in the brain, the metacerebral giant cell (MCC). This innervates a muscle associated with feeding, whose mechanical response is enhanced by up to several-fold following a brief burst in MCC. This is attributable to a combination of EJP facilitation and an independent, slower action on either E-C coupling or the contractile mechanism itself.

Many invertebrate muscles are known to contain nerve terminals with large vesicles having dense cores, dense-cored vesicles (DCVs), which probably contain 5HT, and although 5HT promotes relaxation, not contraction, in the *Mytilus* ABRM (see p. 474), in others it may be acting as it does in the *Aplysia* feeding muscle.

Another natural modulatory substance in insects is the pentapeptide proctolin (see p. 353). Although proctolin is associated with the gut, which it excites, it also has actions on skeletal muscle cells (Piek and Mantel, 1977; O'Shea and Adams, 1981).

The most probable natural vertebrate modulator of skeletal musculature is, of course, adrenaline. It was for a long time after its discovery considered to alter skeletal muscle function in a hormonelike manner, although belief in such involvement eventually waned to the point of extinction. But recently tests have been published on the effects of Ad on single frog semitendinosus fast-twitch cells, with and without skinning or splitting (Gonzales-Serratos et al., 1981). There was no change in properties with fixed amounts of calcium and Ad in the absence of cAMP. But in the presence of cAMP, 10^{-6} *M* Ad dramatically increased the rate of rise of force in the twitch, peak tension being about doubled (Figure 11.11.*1*). The catecholamine presumably promotes cAMP stimulation of the SR calcium pump so that a greater amount of Ca^{2+} is accumulated. When a stimulus arrives, a greater amount of calcium than normal is released thereby effectively prolonging the SA (see Figure 7.4). This mechanism is probably a basic one for hormonal and related potentiations, and is possibly going in the opposite direction for some nutritional and disease defects.

12

Survey of Neuromuscular Mechanisms

When we consider the immense number and variety of animal organisms that populate this planet earth, the small group of mammals that has dominated the scene of physiology comprises less than 1% of the total number of animal species and is really not more than an appendix to the vast number of invertebrates. Ernst Florey†

12.1. Coelenterata

Two principal structural forms occur in coelenterates, the *medusoid* and the *polypoid,* both of which are radially symmetrical. The medusa, a flattened saucer-shaped free-swimming disc, and the polyp, an attached, tentaculate, elongated cylinder, are simple transforms of the same basic shape. In some species they are directly interchangeable as reproductive/growth elements. All possess independent nerve cells, most of which have axons less than 1 μm in diameter and also small (2–3 μm) cell bodies. The cells are either bipolar or multipolar, never unipolar as is characteristic of a majority of the larger cells of nervous systems of other phyla. The largest known coelenterate neurons occur in the subumbrellar circular nerve ring of larger jellyfish, and a few of these are truly of "giant" size, reaching 25 μm in diameter (Anderson and Mackie, 1977) and permitting direct intracellular recording. These axons are to some extent electronically coupled and form a specialized rapidly conducting tract for coordination of the quick contractions brought about by the cross-striated cells of the subumbrellar ring that keeps the jellyfish float and moving through the water.

Special tracts, though with many axons of smaller diameters, also occur in the major mesenteries of sea anemones (Batham et al., 1960). But the majority of neurons in coelenterates are only very loosely arranged, making synapses with each other that are of similar ultrastructure to the average central nervous system synapses of other nervous systems including those of vertebrates (Westfall et al., 1971). Coelenterate neuromuscular junctions are simple but have no unique aspects that might earn the label *primitive* (Figure 12.1 – Westfall, 1973). There are no known three-dimensional aggregates of the neurons into ganglia in polypoid forms, and only rather loosely knit close collections at a few marginal sites on medusae. Evidence for an orderly arrangement of neurons that would indicate the existence of prescribed neural circuits is so far totally lacking. Yet not only is there a large repertoire of subtle reflex actions, but complex coordinated locomo-

†E. Florey, "Introductory Remarks," in *Invertebrate Nervous Systems,* C. A. G. Wiersma, Ed., University of Chicago Press, 1967.

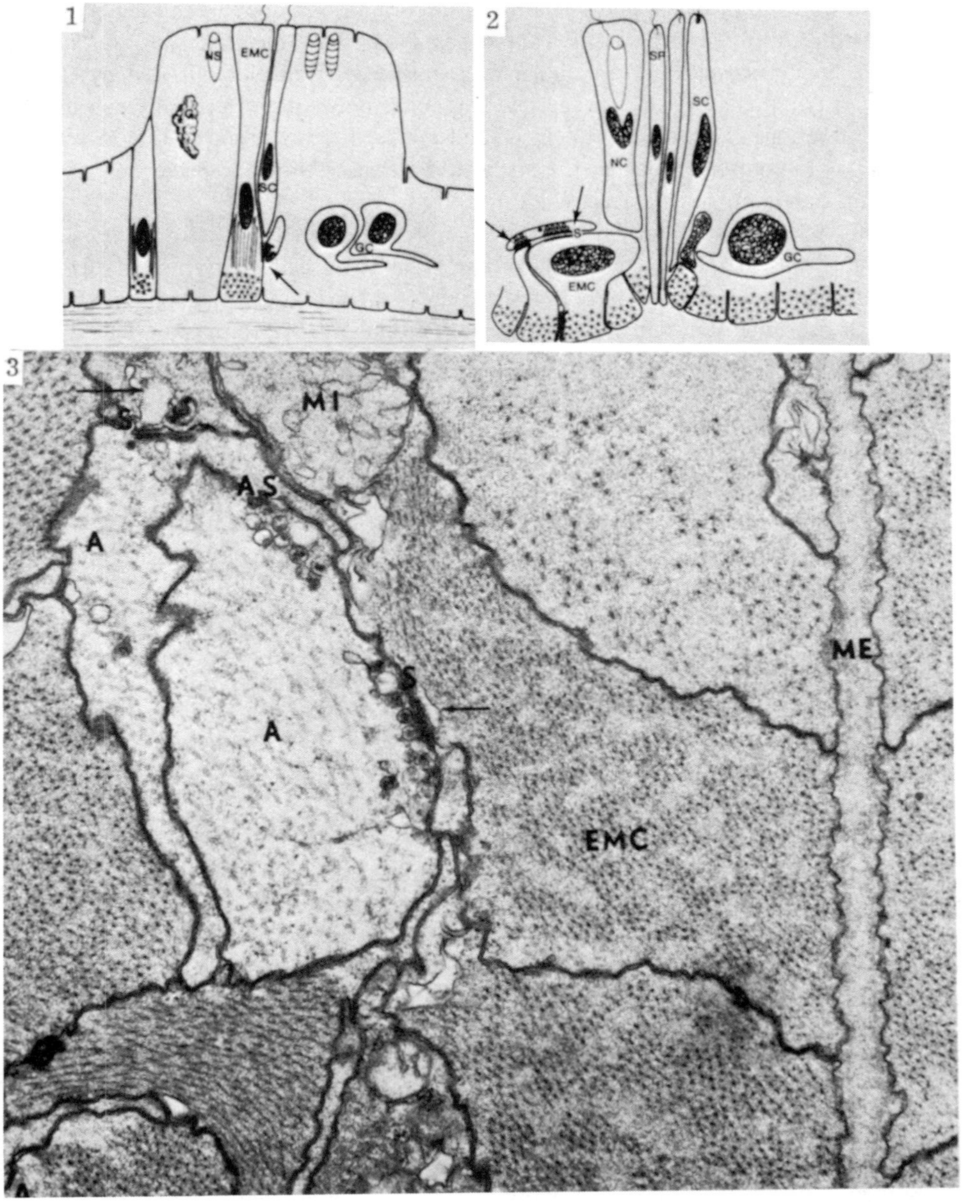

Figure 12.1. Innervation of coelenterate muscle. (*1*) Drawing to show location of neuromuscular junction (*arrow*), with synapse (S), on a musculoepithelial cell (EMC) of a coral. Also present are nematocysts (NS), nerve ganglion cells (GC), sensory receptor regions (SR), and presumed sensory cell (SC). (*2*) Location of junctions on a sea anemone (*Metridium*). (*3*) EM of TS of a tentacle of jellyfish *Haliclystus* showing muscular regions of musculoepithelial cells (*EMC*) and two motor axons (*A*), one of which is making a neuromuscular junction (*S, arrow*). Note that a small SR cisternal element is present at the contact region. Mesoglea (*ME*) is present at right. (All reprinted with permission from Westfall, J. (1973) *Am. Zool.* **13**, 237-246. Copyright the American Society of Zoologists.)

tion is known from the small polyp Hydra, in the form of both looping and somersaulting. In at least three different genera of sea anemones, *Stomphia, Gonactinia,* and *Actinostola,* there is a highly complex fixed-action pattern of escape behavior initiated by contact with one or two predatory starfish species. This comprises scores of perfectly coordinated quick contractions (Figure 3.1 – Yentsch and Pierce, 1955; Robson, 1971) that take a few minutes to complete; each of the many subroutines is known to occur only as a part of swimming and only in its correct place in the sequence.

Very little is known about the details of neuromuscular transmission or neural control of individual muscle components, and nothing at all about cellular events. The speeds of contraction and relaxation range from fast twitch of the medusa's circular striated muscle band, with rise time of about 0.6 s and relaxation time of 0.9 s, to the parietal muscle of a sea anemone body, with a rise time of the order of 5 min and relaxation time of twice that. Circular muscle of the column is just a little faster. Unfortunately, these values are not necessarily equivalent to simple twitch times, since uncontrolled nerve-net cells having excitatory inputs to the muscle are probably active at the time. Also, each part of the animal interacts with every other one, and there is a central hydrostatic "skeleton." The longitudinal retractors of sea anemones come as close to being equivalent to the discrete anatomical muscles of animals with skeletons as any of coelenterates. These have an intermediate rise time of about 3 s but a very slow relaxation time of 15–20 s, so that the "twitch" is markedly asymmetrical as measured in the whole or partially dissected animal. In an isolated mesentery, the times may be reduced, with a minimum rate of shortening of about 0.1 cm/s · cm in a *fast contraction.* The length range for a retractor is about fivefold. Force capability is impossible to determine accurately but is certainly high and impressive; Robson and Josephson (1969) gave a value of 40 kg/cm^2. A second type, *slow contraction,* occurs in the isolated mesentery without a preceding fast one in response to similar shocks spaced 1 s or more apart.

The retractor of the common sea anemone *Metridium* was recognized by Carl Pantin in 1935 to have an interesting response mechanism. A single supramaximal electric shock, no matter to what part of the column it is applied, does not elicit a response. But a close (0.5-s separation or less) suprathreshold pair always gives a fast contraction no matter where on the animal either of the two shocks is applied. Each excites a part of the nerve net in a unitary manner, as if it were a single excitable element that reaches the neuromuscular juntions and sets up a state of excitation there that just fails to exceed the threshold for a mechanical response but sets up a decaying remainder. If a second similar event arrives before the extensive decay, a mechanical response occurs whose magnitude is proportional to the interval. A very small, brief 80-ms (depolarizing) potential is detected by extracellular suction electrodes applied to the mesentery (Figure 12.2 – Robson and Josephson, 1969) in response to the first stimulus, and a much larger one to the second. An even briefer (1-3 ms) miniscule sharp potential precedes the 80-ms one:

Figure 12.2. (*1*) Nerve net of a sea anemone, *Metridium senile,* overlying longitudinal retractor muscle cells, which run vertically. Neurons are bipolar or multipolar. (From Batham et al. 1960.) (*2*) Suction electrical records (*top two traces*) and tension from a preparation of (*1*), electrically excited at 1 Hz. Each electrical response consists of, first a stimulus artifact (*a*), then a muscle-cell response (*m*). The former is of constant height, the latter grows stepwise with each repetition. There was no mechanical response to the first shock, *1*, progressively larger ones (facilitation) to the next four. These were associated with progressively larger muscle electrical events. (From Robson and Josephson, 1969.)

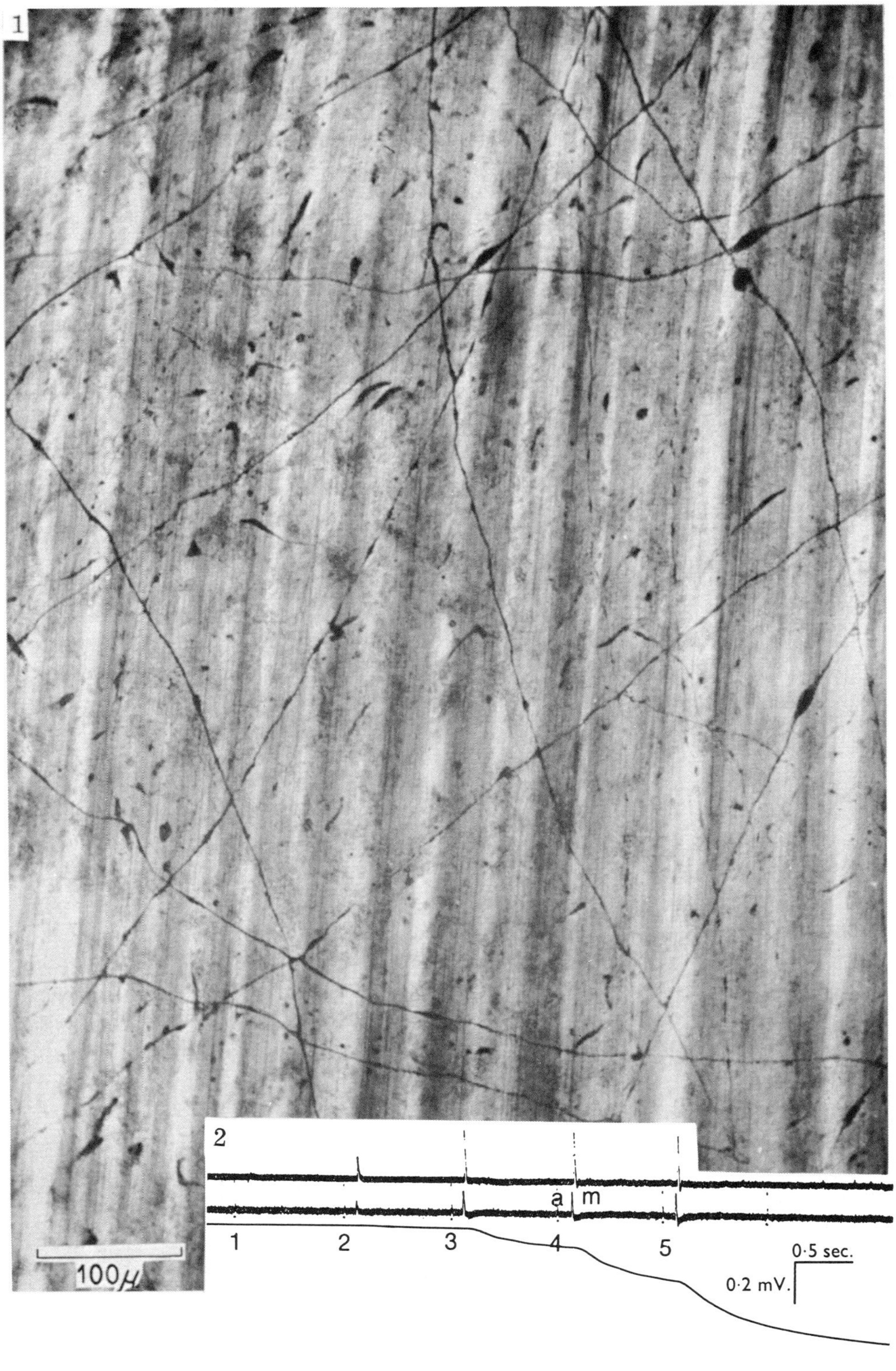
1
100μ
2
a
m
1
2
3
4
5
0·5 sec.
0·2 mV.

this is the nerve-net-conducted potential. By comparison with extracellularly recorded events in many arthropod neuromuscular systems, where the situation is well understood (see p. 596), the longer response is considered to be an EJP. This requires multiterminal innervation of several muscle cells by one or a few common excitatory neurons.

The positive response to the second stimulus indicates not only that a summation may have occurred, but also a growth or *facilitation* of the neuromuscular event. The first increment represents infinitely large facilitation in tension. The succeeding four to six steps show further mechanical facilitation.

The mechanical responses of the retractors of *Metridium* are not matched by those of its circular muscles, which are much slower. By contrast, the retractors of the closely related species *Calliactis* respond like the circular muscles of *Metridium,* but a single pronounced ring of circular muscle at the oral end of this genus responds exactly like the *Metridium* retractors. The ways of life of the two species are quite different, *Metridium* being a tide-pool organism, whereas *Calliactis* lives a commensal, highly mobile life attached to the back of a shell occupied by a hermit crab. For each species a defensive, protective withdrawal is desirable, but each is likely to be buffeted so frequently that a direct reflex response to every sharp disturbance would leave them exhausted and with little time to feed. Each has used neuromuscular facilitation as a device to optimize the situation, responding only to the kind of strong, persistent irritation that sets up successive neural waves. The same principle has, however, been adopted by two different sets of muscles in the two, the retractors for *Metridium* and the sphincter surrounding the mouth for *Calliactis.* This provides us with an early evolutionary example of the remarkable variety and adaptability of neuromuscular control mechanisms and the extent to which these have been utilized in adaptive radiation.

It is presumed, also by analogy with arthropod muscle control mechanisms, that the slow-response capabilities of the same coelenterate muscles that show the facilitated fast responses are determined by a second set of nerve cells. These may or may not innervate the same muscle cells. The contraction characteristics are so different that it seems more likely that a parallel, independent set of muscle cells with very slow contraction and relaxation are responsible.

There is a minimum delay of about 50 ms between the effective stimulus and fast response; for the slow response it is a few hundreds of milliseconds. The former can be attributed to nerve-conduction delay; the latter is probably due to a very slow rise in concentration of intra-muscle-cell activator. The swimming movements brought about by the longitudinal retractors of the swimming sea anemones could not be accomplished either if a long delay occurred or if their neuromuscular junctions required a high degree of facilitation. Large, quick, brief-delay contractions are elicited by single shocks (Hoyle, 1960b).

There is no really satisfactory evidence for peripheral inhibitory neurons in coelenterates, but some results have hinted that inhibition is present at the *Calliactis* sphincter (Ewer, 1960). The excitatory transmitter substance(s) is still unknown; a lot of negative evidence argues against ACh. Tryptamine is a candidate.

The related, also solitary but tube-living, *Cerianthus* is more protected from constant intertidal buffetings but needs to withdraw rapidly into its protective tube when prodded. It responds to a single shock by a quick contraction of uniform size, indicating that some neuromuscular synapses do not require facilitation (Ross, 1957).

12.1.1. Neuroid Conduction

A subsidiary role in coordination and control has recently been recognized for the epithelial cell layer of jellyfish. The cells are electrotonically connected and excitable and support propagated electrical waves (Mackie and Passano, 1968). The epithelium conveys the excitation, which is called *neuroid conduction,* either directly to musculoepithelial cells or indirectly, via nerve-net components (reviewed by Mackie, 1970). Specific reflex behavior can be elicited, but the details of transmission to muscle elements are unknown.

The neuroid-conduction system has acquired a special interest since it was found that a similar mechanism exists in a protochordate, the salp (Figure 12.57). In spite of their lofty phylogenetic status, salps, like their close relative the ascidians (sea squirts), resemble coelenterates in many respects. In pelagic colonial forms, complex coordination occurs in locomotion, requiring accurate timing of the valve and body msucles between succeeding zooids. Yet there are no neural connections between them. However, their outer epithelia are in contact. These are electrically excitable and receive and propagate patterns of impulses that demonstrably regulate their neighbors' behavior (Mackie and Bone, 1977). Comparable waves are now being sought in developing organs of embryos.

Long-term movement and extracellular electrical recordings from a variety of coelenterates have revealed the periodic generation of complex electrical activity, emanating from several different foci in an individual. Unfortunately the findings have been followed by a lot of uncritical jargon about "loose hierarchies of interlocked pacemakers." But the reality of diffusely distributed sites from which trains and brief bursts of impulses originate is a fact. There are two major types of generator, one of which is the hydrozoan *Tubularia,* a polyp, concentrated in the tentacle-bearing region, the other is in the neck, connecting it to a long columnar stalk. Each is itself independently affected by two or three different nerve-net inputs, and the two interact with each other (Josephson, 1965).

The importance of built-in neural mechanisms whose sole function is generating active behavior spontaneously instead of central nervous systems wait for reflex actions to initiate it, escaped two generations of neurobehavioral scientists. A battle of views on this score was handily won by a small group of experimentalists, most of whom studied invertebrates, in the 1960s, but the message has still not been sufficiently widely or fully appreciated in spite of having been repeated *ad nauseam.* The case was eloquently expounded by Delcomyn (1980). Coelenterate spontaneous-activity generators are important in indicating how elementary evolving nervous systems, which must be presumed to have all come from nerve nets, may have begun to advance to the stage of being able to generate complex behavioral sequences like escape swimming.

12.1.2. Colonial Forms

The siphonophores are represented solely by colonial forms with prominent feeding or gastrozooid components that capture as well as digest food for the whole colony. Intracellular recordings have been made in the myoepithelial cells of *Agalma* (Chain, 1981). Overshooting large, probably all-or-none, but relatively slow 20–40-ms spikes occur when the animal is disturbed, in short bursts that evoke quick contractions. There is also a much slower 2–4-s depolarizing response with the shape of a JP that is graded in magnitude but can be 30–40 mV at peak and is associated with slow contraction. Either a sodium-free solution or one containing 20 m*M* manganese abolished the former and left only the slow events. The latter therefore may be a slow synaptic current, but the former seems to be a sodium-dependent event.

The lack of precise understanding of the sources of these potential changes is most unfortunate, since they might, if better understood, afford significant insight into coelenterate neuromuscular mechanisms. Epithelial conduction in such an animal, in which epithelial cells are also the muscle cells, is equivalent to propagation in vegetative muscle in general, and sufficient depolarization can be expected to cause contraction. There is also a subepidermal nerve-net that can be expected to be excited by an external electrical stimulus. Thus it is tempting to suppose that the slow waves are due to inputs from the net and the spikes to an epithelial-propagation wave. An analysis of spontaneous and reflex responses should resolve these questions.

12.1.3. Ctenophora

Ctenophores, or comb jellies, are planktonic free-swimming, relatively advanced coelenterates that have well-marked longitudinal muscle bands and also radial muscle cells. In *Beroë* the component muscle cells have a wide range of sizes but include many that are 20–40 μm in diameter (Figure 12.3.1), extraordinarily large for a coelenterate, and are up to more than 6 cm long, making them easy to study intracellularly (Hernandez-Nicaise et al., 1980). Such cells are commonly fairly uniform cylinders, but some have thick or thin regions, and a few are forked: each has multibranched terminations at its ends, whose tips are attached to epithelial cells, and is multinucleate. The contractile material, as in so many muscle cells, forms a peripheral ring around a core filled with mitochondria that also includes the nuclei. The contractile material comprises both thick filaments, 12–20 nm in diameter, and 6-nm thin ones in orbits of 5–9 (Figure 12.3.2). The array of thick filaments is not regular, but they are fairly evenly dispersed, as are thin filaments around them. The ratio of thin to thick is about 6 : 1 overall, but local variations are higher or lower. A few complete orbits of thin around thick occur, with about 12 in the ring. No Z bodies or clearly marked densities that might be thin-filament attachment sites occur. The thin filaments appear to attach directly to the surface membrane and to be dispersed rather than collected together. They spiral inward helicoidally. Hernandez-Nicaise et al., were unable to assess the lengths of either thin or thick filaments. The SR is sparse, about 2% of contractile-material volume, and there are no ITs. The surface membrane is heavily convoluted, and some thin projections representing evaginations up to a few micrometers long occur. These may serve a structural role, since they are embedded in mesoglea.

Motor axons make simple, vesicle-filled synaptic contacts of one kind with the muscle cells; some of them appear to be *en passant* junctions, the nerve branch continuing to make other junctions.

The resting potentials of radial cells are almost 70 mV, those of longitudinal cells about 10 mV lower. Each is electrogenic, giving all-or-none propagating action potentials

Figure 12.3. (*1*) Giant muscle cell of a ctenophore coelenterate (comb jelly) *Beroë*. Arrowheads point to nuclei. Small multiple evaginations (*SE*) of the sarcolemma occur. (*2*) TEM of TS of (*1*), showing an SE, central nucleus (*N*), mitochondria (*M*), and SR. Contractile material (*upper right*) consists of thick filaments closely similar to those of cross-striated muscle, with numerous loosely related I filaments. (Both from Hernandez-Nicaise et al, 1980.)

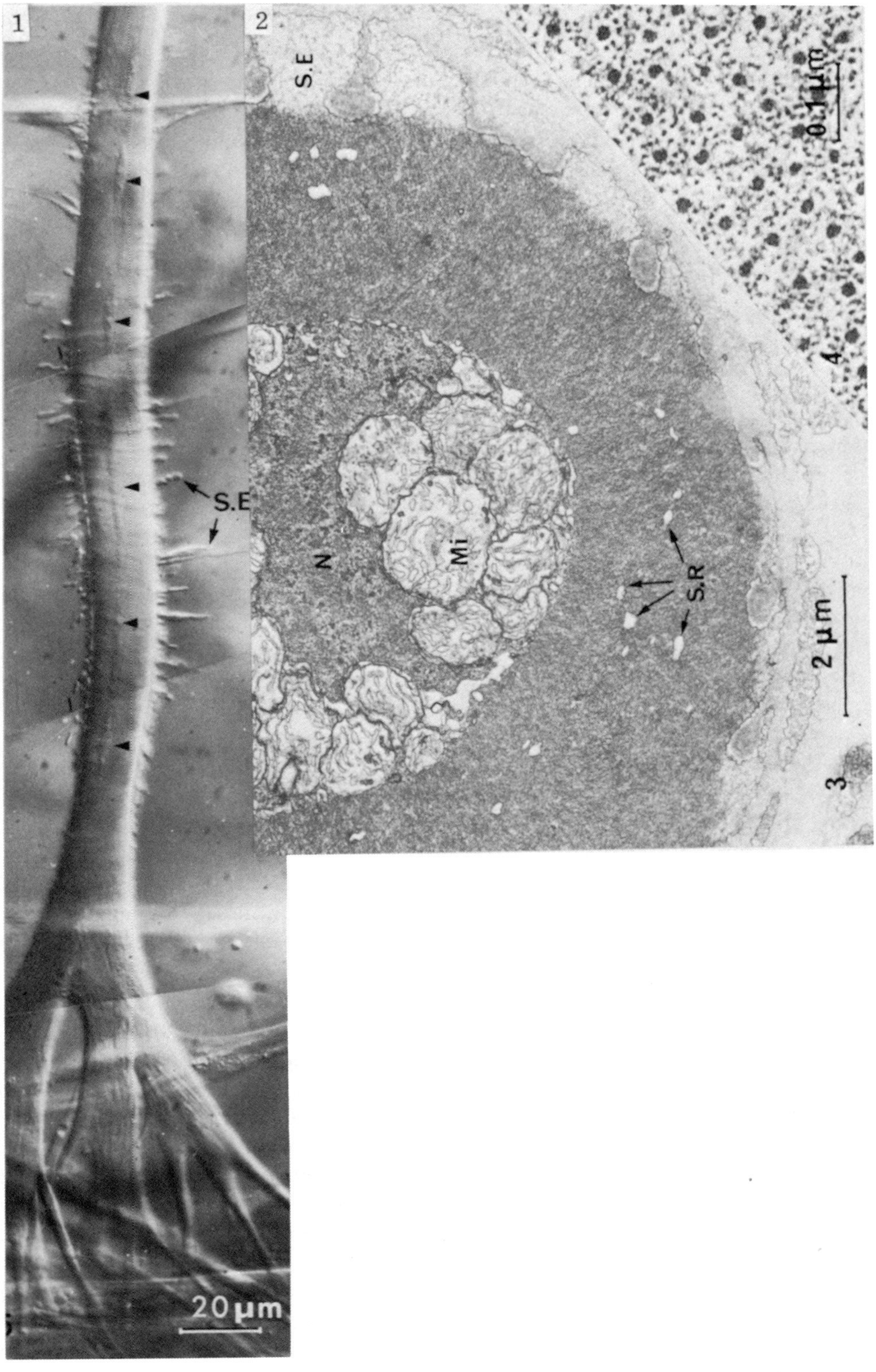
1
2
S.E
S.E
N
Mi
S.R
20 μm
2 μm
0.1 μm
3
4

of 5-ms duration. These are not only TTX insensitive, but they are also unaffected by TEA and manganese. They are also of full size in sodium- or calcium-deficient salines, but reduced when both ions are low. Therefore, the ion channels supporting them permit both sodium and calcium to enter.

The nerve supply is from a nerve net that bombards the cells with excitation constantly, apparently in the form of small summating JPs, but it was not possible to resolve the nature of individual contributing neuromuscular events. When stretched the cells contract, but action potentials do not add to the contraction.

12.2. Platyhelminthes

A majority of this phylum, the most primitive having triplobastic cellular development, is parasitic and lacks complex locomotory behavior. But the phylum is bilaterally symmetrical and also includes interesting forms with "cerebral" ganglia and complex eyes. The members of the phylum have truly amazing capability for the regeneration of missing parts; and they can regress or advance in individual size and structural complexity depending on availability of food. It has been frequently claimed that they have entrainable discriminatory abilities, and filmed evidence of this has been presented by J. B. Best. For a while they were graced with a scientific magazine, *The Worm Runners Digest,* devoted to reports on their training. Unfortunately though, they are not highly amenable to the techniques of neurobiologists; their musculature has advanced but little beyond the coelenterate stage. They have evolved two discrete muscles, both retractors, of the pharynx and penis respectively, but nobody has succeeded in harnessing either one for research.

The brain, though having a highly complex neural organization and capable of exerting some locomotory control, is dispensable. The animals survive well following brain extirpation, until eventually they grow a new one. The rest of the nervous system is submuscular and *plexiform,* full of anastomosing, conducting branches, and this marks a major difference from coelenterates. In addition, there is a subepithelial feltwork of neurons situated between the epithelial surface and deeper muscle-cell layers, and also among the muscle cells, that is closely comparable to the coelenterate nerve net. Synapses between neurons occur in both the net and the plexus and are diverse but conventional (Koopowitz and Chien, 1975).

The muscle cells of all the musculature, whether diffuse sheets or discrete, are small cells with nonstriated myofilament organization not unlike musculoepithelia cells of small polypoid coelenterates. However, the noncontractile portions are not restricted to epithilial-layer locations. A nucleus is situated in a separate cytoplasmic sac (Chien and Koopowitz, 1972). The contractile portion is usually 2-3 μm in diameter by up to 10 μm long, and its noncontractile sarcoplasmic extension or *muscle arm* is also about 10 μm long, running laterally out from it (Figure 12.4). Myofilaments comprise very loosely ordered 18-nm-thick filaments with thin 5.5-nm filaments interspersed, also irregularly.

Motor axons, containing mainly clear-core vesicles of irregular shape, but also some DCVs, are interspersed between muscle cells. Neuromuscular synapses are not made on the contractile portions but on the stalks. About 10–12 axons make contact with each single stalk, making *sarconeural* nerve-muscle-cell junctions. Remarkably similar muscle arms and junctions to those on flatworm muscle cells have been formed in the onychopho-

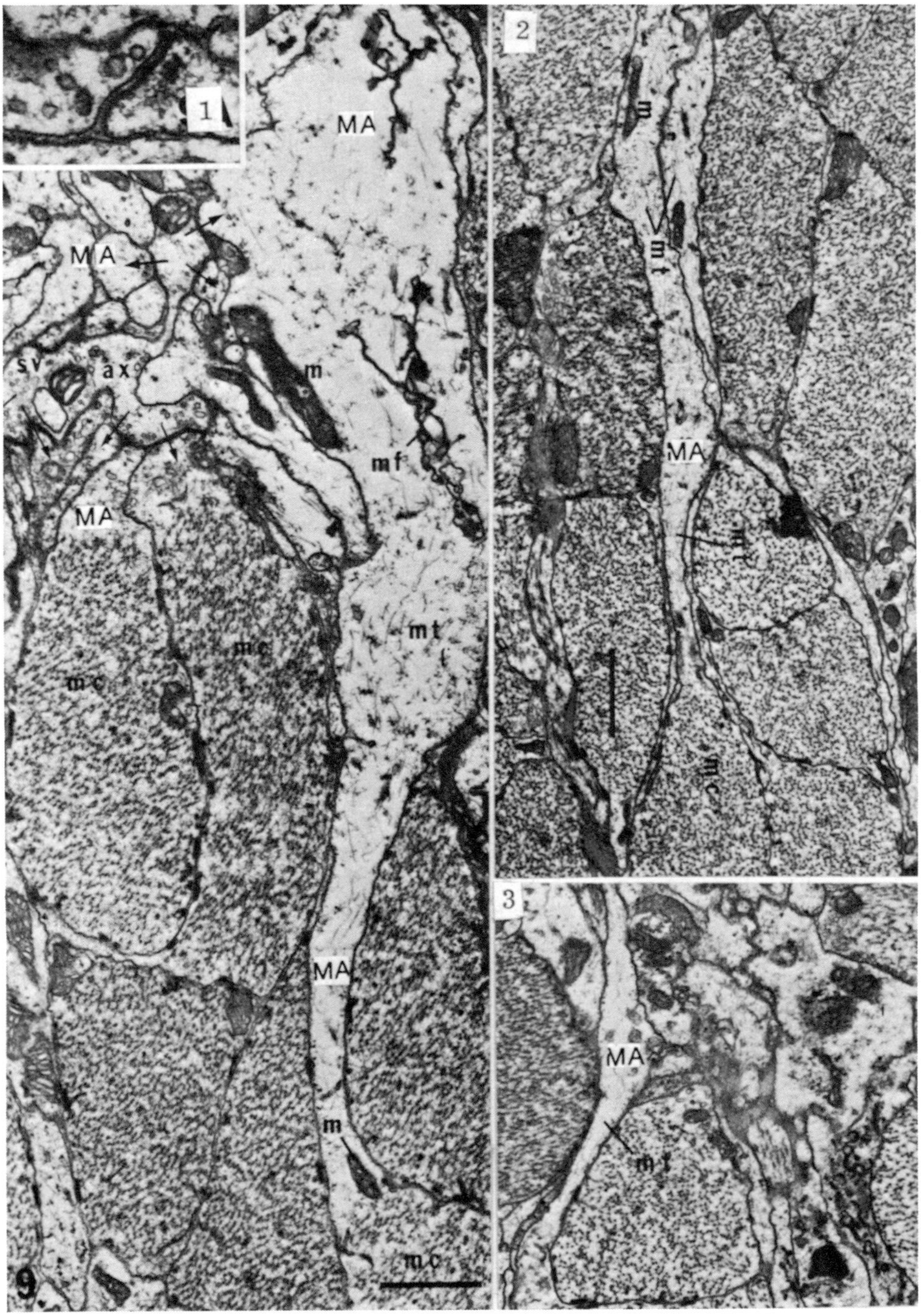

Figure 12.4. Ultrastructure of muscle cells (MC) of the platyhelminth flatworm *Notoplana acticola*. Conspicuous are muscle arms (*MA*) that extend toward motor neurons and on which neuromuscular junctions are formed (*1*). Also present are axons (ax), myofilaments (mf), and mitochondria (M). (From Chien and Koopowitz, 1972.)

ran *Peripatus* (see p. 483). Roughly similar ones occur in the nematode *Ascaris* (see p. 437), the polychaete annelid *Syllis* (see p. 444), the echinoderm starfish, and the phylogenetically important protochordate *Amphioxus* (see p. 562).

The contractile responses of flatworms to neural stimulation are very sensitive to stimulus strength and pattern, as expected from the multiple innervation of single contractile cells. Both slow and quick contractions occur naturally and can be evoked electrically. Partial reduction in strength of directly elicited responses occurs by raising only frequency or more extensively by also increasing strength (Koopowitz and Ewer, 1970). Thus peripheral inhibition may be present, but needs authentication. As with coelenterates, there is no sensitivity to ACh or anticholinesterases; the neuromuscular excitatory transmitter is unknown.

The muscles of parasitic Trematoda have not attracted any attention, but those of their free-swimming larvae, termed *cercaria,* have. The tails show three kinds of movement, a fast tail movement used in swimming, and two different slow movements. One is associated with the body-to-tail juncture and results in changes in the shapes of both. The other is in the two tail forks, and it maintains their rigidity during backward swimming. Each is associated with a different type of muscle-cell ultrastructure. The fast kind is due to transversely striated cells, the only ones known in the phylum, with a remarkably short, 1.3 μm, SL (Figure 12.5). The short thick filaments are present in a regular hexagonal array, and there is prominent H zone, but the thin filaments, present in a 5 : 1 ratio, are only loosely arrayed relative to them (Reger, 1976; Sundararaman and Nadakal, 1979). Invaginated tubules are not present, the surface membrane making many contacts with SR, which are functionally equivalent to dyads. Where these appear on both sides of the membrane at the same level, they have a triadic appearance (Rees, 1975). Slow-type body-tail-junction cells have 6-nm filaments attached to irregularly arranged dense Z bodies: no regularity in myofilament alignment can be discerned. There is only a small amount of SR in these cells. The tail-fork muscle contains cells that also contain an irregular array of filaments. The thick ones are 30–70 nm thick and contain elements with the regular 35–40 nm repeat and subrepeats characteristic of Pm.

The presence of such a range of types of muscle cell in the very small larva of a platyhelminth is further evidence of the extreme plastic capability of ultrastructure of muscles of all phyla in relation to function.

12.3. Nematoda

Nematodes are wormlike parasitic animals unique in many ways and conspicuously so in having a precise number of cell divisions before division stops completely at maturity. The total number of cells in the adult is prescribed, and with diligence each cell can be enumerated and its development traced. Accordingly, a great deal of interest lies with members of this phylum because of what they potentially offer for genetic manipulation.

The muscle cells, however, have features not seen in any other phyla. They are fusi-

Figure 12.5. Ultrastructure of muscle cells of tail of swimming cercaria larva of platyhelminth Digenea trematode larva *Cryptocotyle lingua.* TEM of LS (*1*), TEM of TS (*2*), and three-dimensional drawing (*3*) of overall structure showing surface dyads (*D*), SR, and an arrangement of Z bands. Both lipid droplets (*lip*) and glycogen (*gly*) are present. Also indicated are: cysternal element of SR (C), X bridges (xb), PM, I, A, and Z regions. (From G. Rees, 1975.)

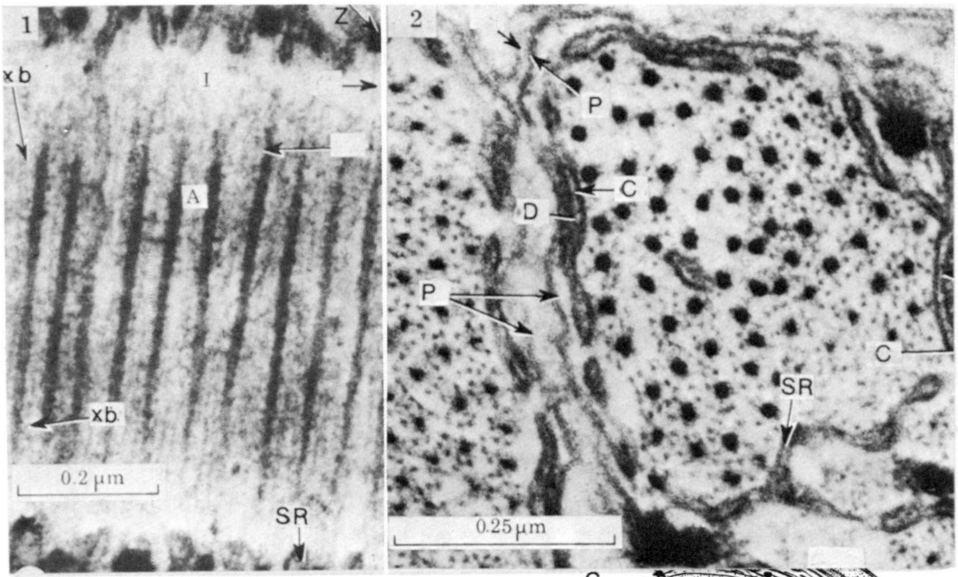

3

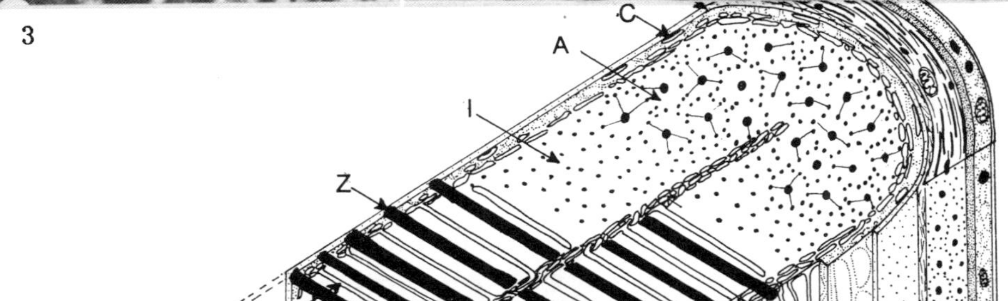

form, entirely longitudinal, parallel, running nearly the entire length of the animal, tapering at their ends. Their size is necessarily directly proportional to that of the animal, several inches long by 0.2 × 0.3 mm in a large specimen of *Ascaris equorum* from horse small intestine. Contractile material forms a peripheral layer surrounding a thick, central noncontractile core. When an opened-out worm slit longitudinally is viewed from the inside, all that can be seen, for the entire length of the animal, are irregular rows of loose cellular blobs, almost spherical in surface view, each 0.2–0.6 mm across. These each contain a muscle-cell nucleus and are the cell bodies. Gently pushing aside some of them reveals that each gives off a slender branch or *muscle arm*, each of which travels directly over the muscle cells transversely toward the nearest nerve cord, of which there are two, one in the dorsal midline, the other in the ventral midline (Figure 12.6). Muscle cells situated close to a nerve have very short arms, ones located laterally have long ones.

The motor axons run in the two nerve cords, and none make any branches. Instead, the tips of the muscle arms make contact with the axons in the cord. Each tip makes many fine branches that pass to a few of the axons located at the nerve surface. The finer branches of several bellies are intermingled, making multiple small synaptic contacts with the same axons. The details of muscle to nerve contact are necessarily difficult to discern, but are effectively equivalent to polyneural excitatory innervation.

Neuromuscular transmission can be studied by an electrode fortuitously inserted directly into an arm close to the junctions, or easily into a belly or muscle cell. The former reveal the presence of many small EJPs occurring sporadically. Most of these are very small and probably represent miniature JPs occurring spontaneously, as do MEPPs at vertebrate junctions. But it is also possible that some are just very small nerve-impulse-evoked synaptic potentials of small size. The "noise" recorded from the belly is an uninterpretable continuous hash with many more components per unit time than has a recording from a presumed single synaptic foot. This indicates that each foot receives independent inputs and that these summate in the belly. Some quite large polarizing transients occur, showing that IJPs as well as EJPs occur, and strong nerve-cord stimulation produces strong, relatively long lasting inhibition.

The records from within the cell body from time to time show much larger electrical events, but in a wide range of sizes from about 12 to 50 mV. These are compound, with a prepotential inflexion on the rising phase leading into a fast-rising graded spike. Both prepotential and spike height are variable. The membrane is electrically gradedly excitable, spikes are initiated by depolarizing currents and spontaneous ones inhibited by polarizing current. Some cell bellies show rhythmic action potentials of fairly constant amplitude at a frequency of about 2–3 Hz (Figure 12.7*A*). Their origin is uncertain: they could be electrode-damage injury discharges. But the worms do have stiffness, so this constant bombardment may be a peripheral tonus-generating mechanism. If so, muscle-cell pacemaking in the critical cells is implicated, since in them there is no inflection on the rising phase of the spike. The cells have a low resting potential, only 30 mV average, so are probably poised below the E-C coupling threshold. Such cells will have strong BT (see p. 255). There is very low gK, and the resting potential is dominated by anion conductances, not conductance to cations as in a majority of other skeletal muscle cells.

Figure 12.6. Drawings of basic structure of Nematoda *Ascaris* showing dorsal (*DNC*) and ventral (*VNC*) nerve cords, with muscle cells arrayed in parallel. Each muscle cell has a cell body (*CB*) and a contractile portion containing myofilaments (*MF*). Muscle arms (*MA*) extend from the cell-body region to one of the nerve cords, where they make muscle-on-nerve synapses with motor neurons (*MN*). [(*1*) From del Castillo et al., 1967; (*2*) from J. Rosenbluth, 1969.]

1

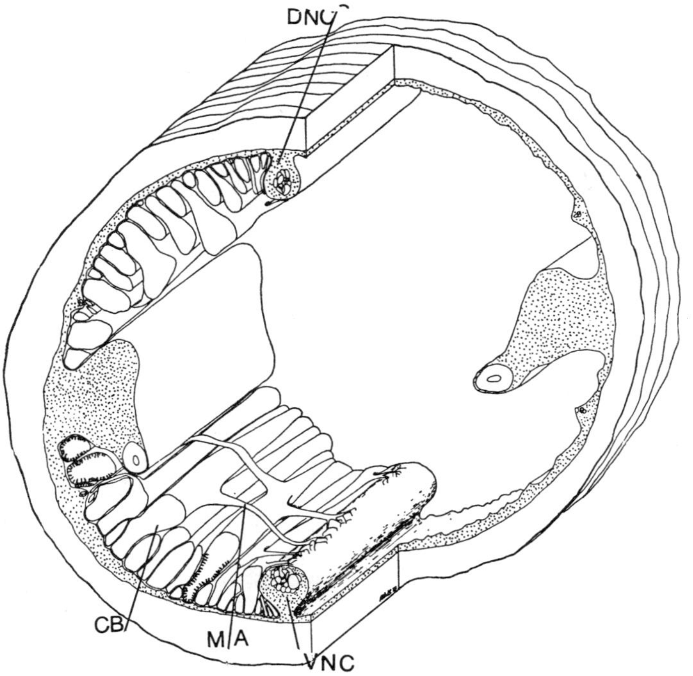

2

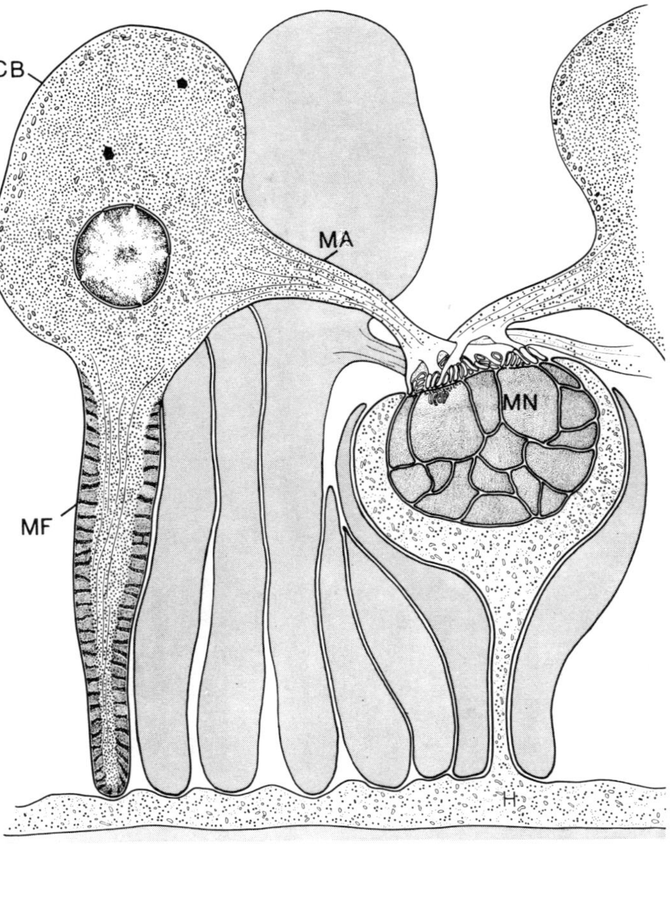

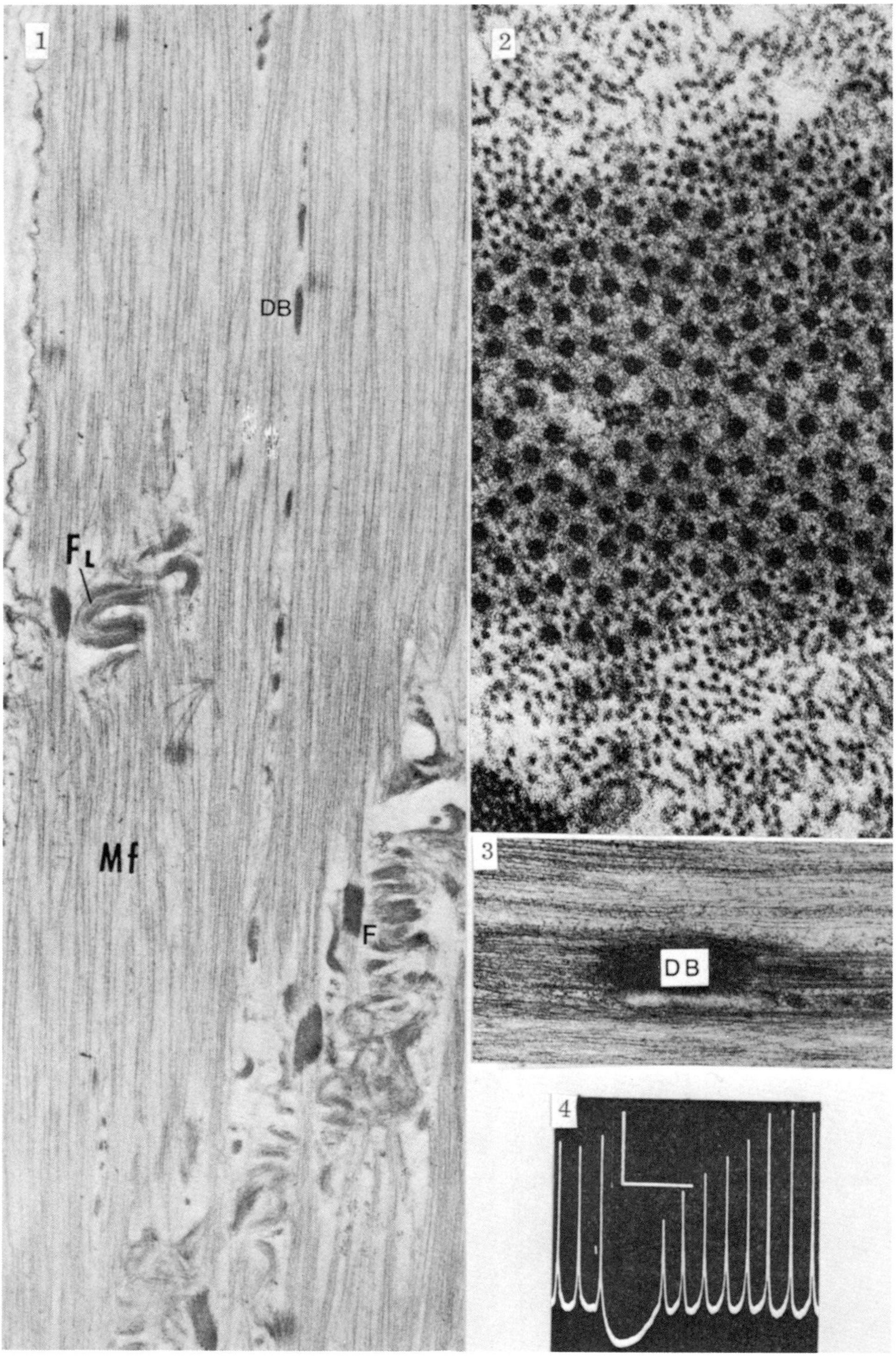
1
DB
FL
Mf
F
2
3
DB
4

In spite of researchers' ability to record intracellularly, the neural control mechanisms have not been worked out. *Ascaris* swims with sine-wave undulations in the dorsoventral plane. These can only be produced by the passage of a highly coordinated wave of excitation along both nerve cords, causing progressing local contraction/relaxation (inhibition?) of the giant longitudinal muscle cells. There seems no reason why contraction responses should not be studied in detail, but they have not been.

The neuromuscular junctions are depolarized by ACh (del Castillo et al., 1967), resulting in the discharge of rhythmic spikes from the cell body. The response is potentiated by neostigmine and blocked by *d*-tubocurarine. Thus if nematodes are placed next above Platyhelminthes in evolutionary sequence, they may be considered the first to have evolved cholinergic neuromuscular synaptic transmission.

There are low resistance pathways between the bellies of adjacent cells, sufficient for the spread of excitation or inhibition from one cell to a few neighboring cells (de Bell et al., 1963). This must result in there being contractile coordination of local *muscle units* comprising clusters of adjacent regions of a few muscle cells. This is a similar mechanism to that used by vegetative organs of vertebrates (see p. 330). However, the longitudinal cells are skeletal, not vegetative.

Vegetative muscle cells occur in the esophagus of *Ascaris*, and these cells are also quite large. Contractile elements within the cells are directed from outside to inside, so their contraction opens up the lumen, causing a sucking action. There are three pairs of cells, one at the end of each of three arms, so the lumen is triangular (Figure 12.8). These cells have large internal spaces from which an intracellular electrode records a large extracellular-type action potential (del Castillo and Morales, 1967). These spaces, though very large, are considered probably to represent a very primitive equivalent to ITs, serving to transmit excitation into the interior to initiate coupling activation. The resting potential is low, about 35 mV, as in the skeletal cells, and likewise dominated by anion permeabilities because *g*K is very low. A pump action may be making a major contribution (Brading and Caldwell, 1971).

Intracellular recording of natural neuromuscular activity and synaptic activity has not been achieved, but the cells are electrically excitable, and they show spontaneous small JPs that may be part of a tonus-determining mechanism. Strangely shaped, very long lasting (up to 2.5 s) large depolarizations also occur spontaneously, and superimposed on their plateaus there are sometimes polarizing IJPs. A more typical action potential first overshoots by about 18 mV, rather slowly, the maximum rate of depolarization being only about 9 V/s, then starts a slow initial recovery leading after about 30 ms to a fast repolarization. This has a very large undershoot of up to 40 mV so that the total peak-to-peak spike height is about 130 mV, the spike portion being due to calcium-conductance activation. There is finally a slow return after the undershoot, so that the total duration is about 100 ms.

Figure 12.7. (*1*) Ultrastructure of *Ascaris lumbricoides* myofilamentar (Mf) contractile material seen in TEM of an LS. The basic ultrastructure is obliquely striated, with thin filaments attached to dense bodies (*DB*). A part of the fibrillar network that supports the cell is cut across longitidinally (F_L) and in various planes at right (*F*). (Reprinted with permission from Reger, J. F. (1964) *J. Ultrastruct. Res.* **10**, 48–57 Copyright Academic Press.) (*2*) High-power EM of TS of extended cell showing effective I, A, and H zones, with DB at lower left. (From J. Rosenbluth, 1967.) (*3*) An LS, showing DB detail. (From J. Rosenbluth, 1967.) (*4*) Typical electrical activity. Large graded spikes recur spontaneously at regular intervals. Electrical stimulation of the nerve cord produces inhibition, with hyperpolarizing potential, and reduced spikes during recovery. Calibration: 10 mV, 1 s; resting potential 30 mV. (From del Castillo et al, 1967.)

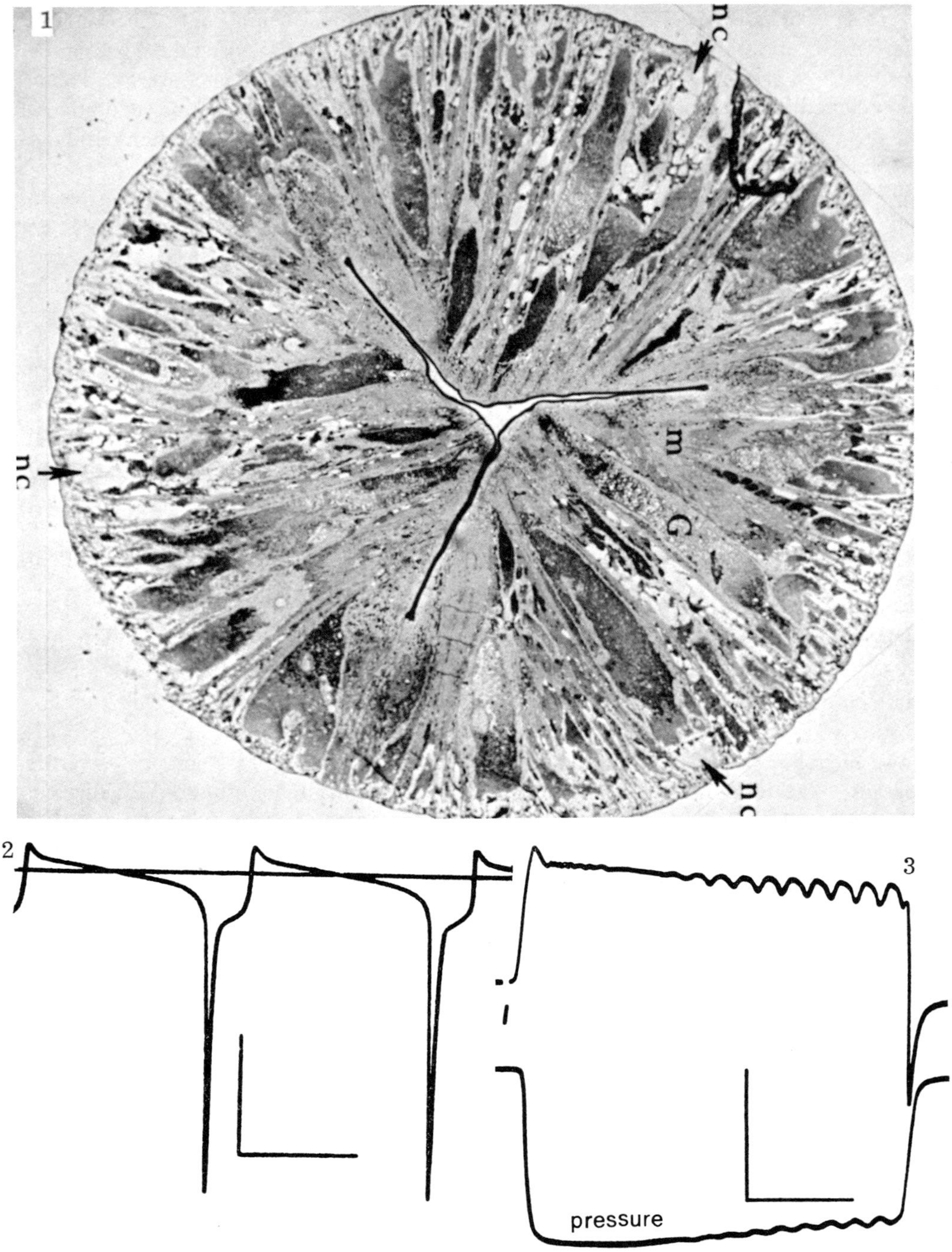

Figure 12.8. (*1*) Transverse section through esophagus of nematode *Ascaris lumbricoides,* showing triangular lumen, triple nerve cords (*nc*), glandular cells (*G*), and giant muscle cells (*m*). (*2*) Spontaneous upside-down spikes recorded intracellularly from a depolarized cell. Calibration: 50 mV, 1 s. (*3*) Simultaneous intracellular record of touch-evoked action potential (*above*) and negative pressure in lumen (*below*). Calibration: 25 mV/1.4 cm Hg, 1 s. (All from del Castillo and Morales, 1967.)

If a prolonged depolarization is applied, or during a very long naturally occurring action potential, tension, as indicated by reduced lumen pressure, is developed, exactly following in amount any small fluctuations of potential superimposed on the plateau (Figure 12.8.*3*).

The rate of relaxation during the fast repolarizing phase is very fast, faster than the rising tension. This is clearly a very interesting adaptation to the sucking function, in which rapid relaxation must necessarily prepare for the next sucking intake. Since the contractile mechanism is of a basically slow, tonic kind, an intensive *activation of relaxation* has been favored. Inward current applied during the depolarization plateau of an action potential produces an immediate rapid repolarization with full undershoot. If the cell is maintained in the depolarized state, large spike-shaped repolarizations occur rhythmically. These are true upside-down action potentials (Figure 12.8.*2* and 12.8.*3*).

The finding of this strong activation of a mechanism for causing relaxation strongly supports the contention made on p. 151 that relaxation is not fundamentally a passive process as earlier supposed. One other instance in which contraction in the unstimulated state is the rule is in eyestalk levator muscles of a crab, and this also shows very fast activation of relaxation (see p. 206).

The interest in nematodes arising largely out of their peculiar form of development (see account of muscle development by Stretton, 1976) has led to intensive studies of their neuroanatomy, muscle-cell structure and innervation, and even overall behavior. The behavior of legless, eyeless, vermiform parasites is hardly inspiring, but any paradigm promoting understanding of how genes affect nerve and muscle function to in turn determine behavior is potentially of fundamental significance. This has been recognized for the minute nematode *Caenorhabdytis elegans* (Brenner, 1974), although physiological understanding cannot be obtained and must be "transferred" from the larger *Ascaris*. There are altogether only about 250 neurons in *C. elegans*, and both these and many of the synapses between them have been determined by serial section/reconstruction methods.

"Behavioral" mutants recognized include slow twitch, roller uncoordinated, and paralyzed (e.g., Epstein et al., 1976). One of the latter, unc-52, results from dystrophy-like effect; another, unc-54, has been shown to be due to muscle-cell defects (Brenner, 1974). The kinetics of normal movements are initiated by a slow myogenic rhythmic depolarization (Weisblat et al., 1976). One *Ascaris* mutant has been shown to be associated with a heavy meromyosin (HMM) defect (Epstein et al., 1974). The thick filaments of all nematode muscle cells, including those of the pharynx, are tapered and contain Pm, but compared with very thick filaments of annelids and molluscs they have slim centers, not more than 25 nm, and they appear to be hollow. Each is 10–12 μm long (Waterston et al., 1974).

12.4. Annelida and Minor Wormlike Phyla

The annelids were probably the first of the evolving phyla to achieve metameric segmentation and enjoy the advantages this provided for developing coordinated locomotion. They successfully adapted to fresh water and terrestrial environments that are sufficiently mosit. At one time it was believed that arthropods had annelidan ancestors, but their neural organization and muscle structures are so very different that nerve/muscle scientists must take the view that they cannot be related. The Onychophora, long regarded as a living-fossil "missing link" between the two phyla, cannot now be regarded as related to either annelids or insects since they have even greater differences from either in nerve/muscle anatomy and physiology (see p. 484).

Annelids, therefore, may represent only their own individual cul-de-sac in evolution. Their skeletal musculature is principally in the form of single outer circular and inner longitudinal body-wall sheets. Discrete anatomical muscles operate the jaws of some marine polychaetes, and other muscles the paired legs of polychaetes or chaetae when present. The former muscles have not been studied, but the latter have been the objects of preliminary studies. Annelidan guts are rhythmically contractile; some show complex, interesting movement patterns and polyrhythms of the anterior part that are clearly related to natural feeding behaviors (see p. 349).

The skeletal cells of each of the major classes – polychaetes, earthworms, and leeches – share major features that are peculiar to the phylum, supporting traditional views of intraphylum relationships. Annelid muscles were long considered to be examples of SM, but electron microscopy revealed that they in fact comprise highly regularly ordered arrays of myofilaments: in the appropriate plane a section has the same appearance as do cross-striated cells in section (Figure 12.9). The arrays are in fact markedly obliquely staggered, but clearly should be classified with striated not smooth cells. The individual cells of longitudinal muscle are attached to radial lamellae in a pinnate manner. The cells are flattened and leaflike rather than rounded as with most skeletal cells. They average about 4 X 30 X 2–4 mm. At right angles to the margins of the broad surfaces, evenly spaced regular sheets of tubules run in parallel from surface to surface. These represent a unique form of nonperforated SR. They terminate in large cisterns lying just below the Sl, but do not open to the exterior. There are no ITs. The tubules parallel Z regions and are close to them. The muscle cell attachments are to connective tissue via hemidesmosomes and simple collagen strands. The subsarcolemmal cisternae closely abut the surface PM, forming functional equivalents to SR cisternal elements of other skeletal cells and serving E-C coupling in a similar manner. Calcium is probably released directly from them, diffusing to the contractile elements, none of which are more than 2 μm away from one. Return is via the parallel tubules, which are only a few tens of nanometers distant from the contractile elements. This is an excellent ultrasturctural design, but it is unique to the phylum.

In the polychaete "bloodworm" *Glycera,* there are two types of muscle cell as based on the amounts of SR. The majority have an orderly arrangement of myofilaments and a large number of SR tubules. The minority have fewer tubules and less-orderly array, with longer and also thicker thick filaments (Rosenbluth, 1968) to form an independent set of slow-type muscle cells, the others being relatively fast by comparison. Thin filaments are fairly evenly dispersed between thick ones, in a ratio of about 12 : 1, but they form general multilayered surrounds, not simple orbits, in annelids generally. The Z region is unique in ultrastructure. As seen in an extended cell sectioned longitudinally, which for annelid skeletal muscle cells is perpendicular to their long transverse axis, Z regions appear as continuous elements reaching almost across the entire cell. They comprise fine, but densely staining, Z filaments not unlike a vertebrate Z band except for the obliquity, to which I filaments are attached with regular separation. The clustering of thin filaments and the breaking up of the Z material into large ovoid DBs that are characteristic of molluscan and nematode obliquely striated cells is not seen in annelids.

Annelid thick filaments are long, generally about 3 μm, tapered at their ends, with thick central regions of about 40 nm. Thin filaments are relatively independent of thick ones in regions of overlap, with few complete orbits. There are wide H zones, since in

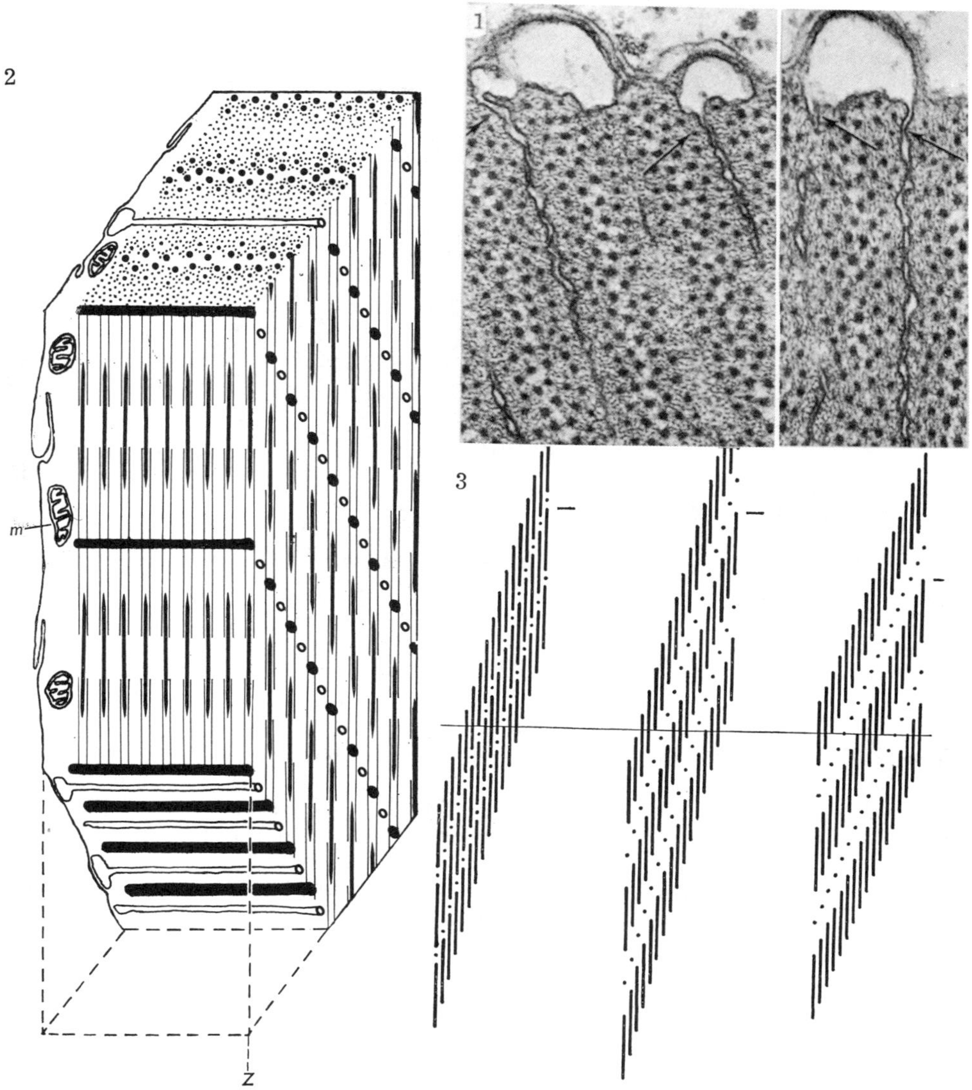

Figure 12.9. Ultrastructure of body muscle of the annelid polychaete bloodworm *Glycera*. (*1*) TEM of TS near the periphery of a muscle cell showing the large cisternal elements of SR forming surface dyads. The SR emerges from them as simple tubes (*arrows*) whose positions alternate on the oblique axis, with Z bands. (From J. Rosenbluth, 1968.) (*2*) Three-dimensional drawing of the myofilamentar array showing the oblique striation. All annelid body muscle conforms to this pattern. (From Mill and Knapp, 1970.) (*3*) Illustrates the two ways in which length changes may be brought about: sliding (*left*) and altered degree of obliquity (*right*). The latter permits obliquely striated muscle to adjust to a wide range of lengths passively, yet still be able to develop active tension equally at each length. (From Knapp and Mill, 1971.)

slightly stretched cells, many thick filaments have no thin ones near them. Orbits vary from 3-4 to 16 or more thin filaments surrounding each thick one. In several annelids, two different sizes of thick filament have been noted, in different cells. In the serpulid polychaete *Mercierella*, these were clearly associated with different cell populations, with which different electrical properties were also associated (Figure 12.10.*1* – Skaer, 1974).

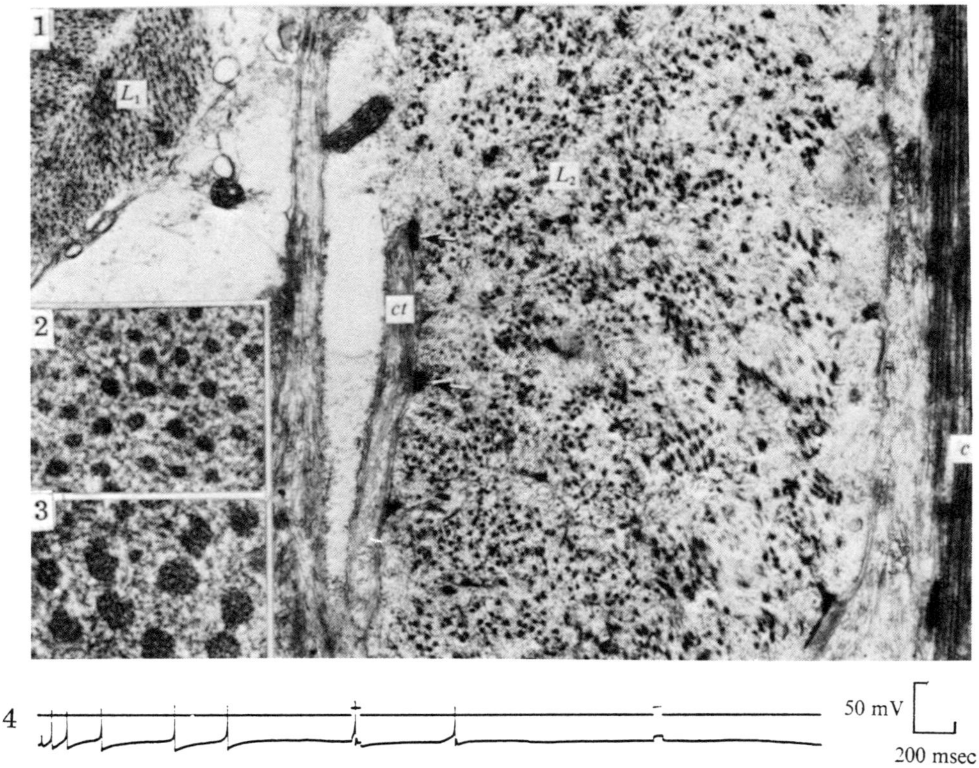

Figure 12.10. (*1*) Three types of muscle cell found in annelid serpulid polychaete *Mercierella:* small longitudinal (L_1), larger longitudinal (L_2), and circular (*c*). The L_2 is anchored to connective tissue (*ct*) by hemidesmosones (*arrow*). (*2*) High-power EM of TS of L_1. (*3*) High-power EM of TS of L_2. (*4*) Intracellular recording of spontaneous and evoked activity from L_2. (All from H. L. B. Skaer, 1974.)

The large cells are electrically excitable and show naturally occurring spikes and are associated with very quick contractions, so are fast cells. Their spikes are all-or-none overshooting and insensitive to sodium-channel blockers and sodium concentration but sensitive to both Ca^{2+} and Cl^-, especially the ratio of these ion in the medium. The ratio is much more important than their absolute concentrations.

A remarkable type of striated cell occurs in the foregut or proventriculus of small syllids, as first discovered almost a century ago by Haswell (1890). The lumen is about 100 μm distant from the epithelium, the distance being bridged by radial myoepithelial cells, each of which in effect comprises two sarcomeres of such length, about 40 μm (Figure 12.11), that constitute the longest ones currently known in the entire animal kingdom. There is a single central Z line of zigzag from between the two. The outer sets of I filaments are attached directly to epithelium/body wall at their outer margins. Thick filaments are extremely thick, up to 90 nm in thickness, though many are ovoid rather than circular in cross section. They have a 14-nm longitudinal repeat period (Figure

Figure 12.11. Ultrastructure and electrophysiology of cross-striated cells with the longest sarcomeres but shortest total length (two sarcomeres total): from the proventriculus of the annelid syllid *Syllis.* (*1*) EM of LS of an entire cell. (*2*) TS of an entire proventriculus, from which the cells radiate. Positions of Z bands are indicated by arrows. (*3*)–(*5*) TEM of TS of I, A, and H regions respectively. (*6*) Paramyosin crystal from thick filaments. (*7*)–(*8*) Intracellular recordings from (*2*) showing EJPs of assorted sizes in single cell, the larger ones eliciting graded spikes, small ones summating, and IJPs. (All from del Castillo et al. 1972.)

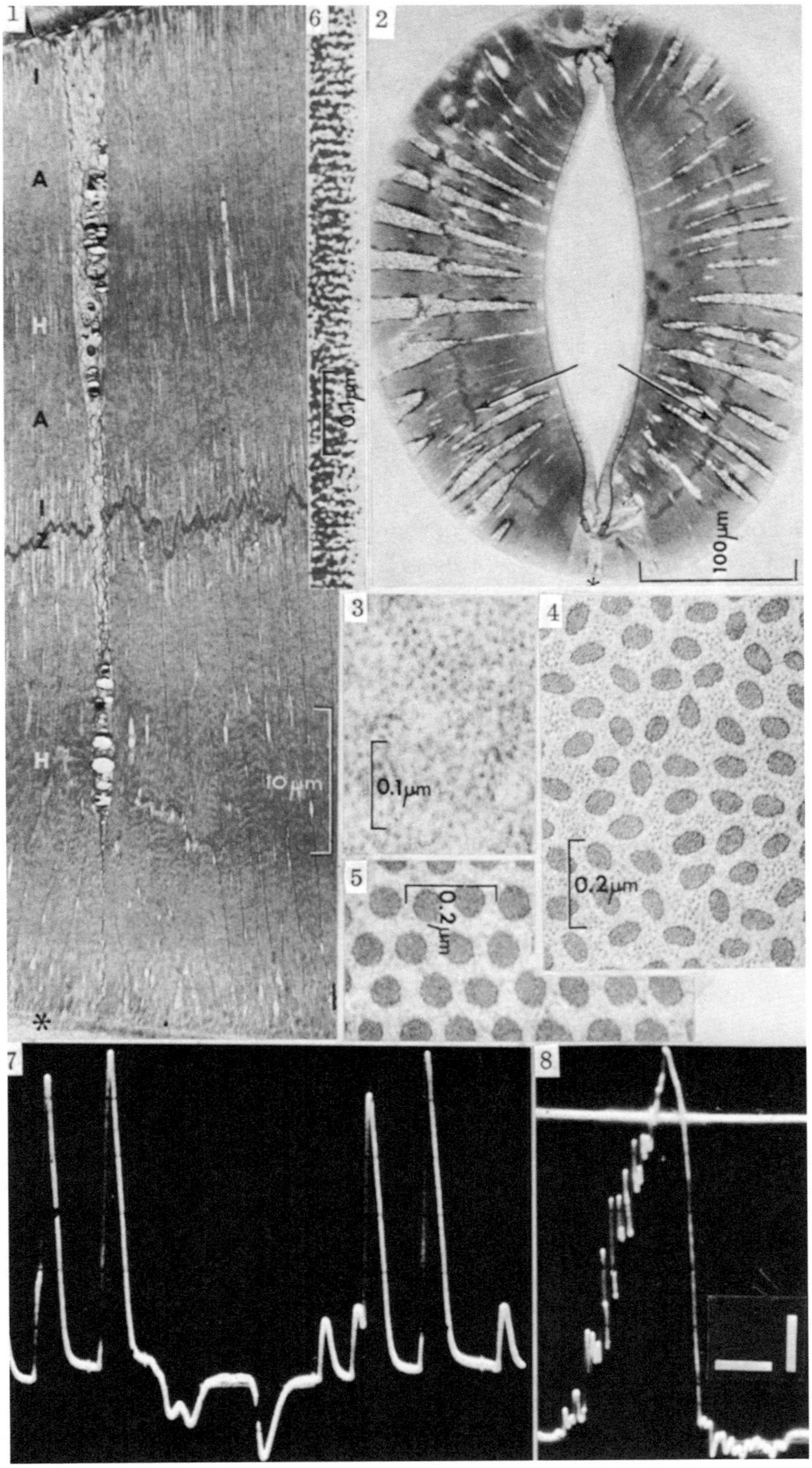
1
I
A
H
A
I
Z
H
10 µm
*
6
0.1µm
2
100µm
3
0.1µm
4
0.2µm
5
0.2µm
7
8

12.11.*6*), corresponding to that of molluscan Pm. This is the only known example of paramyosin-containing very thick filaments located in cross-striated muscle cells. Some Pm occurs in thick filaments of *Limulus* and in those of many other cross-striated cells, but the overall thickness of these thick filaments does not exceed 23 nm in thickness (del Castillo et al., 1972). The *Syllis* proventriculus muscle cells each show spontaneous EJPs of 10-20 mV and large IJPs of 8-16 mV. The larger EJPs initiate large graded spikes, up to 40 mV high. From time to time a strong peristaltic-contraction wave spreads over the muscle. Intracellular recording during this wave shows accompanying bursts of EJPs of different sizes at a single site, facilitating, summating, and eventually exceeding the 56-mV resting potential. EJPs of up to three sizes, and also IJPs, can be evoked by indirect electrical stimulation (Anderson and del Castillo, 1976).

Many polychaete annelids are sedentary, living in a variety of tubes they secrete, manufacture, or simply form by burrowing. The same animals reproduce by making segments that break away filled with gametes and swim around vigorously before bursting. A difference has been noted in the fine structure of some of the muscle cells in the sexual form compared with asexual stages of *Syllis* (Wissocq, 1970). The muscle cells are uninucleate and ribbon shaped, with peripheral myofilaments helicoidally arranged around a thick core crammed with mitochondria in the sexual stage but with few in the asexual stage, which also has much longer cells (Figure 12.12). There is also much more glycogen in the core. The contractions speed is markedly faster in the sexual stage though the myofilaments are similar, with a 7 : 1 thin-thick ratio. The difference in the faster sexual stage is associated with more SR and cisterns.

There are still no really satisfactory published studies on innervation and neuromuscular transmission in annelid muscles. The best are on leech body muscle. There were many studies on neuromuscular transmission in earthworms, but studies were apparently hampered for decades by use of an unsuitable saline that didn't match the ionic composition of the body fluid (Drewes and Pax, 1974a). This has led to low resting potentials, high excitability, and injury discharge. However, optimal resting potentials in a "natural" saline are still below 50 mV. Single cells are innervated by either a slow axon, a fast axon, or both. The former gives rise to small summating and facilitating EJPs and slow sustained contraction that is strongly frequency dependent, so these cells clearly constitute a distinct population of slow muscle cells. The fast-axon gives larger EJPs that decline upon repetition and a rapidly declining small twitch (Drewes and Pax, 1974b), and these cells are, by contrast, fast ones. Dually responding ones are probably intermediate.

It is well known that giant nerve cells occur in earthworms, responsible for evoking rapid contraction of the entire longitudinal musculature. The contraction is mediated by the fast motor axons; all other movements are presumably mediated by the slow ones. Both neuromuscular excitatory systems appear to be cholinergic. The leech dorsal longitudinal body-wall muscle provides the classical bioassay material for ACh (Minz, 1932), and longitudinal muscle of *Branchiomma, Arenicola, Lumbricus, Pontobdella,* and

Figure 12.12. Structure of tubular, obliquely striated, cells of stolon muscle of sexual stage of annelid *Syllis.* (*1*) Isolated entire cell with prominent single nucleus (*N*). (*2*) High-power light micrograph of (*1*) showing "double-oblique" striation pattern. Indicated are myofilament (*My*) clusters, appearing as spiralling lines. (*3*) TEM of LS showing ultrastructural organization, especially the regularly-arrayed dense bodies that effectively form Z bands (Z). (*4*) TEM of TS, as (*3*). (*5*) Three-dimensional drawing of ultrastructural organization: there is a large core area occupied by mitochondria (*m*). (From J. C. Wissocq, 1970.)

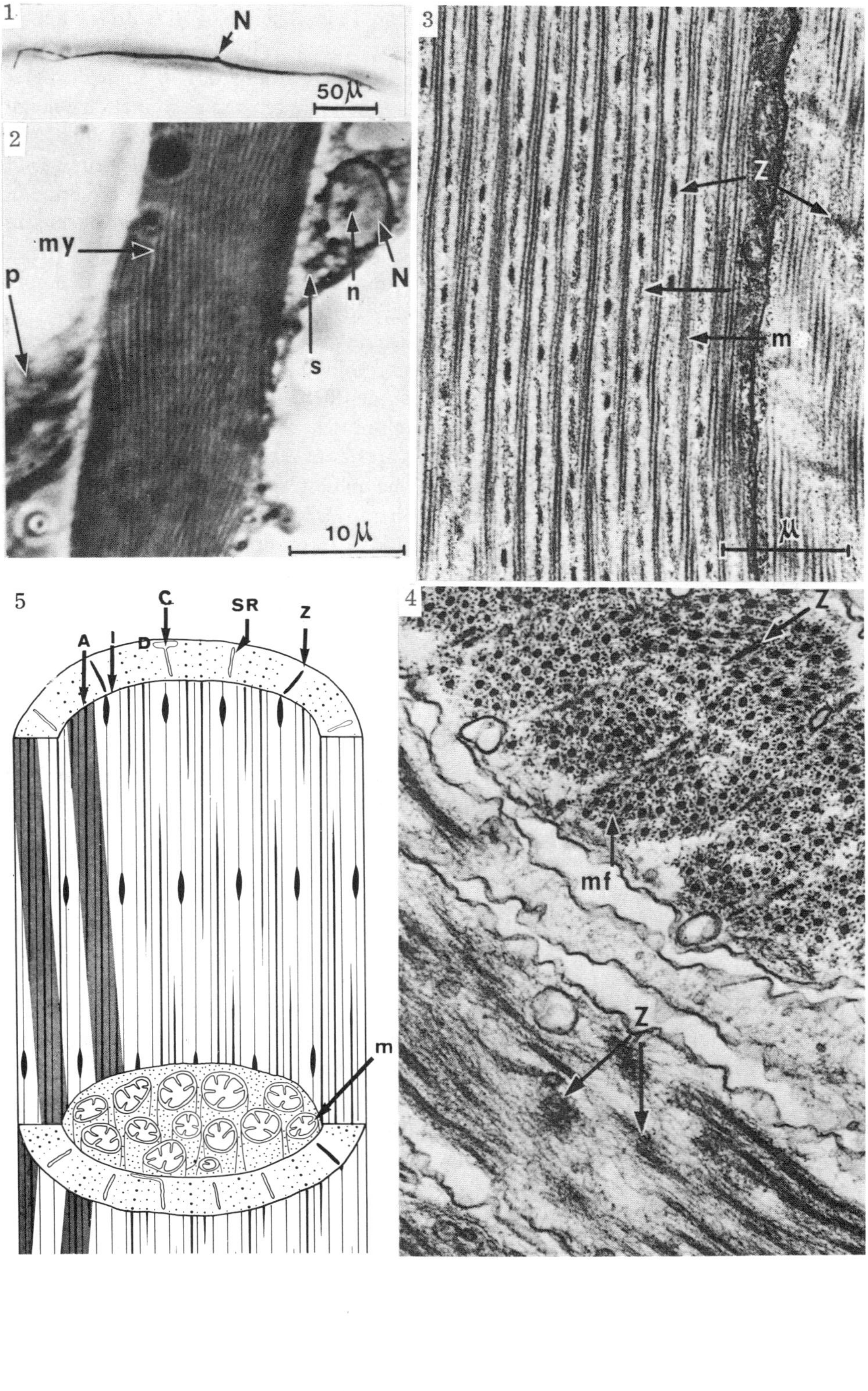
1
N
50μ
2
my
p
n
N
s
10μ
3
Z
m
μ
5
C
SR
Z
A
I
D
m
4
Z
mf
Z

Aphrodite (Wu, 1939b; Bacq and Coppee, 1937) are all cholinergic. At the ultrastructural level, the nerves contain two different sets of motor axons that in turn make two types of junction (Figure 12.13 – Rosenbluth, 1972), one containing numerous 50 nm in diameter spherical vesicles plus a few DCVs, with a 90-nm junctional gap and postjunctional membrane characterized by a complex substructure. The other kind contains mainly large DCVs, of irregular to spherical shape, up to 12 nm in diameter, plus a smaller number of round clear vesicles. These have a narrow 15-nm gap, and there is very little structure in the postjunctional region. The distribution of the two types of ending along the lengths of individual cells is not known.

Leech muscle cells are not leaflike as are those of earthworms, but circular, with a noncontractile sarcoplasmic core surrounded by an orderly array of myofilaments in an obliquely striated arrangement (Figure 12.14). The myofilaments are similar to those of both other annelids and molluscs (Lanzavecchia and DeEquileor, 1976). The leech nervous system permits intracellular studies on individual identified neurons. Motor neurons that are excitors or peripheral inhibitors of body-wall muscle cells have been found (Figure 12.15 – Stuart, 1970). Longitudinal muscle cells are about 20 μm in diameter and receive innervation from two or three excitors and one inhibitor. Each provides distributed terminals along the entire length of the cell. There is a lot of fluctuation in amplitude between successive JPs, each of which is small, 1-6 mV, for most of the neurons. A large cell was found that caused a small twitch in response to a single impulse and showed both facilitation and mechanical summation. Leech longitudinal dorsal body-wall cells are capable of producing overshooting propagated spikes (Washizu, 1967).

12.4.1. Giant Axon Responses of Annelids

Many annelids of several orders have giant axons in their nerve cords. Those of *Myxicola* have yielded the largest axon yet known in the entire animal kingdom; the axon is 1.7 mm in diameter, 0.5-1.0 mm being common. The purpose of all the giant nerve cells is to conduct rapidly in both directions from the region excited to the motor neurons innervating longitudinal musculature. Several examples of this type of response have been studied. The motor-nerve-to-muscle-cell synapses are all cholinergic, those of the sensory nerves to giant axons and giant axons to motor nerves are not. Both these sets of synapses rapidly become refractory to continued stimulation, one or the other accounting for the overall habituation of the response (e.g., Horridge, 1959).

The annelid giant-axon-mediated motor systems are working through typical fast motor axons as judged by extracellular recording. The same muscle field that is excited by a motor axon in the escape response can also be excited to make slow contractions. Unfortunately, it is not known whether or not there are separate populations of muscle-cell types as in arthropods, or if individual muscle cells are innervated by motor neurons producing physiologically different junctional responses.

12.4.2. Nematomorpha

Threadlike and called "horsehair worms," the nematomorpha are parasitic in their asexual stage but free-living in the sexual one. The best known is *Gordius*, which grows to 150 × 0.5 mm and has a single layer of longitudinally running muscle cells just below the cuticle. These are ribbon shaped, with numerous mitochondria arranged end to end in a single row in the core. Thick Pm-containing filaments range at their centers to a gigantic 140-nm

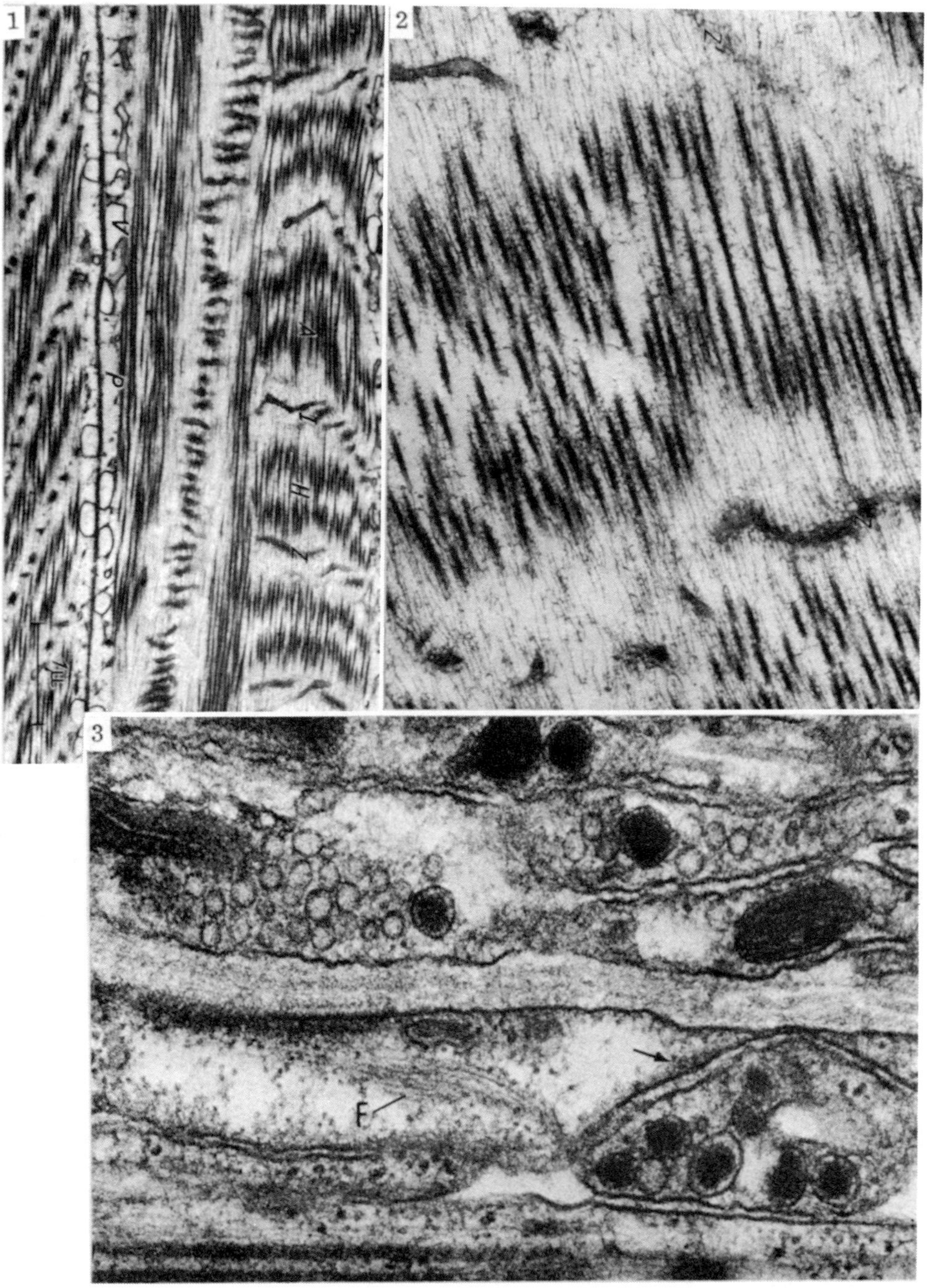

Figure 12.13. (*1*), (*2*) Ultrastructure of earthworm *Lumbricus* body muscle cells, which are very short sarcomere, obliquely striated ones. As with other annelid muscle cells, Z regions alternate with SR. (From Heumann and Zebe, 1967.) (*3*) Two types of motor-nerve endings on earthworm muscle cells. Fine myofilaments (F) occur near a junction, and there is a postsynaptic density (arrow) that may indicate a transmission site. (From J. Rosenbluth, 1972.)

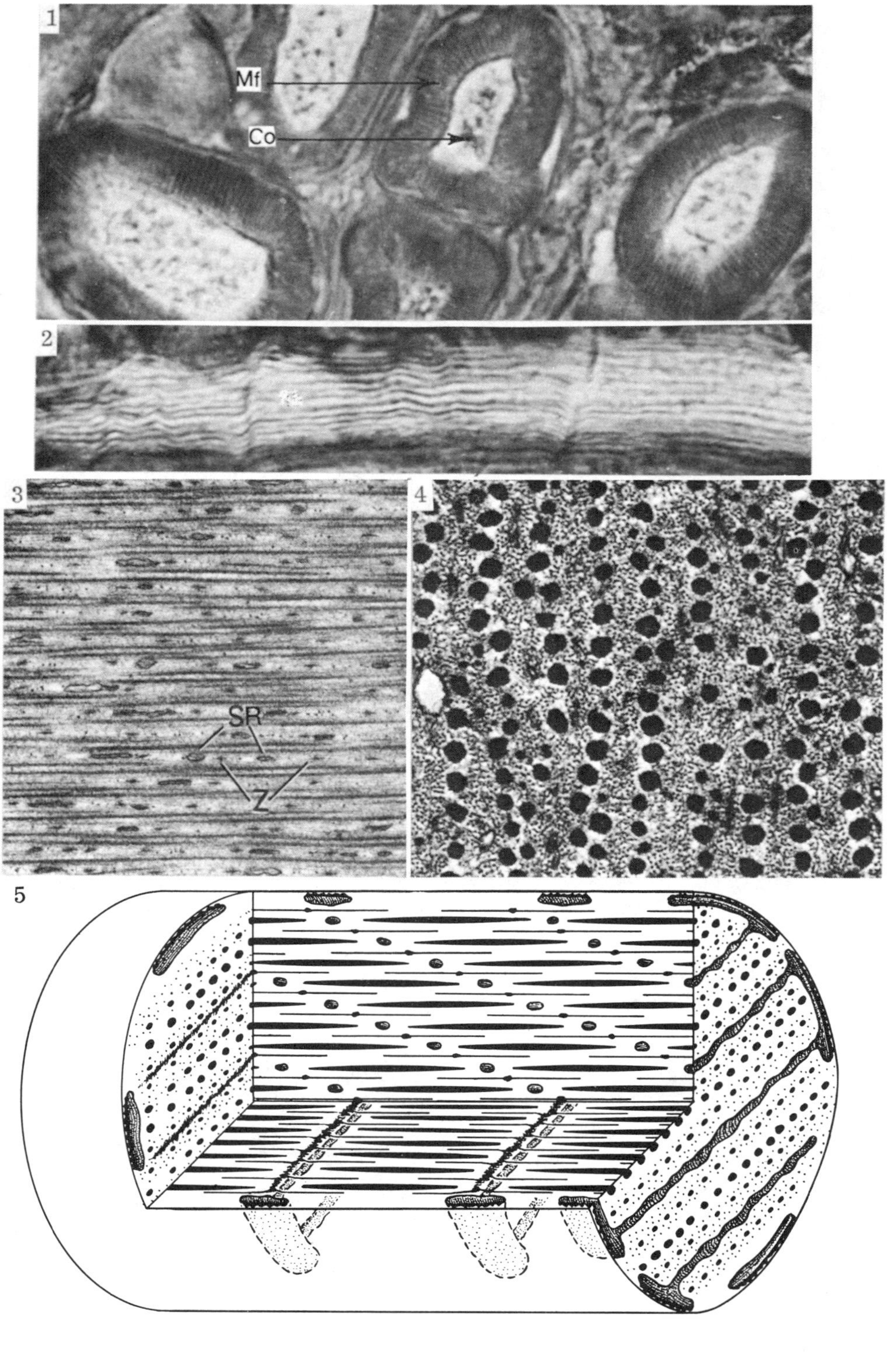
1
Mf
Co
2
3
SR
Z
4
5

thickness, making them the thickest known (Lanzavecchia et al., 1979). The muscle cells are cholinergic (Swanson, 1971). A characteristic behavior of the worms is to get into a tight coil and to maintain it, presumably using a paramyosin "catch" mechanism. Like the molluscan *Mytilus* ABRM (see p. 477) these muscles contract and relax phasically when stimulated with short pulses or AC, but go into "catch" state with DC.

12.4.3. Priapula

A small eight-species phylum members of which are nevertheless worldwide, the type genus *Priapulus* has interesting circular and longitudinal body-wall muscles. Contractile material forms a peripheral ring in the 10–12 μm in diameter cylindrical cells, surrounding a central core containing sarcoplasm, mitochondria, and nuclei. Thin filaments are attached to Z bodies arrayed in rows with a regular radial orientation. There are so few Z bodies, however, that they appear as radial lines rather than as a Z disc. There are numerous relatively thick filaments, but these have no regular array. Thin filaments cluster between thick filaments, and few can be said to be present in orbits. The overall thin-thick number ratio is about 8 : 1. Sarcoplasmic reticulum is sparse and there are no ITs (Carnevali and Fernaguti, 1979).

12.4.4. Archiannelida

A primitive, entirely marine, group of small annelids with few body segments, the Archiannilida have a small powerful muscular pharynx, some 50 × 35 μm. This comprises uninucleate cells, each with a beautifully regular oblique striation (Rieger and Rieger, 1975). Thin filaments are attached to a staggered array of DBs and extend longitudinally to intermingle with oblique arrays of tapered thick filaments with 30 nm in diameter centers. Astonishingly, the myofilaments of one cell occur in two sets of plates that are exactly at right angles with respect to each other (Figure 12.16). This is presumably brought about by progressive flattening of an otherwise-cylindrical cell, with only peripherally located myofilaments in which the obliquity spirals around to make the so-called *double-oblique* striation as seen in light microscopy. The function of these cells is to suddenly increase the volume of the cavity of the foregut to ingest food. The axis of contraction is at 45° with respect to the two filaments axes.

12.4.5. Sipunculoidea

The wormlike sipunculid *Golfingia* (=*Phascolosoma*) has a discrete proboscis retractor muscle that has been studied mechanically, pharmacologically, and with intracellular electrodes (Prosser and Sperelakis, 1959). The muscle cells are of small diameter and length, with resting potentials of no more than 49 mV, but most are capable of producing

Figure 12.14. Ultrastructure of leech body muscle cells. (*1*) Light micrograph of TS showing central noncontractile cores (*Co*) surrounded by regular arrays of myofilament clusters (*Mf*). (*2*) Light micrograph of LS showing parallel arrays of myofilament clusters. (*3*) EM of LS showing alternating small Z bodies and SR tubules. (*4*) EM of TS at high magnification showing tapered thick filaments with very thick centers surrounded by loose arrays of numerous thin filaments. (*5*) Three-dimensional reconstruction showing similarity of contractile unit to that of other annelids. [All reprinted with permission from Röhlich, P. (1962) *J. Ultrastruct. Res.* 7, 399–408. Copyright Academic Press.]

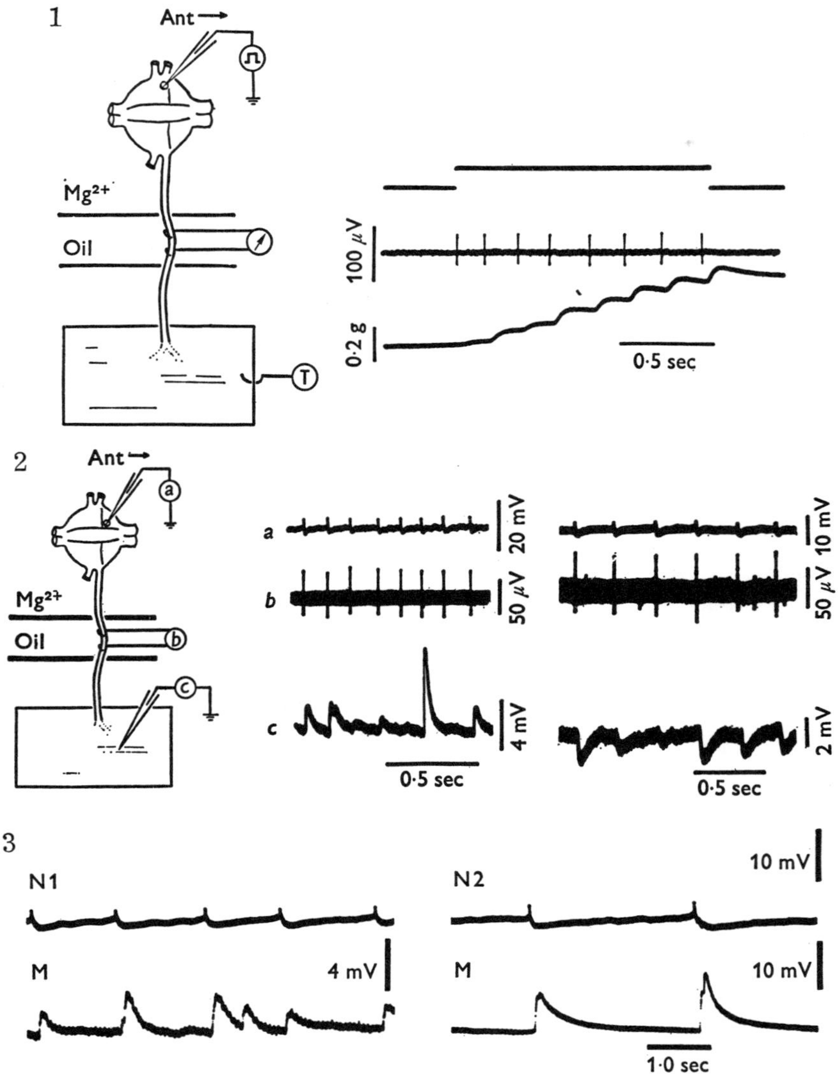

Figure 12.15. Physiology of leech body wall muscle (*1*) Mechanical responses to contralateral motorneuron stimulation. Diagram of preparation at left. (*2*) EJPs and IJPs of very variable amplitude during intracellular recording from muscle cell and direct intracellular neuron stimulation. (*3*) Non-driven EJPs (lower traces) correlated with intracellular recordings of the nerve impulses causing them. (All from A. E. Stuart, 1970.)

large graded spikes up to overshooting. Indirectly evoked electrical responses range from very small EJPs to fairly large ones in the same cells. The former lead to slow sustained contraction when repeated and the latter to faster but rapidly fatiguing ones. The mechanical responses are blocked by curare and potentiated by eserine, so the transmission is probably by ACh. The muscle is of discrete, skeletal type, with complete neural control of muscle cells, and it is not excited by stretch.

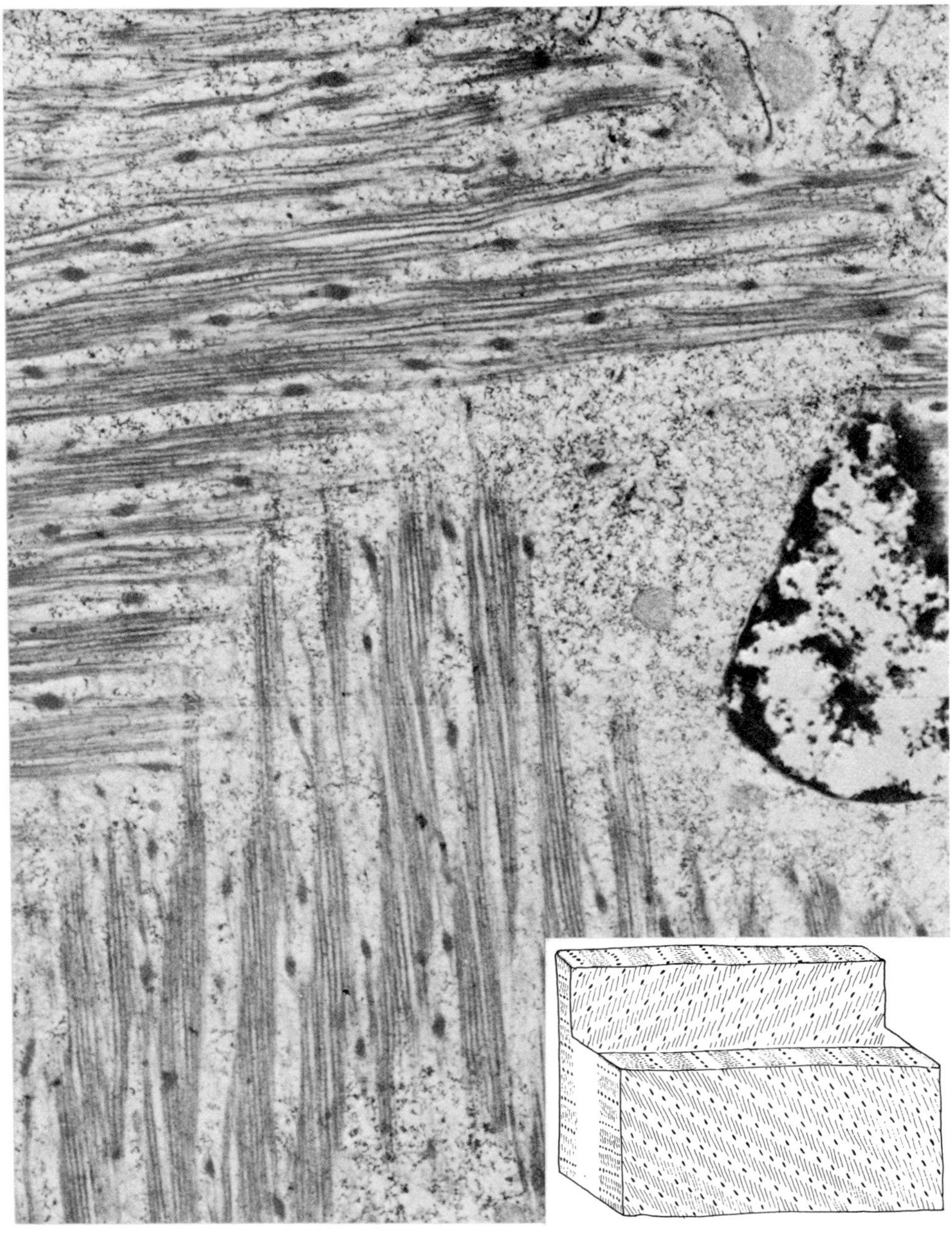

Figure 12.16. Ultrastructure of muscle cell of the pharyngeal bulb of archiannelid *Trilobodrilus*. This seemingly absurd arrangement of striated myofilaments at right angles to each other can be explained by the compression together of two parts of an obliquely striated cell lacking a core, as shown in the inset diagram. (From Rieger and Rieger, 1975.)

12.4.6. Acanthocephala

Acanthocephala enjoy the status of unique phylum. They possess an eversible proboscis that permits attachment, and a parasitic life, as well as thin circular and longitudinal body-wall muscles. Spontaneous EJPs of uniform size, 5-8 mV, occur in the longitudinal body-wall muscles of *Macracanthorhyncus*, the larger of which evoke overshooting spikes (Wong et al., 1979) that are eliminated by TTX. Excitatory transmission appears to be cholinergic.

12.4.7. Chaetognatha

The Chaetognatha are interesting, relatively complex small coelomate animals that include the free-swimming arrowworm *Sagitta,* which is widespread in plankton populations. Their longitudinal body muscle cells, used in swimming, are cross-striated, of very short SL, with an extremely regular array of undoubtedly hollow (tubular) 18-nm-thick filaments (Figure 12.17 – Halvarson and Afzelius, 1969; Duvert, 1969; Duvert and Salat, 1979). Thin filaments are present in a 1 : 3 ratio in the same array found in insect flight muscle cells (Figure 12.17.*2*). There can be no doubt that this arrangement is associated with the production of fast twitches used in swimming. These muscles have another structure in common with insect flight muscles: thin connecting filaments between the thick filaments across the Z bands. The entire Z region is unique. Thin filaments of adjacent sarcomeres also appear to be connected to each other across the Z region by connecting regions that are slightly thicker and more densely staining than in the I region. There is a fine dense Z band in the middle of the Z region, and this band is flanked by broader, denser bands, also on either side. This extraordinary arrangement is unique in the entire animal kingdom.

The cells have an average diameter of about 10 μm. There are no ITs, and the SR consists of perforated material with cisterns located just below the surface membrane. The outer cistern membrane makes close dyad-type juxtaposition with the sarcolemmal membrane. Large mitochondria occur between, but not in, the small cells.

12.4.8. Pentastomida

The pentastomids or linguatulids are wormlike, limbless parasites with superficial annulation that have been considered possibly to have arthropod phyletic connections, but which have an annelidlike gut. The body wall contains distinct circular and longitudinal layers of muscle cells, both sets of which are cross-striated. Electron microscopy has been done on these cells obtained from *Reighardia,* which lives in the airsacs of gulls (Mill and Riley, 1972). The sarcomere length is 3.5-5.5 μm; no distinct H zone can be seen. The Z bands are wide, 0.4 μm, and formed of elongate DBs.

Figure 12.17. Ultrastructure of fast-swimming body muscle of chaetognath *Sagitta.* (*1*) TEM of LS showing spirally aligned SR, unique Z region, and dense N bands (arrows). H, A, and I regions are evident. A higher magnification picture of the Z region and the N bands is shown in the inset; X bridges are clear (arrow). (*2*) TEM of TS in overlap region, showing tubular thick filaments and same 3 : 1 filament array as seen in insect flight muscles. (*3*) The H zone, showing triangular, tubular profiles of thick filaments, X bridges between thick filaments, and ultrathin filaments (*UT*) as continuations of thin ones. (All from Duvert, 1969.)

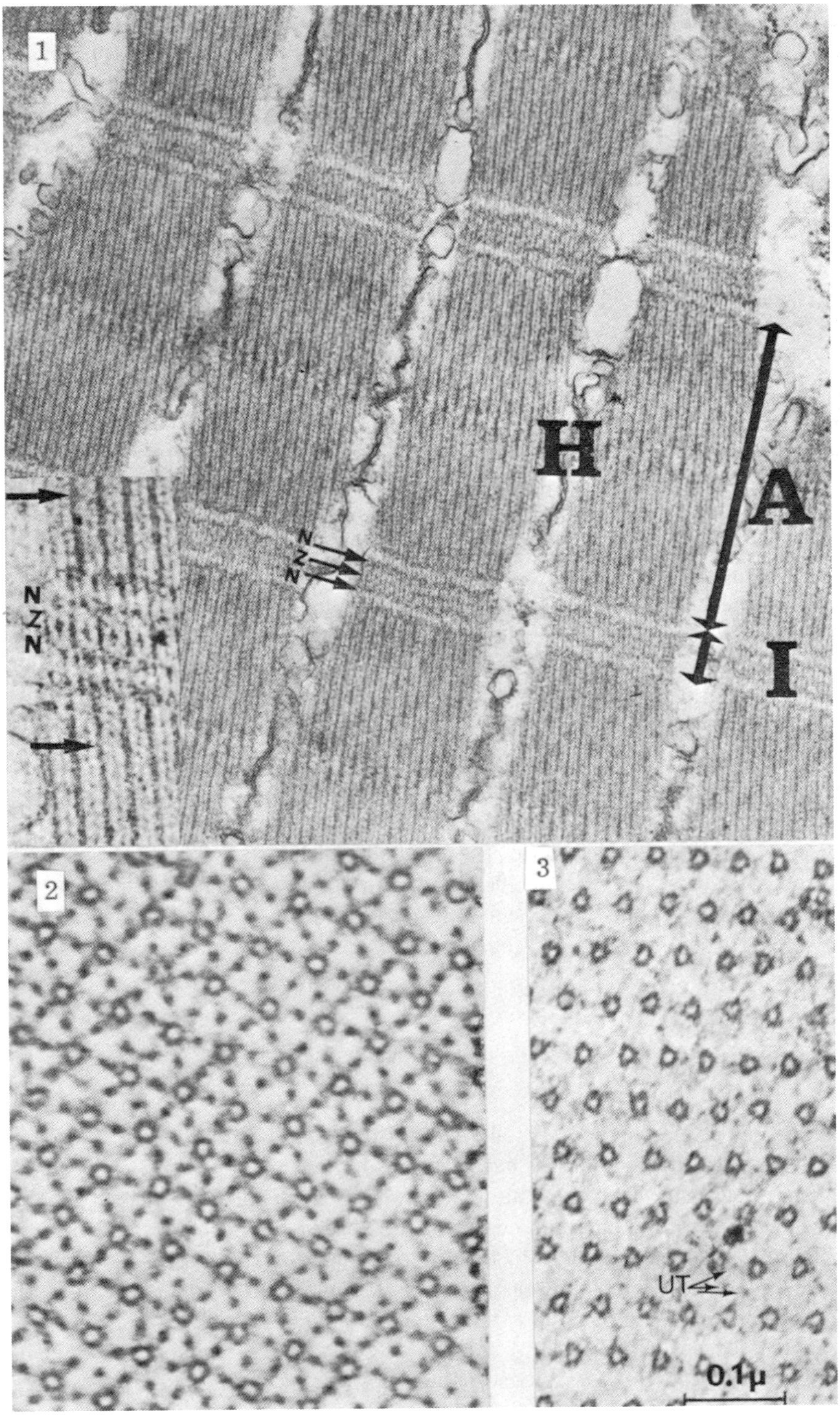
1
N
Z
N
N
Z
N
H
A
I
2
3
UT
0.1μ

Invaginated tubules occur although the diameters of the cells are only 3-5 μm, small enough not to need them, and myofilaments occur throughout the myoplasm. The ITs are more like short grooves than true tubules, running radially, so that the SR contacts both sides to make triads as seen in transverse section. Small mitochondria are scattered throughout the cells, which include both glycogen granules and very large (1 μm in diameter) lipid drops. Thick filaments are about 18 nm in diameter and are surrounded by orbits of about 11 thin filaments.

12.5. Entoprocta

Entoprocta is a phylum of small solitary or colonial marine organisms of unique structure that are interesting because they make jerky bending movements of the calyx with respect to the stalk, giving the popular name "nodding heads." The movement is relatively slow, and the bent position is maintained for a time without further movement. The contractile cells are radially arranged in the stalk, with bulbous noncontractile expansion that extent into the lumen. The cells are obliquely striated (De Equileor and Valvassori, 1977), with tapered 100-nm-thick (at center) thick filaments. There are no ITs, but there are numerous deep small cleft invaginations that serve inward conduction. When strongly shortened, the filaments and Z bodies become so well aligned laterally that the muscle looks cross-striated. The cells of *Barentsia* contain prominent crystalloids comprising linear arrays of 3-7-μm thin filaments that have a helical shape with a 13-15 nm pitch and 34-45° tilt angle (Figure 12.18 – Reger, 1969). These are crosslinked by bridges into regular hexagonal packing. Some of the 3-7-nm filaments branch from the crystalloids, passing into the myofilamentar area. Reger suggested that they represent a storage form of myofilament precursor protein. However, they bear a remarkable resemblance to the paracrystals of actin depicted by Hanson (1973).

The myofilaments of the contractile portion, which is 0.5-1.0 μm across, comprise 6-nm thin filaments and tapered thick ones with 30-nm centers which possibly contain Pm. They have an oblique alignment similar to that of platyhelminths, nematodes, and annelids. Thin filaments are attached to Z bodies, and to the surface membrane at hemidesmosomelike bodies.

12.6. Brachiopoda

The brachiopods are marine animals with a bivalve shell and therefore superficially resemble lamellibranch molluscs, but differ in having a unique ciliated food-current-producing organ with tentacles, the lophophore. These tentacles in the subtidal species *Terebratalia transversa* contain myoepithelial cells, some of which are cross-striated, with short 2.5-μm SL, others adjacent to them having long, thick paramyosin-containing filaments and no obvious lateral myofilamentar alignment (Figure 12.19 – Reed and Cloney, 1977). The cross-striated banding is associated with thick, "blobby" Z lines of the kind commonly seen in cells capable of supercontraction (see p. 261). Thick filaments are 22 nm in diameter, which is thick for such short sarcomeres; thin filaments are 7 nm. The thin filaments are present in orbits of up to 12, with a ratio of thin to thick of about 5 : 1.

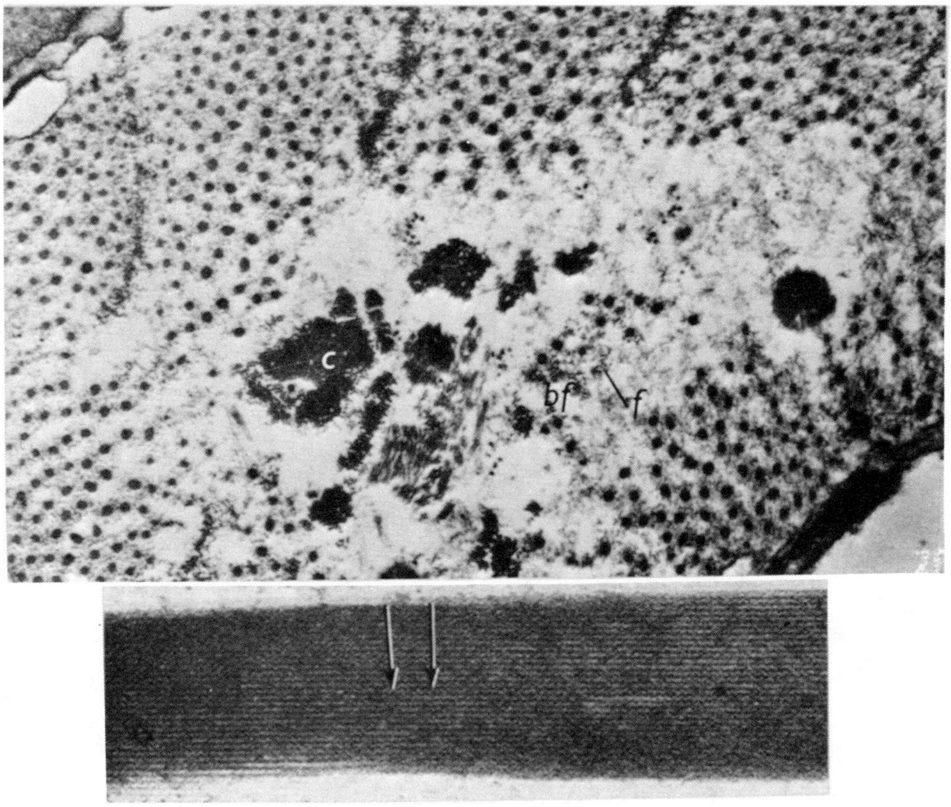

Figure 12.18. EM of TS to show ultrastructure of muscle cell of entoproct basal socket. The cell contains interesting crystalloid inclusions (*C*), shown in higher power below, that closely resemble paracrystals of actin. (From J. F. Reger, 1969.)

The slow cells contain paramyosin thick filaments over 120 nm long and large numbers of thin 7-nm filaments not in regular relation to thick ones. The thick filaments are staggered with respect to each other, but not in a regular manner, so they cannot be regarded as either obliquely or spirally striated and should be classified with other disorderly striated muscle cells. The entire section of the thick filaments shows paramyosin banding.

The two sets of muscle cells must have independent functions in tentacle movements, since the cross-striated ones should be quite fast, the paramyosin-containing ones slow and "holding." The tentacles make flicking motions that are probably exclusively generated by the cross-striated cells. The tentacles also are held down in the flexed position for long periods, which is presumably the role of the slow cells. Interesting differences occur in closely related or even the same species with different habitats and behavior patterns. *Terebratalia transvera* also has an intertidal form and this very interestingly lacks the striated cells. It is argued that the deeper-water form can afford the luxury of quick flicks but that the often-buffeted intertidal one cannot, because it has to shut down its valves frequently and for long periods.

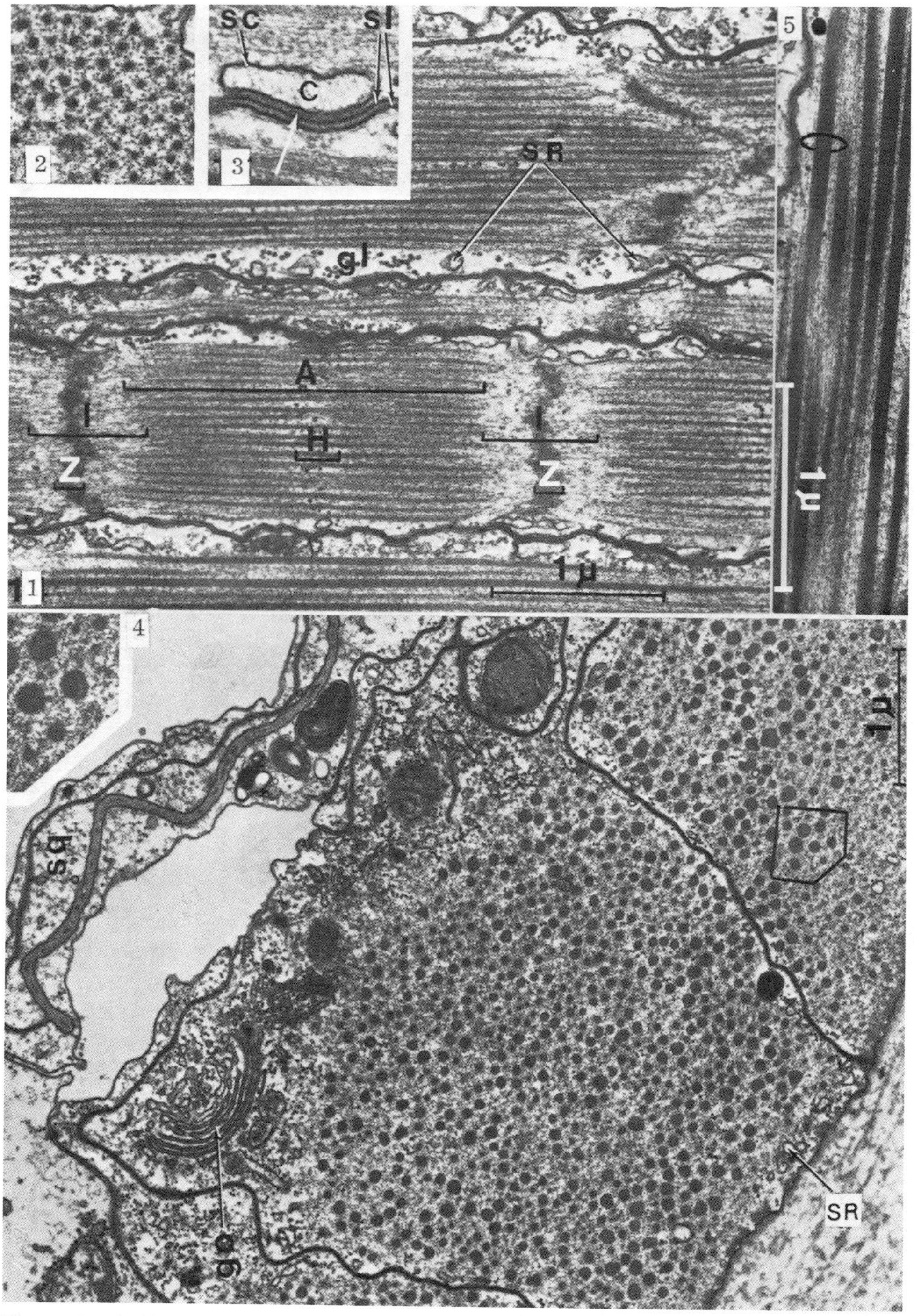

Figure 12.19. Ultrastructure of brachiopod *Terebratalia* tentacle muscle cells. (*1*) EM of LS showing side-by-side short-sarcomere cross-striated fast cells and disorderly striated slow ones. (*2*) TS of overlap region of fast cells. (*3*) Surface "dyad" cisternal element of SR of a fast cell whose Sl lies very close to that of its neighbor. (*4*) TS of slow cells, showing tapered very thick filaments SR near surface membrane. (*5*) Thick filaments showing paramyosin banding. (All from Reed and Cloney, 1977.)

12.7. Echinodermata

12.7.1. General Features

Evolutionists are convinced that the echinoderms are the invertebrate stock from which the chordates were derived. Therefore, it would seem most appropriate to deal with them as if they are at the pinnacle of the invertebrates. However, they have been very poorly researched by comparative nerve and muscle physiologists, in part because both structurally and functionally even the most advanced echinoderms are crude in comparison with some molluscs and all arthropods. Both their nerve and their muscle cells are small and difficult to deal with technically. The radial five-part symmetry and heavy skeletal framework of sea urchins, brittle stars, and starfish have led to the evolution of only slow musculature and a paucity of interesting overall behavior. But discrete muscles occur in the highly complex feeding mechanism of Aristotle's lantern, there are nice retractors of respiratory gills in holothurians (sea cucumbers), and the numerous spines and pedicellariae of all echinoderms are operated by unusual muscle cells.

Investigators of echinoderm nerve/muscle are agreed that all echinoderm muscles must be innervated by a large number of very thin axons. Pople and Ewer (1954) estimated that the holothurian *Cucumaria* pharyngeal retractor comprises as many as 700 motor units, each with about 15 muscle cells, in a muscle about 100 × 6.0 × 2.0 mm. This should be contrasted with arthropod muscles of much larger volume that receive only two excitatory axons for the entire muscle mass (see p. 489). Echinoderms have followed very different paths in the evolution of their muscles and of their neural control than have other invertebrates. They appear to have taken a similar route in many respects to that of the vertebrates, but much more knowledge about them than currently exists will be needed before this can be ascertained with confidence. Careful attention to stimulus strength supports an early observation by Du Buy (1936) that response magnitude can be varied over a wide range by altering stimulus strength alone. No clear evidence has been obtained for dual slow/fast response mechanisms, either associated with different types of muscle cell or with different types of junctional transmission. The nervous system is able to promote very slow contractions taking 2 min to reach peak in addition to ones complete in a few seconds (Pople and Ewer, 1955), but nerve-impulse pattern and delaying of impulses in neural networks could be responsible. The retractor of the lantern likewise shows both a rapid response capability and a slow long-lasting one (Boltt and Ewer, 1963) and even inhibition, but again, all these could be dependent only on central mechanisms. In spite of the fact that no fast-contracting striated cells are known in the phylum, only very small spindle-shaped smooth ones, to control tension, the echinoderms, like the chordates, appear to utilize variation in numbers and clusters from among a large number of motor neurons innervating small groups of muscle cells. We could wish for much more research on echinoderm muscles to firmly establish the principles of motor control, if only because the principles of control of muscles with a large number of component units are still obscure in chordates, which are considered to be phylogenetically related to echinoderms.

Perhaps the most conspicuous mobile elements of echinoderms, present in all, are their hydraulic/muscle-operated tube feet. The muscle cells operating them were originally thought to be innervated by "ribbon axons" (J. E. Smith, 1950). However, a closer study showed that these cells contain some myofilaments (Figure 12.20). They are in fact five muscle arms, and they extend down the tube foot as far as the radial nerve (Cobb, 1967a).

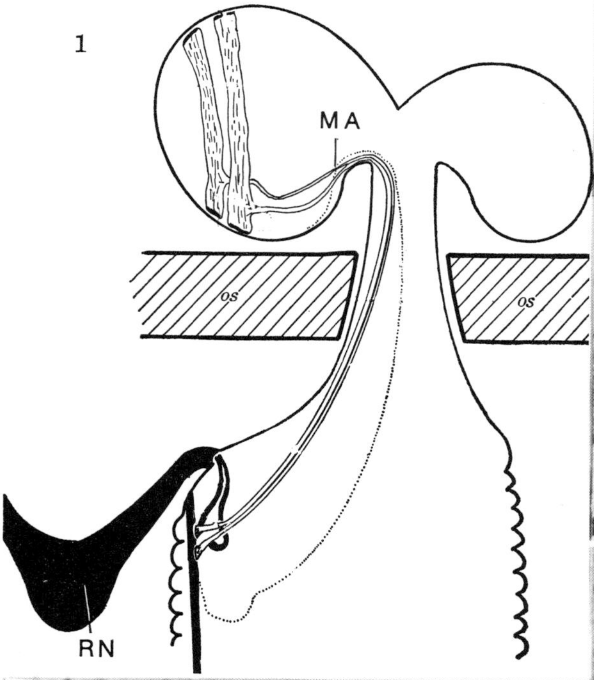
1
M A
os
os
RN

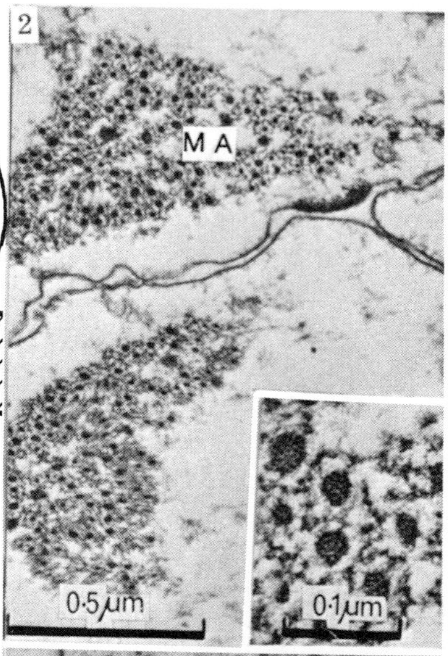
2
M A
0·5 μm
0·1 μm

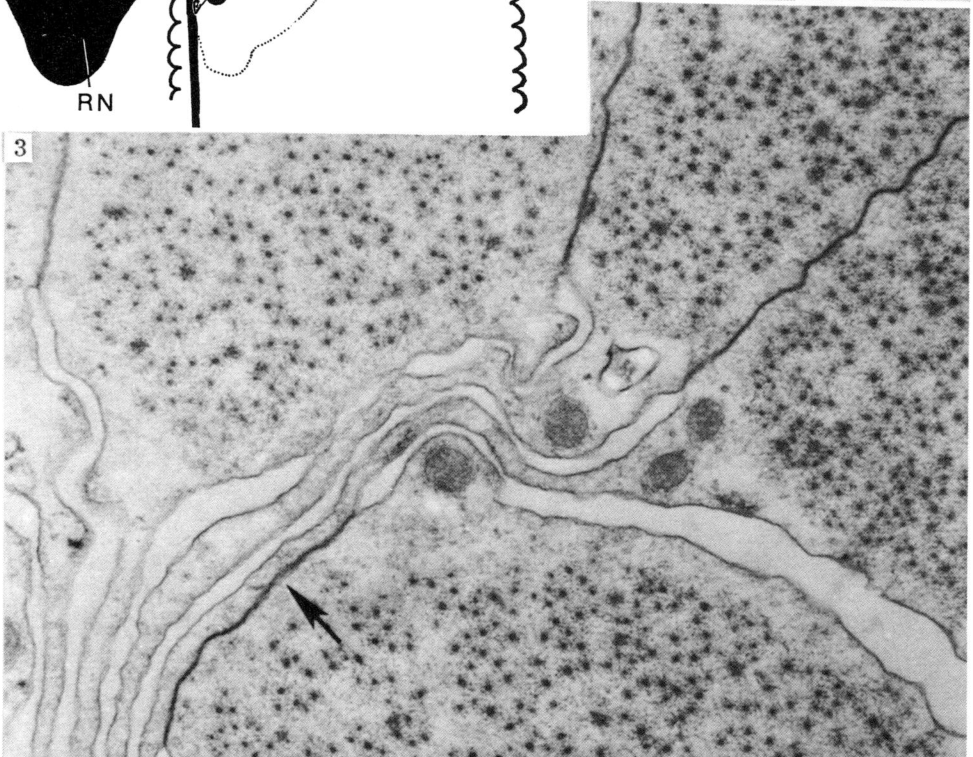
3

This very unexpected finding of muscle cells' sending out processes to nerves rather than the other way around is present in platyhelminths, making one wonder how many other places in the animal kingdom muscle arms are lurking. Is this a primitive form of nerve-to-muscle contact, or a specialized, advanced one?

The muscle cells of the ampullae are small in diameter but contain very long thick filaments, up to at least 15 μm. The numerous ultrathin filaments are attached either to a few scattered, rather small DBs or to the cell membrane.

12.7.2. Echinoidea (Sea Urchins)

Sea urchins are able to move about in any direction using long movable spines that cover the entire dorsal body surface. In some species there are giant spines that are very heavy, several inches long and several millimeters in diameter; in others they are thin and delicate. There is a socket at the base that fits over a rounded basal knob, and a ring of muscle attaches very close to the base. The ring can not only move the spine quickly in any direction, but can also hold the spine in any given position. When in the "hold" mode, a strong force applied to the spine ruptures the holding element beyond capability of recovery, but without affecting the capability of the faster movement. The holding is due to a rapid change of state of an inner ring of material that is heavily impregnated with collagen but is not muscle and cannot itself move the spine (e.g., Takahashi, 1967a, b). The outer layer comprises small, 1–3 μm in diameter, muscle cells running from the base of the spine to the test. These cells make desmosomelike attachments with their neighbors. Each contains numerous 6-nm thin filaments emanating from unusually long DBs, not unlike those that occur apparently randomly within the body of the cell, though with a few at the ends and at the periphery. Small mitochondria occur inside the cells at their periphery; there is no SR and there are no ITs.

Aristotle's lantern is the structurally complex feeding apparatus of sea urchins. It operates five large teeth that are part of a jaw apparatus that can be withdrawn as a whole by retractor muscles or protruded preparatory to a bite by protractors. The teeth can be opened or closed when in the protruded position. When the lantern is first dissected, rhythmic movements occur spontaneously, mainly of opening and closing at about 1/min, superimposed on a slower retraction/protrusion cycle.

Figure 12.20. Muscle arms and myofilaments of echinoderms. (*1*) Drawing of sea urchin *Echinus* tube foot, as it passes through ossicles (*os*) of the test, showing two retractor muscle cells, at upper left, and the long, thin muscle arms (*MA*) they send to the nearest branch from the radial nerve (*RN*). (From Cobb, 1967a.) (*2*) EM of TS through muscle arms showing small clusters of myofilaments, at high magnification in the inset, with PM-containing thick filaments. (From J. L. S Cobb, 1967a.) (*3*) EM of TS of holothurian *Isostichopus* retractor muscle cells, showing muscle arms emerging from them and tight junction between neighboring cells (*arrow*). In this, as in all echinoderm muscles examined, the tapered PM-containing thick filaments lack organization. There is a high thin-thick filament ratio, and although some orbits are made, the majority of myofilaments are scattered loosely. This species lacks SR, other fast moving ones have a little at the periphery. [Hill et al. (1978) Contraction in a muscle with negligible sarcoplasmic reticulum: the longitudinal retractor of the sea cucumber *Isostichopus badionotus* (Selenke). *J. Exp. Zool.* **206**, 143. Reprinted with permission from Alan R. Liss, Inc., New York.]

Motor-nerve stimulation with a single minimal shock produces a weak, nonfacilitating twitch, the size of which increases several-fold, with many steps, as stimulus strength is increased progressively. Thus this muscle is similar to that of other echinoderms tested, with a vertebratelike neuromuscular system. The cells are 8–12 μm in diameter and contain tapered thick filaments with 45-nm centers (Cobb and Laverack, 1966). There is a single motor terminal on each muscle cell, located at its proximal end. The terminals are simple dilations held in place by expansions over them of adjacent muscle.

Another echinoderm muscle that has been studied is the closer of the tridentate pedicellaria of the sea urchin. This provided a surprise: a mixture of cross-striated cells and ones with irregular myofilament array occurs. The SL of the former is remarkably short, rivaling that of coelenterates, at 0.85 μm (Cobb, 1967b). The striated cells, averaging 1.3 μm in diameter, occur in a discrete bundle underlying an outer bundle containing the other muscle cells, which average 2.5-μm diameters.

Thick filaments in the cross-striated cells are 15 nm in diameter, thin ones 5 nm. There is very little SR and no membrane invaginations, but these are not needed for inward spread of excitation since the cells are so thin. The exact lengths of thick filaments in the irregularly arrayed cells is unknown, but one at least 25 μm. The thick filaments are surrounded by 5-nm filaments, but neither is present in an orderly array.

Neural excitation of the cross-striated cells is extremely interesting, since this is achieved by each muscle cell's sending a thin muscle arm to the nearest nerve trunk. The arm comes off the last sarcomere just before its attachment to an ossicle. It then makes a synaptic contact with a single nerve cell. The muscle-arm-on-nerve pattern closely resembles that found in *Peripatus* (see p. 484), except that in *Peripatus* the tip of the muscle arm contacts a few motor axons, as do muscle arms in nematodes, not just one. Thus the tendency for echinoderms to adhere to vertebrate "principles" of nerve/muscle function is upheld even when making a muscle arm on nerve junction!

12.7.3. Holothuria (Sea Cucumbers)

Holothurian muscle cells are obliquely striated, with PM-containing tapered thick filaments, having 72-nm centers (Baccetti and Rosati, 1968), which allies them closely with many invertebrates, notably the annelids and molluscs, but not at all with vertebrates. The holothurian *Isostichopus,* which is very slow moving, has small cells that are organized into discrete bundles of four to six cells each, with extensive surface contacts and also contacts between projections into the centers of the clusters (Figure 12.20 – R. B. Hill et al., 1978). These muscles are notoriously able to elongate, even under very small loads, though paradoxically they strongly resist suddenly applied extending force. The whole animal has a several-fold length range naturally, being soft and pliable when extended, but tough when very short. Extensibility may be due to muscle-cell clusters' slipping past one another.

The cells contain virtually no SR, only a few small mitochondria, and a single nucleus each. Their entire volume is occupied by tapered thick filaments with large arrays of thin ones between. The thin filaments are attached to Z bodies that are in turn attached to the surface membrane and do not occur in the body of the cell. The slightly faster moving *Holothuria tubulosa* has much more SR, lying just below the cell membrane (Baccetti and Rosati, 1968).

12.8. Mollusca

12.8.1. General Features

Molluscs constitute the phylum with the greatest range of diversity in gross anatomical forms, organization of nervous systems, motility, and behavior. At one extreme are simple, extremely slowly moving chitons, limpets, and clams; at the other are the exquisitely subtle, fast-moving, intelligent octopuses. However, of all major invertebrate phyla, they have retained to the largest degree the more primitive characteristics of muscular organization and neural control of muscle. After removal of the entire brain and central nervous ganglia of a nudibranch, what is left moves about in a completely normal looking manner, although it does not seek food. The nudibranch *Tritonia,* deprived of its brain, can feed on food applied to its mouth, copulate, and lay eggs, its only obvious missing capability is that of escape swimming, which is therefore a centrally driven and programmed behavior. Even in the most advanced mullusc, the severed arm of an octopus moves, writhes, and attaches its suckers in much the same way as an attached one. All of these movements are possible because a peripheral nerve net and peripheral ganglia are present, and these control the musculature. Reflex actions initiate and control local movements and the net provides for their coordination.

A very wide range of muscle cell types occurs in mulluscs, from ones that are the slowest and most powerful in the entire animal kingdom to ones among the fastest. The muscles also range from the most economical of energy utilization to among the most wasteful. Many interesting muscles of mulluscs such as those of chitons await even initial studies, but a fairly representative sampling of other types is available. Cross-striations are present in two very different locations: the mantle of squids and a part of the adductor of a few bivalves, such as the scallop *Pecten.* Except for these two disparate sources, molluscs appear to be devoid of the simple cross-striated structural organization favored by skeletal muscles of all insects, most crustaceans and arachnids, and all vertebrates. The favored molluscan cell type is striated, but always with a markedly oblique alignment of the myofilaments as favored also by annelids. At one extreme, such cells are quite fast, and some are used for the faster adductions of lamellibranchs other than *Pecten.* The thick filaments of a majority of molluscan muscle cells are very thick because they contain a lot of paramyosin in ribbon form. The presence of paramyosin is always associated with a relatively slow rate of contraction and an even slower rate of relaxation. In its extreme form it shows the highest holding economy in the animal kingdom, leading von Uexküll (1912) to use the term *Sperrung,* catch mechanism, to describe their contraction.

12.8.2. Cross-Striated Cells

The two sets of cross-striated muscles have each been evolved to permit a high-speed capability. In the bivalve it was to produce repetitive, quick shell closures with such a rapid water ejection that the scallop is literally jet propelled through the water. Jet propulsion is also the aim in the squid mantle, although this is produced very differently, by the nearly synchronous contraction of the entire mantle musculature, which is controlled by the stellate-ganglion/giant-axon excitatory system.

Squids jet propel themselves backward through the water by ejecting water from their mantle cavity. It has entered the cavity to permit respiratory exchange. The forceful ejection is possible because there is synchronous fast contraction of the entire mantle circular musculature. This is achieved by the 10 giant axons. The unitary response is a large twitch. Smooth tetanus starts at 30 Hz, but the tension increment is no more than 5-10% (Prosser and Young, 1937). This indicates that there is little internal or attachment elasticity in series with the contractile elements and that activation is 100% following a single excitation. Many persons have attempted to obtain intracellular recordings from the small cells, but without success. Extracellular records show that the response to the giant axons comprises a large EJP, which in turn evokes a large graded spike. When repeated, the response declines rapidly in size. Even the second in a train is markedly reduced, (D. M. Wilson, 1960).

The same muscle is innervated in addition by many small motor axons that are active in respiration. When several of these are stimulated together with a strong shock, a small twitch follows. Successive stimuli evoke facilitating, summating responses that fuse to a smooth tetanus at about 20 Hz. The rate of rise of tension and plateau tension continue to increase with increasing frequency up to about 100 Hz, when the tetanus-twitch ratio is about 100 : 1. The increments in tension with increasing stimulus strength above threshold are several-fold, showing that many motor axons are present. This shows conclusively that a molluscan muscle can have a dual fast/slow innervation, but it still leaves open the question whether or not there is dual innervation of any individual muscle cells.

The muscle cells of squid mantle circular muscle are thin, 4-7 μm in diameter, with a unique fine structure. A layer only 0.6-μm deep forms a ring of contractile material surrounding a core of very large mitochondria (Figure 3.15). These occupy 60% or more of the entire lumen and are so fat there are at most three, and usually only one or two, abreast, each packed with complex cristae. The contractile material is cross-striated, with the short sarcomere length of 1.8-2.0 μm. Thick filaments are 18 nm thick × 1.2 μm long, with no Pm. Thin filaments are 6 nm in diameter, attached to laterally aligned Z bodies. The thick filaments appear hollow. Thin ones are extremely numerous, occurring in tight clusters with a square array as well as surrounding the thick ones in orbits. There is a large number of thin filaments relative to thick ones – the ratio is about 6 : 1 – and the thick filaments are not in a regular hexagonal array.

Bivalves with fast-contracting, cross-striated adductor muscles include *Pecten, Cuspidaria, Anomia,* and *Lima.* The cells of the first two are small, and complete lateral division is seen, but in the others a spiral arrangement is apparent (Figure 3.6.*14*). However, this is more likely to be the result of small, regular inequalities in sarcomere length; it is quite different from spiraling associated with obliquity (see p. 000). The sarcomere length is least in *Cuspidaria,* at < 2 μm, next in *Pecten,* 2.5 μm, and is 3.5 μm in *Anomia* and *Lima.* Somewhat-longer sarcomeres, 7 μm, have been described in the pedal retractor of *Pecten* (Bowden, 1958).

Thick filaments of the scallop *Aequipecten,* in which the SL is 3 μm, although not especially thick, have been shown to contain PM (Elfvin et al., 1976). These cells respond to graded motor-nerve stimulation with JPs of various sizes, in six or even more steps in single muscle cells (Figure 12.21 – Mellon, 1968). The larger JPs are compound, exceeding 30 mV, the smaller ones being about 5 mV. In some cells graded electrogenic responses with magnitudes up to 45 mV are initiated by summed EJPs. None of the EJPs, not even the smallest ones, show facilitation; the mechanical responses they initiate are all twitchlike.

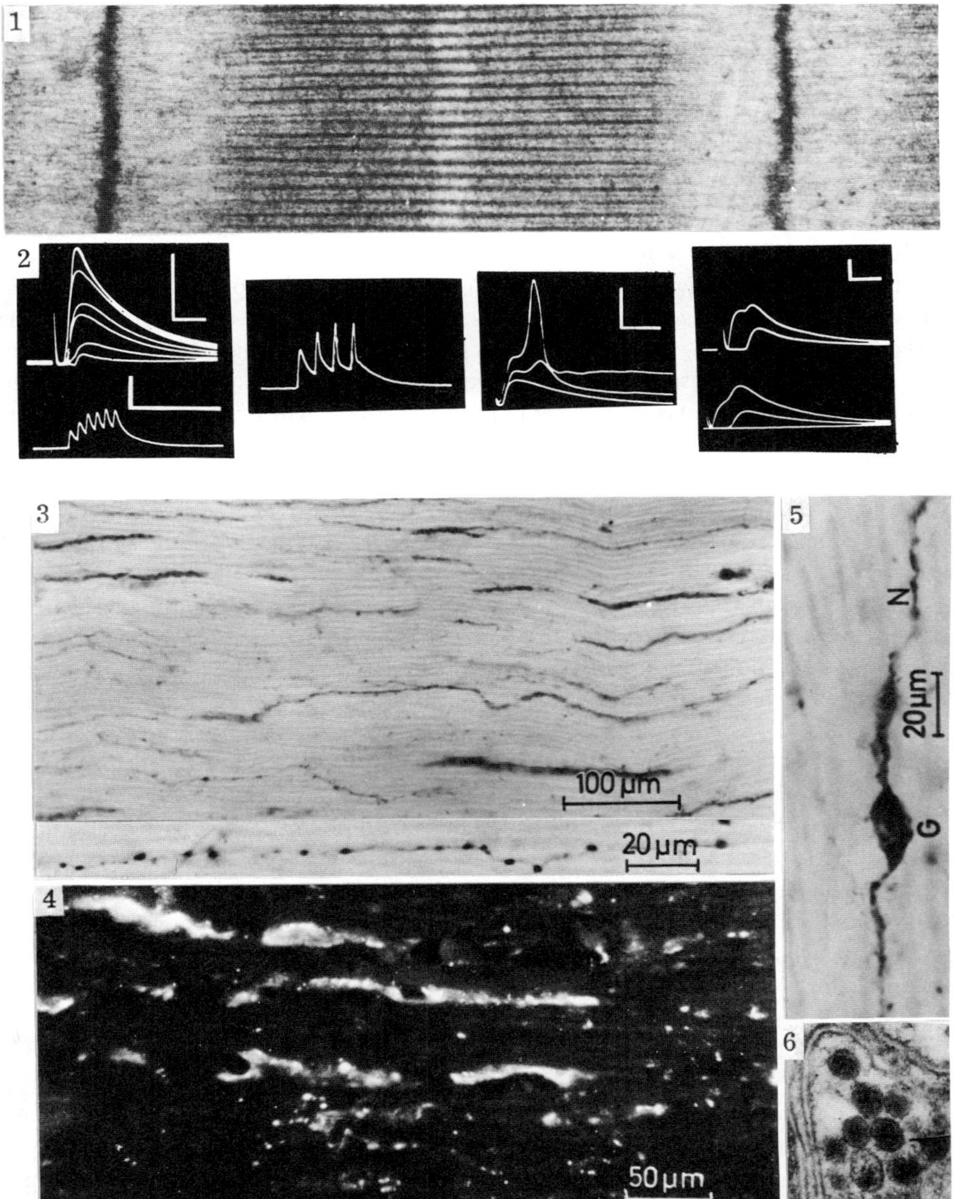

Figure 12.21. (*1*) EM of LS of scallop (*Pecten*) adductor, with 2.6-µm SL. (With permission from *J. Mol. Biol.* **103**, 439–467. (1976) Copyright Academic Press. (London) Ltd.) (*2*) Variety of sizes of EJPs intracellularly recorded from cross-striated adductor of scallop *Aequipecten* to increasing strengths of single shocks or brief trains of stimuli. In the more excitable cells, graded spikes were initiated. Calibrations in first record: 20 mV, 20 ms. (Reprinted with permission from Mellon, D., Jr. (1968) *Science* **160**, 1018–1020 Copyright by the American Association for the Advancement of Science.) (*3*) Innervation of penis retractor of *Helix* seen with methylene blue stain. (From Foh and Bogusch, 1969b.) (*4*) Catecholaminergic innervation of (*3*) revealed with Falck/Owman method. (From Foh and Bogusch, 1969b.) (*5*) Ganglion cell in (*3*). (From Foh and Bogusch, 1969b.) (*6*) A DCV-containing nerve ending in (*3*). (From Foh and Bogusch, 1969b.)

12.8.3. Obliquely Striated Cells

The scallop has a second bundle of muscle cells that contains only obliquely striated cells with very thick Pm-containing thick filaments. When this bundle is destroyed, the scallop looses its ability to remain tightly closed for very long periods and can be forced open much more readily. But both quick and slow contractions still occur. Although not yet directly studied, these results imply that there is a population of cells in the striated adductor, just as there is in most arthropod muscles, ranging from electrically inexcitable slow ones to ones giving strong graded electrogenesis, all of which are polyneuronally innervated. We may presume that the slower cells are responsible for tone and for all closures other than the vigorous ones used in swimming. They probably show at least a weak "catch," since cutting the nerve to a fast adductor immediately following a burst of excitation does not lead to relaxation. The slow adductor muscle that is heavily endowed with Pm probably contracts and relaxes very slowly and probably shows catch. Catch is probably relaxed fully except when the clam is strongly disturbed or out of water.

Behaviorally, the muscle is known only to produce sequences of all-or-none twitches used in swimming escape and prolonged adduction. More research is needed to unravel innervation and muscle-cell-response details for the fast part of the adductor. Since only simultaneous activation of many excitatory neurons to the fast adductor initiates the larger spikes, the swimming neural program must be presumed to consist of synchronous activation of three or four motor neurons. It may be that when these fatigue, others can be recruited to take their place.

Obliquely striated muscles of molluscs from many different sources have been studied to various extents both structurally and physiologically. Each has some unique aspect or other. All the cells are small, and the thick filaments all contain Pm. Of special interest for most is their ability to hold tension with little energy utilization – the phenomenon of catch – which will be dealt with independently (see p. 477). Each major molluscan nerve/muscle system will be described briefly individually.

12.8.3.1. Radula Retractor

Many gastropods are herbivorous and feed by rasping algae and other plant tissue using a muscular toothed tonguelike radula. The principal power stroke is produced by a discrete radula-retractor muscle. As isolated preparations, they are relaxed and give single phasic responses to electrical stimuli, but in the animal they operate highly rhythmically. Rhythmicity can be induced by a combination of ACh and 5HT (R. B. Hill, 1958). This is also true of some other molluscan muscles, including the *Mytilus* ABRM, which is never called upon to contract rhythmically or even phasically *in vivo*. The membrane potential shows oscillations and depolarizes in potassium-free seawater (Figure 12.22.*2*), and a transient hyperpolarization with raised $[K^+]_o$ that is erased by ouabain, which blocks the sodium pump. These anomalies are explained on the basis that K_o activates the sodium pump, whose action itself contributes to the resting potential (Kobayashi and Ando, 1979). Neural excitation evokes EJPs that give rise to graded spikes and twitches of 0.5-s duration. Both ACh and L-glutamate depolarize the membrane and cause contractures, the L-glutamate being the more potent, making it a candidate for excitatory transmitter (Muneoka and Kobayashi, 1980). Both 5HT and OCT suppress excitatory neural responses (Figure 12.22.*3, 4*.).

In support of the rhythmic operation of the retractor as the power-stroke supplier of feeding, the muscle cells have an interesting ultrastructure. Although obliquely striated, the effective SL is small, only a little over 1 μm. The whole substructure is orderly. The contractile material surrounds a wide core that is crammed with small mitochondria (Figure 12.22.*1*).

12.8.3.2. Radula Protractor

The Whelk (*Busycon*) protractor, like the retractor, is rhythmically active following isolation when treated first with ACh, then with 5HT (Hill and McDonald-Ordzie, 1979). The ACh depolarizes the cells and leads to calcium entry and contraction. Then 5HT probably promotes calcium binding within the cell, lowering tension and increasing membrane potential until eventually an unbalanced state exists, in which the potential swings between polarized and depolarized states. As with the retractor, a major active involvement of sodium pumping is implicated. However, there is a major difference, compared with the retractor, in the effects on the protractor of OA and 5HT. In complete contrast to the effect on the retractor, both greatly enhance protractor contraction to neural excitation and ACh (Kobayashi and Muneoka, 1980).

12.8.3.3. Pharynx Retractor

An old classic preparation, the pharynx of the snail, forms a nice muscle or nerve/muscle preparation, since the motor nerve can be dissected out with the muscle quite easily and set up for studies. The magnitude of the mechanical response is a function of the stimulus strength over a wide range, showing that the 40–50-mg muscle receives many motor axons. The electrical responses of individual cells are EJPs of various sizes, from 1 to 20 mV, the larger of which evoke graded spikes (Lloyd, 1980), but these are small, even to the largest summated EJPs.

The electrical response to the second of a pair of submaximal shocks is larger than that to the first, with maximum facilitation occurring at about 40 ms. The mechanical responses are small twitches, weaker ones showing marked facilitation, rising to a maximum at 50 s and decaying slowly. By contrast the large spikes produced by maximal stimulation antifacilitate, declining rapidly at a frequency of 10 Hz or above. There is clearly a very diffuse, multiple-axon innervation quite different from that found in most arthropod skeletal muscles. There is evidence, though inconclusive, for two axon sizes (Ramsay, 1940), which would indicate a duality: slow/fast mechanisms. Most of the responses can be explained on the basis of recruitment of spiking units rather than different types of motor axon or muscle cell, but on available evidence both could be present. The rate of relaxation is increased by 5HT, which enhances the electrogenic response slightly below about 10^{-6} *M*, but depresses it at higher concentration.

12.8.3.4. Mantle Muscle

The mantle of most molluscs is itself contractile and in some shows the interesting feature of giving two contractile responses, with different characteristics, to a single brief electric shock (Wilson, 1969). There is a first rapid component that starts to relax immediately, which very soon leads into a second slower contraction that summates with the relaxing faster one (Figure 12.22.*5*). The muscle cells are 2–7 μm in diameter and contain the usual very thick (up to 90-nm center) mulluscan thick filaments. Although the results suggest dual nerve/muscle mechanisms, possibly with two types of muscle cell, there is also the possibility that the delayed slow response was reflexly initiated by the first via local ganglion cells.

1

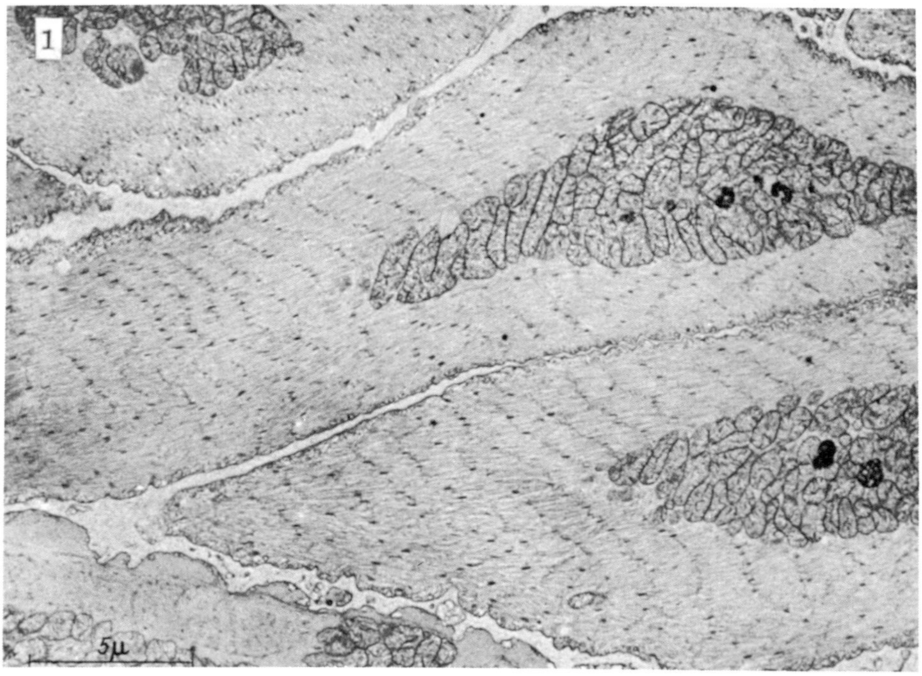

2

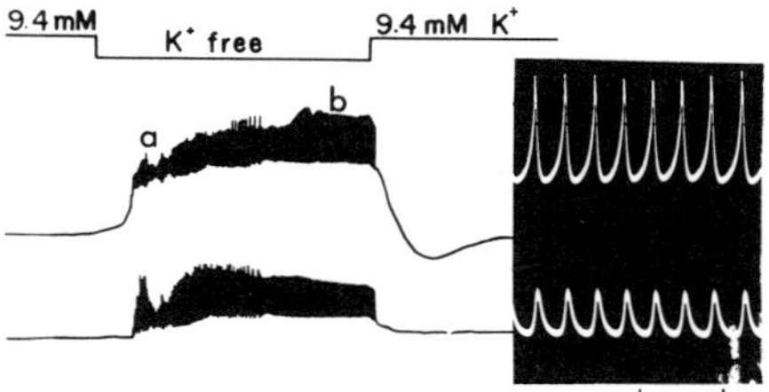

3

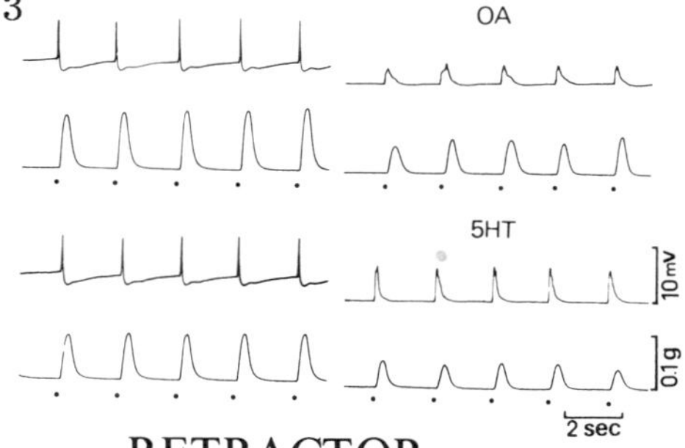

RETRACTOR

4

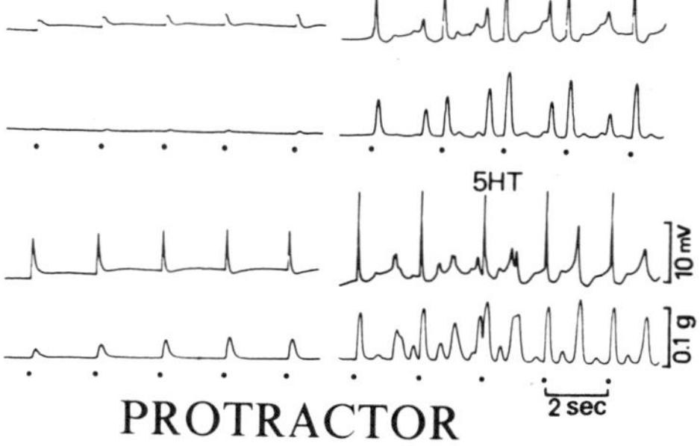

PROTRACTOR

5

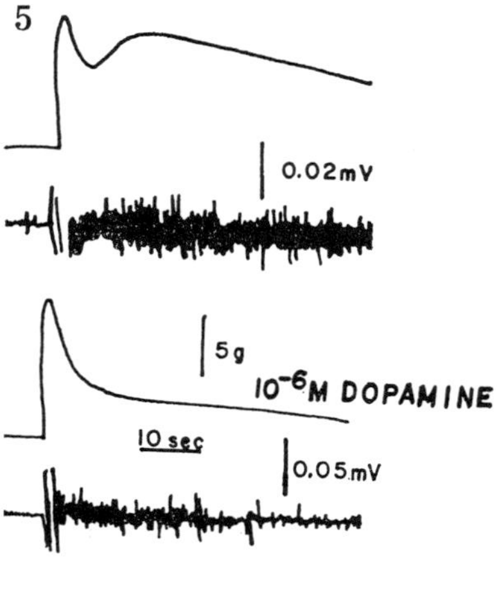

12.8.3.5. Pharynx Levator of Snail

A nerve/muscle preparation with intracellular recording from muscle cells has been made of the *Helix* pharynx levator (Peters, 1979). Each cell is weakly gradedly electrogenic. Innervation is polyneuronal, commonly by six excitatory axons, three of which are from an internal lip nerve, and three from an external counterpart. The EJPs can be divided into two classes, one of which is very small and summates extensively, but with little facilitation. The other is only a little larger initially, but it grows very much larger with repetition. Either type will initiate a large graded spike after combined summation and facilitation. A histological study showed a rich network of nerve branches making junctions on muscle cells but failed to delineate two types.

12.8.3.6. Accessory Radula Closer (ARC)

This muscle is part of the extremely complex buccal mass of opisthobranchs, which comprises more than 20 different muscles (Longley, 1976); it has been intensively studied in *Aplysia* (Cohen et al., 1978). These muscles help to hold food between the radula surfaces as the odontophore rotates back to bring it to the esophagus during a bite. The 7 μm in diameter cells have 75-mV resting potentials and are polyneuronally innervated, EJPs of various sizes occurring in the same cell. Two of the neurons have been identified: B15 and B16. The EJPs produced by B15 facilitate to a large size, about 30 mV, those of B16 either do not facilitate or do so to a much lesser extent. At least one other unidentified neuron gives large facilitating EJPs in the same muscle cell, and another gives very small 3-mV ones. Muscle electrogenic responses have not been seen, even when large EJPs were superimposed on a background depolarization.

The larger EJPs each give rise to small twitches individually when superimposed on a background of depolarization from summed smaller ones. Individual mechnical responses show marked facilitation. There is strong progressive post-tetanic potentiation when brief bursts of stimuli are repeated at intervals. Both these mechanical effects are attributable to continued growth in size of the EJPs. Transmission at all the different junctions appears to be cholinergic. There is no evidence of inhibitory junctions. In addition to the ordinary EJPs, there is a slowly rising, slowly decaying depolarizing potential in response to a large identified neuron of the cerebral ganglion, the so-called metacerebral cell (MCC), which appears to be active during feeding (Weiss et al., 1978). This neuron produces 5HT. Both the neuron and 5HT applied directly produce marked potentiation, up to threefold, of the tension evoked by a standard burst of neural excitation from one of the excitatory neurons (Figure 12.23). This effect peaks in about 1 min following a burst of activity in MCC and decays to normal in about 6 min. At 10^{-8} *M*, serotonin enhances the response to B16 eightfold. The MCC also has central actions that lead to stronger motor-neuron bursts. The 5HT released by MCC can therefore be thought of as serving an arousal func-

Figure 12.22. Ultrastructure and physiology of retractor and protractor muscles of gastropods and dual response of pelycypod mantle. (*1*) EM of LS of retractor of *Buccinum.* (From Mattisson and Arvidsson, 1966.) (*2*) Depolarizing and stimulatory actions of K^+ -free solution (9 min) on retractor of *Rapana.* Portion at higher speed at right: bar = 5 s. (From Kobayashi and Ando, 1979.) (*3*) Depressing actions of both OCT and 5HT at 10^{-6} *M* on neurally evoked responses of *Rapana* retractor. Electrical *above,* tension *below.* (From Muneoka and Kobayashi, 1980.) (*4*) Facilitatory actions of both OCT and 5HT on neurally evoked responses of *Rapana* protractor. Spontaneous rhythmic activity was also stimulated. (From Kobayashi and Muneoka, 1980.) (*5*) Dual response of pelycypod mantle. Tension *above,* extracellular electrical record *below.* At 10^{-6} *M*, DA suppresses the slow phase of the response while leaving the fast one unchanged. (From D. F. Wilson, 1969.)

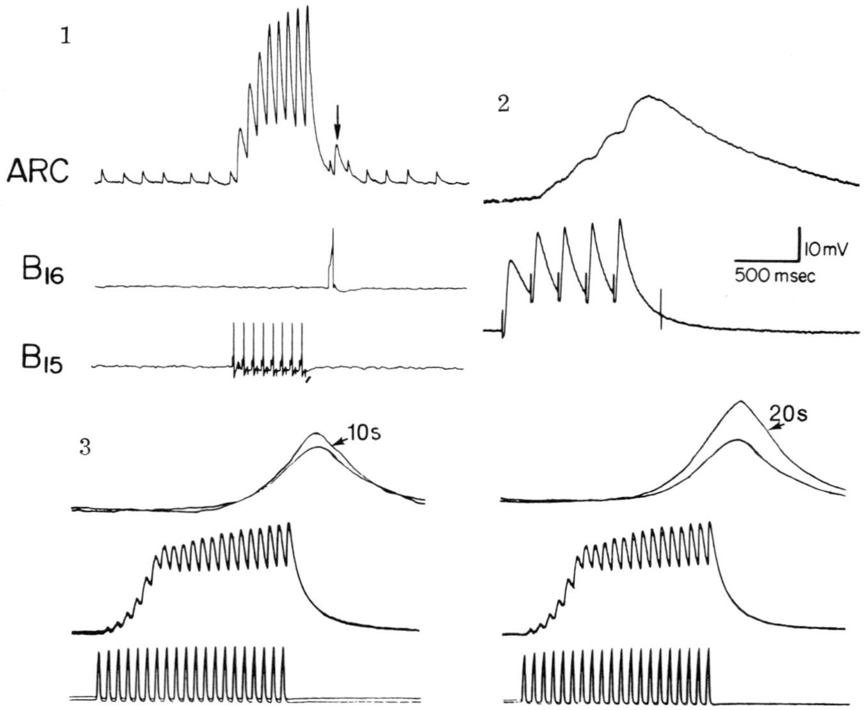

Figure 12.23. Neuromuscular transmission and tension modulation in accessory radula closer (*ARC*) muscle of opisthobranch mollusc *Aplysia*. (*1*) Multiple innervation demonstrated during intracellular recording from muscle cell. Spontaneous small EJPs occurring from unknown spontaneously active input; large facilitating, summating EJP to burst directly applied to motorneuron B_{15}; intermediate size EJP to single impulse in B_{16}. (*2*) Mechanical responses of whole muscle (*above*) to large evoked EJPs. [(*1*) and (*2*) reprinted with permission from Cohen et al., *J. Neurophysiol.* **41**, 157–180 (1978). Copyright 1978 The American Physiological Society.] (*3*) Muscle EJPs (*middle traces*) and tension (*upper trace*) to train of stimuli applied to B_{15} (*lower*) repeated twice, before and again after, stimulating MCC. (*Left*) second trace 10 s after MCC was fired; (*right*) 20 s after MCC was fired. The effect must be within the muscle cells. [Reprinted with permission from Weiss et al., *J. Neurophysiol.* **41**, 181–203 (1978). Copyright The American Physiological Society.]

tion, though why the buccal neuromuscular systems couldn't achieve sufficiently powerful mechanisms on their own, dispensing with the need for this potentiaton, is an interesting question. The mechanisms of potentiation at the muscle level is dual: there is a facilitation of EJP size, but there is also an independent enhancement of contraction, which has a slower time course.

12.8.3.7. Penis Retractor

The penis retractor of *Helix* has been studied electronmicroscopically (Schlote, 1963). It comprises discrete parallel cells not more than about 5 μm in diameter, like all molluscan muscle cells. They contain numerous 5 nm in diameter thin filaments mingled irregularly between roughly obliquely staggered arrays of tapered Pm-containing thick filaments with very thick centers. The very numerous thin filaments are attached to large (∿ 0.15 μm) DBs, most of which are within the myoplasm, but some of which are attached to the surface membrane. There is little SR, most of it lying just below the membrane and making dyadic contacts with it around the entire circumference.

Motor nerves make simple large synaptic endings on the cells that contain three types of vesicle: small clear ones, very large ones with granular cores, and moderately large ones with dense cores (Foh and Bogusch, 1969a, b). Treatment with formaldehyde vapor (Falck and Owman, 1965) reveals the presence of biogenic amine(s) in the nerves, with local flares up to 100 × 10 μm. The terminals take the form of long strings of varicosities along axons running mainly parallel to the muscle cells. They branch and anastomose with each other in the manner expected of a syncytial nerve net. The amine is likely to be 5HT. For its probable role, one should consult the more extensively studied *Mytilus* ABRM (see p. 477). These three types of vesicles are common in all obliquely striated molluscan skeletal muscle (Figure 12.24). The small, clear vesicles are mainly cholinergic, with perhaps some that are L-glutaminergic. The others are either 5HTergic or DAergic.

The penis retractor may serve as a model ultrastructurally for many other molluscan retractors such as the buccal retractor and columella retractors of snails (Heyer et al., 1973).

12.8.3.8. Buccal Retractor

The buccal retractors of the gastropod *Philine* have received attention because combined central/peripheral intracellular recordings should be possible in them. An interesting state of development of ITs was discovered, representing a little greater development than that seen in other retractors. This is related to the fast that the muscle cells are a little thicker;

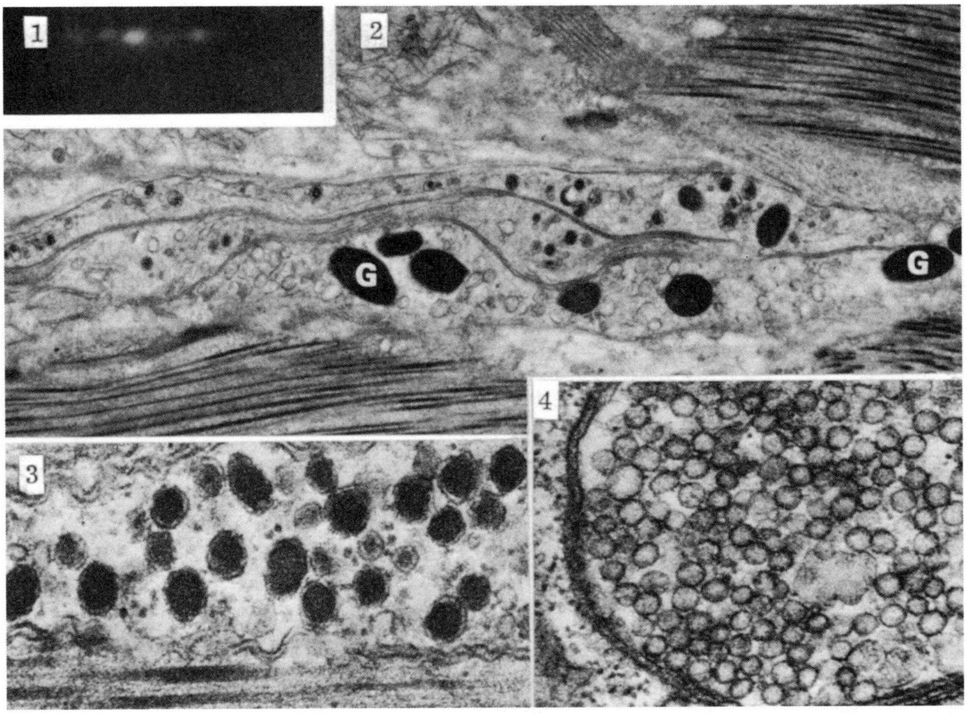

Figure 12.24. Innervation of *Mytilus* ABR. (*1*) Fluorescence after nialamide treatment shows beaded catecholaminergic terminals. (*2*) EM of LS showing beaded DCV-containing axons. The large dark bodies outside the axons are glial processes (*G*). (*3*) Characteristic junctional types of molluscan muscle, full of DCVs. (*4*) Ending containing only clear vesicles. [All reprinted with permission from McKenna and Rosenbluth *J. Ultrastruc. Res.* **42**, 434–450 (1953). Copyright 1973 Academic Press.]

although averaging only 11.4 μm, the range is up to 30 μm (Dorsett and Roberts, 1980). Penetrations by ITs are up to 9 μm deep radially. Cisternal elements of SR make close contacts with the ITs along their entire length.

12.8.3.9. Columellar Muscle

The responses of the muscle that withdraws the entire animal into its shell, the columellar muscle, of the snail *Helisoma* have been studied using intracellular recordings (Kater et al., 1971; Heyer et al., 1973). The cells are parallel, < 10 μm in diameter, embedded in connective tissue, and yield membrane potentials of less than 50 mV. Two distinct types of electrical responses were obtained in different cells (Figures 12.25.*3*, and 12.25.*4*), one consisting of all-or-none overshooting spikes, the other of graded responses. These were atypical, in that the smaller responses had a delay similar to the larger ones; they may therefore have been from damaged all-or-none-type cells or ones with an early abrupt rectification. The mechanical response to a single large spike was a small 0.2-gm twitch, rising to peak in 0.25 s and decaying to baseline in about 1 s. With repetitive stimulation the twitches summated extensively, with no facilitation.

12.8.3.10. Jugalis Muscle

This muscle operates a feeding device called an odontophore, which rocks back and forth to achieve algae scrapings as do rasps by a radula. In the pond snail *Helisoma,* it is innervated by an identified motor neuron whose cell body is in the buccal ganglion, where it has been located and recorded intracellularly. This cell evokes a large spike in each muscle cell it innervates. There is an inflexion on the rising phase that is presumably the transition between a large EJP and the all-or-none spike. So far, this is the only example from molluscs in which the precision associated with direct stimulation of an identified motor neuron combined with intracellular muscle-cell recording has been achieved, and it is especially important because it shows that the mechanical response is slow and weak and subject to much summation in spite of the spike.

12.8.3.11. Mantle Retractor

The retractor of the mantle was first studied in the clam *Mya arenaria* (Pumphrey, 1938). It is innervated by two groups of motor axons, one of which yields large extracellular action potentials and rapid contraction, the other smaller, probably junctional, potentials and slow weak contractions.

12.8.3.12. Longitudinal Body Wall

A preparation of longitudinal body wall of the opisthobranch *Dolabella* (Sugi and Suzuki, 1978) has been examined physiologically and correlated with ultrastructure. The cells are uninucleate, 10–20 × ∿ 200 μm with tapered ends. The end-to-end contacts are markedly zigzag, somewhat resembling intercalated discs of heart cells. Lateral connections between adjacent cells are made via tight junctions and desmosomes. A unique wavy junction also occurs laterally between cells. The contents are tapered very thick PM-containing thick filaments with 50-nm centers and numerous 7-nm thin ones attached to dense bodies that occur either in the body of the cell or attached to the Pm. There is no discernible regularity to the filament arrays. This is the most widespread type of muscle cell found in molluscs, but there are some interesting differences, at least in some of the *Dolabella* cells. There is a little more SR than in the retractors. Not only is it extensive just below the surface membrane, but some occurs throughout the body of the cell. Of

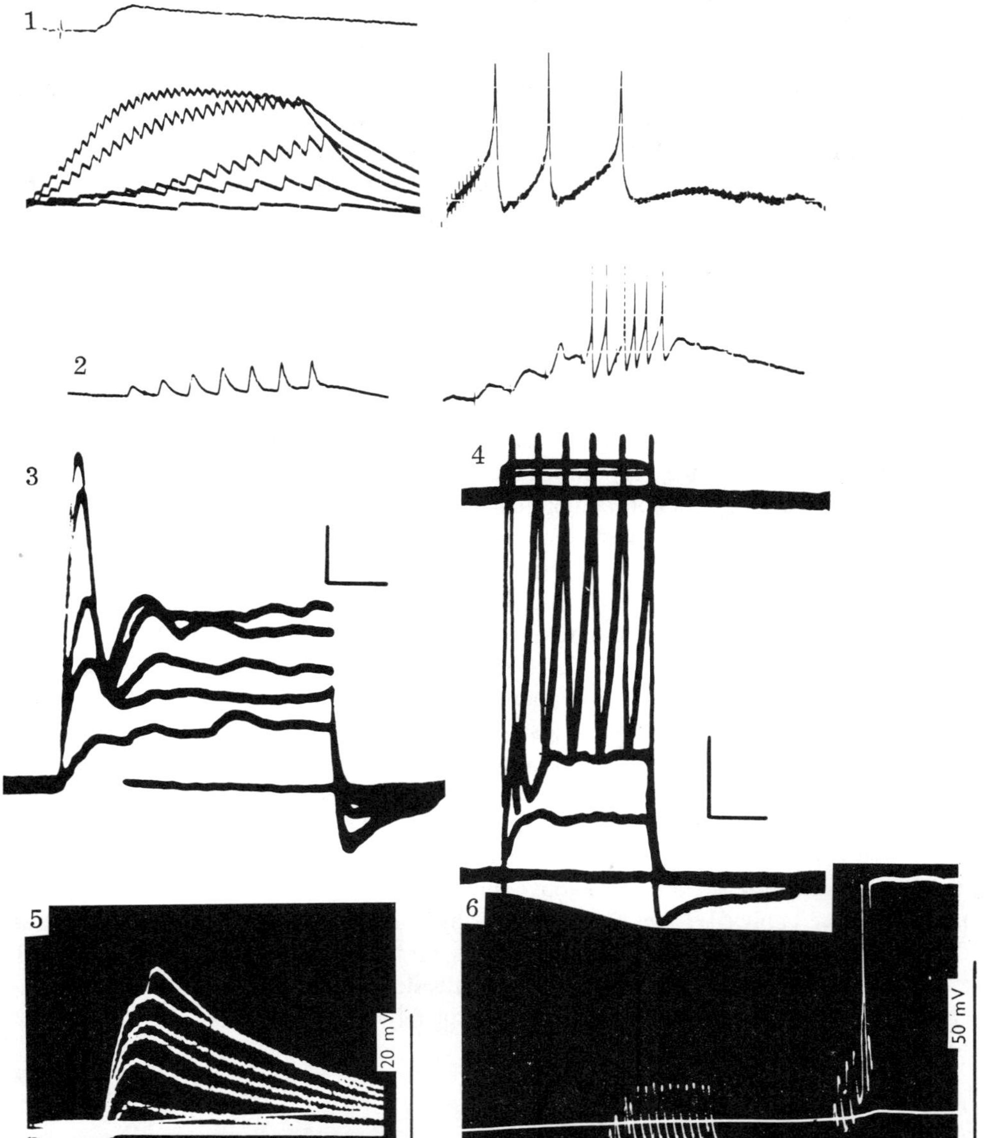

Figure 12.25. Electrical responses, intracellularly recorded, of obliquely striated molluscan muscle cells in response to neural or direct excitation. (*1*) Single EJP and trains of small type EJPs at different frequencies that show only a small extent of facilitation, from mantle muscle of pelecypod *Spisula.* At a very high frequency (*right*), spikes are evoked. (From D. F. Wilson, 1969.) (*2*) Larger, strongly facilitating EJPs that evoke graded electrogenic responses at low levels of summation and large spikes at higher ones. (From D. F. Wilson, 1969.) (*3*) Graded spikes, with heavy damping or perhaps rectification, in columellar muscle of gastropod *Helisoma.* (*4*) All-or-none spikes from another cell of (*3*). [(*3*) and (*4*) reprinted with permission from Heyer et al. (1973) *Am. Zool.* **13**, 247–270. Copyright the American Society of Zoologists.] (*5*) Different sizes of EJP in a cell of the lamellibranch *Mytilus* anterior byssal retractor. The largest evoked a small graded response. (From Twarog, 1967.) (*6*) Facilitation and summation of (*5*) during trains at different strengths. The upper trace registers tension. A large spike evoked by the stronger train elicited a twitch. (From Twarog, 1967.)

special interest are short ITs that make dyadic or triadic contacts with cisternlike expansions of SR. These are only about 2 μm long, so may be thought of as representing an initial stage in evolution of ITs by cells headed in the direction of increased diameter.

Using a calcium-precipitating method (Atsumi and Sugi, 1976) and electron-probe microanalysis (Suzuki and Sugi, 1978), it has been shown that there is a high intracellular calcium content that is concentrated in the cortical region just below the surface membrane in resting molluscan obliquely striated cells (Figure 12.26). In stimulated cells the calcium is scattered throughout. Presumably the short tubules, increased amount of SR, and larger numbers of SR/tubule contacts ensure a little faster contraction and relaxation than in the more typical retractor cells.

The neuromuscular junctions are simple and contain small clear vesicles, some of which are markedly flattened like those of arthropod inhibitory terminals, and some DCVs.

Body-wall muscle of the very slowly moving primitive polyplacophore *Cryptochiton* shows not only a very slow contraction taking more than 10 s to reach peak and 3–4 min to relax fully, but also a quicker, though still slow, one that is complete in about 50 s (Harrison, 1975). No differention of the obliquely striated muscle cells was detected, nor were any grounds found for supposing there are different neuromuscular-transmission details. The slow responses were not equivalent to those shown by some Pm-containing molluscan muscles that go into the catch state; they may be driven entirely by calcium's entering the cell directly during stimulation.

On muscle cells in the tentacle of a slug, which are of a common mulluscan kind with disorderly myofilaments, two distinct populations of nerve endings containing two different kinds of large granulated vesicles, 85 and 119 nm in diameter respectively, have been found (Barrantes, 1970). These occur in addition to clear vesicles. Unfortunately no precise association has yet been made for any molluscan efferent neurons between vesicles, as seen by electron microscopy, and transmitter/modulator substances. One of the latter is certainly 5HT (see p. 000), but which vesicle type this is associated with is disputed. Another is dopamine (DA), which is present in the pedal ganglion. The 5HT and DA very likely account for the large, filled vesicles. Terminals containing almost exclusively clear vesicles, in large numbers, are probably ones releasing ACh or L-glutamate for excitatory transmission.

12.8.4. Sheathed Muscle Cells

Many molluscs are sedentary and shell-less, raising questions about how they can produce extension of the softest parts of their bodies. For example, *Placobranchus* extends parapodia to expose them maximally to sunlight, because it feeds symbiotically on the products of blue-green algae contained in them. There is no hydraulic system to cause the extension: the muscle cells that withdraw the parapodia provide the extensor thrust as they relax. Each has a thick coating, or sheath, of about 0.3 μm, comprising numerous strands including much collagen (Kawaguti, 1968). It is this sheath that provides the necessary restoring force when the contained contractile material relaxes. The contractile material comprises the disarrayed filaments characteristic of disorderly striated muscle, with thin ones attached to DBs. There is a moderate amount of SR.

12.8.5. The Mytilus ABR

Ease of making the preparation, combined with abundance and worldwide easy availability of the animal, have rendered the anterior byssal retractor muscle of the common

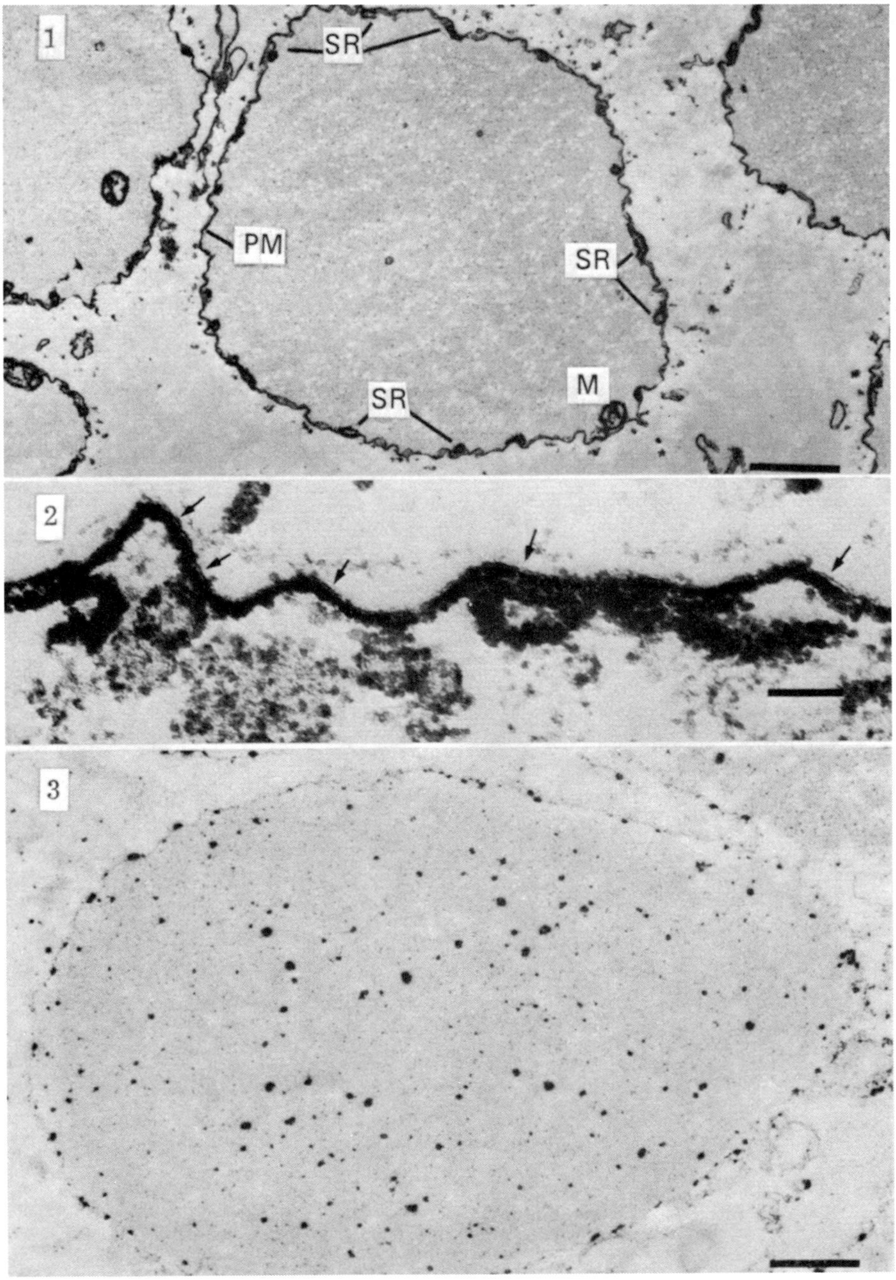

Figure 12.26. Calcium-sequestering surface membrane and surface vesicles of lamellibranch *Mytilus* longitudinal body-wall muscle cell. Electron-opaque precipitate with calcium produced by treatment with pyroantimonate. (*1*) Low-power EM of TS. (*2*) High-power EM section of membrane (PM) and surface SR (arrows) contacting it. The precipitate is localized on the inner surface. (*3*) Cell fixed at the peak of contraction. The calcium is evenly distributed throughout the cell. (All from Atsumi and Sugi, 1976.)

mussel *Mytilus,* of which there are many species, by far the most extensively studied molluscan muscle. Before its complex ultrastructure had been unraveled by electron microscopy, it was thought to be an appropriate "model" for vertebrate SM. The distinguished vertebrate pharmacologist F. R. Winton introduced the preparation in 1925, but the paper that "immortalized" it was published more than a decade later (Figure 12.27 – Winton, 1937). He found that simply to excite the isolated muscle with direct current (DC) was to evoke a very slowly relaxing contraction. The muscle goes into a "semicatch" state, from which full relaxation takes an hour or more. By contrast, alternating current (AC), or a burst of brief DC pulses, though causing similar contraction onsets, is followed by quite fast relaxation that is complete in about 1 min.

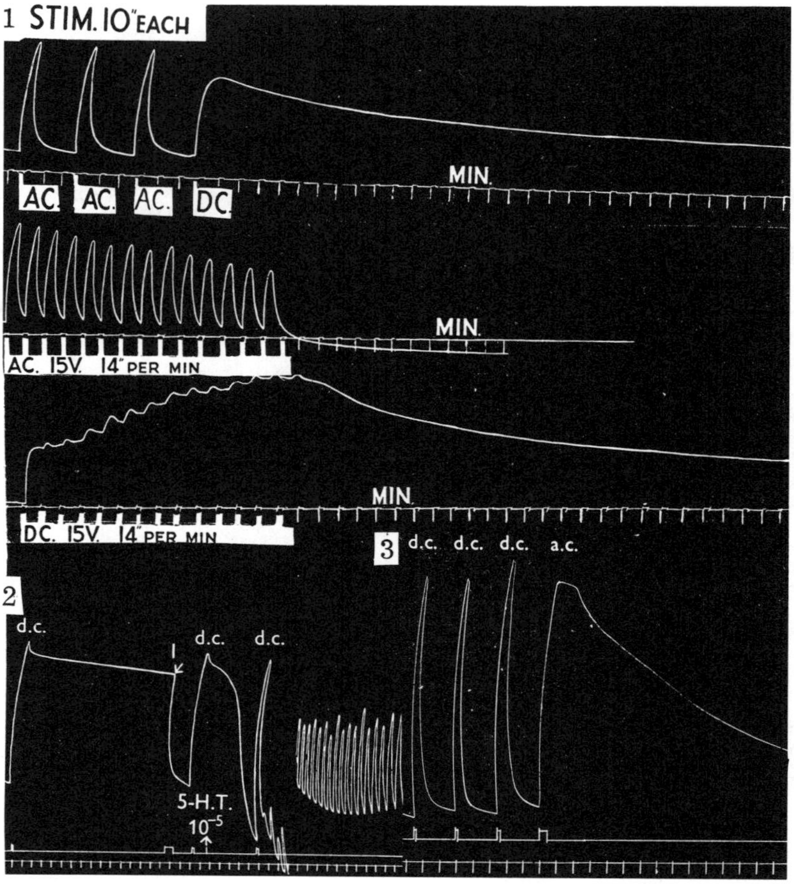

Figure 12.27. Experiments on *Mytilus* ABR tension responses to AC, DC, and effects of 5HT. (*1*) Classical results of Winton (1937). (*2*) Relaxation of strong catch response to DC by precise excitation of relaxing nerve and 10^{-5} *M* 5HT. After treatment with 5HT, development of tension is unaffected, but catch response has been abolished and relaxation is rapid. Spontaneous rhythmic contractions follow. (From Hoyle and Lowy, 1956.) (*3*) Anomalous responses sometimes occur, quick relaxations following DC and slow ones AC. These show that AC and DC stimuli do not have a direct action on the contractile mechanism as Winton supposed. The setting of the catch occurs when excitation is not also accompanied by coincidental stimulation of the relaxing nerves that run in parallel with the excitors. This happens frequently with DC stimulation, rarely with AC. (From Hoyle and Lowy, 1956.)

The interposition of either brief AC or a burst of pulses during the prolonged contraction that follows DC causes only a small additional contraction but is followed immediately by the more rapid relaxation characteristic of either stimulus mode presented in place of DC. Thus these types of stimuli appear to unlock the catch. For many years the effects were attributed to intra-muscle-cell actions, but many anomalies kept surfacing. Sometimes exactly opposite results were obtained, with AC leading to catch and DC relaxing it (Figure 12.27.*3* – Hoyle and Lowy, 1956). The probable explanation was that the stimuli were acting indirectly, via neural elements. The pathway for a plausible explanation was paved by the discovery that the catch state is quickly relaxed by a very low concentration of 5HT and cannot be recovered until this is completely washed away (Twarog, 1954). The 5HT is actually contained in nerves innervating the muscle from the pedal ganglion. Also contained in the ganglion are neurons containing DA, which also relaxes catch, in a concentration of $10^{-7}M$. The two agents may act alternately or synergistically.

Excitatory transmission is cholinergic; a powerful contraction is caused by $10^{-7}M$ ACh, and the indirect response is potentiated by eserine and blocked by ACh antagonists. Acetylcholine invariably leads to a catch-type contraction. Thus contractions of more phasic type must be due to the excitation of another neural element. This does not act like an inhibitor, opposing the excitatory action. But it does drastically promote relaxation from catch, which otherwise is very slow, changing it from an hour or more to a few seconds. Hence the 5HT-, or serotonin-mediated action, is properly described as a *relaxing action* and the mediating nerves as *relaxing nerves.*

The component cells of ABRs are homogeneous, uninucleate, about 5 μm in diameter X 1.2-1.8 nm long (Twarog et al., 1973). They are organized into 100-200 μm in diameter bundles, but with some exchanges of cells between them. They contain oblique arrays of thick filaments containing Pm, with central regions up to 80 nm thick and tapered ends (Figure 12.28). The exact length of the filaments is not known and is difficult to judge because of the lack of orderliness, but is probably about 10 μm for the thick ones. Between thick ones filaments, 5-nm thin filaments occur in profusion. A small proportion make orbits of about 12 thin ones around a thick one in contracted ABR. The relatively small amount of myosin is entirely located as a kind of coat around the Pm (Sobieszek, 1973). Numerous examples have shown apparent bridges between this coat and thin filaments. However, the number of bridges, although sufficient to be compatible with a role in causing shortening, is quite small. There are far too few to explain the maximum tension developed by the ABR, which is larger than an equivalent amount of frog skeletal muscle of similar length. The latter would have at least 100 times more X bridges in their "made" positions at peak contraction than could the ABR. This presents a major problem no matter what view of the *Mytilus* contraction mechanism one considers. The small number of bridges is compatible with the slow shortening rate, but another mechanism is needed to explain force maintenance. Several investigators have reported clumping of thick filaments in ABR fixed in the catch condition (Figure 12.29.*2*). This has been incorporated in one theory of the catch mechanism, illustrated in Figure 12.29.*3*. In ABR fixed in a relaxed state, very few, if any, of the thin filaments are in orbits.

Numerous tight junctions occur laterally between adjacent cells. The passive electrical properties examined with intracellular electrodes suggest that these represent low-resistance pathways. Accordingly, cell clusters are likely to function as wholes as they do in vertebrate vegetative muscle (see p. 340).

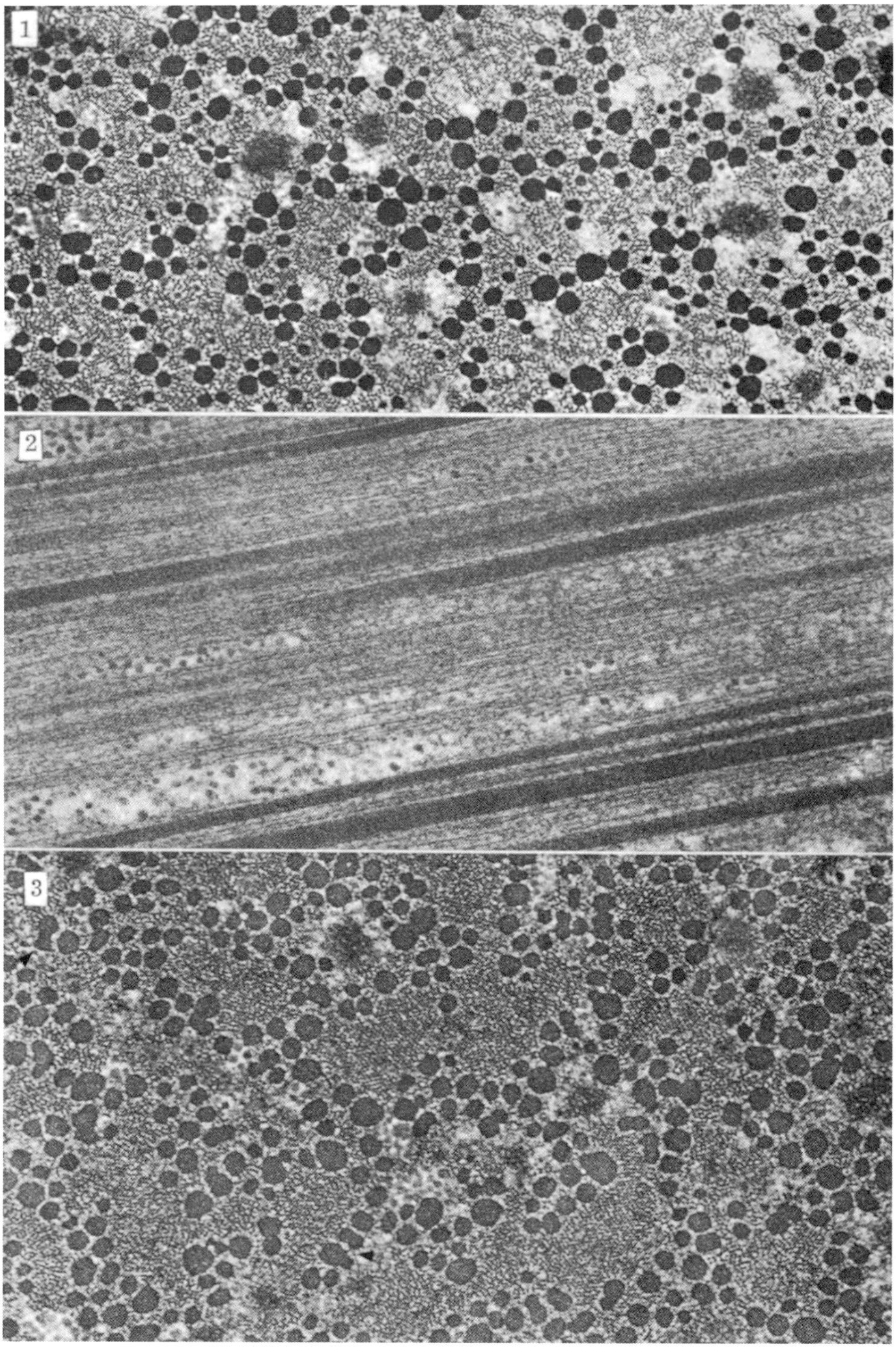

Figure 12.28. Ultrastructure of ABRM muscle cell of *Mytilus edulis.* (*1*) EM of TS of fresh muscle fixed in glutaraldehyde after incubation for 30 min in dimethylsulfoxide (DMSO) to facilitate penetration. (*2*) EM of LS as (*1*). Many thin filaments are not associated with thick ones. (*3*) EM of TS after glycerol extraction 24 h without DMSO. The thin filaments are clustered together in semicrystalline arrays. Some thick-filament clumping. [All reprinted with permission from Sobieszek. *J. Ultrastruct. Res.* **43**, 313–343 (1973). Copyright 1973 Academic Press]

The cells are electrically excitable and most give large nonovershooting graded spikes that are insensitive to TTX and due to calcium activation (Twarog and Muneoka, 1972). The resting potential average is 65 mV, but a depolarization of 35-40 mV is required to elicit the spike. Spike height is proportional to voltage, so it is a graded event (Twarog, 1967). Neural stimulation produces EJPs of at least five different sizes, up to a maximum of about 20 mV, as stimulus strength is raised progressively (Figure 12.25). These show marked facilitation as well as summation, eventually producing graded responses. The mechanical response to a single supramaximal shock is a small slow twitch, that to the second of a close pair a larger slow twitch. A summated, facilitated burst of EJPs at 2.4 Hz failed to give a contraction, or gave only a very weak one, whereas a briefer train at 2.8 Hz, the minimum frequency at which a spike occurred, caused a fairly strong one. The slow-twitch contraction was observed by Twarog only when a large graded spike occurred, so she concluded that the E-C coupling threshold is high and commonly exceeded only by the spikes.

12.8.5.1. The "Catch" Mechanism

The simplest possible explanation offerred for the "catch" phenomenon was that there is really no such thing, but that it represents a form of low-frequency tetanus of a naturally very slow muscle (Hoyle and Lowy, 1956). Continuous spiking occurs in the intact animal during catch, and also in isolated preparations, supporting this view. But ACh-stimulated muscle bundles always go into catch contraction and remain in catch state, although no electrical activity can be detected in them. Even if the neuromuscular junctions are so widely distributed that when depolarized by ACh the mean membrane potential is below E_C, it is difficult to explain why tension remains as high as it does unless there is a capability for a molecular catch state in the contractile machinery in the absence of continued input.

Therefore, it is likely that there truly is a catch state and that it is developed by a normal, though exceptionally slow, contractile mechanism presumably involving myosin and actin. This is activated, as in other muscles, by a rise in calcium concentration. As in other SR-containing muscles, caffeine directly activates calcium release (Twarog and Muneoka, 1972) and causes contraction. However, instead of immediately being sequestered by SR, as it is in other muscles, in the ABRM the calcium just stays around. It can be shown that it is not released continually, because procaine does not relax catch. The relaxing action brought about. the relaxing nerves then, is presumably due to a stimulation of the sluggish calcium-uptake mechanism by 5HT and/or DA, for which there is evidence of release also (Satchell and Twarog, 1979). The 5HT is itself a calcium chelator, which complicates the issue, but insufficient is released for it to be directly responsible for removing the calcium. Both 5HT and DA are now heavily implicated as stimulators of the rapid synthesis of cAMP (e.g., Baguet, 1977). This might in turn activate the dormant calcium-pumping mechanism of the SR. Dopamine also causes hyperpolarization, so would counteract any residual depolarization: it probably affects both gK and the electrogenic pump. Nerve staining has indicated two different kinds of nerve innervating the ABR (Gilloteaux, 1972), and this may be equated with motor axons vs. aminergic relaxors.

There is no evidence for fast as well as slow excitors supplying the ABR. Nothing is yet known or seriously proposed regarding the molecular mechanism of catch or of the role of paramyosin in it. The shortening and some tension development could be due to normal actin/myosin X-bridge interaction, since myosin appears to be present and arrayed

1

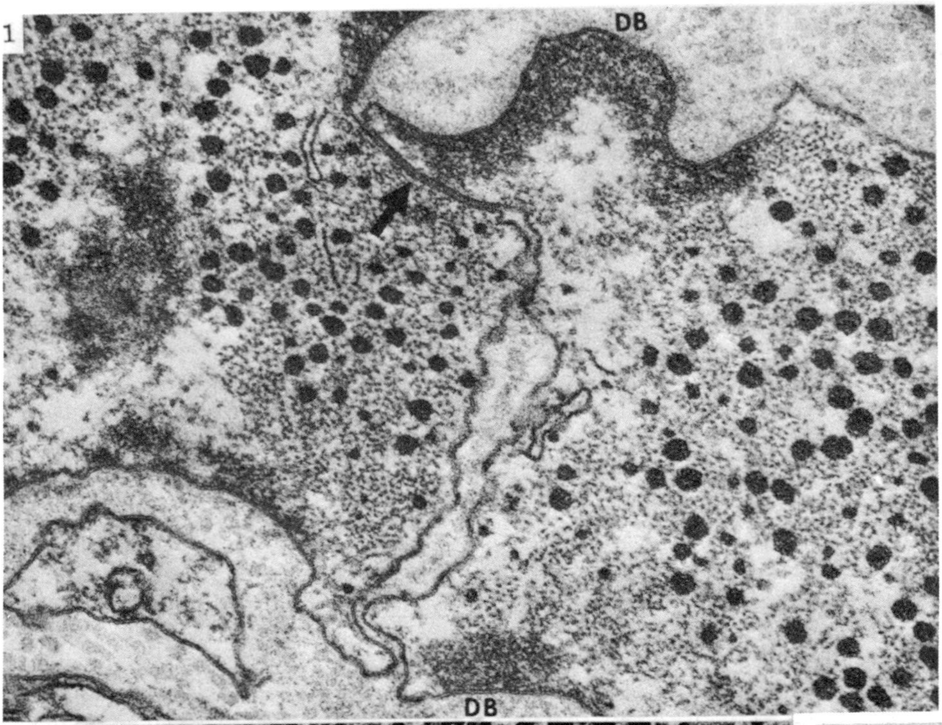

2

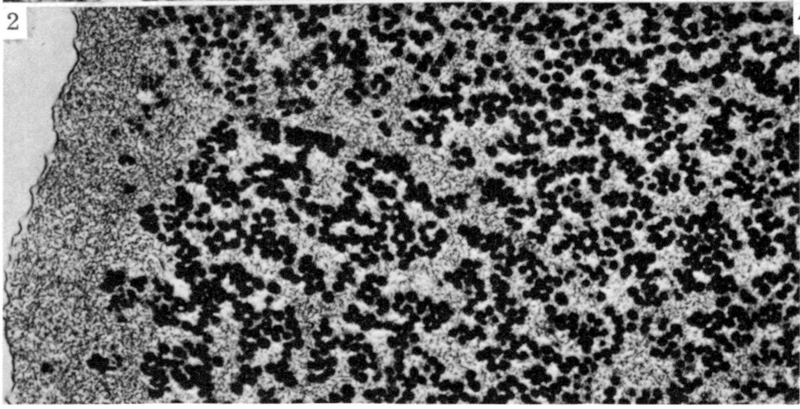

4

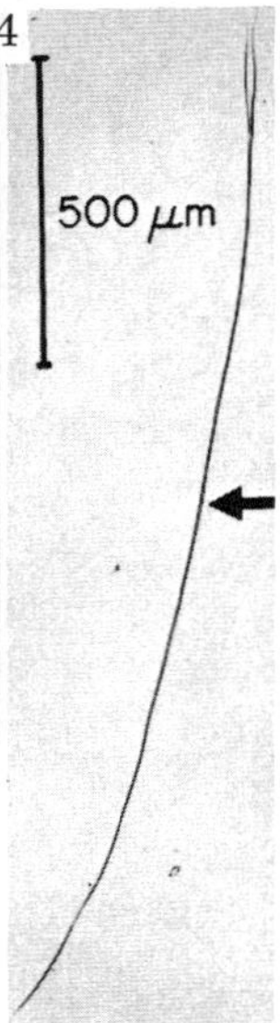

3

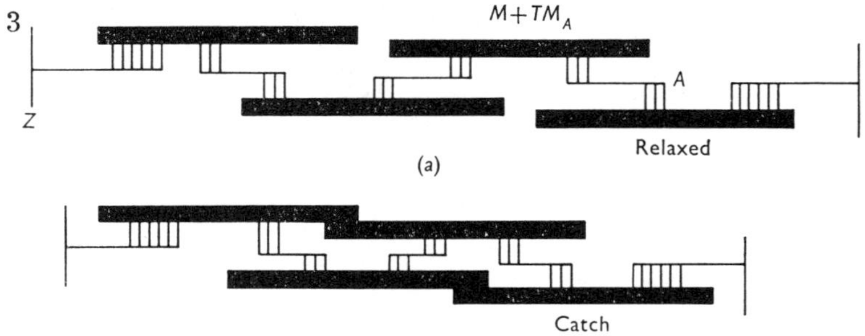

around the periphery of the Pm, but there must be an additional tension-generating mechanism. Various hypotheses have been put forward that invoke crystallization of Pm, or stabilizing crosslink formation between Pm elements of neighboring thick filaments (e.g., Rüegg, 1968b – Figure 12.29.*3*), supported by convincing TEM evidence (e.g., Gilloteaux and Baguet, 1977; Johnson et al., 1979). Paramyosin occurs in thick filaments in a wide variety of muscle cells, all of which cells are characterized by being relatively slowly contracting and relaxing, but only a few of them show the catch phenomenon, and fewer still also are relaxed from catch by 5HT. All hypotheses should take into account the large clusters of thin filaments occurring between distant thick filaments in both relaxed and contracted ABR.

12.8.5.2. Dynamic Properties of ABR

Although there have been numerous studies aimed at understanding something of the kinetic properties of *Mytilus* ABR, only one has been made using modern methods since the ascendancy of the X-bridge hypothesis. This was done on the effects of small changes of load and of length (Sugi and Tsuchiya, 1979). The experiments were somewhat confounded, in relation to their objective of testing X-bridge theory, by the existence of a long compliance of about 5% average length. This is apparently evenly distributed along the length of the cell: it cannot be written off as some effects of elements at the ends of the cells and is therefore like that of barnacle giant cells. It is at least 10 times too long to be attributable to X bridges. The investigators had no idea what might constitute the long effective SEC, but concluded, wistfully, that it "may be essential for the understanding of the contraction mechanism." Their experimental results are not compatible with the X-bridge hypothesis, but they were reluctant to say so after starting their account with, "It is generally accepted that contraction . . . is caused by the alternate formation and breaking of cross-bridges . . . ."

12.8.6. Adductors of Bivalves

Bivalve adductors are possessed of legendary strength, as anyone who has tried to force open an oyster without cutting them will attest. Values as high as 30 kg/cm^2 have often been reported for them: they are certainly the most powerful muscles in the animal kingdom. They can contract very quickly, especially in those bivalves in which many of the muscle cells are cross-striated, but in a majority, the cells are obliquely striated. The muscle is divided by connective tissue into separate parts in some species, one of which parts is for fast contractions, the other for slow ones. In other species two types of muscle cell, presumed to be fast and slow, are mixed together in a single bundle but appear to be separately activated (reviewed by Hoyle, 1964b).

The adductor of the freshwater clam *Anodonta* shows both slow and fast closing. It may remain tonically closed, or it may open and close rhythmically, with either a slow,

Figure 12.29. Ultrastructure of muscle cell of *Mytilus* ABR, a "catch" muscle. (*1*) EM of TS of relaxed cell showing irregularly scattered profiles of Pm-containing very thick filaments, surface DBs to which thin filaments are attached, and numerous thin filaments not obviously related to the thick ones. (From Twarog et al. 1973.) (*2*) As (*1*), contracted. No thick filaments near surface, many are clumped together. (From Rüegg, 1968.) (*3*) Hypothesis to explain (2) while retaining X-bridge hypothesis. (From Rüegg, 1968.) (*4*) Single isolated muscle cell: the fork at one end is genuine and common. (From Twarog et al., 1973.)

low-frequency rhythm or a faster one (Barnes, 1955). The two rhythms are independent and occur in the slow and fast portions of the muscle respectively. The slow one is used in feeding and the faster one in respiration. There is a relaxing nerve to the adductor, as first shown long ago by Pavlov (1885).

The ultrastructure of *Anodonta* adductor cells is essentially the same as that of *Mytilus* ABR (Figure 12.30), and understanding it is accompanied by the same problems. There are large clusters of thin filaments that are not associated with the Pm-containing very thick ones, all running parallel to the long axis. The thick filaments have a preponderant orientation in the same direction, but several run obliquely and some even at right angles to the thin ones.

12.8.7. Molluscan Vegetative Musculature

The muscles of molluscan guts and vegetative organs are worth considering here because their skeletal muscles are so vegetative in many respects. They have been very little studied, but the structure of muscle cells of the *Sepia* (cuttlefish) hindgut has been examined by electron microscopy (Kawaguti, 1964). Perhaps not surprisingly they possess similar basic features to those found in more discrete "skeletal" muscles of molluscs, with loose arrays of probably Pm-containing thick filaments and large numbers of thin ones attached to DBs. The cells are very thin, and there is very little SR, but they contain large and numerous mitochondria. Kawaguti was impressed by their close resemblance to some molluscan heart cells, because these are very different indeed functionally. One is in constant tonus and shows very slow contractile waves: the other gives rhythmic fast phasic contractions only.

12.8.8. Molluscan Heart Muscle Cells

For similar reasons to those expressed in 12.8.7, it is worth paying close attention to molluscan hearts. There is fairly extensive literature on the exclusively myogenic molluscan hearts, both anatomical and physiological/pharmacological. In recent years this literature has expanded since it has been found that both heart and blood vessels are under modulatory neural control from identified, intracellularly addressable cells (Mayeri et al., 1974). All are two-chambered, a thin-walled atrium leading into a 0.5-mm thick ventricle.

The muscle cells are small, 4–10 μm in diameter, connected at their ends by thick wavy intercalated discs, to which the thin filaments are attached, and laterally by typical tight junctions. There is a moderate amount of fenestrated SR, but no ITs. The SR at the surface of the cell makes close juxtapositions with it that presumably serve as coupling sites. Contacts are aided by small, short, 0.5 μm invaginations of the surface membrane

Figure 12.30. Ultrastructure of relaxed and tonically (catch) contracted cells of adductor of lamellibranch *Anodonta.* (*1*) EM of LS of relaxed cell shows Pm-containing very thick filaments arrayed in many directions. A majority of the thin filaments are parallel to each other and to the long axis of the cell even where they cross a thick filament obliquely. (*2*) EM of TS of relaxed cell showing DBs within myoplasm, and large areas (F_2) containing only thin filaments. There are only a very few orbits (*arrows*). (*3*) Same as (*2*) but contracted. There are still large areas containing only thin filaments and no more orbits are present than in (*2*). A few close clusters of thin filaments are present (*large arrow*) and some clusters of thick filaments. [All reprinted with permission from Zs.-Nagy et al., *J. Ultrastruct. Res.* 37, 1–16 (1971). Copyright 1971 Academic Press]

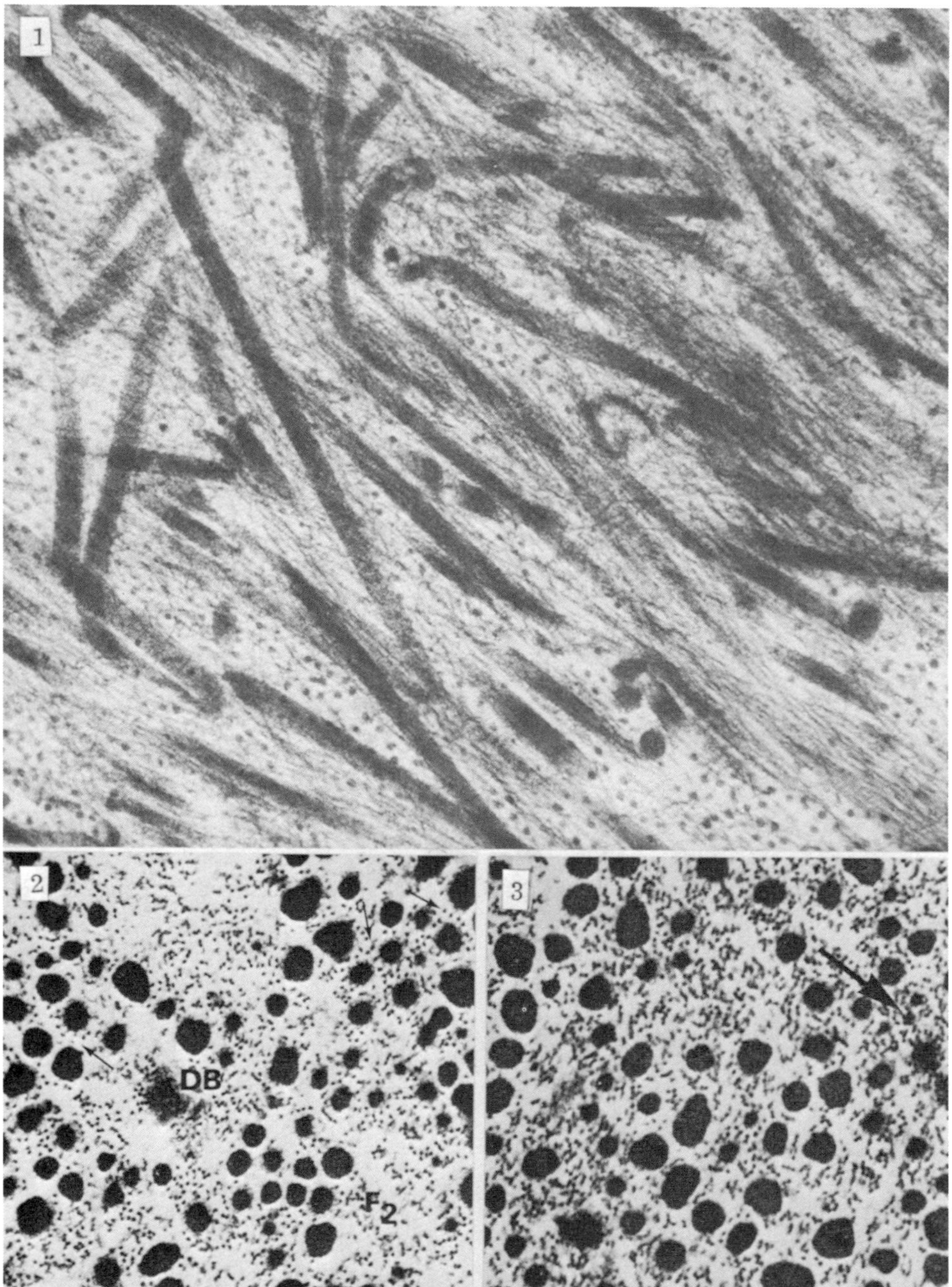
1
2
3
DB
F_2
DB
F_2

that might be thought of as a primitive form of IT (Sanger, 1979). The contractile material comprises a small number of very thick, markedly tapered thick filaments, with centers up to 80 nm thick. These are not regularly arrayed as seen in either transverse or longitudinal sections. The space between them is filled with 7 nm in diameter thin filaments. These heart muscles contain Pm in their thick filaments; since its presence is otherwise thought to be characteristic of slow and catch-type cells, its occurrence in phasic heart cells of molluscs is surprising. It shows that such an organization of itself does not bestow catch-type properties.

Nerve endings occur directly on the heart cells of *Busycon* and are packed with round clear vesicles. Not far away are other cells containing both clear vesicles and many DCVs. Molluscan hearts are very sensitive to ACh, which inhibits the beat. The *Aplysia* heart receives innervation from two cholinergic inhibitory cells, each of which is itself inhibited by the large interneuron L10 (Koester et al., 1974). The *Aplysia* heart is also innervated by two excitatory cells, R17 (also termed RB_{HE}) and its left equivalent LD_{HE}. These may be 5HTergic. Molluscan hearts are depolarized by 5HT, which accelerates the rate in very low concentration.

The heart action potential very closely resembles that of vertebrates in shape and time course. This shows the independent evolutionary capabilities of muscle cell membranes. When function is served by pacemaking or by a postspike plateau, these seem always to emerge, even though structural features and innervation patterns remain conservative.

12.8.9. General Comments on Molluscan Muscles

Except for the few cross-striated adductor cells, molluscan muscles or cells are characterized by remarkable extensibility. They cannot be said to have a characteristic "rest length" because even a vey small load produces an extension, which may be up to a few times initial length, that is not followed by shortening when the load is removed. The original length can only be restored by stimulation. If, after the muscle has been extended fully under a small load, an additional one is added, further extension occurs, but this time there is at least a considerable restoration, but only of the additional extension. Perhaps it is the length under moderate load that should be regarded as the resting length of these muscle cells or muscles. The condition in which they are generally found is then to be regarded as a weakly contracted one. However, it is indeed likely that these cells may not be comparable with those of vertebrate skeletal cells. They are, instead, somewhat similar in this regard to some involuntary unstriated cells or muscles of vertebrates.

This capability of working over a wide range of lengths, together with the lack of a characteristic rest length, must reflect a folded or very loosely arrayed structural/elastic framework for the cells and the connective tissue in which they are embedded. In addition though, the contractile material itself must be extensible. Even though this comprises fairly long thin and thick filaments, relative sliding of these alone cannot account for the extensibility. Clusters of contractile elements must also be able to slide relative to each and yet to become crosslinked following excitation of the cell. This strongly suggests a role for Pm, which is present in cells with these properties, as a lateral link maker, as illustrated in Figure 12.29.*3*.

It is almost always impossible to be sure of the extent to which the mechanical responses of a molluscan muscle are strictly due to the applied stimulation. This is because there are indirect and reflex neural involvements. All large molluscan nerve trunks contain nerve cell bodies, and nerve cells have also been found in all of the molluscan muscles that have been examined carefully. Molluscs can therefore be thought of as having developed only to a small extent in the directions exemplified by both arthropods and vertebrates, in which all the skeletal muscles are controlled by central-nervous elements. Instead they retain characteristics associated with the most primitive metazoa, the coelenterates. Their innervation reflects the looseness of intrinsic organization of the muscles and the modifiability of mechanical responses. No molluscan muscle can yet be said to be reasonably well understood in terms of neural control. Even the simplest ones, the cross-striated adductors of scallops, have multiply innervated muscle cells, and the roles of individual motor neurons are totally unknown. As a whole, each molluscan skeletal muscle resembles a piece of mammalian gut or sea anemone body wall more closely than a vertebrate or arthropod skeletal muscle.

I was forcibly struck by the very close similarity there is between molluscan and annelidan muscle cell untrastructures, and also their innervation. These are so similar that I managed to get illustrations of the junctions (Figures 12.13.*3* and 12.24.*3*) mixed up during initial sorting. These pictures cannot be similarly switched with those of any other phylum, though.

There is still much to be learned about molluscan muscle, but most notably the role(s) played by paramyosin in such diverse muscles as the heart and the ABR.

12.9. Onychophora

The onychophorans are an ancient, geographically widespread group of very similar wormlike animals currently confined to a few genera. They now live entirely in or near moist subtropical forests, but marine fossil forms are also known. Much has been written about them, including two books claiming that they are "living fossils" or "missing links" bridging the "gap" between arthropods and annelids. With insects, and no other animals, they share in common possession of an air-filled tracheal system. With arthropods in general they share a perivisceral hemocoele. With annelids they share possession of repeating, paired excretory tubules. They have several stumpy paired legs, but neither the tracheal system nor the musculature shows any sign of segmentation. There are paired sets of nerves to each pair of legs, but otherwise the nervous system is not segmented either. Their muscle structure and innervation are both unique, indicating the Onychophora are totally unrelated either to annelids or to arthropods. They may represent a unique, independently evolved phylum, or their starting point could have been an advanced platyhelminth, since both the nervous-system organization and the fine structure of muscle cells and neuromuscular junctions have features in common with corresponding ones in this phylum.

Peripatus muscle cells are either directly or indirectly attached to the body wall by collagen. Eight discrete anatomical muscles operate each leg and the claw of each foot, and four operate each of the two jaws. The fastest of these muscles are several tens of times faster in both contraction and relaxation times than the slower circular body-wall muscles (Hoyle and del Castillo, 1979). But the ultrastructures of the two kinds are similar except in amounts of SR (Hoyle and Williams, 1980). Velocity differences appear to be determined in part by differences in neuromuscular transmission and in part by different amounts of SR.

All *Peripatus* muscles contain cells within a wide size range, commonly from 2 to 10 μm in diameter, but larger ones, up to 20 μm, are found in dorsal longitudinal muscles and some even larger ones up to 40 μm occur in the inner circular muscle of the body wall. The longest of these circular cells are about 1 cm long, and dorsal longitudinal cells are even longer, but most are branched or even anastomosing. Most longitudinal muscle cells send off fine branches that pass directly to the skin to become progressively more and more collagenous until they merge with collagen of the subepidermal skin. All of the cells branch and anastomose.

The muscle cells have two independent, unique sets of wide ITs, longitudinal and transverse (Figures 12.31 and 12.32), that cross each other, but nowhere meet or fuse. Their mean diameter is 0.13 μm, 10 times that of vertebrate ITs. However, although these wide tubules must serve to conduct excitation inward by passive electronic spread, there are, in addition, narrow conventional ITs with a lumen less than 0.02 μm. The total numbers of these is small, fewer than in typical arthropod slow-muscle cells, though more than in very slow snake muscle cells (see p. 49). The SR is also intermediate in amount. Dyadic contacts with small bridging feet occur between SR and both wide tubules and narrow ones.

The wide tubules are lined, and near the periphery contain fine collagen strands. Mitochondria are small, scattered throughout the cell, with concentrations just below the Sl. The myofilaments comprise 7-nm diameter thin ones and 23-nm diameter in the center, very long thick ones, in 4 : 1 overall ratio, but only a few orbits of thin around thick ones occur. Both thin and thick filaments are very long, at least a few micrometers, but no semblance of cross-striation occurs in their organization. This is because the thin filaments are attached in clusters to small elongated ovoid dense bodies that have no regular array. Some are attached to wide tubules, others to the surface membrane, and yet others occur scattered within the myoplasm.

The innervation and neuromuscular junctions are also unique in some respects. There are two anatomically different forms, one of which is conventional, with motor axons making synaptic contact with the muscle cell surface. The other partly resembles the muscle arms seen in Platyhelminthes and partly those present in Nematoda. The muscle cells of *Peripatus* send long thin arms not to the nerve cord, but to a major motor-nerve branch. There the arms from a few muscle cells mingle close together before entering the nerve sheath and making synaptic contacts with the motor axons contained within it (Figure 12.31). The nerve-on-muscle-cell synapses are both multiterminal and polyneuronal. From three to seven motor axons situated side by side share a single junctional region (Figure 12.32).

Each muscle arm that extends toward a motor-nerve branch forms near its junction with the nerve a club-shaped expansion or head-section complex. The region just antecedent to the head is packed with mitochondria, and the head itself is filled with complex infoldings resembling ones present also in the subsynaptic region of junctions on muscle cells, but no myofilaments.

Intracellular recording from muscle cells during spontaneous, reflexly evoked and electrical stimulation of the nerve cord or motor nerves shows that most of the JPs are local and small, 1-6 mV. This confirms the presence of endings of several different motor axons on a cell. Many of the cells are electrically excitable and capable of generating large action potentials. From time to time in reflex excitation, spikes appear that have no junctional prepotential. These are followed by small twitches. These spikes may originate at the muscle-on-nerve junctions and propagate as all-or-none waves into the recording region from them. Such dual-excitation processes explain the need for two different

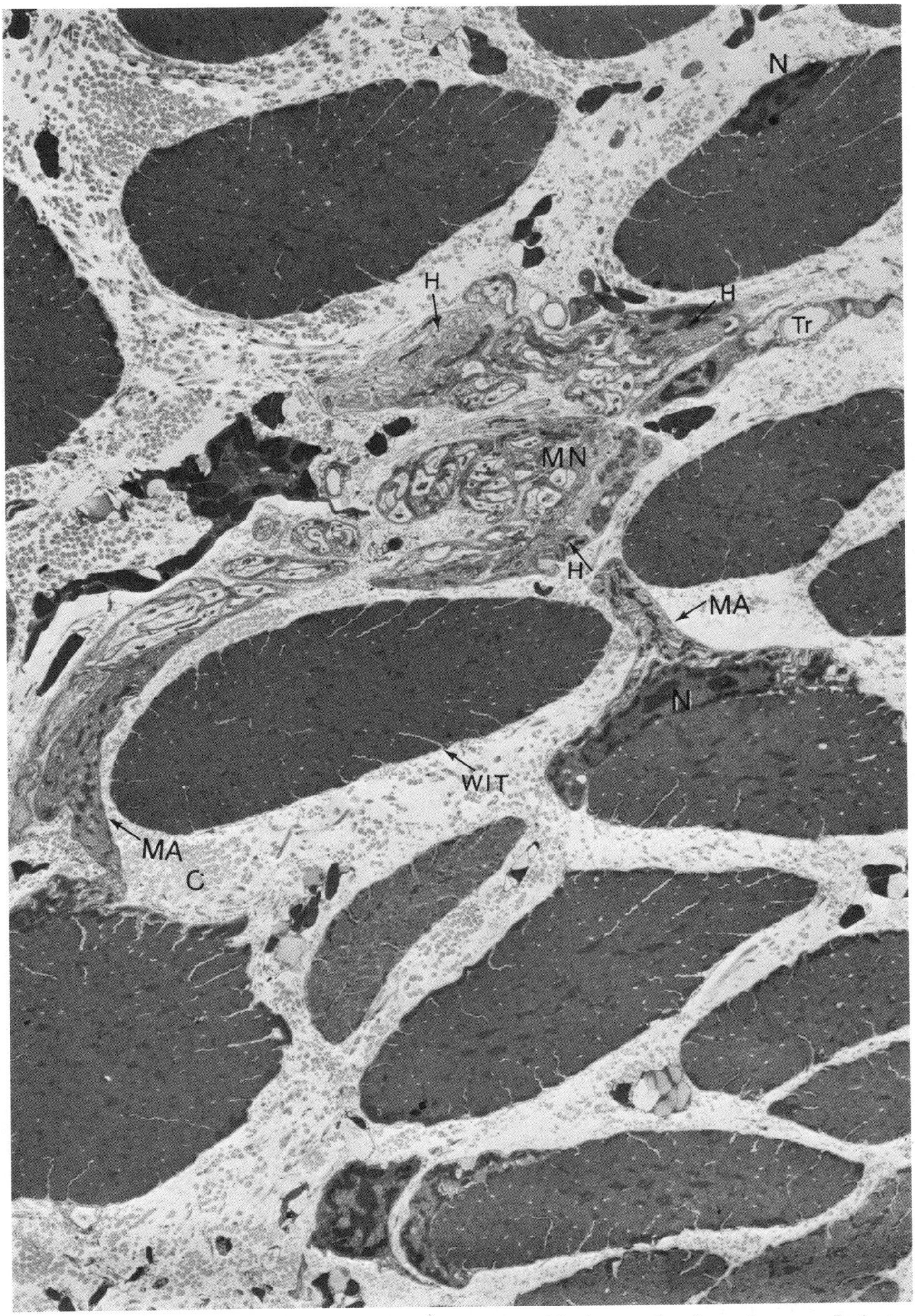

Figure 12.31. Details of muscle cells of dorsal longitudinal body muscle of onychoporan *Peripatus,* showing remarkable muscle arms (*MA*) that extend toward the motor-nerve branch (*MN*). Each arm terminates in a large, complex head section (*H*), which makes synaptic contact with as many as nine motor axons. Also to be seen are tracheae (*Tr*), wide invaginated tubules (*WIT*) lined with basement material plus very fine collagen strands, thick clusters of collagen strands (*C*) between the muscle cells, and large peripheral nuclei (*N*). (From Hoyle and Williams, 1980.)

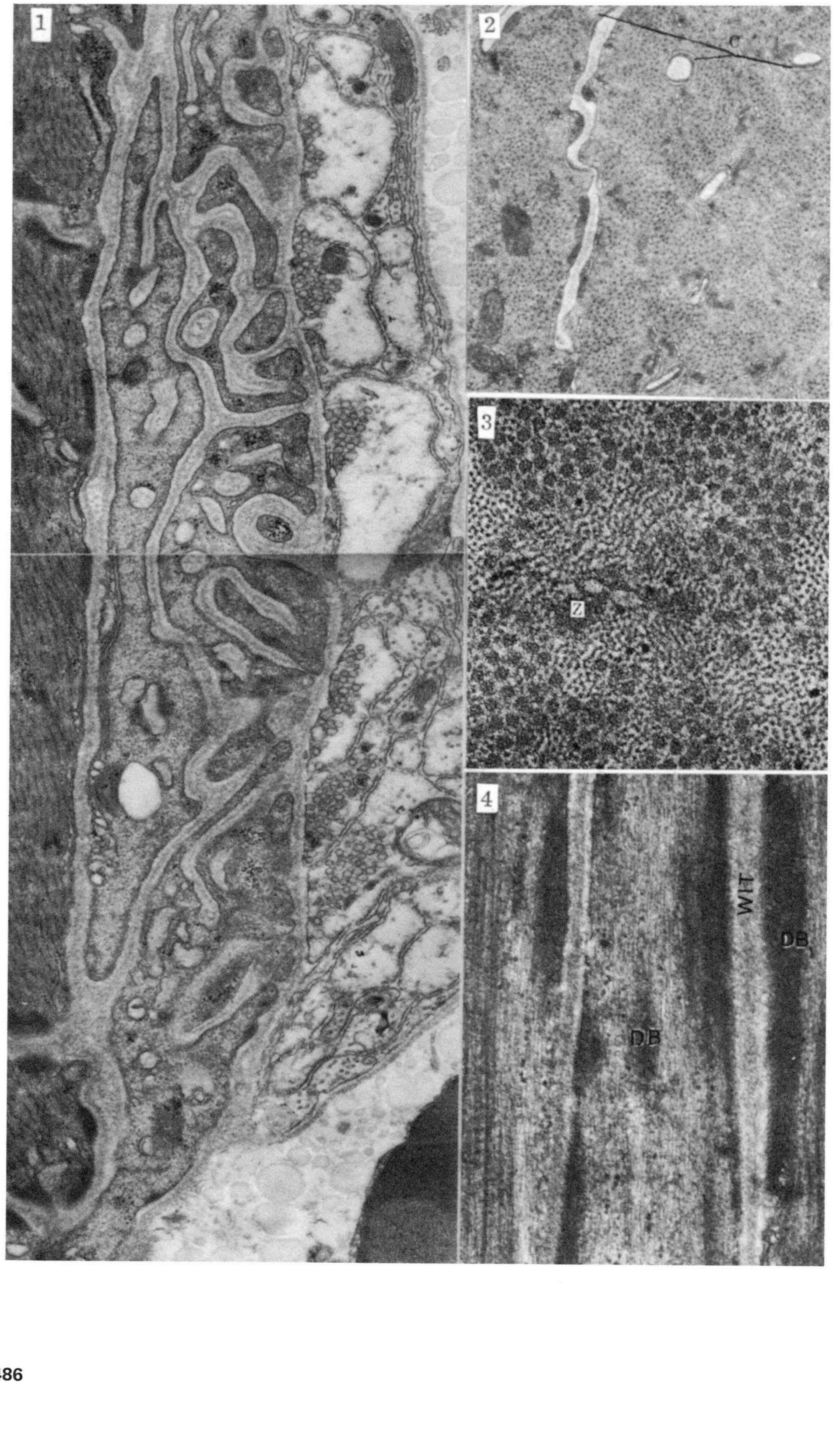
1
2
C
3
Z
4
WIT
DB
DB

locales of neuromuscular excitation. Graded, local, frequency-, summation-, and motor-axon-number-sensitive local events cause tonus and slow contractions. All-or-none impulses originating at the muscle-arm-on-nerve synapses cause fast contractions. These mechanisms represent unique solutions to the problem of tension control and are not known in any other phylum.

No other muscle cells are known in the entire animal kingdom that closely match those of *Peripatus*. In no way can they be regarded as simple or primitive: they are subtler and more complex than any vertebrate skeletal cells. Presumably they have evolved their special structural features in association with their own particular life-style. They permit operation over a threefold body-length range, total flexibility, and a very wide range of speeds.

12.10. Arthropoda

12.10.1. General Features of Neuromuscular Mechanisms of Arthropods

Arthropods, alone among invertebrate phyla, have adaptively radiated to the fullest extent possible, conquering all of the earth's major environments. By using jointed legs, they have mastered terrestrial locomotion, and the most advanced class, insects, have added flight, in association with which they have evolved some interesting unique forms of muscle structure and biophysical properties. Nevertheless they show much less diversity in both bodily and nerve muscle forms than do molluscs, probably because they have an external skeleton. Early in their evolution, they produced a simple method of controlling force development and length change using only cross-striated muscle cells, even in their vegetative organs. This utilizes multiterminal innervation of each single muscle cell by axons ranging from a single excitor to several excitors plus inhibitors to produce local depolarizations of sufficient size range to cover the entire E-C-coupling voltage range in infinitesimal steps. This method is present, with relatively minor variations, in all arthropod musculature. Unlike many other invertebrate phyla, the arthropods either did not invent, or invented and later abandoned, the use of thick strands of paramyosin. They were able to maintain tonic contractions indefinitely, with relatively little energy utilization, by combining the use of this local excitation process with various amounts of BT and other intra-muscle-cell regulation.

Historically, crustaceans were the first to be investigated, but an understanding of the neuromuscular mechanisms was achieved a little earlier in insects. Considerable diversity in detail exists in crustaceans, and to a somewhat lesser extent in insects, in these mechanisms. The fundamental aspect of control is innervation of single muscle cells at many points over the entire surface of the cell no matter how long, termed *multiterminal* innervation (Figure 12.33). The same feature is present for each axon innervating the cell. When there are two or more innervations, it is termed *polyneuronal*. When there are two excitors, one produces larger EJPs than the other and therefore faster contractions. The

Figure 12.32. Further details of *Peripatus* nerve muscle ultrastructure. (*1*) Nerve on muscle-cell synaptic complex. At least six motor axons make synapses side by side on the large, complex end-plate region. (*2*) EM of TS showing relatively independent clusters of thin and thick filaments, transverse WITs, longitudinal WITs cut transversely with C elements of SR (*C*) around them, mitochondria, dense Z bodies, and SR. (*3*) High-power TEM of TS to show independence of thick and thin filament clusters, Z body. (*4*) EM of LS showing longitudinal WITs, dense Z bodies with thin filaments emerging, thin- and thick-filament arrays. (All from Hoyle and Williams, 1980.)

smaller EJPs are highly sensitive to frequency, commonly facilitating extensively as well as summating. Contractions evoked by the axon producing them are always relatively slow, especially at the lowest effective frequencies. Accordingly this axon is termed the *slow* axon. Its partner, producing larger EJPs and faster contraction at the same frequency of excitation, is called the *fast* axon. This is the case even when both responses are actually quite slow. In the simplest instances, each muscle cell of an entire muscle is similarly innervated, and the whole receives but a single, two, or a few axons (Figure 12.33.*1*). There are exceptions to this common, simple scheme, however, in which there is partial, or various degrees up to complete, separation of component muscle cells into independently innervated clusters. The classification scheme takes no account of the latter possibility unless the full details of innervation are known. For three excitatory axons, the extremes in regard to speed of contraction at a given frequency are termed slow and fast, and the other *intermediate.* When there are more than three, they are grouped into one or another of the three classes, which are meant to be relative only, not absolute, time-related events. Since in an increasing number of instances, the locations of the motor-neuron somata are known from direct dye injection or backfilling, an aspect of somata location, anterior, posterior, lateral, and so forth, is used as a secondary descriptive term. Eventually, dynamic synaptic characteristics are likely to be the principal designations.

A second distinctive feature present in many arthropod neuromuscular systems is *peripheral inhibition.* Most commonly there is only a single inhibitory axon. This accompanies the excitor or excitors very closely indeed, branching when they do and making synapses on the muscle cells that are side by side with excitatory ones. Inhibitory axons are effective in reducing the tension produced by slow excitors, but only when discharging repetitively, at frequencies in about the same range as that used by slow excitors. They are less effective against intermediate and fast excitors, and against the fastest excitors, they are totally ineffective. Some inhibitory axons innervate but a single muscle: these are termed *specific inhibitors.* Others innervate more than one muscle and are termed *common inhibitors.* Facilitation is rare in inhibitory junctional potentials (IJPs), and weak when it does occur. The effectiveness of inhibitory action is judged by the ratio inhibitory frequency-excitatory frequency that just completely suppresses tension, or reduces it to the least possible, given optimal timing if relevant. The numbers of each kind of axon innervating an arthropod muscle are subject to extensive variation, as also are the details for different regions, or even individual cells, within a muscle. Some examples are shown in Figure 12.34, and others will be presented in sections on major classes.

For some muscles innervated by an inhibitory axon (I), the mean frequency of I at fixed frequency in the excitatory axon (E), which it opposes, alone determines inhibitory effectiveness regardless of timing. This is termed *simple inhibition.* But for others, especially when the frequencies are the same or closely similar, there is a strong additional effect of timing of arrival of I with respect to E. This extra effect of timing is termed *supplemented inhibition.* It has been conclusively shown that simple inhibition is due to a conductance increase of the muscle cell membrane to chloride, and sometimes also potassium, that is, a *postsynaptic inhibition.* When supplemented inhibition occurs, it is because I releases inhibitory transmitter substance not only at its synapses with the muscle cell membrane, but also in the vicinity of E, probably at I-on-E synapses. There the amount of E transmitter substance released is reduced by *presynaptic inhibition,* which supplements the postsynaptic action. But evidence for natural utilization of this phenom-

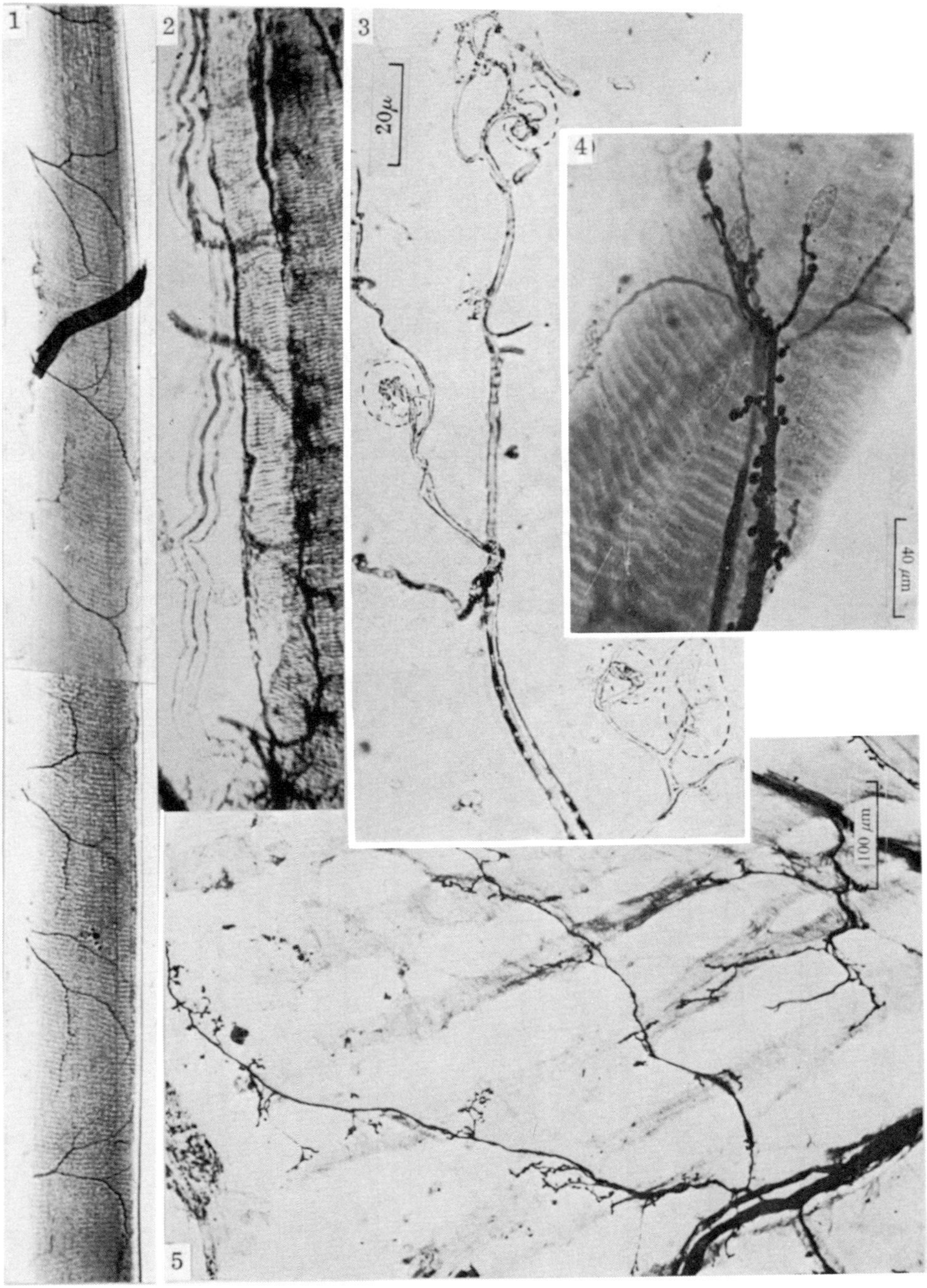

Figure 12.33. Innervation of insect muscle cells. (*1*) Multiterminal innervation of single muscle cell of tergotrochanteral muscle cell of blowfly *Calliphora,* stained to show motor-nerve terminals. (From Auber, 1960.) (*2*) Single cell of locust *Locusta* extensor tibiae with slow and fast axons innervating it at intervals of about 60 μm. (From Hoyle, 1955b.) (*3*) Terminals (*circled*) removed from (*2*). (From Hoyle, 1955b.) (*4*) Terminal on single muscle cell of ventral diaphragm of *Locusta.* (From M. Peters, 1977.) (*5*) Portion of ventral diaphragm muscle of *Locusta* stained to show nerve terminals: six muscle cells. (From M. Peters, 1977.)

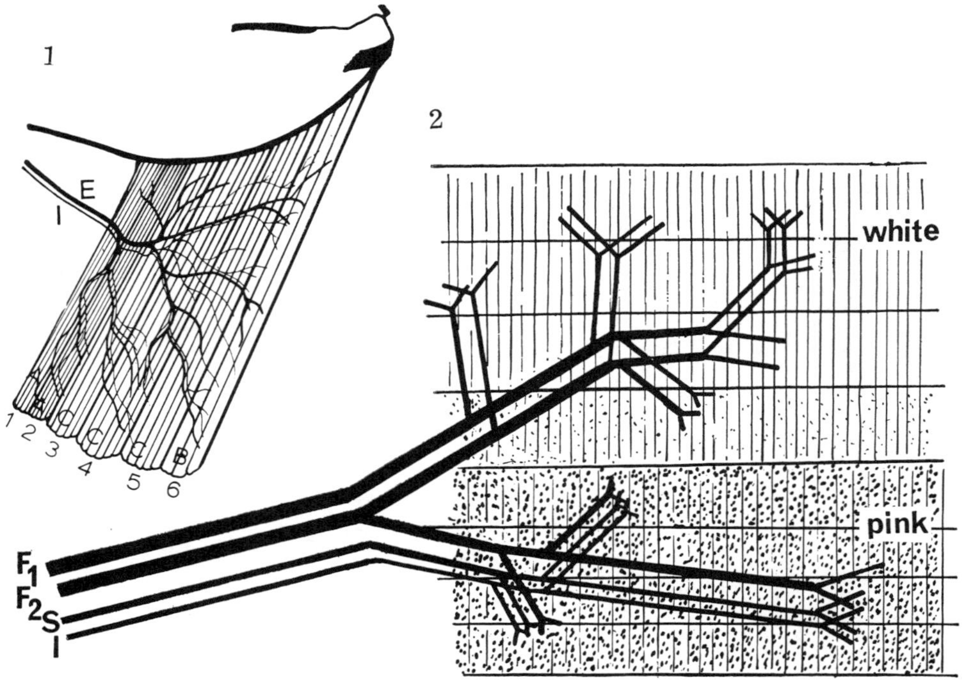
1
E
I
1
2
3
4
5
6
A
C
C
C
B
2
white
pink
F1
F2
S
I

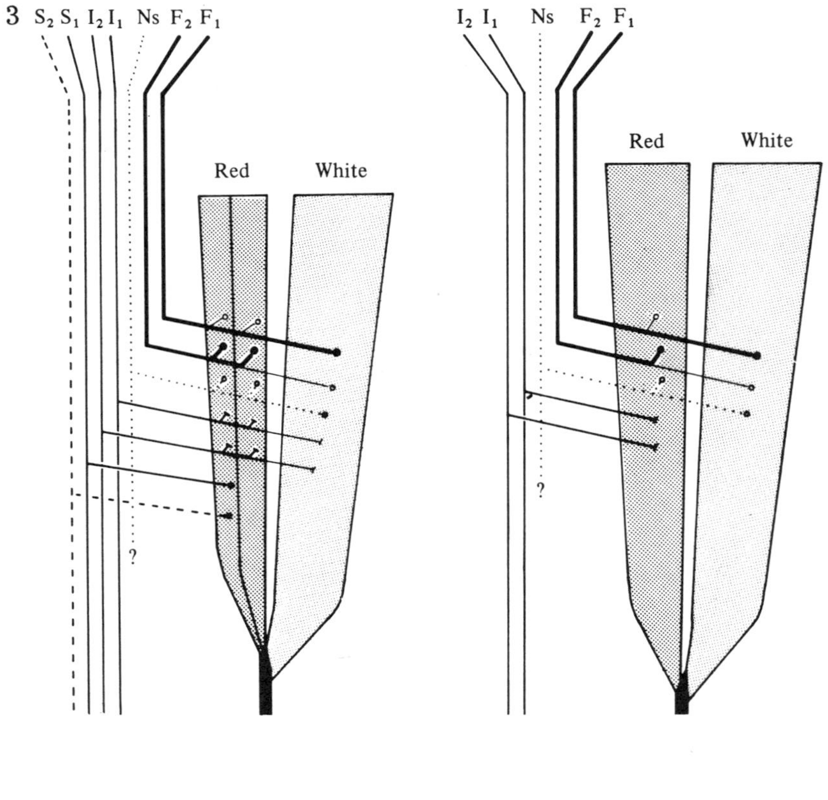
3
S2 S1 I2 I1 Ns F2 F1
Red
White
?
I2 I1 Ns F2 F1
Red
White
?

enon is lacking. It is easy for the experimenter armed with a pair of modern stimulators to demonstrate presynaptic inhibition. But it seems unlikely that the relevant ganglia of the crab could possess the subtle information and executive skills to take advantage of this form of inhibition. It may be there on account of anatomical accidents rather than evolutionary "design."

12.10.2. Transmitter Substances

Throughout the arthropods the same substance appears to be the excitatory transmitter: *L-glutamate.* There are some subtle differences in the actions of L-glutamic acid compared with natural transmitters (Shinozaki and Ishida, 1979), but the concomitant release of other materials from the terminals could account for them. There is direct evidence for its release from some crayfish excitatory junctions (Kowagoe et al., 1981). The only other substance implicated as a possible transmitter at some slow junctions is another amino acid, *L-aspartate.* Although aspartate is an agonist of glutamate at some excitatory synapses, it is an antagonist at others, causing hyperpolarization. Which anion is combined with L-glutamate at the release site is unknown. There is no specific enzyme concerned with destroying L-glutamate that is comparable to acetylcholinesterase. After release and combination with receptors, released transmitter is lost simply by a combination of diffusion together with pinocytotic uptake by the pre-synaptic nerve terminals and normal deamination by ubiquitous enzymes. The transmitter receptors are as sharply localized as are those for ACh in vertebrates (Figure 12.35.*2*).

The inhibitory transmitter substance throughout arthropods is *γ-aminobutyric acid* (GABA). Also for GABA there is no specific destructive enzyme. It may be wondered why ACh, which is otherwise the almost universal neuromuscular transmitter substance, is not used by arthropods. This is most likely because L-glutamate, a natural amino acid, is more readily synthesized in the amounts needed for continuous, unfatiguing production and release. All known cholinergic junctions are subject to fatigue: many L-glutaminergic ones are not. Also, L-glutamate action is a little slower, making it more useful for generating slow movements and tone, yet it is still fast enough for causing quick escape movements when delivered at a high concentration. This is done either at special fast neuromuscular junctions that do fatigue rapidly, producing giant EJPs, or by brief high-frequency bursts at facilitating ones. For more-specific details, the arthropod classes need to be described separately, since there are significant differences between them.

Figure 12.34. Examples of innervation patterns of arthropod muscles. (*1*) Simple. A single excitatory axon (*E*) and a single inhibitor (*I*) branch together to supply each muscle cell in the muscle with *multiterminal* endings. There are slow, *1, 2,* fast, *6,* and intermediate, *3, 4, 5,* muscle cells in parallel. Accessory flexor meropodite in walking leg of crab *Cancer.* [Dorai-Raj, Diversity of crab muscle fibers innervated by a single motor axon. *J. Cell. Comp. Physiol.* 64, Fig. 1, p. 42 (1964). Reprinted with permission from Alan R. Liss, Inc. New York.] (*2*) Moderately complex. Two kinds of muscle cell, slow pink and fast white, occur side by side, and muscle cells are innervated *polyneuronally* from a total of three excitatory axons and one inhibitor. One fast axon F_1 supplies only the white cells, another F_2 supplies both pink and white. The single slow axon and inhibitor each supply only the pink cells. Levator of the eyestalk of the crab *Podophthalmus.* [G. Hoyle, and P. A. McNeill, Correlated physiological and ultrastructural studies on specialised muscles. Ic. Neuromuscular junctions in the eyestalk levator muscles of *Podophthalmus vigil. J. Exp. Zool.* **167,** 525 (1968b). Reprinted with permission from Alan R. Liss, Inc., New York.] (*3*) Complex, with differences in related species of locusts, *Locusta* (*left*), *Schistocera* (*right*). The two kinds of muscle cell are in separate bundles differentially innervated by two fast and two inhibitory axons. Details of slow axon innervation are still uncertain. (From Walther, 1980.)

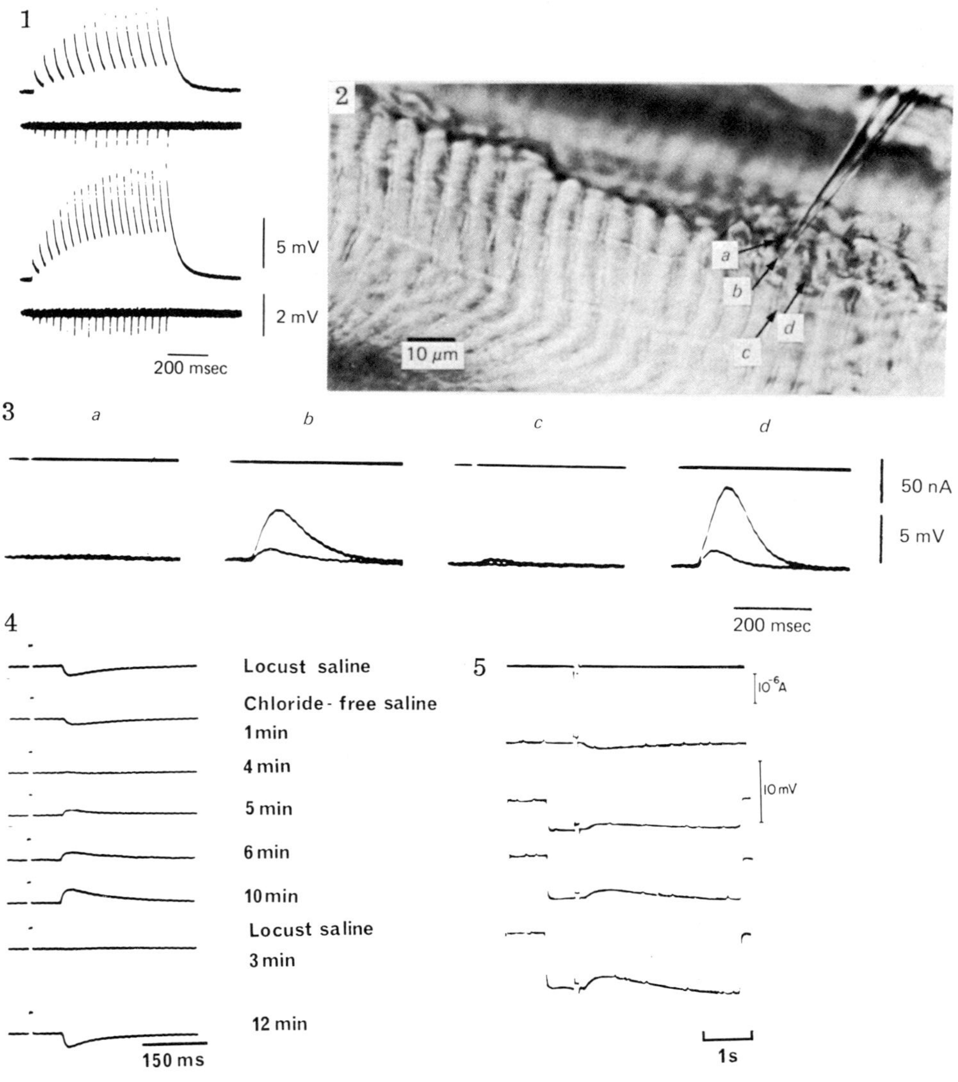

Figure 12.35. Arthropod excitatory and inhibitory transmitters. (*1*) Facilitating EJPs in crayfish *Cambarus* slow flexor muscle cell. (From Kawagoe et al., 1981.) (*2*) Nomarski interference image of junction of (*1*), with iontophoresis electrode, indicates sites where pulses of L-glutamate were applied (*tips of arrows*). (From Kawagoe et al., 1981.) (*3*) Responses at sites indicated in (*2*): *a* and *c* were insensitive; at *b* and *d*, a nerve-terminal bouton is visible. (From Kawagoe et al., 1981.) (*4*) Demonstration that IJP is due to activation of *g*Cl: the potential reverses in Cl^--free medium. *Romalea.* (*5*) Response of (*4*) to iontopheres GABA and reversal by polarizing pulse. [(*4*) and (*5*) reprinted with permission from "*Insect Muscle*," pp. 207–280. Copyright 1975 Academic Press, London]

A comparative study, using intracellular muscle-cell electrodes in several species of crab and lobster, published by Hoyle and Wiersma (1958) was the first to reveal that remarkably extensive variations occur in the detailed events of excitatory neuromuscular transmission. The results provided the first hints of an astonishing range of type of events at individual muscle cells of crustaceans. Different types of cell are in some muscles present in discrete bundles, in others mixed together. The range of membrane properties of the different muscle cells is extensive: from very fast ones whose electrogenicity is to produce only all-or-none overshooting spikes, through ones with all degrees of graded responsiveness (see p. 170) from weak to large spike, to completely electrically inexcitable. Even in a single muscle, the closer of a walking leg of the crab *Cancer magister*, all these types were found (Atwood et al., 1965). Not all of the physiological diversity in transmission properties is to be attributed to local release mechanisms. The final nerve terminals are profusely branched and in addition have both very thin and thick regions. The latter are, or correspond to, varicosities seen on many different kinds of vertebrate nerve terminals. The transition from very thin to thick forms a "bottleneck." At such a region there is either a drastic concentration of current flow or a dispersion, depending on the direction of the impulse relative to the structure. When the latter occurs, the probability of passage of an impulse becomes reduced, and many fail at this point. Transmitter-release sites subsequent to the neck are excited only sporadically compared with ones invaded from thicker axonal regions. These matters have been extensively discussed by Atwood (1976).

12.10.3. Principles of Organization of Muscle Cells in Arthropod Muscles

The larger arthropod muscles may be divided into the two principal kinds, familiar also in vertebrates in spite of the internal skeleton. In one the principal functional objective is a large length change; in the other it is production of large force. For the former the muscle cells are few in number, long, parallel, and attached to a short tendon (apodeme) at the moving end. Their attachment to the skeleton is compact. The large-force producers comprise a large number of shorter muscle cells attached to an extended, even branched, tendon and to broad areas of skeleton. They generally make either a fan or a herringbone pattern if they are attached to both sides of the tendon. Arthropods are endowed with a tremendous advantage over vertebrates in developing great muscular strength because of the tubular exoskeleton, which provides a much greater area for muscle-cell attachment than does an internal one. The claims of enormous tension capability for the extensor of a grasshopper leg or closer of a lobster or crab claw are not exaggerated. Each has a tendon extending the entire length of the joint, and muscle cells are attached to the exoskeleton whole joint. More than 10 times as many muscle cells are available as could be used if the cells ran in parallel from end to end. There are many variations on these plans for muscles in the body, because the endoskeleton has complex shapes. A single tendon may have muscles of both extreme types attached to it. For example, the accessory muscle in crab meropodites is a flat muscle flag with only a single excitatory axon that supplies all the cells. Yet the distal cells are thick, 0.2 mm in diameter, small SL 3–4 μm, all-or-none spiking, and fast, whereas the proximal ones are slow, with 12-μm sarcomeres, 0.1-mm thickness, and electrical inexcitability (Dorai-Raj, 1964). There is a progressive gradient of intermediate forms between the proximal and distal extremes.

For muscles innervated by more than one excitor, it seems likely that each axon has a genetic predisposition to form one or another type of neuromuscular junction, with potential for one or another type of synaptic transmission. There are well-marked differences between excitatory and inhibitory endings that may reflect different transmitter substances, vesicles of excitatory terminals being spherical, those of inhibitors ovoid (see p. 166). All arthropods have a range of SLs, that of crustaceans being by far the largest of any phylum, from 0.6 to about 26 μm. Twitch times have a 1000-fold range, from 0.1 s to more than a minute. By combining different muscle-cell populations with different synaptic properties, an enormous range of control types has been encompassed. The variations are to be seen as much within a single animal, in muscles serving particular functions such as eye-raising, jumping, or fast swimming, as they are between species. The variety, combined with the unsystematic way in which researchers in this field have chosen the subjects for their investigations, makes it difficult to cover the field systematically.

12.11. Crustacea

12.11.1. History

Crustaceans were the first arthropods and indeed the first invertebrates to be studied by mainstream physiologists. Consequently, much of the early history of comparative nerve/muscle physiology is exclusively concerned with one or two large crustaceans. It was natural for scientists studying crustacean myoneural mechanisms for the first time to expect to find features familiar from studies on frogs, so they interpreted their findings in ways compatible with frog experience, even when experimental logic indicated major distinctions. But when Richet (1879) found that tension development by the claw-closer muscle of lobster in response to faradic (repetitive, leading to tetanus) stimulation dropped almost to zero when he raised the stimulus strength only minutely above the threshold for obtaining a contraction, it was obvious that at least one difference stood out above the others. He had discovered peripheral inhibition, and his claw preparation was to be used extensively for more than half a century as a model for central-nervous inhibition. An apparent functional nicety of inhibition was revealed when Biedermann (1887) found that weak stimulation of the nerve *in situ* caused a contraction of the opener while at the same time its antagonist was inhibited. Stronger stimulation caused the closer to contract and the opener to relax: that is, the opener also was peripherally inhibited. The functional importance of this curiosity was realized later when Hoffman (1914) found than an excitatory axon (it is in fact the only one) that innervates the extensor carpopodite (or stretcher) doesn't end there but continues into the next joint also to innervate the opener muscle. When this axon is stimulated, both the opener and the stretcher are excited together. This is the basis of the familiar startle response of crayfish, crabs, and clawed lobsters, which serves as a threat posture and also prepares them for defense. But the claws must be used in other ways too, especially in feeding, which requires opening without concomitant stretching. This can obviously be accomplished only by inhibiting the stretcher selectively at the same time the common excitor is active, by a specific stretcher inhibitor. Central excitation of the stretcher-inhibitory axon must be built into many reflex and self-initiated behaviors.

The first realization of the unique arthropod dual fast/slow excitation mechanisms came when a major figure in early biophysics, Keith Lucas, was examining the closer of the lobster claw (Lucas, 1907). He found that he could obtain either a rapidly fatiguing, powerful fast closure or a slower, weaker prolonged one that did not fatigue. He later (Lucas, 1917) obtained a bimodal curve for the crayfish *Astacus* when plotting contraction height against interval between a pair of stimuli. Lucas concluded correctly that there must be two kinds of nerve, but this view was not accepted by physiologists. It was believed instead, by extrapolation from vertebrate results, that there is only one kind of nerve element and that facilitation and recruitment of muscle cells must somehow account for the bimodality (Pantin, 1934; Katz, 1936).

Eventually Van Harreveld and Wiersma (1937) isolated single motor axons supplying a crayfish claw and stimulated them separately. They found a single inhibitory axon (which is the common inhibitor) and only two excitors, one thicker than the other, each evoking a characteristic contraction. They called the thick one the *fast axon,* because it produced a large twitch in response to only a single brief shock. The thinner one they termed the *slow axon,* because it evoked a slow contraction but only when stimulated repetitively. Clearly there were two different response mechanisms and these accounted for the bimodal curve of Lucas. But the results did not rule out the recruitment hypothesis.

This type of study has been extended to other decapod crustacean muscles and it is apparent that there are some significant variations (Figure 12.36), principally in inhibitory innervation. The functional significance, if any, of the differences is still not apparent. Van Harreveld (1939) showed histologically that the three axons branch together and that they appear to innervate the same single muscle cells, an observation that has been shown by intracellular recording to be only roughly true. Extracellular myograms revealed a progressive slow growth of action currents associated with facilitation in the response to

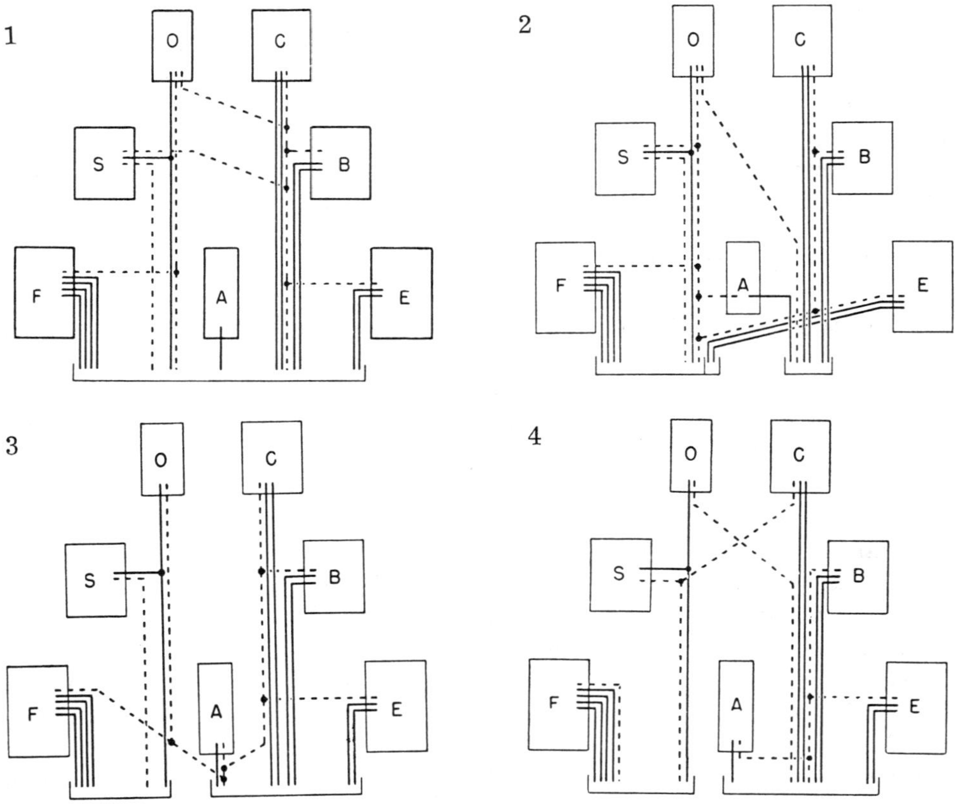

Figure 12.36. Variation in motor-axon innervation patterns in the four tribes of Crustacea: (*1*) Brachyura (crabs), (*2*) Anomura (hermit crabs), (*3*) Palinura (rock lobsters), (*4*) Astacura (crayfish and lobsters). Solid lines represent excitatory axons of all kinds, broken lines with dot connections common inhibitory axons, and unconnected broken lines specific inhibitors. (From Wiersma and Ripley, 1952. Reproduced by permission of Dr. W. Junk Publishers)

the slow axon, and a large declining one for the fast axon (Marmont and Wiersma, 1938). The size of the current was reduced by inhibition. Marmont and Wiersma found that in some of the muscles, there was good inhibition without any attenuations of excitatory action current, regardless of the timing of arrival of inhibitory impulses with respect to excitatory ones, and no change as the timing was altered. This they termed simple inhibition. Other junctions, notably those of the crayfish opener, by contrast showed a marked effect of relative timing, which they termed supplemented inhibition, reflected in an enhanced mechanical effectiveness when the inhibitory impulse arrived at the junctions just slightly ahead of the excitatory ones. Dudel and Kuffler (1961b) showed that supplemented inhibition is due to a reduction in the amount of synaptic-transmitter substance released by excitation.

However, results obtained prior to 1953 still did not reveal the dual excitatory innervation of individual muscle cells. This might have been demonstrated before the introduction of intracellular electrodes, by the feasible experiment of whittling preparations down to single muscle cells, but it was not. It was actually first demonstrated using intracellular electrodes by Fatt and Katz (1953) in the closer of a crayfish. The electrical responses were both very small, the fast being 2 mV after facilitation and summation and the slow, 1 mV. At the same time I found a dual innervation of locust jumping muscle (Hoyle, 1953a), in which the electrical responses are dramatically different, the fast axon giving rise to very large JPs of over 50 mV in response to single shocks, and these in turn to graded electrogenic spikes that overshoot zero. In some of the same muscle cells, the slow axon gives responses, which are smaller junctional potentials, of various initial sizes and extents of facilitation, as well as various sizes of graded response.

12.11.2. Contemporary Studies

Since the early 1960s, a bewildering variety of JP sizes, facilitation rates, electrogenic responses, and mechanical events has been recorded, from many different crustacean muscle cells. When considered in relation to the E-C coupling characteristics of the various kinds of muscle cell present in crustacean muscles, these all translate into exquisitely sensitive functional adaptations. Perfection of control of all kinds of behaviors has been achieved in large part by peripheral variation. Various aspects have been covered in thorough reviews, notably by Atwood (1973, 1976, 1977), to which the reader is referred for additional data. Many studies of the 1960s and early 1970s were aimed at elucidating possible correlations between the ultrastructure and function. Some researchers were intent upon obtaining a better understanding of the details of events at the functionally rich variety of synapses, especially the basis for facilitation, which is the major variant. Others were concerned with understanding the basic features of E-C coupling. Almost all the work has been done on large decapods, and there is as yet no possibility of seeking possible trends that were associated with adaptive radiation and evolution within the class.

Numerous correlated ultrastucture/function studies have been made on a variety of decapod crustaceans. It has transpired that every muscle of every crustacean examined intensively is unique, and investigators have been hard-pressed to perceive general principles when the full details of muscle-cell population and innervation characteristics could be compared. There is no way of predicting structure from function or vice versa. Exceptions have always been found whenever a generalization seemed to be emerging. The best that can be done is to point out some common tendencies in ultrastructural features that turned up and some tentative general rules that are known about innervation.

Slow-type muscle cells are likely to be electrically inexcitable, to have long sarcomeres, little SR, few ITs, and IT/SR contacts, and to lack fast-axon innervation. Fast-type muscle cells are likely to have short SL, abundant SR, numerous ITs and IT/SR contacts, and membranes giving large graded electrogenic responses, and to lack slow-axon and inhibitory innervation. Intermediate cells are likely to have intermediate SL and other structural features, and to receive multiple excitatory and inhibitory axons.

The consistency of relationships between innervation and cell properties is strongly suggestive of trophic influences. But these are in turn following clues that are in part genetically determined, in part developmentally guided, and in part functionally conditioned. An embryonic gradient seems to determine distribution of muscle-cell type in some crustacean muscles independently of nerve influence. Inhibitory terminals have been unequivocally identified by serial electron microscopy and reconstruction (Figure 12.37 – Jahromi and Atwood, 1974) of crayfish cheliped opener. The final inhibitory branches making synaptic contacts with muscle cell membrane are of slightly larger diameter than excitatory ones, but a larger number of branches with presumed functional contact, with larger surface area, occur for the excitors. Both types of axon have many side branches, some joined to the main terminal by only a very narrow neck. These axons also make synapses on the muscle cells, and a greater portion of their total area is devoted to contact than is the case for the main branches.

The cheliped opener is one of the few muscles in which the innervation includes inhibitory-axon synapses on the excitatory axon that are responsible for prejunctional inhibition. Such junctions were located and seen to resemble those on muscle cells (Figure 12.37.*2*).

The same techniques used on cheliped opener enable comparison between the morphologies of physiologically different excitatory axons and of different junctional sites on the same muscle cell. Much of the rich supply of information on this fascinating aspect of transmission has been reviewed by Atwood (1976). There are from about 20 to several hundred active synaptic sites from a single excitatory axon on each muscle cell, depending principally on the size of the cell. The individual synaptic sites on one muscle cell do not necessarily all have similar physiological characteristics, although this can be the case. When it is not so, the individual elements may differ either randomly, or systematically along the length of the cell (Sherman and Atwood, 1972). Inconsistencies of a random nature are likely to be associated with random structural features such as the diameter of connections to the main terminal branch. Smooth trends in response magnitude are probably associated wtih gradients of influence present during development.

The functional characteristics are the initial and final magnitudes of the JPs as seen in trains of different frequencies, and their rates of growth and decline. Whether or not such gradients or local variations are significant in relation to the final criterion of value, the animal's behavior, is problematical, but unlikely. Effects are known that build up only over a period of a few hours and that persist even longer (Sherman and Atwood, 1971).

Junctions that show extensive, slow facilitation and that are highly resistant to fatigue have large axonal diameter, are crammed with very large numbers of vesicles, and make a few small synaptic juxtapositions with the muscle cell membrane. Junctions that produce medium-size EJPs with high facilitation and moderately fast fatigue also contain many vesicles but have extensive, usually continuous, juxtaposition of axonal and muscle cell membranes occupying at least a third of the axon perimeter. The larger, nonfacilitating, rapidly fatiguing junctions have relatively few, clustered vesicles and also a larger area of contact with the muscle cell.

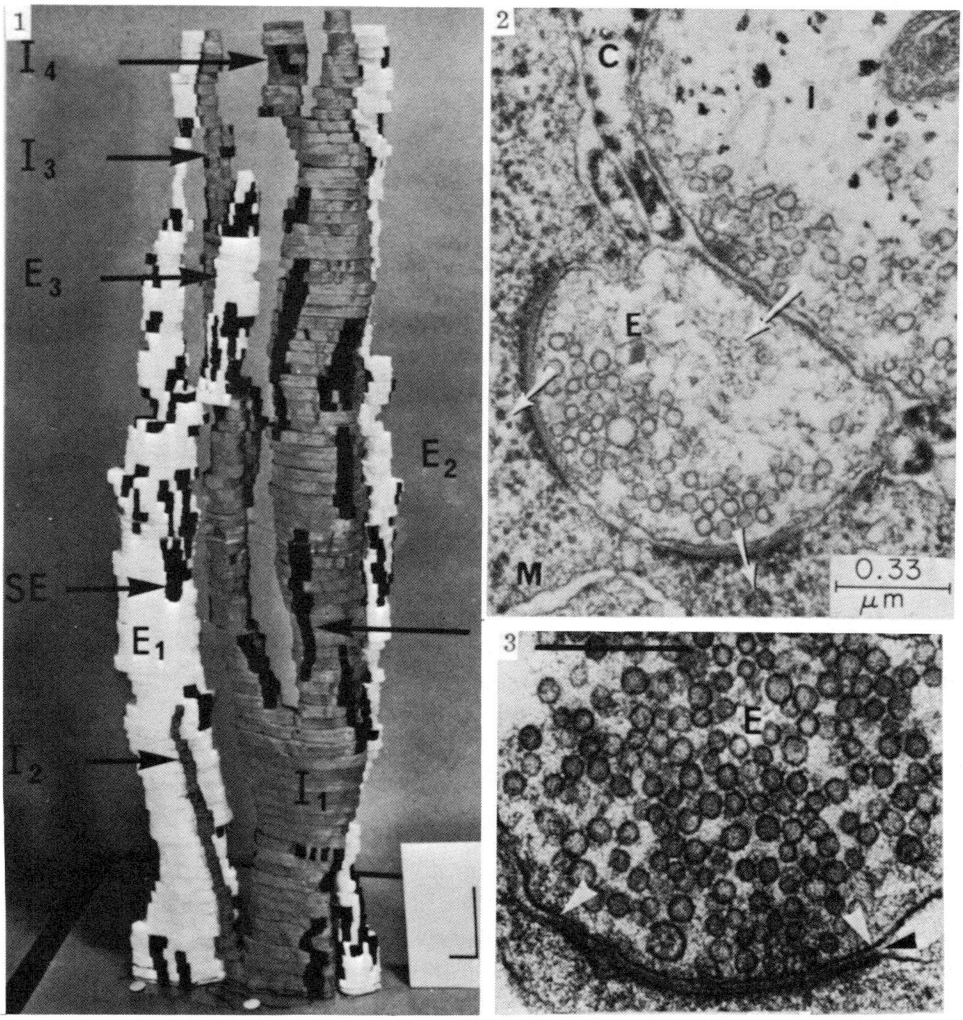

Figure 12.37. (*1*) Details of inhibitory/excitatory axon relationships for crayfish *Cambarus* opener revealed by TEM serial sectioning and reconstruction. There are two excitatory branches: areas of probable synaptic function are indicated in (*1*) by arrows. (*2*) Ultrastructure of a presynaptic inhibitory site. The E axon (*E*) is synapsing with the muscle cell. Directions of transmission indicated by arrows. (*3*) Details of E neuromuscular junction. (All from Jahromi and Atwood, 1974.)

The differences in properties of transmission at individual junctions, and their average for a particular cell, combined with the extensive range encountered even in a single muscle between cells, are often truly remarkable. Yet different individual animals have closely similar junctional varieties with distributions that are astonishingly alike. Therefore, it may be presumed that they are in no way random but rather under genetic control, and that they are the products of variation and natural selection, as are the more familiar peculiarities of external morphology. Taxonomists count the relative numbers of bristles in different locations to determine species; why not also have biophysicists do it by examining JPs?

The range of details in crustacean-muscle organization and innervation is perhaps best illustrated with a few specific examples that happen to have been studied with both intracellular recording and electron microscopy.

12.11.2.1. The Strike Mechanism of Mantis Shrimps

The second thoracic appendage of mantis shrimps is a large raptorial organ on the sharp tip of which it can impale a passing shrimp. The extension lasts as little as 5 ms, but the muscles involved, medial and lateral extensors, are both of slow type (McNeill et al., 1972). The explanation of the high speed of the strike is that it is the result of sudden elastic release of tension accumulated in the extensors. At the same time the powerful extensors contract their weak antagonists are aided in holding the dactyl closed by a "click-joint" locking mechanism (Burrows and Hoyle, 1972). The innervation, neuromuscular transmissions, muscle cell types, and muscle cell ultrastructures all indicate that the evolution of their particular sets of features occurred specifically in association with the operation of the click-joint. Equivalent mechanisms have been evolved by a few different insects in connection mainly with jumping (see p. 551) or with sound production. The innervation and neuromuscular physiology of each of the muscles involved has been studied (Figure 12.38), as well as correlated with ultrastructural details. The results are summarized in Table 12.1. The muscle cells of the power muscles, the extensors, possess some interesting features that may be related to their role. They have a very broad "double" Z region, and also a clearly double-banded M region in which the center of the sarcomere is very thick and devoid of M bridges (see Figures 12.42.*1* and 12.42.*2*).

The principal flexor muscle, which cocks the joint, has an interesting special feature that is clearly adaptive, a specifically slow region (Figure 12.38.*2*) which keeps the dactyl ready for action for a prolonged period when necessary. In keeping with the need to respond rapidly following the appearance of prey, it also has a fast portion. The extensors use fatiguing fast axons to start the development of power as rapidly as possible, then back this up by discharge of slow ones (Figure 12.38.*4*).

12.11.2.2. The Power-Stroke Muscle of a Swimming Crab

The power-stroke muscle of *Portunidae* swimming crabs, the anterior rotator of the coxopodite of the fifth legs, of which the last segment is paddle shaped, has a tripartite substructure (Hoyle, 1973, Figure 12.39). In the Hawaiian species *Portunus sanguinolentus,* the muscle cells attached to one side of the tendon are quite different from those on the other. Not only are they shorter, but they are different in color, the shorter ones being pink and the longer ones white. The pink cells are colored because they contain many mitochondria, up to about 50% total volume, largely in a peripheral ring. There are very few mitochondria in the clear cells. Basic ultrastructures of the contractile materials of the two extremes are similar and so are the SLs, but the pink cells have very much slower contraction and relaxation rates.

Also attached to the tendon, at its proximal margin, are much shorter cells, barely one third the length of the white ones and half the length of the pink ones. These cells are a deep red in color: about half their total volume, including the entire periphery, is occupied by mitochondria. Long ITs pass between the packed ranks of mitochondria to enter the core, where numerous dyadic contacts are made between ITs and an extensive SR. These are features commonly associated with fast cells, but the red ones are slow and unfatigable. White cells fatigue very rapidly. However, very little depolarization is received at the

1

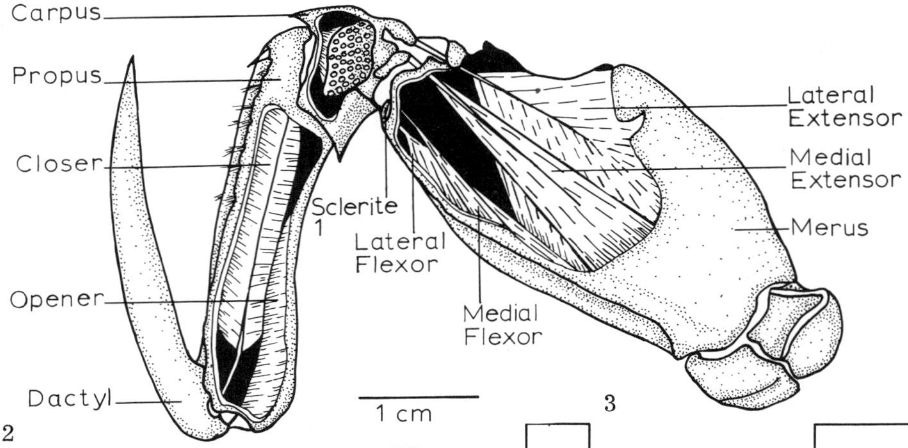
Carpus
Propus
Closer
Opener
Dactyl
Sclerite 1
Lateral Flexor
Medial Flexor
Lateral Extensor
Medial Extensor
Merus
1 cm

2

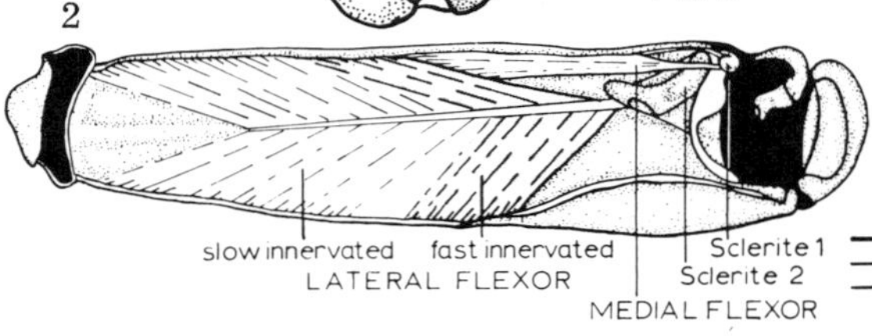
slow innervated
fast innervated
LATERAL FLEXOR
Sclerite 1
Sclerite 2
MEDIAL FLEXOR

3

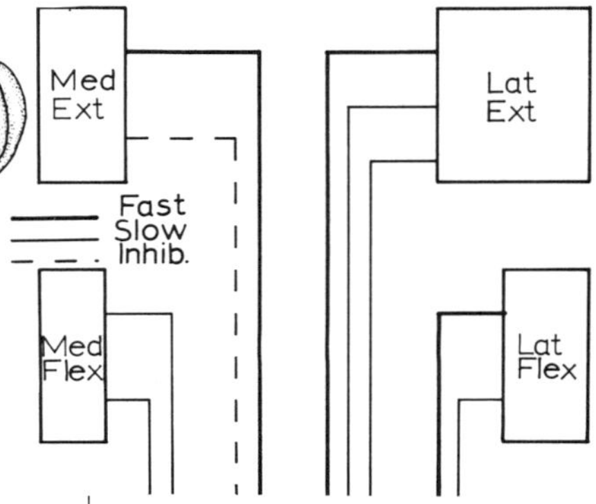
Med Ext
Lat Ext
Fast
Slow
Inhib.
Med Flex
Lat Flex

4

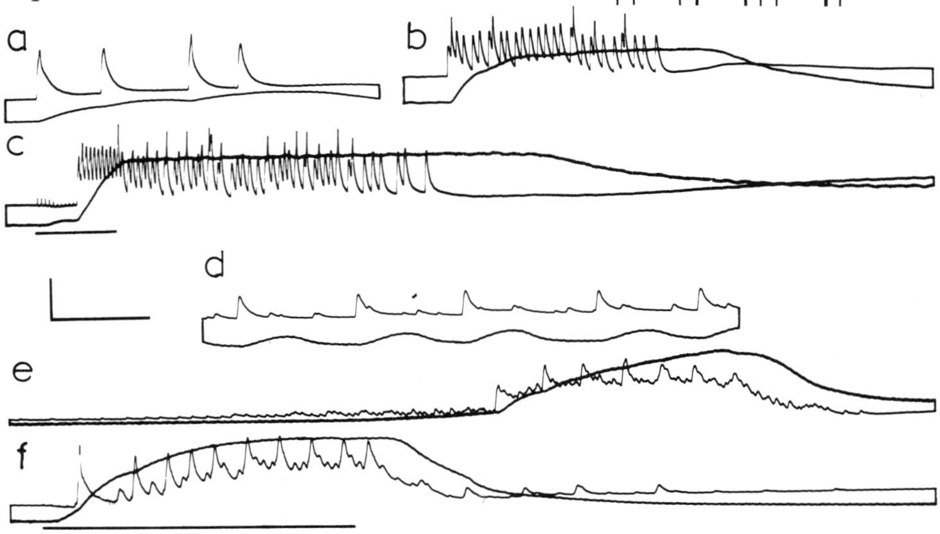
a
b
c
d
e
f

ends of the long ITs in response to slow-axon excitation, so in the red cells the extensive SR may have the function of enhancing the amount of releasable calcium rather than fast mopping up for quick relaxation. Their mean SL is 4.4 μm compared with 4.8 μm for the white ones, another feature normally associated with greater speed, but even the fastest red cells are less than half as fast as slower white ones, for which the whole twitch time is only 100 ms: both are fast by comparison with most crustacean muscle cells, as might be expected in a fast swimmer. There is a major functional exception at mating time for mature males, which hold their paddles aloft as a signal of readiness to mate. The paddles are not brought down until a willing female is encountered, which may not be for a few days.

In this muscle we can see remarkable functional adaptations of component muscle cells. All are fast contracting and relaxing to permit them to contribute to fast swimming. About 40% contain almost 100% contractile material and are the principal producers of power in fast swimming, but they fatigue very rapidly and are useful only for producing brief bursts of extreme power. About 58% of the cells, all of the pink ones, are capable of sustained phasic activity at moderate rates and are responsible for long-duration bouts of swimming. The remaining 10% are unique in that they can maintain tonic contraction indefinitely, yet are still able to contribute extra tension to fast-phasic-swimming activity. It is possible that different parts of these muscle cells are independently addressed by the different axons, the core being brought into the act only with very prolonged, continuous stimulation, the outer region by phasic fast-axon bursts, and that the two regions have different relaxation rates. This has been shown to be likely for some insect slow skeletal cells (see p. 541).

The white cells are electrically homogeneous and produce propagated all-or-none action potentials (Hoyle and Burrows, 1973a). Most pink cells give large graded spikes, the biggest ones overshooting, but a few give all-or-none spikes. The red cells range from electrically inexcitable cells to ones giving large graded responses that fatigue rapidly. The three parts of the muscle each receive a separate innervation, none of which are inhibitory: the white and pink portions share common innervation by three axons and three other axons supply the red portion. The axons supplying the red portion may not be involved in swimming at all. Two of the axons innervating the pink portion are used most often, in all slow swimming and also in fast swimming. The other two are recruited only during fast swimming. Only when swimming is at its fastest are two of the three axons innervating the white cells added. This leaves one final axon to the white portion, and this is active only when the crab is making its most violent motion, to escape from an attack. When this axon alone is stimulated electrically, the muscles' briefest twitch time occurs.

Figure 12.38. Mechanism of the strike of the mantis shrimp *Hemisquilla*. (*1*) Muscles responsible for the strike are the lateral and medial extensors, antagonized by lateral and medial flexors. Two small movable sclerites serves as a click-joint device, locking when flexion is complete. Extension then cannot be achieved until the flexors relax. (*2*) Details of the lateral flexor: it has two separately innervated, physiologically distinct portions, a proximal slow-axon-innervated slow region and a 10 times faster, fast-axon-innervated distal portion. (*3*) Plan of innervations of the four muscles. (*4*) Intracellular recordings (*above* with tension below) during cocking and striking correlated with tension. (*a*) Medial flexor – fast portion, spontaneous. (*b*) Medial flexor during natural strike. (*c*) Medial flexor during electrically driven strike. (*d*) Lateral extensor – spontaneous activity showing three sizes of JPs. (*e*) Lateral extensor during natural strike. (*f*) Lateral extensor during evoked strike. [All from M. Burrows and G. Hoyle, Neuromuscular physiology of the strike mechanism of the mantis shrimp, *Hemisquilla*. *J. Exp. Zool.* **179**, Fig 1, p. 381, Fig 6A, p. 386, Fig 13, p. 391 (1972). Reprinted with permission from Alan R. Liss, Inc., New York.]

Table 12.1 Summary of Muscle Cell Types Found in Functionally Specialized Crustacean Leg Muscles Involved in the Strike of the Mantis Shrimp[a,b]

				Lateral Flexor	
	Medial Extensor	Lateral Extensor	Medial Flexor	Distal Portion	Proximal Portion
Function	Movements other than strike; accessory in strike	Provides power for strike	Movements other than strike	Locks sclerite to "cock" strike mechanism	Holds sclerite in "cocked" position. Relaxation releases strike
Innervation (axons)	1 "fast" 1 inhibitor	1 "fast" 2 "slow"	2 "slow"	1 "fast"	1 "slow"
Twitch time	200 ms	400 ms	Does not twitch	80 ms	Does not twitch
Rise time to peak tetanic contraction under maximum stimulation	300 ms	700 ms	500 ms	50ms	500 ms
Decay time from peak contraction to rest	500 ms	1000 ms	750 ms	100 ms	700 ms
Electrical excitability	Graded	Graded	Weakly graded	Large graded to spike	Graded
Sarcomere length	Variable 6.0–10.2 μm	Constant 7.8 μm	Constant 5.3 μm	Constant 4.3 μm	Constant 4.5 μm
A-band length	4.4–7.2 μm	5.1 μm	3.4 μm	2.8 μm	3.3 μm
Half I-band length	2.7–4.2 μm	3.1 μm	1.9 μm	1.8 μm	1.7 μm
"Fibril" diameter (average)	2.2 μm	3.5 μm	1.4 μm	0.7 μm	1.3 μm
Ratio of thin to thick filaments	3.5 : 1	4 : 1	3 : 1	3 : 1	3 : 1

Z-disc thickness	Variable 0.12–0.32 μm	Very variable 0.11–0.38 μm	Fairly uniform 0.12 μm	Uniform 0.12 μm	Uniform 0.1 μm
Sarcoplasmic reticulum	Moderate development; mostly single layer	Well developed; mostly double layer and wider	Moderately well developed; mostly single layer, some double	Very extensive and well developed; usually two or more layers	Moderately well developed; mostly single layer, some double
Mitochondria	Very few	A well-developed peripheral ring	Very few	Very few	Very few
Dyads (number per unit cross-sectional area of 32 μm^2)	5	9	7	24	12

[a]McNeill, P., Burrows, M. and Hoyle, G. (1972). Fine structure of muscles controlling the strike of the mantis. *Hemisquilla. J. Exp. Zool.* **179**, 395–416. Reprinted with permission from ALAN R. LISS, INC., N.Y.
[b]See Figure 12.38.

1

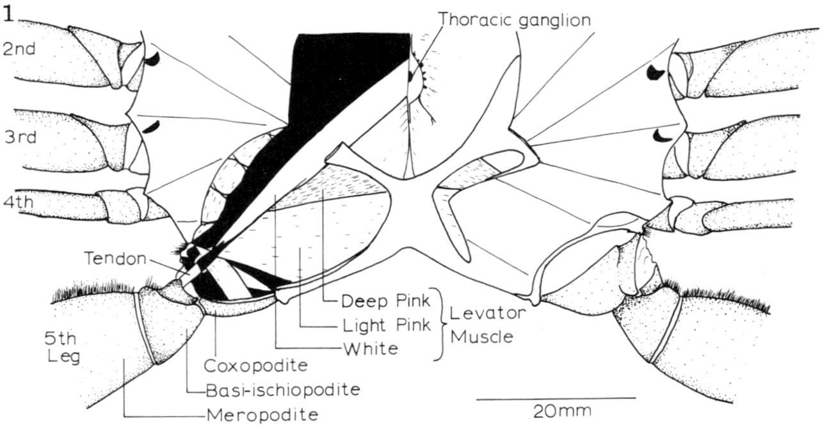
Thoracic ganglion
2nd
3rd
4th
Tendon
5th Leg
Deep Pink
Light Pink
White
Levator Muscle
Coxopodite
Basi-ischiopodite
Meropodite
20mm

2

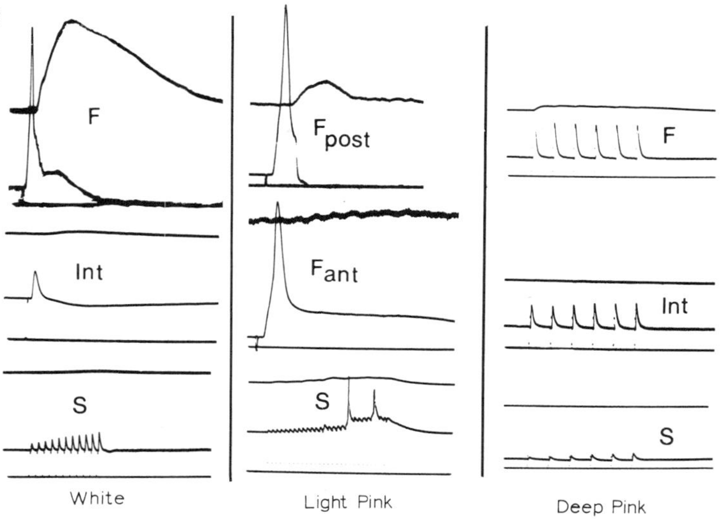
F
Int
S
White
F post
F ant
S
Light Pink
F
Int
S
Deep Pink

3

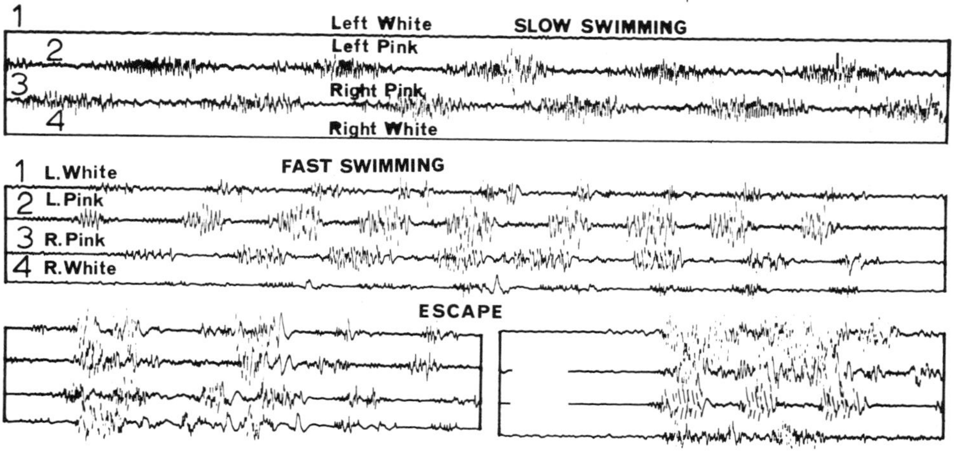
1
Left White
SLOW SWIMMING
2
Left Pink
3
Right Pink
4
Right White
1 L.White
FAST SWIMMING
2 L.Pink
3 R.Pink
4 R.White
ESCAPE

12.11.2.3. Dactylopodite Adductor (Closer) of Walking Leg

This muscle was examined in a crab (Atwood et al., 1965) using intracellular recording and tension recording from only the single muscle cell the electrode was in. The innervation was left intact and the slow, fast, and common inhibitor axons each stimulated separately. A second intracellular electrode was used to pass current into the same cell to study its electrical excitability and related mechanical responses. It is desirable, in principle, to study all muscles of interest in this manner, but unfortunately even for the dactylopodite adductor of this large crab, in which the anatomical layout is favorable and the axons are large and separable, dissection is extremely tedious and preparations easily damaged. The closer remains the only muscle examined intensively in the desirable manner: it may serve as a model of the extent to which complexity of muscle-cell composition and innervation have evolved in crustaceans.

Major cell types range from fast strongly electrically excitable short SL cells, giving all-or-none spikes, to ones that are slow, have long SL, and are electrically completely unresponsive. These may be subdivided on the basis of the proximity of membrane potential to E-C threshold. Some are tonically contracted at their resting potential. The majority show some degree of electrogenesis, ranging from strong delayed rectification without depolarizing responsiveness to near all-or-none spiking. Some examples of the different types are illustrated in Figure 12.40.

The summary table of results of this work is reproduced in Table 12.2. Muscle cells of different physiological type are mixed together apparently randomly even though their responsiveness ranges from totally inexcitable to large all-or-none spiking. Innervation is from a slow excitor, a fast excitor, and an inhibitor, with a strong tendency for single excitatory innervation only of individual cells. Thus specialized slow cells are innervated by the slow axon and the inhibitor and fast ones only by the fast axon. Intermediates are likely to have dual excitor innervation plus the inhibitor. At the time this work was carried out, all arthropod research had relied only on sampling electrical activity in single muscle cells while recording tension from either the entire muscle or a major fraction. Because of the heterogeneity of the cell population, there were many anomalies, including the possibility that a direct action of some inhibitors could occur, that is bypassing membrane-potential effects. Correlations of neural inputs made with tension in individual cells (Figure 12.41) showed unequivocally that all neural actions in crustaceans are mediated by way of mean membrane potential.

Figure 12.39. Neuromuscular physiology of the power-stroke muscle of a swimming crab, *Portunus.* (*1*) Drawing to show the location and anatomy of the large, triangular muscle, which has three separate flags attached to the same tendon: long, narrow fast white, broad light pink, short triangular deep pink. (*2*) Intracellular (middle traces) and tension (top traces) responses to stimulation of nine axons innervating different flags as indicated: fast (*F*) intermediate (*Int*) slow (*S*). Stimuli monitored on lower trace. The two fast axons innervating the light-pink portion supply the posterior (F_{post}) and anterior (F_{ant}) portions respectively. The slow deep-pink portion requires high-frequency, long-sustained excitation for development of appreciable amounts of tension: it is unfatigable. (*3*) Electromyograms from left and right pink and white portions of the power-stroke muscles during slow, fast, and escape (fastest) swimming behavior. [All from G. Hoyle and M. Burrows, Correlated physiological and ultrastructural studies on specialized muscles. IIIa. Neuromuscular physiology of the power-stroke muscle of the swimming leg of *Portunus sanguinolentus, J. Exp. Zool.* **185,** Fig 1, p. 85, Fig 2, p. 87, parts of Figs. 4, 6, 8, 9 on pp. 88, 91, 93 (1973). Reprinted with permission from Alan R. Liss, Inc., New York.]

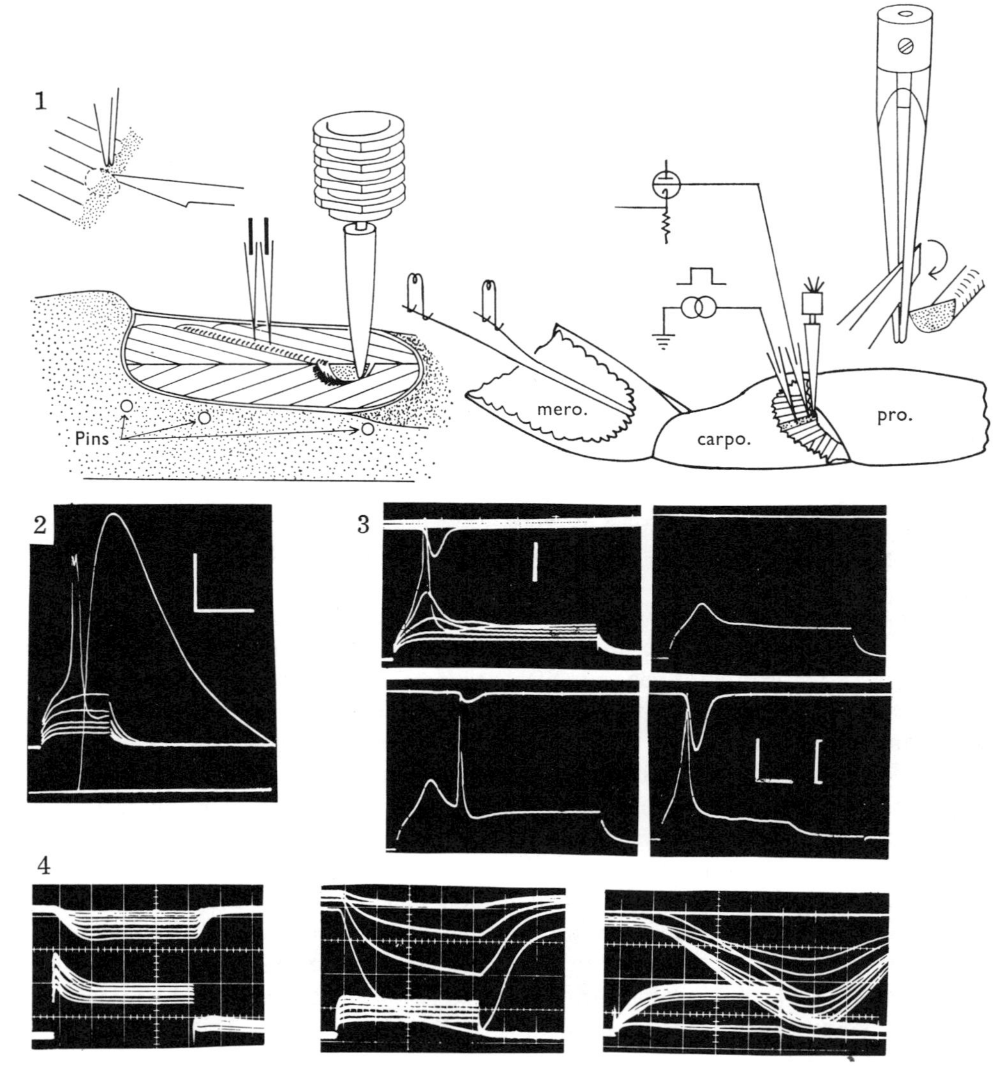

Figure 12.40. Responses of single muscle cells sampled *in situ* from crab *Cancer* dactylopodite flexor. (*1*) Diagrams illustrating technique for preparation of single cells for recording tension and direct and neural electrical events. (*2*) Example of a fast-type cell with all-or-none propagated spike electrogenesis and brisk twitch (increasing tension upward). (*3*) Example of electrogenically gradedly responding cells. In each case the E-C coupling threshold was high and exceeded only be large spikes. (*4*) Examples of electrically inexcitable cells. In each the membrane resting potential was very close to E-C coupling threshold and even a small depolarization evoked a contraction. The cell at left is not giving graded responses but rather an early onset of rectification, which was also present, but much weaker, in the middle example and not at all in the very slowly, though strongly responding cell at right. (All from Atwood et al., 1965.)

Table 12.2 Summary of Muscle Cell Types Found in the Power-Stroke (Swimming) Muscle of the Swimming Crab *Portunus sanguinolentus*[a,b]

	White Axons			Light Pink Axons				Deep Pink Axons		
Innervation	1	2	3	1	2	3	4	1	2	3
Twitch times(ms)	110	160	200	–	–	160	160	160	180	–
Rise time to peak contraction(ms)	Twitch 30	Twitch 40	(Tetanus) 120	Tetanus only 1000	1000	Twitch 20	40	80	100	700
Decay from peak(ms)	Twitch 80	Twitch 120	Tetanus 200	Tetanus 750	750	Twitch 40	120	800	800	800
Electrical excitability	Propagated action potential			Graded – propagated action potential				Inexcitable – variously graded		
Behavioral role	Escape swimming only	Fast swimming and escape response		Slow swimming – active also in fast swimming and escape swimming		Recruited during fast swimming and escape		Probably not involved in swimming		

[a]From Hoyle, G., Burrows, M. (1973). Correlated physiological and ultrastructural studies on specialized muscles. 1973a. Neuromuscular physiology of the power-stroke muscle of the swimming leg of *Portunus sanguinolentus*. *J. Exp. Zool.* **185**, Table 1, p. 86. Reprinted with permission from ALAN R. LISS, INC. N.Y.
[b]See Figure 12.39.

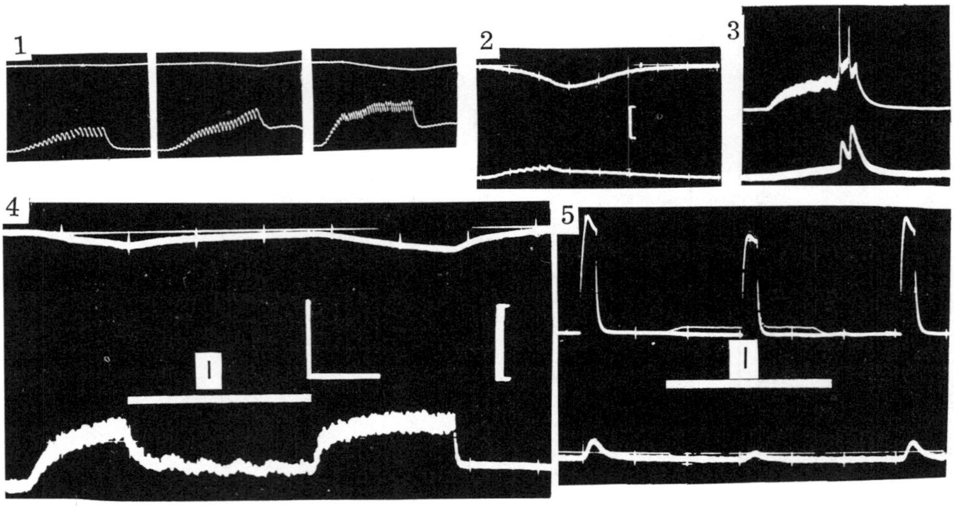

Figure 12.41. Intracellularly-recorded response of innervated crab *Cancer* muscle cells to neural stimulation, with tension recorded from the same cell only (see Figure 12.40). (*1*) Responses of an electrically inexcitable cell to slow-axon stimulation at increasing frequencies. Increasing tension (*top traces*) downward. (*2*) Response of a slow-type cell whose resting potential was poised at E-C threshold. (*3*) Responses of an intermediate-type cell with graded electrogenic responses. Tension (*on lower trace*) increasing upward. Very little depolarization was required from the facilitating, summating EJPs to initiate tension. The spikes promoted much faster, strong twitches superimposed on the slow tension. (*4*) Peripheral inhibition (*I*) applied, as indicated by the bar, of a slow-axon-excited cell. As (*1*). (*5*) Test of effectiveness of peripheral inhibition of a cell with a high E-C threshold. Three constant-current pulses were applied intracellularly. Inhibition (*I*) was applied as indicated by the bar, so that the second of the three occurred during the inhibition. The tension response was reduced. (All from Atwood et al., 1965.)

12.11.2.4. Levator of an Eyestalk

The eyes of crustaceans are located on articulated, movable stalks. Although this gives them maneuverability, it renders it necessary for the muscles operating them to possess subtle control mechanisms. The eyestalks must be kept raised and still for hours at a time by unfatigable cells, whereas at other times the eyes are required to move over a wide range of speeds with extreme angular accuracy in following objects moving in the visual field. Also, for their protection there are grooves in the front of the carapace into which the stalks fit, and sockets for the eyes. When the animal is disturbed, the stalks are rapidly depressed into the grooves. In the Hawaiian genus *Podophthalmus,* the eyestalks are very long, about 5 cm in a large specimen, operating like periscopes, permitting the body of the crab to remain hidden in mud while watching for food.

The levator muscle comprises separate bundles of muscle cells, one comprising short-SL, fast, white cells, the others long-SL, slow, pink-colored cells. The white portion receives one fast axon exclusive to itself plus another that it shares with the pink portion (Hoyle and McNeill, 1968a). The pink portion receives an exclusive slow axon and also an inhibitor. Some of the cells in the pink portion present an unusual but very interesting situation: their resting potential is so far above E-C threshold that they are in a constant state of mild excitation, which is the same as having strong BT (see p. 530). The tension is sufficient to cause the eyestalks to be normally in the raised position even in the absence of excitatory neural input (Hoyle, 1968a). Either a strong contraction of the antagonist, action of the inhibitor, or both, are needed to lower the eyestalk into its protective groove. The muscle can relax more quickly than it can contract (see p. 206). The pink color, like that of the deep pink cells used in raising of the swimming paddle of *Portunus* (Figure 12.42) is due to the presence of a thick peripheral layer of mitochondria (Hoyle and McNeill, 1968a).

12.11.3. Effects of Genetically Determined Disuse

A very interesting specific comparison has been made between eye muscles of a cave-dwelling species of crayfish. *Procambarus erythrops,* with *P. clarkii.* The cave dweller has fewer muscle cells, with smaller diameter (causing a high input resistance) in homologous muscles. But these differences are slight compared with the loss of the retina and most of the optic ganglia. The eye reflexes are completely intact (Mellon and Lnenicka, 1980). It is assumed that the retina/ganglion losses are associated with lack of use, but perhaps it was their loss genetically that left this species no alternative but to make a living in dark places! It does not follow that a genetically determined organ system will be lost just because it is not used.

12.11.4. Transmitter Substances

The predominant excitatory transmitter substance onto skeletal muscles of decapod crustaceans was first recognized to be L-glutamate during studies on the effects of amino acids on crayfish muscles (Robbins, 1959). L-Glutamate depolarizes many different crustacean muscle cells when applied iontophoretically in low concentration. The cell body of a major motor neuron of the lobster can be first identified functionally by direct intracellular stimulation in association with muscle recording and then removed by dissection followed by suction into a fine pipette (Otsuka et al., 1967). Clusters of like cell bodies, or even single large ones, can then be analyzed chemically. Excitatory cell bodies contain sufficient glutamate, 10^{-10} mol, to indicate that specific synthesis is occurring in them. Similar treatment of inhibitory cells shows a similar concentration of GABA, but only $\frac{1}{10}$ as much L-glutamate as in excitatory cell bodies.

Neither amino acid is destroyed by a specific enzyme equivalent to acetylcholinesterase in cholinergic transmission. They are presumed to be lost from the junctions by diffusion combined with deamination. The trick of reducing the loss of the transmitter at the synapse by blocking the local enzyme, as with the use of eserine in vertebrate cholinergic systems, is simply not available for any arthropod. Indeed, there is apparently no equivalent enzyme in arthropods. Nor is there a known competitive antagonist such as curare, or even suitable agonists. This means that the evidence in support of L-glutamate and GABA's being utilized is less complete than in the case for ACh at vertebrate junctions.

12.11.5. Crustacean Neuromuscular Transmission as a "Model" System

Advancing techniques and especially the widespread use of multiple intracellular recording from small cells in association with their identification and mapping are rapidly rendering obsolete the earlier approach to understanding based upon the use of "model" systems. Peripheral inhibition of crustacean muscles was early used as a model of central inhibition processes of vertebrates, which for a long time could not be examined directly. Later, the subtle facilitation and summation events were regarded as models for central-

1
3
2
Br
UT

nervous integration in vertebrates, the crab or lobster being said to "think in its claws." Much crustacean research was "justified," or even naively pursued, with this concept in mind.

It is probably fair to expect that every process known in crustacean neuromuscular synapses will be found to have been utilized by some vertebrate central synapse or other. But it is no longer necessary to pretend that the information may be directly transferred: vertebrate central-integration mechanisms are amenable to similar analyses, particularly by the use of tissue slices, so they can be examined directly. However, vertebrate brains are enormously complex structurally, and it is already doubtful if they will ever be truly comprehended by the human mind. Has the crustacean research aided in achieving the current level of understanding of vertebrate central-nervous function? It is possible that the specific knowledge derived from crustaceans in the past, which is lodged in the minds of the current investigators, will guide them to seek, or be alert to, similar phenomenona in vertebrates, but probably the usefulness ends there.

The true significance of crustacean studies is for biologists interested in muscle, neurobiology, behavior and its neural control, and evolution. It really is impossible to provide a simple statement that fairly covers "the general crustacean muscle." Crustacean muscles seem to offer every conceivable alternative to any generalization one may make about them beyond those mentioned at the beginning of this chapter that also apply to arthropods as a whole. For example, just when it seemed that long sarcomeres (10 μm and above) were specifically associated with slow contraction, and all such cells had only weakly electrically excitable or inexcitable membranes, a bundle was found in the extensor of the carpopodite of the crayfish *Procambarus* that gave overshooting, propagated spikes (Ozeki, 1969).

It should be borne in mind in evaluating these complexities that evolution cannot reverse itself: it must be additive. Suppose a muscle has acquired some slow-type muscle cells structurally, by way of increased SL, that enabled it to show a slow behavior of adaptive value. Later, as another aspect of its life changed, it needed to have speed in the muscle rather than slowness. It may not have had any mutations reverting the long-sarcomere cells to shorter ones, but some may have occurred that endowed their membranes with all-or-none spiking. This would at least partly counteract the intrinsic slowness associated with long SL.

A feature that a particular set of muscle cells may be required to live with is a small number of fixed motor programs. If these are all fast phasic, it would be to the animal's advantage to have muscle-cell structures, membrane excitability, and excitatory

Figure 12.42. Ultrastructural features of interest in two specialized crustacean muscle cells. (*1*) EM of LS of fast white cell from medial extensor of mantis shrimp *Hemisquilla.* Note wide Z band with dual denser bands, double M band. (*2*) EM of TS of (*1*), crossing one set of M bridges (*arrows*). Note abrupt transition to thick central bare zone. Ultrathin filaments (*UT*) are conspicuous around the thick filaments in the H zone. Bridges (*Br*) occur between thick filaments. (*1*) and (*2*) from P. A. McNeill et al. Fine structure of muscles controlling the strike of the mantis shrimp. *J. Exp. Zool.* **179**, part of plate 4, p. 402, plate 5c, p. 402, (1972). Reprinted with permission from Alan R. Liss, Inc., New York. (*3*) EM of TS of slow unfatigable deep-pink cell from power-stroke muscle of Hawaiian swimming crab *Portunus sanguinolentus.* Note thick layer of mitochondria surrounding contractile material, less SR in core. [From G. Hoyle, Correlated physiological and ultrastructural studies on specialized muscles. IIIb. Fine structure of the power-stroke muscle of the swimming leg of *Portunus sanguinolentus. J. Exp. Zool.* **185**, Fig 5, p. 104 (1973). Reprinted with permission from Alan R. Liss, Inc., New York.]

neuromuscular-transmission mechanisms to match. Conversely, slow programs could not be executed with only twitchy cells. There are examples of good matching of cells to programs, for example, in the three muscles of the second maxilla of crayfish (Moody-Corbett and Pasztor, 1980). But such matching can only be at the expense of versatility in the adoption of additional motor programs or new varieties thereof in the future.

12.11.6. Principles of Control of Crustacean Muscles

The innervation patterns and neuromuscular-transmission properties of crustaceans, in common with those of other arthropods (see p. 488), although very diverse in details, nevertheless possess shared features that distinguish them from all other major phyla. These are the following:

1. Small number of excitatory axons per whole muscle, commonly one or two and not more than about 8.
2. Peripheral inhibitory axon or axons often present in addition to excitor(s).
3. Shared, polyneuronal innervation of individual muscle cells.
4. Multiterminal distributed synapses on muscle cells from the axons.
5. Different characteristics of neuromuscular transmission by different excitatory axons, particularly fast compared with slow.

The above properties dictate certain aspects of control, especially the following:

1. Selection of the appropriate excitatory axon(s) to be activated in order to achieve the relevant movement.
2. Achieving a neuron's excitation at a frequency appropriate to the task in hand.
3. Adjusting the frequency with time to produce the needed tension profile.
4. Setting the timing and frequency of inhibitory axon discharge, for those actions that require peripheral inhibition.
5. Planning orderly recruitment of excitatory axons for muscles in which several excitors act synergistically.

Items 1 to 3 in control are general. Only one case of item 4 is known, but it is very important one: the stretcher/opener of the claws of crayfish and lobsters, in which a single excitatory axon supplies both the opener and the stretcher (see p. 495). Opening without stretching, which is essential in feeding, can only occur when the stretcher inhibitor is also activated.

Several crustacean muscles fall into the category requiring operation of the fifth principle. One way this is done is for the nervous system to use a size principle. The latter is the simplest, in principle, to "program" and will be dealt with first, but it applies to only a few of the muscles.

12.11.6.1. Operation of a "Size Principle"

Swimmerets of lobsters are operated by 12 small muscles; the principal power stroke is provided by a cluster of four muscle-cell bundles. Innervation of the muscles is by at least three shared excitatory axons that produce EJPs of different sizes. These have been related to particular somas that average 55 μm (small), 85 μm (medium), and 125 μm (large) respectively (Davis, 1971). The small one evokes the smaller EJPs, which facilitate very extensively and fatigue little. The medium one evokes initially larger EJPs that facilitate less than those of the slow cell. And the largest one evokes larger EJPs that facilitate little; they evoke, respectively, slow, medium, and fast contractions and may themselves be so labeled. Davis was able to cause natural swimmeret beating by continuous electric stimulation of an interneuron. This was better than reflex excitation because it removed the ambiguity associated with uncontrolled input: in reflexes, there is a sequence in the order of requirement of sensory axons, the smaller ones being excited first. When the interneuron threshold was exceeded, at a suitable frequency, beating always followed. When the power stroke occurred, there was always excitation of all of three motor neurons but the onset of their discharge followed an orderly sequence: slow → medium → fast.

A recruitment in order of increasing soma size of motor neurons in a common motor neuron "pool" had been recognized for mammals by Henneman et al. (1965) and termed by them *the size principle* (see p. 606). It appears that a similar principle is operating in the swimmeret system. The underlying mechanism is a very simple one, depending only on shared common inputs from the driver interneuron(s). Assuming qualitatively and quantitatively similar inputs, the order of discharge is then solely dependent on the passive electrical properties of the follower motor neurons, which are directly related to their size.

There are doubtless other crustacean muscles whose control follows the same pattern. This is particularly likely to be the case in walking and has already been shown for a swimming crab (Hoyle and Burrows, 1973a). Slow walking and swimming involve only the slow motor axons. As stepping speeds up, fast neurons acting in parallel with the slow ones are automatically recruited. But this can only be applicable when there are several excitatory motor neurons supplying the same muscle or group of synergists. Even there, operation according to the size principle is obligatory only in instances in which an excitatory interneuron innervates all the excitatory motor neurons equally. In some phasic movements such as eyestalk flicks of crabs, the crustacean needs (and functionally is able) to activate a fast motor neuron without at the same time exciting slow and intermediate ones.

12.11.6.2. Motor-Program/Motor-Pattern Generation (MPG)

Behavior is produced by a sequence of contractions, the overall *motor program.* Within each burst of excitation is a detailed frequency/time profile for each motor neuron involved, which is termed *motor pattern generation* (MPG). Since some motor patterns are generated by sequences of reflex actions the expression *central pattern generator* (CPG) has often been used to distinguish that contribution to a motor pattern that is generated by the central nervous sytem independently of reflex sensory neural input information. The first step in understanding MPG is to determine quantitatively the output associated with a particular behavior. This is not an easy task for any animal. Studies that have attempted to quantify motor output to crustacean leg muscles have been frustrated in spite of the simplicity of the motor innervation. It is clear that this is in part due to there being nine other legs operating at the same time, each making simultaneous inputs that influence the one being examined in detail (e.g., Evoy and Fourtner, 1973). There are disappointingly few studies on the actual patterns of motor-axon outputs used by crustaceans in natural movements, though this information is not difficult to obtain.

The ideal approach to this is to first study innervation details using axon isolation and stimulation, with extracellular nerve or muscle leads. For recording from axons *in situ,* a cuff containing recording leads is first implanted round the motor nerve. Both of these techniques were first developed for use with insects (see p. 550). Since the leads can simply be long, fine trailing wires, a great deal of freedom of movement may be permitted, especially for aquatic animals that can be confined to a tank. But even for the fastest-running terrestrial crab known, *Ocypode ceratophthalma,* it has proved to be possible to get good recordings by placing the crab in a narrow runway and running alongside it while hand-carrying the lead holder and cable (Burrows and Hoyle, 1973). A less satisfactory alternative is to restrain the animal and let it run on the spot, as it were, or walk on a treadmill, wheel, or disc. Movements studied in these ways include the mantis shrimp strike (see p. 500, Burrows and Hoyle, 1972), swimming by a portunid crab (see p. 504, Hoyle and Burrows, 1973a), walking by a crab (Barnes et al., 1972; Barnes, 1975), and walking by a crayfish (Barnes, 1977; Evoy, 1977). There is a larger body of pertinent information derived from semiintact, but much more restrained, preparations, for example Clarac (1977a). Had the same preparations in addition been recorded as freely moving ones, valuable information for understanding motor programming would doubtless have been obtained.

Often the investigators had a restricted area of interest in mind during their experiments, the commonest being to study the role of reflexes in the generation of locomotion. Earlier, most interest was in trying to determine if any complex behaviors can be determined solely by CPG without reference to proprioception. No hard evidence was obtained for this in crustaceans. However, many relatively complex movements, some of which are classifiable as significant behavioral acts, such as the "startle/defensive" posture of crabs, crayfish, and lobsters, can be elicited by unpatterned stimulation of single interneurons, so some pure central programming can occur.

12.11.6.3. Single Premotor Interneurons that Drive Coordinated Behaviors

Interneurons that evoke coordinated movements recognizable as natural behavioral acts could, perhaps, be turned on by the "will" of the animal and so be classed as centrally programmed, though there is no evidence yet that they are. Interneurons of this type have been the subject of much discussion and disagreement. In a count of expert opinion made in 1978 (see commentary by Hoyle in Kupferman and Weiss, 1978) opinions were about equally divided as to whether or not they should rationally be called *command interneurons,* as they often are. This flamboyant descriptive term was first applied to them by Wiersma and Ikeda (1964), and it was unfortunate because *command,* used adjectivally, conveys a behaviorally highly significant implication, that is, that these are the neurons used by the organism to evoke this behavior when it has integrated sensory inputs at its "highest" neural center and made the (pontifical) decision to require this particular one to be put into operation. Such interneurons may well exist in many nervous systems, but the criterion for their identification is the demonstration experimentally that the animal actually uses them to initiate the behavior when it is evoked by "will," that is, other than reflexly. Some of the neurons that have been labeled "command" cells may be such, but none have yet been proved to be. The cells to which the label command has been applied undoubtedly excite follower interneurons and motor neurons in such a manner that a discrete, coordinated behavior results. Other terms have been used for such neurons, including *driver interneurons,* which I proposed at a meeting held in 1962 (Hoyle, 1964b). Neurons such as the crayfish medial and lateral giant axons that mediate fast escape swimming responses to sudden stimuli have been regarded by some as prime examples of "command" interneurons (Evoy and Kennedy, 1967), as have the giant axons of the cockroach abdomen. However, the escape behaviors that electrical stimulation of them evokes still occur when giant axons are absent (Dagan and Parnas, 1970; Krasne and Wine, 1977). Also, ascending impulses, as in both of these examples, are clearly going the wrong way to match the meaning of the term *command*. The complexity of follower neuron responses to driver-interneuron action is readily seen, even in a simple movement uncomplicated by limb coordination requirement – flexion of the crayfish abdomen. Interneurons exist that drive slow abdominal flexor muscles, and others drive the extensors. When either of these is stimulated at a constant, adequate frequency after a short delay, the antagonist, as well as the agonist, discharges, though at a lesser frequency (Evoy and Kennedy, 1967). The antagonist discharge is accompanied by discharge of the inhibitor to that muscle. Discharge of the slowly adapting muscle receptor organ (MRO) reflexly excites only the slow tonic system of the tail – the superficial flexor muscle cells that are themselves slow, associated with a long SL (Fields and Kennedy, 1965). Their contraction is unlikely to make a significant contribution to the behavior.

12.11.6.4. Proprioceptive Sense Organs

Crustaceans are richly endowed with sense organs registering joint angle and/or movement. These include strain-type organs, in which a relatively large length change of the organ itself occurs, and stress-type organs, in which there is deformation but little length change. Doubtless they all play major roles in the control of movements, although they provide too much information for more than a small fraction to be continually assessed during the production of behavior. The ratio of sensory to motor neurons is several thousand to one, but each is capable of initiating or modulating movements or adjusting posture. An extensive description of the various types has been presented in the book edited by Mill (1976). An incomplete list includes *stress* – slit sensilla, campaniform sensilla, hair plates, cuticular articulated pegs (CAP organs), hair sensilla; *strain* – MROs and chordotonal organs. Although a great deal is known about the sense organs themselves and their properties, little is known about their involvement in posture and movement control. In an animal such as a lobster, in which many behaviors are generated by driver interneurons, whose functioning is necessarily independent of proprioception, it is almost surprising that there is such a rich variety of complex postural receptors. One possible explanation (Hoyle, 1975b) is that they are not much used during the actual generation of behavior but rather during development and practicing, to perfect the functioning of inherited CPG neuronal circuitry, that is, during the initial development of the neural connections and the later maintenance of details of transmission at the synapses: although much of the elemental circuitry is genetically determined, the finer details are left to be set and reset if necessary as a result of experience. Some major organs will be considered individually.

Muscle receptor organs (MROs). The most elaborate movement/position-sensitive organs are MROs, located in pairs, one slowly adapting, the other rapidly adapting, in both left and right halves of each segment of the abdomens of crayfish and lobsters. Each comprises a single muscle cell with SL appropriate to function, long for the slowly adapting and short for the rapidly adapting. A central cell body gives off a single axon to the central nervous system (CNS) and numerous small dendrites into the muscle cell. The muscular portion of the receptor receives typical distributed excitatory innervation and presumed inhibitory innervation also to the slow one. The sensory cell body and nearby dendrites are also independently innervated, by an accessory inhibitory axon. This provides the animal with a peripheral means of annulling the reflex effects initiated by stretch. The excitatory innervation of the muscle element provides the animal with an indirect means of controlling abdominal posture, since contraction must have the same effect as abdominal flexion that stretches the organ.

The overall anatomy of the arrangement provides a closed-loop control system, since the sensory cell excites motor neurons supplying the principal extensor muscle (Figure 12.43). Provided the organism could send posture-setting instructions to the muscular part of the MRO, reflex load compensation would be introduced automatically. However, when the abdomen flexes, one reflex consequence is activation of the accessory nerve, causing the organ to inhibit itself (Eckert, 1961a). Therefore, the organ can only be used in phasic activity, notably in the repeated tail flips of the escape swim (Eckert, 1961b), unless the inhibitory output to the slow MRO can be suppressed, as it might be during a centrally programmed slow movement sequence. That this can happen was demonstrated in crayfish by Fields et al. (1967), so the load-compensation servo device is available to the organism. The reflex excitation evoked is of superficial extensor muscle cells, which are of slow type. Position control is thereby possible in the face of variable load, although this system works only during behavior that is centrally driven, indirectly determined via proprioceptor setting. It is not part of the basic reflex mechanism, which is aimed solely at supporting fast tail flips.

Chordotonal organs. The most widespread of all arthropod strain receptors are called *chordotonal organs.* These commonly contain a cluster of, though some but a single, sensory element(s) characterized by the presence of an internal dense lattice associated with microtubules, called a *scolopale.* Similar cellular structures are associated with a variety of surface sensillae, but these are stress or deformation receptors. The length change in a chordotonal organ is not, however, within these cells, but in an elastic strand to which they are attached. This is in turn attached to a stiff, thin tendinous extension attached to a joint. In some, there are one or even two small muscles attached. When a muscle is involved, the organ is termed a *myochordotonal organ.*

A myochordotonal organ that has been extensively studied is located in the proximal region of the principal leg joint of crabs, the meropodite, and registers the angle this makes with the next more distal joint, the coxopodite – hence termed the MC joint. Attached to the tendoextension are a larger proximal muscle and a smaller distal one

Figure 12.43. Stretch receptors of decapod Crustacea. (*1*) Sketch of lateral longitudinal section of one side of a lobster abdomen to show the insertions of the two thin muscle cells associated with the slowly adapting (RM_1) and rapidly adapting (RM_2) stretch-receptor cells. (From Alexandrowicz, 1951.) (*2*) Photomicrograph of a pair of stained stretch receptors of the lobster *Homarus vulgaris.* At left is a sketch to show the shared excitatory axon (*E*) supplying the muscle cells and the inhibitor (*I*) that directly inhibits the sensory cell. The muscle cell on which RM_1 sits has long sarcomeres and is a slow cell; that on which RM_2 sits has short sarcomeres and is a fast cell. (From Alexandrowicz, 1951.) (*3*) Drawing of a part of a dissected thorax of the crab *Carcinus maenas* to show the location of the several stretch receptors serving a single leg. (From Alexandrowicz and Whitear, 1957.) (*4*) Enlarged view of proximal region of (*3*). The two larger receptors termed, T and S respectively, are each a single cell of giant size, up to more than 0.1 mm in diameter, are electrically inexcitable, and transmit electrotonically. (From Alexandrowicz and Whitear, 1957.)

1

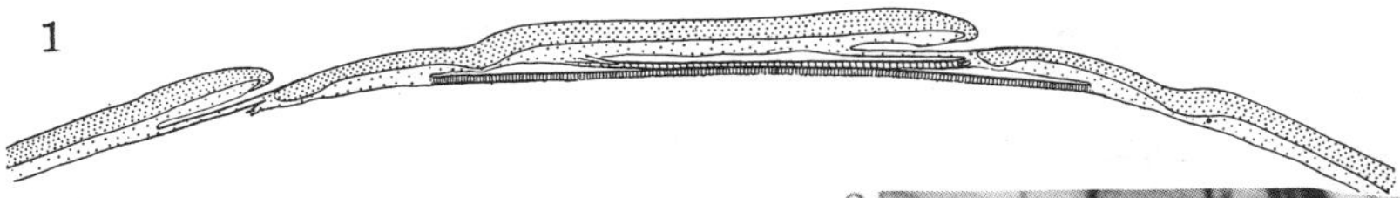

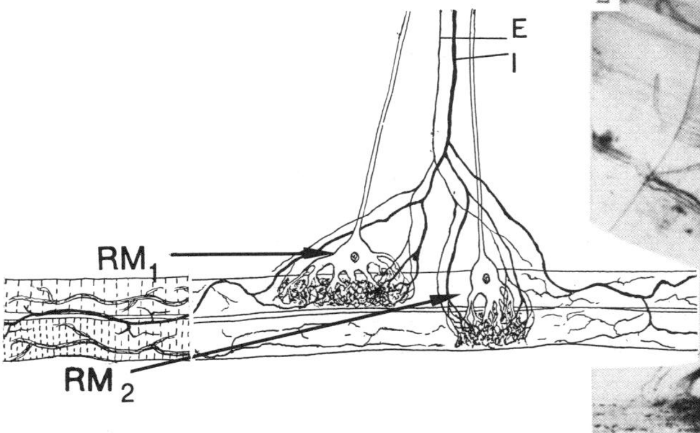
E
I
RM1
RM2

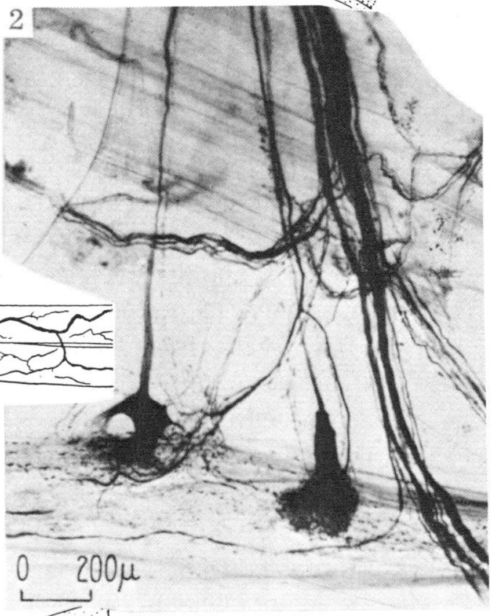
2
0 200μ

3

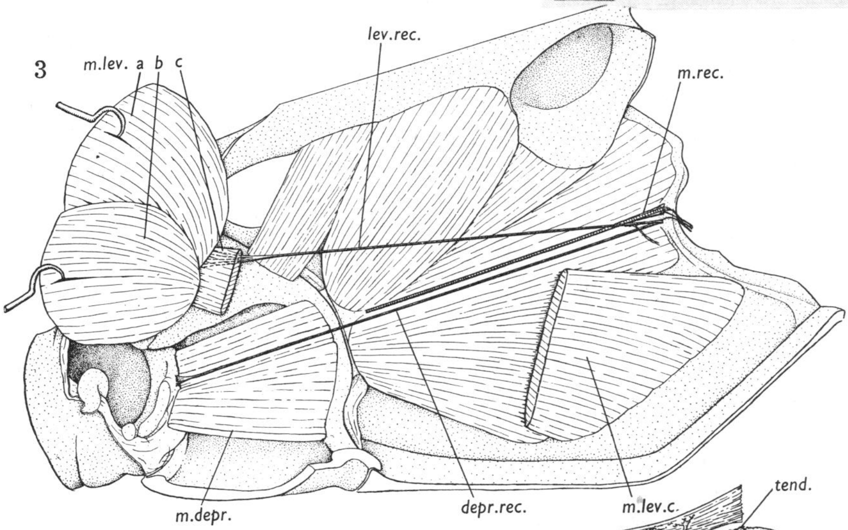
m.lev. a b c
lev.rec.
m.rec.
m.depr.
depr.rec.
m.lev.c.

4

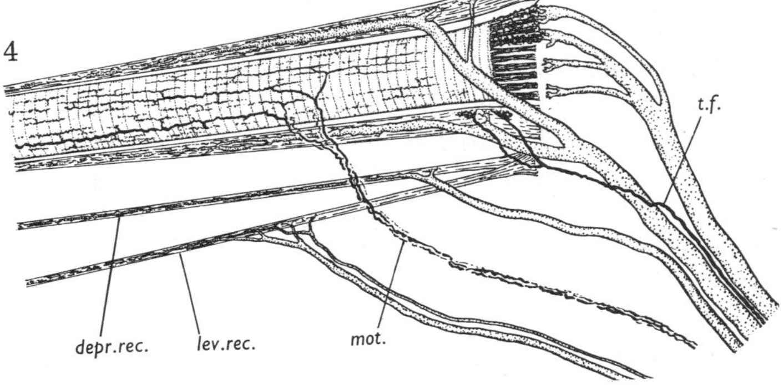
tend.
t.f.
depr.rec.
lev.rec.
mot.

(Figure 12.44). Therefore, although very different structurally from the MROs, the myochordotonal organ has exactly similar functional capabilities. The properties of component sense organs are such that they divide the angular movement range between them, some responding in one part, others at another, to produce *range fractionation.* Some elements are rapidly adapting, others slowly adapting, covering the entire range both phasically and tonically (Cohen, 1963).

The most prominent action so far established to be associated with the organ is that of initiating a *resistance reflex.* The myochordotonal organ is itself attached to the same place on the carpopodite as the flexor and was originally thought, because of its major muscular portion, to be an accessory flexor muscle. When the organ is stretched, as it would be by carpopodite extension, the slow extensor motor neuron is reflexly excited (Cohen and Evoy, 1969). The control system is further complicated, however, because there are also two distinct additional sets of proprioceptive sense organs in the distal region of the meropodite that are also affected by the same movements, and yet another set associated with both the flexor and the extensor tendons (Macmillan and Dando, 1972). These are spontaneously active, and increase in isometric tension increases the rate of discharge.

The distal accessory flexor, the relatively large distal muscle associated with the MC myochordotonal organ, is innervated by a single excitatory axon and a single inhibitor. This is the muscle that has a progression of muscle cell types from proximal to distal (see p. 490), so is capable of phasic or tonic response depending on the input frequency and duration. In a heroic effort to determine the involvement of myochordotonal organs in crab behavior, Evoy and Cohen (1971) removed them from the entire set of walking legs of a large *Cancer magister.* The deprived crab sat around after the operation for more than a week. But just when it began to look as though walking were no longer possible, the crab gave a fine display of perfectly normal looking walking.

The crab was doubtless partly sick from the effects of the operation, but possibly also suffering from lack of the stimulatory effect on behavior initiation engendered by inputs from the organs: it spent all of its time in a slumped posture. The latter is characteristic of normal crabs a lot of the time, and in this condition they are rather unresponsive and also spontaneously inactive. In an alternative "alert" condition, the crabs have some posture, the legs being slightly extended and both active spontaneously and readily responsive. Transition from slumped to alert, or vice versa, could therefore be brought about by altered central output to the proprioceptors. The descriptive term applied to changing states, borrowed from psychology, is *mood setting,* and it applies to most animals. It is important to bear in mind that the *central excitatory state,* as Sherrington (1906) termed it, of vertebrates, fluctuates for both known and unknown reasons. The former include daily and seasonal rhythms, sex-hormone levels, and hunger/satiation. In insects (see p. 533) and man (see p. 418), and possibly in crustaceans also, there are sufficiently large fluctuations in blood composition, including mineral-ion concentrations, associated with feeding to affect neuromuscular transmission and thereby behavior also, even if the blood-brain barrier and nerve sheaths render the CNSs immune to them. All physiological mechanisms associated with posture and movement are subject to such modulatory influences.

Electrically inexcitable stretch receptors. In the thorax of a crab is located a set of receptors, one for each leg, that register movement of the base of the leg, the coxopodite. These have attracted attention because they are not electrically excitable and rely upon

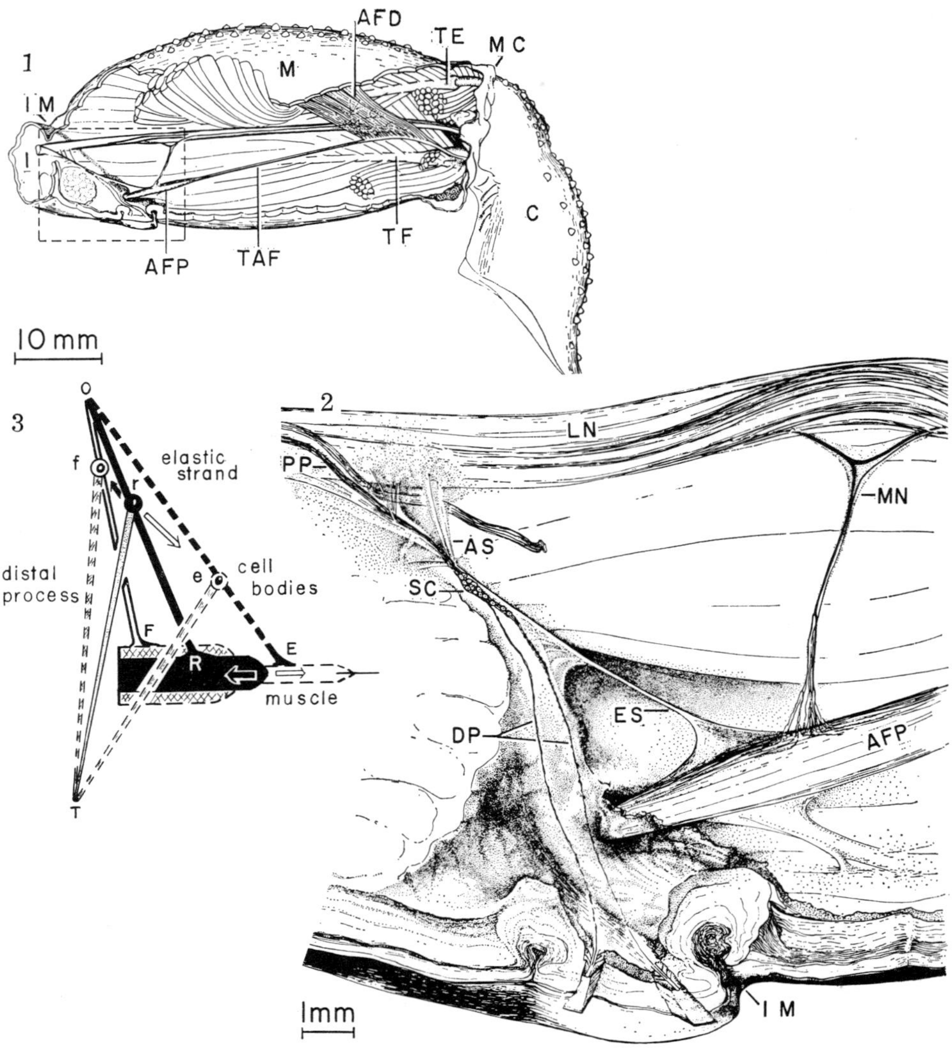

Figure 12.44. Chordotonal-organ type stretch receptors of crab *Cancer magister* leg. A cluster of sensory cells (*SC*) is stressed when the joint is extended, stretching an elastic strand (*ES*) to which they are attached. The strand in turn attaches to a small muscle, the proximal accessory flexor (*AFP*). A long, thin tendon (*TAF*) extends from this to the joint. A somewhat larger muscle, the distal accessory flexor (*AFD*) attaches to the same tendon directly. Contraction of either muscle reduces the stress caused by extension. (*1*) Drawing of dissected leg to show location of the organ. (*2*) Enlarged view of proximal region outlined by broken line in (*1*). (*3*) Diagram to show mechanics of the receptor. [All reprinted with permission from M. J. Cohen, The crustacean myochordotonal organ as a proprioceptive system. *Comp. Biochem. Physiol.* 8, (1963) Copyright 1963 Pergamon Press, Ltd.]

electrotonic propagation to transmit to the CNS information about the stretched condition. Each is no more than a branched sensory-nerve terminal whose dendrites run along either a special muscle cell, or one or another elastic strands. There is a pair, S and T, for each organ, both giant cells with diameters exceeding 50 μm. Contraction of the muscle must affect them. The nature of their action has been confirmed by directly depolarizing them via an intracellular electrode (Cannone and Bush, 1980). Graded depolarization evokes progressively faster discharge in three synergists.

When stretched rapidly these receptors initiate excitatory discharge to the extensor of the MC joint, and when released suddenly, the flexor of the carpopodite and also the distal accessory flexor are excited. The frequency is proportional to the rate of movement (Bush et al., 1978). Reduced stretch of the MC receptors alone would be likely in turn to initiate extension, but this tendency is counteracted by contraction of the accessory muscle, which produces a contrary signal. Modulation of a proprioceptor by one of a different kind located in another segment is a unique, significant process suggesting that proprioceptive sensory inputs are not acting independently of each other, thereby requiring extensive central coordination to collate their messages.

Also acting in parallel with the above stretch receptor are two chordotonal organs. One, the coxobasal organ (CB), mediates an intrasegmental reflex involving the two levator and two depressor muscles of the same leg segment, as well as intersegmental reflexes involving all eight muscles of the more distal leg segments and also the muscles operating the tergal/coxal joint proximal to it (Clarac et al., 1978).

Two principal kinds of stress detectors occur in crustaceans: cuticular stress detectors (CSD) located within the exoskeleton, and tendon organs. The former are affected by gravitational or muscular forces that cause the exoskeleton to bend; the latter react to slight deformation of the tendon when a muscle attached to it contracts. The CSDs of crustaceans, unlike campaniform sensillae of insects and their equivalent in arachnids, lyriform organs, are not actually built into the cuticle. They merely have sensory filaments that are embedded in it. These occur together in clusters (Wales et al., 1970). The most dramatic known function of this type of sense organ is mediation of autotomy. Each limb has a fracture line of structural weakness in the segment termed the basiischiopodite, located between the coxopodite and the meropodite. Specific muscles are attached near this line of weakness, whose sole function is to cause a break there when it is very strongly stressed. When the leg breaks off, the opening to the body, which is narrow at this point, is closed neatly by further contraction of the autotomy muscle. Strong stress on the CSDs in the region directly initiate autotomy as a reflex (Moffett, 1975).

The CSDs tend to be associated with soft, relatively flexible areas of cuticle. Prodding these from the outside stimulates the organs. In both a crab, *Carcinus,* and a crayfish, such prodding initiates a discharge and also reflex actions on levator and flexor muscles, including the crab accessory flexor. Passive rhythmic extension of the carpopodite in the crayfish evokes matching discharges in the flexors with every extension (Clarac, 1977b).

Vertebrate tendon organs, named after Golgi, produce strong negative feedback onto their own muscle, protecting the tendon from damage due to too great force's being developed. It was natural to expect a similar function in crustaceans, but ambiguous results were obtained, including different effects from the same movement when repeated (Macmillan, 1976). Isometric contractions of the flexor of *Cancer* cause fairly strong discharges in the tendon-organ nerve that increase with increasing force. When the joint is passively flexed only very weak discharges occur. However, the receptor could serve to some extent as a movement indicator as well as a tendon-stress indicator.

It has been found that input from the organ in *Cancer* is inhibitory, but of the two strong resistance reflexes – flexor and extensor – involved in carpopodite movement. This suggests a role in preventing oscillation at the joint.

12.11.6.5. Utilizing Proprioceptive Input During Behavior

As pointed out earlier, MPG and detailed patterning of the individual neuron output would be difficult enough for a simple nervous system without the complications associated with complex and interactive feedback that have evolved to assist and improve their operation. A list of available options is presented on p. 596. One of the special interests associated with studies of crustaceans, with their relatively simple nervous systems, is that because they are amenable to analysis, it should be possible to determine experimentally for a particular piece of behavior which method is being used.

The crayfish tail-flip escape behavior is an example of the first type, simple reflex, though it approaches the second option, combined CPG plus reflex, when operated "voluntarily." There is good evidence that when certain kinds of walking occur in the crab *Cardisoma,* the resistance reflexes are shut off centrally, meeting option 3 (Barnes et al., 1972). Option 4, patterned output to muscular parts of MROs, is very attractive, as it is also for vertebrates (see p. 578), but unfortunately there is not a shred of evidence for it. Option 5, proprioceptively modulated central generation, is likely to be used in relatively simple, cyclical movements such as swimming. In the walking crayfish, there is direct evidence of its using Option 6, in which proprioceptive inputs are incorporated into the MPG (Barnes, 1977).

12.11.6.6. Is Peripheral Presynaptic Inhibition Used in Behavior?

Dudel and Kuffler (1961b) elegantly demonstrated that the phenomenon of supplemented inhibition discovered by Marmont and Wiersma (1938) is due to a presynaptic action; the same phenomenon has also been found at many vertebrate central-nervous synapses. Specifically, it appears to play a major role at sensory inputs to the spinal cord (reviewed by Schmidt, 1971). There, its persistence is certainly no accident. In those crustacean neuromuscular junctions where the phenomenon exists, can there be some action that is independent of the greater mechanical-suppression capability, which seems so unlikely to be feasible? One possible function, suggested by Kennedy (1977), is to mask progressive facilitation occurring in the excitor. This effect is not seen until the inhibitor stops firing; then there is a rebound contraction of greater speed and strength than the excitor can produce acting alone. This is an interesting notion and it is to be hoped that such a possibility will be sought by recording during natural behavior. However, it is the kind of thing one might expect from an animal that has been forced to live with an irreversible genetically determined factor that may have lost its original significance. I believe that the primeval function of peripheral inhibitory-type axons was in the regulation of basic tonus rather than any of these alternatives.

Among the many varieties of synaptic transmission that have been found in crustaceans is presynaptic inhibition of very long duration, which occurs in the closer of walking legs of the crab *Eriphia.* An inhibitory burst is not only almost completely effective against an excitatory one following after an inhibitory burst has ended, it is not until after 5 min that normal contraction strength is resumed (Rathmayer and Florey, 1974). This effect is due to persistence of the presynaptic action, not a postsynaptic one.

A probable value of presynaptic inhibition is in improving efficacy of low-frequency inhibitory discharge. While this will not exert a significant effect on fast phasic activity, it will reduce or eliminate contributions from specifically slow muscle cells, which would otherwise impede quick movements. Postural reflexes would be temporarily overridden by a background of inhibition, such as often occurs during phasic movement. The specific timing of individual IJPs would not be significant, only the overall inhibition.

12.11.7. Myographically Analyzed Behaviors

Very few specific behaviors of crustaceans have been analyzed in detail and only one, the crayfish tail flip, extensively. Walking has been a major general target, but the objective has been understanding patterning rather than principles used in neural control of muscles. For the tail flip, two forms of the behavior are recognized (Krasne and Wine, 1977), one mediated by the medial giant axon(s), the other by lateral giant axon(s). In the former, all the abdominal segments flex, so that the force is directed in line with its body. In the latter, only the anterior segments flex so that the force is partly directed downward, causing pitching. Each of these movements can be initiated by other unidentified interneurons, and so can slightly different overall movements.

The motor neurons involved are fast flexors innervating the large, deep abdominal muscles that provide the power stroke, and weaker dorsal extensors that restore the initial posture. Basic circuitry consists of common motor neurons synaptically excited by both giants, some that supply anterior muscles only, and other supplying the posterior muscles. Anterior muscles are excited by the lateral giant, and posterior ones by the medial but not the lateral giant. The lateral giant excites the flexor motor neurons at electrotonic synapses and at the same time directly inhibits the excitatory motor neurons and excites a peripheral inhibitor of the muscle cells. The extensors are then reflexly excited by dorsal stretch receptors extended by the flexion, aided by peripheral inhibition of the flexor muscle cells. Continued output from the giant axon now excites the flexor motor neurons again, and the cycle repeats. This behavior is only partially centrally programmed, by a mechanism that is dependent on a combination of anatomical connections and characteristics of inhibitory synaptic events (Wine and Hagiwara, 1978). It is basically a simple classical reflex chain.

Recordings have been made of the electromyograms of flexor and extensor, opener and closer, levator and depressor muscles of the legs of *Ocypode ceratophthalma* during natural running at various speeds (Burrows and Hoyle, 1973). The innervation and neuromuscular responses were determined independently, permitting a reasonably accurate assessment of the roles of the various motor axons supplying each muscle. There is a remarkable shift in behavior as higher speeds are reached, the greatest of which is 2.1 m/s, which theoretically requires leg movements at frequencies greater than 40 Hz. But it was found that the crab actually never swings its legs above 20 Hz and is physically incapable of operating above about 25 Hz. This apparent paradox is explained by a change from typical crablike gait, with both leading and trailing legs being on the ground, to one in which the trailing legs alone provide the power during fast running. The leading legs are lifted up all the time, and the crab advances using only the two most powerful legs by a form of bipedal leaping through the air, as seen in human running, from one leg to the other (Figure 12.45). The motor-output requirements are sufficiently great at the extremes to suggest that different central programs may be being utilized, but this need not be so. At moderate speeds, legs 2 and 4 of one side move together with leg 3 of the opposite side, in alternating tripods of supports, so 2 and 3 of one side are already providing thrust alternately. It only requires them to take over totally by upward tilting of the leading margin to achieve bipedal running.

Twitch times of the muscles involved range from 250 ms for dactyl to 70 ms for the

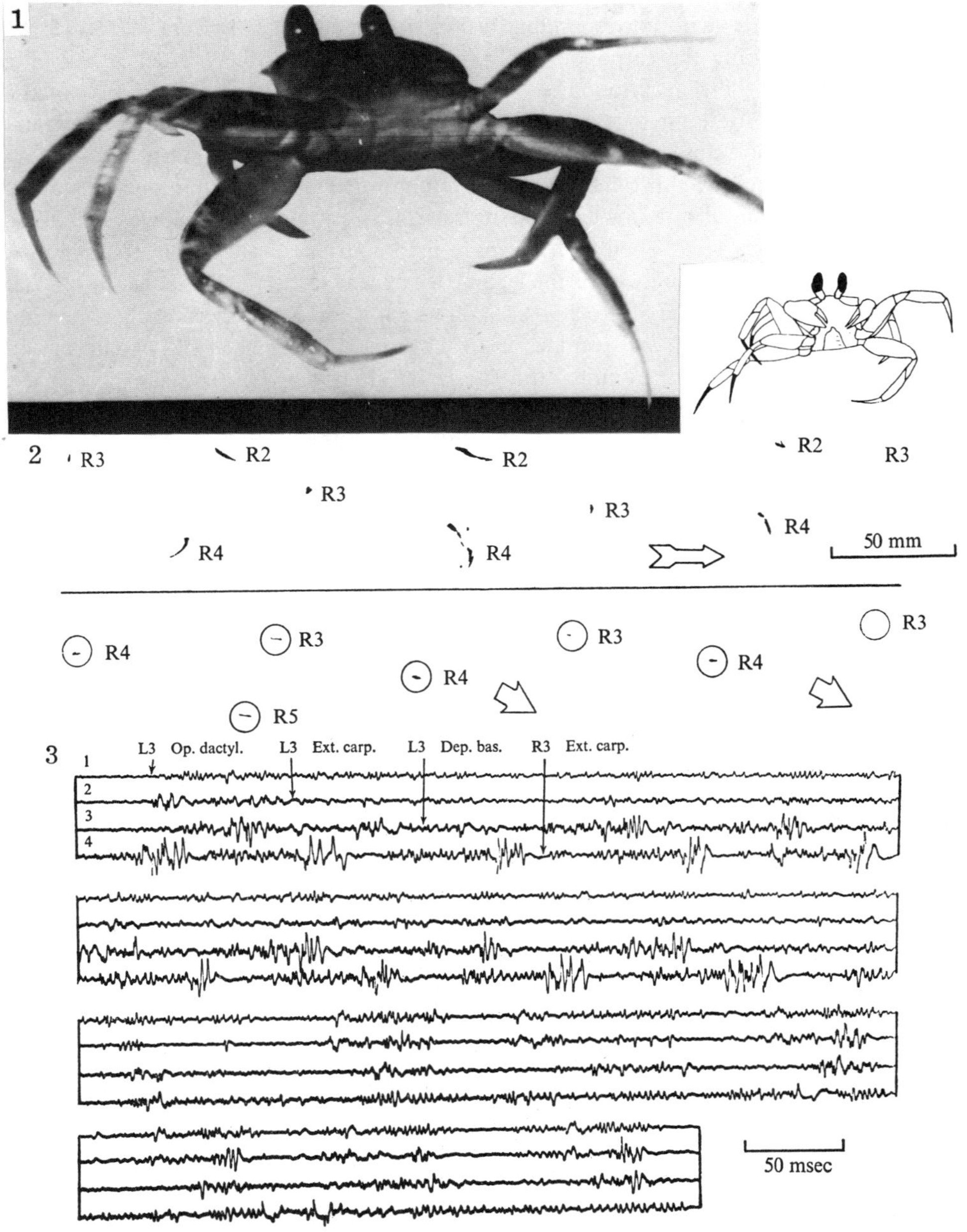

Figure 12.45. Rapid running by the ghost crab *Ocypode ceratophthalma.* (*1*) Flash photograph of crab during high-speed run. The crab is in the air during this "ballet leap," which is just one step in the running sequence. (*2*) Marks made on paper while the crab is running from left to right with all the legs painted (see drawing). Only the two middle trailing legs touched the ground. (*3*) Continuous myogram made from key muscles during running. *1*, left extensor/opener of the dactylopodite; *2*, left extensor of the carpopodite; *3*, left depressor of the basipodite; *4*, right extensor of the carpopodite. The crab ran to its right first, then to its left. There is only continuous weak tonic activity in the muscles of the left legs when they are leading, but large spikes in those of the right, and vice versa. (All from Burrows and Hoyle, 1973.)

extensor carpopodite, with corresponding fusion frequencies ranging from 100 to 15 Hz. The muscles operating the carpopodite always were the first to be excited, usually the flexor, followed by the dactylopodite "opener" and levator. On the return part of the cycle, the carpopodite is first flexed about the meropodite, then the leg is depressed, and finally the dactyl is "closed." There is a 25–30 ms gap before activity starts and the contralateral flexor fires. In ghost crab leg muscles innervated by more than one excitatory axon, which includes all the power-stroke muscles, a size principle is followed in making each burst causing a step. There is little more than continuous slow-axon tonic levator and extensor discharge in muscles of the leading leg except for leg 3. This is depressed and flexed so as to bring its dactyl into a position in which it can serve occasionally, when unbalanced or during slowing and turning, as a skid support. Even when the leading legs do touch the ground in slower running, the discharges are much weaker than those of the trailing legs. The bursts include fewer, or no, fast-axon spikes. Leading legs are probably doing little more than providing support, not pulling the animal.

During running by the ghost crab, the last pair of legs is kept raised off the ground throughout and does not contribute thrust. It is interesting to compare this situation with that found in control of the power-stroke muscle of the swimming crab *Portunus sanguinolentus* (see p. 504), in which the last pair of legs are the only ones used in its common form of locomotion. These legs are greatly enlarged and modified structurally to form a paddle, which has great flexibility and can be either "feathered" or swept powerfully. The range is accomplished mainly by motor-neuron recruitment, not by frequency adjustment.

12.11.8. Vegetative Musculature

With one exception, the stomach of spiny lobsters, there has been almost no modern interest in the vegetative musculature of crustaceans and its neural control. The stomach has been accorded extensive attention because of the accessibility of the approximately 30 neurons of the stomatogastric ganglion that control its operation. This ganglion exerts complete control, apart from generation of peristaltic waves, of the entire large stomach region. In general, crustacean guts seem to be to a large extent independent of the local nerve nets that characterize most guts. They have apparently advanced much farther in the direction of abandoning local control in favor of higher-level, centralized programming and control of gut movements than is the case with any other class of animals. They even appear to have eliminated much of the spontaneous activity intrinsic to the musculature of most guts. These factors may or may not have been correlated with the use of cross-striation in the contractile cells. All the known gut muscles of crustaceans are striated.

The decapod foregut and stomach have come to be operated more by muscles attached to them and to internal skeletal elements (ossicles) than by intrinsic muscle cells. Although internal and associated with vegetative function, they function like skeletal muscles. Several have been examined physiologically, notably foregut muscles of *Callinectes* and *Panulirus* (Govind et al., 1975) and pyloric and gastric mill muscles of *Callinectes*. Except for the dorsal dilator muscles of the crab pyloric region, all appear to be glutaminergic: the dilator seems to be cholinergic (Marder, 1976; Atwood et al., 1977). The excitatory neurons innervating gastric mill muscles have their cell bodies in the stomatogastric ganglion; they make inhibitory synaptic connections with each other within the ganglion.

There is a wide range of neuromuscular-transmission properties onto these various muscles, from muscles in which the transmission to individual cells is rather homogeneous to others that have a wide range comparable to that which occurs in leg muscles. The differences provide further emphasis on the notion of genetically/developmentally controlled plasticity in the neuromuscular apparatus. The stomatogastric ganglion produces two characteristic MPGs, one for the anterior portion of the stomach, the *gastric mill cycle,* and the other for the posterior region, the *pyloric cycle.* The former has been extremely intensively studied (reviewed by Selverston et al., 1976) and continues to be examined by refined techniques that permit assessment of the roles not only of individual neurons but also of local regions of these. It has become apparent that it is a combination of cell properties and connectivity that determine the program and the patterns of output by individual neurons. The program is, however, subject to overall *modulation* of cycle time and amplitude by neural inputs to the ganglion, particular ones mediated by a large dopaminergic cell in the commissural ganglion (Kushner, 1979). This ganglion receives inputs from a variety of groups of proprioceptive cells whose axons travel in stomach nerves. There are both phasic and tonic components to the discharges. Electrical stimulation of one of these nerves causes changes in the output program and patterns. The proprioceptive feedback does not alter the details of individual outputs, only those of several or all. The action is probably via the DA-releasing cell. The average overall level of DA within the ganglion determines first, whether or not the rhythm is produced at all and second, some aspects of the timing of the neural program and some details of neural bursts.

When observed *in situ,* crustacean intestines show relatively fast rhythmic waves of contractions; these waves are associated with action potentials and are thought to be neurally generated (Prosser et al., 1965).

12.12. Arachnida

12.12.1. Xiphosura

The most primitive living arachnid (or merastomate chelicerate) is the horseshoe crab *Limulus,* so there is special interest in its neuromuscular mechanisms. The commonest arachnids, spiders, are of interest in comparison, partly because they are very unusual in that they lack extensor musculature, blood pressure being used to extend the legs.

Limulus leg "closer," extensor, and bender muscles are innervated by either two or three excitatory axons (Hoyle, 1958; Parnas et al., 1968), which can be considered to correspond to slow, fast, and intermediate of crustacean and insect muscles. However, there is relatively much less differentiation, both in the synaptic properties and in the mechanical responses to the axons sharing a muscle, than occurs in most crustacean and insect neuromuscular systems. The *Limulus* system is polyneuronal, but by functionally relatively equal motor axons (Figure 12.46). The slow- and intermediate-axon JPs facilitate to give rise to small graded responses; those of the fast axons are larger and in a few cells initiate large spikes of all-or-none character. All appear to be glutaminergic. By contrast, the flexor, which moves the merocarpopodite joint, is innervated by a large number of motor axons, perhaps as many as 12 (Fourtner and Pax, 1972). This is especially interesting in view of the finding that insect leg flexors can have as many as nine excitors and two inhibitors (see p. 540). The muscle operating the corresponding

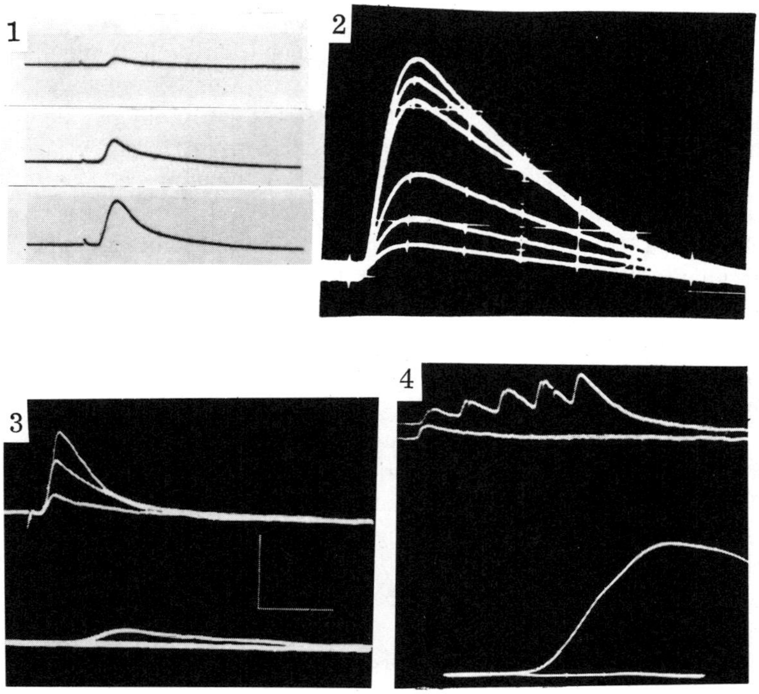

Figure 12.46. Neuromuscular transmission in arachnid skeletal muscles. (*1*) Intracellularly recorded EJPs from a single *Limulus* claw-closer muscle cell to each of three axons. (From Parnas et al., 1968.) (*2*) Tension steps, in two clusters of three each, obtained from pedipalp claw of scoprion *Leiurus.* (From Gilai and Parnas, 1970.) (*3*) Intracellular EJPs from a spider *Eurypelma* flexor, with increasing stimulus strength applied to the motor nerve. Tension from whole muscle *below*. (From Rathmeyer, 1965.) (*4*) Single EJP and a brief train, to show summation, with slowly rising tension, from spider flexor. (From Rathmeyer, 1965.)

joint in crustaceans is innervated by four excitors and an inhibitor. Thus flexion of this particular joint has more-complex control than have the majority of muscles throughout the arthropods.

Peripheral inhibitory axons have not been satisfactorily demonstrated in *Limulus,* but GABA does inhibit the excitatory action, and its effects are in turn blocked by picrotoxin, which blocks inhibitory transmission in crustacean muscles, so it is likely that peripheral inhibitory axons are present.

There is a notable uniformity in ultrastructure of *Limulus* muscle cells compared with those of most other arthropods. All the *Limulus* cells are cross-striated, with relatively long sarcomeres, in the range 6.5–8.0 μm. An H zone can be discerned, though it is not pronounced, and Z bands are wide but solid, that is, not broken up into Z bodies as they are in many crustacean and insect muscles. An extensive system of ITs is present, with abundant SR and SR/IT abutments located at the A/I overlap regions.

Special attention is being focused on *Limulus* muscle cells since the thick A filaments, which contain both myosin and paramyosin, shorten during contraction (see p. 152). Using laser diffraction images, it has been found that as the SL shortens in supercontraction from 7 to 3 μm, the A band shortens from 4.9 to 2.0 μm (Walcott and Dewey, 1980). This, like so many aspects of muscle-cell structure and function, appears to be unique, confined to a single organism, and therefore can tell us little about the evolution of arthropod cells and neuromuscular transmission. However, an interesting relevant feature does occur in innervation of *Limulus* muscle. The muscle cells send short muscle arms toward the motor axons, on which the synapses are made. Arms from more than one muscle cell can occur at the same region of a motor axon, and three axons can make contact side by side on the arm. Vesicles are largely of small, round clear type. These muscle arms could be taken as linking the phylum with the Onychophora, where corresponding ones are prominent (see p. 484), but in other aspects of ultrastructure there are marked differences.

No division of the muscle cells into fast/slow types, with attendant ultrastructural, physiological, and biochemical differentiation has yet been recognized in *Limulus*. In so far as the state of affairs in *Limulus* can be taken to indicate the primitive arthropod condition, it would be as follows. Polyneuronal innervation by a few motor axons only, with multiterminal innervation permitting graded local contractions, has evolved, but there is as yet little separation of function of these axons. Development of fast and slow extremes in junctional transmission has failed to occur: likewise with the muscle-cell population.

12.12.2. Scorpion

By contrast to the conventional appearance of *Limulus* muscle cells, those of scorpion claw-closer muscles have some interesting peculiarities. Both nuclei and mitochondria are located in a central core, which lacks contractile material, of the 40–60 μm in diameter cylindrical cells. The contractile material is separated into lamellae, 0.5–1.5 μm wide × 10–20 μm radially, by extensive radial strips of SR (Figure 12.47 – Gilai and Parnas, 1972). An IT occurs between each lamella and passes inward almost as far as the central core. Each IT makes numerous dyadic or triadic contacts with SR along its length. The SR is narrow except at these contacts, where wide cisternal sacs are formed. The contacts are all at A/I overlap level: expansions of ITs match the cisternal elements in shape. The Z bands are of medium thickness except at their abutments with the cell membrane, where they broaden into plaques. A-band filaments are 20–22 nm thick and surrounded by 12 thin ones, a 5 : 1 ratio.

The fine structure of scorpion muscle cells very closely resembles that of cockroach leg muscles. This is also true of the neuromuscular junctions, which are very simple points of contact, many per axon-terminal branch, between the terminal, which is not structurally specialized at the point of contact (D. S. Smith, 1972), and the muscle cells.

12.12.3. Spider

Spider muscle cells are thicker than those of *Limulus* or scorpion, 100–250 μm in diameter, and the SL range is much greater. Sarcomere lengths for a particular muscle are generally clustered around either 3–4 or 6–7.3 μm, but some intermediates also occur. An entire muscle such as the levator praetarsus of the tarantula *Eurypelma* may have only the short type, and its antagonist only the long type (Sherman and Luff, 1971). Both have abundant SR and numerous ITs, with contacts at A/I-overlap level.

The tarantula spider definitely has an inhibitory axon, to the promotor of the tibia (H. R. Brenner, 1972). There are commonly three excitors, giving rather similar JPs (Figure 12.46.*3*). Orthosummations or heterosummations lead to weak graded electrogenic responses (Rathmayer, 1965). The rates of facilitation in the axons to the basitarsus

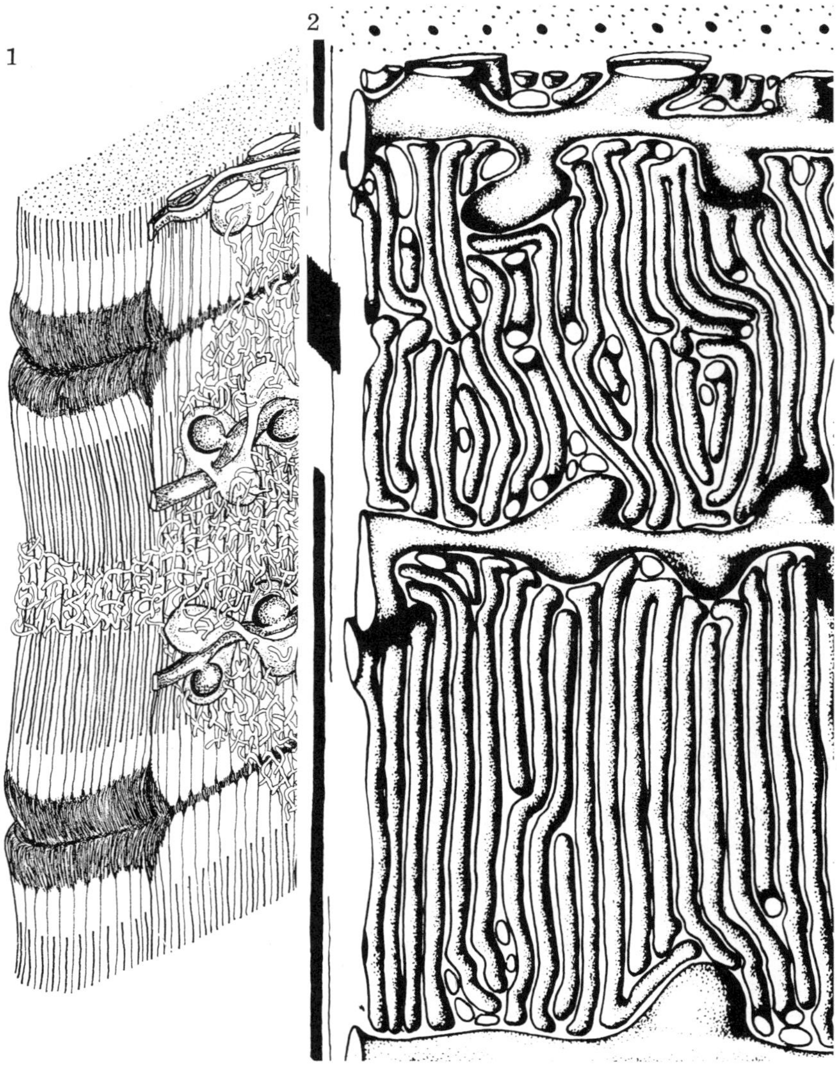

Figure 12.47. Ultrastructure of an arachnid and an insect skeletal cell. (*1*) Reconstruction of lamellar unit of muscle cell of the pedipalp claw of the scorpion *Leiurus.* The Z region is greatly expanded at the surface. The ITs cross each sarcomere at the mid-I/A-overlap regions. Each has pockets alternating in location, which contact corresponding swellings on the otherwise tubular network of SR, to make large dyads. (From Gilai and Parnas, 1972.) (*2*) Reconstruction of a housefly femoral muscle SR, with ITs. These have alternating pockets closely resembling those of the scorpion. (From I. Pasquali-Ronchetti, 1970.)

are, however, different, and contractions initiated by two of the axons are faster than those of the other. This muscle does not receive an inhibitory axon.

12.13. Insects

12.13.1. General Features

Considering that insects are by far the single most successful animal class and that they are of enormous economic importance both as pests and producers, relatively little data exists on their nerve/muscle systems. Only two kinds of insects, both orthopterans – locusts and cockroaches – have received almost all of the attention. Partly, this lack of interest may have been associated with the similarity in details to those found in crustaceans. The cynic will relate the difference in research effort to the beauty of the marine-station locales where much of the crustacean research is carried out, combined with the edibility of the parts left over after completion of the experiment. Greater size of lobsters and crabs, with concomitant ease of handling, and widespread physical revulsion to insects have played secondary roles.

The innervations of homologous muscles of locusts and cockroaches are similar, providing grounds for the belief that they shared a common ancestral motor-neuron plan. The extensor tibiae (ETi) of all the legs of both receive a single slow axon (SETi), a single fast axon (FETi), and a single common inhibitor (CI) shared with a still undetermined number of other leg muscles. Its antagonist, the flexor tibiae (FITi) is, by contrast, innervated by 11 axons (Phillips, 1980). There are three anatomically slightly separate groups of three excitatory axons each, and two inhibitors. The groups of excitors each comprise a fast, a slow, and an intermediate, and are classed as anterior, lateral, and posterior.

There is considerable overlap in the excitatory innervation of the flexors, and a single muscle cell may receive as many as seven excitors. There is yet no explanation for the flexor complexity, compared with that of the extensor. It does permit, in principle, a combination of some of the advantages offered by the typical vertebrate motor-innervation pattern with those offered by multiterminal, polyneuronal innervation combined with summation/facilitation, frequency-operated control. However, the actual ways it is operated naturally in the animal are not known. There is a great deal of shared common input to the flexor motor neurons (Hoyle and Burrows, 1973b), resulting in general synergism. To a small extent, a size principle operates: slow excitors are recruited first in any general excitation, followed by the intermediate and fast axons. However, a change in overall posture or orientation can produce an abrupt change in flexor motor pattern even though the type of behavior, walking on a round stick, remains the same (Hoyle, 1964a). This was seen when the stick was slowly rotated during walking.

12.13.2. Basic Tonus and Catch

Locust hind-leg ETis have at least a small amount of tonus in the absence of excitatory neural input; this is termed basic tonus (BT – see p. 255). The BT fluctuates in amount and is under neural control. A moderate change in posture is frequently observed to occur by way of change in BT alone, for example, during operant-conditioning experiments in which tibial position is linked to a reinforcement (Hoyle, 1980). A candidate for altering

BT naturally is the pentapeptide proctolin (Piek and Mantel, 1977): each ganglion contains a pair of proctolinergic cells (O'Shea and Adams, 1981) that each have an axon extending into a leg, and which presumably end in the flexors; they are not associated anatomically with ETi (Hoyle and Williams, 1980). The BT, when strong, is relaxed by OCT, which is released directly among ETi muscle cells (Figure 12.48), but when BT is low, OCT may cause an increase rather than a decrease.

The assessment of control of BT in locusts and grasshoppers is complicated by the presence of the slow intrinsic rhythmic contraction (IR – see p. 531) whose peak-tension amplitudes about equal strong BT. As BT increases, it does so at the expense of IR amplitude, and when BT is very strong, IR is masked completely (Figure 12.48). Therefore, the same cells must be involved in the production of both. Sometimes, an IR contraction may fail to relax after reaching its peak. Such a contraction is indistinguishable from increase of BT. It is relaxed by OCT or by activity in the single dorsal unpaired median nerve that innervates ETi (DUMETi). All neural inputs, FETi, SETi, and CI affect the timing, duration, and amplitude of IR, mainly be reducing them.

I have recently found that New Zealand "Wetas," members of a primitive orthopteran family, the stenopelmatidae, adopt limb postures requiring considerable force that last for several minutes without maintained neural output. The tension is therefore functionally comparable to the catch contractions of some clam muscles (p. 475). In the insect a catch contraction is brought about by a strong release of OCT by DUMETi and associated neurons, prior to a brief excitatory burst in a slow excitatory motor neuron supplying the relevant muscle(s). The catch state is always released by a single IJP, and often even by a single EJP, catch tension falling very quickly to zero upon release.

12.13.3. Intrinsic Rhythmicity (IR)

Intrinsic rhythmicity is a prominent feature of all locust and grasshopper ETis examined. It has a variable amplitude and frequency, all of which fluctuate erratically though slowly. Most commonly the frequency is about 3/min, in the range 8/min to 1/4 h. The duration of individual contractions is in the range 1 s to 4 h (Hoyle, 1978). Amplitude maximum is about 2.0 g. Such a contraction extends the tibia to about 120° with respect to the femur. No function is known for IR, but it is strong enough to initiate both flexion- and extension-resistence reflex, so could be serving as an automatic exercise mechanism or in the capacity envisioned for some sense organs, as a general background stimulator of the animal. A similar rhythm has been noted occasionally in locust flexor tibiae but not, so far, in any other insect muscle.

Activity in CI at low frequencies causes amplitude depression and reduced duration of individual IR contractions (Figure 12.49*.3*), but increased frequency. Above about 30

Figure 12.48. Effects of neural modulators, the pentapeptide proctolin and the phenolamine octopamine on IR and BT in locust extensor tibiae muscles. (*1*) The BT was high in 10^{-8} *M* proctolin, obscuring IR, but BT shifted to a lower value in 10^{-9} *M* proctolin revealing IR. The SETi axon was stimulated briefly at the times and frequencies stated. When BT is high, only a poststimulus relaxation of BT is seen at the lower frequencies: *Locusta migratoria.* (From Piek and Mantel, 1977.) (*2*) Complex effects of stimulus number and frequency applied intracellularly to a presumed proctolinergic neuron of the metathoracic ganglion on IR and BT: *Schistocerca gregaria.* The BT was relaxed for a short time, then recovered spontaneously (*at arrows*). The IR was attennuated at the higher frequencies only. Calibrations: 0.1 g, 10 s. (From Hoyle, 1978a.) (*3*) Inhibition of IR and enhancement of BT by one drop of 10^{-8} *M* OCT. Same as (*2*). (From Hoyle, 1981a.)

1

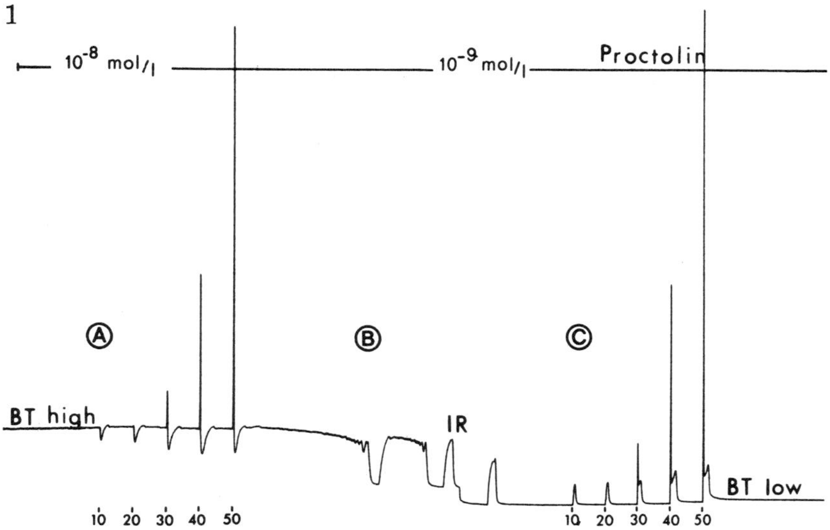
10^-8 mol/l
10^-9 mol/l
Proctolin
Ⓐ
Ⓑ
Ⓒ
BT high
IR
BT low
10 20 30 40 50
10 20 30 40 50

2

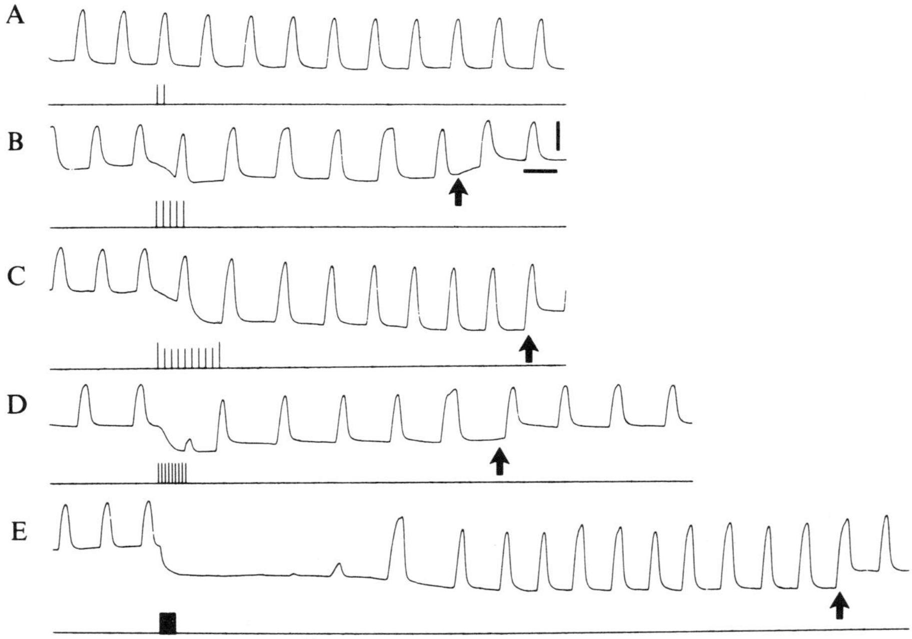
A
B
C
D
E

3

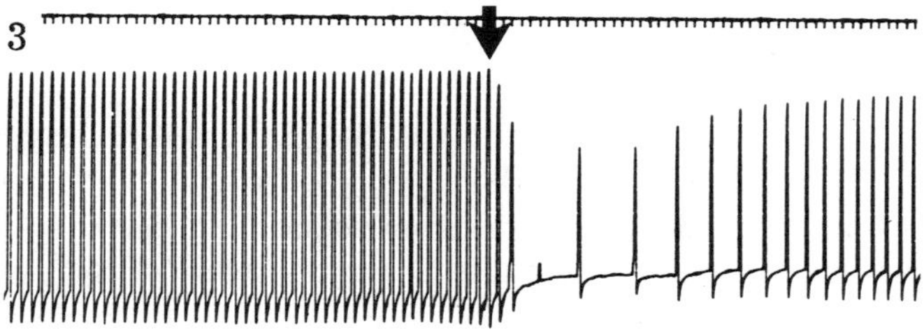

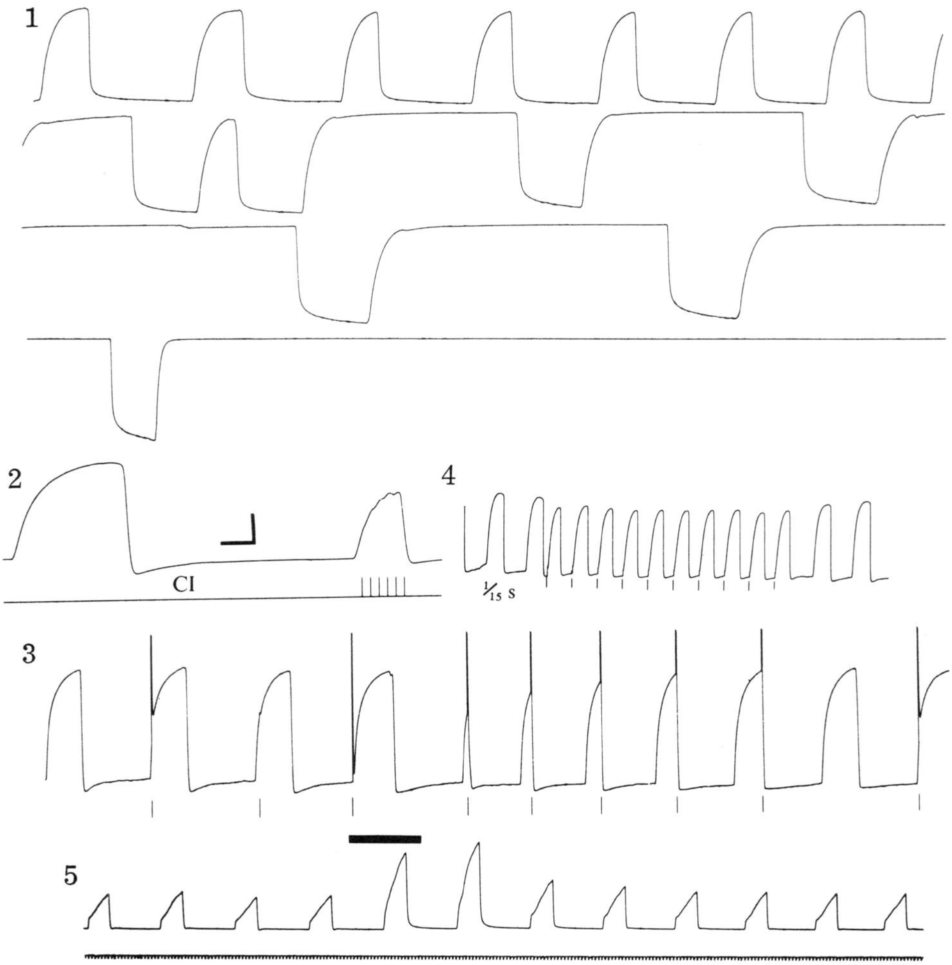

Figure 12.49. Modulation of IR of the extensor tibiae spontaneously and by the motor neurons innervating the muscle: tension records from ETi of grasshopper and locust. (*1*) Spontaneously occurring alteration of the durations of individual IR contractions in a nearly intact preparation of *Brachystola.* (*2*) Partial inhibition and early termination of IR by CI activity – six stimuli (*at right*) in *S. gregaria.* (*3*) Critical timing period evidenced for abrupt termination of IR contractions by FETi impulses. Note also that the interval is decreased in proportion to the amount of abbreviation of IT. (*4*) Acceleration and synchronization of IR by single impulses in SETi at one per 15 s, which was faster than the natural IR frequency. (*5*) Neuronal enhancement of IR amplitude caused by another metathoracic neuron stimulated intracellularly (*bar*). Time 1/s. (All from Hoyle, 1978a.)

Hz, IR is inhibited completely by CI. At frequencies below 10 Hz, SETi causes increased frequency and decreased amplitude of IR. Above 15 Hz, the evoked contraction swamps IR, though this is still seen as a rhythmically occuring contraction supplement. In a quiescent preparation not showing IR, a SETi burst often starts up IR. Single SETi or FETi impulses can initiate an early IR contraction, causing altered phasing; when either is stimulated repeatedly at a low frequency, up to a quintupling of IR frequency can be obtained. Single impulses in either CI, SETi, or FETi cause an abrupt early termination of an IR contraction if applied following a critical period (Figure 12.49.*3*). Such release of an IR contraction closely parallels release of catch tension in the Weta, in time-course of relaxation. Early termination of an IR contraction is followed by a reduced interval, which is proportional to the reduced IR contraction time. The rhythm of accumulated readiness to go into an IR contraction is independent of the pacemaker rhythm that initiates the contraction.

The IR is affected by the ionic concentration of the bathing medium. In standard locust saline (Hoyle, 1953), it sometimes stops completely, though it can then easily be started again by adding a little ACh ($10^{-6} M$) to the saline. The muscles are not otherwise sensitive to ACh, even in high concentration.

12.13.4. Effects of Hemolymph Composition on Contraction

The blood, or hemolymph, of insects commonly has a bizarre average ionic composition compared with that present in vertebrates and most invertebrates. It also is much less drastically regulated, some ion levels rising and falling daily in association with feeding. The potassium concentration in many is sufficiently high, 10–70 mM/1, for depolarization of muscle cells, both skeletal and vegetative, to an E_M near or beyond the contraction threshold. In this state the contractile system is highly susceptible to hormonal action. This could occur directly, or indirectly via synaptic sites. After a cockroach hindgut is depolarized by isosmotic potassium, 10^{-8} M proctolin still causes a substantial contraction. An agent such as this peptide is likely to be acting at a nonsynaptic site, possibly intracellularly and directly or indirectly on calcium mobilization. An action of OCT can, however, be detected in as little as 50 ms after release (Hoyle, 1975c), so it is likely to be at least partly having a direct membrane action.

Phytophagous insects not only have a high hemolymph potassium concentration, but they also have large quantities of magnesium, up to 80 mM (replacing sodium), which is an almost universal blocker of synaptic transmission. Magnesium action is presynaptic, preventing the release of transmitter substance. The synapses in some cases are permanently depressed by magnesium. But for others there must be antagonism of the blocking or some form of adaptation. The excitatory axons are tightly wrapped by sheath cells up to the point where they must emerge to make synaptic contacts with the muscle cell. The side away from the synapse still has this covering, tightly linked to the sarcolemmal complex. Thus it is possible that Mg^{2+} ions are kept away from the junction by the sheath.

Insect muscle cells are surrounded by a *basement lamina* outside the muscle cell membrane. This is composed of mucopolysaccharide and is extremely variable in thickness within any given muscle, from as little as 0.03 to 10 μm with extreme thickenings up to 30 μm at the corners of irregularly shaped cells. There also appear to be some general differences between cell types. The thickness is greater around slow-type cells than fast ones. Large amoebocyte cells wander between the muscle cells, constantly laying down new basement lamina material (Figure 3.10). It is possible that this serves as a lubricant between cells: since no two adjacent muscle cells of an insect are excited in exactly the same way, there must be sliding of local regions relative to each other. Alternatively, the material may serve as a local reservoir of metabolic substrate.

Desmosomelike structures occur at a few spots between the lamina and the membrane. These are particularly prominent near Z regions but are not confined to them. Membranes, even multiple layers of them up to 10 μm thick, occur in some places embedded in the lamina, especially where this is thick and near a nerve terminal. These membranes are derived from nerve-sheath cells and are in part anchoring the nerve to the muscle cell. It is this covering that shields the neuromuscular junctions from many substances present in the blood such as L-glutamate and magnesium, which would otherwise affect transmission. Little or no sodium is present in the hemolymph of lepidopterans. The nerves have adjusted to this to support conventional axonal conduction by developing the cells lining the inside of the sheath as storage sacs for sodium. Sodium is abundantly available to the animal in the larva or caterpillar stage, where it is selectively accumulated by these perineurium cells. It is presumably released slowly as needed, to keep a fluid with high sodium content around the axons in the adults, which live exclusively on sodium-free sugar water (nectar). There is no such device for the muscle cell membranes, which must therefore use ionic mechanisms appropriate to being bathed in such a medium. Insect muscle cells are gradedly electrogenic (see p. 170), and in the virtual absence of sodium, they must use conductance increase to divalent cations, principally calcium but also magnesium in some, to generate secondary responses. These range from very weak through all degrees of gradedness to rare all-or-none spikes (Washio, 1972a; Piek and Njio, 1979; Ashcroft et al., 1979).

The membrane potential of most insect muscle cells is dominated, as of other classes, by K^+. Insect muscle cell membranes are less permeable to Cl^- than are those of vertebrates, but their IT walls are especially permeable to Cl^- (Usherwood and Grundfest, 1965).

12.13.5. Ultrastructure

Insect skeletal muscle cells occur in a wide range of gross substructures overall, especially flight muscles (Figure 8.1). The variety is associated mainly with differences in the amounts and arrangements of SR, and in sizes, numbers, and locations of mitochondria and nuclei. The core is totally filled with contractile material in some, even ones with diameters of over 100 μm, but in others a part of the core, up to 50% of the total volume, contains no contractile material. The core contains the nuclei in some cells, mitochondria in others, or it may contain only clear cytoplasm. There is a strong tendency for the SR to be arranged radially, particularly in cells with hollow cores, giving rise to a lamellate appearance of the contractile material. In such cells, which are predominant for example in cockroach leg and dragonfly flight muscles, ITs run in each lamella, making mainly dyadic and some triadic connections with the SR. If it were not for the much larger mitochondria occurring in these flight cells, together with nuclei in their cores, it would be difficult to distinguish them at low magnification.

Locust fast-type leg muscle cell structures range from cells with typical vertebrate *Fibrillenstruktur,* in which the SR is abundant but evenly dispersed, almost completely invading all the contractile material, making cylindrical surrounds of the contractile material, to ones with radial lamellae that resemble cockroach leg muscle cells. There are no differences in the contractile characteristics of the different types provided excitabilities are similar. Nor can any particular general form be assigned to flight or leg muscles that could be related to the very different functions they perform. Longitudinal alignment of sarcomeres of fast-type cells is virtually nonexistent, and most do not even look cross-striated in the light microscope. This is surprising because it has long been assumed that speed is associated with orderly alignment.

Slow-type leg muscle cells have a variety of appearances also, in transverse sections. Most have a coarsely fibrillar appearance with a partially radial alignment. The cores may either be without myofilaments or have a similar appearance to the periphery. The longitudinal alignment of sarcomeres is generally good, with only a few vernier patterns (see p. 46), and in some it is almost perfect.

Cross-striated muscle cells of insect appendages and body walls have a wide range of fine structures. This is also true of flight muscle cells, but all fast-flapping (usually asynchronous) direct flight muscle cells have certain features in common that distinguish them from all the others. Indirect flight muscles that are generally bifunctional, serving to operate leg muscles as well as flight muscles, and synchronous flight muscles that have a slow beat, such as those of most large-winged moths and butterflies, all have features characteristic of the latter, not of the direct flight muscles. In locust ETi the faster muscle cells have SLs of 5.8 μm; those of the slower ones are but little different, averaging 7.2 μm. The longest sarcomere average yet described in an insect was 8.3 μm, in dorsal intersegmental muscles of a cockroach (Smith, 1966a). Flight muscles, both direct and indirect, are the only ones in which relatively short sarcomeres are encountered. The range described is from 2.4 to 4.0 μm.

12.13.5.1. Flight Muscle Cells

Direct flight muscle cells, though showing great variations in overall plan, principally because of differences in distribution of SR, all have one characteristic in common, a remarkably regular array of thick filaments and thin ones, at least in all the in-line positions. This distinguishes all strictly flight muscle cells, both fibrillar and afibrillar, from all skeletal ones, suggesting either that flight muscle cells had an ancient common origin or that these features are in some way associated with the making of fast, repetitive equal motions. I filaments are in line with thick ones as seen in transverse sections, in orbits of six, but in a 3 : 1 ratio compared with the 2 : 1 ratio for vertebrates. The 3 : 1 ratio is associated with their occupying the trigonal positions. The I region is extremely short in the relaxed condition. Ultrathin filaments, sometimes termed C (connecting) filaments, extend from the tips of the thick filaments into the Z lines. The I filaments enter the Z region in hexagonal array. Exactly how they enter the Z region is a matter of controversy (see p. 72) and may be different for different kinds of flight muscle. The 3 : 1 filament ratio, with thin filaments in the same planes as thick ones, is seen in a few other arthropod muscle cells including antennal muscles of copepods (see p. 54), where the function, as in flight, calls for repeated quick twitches. But the organization of Z regions is quite different in arthropod nonflight muscles, where there is no sign of C filaments. Flight muscle cells have a high stiffness, strongly resisting extension even of only a few percent. Yet they can be extended to beyond I/A-filament overlap as with other skeletal cells. It is assumed that special properties of C filaments determine much of the elastic stiffness, in parallel with sarcolemmal elastic elements. The mode of attachment of C filaments to A filaments is not known, and it has been proposed that they may pass down the cores of A filaments, either as far as the M region, or even the next Z region. The chemical nature of C filaments is unknown. No details of flight muscle M-band structure have been published. It is broad, about 0.3 μm, and a total of five or six rows of M bridges can be discerned in longitudinal TEM sections. The thick filaments have a solid, oval profile in the M region, compared with a hollow, circular one elsewhere. Profiles of six very thin dense filaments, which may correspond to the M filaments of vertebrate M regions (see p. 76), occur. Unlike the I filaments, these are present in trigonal positions in a 2 : 1 ratio, as are longitudinal M filaments of vertebrates. Two rings of dense material, presumably LMM molecules, can be discerned in each thick filament in the A/I overlap regions. The X bridges between thick filaments can be discerned unusually clearly (Figure 12.50). When

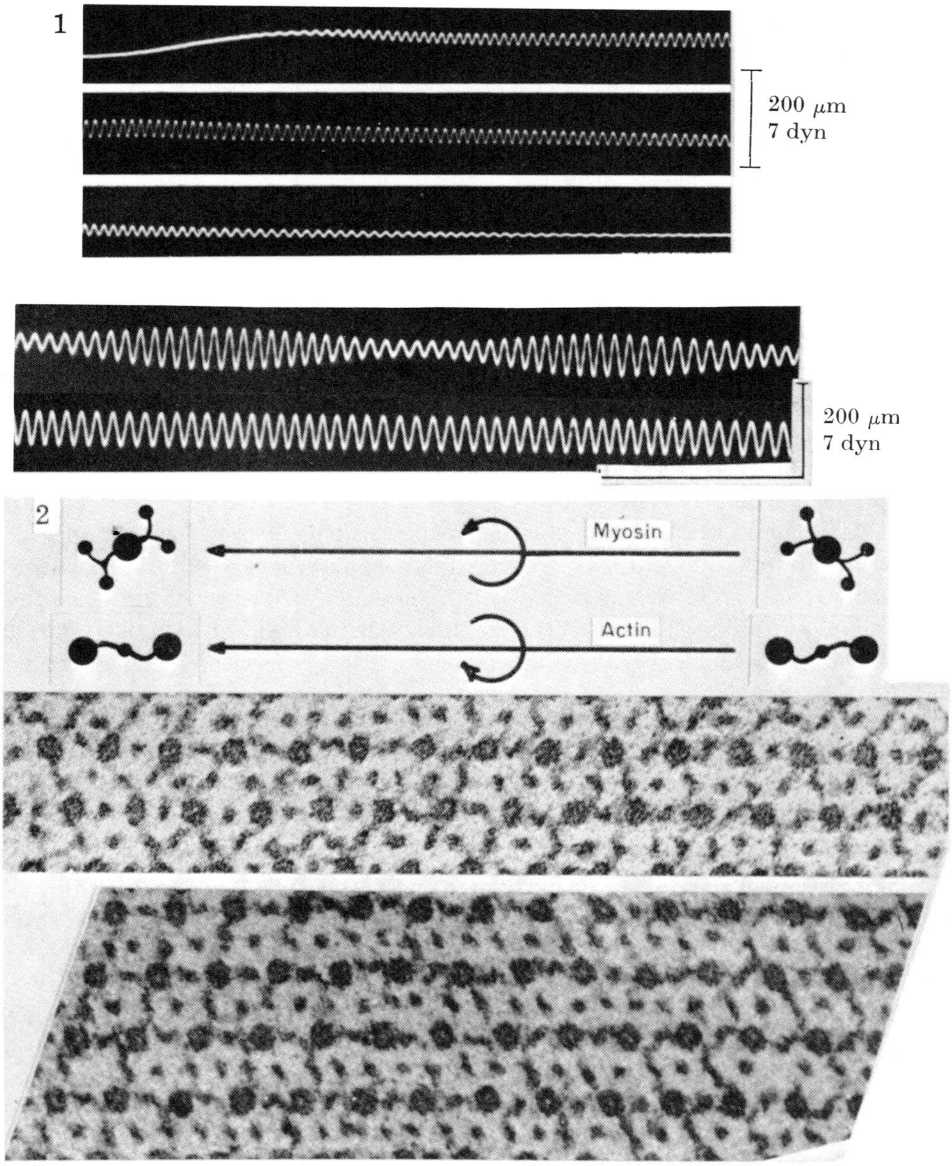

Figure 12.50. (*1*) Oscillatory contractions of isolated glycerinated insect (*Lethocerus*) flight muscle initiated by the addition of ATP and calcium (1.2×10^{-6} *M, above*, 2.9×10^{-7} *M, below*). Amplitude modulation (*below*) was spontaneous. (From Jewell and Rüegg, 1964.) (*2*) Details of X bridges as seen in very thin TEM of TS of (*1*). The arrows indicate the direction of traverse along myosin and actin helices respectively. Anticlockwise rotation of myosin-centered groupings (flared X) is accompanied by clockwise rotation within I-centered groupings (S-shaped). Attachment to I filaments proceeds along a right-handed helix. The upper micrograph shows the flared X exceptionally well, the lower one the S shape. [With permission, from M. K. Ready, *J. Molecular Biol.* **31,** 155–176 (1968). Copyright 1968 Academic Press, London.]

viewed from in front or behind there is a strong twist to each bridge, associated with the opposite screw senses of the A and I filaments. When a single pair of bridges are seen they have an S shape, but when two pairs occur together, the appearance may be termed a "flared X" (Reedy, 1968).

A drawing of the details of the M region in *Lethocerus* flight muscle was shown in Figure 3.24. The functional requirement for flight muscles is for synchronous, powerful unitary contractions, each caused by single or close pairs, triplets, or quadruplets of impulses repeated exactly, at a moderate to high frequency (20–150 Hz) depending on wing size. There is good evidence that even the X bridges make their individual strokes together in flight muscles. The ultrastructural features were presumably all evolved to achieve X-bridge stroke synchrony, by optimizing X-bridge angles with respect to interaction sites on the actin. Various nonflight skeletal muscles of many species have now been described, for example, by Hoyle (1967a), D. S. Smith (1968), Toselli and Pepe (1968), Cochrane et al. (1972), Elder (1975), and Hoyle (1978b). We also have a portfolio of unpublished micrographs on which this account is partly based.

12.13.5.2. Leg Muscle Cells

No insect leg muscle cells so far examined have the same characteristic features as those seen in direct-flight-muscle cells. There are a few fast-type cells that have a 3 : 1 I–A filament ratio overall, but these do not have very orderly arrays of thick filaments, and I-filament orbits are often slightly misplaced or incomplete. The hexagonal array of I filaments does not persist into the Z region, instead becoming highly irregular. When any order can be discerned, it is essentially square as the Z region is approached, as it is in vertebrates, but with closer packing.

The commonest I–A ratio overall is 4 : 1, an average corresponding to orbits of 10, some I filaments being in trigonal positions, but the ratio is not adhered to strictly, some of the sites not being occupied. The approach of thin filaments to the Z line, in contrast to that in flight muscle, but as in vertebrates, is highly irregular, with a slight tendency toward making a close-packed square array. Many fast- and intermediate-type leg- and body-muscle cells have this array. Others, and some slow-type cells, have a 5 : 1 ratio in fairly regular orbits of 12 I filaments. In places, the sets of orbits are highly regular, with each of the six trigonal positions occupied, and likewise each of the six positions in line with thick filaments. However, nonconformities, both in thick-filament arrays and orbits, are common.

Slow-type cells have an overall thin–thick filament ration of 7 : 1, rarely 6 : 1. The thick filaments are fairly evenly spaced with respect to each other, but almost twice as far apart as in fast cells and not in a strictly hexagonal array. The thin filaments are only loosely collected into orbits. In places, clusters of thin filaments stand apart from those making orbits. Mitochondria are almost always located rather strictly at I regions and are commonly about equal in length to the half I-band length in the resting muscle. However, the sizes of these mitochondria range from very small to as large as $2 \times 2 \times 4$ μm, depending on the muscle. The light microscopic appearance is dominated not so much by A and I alternation as by the two transverse rows of mitochondria present on each side of the Z lines. The A/I overlap is the site for most dyads, with a few occurring near the center of the sarcomere.

Unlike wing muscle cells, where there is always a conspicuous M band, cells of legs and body show no sign of one. This feature alone could serve to distinguish flight cells. Transverse M bridges linking thick filaments are not localized but occur, as in many crustacean muscles, along their entire length.

Thick filaments are uniform in diameter throughout the insects, at 22–23 nm. This means that more myosin molecules are packed together per unit length than in vertebrates, possibly 18 compared with 12. The cores are markedly hollow in flight muscles, much less so or not at all in the slower leg muscle cells. The hollow ones may contain fewer myosin molecules per unit length.

12.13.6. Muscle Cell Types Within a Muscle

Few insect muscles have yet been examined with regard to component cell types. It was early recognized that cockroach thoracic muscles operating the legs could be distinguished on the basis of a slight color tint into "pink" (or "red"), "white" and "intermediate" (Becht and Dresden, 1956). The pink ones (muscles 135b,d,e) are more slowly contracting and more resistant to fatigue. Their cell structure is of *Felderstruktur* type (see p. 49), that is, there is relatively little SR. The white ones (muscles 136, 137, 135a,c) are *Fibrillenstruktur* type (see p. 50), that is, extensively invaded by SR. The color tint is associated with absorption of green by cytochrome *b* in the relatively large number of mitochondria present in the slow-type cells. This is an incidental, not an obligatory, accompaniment of slowness in the cells of these particular cockroach muscles. A color difference is not a common feature, either of cockroach muscles or of insect muscles in general. However, a similar color tint characterizes a cluster of muscle cells that form part of the retractor unguis muscle of locusts. This is a small muscle comprising only 16 cells in *Schistocerca gregaria* and 15 in *Locusta migratoria.* These cells are divided into pink and white clusters of 5 or 6 pink and 10 white cells. The pink ones tend to be slightly smaller, have more prominent mitochondria, a very slightly longer SL, and more succinate dehydrogenase than white ones. Contents of both glycogen and aerobic metabolic enzymes are higher in the former type.

However, many other factors affect muscle-cell-contraction characteristics, notably membrane excitability, amount of SR, numbers of SR/IT contacts, and so forth, which can be mixed together in different combinations, and slow/fast-cell differences occur that are not reflected in mitochondria or oxidative enzymes. These latter are directly related to the work load rather than to genetically determined intrinsic factors.

12.13.6.1. Retractor Unguis

The *S. gregaria* retractor unguis, a tiny muscle attached to the proximal margin of the femur, operates the double claw at the tip of the tarsus by way of a very long apodeme that spans femur, tibia, and tarsus. Pink and white muscle cells are innervated by the same two major excitatory axons, both of which evoke large EJPs and twitches and can be classified as fast. Each axon innervates both sets of muscle cells. Therefore, it had to be assumed that the determination of the pink/white color is genetically/developmentally controlled rather than trophically. However, an intensive study using electron microscopy of the nerve revealed additional, very small axons. Both of these are inhibitory and in *S. gregaria* innervate only the pink cells (Figure 12.34.3 – Walther, 1980). In *Locusta,* a slow excitor innervating only pink cells was found. Thus a trophic determination of the cell characteristics is possible. A small axon containing DCVs that closely resemble those found in DUMETi (see p. 420) also occurs in the retractor unguis nerve.

12.13.6.2. Extensor Tibiae

The extensor tibiae of *S. gregaria* has a complex cell composition and innervation even though there are only four axons (Hoyle, 1978b). There is a single fast excitor, a single slow excitor, a single common inhibitor, and the single octopaminergic modulator cell DUMETi (Figure 12.51). The muscle cells have no color differences, but there is a two-fold range in twitch times and a wide range of electric responsiveness. The property of IR is confined to a small cluster of proximal cells. Ultrastructures range from fast type, as shown by abundant SR, ITs, and SR/IT contact, to slow type, which in the extreme has very little SR in the core and into which the ITs do not enter. These slow cells show an interesting, functionally significant response variation. The quality of contraction is markedly dependent not only on the frequency but also on the duration of stimulation, expecially at low frequencies. This is because brief bursts of stimuli cause contraction, which is relatively fast, only of the periphery of the cell, which is well supplied with ITs. The core of the cell gets excited only when the stimuli are repeated for a long time, presumably building up a sufficient excess of calcium for some to diffuse into the core and activate it; a very low frequency eventually recruits the core fully. The core can relax only very slowly, presumably because there are no SR cisternae there to which the calcium can be returned.

The greatest complexity is seen in the details of axon distribution and muscle-cell innervation. In general, CI innervates only cells innervated by SETi. About 60% of the cells receive only FETi, 15% receive FETi, SETi, and CI (triple), and 20% receive SETi and CI (double). An undetermined number, about 5%, receive FETi and DUMETi. The muscle cells receiving only SETi and CI are confined to three bundles, the most distal and the two most proximal ones. The proportion of cells receiving both of the excitors and the inhibitor is greatest toward the two ends near these bundles (Figure 12.52).

Somewhat similar distributions of innervation occur in the homologous muscle of the cockroach, but the ETIs of the prothoracic and mesothoracic legs appear to have a simple gradient for triple innervation, which is greatest at the proximal end.

12.13.6.3. Dorsal Longitudinal (DL) Flight Muscle

The dorsal longitudinal flight muscles of a locust receive only fast axons, five in all, one going to each of five separate bundles of muscle cells (Neville, 1963), and branches of a single octopaminergic unpaired median neuron (DUMDL) – Hoyle et al. (1980). The mesothoracic DL muscle of a katydid is composed of only three bundles and receives four motor neurons, three being fast the other intermediate, giving a slower twitch (Stokes et al., 1900). The ventral and medial bands comprise 4.0- and 4.3-μm SL cells with 24% SR, innervated only by fast axons, some medials being doubly innervated. The medial bands are crammed with mitochondria, both distributed and clumped, taking up about half the cell volume. These must represent the workhorse region of the muscle. The dorsal band is innervated by one fast axon and also the intermediate, with some double innervation. Its cells are distinguished by having much longer, 9.5 μm, SL, and only 10% SR. Clearly these cells must be responsible for the slow twitch evoked by the intermediate axon.

The excitatory junctions onto flight muscle cells from different orders of insects are recognizably different from those on their leg-muscle counterparts because they have a much wider range of vesicle sizes (Figure 12.53). What the functional significance may be is unknown, but it may be related to the need to produce large EJPs to ensure spikes and fast twitches.

12.13.6.4. Flexor Tibiae

The largest number of excitatory axons known to innervate an insect muscle is nine, for the flexor tibiae of the metathoracic leg of locusts. There is also a relatively large number for the prothoracic and mesothoracic flexor tibiae, but the full details are not yet known.

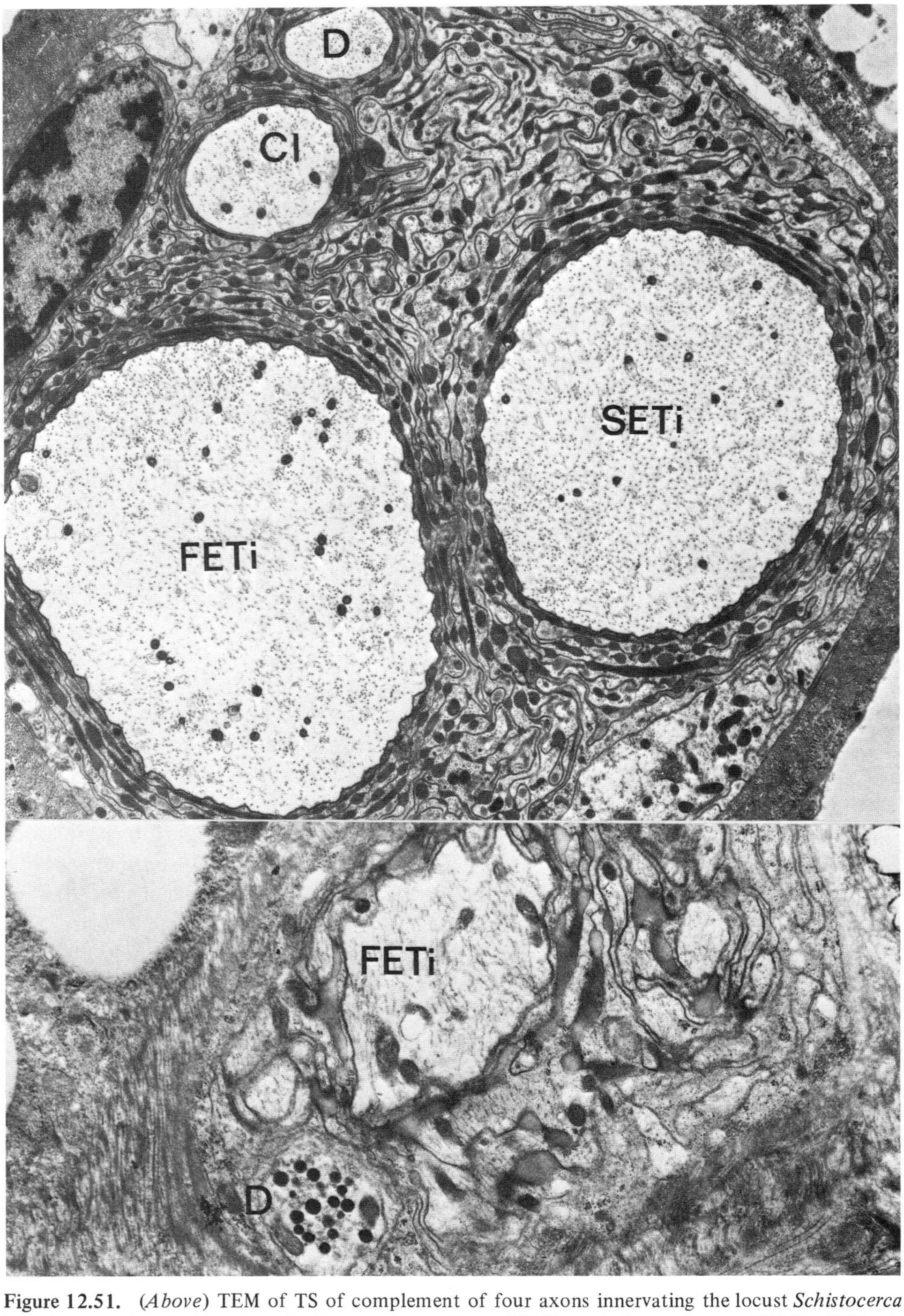

Figure 12.51. (*Above*) TEM of TS of complement of four axons innervating the locust *Schistocerca* jumping muscle: L-glutaminergic fast extensor tibiae (*FETi*) and slow extensor (*SETi*), 5HTergic common inhibitor (*CI*), and octopaminergic dorsal unpaired medium neuron DUMETi (*D*). (*Below*) DUMETi accompanying FETi in surface (Doyère cone) complex on single extensor muscle cell. (From Hoyle et al., 1980.)

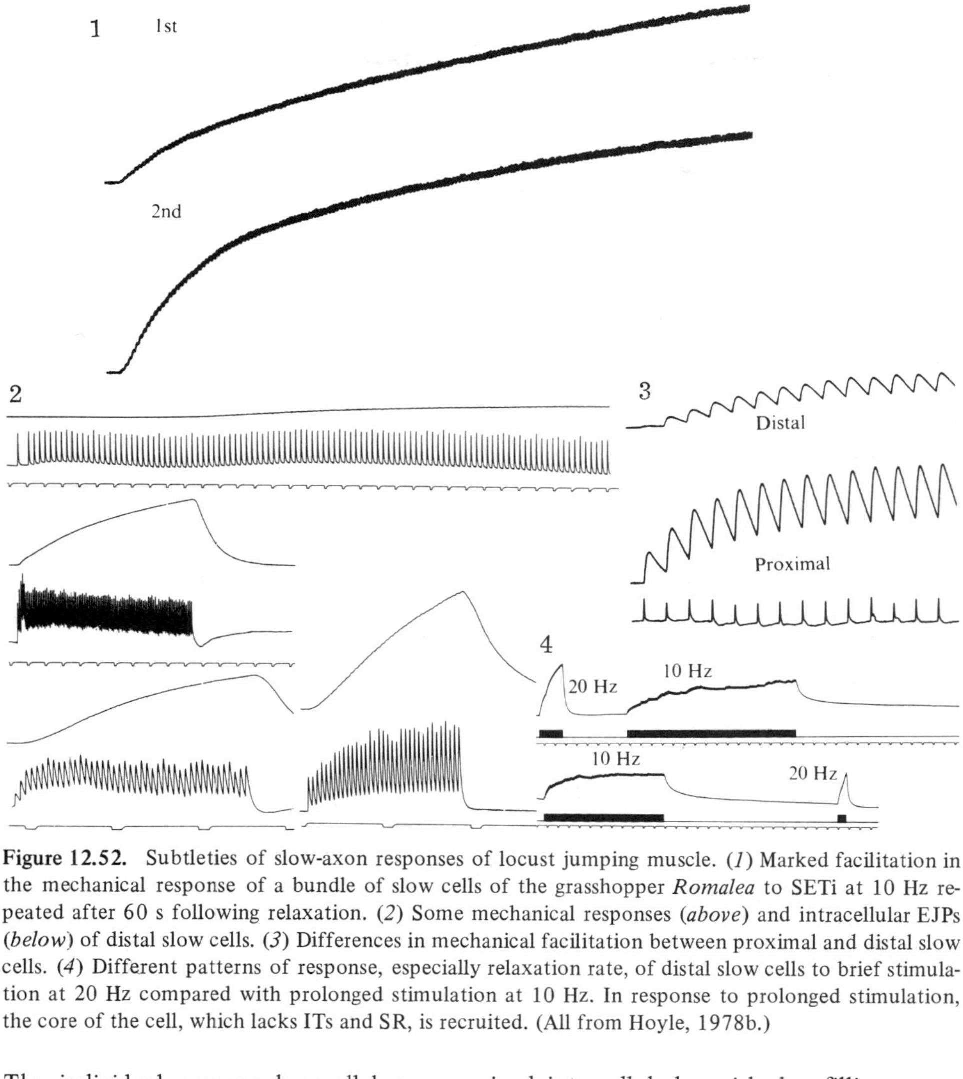

Figure 12.52. Subtleties of slow-axon responses of locust jumping muscle. (*1*) Marked facilitation in the mechanical response of a bundle of slow cells of the grasshopper *Romalea* to SETi at 10 Hz repeated after 60 s following relaxation. (*2*) Some mechanical responses (*above*) and intracellular EJPs (*below*) of distal slow cells. (*3*) Differences in mechanical facilitation between proximal and distal slow cells. (*4*) Different patterns of response, especially relaxation rate, of distal slow cells to brief stimulation at 20 Hz compared with prolonged stimulation at 10 Hz. In response to prolonged stimulation, the core of the cell, which lacks ITs and SR, is recruited. (All from Hoyle, 1978b.)

The individual neurons have all been examined intracellularly, with dye filling, many times. Two groups of three were first recognized on the basis of a combination of location properties, contractile response, and size of soma. These are the anterior and posterior groups, each comprising a fast, a slow, and an intermediate axon (Hoyle and Burrows, 1973b). Later Phillips (1980) found a large lateral fast flexor motor neuron some (La FFl) that was undoubtedly present in addition to the other two fast cells. This led to an intensive search based upon a combination of microscopic study of the flexor muscle and its nerve supply, backfilling of neurons, and additional probing in the ganglion. It now seems probable that an intermediate and a slow cell may be considered to be grouped with La FFl. There are also two inhibitory axons.

The flexor-muscle-cell population ranges from mainly fast type proximally to mainly slow type distally. The innervation, which is not fully worked out, is complex, with as many as seven of the probable excitatory axons innervating a single proximal muscle cell. This rules out the possibility of separate innervation of independent muscle units. In spite

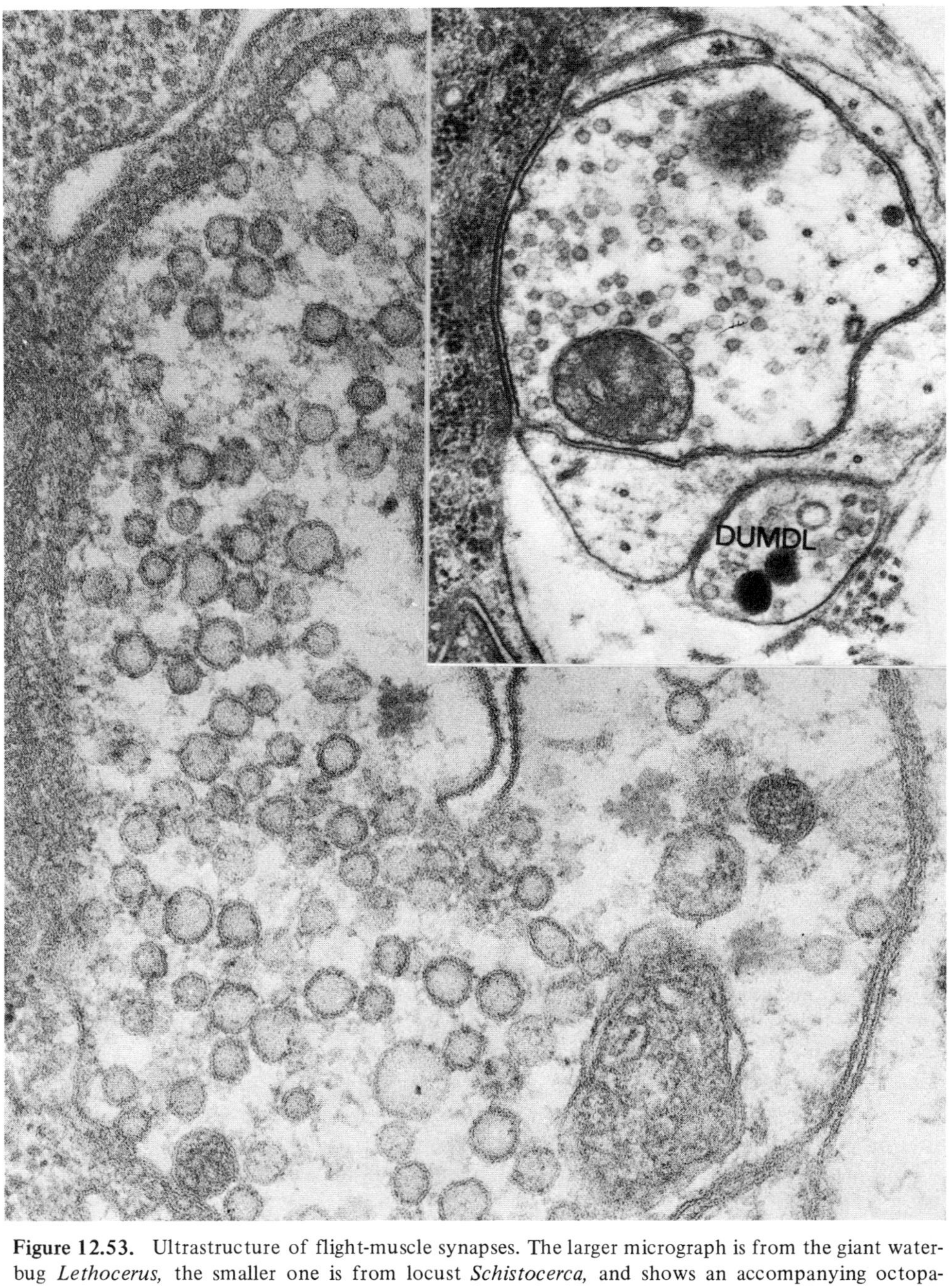

Figure 12.53. Ultrastructure of flight-muscle synapses. The larger micrograph is from the giant waterbug *Lethocerus,* the smaller one is from locust *Schistocerca,* and shows an accompanying octopaminergic neuron, DUMDL. These excitatory junctions have an unusually wide range of vesicle size. (*Lethocerus,* unpublished, by B. J. Walcott; *Schistocerca,* from Hoyle et al., 1980.)

of intensive recent studies, the rules governing the operation of the flexor are not understood. All the flexor motor neurons receive a lot of common excitatory inputs. There is a strong tendency for a size principle to operate in both spontaneous and reflex actions. First the slow axons discharge tonically, then one or another intermediate is added (or sometimes both discharge), and finally one or two fast ones, in brief phasic bursts only. In a prolonged contraction the fast axon(s) cease(s) to fire first, followed by the intermediate(s), and finally the slow. But a fast axon is often fired in a brief burst to flex the tibia fully even when none of the other axons are active. Independent direct excitatory central inputs specifically to one or another of the fast flexors therefore occur in addition to the common inputs whose actions are governed by correlates of size. The fast flexor that is used independently most often is the anterior fast flexor of the tibia (AFFITi).

12.13.7. Patterns of Innervation

All insect muscle cells have more than one motor nerve terminal on them from each motor axon that innervates them, the number of which is from one to nine. These axons, both excitatory and inhibitory, travel together in a common sheath, and they leave this covering together. There may be only one nerve branch going to a muscle cell, or a few, but when there are more than one they are all derived from the same trunk and contain branches of the same axons. After contacting a muscle cell, the axon(s) may simply run straight along its parallel to its margin or branch at least a few times, perhaps profusely. There are species differences and also differences between muscles and even different muscle cells of the same muscle, in the final branching patterns. But no correlations have yet been discerned between the different structures and physiological features, that is, very different anatomical features can end up having similar effects. When one axon branches, the others do also, at almost all the forks. Beyond each secondary fork there may be no more branching, giving rise to a series of end-plate claws, each of which resembles an amphibian end plate. Or branching can go on almost indefinitely, covering the muscle cell in a web of progressively finer nerve processes.

Numerous types of contact are formed. In some, the muscle cell gives off a series of short pillars resembling truncated muscle arms (see p. 158) in the direction of the passing axon. At each contact point a synapse is formed. If there is more than one axon, multiple synapses may be formed side by side on the same pillar, or there may be more than one pillar but still with the junctions quite close. In others the axon simply flattens itself against the unyielding muscle cell, whereas in others the axon(s) fall(s) into (a) groove(s) indented into the muscle cell surface. The grooves range from shallow depressions to deep invaginations. Sometimes a nerve terminal will follow an invagination, accompanied by a tracheole, before making (a) synaptic juxtaposition(s). In this way junctions occur, albeit rarely, that are inside, even near the middle of, a muscle cell. These are probably chance events arising from the vagaries of developmental processes and have no special functional significance.

The terminal axons may run straight, sides parallel, without bulging, or they may make varicosities. These can take the form of a series of lateral pockets, or a sequence of widenings along the length, without branchlets. There may be only two or three bulges or scores of them. Clusters of clear vesicles occur at intervals in the parallel-sided ones, in the bulges in the others, that are presumably the release sites. There are from about 20 to several hundred such sites on a single muscle cell. They may be confined to a single line or be around the entire circumference.

12.13.8. Innervation and Neuromuscular-Transmission Plasticity in Insect Evolution

It was early found that the SETi and FETi axons of locusts and grasshoppers leave the ganglion via separate nerve trunks, a remarkable fact that is difficult to understand from an evolutionary or developmental point of view. The two axons come together when the small nerve branch 3b carrying SETi joins the large crural nerve N5 carrying FETi just before the slender branch supplying ETi leaves it. And CI travels by the side of SETi throughout. This final branch contains only FETi, SETi, CI, and DUMETi. In the cockroach, although the nerve trunks have a similar configuration and the axonal supplies are similar, FETi, SETi, and CI leave together in N5.

What is especially interesting is that the pathways of FETi and SETi from the prothoracic and mesothoracic ganglia of locusts and grasshoppers are reversed (Hoyle, 1957). Furthermore, both the cell-body locations and the branching patterns of these cells in prothoracic and mesothoracic ganglia are exactly similar to those of their opposites in the metathoracic ganglion (Figure 8.16.2). Since the ganglia are serial homologues, they should contain similar sets of neurons. It may therefore be deduced that as the hind leg evolved into a jumping organ, the original slow axon came to serve the function of exciting the extensor muscle in jumping, mainly by exclusively innervating the additional muscle cells added when the femur enlarged (see p. 309 – Wilson and Hoyle, 1978). These cells must have been added to the center of the muscle, the original parts being preserved as proximal and distal regions. Secondarily, changes in neuromuscular-junctional transmission completed the transformation from slow to fast and vice versa. Central-neuronal anatomy is thereby shown to be more conservative than peripheral anatomy, muscle-cell innervation, and function. Indeed, the essentially infinite variety of details found in muscle-cell contents, innervation pattern, and synaptic function point to the periphery as a major source of variation in adaptive radiation and evolution, not only of insects, but also of all metazoan animal species.

However, a major source of functional variation used by crustacean muscles, variation of SL, has apparently not been used by insects as a major variable. In crustaceans the known range is from 0.6 to 28 μm, whereas in insects it is only 2.0–9.8 μm, and so far there is only one known example of the longer one (see p. 539). The majority of insect skeletal cells have 5–7-μm SL.

12.13.9. Electrogenesis

All larger insect muscles that have been extensively examined, from the dorsal longitudinal body-wall cells of *Tenebrio* larva (Yamamoto and Washio, 1979) through coxal adductors (Washio, 1972a) to locust ETi (Hoyle, 1978b), have a population of muscle cells in terms of electrogenesis that ranges from passive inexcitable through all shades of graded electrogenesis to an occasional all-or-none spiker. The inward current is blocked by 15 n*M* cobalt and is unaffected by glucose-containing saline, just as inward current is blocked in barnacle giant cells (Maruyama and Lakshminarayanaiah, 1977) where the evidence strongly favors Ca^{2+} as the carrier. The ionic strength as such is a critical variable (Yamamoto et al., 1981). The transient outward current in graded responses rectifies considerably, reversing at –46 mV. The slow component has a 5 ms time constant, so a hump occurs in the rising phase. The delayed outward current is the only one in the presence of cobalt, and it is lost when calcium is absent externally. It is therefore due to (V, t) Ca^{2+}; sodium probably uses the same channels. The gradedness is dependent on the number of these channels compared with the number of (V, t) K^+ channels. When these are blocked with TEA, all gradedly responding cells give spikes whose overshoot, if any, is proportional to external Ca^{2+}. Inactivation of calcium current is not simply related to voltage and time as it is for squid giant axon. It is dependent on the calcium concentration (Ashcroft and Stanfield, 1981).

12.13.10. Transmitter Substances

Acetylcholine has failed to cause contraction in all types of insect muscle cells so far tested, even after treatment with eserine, so it cannot be a peripheral excitatory-transmitter substance. Following the indications that L-glutamate is the excitatory transmitter at most crustacean neuromuscular synapses, this substance was applied grossly and iontophoretically at junctions to a variety of insect muscles, and caused most to be depolarized and to contract (reviewed by Usherwood and Cull-Candy, 1975; Usherwood, 1978) at reasonably low concentrations (10^{-5}–10^{-6} M). Both fast and slow excitatory actions are lost following both applications, suggesting that both types of axon use the same transmitter substances. There is no enzyme to destroy it apart from deaminases, but it is probably actively taken up and recycled by the motor axons themselves. There is corresponding evidence that GABA is the peripheral inhibitory-transmitter substance.

One exception has been claimed. In the ventrolateral body wall of larval flies, bath-applied L-glutamate failed to abolish slow EJPs, though abolishing the fast ones (Irving and Miller, 1980). However, L-aspartate was effective in abolishing the slow EJPs, suggesting that here the slow-axon transmitter substance is L-aspartate, and only the fast-axon transmitter is L-glutamate.

12.13.11. Quantitative Aspects of Excitatory and Inhibitory Junctions

The ultrastructural details of synapses and their vesicle contents have been experimentally correlated with their axon types in the *Schistocerca* ETi (Figure 5.5 – Titmus and Hoyle, 1977; Titmus, 1981). The neurons were electrically stimulated after individual identification while horseradish peroxidase was added to the fluid bathing the muscle. After gluteraldehyde fixation, incubation with 3-3′-diaminobenzidine-HCl and H_2O_2, followed by preparation for electron microscopy, the material was examined. Small dense particles occur in vesicles of terminals that have been stimulated because they have picked up peroxidase. By thus labeling the individual terminals, each kind could be examined and compared. At excitatory junctions there is granular material below the muscle membrane that is lacking or sparse at CI junctions. Dense material occurs in the synaptic cleft at excitatory junctions but not at inhibitory ones.

The CI vesicles are irregular in shape, with a maximum-minimum diameter ratio of 1.4. The SETi and FETi vesicles are both nearly spherical, but those of FETi are larger, mean cross-sectional area being 45% greater. The overall volume of a CI vesicle is also larger than that of a SETi vesicle and about the same as that of a FETi vesicle.

12.13.12. Modulation of Neuromuscular Transmission

12.13.12.1. Octopamine Action

A small cluster of dorsal, unpaired, median (DUM) neurons is a conspicuous feature of many insect ganglia. The DUM nerrons (see p. 530) are probably all, like DUMETi in the locust, octopaminergic; they secrete the OCT from release sites present in muscle cells and resembling neuromuscular synapses except for a lack of apposition with the muscle cell membranes. From the terminals the OCT diffuses to both nearby and distant synaptic sites. The functions of OCT are by no means clear yet, and they may be different in different insects. In Wetas, OCT release prepares a muscle for entering the catch state. In fireflies it triggers the light flood. For insects in general, it serves a role similar to adrenaline released from the adrenal glands in vertebrates, that is, as a multiple-site activator, the overall purpose of which is to improve escape performance. Neuromuscular-junctional transmission for slow axons is briefly enhanced in locusts up to a maximum of threefold by 10^{-8} *M* OCT (see p. 420). In Wetas, I have found long-lasting potentiation, up to a few hours, of SETi EJPs, in amounts up to 12-fold enhancement by 10^{-8} *M* OCT. SETi-evoked tension is even more dramatically increased, especially in meso/thoracic extensors, which develop more tension to SETi under OCT potentiation than to FETi stimulated maximally.

An important known biochemical effect of OCT is to promote the formation/release of cAMP and activation of phosphorylase. Glycogenolysis, fat utilization, and formation of the insect-blood-sugar trehalose, which insects use in place of glucose, are all promoted also (Robertson and Steele, 1972).

No effect of OCT on fast-axon neuromuscular transmission to leg muscles of locusts has been detected, but in Wetas an enhancement of from 15–40% occurs with 10^{-8} *M* OCT. In the moth *Manduca sexta*, it has been found that 10^{-6} *M* OCT greatly enhances fast EJPs in DL of the developing adult. When OCT is applied four days before molting, the EJP grows from 10–20 to 30–35 mV, sufficient to elicit an action potential (Klaassen and Kammer, 1980). There was no comparable enhancement in adult muscle, perhaps because there is so much OCT around normally after handling that the response is already fully potentiated.

12.13.12.2. Postexcitation Enhancement

In *Drosophila* flight muscle, it has been found that prior electrical activity has a short-lived (1.5 s) but perhaps functionally significant facilitatory action on subsequent activation (Salkoof and Wyman, 1980). This action is via a reduced threshold for the calcium-dependent spike.

Long-term facilitation is present in locust slow excitatory axon transmission, but it is less dramatic than that seen in some crustacean muscles and has been but little studied.

12.13.13. Supercontraction

The process of smooth reversible contraction beyond the point at which I filaments have passed each other and A-filament tips have reached the Z line was first found in giant barnacle cells (see p. 261). It has since turned up in a variety of muscles including some from insects, notably body wall (Osborne, 1967) and viscera (Rice, 1970). It will be recalled that the latter are cross-striated. The capability for supercontraction is not, however, universal in these visceral muscles. Those extrinsic to the canal, such as dilators, more closely resemble skeletal muscle cells and are not supercontractile. The difference is also expressed in the Z regions, which are continuous and relatively thin in the nonsupercontractors. The others lack continuous Z material: their I filaments are attached to elongated ovoid Z bodies, which are in line laterally and attached to each other, the whole forming an irregularly perforated disc.

12.13.14. Control of Spiracular Muscles: CO_2-Sensitive Neuromuscular Transmission

In order to conserve body water, the cuticle is covered with wax, and the spiracular openings with a valve or valves. The resting position of these valves is the open one, and the insect keeps them closed by tetanically exciting a small closer muscle. In the locust mesothoracic segment, the spiracular closer is innervated by two excitatory axons of nearly equal size, each of which produces a large EJP in each muscle cell (Hoyle, 1959). One axon is, however, larger than the other and produces a larger twitch. Both axons discharge at about the same frequency and with similar burst characteristics most of the time, though sometimes the one giving the smaller EJPs fires continually. There are two groups of muscle cells, an outer thicker set, and an inner thinner one, that may be of fast and slow types respectively.

Special interest was focused on this muscle because a peripheral sensory-detection/inhibition mechanism was suspected for CO_2 regulation; CO_2 causes the spiracles to open immediately. Rather than a reflex mechanism, it was found that CO_2 acts entirely peripherally (Hoyle, 1960a). The neural-output pattern remains unchanged, but the EJPs are drastically reduced. The spiracular neuromuscular junctions are enormously more sensitive to CO_2 than are those on other muscles. The mechanism has not been worked out, but one possibility is that a carbamate compound is formed at the junction. Carbamates are short lived, but they are potent inhibitors of excitory-junctional transmission.

12.13.15. "Fibrillar" Asynchronous and "Afibrillar" Synchronous Flight Muscles

Insect direct flight muscles include examples having the most regular arrangements of myofilamentar components in the whole animal kingdom. They include the only muscle cells that can be truly said to be composed of myofibrils (defined as discrete bundles of myofilaments totally surrounded by SR and not branching or anastomosing). They are also often close to being perfectly aligned longitudinally. Flight-muscle cells of this kind are specifically labeled *fibrillar* not only because they have this appearance, but also because a major physiological distinction is associated with this appearance compared with other flight muscles in which there is an *afibrillar* appearance. Fibrillar flight muscles contract rhythmically *in situ* or in an external mechanical system with suitable reverberatory characteristics, at a frequency that is much greater than that of the action potentials driving it: these need not even be applied at a regular rate. The muscles are therefore also called *asynchronous:* they are found in flies (Diptera), bees and wasps (Hymenoptera), beetles (Coleoptera), and many, but not all, bugs (Hemiptera), especially cicadas. When detached, such muscles show normal smooth tetanus at a low frequency, generally below 20 Hz. But in the thorax they can reach frequencies of 100–300 Hz depending on the species. The special properties that permit this apparent paradox will be discussed in detail below.

Other insect flight and sound-producing muscles are dependent on a single contraction/relaxation cycle for each excitation. These are therefore called *synchronous,* and it is a curious fact that their contractile material is also less clearly subdivided into bundles of myofilaments, so they are afibrillar. But this does not mean that they are also intrinsically slower – quite the contrary. They include the fastest invertebrate muscles yet known, ones that are used in song production by bush crickets and katydids (Josephson, 1973). The song frequencies are up to 280 Hz. The thoracic temperature of the katydid *Neoconocephalus* rises to 35°C during singing. At this temperature complete fusion of mechanical responses requires 400 Hz: twitch rise time is 3 ms and relaxation time about 15 ms (Josephson et al., 1975). The membrane is only weakly gradedly responsive, most of the neurally evoked depolarization being due to large EJPs. These are unusually short-lasting, due largely to a low membrane resistance (162 Ωcm^2), together with a delayed rectification that develops rapidly following depolarization but is not present at rest.

12.13.16. Mechanism Underlying Oscillatory Asynchronous Contractions

The physiological molecular mechanisms thought to underlie the rapid oscillatory contractions given by asynchronous flight muscles when suitably loaded was discussed on p. 145. The events are dependent upon the muscle's being attached to a load with oscillatory properties such as the cicada's sound-producing tympanic membrane or the wing/wing hinge of short-winged insects. At first it was thought that the basis is deactivation of the contractile mechanism following the abrupt shortening that occurs when the tympanum "clicks" to its inward position as the tension developed passes a threshold, or the wing flips to its up position. This subject was the subject of intensive research, which revealed a different type of mechanism from the one originally envisaged, essentially dependent on a high degree of synchrony of X bridges. It is probable that only in these types of muscle do X bridges stroke together synchronously.

The fundamental event of significance occurs during the time the muscle is being stretched, not when it is shortening. Stretch causes an activation, at the molecular level, either via the X bridges or T filaments, of the energy-liberation process. But the enhancing effect is delayed a little, coming during the ensuing shortening period that powers wing raising or inward click of the tympanum. A suitably loaded isolated glycerinated flight muscle will produce oscillations spontaneously when ATP is added (Figure 12.50). In katydids that use the forewing to make stridulatory sounds by beating at frequencies above 200 Hz, an interesting comparison can be made between the wing muscles and the corresponding ones of the metathoracic segment, which are called upon to produce only 20 Hz for flight. The critical element in determining oscillatory properties may well be properties of C filaments.

12.13.17. Motor Control

12.13.17.1. Motor Control During Flight

Tethered flight is easy to obtain by lifting the tarsi off the ground and blowing on the head or thorax, and recording of the motor output is also easy. Consequently the outputs during weak-to-powerful flight are known. A single weak wingbeat occurs in association with a single impulse in each synergistic motor neuron (Wilson, 1968). More-powerful flight is associated first with a slightly higher frequency. Then each muscle is excited not by one EJP but by a close pair. A further stage is produced by triplets of EJPs, until in the strongest flight, most individual contractions are caused by quadruplets. These changes are not neurally programmed in some complex manner but occur automatically as excitatory synaptic input bursts driving the flight motor neurons become progressively larger. Each larger potential swing initiates more impulses, at shorter intervals, in the impulse initiation region of the motor neurons themselves (Figure 12.54.*3* – Hoyle and Burrows, 1973b).

The flight MPG, often called the flight motor, is located in the mesothoracic/metathoracic ganglia and comprises an unknown number of interneurons together with the motor neurons themselves and interconnections between them (Burrows, 1977). The basic oscillatory mechanism is thought to involve few neurons and to be strongly dependent on reciprocal inhibitory connections between the antagonistic motor neurons. The flight motor is not the only motor program to be turned on during flight. In addition, there is coordinated stereotyped output to muscles controlling legs, antennae, and abdomen, to set them in a flight posture. This program is still present and turned on during a jump in flightless grasshoppers (Arbas, 1980).

12.13.17.2. Motor Output During Walking and Courtship Behavior

The motor output for a majority of the muscles used in walking can be assessed from simple whole-muscle electromyograms recorded with long implanted fine wire leads (Hoyle, 1964b). It is also possible to record the motor-nerve impulses with the aid of an implanted cuff electrode (Runion and Usherwood, 1966). With each of these techniques, the insect can be allowed to make natural movements and walk relatively freely around a volume many hundreds of times its own. This method has been refined to the point where as many as 30 muscles have been recorded simultaneously in grasshoppers *Gomphocerippus* and *Stenobothrus* only an inch long (Figure 12.54.*1*).

The locust hind-leg extensor is operated exclusively by SETi for all movements except jumping. The prothoracic and mesothoracic legs use FETi in addition, as does the coxal levator of cockroach in ordinary walking. All recordings from these muscles that commonly use both slow and fast axons follow a size principle: the slow axon always fires first and last, causing the basic contraction that is simply boosted by a brief burst of fast-axon recruitment at the peak of excitation only. The flexors, in spite of their larger number of axons (see p. 539), nevertheless follow the same principle, with slow first, then intermediate, and finally fast acting as a booster. However, in details such as total number of impulses per burst, mean frequency, rates of rise and fall of frequency, and interval between bursts, there is a great deal of variation, even when the movements look similar.

In the more stereotyped, but highly complex movements involved in the generation of calling and courtship songs of crickets (Bentley, 1969, 1970, 1971) and grasshoppers (Elsner, 1968), a truly remarkable finding has emerged that is in stark contrast to the situation in walking insects. The output patterns in particular motor neurons playing roles in these sex-related behaviors are highly prescribed (Figure 12.54.*4*). This is the case to the ultimate possible limit in a burst: a simple impulse. Furthermore, for two related species of the cricket *Teleogryllus,* which have different song patterns, Bentley (1971) was able to prove, by interbreeding combined with backcrossing, that there is accurate genetic regulation of the impulse number. In this case the item studied was the number of pulses in a single trill of the species – specific calling song.

12.13.17.3. Walking and Jumping

The first two pairs of legs of the best-known insects, locusts and cockroaches, are similar in innervation and motor control. There is an intrinsic tonic mechanism, BT (see p. 531), that is supplemented by low-frequency discharges in slow axons. Slow movements are

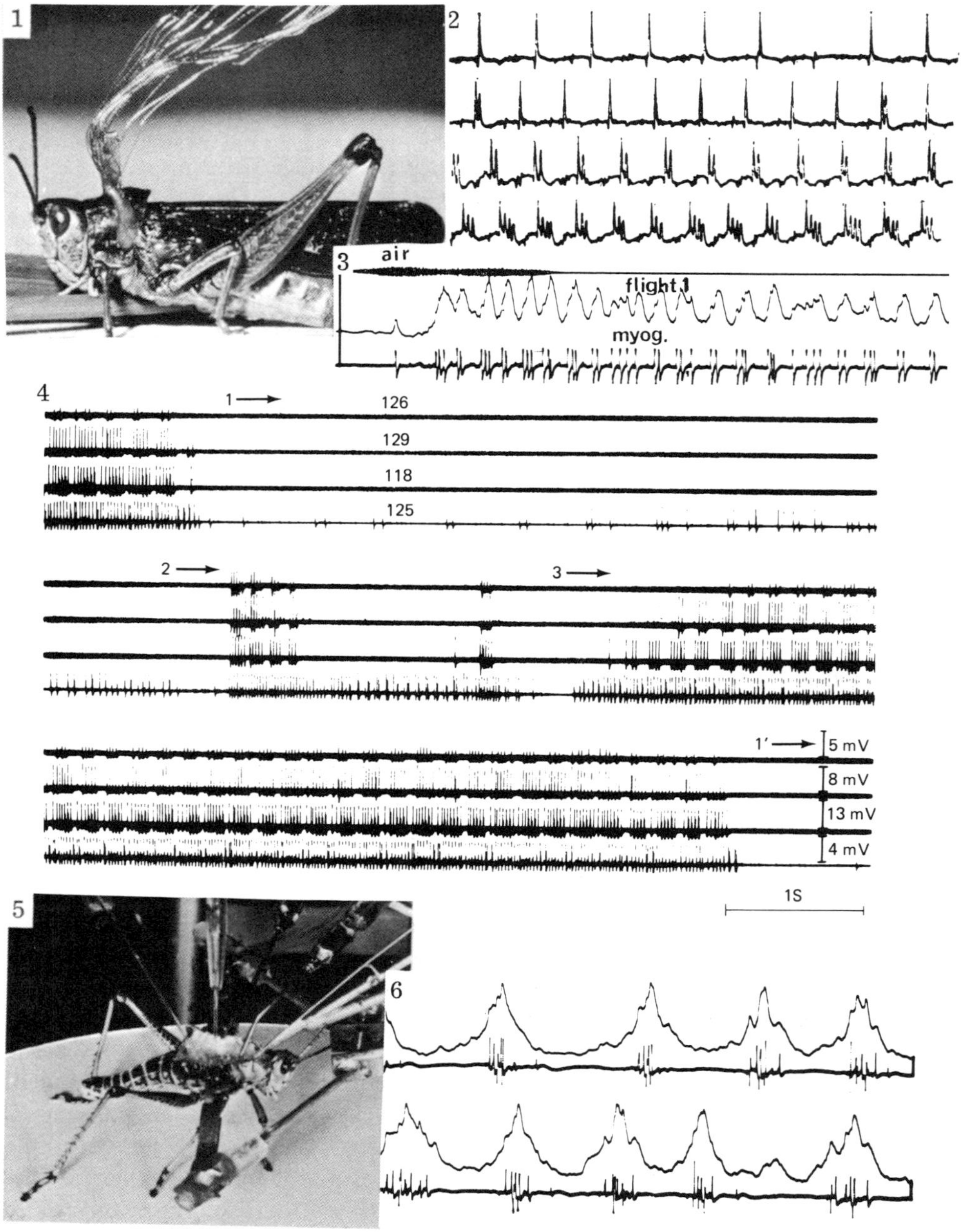

Figure 12.54. Techniques and some examples of recordings in neuroethology of insects. (*1*) Extracellular myography during courtship behavior of male grasshopper, with a total of 30 leads. (Photo by N. Elsner.) (*2*) Extracellular myograms during tethered flight of locust. The beat first increases in frequency, with single impulses per contraction. Increased power is produced first by doublets, then by triplets of closely spaced impulses. [With permission from *Adv. Insect Physiol.* **5**, 289–338 (1968). Copyright 1968 Academic Press, London.] (*3*) Intracellular recording from identified flight motor neuron during tethered "flight," showing strong depolarizing waves producing doublets and triplets. (From Hoyle and Burrows, 1973.) (*4*) Myograms from four muscles of (*1*) during courtship sequence. Each contraction is caused by a brief burst of impulse with fixed frequency and repetition rate within the behavior. (From Elsner, 1969.) (*5*) Grasshopper walking on a pivoted disc during intracellular recording. [G. Hoyle, Identified neurons and the future of neuroethology. *J. Exp. Zool.* **194**, Fig 4, p. 57 (1975). Reprinted with permission from Alan R. Liss, Inc., New York.] (*6*) Intracellular recording from identified locust flexor motor neuron during walking-type discharge, correlated with myogram (*lower trace*). (From Hoyle and Burrows, 1973b.)

produced by increases in frequency, often at a progressively increasing rate during the movement. Originally thought to be entirely reflexly generated, it is now firmly established that NPGs for each behavior are the primary determinants, with reflex action supplementing the basic oscillatory outputs. Neuromuscular facilitation plays a major role in achieving the correct tension-time profile. A minor role seems to be played by the timing of impulses in the "inhibitor" axon with respect to the excitatory ones (see p. 175). A constant low-frequency background inhibitory discharge is turned on during walking in order to suppress reflex and active tonus (Zill and Moran, 1982) and also BT.

As movements speed up, requiring greater strength to be developed in a shorter time, the extensors add brief phasic bursts of fast-axon discharges. These arise as the slow-axon frequency is reaching its peak. The flexors, which receive many more excitatory axons than the extensors, up to nine (see p. 539) and two "inhibitors", are in principle capable of complex control, and little is yet known about the rules by which the multiple axons are regulated. Basically similar sequences occur, with slow axons discharging for tonus and alone being responsible for slow movements. Faster movements are produced by adding intermediate-axon bursts, and the fastest movements by further supplementation from fast axons. The rules governing which of the slow, intermediate, or fast axons are selected are unknown.

The third pair of legs of locusts and grasshoppers is greatly enlarged for jumping, and it was natural to think that the jump could be explained on the basis of a brief, quick contraction evoked by a short burst in the fast axon. However, although this provides a sufficiently fast contraction to explain a jump of modest proportions, it became evident that some other mechanism is at work, because the more prodigious leaps required greater speeds and forces than those developed maximally by the muscle's operating directly. The disparity was resolved by the discovery that a simple structural internal knob due to invaginated ventral cuticle, located on the ventral margin just posterior to the knee, serves to lock the flexor tendon in place when the tibia is fully flexed (Figure 12.55.*2* – Heitler, 1974). There is a broad forked region in the distal part of the tendon that has a soft center that fits neatly over the knob. Coupled with a reduced torque for the extensor in this position, extension cannot occur after full flexion until the flexor muscle relaxes. Thus the extensor contraction needs to be powerful but need not be ultrafast – and it is not. The gross structure is pinnate (see p. 581), providing the largest possible power, and the ultrastructure has been able to make a compromise between speed and strength. There is one additional trick; as the extensor co-contracts with the flexor, great strain is placed on the knee. This has two thick arcs of the proteinaceous elastic material resilin, which is unique to insects, built into it; the arcs are bent, storing energy elastically. This extra energy is released during the initial extension after relaxation of the flexor, that is, during the takeoff. The height and distance in a jump are about doubled by the combined use of the knob and elastic elements.

Another example of leaping to a great height by an insect is the jump of the flea. In a typical jump, 2.25 ergs are delivered in 0.75 ms, much more than flea muscles can manage. Here also the difference is due to elastic storage by a resilin pad situated at the base of the leg between notum and pleuron. This energy is released by a second muscle that pulls the depressor femur tendon away from a position in which it is otherwise locked (Figure 12.55.*1* – Bennet-Clark and Lucy, 1967).

The evolution of the diverse, complex accessory structures associated with these and other tricks such as clicking of click beetles shows remarkable plasticity and cooperativity in central programming and muscle-cell ultrastructure, as well as skeletal morphology. It was pointed out on p. 309 that FETi was primevally a slow axon and that it has changed

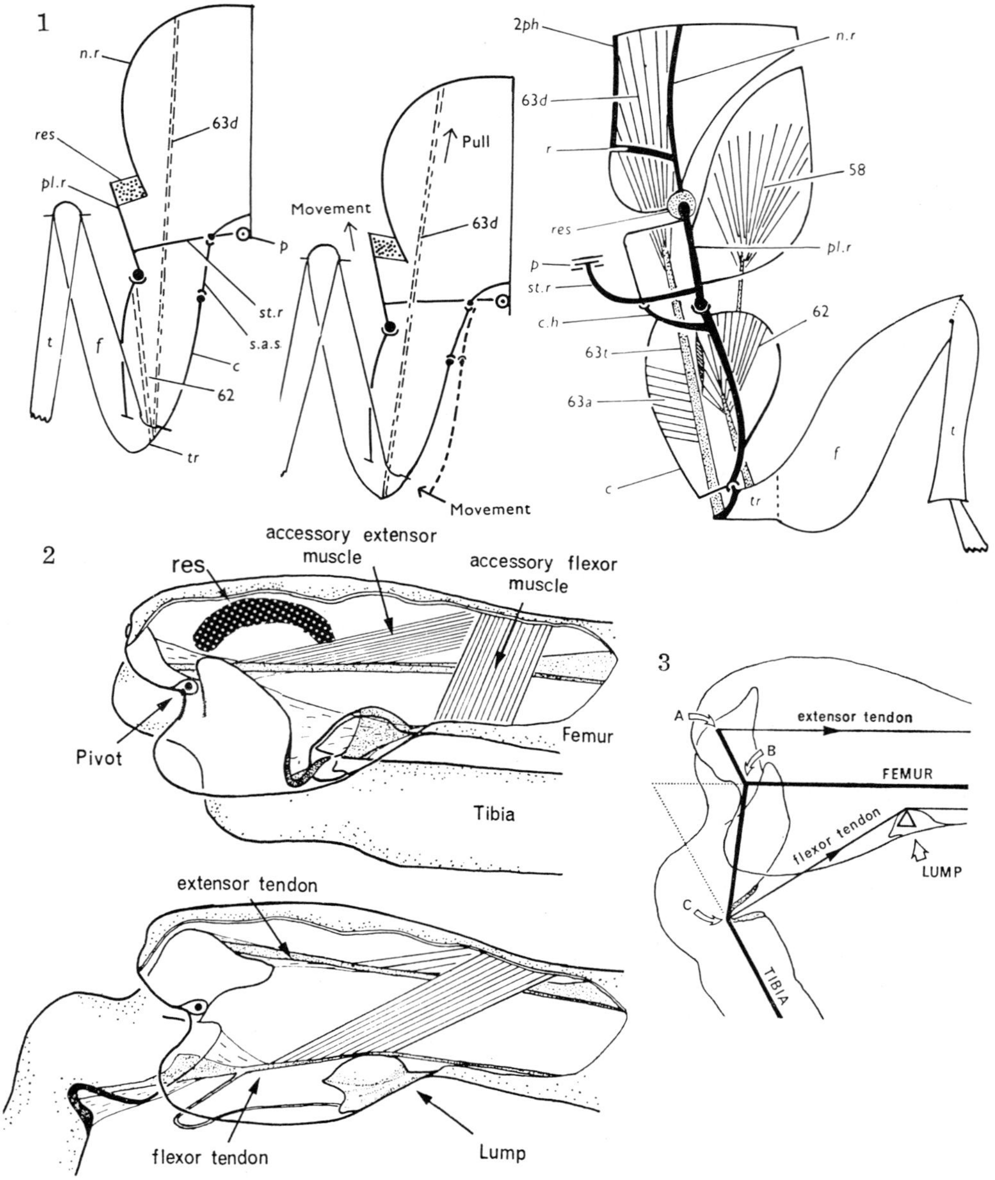

Figure 12.55. (*1*) Drawings to illustrate the mechanism underlying operation of the jump of the flea; left jumping leg as viewed from behind (*left*) and in front (*right*). Preparation for the jump is illustrated at left, in which the femur (*f*) is fully lifted by contraction of muscle 62. Muscle 63d, running from the notal ridge (*n.r*) of the thorax to the trochanter (*tr*) of the leg, then contracts (*center*), pulling the whole coxa (*c*) upward, and stressing the elastic resilin pad (*res*) linking the base of the leg to the thorax. As the coxa now rotates outward, it also rotates the sternal articular sclerite (*s.a.s*) past the pivot point (*p*) of the sternal ridge (*st. r*). A position is reached in which the femur is free to extend abruptly when muscle 62 relaxes, and does so aided by the released energy stored in the distorted resilin pad, which returns to its original shape as the bent base of the leg (*below*) straightens. (From Bennett-Clark and Lucey, 1967.) (*2*) Tricks used by grasshoppers and locusts to improve jumping performance. (*Above*) the fully flexed position in which the distal fork of the flexor tendon is locked over the invaginated cuticular lump. The positions of the two resilin pads are indicated, one is on the inside, the other on the outside. (*Below*) the extended position in which the flat portion of the flexor tendon slides over the lump. (Modified, from Heitler, 1974.) (*3*) Diagram to show lever actions used in jumping (*2*). (From Heitler, 1974.)

its entire peripheral physiology in coming to subserve jumping but not its own location, morphology or pathway in the ganglion. At least some aspects of neural organization are conservative.

Locusts and grasshoppers can swim quite well, even on their first encounter with water. They achieve this by synchronous tibial extension following flexor release after co-contraction, as in jumping (Pflüger and Burrows, 1978). After a swim, kicking is used to aid in drying the insect.

12.13.17.4. Roles of Proprioceptors

Direct study of the roles of the various proprioceptors of insects in the control of locomotion and more-specific behaviors is still in its infancy, but beginning to attract attention. There are so many sense organs serving every joint and every type of movement that redundancy is inevitable and substitution of information always possible. Accordingly, results of ablations (e.g., Wendler, 1971) have tended to support the view that central pattern generators (CPGs) can drive detailed behavior and that the proprioceptors are mainly serving trimming roles.

However, when the more effective method of testing, by cross-connecting a proprioceptive input, was used in stick insects, walking was drastically altered (Graham and Bässler, 1981). The chordotonal organ was switched so that it was excited by flexion rather than the normal extension. The operated leg could indeed still walk normally as if using a CPG. But from time to time, the operated leg would remain raised in the "salute" position. When this happened, the pattern of all the legs changed to one similar to that occurring when the leg has been amputated. However, regular depressor (extensor) contractions occur during a "salute," showing continued operation of the CPG and less dependence of the extensor than the flexor musculature on the proprioception.

12.13.17.5. Control of Head Movements

Fourteen muscles on each side operate the locust head. They are innervated by a total of about 60 axons each originating in the subesophageal, prothoracic, and mesothoracic ganglia (Shepheard, 1974). The most dramatic movements they produce are similar to those seen in a wide range of vertebrates also, namely tracking of an object moving in the visual field, followed by rapid return and repeat (nystagmus). Each major muscle receives one slow, one intermediate, and one fast excitatory axon plus an inhibitory one.

Stationary head position is maintained by tonic discharges of slow axons only. These axons, firing alone, also cause the slower head movements by temporary increases in their frequency. The intermediates are turned on and become progressively more active when the eye must make faster following movements. Fast-axon discharges occur in only the fastest movements, when they are superimposed on the slow axon background. The fast phasic return is produced by the discharge of fast axons to appropriate muscles alone. Thus we see here a size principle operating when the eye is making a following movement. There are specific direct-to-fast-axon "commands" in nystagmus, and they are proprioceptively driven. The slower-movement sequence of excitation must be "planned," following assessment of the speed of the object being watched.

12.13.17.6. Use of Inhibitors

There has been only a single study of the use made by an insect of one of its inhibitors. This was the common inhibitor of the *S. gregaria* hind leg, in relation to walking and SETi (Usherwood and Runion, 1970). Electrodes were first implanted to record from the ETi nerve, then the insect was freed to walk around while the impulses in both axons were tape-recorded and its behavior was filmed. Next a physiological preparation of the same animal was made. The tape recording of the impulses was then played back, amplified to a strength sufficient for the impulses to stimulate the axons. First, the excitatory pattern alone was applied and the resulting tension recorded. Then the pattern was repeated adding the CI train. After the addition, the tension peaks were reduced, but the effects were slight, and it is debatable whether reduced peaks have provided convincing evidence that the animal was seriously utilizing peripheral inhibition, at least of ETi, during the particular walking bout. Nevertheless they showed that some use of CI is built into either the walking program or reflexes associated with it. These experimenters succeeded in exemplary fashion to demonstrate how to set about determining what it is that any arthropod does use its inhibitors for. More research along the same lines is needed.

During head movements, inhibitors fire just before each fast turn and also during the turn. Clearly they don't inhibit the contractions causing the turn. Instead, they seem to make them brisker. A similar use has been seen in the anterior coxal adductor, a muscle innervated by a single excitatory axon that is always tonically active but also makes fast phasic movements. A hyperpolarization is caused by the inhibitor just before the onset of each fast movement. This ensures that the burst, when it comes, will give large EJPs leading to large graded responses, thereby promoting quick tension rise. Because the inhibitor is active during the burst, when this ceases the membrane potential returns rapidly to baseline or above, ensuring rapid relaxation. The "inhibitor" is not used to cause inhibition, which indeed it cannot, for most muscles, but rather to sharpen up phasic movements, both their rising and relaxing phases being enhanced, especially when these must be superimposed on ongoing tonic activity.

12.13.18. Vegetative Muscles

There are many insect muscles whose functions ally them with those collectively termed vegetative in vertebrates (see p. 345). Unfortunately, very few have been studied in any depth physiologically. Those that have constitute mainly parts of the alimentary canal, especially the anterior and posterior specializations, esophagus and proctodeum, and the diaphragm. The latter is neurogenic, but all parts of the gut seem to be myogenic, though with local neural control of both direct and modulatory kinds. The cells of insect vegetative muscles are all cross-striated and very small, commonly about $5 \times 5 \times 20$ μm, with tapered ends, a single nucleus placed either centrally or at the margin, and only a few small mitochondria. The sarcomeres are generally fairly long, 7–10 μm, but some are as short as 2.2–3.5 μm, shorter than those of fast-contracting leg muscles of the same animal. The Z regions are formed of elongated ovoid DBs to which I filaments are attached in close clusters, permitting easily reversible supercontraction. However, in addition to these attachments, some I filaments are attached to desmosomelike DBs at the surfaces of adjacent cells. The desmosome DBs also serve to link the cells laterally and at their ends. There is some basement lamina material between the cells, with collagen fibers imbedded. There are no M bands, and H regions are at best poorly defined. There is a small amount of SR and a few obliquely running ITs, which make dyadic contacts with sparse SR cisternae. Some effectively dyadic regions occur by SR's contacting the surface membrane. A 7 : 1 thin-thick filament ratio is common.

Not only the intrinsic gut muscle cells but also those of extrinsic muscles operating on parts of the gut are cross-striated in insects. Most show rhythmic myogenic contractions at a relatively fast rate, 1–4/s.

12.13.18.1. Diaphragm

The diaphragm is a sheet of transverse muscle cells present in many insects from the metathorax through most of the abdomen; the diaphragm serves to assist the heart. The cells in *S. gregaria* are cross-striated, and polyneuronally innervated, but contraction appears to be primarily myogenic (Guthrie, 1962). The cells not only anastomose, but they are interconnected by very thin bridges composed of strands of cross-striated muscle (Peters, 1977). The cells are sensitive to stretch, but local sensory cells occur in them, so reflex mediation of contraction may be occurring as well as a direct action. A pair of excitatory neurons to diaphragm muscle cells exists in each abdominal ganglion, innervating cells of both its own segment and the immediately antecedent one, so that innervation fields overlap.

12.13.18.2. Hyperneural Muscle

This is present in *Periplanata* as a sheet of longitudinal muscle cells that probably also aids blood flow. The anteriormost cells are innervated from medial nerves, the posterior ones by a branch of the sixth abdominal ganglion. All contraction ceases when the nerves are cut, so the muscle is not myogenic (Miller and James, 1976).

12.13.18.3. Esophagus

The foregut of insects is spontaneously contractile but is capable of being driven and modulated by neural input from the ingluvial ganglion, which is itself driven from the frontal ganglion. A number of possible candidates as transmitter/modulator substances involved have been proposed, most notable and convincing of which are 5HT, which is present in the corpus cardiacum, and NAd (Cook et al., 1969).

12.13.18.4. Proctodeum

Superior rectal longitudinal cells of *Periplaneta* have a 40–64-mV resting potential and give spontaneous action potentials at a frequency of about $\frac{1}{2}$ s (Nagai and Brown, 1969). These are preceded by a slow depolarization, and have a shape resembling that of heart muscle cells, with a plateau. The total duration of the action potential is about 0.6 s. Also, EJPs occur sporadically and can be evoked by nerve stimulation. Three sizes of weakly facilitating EJPs can be obtained from a single muscle cell. There is a large slow-twitch contraction in the entire muscle when each action potential occurs, with a total duration of about 5 s. Another, more slowly conducting type of axon innervates the muscle, and this does not evoke a JP.

Stretch of the preparation initiates early depolarizations leading to repeating action potentials and contractions when maintained (Nagai, 1970). The action potentials are both sodium and calcium dependent but are TTX insensitive (Nagai, 1972).

The entire hindgut is enveloped by a periproctodeal net, comprising a syncytium of cells that is itself innervated (Brown and Nagai, 1969). What the function of this net may be is not known, but it may well be the site of storage of a modulatory substance. Brown (1975) discovered that the proctodeal muscle cells are excited by an agent, present in the ganglia, eventually shown to be active at concentrations as low as 10^{-9} M, to which he gave the name *proctolin.* Starratt and Brown (1975) soon showed that this is the pentapeptide Arg-Tyr-Leu-Pro-Thr. A pair of (probably) proctolinergic cells occurs in each abdominal ganglion, and also in the metathoracic ganglion (O'Shea and Adams, 1980, 1981 and personal communication).

A net of some kind has also been demonstrated in another cockroach, *Leucophaea* (Cook and Reinecke, 1973). The gut shows peristaltic and reverse peristaltic waves.

Recordings were made from circular as well as longitudinal muscles. The former show rhythmic action potentials at 2/s that are TTX insensitive.

12.14. Chordata

The phylum chordata, which encompasses animals with the most highly developed nervous systems, notably man, also includes some forms that have invertebratelike adult stages. The range of structure and function encompassed is so great that these organisms provide a special interest in regard to muscle-cell diversity and control. The invertebrate-like forms are classed as Protochordata.

12.14.1. Protochordata

The protochordates are phylogenetically important but have received much less attention than they deserve from nerve/muscle anatomists, ultrastructurists, and physiologists. Some data are available for a few tunicate ascidians and salps, from the hemichordates *Balanoglossus* and *Ptychodera,* and structural, but not physiological, information for the advanced free-swimming lancelot *Amphioxus.*

The hemichordate worm *Ptychodera* shows a strong startle response. As few as two impulses in the ventral nerve cord trigger this. Their final effect, as recorded extracellularly from the muscle, is to initiate a massive, complex, high-frequency discharge (Pickens, 1973), such as could only be produced by a large number of motor neurons acting asynchronously but producing similar muscle effects. It is a sad reflection on the state of our knowledge of both echinoderms and protochordates that we do not know what these muscle-cell events may be.

12.14.1.1. Ascidiacea

The ascidians, sessile tunicates, provide a wide variety of contractile cells. As adults they are sessile filter feeders known as sea squirts that are so much like invertebrates in character that standard textbooks of invertebrates include them, whereas chordate texts ignore them completely. Yet the larval stage is a fast-moving tadpole with a dorsal nerve cord, an organism that is very much the prototype for evolutionary images of early chordate stock. The transformation from larva to adult is itself a fascinating process that has been extensively studied. The muscles of the adult, however, are completely new elements, not transformation of those of the tadpole, which are resorbed.

The larval tail muscle cells are cross-straited, but they are small and squat, about $14 \times 9 \times 21$ μm, and mononucleate. A total of about 800 such cells forms a single muscle band, and only a portion of the cell contains myofibrils, the remaining space being occupied by numerous mitochondria, perforated sheets of longitudinal SR, and myoplasm containing glycogen rosettes (Cavey and Cloney, 1976). The sarcomeres are very vertebratelike in some respects, SL is short, 1.8 μm, with strongly marked Z band, H zone, and M bridges, but there are also differences. The SR has flattened cisternae at the Z regions, and these make dyadic connections with a sparse system of ITs. The Z region is formed of fine filaments in a square array essentially indistinguishable from that of a typical vertebrate skeletal cell. The tiny cells are connected to each other by tight junctions. Motor axons terminate on only a few of the muscle cells, so most must be excited by spread of electric excitation via the tight junctions from the innervated ones.

Of special interest is contractility in epithelial cells of the tail that is responsible for intrusion of the tail at metamorphosis. Before contraction, 6-nm filaments are scattered in all directions, but they become oriented in the future contraction axis before the event. No thick filaments appear, but the cells proceed to shorten down to about 6% of their initial length (Cloney, 1966). The length of the thin filaments is not known. Most of their central portions disappear when the cells are treated with cytochalasin B, an agent that disrupts microtubules, reappearing very quickly when this is washed out, but thin filaments close to the end membranes remain in its presence. These residual filaments are attached directly to the membrane and are dispersed evenly along it rather than being clustered. Their lengths could not be ascertained.

The larvae have motile adhesive papillae that are used in attachments. The contractile elements are myoepithelial cells, of which a portion contains both thick and thin filaments (Cavey and Cloney, 1976). However, no longitudinal organization occurs: even DBs to which the thin filaments might be attached are absent.

There also is a heart in the larva, as in the adult (see p. 371), comprising short, cross-striated cells with about 4 × 10 μm transverse axes, each with numerous mitochondria and up to 50% cytoplasm that lacks myofilaments (Oliphant and Cloney, 1972). There are no ITs, but SR makes close contacts with the surface membrane in the form of collars situated at the M regions (Figure 12.56). These are a unique form of E-C coupling device, like a giant dyad encircling the cell. The cells are connected end to end by simple, slightly wavy regions that may be regarded as primitive intercalated discs, with a wide separation of the membranes. The Z bands are situated a short distance inside each, abutting cell membranes. The hearts of the tadpoles, unlike those of the adult, show local, asynchronous "fibrillations" rather than coordinated contractions.

Adult tunicates behave more like sea anemones than chordates both in what they do and how they respond to stimuli. A single strong shock applied to the stiff external "skeletal" test of *Phallusia mammillata* or *Ascidia aspersa* evokes a response that is independent of either stimulus strength or location (Hoyle, 1952), taking the form of a weak siphon closure. But if a second shock is given soon after the first, a contraction of the entire branchial and atrial musculature occurs. This produces the powerful familiar ejection of water that gives them their popular name. The similarity of the squirt mechanism to the total response of a sea anemone (see p. 424) is unmistakable. The state of readiness for facilitation set up by the first shock lasts even longer in the ascidian than it does in the anemone, and it has a significant consequence after as long an interval as 16 s.

Local contractions of individual muscle cells can be obtained by local stimulation, but strong shocks invoke whole-bundle quick responses, probably by exciting neural elements innervating many cells. The simplicity of the responses may be due to a major departure from the characteristic chordate innervation; this is an area that still needs much more study. The brain/nerve cord of the tadpole is replaced by a simple ganglion in the sea squirt.

Aside from these marked divergences from the vertebrate pattern, the muscles of ascidians are typical for chordates: they are highly sensitive to ACh, and transmission to them is blocked by curare (Florey, 1967). Acetylcholine is also present in the ganglion from which the motor axons emanate.

12.14.1.2. Larvacea

One class of tunicate, the Larvacea, attains maturity without metamorphosing. It does

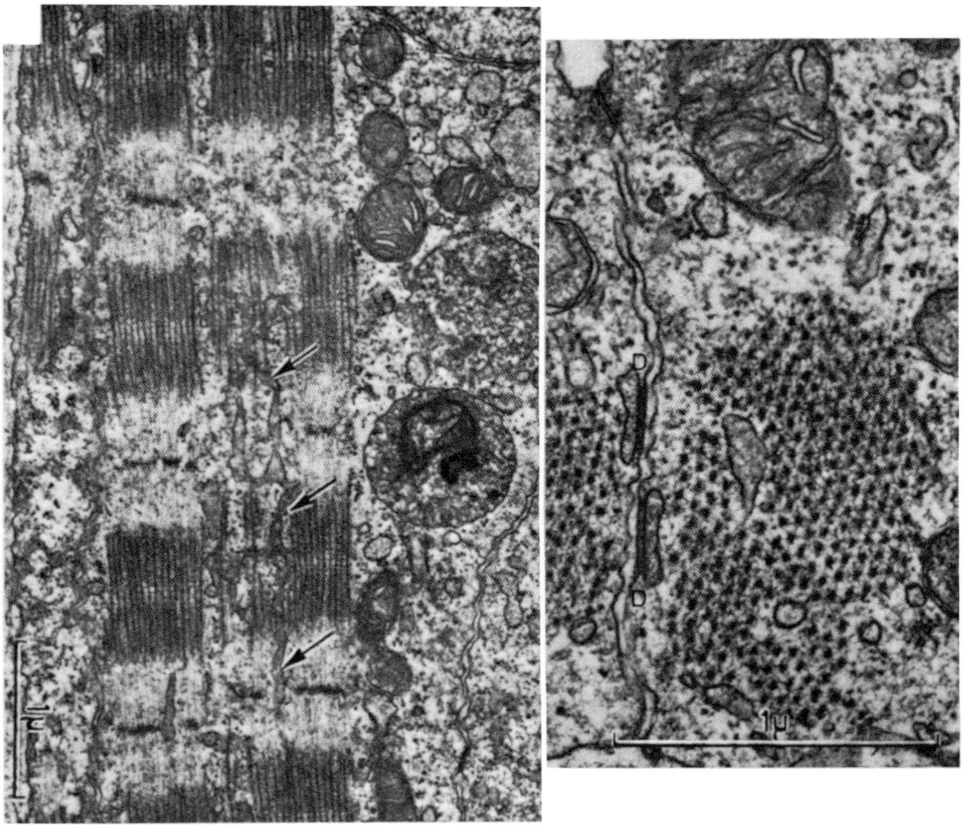

Figure 12.56. Ultrastructure of cardiac cells of ascidian *Corella.* (*Left*) TEM of LS showing very short SL and sparse, mainly longitudinal SR (*arrows*). (*Right*) TEM of TS showing margins of two adjacent cells, with dyadic-type contacts between SR and cell membranes. These are at the M region, in the form of a collar around the entire cell. (From Oliphant and Cloney, 1972.)

make temporary coverings for itself, but not like the tough tunican tests of ascidians. Examples are *Oikopleura* and *Dendrodoa.* The epithelial cells of these tadpoles are interconnected and electrogenic like the surface cells of jellyfish (see p. 427); the wave is able to enter the animal in unknown ways and affect the nervous system. Curiously, a wave of excitation in the epithelium is excitatory to the motor system in *Oikopleura* but inhibitory in *Dendrodoa* (Bone and Mackie, 1975; Mackie and Bone, 1976). Such events were first found in young amphibian tadpoles (Roberts and Stirling, 1971), and it is truly remarkable that they are so widespread, hinting at commonality and a very primitive precoelenterate evolutionary origin of epithelian conduction.

The electrical activity of muscle cells of ascidian tadpoles has been compared in presumptive cells of the gastrula stage, as well as in prehatching and posthatching tadpoles (Takahashi et al., 1971). At first the membrane is electrically unresponsive, then it develops a slowly decaying inward-current response. At the time of hatching, the response is an all-or-none overshooting plateau resembling that of a cardiac action potential, but with a very long, almost 3 s, duration before abrupt repolarization occurs. After hatching the plateau has gone and the spike peak is immediately followed by a large undershoot.

12.14.1.3. Thaliacea

The salps, members of the protochordate class Thaliacea, are free-swimming pelagic organisms that may occur singly or in colonies of from a few individuals linked as a chain, to long cylindrical columns comprising hundreds of members. They therefore offer interesting questions, both of muscular contraction and control in the individual and of coordination between individuals. Salps keep afloat and motile by making quick contractions in the manner of jellyfish, by propelling water from their internal cavity. Like larvaceans (see p. 558), they use the entire skin as a receptor/conducting/stimulating element (Figure 12.57). In addition, colonial forms use epithelial spikes to transmit excitation from one member to another in order to ensure coordination (Mackie and Bone, 1977). The events of transmission from the epithelium to the musculature are unknown.

The muscles of salps are in the form of circular bands whose component cells in large specimens are as large as those of vertebrates, commonly 50–100 μm in diameter and 1–2 cm long. They are cross-striated, with a longer SL than that of ascidian tadpoles, 2.5–3.5 μm (Bone and Ryan, 1973). Some other features are also different, the Z lines being irregular; there are well-marked H zones and no ITs. The cells are linked to each other at desmosomelike spots. As in ascidian hearts, E-C coupling is achieved by SR contacts with the surface membrane, but SR is largely longitudinal.

The neuromuscular junctions are simple flattened naked axonal terminals embedded in the sarcolemmal matrix. They contain uniform spherical 50-nm clear vesicles and are probably cholinergic.

The central core of the muscle cells is crammed with large mitochondria, three or four abreast, plus several nuclei, so that only about half the cell contents comprises contractile material, the remainder being devoted to a continuous strong energy supply for the sustained swimming.

12.14.1.4. Cephalochordata

The highest group of the Protochordata, and the one most nearly resembling a vertebrate in appearance, is represented by the marine lancelots *Amphioxus* and *Branchiostoma.* They are therefore of great theoretical significance in relation to understanding the form and function of ancestral vertebrates. The musculature consists of lamellae, which are cells each 0.7–2 μm thick, 10–300 μm wide, and about 600-μm long, termed myotomes because of their resemblance to fish musculature. A longitudinal section reveals cross-striated contractile material in which H zones and M bands are well marked (Flood, 1968). The Z lines are clear, zigzag structures as seen in longitudinal sections, closely resembling those of fish. However, they are not laterally aligned, although their lengths are similar. The SL is short, 2.0 μm, with 1–8-μm A bands.

Two kinds of lamella occur, one is pink, containing many mitochondria and glycogen particles, the other, lying deeper and being white, has few of these. There are also lamellae that are intermediate between the two. Thus differentiation of cell types into slow, fast, and intermediate is present. The faster cells have contraction times of 70 ms, compared with 250 ms for the slowest ones (Guthrie and Banks, 1970).

At the inner margin, fine tapering strands of the cross-striated contractile material pass laterally into a broad process that is largely noncontractile cytoplasm (Figure 12.58). This is of considerable interest because it is a muscle arm that narrows and passes to the nerve cord to form neuromuscular junctions and receive excitation. This pattern is similar to that seen in a number of invertebrates, notably platyhelminths, nematodes, echino-

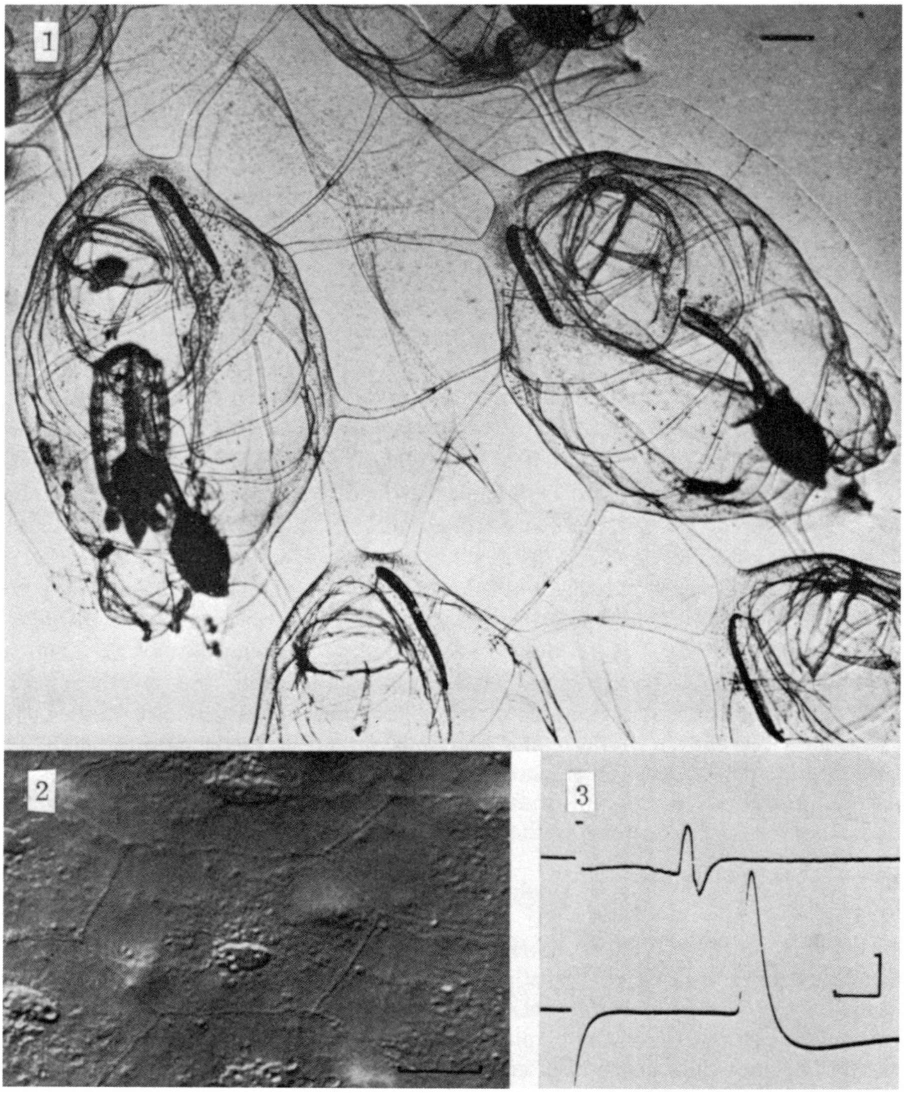

Figure 12.57. Electrical conduction leading to movement control occurs not only in nerve cells but also in epithelia of coelenterates and tunicates such as this salp *Thalia,* which lives colonially. Parts of six connected individuals of *Salpa fusiformis* are seen in (*1*), epithelial cells in (*2*), and electrical recordings from (*2*) in (*3*). The upper trace is extracellular from the epithelium, the lower one intracellular from a single epithelial cell. (From Mackie and Bone, 1977.)

Figure 12.58. Innervation of *Amphioxus* muscle. This is entirely by muscle arms extending to the nerve cord. There are three kinds of muscle cell, each lamellar, one pink, containing numerous mitochondria and presumably slow (*S*), one white and fast (*D*), and one intermediate (*I*). The S lamellae make fine muscle arms (*t*) reaching a dorsal location (*Do*). The fast and intermediate myotomes make broad extensions to a ventral location (*Ve*). (*1*) Three-dimensional diagram of myotomes and muscle arm. (*2*) Fine (*t*) and broad (*T*) muscle arms viewed in low-power micrograph. (*3*) Broad muscle arms from a single cell. [All from P. R. Flood, A peculiar mode of muscle innervation in *Amphioxus.* Light and electron microscopic studies of the so-called ventral roots. *J. Comp. Neurol.* **126**, Plate 1, p 195 (1966). Reprinted with permission from Alan R. Liss, Inc., New York.]

1

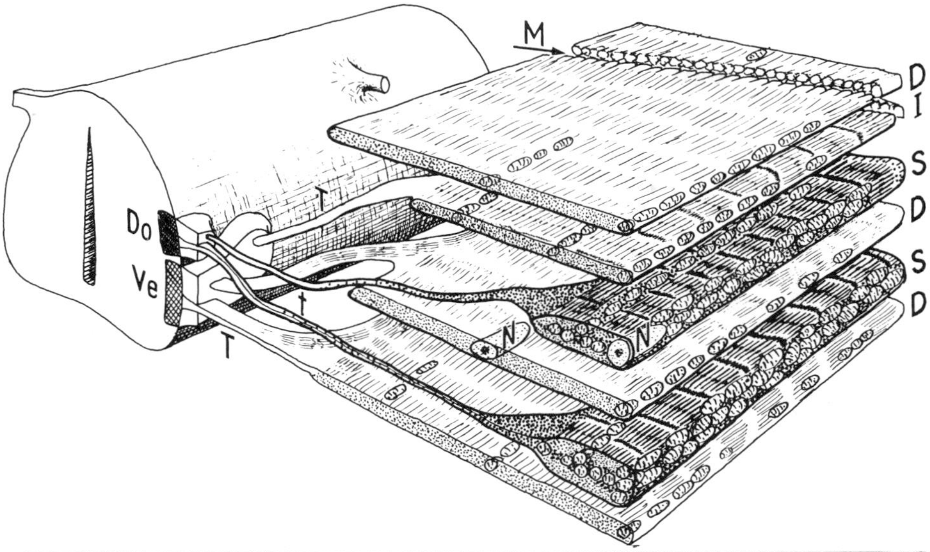

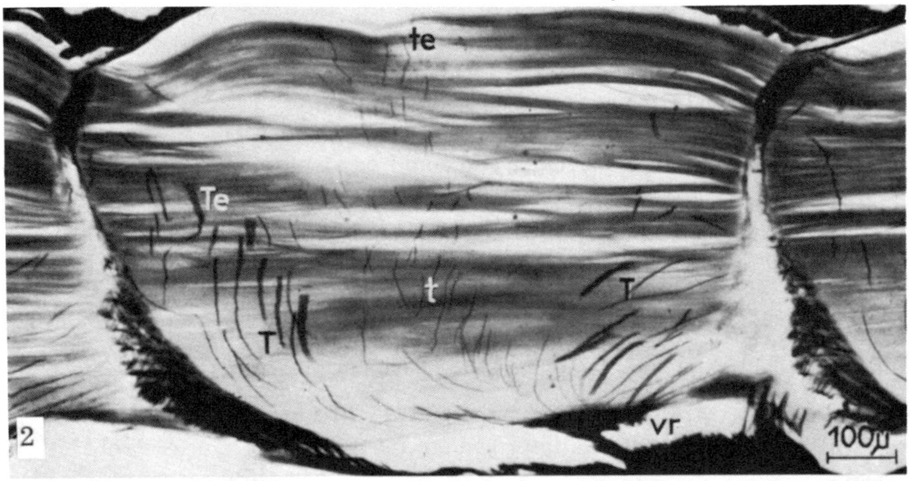

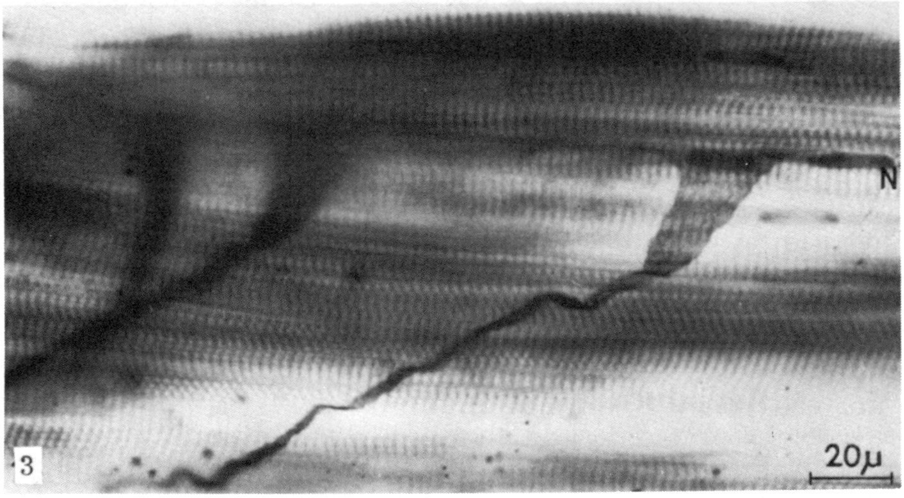

derms, and onychophorans. The echinoderms, because of their type of development, are considered to be the specific ancestral phylum. There is no evidence of any nerve-on-muscle cell junctions in *Amphioxus,* although this more advanced feature, compared with muscle-on-nerve synapses, does occur in echinoderms in addition to muscle arms (see p. 459).

Amphioxus is capable of long-sustained gentle swimming and also short-lasting intense fast bursts that are doubtless associated with the two extremes of muscle cell type. Its muscle cells lack both longitudinal SR and ITs. Therefore, it seemed likely that they must be using directly entering calcium to activate the contractile machinery. Since the cells are so thin, diffusion from the surface could account for E-C coupling. The related *Branchiostoma* has a resting potential of -53 mV and gives spikes that overshoot by 29 mV when excited (Hagiwara and Kidokoro, 1971). The action potential is, however, TTX sensitive, but a full spike is restored after TTX block in the presence of 7.3 m*M* procaine. The size of the spike is directly related to external calcium concentration. Thus the membrane has dual depolarizing electrogenesis mechanisms, a sodium-dependent one for propagation along the muscle arm and lamella, and a Ca^{2+}- dependent one for activation of contraction.

The lack of SR and dependence on directly entering Ca^{2+} could have been secondary to evolution of the thin lamellar structure. Alternatively, it must be assumed that the *Amphioxus* muscle cells represent an extremely primitive state, beyond which almost all invertebrates have advanced, in which no Ca^{2+}-releasing/sequestering internal membranes have yet evolved. Even the presumed precursors of protochordates, echinoderms, have evolved muscle SR that probably functions in this manner. However, *Branchistoma* has a caffeine-activatable Ca^{2+} source (Hagiwara et al., 1971a), so may have a little SR equivalent that could represent the starting point for evolution of SR of vertebrate muscle. That this source is not necessary or normally used in coupling was shown by first depleting it fully by caffeine address. A full-size twitch is obtained before the caffeine response has recovered. Thus calcium-sequestering/releasing SR is probably not an ancient device of common cellular origin but one that has evolved independently in several phyla.

12.14.2. Pisces

For our present purposes, it seemed better to consider jawless, cartilaginous, and bony fishes together because of the basic similarity of their forms, habitats, and locomotion. This permits comparison of their musculature, innervation, and neural control collectively with those of the other major vertebrate subdivisions: amphibia, reptiles, birds, marsupials, and mammals. All forms of fish have as their principal musculature that of the back/body used in swimming. This is considerably different from that present in amphibia and all higher forms that have evolved limbs. The muscle cells of fish myotomes are markedly different from those of all back and leg muscle cells of amphibia and higher forms. Few transition forms of muscle cell are known except, perhaps, in the jawless lamprey. So how, or even why, the switch to characteristic vertebrate cylindrical muscle cells occurred needs to be considered.

12.14.2.1. Hagfish and Lampreys

In ammocoetes larvae of lampreys, the myotomes are lamellate, with five or six to a segment. Dorsal and ventral lamellae are rich in mitochondria, fat droplets, and glycogen, whereas the inner ones have little. In the adult, however, the inner lamellae are each broken up into several cells, though each of these is flattened. But the outer, reddish ones form rounded cells that are transitional to typical vertebrate ones.

The hagfish *Myxine* has muscle cells that in section are shaped just like the typical vertebrate example. White, fast-type cells, though of relatively small diameter (Figure 12.59) form the majority, occurring throughout the myotome except for the ventral and outside margins, which comprise yet-smaller-diameter reddish ones. These are considered to be slow-type cells. A few intermediates occur between the two sets. There are fast- and slow-type cells in lamprey body muscle, with slow ones being a majority (Teräväinen and Rovainen, 1971). The former have a 7-Hz fusion frequency, and tension is maintained, after slowly increasing, at this frequency. The latter fuse at 12 Hz, which is also low, but start to fatigue immediately. The contraction times range from 150 ms for fast-type cells to 500 ms for slow ones (Anderson et al., 1963).

12.14.2.2. Cartilaginous Fish

The myotomes of cartilaginous fish comprise standard vertebrate white cells except at the outer margins, where smaller red ones exist, with pink intermediates between them. Such amphibia as possess tails have muscles that are organized similarly.

12.14.2.3. Bony Fish

In bony fish, the red, slow-type muscle-cell population is reduced to a narrow band running the whole length of the fish and confined to the lateral regions near the lateral line. Between this and the massive bulk of the white muscle is a small region of pinkish intermediate-type cells. No links between *Amphioxus* and fish are known in which muscle-arm–on–nerve junctions coexist with ones of nerve-on-muscle cells. All fish have a characteristic vertebrate-type innervation. The motor nerves of all fish travel directly to the muscle cells and make synaptic contacts on them. The nerves travel down the septae separating cell clusters longitudinally, thereby first contacting the cells at their ends. On some white fast cells, the nerves make single clawlike or basket endings at one end only on each cell, or at both ends on others, depending on the species; uncommonly, multiple endings are made. Intermediate-type cells are innervated in their centers in some, or at their ends in others. Motor axons innervate red cells near both their ends but do not terminate in single endings. They continue inward toward the centers and make additional junctions en route (Figure 12.60).

The individual junctions range from ones with multiple long, thin, wavy filamentous branches running mainly parallel to the long axis of the cell to ones that have compact, grape-cluster form.

There seems to have been a partial transition from distributed to focal innervation in overall evolution, and division into white fast-type cells as the majority, compared with red slow-type ones as the hard-working minority. Associated with the transition, the white cells are of larger diameter and produce very large, rapidly fatiguing JPs with a high safety factor for action-potential initiation. These are all fast-twitch cells with overshooting, propagated action potentials. The red cells are either slow-twitch or gradedly responding and slowly contracting, depending on the species. Intermediate types are twitch-type, with propagated potentials, but they fatigue less rapidly than do white cells.

When innervated at both ends, fish muscle cells could be receiving two different axons; this question has not been resolved. But polyneuronal innervation of single muscle cells is at most a rarity, even for those red cells that are gradedly responding.

A division of function in fish tail muscle has been firmly established for a wide range

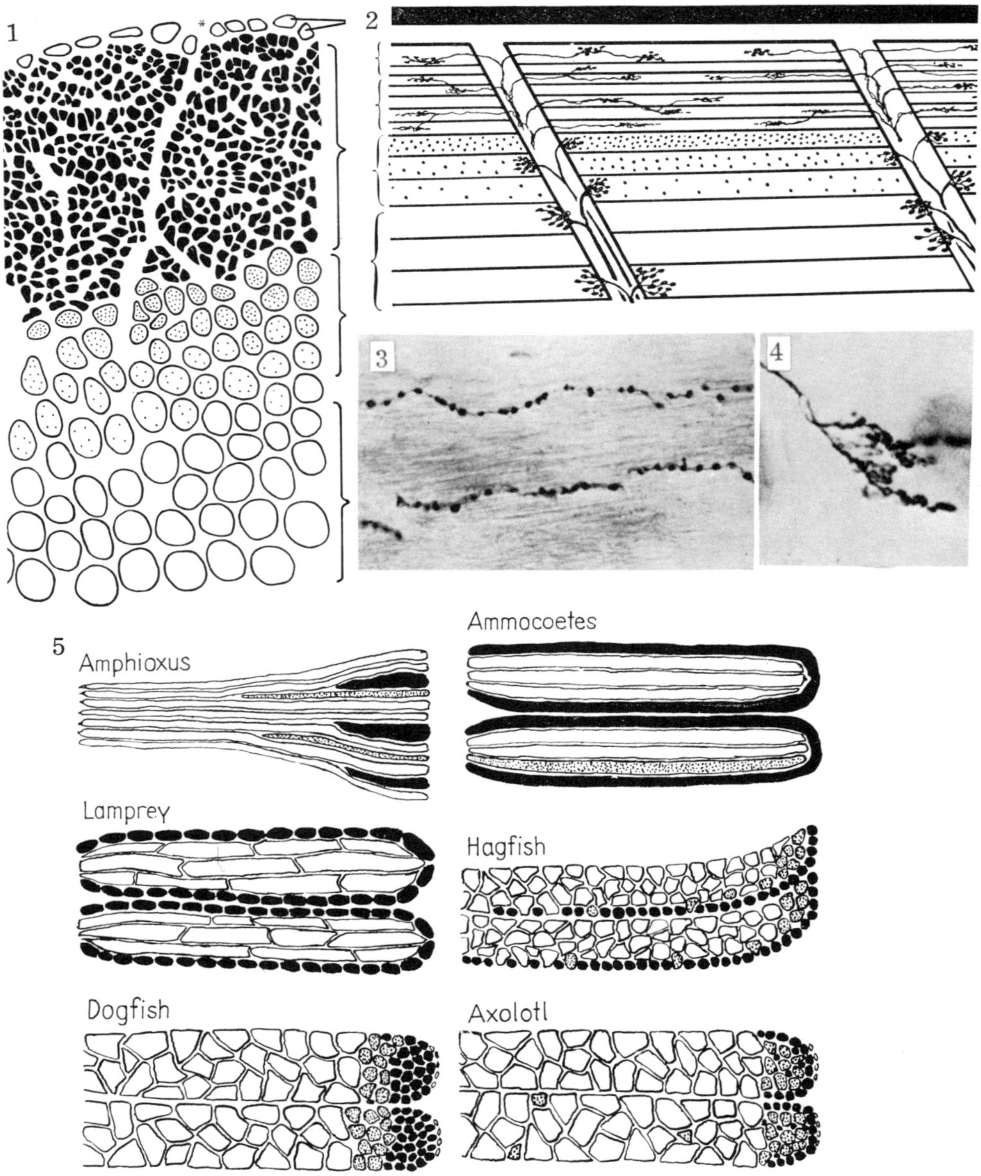

Figure 12.59. Structure and innervation of hagfish (*Myxine*) muscle and possible evolution of vertebrate muscle cells. (*1*) Diagram of different types of muscle cell in a myotome, in transverse section, skin at top. First are small-diameter, slow, red cells; next are pink intermediates; innermost are large, fast, white cells. (*2*) Diagram of innervations of the three cell types, in longitudinal section. (*3*) Varicosities in stained distributed slow-type terminals. (*4*) Typical compact terminal at end of fast-type cell. (*5*) Diagram of possible evolution of vertebrate muscle cells from *Amphioxus:* slow (*dark*) intermediate (*dotted*) and fast (*clear*) lamellar muscle cells progressing to comparable cylindrical ones. (All from Jansen et al., 1962.)

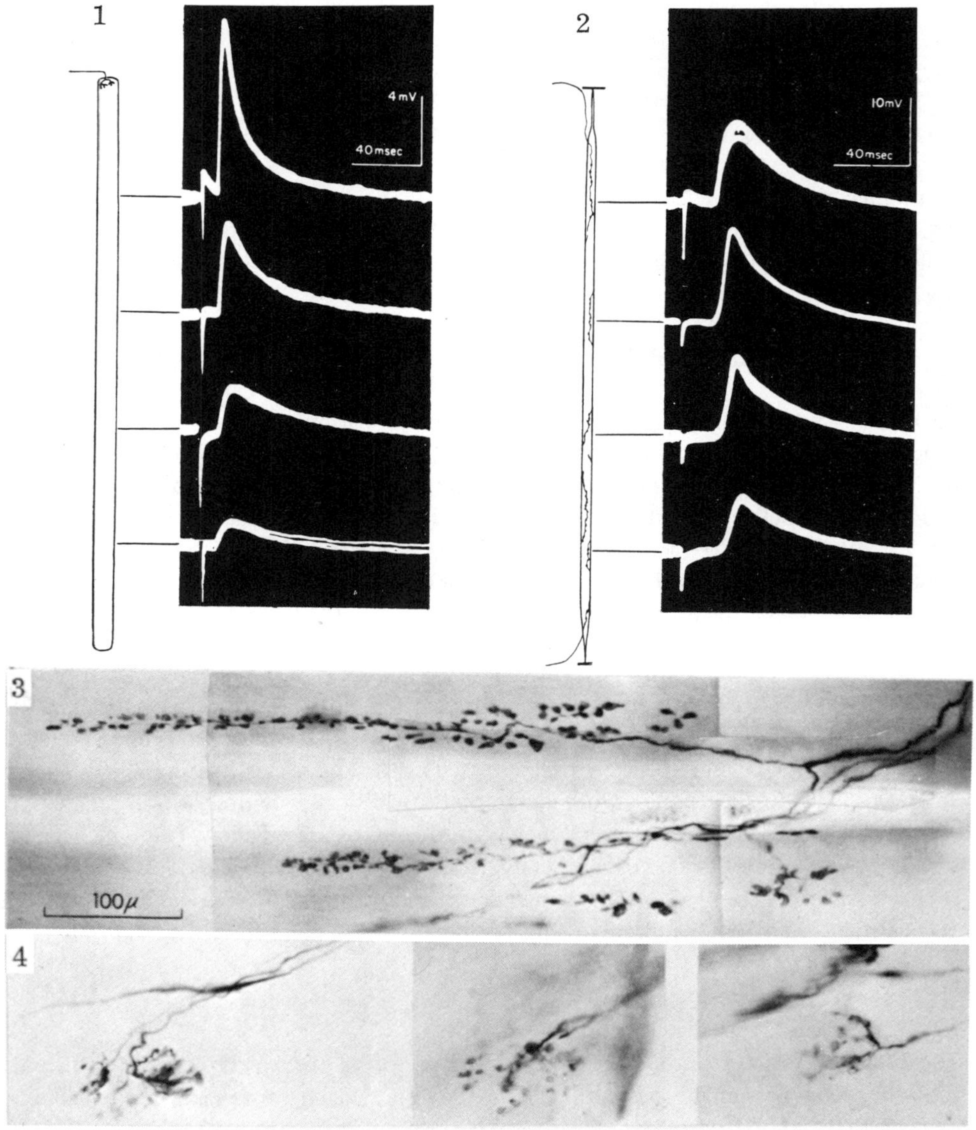

Figure 12.60. Neuromuscular transmission to fish muscle cells and typical motor nerve terminals. (*1*) EJPs intracellularly recorded, at increasing distances indicated, from single fast-type *Myxine* cell. (From Andersen et al., 1963.) (*2*) Distributed EJPs intracellularly recorded, at sites indicated, from slow-type *Myxine* cell. (From Andersen et al., 1963.) (*3*) Stained terminal on slow cell. (From Jansen and Andersen, 1963.) (*4*) Stained terminals on fast cells (From Jansen and Andersen, 1963.)

of species and is quite dramatic (Bone, 1966; Rayner and Keenan, 1967). Barely 5% of the musculature, the two thin bands of "slow"-type red cells, does more than 95% of the work (Bone, 1966). The pinker band outside it is recruited for the more vigorous bouts of normal swimming. But the bulk of the musculature is utilized only when either an escape response or a brief power surge against a current is called for. There is here operating a rudimentary size principle (see p. 606) for the motor neurons involved. Small motor neurons innervating small-size red cells start all movements. These are supplemented first by the larger ones innervating the larger pink cells, and finally by the largest ones that supply the largest muscle cells.

There is regrettably little information on the innervation, control, and output to pectoral and pelvic fin musculature. This might provide clues about the evolution of vertebrate limb muscle control. Basic quadrupedal locomotion is clearly derived from fish swimming: the primeval sinusoidal ripple can easily be seen traveling along the back of a walking horse. During swimming of carp, the fins only are used at speeds of 0.3-0.5 body lengths per second (BL/s). Small red cells are primarily responsible for the fin movements. Myotomal red muscle is recruited first, as speed rises above 0.5 BL/s, and pink muscle, lying between the red and white muscle, is recruited next, between 0.5 and 1.5 BL/s. Finally, the large white myotomal cells are added, at 1.5-2.5 BL/s (Davison et al., 1976).

12.14.2.4. A Major Shift in Resting-Potential Determination

A major physiological switch apparently occurred in evolution of teleost fish. Elasmobranchs determine their resting muscle-cell membrane potential mainly by Cl^- concentration difference between inside and outside and Cl^- conductance (Hagiwara and Takahashi, 1967a): the value is 70-75 mV. A 25-30-mV overshoot occurs in the action potential, as it does also in hagfish. By contrast, in teleost muscle cells, as in those of all other vertebrates, E_m is determined by K^+ distribution and gK^+. In the teleost cells, there is little or no overshoot of the action potentials. The white muscle cells of silver carp have about a 10-mV higher resting potential than the red ones (Hidaka and Toida, 1969). Both general ultrastructures and extents of IT invaginations are about the same. The only dramatic difference is in fatigability, which is fast for white cells, slow for red ones. The latter produce very much more tension when treated with similar depolarizing K^+-containing solutions than do the former. Both kinds of fish cells produce progressively increasing graded mechanical responses as frequency of stimulation is raised above that which causes smooth contraction, distinguishing them from frog and mammalian fast-twitch cells (Flitney and Johnston, 1979). This means that the amount of calcium released in a twitch is much less than that required to saturate the contractile machinery in fish muscle of either type.

12.14.2.5. Neuromuscular Junctions

The junctions have a smooth lining, unlike those of all higher vertebrates. Junctional transmission is presumably less efficient in the fish. This in turn makes it amenable to enhancement, and it has indeed been found that Ad facilitates it. However, the effect appears to be a presynaptic one, on the amount of transmitter released (Hidaka and Kuriyama, 1969).

12.14.3. Amphibia

Much of the classical muscle and neuromuscular literature, some of which has been reviewed above in various contexts, is in fact on amphibian, frog, muscles. The majority of frog muscles comprise largely fast-twitch cells. These cells are innervated at a single point by a branched ending of a myelinated axon. The terminals are grooves, more or less parallel to the long axis of the cell, invaginated at regularly spaced intervals with simple, parallel shallow grooves in transverse planes. The region just below the groove is occupied by nucleated granular cytoplasm containing cholinesterase. The synaptic gap is wide, 90 nm, and a line of densely staining material of unknown significance lies in its center. A majority of fast-type muscle cells of amphibians receive a single end plate, but some, particularly on sartorius, receive two, or rarely more.

The speeds of individual cells of the fast-twitch type vary between muscles, as does the mixture within a given muscle. Some, such as the sartorius, are fairly homogeneous. A majority, such as rectus abdominis and gastrocnemius, are strongly heterogeneous, with a greater than fivefold range in twitch times of component cells. Associated variables are the extent and characteristics of latency relaxation if present (see p. 257), duration of latency to tension onset, ratio of relaxation time to rise times, E-C threshold, E_m, and time to relaxation under constant depolarization, for which there is an especially wide range.

A second characteristic class of cell occurs in amphibia, the slow-type cell (see p. 296). These are located in a specific bundle in some frog muscles such as the iliofibularis (see p. 265), or are otherwise mixed with the fast-type cells. They appear to be totally absent from some fast muscles such as sartorius. The slow-type cells are multiterminally innervated. Their resting potential is less than that of fast-twitch cells, averaging -80 mV compared with -93 mV. They lack the anomalous rectification (see p. 187) seen in fast-twitch cells, and their membrane resistance is about 20 times higher (Stefani and Steinbach, 1969), but they also show delayed rectification, which is at least partially due to their having a much lower permeability to Cl^-, which is high in the fast-twitch cells. The ultrastructure of the slow-twitch cells is somewhat different (Page, 1965), with fewer ITs, less SR, thicker Z bands, and poorly defined M bands.

The slow cells are visually recognized by the presence of distributed JPs of about 10 mV rather than overshooting action potentials. Their motor axons, being thinner, have longer latencies and require higher threshold current than do axons to fast cells. Even after JP summation during repetitive stimulation, they do not spike, but following denervation they give overshooting action potentials, showing that the lack of action potentials is a trophically determined, as well as functionally significant, phenomenon (Schmidt and Stefani, 1977). This in turn suggests that the amphibian slow cells are either a relatively recent addition, not arising from fish slow muscle cells, or that the ionic mechanisms for producing spikes, once evolved, had to be suppressed in these cells. The maximum shortening velocity of the fast cells is six times greater than that of the slow ones. Calculations by Lännergren (1978) show that according to X-bridge theory, this requires about 15 times slower bridge-cycle time for the slow cells than fast ones.

In spite of the frog's popularity as a source of muscles, there has been almost no research on neural control of any muscles in the intact animal. It is not known under what circumstances the slow-muscle-cell system is utilized. The frog has a sitting posture that requires continuous weak extensor tone in forelimbs. Also, the male has a strong clasping reflex associated with mating that persists without fatigue for up to a few days. These probably both utilize specific activation of slow cells receiving a separate innervation via small-diameter motor neurons, that can be excited independently of the fast-twitch musculature.

12.14.4. Reptiles

Surprisingly, and disappointingly, reptiles have been but little studied in regard to cellular aspects of musculature and its neural control in spite of having some of the most remarkable of mechanisms, such as the tongue strike of chameleons. A few preparations have been exploited as test objects for studies on biophysics of junctional transmission, muscle-cell repriming, and synaptic and cellular ultrastructure, but nothing systematic has been attempted. Accordingly, the desirable objectives of assessing extents of transition from lower to higher form, or specific adaptive radiation within the class, is not possible.

The most conspicuous known feature is a change in the forms of the neuromuscular-junctional grooves that accommodate the end-plate nerve terminals. The major longitudinal grooves not only have transverse grooves as do those of amphibia, but also fine longitudinal clefts (Robertson, 1956). The endings are also quite different in shape, being highly compact compared with the multiple, long fingerlike processes of amphibians. Both these developments continued into the mammals. These compact endings comprise clusters of a hundred or so varicosities containing synaptic vesicles. After treatment with weak protein/connective-tissue-dissolving enzyme solution, the terminals can be pulled away. The postsynaptic membrane is thereby exposed, permitting the most accurate studies yet possible on junctional transmission (see p. 160). A sound estimate of the number of transmitter molecules contained in a single vesicle was obtained using this preparation (Kuffler and Yoshikami, 1975a, b).

The ultrastructures of reptilian muscle cells indicate a very wide range of intrinsic speeds. There is also a full spectrum both between slow species such as tortoise, and fast ones such as certain lizards. But an equally full range occurs in a single animal such as the snake, which can maintain a motionless extended posture for hours yet also make a dramatically rapid extension strike. A marked SL difference is associated with the two extremes, 2.0 μm for the fast types and 3.0 μm for the slow (Hess, 1965). However, the principal difference is in the amounts of SR and the numbers of ITs: there is a large amount of SR in the fast cells and numerous ITs, whereas SR is sparse and ITs very few in the slow ones (Hoyle, 1965). A wide range of numbers of mitochondria occurs, being small in seldom-used cells, large in the heavily used tonic ones. In slow cells, M bands are not present, but they are well marked in the fast ones: this is a common feature in birds and mammals also and must be significant, but there is yet no explanation of it.

The slow-type cells have markedly different neuromuscular junctions from fast ones, which are less compact, resembling a bunch of grapes with few grapes, several bunches being distributed along the length of the muscle cell. The postjunctional region has both transverse and longitudinal folds, but they are far fewer than those of fast cells.

In the iguana *Anolis* two kinds of SR occur – one of which is relatively common in invertebrates, the other unique – in addition to normal SR (Forbes and Sperelakis, 1980). The first is subsarcolemmal SR that is closely coupled to the surface muscle membrane. The other takes the form of a massive cylindrical body that is wrapped around myofilament clusters.

The characteristic fast-type cells of the garter snake occur intermingled with the polyneuronally, multiterminally innervated ones. They can themselves be subdivided by their individual responses into fast-twitch, slow-twitch, and intermediate (Ridge, 1971). The nerve supply to the slow cells is separate, as a final branch, from that supplying the three fast groups. The faster twitch has a rise time of 40 ms compared with 60 ms for the slower one.

The concentration of research interest on vertebrate muscles of a presumed basic representative, the frog, has led to the neglect of what is truly a rich diversity within all the higher classes. One exception in reptiles concerns the truly remarkable operation of the chameleon tongue. A chameleon can strike a fly at a distance greater than its entire body length. The extension occupies less than 40 ms, so fast that the hyoid horn that forms its base can be playing no more significant a role than bringing the tongue out of the mouth to position it for the strike.

The entire tongue is shot out by an *accelerator muscle,* which is a hollow one comprising longitudinal cross-striated cells that forms the outer head of the tongue (Gnanamuthu, 1930; Zoond, 1932). Its central lumen fits over the tip of the hyoid horn (Figure 12.61). Contraction of these cells raises the pressure in the small space between the muscle and the hyoid to such an extent that the tongue and the contracted accelerator are shot forward to the target by a hydraulic action. Outward movement occurs at the expense of extension of its antagonist, the *retractor*, also hollow, which is folded along the hyoid horn in its rest position. A long, thin tendon links the tongue to the tip of the hyoid horn, and when this is stretched out fully, the tongue has reached its greatest possible extent. The extent of the extrusion is varied to match the accurate assessment of the distance, apparently by adjustment of the timing of onset of contraction in the retractor. Withdrawal of the extended tongue is brought about by contraction of the retractor. In spite of its being a very fast muscle, the retractor is able to supercontract, shortening to less than $\frac{1}{6}$ its fully extended length. Judging by the speed at which extension occurs, it cannot offer much resistance, so is not greatly stretched in the extended condition. Therefore, it must be capable of supercontraction comparable to that shown by giant barnacle cells and must be contracted continually when the tongue is in its rest position with the hyoid horn bent back. Striking must require a combination of relaxation of the retractor with activation of the extensor.

Electron micrographs of longitudinal sections of the retractor show a typical cross-striated vertebrate pattern with SL about 2.3 μm. However, the Z regions, like those of the barnacle cells, are perforated (Rice, 1973): the details are different because the chameleon Z region is banded, but the principle is the same.

12.14.5. Birds

Birds are little better known than reptiles in the context of our interest, though some attention has been focused on flight muscles, especially of humming birds, on account of their speed. Humming bird wing muscle contraction time is only 8 ms. The contraction times of other birds are from a few to several times longer, about in proportion to the birds' size and wingspan. Slow- and fast-type muscles of birds have been of interest in the context of possible trophic influences (see p. 407). The different color of chicken white, compared with red (dark), muscle is the best-known and one of the most conspicuous of basic muscle-cell-type differences. The most experimentally convenient separation of cell types is seen in the two latissimi dorsi, anterior (ALD) and posterior (PLD), of the chick (Figure 12.62). The ALD comprises entirely slow-type cells. These each receive several terminals, distributed at intervals from 200 to 2000 μm (790 μm average) at age 15 weeks along the length of the cell (Ginsborg and MacKay, 1961), but there is a major difference compared with distributed junctions on typical arthropod skeletal cells: the terminals arise not from the same axon but from different ones. The PLD cells, by contrast, are of fast type and are rarely innervated at more than one site.

When MEPPs (see p. 157) are recorded intracellularly from an ALD cell, they occur with a wide range of sizes and rise times, as expected on the basis of independent release from several different terminals on the same cell. These cells go into prolonged contracture in the presence of 10^{-5} *M* ACh, whereas the fast-type PLD cells give only a brief burst of partly fused contractions. The summed local contractions of ALD cells resemble

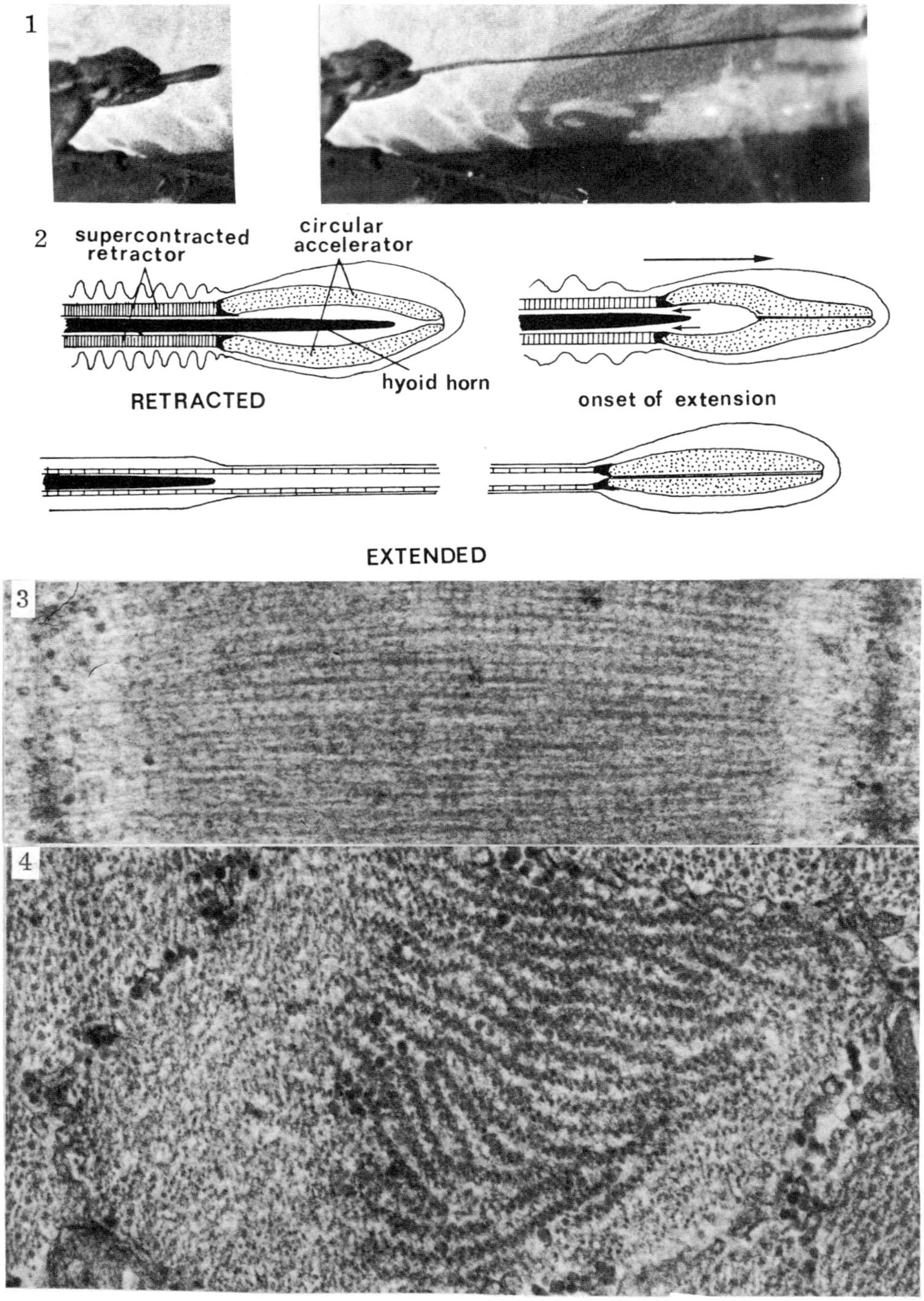

Figure 12.61. Strike of the chameleon tongue. (*1*) Start and finish of a long strike. Micrographs courtesy of Dr. Martin Rice.) (*2*) Diagrams to show the strike mechanism. Fluid pressure builds up around the hyoid horn as a result of contraction in the circular accelerator muscle within the tongue. Relaxation of the retractor permits the entire tongue to shoot forward. (ORIG.) (*3*) TEM of LS of retractor, SL = 1.5 μm. (*4*) TEM of TS, same as (*3*), through Z region. The spaces between the layers of Z material permit thick filaments to pass across. [By M. Rice. Reprinted by permission from *Nature* **243**, 238–240 (1973). Copyright 1973 Macmillan Journals Ltd.]

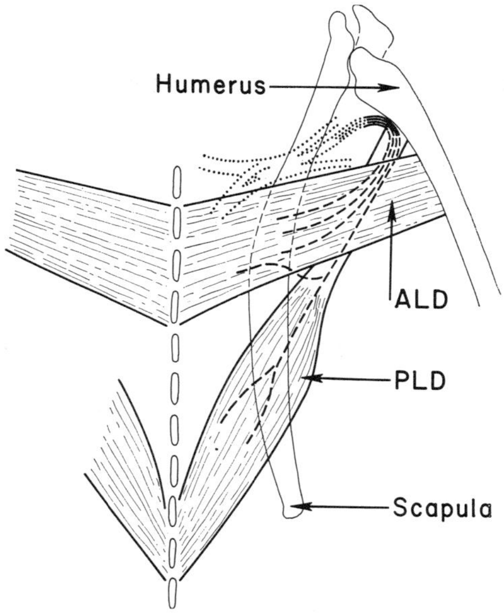

Figure 12.62. Diagram of locations of anterior latissimus dorsi (ALD) and posterior latissimus dorsi (PLD) in pigeon. (From B. L. Ginsborg, 1960.)

normal neuromuscular activation in arthropods and slow cells of frog. However, most of the avian slow muscle cells that have been encountered by microelectrodes have been able to give overshooting spikes, but the junctional responses from more than one motor neuron may be needed to reach spike threshold.

In many other muscles of the chick, such as neck semispinalis biventer, gastrocnemius, sartorius, and extensor digiti, both slow and fast cell types occur, but they are mixed together rather than located in separate bundles. It would be interesting to know the recruitment patterns of the various motor neurons that innervate the slow cells, in relation to behavioral use, and the frequency range over which they operate naturally.

12.14.5.1. Iris Muscle

The iris is unquestionably a vegetative organ because it is completely involuntary, tonic, and circular. It is not attached to a movable skeletal structure, and it is innervated by the parasympathetic nervous system. However, curiously and paradoxically in birds and reptiles, iris muscles are not smooth but cross-striated (see p. 64). Furthermore, to make the situation even more strange, they have the shortest known SLs of any vertebrate striated cell: 1.3 μm. Triads are abundant and not precisely located. If SL alone were a determinant of speed, iris cells would also be the fastest, which they are not. It is, therefore, on many counts a very interesting muscle: its cells are very thin and they anastomose (Hess et al., 1969). In frog, the membrane contains rhodopsin and a very strong light actually triggers contraction directly (Barr and Laverack, 1981). There are numerous small neuromuscular junctions that are similar to ones seen on extraocular slow cells. The Z regions are thin and the M regions clearly marked as in twitch, rather than slow, cells. Curiously, when excited via the nerve that innervates them, they do indeed respond with a fast 50-ms twitch (Pilar and Vaughan, 1969b), which is not exactly the characteristic "behavior" of iris muscles. This is probably associated with the synchronous excitation of many motor neurons innervating the same muscle cells. In natural excitation this does not occur, only a few cells discharging, asynchronously, and giving rise to small

EJPs each less than 5 mV. The cells, or some of them, do have a spike mechanism, though, which is somewhat surprising since they probably do not use it in normal functioning. At intensities below spiking, and at a moderate frequency, the distributed small EJPs cause weak tonic contractions. This mechanism is probably the normal one in iris-muscle excitation. It is strongly reminiscent of the basic excitatory mechanism of arthropods.

12.14.6. Mammals

The literature on mammalian muscles and their control is vast, because of medical relevance. Only a very brief survey is possible or needed here; just sufficient material will be covered to enable comparisons to be made with the better-known invertebrates and with other vertebrate classes. Mammals alone among vertebrates have abandoned true slow-type, multiply innervated cells and distributed junctions for their skeletal musculature except for some of those moving the eyes. In the extrinsic eye muscles of the cat, both true slow-type and fast-type muscle cells occur, mixed together. There is a predominance of fast ones in the cores and slow ones peripherally (Hess and Pilar, 1963; Pilar, 1967). The slow type receive distributed polyneuronal innervation, which promotes local JPs of about 20 mV (Figure 12.63). Most of these cells are electrically inexcitable, but there are some exceptions (Bach-y-Rita and Ito, 1966), which may represent mistakes in innervation. The fast-type cells of the eye muscles are typical fast cells receiving a single end plate.

This mixture of rapidly fatiguing fast cells with slow ones that are highly resistant to fatigue is eminently suited to the needs of controlling eye position. It is not to be regarded as an evolutionary relic but rather represents an advanced state permitting much more efficient control than can be achieved with slow-twitch (a variant of fast-type) cells. Another function that might be expected to be served by slow muscles is maintenance of the erect posture of apes and man. A place to look for slow cells would be in back muscles, which are connected to vertebrae, which are distinctive in snakes. They do not seem to have been examined.

The bulk of the skeletal musculature of mammals is all of fast type by comparison with the true slow-type muscle cells that form the bulk of the musculature of most invertebrates and that constitute up to 10% or so of the musculature of fish, amphibia, reptiles, and birds. The cells have, however, been divided into classes in a variety of schemes depending on whether histological or physiological criteria were used. Attempts to reconcile different sets of observations have resulted in fairly widespread acceptance of the scheme set out on p. 298. Inevitably the extremes within any given muscle have been termed *fast* and *slow* without reference to muscle-cell features in general, of other species and classes.

The fastest twitches of mammals occur in the sound-producing cricothyroid muscle of bats; they were examined ultrastructurally in the expectation that they would be unusual in ultrastructure in association with specialization for high speed, as indeed they are (see p. 273). Their contraction time is only 4 ms, the shortest known (Revel, 1962). Sloth muscle was expected to be the slowest, and its gastrocnemius, normally a fast muscle because it is used in running and jumping, has a 174-ms contraction time (Göffart, 1971), which is quite slow in comparison to the majority of vertebrate skeletal cells, though still fast relative to some other muscles of the sloth.

The commonest mammalian muscles compared by physiologists to which the terms *fast* and *slow* are applied are the slow soleus (SOL) and fast extensor digitorum longus

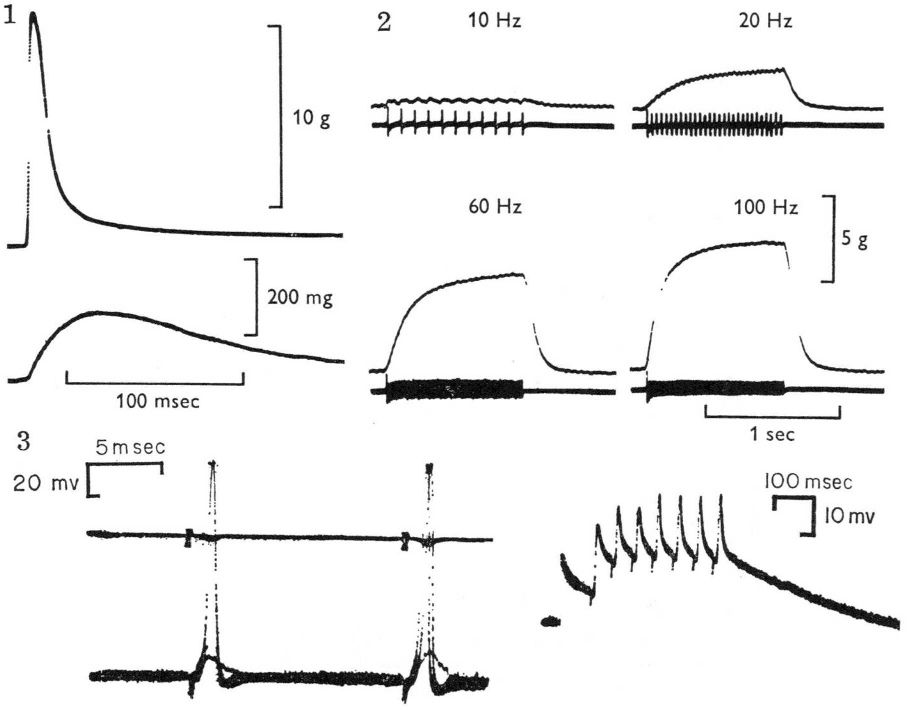

Figure 12.63. Mechanical and electrical responses of true slow and fast muscle cells of mammalian extraocular muscles. (*1*) Fast response of whole superior oblique of sheep (*above*) and slow response (*below*) evoked by selective stimulation of trochlear nerve. (From J. S. Browne, 1976.) (*2*) Slow responses of (*1*) at increasing frequencies of stimulation. (From J. S. Browne, 1976.) (*3*) (*Left*) intracellular responses of fast cell of cat superior oblique. The EPP commonly elicits overshooting spike. (*Right*) summation of EJPs, with no electrogenic response, in slow cell of cat superior oblique. (From G. Pilar, 1967.)

(EDL) (see p. 290). Contraction times of SOL range from about 40 ms for an average rat to about 90 ms for a cat. For EDL they are 13 and 20 ms respectively, three to four times faster. But because the variable underlying speeds in different cells are so numerous, such distinctions are somewhat spurious. As pointed out elsewhere (see p. 260), ultrastructures, enzyme contents, and E-C coupling are each subject to variation independently of the other factors. In the case of SOL compared with EDL, the principal recognized differences are in myosin ATPase activity, but numbers of mitochondria and amounts of other enzymes enter the picture also.

Most mammalian muscles are not homogeneous in cell population but comprise a mixture of cell types as determined by histological treatments. The different types of cells are scattered apparently randomly within a muscle. A clue to their different natures in relation to distribution is provided by experiments in which a single motor neuron supplying a muscle is directly excited to the point of fatigue. The total muscle cells innervated by a single motor neuron constitues a *motor unit* (MU). Continued stimulation depletes the glycogen content of all the cells in the motor unit; hence by quickly fixing after stimulation and staining for glycogen, the cells constituting the motor unit can be picked out. If the tension response was followed during the stimulation, the character of

1

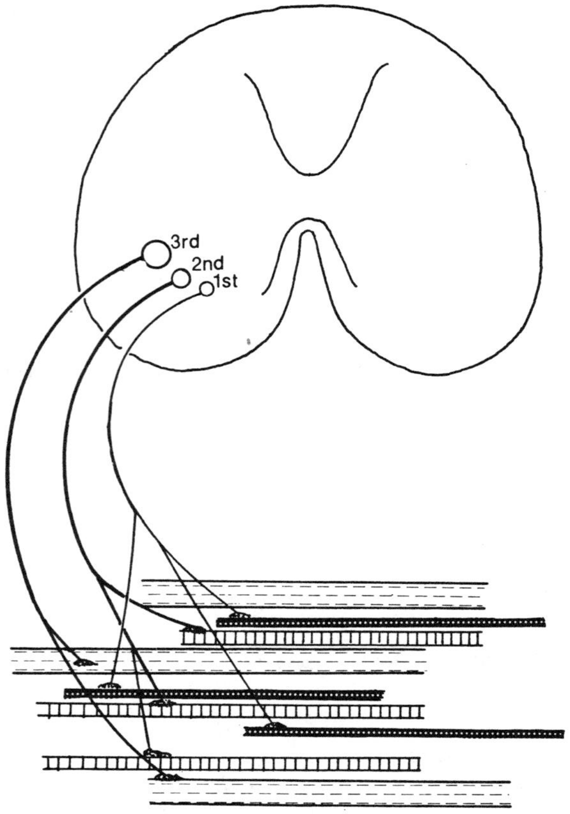
3rd
2nd
1st

2

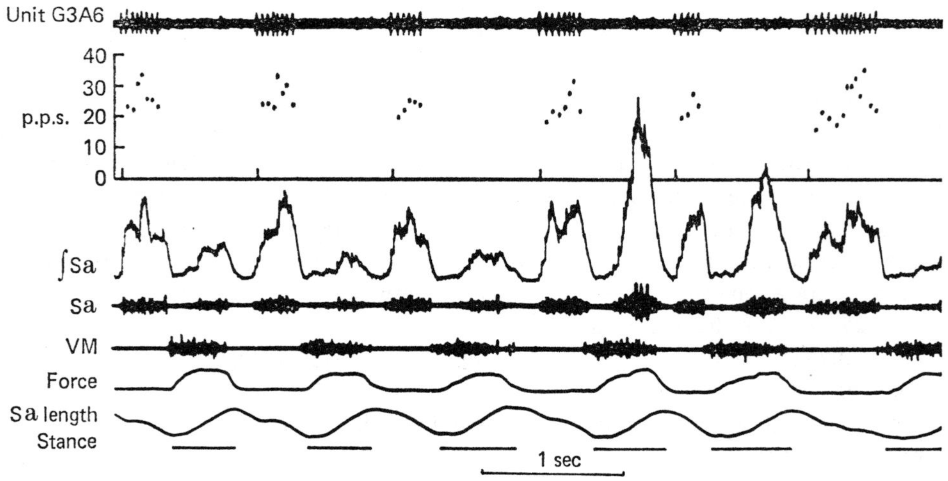
Unit G3A6
40
30
p.p.s. 20
10
0
∫Sa
Sa
VM
Force
Sa length
Stance
1 sec

the unit as a whole can be assessed. By serially sectioning the same area and applying a variety of staining techniques to different sections, an entire set of enzyme-content studies can be carried out on the same cells of the unit. In this way it has transpired that the cells constituting a unit are of like kind in regard to their enzyme and enzyme-substrate contents. This is presumably associated with trophic interaction between the neuron and the muscle cells it innervates (see p. 408), following the initial genetic determination. This may be stated, as the first principle underlying mammalian muscle control, as: *each MU is composed of cells of like type.*

The muscle cells in a mammalian muscle differ somewhat in size, and their sizes go along with their histological/physiological type. Thus small-diameter cells are also relatively slowly contracting and much more fatigue-resistant than the faster ones. They also contain more mitochondria, oxidative enzymes, and glycogen reserves. But that is only a part of the story. The sizes and properties of muscle cells in a unit are directly related to the sizes and properties of electrical characteristics of the cell bodies of their motor neurons (MN).

12.14.6.1. The Size Principles

There are two basic principles that guide the excitation of mammalian MNs, illustrated diagrammatically in Figure 12.64. Neither was originally proposed on the basis of a reasonably adequate sample of muscles or species, but it is fortuitously transpiring, as data accumulates, that both are probably true, allowing for some exceptions. Nor has there been nearly enough comparative testing since the principles were enunciated, but continuing interest in the possible existence of a principle in a field saturated with apparently unrelated diversity is likely to change that. The first of the two size principles to be discovered, which is really the second principle because it follows the other one logically, was discerned in studies on cats by Henneman (1957) and is now named after him the *Henneman size principle.* Strictly, it is the *motor-neuron cell-body size principle,* and it will be convenient to refer to the Henneman principle as the MN size principle. This was elaborated by Henneman and Olson (1965) and Henneman et al. (1965; 1969). From the start of our description and discussion, though, it should be emphasized that there are many exceptions, especially when learned, complex, voluntary behaviors are involved. The Henneman size principle states that *MNs innervating a muscle are recruited in an orderly manner, starting with those with the smallest cell bodies and progressively adding neurons in ascending order of soma size.*

The other principle, attributed independently to several authors, is the *motor-unit speed/strength rule.* This states that *contraction time of a given MU is inversely related to the fraction it contributes to the total tension of the whole muscle* (Bagust et al., 1973).

Figure 12.64. Control of mammalian skeletal muscles. (*1*) Diagram illustrating the three principles of mammalian muscle control. Each MU comprises like muscle cell types (*principle of MU homogeneity*). The size of the soma of the MU is directly proportional to the maximum-force capability of its muscle cells (*MU size principle*). The order of recruitment in most reflex actions is in order of ascending cell-body size *recruitment by Henneman size principle*). (ORIG.) (*2*) Example of a modern analysis of the use of a motor unit during walking in cat. The unit, of the sartorius, was recorded using flexible wire electrodes chronically implanted in lumbar 5 ventral rootlets. There was a double burst in each step in the rectified, filtered (20 Hz) electromyogram (*Sa*). Also shown are extensor vastus medialias (*VM*), force from a strain gauge implanted on the patellar ligament, and length from an implanted gauge spanning the iliac crest to the proximal patellar margin. The stance phase is indicated by bars. (From Hoffer et al., 1980.)

Powerful units are fast, weak units are slow. But the weak units of most muscles are also the ones most often used. This is because their MN cell bodies are smaller: a small cell body is depolarized to a greater extent by a unitary excitatory synaptic input than is a larger one (Luscher et al., 1979), likewise with hyperpolarization produced by inhibitory input. To contrast this principle with the Henneman one, it will be referred to as the motor unit (MU) size principle.

The Henneman size principle is a direct consequence of the biophysical properties of the cells that receive similar shared, physiologically equally weighted, excitatory synaptic inputs. The MU size principle is a consequence of the combination of genetic determination of MN properties and in turn the MN's regulation of the muscle cells it innervates. During a period of rising excitatory state, especially one brought about reflexly by progressively increasing stimulus strength, the smaller MNs fire first, followed by the next larger, and so on to the largest. The initial contraction is necessarily weak; it is supplemented progressively. This process can be thought of as conserving energy by keeping the most powerful MUs in reserve until it is certain that they are needed rather than risking their being triggered by stray, irrelevant stimuli.

There are, of course, many MNs of each size class, not all of which are needed for many tasks, particularly manipulations of small objects. How does the body choose which ones to use; is it random or are there more-likely or preferred pathways? Data is beginning to accumulate on this matter in humans, including comparisons of identical versus nonidentical twins. In these experiments, very small syringe-needle electrodes are used for long periods during everyday activities such as drinking tea, sewing, and so forth. These electrodes sample from only a very few units and are usually moved from place to place until a commonly used MU is encountered.

It seems that most persons have one or two MUs in a particular muscle that they use most often, almost all their minor skilled movements utilizing the same ones. These tend to be similarly located and therefore possibly homologous MUs in identical twins but not in genetically dissimilar individuals. So there is probably a genetic component involved in determining which ones are favored. This may of course be related to size and concomitant ease of activation in general reflex actions. But ease of address from willpower/learning centers may be equally important. There is ample evidence that a person can learn to cause an arbitrarily selected MU to fire by referring to the recorded output of the unit visually or auditorily (Basmajian, 1963). This ability is retained even when the motor nerve is shocked repeatedly (Scully and Basmajian, 1969). Neighboring units progressively cease firing as the subject concentrates on one. The size principle in MU selection is seen in untrained humans (Milner-Brown et al., 1973a, b) and in persons trained to respond with the minimum possible discharge: a single impulse in a single MU in response to a visual or auditory signal. The mean reaction time for 100 subjects was 300 ms (Kosarov, 1979). For a response consisting of a train, the value was 260 ms, and for a phasic burst only 200 ms. Each type of response uses units, which have intrinsically different thresholds, that are progressively larger and faster.

The concept of selective use of one or two MUs to execute skill-developed motor programs is important and much more satisfactory than the random-selection model first developed by Sherrington, which has presented too many difficulties for circuitry modeling. The newer concept, by contrast, fits easily with the size principle and permits modeling. It also reduces the otherwise enormous gap between requirements for neural programming in arthropods, in which single clusters of muscle cells called muscle units are necessarily used because of the small number of MNs, and vertebrates, which have such a large number of MNs available for every muscle so they appear to be "spoilt for choice." In fact, since a majority, and certainly all the larger ones, of vertebrate MUs are reserved strictly for powerful, mainly ballistic acts, the numbers available for skilled movements is not large. Also, these smaller units are much more sensitive to frequency modulation than are the larger ones, bringing them closer into line with the situation in arthropods, in which all the muscles are dependent upon frequency of MN discharge.

There are conflicting data in the literature, particularly on mammals, but the above is apparently emerging as a general principle at all major levels above fish. It is supported in amphibia (frog – Luff and Praska, 1976), reptiles (snake – Hammond and Ridge, 1978; Ridge, 1980) and lower mammals (cat – Bagust, 1974) as well as humans (Monster and Chan, 1977; Monster et al., 1978). But research on this important topic is in its infancy, and much more recording from units in freely moving animals is needed. Experiences – and therefore opinions – differ regarding the extent to which the MN size principle can be bypassed in human voluntary ballistic movements. Some authors have considered that man is unable to voluntarily miss out the smaller units (e.g., Desmedt and Godeaux, 1977). However, the operation of the principle depends upon the movement being made. When the human interosseus is acting as a prime mover of the index finger, recruitment follows size, but when it is acting as a synergist, the recruitment order is reversed (Desmedt and Godeaux, 1981).

Whether or not the MU size principle operates in any given situation is dictated by the extent to which the driving innervation is shared and equally weighed. At best this is likely to apply to MNs whose cell bodies are closely clustered together in the cord. For ones innervating the same muscle that are not close together, as does occur (Wyman et al., 1974), different qualitative and quantitative inputs are likely, and independent operation is inevitable. Numerous exceptions to the operation of the Henneman size principle have been documented, especially during voluntary movements (e.g., Bawa and Tatton, 1979 – monkey wrist).

But in a majority of movements that are reflexly generated, certainly all in which the skin is stimulated, the recruitment order follows the MN size quite strictly. An example of a recent study that illustrates the orderliness of recruitment is shown in Figure 12.64. This recording was made from a cat during normal walking, using flexible wire chronically implanted in an L5 ventral root (Hoffer et al., 1980), which may be taken as exemplary of how best to determine natural usage of units. Cat slow units have twitch times in the approximate range 130–200 ms; intermediates, 60–130 ms; and fast ones, 15–60 ms.

It is not possible to express the total involvement of the many different MUs by a simple graph. Burke and his colleagues have attempted to do it, specifically for cat hindlimb muscles, using a three-dimensional grid relating force as ordinate to cell type and MU type respectively, as the two abscissas (see p. 295). The same thing can be done, alternatively, with other parameters such as enzyme content. These diagrams enable a graphic assessment of recruitment of cell types in relation to movement type.

12.14.6.2. Cortical Involvement

The motor cortex of the cerebral hemispheres is recognized because direct electrical stimulation of discrete regions causes discrete movements. Mapping produces the well-known prescribed "motor homunculus." Contraction of a particular small muscle can be evoked by focal stimulation, with a precision of about ∓0.3-mm radius. But if the same animal is tested over a long period, the effective site is found not to be constant but to wander around. It is in the same general area, but not precisely so, and may move by a few millimeters (Penfield and Roberts, 1959). If the entire area within which the site wanders is removed, the movement cannot be evoked by cortical stimulation for several days or a few weeks. But eventually the capability returns and is now associated with another nearby region, previously ineffective.

The significance of the wandering and the plasticity are unknown, but they clearly argue against there being hard wiring for cortical control. Specific cells in the region can be brought into action by operant conditioning. When an isometric contraction of an arm muscle is reinforced, correlations between a specific cortical cell and a specific muscle appear consistently (Fetz and Finocchio, 1971). The link can then be broken by reinforcement correlated with the cell's activity while restraining the movement. Thus there are temporary hookups between specific cortical neurons and MNs, but the overall characteristics of an established relation, while being retained, must be shared with other cells and transferred back and forth among them. The significance of this in cellular events is currently beyond our comprehension.

12.14.6.3. Role of the Cerebellum

Damage to the cerebellum generally produces aberrant locomotion. Certain mouse mutants with abnormal locomotion such as "reelers" are associated with undoubted cerebellar malformation. However, efforts to assign very precise motor functions to the cerebellum are confounded by the fact that in a few humans the entire organ is vestigial, yet this is only discovered accidentally and incidentally, since motor behavior often seems quite normal. In these persons other parts of the brain have somehow undertaken the cerebellum's roles. Possibly certain motor tasks are assigned to the cerebellum if it is there, but can be handled by other parts of the brain if it is not. The cerebellum may thus be a general-purpose computer that is built to handle motor and related sensory information processing. It is not itself a memory device, nor is it a place where any basic motor command functions originate. But neither occurs in the normal animal or person without the active participation of the cerebellar circuits.

The cerebellum has provided brain neuronal modelers with their supreme source of inspiration since it is a highly organized structure with regularly repeated neuronal architecture. The neuronal links provide for both the feedback and feedforward of information beloved of cyberneticists. Some years ago Sir John Eccles informed me with great confidence – and his characteristic enthusiasm – that he would soon be able to state precisely how central nervous systems work; his expectation was based largely upon his studies on cerebellar cortex.

The historically oldest part, the medial vermis, is concerned directly with the semicircular canals and hence with equilibrium in posture. The intermediate region has long been associated with coordination of complex movements, perhaps with preplanning of detailed commands to be issued to the γ-efferent system. The lateral parts have been considered to be involved in voluntary movements. There are no direct pathways from the cerebellum to the cord, only routes that pass via the motor areas of the cortex. Many sensory inputs project to the cerebellum, notably from proprioceptors of the eye muscles, so a role normally in control of eye movements is assumed. The convergence of eye afferents and vestibular inputs point to major roles in overall posture control, which commonly uses a combination of visual, proprioceptive, and vestibular signals.

13

Evolution of Muscles and Their Control Mechanisms

The unit reaction in nervous integration is the reflex . . . . Co-ordination, therefore, is in part the compounding of reflexes. C. S. Sherrington†

Can an animal initiate and maintain patterns of co-ordinated muscular movement without reference to the outside world? J. Gray‡

Is the sensory influx a constructive agent, instrumental in building up the motor patterns, or is it a regulatory agent, merely controlling the expression of autonomous patterns without contributing to their differentiation? P. Weiss§

The purpose of muscles is to produce behavior. There is no sharp dividing line between a muscular contraction, even tonus, as a physiological process and the movement-oriented activity termed behavior. Behavior may be defined, as it is in Webster's dictionary, as movement. Initially movements were entirely random but still functionally useful because they promoted dispersion, reducing competition for energy sources and cellular building components and introducing the primeval organism to new supplies. The evolutionary process produced more-complex organisms and specialized cells. Once cellular differentiation was under way, the setting aside of some cells as presumptive controllers of contraction was possible. It is quite likely that the first nerve cells were inhibitors developed to regulate the actions of otherwise randomly contracting presumptive muscle cells. The processes of regulation of contraction by inhibition of an intrinsic cellular action, and by release from inhibition, are emerging as perhaps the two most important common principles governing how neural circuits in general operate (e.g., Pinsker and Willis, 1980). Because of the occurrence of musculoepithelial cells in existing coelenterates, these cells have usually been made the starting point for evolutionary models (e.g., Parker, 1919; Prosser, 1979). Let us say only that this is another possibility.

Beyond the stage of either of these alternatives – or perhaps a combination of the two – the controlling nerve cells and the muscle cells they control progressively developed

†C. S. Sherrington, *The Integrative Action of the Nervous System,* Yale University Press, New Haven, CT, 1906.
‡J. Gray, *Symp. Soc. Exp. Biol.* **4**, 112–126 (1950).
§P. Weiss, *Schweiz. Med. Wschr.* 71, 406–407 (1941).

increasing cooperative organization in the diverse ways I have outlined. The requirements for neural control and the capability of meeting them were dictated by both the form of a muscle, the combined properties of all the component cells, and the neuromuscular synaptic properties. All of these have been described individually in previous chapters, with little reference to overall control principles. Our task now is to consider the functional requirements of muscle control in a general way, in relation to what the largely haphazard, certainly unguided, additive only, evolutionary process has provided.

First we shall survey the various forms of musculature that have evolved, which are strictly related to the skeletal attachments, if any, that were available in the organism. The most primitive organism with muscles was undoubtedly totally lacking in hard parts: a soft elongated tube. Ths form is, however, so flexible and useful that it has been retained as the major body form by some phyla and for many internal organs by all animals.

13.1. Skeletal Muscle Design

13.1.1. Muscles Associated With Hard Skeletons

In spite of the great differences in their skeletal hard structures, vertebrates and invertebrates with hard-skeletal attachment sites for muscle cells have come up with precisely similar designs, which are diagramed below. There are two major types depending on whether the principal purpose of the muscle has been to achieve length change – whether for amount, precision, or high speed/frequency – or large force or power output. Alternatively, there has been a compromise between these conflicting force and movement options. *Movement* muscles have their cells in *parallel,* provided there is the possibility of equal areas of attachment at both ends. If one end is restricted by a narrow tendon, the muscle is *fusiform* instead. In *power* muscles, a greater number of cells are packed into a similar volume by arranging many cells of shorter length side by side in a *pinnate* arrangement. When only one major basal attachment area is available, the configuration is unipinnate; with dual attachment areas it is bipinnate. In either case there is a very long tendon parallel to the basal attachment so that the cells can lie in parallel without the need for tapering or lateral bowing. A further development of the pinnate arrangement is the multipinnate, in which there are a few subsidiary tendons radiating from the principal one (Figure 13.1).

Invertebrates, with their complex exoskeletons, have been able to evolve combinations of parallel and pinnate muscles converging on a single tendon. Another invertebrate specialty is a radial arrangement around a central cavity that is expanded as a result of the contraction to provide suction for food intake.

Both vertebrates and invertebrates have many muscles in which the muscle cells must taper markedly at one or both ends in order to be able to fit onto a small tendon. Such muscles are *fan shaped.* Some invertebrate cells, such as those of the giant barnacle, have a separate tendon for each muscle cell. These taper progressively toward their distal ends. The tendons are all attached to rock-hard calcified shell or to bone; at their other ends they may attach directly to shell or bone. A few have an alternation between tendinous and direct attachment at either end.

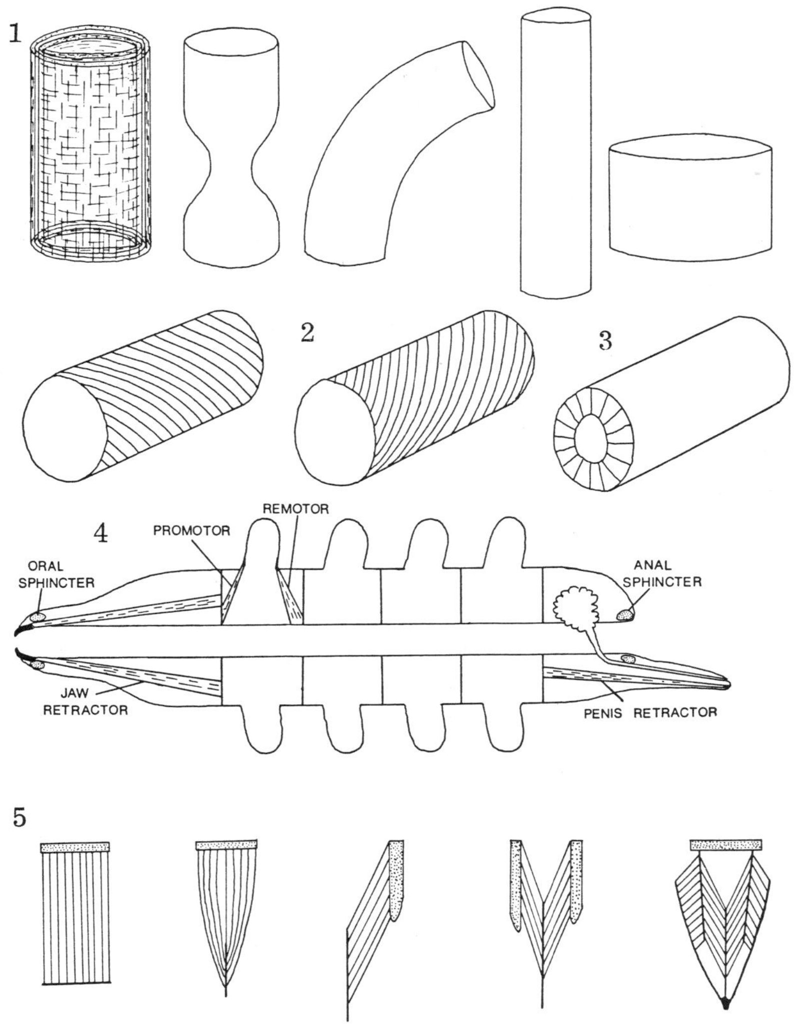

Figure 13.1. Basic forms of muscles. (*1*) Antagonistic cylinders of inner circular and outer longitudinal sheets of muscle. This was the most primitive type, retained by coelenterates, in which the contractile elements were derived from musculoendothelial and musculoepithelial cells respectively. Some examples of ways in which the shape can be altered by local contractions are shown at right. All vegetative muscles of all animals essentially retain this basic configuration. (*2*) Oblique or spirally aligned sheets are added outside the basic cylinders in the bodies of soft-bodied triplobastic animals. (*3*) A variant in which cells are radially arranged around the inside of the anterior portions of guts to create suction used in feeding. (*4*) The first discrete muscles to evolve depended on the prior evolution of hard parts, metameric segmentation, and discrete organs. (*5*) The commonest forms of arrangement of muscle cells in discrete muscles of both invertebrates and vertebrates. These are respectively: parallel, fusiform, pinnate, bipinnate, and multipinnate. The first two are optimized for speed, the others for force.

13.1.2. Muscles Associated With Soft Skeletons

The overall organization of the musculature of the soft organs of vertebrates closely resembles that, not only of the corresponding organs of invertebrates, but also of the entire invertebrate body with the exception of arthropods. The principal types are best seen in tubular structures: rounded bodies generally have very diffuse musculature aimed only at actively reducing the volume of the body or providing it with overall stiffness.

The most significant types of whole-body or whole-organ muscles are the *longitudinal* and the *circular*. The former generally consist of long parallel cells with a small amount of anastomosis and with end-to-end as well as lateral attachments. The commonest binding material for these cells is collagen. Their principal function is to cause extensive shortening of the whole body for protective purposes or local shortening for locomotion. Circular muscles consist of parallel cells, also with some anastomosis; it is possible that some are truly endless ring cells, but I know of no single instance where this has been shown conclusively. Circular muscles are used to cause elongation of the whole, to raise body pressure and increase stiffness, and in coordinated waves to produce propulsion internally or of the whole. Common in invertebrate body walls in addition to longitudinal and circular muscles are *oblique* muscles. These are oriented approximately at 45° to the longitudinal body axis. There are usually two sets, which generally cross over and under each other in the dorsal and ventral midlines.

Dorsoventral sheets comprising parallel muscle cells running at right angles to the body axis are common. Muscles operating small skeletal hard parts such as jaws, bristles, claws, and so forth are quite similar to their counterparts in animals with hard skeletons, being commonly parallel, fusiform, and fan shaped, with cells that are not of uniform thickness but taper toward the part to be moved. Pinnate forms are rare. At the anterior end, *radial* arrangements are common.

13.1.3. Cell Types Within a Muscle

The question of homogeneity vs. heterogeneity of cell-type composition of a muscle in relation to function has been dealt with throughout the book, especially in Chapter 8. There is virtually infinite variation in this, and no rules can be discerned. Even the very fundamental question of whether the cells in undoubtedly homologous muscles of closely related species of some invertebrates will be cross-striated or not cannot be answered *a priori,* without direct examination. As shown for brachiopod tentacles (see p. 454), a small difference in habitat, with accompanying life-style variation, can be associated with extreme differences in cell type (Figure 13.2.*1*). There has been a tendency to think of cross-striation as being an "advanced" evolutionary state compared with nonstriation, likewise with SR, ITs, and coupling dyads or triads (Figure 13.2.*2*). Yet the actual evolutionary sequence is just as likely to have been in the reverse direction. The breakup of highly ordered structure favors many functions including slow-speed operations. It offers a much greater speed range, an enormously greater length range, more precise control of length, and economy in metabolism and energetics.

13.2. Reflexes Versus Centrally Generated Behaviors

All "explanations" of movement prior to about 1950 assumed that each is associated with linear information flow. Mathematician Norbert Wiener, who first clearly enunciated

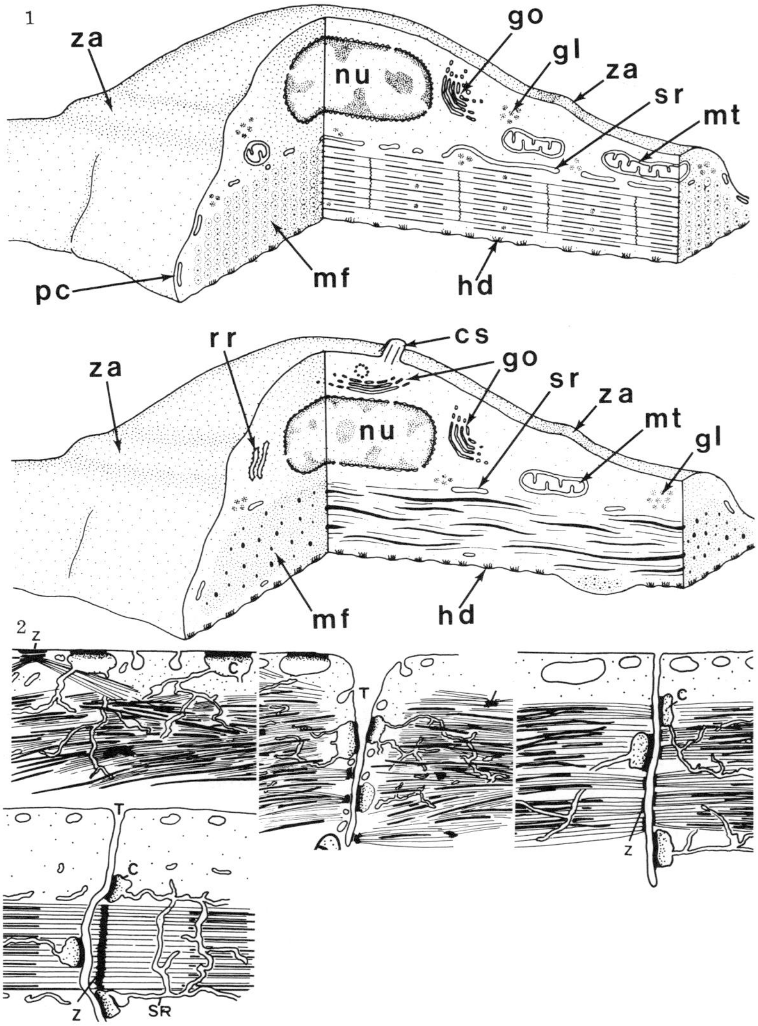

Figure 13.2. Stages in adaptive radiation of contractile material and associated tubules, showing course of cross-striated-smooth transition. (*1*) Orderly cross-striated and disorderly ("smooth"-type) homologous muscle cells of tentacles of two closely related brachiopods from the same geographical area but living in different habitats. (From Reed and Cloney, 1977.) (*2*) Hypothetical sequence of transition from "smooth" to cross-striated (or vice versa?) in molluscan muscle. (From Dorsett and Roberts, 1980.)

the principles of automatic control in his book *Cybernetics* (Wiener, 1948), was horrified in his first incursions into biological literature of movement control to find a total unawareness of the significance of the information feedback loops actually known to be present. Prior to publication of Wiener's book, the fundamental basis of behavior was sought in terms of a specific stimulus's initiating a *reflex response*. The movement might be complete in itself and as simple as a knee jerk in response to a tap below the petalla or as complex as a cough in response to a hair in the throat. The initial response could lead to a second, and this in turn to third, fourth, and so on, reflexes, the whole forming *a chain reflex*. In this way, it was thought, behavior as complex as that of a spider building its intricate web, or a bird going through mating rituals or nest building, could be simply explained by breaking the behavior down into successive reflex components. This view, which in retrospect seems extraordinarily naive, was shattered only after a generation of European students of animal behavior, who came to call themselves ethologists, made their first attempts at synthesizing their findings and publicizing them. This was done especially by Lorenz (1950) and Tinbergen (1951), just one year after Wiener's book had paved the way for some new thinking.

Regrettably, ethologists called their subject "the *scientific* study of behavior," thereby antagonizing many careful, competent naturalists and all experimental psychologists. But the ethologists succeeded in convincing many zoologically inclined physiologists that much complex animal behavior is capable of occurring even in the absence of specific initial stimuli. Spontaneous behavioral acts, for which neither an external nor internal stimulus is apparent, are a major part of the repertoire of all animals. Furthermore, in an ambiguous or conflicting situation, such as when it is not obvious whether to fight or flee, animals often do something that is irrelevant, such as digging, which is then referred to as a *displacement activity*. Ethologists also showed that many complex behavioral acts, involving multiple-component muscular-contraction sequences by different muscles, are under genetic control as effective wholes that are subject to familiar Mendelian rules of inheritance. As researchers tried to come to grips with the problems these new views raised about underlying neuroanatomical organization and neurophysiological events, many new concepts arose. Animals act as if they inherit sets of fixed instruction sequences, each restricted to the control of a single, though often complex, "piece" of behavior in their nervous systems. The ethologists relied largely on observations on birds among vertebrates and insects among invertebrates. In view of the very different organizations of the nervous systems of birds, compared with those of insects, the similarities were striking and suggested that common general principles have evolved in neural control of movement. Selected invertebrates offering ease of experimental manipulation soon began to be studied for the light they might throw on this. It was found, for example, that a semiintact locust can produce flight motor patterns even though proprioceptive inputs have been removed (Wilson, 1961). More dramatically, the completely isolated brain of the mollusc *Tritonia* produces the same neural program appropriate to swimming that is seen in the same cells in the intact animal (Dorsett et al., 1973). A nonspecific electrical stimulus applied to an anterior sensory nerve triggers the pattern. Six identified nerve cells involved in swimming were first recorded intracellularly during tethered swimming in an animal that was intact except for a small slit above its brain. The brain was then removed and some of the same neurons again recorded (Figure 13.3). Both the neural program and its timing and the patterns of impulses in individual cells were virtually identical to those obtained previously from the intact animal. These observations and others along similar lines shattered the earlier concepts and paved the way for studies on the generation of motor patterns as the fundamental basis of behavior.

13.3. Generation of Motor Patterns

In examples such as those quoted, when an inherited neuronal instruction sequence has been selected and is "played" back, a *fixed action pattern* or complex, but stereotyped, movement sequence occurs. This cannot be started anywhere but at its beginning. It cannot be run backward or have its components appear out of their correct sequence. Components are never naturally dropped from a sequence, and if one is forcibly prevented

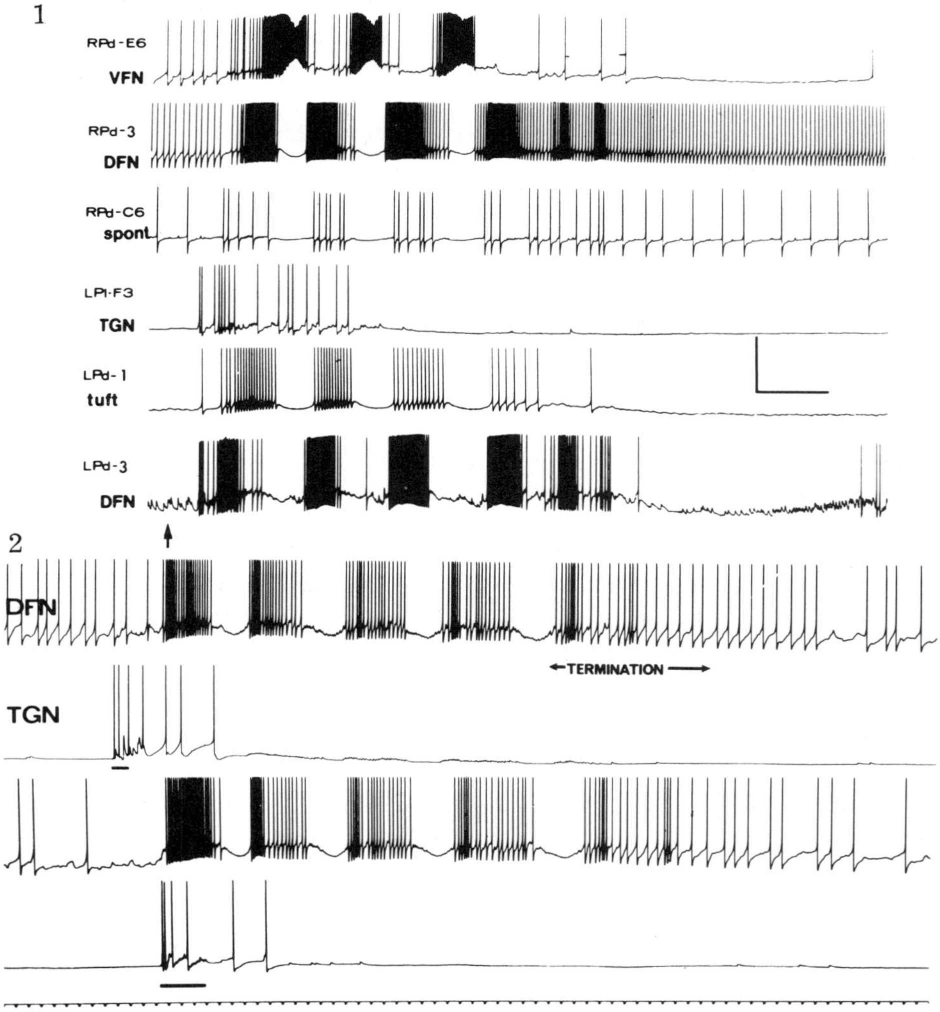

Figure 13.3. Generation of a fixed action pattern by a "motor tape" mechanism. (*1*) Simultaneous intracellular recordings from six brain neurons of the mollusc *Tritonia* during an escape swim evoked at the arrow. The labels identify the neurons functionally, and give their locations according to a computer-standardized grid/ganglion map. (From Willows et al., 1973.) (*2*) Two "escape swim" neural sequences evoked in an isolated brain by electrical stimulation of a sensory nerve (*bar*). The details of burst structure are closely similar to each other and to ones seen in the nearly intact animal in (*1*). G. Hoyle, Identified neurons and the future of neuroethology. *J. Exp. Zool.* **194**, Fig 5, p. 58 (1975). Reprinted with permission from Alan R. Liss, Inc., New York.]

from being expressed, the next expected and feasible movement occurs at its normal time. The delay in its appearance is the same as that which would have occurred had normal movement been permitted in spite of the altered proprioceptive input. The amount of fixity in patterns comes in a range from ones in which it is precise, even to the level of each underlying motor impulse (courtship song of some crickets – Bentley, 1971; and courtship song of a grasshopper – Elsner, 1968), to others in which components can be emphasized to various extents, or even left out (facial grooming in mice – Fentress, 1973).

If backward walking can occur, it is because there is a separate program for its production, not because the control components of forward walking can be reversed. Such behaviors will occur once the appropriate MPG has been "switched on." Some may be on continually as CNS events, but not linked to motor output, in the manner of a car idling at a STOP sign. When the clutch is engaged to connect the idling motor to the drive shaft, the vehicle moves. Switching on, or engaging an act of behavior in an animal, is commonly the result of a specific *trigger* or "*releaser*" stimulus or situation. But some acts occasionally start up merely because the general level of central excitation is high or that of central inhibition is low. The behavior would then appear to be spontaneous or occurring as a displacement activity. At the other extreme there is the classical built-in unconditioned reflex and its attendants, conditioned reflexes. Some of the former are proprioceptively initiated, by position-sensitive sense organs. Common ones are the stretch reflex and the resistance reflexes that oppose limb displacement in either direction. If the animal has eyes, its own movement may generate visual signals and gravity receptors will also be affected, providing secondary inputs. When a movement is initiated by an MPG, these secondary inputs are necessarily activated. Thereafter, they must interact with the program generator unless, of course, operation of the generator includes automatic blockage of the secondary input by way of central inhibition. Alternatively, there can be a variety of extents of interaction. It is also possible that a program takes into account the expected input and either allows for it, while remaining dominant, or else incorporates the reflex input, letting it take over temporarily. All of these various options are available in principle, and the alert student of the nervous control of behavior should expect sooner or later to come across examples of them all.

There is another alternative, though, that at least the more complex nervous systems can use to generate behavioral sequences. This is to utilize information stored as memory of the desired objective, to guide the formation of a motor program. For example, the form of a nest is stored as a visual image of what a nest looks like, not as a series of motor instructions on how to make one. The memory may be only a temporary one, acquired as a result of experience, or it may be inherited. A bird that can build a nest without ever seeing one must have guidance built into its developed nervous system, what we may call a *species sensory memory*. With either type of memory the program generation is a continuous process dependent upon comparing sequences of sensory inputs with the sensory memory.

Neuroscientists appear to have difficulty grasping this "sensory tape" concept. But let them consider the hermit crab's shell-selection behavior. In the absence of a shell, intense exploratory behavior is automatically initiated. This goes on until, as has recently been determined (Mesce, 1982), a body is encountered that is emitting calcium salt but not also the odor of another crab. In the natural environment this is likely to be an unoccupied shell, ones with the mollusc present being covered by a thin living tissue that prevents calcium loss. When such a body is encountered, it is tested for size, not by the crab's actually inserting its abdomen, but by touching all round the shell with its chelipeds. If the size is "judged" right and calcium is being emitted very fast, a fixed action pattern of movements is released that causes insertion of the abdomen into the shell. That is, if there is an aperture there! When the experimenter has blocked the aperture or made a solid, instead of hollow, shell replica, the abdomen-entering movements keep being repeated until the tender abdomen has received bruising treatment. The crab was judging where the "expected" hole of the shell should be, based not on the actual presence of a cavity of suitable size but upon the overall feel of the object of which it has sensory memory. Both emanating calcium and an object of about the right size are needed to release this fixed action pattern. A lump of calcite or of calcium sulphate, which is more soluble than the calcium carbonate of shells, of about the same overall size as a suitable shell is so attractive that it is preferred to real shells regardless of whether or not it has an aperture of the right size for the hermit crab. Once initiated, shell-searching behavior stops only when sensory input tells the crab that there is now an object it can move, of about the right size and weight, installed on its back. The crab frustrated by the lump of calcite will eventually move away and try a shell. The requirement for the correct final sensory input is inherited by most hermit crab species, but has been secondarily lost by the coconut hermit crab *Birgus.* Another example of released behavior that is not appropriate is shown by nesting birds, which show a preference for artificial eggs of extremely large size over normal ones, provided they have appropriate markings (Tinbergen, 1951). Numerous other examples show clearly that similar mechanisms are widespread.

At this time there is no obvious way in which to tackle the cellular circuits and the associated neural mechanisms involved in the MPGs of vertebrates. But for some molluscs such as *Tritonia* (see p. 585) and large arthropods such as locusts, it is feasible and a start has been made. This is in regard to behaviors in which initiation occurs obligatorily in the ultimately simplest possible manner: when a single interneuron, either a specific individual or one of a small class, is excited with a train of stimuli. The number of complex movements currently known to be so initiated is at least 20, and includes both forward and backward walking in lobsters (Larimer, 1976). There are only small numbers of MNs in crustacean ganglia, of the order of 2-300 per ganglion. The total number of neurons is much larger, 5-10,000 per ganglion, and there are about 100,000 cell bodies of neurons in the entire central nervous system of a crayfish or lobster. These figures do not include sensory neurons, whose cell bodies are peripheral. The total number of sensory neurons is unknown, but a single optic tract of a crayfish has 20,000 sensory axons (McAlear, 1967), and the total for the entire animal may be as high as 0.5×10^6. It is certainly much greater than the number of central nerve cells. The number of MNs is also unknown, but in view of the small number of motor axons innervating all crustacean muscles, it cannot be more than about 2000. Therefore, the central nervous system has first to scale down the input-to-output signal ratio from about 2500 to 1. A large number of the interneurons must be solely concerned with reducing the enormous barrage of incoming sensory information to a small enough number of channels for practical handling.

Many of the sensory inputs infringe directly onto MNs so a lot of the integration and motor coordination is handled directly by them. But overall direction of movement must be superimposed. The central nervous system must be able to suppress or ignore the confusing morass of sensory inputs, which must change drastically with every move the animal makes, when it commences purposive complex behaviors. There is direct evidence for the occurrence of such inhibition is crustaceans (Evoy, 1977). When a tail flip is occurring in the crayfish, there is a series of new sensory inputs onto the lateral giant interneurons. But these interneurons are automatically strongly inhibited during the entire period of execution of the flip (Roberts, 1968a,b). Inhibition of interneurons on which the sensory inputs converge occurs whenever coordinated major movements commence.

The setting aside of many central interneurons to reduce sensory information to manageable levels, and the delegation of many others to coordinate MNs ipsilaterally and and contralaterally within a segment and also between segments, leaves relatively few to handle decision making and neural programming. Therefore, it should not have come as much of a surprise, as apparently it did, that when stimulated, some single driver interneurons evoke complex movement sequences of behavioral relevance.

Technical improvements permit working out the details of cellular organization and synaptic properties associated with the actions of driver interneurons. However, except for crayfish tail flips and swimmeret beating, this has not yet been achieved.

Swimmerets are the simpler, beating rhythmically for long periods to promote forward swimming only. The tail flip produces rapid backward swimming, but is rapidly fatigued. The swimmeret motor program consists of bursts of impulses occurring alternately in clusters of power stroke and return MNs. Each crayfish ganglion contains its own MPG for swimmeret beating, but the overall activity is a metachronal wave that starts in the most posterior ganglion. In each ganglion there are matched left and right neural elements: the MNs and three (or possibly three classes of) electrically inexcitable interneurons (Heitler and Pearson, 1980). One of the latter is able to reset the period of the swimmeret rhythm when this is going, but not to initiate it. Another can turn it on. The third can neither start it nor alter the period, but it can alter the motor-output balance. One is certainly, and the others may be, local, that is, without axonal connections to other ganglia. Interganglionic coordination neurons have not been discovered. As with most arthropod motor activities, the MNs are not simply passive elements responding to the interneurons but are themselves involved in the integration and programming of their own output.

Although the finding of these interneurons represents no more than a beginning of detailed understanding, it already gives us a good grasp of the fundamental simplicity of the circuitry and its operation. There might have been many thousands of cells involved, as appears to be the case with vertebrates (see p. 578), but in an invertebrate there are only a small number, most or all of which are identifiable and repeatedly addressable in different individuals.

Exact recording of all the motor output associated with movements of a vertebrate is impossible, even for a single joint, because so many units are involved. But for some joints of arthropods, it is possible accurately to determine the entire output, since the innervation is so simple (see p. 488). I have made thousands of myographic recordings of activity in antagonistic muscles of insects and crustaceans during visually/photographically identical movements, but when the underlying impulse patterns were examined, they were all different in regard to detailed patterns of impulses in the MNs. This was clearly not the case for the recordings made by Norbert Elsner from the male grasshopper *Gomphocerripus* during its courtship song. But comparable stereotypic fixed action patterns are not common.

My own experience shows conclusively that MPG action is not simply brought about by a central mechanism acting indepently of ongoing proprioceptive inputs. But neither does it indicate that the latter reflexly drive motor output. Nor does the truth necessarily lie somewhere in between, as has been commonly assumed in recent literature. A mixture of motor-program generation based upon neuronal-network properties with some proprioceptive input superimposed in modulatory actions is insufficient to explain the perfection of the movements as behavioral events. Furthermore, it ignores the fact that a set of arthropod legs can make many different motor programs though using exactly the same muscles: two or three forward walking gaits, backward walking, sideway walking, running, swimming, righting, turning, tilting, climbing, and so forth, and in some species also hopping and jumping, and a variety of maneuvers specifically associated with courtship, mating, egg laying, and care of eggs.

Intensive study of these movements shows that each has an element of fixed motor-pattern generation as well as modification in regard to detail. The animal therefore must have a collection of motor programs, one for each behavior, which may be thought of as a series of tape-recorded instructions that I have long referred to as *motor tapes* (Hoyle, 1964b), to be called upon as occasion demands. Each is played back in a continuous sequence, and each sequence can be repeated endlessly, But that is not all. The details of these movements can be subtly varied according to the specific immediate requirement. This is especially clearly seen in righting, climbing, and mating, where, unlike steps in walking, the individual leg movements are patently different on different occasions and are related to inconsistencies in the environment. Such movements can only be brought about by guidance based upon a sensory model. It may be deduced that the nervous system has an information store not only of the major sensory inputs to be achieved, but also of their correct sequence. It can program motor output by reference to this store, comparing it with recent inputs. Such an information store I call a *sensory tape.* Many insects, crustaceans, cephalopod molluscs, and vertebrates inherit a set of such sensory tapes. Proprioceptive, tactile, visual, and sometimes also chemical sensory inputs all are used.

The more stereotyped the behavior to be produced, the more likely it is that the animal will resort to programming that is independent of the complicating sensory feedback. This is particularly true of movements that must be produced quickly, which is to say all aggressive, defensive and escape movements. For these the information is direct: a straightforward instruction sequence or "how-to" series, rather than "this is what it feels like." This is the type of open loop information store we called motor tape. Most animals inherit one or more of these by way of neural-circuit connectivity and properties.

In addition to the inherited "tapes," some are doubtless accrued through experience – learned tapes. At least some of the tapes are somewhat plastic in regard to details. As the organism grows it changes in size and often also in shape with maturation. Proprioceptive/visual/tactile experiences lead to long-lasting modifications aimed at keeping the behavior optimal. Diagrams illustrating the tape concept in MPG are presented in Figure 13.4.

13.3.1. Forming New Tapes by Learning

Animals with complex nervous systems, certainly many insects and vertebrates, are able to learn to make behavioral acts they did not inherit the ability to execute. Non-inherited behavioral acts are acquired by conditioning, both associative and operant. Little is yet known about the sequence of events leading to the establishment of such memories. But it is reasonable to expect that they are often started by relating motor output to sensory memory, that is, by first establishing a sensory tape. As each component movement is repeated and practiced, the motor output required to cause it is committed to a different store – of motor information – and placed in its correct sequence in time. By degrees a learned motor tape is thereby acquired.

When the tape is sufficiently detailed and precise to operate in the absence of sensory input, it will generally be much more efficient in use than the sensory tape; the latter can then be dispensed with. As complex behaviors, such as decapping of milk bottles by birds, are learned, a progression from the sensory-tape trial/error comparison method to direct motor-tape control of the appropriate movements is envisaged.

Although the concepts outlined above apply to all animals with sufficiently complex nervous systems, they are most readily understood in humans, where they are part of

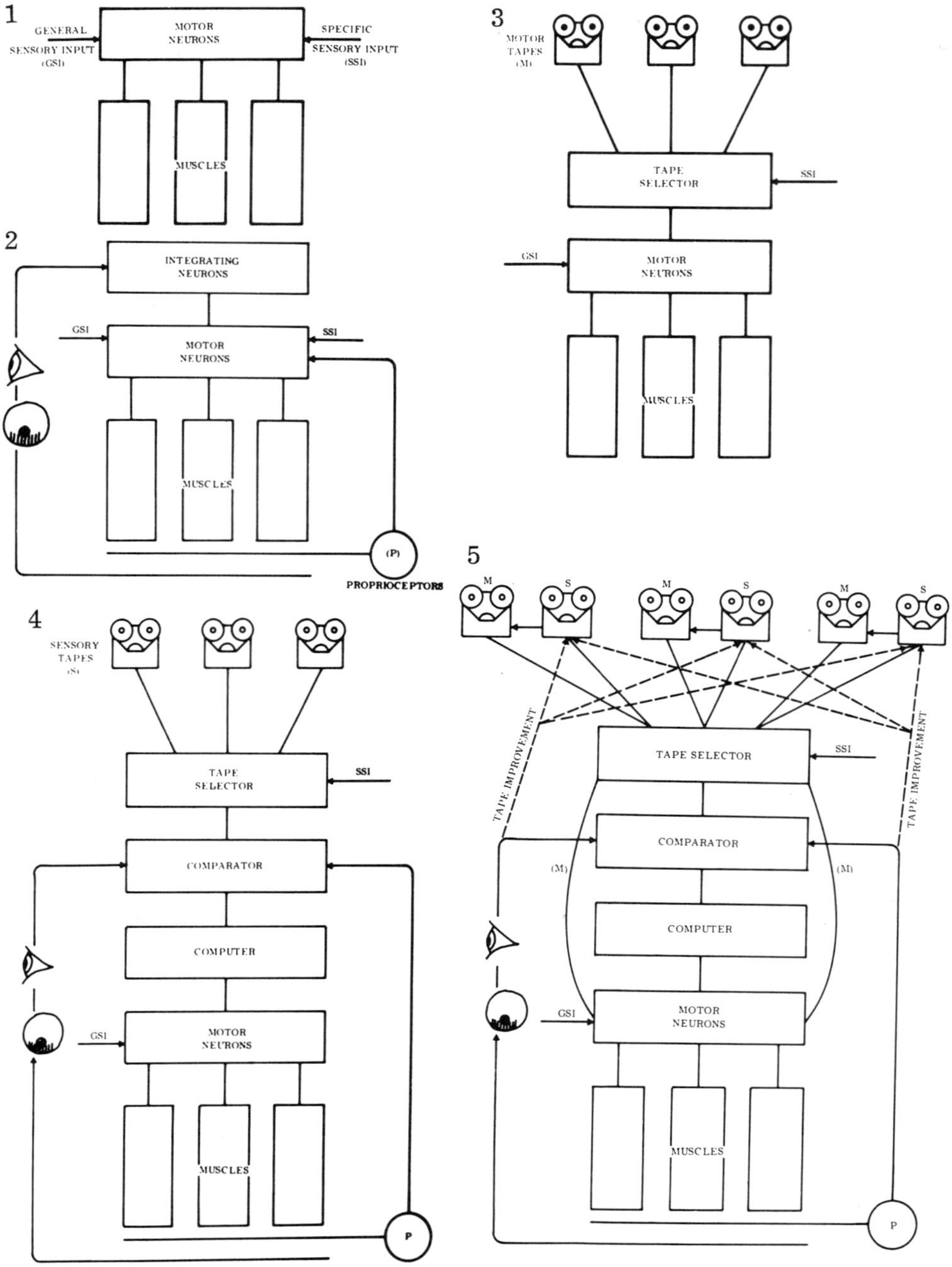
1
GENERAL SENSORY INPUT (GSI)
MOTOR NEURONS
SPECIFIC SENSORY INPUT (SSI)
MUSCLES
2
INTEGRATING NEURONS
GSI
MOTOR NEURONS
SSI
MUSCLES
(P)
PROPRIOCEPTORS
3
MOTOR TAPES (M)
TAPE SELECTOR
SSI
GSI
MOTOR NEURONS
MUSCLES
4
SENSORY TAPES (S)
TAPE SELECTOR
SSI
COMPARATOR
COMPUTER
GSI
MOTOR NEURONS
MUSCLES
P
5
M
S
M
S
M
S
TAPE IMPROVEMENT
TAPE SELECTOR
SSI
TAPE IMPROVEMENT
COMPARATOR
(M)
(M)
COMPUTER
GSI
MOTOR NEURONS
MUSCLES
P

everyday experience. A golf or tennis swing is not a natural act, but once learned and perfected, it can be executed as readily as a simple reflex. While a human learns to make it, the emphasis is on acquiring a "feel" for the movement, that is, building up the sensory tape of the relevant proprioceptive input sequences. But this is only the first stage. As performance improves, the ideal becomes to have a "senseless" motor tape of what to do, which is subject only to minor modulation from visual and proprioceptive inputs.

In animal studies, the best experimental evidence for the acquisition of motor tapes comes from the research of Emilio Bizzi and his collaborators (Bizzi et al., 1978). A monkey is first trained to point quickly (2 s maximum is allowed before the automatic apple-juice reward is disallowed) to a small light when this is switched on. The light is one of up to 12 set in an arc horizontally in front of the monkey. The arm is strapped to a platform pivoted under the elbow that automatically returns to the same fixed point. A large collar prevents the monkey from seeing its arm or hand when the light is on, and the experiment is done in the dark. Eventually the animal is able to point inside 1 s to any light that is turned on. Then its nerve roots are sectioned so that there are no proprioceptive inputs from the arm. Once it has recovered from the surgery, the monkey is able to perform the task as well as it was before.

Electromyography of the forearm muscles during the ballistic arm swing shows that each swing is made by a strong extensor excitation followed by a flexor burst timed to initiate a braking action. Either altering the rest position or loading the arm leads to error, because the monkey has in effect learned a number of different precise motor programs corresponding to each one of the lights used, all of which function in the absence of proprioceptive feedback. The releasing stimulus is a combination of light on and its position relative to the monkey's sagittal head axis.

Although movements of human fingers are significantly impaired following dorsal root section (to eliminate chronic pain), many movements, especially skilled ones associated with prior training, remain (Foerster, 1927). The statement: "Patients should be trained to execute prescribed finger movements before rhizotomy," attributed to C. G. Phills (in Evarts et al., 1971), very clearly emphasizes the importance of establishing motor tapes by experience. The various forms of control system, from simple reflex to sensory tape control, are illustrated schematically in Figure 13.4.

Figure 13.4. Diagrams illustrating the principal forms of behavioral-control systems. (*1*) Simple reflex: open loop, operating without either proprioceptive or visual feedback, or programmed instructions from nerve centers. (*2*) Simple reflex with servoadditions: the motor output is continually modified during the reflex by feedback arising out of the animal's own movements. (*3*) Control by stored information programs of motor output based upon neuronal circuitry and properties, symbolized by magnetic tape recorders (= *motor tapes*); when the level of general sensory input (*GSI*) is sufficiently high, i.e., the central excitatory state is suprathreshold for action, an appropriate motor program is selected on the basis of specific sensory input (SSI) and played out, causing behavior. No feedback is involved in programming. (*4*) Control by stored sequences of sensory information (= *sensory tape*), for example, consecutive notes of a song. In operation, the instantaneous sensory input is compared with the stored sequence. From the difference an appropriate motor program is computed. (*5*) Combined motor tape and sensory tape, with improvement based upon experience (learning). The animal first uses the sensory tape as the movement is being learned and perfected by trial and error. As the movements become improved, the information is transferred to a parallel motor tape. This bypasses the comparator and computer elements and, hence, is more automatic and reliable than sensory-controlled movement. (All from Hoyle, 1975b.)

13.3.2. The Early Origin of Motor Tapes

Coelenterates represent the lowliest level of organization of neural elements known, and the animals themselves have an elemental, diploblastic structure. The nervous system lacks "central" elements: there is not even a ganglion in which neurons are concentrated three-dimensionally. Yet behavior of remarkable subtlety is seen in them. The most dramatic is the fixed action pattern of escape-swimming behavior shown by the sea anemones *Stomphia, Gonoactinia,* and *Actinostola* (see p. 38). Very rarely indeed, this escape swim occurs spontaneously. It can be triggered by a single very strong electric shock or a touch by one of a few species of starfish. Involved are several subroutines, always performed in the same sequence, starting with filling with water and elongation. Next the oral disc expands enormously, and at the same time, the anemone detaches itself from its substrate by a rapid ungluing that itself is a remarkable event. It now pushes itself off by making a large bulge in its base, requiring general contraction of body-wall muscles but relaxation of muscles in the center of the basal disc. Thereafter, there is a sequence of violent fast contractions of longitudinal muscles, starting on one side, moving next to 180° around the animal, then to about 370° from the starting point, and so forth in 190° increments until all the longitudinal muscles have had a turn. Pinning the anemone down does not either prevent its making an attempt to execute all of the maneuvers or affect the timing. The entire sequence lasts for more than a minute. Many evolutionarily more "advanced" animals with "brains" do not show a behavior that even nearly matches the above for subtlety or complexity. No comparably complex behavior is known to occur in annelids or platyhelminths. No more-complex behavior occurs in molluscs except octopus, or in "lower" orders of crustaceans.

These sequences represent a major challenge to students of the neural mechanisms underlying behavior. At the present time there is not even the beginning of speculation on the neural circuitry and physiology of swimming MPG in these sea anemones.

13.3.3. Is a MPG Equivalent to a Motor Tape?

I have always emphasized the term *motor tape,* partly in order to draw attention to its neglected, but equally important counterpart, the sensory tape. However, the reader will recognize that the concept of MPG embodies the same features. An MPG is indeed a motor tape, and since a few of these are now fairly well worked out in invertebrates at the neuronal level, we can seek their common features at the neuronal level. The best current example overall is the triggered escape-swimming behavior of *Tritonia.* Swimming behavior has also been analyzed in two other molluscs, *Melabe* and *Aplysia,* but these are not complete acts with a precise end as is the *Tritonia* swim. Much central synaptic efficacy is directly affected by use, either decreasing or increasing (Gallego et al., 1979b), within a few days of sustained change in input. Each MPG is based upon a combination of genetically programmed fixed neural circuitry – the hard wiring – and early and continuing experiences that affect modifiable properties of the cells within the circuits. Which of the properties are modifiable has not yet been determined. Earlier thinking (e.g., Eccles, 1953) placed it squarely with the synapses. Two kinds of modification were envisaged: changes in morphology, especially sizes of individual synapses and their total numbers, and changes in efficacy, promoting or depressing quantitative aspects. Following direct actions, there are also indirect ones that may occur during the partial transfer of learning from ipsilateral to contralateral side, or to the other ipsilateral limb. Such transfers are not the exclusive attribute of higher vertebrates. Learning transfer affecting function of reflex pathways not themselves involved in the training has been demonstrated in a cockroach (Reep et al., 1980). To these various candidates that may be invoked as underlying events in learning we must add all membrane properties affecting integration of synaptic events by way of time constant, and all biophysical factors affecting the firing pattern of the follower neuron (Hoyle, 1982a), which include all of its many ion channels (see p. 184). It is important to stress that these are *postsynaptic,* and that many are at sites distant from the input synapses.

When these sources of variation are combined with the capability for diversity at the level of the muscle-cell composition of each muscle involved in a particular behavior, we are forced to realize that it will take a much broader base of knowledge than currently exists before it becomes possible to establish what general principles have been utilized in the complex interactive relationship between MPGs and muscles. Since muscles must perform many different tasks, which can be mutually antagonistic, often for not one but several different MPGs, compromises must have been achieved, at best for expediency, given the constraints of the genetic/evolutionary-pressure interactions.

It is the use to which a muscle has been put based upon the animal's entire evolutionary history that determines all these features in both vertebrates and invertebrates. In the latter we have noted as an example the extraordinary switches in muscle form, cell population, innervation, synaptic physiology, and motor-pattern generation that accompanied jumping by the locust. In a vertebrate we may compare two synergistic extensors of the foot, the soleus (SOL) and gastrocnemius (GAST), as they came to be used both independently and cooperatively in standing, and jumping or running, respectively. The SOL contains a relatively homogenous cell population of intermediate-type cells, though they can be subdivided. But there is no difficulty in dividing those of GAST into fast, intermediate, and slow, the extremes being well beyond the limits found in SOL. The division of labor between them is said to be very precisely matched to the characteristics of their different MU populations (Walmsley et al., 1978). The SOL develops a fixed peak force over a wide range of walking to running speeds. The GAST, by contrast, shows a threefold range.

The most compelling evidence for the involvement of experience in setting MN properties is to be seen in experiments in which posture is shifted by operant conditioning. Examples are known from both invertebrates (locust/grasshopper – Hoyle, 1980; crab – Hoyle, 1976) and vertebrates (frog – Farel and Buerger, 1972; rat – McLoon and Bueger, 1974). When tonic leg position is coupled to a negative (punishment) reinforcement, or as has been shown for the locust/grasshopper a positive (reward) reinforcement, after from one to several trials, the position changes to one in which the punishment is avoided or reward obtained. The change can also be accomplished after removing the proprioceptive feedback by deafferentation. There must then be an association between a central memory of the motor output (efference memory) and its consequences in regard to receipt of reinforcement. This automatically leads to long-term resetting of motor output. In the case of the locust it has been shown that MN pacemaker frequency is shifted (Woollacott and Hoyle, 1977), undoubtedly a postsynaptic process. A probable biophysical basis for the underlying event is altered K^+ conductance of the cell near its pacemaker region (Hoyle, 1982a). As would be expected, this type of learning is accompanied by altered reflex action involving the motor neuron, as well as altered posture (Forman and Zill, 1981).

13.3.4. Pattern of Impulses as a Significant Variable

Many invertebrate muscles, especially of arthropods, and possibly a few vertebrate ones also – though this possibility does not seem to have been seriously tested – are extremely sensitive to motor-input pattern. The same total number of impulses may give markedly increased tension when delivered as close pairs or in triplets compared with tension when impulses are evenly spaced. This quality resides in part in neuromuscular events, notably facilitation and posttetanic potentiation, and in part also in E-C and in contractile-mechanism events. The effect may be a phasic, short-lived one, or it may last for many minutes. Associated with the latter, some muscle cells possess a paramyosin-related catch mechanism. A few others show a strong BT that can be reset to a higher or lower level. In all of these, the event that causes the drastic change in muscular responses is an abrupt decrease in interval between MN impulses. The addition of as little as only a single extra impulse, or at most a brief burst, greatly enhances the tension, in the more dramatic cases by a fewfold. In a train, the extra tension is maintained at the background stimulation frequency, that would not, by itself elicit nearly so much tension. In a succession of phasic responses, the enhancement declines gradually.

These facilitations and potentiations are often so potent that if they were to occur naturally, they would have to lead to significant behavioral effects. It has been wondered now for half a century whether or not the animal's MPGs take advantage of them. I believe this to be true, and to be the purpose, for example, of a brief tetanus before the start of a run of the ghost crab *Ocypode.* The prestart provides its leg muscles with the opportunity to develop the maximum thrust its muscles are capable of developing from the first steps. A large facilitation, up to 10 times, occurs and lasts for a minute or so, long enough for escape. In similar vein, though to raise muscle temperature rather than to initiate facilitation, human athletes "warm-up" by co-contracting antagonists and by making cyclical flex-extend sequences. In the case of many flying insects, especially moths, a period of warm-up activity to raise the body temperature is essential for flight to be possible.

13.4. Proprioceptive Involvement in Generation and Control of Movements

Vertebrates are at least as richly endowed with proprioceptors as are arthorpods: both qualitatively and quantitatively the two sets of proprioceptors are remarkable similar except for the existence of sensilla within the arthropod cuticle, which have no equivalent in vertebrates. Spindles (intrafusal cells) have a mixture of sensory endings with different rates of adaptation and different muscular regions. Between them they cover registration of rapid movements (phasic receptors) to maintained position (tonic receptors). In addition, there are receptors in the tendons that were once thought to mediate only inhibition of motor output following development of a very large force. Now it is realized that some of these also respond to both movement and posture, perhaps even with range fractionation (e.g., Desmedt and Godeaux, 1978). In the joints are Paccinan corpuscles responding to stresses associated with change in angle. In the skin are touch receptors affected directly and indirectly by movements.

With this plethora of sources, the central nervous system is well provided with information at all times about the position and movement of all limbs. Each of the sense organs can be semi-isolated individually and shown to be able to modify motor output by reflex action. But when the multiplicity of sense organs is put together, the demands of all the sense organs are overlapping, even conflicting – a babel of excess – as they are also for invertebrates (see p. 551).

The stretch receptors mediate monosynaptic reflex arcs onto the extrafusal muscle cells of the muscle that contains them. Each has muscle of its own that is innervated by small motor cells – the γ-efferent system. It was not until following the initial statement of the principles of control theory – cybernetics – by Norbert Wiener, Claude Shannon, and others that the possible significance of this servoloop system, the equivalent of which is also present in many invertebrates, was appreciated. It is possible in principle for a MPG to control movements not only by sending motor output directly to the muscle cells but also indirectly, via the γ-efferent output. By causing alternate stretch and relaxation of the spindle's sensory element, the servoloop must cause reflex following of motor output. The disadvantage of this route would be that it introduces a quite large delay in execution, but this might be more than outweighed by a marked advantage: this secondary route automatically provides compensation for load fluctuations, so it should be much more accurate and sensitive (Eldred et al., 1953) than the other. Early research suggested that the γ-route might be preferentially used, via cerebellar output, for all slower, skilled voluntary movements. However, 30 years later, the matter is still very much in doubt, as is the whole question of the detailed roles of proprioceptors.

Previously learned skilled movements can still be performed quite well by monkeys subjected to cervical/spinal-dorsal-root section of nerves carrying inputs from the limb used to make the movement (Bizzi et al., 1978). The proprioception is essential during the learning stage of a complex movement sequence, though not of simple posture (Taub et al., 1965). Proprioception may well be used (when still available) in execution, but it is no longer essential. A *memory* of the *motor output* required to execute the learned movement has been acquired and is utilized without benefit of sensory inputs beyond those triggering the initial release of the action.

Also, as we have seen, some motor learning can occur by way of operant (instrumental) conditioning even after deafferentation. The memory of motor outputs made spontaneously is used, in relation to limited indirect sensory return from punishment or reward, to adjust later output until satisfactory posture or movement pattern is achieved.

In invertebrates it has been firmly established that there is a full range of involvements of proprioceptors, depending on the behavior and how it is generated. At one extreme complex, long-lasting behaviors are driven from motor tapes: in these, proprioceptive input is totally shut off and not used at all. Apart from very simple reflex actions, it is unlikely that any behavior is totally generated reflexly, as once thought to be true for all. For several of the thoroughly studied examples, it is evident that motor output is the result of a combination of MPG with proprioceptive modulation. Movements are enhanced or reduced in amplitude or velocity by the feedback, generally in useful ways. Some behaviors are doubtless run from sensory tapes, in which case the proprioceptive input is absolutely critical for the generation, as well as control, of the movements.

13.5. Cerebellar Involvement in Vertebrates

It has long been assumed that the coordination of movements, especially control of smooth operation by continuous correction, is a primary role of the cerebellum (e.g., Eccles, 1973). However, although it has also been assumed that a memory capability is associated with this control (e.g., Gilbert, 1975), none have ever been satisfactorily demonstrated. Cerebellar lesions and abnormalities are associated with erroneous movements in a number of vertebrates. A well-known example is the mouse "reeler" mutant. However, some humans totally lack a cerebellum, and this fact is not ascertained because their movements are in some way abnormal. If a cerebellum is present, motor coordination and control tasks are likely to be assigned to it, but other parts of the CNS are apparently equally capable of performing the same tasks.

13.6. Motor Control Options

According to the principles outlined above, the neurally operated organism has seven basic options for control of any particular behavior:

1. Simple reflex action.
2. An MPG control of parts of the action, with reflex action filling in the gaps.
3. Pure direct MPG control, perhaps with central inhibition used to cut off all sensory input that could interfere.
4. Indirect MPG control operating via proprioceptors, by sending motor patterns to the muscular parts of strain receptors. Because of its servo-operation, this route takes care of load variation automatically. Theoretically excellent and certainly available, there is little evidence that any organisms actually utilize it.
5. "Sloppy" MPG with intermediate output stages sensitive to immediate sensory modulation.
6. An MPG incorporating anticipated proprioceptive inputs as part of the mechanism.
7. Sensory-tape control.

None of these options need be exclusive for a given behavior. If the behavior is long and complex, some sequences could use one option, others different ones.

13.7. Has There Been an Evolutionary Sequence in Muscles and Their Control?

It is customary to take it for granted that as organisms diversified and some became larger, there were trends toward "more advanced" or "improved" states. It is easy to read "progress" into a transition from cyclostomes through elasmobranchs, bony fish, amphibia, reptiles, and protomammals to mammals, and within the latter a narrowing pyramid having humans at its pinnacle. Along with this overall trend, it is assumed that cellular tricks were "discovered" and perpetuated by each "enriched" branch. It is supposed to be the principal task of comparative physiologists to elucidate points of branching based on physiological events and in general to illuminate the evolutionary tree concocted by morphologists.

I have diligently sought evidence of trends both within phyla and between phyla. Here and there a glimmer of light appears: the neuromuscular synaptic gutters of vertebrates shift from the smooth ones of fish through the transversely grooved ones of amphibia to the multiply indented ones of reptiles. But there is not a shred of evidence that any improvement was associated with this. Neuromuscular transmission is apparently just as efficient in a fish or frog as it is in mammals. A musculoepithelial cell has been seen as offering no more than a primitive starting point, but for an aquatic organism with a thin body wall, such a cell is entirely satisfactory. It could just as readily represent the culmination of evolutionary progress toward an optimal functional solution to the problems of being a coelenterate as an elemental starting point.

It is too easy to think of a fast-twitch cross-striated muscle cell as the ultimate contractile machine, developed after countless experiments from the disorderly mix of myofilaments sloshing about in some primeval cell. But the typical unstriated muscle cell of a "lowly" invertebrate is much more versatile and controllable than any twitchy striated one. It can be stretched to great lengths; shorten reversibly to a small fraction of mean body length; maintain steady tension without fatigue, using little energy; and be caused to contract at rates from near zero to almost as fast as its admired partner. Cross-striation equivalent to that of the most advanced mammal is present in some cells of the lowliest organisms that possess muscles. It is entirely possible that the sequence was the reverse of what is usually supposed, and that cross-striation is the primitive, not the advanced feature. The seeming disorder of the "smooth" cells could be the culmination of an evolutionary improvement sequence aimed at overcoming the constraints associated with striation in regard to speed and control. Likewise, ITs have been thought of as a significant advance in muscle fine structure. Certainly they permit, and are essential for, functioning of cells of larger diameter. However, no known advantage accrues from large cell diameter. This is only one of many possible ways of making a larger organism, and it is attended by problems in metabolic exchange as well as E-C coupling. The largest muscle cells occur in the largest crustaceans, which have a fixed number per muscle. A giant barnacle or king crab of exceptional size ends up with 4 mm in diameter muscle cells!

Another admired and presumed advanced feature of fast muscle cells is the presence of an extensive array of ITs. Yet ITs occur in many cells of invertebrates, such as those of the *Peripatus* body wall (see p. 484), that have certainly not the slightest semblance of longitudinal organization. When Dorsett and Roberts (1980) found ITs in a molluscan unstriated cell, it prompted them to propose a "sequence" for the evolution of cross-striated cells from unstriated ones including many regularly arrayed ITs from a "primitive" nonstriated cell (Figure 13.2.*2*). Their sequence could equally well have been the other way round. Slow cells of a snake body (see p. 49) have certainly been improved for the role they must play, by secondarily losing the ITs they had to start with.

Certainly muscle cells are enormously plastic and can change in many respects within the lifetime of a single animal; most must have been subjected to drastic changes during evolution. It seems likely that the organization of the components of a muscle cell has been very flexible: it probably changed in various directions, from disorganized (smooth) to cross-striated or obliquely striated, and back again as functional needs changed. A muscle that is now functionally very fast may have been slow at some stage in its history and vice versa. Fast flight muscles did not appear *de novo* but were converted from ones that originally served to perform a slow tonic or multiple locomotory function. In insects, some muscles even retained the original function as well as became involved in flight, so are now bifunctional (Wilson, 1962). The major indirect insect flight muslces of the otherwise very different orders have myofilamentary organizations that are very similar and unique as a class. This homogeneity could have been achieved by chance convergence, each insect flight muscle finding the same structural solution, or be an indication of common origin, flight having perhaps been evolved but once by insects. The indirect flight muscle cells do all have very different organizations at the more gross subcellular level, and their mitochondrial types are markedly different from each other, so there has at least been extensive radiation.

The differences seen at the level of cellular cytoarchitecture: forms, amounts, and distribution patterns of SR, forms and details of ITs and their contacts with SR, and so forth are all different for muscle cells from different animal types. A specialist can tell from a glance at an electron micrograph whether it is from an insect rather than some other arthropod class or even another phylum. The expert can go even further, distinguishing a TEM picture of a cockroach muscle from that of its homologue in another orthopteran, the locust. Some differences are superficial: SR length, amounts of SR, numbers of ITs and IT/SR contacts, mitochondria, and so forth can be essentially identical and yet the cells be very different in appearance internally because in one the SR is arranged in lamellae and in the other in cylinders. No functional difference can be related to many of the structural differences observed. This is true in a general way of the majority of the fine differences seen in cellular morphology. They are a result of happenstance in genetic/developmental/evolutionary factors. Many are totally without functional significance, and others have combined positive and negative contributions canceling, so that the end products match functionally.

13.8. Evolution of the Contractile Machinery

The arguments supporting the X-bridge theory have recently been eruditely stated by A. F. Huxley (1980). However, as he is at pains to point out throughout his book, they are often specious. There is yet no absolute proof that X bridges alone provide the motile force: the way remains open for consideration of alternatives. In fact, there is no agreement, because of a lack of knowledge, on the causality of any of the myriad forms of intracellular and whole-cell movements known in animal cells, in spite of the insistent attentions of hundreds of major scientists during the last half century. Does this not suggest the possible pursuit of false trails?

A. F. Huxley proposes that progress in understanding muscle started when attention was paid to a specialized form, vertebrate fast-twitch cross-striated cells, and away from an earlier search from a presumed primeval universal contractile machine for all types of cellular movement. Now, however, there is a strong need to at least attempt to fit the X-bridge model quantitatively to other muscles, especially ones with few myosin molecules, lacking filamentar organization (see p. 314), and ones with random distribution of myofilaments (see p. 481). Some of the latter can develop much more maximal force than vertebrate twitch cells, as well as functioning over a 10-fold greater length range. The usual argument given by high priests of X-bridge theory – who, incidentally, support their own ideas with an ardor that matches that of any religious fanatic – are that these other materials are "unsuited to analysis," or are "to atypical to merit serious attention." Their passion is matched by a disdainful intolerance of, and impatience with, any alternatives to X-bridge theory, an intolerance that must match the fire of those who "burnt out heresies" at the peak of the Spanish Inquisition.

A. F. Huxley's first X-bridge theory (A. F. Huxley, 1957a) relied solely upon different attachment, compared with detachment, kinetics for a bridge that oscillated in position longitudinally (Figure 4.8.*1*). Later ideas, such as that of R. E. Davies (1963), proposed shortening in the bridge itself (Figure 4.8.*2*). This idea was widely supported until ultrastructural evidence (see p. 112 – Reedy et al., 1965; Reedy, 1968) combined with antibody staining methods (see p. 111 – Pepe, 1968), together with studies on self-assembly of myosin (H. E. Huxley, 1963) and realistic modeling of thick-filament molecular architecture (see p. 114 – Pepe, 1971) made it clear that the bridges do not point outward from the myosin to the actin. Instead they curl around so that the heads point inward. This means that the only way they can possibly develop force is by local movement of the head, either by a kind of wristy twist or by local movement at the head. These realizations forced a new generation of hypotheses. H. E. Huxley (1969) proposed a relative sliding of the two head sections (Figure 4.8.*3*). A. F. Huxley and Simmons (1971) proposed rotation of the entire head (Figure 4.8.*4*). Two forms of this hypothesis appeared: rotation of the head about a single attachment (Figure 4.8.5) and progressive rolling of the head around a sequence of different attachment points (the "rocker" hypotheses – see p. 139; Figure 4.8.*6*). Short-range elasticity, empirically shown to be present in cross-striated cells, has been extremely carefully measured for the light it might throw on the molecular mechanism. It is presumed to be associated with either the hinge or arm of the myosin molecule, or it could be associated in part with the I filament near the attachment of the myosin head.

Let us next look at a realistic model that adds a third major filament to muscle, the ultrathin (T) filament, as required by experience with stretch and extraction. A model of the fundamental elements present in muscle cells is presented in Figure 13.5. Unfortunately, we know neither the chemical nature of the T filament nor its reactions, if any, with the major muscle proteins actin, myosin, tropomyosin (Tm), and the troponin (TN) complex. We do not even know its attachment sites, but they could be directly onto the cell membrane at its outer margins. It seems unlikely that the T filaments are attached to Z lines, as in each of the alternatives suggested by A. F. Huxley (1980). They must closely parallel I filaments. In some cells at least, some of them may pass down the cores of A filaments. The prominent C filaments of insect flight muscle cells probably evolved out of such core T filaments. It is likely that the N material is attached to T filaments. Data that bear on the question of T-filament attachment sites are provided by the shift in N position associated with stretch (see p. 97). The shift away from Z bands is less than it would be on the basis of the N materials being attached to a free elastic T filament

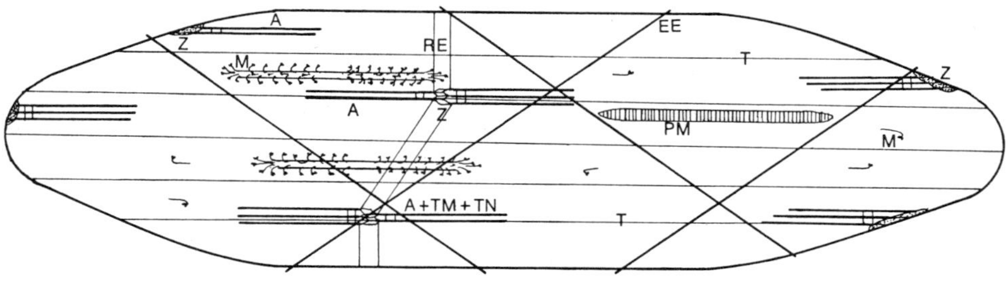

Figure 13.5. Fundamental common components of muscle cells. *Universal.* External elastic elements, *EE,* with radial and longitudinal components. Internal ultrathin longitudinal elastic components, *T.* Z Bodies, *Z,* both attached to the surface membrane and loosely attached internally, to which actin filaments are attached. Radial elastic elements, *RE,* linking Z bodies to the surface membrane. Four polarized actin filaments attached to Z regions. Myosin, *M,* either as loose molecules or with the molecules organized into rods. *Not universal.* Paramyosin, *PM,* organized into crystalline rods. Troponin, *TN,* and tropomyosin, *TM,* present in association with the actin rods.

running from Z band to Z band with movement of N's being unimpeded. But either if there is a resistance to the movement of N or if T attaches to A close to its outer margin, the observed relative shift in N could be explained. The T-filament relationships with thick ones might be different in different cells. The data on shift with stretch require N to be attached to an elastic filament that runs at least for some distance along the thick filament before making a firm attachment. In a cell at less than rest length, N takes up a position only 0.12 μm distant from the Z center, compared with 0.3 μm at rest length. Therefore the elastic element to which it is attached must be appreciably stretched at rest length. A force is needed to counteract the elastic shortening tendency, which could be either longitudinal electrostatic repulsion (solenoid), or inwardly directed (compressive) radial elastic forces. In those cross-striated cells having an M band, T filaments seem likely to be attached to the M bridges.

I filaments are almost universally attached to DBs and run parallel to the axis of contraction or nearly so. Except near the ends of the cell, they are attached to both sides of the bodies. These are clearly homologous wherever they occur, being equivalent to the dense material of Z regions: they should therefore be referred to as Z bodies. The Z filaments are attached to the ends of I filaments to form a lateral Z-filament lattice that is the structural underpinning of Z regions.

Another important aspect of muscle cells that has been ignored is their radial elasticity. There is a balanced system of radial forces that matches the elastic longitudinal ones. This contributes to resting elasticity, restoration of length after shortening, negative contractions in response to hyperpolarization of cells showing strong BT (see p. 206), and the immediate restoration of normal ultrastructure following stretch beyond overlap. But the identity of the radial elasticity has not been established. In the simple model currently favored by both A. F. and H. E. Huxley and most muscle biophysicists, radial elastic elements cannot be explained, except for the contributions made by Z bands.

There has long been incontrovertible evidence for the existence of T filaments (see p. 90). Regarding their nature, there is certainly the major difficulty that the current balance sheet of proteins known to be present seems to have little room for another major one. Numerous attempts to obtain T filaments structurally in our laboratory have left us with a set of fine filaments of uniform appearance, some of which were greater than 2 μm long even from rabbit psoas residuum after classical myosin and actin extraction procedures have been applied (Figures 3.34.*4* and 3.34.*5*). These filaments are made up of uniform club-shaped elements arranged helically. The average repeat periods of component length and pitch closely resemble those of actin, but with a range of both shorter and longer repeats compatible with their elastic nature. An impression of their structure is shown in Figure 4.5. The strands tend to pair up, in the ways shown, to make structures closely resembling I filaments in appearance, and this would account for their having been so obscure. If the H. E. Huxley model of muscle were valid, bridges would indeed have to also be responsible for generating movement, unless the operation is electrostatic. But the presumption of this role depended directly on his belief that no other filaments are present in parallel.

The T filaments alone, without any I/A-filament links, are able to bear forces greater than those developed by a fully activated sarcomere at rest length. When a muscle cell breaks, it is because the T filaments either break or detach. There is no doubt in my mind that T filaments overlap A filaments in most forms of cross-striated muscle cell. This automatically raises the question of their possibly being actively involved in contraction.

I have never felt comfortable with X bridges as generators of movement. As elements that cause stiffness and resist stretch they are fine. As undoubted devices for bringing an enzyme molecule close to its site of action and as carriers of the regulatory molecules governing this enzyme's activation, they are a wonderful invention, with plenty for them to do. But to attribute to them all these actions and in addition movement generation is a bit much, especially in view of their curling toward the M region. Curling is understandable as a desirable feature aiding in finding attachment sites, especially during lateral separation of the filaments as shortening proceeds. But for the process of shortening, I prefer the notion of local longitudinal shortening of either the T filament or of the actin molecule or both, following ATP splitting. Except for tubulin-based motions of ciliae flagella and microtubule assembly/disassembly, actin is the one substance known to be always present when there is movement, so local shortening of actin should be entertained as a serious possibility also.

There is at the present time a belief that actin in muscle is a rigid rod that doesn't shorten, but local length changes could occur without significantly altering overall length.

In no current molecular theory is it easy to discern any major trends that could be called evolutionary. All known specialized features appear to have been aimed at improving the performance of a particular muscle. They operate in terms of the contributions of individual cells within the muscle, not the muscle as a whole, in close conjunction with the properties of the MN or neurons that innervate them. Variation and adaptive radiation there have been, including the development of the TM/TN- and leiotonin-regulator systems, which almost certainly superceded a primeval phosphorylation of myosin light chains (see p. 119). But what advantages may have ensued from this step are not clear. They may have permitted greater speed of activation, and faster is better for prey and predator alike without secondary attack/defence measures such as poisons and armor. But speed and economy had to be traded off, so the fleet of foot had to eat more – if the food could be found. Slowness has its advantages too! There have been systematic changes in other molecules including actin, which is superficially the most conservative, as determined by progressive changes in its amino acid composition (Ogievetskaya, 1976). But no possible functional changes associated with the progression have been proposed.

13.9. Are All Muscles Phylogenetically Related?

A major obstacle for me in accepting the "muscle is a muscle is a muscle" assumption is the extraordinary range of intrinsic speeds of shortening, leading to contraction rise times that vary over a 100,000-fold range, and of relaxation times that are even greater, up to 1,000,000-fold, even if catch muscles are excluded. The force range is so much less, being only about 30-fold, that most theories of contraction can easily accommodate it. The probable range in X-bridge recycling rates is miniscule by comparison, as is that for myosin ATPase activities. There are also different rates of preparing or uncovering the enzyme interaction sites and different intrinsic viscosities of myoplasm, but even collectively they cannot cover the entire range.

Stiffness as such is another major but neglected variable: some striated muscles strongly resist even a few percent stretch, others can be stretched over 50% before there is more than a slight resistance. Many muscle cells, especially smooth ones, but also the cross-striated giant barnacle cells, lack a precise "rest" length; their unloaded length is altered by recent performance. There is a marked hysteresis in the force–length relation after strong stretch, the cell returning to a longer "rest" length than its original one. The extent of the new rest length is a function of the amount of the stretch (Hoyle, 1982b). But if as little as one strong excitation is applied and the cell is allowed to shorten under load, it resumes its original length upon release. From this point the original elastic relations hold again. For this and related reasons, giant barnacle cells, although cross-striated, could be considered to embody some of the features of classical smooth muscles, for which it is virtually impossible to define rest length. But what does such a compromise mean in terms of myofilaments? Since the barnacle A and I filaments appear to be just as inelastic as are those of vertebrates, the changes associated with stretch can only be brought about by decreasing the overlap between them. The properties of the elastic material must undergo a change under strong stretch, as happens in some plastics or even metal springs. Unlike the metal spring, however, the original properties are restorable by excitation. The elastic elements include the Sl, but in many cross-striated cells, this is folded at rest, looping from peripheral Z element to Z element where it is attached, and starts to contribute to the force during stretch only at about 25% above rest length, the same length as that to which the greatly stretched barnacle cell returns when released under zero load. The recovery of the original properties in response to an excitation cannot be explained in terms of any known muscle properties.

Many invertebrate cross-striated cells such as those of *Limulus* telson and giant barnacle retractors develop very large active tensions when stimulated after being stretched very slowly to well beyond the point of I- and A-filament overlap. It is possible to find regions in TEM section in which adjacent I and A filaments from neighboring myofilament clusters have come into contact with one another because of skewing in both sarcomeres. These contact points meet the enzymatic requirement for actomyosin, but they are too few and too small in area to account for the large force developed (Hoyle, 1982b).

Set aside from all other muscle cells in form and ultrastructure are vertebrate smooth-muscle cells. At the present time it is abundantly evident that we do not understand the basis of contraction in any vertebrate SM cells. We cannot explain the large forces they can develop with so little myosin, which in some is barely one one-hundredth that present in cross-striated ones. Even their structural organization is completely obscure. Stretchability of some smooth cells is, of course, very much greater than that of even barnacle cross-striated ones, which break at about three times rest length even when stretched very slowly, and at shorter lengths when stretched quickly. The maximum for a non-cross-striated whole muscle is about 70-fold, but this is associated with movement of cells relative to each other as well as cell stretching. However, individual cell-length changes up to 10-fold have been seen. The details of redistribution of contractile material are as obscure as are those of their basic organization in such cells.

The differences between muscle cells are enormous and are such as to make it quite likely that Carl Pantin and the others have indulged only a speculative fancy. Until much more is known about the comparative ultrastructure and molecular components of a wide range of both cross-striated cells and smooth ones alike, the possibility of there being more than one molecular element generating contraction must be considered. Likewise, the possibility that they might have had independent, polyphyletic evolutionary origins is at least as good as an assumed homogeneity and common ancestry.

13.10. Evolution of Innervation of Individual Muscle Cells

13.10.1. Skeletal Muscles

As with muscle-cell structure, it is not possible to point with conviction to any evolutionary trends that have occurred in innervation patterns. There has been a tendency to regard slow-type cells with multiterminal and often also polyneuronal innervation as representing the primitive condition. It is true that this was evolved early and is more abundant in invertebrates and lower vertebrates than in mammals. However, since it offers a much more efficient way of controlling contraction than is possible using all-or-none action potentials, it might just as well be argued that it represents the more evolved condition. It is just as likely to have been evolved secondarily, for example for use in controlling eye movements (of vertebrates), as to have been retained from an early widespread distribution while skeletal cells developed unitary end plates. There is no reason even why innervation of a muscle may not have changed from one form to the other and then to a different one that was effectively back to the earlier one again.

We should certainly address the question whether muscle arm-to-nerve junctions came before nerve-on-muscle synapses. The latter are by far the commonest and clearly the only kind suitable for larger animals. The former are widespread in invertebrates that on general morphological grounds are considered "primitive," but they do not occur at all in the ones presumed to be the most primitive of all, coelenterates. Both kinds occur in *Peripatus,* which in many respects is an advanced, sophisticated animal, certainly the equal in behavioral complexity of many insects and most crustaceans. The answer may be that both types of junction were evolved at about the same time, totally independently.

The next question is what the initial pattern of innervation in the nerve-cell-on-muscle mode may have been. I think it likely that multiterminal innervation was the primitive condition. Contractile activation was likely to have been brought about primitively by directly entering calcium. The primeval conductance increase is likely to have been to Ca^{2+}, because this is the contraction activator. At the membrane level, gCa^{2+} increases lead to graded responses that spread decrementally. In this view, Na^{+} spikes were a later invention, adopted by larger organisms, that enabled further increase in size because of the faster propagation along, and also into, the longer, fatter muscle cells. But this could only be achieved by organisms with nerve cells to spare and plenty of space to put them. Decremental propagation permits contraction control by way of frequency and the whole arthropod route of a single excitatory axon's supplying an entire muscle. Decremental propagation is also compatible with multiple excitatory axons; its only disadvantage is that multiterminal innervation is obligatory. Large size is not incompatible with this mode, as witnessed by the vigor of giant king crabs with 6-ft leg span, and by 70-lb lobsters. All-or-none sodium spikes demand tension control by a large number of axons, except for hearts, in which little control of either the amount or the rate of change of tension is needed.

I think it likely that the precursors of both vertebrates and modern invertebrates had multiterminal innervation of each muscle cell by a few excitatory axons each. This was whittled down in most (at least in the mature animal), but retained by some cells, presumably because of its superiority for control of tonus and very slow movements. The multiple motor-unit/single-terminal/propagated-spike mode favors ballistic movements.

13.10.2. Vegetative Muscles

Vegetative functions cover a wide range, not only between organs, from urine storage to iris-aperture control, but also within an organ, from grinding up food to voiding waste. Vegetative musculature is basically very ancient, and because it retained elemental innervation while becoming involved also with evolving central nervous systems, it comes in such an immense variety that it is difficult to discern possible trends. Some of its features are unique, leaving special problems to be resolved, including the origin of sensitivity to stretch, interconnectedness and cooperativity between cells, myogenic rhythms, and propagated spikes.

An origin of muscle from myoepithelial cells as found in coelenterates has often been proposed, most recently by Prosser (1977); these hypotheses imagine loss of the epithelial part to be the starting point. In this view, these cells are not myogenic, requiring this quality to have evolved later and especially in vertebrate guts. Stretch sensitivity would have evolved independently in various guts. According to the view that the earliest contractile cells were intrinsically active and come to be controlled by contact with other cells, myogenicity is their most ancient attribute. Interconnectedness may have been very early, and likewise stretch sensitivity, because of the advantages offered by coordination of the cells. Both would have preceded innervation. The earliest electrical activity would have been graded Ca^{2+}-conductance increase occurring spontaneously, perhaps rhythmically, and this would have been sensitive to stretch.

13.10.3. Evolution of Central Control

Most evolutionary neuroscientists agree that the primeval motor-nerve cells were diffusely scattered and situated near the muscle cells as they are in present-day coelenterates. These cells were loosely connected with each other in the form of a nerve net, much as they are in *Hydra.* This condition has been retained for parts, at least, of all organisms including man. Its advantage is that it permits rapid, direct, local control. This is especially necessary in an alimentary canal, in which the various regions are required to perform very different functions at any given time. As evolution proceeded, the vegetative musculature added extra innervation, but without losing the old. The new innervation served only to modulate the actions of the old, not to act as a substitute.

The story of muscles operating skeletal parts or the body as a whole is quite different. The local nerve-net control was gradually replaced by evolving central-nervous machinery. Molluscs have progressed the least among major invertebrate phyla in the direction of switching to central control. Even the most advanced, the cephalopods, squids, and octopuses, have an extensive peripheral net. An excised arm of an octopus moves continually in a complex manner: its suckers attach reflexly to objects. The foot of a snail or nudibranch creeps around after being severed from the rest of the body. By contrast, arthropods and vertebrates have each totally lost all semblance of peripheral control of their skeletal muscles.

For many arthropod muscles, it has been firmly established that the frequency of discharge of a MN is the primary variable determining speed and force for all contractions except brief twitches. The situation is more complex in all vertebrates. Even in fast phasic contractions, MUs are commonly activated by brief bursts, and these tend to be of a fairly fixed frequency, just sufficient to develop a smooth tetanus. But a single unit can seldom generate sufficient force to be useful, so more than one must be activated.

Recruitment of units has thus become the major single variable. The extent to which force and speed can be varied during a sustained contraction by frequency alteration is very dependent on the muscle-cell composition of the particular muscle, being greatest for those with a lot of slow red cells and least for ones that are predominantly fast white. Examples of the latter in man are the adductor of the thumb and the finger muscles. These are all relatively small, so contain few MUs, yet must be controlled very precisely.

In some vertebrate muscles, there are complex relationships between performance and frequency of excitation. When an oscillatory movement is superimposed on a contracting cat soleus muscle stimulated at a constant low frequency (7 Hz), the steady tension fades to nothing, but at a higher frequency (35 Hz), it is maintained (Joyce et al., 1969). There is a very steep section in the curve relating tension to frequency. This steep region occurs over different ranges of stimulus rate at different muscle lengths. There is a reciprocal relation between frequency and length. At low frequencies the maximum force is developed at great length, but at high frequencies it occurs at short lengths. So far these facts have not been interpreted, but this complexity may be one reason why some vertebrate muscles do not seem to use frequency at all as a control variable. Because the larger phasic muscles have been the more popular research objects until recently, the impression was spread that contraction strength/speed of vertebrate muscles in general is determined by the number of MUs active and hardly at all by frequency. The truth is probably that the muscles used most often, and also those used most delicately, use frequency at least as much as they use recruitment and derecruitment. For some tonic units and for slow extrinsic eye-muscle movements, frequency is the principal variable. Also, in muscles of the fingers of man, change from a low frequency base upward threefold to fivefold is utilized and is at least as important as recruitment (e.g., Milner-Brown et al., 1973b). The maximum frequencies used are relatively low, generally below 90 Hz. This maximum frequency seems to be independent of type of muscle cell – and therefore also of motor-neuron·cell body. The limitation to this frequency certainly cannot be attributed to the neurons or even the synapses. It is possibly associated with the muscle-cell action-potential mechanism, which cannot sustain higher rates.

In many classifications of muscle cell types, fatigability has been correctly regarded as a major variable. Certainly this must play a major role in the survival of all smaller animals likely to be preyed upon. It is interesting that there is such a trade-off between fatigability and speed in vertebrate muscle cells. In invertebrates the same muscle cell can function indefinitely when excited via its slow axon but must rapidly fade when excited by the fast one. There is an exponential increase in the energy required to push the cell toward the limits of its capability to either shorten at its fastest or to maintain its greatest force.

It is surprising, in light of the probable X-bridge contribution to force generation, that vertebrates have not evolved a controllable low-energy-cost method of maintaining steady tension, by keeping bridges locked in the attached configuration. Some invertebrates, by way of the not-yet-understood use of paramyosin, have discovered a means of doing just this in "catch" (see p. 477). A few vertebrates, notably tortoises and sloths, have taken a step in this direction, but only by slowing down all their muscles. There is no reason in principle why a vertebrate animal should not be able to be able to make very fast and also very strong sustained contraction in the same muscle by way of utilizing a mixture of two extreme kinds of muscle cell, but such didn't evolve! Several invertebrates of various phyla not only succeeded in achieving this, but they also evolved, in insect flight muscles, one that are both extremely fast and very resistant to fatigue.

13.10.4. Evolution of the Size Principles

The only accepted principles we have regarding the control of muscles at the present time are the two, described above, that prescribe the order of recruitment of MUs: the cell-body size and the relation this bears to homogeneity of properties of muscle cells innervated. The former principle has been utilized independently by both vertebrates and invertebrates. The phenomenon is to be expected, since it is a simple biophysical consequence of cell-body size provided there is shared, equally weighted, innervation. The input resistance of a cell increases with decreasing size, so that like synaptic inputs exert relatively large potential changes in smaller cells, and smaller ones elicit action potentials only in small somas. Cell-body size is principally related to the amount of peripheral work the MN has to do. There are three aspects to this: the number of muscle cells supplied, the sizes of these cells, and the size of transmitter package delivered to them per impulse. For invertebrate slow-, intermediate-, or fast-type junctions, the latter, the size of transmitter package, is by far the more important.

In vertebrates it is the size of the MU, which is in turn related to the type of cell innervated: slow-, intermediate- or fast-twitch, that dominates. The maximum force developed by MUs generally ranges about 10-fold in each muscle. The cat soleus red slow-twitch unit, for example, has a maximum twitch force of 4 g, whereas the fastest white ones develop a force of about 40 g, the average being 14.8 g (Henneman et al., 1965), reflecting summation effects of sizes of individual cells with numbers of cells. The MU size principle has not been used by invertebrates. In arthropods this is because there are few (often only one) MUs and because each muscle cell is operated by more than one size of nerve cell. It has long been known that the recruitment order in invertebrates about which we have detailed knowledge, when there are two or more excitors, is in order of their ascending size (Henneman size principle). There has, however, been no simple, slavish use of this "rule," and various mixtures are known. One is to have a few clusters of neurons each containing slow, intermediate, and fast neurons. The principle then operates within, but not between, clusters. For example, the locust flexor tibiae myograms show a slow-intermediate-fast recruitment sequence that is also related to cell size. However, the muscle – and individual cells – receives inputs from three similar, but differently located, sets of three cells each. Each set has the same range of cell sizes. The principle applies here, as elsewhere, only to neurons with quantitatively similar, shared inputs. However, it is probably not by chance that there are three anatomically separate sets of three sizes in the locust. Each set receives some separate inputs as well as some inputs in common. These permit there to be different control pathways, as there are, perhaps to enable more MPGs to include the same muscle in their operations, each of which can use the cell-body size principle.

The evolution of the MU size principle in vertebrates presumably started with fishes, but owing to a dearth of information on the precursors of tetrapod limb muscles – fish fin musculature – the early stages cannot be assessed. The well-studied back muscles of fish afford no clues because of the complete separation of the tiny red lateral strip that does most of the work. But even here the principle operates. The red cells are slower, and they are innervated by smaller, slower-conducting axons, than the white ones. In some fish an intermediate, pink, layer occurs between the two, and this is recruited as swimming speed picks up (Johnston et al., 1975), followed by the white fast majority. The MU size principle, at least, certainly works in fish. However, this fact shows that the relationship to size may be arbitrary. It is the physiological/ultrastructural/biochemical-combination type that is the real determinant. *A priori,* I see no reason why the cells should have a unique size because of the particular combination of these factors that determines their speed and endurance. It is therefore possible that the size differences arose from random variants via natural selection. The feature that would reinforce such selection is clearly the potential behavioral/survival value of using the two size principles in combination. In vertebrate animals one would have little or no value without the other.

It might have been better to eliminate the muscle-cell size/MU-strength elements from the consideration of the contribution a particular MU makes. Instead, perhaps we should concentrate on the really significant aspects – speed and endurance – which are simply inversely related. Fortunately, sizes – of muscle cells, MU, and cell body – are all directly related to these cell properties, so provided one is not misled into thinking that muscle-cell size as such is functionally significant, size can be retained as a useful and simple summary term.

13.11. Interpreting Muscle Function During Behavior

13.11.1. Fluctuations During Operation

The uses to which any particular muscle are put can change even in a single behavior, as speed changes, mainly because there are progressive changes in gait, with transitions between one gait and another. These can in turn affect the MU-recruitment pattern by altered reflex action. Allowances for the need to alter MU-recruitment are presumably built into the MPG mechanisms. The stretched antagonist in locomotion stores energy elastically that is utilized when its turn comes to shorten. But as the pattern changes when the gait shifts from walking through trotting or hopping to jumping or galloping; synergistic muscles are recruited, even whole movements such as those of the back being added. The forces exerted change drastically within each cycle so that proprioceptive factors alter in a very complex manner. Interpretations using only reflexology, even after the discovery of the size principles, were not able to explain the events quantitatively. There has been a recent shift toward an explanation based on MPG (e.g., Grillner, 1976), which, although still incomplete, has largely eliminated the difficulties. But of course it has not made detailed description of the events any easier. Now it is realized that the details underlying a particular movement are never likely to be understood on the basis of any *general* principles. Each must be determined empirically, since the MPGs or "tapes" (see p. 588) are themselves widely different from each other in details.

Long-term changes in MU properties are now recognized as major factors in determining performance, demanding corresponding stable shifts in properties of the MPGs if vertebrate MU effectiveness is to be retained, except in movements that can be compensated for by servo-operating subsystems. Component muscle cells of an MU, resemble each other very closely indeed, which may be taken as evidence that their MN trophically specifies their properties within very narrow limits. Neverthless, the range of properties of any given unit are very wide indeed, even to the point of overlap with those of otherwise different units. This means that two initially dissimilar units could become indistinguishable if some properties of each shift in the direction of the other's characteristics. Such changes undoubtedly occur during effective training (see p. 405). They also shift in uncontrollable and usually undesirable ways in association with many diseases (see p. 414). Change in properties occur as a consequence of the type of work the muscle is called upon to perform, hence the successes of training.

13.11.2. Effects of Heterogeneity of Form Within a Muscle

In addition to the many differences in the properties of individual muscle cells, there are many gross factors that affect overall performance. The length ranges of individual muscle cells in vertebrate bipinnate muscles is more than threefold, and the angles of attachment vary by as much as 40°. Thus no simple relationship can be found that expresses the relation of design to performance exactly for these more-complex muscle arrangements. The situation is even more complex in arthropods among invertebrates. Here a very tiny muscle comprising but three cells may have a slow one, with 12-μm sarcomeres; an intermediate, with 6-μm sarcomeres; and a fast one, with 2.0-μm sarcomeres. Or, with several cells, they may range in a continuum from slow to fast, each of which is supplied by the same single excitatory axon. So far no rules are recognized that relate cell-type content to function in arthropods, and no two muscles have been found to be strictly alike among the about 200 so far examined. Each has its own unique characteristics. Control is highly dependent on nerve-impulse pattern, which in turn determines the contraction pattern. The evolution of a MPG is a subtle two-way interaction with the muscles to be controlled, which are themselves plastic and capable of shifting in the direction of making the movements more efficiently.

Conclusion

For the compulsive wielder of Occam's razor, muscles and their control provide a field that must promote wild despair. By contrast, for comparative physiologists, it is an incredible happy hunting ground, for these scientists are seekers of examples of the richness of animal diversity, and in muscles, adaptive radiation has successfully run rampant. The reasons, apart from the intrinsic molecular looseness of the components, are the immediacy of muscular actions and their consequences. There is no possibility in muscle for slow, long-term trends, each trivial and inconsequential in itself, which after many millenia will begin to confer a selective advantage. Within its lifespan, an animal's muscles can be lost, gained or regained, as well as changed almost beyond recognition: they are plastic in a way that no other tissue is.

For the forseeable future, there is no shortage of major problems in muscle science, even at the level of molecular events of contractions, and there is an infinity of minor problems, any of which might yield clues about where to look for solutions to the bigger ones. Research is particularly needed on the biochemistry and properties of the structural components of arthropod skeletal cross-striated cells, on the role(s) of paramyosin wherever it occurs, and of almost any aspect of vertebrate SM cells. Muscle is not just a matter of the interaction of actin and myosin in water, as Nobel laureate Albert Szent-Györgi, physiologist W. H. F. Mommaerts, and many others were staunchly claiming but a few years ago. Soon afterwards, S. Ebashi discovered TN, and the scene changed dramatically. But perhaps not sufficiently! I anticipate that other surprises lie in store for us.

These surprises will not come quickly though, as long as 80% of the effort continues to be devoted to the ardent pursuit of the question of what exactly happens at the head of the myosin molecules, as it has for the last 15 years! The seriousness of this search for muscle's presumed "Holy Grail," the X-bridge cycle, resembles nothing so much as that evinced in medieval times concerning how many angels may stand on the head of a pin. Now, however, the question concerns the type of dance they are performing on it.

There is no longer any room in the study of muscle for the kind of idealistic wishful thinking that has so often marred progress in the past. Even Sir Andrew Huxley (1980) has complained about the biochemists and biophysicists who periodically enter the field and work in it for a lifetime without ever doing their homework on its rich history. I hope his book will have directed their attention to past follies. I can only wish that the newcomers will at least glance through the present volume so that they may quickly see for themselves what manner of murky puddle this is into which they would step. We have great need of their contributions at many levels and on a great many different kinds of cell, from many different species. I hope they will neither feel inferior nor see their research go unsupported if the source of their cells lacks a backbone. But if it has one, may they not succumb to the temptation of feeling superior that in the past led so many into alleys that had blind endings.

References

Abbott, B. C. (1951) The heat production associated with the maintenance of a prolonged contraction and the extra heat produced during large shortening. *J. Physiol.* **112**, 438–445.

Abbott, B. C., and Baskin, R. J. (1962) Volume changes in frog muscle during contraction. *J. Physiol.* **161**, 379–391.

Abbott, B. C., and Lowy, J. (1956a) Resting tension in snail muscle. *Nature* **178**, 147–148.

Abbott, B. C., and Lowy, J. (1956b) Mechanical properties of *Pinna* adductor muscle. *J. Mar. Biol. Assoc.* **35**, 521–530.

Abbott, B. C., and Lowy, J. (1957) Stress relaxation in muscle. *Proc. R. Soc. Lond. Ser. B* **146**, 280–288.

Abbott, B. C., Aubert, X. M., and Hill, A. V. (1951) The absorption of work by a muscle stretched during a single twitch or a short tetanus. *Proc. R. Soc. Land. Ser. B* **139**, 86–117.

Abbott, B. C., Bigland, B., and Ritchie, J. M. (1952) The physiological cost of negative work. *J. Physiol.* **117**, 380–390.

Abbott, B. C., Lang, F., and Parnas, I. (1969) Physiological properties of the heart and cardiac ganglion of *Limulus polyphemus. Comp. Biochem. Physiol.* **28**, 149–158.

Abbott, R. H. (1973) The effects of fibre length and calcium ion concentration on the dynamic response of glycerol extracted insect fibrillar muscle. *J. Physiol.* **231**, 195–208.

Abbott, R. H. and Cage, P. E. (1978) Periodicity in insect flight muscle stretch activation. *J. Physiol.* **285**, 47P.

Abbott, R. H. and Leech, A. R. (1973) Persistence of adenylate kinase and other enzymes in glycerol extracted muscle. *Pflüg. Arch.* **344**, 233–243.

Acosta-Urquidi, J. (1978) Dormant synapses in a crustacean neuromuscular system. *Canad. J. Zool.* **56**, 463–468.

Adachihara, K., and Takagi, Y. (1974) A case of hypokalemic and hyperkalemic periodic paralysis with hyperthyroidism. *Clin. Neurol.* **14**, 369–375.

Adelman, W. J., Palti, Y., and Senft, J. P. (1973) Potassium ion accumulation in a periaxonal space and its effect on the measurement of membrane potassium conductance. *J. Memb. Biol.* **13**, 387–410.

Adrian, R. H., and Freygang, W. H. (1962a) The potassium and chloride conductance of frog muscle membrane. *J. Physiol.* **163**, 61–103.

Adrian, R. H., and Freygang, W. H. (1962b) Potassium conductance of frog muscle membrane under controlled voltage. *J. Physiol.* **163**, 104–114.

Adrian, R. H., and Peachey, L. D. (1965) The membrane capacity of frog twitch and slow muscle fibres. *J. Physiol.* **181**, 324–336.

Adrian, R. H., Chandler, W. K., and Hodgkin, A. L. (1969) The kinetics of mechanical activation in frog muscle. *J. Physiol.* **204**, 207–230.

Agarwal, R. A., Ligon, P. J. B., and Greenberg, M. J. (1972) The distribution of cardioactive agents among molluscan species and tissues. *Comp. Gen. Pharmacol.* **3**, 249–260.

Aidley, D. J. (1965) Transient changes in isotonic shortening velocity of frog rectus abdominis muscles in potassium contracture. *Proc. R. Soc. Lond. Ser. B* **163**, 215–223.

Albuquerque, E. X., Warnick, J. E., Tasse, J. R., and Sansone, F. M. (1972) Effects of vinblastine and colchicine on neural regulation of the fast and slow skeletal muscles of the rat. *Exp. Neurol.* **37**, 607–634.

Alexandrowicz, J. S. (1951) Muscle receptor organs in the abdomen of *Homarus vulgaris* and *Palinurus vulgaris. Q. J. Microsc. Sci.* **92**, 163–199.

Alexandrowicz, J. S., and Whitear, M. (1957) Receptor elements in the coxal region of *Palinurus vulgaris*. *J. Mar. Biol. Assoc.* **47**, 415–432.

Aljure, E. F., and Borrero, L. M. (1968) The influence of muscle length on the development of fatigue in toad sartorius. *J. Physiol.* **199**, 241-252.

Allen, D. G., and Blinks, J. R. (1978) Calcium transients in aequorin-injected frog cardiac muscle *Nature* **273**, 509–513.

Allen, D. G., and Kurihara, S. (1980) Intracellular calcium transients in mammalian cardiac muscle following a change in length. *J. Physiol.* **305**, 29P.

Allen, D. G., Blinks, J. R., and Prendergast, F. G. (1977) Aequorin luminescence: relation of light emission to calcium concentration – a calcium independent component. *Nature* **195**, 996–998.

Allen, E. R. (1978) Development of vertebrate skeletal muscle. *Am. Zool.* **18**, 101–111.

Allen, R. D., Cooledge, J. W., and Hall, P. J. (1960) Streaming in cytoplasm dissociated from the giant ameba *Chaos chaos*. *Nature* **187**, 896–897.

Almers, W. (1972) The decline of potassium permeability during extreme hyperpolarization in frog skeletal muscle. *J. Physiol.* **225**, 57–83.

Alvarado-Mallart, R-M., and Pincon-Raymond, M. (1979) The palisade endings of cat extraocular muscles: a light and electron microscope study. *Tissue and Cell* **11**, 567–584.

Alzaid, N., Goldspink, G., and Williams, P. E. (1980) Increase in collagen concentration in overloaded skeletal muscle. *J. Physiol.* **308**, 73P.

Amos, W. B. (1972) Structure and coiling of the stalk in the peritrich ciliates *Vorticella* and *Carchesium*. *J. Cell Sci.* **10**, 95–122.

Amphlett, G. (1975) Cross innervation and the regulatory protein system of rabbit soleus muscle. *Nature* **257**, 602–604.

Amsellem, J., and Clement, P. (1977) Correlations between ultrastructural features and contraction rates in rotiferan muscle. I. Preliminary observations on longitudinal retractor muscles in *Trichocerca rattus*. *Cell Tiss. Res.* **181**, 81–90.

Amsellem, J., and Nicaise, G. (1980) Ultrastructural study of muscle cells and their connections in the digestive tract of *Sepia officinalis*. *J. Submicrosc. Cytol.* **12**, 219–231.

Andersen, P., Jensen, J. K. S. and Løyning, Y. (1963) Slow and fast muscle fibers of hagfish. *Acta Physiol. Scand.* **57**, 167–179.

Anderson, M. (1968) Electrophysiological studies on initiation and reversal of the heart beat in *Ciona intestinalis*. *J. Exp. Biol.* **49**, 363–386.

Anderson, M. (1979) Ultrastructural evidence for disuse atrophy in insect skeletal muscle. *Cell Tiss. Res.* **203**, 503–506.

Anderson, M., and Castillo, J. del (1976) Electrical activity of the proventriculus of the polychaete worm *Syllis spongiphila*. *J. Exp. Biol.* **64**, 691–710.

Anderson, M. E., and Smith, D. S. (1971) Electrophysiological and structural studies on the heart muscle of the lobster *Homarus americanus*. *Tissue and Cell* **3**, 191–205.

Anderson, N. C., Ramon, P., and Snyder, A. (1971) Voltage clamp of uterine smooth muscle. *J. Gen. Physiol.* **58**, 322–339.

Anderson, P., and Henriksson, J. (1977) Training induced changes in the subgroups of human type II skeletal muscle fibres. *Acta Physiol. Scand.* **99**, 123–125.

Anderson, P. A. V., and Mackie, G. O. (1977) Electrically coupled, photosensitive neurons control swimming in a jellyfish. *Science* **197**, 186–188.

Anderson, W. W. (1980) Synaptic mechanisms generating nonspiking network oscillations in the stomatogastric ganglion of the lobster, *Panulirus interruptus*. Ph.D. thesis, University of Oregon.

Andersson-Cedergren, E. (1951) Ultrastructure of motor endplate and sarcoplasmic components of mouse skeletal muscle fiber as revealed by three-dimensional reconstruction from serial sections. *J. Ultrastruct. Res.* **1 (Suppl.)**, 1–191.

Andersson, K. E., and Edman, K. A. P. (1974a) Effects of lanthanum on the coupling between membrane excitation and contraction of isolated frog muscle fibres. *Acta Physiol. Scand.* **90**, 113–123. 113–123.

Andersson, K. E., and Edman, K. A. P. (1974b) Effects of lanthanum on potassium contractures of isolated twitch muscle fibres of the frog. *Acta Physiol. Scand.* **90**, 124-131.

April, E., Brandt, P. W., Reuben, J. P., and Grundest, H. (1968) Muscle contraction: The effect of ionic strength. *Nature* **220**, 182-184.

Arbas, E. A. (1980). A neuroethological study of flight loss and aerial maneuvering in insects. Ph.D. thesis, University of Oregon.

Armstrong, R. B., Ianuzzo, C. D., and Kunz, T. H. (1977) Histochemical and biochemical properties of flight muscle fibers in the little brown bat, *Myotis lucifugus J. Comp. Physiol.* **119**, 141-154.

Armitage, P. M., Tregear, R. T., and Miller, A. (1975) Effect of activation by calcium on the X-ray diffraction pattern from insect flight muscle. *J. Mol. Biol.* **92**, 39-53.

Arndt, I., and Pepe, F. A. (1975) Antigenic specificity of red and white muscle myosin. *J. Histochem. Cytochem.* **23**, 159-168.

Aronson, J. (1963) Observations on the variations in size of the A region of arthropod muscle. *J. Cell Biol.* **19**, 359-367.

Ashcroft, F. M., and Stanfield, P. R. (1980) Inactivation of calcium currents in skeletal muscle fibres of an insect depends on calcium entry. *J. Physiol.* **308**, 36P.

Ashcroft, F. M., and Stanfield, P. R. (1981) Calcium dependence of the inactivation of calcium currents in skeletal muscle fibres of an insect. *Science* **213**, 224-226.

Ashcroft, F. M., Standen, N. B., and Stanfield, P. R. (1979) Calcium currents in insect muscle. *J. Physiol.* **291**, 51P.

Ashley, C. C. (1967) The role of cell calcium in the contraction of single cannulated muscle fibers. *Am. Zool.* **7**, 647-659.

Ashley, C. C. (1970) Ion movements in skeletal muscle. In: *Membranes and Ion Transport*, Vol. 2, E. Biltov, Ed., pp. 1-316 Wiley, New York.

Ashley, C. C., and Caldwell, P. C. (1974) Calcium movements in relation to contraction. *Biochem. Soc. Symp.* **39**, 29-50.

Ashley, C. C., and Lignon, J. (1981) Aequorin responses during relaxation of tension of single muscle fibres stimulated by voltage clamp. *J. Physiol.* **311**, 19P.

Ashley, C. C., and Moisescu, D. G. (1977) Effect of changing the composition of the bathing solutions upon the isometric tension-pCa relationship in bundles of crustacean myofibrils. *J. Physiol.* **270**, 627-652.

Ashley, C. C., and Ridgway, E B. (1970) On the relationships between membrane potential, calcium transient and tension in single barnacle muscle fibres. *J. Physiol.* **209**, 105-130.

Ashley, C. C., Ellory, J. C., and Griffiths, P. J. (1977) Caffeine and the contractility of single muscle fibres from the barnacle *Balanus nubilus. J. Physiol.* **269**, 421-439.

Ashurst, D. E. (1971) The Z-line in insect flight muscle. *J. Mol. Biol.* **55**, 283-285.

Ashurst, D. E., and Vrbová, G. (1979) Experimentally induced differenciation of slow tonic and fast twitch muscles in the chick. *J. Cell. Sci.* **36**, 137-154.

Askanas, V., Engel, W. K., Reddy, N. B., Bath, P. G., Bethlem, J., Krauss, D. R., Hibberd, M. E., Lawrence, J. V., and Carter, L. S. (1979) X-linked recessive congenital muscle fiber hypotrophy with central nuclei. Abnormalities of growth and adenylate cyclase in muscle tissue culture. *Arch. Neurol.* **36**, 604-609.

Astbury, W. T. (1946) On the structure of biological fibres and the problem of muscle. *Proc. R. Soc. Lond. Ser. B* **134**, 303-328.

Atsumi, S. (1978) Sarcoplasmic reticulum and intracellular calcium localization at rest and during contraction in *Mytilus* smooth muscle. *Arch. Histol. Jap.* **41**, 239-258.

Atsumi, S., and Sugi, H. (1976) Localization of calcium-accumulating structures in the anterior byssal retractor muscle of *Mytilus edulis* and their role in the regulation of active and catch contractions. *J. Physiol.* **257**, 549-560.

Atwater, I., Rojas, E., and Vergara, J. (1974) Calcium influxes and tension development in perfused single barnacle muscle fibres under membrane potential control. *J. Physiol.* **243**, 523-551.

Atwood, H. L. (1973) An attempt to account for the diversity of crustacean muscles. *Am. Zool.* **13**, 357-378.

Atwood, H. L. (1976) Organization and synaptic physiology of crustacean neuromuscular systems. *Prog. Neurobiol.* 7, 291–391.

Atwood, H. L. (1977) Crustacean neuromuscular systems: past, present, and future. In: *Identified Neurons and Behavior of Arthropods.* G. Hoyle, Ed., pp. 9–29. Plenum, New York.

Atwood, H. L., and Hoyle, G. (1965) A further study of the paradox phenomenon of crustacean muscle. *J. Physiol.* **181**, 225–235.

Atwood, H. L., Hoyle, G., and Smyth, Jr., T. (1965) Mechanical and electrical responses of single innervated crab-muscle fibres. *J. Physiol.* **180**, 449–482.

Atwood, H. L., Swenarchuk, L. E., and Gruenwald, C. R. (1975) Long-term synaptic facilitation during sodium accumulation in nerve terminals. *Brain Res.* **100**, 198–204.

Atwood, H. L., Govind, C. K., and Jahromi, S. S. (1977) Excitatory synapses of blue crab gastric mill muscles. *Cell Tiss. Res.* **177**, 145–158.

Auber, J. (1960) Observations sur l'innervation motrice des muscles des Insectes. *Z. Zellforsch.* **51**, 705–724.

Auber, J. (1967a) Particularitiés ultrastructurales des myofibrilles des muscles du vol chez des lepidoptères. *C. R. Acad. Sci.* **264**, 621–624.

Auber, J. (1967b) Remarques sur la structure des fibrilles des muscles du vol d'insectes au niveau de la strie M. *C. R. Acad. Sci.* **264**, 2916–2918.

Auber, J. (1969) La myofibrillogenèse du muscle strié. I. Insectes. *J. Microscopie* **8**, 196–232.

Auber, J., and Couteaux, R. (1962) L'attache des myofilaments secondaires au niveau de la strie Z dans les muscles de Diptères. *C. R. Acad. Sci.* **254**, 3425–3426.

Auber, J., and Couteaux, R. (1963) Ultrastructure de la strie Z dans des muscles de Diptères. *J. Microscopie* **2**, 309–324.

Auber-Thomay, M. (1967) Modifications ultrastructurales au cours de la degenerescence et de la croissance de fibres musculaires chez un Insecte. *J. Microscopie.* **6**, 627–638.

Aubert, X., and Gilbert, S. H. (1980) Variation in the isometric maintenance heat rate with muscle length near that of maximum tension in frog striated muscle. *J. Physiol.* **303**, 1–8.

Axelsson, J., and Thesleff, S. (1957) A study of supersensitivity in denervated mammalian skeletal muscle. *J. Physiol.* **149**, 178–193.

Baccetti, B., and Rosati, F. (1968) On the thick filaments of Holothurian muscles. *J. Microscopie* **7**, 455–458.

Bach-y-Rita, P., and Ito, F. (1966) In vivo studies on fast and slow muscle fibers in cat extraocular muscles. *J. Gen. Physiol.* **49**, 1177–1198.

Bacq, Z. M., and Coppée, G. (1937) Réaction des vers et des mollusques à l'ésérine. Existence de nerfs cholinergiques chez les vers. *Arch. Int. Physiol.* **45**, 310–324.

Badley, R. A., Couchman, J. R., and Rees, D. A. (1980) Comparison of the cell cytoskeleton in migratory and stationary chick fibroblasts. *J. Muscle Res. Cell Motil.* **1**, 5–14.

Bagby, R. M. (1966) The fine structure of myocytes in sponges *Microciona prolifera* and *Tedania ignis. J. Morph.* **118**, 167–181.

Bagby, R. M., and Abercrombie, R. K. (1979) Observations on the three-dimensional organization of myofibrils in vertebrate smooth muscle. In: *Motility in Cell Function,* F. A. Pepe, J. W. Sanger, and V. T. Nachmias, Eds., pp. 427–431. Academic, New York.

Baguet, F. (1977) Cyclic-AMP and relaxation mechanism in molluscan catch muscle (ABRM). In: *Excitation-contraction coupling in smooth muscle,* R. Casteels, T. Godfraind, and J. C. Rüegg, Eds., pp. 407–414. Elsevier/North Holland, Amsterdam, New York and Oxford.

Bagust, J. (1974) Relationships between motornerve conduction velocities and motor unit contraction characteristics in a slow twitch muscle of the cat. *J. Physiol.* **238**, 269–278.

Bagust, J. (1979) The effects of tenotomy upon the contraction characteristics of motor units in rabbit soleus muscle. *J. Physiol.* **290**, 1–10.

Bagust, J., Knot, S., Lewis, D. M., Luck, J. C., and Westerman, R. A. (1973) Isometric contractions of motor units in a fast twitch muscle of the cat. *J. Physiol.* **231**. 87–104.

Bahler, A. S., Fales, J. T., and Zierler, K. L. (1967) The active state of mammalian skeletal muscle. *J. Gen. Physiol.* **50**, 2239–2253.

Bailey, K. (1946) Tropomyosin: a new asymmetric protein component of muscle. *Nature* **157**, 368–370.

Bailey, K. (1956) The proteins of adductor muscles. *Publ. Staz. Zool. Napoli* **29**, 96–105.

Bainbridge, F. A. (1915) The influence of venous filling upon the rate of the heart. *J. Physiol.* **50**, 65–84.

Bajer, A. (1973) Interaction of microtubules and the mechanism of chromosome movement. The "zipper hypothesis." *Cytobios* 8, 139–160.

Bannister, L. H., and Tatchell, E. C. (1968) Contractility and the fibre systems of *Stentor coeruleus*. *J. Cell Sci.* **3**, 295–308.

Bárány, M. (1967) ATPase activity of myosin correlated with speed of muscle shortening. *J. Gen. Physiol.* **50**, 197–218.

Bárány, K., Bárány, M., Gillis, J. M., and Kushmerick, M. J. (1980) Myosin light chain phosphorylation during the contraction cycle of frog muscle. *Fed. Proc.* **39**, 1547–1551.

Barden, J. A., and Mason, P. (1978) Muscle crossbridge stroke and activity revealed by optical diffraction. *Science* **199**, 1212–1213.

Barnes, G. E. (1955) The behaviour of *Anodonta cygnea* L. and its neurophysiological basis. *J. Exp. Biol.* **32**, 158–174.

Barnes, W. J. P. (1975) Nervous control of locomotion in Crustacea. In: *"Simple" Nervous Systems*, P. N. R. Usherwood and D. R. Newth, Eds., pp. 415–441. Arnold, London.

Barnes, W. J. P. (1977) Proprioceptive influences on motor output during walking in the crayfish. *J. Physiol. Paris.* **73**, 543–564.

Barnes, W. J. P., Spirito, C. P., and Evoy, W. H. (1972) Nervous control of walking in a crab, *Cardiosoma guanhumi*. II. Role of resistance reflexes in walking. *Z. Verg. Physiol.* **76**, 16–31.

Barr, L., and Laverack, E. (1981) Responsivity of the amphibian iris to light. *J. Gen. Physiol.* **78**, 12a.

Barrantes, F. J. (1970) The neuromuscular junctions of a pulmonate mollusc. *Z. Zellforsch.* **104**, 205–212.

Barrett, J. N., and Barrett, E. F. (1978) Excitation-contraction coupling in skeletal muscle – blockade by high extracellular concentrations of calcium buffers. *Science* **200**, 1270–1272.

Basar, E., and Eroglu, C. (1976) Spectral analysis of spontaneous activity in smooth muscles. In: *Physiology of Smooth Muscle* E. Bülbring and M. F. Shuba, Eds., pp. 137–146. Raven, New York.

Baskin, R. J., and Paolini, P. J. (1966) Muscle volume changes. *J. Gen. Physiol.* **49**, 387–404.

Basmajian, J. V. (1963) Control and training of individual motor units. *Science* **141**, 440–441.

Bass, A., Gutmann, E., Melichna, J., and Syrový, I. (1973) Contractile and enzymatic properties of fast and slow muscles of rabbit and hare. *Physiol. Bohemoslov.* **22**, 477–486.

Bassingthwaite, J. B., Beeler, G. W., Sidell, P. M., Reuter, H., and Safford, R. E. (1973). In: *Regulation and Control in Physiological Systems*, pp. 36–38. ISA, Pittsburgh.

Bassingthwaite, J. B., Fry, C. H., and McGuigan, J. A. S. (1976) Relationship between internal calcium and outward current in mammalian ventricular muscle; A mechanism for the control of the action potential duration. *J. Physiol.* **262**, 15–37.

Bastian, J., and Nakajima, S. (1974) Action potential in the transverse tubules and its role in the activation of skeletal muscle. *J. Gen. Physiol.* **63**, 257–278.

Batham, E. J., Pantin, C. F. A., and Robson, E. A. (1960) Nerve-net of the sea-anemone *Metridium senile:* the mesenteries and the column. *Q. J. Micr. Sci.* **101**, 487–510.

Baumgarten, H., Bjorklund, L., Lachenmayer, L., Nobin. A., and Rosengren, E. (1973) Evidence for existence of serotonin, dopamine and noradrenaline-containing neurons in the gut of *Lampetra fluviatilis*. *Z. Zellforsch.* **141**, 33–54.

Bawa, P., and Tatton, W. G. (1979) Motor unit response in muscles stretched by imposed displacements of the monkey wrist. *Exp. Brain Res.* **37**, 417–437.

Baylor, S. M., and Oetliker, H. (1977a) A large birefringence signal preceding contraction in single twitch fibres of the frog. *J. Physiol.* **264**, 141–162.

Baylor, S. M., and Oetliker, H. (1977b) Birefringence signals from surface and T-system membranes of frog single muscle fibres. *J. Physiol.* **264**, 199–213.

Becht, G. (1959) Studies on insect muscles. *Bijdr. Dierk.* **29**, 1-36.

Becht, G., and Dresden, D. (1956) Physiology of the locomotory muscles in the cockroach. *Nature* **177**, 836–837.

Becht, G., Hoyle, G., and Usherwood, P. N. R. (1960) Neuromuscular transmission in the coxal muscles of the cockroach. *J. Ins. Physiol.* **4**, 191–201.

Becht, H. (1969) Seasonal changes in the contents of inorganic ions in the blood of *Rana esculenta. Neth. J. Zool.* **19**, 610–620.

Becker, R. P., and Lombard, R. E. (1977) Structural correlates of function in the "opercularis" muscle of amphibians. *Cell Tiss. Res.* **175**, 499-522.

Beeler, G. W., and Reuter, H. (1970) The relation between membrane potential, membrane currents and activation of contraction in ventricular myocardial fibres. *J. Physiol.* **207**, 211–229.

Begenisich, T. (1975) Magnitude and location of surface charges on *Myxicola* giant axons. *J. Gen. Physiol.* **66**, 47–65.

Behrendt, H. (1977) Effects of anabolic steroids on rat heart muscle cell. *Cell Tiss. Res.* **180**, 303–315.

Beinbrecht, G., Kuhn, H. J., and Rüegg, J. C. (1972) Electron microscope and optical diffraction studies on glycerol-extracted insect flight muscle fibres relaxed by pyrophospate. *Experientia* **28**, 511-513.

Bekoff, A., and Betz, W. J. (1976) Acetylcholine hot spots: development on myotubes cultivated from aneural limb buds. *Science* **193**, 915–918.

Bekoff, A., and Betz, W. J. (1977) Physiological properties of dissociated muscle fibres obtained from innervated and denervated adult rat muscle. *J. Physiol.* **271**, 25–40.

Belton, P., and Grundfest, H. (1962) Potassium activation and K spikes in muscle fibers of the mealworm larva (*Tenebrio molitor*). *Am. J. Physiol.* **203**, 588–594.

Bendall, J. R. (1958) Muscle-relaxing factors. *Nature* **181**, 1188–1190.

Bendel, U. (1979) New system theory of muscle mechanics. *Stud. Biophys.* **75**, 31–40.

Bennett, T. (1970) Interaction of nerve-mediated excitation and inhibition of single smooth muscle cells of the avian gizzard. *Comp. Biochem. Physiol.* **32**, 669–680.

Bennett-Clark, H. C., and Lucey, E. C. A. (1967) The jump of the flea: a study of the energetics and a model of the mechanism. *J. Exp. Biol.* **47**, 59–76.

Bentley, D. R. (1969) Intracellular activity in cricket neurons during generation of song patterns. *Z. Vergl. Physiol.* **62**, 267–283.

Bentley, D. R. (1970) A topological map of the locust flight system motor neurons. *J. Ins. Physiol.* **16**, 905–918.

Bentley, D. R. (1971) Genetic control of an insect neuronal network. *Science* **174**, 1139–1141.

Berl, S., Puszkin, S., and Nicklas, W. J. (1973) Actomyosin-like protein in brain: function in the release of transmitter at synaptic endings. *Science* **179**, 441–446.

Berridge, M. J., and Rapp, P. E. (1979) A comparative survey of the function, mechanism and control of cellular oscillations. *J. Exp. Biol.* **81**, 217–280.

Betz, W. J. (1976) The formation of synapses between chick embryo skeletal muscle and ciliary ganglia grown *in vitro. J. Physiol.* **254**, 63–73.

Betz, W. J., Caldwell, J. H., and Ribchester, R. R. (1980a) The effects of partial denervation at birth on the development of muscle fibres and motor units in rat lumbrical muscle. *J. Physiol.* **303**, 265–279.

Betz, W. J., Caldwell, J. H., and Ribchester, R. R. (1980b) Sprouting of active nerve terminals in partially inactive muscles of the rat. *J. Physiol.* **303**, 281–297.

Bezanilla, F., Caputo, C., Gonzalez-Serratos, H., and Venosa, R. A. (1972) Sodium dependence of the inward spread of activation in isolated twitch muscle fibres of the frog. *J. Physiol.* **223**, 507–523.

Bianchi, P. (1960) The effect of caffeine on radiocalcium movement in frog sartorius. *J. Gen. Physiol.* **44**, 845–855.

Biedermann, W. (1887) Beitrage zur allgemeinen Nerven and Muskel Physiologie. XX. Über die Innervation der Krebsschere. *Akad. Wiss. Wien Abt. 3* **95**, 7–40.

Bigland-Ritchie, B., and Woods, J. J. (1976) Integrated electromyogram and oxygen uptake during positive and negative work. *J. Physiol.* **260**, 267–278.

Birks, R. I., and Davey, D. F. (1972) An analysis of volume changes in the T-tubes of frog skeletal muscle exposed to sucrose. *J. Physiol.* **222**, 95–111.

Bittner, G. D. (1968) The differentiation of crayfish muscle fibers during development. *J. Exp. Zool.* **167**, 439–456.

Bittner, G. D. (1973) Trophic dependence of fiber diameter in a crustacean muscle. *Exp. Neurol.* **41**, 38–53.

Bizzi, E., Dev, P., Morasso, P., and Polit, A. (1978) Effect of load disturbances during centrally initiated movements. *J. Neurophysiol.* **41**, 542–556.

Blaschko, H., Cattell, M., and Kahn, J. L. (1931) On the nature of the two types of response in the neuromuscular system of the crustacean claw. *J. Physiol.* **73**, 25–35.

Blinks, J. R., Rüdel, R., and Taylor, S. R. (1978) Calcium transients in isolated amphibian skeletal muscle fibres: detection with aequorin. *J. Physiol.* **277**, 291–323.

Blum, J. J., Creese, R., Jenden, D. J., and Scholes, N. W. (1957) The mechanism of action of ryanodine on skeletal muscle. *J. Pharmacol. Exp. Ther.* **121**, 447–486.

Bock, W. J., and Hikada, R. S. (1968) An analysis of twitch and tonus fibers in the hatching muscle. *Condor.* **70**, 211–222.

Boeke, J. (1949) The sympathetic endformation, its synaptology, the interstitial cells, the periterminal network, and its bearing on the neurone theory. Discussion and critique. *Acta Anat.* **8**, 18–61.

Boesiger, B. (1951) Comparaison histochimique et biometrique du muscle peroneus longus et de son innervation chez *Gallus gallus* et *Coturnix c. japonica. Acta Anat.* **93**, 100–114.

Bolton, T. B. (1979) Mechanisms of action of transmitters and other substances on smooth muscles. *Physiol. Rev.* **59**, 606–718.

Boltt, R. E., and Ewer, D. W. (1963) Studies on the myoneural physiology of Echinodermata. V. The lantern retractor muscle of *Parechinus*: responses to drugs. *J. Exp. Biol.* **40**, 727–733.

Bone, Q. (1966) On the function of the two types of myotomal muscle fibre in elasmobranch fish. *J. Mar. Biol. Assoc. U. K.* **46**, 321–349.

Bone, Q., and Mackie, G. O. (1975) Skin impulses and locomotion in *Oikopleura* (Tunicata: Larvacea). *Biol. Bull.* **149**, 267–286.

Bone, Q., and Ryan, K. P. (1973) The structure and innervation of the locomotor muscles of salps (Tunicata: Thaliacea). *J. Mar. Biol. Assoc. U. K.* **53**, 873–883.

Bonke, F. I. M. (1973) Electrotonic spread in the sinoatrial node of the rabbit heart. *Pflüg. Arch.* **339**, 17–23.

Bonner, R. F., and Carlson, F. D. (1975) Structural dynamics of frog muscle during isometric contraction. *J. Gen. Physiol.* **65**, 555–581.

Boone, L. P., and Bittner, G. D. (1974) Morphological and physiological measures of trophic dependence in a crustacean muscle. *J. Comp. Physiol.* **89**. 123–144.

Booth, F. W., and Kelso, J. R. (1973) Effect of hind-limb immobilization on contractile and histochemical properties of skeletal muscle. *Pflüg. Arch.* **342**, 231–238.

Borejdo, J., and Morales, M. F. (1977) Fluctuations in tension during contraction of single muscle fibers. *Biophys. J.* **20**, 315–334.

Borg, T. K., and Caulfield, J. B. (1980) Morphology of connective tissue in skeletal muscle. *Tissue & Cell* **12**, 197–208.

Borovikov, Y. S., and Chernogriadskaie, N. A. (1979) Studies on conformational changes in F-actin of glycerinated muscle fibers during relaxation by means of polarized ultraviolet fluorescence microscopy. *Microsc. Acta* **81**, 383–392.

Bossen, E. H., Sommer, J. R., and Waugh, R. A. (1978) Comparative stereology of the mouse and finch left ventricle. *Tissue & Cell* **10**, 773–784.

Botham, R. P., Beadle, D. J., Hart, R. J., Potter, C., and Wilson, R. G. (1979a) Effects of stimulation and rest on the ultrastructure of the excitatory neuromuscular junctions of *Locusta migratoria* L. *Cell Tiss. Res.* **203**, 367–371.

Botham, R. P., Beadle, D. J., Hart, R. J., Potter, C., and Wilson, R. G. (1979b) Changes in the distribution and size of synaptic vesicles in neuromuscular junctions of *Locusta migratoria* after stimulation and rest. *Cell Tiss. Res.* **203**, 373–378.

Bouligand, Y. (1962) Les ultrastructures du muscles strié et des ses attaches au squelette chez les *Cyclops* (Crustacés copépodes). *J. Microscopie* **1**, 377–394.

Bouligand, Y. (1963) Les ultrastructures musculaires des *Copépodes*. II. Membrane sarcoplasmique, reticulum sarcoplasmique et jonction neuro-musculaire chez les *Cyclops*. *J. Microscopie* **2**, 197–212.

Bowden, J. (1958) The structure and innervation of Lamellibranch muscle. *Int. Rev. Cytol.* **7**, 295–335.

Bowerman, R. F., and Burrows, M. (1980) The morphology and physiology of some walking leg motor neurons in a scorpion. *J. Comp. Physiol.* **140**, 31–42.

Bowman, W. (1840) On the minute structure and movements of voluntary muscle. *Phil. Trans. R. Soc. Lond. Ser. B* **130**, 457-501.

Boyd, I. A., and Palhak, C. L. (1965) The response of perfused frog hearts to minute quantities of acetylcholine, and variation in sensitivity with seasons. *J. Physiol.* **176**, 191-204.

Bozler, E. (1947) The response of smooth muscle to stretch. *Am. J. Physiol.* **149**, 299-301.

Brading, A., Bülbring, E., and Tomita, T. (1969) The effect of sodium and calcium on the action potential of the smooth muscle of the guinea-pig taenia coli. *J. Physiol.* **200**, 637–654.

Brading, A. F., and Caldwell, P. C. (1971) The resting membrane potential of the somatic muscle cells of *Ascaris lumbricoides*. *J. Physiol.* **217**, 605–624.

Bradley, W. G., and Fulthorpe, J. J. (1978) Studies of sarcolemmal integrity in myopathic muscle. *Neurology* **28**, 670–677.

Brady, A. J. (1967) Length–tension relations in cardiac muscle. *Am. Zool.* **7**, 603–610.

Brady, A. J. (1968) Active state in cardiac muscle. *Physiol. Rev.* **48**, 570–600.

Brandt, P. W., and Orentlicher, M. (1972) Muscle energetics and the Fenn effect. *Biophys. J.* **12**, 512-527.

Brandt, P. W., Reuben, J. P., and Grundfest, H. (1968) Correlated morphological and physiological studies on isolated single muscle fibers. II. The properties of the crayfish transverse tubular system: localization of the sites of reversible swelling. *J. Cell Biol.* **39**, 451-467.

Brann, L., Dewey, M. M., Baldwin, E. A., Brink, P., and Walcott, B. (1979) Requirements for *in vitro* shortening and lengthening of isolated thick filaments of *Limulus* striated muscle. *Nature* **279**, 256-257.

Branton, D., Bullivant, S., Gilula, N. B., Karnovsky, M. J., Moor, H., Mullenthaler, K., Northcote, D. H., Packer, L., Sater, P., Speth, V., Staehelin, L. A., Steere, R. L., and Weinstein, R. S. (1975) Freeze-etching nomenclature. *Science* **190**, 54-56.

Bray, D. (1973) Model for membrane movements in the neural growth cone. *Nature* **244**, 93–95.

Bray, J. J., Hawken, M., Hubbard, J. I., Pockett, S., and Wilson, L. (1976) The membrane potential of rat diaphragm muscle fibres and the effect of denervation. *J. Physiol.* **255**, 651-667.

Bray, J. J., Hubbard, J. I. and Milles, R. G. (1979) The trophic influence of tetrodotoxin-inactive nerves on normal and reinnervated rat skeletal muscles. *J. Physiol.* **297**, 479–491.

Bremel, R. D., Murray, J. M., and Weber, A. (1973) Manifestations of cooperative behavior in the regulated actin filament during actin-activated ATP hydrolysis in the presence of calcium. *C. Spr. Harb. Symp. Quant. Biol.* **37**, 267–275.

Brenner, B. (1979) An indirect proof of stretch-induced Ca^{2+} release from the sarcoplasmic reticulum in glycerinated skeletal and heart muscle preparations. *Basic Res. Cardiol.* **74**, 177–202.

Brenner, H. R. (1972) Evidence for peripheral inhibition in an arachnid muscle. *J. Comp. Physiol.* **80**, 227–231.

Brenner, S. (1974) The genetics of *Caenorhabditis elegans*. *Genetics* **77**, 71-94.

Bressler, B. H., Phillis, J. W., and Kozachuk, W. (1975) Noradrenaline stimulation of a membrane pump in frog skeletal muscle. *Europ. J. Pharmacol. (Amst.)* **33**, 201-204.

Brevet, A., Pinto, E., Peacock, J., and Stockdale, F. E. (1976) Myosin synthesis increased by electrical stimulation of skeletal muscle cell cultures. *Science* **193**, 1152-1154.

Briden, K. L., and Alpert, N. R. (1972) The effect of shortening on the time-course of active state decay. *J. Gen Physiol.* **60**, 202-220.

Brinley, Jr., F. J. (1968) Sodium and potassium fluxes in isolated barnacle muscle fibers. *J. Gen Physiol.* **51**, 445-447.

Briskey, E. J. (1966) The modification of actomyosin by α-actinin. III. The interaction between α-actinin and actin. *Biochim. Biophys. Acta* **133**, 424-434.

Brooks, M. A. (1957) Succinoxidase in the american cockroach. *Ann. Ent. Soc. Am.* **50**, 122-125.

Brown, A. M. (1967) Cardiac sympathetic adrenergic pathways in which synaptic transmission is blocked by atropine sulphate. *J. Physiol.* **191**, 271-288.

Brown, B. E. (1967) Neuromuscular transmitter substance in insect visceral muscle. *Science* **155**, 595-597.

Brown, B. E. (1975) Proctolin: a peptide transmitter candidate in insects. *Life Sci.* **17**, 1241-1252.

Brown, B. E., and Nagai, T. (1969) Insect visceral muscle: Neural relations of the proctodeal muscles of the cockroach. *J. Ins. Physiol.* **15**, 1767-1783.

Brown, H. F. (1964) Electrophysiological investigations of the heart of *Squilla mantis.* II. The heart muscle. *J. Exp. Biol.* **41**, 701-722.

Brown, H., and Di Francesco, D. (1980) Voltage-clamp investigations of membrane currents underlying pace-maker activity in rabbit sino-atrial node. *J. Physiol.* **308**, 331-351.

Brown, H., Di Francesco, D., and Noble, S. (1979) Cardiac pacemaker oscillation and its modulation by autonomic transmitters. *J. Exp. Biol.* **81**, 175-204.

Brown, T., Chance, E. M., Dawson, M. J., Gadian, D. G., Radda, G. K., and Wilkie, D. R. (1980) The activity of creatine kinase in frog skeletal muscle studied by saturation transfer nuclear magnetic resonance. *J. Physiol.* **305**, 84P.

Browne, J. S. (1976) The contractile properties of slow muscle fibres in sheep extraocular muscle. *J. Physiol.* **254**, 535-550.

Bruce, D. S., and Nicol, C. J. M. (1981) Contractile and fibre population changes induced by hyperthryroidism in rat skeletal muscle. *J. Physiol.* **310**, 75P.

Brulé, G., Guilbault, P., and Haudecoeur, G. (1976) Delayed rectification in the skeletal muscle fibre of the crab. *J. Physiol.* **263**, 21P.

Bryan, G. W. (1967) Zinc concentrations of fast and slow contracting muscles in the lobster. *Nature* **213**, 1043-1044.

Bryant, S. H. (1973) The electrophysiology of myotonia with a review of congenital myotonia of goats. In: *New Developments in Electromyography and Clinical Neurophysiology,* Vol. I., J. E. Desmedt, Ed., pp. 420-450. Karger, Basel.

Buchthal, F., and Rosenfalck, P. (1960) Dynamic elasticity in the initial phase of an isotonic twitch. *Acta Physiol. Scand.* **49**, 198-210.

Buchthal, F., and Schmalbruch, H. (1980) Motor unit of mammalian muscle. *Physiol. Rev.* **60**, 90-142.

Buchthal, F., Kaiser, E., and Rosenfalck, P. (1951) The rheology of the cross-striated muscle fibre. *Biol. Medd. Kbh.* **21**, 1-318.

Buchthal, F., Weis-Fogh, T., and Rosenfalck, P. (1957) Twitch contractions of isolated flight muscle of locusts. *Acta Physiol. Scand.* **39**, 246-276.

Bülbring, E. (1954) Membrane potentials of smooth muscle fibres of the taenia coli of the guinea-pig. *J. Physiol.* **125**, 302-315.

Bülbring, E. (1955) Correlation between membrane potential, spike discharge and tension in smooth muscle. *J. Physiol.* **128**, 200-221.

Bülbring, E., and Hooton, I. N. (1954) Membrane potentials of smooth muscle fibres in the rabbits sphincter pupillae. *J. Physiol.* **125**, 292-301.

Bullard, B., Luke, B., and Winkelman, L. (1973) The paramyosin of insect flight muscle. *J. Mol. Biol.* **75**, 359-367.

Bullard, B., Pringle, J. W. S., and Sainsbury, G. (1981) Localization of a new protein in the Z discs of insect flight and vertebrate striated muscle. *J. Physiol.* **317**, 19P.

Buller, A. J., Eccles, J. C., and Eccles, R. M. (1960) Differentiation of fast and slow muscles in the cat hind limb. *J. Physiol.* **150**, 399-416.

Bullock, T. H. (1943) Neuromuscular facilitation in Scyphomedusa. *J. Cellular Comp. Physiol.* **22**, 251-272.

Burden, S. J., Hartzell, H. C., and Yoshikami, D. (1975) Acetylcholine receptors at neuromuscular synapses: phylogenetic differences detected by snake α-neurotoxins, *Proc. Nat. Acad. Sci.* **72**, 3245-3249.

Burke, R. E., Levine, D. N., Tsairis, P., and Zajac, III, F. E., (1973) Physiological types and histochemical profiles in motor units of the cat gastrocnemius. *J. Physiol.* **234**, 723-748.

Burke, R. E., Levine, D. N., Salcman, M., and Tsairis, P. (1974) Motor units in cat soleus muscle: Physiological, histochemical and morphological characteristics. *J. Physiol.* **238**, 503-514.

Burke, W., and Ginsborg, B. L. (1956) The action of the neuromuscular transmitter on the slow fibre membrane. *J. Physiol.* **132**, 599-612.

Burnstock, G. (1958) Reversible inactivation of nervous activity in a fish gut. *J. Physiol.* **141**, 35-45.

Burnstock, G. (1972) Purinergic nerves. *Pharmacol. Rev.* **24**, 509-581.

Burnstock, G., and Holman, M. (1961) The transmission of excitation from autonomic nerve to smooth muscle. *J. Physiol.* **155**, 115-133.

Burnstock, G., and Prosser, C. L. (1960a) Conduction in smooth muscles: comparative electrical properties. *Am. J. Physiol.* **199**, 553-559.

Burnstock, G., and Prosser, C. L. (1960b) Delayed repolarization in smooth muscles. *Proc. Soc. Exp. Biol. Med.* **103**, 269-270.

Burnstock, G., Holman, M. E., and Prosser, C. L. (1963a) Electrophysiology of smooth muscle. *Physiol. Rev.* **43**, 482-527.

Burnstock, G., Campbell, G., Bennett, M., and Holman, M. E. (1963b) Inhibition of the smooth muscle of the taenia coli. *Nature* **200**, 581-582.

Burnstock, G., Holman, M. E., and Kuriyama, H. (1964) Facilitation of transmission from autonomic nerve to smooth muscle of guinea-pig vas deferens. *J. Physiol.* **172**, 31-49.

Burridge, K., and Phillips, J. H. (1975) Association of actin and myosin with secretory granule membranes. *Nature* **254**, 526-529.

Burrini, A. G., Baccetti, B., Campanella, C., Runger-Brandle, E., and Gabbiani, G. (1980) Pericentriolar actin in spermatozoa. *J. Submicrosc. Cytol.* **12**, 161-164.

Burrows, M. (1977) Flight mechanisms of the locust. In: *Identified Neurons and Behavior of Arthropods*, G. Hoyle, Ed., pp. 339-356. Plenum, New York.

Burrows, M., and Hoyle, G. (1972) Neuromuscular physiology of the strike mechanism of the mantis shrimp, *Hemisquilla*, *J. Exp. Zool.* **179**, 379-394.

Burrows, M., and Hoyle, G. (1973) The mechanism of rapid running in the ghost crab, *Ocypode ceratophthalma*. *J. Exp. Biol.* **58**, 327-349.

Busch, W. A., Stromer, M. H., Goll, D. E., and Suzuki, A. (1972) Ca^{2+}-specific removal of Z lines from rabbit skeletal muscle. *J. Cell Biol.* **52**, 367-381.

Bush, B. M. H., Vedel, J. P., and Clarac, F. (1978) Intersegmental reflex actions from a joint sensory organ (CB) to a muscle receptor (MCO) in decapod crustacean limbs. *J. Exp. Biol.* **73**, 47-63.

Butler, T. M., Siegman, M. J., Mooers, S. U., and Davies, R. E. (1977) High energy phosphate utilization during force development and force maintenance in mammalian smooth muscle. In: *Excitation-Contraction Coupling in Smooth Muscle*, R. Casteels, T. Godfraind, and J. C. Rüegg, Eds., pp. 463-469. Elsevier, Amsterdám.

Buy, Du H. G. (1936) The physiology of an invertebrate smooth muscle (retractor of *Thyone briareus*). *Am. J. Physiol.* **116**, 22-30.

Cachon, J., Cachon, M., Tilney, L. G., and Tilney, M. S. (1977) Movement generated by interactions between the dense material at the ends of microtubules and non-actin-containing microfilaments in *Sticholonche zanclea*. *J. Cell Biol.* **72**, 314–338.

Cain, D. F., Infante, A., and Davies, R. E. (1962) Adenosine triphosphate and phosphorylcreatine as energy suppliers for single contractions of working muscle *Nature* **296**, 214–216.

Calabrese, R. L. (1979) Neural generation of the peristaltic and non-peristaltic heartbeat coordination modes in the leech, *Hirudo medicinalis*. *Am. Zool.* **19**, 87–102.

Caldwell, P. C. (1958) Studies on the internal pH of large muscle and nerve fibres. *J. Physiol.* **142**, 22–62.

Campbell, G. (1966) Nerve-mediated excitation of the taenia of the guinea-pig caecum. *J. Physiol.* **185**, 148–159.

Campbell, G. R., Uehara, Y., Mark, G., and Burnstock, G. (1971) Fine structure of smooth muscle cells grown in tissue culture. *J. Cell Biol.* **49**, 21–34.

Campos, H. A., and Urquilla, P. R. (1969) Action of cocaine and chronic sympathetic denervation on vagal escape. *J. Physiol.* **200**, 311–320.

Cannon, W. B., and Rosenblueth, A. (1937) *Autonomic Neuro-effector Systems*. Macmillan, New York.

Cannon, W. B., and Rosenblueth, A. (1949) *The Supersensitivity of Denervated Structures*. Macmillan, New York.

Cannone, A. J., and Bush, B. M. H. (1980) Reflexes mediated by non-impulsive afferent neurones of thoracic-coxal muscle receptor organs in the crab, *Carcinus maenas*. II. Reflex discharge evoked by current injection. *J. Exp. Biol.* **86**, 305–331.

Caputo, C. (1972a) The effect of low temperatures on the excitation-contraction coupling phenomena of frog single muscle fibres. *J. Physiol.* **223**, 461–482.

Caputo, C. (1972b) The time course of potassium contractures of single muscle fibres. *J. Physiol.* **223**, 483–505.

Caputo, C., (1976) The effect of caffeine and tetracaine on the time course of potassium contractures of single muscle fibres. *J. Physiol.* **255**, 191–27.

Caputo, C., and Gimenez, M. (1967) Effects of external calcium deprivation on single muscle fibers. *J. Gen. Physiol.* **50**, 2177–2195.

Carew, T. J., Pinsker, H., Rubison, K., and Kandel, E. R. (1974) Physiological and biochemical properties of neuromuscular transmission between identified motoneurons and gill muscle in *Aplysia*. *J. Neurophysiol.* **37**, 1020–1040.

Carlson, F. D., and Wilkie, D. R. (1974) *Muscle Physiology*. Prentice-Hall, Englewood Cliffs, N. J.

Carnevali, M. D. C., and Ferraguti, M. (1979) Structure and ultrastructure of muscles in the Priapulid *Halicryptus spinulosus*: functional and phylogenetic remarks. *J. Mar. Biol. Ass. U. K.* **59**, 737–747.

Carvalho, A. P. (1968) Effects of potentaitors of muscular contraction on binding of cations by sarcoplasmic reticulum. *J. Gen. Physiol.* **51**, 427–442.

Casella, C. (1951) Tensile force in total striated muscle, isolated fibre and sarcolemma. *Acts Physiol. Scand.* **21**, 380–401.

Caspar, D. L. D., Cohen, C., and Longley, W. (1969) Tropomyosin: crystal structure, polymorphism and molecular interactions. *J. Mol. Biol.* **41**, 87–107.

Cassens, R. G., Hockstra, W. G., Faltin, E. C., and Briskey, E. J. (1967) Zinc content and subcellular distribution in red vs. white porcine skeletal muscle. *Am. J. Physiol.* **212**, 688–692.

Castellani, L. C., Alberici, A., Bezzi, E., and Landoni, C. (1978) Structure and biochemical analysis of paramyosin from annelid muscles. *J. Submicrosc Cytol.* **10**, 315–325.

Castillo, J. del, and Katz, B. (1953) Statistical nature of "facilitation" at a single nerve-muscle junction. *Nature* **171**, 1016–1017.

Castillo, J. del, and Katz, B. (1954) Quantal components of the endplate potential. *J. Physiol.* **124**, 560–573.

Castillo, J. del, and Katz, B. (1955a) Production of membrane potential changes in the frog's heart by inhibitory nerve impulses. *Nature* **175**, 1035.

Castillo, J. del, Katz, B. (1955b) On the localization of acetylcholine receptors. *J. Physiol.* **128**, 157–181.

Castillo, J. del, Katz, B. (1955c) Production of membrane potential changes in the frog's heart by inhibitory nerve impulses. *Nature* **175**, 1035–1036.

Castillo, J. del, and Morales, T. (1967) The electrical and mechanical activity of the esophageal cell of *Ascaris lumbricoides. J. Gen Physiol.* **50**, 603–629.

Castillo, J. del, Morales, T. A., and Sanchez, V. (1963) Action of piperazine on the neuromuscular system of *Ascaris lumbricoides. Nature* **200**, 706-707.

Castillo, J. del, de Mello, W. C., and Morales, T. (1967) The initiation of action potentials in the somatic musculature of *Ascaris lumbricoides. J. Exp. Biol.* **46**, 263-279.

Castillo, J. del, Anderson, M., and Smith, D. S. (1972) Proventriculus of a marine annelid: Muscle preparation with the longest recorded sarcomere. *Proc. Nat. Acad. Sci.* **69**, 1669–1672.

Cavey, M. J., and Cloney, R. A. (1972) Fine structure and differentiation of ascidian muscle. I. Differentiated caudal musculature of *Distaplia occidentalis* tadpoles. *J. Morph.* **138**, 349–374.

Cavey, M. J., and Cloney, R. A. (1976) Ultrastructure and differentiation of Ascidian muscle. *Cell Tiss. Res.* **174**, 289–313.

Cecchi, C., Colomo, F., Lambardi, V. and Piazzesi, G. (1979) Development of activation and rise of tension in an isometric tetanus. *Pflüg. Arch.* **381**, 71–74.

Cederblad, G., Bylund, A. -C., Holm, J., and Schersten, T. (1976) Carnitine concentration in relation to enzyme activities and substrate utilization in human skeletal muscles *J. Clin. Lab. Invest.* **36**, 547–552.

Chain, B. M. (1981) A sodium-dependent twitch muscle in a coelenterate. The ectodermal myoepithelium of the gastrozooids in *Algama* sp. (Siphonophora). *J. Exp. Biol.* **90**, 101-108.

Chambers, R., and Hale, H. P. (1932) The formation of ice in protoplasm. *Proc. R. Soc. Lond. Ser. B* **110**, 336-352.

Chandler, W. K., Rakowski, R. F., and Schneider, M. F. (1976) Effects of glycerol treatment and maintained depolarization on charge movement in skeletal muscle. *J. Physiol.* **254**, 286–316.

Chapman, D. M., Pantin, C. F. A., and Robson, E. A. (1962) Muscle in coelenterates. *Rev. Canad. Biol.* **21**, 267–278.

Chapman, J. B., and Gibbs, C. L. (1972) Energetics of isometric and isotomic twitches in toad sartorius. *Biophys. J.* **12**, 215–226.

Chapman, R. A. (1979) Excitation-contraction coupling in cardiac muscle. *Prog. Biophys. Mol. Biol.* **35**, 1-52.

Chaplain, R. A. (1966) Indication for an allosteric effect of adenosine diphosphate in actomyosin gels from insect fibrillar flight muscle. *Arch. Biochem. Biophys.* **115**, 450–461.

Chari, N., Bass, A., and Hajek, I. (1973) Biochemical differentiation of fast and slow coxal muscles of male and female *Periplaneta americana. Current Sci.* **43**, 541–542.

Charlton, M. P., Silverman, H., and Atwood, H. L. (1981) Intracellular potassium activities in muscles of normal and dystrophic mice: an *in vivo* electrometric study. *Exp. Neurol.* **7**, 203–219.

Cheng-Minoda, K., Davidowitz, J., Liebowitz, A., and Breinin, G. M. (1968) Fine structure of extraocular muscle in rabbit. *J. Cell Biol.* **39**, 193–197.

Chevreul, E. C. (1835) Untersuchungen über die chemische Zusammentsetzung des Fleischbrühe. *J. Prakt. Chem.* **6**, 120–150.

Chiarandini, D. J., and Stefani, E. (1974) Twitch potentiation by potassium contractures in single muscle fibres of the frog. *J. Physiol.* **240**, 1-14.

Chiarandini, D. J., Reuben, J. P., Girardier, L., Katz, G. M., and Grundfest, H. (1970) Effects of caffeine on crayfish muscle fibers. *J. Gen. Physiol.* **55**, 640–664.

Chien, P., and Koopowitz, H. (1972) The ultrastructure of neuromuscular systems in *Notoplana acticola,* a free-living polyclad flatworm. *Z. Zellforsch.* **133**, 277–288.

Cho, Y., Sidie, J. M., and DeBruyn, P. P. H. (1972) Electron microscopic studies on a tubulofilamentous fasciculus in the bat cricothyroid muscle. *J. Ultrastruct. Res.* **41**, 344-357.

Civan, M. M., and Podolsky, R. J. (1966) Contraction kinetics of striated muscle fibres following quick changes in load. *J. Physiol.* **184**, 511-534.

Claassen, H., Hoppeler, H., Straub, R., and Weibel, E. R. (1981) Correlation of histochemistry and quantitative ultrastructure of muscle fibres in skeletal muscles. *J. Physiol.* **318**, 3–4P.

Clarac, F. (1977a) Motor coordination in crustacean limbs. In: *Identified Neurons and Behavior of Arthropods,* G. Hoyle, Ed., pp. 167-186. Plenum, New York.

Clarac, F. (1977b) Crustacean cuticular stress detectors. In: *Structure and Function of Proprioceptors in the Invertebrates,* P. J. Mill, Ed., pp. 299-321. Chapman and Hall, London.

Clarac, F., Vedel, J. P., and Bush, B. M. H. (1978) Intersegmental reflex coordination by a single joint receptor organ (CB) in rock lobster walking legs. *J. Exp. Biol.* **73**, 29-46.

Clark, A. W., Mauro, A., Longenecker, Jr., H. E., and Hurlbut, W. P. (1970) Effects on the fine structure of the frog neuromuscular junction. *Nature* **228**, 705-706.

Clark, R. B., Gration, K. A. F., and Usherwood, P. N. R. (1979) Relative trophic influence of excitatory and inhibitory innervation of locust skeletal muscle fibres. *Nature* **280**, 679-682.

Clarke, M., and Spudich, J. A. (1974) Biochemical and structural studies of actomyosin-like proteins from non-muscle cells. *J. Mol. Biol.* **86**, 209-222.

Cloney, R. A. (1966) Cytoplasmic filaments and cell movements: epidermal cells during ascidian metamorphosis. *J. Ultrastruct. Res.* **14**, 300-328.

Close, R. (1964) Dynamic properties of fast and slow skeletal muscles of the rat during development. *J. Physiol.* **173**, 74-95.

Close, R. (1965) Force: velocity properties of mouse muscles. *Nature* **206**, 718-719.

Close, R. (1968) Effects of nerve cross-union on fast-twitch and slow-graded muscle fibres in the toad. *J. Physiol.* **198**, 103-125.

Close, R. (1969) Dynamic properties of fast and slow skeletal muscles of the rat after nerve cross-union. *J. Physiol.* **204**, 331-346.

Close, R., and Hoh, J. F. Y. (1968) The after-effects of repetitive stimulation on the isometric twitch contraction of rat fast skeletal muscle. *J. Physiol.* **197**, 461-477.

Cobb, J. L. S. (1967a) The innervation of the ampulla of the tube foot in the starfish *Astropecten irregularis. Proc. R. Soc. Lond. Ser. B* **168**, 91-99.

Cobb, J. L. S. (1967b) The fine structure of the pedicellariae of *Echinus esculentus* (L.). I. The innervation of the muscles. *J. Roy. Micr. Soc.* **88**, 211-221.

Cobb, J. L. S., and Laverack, M. S. (1966) The lantern of *Echinus esculentus* (L.) III. The fine structure of the lantern retractor muscle and its innervation. *Proc. R. Soc. Lond. Ser. B* **164**, 651-658.

Cobb, S. (1925) Review on the tonus of skeletal muscle. *Physiol. Rev.* **5**, 518-550.

Coburn, R. F. (1979) Electromechanical coupling in canine trachealis muscle; acetylcholine contractions. *Am. J. Physiol.* **5**, 177-183.

Cochrane, D. G., Elder, H. Y., and Usherwood, P. N. R. (1972) Physiology and ultrastructure of phasic and tonic skeletal muscle fibres in the locust *Schistocerca gregaria. J. Cell Sci.* **10**, 419-441.

Cohen, C. (1975) The protein switch of muscle contraction. *Sci. Am.* **233**, 35-46.

Cohen, C. (1979) Identification of calmodulin as the fourth subunit of phosphorylase kinase. *Fed. Proc.* **38**, Abst. 2948.

Cohen, J. L., Weiss, K. R., and Kupferman, I. (1978) Motor control of buccal muscles in *Aplysia. J. Neurophysiol.* **41**, 157-180.

Cohen, M. J. (1963) The crustacean myochordotonal organ as a proprioceptive system. *Comp. Biochm. Physiol.* **8**, 223-243.

Cohen, M. J., and Evoy, W. H. (1969) Sensory and motor interaction in the locomotor reflexes of crabs. *J. Exp. Biol.* **51**, 151-169.

Cole. W. C. (1955) Motor endings in the striated muscle of vertebrates. *J. Comp. Neurol.* **102**, 671-716.

Collins, C., and Miller, T. (1977) Studies on the action of biogenic amines on cockroach heart. *J Exp. Biol.* **67**, 1-15.

Connor, J. A. (1979) On exploring the basis for slow potential oscillations in the mammalian stomach and intestine. *J. Exp. Biol.* **81**, 153-173.

Connor, J. A., and Prosser, C. L. (1974) Comparison of ionic effects on longitudinal and circular muscle of cat jejunum. *Am. J. Physiol.* **226**, 1212-1218.

Connor, J. A., Prosser, C. L., and Weems, W. A. (1974) A study of pace-maker activity in intestinal smooth muscle. *J. Physiol.* **240**, 671-701.

Conti, M. A., and Adelstein, R. S. (1980) Phosphorylation by cyclic adenosine 3′ : 5′-monophosphate-dependent protein kinase regulates myosin light chain kinase. *Fed. Proc.* **39**, 1569–1573.

Cook, B. J., and Holman, G. M. (1978) Comparative pharmacological properties of muscle function in the foregut and the hindgut of the cockroach *Leucophaea maderae. Comp. Biochem. Physiol.* **61**, 291–295.

Cook, B. J., and Holman, G. M. (1979a) The action of proctolin and L-glutamic acid on the visceral muscles of the hindgut of the cockroach *Leucophaea maderae. Comp. Biochem. Physiol.* **64**, 21–28.

Cook, B. J., and Holman, G. M. (1979b) The pharmacology of insect visceral muscle. *Comp. Biochem. Physiol.* **64**, 183–190.

Cook, B. J., and Reinecke. J. P. (1973) Visceral muscles and myogenic activity hindgut of the cockroach, *Leucophaea maderae. J. Comp. Physiol.* **84**, 95–118.

Cook, B. J., and Erakes, J., and Anderson, G. R. (1969) The effect of various amines on the activity of the foregut of the cockroach, *Blaberus giganteus. J. Ins. Physiol.* **15**, 445–455.

Cook, B. J., Thompson, J. M. and Shelton, W. D. (1980) Some structural and functional properties of nerves and muscles in the oviduct of the pink bollworm moth. *Int. J. Invert.* **2**, 351–362.

Cooke, P. (1976) A filamentous cytoskeleton in vertebrate smooth muscle fibers. *J. Cell Biol.* **68**, 539–556.

Cooke, P. H., and Chase, R. H. (1971) Potassium chloride-insoluble myofilaments in vertebrate smooth muscle cells. *Exp. Cell Res.* **66**, 417–425.

Cooke, P. H., and Fay, F. S. (1972a) Correlation between fiber length, ultrastructure, and the length–tension relationship of mammalian smooth muscle. *J. Cell Biol.* **52**, 105–116.

Cooke, P. H., and Fay, F. S. (1972b) Thick myofilaments in contracted and relaxed mammalian smooth muscle cells. *Exp. Cell Res.* **71**, 265–272.

Cooke, R. (1975) The bound nucleotide of actin. *J. Supramolec. Struct.* **3**, 146–153.

Costa, M., Patel, Y., Furness, J. B., and Arimura, A. (1977) Evidence that some intrinsic neurons of the intestine contain somatostatin. *Neurosci. Let.* **6**, 215–222.

Costantin, L. L., and Podolsky, R. J. (1967) Depolarization of the internal membrane system in the activation of frog skeletal muscle. *J. Gen. Physiol.* **50**, 1101–1124.

La Course, J. R., and Northrop, R. B. (1978) A preliminary study of mechanoreceptors within the anterior byssus retractor muscle of *Mytilus edulis* L. *Biol. Bull.* **155**, 161–168.

Craig, R., and Megerman, J. (1977) Assembly of smooth muscle myosin into side-polar filaments. *J. Cell Biol.* **75**, 990–996.

Craig, R., and Megerman, J. (1979) Electron microscope studies on muscle. Thick filaments. In: *Motility in Cell Function,* F. A. Pepe, J. W. Sanger, and V. T. Nachmias, Eds., pp. 91–102. Academic, New York.

Crass, M. F., and Pang, P. K. T. (1980) Parathyroid hormone: a coronary artery vasodilator. *Science* **207**, 1087–1089.

Creighton, T. E. (1978) Experimental studies of protein folding and unfolding. *Prog. Biophys. Mol. Biol.* **33**, 231–297.

Creuzfeldt, O. D., Abbott, B. C., Fowler, W. M., and Pearson, C. M. (1963) Muscle membrane potentials in episodic adynamia. *Electroenceph. Clin. Neurophys.* **15**, 508–519.

Crowe, L. M., and Baskin, R. J. (1978) Freeze-fracture of intact sarcotubular membranes. *J. Ultrastruct. Res.* **62**, 147–154.

Cull-Candy, S. G. (1975) Effects of denervation and local damage on extrajunctional L-glutamate receptors in locust muscle. *Nature* **258**, 530–531.

Cull-Candy, S. G. (1976a) Two types of extrajunctional L-glutamate receptors in locust muscle fibres. *J. Physiol.* **255**, 449–464.

Cull-Candy, S. G. (1976b) L-glutamate receptors in locust muscle fibres. In: *Motor Innervation of Muscle.* S. Thesleff, Ed., pp. 263–288. Academic, London and New York.

Cull-Candy, S. G., Miledi, R., and Trautmann, A. (1979) End-plate currents and acetylcholine noise at normal and myasthenic human end-plates. *J. Physiol.* **287**, 247–266.

Cull-Candy, S. G., Miledi, R., Trautmann, A., and Uchitel, O. D. (1980) On the release of transmitter at normal, myasthenia gravis and myasthenic syndrome affected human end-plates. *J. Physiol.* **299**, 621–638.

Curry, O. B., Basten, J. F., Francis, M. J. O., and Smith, R. (1974) Calcium uptake by sarcoplasmic reticulum of muscle from Vitamin D-deficient rabbits. *Nature* **249**, 83–84.

Curtin, N. A., and Woledge, R. C. (1974) Energetics of relaxation in frog muscle. *J. Physiol.* **238**, 437–446.

Curtin, N. A., and Woledge, R. C. (1979) Chemical change, production of tension and energy following stretch of active muscle of frog. *J. Physiol.* **297**, 539–550.

Curtin, N. A., Gilbert, C., Kretzschmar, K. M., and Wilkie, D. R. (1974) The effect of the performance of work on total energy output and metabolism during muscular contraction. *J. Physiol.* **238**, 455–472.

Czeh, G., Gallego, R., Kudo, N., and Kuno, M. (1978) Evidence for maintenance of motoneuron properties by muscle activity. *J. Physiol.* **281**, 239–252.

Dabrowska, R., Sherry, J. M. F., and Hartshorne, D. J. (1979) Phosphorylation of myosin: a possible regulatory mechanism in smooth muscle. In: *Motility in Cell Function.* F. A. Pepe, J. W. Sanger, and V. T. Nachmias, Eds., pp. 147–160. Academic, New York.

Dagan, D., and Parnas, I. (1970) Giant fibre and small fibre pathways involved in the evasive response of the cockroach, *Periplaneta americana. J. Exp. Biol.* **52**, 313–324.

Dahlbäck, O., Elmquist, D., Johns, T. R., Radner, S., and Thesleff, S. (1961) An electrophysiological study of the neuromuscular junction in myasthenia gravis. *J. Physiol.* **156**, 336–343.

Davey, D. F., Gibbs, C. L., and McKirdy, H. C. (1975) Structural, mechanical and myothermic properties of rabbit rectococcygeus muscle. *J. Physiol.* **248**, 207–230.

Davidheiser, S., O'Keefe, K., and Davies, R. E. (1980) Shortening of A bands in living crayfish single muscle fiber. *Fed. Proc.* **39**, Abstr. 671.

Davies, R. E. (1963) A molecular theory of muscle contraction: calcium-dependent contractions with hydrogen bond formation plus ATP-dependent extensions of part of the myosin-action cross-bridges. *Nature* **199**, 1068–1074.

Davies, R. E., Goldspink, G., and Larson, R. E. (1969) ATP utilization by fast and slow muscles during the development and maintenance of isometric tension. *J. Physiol.* **206**, 28–29P.

Davis, W. J. (1971) Functional significance of motoneuron size and soma position in swimmeret system of the lobster. *J. Neurophysiol.* **34**, 274–288.

Davison, W., Goldspink, G., and Johston, I. A. (1976) Division of labor between fish myotomal muscles during swimming. *J. Physiol.* **263**, 185P.

Dawson, M. J., Gadian, D. G., and Wilkie, D. R. (1977) Contraction and recovery of living muscles studied by ^{31}P nuclear magnetic resonance. *J. Physiol.* **267**, 703–735.

Dawson, M. J., Gadian, D. G., and Wilkie, D. R. (1980) Mechanical relaxation rate and metabolism studied in fatiguing muscle by phosphorus nuclear magnetic resonance. *J. Physiol.* **299**, 465–484.

DeBell, J. T., Castillo, J. del, and Sanchez, V. (1963) Electrophysiology of the somatic muscle cells of *Ascaris lumbricoides. J. Cell Comp. Physiol.* **62**, 159–178.

Dedman, J. R., Potter, J. D., and Means, A. R. (1977) Biological cross-reactivity of rat testis phosphodiesterase activator protein and rabbit skeletal muscle troponin-C. *J. Biol. Chem.* **252**, 2437–2440.

Deitmer, J. W. (1977a) Electrical properties of skeletal muscle fibres of the flour moth larva *Ephistia kühniella. J. Ins. Physiol.* **23**, 33–38.

Deitmer, J. W. (1977b) Effects of cobalt and manganese on the calcium-action potentials in larval insect muscle fibres (*Ephestia kühniella*). *Comp. Biochem. Physiol.* **58A**, 1–4.

Delaleu, J. C., Blondeau, A., and Holley, A. (1972) Electrophysiology of the heart of an isopod crustacean: *Porcellio dilatatus.* II. Effects of ions and membrane permeability inhibitors. *J. Exp. Biol.* **57**, 609–632.

Delcomyn, F. (1980) Neural basis of rhythmic behavior in animals. *Science* **210**, 492–498.

Depue, R. H., and Rice, R. V. (1965) F-actin is a right-handed helix. *J. Mol. Biol.* **12**, 302–303.

De Rosier, D. J., Tilney, L. G., and Egelman, E. (1980) Actin in the inner ear – the remarkable structure of the stereocilium. *Nature* **287**, 291–296.

Desmedt, J. E. (1957) Nature of the defect of neuromuscular transmission in myasthenic patients: "post-tetanic exhaustion." *Nature* **179**, 156–157.

Desmedt, J. E., and Godeaux, E. (1977) Fast motor units are not preferentially activated in rapid voluntary contractions in man. *Nature* **267**, 717–719.

Desmedt, J. E., and Godeaux, E. (1978) Mechanism of the vibration paradox: excitatory and inhibitory effects of tendon vibration on single soleus muscle motor units in man. *J. Physiol.,* **285**, 197–207.

Desmedt, J. E., and Godeaux, E. (1981) Spinal motoneuron recruitment in man: rank deordering with direction but not with speed of voluntary movement. *Science* **214**, 933–955.

Desmedt, J. E., and Hainaut, K. (1976) The effect of A23187 ionophore on calcium movements and contraction processes in single barnacle muscle fibers. *J. Physiol.* **257**, 87–107.

Desmedt, J. E. and Hainaut, K. (1977) Inhibition of intracellular release of calcium by dantrolene in barnacle muscle fibers. *J. Physiol.* **265**, 565–585.

Dewey, M. M., Walcott, B., Colflesh, D. E., Terry, H., and Levine, R. J. C. (1977) The changes in thick filament length in *Limulus* striated muscle. *J. Cell Biol.* **75**, 366–380.

Dhoot, G. K., and Perry, S. V. (1979) Distribution of polymorphic forms of troponin components and tropomyosion in skeletal muscle. *Nature* **278**, 714–718.

Dhoot, G. K., Gell, P. G. H., and Perry, S. V. (1978) The localization of the different forms of troponin I in skeletal and cardiac muscle cells. *Exp. Cell Res.* **117**, 357–370.

Dickinson, V., and Woledge, R. C. (1973) The thermal effects of shortening in tetanic contractions of frog muscle. *J. Physiol.* **233**, 659–671.

DiFransesco, D. (1980) The pace-maker current i_{K_2} in Purkinje fibres is carried by sodium and potassium. *J. Physiol.* **308**, 32P.

DiFrancesco, D., and Ojeda, C. (1980) Properties of the current i_f in the SA node of the rabbit compared with those of the current i_{K_2} in Purkinje fibres. *J. Physiol.* **308**, 353–368.

Dillon, P. F., Aksoy, M. O., Driska, S. P., and Murphy, R. A. (1981) Myosin phosphorylation and the cross-bridge cycle in arterial smooth muscle. *Science* **211**, 495–497.

Dorai-Raj, B. S. (1964) Diversity of crab muscle fibers innervated by a single motor axon. *J. Cell Comp. Physiol.* **64**, 41–54.

Dorsett, D. A., and Roberts, J. B. (1980) A transverse tubular system and neuromuscular junctions in a molluscan unstriated muscle. *Cell Tiss. Res.* **206**, 251–260.

Dorsett, D. A., Willows, A. O. D., and Hoyle, G. (1973) The neuronal basis of behavior in *Tritonia.* IV. The central origin of a fixed action pattern demonstrated in the isolated brain. *J. Neurobiol.* **4**, 287–300.

Dörrscheidt-Käfer, M. (1979) Excitation-contraction coupling in frog sartorius and the role of the surface charge due to the carboxyl group of sialic acid. *Pflüg Arch.* **380**, 171–179.

Douglas, W. B., and Cosmos, E. (1975) *Exploratory Concepts in Muscular Dystrophy,* A. J. Milhorat, Ed. Excerpta Medica, Amsterdam.

Drachman, D. B. (1979) Acetylcholine receptors and myasthenia gravis. *Proc. Soc. Exp. Biol. Med.* **162**, 26–30.

Drachman, D. B., and Witzke, F. (1972) Trophic regulation of acetylcholine sensitivity of muscle; effect of electrical stimulation. *Science* **176**, 514–516.

Dreizen, P., and Kim, H. D. (1971) Contractile proteins of a benthic fish. I. Effects of temperature and pressure on myosin ATPase. *Am. Zool.* **11**, 513–521.

Drewes, C. D., and Pax, R. A. (1974a) Neuromuscular physiology of the longitudinal muscle of the earthworm, *Lumbricus terrestris.* I. Effects of different physiological salines. *J. Exp. Biol.* **60**, 445–452.

Drewes, C. D., and Pax, R. A. (1974b) Neuromuscular physiology of the longitudinal muscle of the earthworm, *Lumbricus terrestris.* II. Patterns of innervation. *J. Exp. Biol.* **60**, 453–467.

Duance, V. C., Stephens, H. R., Dunn, M., Bailey, A. J., and Dubowitz, V. (1980) A role for collagen in the pathogenesis of muscular dystrophy? *Nature* **284**, 470–472.

Dubowitz, V. (1978) *Muscle Disorders in Childhood.* Saunders, Philadelphia.

Dudel, J., and Kuffler, S. W. (1961a) Mechanisms of facilitation at the crayfish neuromuscular junction, *J. Physiol.* **155**, 530–542.

Dudel, J., and Kuffler, S. W. (1961b) Presynaptic inhibition at the crayfish neuromuscular junction. *J. Physiol.* **155**, 543–562.

Dudel, J., and Kuffler, S. W. (1961c) The quantal nature of transmission and spontaneous miniative potentials at the crayfish neuromuscular junction. *J. Physiol.* **155**, 514–529.

Dudel, J., and Rüdel, R. (1969) Voltage controlled contractions and current voltage relations of crayfish muscle fibres in chloride-free solutions. *Pflüg. Arch.* **308**, 291–314.

Dulhunty, A. F., and Franzini-Armstrong, C. (1975) The relative contributions of the folds and caveolae to the surface membrane of frog skeletal muscle fibres of different sarcomere lengths. *J. Physiol.* **250**, 513–539.

Dulhunty, A. F., and Gage, P. W. (1973) Electrical properties of toad sartorius muscle fibres in summer and winter. *J. Physiol.* **230**, 619–641.

Dunlap, K. (1977) Localization of calcium channels in *Paramecium caudatum. J. Physiol.* **271**, 119–133.

Duvert, M. (1969) Ultrastructure des myofibrilles dans les muscles longitudinaux du tronc de *Sagitta setosa* (Chaetognathes). *C. R. Acad. Sci. Paris.* **268**, 2452–2454.

Duvert, M., and Salat, C. (1979) Fine structure of muscle and other components of the trunk of *Sagitta setosa* (Chaetognath). *Tissue & Cell* **11**, 217–230.

Dydynska, M., and Wilkie, D. R. (1966) The chemical and energetic properties of muscles poisoned with FDNB. *J. Physiol.* **184**, 751–769.

Eastwood, A., Zollman, J., Katz, G. M., and Reuben, J. P. (1973) Disruption of the TTS in crayfish muscle fibers. *Fed. Proc.* **32**, Abst. 876.

Eastwood, A. B., Wood, D. S., and Reuben, J. P. (1978) Unusual thick and thin filaments packing in a crustacean muscle. *J. Cell Biol.* **77**, 48–58.

Eastwood, A. B., Wood, D. S., and Bock, K. L., and Sorenson, M. M. (1979) Chemically skinned mammalian skeletal muscle. I. The structure of skinned rabbit psoas. *Tissue & Cell.* **11**, 553–566.

Ebashi, S. (1961) The role of the "relaxing factor" in contraction–relaxation cycle of muscle. *Prog. Theor. Phys.* **Suppl. 17**, 35–41.

Ebashi, S. (1979) Calcium ion and muscle contraction. *Adv. Pharmacol. Therap.* **3**, 81–98.

Ebashi, S., and Ebashi, F. (1965) α-actinin, a new structural protein from striated muscle. I. Preparation and action on actomyosin-ATP interaction. *J. Biochem.* **58**, 7–12.

Ebashi, S., and Endo, M. (1969) Calcium ion and muscle contraction. *Prog. Biophys. Mol. Biol.* **18**, 123–183.

Ebashi, S., and Kodama, A. (1965) A new protein factor promoting aggregation of tropomyosin. *J. Biochem.* **58**, 107–108.

Ebashi, S., Endo, M., and Ohtsuki, I. (1969) Control of muscle contraction. *Q. Rev. Biophys.* **2**, 351–384.

Ebashi, S., Ohnishi, S., Abe, S., and Maruyama, K. (1975) A spin-label study on calcium-induced conformational changes of troponin components. *J. Biochem.* **75**, 211–213.

Eberstein, A., and Goodgold, J. (1968) Slow and fast twitch fibers in human skeletal muscle. *Am. J. Physiol.* **215**, 535–541.

Eccles, J. C. (1953) *The Neurophysiological Basis of Mind.* Oxford University Press, Oxford.

Eccles, J. C. (1973) The cerebellum as a computer. *J. Physiol.* **229**, 1–32.

Eckert, R. O. (1961a) Reflex relationships of the abdominal stretch receptors of the crayfish. I. Feedback inhibition of the receptors. *J. Cell Comp. Physiol.* **57**, 149–162.

Eckert, R. O. (1961b) Reflex relationships of the abdominal stretch receptors of the crayfish. II. Stretch receptor involvment during the swimming reflex. *J. Cell Comp. Physiol.* **57**, 163–197.

Eckert, R., and Naitoh, Y. (1972) Bioelectric control of locomotion in the ciliates. *J. Protozool.* **19**, 237–243.

Eckert, R., and Randall, D. (1978) *Animal Physiology.* Freeman, San Francisco.

Edman, K. A. P. (1970) The rising phase of the active state in single skeletal fibres of the frog. *Acta Physiol. Scand.* **79**, 167–173.

Edman, K. A. P. (1979) The velocity of unloaded shortening and its relation to sarcomere length and isometric force in vertebrate muscle fibres. *J. Physiol.* **291**, 143–159.

Edman, K. A. P., and Höglund, O. (1981) A technique for measuring length changes of individual segments of an isolated muscle fibre. *J. Physiol.* **317**, 8P.

Edman, K. A. P., and Hwang, J. C. (1977) The force-velocity relationship in vertebrate muscle fibres at varied tonicity by the external medium. *J. Physiol.* **269**, 255–272.

Edman, K. A. P., and Kiessling, A. (1971) The time course of the active state in relation to sarcomere length and movement studied in single muscle fibres of the frog. *Acta Physiol. Scand.* **81**, 182–196.

Edman, K. A. P., and Nilsson, E. (1969) The dynamics of the inotropic change produced by altered pacing of rabbit papillary muscle. *Acta Physiol. Scand.* **76**, 236–247.

Edman, K. A. P., and Nilsson, E. (1972) Relationships between force and velocity of shortening in rabbit papillary muscle. *Acta Physiol. Scand.* **85**, 488–500.

Edman, K. A. P., Grieve, D. W., and Nilsson, E. (1966) Studies of the excitation-contraction mechanism in the skeletal muscle and the myocardium. *Pflüg. Arch.* **290**, 320–334.

Edman, K. A. P., Elzinga, G., and Noble, M. I. M. (1978) Enhancement of mechanical performance by stretch during tetanic contractions of vertebrate skeletal muscle fibres. *J. Physiol.* **281**, 139–155.

Edman, K. A. P., Elzinga, G., and Noble, M. I. M. (1980) Critical release required to remove a stiff component of the force enhancement by stretch during activity in single muscle fibres in the frog. *J. Physiol.* **300**, 19P.

Edström, A. (1974) Effects of Ca^{2+} and Mg^{2+} on rapid axonal transport of proteins in vitro in frog sciatic nerves. *J. Cell Biol.* **61**, 812–818.

Edwards, C., Chichibu, S., and Hagiwara, S. (1964) Relation between membrane potential changes and tension in barnacle muscle fibers. *J. Gen. Physiol.* **48**, 225–234.

Edwards, R. H. T., Hill, D. K., and Jones, D. A. (1975) Heat production and chemical changes during isometric contractions of the human quadriceps muscle. *J. Physiol.* **251**, 303–315.

Eggleton, D., and Eggleton, G. P. (1927) The inorganic phosphate and a labile form of organic phosphate in the gastrocnemius of the frog. *Biochem. J.* **21**, 190–195.

Eichelberg, H. (1976) The fine structure of the drum muscles of the tigerfish, *Therapon jarbina* as compared with the trunk musculature. *Cell Tiss. Res.* **174**, 453–463.

Eichelberg, H. (1977) Fine structure of the drum muscles of piranha (Serrasalminae, characidae). *Cell Tiss. Res.* **185**, 547–555.

Einwächter, H. M., Haas, H. G., and Kern, R. (1972) Membrane current and contraction in frog atrial fibres. *J. Physiol.* **227**, 141–171.

Eisenberg, E., and Hill, T. L. (1978) A cross-bridge model of muscle contraction. *Prog. Biophys. Mol. Biol.* **33**, 55–82.

Eisenberg, R. S., and Gage, P. W. (1967) Frog skeletal muscle fibers: changes in electrical properties after disruption of transverse tubular system. *Science* **158**, 1700–1701.

Eisenberg, R. S., and Gage, P. W. (1969) Ionic cunductances of the surface and transverse tubular membranes of frog sartorius fibers. *J. Gen. Physiol.* **53**, 279–297.

Eisenberg, B. R., and Salmons, S. (1980) Ultrastructural correlates of fibre type transformation in rabbit skeletal muscle. *J. Physiol.* **300**, 16P.

Eisenberg, B. R., Kuda, A. M., and Peter, J. B. (1974) Stereological analysis of mammalian skeletal muscle. I. Soleus muscle of the adult guinea pig. *J. Cell Biol.* **60**, 732–754.

Elder, H. Y. (1973) Distribution and functions of elastic fibers in the invertebrates. *Biol. Bull.* **144**, 43–63.

Elder, H. Y. (1975) Muscle structure. In: *Insect Muscle,* P. N. R. Usherwood, Ed., pp. 1–74. Academic, New York.

Eldred, E., Granit, R., and Merton, D. (1953) Supraspinal control at the muscle spindles and its significance. *J. Physiol.* **122**, 498–523.

Elfvin, M., Levine, R. J. C., and Dewey, M. M. (1976) Paramyosin in invertebrate muscles. I. Identification and localization. *J. Cell Biol.* **71**, 261–272.

Elliott, G. F., Hanson, J., and Lowy, J. (1957) Paramyosin elements in lamellibranch muscles. *Nature* **180**, 1291–1292.

Elliott, G. F., Rome, E. M., and Spencer, M. (1970) A type of contraction hypothesis applicable to all muscles. *Nature* **226**, 417–420.

Ellisman, M. H. (1978) Appearance in slow muscle sarcolemma of specializations characteristic of fast muscle after reinnervation by a fast muscle nerve. *Exp. Neurol.* **58**, 59–67.

Ellisman, M. H., Brooke, M. H., Kaiser, K. K., and Rash, J. E. (1978) Appearance in slow muscle sarcolemma of specializations characteristic of fast muscle after reinnervation by a fast muscle nerve. *Exp. Neurol.* **58**, 59–67.

Ellisman, M. H., Rash, J. E., Staehlin, L. A., and Porter, K. R. (1976) Studies on excitable membranes. II. A comparison of specializations at neuromuscular junctions and nonjunctional sarcolemmas of mammalian fast and slow twitch muscle fibers. *J. Cell Biol.* **68**, 752–774.

Elsner, N. (1968) Die neuromuskularen Grundlagen des Werbeverhaltens der Roten Keulenheuschrecke *Gomphocerippus rufus* (L.). *Z. Verg. Physiol.* **60**, 308–350.

Embden, G., and Adler, E. (1921) Ueber die Phosphorsaureverteilung in den weissen und roten Muskulatur des Kaninchens. *Z. Physiol. Chem.* **113**, 201–210.

Endo, M. (1977) Calcium release from the sarcoplasmic reticulum. *Physiol. Rev.* **57**, 71–108.

Endo, M., and Nakajima, Y. (1973) Release of calcium induced by "depolarisation" of the sarcoplasmic reticulum membrane. *Nature New Biol.* **246**, 216–218.

Endo, M., Tanaka, M., and Ogawa, Y. (1970) Calcium induced release of calcium from the sarcoplasmic reticulum of skinned skeletal muscle fibres. *Nature* **228**, 34–36.

Engelmann, T. W. (1878) Neue Untersuchungen über die mikroskipischen Vorgänge bei der Muskelkontraction. *Pflüg. Arch.* **18**, 1–25.

Epstein, H. F., Isachsen, M. M., and Suddleson, E. A. (1976) Kinetics of movement of normal and mutant nematodes. *J. Comp. Physiol.* **110**, 317–322.

Epstein, H. F., Waterston, R. H., and Brenner, S. (1974) A mutant affecting the heavy chains of myosin in *Caenorhabditis elegans. J. Mol. Biol.* **90**, 291–300.

Equileor, M. De, and Valvassori, R. (1977) Studies on the helical and paramyosinic muscles. VII. Fine structure of body wall muscles in *Sipunculus nudus. J. Submicrosc. Cytol.* **9**, 363–372.

Ernst, E. (1925) Untersuchungen über Muskelkontraktion. I. Volumanderung bei der Muskelkontraktion. *Pflüg. Arch.* **209**, 613–622.

Ernst, E., Kovacs, K., Metzger-Torok, G., and Trombitas, C. (1969a) Longitudinal structure of the striated fibril. *Acta Biochim. Biophys.* **4**, 177–186.

Ernst, E., Kovacs, K., Metzger-Torok, G., and Trombitas, C. (1969b) Transversal structure of the striated fibril. *Acta. Biochim. Biophys.* **4**, 187–194.

Erspamer, V., and Boretti, G. (1951) Identification and characterization, by paper chromatography, of enteramine, octopamine, tyramine, histamine, and allied substances in extracts of posterior salivary glands of *Octopoda* and in other extracts of vertebrates and invertebrates. *Arch. Int. Pharmacodyn.* **88**, 296–332.

Euler, U. S. von (1961) Occurrence of catecholamines in Acrania and invertebrates. *Nature* **190**, 170–171.

Evans, P. D., Talamo, B. R., and Kravitz, E. A. (1975) Octopamine neurons: morphology, release of octopamine and possible physiological role. *Brain Res.* **90**, 340–347.

Evarts, E. V., Bizzi, E., Burke, R. E., DeLong, M., and Thach, W. T. (1971) *Central Control of Movement.* N. R. P. Bulletin 9 (1). MIT Press, Boston.

Evoy, W. H. (1977) Crustacean motor units. In: *Identified Neurons and Behavior of Arthropods,* G. Hoyle, Ed., pp. 67–86. Plenum, New York.

Evoy, W, H., and Cohen, M. J. (1971) Central and peripheral control of arthropod movements. *Adv. Comp. Physiol. Biochem.* **4**, 225–266.

Evoy, W. H., and Fourtner, C. R. (1973) Nervous control of walking in the crab, *Cardisoma guanhumi.* III. Proprioceptive influences on intra- and inter-segmental coordination. *J. Comp. Physiol.* **83**, 303–318.

Evoy, W. H., and Kennedy, D. (1967) The central nervous organization underlying control of antagonistic muscles in the crayfish. I. Types of command fibres. *J. Exp. Zool.* **165**, 223–238.

Ewer, D. W. (1960) Inhibition and rhythmic activity of the circular muscles of *Calliactis parasitica* (Couch). *J. Exp. Biol.* **37**, 812–831.

Fabiato, A., and Fabiato, F. (1975) Contractions induced by a calcium-triggered release of calcium from the sarcoplasmic reticulum of single skinned cardiac cells. *J. Physiol.* **249**, 469–495.

Fahrenbach, W. H. (1964) A new configuration of the sarcoplasmic reticulum. *J. Cell Biol.* **22**, 477–481.

Fahrenbach, W. H. (1967) The fine structure of fast and slow crustacean muscles. *J. Cell Biol.* **35**, 69–97.

Falck, B., and Hillarp, N. A. (1962) Fluorescence of catecholamines and related compounds condensed with formaldehyde. *J. Histochem. Cytochem.* **10**, 348–354.

Falck, B., and Owman, C. (1965) A detailed methodological description of the fluorescence method for the cellular demonstration of biogenic monoamines. *Acta Univ. Lund.* Sect. III. **7**, 1–24.

Falk, G. (1968) Predicted delays in the activation of the contractile system. *Biophys. J.* **8**, 608–625.

Falk, G., and Fatt, P. (1964) Linear electrical properties of striated muscle fibres observed with intracellular electrodes. *Proc. R. Soc. Lond. Ser. B* **160**, 69–123.

Fambrough, D. M. (1969) Acetylcholine sensitivity of muscle fiber membranes: mechanism of regulation by motoneurons. *Science* **168**, 372–373.

Fambrough, D. M. (1979) Control of acetylcholine receptors in skeletal muscle. *Physiol. Rev.* **59**, 165–227.

Fambrough, D. M., Drachman, D. B., and Satyamurti, S. (1973) Neuromuscular junction in myasthenia gravis: decreased acetylcholine receptors. *Science* **182**, 293–295.

Fardeau, M., Harpey, J. P., and Caille, B. (1975) Congenital fibre type disproportion, with relative smallness of Type 1 fibers. *Rev. Neurol.* **131**, 745–766.

Farel, P. B., and Buerger, A. A. (1972) Instrumental conditioning of leg position in chronic spinal frog: before and after sciatic section. *Brain Res.* **47**, 345–351.

Fatt, P., and Ginsborg, B. L. (1958) The ionic requirements for production of action potentials in crustacean muscle fibres. *J. Physiol.* **142**, 516–543.

Fatt, P., and Katz, B. (1951) An analysis of the end-plate potential recorded with an intra-cellular electrode. *J. Physiol.* **115**, 320–370.

Fatt, P., and Katz, B. (1953) Distributed "end-plate potentials" of crustacean muscle fibres. *J. Exp. Biol.* **30**, 433–439.

Favard, P., and Carasso, N. (1965) Mise en evidence d'un reticulum endoplasmique dans le spasmoneme de cilies peritriches. *J. Microscopie* **4**, 567–572.

Fawcett, D. W. (1968) The sporadic occurrence in cardiac muscle of anomalous Z bands exhibiting a periodic structure suggestive of tropomyosin. *J. Cell Biol.* **36**, 266–270.

Fawcett, D. W., and Revel, J. P. (1961) The sarcoplasmic reticulum of fast-acting fish muscle. *J. Biophys. Biochem. Cytol.* **10**, 89–109.

Fay, F. S. (1977) Isometric contractile properties of single isolated smooth muscle cells. *Nature* **265**, 553–556.

Fay, F. S., Shlevin, H. H., Granger, W. C., and Taylor, S. R. (1979) Aequorin luminescence during activation of single isolated smooth muscle cells. *Nature* **280**, 124–127.

Fenn, W. O. (1923) A quantitive comparison between the energy liberated and the work performed by the isolated sartorius of the frog. *J. Physiol.* **58**, 175–203.

Fenn, W. O. (1924) The relationships between the work performed and the energy liberated in muscular contraction. *J. Physiol.* **58**, 378–395.

Fentress, J. C. (1973) Development of grooming in mice with amputated forelimbs. *Science* **179**, 704–705.

Ferenczi, M. A., Goldman, Y. E., and Simmons, R. M. (1979) The relation between maximum shortening velocity (V_{max}) and (Mg-ATP) in frog skinned muscle fibres. *J. Physiol.* **288**, 89P.

Fernand, V. S. V., and Hess, A. (1969) The occurrence, structure and innervation of slow and twitch muscle fibres in the tensor tympani and stapedius of the cat. *J. Physiol.* **200**, 547–554.

Fetz, E., and Finocchio, D. V. (1971) Operant conditioning of specific patterns of neural and muscular activity. *Science* **174**, 431–435.

Fields, H. L. and Kennedy, D. (1965) Functional role of muscle receptor organs in crayfish. *Nature* **206**, 1232–1237.

Fields, H. L., Evoy, W. H., and Kennedy, D. M. (1967) Reflex role played by efferent control of an invertebrate stretch receptor. *J. Neurophysiol.* **30**, 859–874.

Fields, R. W. (1970) Mechanical properties of the frog's sarcolemma. *Biophys. J.* **10**, 462–480.

Fischer, E. (1944) Vertebrate smooth muscle. *Physiol. Rev.* **24**, 467–490.

Fischman, D. A. (1967) An electron microscope study of myofibril formation in embryonic chick skeletal muscle. *J. Cell Biol.* **32**, 557–575.

Fiske, C. H., and Subbarow, Y. (1929) Phosphocreatine. *J. Biol. Chem.* **81**, 629–679.

Fitts, R., Campion, D., Nagle, F., and Grissens, R. (1973) Contractile properties of skeletal muscle from trained miniature pig. *Pflüg. Arch.* **343**, 133–141.

Fletcher, W. M., and Hopkins, F. G. (1907) Lactic acid in amphibian muscle. *J. Physiol.* **35**, 247–309.

Flitney, F. W., and Hirst, D. G. (1974) Rapid "give" of sarcomeres and tension changes during servo-controlled stretches applied to contracting frog's muscle. *J. Physiol.* **239**, 72–73P.

Flitney, F. W., and Johnston, I. A. (1979) Mechanical properties of isolated fish red and white muscle fibers. *J. Physiol.* **295**, 67P.

Flood, P. R. (1966) A peculiar mode of muscular innervation in *Amphioxus.* Light and electron microscopic studies of the so-called ventral roots. *J. Comp. Neurol.* **126**, 181–218.

Flood, P. R. (1968) Structure of the segmental trunk muscle in *Amphioxus. Z. Zellforsch.* **84**, 389–416.

Flood, P. R., Guthrie, D. M., and Banks, J. R. (1969) Paramyosin muscle in the notochord of *Amphioxus. Nature* **222**, 87–88.

Foerster, O. (1927) Schteffe und spastische Hahmung. In: *Handbuch der normalen und pathologischen Physiologia.* A. Bethe, G. V. Bergmann, G. Embden, and A. Ellinger, Eds., pp. 900–901. Springer-Verlag, Berlin.

Florey, E. (1967) Newstransmitters and modulators in the animal kingdom. *Fed. Proc.* **26**, 1164–1178.

Foh, E., and Bogusch, G. (1969a) Die Auswirkung passiver Dehnungen auf die Struktur des glatten Musculus retractor penis von *Helix pomatia* L. *Z. Zellforsch.* **93**, 414–433.

Foh, E., and Bogusch, G. (1969b) Die Nervennetze im Penisretraktor von *Helix pomatia. Z. Zellforsch.* **93**, 439–446.

Forbes, M. S., and Sperelakis, N. (1980) Membrane systems in skeletal muscle of the lizard *Anolis carolinensis. J. Ultrastruct. Res.* **73**, 245–261.

Ford, L. E., and Podolsky, R. J. (1970) Regenerative calcium release within muscle cells. *Science* **167**, 58–59.

Ford, L. E., and Podolsky, R. J. (1972) Calcium uptake and force development by skinned muscle fibres in EGTA buffered solutions. *J. Physiol.* **223**, 1–20.

Ford, L. E., Huxley, A. F., and Simmons, R. M (1977) Tension responses to sudden length change in stimulated frog muscle fibres near slack length. *J. Physiol.* **269**, 441–515.

Ford, L. E., Huxley, A. F., and Simmons, R. M. (1981) The relation between stiffness and filament overlap in stimulated frog muscle fibres. *J. Physiol.* **311**, 219–249.

Forman, R. R., and Zill, S. N. (1981) Use of a learning paradigm to define behavioral state in an insect. In: *11th Ann. Mtg., Soc. Neurosci.* Abstr. 212.7.

Fourtner, C. R., and Pax, R. A. (1972) Chelicerate neuromuscular systems: the distal flexor of the mero-carpopodite of *Limulus polyphemus* (L). *Comp. Biochem. Physiol.* **41**, 617–627.

Frank, E. (1973) Matching of facilitation at the neuromuscular junction of the lobster: a possible case for influence of muscle on nerve. *J. Physiol.* **233**, 635–658.

Franzini-Armstrong, C. (1970a) Studies of the triad. I. Structure of the junction in frog twitch fibers. *J. Cell Biol.* **47**, 488–499.

Franzini-Armstrong, C. (1970b) Details of the I band structure as revealed by the localization of ferritin. *Tissue & Cell* **2**, 327–338.

Franzini-Armstrong, C. (1971) Studies of the triad. II. Penetration of tracers into the junctional gap. *J. Cell Biol.* **49**, 196–203.

Franzini-Armstrong, C. (1973) Membranous systems in muscle fibers. In: *The Structure and Function of Muscle,* Vol. 2, G. H. Bourne, Ed., pp. 531–619. Academic, New York and London.

Franzini-Armstrong, C., and Porter, K. R. (1964) Sarcolemmal invaginations constituting the T system in fish muscle fibers. *J. Cell Biol.* **22**, 675–696.

Franzini-Armstrong, C., Venosa, R. A., and Horowicz, P. (1973) Morphology and accessibility of the transverse tubular system in frog sartorius muscle after glycerol treatment. *J. Mem. Biol.* **14**, 197–212.

Freeman, M. A. (1966) The effect of drugs on the alimentary canal of the African migratory locust, *Locusta migratoria. Comp. Biochem. Physiol.* **17**, 755–764.

Freygang, W. H., Goldstein, D. A., Hellam, D. C., and Peachey, L. D. (1964) The relation between the late after-potential and the size of the transverse tubular system of frog muscle. *J. Gen. Physiol.* **48**, 235–263.

Furness, J. B. (1969) An electrophysiological study of the innervation of the smooth muscle of the colon. *J. Physiol.* **205**, 549–562.

Furness, J. B., and Costa, M. (1979) Types of nerves in the enteric nervous system. *Neuroscience* **5**, 1–21.

Furness, J. B., Costa, M., and Howe, P. R. C. (1980) Intrinsic amine-handling neurons in the intestine. In: *Histochemistry and Cell Biology of Autonomic Neurons, SIF Cells, and Paraneurons.* Eränkö, O., Soinila, S., and Päivärinta, H., Eds., pp. 367–372. Raven, New York.

Gabella, G. (1976) The force generated by a visceral smooth muscle. *J. Physiol.* **263**, 199–214.

Gage, P. W., and Eisenberg, R. S. (1967) Action potentials without contraction in frog skeletal muscle fibers with disrupted transverse tubules. *Science* **158**, 1702–1703.

Gainer, H. (1968) The role of calcium in excitation-contraction coupling of lobster muscle. *J. Gen. Physiol.* **52**, 88–110.

Galey, F. R. (1969) Elastic properties of fixed and fresh muscle. *J. Ultrastruct. Res.* **26**, 424–441.

Gallego, R., Kuno, M., Núñez, R., and Snider, W. D. (1979a) Dependence of motoneurone properties on the length of immobilized muscle. *J. Physiol.* **291**, 179–189.

Gallego, R., Kuno, M., Núñez, R., and Snider, W. D. (1979a) Dependence of motoneurone properties spinal motoneurons. *J. Physiol.* **291**, 191–205.

Gannon, J., and Burnstock, G. (1969) Excitatory adrenergic innervation of the fish heart. *Comp. Biochem. Physiol.* **29**, 765–773.

Garamvölgyi, N. (1963) Observations preliminaires sur la structure de las strie Z dans le muscle alaire de l'Abeille. *J. Microscopie* **2**, 107–112.

Garamvölgyi, N. (1966a) The ultrastructure of the insect flight muscle fibril ghost. *Biochem. Biophys. Acta* **1**, 81–88.

Garamvölgyi, N. (1966b) Elongation of the primary myofilaments in highly stretched insect flight muscle fibrils. *Biochem. Biophys. Acta* **1**, 89–100.

Garamvölgyi, N., and Kerner, J. (1966) The ultrastructure of the insect flight muscle fibril ghost. *Acta Biochim. Biophys. Hung.* **1**, 81–88.

Garamvölgyi, N., Metger-Torok, G., and Tigyi-Sebes, A. (1962) The Z- and M-formations of striated muscle. *Acta Physiol. Hung.* **22**, 223–233.

Garrels, J. I., and Hunter, T. (1979) Post-translational modification of actins synthesized *in vitro. Biochim. Biophys. Acta* **564**, 517–525.

Gauthier, G. F., and Dunn, R. A. (1973) Ultrastructural and cytochemical features of mammalian skeletal muscle fibres following denervation. *J. Cell Sci.* **12**, 525–547.

Gauthier, G. F., and Lowey, S. (1977) Polymorphism of myosin among skeletal muscle fiber types. *J. Cell Biol.* **74**, 760–779.

Gauthier, G. F., and Padyulka, H. A. (1966) Cytological studies of fiber types in skeletal muscle. *J. Cell Biol.* **28**, 333–354.

Gauthier, G. F., Lowey, S., and Hobbs, A. W. (1978) Fast and slow myosin in developing muscle fibers. *Nature* **274**, 25–29.

Geddes, L. E., and Mailman, D. S. (1973) Contrasting twitch and latency relaxation in skeletal muscle. *Comp. Biochem. Physiol.* **44A**, 693–710.

George, J. C., and Bhakthan, N. M. G. (1961) Lipase activity in the slow and fast-contracting leg muscles of the cockroach. *Nature* **192**, 356.

George, J. C., and Scaria, K. S. (1958) A histochemical study of dehydrogenase activity in the pectoralis major muscle of the pigeon and certain other vertebrate skeletal muscles. *Q. J. Microsc. Sci.* **99**, 469–473.

Gergely, J. (1973) Muscle contraction, conformational changes and energy transduction. *Ann. N. Y. Acad. Sci.* **227**, 587–593.

Gergely, J. (1974) *Circulation Research.* **34 Suppl. III**, 74–82.

Giacomelli, F., Wiener, J., and Spiro, D. (1970) Cross-striated arrays of filaments in endothelium. *J. Cell Biol.* **45**, 188–192.

Gibbs, C. L., and Gibson, W. R. (1970) Energy production in cardiac isotonic contractions. *J. Gen. Physiol.* **56**, 732–750.

Gibbs, C. L., Ricchiuti, N. V., and Mommaerts, W. F. H. M. (1966) Activation heat in frog sartorius muscle. *J. Gen. Physiol.* **49**, 517–535.

Gilai, A., and Parnas, I. (1970) Neuromuscular physiology of the closer muscles in the pedipalp of the scorpion *Leiurus quinquestriatus. J. Exp. Biol.* **52**, 325–344.

Gilai, A., and Parnas, I. (1972) Electromechanical coupling in tubular muscle fibers. I. The organization of tubular muscle fibers in the scorpion *Leiurus quinquestriatus. J. Cell Biol.* **52**, 626–638.

Gilbert, C., Kretzschmar, K. M., Wilkie, D. R., and Woledge, R. C. (1971) Chemical change and energy output during muscular contraction. *J. Physiol.* **218**, 163–193.

Gilbert, P. (1975) How the cerebellum could memorise movements. *Nature* **254**, 688–689.

Gilëv, V. P., Perov, N. A., and Melnikova, E. Y. (1968) Ultrastructure of thick filaments of striated muscle fibre in studies of sections from unembedded objects (in Russian). *Arch. Anat. Histol.* **54**, 40–53.

Gillespie, J. S. (1962) Spontaneous mechanical and electrical activity of stretched and unstretched intestinal smooth muscle cells and their response to sympathetic nerve stimulation. *J. Physiol.* **162**, 54–75.

Gillis, J. M. (1969) The site of action of calcium in producing contraction in striated muscle. *J. Physiol.* **200**, 849–864.

Gilloteaux, J. (1972) Innervation of the anterior byssal retractor muscle in *Mytilus edulis* L. I. Histology. *Z. Zellforsch.* **124**, 204–216.

Gilloteaux, J., and Baguet, F. (1977) Contractile filaments organization in functional states of the anterior byssus retractor muscle (ABRM) of *Mytilus edulis* L. *Cytobiol.* **15**, 192–220.

Gilly, W. F., and Hui, C. S. (1980a) Mechanical activation in slow and twitch skeletal muscle fibres of the frog. *J. Physiol.* **301**, 137–156.

Gilly, W. F., and Hui, C. S. (1980b) Membrane electrical properties of frog slow muscle fibres. *J. Physiol.* **301**, 157–173.

Gilly, W. F., and Hui, C. S. (1980c) Voltage-dependent charge movement in frog slow muscle fibres. *J. Physiol.* **301**, 175–190.

Ginsborg, B. L. (1960) Some properties of avian skeletal muscle fibres with multiple neuromuscular junctions. *J. Physiol.* **154**, 581–598.

Ginsborg, B. L., and Mackay, B. (1961) A histochemical demonstration of two types of motor innervation in avian skeletal muscle. *Bibl. Anat.* **2**, 174–181.

Gnanamuthu, C. P. (1930) The anatomy and mechanism of the tongue of *Chamaeleon carcaratus. Proc. Zool. Soc. Lond.* **31**, 467–486.

Goffart, M. (1971) *Function and Form in the Sloth.* Pergamon, New York.

Goldberg, A. L., Etlinger, J. D., Goldspink, D. F., and Jablecki, C. (1975) Mechanism of work-induced hypertrophy of skeletal muscle. *Med. Sci. Sports* 7, 248–261.

Goldstein, L., Ed. (1977) *Introduction to Comparative Physiology*. Holt, Rinehart and Winston, New York.

Goldstein, M. A., Schroeter, J. P., and Sass, R. L. (1979) The Z lattice in canine cardiac muscle. *J. Cell Biol.* 83, 187–204.

Golenhofen, K., and Loh, D. (1970) Spontaneous activity of isolated smooth muscle preparation from different regions of the guinea pig stomach. *Pflüg. Arch.* **315**, 336–356.

Goll, D. E., Mommaerts, W. F. H., Reedy, M. K., and Seraydarian, K. (1969) Studies on α-actinin-like proteins liberated during trypsin digestion of α-actinin and of myofibrils. *Biochim. Biophys. Acta* **175**, 174–194.

Gonzalez-Serratos, H. (1971) Inward spread of excitation in vertebrate muscle fibers. *J. Physiol.* **212**, 777–800.

Gonzalez-Serratos, H. (1975) Graded activation of myofibrils and the effect of diameter on tension development during contractures in isolated skeletal muscle fibres. *J. Physiol.* **253**, 321–339.

Gonzales-Serratos, H., Hill, L., and Valle-Aguilera, R. (1981) Effects of catecholamines and cyclic AMP on excitation-contraction coupling in isolated skeletal muscle fibres of the frog. *J. Physiol.* **315**, 267–282.

Gordon, A. M., Huxley, A. F., and Julian, F. J. (1966a) Tension development in highly stretched vertebrate muscle fibers. *J. Physiol.* **184**, 143–169.

Gordon, A. M., Huxley, A. F., and Julian, F. J. (1966b) The variation in isometric tension with sarcomere length in vertebrate muscle fibres. *J. Physiol.* **184**, 170–192.

Gordon, D. J., and Korn, E. D. (1976) Biochemical comparison of actins from *Acanthameba castellani* and rabbit skeletal muscle. *Fed. Proc.* **35**, 1359.

Gordon, T., Perry, R., Sriharai, T., and Vrbova, G. (1977) Differentiation of slow and fast muscles in chickens. *Cell Tiss. Res.* **180**, 211–222.

Govind, C. K., Atwood, H. L., and Maynard, D. M. (1975) Innervation and neuromuscular physiology of intrinsic foregut muslces in the blue crab and spiny lobster. *J. Comp. Physiol.* **96**, 185–204.

Gower, D. and Kretschmer, K. M. (1976) Hear production and chemical change during isometric contraction of rat soleus muscle. *J. Physiol.* **258**, 659–671.

Gradwell, N., and Walcott, B. (1971) Dual functional and structural properties of the interhyoideus muscle of the bullfrog tadpole (*Rana catesbeiana*). *J. Exp. Zool.* **176**, 193–218.

Graham, D., and Bässler, U. (1981) Effects of afference sign reversal on motor activity in walking stick insects (*Carausius morosus*). *J. Exp. Biol.* **91**, 179–193.

Graham, J. A., and Lamb, J. F. (1968) The effect of adrenaline on the tension developed in contractures and twitches of the ventricle of the frog. *J. Physiol.* **197**, 479–509.

Gray, E. G. (1958) The structures of fast and slow muscle fibres in the frog. *J. Anat.* **92**, 559–562.

Greenberg, M. J., and Price, D. A. (1979) FMRamide, a cardioexcitatory neuropeptide of molluscs: an agent in search of a mission. *Am. Zool.* **19**, 163–174.

Greeves, M. A., Midelfort, F., Trentham, D. R., and Boyer, P. D. (1979) Recent advances in the myosin and actomyosin ATPase mechanisms. In: *Motility in Cell Function*. F. A. Pepe, J. W. Sanger, and V. T. Nachmias, Eds., pp. 27–52. Academic, New York.

Greven, K. (1953) Ruhe- und Aktionspotentiale des glatten Muskulatur nach Untersuchungen mit Glaskepillar-electroden. *Z. Biol.* **106**, 1–15.

Grigor'Eva, T. A. (1962) *The Innervation of Blood Vessels*. Pergamon, Oxford.

Grillner, S. (1976) Some aspects of the descending control of the spinal circuits generating locomotor movements. In: *Neural Control of Locomotion*. R. M Herman, S. Grillner, P. S. G. Stein, and D. G. Stuart, Eds., pp. 351–376. Plenum, New York.

Grimstone, A. V., and Cleveland, L. R. (1965) The fine structure and function of the contractile axostyles of certain flagellates. *J. Cell Biol.* **24**, 387–400.

Grinyer, I., and George, J. C. (1969) Some observations on the ultrastructure of the hummingbird pectoral muscles. *Canad. J. Zool.* **47**, 771–774.

Gruber, H. (1978) Motor innervation of the striated oesophagus muscle. I. Intramural distribution of the right and left vagus nerve in the rat oesophagus as revealed by the glycogen depletion technique. *J. Neurol. Sci.* **36**, 41–53.

Gruener, R., McArdle, B., Ryman, B. E., and Weller, R. O. (1968) Contracture of phosphorylase deficient muscle. *J. Neurol. Neurosurg. Psychiatry* **31**, 268–283.

Guba, F., Harsányi, V., and Vajdae, E. (1968) The muscle protein fibrillin. *Acta Biochim. Biophys. Hung.* **3**, 353–363.

Gulati, J., and Palmer, L. G. (1977) Untitled. *Science* **198**, 1283–1284.

Gunn, M. (1951) A study of the enteric plexuses in some amphibians. *Q. J. Microsc. Sci.* **92**, 55–78.

Guth, K. (1980) Polarization of trytophan fluorescence measurements in muscle. A re-evaluation. *Biophys. Str.* **6**, 81–93.

Guth, K., and Kuhn, H. J. (1978) Stiffness and tension during and after sudden length changes of glycerinated rabbit psoas muscle fibers. *Biophys. Str.* **4**, 223–236.

Guth, L. (1968) "Trophic" influences of nerve on muscle. *Physiol. Rev.* **48**, 645–687.

Guthrie, D. M. (1962) Control of the ventral diaphragm in an insect. *Nature* **196**, 1010–1012.

Guthrie, D. M., and Banks, J. R. (1970) Observations on the function and physiological properties of a fast paramyosin muscle – the notochord of *Amphioxus* (*Branchiostoma lanceolatum*). *J. Exp. Biol.* **52**, 125–138.

Gutmann, E., and Hanzlíková, V. (1973) Basic mechanisms of aging in the neuromuscular system. *Mech. Age Dev.* **1**, 327–349.

Guy, P. S., and Snow, D. H. (1977) Effect of training and detraining on muscle composition in horse. *J. Physiol.* **269**, 33–51.

Hagiwara, S., and Kidokoro, Y. (1971) Na and Ca components of action potential in *Amphioxus* muscle cells. *J. Physiol.* **219**, 217–232.

Hagiwara, S., and Naka, K. (1964) The initiation of spike potential in barnacle muscle fibers under low intracellular Ca^{2+}. *J. Gen. Physiol.* **48**, 141–162.

Hagiwara, S., and Nakajima, S. (1966) Effects of the intracellular Ca ion concentration upon the excitability of the muscle fiber membrane of a barnacle. *J. Gen. Physiol.* **49**, 807–818.

Hagiwara, S., and Takahashi, K. (1967a) Resting and spike potentials of skeletal muscle fibres of salt-water elasmobranch and teleost fish. *J. Physiol.* **190**, 499–518.

Hagiwara, S., and Takahashi, K. (1967b) Surface density of calcium ions and calcium spikes in the barnacle muscle fiber membrane. *J. Gen. Physiol.* **50**, 583–601.

Hagiwara, S., and Takahashi, K. (1974) Mechanism of anion permeation through the muscle fibre membrane of an elasmobranch fish, *Taeniura lymma*. *J. Physiol.* **238**, 109–127.

Hagiwara, S., Chichibu, S., and Simpson, N. (1968) Neuromuscular mechanisms of wing beat in hummingbirds. *Z. Verg. Physiol.* **60**, 209–218.

Hagiwara, S., Henkart, M. P., and Kidokoro, Y. (1971a) Excitation-contraction coupling in *Amphioxus* muscle cells. *J. Physiol.* **219**, 233–251.

Hagiwara, S., Takahashi, K., and Junge, D. (1968) Excitation-contraction coupling in a barnacle muscle fiber as examined with voltage clamp technique. *J. Gen. Physiol.* **51**, 157–175.

Hagiwara, S., Toyama, K., and Hayashi, H. (1971b) Mechanisms of anion and cation permeation in the resting membrane of a barnacle muscle fiber. *J. Gen. Physiol.* **57**, 408–434.

Hagopian, M. (1967) Three shapes of mitochondria in femoral muscles of the cockroach *Leucophaea maderae* Fabricius. *J. Morph.* **122**, 147–168.

Hall, C. E., Jakus, M. A., and Schmitt, F. O. (1945) The structure of certain muscle fibrils as revealed by the use of electron stains. *J. Appl. Phys.* **16**, 459–472.

Halpern, W., Mulvany, M. J., and Warshaw, D. M. (1978) Mechanical properties of smooth muscle cells in the walls of arterial resistance vessels. *J. Physiol.* **275**, 85–101.

Halverson, M., and Afzelius, B. A. (1969) Filament organization in the body muscles of the arrow-worm. *J. Ultrastruct. Res.* **26**, 289–295.

Hammond, G. R., and Ridge, R. M. A. P. (1978) Properties of twitch motor units in snake costocutaneous muscle. *J. Physiol.* **276**, 525–534.

Hanson, J. (1967) Axial period of actin filaments. *Nature* **213**, 353-356.

Hanson, J. (1973) Evidence from electron microscope studies on actin paracrystals concerning the origin of the cross-striation in the thin filaments of vertebrate skeletal muscle. *Proc. R. Soc. Lond. Ser. B* **183**, 39–58.

Hanson, J., and Huxley, H. E. (1956) The structural basis of contraction in striated muscle. *Symp. Soc. Exp. Biol.* **9**, 228–264.

Hanson, J., and Huxley, H. E. (1957) Quantitative studies on the structure of cross-striated myofibrils. II. Investigation by biochemical techniques. *Biochim. Biophys. Acta* **23**, 250–260.

Hanson, J., and Lowy, J. (1953) Structure of smooth muscles. *Nature* **180**, 906–909.

Hanson, J., and Lowy, J. (1961) The structure of the muscle fibres in the translucent part of the adductor of the oyster *Crassostrea angulata*. *Proc. R. Lond. Soc. Ser. B* **154**, 173–196.

Hanson, J., and Lowy, J. (1963) The structure of F-actin and of actin filaments isolated from muscle. *J. Mol. Biol.* **6**, 46–60.

Hanson, J., and Lowy, J. (1964) Proteins of the contractile mechanism of mammalian smooth muscle and their possible location in the cell. *Proc. R. Soc. Lond. Ser. B* **160**, 523–524.

Hanson, J., and Lowy, J. (1965) Molecular basis of contractility in muscle. *Brit. Med. Bull.* **21**, 264-271.

Hanson, J., Lowy, J., Huxley, H. E., Bailey, K., Kay, C. M., and Rüegg, J. C. (1957) Structure of molluscan tropomyosin. *Nature* **180**, 1134–1135.

Harpey, J. P., and Caille, B. (1975) Congenital fiber type disproportion, with relative smallness of type I fibers – muscle biopsy study of three cases from the same family. *Rev. Neurol. (Paris)*. **131**, 745–766.

Harreveld, A. van (1939) The motor innervation of a triply innervated crustacean muscle. *J. Exp. Biol.* **16**, 398–402.

Harreveld, A. van, and Wiersma, C. A. G. (1937) The triple innervation of crayfish muscle and its function in contraction and inhibition. *J. Exp. Biol.* **14**, 448–461.

Harrington, W. F. (1979) On the origin of the contractile force in skeletal muscle. *Proc. Nat. Acad. Sci.* **76**, 5066–5070.

Harrington, W. F., Sutoh, K., and Chiao, Y. C. (1979) Myosin filaments and cross-bridge movement. In: *Motility in Cell Function.* F. A. Pepe, J. W. Sanger, and V. T. Nachmias, Eds., pp. 69–85. Academic, New York.

Harrison, J. T. (1975) Isometric responses of the somatic musculature of *Cryptochiton stelleri. Veliger* **18 (Suppl.)**, 79–82.

Hartree, W., and Hill, A. V. (1924) The heat production of muscles treated with caffeine or subjected to prolonged discontinuous stimulation. *J. Physiol.* **58**, 441–454.

Haselgrove, J. C. (1975) X-ray evidence for conformational changes in the myosin filaments of vertebrate striated muscle. *J. Mol. Biol.* **92**, 113–143.

Haselgrove, J. C., and Huxley, H. E. (1973) X-ray evidence for radial cross bridge movement and for the sliding filament model in actively contracting skeletal muscle. *J. Mol. Biol.* **77**, 549–568.

Haselgrove, J. C., and Reedy, M. K. (1978) Modeling rigor cross-bridge patterns in muscle. I. initial studies of the rigor lattice of insect flight muscle. *Biophys. J.* **24**, 713–728.

Hasselbach, W. (1964) Relaxing factor and the relaxation of muscle. *Prog. Biophys. Mol. Biol.* **14**, 167–222.

Hasselbach, W., and Makinose, M. (1961) Die Calciumpumpe der "Erschlaff-ungsgrana" des Muskels und ihre Abhängigkeit von der ATP-Spaltung. *Biochem. Z.* **333**, 518–523.

Haswell, W. A. (1890) A comparative study of striated muscle. *Q. J. Microsc. Sci.* **30**, 31–50.

Hawkins, R. D., and Bruner, J. (1979) Maintained contraction of the crayfish claw opener muscle in the absence of motor neuron activity. *Brain Res.* **162**, 129–136.

Hayat, A., Tradieu, C., Tabary, J. C., and Tabary, C. (1978) Effects of denervation on the reduction of sarcomere number in cat soleus muscle immobilized in shortened position during seven days. *J. de Physiol.* **74**, 563–567.

Heidenhain, M. (1898) Struktur der contractilen Materie. *Erg. Anat.* **8**, 71–111.

Heilbrunn, L. V., and Wierczinski, F. J. (1947) The action of various cations on muscle protoplasm. *J. Cell Comp. Physiol.* **29**, 15–32.

Heilmeyer, Jr., L. M. G., Rüegg, J. C., and Wieland, Th., Eds. (1976) *Molecular Basis of Motility.* Springer-Verlay, Berlin, Heidelberg and New York.

Heinl, P. (1972) Mechanische Aktivierung und Deaktivierung der isolierten contractilen Struktur des Froschsartorius durch rechteckformige und sinusformige Langenaderungen. *Pflüg. Arch.* **333**, 213–226.

Heistracher, P., and Hunt, C. C. (1969a) The relation of membrane change to contraction in twitch muscle fibres. *J. Physiol.* **201**, 589–612.

Heistracher, P., and Hunt, C. C. (1969b) Contractile repriming in snake twitch muscle fibres. *J. Physiol.* **201**, 613–626.

Heitler, W. J. (1974) The locust jump. Specialization of the metathoracic femero-tibial joint. *J. Comp. Physiol.* **89**, 93–104.

Heitler, W. J., and Pearson, K. G. (1980) Non-spiking interactions and local interneurones in the central pattern generator of the crayfish swimmeret system. *Brain Res.* **187**, 206–211.

Hellam, D. C., and Podolsky, R. J. (1969) Force measurements in skinned muscle fibres. *J. Physiol.* **200**, 807–819.

Hellewell, S. B., and Taylor, D. L. (1979) The contractile basis of ameboid movement. VI. The solation-contraction coupling hypothesis. *J. Cell Biol.* **83**, 633–648.

Henderson, E. G. (1970) Potassium exchange and afterpotentials in frog sartorius muscles treated with glycerol. *J. Gen. Physiol.* **56**, 692–715.

Henon, B. K., and Ikeda, K. (1981) Changes in membrane properties of the *Drosophila* dorsal longitudinal flight muscle induced by sodium pomp inhibitors. *J. Exp. Biol.* **90**, 175–183.

Henneman, E. (1957) Relation between size of neurons and their susceptibility to discharge. *Science* **126**, 1345–1347.

Henneman, E. C., and Olsen, J. (1965) Relations between structure and function in the design of skeletal muscle. *J. Neurophysiol.* **28**, 581–598.

Henneman, E., Somjen, G., and Carpenter, D. O. (1965) Functional significance of cell size in spinal motoneurons. *J. Neurophysiol.* **28**, 560–580.

Henneman, E., Somjen, G., and Carpenter, D. O. (1969) Excitability and inhibitability of motoneurons of different sizes. *J. Neurophysiol.* **28**, 599–620.

Hernandez-Nicaise, M-L., Mackie, G. O., and Meech, R. W. (1980) Giant smooth muscle cells of *Beroë*. Ultrasturcture, innervation, and electrical properties. *J. Gen. Physiol.* **75**, 79–105.

Hertz, R., and Weber, A. (1965) Caffeine inhibition of calcium uptake by muscle reticulum. *Fed. Proc.* **24**, 208–214.

Hess, A. (1961a) Structural differences of fast and slow extrafusal muscle fibers and their nerve endings in chickens. *J. Physiol.* **157**, 221–231.

Hess, A. (1961b) The structure of slow and fast extrafusal muscle fibers in the extraocular muscles and their nerve endings in guinea pigs. *J. Cell. Comp. Physiol.* **58**, 63–80.

Hess, A. (1961c) Two kinds of motor nerve endings on mammalian intrafusal muscle fibers revealed by the cholinesterase technique. *Anat. Rec.* **139**, 173–183.

Hess, A. (1963) Two kinds of extrafusal muscle fibers and their nerve endings in the garter snake. *Am. J. Anat.* **113**, 347–364.

Hess, A. (1965) The sarcoplasmic reticulum, the T system and the motor terminals of slow and twitch muscle fibres in the garter snake. *J. Cell Biol.* **26**, 467–476.

Hess, A. (1967) The structure of vertebrate slow and twitch muscle fibers. *Invest. Ophthalmol.* **6**, 217–228.

Hess, A., and Pilar, G. (1963) Slow fibres in the extraocular muscles of the cat. *J. Physiol.* **169**, 780–798.

Hess, A., Pilar, G., and Weakly, J. N. (1969) Correlation between transmission and structure in avian ciliary ganglion synapses. *J. Physiol.* **202**, 339–354.

Hettinger, Th. (1961) *The Physiology of Strength.* Thomas, Springfield, Ill.

Heumann, H. G., and Zebe, E. (1967) Über Feinbau und Funktionsweise der Fasern aus dem Hautmuskelschlauch des Regenwurms, *Lumbricus terrestria* L. *Z. Zellforsch.* **78**, 131–150.

Heuson-Stiennon, J. A. (1965) Morphogenèse de la cellule musculaire strieé, étudiée au microscope électronique. I. Formation des structures fibrillaires. *J. Microscopie.* **4**, 657–678.

Heyer, C. B., Kater, S. B., and Karlsson, U. L. (1973) Neuromuscular systems in molluscs. *Am. Zool.* **13**, 247–270.

Hibberd, M. G., and Jewell, B. R. (1979) Length-dependence of the sensitivity of the contractile system to calcium in rat ventricular muscle. *J. Physiol.* **290**, 30–31P.

Hidaka, R. S. (1978) Z-line extraction: comparative effects in avian skeletal muscle fiber types. *J. Ultrastruct. Res.* **65**, 266–278.

Hidaka, T., and Kuriyama, H. (1969) Effects of catecholamines on the neuromuscular transmission in fish red muscle. *J. Physiol.* **201**, 61–71.

Hidaka, T., and Toida, N. (1969) Biophysical and mechanical properties of red and white muscle fibres in fish. *J. Physiol.* **201**, 49–59.

Higgins, W. J., and Greenberg, M. J. (1974) Intracellular actions of 5-hydroxytryptamine on the bivalve myocardium. II. Cyclic nucleotide-dependent protein kinases and microsomal calcium uptake. *J. Exp. Zool.* **190**, 305–316.

Hikida, R. S. (1972) Morphological transformation of slow to fast muscle fibres after tenotomy. *Exp. Neurol.* **35**, 265–273.

Hikida, R. S. (1978) Z-line extraction: comparative effects in avian skeletal muscle fiber types. *J. Ultrastruct. Res.* **65**, 266–278.

Hikida, R. S., and Bock, W. J. (1968) An analysis of twitch and tonus fibers in the hatching muscle. *Condor.* **70**, 211–222.

Hildebrand, H. F., and Gilloteaux, J. (1979) Étude des filaments de paramyosine du muscle lisse de *Mytilus edulis* L.: ultrastructure, purification et interaction paramyosine-myosine. *C. R. Acad. Sci.* **288**, 109–112.

Hill, A. V. (1938) The heat of shortening and the dynamic constants of muscle. *Proc. R. Soc. Lond. Ser. B* **126**, 136–195.

Hill, A. V. (1949a) The heat of activation and the heat of shortening in a muscle twitch. *Proc. R. Soc. Lond. Ser. B* **136**, 196–238.

Hill, A. V. (1949b) The abrupt transition from rest to activity in muscle. *Proc. R. Soc. Lond. Ser. B* **136**, 399–420.

Hill, A. V. (1964) The effect of load on the heat of shortening of muscle. *Proc. R. Soc. Lond. Ser. B* **159**, 279–318.

Hill, A. V., and Lupton, H. (1922) Muscular exercise, lactic acid, and the supply and utilization of oxygen. *Q. J. Med.* **16**, 135–171.

Hill, D. K. (1940) The time course of oxygen consumption of stimulated frog's muscle. *J. Physiol.* **98**, 207–227.

Hill, D. K. (1960) The location of adenine nucleotide in frog's striated muscle. *J. Physiol.* **150**, 347–373.

Hill, D. K. (1964) The location of adenine nucleotide in the striated muscle of the toad. *J. Cell Biol.* **20**, 435–458.

Hill, R. B. (1958) The effects of certain neurohumours and of other drugs on the ventricle and radula protractor of *Busycon canaliculatum* and on the ventricle of *Strombus gigae*. *Biol. Bull.* **115**, 471–482.

Hill, R. B. (1974) Effects of acetylcholine on resting and action potentials, and on contractile force in the ventricle of *Dolabella auricularia*. *J. Exp. Biol.* **61**, 629–637.

Hill, R. B., and McDonald-Ordzie, P. E. (1979) Ionic dependence of the response to acetylcholine of a molluscan buccal muscle: The radular protractor of *Busycon canaliculatum*. *Comp. Biochem. Physiol.* **62C**, 19–30.

Hill, R. B., Marantz, E., Beattie, B. A. and Lockhart, J. M. (1968) Mechanical properties of the radula protractor of *Busycon canaliculatum*. *Experientia* **24**, 91.

Hill, R. B., Sanger, J. W., Yautorno, R. E., and Deutsch, C. (1978) Contraction in a muscle with negligible sarcoplasmic reticulum: the longitudinal retractor of the sea cucumber *Isostichopus badionotus* (Selenke), Holothuria Aspidochirota. *J. Exp. Zool.* **206**, 137–150.

Hill, T. L. (1975a) Theoretical formalism for the sliding filament model of contraction of striated muscle, Part 1. *Prog. Biophys. Mol. Biol.* **28**, 267–340.

Hill, T. L. (1975b) Theoretical formalism for the sliding filament model of contraction of striated muscle, Part 2. *Prog. Biophys. Mol. Biol.* **29**, 105–159.

Hille, B. (1972) Permeability of sodium channels to metal cations in myelinated nerve. *J. Gen. Physiol.* **59**, 637–658.

Hinssen, H., D'Haese, J., Small, J. V., and Sobieszek, A. (1978) Mode of filament assembly of myosin from muscle and non-muscle cells. *J. Ultrastruct. Res.* **64**, 282–302.

Hirabayashi, T. (1971) Studies on muscle differentiation. *Dev. Growth Diff.* **12**, 241–252.

Hirakow, R. (1971) The fine structure of the *Necturus* (amphibia) heart. *Am. J. Anat.* **132**, 401–422.

Hiramoto, Y. (1956) Physical state of muscle protoplasm. *Annot. Zool. Jap.* **29**, 63–68.

Hirata, M., and Kuriyama, H. (1980) Does activation of cyclic AMP dependent phosphorylation by β-adrenergic agent control the tone of vascular muscle? *J. Physiol.* **307**, 143–161.

Ho, K. W., Tweedle, C. D., Stephens, K. R., Roy, R. R., Heusner, W. W., and Van Huss, W. (1978) Muscle fiber splitting with exercise in rats. *Fed. Proc.* **37**, 786.

Hoar, P. E., and Kerrick, W. G. L. (1979) Rabbit diaphragm: two types of fibres determined by calcium strontium activation and protein content. *J. Physiol.* **295**, 345–352.

Hodgkin, A. L., and Horrowicz, P. (1959a) Movements of Na and K in single muscle fibres. *J. Physiol.* **145**, 405–432.

Hodgkin, A. L., and Horrowicz, P. (1959b) The influence of potassium and chloride ions on the membrane potential of single muscle fibres. *J. Physiol.* **148**, 127–160.

Hodgkin, A. L., and Horowicz, P. (1960) Potassium contractures in single muscle fibres. *J. Physiol.* **153**, 386–403.

Hoffer, J. A., Loeb, G. E., O'Donovan, M. J., and Pratt, C. A. (1980) Unitary activity patterns during walking confirm the existence of two functionally distinct classes of sartorius motoneurones in cat. *J. Physiol.* **308**, 20P.

Hoffman, B. F., De Carvalho, A. P., De Mello, W. C., and Cranefield, P. F. (1959) Electrical activity of single fibres of the AV node. *Circ. Res.* **7**, 11–20.

Hoffmann, P. (1914) Über die doppelte Innervation der Krebsmuskeln. Zugieich ein Beitrag zur Kenntnis nervoser Hemmungen. *Z. Biol.* **63**, 411–442.

Holman, G. M., and Cook, B. J. (1970) Pharmacological properties of excitatory neuromuscular transmission in the hindgut of the cockroach, *Leucophaea maderae*. *J. Ins. Physiol.* **16**, 1891–1907.

Holman, G. M., and Cook, B. J. (1979) Evidence for proctolin and a second myotropic peptide in the cockroach *Leucophaea maderae*, determined by bioassay and HPLC analysis. *Ins. Biochem.* **9**, 149–154.

Holman, M. E. (1958) Membrane potentials recorded with high resistance microelectrodes, and the effects of changes in ionic environment on the electrical and mechanical activity of the smooth muscle of the taenia coli of the guinea-pig. *J. Physiol.* **141**, 464–488.

Homsher, E., Kean, C. J., Wallner, A., and Garibian-Sarian, V. (1979) The time-course of energy balance in an isometric tetanus. *J. Gen. Physiol.* **73**, 553–567.

Horridge, G. A. (1959) Analysis of the rapid responses of *Nereis* and *Harmothoe* (Annelida). *Proc. R. Soc. Lond. Ser. B* **150**, 245–262.

Horrobin, D. F., and Morgan, R. O. (1980) Myotonic dystrophy – a disease caused by functional zinc deficiency due to an abnormal zinc-binding ligand. *Med. Hypoth.* **6**, 375–388.

Howard, T. H., and Liu, S. (1979) Specific interaction of cytochalasins with muscle and platelet actin filaments in vitro. *J. Supramol. Struct.* **11**, 283–293.

Hoyle, G. (1952) The response mechanism in ascidians. *J. Mar. Biol. Assoc.* **31**, 287–305.

Hoyle, G. (1953a) "Slow" and "fast" nerve fibres in locusts. *Nature* **172**, 165.

Hoyle, G. (1953b) Potassium ions and insect nerve muscle. *J. Exp. Biol.* **30**, 121–135.

Hoyle, G. (1955a) The effects of some common cations on neuromuscular transmission in insects. *J. Physiol.* **127**, 90–103.

Hoyle, G. (1955b) The anatomy and innervation of locust skeletal muscle. *Proc. R. Soc. Lond. Ser. B* **143**, 281-292.

Hoyle, G. (1957) Nervous control of insect muscles. In: *Recent Advances in Invertebrate Physiology,* B. T. Scheer, Ed., pp. 73–98. University of Oregon Press, Eugene, Or.

Hoyle, G. (1958) Studies on neuromuscular transmission in *Limulus. Biol. Bull.* **115**, 209–218.

Hoyle, G. (1959) The neuromuscular mechanism of an insect spiracular muscle. *J. Insect Physiol.* **3**, 378-394.

Hoyle, G. (1960a) The action of carbon dioxide gas on an insect spiracular muscle. *J. Ins. Physiol.* **4**, 63-79.

Hoyle, G. (1960b) Neuromuscular activity in the swimming sea anemone, *Stomphia coccinea* (Müller). *J. Exp. Biol.* **37**, 671–688.

Hoyle, G. (1964a) Muscle and Neuromuscular Physiology. In: *Physiology of Mollusca,* K. M. Wilbur and C. M. Yonge, Eds., pp. 313-351. Academic, New York.

Hoyle, G. (1964b) Exploration of neuronal mechanisms underlying behavior in insects. In: *Neural Theory and Modelling,* R. F. Reiss, Ed., pp. 346-376. Stanford University Press, Stanford, Calif.

Hoyle, G. (1965) Nature of the excitatory sarcoplasmic reticular junction. *Science* **149**, 70–72.

Hoyle, G. (1966) Interpreting muscle function in invertebrates. *Pflüg. Arch.* **291**, 12–27.

Hoyle, G. (1967a) Diversity of striated muscle. *Am. Zool.* **7**, 435–449.

Hoyle, G. (1967b) Specificity of muscle. In: *Invertebrate Nervous Systems,* C. A. G. Wiersma, Ed., pp. 151-167. University of Chicago Press, Chicago.

Hoyle, G. (1968a) Correlated physiological and ultrastructural studies on specialized muscles. Ia. Neuromuscular physiology of the levator of the eyestalk of *Podophthalmus vigil* (Weber). *J. Exp. Zool.* **167**, 471-486.

Hoyle, G. (1968b) Resting tension, "negative" contraction and "break" contraction in specialised crustacean muscle fibers. *J. Exp. Zool.* **167**, 551-566.

Hoyle, G. (1968c) untitled (review of comparative ultrastructure). *Symp. Biol. Hung.* **8**, 34–48.

Hoyle, G. (1970) How is muscle turned on and off? *Sci. Am.* **222**, 84–93.

Hoyle, G. (1973) Correlated physiological and ultrastructural studies on specialized muscles. IIIb. Fine structure of the power-stroke muscle of the swimming leg of *Portunus sanguinolentus. J. Exp. Zool.* **185**, 97–110.

Hoyle, G. (1974a) Neural control of insect muscle. In: *The Physiology of Insects,* Vol. 4, 2nd Ed., M. Rockstein, Ed., pp. 176-236. Academic, New York.

Hoyle, G. (1974b) A function for neurons (DUM) neurosecretory on skeletal muscle of insects. *J. Exp. Zool.* **189**, 401–406.

Hoyle, G. (1975a) Identified neurons and the future of neuroethology. *J. Exp. Zool.* **194**, 51–74.

Hoyle, G. (1975b) Neural mechanisms underlying behavior of invertebrates. In: *Handbook of Psychobiology,* C. Blakemore and M. S. Gazzaniga, Eds., pp. 3–48. Academic, New York.

Hoyle, G. (1975c) Evidence that insect dorsal unpaired median (DUM) neurons are octopaminergic. *J. Exp. Zool.* **193**, 425-431.

Hoyle, G. (1976) Learning of leg position by the ghost crab *Ocypode ceratophthalma. Behav. Biol.* **18**, 147-163.

Hoyle, G. (1978a) Intrinsic rhythm and basic tonus in insect skeletal muscle. *J. Exp. Biol.* **73**, 173–203.

Hoyle, G. (1978b) Distribution of nerve and muscle fibre types in locust jumping muscle. *J. Exp. Biol.* **73**, 205-233.

Hoyle, G. (1980a) Comprative aspects of the onset of muscular contraction with special reference to latency relaxation. *Comp. Biochem. Physiol.* **66A**, 57–68.

Hoyle, G. (1980b) Learning, using natural reinforcements, in insect preparations that permit cellular neuronal analysis. *J. Neurobiol.* **11**, 323-354.

Hoyle, G. (1982a) Pacemaker changes in a learning paradigm. In: *Cellular Pacemakers,* Vol. 2, D. O. Carpenter, Ed., pp. 3-25. Wiley, New York.

Hoyle, G. (1982b) Giant barnacle and *Peripatus* muscles: striated smooth transitions. In: *Basic Biology of Muscles,* B. Twarog, R. Levine, and M. Dewey, Eds., pp. 000–000. Raven, New York (in press).

Hoyle, G., and Abbott, B. C. (1967) Dynamic properties of giant muscle fibers of the barnacle. *Am. Zool.* 7, 611–614.

Hoyle, G., and Burrows, M. (1973a) Correlated physiological and ultrastructural studies on specialized muscles. IIIa. Neuromuscular physiology of the power-stroke muscle of the swimming leg of *Portunus sanguinolentus. J. Exp. Zool.* **185**, 83–96.

Hoyle, G., and Burrows, M. (1973b) Neural mechanisms underlying behavior in the locust *Schistocerca grégaria.* II. Integrative activity in metathoracic neurons. *J. Neurobiol.* **4**, 43–67.

Hoyle, G., and Castillo, J. del (1979) Neuromuscular transmission in *Peripatus. J. Exp. Biol.* **83**, 13-29.

Hoyle, G., and Dagan, D. (1978) Physiological characteristics and reflex activation of DUM (octopaminergic) neurons of locust metathoracic ganglion. *J. Neurobiol.* **9**, 59–79.

Hoyle, G., and Lowy, J. (1956) The paradox of *Mytilus* muscle. A new interpretation. *J. Exp. Biol.* **33**, 295-310.

Hoyle, G., and McNeill, P. A. (1968a) Correlated physiological and ultrastructural studies on specialized muscles. 1b. Ultrastructure of white and pink fibers of the levator of the eyestalk of *Podophthalmus vigil* (Weber). *J. Exp. Zool.* **167**, 487–522.

Hoyle, G., and McNeill, P. A. (1968b) Correlated physiological and ultrastructural studies on specialised muscles. Ic. Neuromuscular junctions in the eyestalk levator muscles of *Podophthalmus vigil* (Weber). *J. Exp. Zool.* **167**, 523–550.

Hoyle, G., and Smyth, T. (1963) Neuromuscular physiology of giant muscle fibers of a barnacle, *Balanus nubilus* Darwin. *Comp. Biochem. Physiol.* **10**, 291-314.

Hoyle, G., and Wiersma, C. A. G. (1958) Excitation at neuromuscular junctions in Crustacea. *J. Physiol.* **143**, 403–425.

Hoyle, G., and Williams, M. (1980) The musculature of *Peripatus* and its innervation. *Phil. Trans. R. Soc. Lond. Ser. B* **288**, 481–510.

Hoyle, G., Colquhoun, W., and Williams, M. (1980) Fine structure of an octopaminergic neuron and its terminals. *J. Neurobiol.* **11**, 103–126.

Hoyle, G., McAlear, J. H., and Selverston, A. I. (1965) Mechanism of supercontraction in a striated muscle. *J. Cell Biol.* **26**, 621–640.

Hoyle, G., McNeill, P. A., and Selverston, A. I. (1973) Ultrastructure of barnacle giant muscle fibers. *J. Cell Biol.* **56**, 74–91.

Hoyle, G., McNeill, P. A., and Walcott, B. (1966) Nature of invaginating tubules in *Felderstruktur* muscle fibers of the garter snake. *J. Cell Biol.* **30**, 197–201.

Hoyle, G., Dagan, D., Moberly, B., and Colquhoun, W. (1974) Dorsal unpaired median insect neurons make neurosecretory endings on skeletal muscle. *J. Exp. Zool.* **187**, 159–165.

Huang, B., and Pitelka, D. R. (1973) The contractile process in the ciliate, *Stentor coeruleus.* I. The role of microtubules and filaments. *J. Cell Biol.* **57**, 704–728.

Huddart, H. (1966) Electrical and mechanical responses recorded from lepidopterous skeletal muscle. *J. Insect. Physiol.* **12**, 537–545.

Huddart, H. (1967) The effect of chloride ions on moth skeletal muscle fibres. *Comp. Biochem. Physiol.* **20**, 355–359.

Huddart, H. (1975) *The Comparative Structure and Function of Muscle.* Pergamon, Oxford.

Huddart, H., and Hunt, S. (1976) *Visceral Muscle: Its Structure and Function.* Wiley, New York.

Huddart, H., Hunt, S., and Oates, K. (1977) Calcium movements during contraction in molluscan smooth muscle, and the loci of calcium binding and release. *J. Exp. Biol.* **68**, 45–56.

Hughes, J., Smith, T. W., Kosterlitz, H. W., Fothergill, L. A., Morgan, B. A., and Morris, H. R. (1975) Identification of two related pentapeptides from the brain with potent opiate agonist activity. *Nature* **258**, 577–580.

Hughes, T. D. (1980a) The imaginal ecdysis of the desert of locust, *Schistocerca gregaria*. I. A description of the behavior. *Physiol. Ent.* **5**, 47–54.

Hughes, T. D. (1980b) The imaginal ecdysis of the dessert locust, *Schistocerca gregaria*. II. Motor activity underlying the pre-emergence and emergence behavior. *Physiol. Ent.* **5**, 55–71.

Hui, C. S., and Gilly, W. F. (1979) Mechanical activation and voltage-dependent charge movement in stretched muscle fibres. *Nature* **281**, 223–225.

Hürthle, K. (1909) Über die Struktur der quergestreiften Muskelfasern von *Hydrophilus* im ruhenden und tätigen Zustand. *Pflüg. Arch.* **126**, 1–164.

Hutter, O. F., and Padsha, S. M. (1959) Effect of nitrate and other anions on the membrane resistance of frog skeletal muscle. *J. Physiol.* **146**, 117–131.

Hutter, O. F., and Trautwein, W. (1956) Vagal and sympathetic effects on the pacemaker fibers in the sinus venosus of the heart. *J. Gen. Physiol.* **39**, 715–733.

Huxley, A. F. (1957a) Muscle structure and theories of contraction. *Prog. Biophys. Mol. Biol.* **7**, 257–317.

Huxley, A. F. (1957b) Das Interferenz-Mikroskop und seine Anwendung in der biologischen Forschung. *Naturwissenschaften* **7**, 189–196.

Huxley, A. F. (1974) Muscular Contraction. *J. Physiol.* **243**, 1–43.

Huxley, A. F. (1980) *Reflections on Muscle.* Princeton University Press, Princeton.

Huxley, A. F., and Simmons, R. M. (1971) Mechanical properties of the cross-bridges of frog striated muscle. *J. Physiol.* **218**, 59–60P.

Huxley, A. F., and Taylor, R. E. (1958) Local activation of striated muscle fibres. *J. Physiol.* **144**, 426–441.

Huxley, H. E. (1953) Electron microscope studies of the organisation of the filaments in striated muscle. *Biochim. Biophys. Acta* **12**, 387–394.

Huxley, H. E. (1957) The double array of filaments in cross-striated muscle. *J. Biophys. Biochem. Cytol.* **3**, 631–648.

Huxley, H. E. (1958) The contraction of muscle. *Sci. Am.* **199**, 67–82.

Huxley, H. E. (1963) Electron microscope studies on the structure of natural and synthetic protein filaments from striated muscle. *J. Mol. Biol.* **7**, 281–308.

Huxley, H. E. (1964a) Structural arrangements and the contraction mechanism in striated muscle. *Proc. R. Soc. Lond. Ser. B* **160**, 442–448.

Huxley, H. E. (1964b) Evidence for continuity between the central elements of the triads and extracellular space in frog sartorius muscle. *Nature* **202**, 1067–1071.

Huxley, H. E. (1969) The mechanism of muscular contraction. *Science* **164**, 1356–1360.

Huxley, H. E. (1972) In: *The Structure and Function of Muscle,* Vol. 1, G. H. Bourne, Ed., pp. 301–387. Academic, New York.

Huxley, H. E., and Brown, W. (1967) The low-angle X-ray diagram of vertebrate striated muscle and its behaviour during contraction and rigor. *J. Mol. Biol.* **30**, 383–434.

Ianuzzo, C. D., Gollnick, P. D., and Armstrong, R. B. (1976) Compensatory adaptations of skeletal muscle fiber types to a long-term functional overload. *Life Sci.* **19**, 1517–1524.

Ihemelandu, E. C. (1980) Decrease in fibre numbers of dog pectineus muscle with age. *J. Anat.* **130**, 69–73.

Ikemoto, N., and Kawaguti, S. (1967) Elongating effect of tropomyosin A on the thick myofilaments in the long-sarcomere muscle of the horse-shoe crab. *Proc. Jap. Acad.* **43**, 974–979.

Infante, A. A., Klaupiks, D., and Davies, R. E. (1965) Phosphorylcreative consumption during single-working contractions of isolated muscle. *Biochim. Biophys. Acta* **94**, 504–515.

Ingels, N. P., and Thompson, N. P. (1966) An elektrokinematic theory of muscle contraction. *Nature* **211**, 1032–1035.

Inomata, H., and Kao, C. Y. (1976) Ionic currents in the guinea-pig taenia coli. *J. Physiol.* **255**, 347–378.

Inomata, H., and Kao, C. Y. (1979) Ionic mechanisms of repolarization in the guinea-pig taenia coli as revealed by the actions of strontium. *J. Physiol.* **297**, 443–462.

Irisawa, H., and Hama, K. (1965) Some observations on the fine structure of the mantis shrimp heart. *Z. Zellforsch.* **68**, 674–688.

Irisawa, H., Kobayashi, M., and Matsubayashi, T. (1961) Relaxation of oyster heart through the anodal current pulse. *Jap. J. Physiol.* **11**, 385–392.

Irish, M. J., and Wilson, F. J. (1979) The structure of A segments from glycerinated skeletal muscle. *Tissue and Cell* **11**, 201–207.

Irving, S. N., and Miller, T. A. (1980) L. glutamate and aspartate as possible transmitters at 'fast' and 'slow' neuromuscular junctions in body wall of *Musca* larva. In: *Insect Neurobiology and Pesticide Action,* pp. 227–233. Society of the Chemical Industry, London.

Irwin, R. L., and Hein, M. M. (1963) Contracture of slow striated muscle during calcium deprivation. *J. Gen. Physiol.* **47**, 133–150.

Irwin, R. L., and Hein, M. M. (1966) Ionic requirements for relaxation of slow striated muscle. *Am. J. Physiol.* **211**, 1117–1122.

Isaacson, A., and Sandow, A. (1963) Effects of zinc on responses of skeletal muscle. *J. Gen. Physiol.* **46**, 655–677.

Isenberg, G., and Small, J. V. (1978) Filamentous actin, 100A filaments and microtubules in neuroblastoma cells–their distribution in relation to sites of movement and neuronal transport. *Cytobiology* **16**, 326–344.

Isenberg, G., Rathke, P. C., Hülsmann, N., Franke, W. W., and Wohlfarth-Botterman, K. E. (1976) Transformation of cytoplasmic actin. Importance of the contractile gel reticulum–relaxation cycle of cytoplasmic actomyosin. *Cell Tiss. Res.* **173**, 495–528.

Ishikawa, H., Bischoff, R., and Holtzer, H. (1968) Mitosis and intermediate-sized filaments in developing skeletal muscle. *J. Cell Biol.* **38**, 538–555.

Ishikawa, H., Bischoff, R., and Holtzer, H. (1969) Formation of arrowhead complexes with heavy meromyosin in a variety of cell types. *J. Cell Biol.* **43**, 312–328.

Iwazumi, T. (1970) A new field theory of muscle contraction. Ph. D. thesis University of Pennsylvania, Philadelphia.

Izumi, T., Hattori, A., Higuma, N., and Tamura, K. (1978) Cardiac myofibril disorientation and Z-band abnormalities in idiopathic cardiomyopathy. *Arch. Histol.* **41**, 293–308.

Jahromi, S. S., and Atwood, H. L. (1974) Three-dimensional ultrastructure of the crayfish neuromuscular apparatus. *J. Cell Biol.* **63**, 599–613.

Jahromi, S. S. and Bloom, J. W. (1979) Structural changes in locust leg muscle fibres in response to tenotomy and joint immobilization. *J. Ins. Physiol.* **25**, 767–780.

Jahromi, S. S. and Charlton, M. P. (1979) Transverse sarcomere splitting. A possible means of longitudinal growth in crab muscles. *J. Cell Biol.* **80**, 736–742.

Jaimovich, E., Venose, R. A., Shrager, P., and Horowicz, P. (1976) Density and distribution of tetrodotoxin receptors in normal and detubulated frog sartorius muscle. *J. Gen. Physiol.* **67**, 399–416.

James, N. T., (1979) Studies on the responses of different types of muscle fibre during surgically induced compensatory hypertrophy. *J. Anat.* **129**, 769–776.

Jansen, J., Andersen P., and Jansen, J. K. S. (1962) On the structure and innervation of the parietal muscle of the hagfish (*Myxine glutinosa*). *Acta Morphol. Neerlando-Scandinavica.* **5**, 329–338.

Jansen, J. K. S., and Andersen, P. (1963) Anatomy and physiology of the skeletal muscles. In: *The Biology of Myxine,* pp. 161–194. A. Brodal and R. Fänge, Eds. University of Oslo Press, Oslo.

Jenerick, H. P., and Gerard, R. W. (1953) Membrane potential and threshold of single muscle fibers. *J. Cell. Comp. Physiol.* **42**, 79–102.

Jensen, D. (1965) The aneural heart of the hagfish. *Ann. N. Y. Acad. Sci.* **127**, 443–458.

Jensen, D. (1969) Intrinsic cardiac rate regulation in the sea lamprey, *Petromyzon marinus* and the rainbow trout, *Salmo gairdneri. Comp. Biochem. Physiol.* **30**, 685–690.

Jensen, H. (1977) Ultrastructure of the myocardial cell and its membrane systems in the adult fly *Calliphora erythrocephala* (Insecta: Diptera). *Cell Tiss. Res.* **180**, 293–302.

Jensen, H., and Mykleburst, R. (1975) Ultrastructure of muscle cells in *Siboglinum fiordicum* (Pogonophora). *Cell Tiss. Res.* **163**, 185–197.

Jensen, H., and Tjønneland, A. (1977) Ultrastructure of the heart muscle cells of the cuttlefish *Rossia macrosoma* (Delle chiaje) (Mollusca: Cephalopoda). *Cell Tiss. Res.* **185**, 147–158.

Jewell, B. R., and Rüegg, J. C. (1964) Oscillatory contraction of insect fibrillar muscle after glycerol extraction. *Proc. R. Soc. Lond. Ser. B* **164**, 428–459.

Jewell, B. R., and Wilkie, D. R. (1958) An analysis of the mechanical components in frog's sartorius muscle. *J. Physiol.* **143**, 515–540.

Jewell, B. R., and Wilkie, D. R. (1960) The mechanical properties of relaxing muscle. *J. Physiol.* **152**, 30–47.

Johnson, E. A., and Sommer, J. R. (1967) A strand of cardiac muscle: its ultrastructure and the electrophysiological implications of its geometry. *J. Cell Biol.* **33**, 103–129.

Johnson, J. D., Charlton, S., and Potter, J. D. (1978) Kinetic analysis of Ca^{2+} binding to troponin C. *Biophys. J.* **21**, 105A.

Johnson, W. H., and Van P. Thompson. (1979) Evidence for direct large filament interactions in molluscan catch muscles. In: *Motility in Cell Function.* F. A. Pepe, J. W. Sanger, and V. T. Nachmias, Eds., pp. 467–469. Academic, New York.

Johnston, I. A., and Lucking, M. (1978) Temperature-induced variation in distribution of different types of muscle fiber in goldfish. *J. Comp. Physiol.* **124**, 111–116.

Johnston, I. A., and Moon, T. W. (1980) Exercise training in skeletal muscle of brook trout (*Salvelinus fontinalis*). *J. Exp. Biol.* **87**, 177–194.

Johnston, I. A., Ward, P. S., and Goldspink, G. (1975) Studies on the swimming musculature of the rainbow trout. I. Fibre types. *J. Fish Biol.* **7**, 451–458.

Jones, P. A., Scott-Burden, T., and Gevers, W. (1979) Glycoprotein, elastin, and collagen secretion by rat smooth muscle cells. *Proc. Nat. Acad. Sci.* **76**, 353–357.

Josephson, R. K. (1965) The coordination of potential pacemakers in the hydroid *Tubularia. Am. Zool.* **5**, 483–490.

Josephson, R. K. (1973) Contraction kinetics of the fast muscles used in singing by a katydid. *J. Exp. Biol.* **59**, 781–801.

Josephson, R. K. (1975) Extensive and intensive factors determining the performance of striated muscle. *J. Exp. Zool.* **194**, 135–154.

Josephson, R. K., and Halverson, R. C. (1971) High frequency muscles used in sound production by a katydid. I. Organization of the motor system. *Biol. Bull.* **141**, 411–433.

Josephson, R. K., and Young, D. (1981) Synchronous and asynchronous muscles in cicadas. *J. Exp. Biol.* **91**, 219–237.

Josephson, R. K., Stokes, D. R., and Chen, V. (1975) The neural control of contraction in a fast insect muscle. *J. Exp. Zool.* **193**, 281–300.

Joyce, G. C., Rack, P. M. H., and Westbury, D. R. (1969) The mechanical properties of cat soleus muscle during controlled lengthening and shortening movements. *J. Physiol.* **204**, 461–474.

Julian, F. (1971) The effect of calcium on the force-velocity relation of briefly glycerinated frog muscle fibres. *J. Physiol.* **218**, 117–145.

Julian, F. J., and Morgan, D. L. (1979a) The effect on tension of non-uniform distribution of length changes applied to frog muscle fibres. *J. Physiol.* **293**, 379–392.

Julian, F. J., and Morgan, D. L. (1979b) Intersarcomere dynamics during fixed-end tetanic contractions of frog muscle fibres. *J. Physiol.* **293**, 365–378.

Julian, F. J., and Morgan, D. L. (1980) Comparison of tension transients in frog twitch and slow fibres. *J. Physiol.* **301**, 97P.

Julian, F. J., and Moss, R. L. (1976) The concept of active state in striated muscle. *Circulation Res.* **38**, 53–59.

Julian, F. J., and Sollins, M. R. (1975) Variation of muscle stiffness with force at increasing speeds of shortening. *J. Gen. Physiol.* **66**, 287–302.

Julian, F. J., Moore, J. W., and Goldman, D. E. (1962) Membrane potentials of the lobster giant axon obtained by use of the sucrose gap technique. *J. Gen. Physiol.* **45**, 1195–1216.

Julian, F. J., Sollins, M. R., and Moss, R. L. (1978) Sarcomere length non-uniformity in relation to tetanic responses of stretched skeletal muscle fibres. *Proc. R. Soc. Lond. Ser. B* **200**, 109-116.

Kamada, T., and Kinosita, H. (1943) Disturbances initiated from naked surface of muscle protoplasm. *Jap. J. Zool.* **10**, 469-493.

Kane, R. E. (1980) Induction of either contractile or structural actin-based gels in sea-urchin egg cytoplasmic extract. *J. Cell Biol.* **86**, 803-809.

Kappagoda, C. T., Linden, R. J., and Mary, D. A. S. G. (1975) Gradation of the reflex response from atrial receptors. *J. Physiol.* **251**, 561-567.

Karlsson, J., and Komi, P. V. (1976) Motion and its significance on innervation, fibre recruitment and muscle metabolism in man. *Acta Physiol. Scand.* **98 (Suppl. 440)**, 12.

Karlsson, J., Sjodin, B., Tesch, P., and Larsson, L. (1978) The significance of muscle fibre composition to human performance capacity. *J. Rehab. Med.* **10**, 50-61.

Kater, S. B., Heyer, C., and Hegmann, J. P. (1971) Neuromuscular transmission in the gastropod mollusc *Helisoma trivolvia:* Identification of motoneurons. *Z. Vergl. Physiol.* **74**, 127-139.

Katz, A. M. (1967) Patterns of energy production and energy utilization in cardiac skeletal muscle. In: *Myocardial Contractility,* R. D. Tanz, F. Kavaler, and J. Roberts, Eds., pp. 401-415. Academic, New York and London.

Katz, B. (1936) Neuromuscular transmission in crabs. *J. Physiol.* **87**, 194-221.

Katz, B. (1939) The relation between force and speed in muscular contraction. *J. Physiol.* **96**, 45-64.

Katz, B. (1962) The transmission of impulses from nerve to muscle and the subcellular unit of synaptic action. *Proc. R. Soc. Lond. Ser. B* **155**, 455-479.

Katz, B., and Miledi, R. (1967a) The timing of calcium action during neuromuscular transmission. *J. Physiol.* **189**, 535-544.

Katz, B., and Miledi, R. (1967b) Modification of transmitter release by electrical interference with motor nerve endings. *Proc. R. Soc. Lond. Ser. B* **167**, 1-7.

Kawagoe, R. Onodera, K., and Takeuchi, A. (1981) Release of glutamate from the crayfish neuromuscular junction. *J. Physiol.* **312**, 225-236.

Kawaguti, S. (1964) Electron microscopy on the intestinal wall of the cuttlefish with special references to the muscles and the mucous cells. *Biol. J. Okayama Univ.* **10**, 93-103.

Kawaguti, S. (1968) Electron microscopy on sheathed muscle fibers of an opisthobranchiate mollusc. *Biol. J. Okayama Univ.* **14**, 13-20.

Kawai, M., Brandt, P. W., and Orentlicher, M. (1977) Dependence of energy transduction in intact skeletal muscles on the time in tension. *Biophys. J.* **18**, 161-172.

Kawata, H., and Chen, Y. F. (1972) Post-anodal inhibition of the electrically induced sustained contraction in the skeletal muscle. *Jap. J. Physiol.* **22**, 187-204.

Kean, C. J. C., Lewis, D. M., and McGarrick, J. D. (1974) Dynamic properties of denervated fast and slow twitch muscle of the cat. *J. Physiol.* **237**, 103-113.

Keil, E. M., and Sichel, F. J. M. (1936) The injection of aqueous solutions, including acetylcholine, into the isolated muscle fiber. *Biol. Bull.* **71**, 402A.

Kelly, D. E. (1967) Models of muscle Z-band fine structure based on a looping filament configuration. *J. Cell Biol.* **34**, 827-840.

Kelly, D. E., and Hayes, R. L. (1969) The ultrastructure of smooth cardiac muscle in the clam *Venus mercenaria. J. Morph.* **127**, 163-176.

Kelly, D. E., and Kuda, A. M. (1979) Subunits of the triadic junction in fast skeletal muscle as revealed by freeze-fracture. *J. Ultrastruct. Res.* **68**, 220-233.

Kendrick-Jones, J., and Jakes, R. (1977) Myosin-linked regulation – chemical approach. In: *Myocardial Failure,* G. Riecker, A. Weber, and J. Goodwin, Eds., pp. 28-40. Springer-Verlag, Berlin.

Kendrick-Jones, J., Szentkiralyi, E. M., and Szent-György, A. G. (1973) *C. Spr. Harb. Symp. Quant. Biol.* **37**, 47-54.

Kennedy, D. (1977) Inhibition in the center and the periphery. In: *Identified Neurons and Behavior of Arthropods,* G. Hoyle, Ed., pp. 101-116. Plenum, New York.

Kensler, R., Brink, P., and Dewey, M. M. (1977) The structure of the frog ventricular nexus. *J. Cell Biol.* **73**, 768-781.

Kerkut, G. A., and Leake, L. D. (1966) The effect of drugs on the snail pharyngeal retractor muscle. *Comp. Biochem. Physiol.* **17**, 623–633.

Kerrick, W. G. L., Hoar, P. E., Malencik, D. A., Stamps, L., and Fischer, E. H. (1979) Characterization of Ca^{2+} and Sr^{2+}-activated tension in functionally skinned chicken fibers of normal and dystrophic skeletal and normal cardiac muscle. *Pflüg. Arch.* **381**, 53–62.

Kerrick, W. G. L., Hoar, P. E., and Cassidy, P. S. (1980) Calcium-activated tension: the role of myosin light chain phosphorylation. *Fed. Proc.* **39**, 1558–1563.

Keynes, R. D. (1949) The ionic permeability of nerve and muscle membranes. Ph.D. thesis, Cambridge University.

Khabarova, A. Ya. (1963) *The Afferent Innervation of the Heart.* (Trs) Consultants Bureau, New York.

Kibbelaar. M. A., Ramaekers, F. C. S., Ringens, P. J., Selten-Versteegen, A. M. E., Poels, L. G., Jap, P. H. K., van Rossum, A. L., Feltkamp, T. E. W., and Bloemendal, H. (1980) Is actin in eye lens a possible factor in visual accommodation? *Nature* **285**, 504–508.

Kilarski, W., and Jakubowska, M. (1979) An electron microscope study of myofibril formation in embryonic rabbit skeletal muscle. *Z. Mikr. Anat. Forsch.* **93**, 1159–1181.

Kimizuka, H., and Koketsu, K. (1963) Changes in the membrane permeability of frog's sartorius muscle fibers in Ca-Free EDTA solution. *J. Gen. Physiol.* **47**, 379–392.

King, B. F., McKirdy, H. C., and Wai, S. S. (1977) Innervation of rabbit rectococcygeus muscle and the functional relationship of the muscle to the terminal large intestine. *J. Physiol.* **264**, 607–619.

King, J. A., and Govind, C. K. (1980) Development of excitatory innervation in the lobster claw closer muscle. *J. Comp. Neurol.* **194**, 57–70.

Klaassen, L. W., and Kammer, A. E. (1980) Habituation and effects of an octopamine agonist in the developing moth flight control system. *10th Ann. Mtg. Soc. Neurosci.* Abs. 213.10.

Klemm, N. (1972) Monoamine-containing nervous fibres in foregut and salivary gland of the desert locust, *Schistocerca gregaria* Forskal (Orthoptera, Acrididae). *Comp. Biochem. Physiol.* **43A**, 207–211.

Kluge, A. G. (Ed.) (1977) *Chordate Structure and Function,* 2nd Ed. Macmillan, New York.

Knapp, M. F. (1971) The contractile mechanism in obliquely striated body wall muscle of the earthworm, *Lumbricus terrestris. J. Cell Sci.* **8**, 413–425.

Knappeis, G. G., and Carlsen, F. (1962) The ultrastructure of the Z disc in skeletal muscle. *J. Cell Biol.* **13**, 323–335.

Knappeis, G. G., and Carlsen, F. (1968) The ultrastructure of the M line in skeletal muscle. *J. Cell Biol.* **38**, 202–211.

Kobayashi, M., and Ando, M. (1979) Evidence of an electrogenic sodium pump in molluscan radular muscle cells. *J. Comp. Physiol.* **132**, 141–147.

Kobayashi, M., and Muneoka, Y. (1980) Modulatory actions of octopamine and serotonin on the contraction of buccal muscles in *Rapana thomasiana.* I. Enhancement of contraction in radula protractor. *Comp. Biochem. Physiol.* **65C**, 73–79.

Kobayashi, M., Prosser, C. L., and Nagai, T. (1967) Electrical properties of intestinal muscle as measured intracellularly and extracellularly. *Am. J. Physiol.* **213**, 275–286.

Koester, J., Mayeri, E., Liebeswar, G., and Kandel, E. R. (1974) Neural control of circulation in *Aplysia. J. Neurophysiol.* **37**, 476–496.

Komi, P. V., and Karlsson, J. (1977) Skeletal muscle fibers and muscle enzyme activities in monozygous and dizygous twins of both sexes. *Acta Physiol. Scand.* **100**, 385–392.

Komi, P. V., and Karlsson, J. (1978) Skeletal muscle fiber types, enzyme activities and physical performance in young males and females. *Acta Physiol. Scand.* **103**, 210–218.

Koopowitz, H., and Chien, P. (1975) Ultrastructure of nerve plexus in flatworms. II. Sites of synaptic interactions. *Cell Tiss. Res.* **157**, 207–216.

Koopowitz, H., and Ewer, D. W. (1970) Observations on the myo-neural physiology of a polyclad flatworm: Inhibitory systems. *J. Exp. Biol.* **53**, 1–8.

Koretz, J. F. (1979a) Structural studies of synthetic filaments prepared from column-purified myosin. *Biophys. J.* **27**, 423–432.

Koretz, J. F. (1979b) Effects of C-protein on synthetic myosin filament structure. *Biophys. J.* **27**, 433–446.

Kosarov, D. (1979) The reaction time of single motor units in the human muscle. *Aggressologie* **20**, 279–285.

Kovacs, L., and Schneider, M. F. (1978) Contractile activation by voltage clamp depolarization of cut skeletal muscle fibres. *J. Physiol.* **277**, 483–506.

Kowagoe, R., Onodera, K., and Takeuchi, A. (1981) Release of glutamate from the crayfish neuromuscular junction. *J. Physiol.* **312**, 225–236.

Krasne, F. B., and Wine, J. J. (1977) Control of crayfish escape behavior. In: *Identified Neurons and Behavior of Arthropods,* G. Hoyle, Ed., pp. 275–292. Plenum, New York.

Krause, W. (1868) Mikroskopische Untersuchungen über die quergestreifte Muskelsubstanz. *Nach. Gesell. Göttingen* **17**, 357.

Kriebel, M. E. (1968) Cholinoceptive and adrenoceptive properties of the tunicate heart pacemaker. *Biol. Bull.* **135**, 174–180.

Krolenko, S. A., and Adamyan, S. Y. (1967) Permeability of muscle fibers to nonelectrolytes. *Cytology* **9**, 185–192.

Krueger, J. W., and Pollack, G. H. (1975) Myocardial sarcomere dynamics during isomatic contraction. *J. Physiol.* **251**, 627–643.

Krüger, P. (1952) *Tetanus und Tonus der quergestreiften Skelettmuskeln der Wirbeltiere und des Menschen.* Akademisher Verlag. Leipzig.

Kryvi, H. (1978) Ultrastructure of fiber types in axial muscles of the sharks *Etmopterus spinax* and *Galeus melastomus. J. Ultrastruct. Res.* **63**, 99A.

Kuffler, S. W. (1954) Mechanisms of activation and motor control of stretch receptors in lobster and crayfish. *J. Neurophysiol.* **17**, 558–574.

Kuffler, S. W., and Vaughan Williams, E. M. (1953a) Small-nerve junctional potentials. The distribution of small motor nerves to frog skeletal muscle, and the membrane characteristics of the fibres they innervate. *J. Physiol.* **121**, 289–317.

Kuffler, S. W., and Vaughan Williams, E. M. (1953b) Properties of the 'slow' skeletal muscle fibres of the frog. *J. Physiol.* **121**, 318–340.

Kuffler, S. W., and Yoshikami, D. (1975a) The distribution of acetylcholine sensitivity at the postsynaptic membrane of vertebrate skeletal twitch muscles: Iontophoretic mapping in the micron range. *J. Physiol.* **244**, 703–730.

Kuffler, S. W., and Yoshikami, D. (1975b) The number of transmitter molecules in a quantum: an estimate from iontophoretic application of acetylcholine at the neuromuscular synapse. *J. Physiol.* **251**, 465–482.

Kühne, W. (1859) Untersuchungen über Bewegungen und Veränderungen der contractilen Substanzen. *Arch. Anat. Physiol.* **27**. 784.

Kumari, S., Ojha, J., Saha, M. P., and Datta, J. S. (1979) Cytochemical differentiation of tonic (slow) and phasic (fast) muscle units in the ocular muscles of a fresh-water carp, *Catla catla* (Ham.). *Z. Mikr. Anat. Forsch.* **93**, 559–567.

Kupferman, I., and Weiss, K. R. (1978) The command neuron concept. *Behav. Brain. Res.* **1**, 1–40.

Kuriyama, H., and Csapo, A. (1961) A study of the parturine uterus with the microelectrode technique. *Endicrinology* **68**, 1010–1025.

Kusch, V. T. (1975) Der quantitative Gehalt von 5-hydroxytryptamine und dopamin im Nervensystem von *Periplaneta americana* (L.) *Zool. Jb. Physiol.* **79**, 513–517.

Kushner, P. D. (1979) The presence and function of dopamine in the lobster stomatogastric nervous system; the generation and regulation of motor rhythms. Ph.D. thesis, University of Oregon.

Kushner, P. D., and Ono, J. (1978) Endogenous dopamine in the stomatogastric ganglion and a single, identified neuron of the commissural ganglion. *Neurosci. Soc.* **9**, Abst. 198.

Kuwasawa, K. (1979) Effects of ACh and IJPs on the AV valve and the ventricle of *Dolabella auricularia. Am. Zool.* **19**, 129–143.

Ladanyi, P., and Leray, C. (1967) Caractérisation des fibres musculaires chez un céphalopode (*Octopus vulgaris*) sur la base de quelques reactions histoenzymologiques. *Ann. Histochim.* **12**, 395–403.

Laevis, P. C., and Kraft, E. L. (1978) Calcium binding to cardiac troponin C. *Arch. Biochem. Biophys.* **186**, 411–415.

Lakatta, E. G., and Spurgeon, H. A. (1980) Force staircase kinetics in mammalian cardiac muscle modulation by muscle length. *J. Physiol.* **299**, 337–352.

Lakie, M., Tsementzis, S. T., Walsh, E. G., and Wright, G. (1980) Anaesthesia does not (and cannot) reduce muscle tone? *J. Physiol.* **301**, 33P.

Landowne, D. (1974) Changes in fluorescence of skeletal muscle stained with merocyanine associated with excitation-contraction coupling. *J. Gen. Physiol.* **64**, 5A.

Lane, B. P., and Rhodin, J. A. C. (1964) Cellular interrelationships and electrical activity in two types of smooth muscle. *J. Ultrastruct. Res.* **10**, 470–488.

Lang, F., Costello, W. J., and Govind, C. K. (1977a) Development of the dimorphic claw closer muscles of the lobster Homarus americanus. I. Regional distribution of muscle fiber types in adults. *Biol. Bull.* **152**, 75–83.

Lang, F., Govind, C. K., and She, J. (1977b) Development of the dimorphic claw closer muscles of the lobster, *Homarus americanus.* II. Distribution of muscle fiber types in larval forms. *Biol. Bull.* **152**, 382–391.

Lang, F., Govind, C. K., and Costello, W. J. (1978) Experimental transformation of muscle fiber properties in lobster. *Science* **201**, 1037–1039.

Langelaan, J. W. (1915) On muscle tonus. *Brain* **38**, 235–380.

Langelaan, J. W. (1922) On muscle tonus. De Boer's experiments on the frog. *Brain* **45**, 434–453.

Lännergren, J. (1978) The force-velocity relation of isolated twitch and slow muscle fibres of *Xenopus laevis. J. Physiol.* **283**, 501–521.

Lännergren, J., and Smith, R. S. (1966) Types of muscle fibres in toad skeletal muscle. *Acta Physiol. Scand.* **68**, 263–274.

Lanzavecchia, G., and De Equileor, M. (1976) Studies on the helical and paramyosinic muscles. V. Ultrastructural morphology and contraction speed of muscular fibres of *Erpobdella octoculata* and *Erpobdella testacea* (Annelida, Hirudinea). *J. Submicr. Cytol.* **8**, 69–88.

Lanzavecchia, G., Valvassori, R., De Equileor, M., and Lanzavecchia, P. (1979) Three-dimensional reconstruction of the Nematomorpha muscle fiber. *J. Ultrastruct. Res.* **66**, 201–223.

Larimer, J. L. (1976) Command interneurons and locomotor behavior in crustaceans. In: *Neural Control of Locomotion,* R. M. Herman, S. Grillner, P. S. G. Stein, and D. G. Stuart, Eds., pp. 293–326. Plenum, New York.

Lawrentjen, B. J. (1934) Einige Bemerkungen über Fortschritte und Aüfgaben der Erforschung des autonomen Nervensystems. *Z. Mikr. Anat. Forsch.* **36**, 651–659.

Lehman, W. J., and Rebhun, L. J. (1971) The structural elements responsible for contraction in the ciliate *Spirostomum. Protoplasma* **72**, 153–167.

Lehman, W., and Szent-Györgyi, A. G. (1975) Regulation of muscular contraction. Distribution of actin control and myosin control in the animal kingdom. *J. Gen. Physiol.* **66**, 1–30.

Lehman, W., Kendrick-Jones, J., and Szent-Györgyi, A. G. (1973) Myosin-linked regulatory systems. Comparative studies. *C. Spr. Harb. Symp. Quant. Biol.* **37**, 319–330.

LeHouelleur, J. (1978) Electrical properties of phasis and tonic muscle fibres of the adbominal flexor muscles in crayfish. *J. Physiologie* **74**, 675–686.

Léoty, C. (1974) Membrane currents and activation of contraction in rat ventricular fibres. *J. Physiol.* **239**, 237–250.

Levi, H., and Ussing, H. H. (1948) The exchange of sodium and chloride ions across the fibre membrane of the isolated frog sartorius. *Acta Physiol. Scand.* **16**, 232–240.

Levine, B. A., Mercola, D., Coffman, D., and Thornton, J. M. (1977a) Calcium-binding by troponin C – protein magnetic resonance study. *J. Mol. Biol.* **115**, 743–760.

Levine, B. A., Thornton, J. M., Coffman, D., and Mercola, D. (1977b) High affinity calcium sites of troponin C direct and stabilize folding of almost all of structure. *Biophys. J.* **17**, 119A.

Levine, R. J. C., Dewey, M. M., Elfvin, M., and Walcott, B. (1974) *Lethocerus* flight muscle paramyosin: antibody localization and electrophoretic studies. *Am. J. Anat.* **141**, 453–458.

Levine, R. J. C., Elfvin, M., Dewey, M. M., and Walcott, B. (1976) Paramyosin in invertebrate muscles. II. Content in relation to structure and function. *J. Cell Biol.* **71**, 273–279.

Levitt, T. A., Loring, R. H., and Salpeter, M. M. (1980) Neuronal control of acetylcholine receptor turnover rate at a vertebrate neuromuscular junction. *Science* **210**, 550–551.

Lewis, D. M., Rowlerson, A., and Webb, S. N. (1980) Motor units in cat soleus muscle after long periods of cross-reinnervation. *J. Physiol.* **308**, 23P.

Licht, P. (1964) The temperature dependence of myosin-adenosinetriphosphatase and alkaline phosphatase in lizards. *Comp. Biochem. Physiol.* **12**, 331–340.

Liddell, E. G. T., and Sherrington, C. S. (1924) Reflexes in response to stretch (myotatic reflexes). *Proc. R. Soc. Lond. Ser. B* **96**, 212–242.

Liddell, E. G. T., and Sherrington, C. S. (1925) Further observations on myotatic reflexes. *Proc. R. Soc. Lond. Ser. B* **97**, 267–283.

Ling, G., and Gerard, R. W. (1949) The influence of stretch on the membrane potential of the striated muscle fiber. *J. Cell Comp. Physiol.* **34**, 397–405.

Lloyd, P. E. (1980) Modulation of neuromuscular activity by 5-hydroxytryptamine and endogenous peptides in the snail, *Helix aspersa*. *J. Comp. Physiol.* **139**, 333–339.

Locker, R. H., and Leet, N. B. (1975) Histology of highly-stretched beef muscle. I. The fine structure of grossly stretched single fibers. *J. Ultrastruct. Res.* **52**, 64–75.

Locker, R. H., and Leet, N. G. (1976a) Histology of highly-stretched beef muscle. II. Further evidence on the location and nature of gap filaments. *J. Ultrastruct Res.* **55**, 157–172.

Locker, R. H., and Leet, N. G. (1976b) Histology of highly-stretched beef muscle. IV. Evidence for movement of gap filaments through the Z-line, using the N_2-line and M-line as markers. *J. Ultrastruct. Res.* **56**, 31–38.

Locker, R. H., Daines, G. J., Carse, W. A., and Leet, N. G. (1977) Meat tenderness and the gap filaments. *Meat Science* **1**, 87–104.

Lohmann, K. (1929) Ueber die Pyrophosphatfraktion im Muskel. *Naturwissenschaften* **17**, 624–626.

Lømo, T. (1974) Neurotrophic control of colchicine effects on muscle. *Nature* **249**, 473–474.

Lømo, T. and Rosenthal, J. (1972) Control of ACh sensitivity by muscle activity in the rat. *J. Physiol.* **221**, 493–513.

Lømo, T., Westgaard, R. H., and Dahl. H. A. (1974) Contractile properties of muscle: control by pattern of muscle activity in the rat. *Proc. R. Soc. Lond. Ser. B* **187**, 99–103.

Longley, R. D. (1976) Analysis of the feeding mechanism in the mollusc *Tritonia*. Ph.D. thesis, University of Oregon.

Lopina, O. D. (1976) Calcium transport and ATPase activity of the sarcoplasmic reticulum of normal and denervated rabbit muscles. *Bull. Exp. Biol. Med.* **81**, 669–672.

Lorenz, K. Z. (1950) The comparative method in studying innate behavior patterns. *Symp. Soc. Exp. Biol.* **4**, 221–268.

Lorenzini, S. (1678) *Usservazio ni intorno alle Torpedim.* Onofri, Firenze.

Loewy, A. (1952) An actomyosin-like substance from the plasmodium of a myxomycete. *J. Cell Comp. Physiol.* **40**, 127–137.

Lowey, S., and Risby, D. (1971) Light chains from fast and slow muscle myosins. *Nature* **234**, 81–85.

Lowey, S., Silberstein, L., Gauthier, G. F., and Holt, J. C. (1979) Isolation and distribution of myosin isoenzymes. In: *Motility in Cell Function.* F. A. Pepe, J. W. Sanger, and V. T. Nachmias, Eds., pp. 53–67. Academic, New York.

Lowey, S., Slater, H. S., Weeds, A. G., and Baker, H. (1969) The substructure of the myosin molecule. I. Subfragments of myosin by enzymic degradation. *J. Mol. Biol.* **42**, 1–29.

Lowy, J., and Mulvany, M. J. (1973) Mechanical properties of guinea pig taenia coli muscles. *Acta Physiol. Scand.* **88**, 123–136.

Lowy, J., and Small, J. V. (1970) The organization of myosin and actin in vertebrate smooth muscle. *Nature* **227**, 46–51.

Lowy, J., Vibert, P. J., Haselgrove, J. C., and Poulsen, F. R. (1973) Structure of myosin elements in vertebrate smooth muscle. *Phil. Trans. R. Soc. Ser. B* **265**, 191–196.

Lucas, K. (1907) The analysis of complex excitable tissues by their response to electric currents of short duration. *J. Physiol.* **35**, 310–331.

Lucas, K. (1917) On summation of propagated disturbances in the claw of *Astacus* and on the double neuromuscular system of the adductor. *J. Physiol.* **51**, 1–35.

Luff, A. R., and Atwood, H. L. (1972) Membrane properties and contraction of single muscle fibers in the mouse. *Amer. J. Physiol.* **222**, 1435–1440.

Luff, A. R., and Praska, U. (1976) Properties of motor units of the frog sartorius muscle. *J. Physiol.* **258**, 673–685.

Lukowiak, K. (1979) L_7 modulation of gill withdrawal reflex habituation in *Aplysia. J. Neurobiol.* **10**, 255–271.

Lundsgaard, E. (1931) Ueber die Energetik der anaeroben Muskelkontraction. *Biochem. Z.* **233**, 323–343.

Luscher, H. R., Ruenzel, P., and Henneman, E. (1979) How the size of motoneurones determines their susceptibility to discharge. *Nature* **282**, 859–861.

Luttau, H. C., and Niedergerke, R. (1958) The antagonism between Ca and Na ions on the frog heart. *J. Physiol.* **143**, 486–505.

Luther, P., and Squire, J. (1978) Three-dimensional structure of the vertebrate muscle M-region. *J. Mol. Biol.* **125**, 313–324.

Lüttgau, H. C. (1963) The action of calcium ions on potassium contractures of single muscle fibres. *J. Physiol.* **168**, 679–697.

Lüttgau, H. C. (1965) The effect of metabolic inhibitors on the fatigue of the action potential in single muscle fibres. *J. Physiol.* **178**, 45–67.

Lüttgau, H. C., and Niedergerke, R. (1958) The antagonism between Ca and Na ions on the frog heart. *J. Physiol.* **143**, 486–505.

Lüttgau, H. C., and Spiecker, W. (1979) The effects of calcium deprivation upon mechanical and electrophysiological parameters in skeletal muscle fibres of the frog. *J. Physiol.* **296**, 441–429.

Lüttgau, H. C., Melzer, W., and Spiecker, W. (1977) The effects of Ca^{2+} removal on excitation-contraction coupling. *J. Physiol.* **271**, 45P.

Lutz, H., Weber, H., Billeter, R., and Jenny, E. (1979) Fast and slow myosin within single sketal muscle fibres of adult rabbits. *Nature* **281**, 142–144.

Lymn, R. W. (1975) Equatorial X-ray reflections and cross arm movement in skeletal muscle. *Nature* **258**, 770–772.

Machin, K. E., and Pringle, J. W. S. (1959) The physiology of insect fibrillar muscle. II. Mechanical properties of a beetle flight muscle. *Proc. R. Soc. Lond. Ser. B* **151**, 204–225.

Mackie, G. O. (1960) Studies on *Physalia physalis* L. *Disc. Repts.* **30**, 369–408.

Mackie, G. O. (1970) Neuroid conduction and the evolution of conducting tissues. *Q. Rev. Biol.* **45**, 319–332.

Mackie, G. O., and Bone, Q. (1976) Skin impulses and locomotion in an ascidian tadpole. *J. Mar. Biol. Ass. U.K.* **56**, 751–768.

Mackie, G. O., and Bone, Q. (1977) Locomotion and propagated skin impulses in salps (Tunicata: Thaliacea). *Biol. Bull.* **153**, 180–197.

Mackie, G. O., and Passano, L. M. (1968) Epithelial conduction in hydromedusae. *J. Gen. Physiol.* **52**, 600–621.

Mackie, G. O., and Singla, C. L. (1975) Neurobiology of Stomatoca. I. Action Systems. *J. Neurobiol.* **6**, 339–356.

Macmillan, D. L. (1976) Arthropod apodeme tension receptors. In: *Structure and Function of Proprioceptors in the Invertebrates,* P. J. Mill, Ed., pp. 427–442. Chapman and Hall, London.

Macmillan, D. L., and Dando, M. R. (1972) Tension receptors on the apodemes of muscles in the walking legs of the crab, *Cancer magister. Mar. Behav. Physiol.* **1**, 185–208.

Maeda, Y., Matsubara, I., and Yagi, N. (1979) Structural changes in thin filaments of crab striated muscle. *J. Mol. Biol.* **127**, 191–201.

Mandrino, M. (1977) Voltage-clamp experiments on frog single skeletal muscle fibres: evidence for a tubular sodium current. *J. Physiol.* **269**, 605–625.

Mangel, A. W., Nelson, D. O., Connor, J. A., and Prosser, C. L. (1979) Contractions of cat small intestinal smooth muscle in calcium-free solution. *Nature* **281**, 582–583.

Mannherz, H. G., Leigh, J. B., Holmes, K. C., and Rosenbaum, G. (1973) Identification of the transitory complex myosin-ATP by the use of α-β-methylene ATP. *Nature* **241**, 226–229.

Manton, S. (1977) *The Arthropods.* Oxford University Press, Oxford.

Marder, E. (1976) Cholinergic motor neurones in the stomatogastric system of the lobster. *J. Physiol.* **257**, 63–86.

Marek, M., and MacLennan, D. H. (1980) Assembly of the sarcoplasmic reticulum: biosynthesis of the high affinity calcium-binding protein in rat skeletal muscle cell cultures. *J. Biol. Chem.* **255**, 1327–1334.

Mark, R. F. (1980) Synaptic repression at neuromuscular junctions. *Physiol. Rev.* **60**, 355–395.

Marmont, G., and Wiersma, C. A. G. (1938) On the mechanism of inhibition and excitation of crayfish muscle. *J. Physiol.* **93**, 173–193.

Marston, S. B. (1973) The nucleotide complexes in glycerol-extracted muscle fibres. *Biochim. Biophys. Acta* **305**, 397–412.

Marston, S. B., and Tregear, R. T. (1972) Evidence for a complex between myosin and ADP in relaxed muscle fibres. *Nature New Biol.* **235**, 23–24.

Marston, S. B., and Treagear, R. T. (1974) Calcium binding and the activation of fibrillar insect flight muscle. *Biochim. Biophys. Acta,* **347**, 311–318.

Marston, S. B., Tregear, R. T., Rodger, C. D., and Clarke, M. L. (1979) Coupling between the enzymatic site of myosin and the mechanical output of muscle. *J. Mol. Biol.* **128**, 111–126.

Maruyama, K. (1965) A new protein factor hindering network formation of F actin in solution. *Biochim. Biophys. Acta* **94**, 208–225.

Maruyama, K., and Ebashi, S. (1965) α-Actinin, a new structural protein from striated muscle. II. Action on actin. *J. Biochem.* **58**, 13–19.

Maruyama, K., and Lakshminarayanaiah, N. (1977) Some electrical properties of the membrane of the barnacle muscle fibres under internal perfusion. *J. Memb. Biol.* **35**, 257–283.

Maruyama, K., and Yamashita, Y. (1962) The effects of barium, strontium and TEA ions on the production of action potentials in the cheliped muscle of the crayfish. *J. Fac. Sci. Hokkaido U.* **15**, 123–136.

Maruyama, K., Cage, P. E., and Bell, J. L. (1978a) The role of connectin in elastic properties of insect flight muscle. *Comp. Biochem. Physiol.* **61A**, 623–627.

Maruyama, K., Murakami, F., and Ohashi, K. (1977a) Connectin, an elastic protein of muscle. *J. Biochem.* **82**, 339–345.

Maruyama, K., Natori, R., and Nonomura, Y. (1976) New elastic protein from muscle. *Nature* **262**, 58–60.

Maruyama, K., Kimura, S., Koroda, M., and Handa, S. (1977b) Connectin, an elastic protein of muscle: its abundance in cardiac myofibrils. *J. Biochem.* **82**, 347–350.

Maruyama, K., Kunitomo, S., Kimura, S., and Ohashi, K. (1978b) I-protein: a new regulatory protein from vertebrate skeletal muscle. III. Function. *J. Biochem.* **81**, 243–247.

Maruyama, K., Kimura, S., Ishii, T., Kuroda, M., Ohashi, K., and Muramatsu, S. (1977c) β-Actinin: a regulatory protein of muscle. *J. Biochem.* **81**, 215–232.

Mathias, R. T. (1978) An analysis of the electrical properties of a skeletal muscle fiber containing a helicoidal T system. *Biophys. J.* **23**, 277–284.

Mathews, E. K., and O'Connor, M. D. L. (1979) Dynamic oscillations in the membrane potential of pancreatic islet cells. *J. Exp. Biol.* **81**, 75–91.

Matsubara, I. (1974) Light and X-ray diffraction studies on chick skeletal muscle under controlled physiological conditions. *J. Physiol.* **238**, 473–486.

Matsubara, I., and Millman, B. M. (1974) X-ray diffraction patterns from mammalian heart muscle. *J. Mol. Biol.* **82**, 527–536.

Matsumoto, Y. (1967a) Validity of the force-velocity relation for muscle contraction in the length region $l_{min} - l_o$. *J. Gen. Physiol.* **50**, 1125–1137.

Matsumoto, Y. (1967b) Theoretical series elastic element length in *Rana pipiens* sartorius muscles. *J. Gen. Physiol.* **50**, 1139–1156.

Matsuzaki, Y., Hamasaki, Y., and Said, S. I. (1980) Vasoactive intestinal peptide: a possible transmitter of nonadrenergic relaxation of guinea pig airways. *Science* **210**, 1252–1253.

Matthaei, E., and Tiegs, O. W. (1955) The path of the slow contractile wave in arthropod muscle fibre. *Phil. Trans. R. Soc. Lond. Ser. B* **238**, 349–359.

Mattisson, A. G. M., and Arvidsson, J. A. (1966) Some effects of electrical stimulation and exogenous metabolites on the contractile activity and the ultrastructure of the radula muscle of *Buccinum undatum*. *Z. Zellforsch.* **73**, 37–55.

Mattisson, A. G. M., and Birch-Andersen, A. (1962) On the fine structure of the mitochondria and its relation to oxidative capacity in muscle in various invertebrates. *J. Ultrastruct. Res.* **6**, 205–228.

Mauro, A., and Adams, W. R. (1961) The structure of the sarcolemma of the frog skeletal muscle fiber. *J. Biophys. Biochem. Cytol.* **10**, 177–185.

Mayeri, E. J., Koester, J., Kupferman, I., Liebeswar, G., and Kandel, E. (1974) Neural control of circulation in *Aplysia*. I. Motoneurons. *J. Neurophysiol.* **37**, 458–475.

Maylie, J., Morad, M., and Weiss, J. (1981) A study of pace-maker potential in rabbit sino-atrial node: measurement of potassium activity under voltage-clamp conditions. *J. Physiol.* **311**, 161–178.

Maynard, D. M., and Selverston, A. I. (1975) Organization of the stomatogastric ganglion of the spiny lobster. IV. The Pyloric System. *J. Comp. Physiol.* **100**, 161–182.

Mayr, R. (1971) Structure and distribution of fibre types in the external eye muscles of the rat. *Tissue and Cell* **3**, 433–462.

Meiss, D. E., and Govind, C. K. (1980) Heterogeneity of excitatory synapses at the ends of single muscle fibers in lobsters, *Homarus americanus*. *J. Neurobiol.* **11**, 381–395.

Meketa, F. (1971) Electrophysiological studies of the smooth muscle cell membrane of the rabbit common carotid artery. *J. Gen. Physiol.* **57**, 738–751.

Mello, W. C. de (1976) Sodium pump: its importance to intercellular communication in heart fibres. *Experientia* **32**, 355–356.

Mellon, DeF. (1968) Junctional physiology and motor nerve distribution in the fast adductor muscle of the scallop. *Science* **160**, 1018–1020.

Mellon, DeF., and Lnenicka, G. (1980) Structure and electrical properties of eye muscles in cave- and surface-dwelling crayfishes. *J. Exp. Biol.* **84**, 187–199.

Mellon, DeF., and Stephens, P. J. (1978) Limb morphology and function are transformed by contralateral nerve section in snapping shrimps. *Nature* **272**, 246–248.

Mendell, J. R. (1971) Unusual features of the T system of the pectoralis muscle of the chicken. *J. Ultrastruct. Res.* **37**, 383–387.

Mendelson, M. (1969) Electrical and mechanical characteristics of a very fast lobster muscle. *J. Cell Biol.* **42**, 548–563.

Mendelson, R. A., and Cheung, P. (1976) Muscle crossbridges: absence of direct effect of calcium movement away from the thick filaments. *Science* **194**, 190–191.

Mercola, D., Bullard, B., and Priest, J. (1975) Crystallization of troponin C. *Nature* **254**, 634–635.

Mesce, K. (1982) Calcium-bearing objects elicit shell selection behavior in a hermit crab. *Science.* **215**, 993–995.

Meyer, D. J., and Walcott, B. (1979) Differences in the responsiveness of identified motoneurons in the cockroach: role in the motor program for stepping. *Brain Res.* **178**, 600–605.

Meyerhof, O., and Lohmann, K. (1928) Ueber den Ursprung der Kontraktionswarme. *Naturwissenschaften* **15**, 670–672.

Meyling, H. A. (1953) Structure and significance of the peripheral extension of the autonomic nervous system. *J. Comp. Neurol.* **99**, 495–544.

Meyling, H. A. (1956) Das periphere Nervennetz und sein Zusammenhang mit den ortho-und parasympathischen Nervenfasern. *Acta Neuroveg.* **6**, 35–63.

Mikawa, T., Toyo-oka, T., Nonomura, Y., and Ebashi, S. (1977a) Essential factor of gizzard 'troponin' fraction. *J. Biochem.* **81**, 273–275.

Mikawa, T., Toyo-oka, T., Nonomura, Y., and Ebashi, S. (1977b) Kinetics of conformational change of troponin-C induced by proton binding or removal in the absence of calcium ions. *J. Biochem.* **81**, 277–283.

Miledi, R., Parker, I., and Schalow, G. (1977) Measurement of calcium transients in frog muscle by the use of arsenazo III. *Proc. R. Soc. Lond. Ser. B.* **198**, 201–210.

Milner-Brown, H. S., Stein, R. B., and Yemm, R. (1973a) The orderly recruitment of human motor units during voluntary isometric contractions. *J. Physiol.* **230**, 359–370.

Milner-Brown, H. S., Stein, R. B. and Yemm, R. (1973b) Changes in firing rate of human motor units during linearly changing voluntary contractions. *J. Physiol.* **230**, 371–390.

Mill, P. J. (Ed.) (1976) *Structure and Function of Proprioceptors in the Invertebrates. Chapman and* Hall, London.

Mill, P. J., and Knapp, M. F. (1970) The fine structure of obliquely striated body wall muscles in the earthworm, *Lumbricus terrestris* Linn. *J. Cell Sci.* **7**, 233–261.

Mill, P. J., and Riley J. (1972) An ultrastructural study of the body wall muscles of the pentastomid *Reighardia sternae,* parasitic on the herring gull *Larus argentatus. Z. Zellforsch.* **130**, 12–17.

Miller, A., and Tregear, R. T. (1970) Evidence concerning crossbridge attachment during muscle contraction. *Nature* **226**, 1060–1061.

Miller, T. (1968) Role of cardiac neurons in the cockroach heartbeat. *J. Ins. Physiol.* **14**, 1265–1275.

Miller, T., and James, J. (1976) Chemical sensitivity of the hyperneural nerve-muscle preparation of the american cockroach. *J. Ins. Physiol.* **22**, 981–988.

Millman, B. M., and Bennett, P. M. (1976) Structure of the cross-striated adductor muscle of the scallop. *J. Mol. Biol.* **103**, 439–467.

Millott, N. (1943a) The visceral nervous system of the earthworm. I. Nerves controlling the tone of the alimentary canal. *Proc. R. Soc. Lond. Ser. B* **131**, 271–295.

Millott, N. (1943b) The visceral nervous sytem of the earthworm. II. Evidence of chemical transmission and the action of sympathomimetic and parasympathomimetic drugs on the tone of the alimentary canal. *Proc. R. Soc. Lond. Ser. B* **131**, 362–373.

Minz, B. (1932) Pharmakologische Untersuchungen am Blutegelpräparat. Zugleich eine Methode zum biologischen Nachweis von Acetylcholin bei Anwesenheit anderer pharmakologisch wirksamer korpereigener Stoffe. *Arch. Exp. Path.* **168**, 292–304.

Misaki, N. (1964) Effect of administration of sexual hormones on the inonic fluxes in frog's sartorius muscle. *Kobe J. Med. Sci.* **10**, 69–83.

Mittenthal, J. E. (1975) A sliding filament model for skeletal muscle: dependence of isometric dynamics on temperature and sarcomere length. *J. Theor. Biol.* **52**, 1–16.

Mittenthal, J. E. (1980) On the form and size of crayfish legs regenerated after grafting. *Biol. Bull.* **159**, 700–713.

Miyahara, M., and Noda, H. (1980) Interaction of C-protein with myosin. *J. Biochem.* **87**, 1413–1420.

Miyahara, M., Kishi, K., and Noda, H. (1980) F-protein, a myofibrillar protein interacting with myosin. *J. Biochem.* 87, 1341–1345.

Miyamoto, Y. (1964) Effects of reduction of calcium ions on the electrical properties of an isolated single muscle fibre from the crayfish. *J. Fac. Sci. Hokkaido U.* **15**, 458–469.

Mobley, B. A., and Eisenberg, B. R. (1975) Size of components in frog skeletal muscle measured by methods of stereology. *J. Gen. Physiol.* **66**, 31–45.

Moffett, S. B. (1975) Motor patterns and structural interactions of basiischiopodite levator muscles in routine limb elevation and production of autotomy in the land crab *Cardisoma guanhumi. J. Comp. Physiol.* **96**, 285–305.

Moisescu, D. G., and Ashley, C. C. (1977) The effect of physiologically occurring cations upon aequorin light emission. *Biochim. Biophys. Acta* **460**, 189–205.

Mommaerts, W. F. H. M. (1965) Muscular contraction, a subject of molecular physiology. *Arch.*

Biol. **76**, 355-363.

Monster, A. W., and Chan, H. (1977) Isometric force production by motor units of extensor digitorum communis muscle in man. *J. Neurophys.* **40**, 1432–1443.

Monster, A. W., Chan, H. C., and O'Connor, D. (1978) Activity patterns of human skeletal muscles: relation to muscle fiber type composition. *Science* **200**, 314–317.

Moody-Corbett, F., and Pasztor, V. M. (1980) Innervation, synaptic physiology, and ultrastructure of three muscles of the second maxilla in crayfish. *J. Neurobiol.* **11**, 21–30.

Moon, T. W., and Hulbert, W. C. (1975) The ultrastructure of the mantle musculature of the squid *Symplectoteuthis oualaniensis. Comp. Biochem. Physiol.* **52**, 145–149.

Moore, J. W., and Narahashi, T. (1967) Tetrodotoxin's highly selective blockage of an ionic channel. *Fed. Proc.* **26**, 1655–1663.

Moos, C. (1981) Fluorescence microscope study of the binding of added C protein to skeletal muscle myofibrils. *J. Cell Biol.* **90**, 25–31.

Mooseker, M. S., and Tilney, L. G. (1975) The organization of an actin-filament-membrane complex: Filament polarity and membrane attachment in the microvilli of intestinal epithelial cells. *J. Cell Biol.* **67**, 725–743.

Morad, M., and Orkand, R. K. (1971) Excitation-contraction coupling in frog ventricle: evidence from voltage clamp studies. *J. Physiol.* **219**, 167–189.

Morin, W. J., and McLaughlin, E. (1973) Glycogen in crustacean fast and slow muscles. *Am. Zool.* **13**, 435–445.

Morison, G. D. (1927) The muscles of the adult honey-bee (*Apis mellifera,* L.) *Q. J. Microsc. Sci.* **71**, 563–651.

Moss, R. L., and Halpern, W. (1977) Elastic and viscous properties of resting frog skeletal muscle. *Biophys. J.* **17**, 213–228.

Moss, R. L., Sollins, M. R., and Julian, F. J. (1976) Calcium activation produces a characteristic response to stretch in both skeletal and cardiac muscle. *Nature* **260**, 619–621.

Moulins, M., and Nagy, F. (1981) Participation of an unpaired motor neurone in the bilaterally organized oesophageal rhythm in the lobsters *Jasus lalandii* and *Palinurus vulgaris. J. Exp. Biol.* **90**, 205–230.

Mrwa, U (1979) The inhibitory role of the regulatory light chain in vascular smooth muscle actomyosin ATPase. In: *Motility in Cell Function.* F. A. Pepe, J. W. Sanger and V. T. Nachmias, Eds., pp. 397–398. Academic, New York.

Mrwa, U., and Hartshorne, D. J. (1980) Phosphorylation of smooth muscle myosin and myosin light chains. *Fed. Proc.* **39**, 1564–1568.

Mullins, L. J., and Noda, K. (1963) The influence of sodium-free solutions on the membrane potential of frog muscle fibers. *J. Gen. Physiol.* **47**, 117–132.

Mulloney, B., and Selverston, A. I. (1974) Organization of the stomatogastric ganglion of the spiny lobster. III. Coordination of the two subsets of the gastric system. *J. Comp. Physiol.* **91**, 53–78.

Muneoka, Y., and Kobayashi, M. (1980) Modulatory actions of octopamine and serotonin on the contraction of buccal muscles in *Rapana thomasiana.* II. Inhibition of contraction in radula retractor. *Comp. Biochem. Physiol.* **65**, 81–86.

Murphy, D. B., and Tilney, L. G. (1974) The role of microtubules in the movement of pigment granules in teleost melanophores. *J. Cell Biol.* **61**, 757–779.

Murphy, R. A., Driska, S. P., and Cohen, D. M. (1977). Variations in actin to myosin ratios and cellular force generation in vertebrate smooth muscles. In: *Excitation - Contraction Coupling in Smooth Muscle*, R. Casteels, T. Godfraind, and J. C. Rüegg, Eds., pp. 417–426. Elsevier North Holland, Amsterdam.

Musick, J., and Hubbard, J. I. (1972) Release of protein from mouse motor nerve terminals. *Nature* **237**, 279–281.

Mykleburst, R. (1977) The membrane systems of the cardiac muscle cell of *Meganychtiphanes norvegica* (M. Sars), euphausiacea, crustacea. *Cell Tiss. Res.* **181**, 339–347.

McAlear, J. H. (1967) Comparison of neuron maps of the optic tracts of mouse and crayfish. In: *Invertebrate Nervous Systems,* C. A. G. Wiersma, Ed., pp. 263–268.

McDonald, K. L., Edwards, M. K., and McIntosh, R. J. (1979) Cross-sectional structure of the central mitotic spindle of *Diatoma vulgare.* Evidence for specific interactions between antiparallel microtubules. *J. Cell Biol.* **83**, 443–461.

McDowall, R. J. S. (1924) A vago-pressor reflex. *J. Physiol.* **59**, 41–47.

McKenna, O. C., and Rosenbluth, J. (1973) Myoneural and intermuscular junctions in a molluscan smooth muscle. *J. Ultrastruct. Res.* **42**, 434–450.

McLoon, S. C., and Buerger, A. A. (1974) Learning and retention of a shock avoidance task by rat using deafferented hind limb. *Physiol. Behav.* **12**, 39–43.

McNeill, P. A., and Hoyle, G. (1967) Evidence for superthin filaments. *Am. Zool.* **7**, 483–498.

McNeill, P. A., Burrows, M., and Hoyle, G. (1972) Fine structure of muscles controlling the strike of the mantis shrimp, *Hemisquilla. J. Exp. Zool.* **179**, 395–416.

Nachmias, V. T., and Ingram, W. C. (1970) Actomyosin from *Physarum polycephalum:* electron microscopy of myosin-enriched preparations. *Science* **170**, 743–745.

Nagai, T. (1970) Insect visceral muscle. Responses of the proctodeal muscles to mechanical stretch. *J. Ins. Physiol.* **16**, 437–448.

Nagai, T. (1972) Insect visceral muscle: Ionic dependance of electrical potentials in the proctodeal muscle fibers. *J. Ins. Physiol.* **18**, 2299–2318.

Nagai, T., and Brown, B. E. (1969) Insect visceral muscle. Electrical potentials and contraction in fibres of the cockroach proctodeum. *J. Ins. Physiol.* **15**, 2151-2167.

Naitoh, Y., and Kaneko, H. (1972) Rectivated triton-extracted models of *Paramecium:* modification of ciliary movement by calcium ions. *Science* **176**, 523–524.

Nakajima, Y. (1969) Fine structure of red and white muscle fibers and their neuromuscular junctions in the snake fish (*Ophiocephalus argus*). *Tissue & Cell* **1**, 229–246.

Nakajima, S., and Hodgkin, A. L. (1970) Effect of diameter on the electrical constants of frog skeletal muscle fibres. *Nature* **227**, 1053–1055.

Nakajima, S., Nakajima, Y., and Peachey, L. D. (1968) Speed of repolarization and morphology of glycerol-treated muscle. *J. Physiol.* **200**, 115–116P.

Narahashi, T., Deguchi, T., Urakawa, N., and Ohkubo, Y. (1960) Stabilization and rectification of muscle fiber membrane as manifested by tetrodotoxin. *Am. J. Physiol.* **198**, 934–938.

Nasse, O. (1869) Beitrage zur Physiologie der contractilen Substanz. *Pflüg. Arch.* **2**, 97–114.

Nastuk, W. L. (1953) The electrical activity of the muscle cell membrane at the neuro-muscular junction. *J. Cell. Comp. Physiol.* **42**, 249–272.

Natori, R. (1954) The property and contraction process of isolated myofibrils. *Jikei. Med. J.* **1**, 119–126.

Natori, R., Isojima, C., and Natori, R. (1976) On various types of periodic contractions in skinned fibers of skeletal muscle. *Jikei. Med. J.* **23**, 255–269.

Nayler, W. G., and Chipperfield, D. (1969) Effect of skeletal muscle potentiators on Ca^{2+} in cardiac sarcoplasmic reticulum. *Am. J. Physiol.* **217**, 609–614.

Needham, D. M. (1971) *Machina Carnis.* Cambridge University Press, Cambridge.

Needham, D. M., and Schoenberg, C. F. (1964) Proteins of the contractile mechanism of mammalian smooth muscle and their possible location in the cell. *Proc. R. Soc. Lond. Ser. B* **160**, 517–522.

Neering, I. R., and Morgan, K. G. (1980) Use of aequorin to study excitation-contraction coupling in mammalian smooth muscle. *Nature* **288**, 585–587.

Nelson, D. A., and Benson, E. S. (1963) On the structural continuities of the transverse tubular system of rabbit and human myocardial cells. *J. Cell Biol.* **16**, 297–313.

Nemeth, P., Hofer, H. W., and Pette, D. (1979) Metabolic heterogeneity of muscle fibres classified by myosin ATPase. *Histochemistry* **63**, 191–201.

Neville, A. C. (1963) Motor unit distribution of the dorsal longitudinal flight muscles in locusts. J. *Exp. Biol.* **40**, 123–136.

Newman, E. (1972) Contraction in *Stentor coeruleus*: a cinematic analysis. *Science* **177**, 447–449.

Newman, E. (1974) Scanning electron microscopy of the cortex of the ciliate *Stensor coeruleus.* A view from the inside. J. *Protozool.* **21**, 729–737.

Nicolai, L. (1936) Über das Beugungsspektrum der Quergestreifung des Skeletmuskels und einen direkten Beweis der Diskontinuität der tetanischen Kontraktion. *Pflüg. Arch.* **237**, 399–410.

Nicolaysen, K. (1976) The spread of the action potential through the T-system in hagfish twitch muscle fibers. *Acta Physiol.* **96**, 29–49.

Niedergerke, R. (1963) Movements of Ca in beating ventricles of the frog heart. *J. Physiol.* **167**, 551–580.

Nishihara, H. (1967) Studies on the fine structures of red and white fin muscles of the fish *Carassius auratus. Arch. Histol. Jap.* **28**, 425–447.

Nistal, M. and Paniagua, R. (1977) Variable patterns of actin filaments in mammalian skeletal muscle. *Cell Tiss. Res.* **185**, 285–288.

Nistal, M., Paniagua, R. and Morales, C. (1977) Irregular patterns of actin and myosin filaments in human skeletal muscle cells. *Cell Tiss. Res.* **181**, 403–407.

Njio, K. D., and Piek, T. (1977) Localization of sodium and potassium ions in a flight muscle of *Pieris brassicae. J. Ins. Physiol.* **23**, 919–929.

Noma, A., and Trautwein, W. (1978) Relaxation of the ACh-induced potassium current in the rabbit sinoatrial node. *Pflüg. Arch.* **372**, 43–51.

Nomura, H. (1963) The effect of stretching on the intracellular action potential from the cardiac muscle fibre of the marine mollusc, *Dolabella auricula. Sci. Rep. Tokyo Daig.* **11**, 153–165.

Nonidez, J. F. (1944) The present status of the neurone theory. *Biol. Rev.* **19**, 30–40.

Nonomura, Y. (1980) Leiotonin, regulatory protein of smooth muscle. In: *Muscle Contraction. Its Regulatory Mechanisms,* S. Ebashi, K. Maruyama, and M. Endo, Eds., pp. 329–346. Springer-Verlag, Berlin.

Nuccitelli, R., Poo, M. M., and Jaffe, L. F. (1977) Relations between ameboid movement and membrane-controlled electrical currents. *J. Gen. Physiol.* **69**, 743–763.

Ochs, S. (1974) Systems of material transport in nerve fibers (axoplasmic transport) related to nerve function and trophic control. *Ann. N. Y. Acad. Sci.* **228**, 202–223.

Ockleford, C. D., and Tucker, J. B. (1973) Growth, breakdown, repair, and rapid contraction of microtubule axopodia in the heliozoan *Actinophrys sol. J. Ultrastruct. Res.* **44**, 369–372.

Oetliker, H. (1980) Studies on the mechanism causing optical 'EC-coupling-signals' in skeletal muscle. *J. Physiol.* **305**, 26P.

Offer, G., and Elliot, A. (1978) Can a myosin molecule bind to two active filaments? *Nature* **271**, 325–329.

Ogievetskaya, M. M. (1976) The determination of evolution rates of acin by amino acid composition. *Acad. Sci. USSR.* Special printing.

Ohashi, M., and Nonomura, Y. (1979) Disappearance of smooth muscle thick filaments during K^+ contraction. *Cell. Struct. Funct.* **4**, 325–329.

Ohashi, K., Kimura, S., Deguchi, K., and Maruyama, K. (1977a) I-protein: a new regulatory protein from vertebrate skeletal muscle. I. Purification and characterization. *J. Biochem.* **81**, 233–236.

Ohashi, K., Masaki, T., and Maruyama, K. (1977b) I-protein: a new regulatory protein from vertebrate skeletal muscle. II. Localization. *J. Biochem.* **81**, 237–242.

Ohtsuki, I., Masaki, T., Nonomura, Y., and Ebashi, S. (1967) Periodic distribution of troponin along the thin filament. *J. Biochem.* **61**, 817–904.

Oliphant, L. W., and Cloney, R. A. (1972) The ascidian myocardium: sarcoplasmic reticulum and excitation-contraction coupling. *Z. Zellforsch.* **129**, 395–412.

Oplatka, A., and Tirosh, R. (1973) Active streaming in actomyosin solutions. *Biochim. Biophys. Acta* **305**, 684–688.

Oplatka, A., Gadasi, H., and Borejdo, J. (1974a) The contraction of "ghost" myofibrils and glycerinated muscle fibers irrigated with heavy meromyosin gut fragment-I. *Biochem. Biophys. Res. Comm.* **58**, 905–912.

Oplatka, A., Gadasi, H., Tirosh, R., Lamed, Y., Muhlrad, A., and Liron, V. (1974b) Demonstration of mechanochemical coupling in systems containing actin, ATP and non-aggregating active myosin deviatives. *J. Mechanochem. Cell Mot.* **2**, 295–306.

Orkand, R. K. (1962a) The relation between membrane potential and contraction in single muscle fibres. *J. Physiol.* **161**, 143–159.

Orkand, R. K. (1962b) Chemical inhibition of contraction in directly stimulated crayfish muscle fibres. *J. Physiol.* **164**, 103–115.

Orlov, J. (1924) Die Innervation des Darmes der Insekten (Larven von Lamellicorniern) *Z. Wiss. Zool.* **122**, 425–502.

Osborne, M. P. (1967a) The fine structure of neuromuscular junctions in the segmental muscles of the blowfly larva. *J. Ins. Physiol.* **13**, 827–833.

Osborne, M. P. (1967b) Supercontraction in the muscles of the blowfly larva. An ultrastructural study. *J. Ins. Physiol.* **13**, 827–835.

O'Shea, M., and Adams, M. (1980) Identification of giant peptidergic (Proctolin containing?) neurons in insects. *10th Am. Mtg. Soc. Neurosci.* Abstr. 240.11.

O'Shea, M., and Adams, M. E. (1981) Pentapeptide (proctolin) associated with an identified neuron. *Science* **213**, 567–569.

O'Shea, M., and Evans, P. D. (1979) Potentiation of Neuromuscular Transmission by an Octopaminergic Neurone in the Locust. *J. Exp. Biol.* **79**, 169–190.

Otsuka, M., and Ohtsuki, I. (1965) Mechanism of mescular paralysis by insulin with particular reference to familial periodic paralysis. *Nature* **207**, 300–301.

Otsuka, M., Kravitz, E. A., and Potter, D. D. (1967) Physiological and chemical architecture of a lobster ganglion with particular reference to gamma-aminobutyrate and glutamate. *J. Neurophysiol.* **30**, 725–752.

Otto, J. J. (1977) Orientation and behavior of epithelial cell muscle processes during *Hydra* budding. *J. Exp. Zool.* **202**, 307–322.

Ovalle, W. K. (1971) Fine structure of rat intrafusal muscle fibers. *J. Cell Biol.* **51**, 83–103.

Ozeki, M. (1962) Electrical and mechanical activities of the retractor pharynx muscle of a snail. *Jap. J. Physiol.* **12**, 293–311.

Ozeki, M. (1969) Crayfish muscle fiber: spike electrogenesis in fibers with long sarcomeres. *Science* **163**, 82–83.

Ozeki, M., and Sato, M. (1970) Potentiation of excitatory junctional potentials and glutamate-induced responses in crayfish muscle by 5'-ribonucleotides. *Comp. Biochem. Physiol.* **32**, 203–218.

Padyulka, H. A., and Gauthier, G. F. (1970) The ultrastructure of the neuromuscular junctions of mammalian red, white, and intermediate skeletal muscle fibers. *J. Cell Biol.* **46**, 27–41.

Page, E., and Fozzard, H. A. (1973) Capacitative, resistive, and syncylial properties of heart muscle - ultrastructural and physiological considerations. In: *The Structure and Function of Muscle,* G. H. Bourne, Ed., pp. 91–158. Academic, New York.

Page, S. G. (1965) A comparison of the fine structures of frog slow and twitch muscle fibres. *J. Cell Biol.* **26**, 477–497.

Page, S. G. (1968) Fine structure of tortoise skeletal muscle. *J. Physiol.* **197**, 709–715.

Page, S. (1969) Structure and some contractile properties of fast and slow muscles of the chicken. *J. Physiol.* **205**, 131–145.

Page, S., and Huxley, H. E. (1963) Filament lengths in striated muscle. *J. Cell Biol.* **19**, 369–391.

Painter, R. G., and McIntosh, A. T. (1979) The regional association of actin and myosin with sites of particle phagocytosis. *J. Supramol. Struct.* **12**, 369–384.

Paleček, F. (1965a) Mechanical aspects of muscle contraction during activity. *Physiol. Bohemoslov.* **14**, 23–27.

Paleček, F. (1965b) Mechanical aspects of denervated msucle contractions. *Physiol. Bohemoslov.* **14**, 28–31.

Panner, B. J., and Honig, G. R. (1967) Filament ultrastructure and organization in vertebrate smooth muscle. *J. Cell Biol.* **35**, 303–321.

Pantin, C. F. A. (1934) On the excitation of crustacean muscle. *J. Exp. Biol.* **11**, 11–27.

Pantin, C. F. A. (1935) The nerve net of the Actinozoa. I. Facilitation. *J. Exp. Biol.* **12**, 119–138.

Pantin, C. F. A. (1956) Comparative physiology of muscle. *Brit. Med. Bull.* **12**, 199–202.

Paolini, P. J., Roos, K. P., and Baskin, R. J. (1977) Light diffraction studies of sarcomere dynamics in single skeletal muscle fibers. *Biophys. J.* **20**, 221–232.

Parker, G. H. (1919) *The Elementary Nervous System.* Lippincott, Philadelphia.

Parker, G. H. (1932) Neuromuscular activities of the fishing filaments of *Physalia. J. Cell. Comp. Physiol.* **1**, 53–63.

Parnas, I., Abbott, B. C., Shapiro, B., and Lang, F. (1968) Neuromuscular system of *Limulus* leg closer muscle. *Comp. Biochem. Physiol.* **26**, 467–478.

Parnas, J. K., and Wagner, R. (1914) Ueber den Kohlehydratumsatz isolatier Amphibienmuskeln und über die Beziehungen zwischen Kohlehydratschwund und Milchsaurebildung im Muskel. *Biochem. Z.* **61**, 387–427.

Pasquali-Ronchetti, I. (1970) The ultrastructural organization of femoral muscles in *Musca domestica* (Diptera). *Tissue and Cell* **2**, 339–354.

Paul, R. J., and Peterson, J. W. (1977) The mechanochemistry of smooth muscle. In: *The Biochemistry of Smooth Muscle,* N. L. Stephends, Ed., pp. 15–40. University Park Press, Baltimore.

Pavlov, J. (1885) Wie die Muschel ihre Schaale offnet. *Pflüg. Arch.* **37**, 6–31.

Payne, C. M., Stern, L. Z., Curless, R. G., and Hannapel, L. K. (1975) Ultrastructural fiber typing in normal and diseased human muscle. *J. Neurol. Sci.* **25**, 99–108.

Peachey, L. D. (1961) Structure of the longitudinal body muscles of *Amphioxus. J. Biophys. Biochem. Cytol.* **10**, 159–176.

Peachey, L. D. (1965) The sarcoplasmic reticulum and transverse tubules of the frog's sartorius. *J. Cell Biol.* **25**, 209–232.

Peachey, L. D. (1972) Electrical events in the T-system of frog skeletal muscle. C. S. Harbor *Symp. Quant. Biol.* 37, 479–487.

Peachey, L. D., and Eisenberg, B. R. (1975) The T-system and striations of frog skeletal muscle are spiral. *Biophys. J.* **15**, 253A.

Peachey, L. D., and Eisenberg, B. (1978) Helicoids in the T system and striations of frog skeletal muscle fibers seen by high voltage electron microscopy. *Biophys. J.* **22**, 145–155.

Peachey, L. D., Takeichi, M., and Nagi, A. C. (1973) Muscle fiber types and innervation in adult cat extraocular muscles. In: *Exploratory Concepts in Muscular Dystrophy* II, Ed., pp. 246–254.

Pease, D. C. (1968) Structural features of unfixed mammalian smooth and striated muscle prepared by glycol dehydration. *J. Ultrastruct. Res.* **23**, 280–303.

Peeler, R. H., McGraw, C. P., and Wagoner, R. O. (1979) The effect of thiamine, ascorbic acid and alpha-tocopherol on experimental autoimmune myasthenia gravis (EAMG) in rats and rabbits. *J. Appl. Nutr.* **31**, 34–44.

Penfield, W., and Roberts, L. (1959) *Speech and Brain Mechanisms.* Princeton University Press, Princeton.

Pennycuick, C. J. (1964a) Frog fast muscle. I. Mechanical power output in isotonic twitches. *J. Exp. Biol.* **41**, 91–111.

Pennycuick, C. J. (1964b) Frog fast muscle. III. Twitches with isometric and inertial load. *J. Exp. Biol.* **41**, 273–289.

Pentreath, V. W., Osborne, N. N., and Cottrell, G. A. (1973) Anatomy of giant serotonin-containing neurones in the cerebral ganglia of *Helix pomatia* and *Limax maximus. Z. Zellforsch.* **143**, 1-20.

Pepe, F. A. (1966) Some aspects of the structural organization of the myofibril as revealed by antibody-staining methods. *J. Cell Biol.*. **28**, 505–526.

Pepe, F. A. (1967a) The myosin filament. I. Structural organization from antibody staining observed in electron microscopy. *J. Mol. Biol.* **27**, 203–225.

Pepe, F. A. (1967b) The myosin filament. II. Interaction between myosin and actin filaments observed using antibody staining in fluorescent and electron microscopy. *J. Mol. Biol.* **27**, 227–236.

Pepe, F. A. (1968) Analysis of antibody staining patterns obtained with striated myofibrils in fluorescence microscopy and electron microscopy. *Int. Rev. Cytol.* **24**, 193–231.

Pepe, F. A. (1971) Structure of the myosin filament of striated muscle. *Prog. Biophys. Mol. Biol.* **22**,

77–96.

Pepe, F. A. (1979) The myosin filament: molecular structure. In: *Motility in Cell Function.* F. A. Pepe, J. W. Sanger and V. T. Nachmias, Eds., pp. 103–116. Academic, New York.

Pepe, F. A., and Drucker, B. (1975) The myosin filament. III. C-protein. *J. Mol. Biol.* **99**, 609–617.

Pepe, F. A., and Drucker, B. (1979) The myosin filament. VI. Myosin content. *J. Mol. Biol.* **130**, 379–393.

Peretz, B., Jacklet, J. W., and Lukowiak, K. (1976) Habituation of reflexes in *Aplysia.* Contribution of the peripheral and central nervous systems. *Science* **191**, 396–399.

Perry, S. V., and Grey, T. C. (1956) A study of the effects of substrate concentration and certain relaxing factors on the magnesium-activated myofibrillar adenosine triphosphatase. *Biochem. J.* **64**, 184–192.

Perry, S. V., Cole, H. A., and Dhoot, G. K. (1979) The regulatory proteins of the I filament and the control of contractile activity in different types of striated muscle. In: *Motility in Cell Function.* F. A. Pepe, J. W. Sanger, and V. T. Nachmias, Eds., pp. 129–145. Academic, New York.

Peters, M. (1977) Innervation of the ventral diaphragm of the locust (*Locusta migratoria*). *J. Exp. Biol.* **69**, 23–32.

Peters, M. (1979) Motor innervation of the pharynx levator muscle of the snail, *Helix pomatia:* Physiological and histological properties. *J. Neurobiol.* **10**, 137–152.

Peterson, A. C. (1974) Chimaera mouse study shows absence of disease in genetically dystrophic muscle. *Nature* **248**, 561–564.

Peterson, R. P. (1963) A note on the structure of crayfish myofilaments. *J. Cell Biol.* **18**, 213–218.

Petrofsky, J. S., and Fitch, C. D. (1980) Contractile characteristics of skeletal muscles depleted of phosphocreatine. *Pflüg. Arch.* **384**, 123–129.

Pette, D., Ramirez, B. U., Müller, W., Simon, R., Exner, G. V., and Hildebrand, R. (1975) Influence of intermittent long-term stimulation on contractile, histochemical and metabolic properties of fiber populations in fast and slow rabbit muscles. *Pflüg. Arch.* **361**, 1–7.

Pflüger, H-J., and Burrows, M. (1978) Locusts use the same basic motor pattern in swimming as in jumping and kicking. *J. Exp. Biol.* **75**, 81–93.

Phillips, C. E. (1980) An arthropod muscle innervated by nine excitatory motor neurons. *J. Exp. Biol.* **88**, 249–258.

Pick, E. P., and Tullius, E. J. (1951) Pharmacological studies on the rigidity of striated muscle produced by ryanodine. *Arch. Int. Pharmacodyn.* **86**, 121–136.

Pickens, P. E. (1973) Bioelectric activity during the startle response of a Hemichordate worm. *J. Exp. Biol.* **58**, 295–304.

Pickett, J. B. (1980) The rate at which glucose enters nerve terminals in muscle varies with their histochemical type. *Neurology* **30**, 375A.

Pickett, J. B. (1980) Nerve terminals are as different as the muscle fibers they innervate. *Science* **210**, 927–928.

Piek, T., and Njio, K. D. (1979) Morphology and electrochemistry of insect muscle fibre membrane. *Adv. Ins. Physiol.* **16**, 185–250.

Piek, T., and Mantel, P. (1977) Myogenic contractions in locust muscle induced by proctolin and by wasp, *Philanthus triangulum,* venom. *J. Ins. Physiol.* **23**, 321–325.

Piek, T., Van Marle, J., Mantel, P., and Lind, A. (1979) Peripheral control of basic tonus and myogenic rhythm in the extensor tibiae of *Schistocerca gregaria. Comp. Biochem. Physiol.* **64A**, 565–569.

Pilar, G. (1967) Further studies of the electrical and mechanical responses of slow fibers in cat extraocular muscles. *J. Gen. Physiol.* **50**, 2289–2300.

Pilar, G., and Vaughan, P. C. (1969a) Electrophysiological investigations of the pigeon iris neuromuscular junctions. *Comp. Biochem. Physiol.* **29**, 51–72.

Pilar, G., and Vaughan, P. C. (1969b) Mechanical responses of the pigeon iris muscle fibres. *Comp. Biochem. Physiol.* **29**, 73–87.

Pilar, G., and Vaughan, P. C. (1971) Ultrastructure and contractures of the pigeon iris striated muscle. *J. Physiol.* **219**, 253–266.

Pinsker, H. D., and Willis, W. D. (Eds.) (1980) *Information Processing in the Nervous System.* Raven, New York.

Podlubnaya, Z. A., Tskhovrebova, L. A., Zaalishvili, M. M., and Stefanenko, G. A. (1975) Electron microscopic study of α-actinin. *J. Mol. Biol.* **92**, 357-361.

Podolsky, R. J., and Teichholz, L. E. (1970) The relation between calcium and contraction kinetics in skinned muscle fibres. *J. Physiol.* **211**, 19-35.

Polak, J. M., and Bloom, S. R. (1979) The diffuse neuroendocrine system. Studies of this newly discovered controlling system in health and disease. *J. Histochem. Cytochem.* **27**, 1398-1400.

Pollack, G. H., Iwazumi, T., ter Keurs, H. D. D. J., and Shibata, E. F. (1977) Sarcomere shortening in striated muscle occurs in stepwise fashion. *Nature* **268**, 757-759.

Pollard, T. D. (1977) Introductory remarks: myosin and control mechanisms. In: *Cell Motility,* Vol. 2, R. Goldman, T. Pollard, and J. Rosenbaum, Eds., pp. 685-687. Cold Spring Harbor Labs.

Pollard, T. D., and Ito, S. (1970) Cytoplasmic filaments in *Amoeba proteus.* I. The role of filaments in consistency changes and movement. *J. Cell Biol.* **46**, 267-289.

Pollard, T. D., and Korn, E. D. (1971) Filaments of *Amoeba proteus.* II. Binding of heavy meromyosin by thin filaments in motile cytoplasmic extracts. *J. Cell Biol.* **48**, 216-219.

Pollard, T. D., and Korn, E. D. (1973) *Acanthamoeba* myosin. I. Isolation from *Acanthamoeba castellanii* of an enzyme similar to muscle myosin. *J. Biol. Chem.* **248**, 4682-4690.

Pople, W., and Ewer, D. W. (1954) Studies on the myoneural physiology of Echinodermata. I. The pharyngeal retractor muscle of *Cucumaria. J. Exp. Biol.* **31**, 114-126.

Pople, W., and Ewer, D. W. (1955) Studies on the myoneural physiology of Echinodermata. II. Circumoral conduction in *Cucumaria. J. Exp. Biol.* **32**, 59-69.

Popescu, L. M. (1977) Cytochemical study of the intracellular calcium distribution in smooth muscle. In: *Exctiation-Contraction Coupling in Smooth Muscle,* R. Casteels, T. Godfraind, and J. C. Rüegg, Eds., pp. 13-23. Elsevier/North-Holland, Amsterdam, New York and Oxford.

Porteau, D., and Raymond, G. (1980) Slow inward barium current and contraction on frog single muscle fibres. *J. Physiol.* **303**, 91-109.

Porter, K. R. (1976) Introduction: motility in cells. In: *Cell Motility,* Vol. 1, R. Goldman, T. Pollard, and J. Rosenbaum, Eds., pp. 1-28. Cold Spring Harbor Labs.

Potter, J. D., and Gergely, J. (1975) The calcium and magnesium binding sites on troponin and their role in the regulation of myofibrillar adenosine triphosphatase. *J. Biol. Chem.* **250**, 4628-4633.

Price, H. M., Gordon, G. B., Pearson, C. M., Munsat, T. L., and Blumberg, J. M. (1966) New evidence for excessive accumulation of Z-band material in nemaline myopathy. *Proc. Nat. Acad. Sci.* **54**, 1398-1406.

Pringle, J. W. S. (1954) The mechanism of the myogenic rhythm of certain insect striated muscles. *J. Physiol.* **124**, 269-291.

Pringle, J. W. S. (1957) *Insect Flight.* Cambridge University Press.

Pringle, J. W. S. (1960) Models of muscle. *Symp. Soc. Exp. Biol.* **14**, 41-68.

Pringle, J. W. S. (1967) Evidence from insect fibrillar muscle about the elementary contractile process. *J. Gen. Physiol.* **50**, 139-156.

Prives, J., Christian, C., Penman, S., and Olden, K. (1980) Neuronal regulation of muscle acetylcholine receptors: role of muscle cytoskeleton and receptor carbohydrate. In: *Tissue Culture in Neurobiology,* E. Giccobini, A. Vernadekis, and A. Shahar, Eds., pp. 35-52. Raven, New York.

Prosser, C. L. (1974) Smooth muscle. *Ann. Rev. Physiol.* **36**, 503-535.

Prosser, C. L. (1979) Evolution and diversity of nonstriated cells. In: *Handbook of Physiology. The Cardiovascular System II,* pp. 635-670. The American Physiological Society.

Prosser, C. L. (1980) Evolution and diversity of nonstriated muscles. In: *Handbook of Physiology,* Sec. 2, Vol. 11, pp. 635-670.

Prosser, C. L., and Sperelakis, N. (1959) Electrical evidence for dual innervation of muscle fibers in the sipunculid *Golfingia (=Phascolosoma). J. Cell. Comp. Physiol.* **54**, 129-133.

Prosser, C. L., and Young, J. Z. (1937) Responses of muscles of the squid to repetitive stimulation of the giant nerve fibers. *Biol. Bull.* **73**, 237-241.

Prosser, C. L., Nystrom, R. A., and Nagai, T. (1965) Electrical and mechanical activity in intestinal muscles of several invertebrate animals. *Comp. Biochem. Physiol.* **14**, 53-70.

Pumphrey, R. J. (1938) The double innervation of muscles in the clam *(Mya arenaria)*. *J. Exp. Biol.* **15**, 500-505.

Putney, J., and Bianchi, C. P. (1974) Site of action of Dantrolene in frog sartorius muscle. *J. Pharmacol. Exp. Ther.* **189**, 202-212.

Pybus, J., and Tregear, R. T. (1975) The relationship of adenosine triphosphatase activity to tension and power output in insect flight muscle. *J. Physiol.* **247**, 71-89.

Rack, P. M. H., and Westbury, D. R. (1974) The short range stiffness of active mammalian muscle and its effect on mechanical properties. *J. Physiol.* **240**, 331-350.

Rall, J. A. (1979) Effects of temperature on tension, tension-dependent heat, and activation heat in twitches of frog skeletal muscle. *J. Physiol.* **291**, 265-275.

Rall, J. A. (1980) Effects of previous activity on the energetics of activation in frog skeletal muscle. *J. Gen. Physiol.* **75**, 617-631.

Rall, T. W., and Sutherland, E. W. (1957) Formation of a cyclic adenine ribonucleotide by tissue particles. *J. Biol. Chem.* **232**, 1065-1076.

Ramsay, J. A. (1940) A nerve-muscle preparation from the snail *J. Exp. Biol.* **17**, 96-115.

Ramsey, R. W., and Street, S. F. (1940) The isometric length-tension diagram of isolated skeletal muscle fibers of the frog. *J. Cell. Comp. Physiol.* **15**, 11-34.

Rapoport, S. I. (1972) Mechanical properties of the sarcolemma and myoplasm in frog muscle as a function of sarcomere length. *J. Gen. Physiol.* **59**, 559-585.

Rapoport, S. I., Peachey, L. D., and Goldstein, D. A. (1969) Swelling of the transverse tubular system in frog sartorius. *J. Gen. Physiol.* **54**, 166-177.

Rash, J. E., Shay, J. W., and Biesele, J. J. (1968) Urea extraction of Z bands, intercalated disks, and desmosomes. *J. Ultrastruct. Res.* **24**, 181-189.

Rathmayer, W. (1965) Neuromuscular transmission in a spider and the effect of calcium. *Comp. Biochem. Physiol.* **14**, 673-687.

Rathmayer, W., and Florey, E. (1974) Presynaptic inhibition of long duration at crab neuromuscular junctions. *Pflüg. Arch.* **348**, 77-81.

Rathmayer, W., and Walther, C. (1976) Mode of action and specificity of *Habrobracon* venom (Hymenoptera, Braconidae). In: *Animal, Plant, and Microbial Toxins,* Vol. 2, A. Ohsaka, K. Hayashi, and Y. Sawai, Eds., pp. 299-307. Plenum, New York.

Rathmayer, W., Walther, C., and Zlotkin, E. (1977) The effect of different toxins from scorpion venom on neurosmuclar transmission and nerve action potentials in the crayfish. *Comp. Biochem. Physiol.* **56C**, 35-39.

Rauh, F. (1922) Die Latenzzeit des Muskelelements. *Z. Biol.* **76**, 25-48.

Rayner, M. D., and Keenan, M. J. (1967) Role of red and white muscle in the swimming of the skipjack tuna. *Nature* **214**, 392-393.

Redfern, P., and Thesleff, S. (1971) Action potential generation in denervated rat skeletal muscle. *Acta Physiol. Scand.* **82**, 70-78.

Reed, C. G., and Cloney, R. A. (1977) Brachiopod tentacles: ultrastructure and functional significance of the connective tissue and myoepithelial cells in *Terebratalia*. *Cell Tiss. Res.* **185**, 17-42.

Reedy, M. K. (1964) The structure of actin filaments and the origin of the axial periodicity in the I-substance of vertebrate striated muscle. *Proc. R. Soc. Lond. Ser. B* **160**, 458-460.

Reedy, M. K. (1968) Ultrastructure of insect flight muscle. I. Screw sense and structural grouping in the rigor cross-brige lattice. *J. Mol. Biol.* **31**, 155-176.

Reedy, M. K., Holmes, K. C., and Tregear, R. T. (1965) Induced changes in orientation of the cross-bridges of glycerinated insect flight muscle. *Nature* **207**, 1276-1280.

Reep, R. L., Eisenstein, E. M., and Tweedle, C. D. (1980) Neuronal pathways involved in transfer of information related to leg position learning in the cockroach, *P. americana*. *Physiol. Physiol. Behav.* **24**, 501-513.

Rees, D., and Usherwood, P. N. R. (1972) Effects of denervation on the ultrastructure of insect muscle. *J. Cell Sci.* **10**, 667–682.

Rees, G. (1975) The arrangement and ultrastructure of the musculature, nerves and epidermis, in the tail of the cercaria of *Cryptocotyle lingua* (Creplin) from *Littorina littorea* (L.). *Proc. R. Soc. Lond. Ser. B* **190**, 165–186.

Reger, J. F. (1964) The fine structure of the fibrillar network and sarcoplasmic reticulum in smooth muscle cells of *Ascaris lumbricoides*. *J. Ultrastruct. Res.* **10**, 48–57.

Reger, J. F. (1969) Studies on the fine structure of muscle fibres and contained crystalloids in basal socket muscle of the entoproct, *Barentsia gracilis*. *J. Cell Sci.* **4**, 305–325.

Reger, J. F. (1976) Studies on the fine structure of the cercarial tail muscle of *Schistosoma sp.* (Trematoda). *J. Ultrastruct. Res.* **57**, 77–86.

Reger, J. F. (1978) A comparative study on the fine structure of tongue and cricothyroid muscle of the bat, *Myotis grisescens,* as revealed by thin section and freeze-fracture techniques. *J. Ultrastruct. Res.* **63**, 275–286.

Remedios, C. G. dos, and Gilmour, D. (1978) Is there a third type of filament in striated muscles? *J. Biochem.* **84**, 235–238.

Reuben, J. P., Brandt, P. W., Garcia, H., and Grundfest, H. (1967) Excitation-contraction coupling in crayfish. *Am. Zool.* **7**, 623–645.

Revel, J. P. (1962) The sarcoplasmic reticulum of the bat cricothyroid muscle. *J. Cell Biol.* **12**, 571–588.

Rheuben, M. B., and Reese, T. S. (1978) Three-dimensional structure and membrane specializations of moth excitatory neuromuscular synapse. *J. Ultrastruct. Res.* **65**, 95–111.

Rhodes, Jr., W. R. (1977) Anatomical and physiological correlates of asymmetry and courtship display of male fiddler crabs. Ph. D. thesis, University of Wisconsin.

Rice, M. J. (1970) Supercontracting and non-supercontracting visceral muscles in the tsetse fly, *Glossina austeni*. *J. Ins. Physiol.* **16**, 1109–1122.

Rice, M. J. (1973) Supercontracting striated muscle in a vertebrate. *Nature* **243**, 238–240.

Rice, R. V., Moses, J. A., McManus, G. M., Brady, A. C., and Blasik, L. M. (1970) The organization of contractile filaments in a mammalian smooth muscle. *J. Cell Biol.* **47**, 183–196.

Richardson, J. H., and Drake, P. D. (1979) The effects of zinc on fatigue of striated muscle. *J. Sports Med.* **19**, 133–134.

Richet, C. (1879) Contribution à la physiologie des centres neveux et des muscles de l'écrevisse. *Arch. Physiol. Norm. Path.* **6**, 262–292.

Ridge, R. M. A. (1971) Different types of extrafusal muscle fibres in snake costocutaneous muscles. *J. Physiol.* **217**, 393–418.

Ridge, R. M. A. P. (1980) Contraction times of single muscle fibres in motor units and muscles of the garter snake. *J. Physiol.* **308**, 103P.

Ridge, R. M. A. P., and Thomson, A. M. (1978) Polyneuronal innervation in amphibian skeletal muscle. *J. Physiol.* **277**, 2–4P.

Ridgway, E. B., and Gordon, A. M. (1975) Muscle activation: effects of small length changes on calcium release in single fibers. *Science* **189**, 881–884.

Ridgway, E. B., Ashley, C. C., and Hoyle, G. (1967) Calcium mediated light emission from single muscle fibres. *Fed. Proc.* **26**, 375A.

Rieger, R. M., and Rieger, G. E. (1975) Fine structure of the pharyngeal bulb in *Trilobodrilus* and its phylogenetic significance within Archiannelida. *Tissue and Cell* **7**, 267–279.

Ritchie, J. M. (1954a) The duration of the plateau of full activity in frog muscle. *J. Physiol.* **124**, 605–612.

Ritchie, J. M. (1954b) The effect of nitrate on the active state of muscle. *J. Physiol.* **126**, 155–168.

Ritchie, J. M., and Wilkie, D. R. (1958) The dynamics of muscular contraction. *J. Physiol.* **143**, 104–113.

Robbins, J. (1959) The excitation and inhibition of crustacean muscle by amino acids. *J. Physiol.* **148**, 39–50.

Robbins, J., and Van der Kloot, W. G. (1958) The effect of picrotoxin on peripheral inhibition in the crayfish. *J. Physiol.* **143**, 541–552.

Roberts, A. (1968a) Recurrent inhibition in the giant-fibre system of the crayfish and its effect on the excitability of the escape response. *J. Exp. Biol.* **48**, 545–567.

Roberts, A. (1968b) Some features of the central co-ordination of a fast movement in the crayfish. *J. Exp. Biol.* **49**, 645–656.

Roberts, A., and Stirling, C. A. (1971) The properties and propagation of a cardiac-like impulse in the skin of young tadpoles. *Z. Vergl. Physiol.* **75**, 388–401.

Robertson, H. A., and Steele, J. E. (1972) Activation of insect nerve-cord phosphorylase by octopamine and adenosine 3' 5'-monphosphate. *J. Neurochem.* **19**, 1603–1606.

Robertson, J. D. (1956) The ultrastructure of a reptilian myoneural junction. *J. Biophys. Cytol.* **2**, 381–394.

Robson, E. A. (1971) The behaviour and neuromuscular system of *Gonactinia prolifera,* a swimming sea-anemone. *J. Exp. Biol.* **55**, 611–640.

Robson, E. A., and Josephson, R. K. (1969) Neuromuscular properties of mesenteries from the sea-anemone *Metridium. J. Exp. Biol.* **50**, 151–168.

Röhlich, P. (1962) The fine structure of the muscle fiber of the leech *Hirudo medicinalis. J. Ultrastruct. Res.* **7**, 399–408.

Rollet, A. (1890) Untersuchungen über Kontraction und Doppelbrechung der quergenstreiften Muskelfasern. *Denk. Akad. Wiss. Wien* **58**, 41–98.

Rome, E. (1968) X-ray diffraction studies of the filament lattice of striated muscle in various bathing media. *J. Mol. Biol.* **37**, 331–344.

Rose, S. J., Pappas, G. D., and Kriebel, M. E. (1978) The fine structure of identified frog neuromuscular junctions in relation to synaptic activity. *Brain Res.* **144**, 213–239.

Rosenbluth, J. (1965) Ultrastructural organization of obliquely striated muscle fibers in *Ascaris lumbricoides, J. Cell Biol.* **25**, 495–515.

Rosenbluth, J. (1967) Obliquely striated muscle. III. Contraction mechanism of *Ascaris* body muscle. *J. Cell Biol.* **34**, 15–33.

Rosenbluth, J. (1968) Obliquely striated muscle. IV. Sarcoplasmic reticulum, contractile apparatus, and endomysium of the body muscle of a Polychaete, *Glycera,* in relation to its speed. *J. Cell Biol.* **36**, 245–259.

Rosenbluth, J. (1969) Sarcoplasmic reticulum of an unusually fast-acting crustacean muscle. *J. Cell Biol.* **42**, 534–547.

Rosenbluth, J. (1971) Myosin-like aggregates in trypsin-treated smooth muscle cells. *J. Cell Biol.* **48**, 174–188.

Rosenbluth, J. (1972) Myoneural junctions of two ultrastructurally distinct types in earthworm body wall muscle. *J. Cell Biol.* **54**, 566–579.

Ross, D. M. (1957) Responses of *Cerianthus* (Coelenterata). *Nature* **180**, 1368–1370.

Ross, D. M., and Sutton, L. (1967) Swimming sea anemones of Puget Sound: Swimming of *Actinostola* new species in response to *Stomphia coccinea. Science* **155**, 1419–1421.

Routledge, L. M., Amos, W. B., Yew, F. F., and Weis-Fogh, T. (1976) New calcium-binding contractile proteins. In: *Cell Motility,* Vol. 1., R. Goldman, T. Pollard, and J. Rosenbaum, Eds., pp. 93–113. Cold Spring Harbor Labs.

Rovainen, C. M. (1979) Neurobiology of lampreys. *Physiol. Rev.* **59**, 1007–1077.

Rowe, A. J. (1964) The contractile proteins of skeletal muscle. *Proc. R. Soc. Lond. Ser. B* **160**, 437–441.

Rowe, R. W. D. (1971) Ultrastructure of the Z line of skeletal muscle fibers. *J. Cell Biol.* **51**, 674–685.

Rowe, R. W. D. (1973) The ultrastructure of Z discs from white, intermediate, and red fibers of mammalian striated muscle. *J. Cell Biol.* **51**, 261–277.

Rüdel, R., and Taylor, S. R. (1979) Detection of free calcium transients in single frog muscle fibers using aequorin. In: *Detection and Movement of Free* Ca^{2+} *in Cells.* C. C. Ashley and A. K. Campbell, Eds., pp. 153–158. Elsevier, Amsterdam.

Rubinstein, N. A., and Holtzer, H. (1979) Fast and slow muscles in tissue culture synthesise only fast myosin. *Nature* **280**, 323-325.

Rubinstein, N., Mabuchi, K., Pepe, F., Salmons, S., Gergely, J., and Sreter, F. (1979) Fast-to-slow myosin transformation in existing muscle fibers during chromic stimulation. In: *Motility in Cell Function*. F. A. Pepe, J. W. Sanger, and V. T. Nachmias, Eds., pp. 367-370. Academic, New York.

Rüdel, R., and Taylor, S. R. (1973) Aequorin luminescence during contraction of amphibian skeletal muscle. *J. Physiol.* **233**, 5-6P.

Rüegg, J. C. (1968a) Oscillatory mechanism in fibrillar insect flight muscle. *Experientia* **24**, 529-536.

Rüegg, J. C. (1968b) Contractile mechanisms of smooth muscle. *Symp. Soc. Exp. Biol.* **22**, 45-66.

Runion, H. I., and Usherwood, P. N. R. (1966) A new approach to neuromuscular analysis in the intact free-walking insect preparation. *J. Ins. Physiol.* **12**, 1255-1263.

Saavedra, J. M., Brownstein, M. J., Carpenter, D. O., and Axelrod, J. (1974) Octopamine: presence in singe neurons of *Aplysia* suggests neurotransmitter function. *Science* **185**, 364-365.

Saetersdal, T., Engedal, H., Lie, R., and Myklebust, R. (1980) On the origin of Z-band material and myofilaments in myoblasts from the human atrial wall. *Cell Tiss. Res.* **207**, 21-29.

Saide, J. D., and Ullrick, W. C. (1973) Fine structure of the honeybee Z-disc. *J. Mol. Biol.* **79**, 329-337.

Saiki, M., and Hiramoto, Y. (1975) Control of ciliary activity in *Paramecium* by intracellular injection of calcium buffers. *Cell Str. Funct.* **1**, 33-41.

Salkoff, L., and Wyman, R. (1980) Facilitation of membrane electrical excitability in *Drosophila*. *Proc. Nat. Acad. Sci.* **77**, 6216-6220.

Salmons, S., and Sreter, F. A. (1976) Significance of impulse activity in transformation of skeletal muscle type. *Nature* **263**, 30-34.

Sanchez, J. A., and Stefani, E. (1978) Inward calcium current in twitch muscle fibres of the frog. *J. Physiol.* **283**, 197-209.

Sandow, A. (1965) Excitation-contraction coupling in skeletal muscle. *Pharm. Rev.* **17**, 265-320.

Sandow, A. (1966) Latency relaxation: a brief analytical review. *MCV Quart.* **2**, 82-89.

Sandow, A. and Isaacson, A. (1966) Topochemical factors in potentiation of contraction by heavy metal cations. *J. Gen. Physiol.* **49**, 937-961.

Sandow, A., and Preiser, H. (1964) Muscular contraction as regulated by the action potential. *Science* **146**, 1470-1472.

Sanger, J. W. (1971) Sarcoplasmic reticulum in the cross-striated adductor muscle of the bay scallop, *Aequipecten irridians*. *Z. Zellforsch.* **118**, 156-161.

Sanger, J. W. (1979) Cardiac fine structure in selected arthropods and molluscs. *Am. Zool.* **19**, 9-27.

Sanger, J. W., and McCann, F. V. (1968a) Ultrastructure of the myocardium of the moth, *Hyalophora cecropia*. *J. Ins. Physiol.* **14**, 1105-1111.

Sanger, J. W., and McCann, F. V. (1968b) Ultrastructure of moth alary muscles and their attachment to the heart wall. *J. Ins. Physiol.* **14**, 1539-1544.

Sanger, J. W., Sanger, J. M., and Gwinn, J. (1979) Actin and the mitotic spindle. In: *Motility in Cell Function*, F. A. Pepe, J. W. Sanger, and V. T. Nachmias Eds., pp. 313-323. Academic, New York.

Santer, R. M. (1972) Ultrastructural and histochemical studies on the innervation of the heart of a teleost, *Pleuronectes platessa* L. *Z. Zellforsch.* **131**, 519-528.

Santer, R. M., and Cobb, J. L. S. (1972) Fine structure of heart of teleost *Pleuronectes platessa* L. *Z. Zell. Mikr. Anat.* **131**, 1-14.

Sasaki, F. (1974) Histochemical and ultrastructural studies of tail muscles in the anuran tadpole. *Acta Histochem. Cytochem.* **7**, 239-256.

Satchell, D. G., and Twarog, B. M. (1979) Identification of dopamine release from the anterior byssus retractor muscle of *Mytilus californianus* in response to nerve stimulation. *Comp. Biochem. Physiol.* **64C**, 231-235.

Sato, A. (1960) Electrophysiological studies of the working mechanism of muscle walls of the stomach. *Jap. J. Physiol.* **10**, 359-373.

Schachat, F. H., Harris, H. E., and Epstein, H. F. (1977) Two homogeneous myosins in body-wall muscle of *Caenorhabditis elegans*. *Cell* **10**, 721–728.

Schädler, M., Steiger, G., and Rüegg, J. C. (1971) Mechanical activation and isometric oscillation of insect fibrillar flight muscle. *Pflüg. Arch.* **330**, 217–229.

Schäffer, B., and Örkenyi, J. (1972) Time-relations of initial volume decrease and contraction in frog muscles. *Acta Biochim. Biophys.* 7, 255–266.

Schalow, G., and Schmidt, H. (1979) Local development of action potentials in slow muscle fibres after complete or partial denervation. *Proc. R. Soc. Lond. Ser. B* **203**, 445–457.

Scherlag, B. J., Bartelstone, H. J., Wyte, S. R., and Hoffman, B. F. (1966) Variable diastolic ventricular compliance: general property of mammalian cardiac muscle. *Nature* **209**, 1246.

Schiaffino, S., and Margreth, A. (1969) Coordinated development of the sarcoplasmic reticulum and T system during postnatal differentiation of rat skeletal muscle. *J. Cell Biol.* **41**, 855–875.

Schiaffino, S., Nunzi, M. G., and Burighel, P. (1976) T system in ascidian muscle: organization of the sarcotubular system in the caudal muscle cells of *Botryllus schlosseri* tadpole larvae. *Tissue & Cell* **8**, 101–110.

Schlote, F. W. (1963) Die Muskulatur von Helix pomatia und ihre Innervation. *Proc. 1st Europ. Malac. Congr.* 113–151.

Schultzberg, M., Hökfelt, T., and Lundberg, J. M. (1980) Peptide neurons in the autonomic nervous system. *Adv. Biochem. Psychopharmacol.* **25**, 341–348.

Schmalbruch, H. (1971) "Rote" Muskelfasern. *Z. Zellforsch.* **119**, 120–146.

Schmalbruch, H. (1979) The membrane systems in different fibre types of the triceps surae muscle of cat. *Cell Tiss. Res.* **204**, 187–200.

Schmidt, H., and Stefani, E. (1977) Action potentials in slow muscle fibres of the frog during regeneration of motor nerves. *J. Physiol.* **270**, 507–517.

Schmidt, R. F. (1971) Presynaptic inhibition in the vertebrate central nervous system. *Ergb. Physiol.* **63**, 21–108.

Schoenberg, C. F., and Haselgrove, J. C. (1974) Filaments and ribbons in vertebrate smooth muscle. *Nature* **249**, 152–154.

Schoenberg, C. F., Rüegg, J. C., Needham, D. M., Schirmer, R. H., and Nemetchek-Gansler, H. (1966) A biochemical and electron microscope study of the contractile proteins in vertebrate smooth muscle. *Biochem. Z.* **345**, 255–266.

Schoenberg, M., and Podolsky, R. J. (1972) Length-force relation of calcium activated muscle fibers. *Science* **176**, 52–54.

Schofield, G. C. (1968) Anatomy of muscular and neural tissues in the alimentary canal. In: *Handbook of Physiology,* C. F. Code, Ed., Americn Physiological Society, Washington, D.C.

Scholte, H. R., and Busch, H. F. M. (1980) Early changes of muscle mitochondria in Duchene dystrophy. Partition and activity of unaffected boys and adults and patients. *J. Neurol. Sci.* **45**, 217–234.

Schulman, J. A., and Weight, F. F. (1976) Synaptic transmission: long-lasting potentiation by a postsynaptic mechanism. *Science* **194**, 1437–1439.

Scully, H. E., and Basmajian, J. V. (1969) Effect of nerve stimulation on trained motor unit control. *Arch. Phys. Med. Rehabil.* **50**, 32–33.

Seidel, J. C., and Gergely, J. (1973a) Investigation of conformational changes in spin-labeled myosin: implications for the molecular mechanism of muscle contraction. *C. Spr. Harb. Symp. Quant. Biol.* **37**, 187–193.

Seidel, J. C., and Gergely, J. (1973b) The use of spin labels in the study of muscle proteins. *Ann. N. Y. Acad. Sci.* **222**, 574–584.

Sellin, L. C., and McArdle, J. J. (1977a) Effect of ouabain on reinnervating mammalian skeletal muscle. *Europ. J. Pharmacol.* **41**, 337–340.

Sellin, L. C., and McArdle, J. J. (1977b) Colchicine blocks neurotrophic regulation of the resting membrane potential in reinnervating skeletal muscle. *Exp. Neurol.* **55**, 483–492.

Selverston, A. I. (1967) Structure and function of the transverse tubular system in crustacean muscle fibers. *Am. Zool.* 7, 515–525.

Selverston, A. I., Russel, D. F., Miller, J. P., and King, D. G. (1976) The stomatogastric nervous system: Structure and function of a small neural network. *Prog. Neurobiol.* **7**, 215–290.

Seraydarian, M. W., Jenden, D. J., and Abbott, B. C. (1961) The effect of ryanodine on relaxation of frog sartorius. *J. Pharmacol. Exp. Ther.* **135**, 374–381.

Shimada, Y., Fischman, D. A., and Moscona, A. A. (1967) The fine structure of embryonic chick skeletal muscle cells differentiated in vitro. *J. Cell Biol.* **35**, 445–453.

Shafiq, S. A., Leung, B., and Schitta, H. S. (1979) Freeze-fracture study of fiber types in normal human muscle. *J. Neurol. Sci.* **42**, 129–138.

Sharp, A. P., and Thomas, R C. (1979) The effect of chloride substitutes on pH: in crab muscle fibres. *J. Physiol.* **293**, 75P.

Shaw, J. (1955) Ionic regulation in the muscle fibres of *Carcinus maenas.* I. The electrolyte composition of single fibres. *J. Exp. Biol.* **32**, 383–396.

Shear, D. B. (1969) The electrical double layer, long range forces, and muscle contraction. *Physiol. Chem. Phys.* **1**, 495–508.

Shepheard, P. (1974) Control of head movement in the locust, *Schistocerca gregaria. J. Exp. Biol.* **60**, 735–767.

Sherman, R. G., and Atwood, H. L. (1971) Synaptic facilitation: long-term neuromuscular facilitation in crustaceans. *Science* **171**, 1248–1250.

Sherman, R. G., and Atwood, H. L. (1972) Correlated electrophysiological and ultrastructural studies of a crustacean motor unit. *J. Gen Physiol.* **59**, 586–615.

Sherman, R. G., and Luff, A. R. (1971) Structural features of the tarsal claw muscles of the spider *Eurypelma marxi* Simon. *Canad. J. Zool.* **49**, 1549–1556.

Sherrington, C. S. (1906) *The Integrative Action of the Nervous System.* Yale University Press, New Haven.

Shigeto, N. (1970) Excitatory and inhibitory actions of acetylcholine on hearts of oyster and mussel. *Am. J. Physiol.* **218**, 1773–1779.

Shimada, Y., Fischman, D. A., and Moscona, A. A. (1967) The fine structure of embryonic chick skeletal muscle cells differentiated in vitro. *J. Cell Biol.* **35**, 445–454.

Shinozaki, H., and Ishida, M. (1979) Pharmacological distinction between the excitatory junctional potential and the glutamate potential revealed by concanavalin A at the crayfish neuromuscular junction. *Brain Res.* **161**, 493–501.

Shizuta, Y., Shizuta, H., Gallo, M., Davies, P., Pastan, I., and Lewis, M. S. (1976) Purification and properties of filamin, an actin binding protein from chicken gizzard. *J. Biol. Chem.* **251**, 6563–6567.

Shoenberg, C. F., and Stewart, M. (1980) Filament formation in smooth muscle homogenates. *J. Muscle Res. Cell Motil.* **1**, 117–126.

Siddiqi, O., and Benzer, S. (1976) Neurophysiological defects in temperature-sensitive paralytic mutants of *Drosophila melanogaster. Proc. Nat. Acad. Sci.* **73**, 3253–3257.

Simpson, F. O., Rayns, D. G., and Ledingham, J. M. (1973) The ultrastructure of ventricular and atrial myocardium. In: *Ultrastructure of the Mammalian Heart,* C. E. Challice, and S. Viragh, Eds., pp. 1–41. Academic, New York.

Singh, S. I., and Singh, I. (1951) Active relaxation of unstriated muscle produced by adrenaline. *Nature* **168**, 829.

Sjodin, R. A. (1980) Contribution of Na/Ca transport to the resting membrane potential. *J. Gen. Physiol.* **76**, 99–108.

Sjöstrand, F. S. (1962) The connections between A- and I-band filaments in striated frog muscle. *J. Ultrastruct. Res.* **7**, 225–246.

Sjöstrom, M., Kidman, S., Larsen, K., and Angqist, K. A. (1979) Definition of M-band and Z-band appearances in histochemically-characterized human skeletal muscle fibers. *J. Ultrastruct. Res.* **66**, 86–87.

Skaer, H. L. B. (1974) The water balance of a serpulid polychaete *Mercierella enigmatica* (Fauvel). IV. The excitability of the longitudinal muscle cells. *J. Exp. Biol.* **60**, 351–370.

Skoglund, C. R. (1959) Neuromuscular mechanism of sound production in *Opsanus tan. Biol. Bull.*

117, 438–443.

Skoglund, C. R. (1961) Functional analysis of swim-bladder muscles engaged in sound production of the toadfish. *J. Biophys. Biochem. Cytol.* **10**, 187–200.

Slater, H. S., and Lowey, S. (1967) Substructure of the myosin molecule as visualized by electron microscopy, *Proc. Nat. Acad. Sci.* **58**, 1611–1618.

Sloper, J. C., and Partridge, T. A. (1980) Skeletal muscle: regeneration and transplantation studies. *Brit. Med. Bull.* **36**, 153–158.

Small, J. V. (1974) Contractile units in vertebrate smooth muscle cells. *Nature* **249**, 324–327.

Small, J. V. (1977) The contractile and cytoskeletal elements of vertebrate smooth muscle. In: *Excitation-Contraction Coupling in Smooth Muscle,* R. Casteels, T. Godfraind, and J. C. Rüegg, Eds., pp. 305–315. Elsevier/North Holland, Amsterdam, New York and Oxford.

Smit, W. A. (1958) Muscle fibres of the fast and the slow contracting muscles of the mesothoracic leg of the American cockroach (*Periplaneta americana* L.) *Nature* **181**, 1073–1074.

Smith, D.S. (1961) The structure of insect fibrillar flight muscle. *J. Biophys. Biochem. Cytol.* **10**, 123–158.

Smith, D. S. (1966a) The structure of intersegmental muscle fibers in an insect, *Periplaneta americana. J. Cell Biol.* **29**, 449–460.

Smith, D. S. (1966b) The organization and function of the sarcoplasmic reticulum and T-system of muscle cells. *Prog. Biophys. Mol. Biol.* **16**, 107–142.

Smith, D. S. (1966c) The organization of flight muscle fibers in the Odonata. *J. Cell Biol.* **28**, 109–126.

Smith, D. S. (1971) On the significance of cross-bridges between microtubules and synaptic vesicles. *Phil Trans. R. Soc Ser. B.* **261**, 395–405.

Smith, D. S. (1972) *Muscle.* Academic, New York.

Smith, D. S. (1968) *Insect Cells, Their Structure and Function.* Oliver & Boyd, Edinburgh.

Smith, D. S., and Anderson, M. E. (1972) The disposition of membrane systems in cardiac muscle of a lobster, *Homarus americanus. Tissue & Cell* **4**, 629–645.

Smith, D. S., Castillo, J. del, and Anderson, M. (1973) Fine structure and innervation of an annelid muscle with the longest recorded sarcomere. *Tissue & Cell* **5**, 281–302.

Smith, J. E. (1950) The motor neurons system of the starfish, *Astropecten irregularis* (Pennant), with special reference to the innervation of the tube feet and ampullae. *Phil. Trans. R. Soc. Lond. Ser. B* **234**, 521–558.

Smith, R. S., and Ovalle, Jr., W. K. (1973) Varieties of fast and slow extrafusal muscle fibers in amphibian hind limb muscles. *J. Anat.* **116**, 1–24.

Sobieszek, A. (1973) The fine structure of the contractile apparatus of the anterior byssus retractor muscle of *Mytilus edulis. J. Ultrastruct. Res.* **43**, 313–343.

Solandt, D. Y. (1936) The effect of potassium on the excitability and resting metabolism of frog's muscle. *J. Physiol.* **86**, 162–170.

Somlyo, A. P., Somlyo, A. V., Ashton, F. T., and Vallières, J. (1976a) Vertebrate smooth muscle: ultrastructure and function. In: *Cell Motility,* Vol. 1, R. Goldman, T. Pollard, and J. Rosenbaum, Eds., pp. 165–183. Cold Spring Harbor Labs.

Somlyo, A. P., Somlyo, A. V., Devine, C. E., and Rice, R. V. (1971) Aggregation of thick filaments into ribbons in mammalian smooth muscle. *Nature New Biol.* **231**, 243.

Somlyo, A. V. (1979) Bridging structures spanning the junctional gap at the triad of skeletal muscle. *J. Cell Biol.* **80**, 743–750.

Somlyo, A. V., Shuman, H., and Somlyo, A. P. (1979) The composition of the sarcoplasmic reticulum of striated muscle: electron probe studies. In: *Motility in Cell Function,* F. A. Pepe, J. W. Sanger, and V. T. Nachmias, Eds., pp. 195–209. Academic, New York.

Somlyo, A. V., Ashton, F. T., Lamanski, L., Vallières, J., and Somlyo, A. P. (1976b) Filament organization and dense bodies in vertebrate smooth muscle. In: *Biochemistry of Smooth Muscle,* N. L. Stephens, Ed., pp. 445–472. University Park Press, Baltimore.

Sommer, J. R., and Johnson, E. A. (1979) Ultrastructure of cardiac muscle. In: *Handbook of Physi-*

ology, pp. 113–186.

Sommer, J. R., and Waugh, R. A. (1976) The ultrastructure of the mammalian cardiac muscle cell, with special emphasis on the tubular membrane systems. *Am. J. Path.* **82**, 192–232.

Sommerkamp, H. (1928) Das Substrat der Dauerverkurzung am Froschmuskel. *Arch. Exp. Path. Pharm.* **128**, 99–115.

Sotavalta, O. (1953) Recording of high wing-stroke and thoracic vibration frequency in some midges. *Biol. Bull.* **104**, 439–444.

Spector, S. A., Gardiner, P. F., Zermicke, R. F., Roy, R. R., and Edgerton, V. R. (1980) Muscle architecture and force-velocity characteristics of cat soleus and medial gastrocnemius: implications for motor control. *J. Neurophysiol.* **44**, 951–960.

Spencer, M., and Worthington, C. R. (1960) A hypothesis of contraction in striated muscle. *Nature* **187**, 388–394.

Sperelakis, N. (1970) Ultrastructure of the neurogenic heart of *Limulus polyphemus. Z. Zellforsch* **116**, 443–463.

Sperelakis, N., and Prosser, C. L. (1959) Mechanical and electrical activity in intestinal smooth muscle. *Am. J. Physiol.* **196**, 850–856.

Squire, J. M. (1971) General model for the structure of all myosin-containing filaments. *Nature* **233**, 457–462.

Squire, J. M. (1972) General model of myosin filament structure. II. Myosin filaments and cross-bridge interactions in vertebrate striated and insect flight muscles. *J. Mol. Biol.* **72**, 125–138.

Squire, J. M. (1973) General model of myosin filament structure. III. Molecular packing arrangements in myosin. *J. Mol. Biol.* **77**, 291–323.

Squire, J. M. (1974) Symmetry and three-dimensional arrangement of filaments in vertebrate striated muscle. *J. Mol. Biol.* **90**, 153–160.

Sréter, F. A., Gergely, J., Salmons, S., and Romanul, F. (1973) Synthesis by fast muscle of myosin light chains characteristic of slow muscle in response to long-term stimulation. *Nature* **241**, 17–19.

Sréter, F. A., Luff, A. R., and Gergely, J. (1975) Effect of cross-reinnervation on physiological parameters and on properties of myosin and sarcoplasmic reticulum of fast and slow muscles of the rabbit. *J. Gen. Physiol.* **66**, 811–821.

Staley, N. A., and Benson, E. S. (1969) The ultrastructure of frog reticular cardiac muscle and its relationship to mechanisms of excitation-contraction coupling. *J. Cell Biol.* **38**, 99–114.

Stämpfli, R. (1954) A new method for measuring membrane potentials with external electrodes. *Experientia* **10**, 508–509.

Standaert, F. G. (1964) The mechanism of post-tetanic potentiation in cat soleus and gastrocnemius muscles. *J. Gen. Physiol.* **47**, 987–1001.

Standen, N. B., and Stanfield, P. R. (1979) Potassium depletion and sodium block of potassium currents under hyperpolarization in frog muscle. *J. Physiol.* **294**, 497–520.

Standte, H. W., Exner, G. U., and Pette, D. (1973) Effects of short-term, high intensity (sprint) training on some contractile and metabolic characteristics of fast and slow muscle of the rat. *Pflüg. Arch.* **344**, 159–168.

Stanfield, P. R. (1972) Electrical properties of white and red fibres of the elasmobranch fish *Scyliorhinus canicula. J. Physiol.* **222**, 161–186.

Starling, E. H. (1914) *The Linacre Lecture on the Law of the Heart.* Longmans, Green, London.

Starr, R., and Offer, G. (1971) Polypeptide chains of intermediate molecular weight in myosin preparations. *FEBS Lett.* **15**, 40.

Starratt, A. N., and Brown, B. E. (1975) Structure of the pentapeptide proctolin: proposed neurotransmitter in insects. *Life Sci.* **17**, 1253–1256.

Steedman, W. M. (1966) Microelectrode studies on mammalian vascular muscle. *J. Physiol.* **186**, 382–400.

Stefani, E., and Steinbach, A. B. (1969) Resting potential and electrical properties of frog slow muscle fibres. Effect of different external solutions. *J. Physiol.* **203**, 383–402.

Stein, R.J., Richter, W. R., Znssman, R. A., and Brynjolfsson, G. (1966) Ultrastructural characterization of *Daphnia* heart muscle. *J. Cell Biol.* **29**, 168–170.

Stendahl, O. I., Hartwig, J. H., Brotschli, E. A., and Stössel, T. P. (1980) Distribution of actin-binding protein and myosin in macrophages during spreading and phagocytosis. *J. Cell Biol.* **84**, 215–224.

Sten-Knudsen, O. (1954) The ineffectiveness of the 'windowfield' in the initiation of muscle contraction. *J. Physiol.* **125**, 396–404.

Stephens, R. E. (1965) Analysis of muscle contraction by ultraviolet microbeam disruption of sarcomere structure. *J. Cell Biol.* **25**, 129–139.

Stephenson, E. W. (1978) Properties of chloride-stimulated ^{45}Ca flux in skinned muscle fibers. *J. Gen. Physiol.* **71**, 411–430.

Stephenson, E. W., and Podolsky, R. J. (1977) Influence of magnesium on chloride-induced calcium release in skinned muscle fibers. *J. Gen. Physiol.* **69**, 17–35.

Stevens, C. F. (1979) The neuron. *Sci. Am.* **318**, 54–65.

Stockdale, F. E., and Baden, H. (1979) On the origin of fast and slow muscle. *J. Cell Biol.* **83**, 385A.

Stöhr, P. (1954) Zusammenfassende Ergebnisse über die Endigungsweise des vegetativen Nervensystems. *Acta Neuroveg.* **10**, 21–109.

Stössel, W., and Zebe, E. (1968) Zur intracellulären Regulation der Kontraktionsaktivität. *Pflüg. Arch.* **302**, 38–56.

Straub, F. B. (1943) Actin. In: *Studies on Muscle. Inst. Med. Chem. U. Szeged.* University of Szeged, Szeged, Hungary.

Stretton, A. O. W. (1976) Anatomy and development of the somatic musculature of the nematode *Ascaris. J. Exp. Biol.* **64**, 773–788.

Strickholm, A. (1966) Local sarcomere contraction in fast muscle fibres. *Nature* **212**, 835–836.

Strickholm, A. (1974) Intracellular generated potentials during excitation-contraction coupling in muscle. *J. Neurobiol.* **5**, 161–187.

Stromer, M. H., and Goll, D. E. (1972) Studies on purified α-actinin. II. Electron microscopic studies on the binding of α-actinin and tropomyosin to Z-line extracted myofibrils. *J. Mol. Biol.* **67**, 484–494.

Stromer, M. H., Hartshorne, D. J., and Rice, R. V. (1967) Removal and reconstitution of Z-line material in a striated muscle. *J. Cell Biol.* **35**, C23–C28.

Stromer, M. H., Hartshorne, D. J., Mueller, H., and Rice, R. V. (1969) The effect of various protein fractions on Z- and M-line reconstruction. *J. Cell Biol.* **40**, 167–178.

Stuart, A. E. (1970) Physiological and morphological properties of motoneurones in the central nervous system of the leech. *J. Physiol.* **209**, 627–646.

Stull, J. T., and Mayer, S. E. (1979) Biochemical mechanisms of adrenergic and cholinergic regulation of myocardial contractility. In: *Handbook of Physiology,* Sec. 2, Vol. 1, pp. 741–774. American Physiological Society, Bethesda, Md.

Suarez-Kurtz, G., and Sorenson, A. L. (1977) Inhibition by hypertonic solutions of Ca-dependent electrogenesis in single crab muscle fibers. *J. Gen. Physiol.* **70**, 491–505.

Suarez-Kurtz, R., Reuben, J. P., Brandt, P. W., and Grundfest, H. (1972) Membrane calcium activation in excitation-contraction coupling. *J. Gen. Physiol.* **59**, 676–688.

Sugaya, E., and Onozuka, M. (1978) Intracellular calcium: its release from granules during bursting activity in snail neurons. *Science,* **202**, 1195–1197.

Sugi, H. (1960) Propagation of contraction in the stalk muscle of *Carchesium. J. Fac. Sci. U. Tokyo* **8**, 603–615.

Sugi, H. (1968) Local activation of frog muscle fibres with linearly rising currents. *J. Physiol.* **199**, 549–567.

Sugi, H. (1972) Tension changes during and after stretch in frog muscle fibers. *J. Physiol.* **225**, 237–253.

Sugi, H. (1974) Inward spread of activation in frog muscle fibres investigated by means of high-speed microcinematography. *J. Physiol.* **242**, 219–235.

Sugi, H., and Atsumi, S. (1973) Localization of calcium-accumulating structures in the anterior byssal retractor muscle of *Mytillus edulis* and their role in excitation-contraction coupling. *Proc. Jap.*

Acad. **49**, 638–642.

Sugi, H., and Ochi, R. (1967a) The mode of transverse spread of contraction initiated by local activation in single crayfish muscle fibers. *J. Gen. Physiol.* **50**, 2145–2166.

Sugi, H., and Ochi, R. (1967b) The mode of transverse spread of contraction initiated by local activation in single frog muscle fibers. *J. Gen. Physiol.* **50**, 2167–2176.

Sugi, H., and Suzuki, S. (1978) Ultrastructural and physiological studies on the longitudinal body wall muscle of *Dolabella auricularia.* I. Mechanical response and ultrastructure. *J. Cell Biol.* **79**, 454–466.

Sugi, H., and Tameyasu, T. (1979) Origin of the instantaneous elasticity in single frog muscle fibers. *Experientia* **35**, 227–228.

Sugi, H., and Tsuchiya, T. (1979) The change in load-sustaining ability and in the series elasticity in *Mytilus* smooth muscle during isotonic shortening. *J. Physiol.* **288**, 635–648.

Sulakhe, P. V., and Dhalla, N. S. (1971) Excitation-contraction coupling in heart. VI. Demonstration of calcium-activated ATPase in dog heart sarcolemma. *Life Sci.* **10**, 185–191.

Sundararaman, V., and Nadakal, A. M. (1979) The ultrastructure of the striated muscle of the tail of *Cercaria chackai* Nad. *Cell Tiss. Res.* **201**, 479–486.

Suzuki, S., and Sugi, H. (1978) Ultrastructural and physiological studies on the longitudinal body wall muscle of *Dolabella auricularia.* II. Localization of intracellular calcium and its translocation during mechanical activity. *J. Cell Biol.* **79**, 467–478.

Swanson, C. J. (1971) Occurrence of paramyosin among the Nematomorpha, *Nature New Biol.* **232**, 122–123.

Swatland, H. J. (1979) Allometric radial growth in muscle, comparing fibres with strong and with weak adenosine triphosphatase activity. *J. Anat.* **129**, 591–596.

Szent-Györgyi, A. G. (1944) Actomyosin. In: *Studies on Muscle. Inst. Med. Chem.* University of Szeged, Szeged, Hungary.

Szent-Györgyi, A. (1951) *Chemistry of Muscular Contraction,* 2nd Ed. Academic, New York.

Szent-Györgyi, A. G. (1965) Evidence for conformational changes in actin on contraction. In: *Muscle,* E. Daniel, Ed., pp. 141–151. Pergamon, Oxford.

Takahashi, K. (1967a) The catch apparatus of the sea-urchin spine. I. Gross histology. *J. Fac. Sci. U. Tokyo* **4**, 109–120.

Takahashi, K. (1967b) The catch apparatus of the sea-urchin spine. II. Responses to stimuli. *J. Fac. Sci. U. Tokyo* **4**, 121–130.

Takahashi, K., Miyazaki, S., and Kidokoro, Y. (1971) Development of excitability in embryonic muscle cell membranes in certain tunicates. *Science* **171**, 415–418.

Takeda, K. (1967) Permeability changes associated with the action potential in procaine-treated crayfish abdominal muscle fibers. *J. Gen. Physiol.* **50**, 1049–1074.

Takeuchi, A., and Takeuchi, N. (1959) Active phase of frog's end-plate potential. *J. Neurophysiol.* **22**, 395–411.

Tameyasu, T., and Sugi, H. (1976) Series elastic component and force-velocity relation in anterior byssal retractor muscle of *Mytilus edulis* during active and catch contractions. *J. Exp. Biol.* **64**, 497–510.

Tariguchi, M., Yamazaki, K., and Ohta, J. (1978) Extraction of contractile protein from myxamebae of *Physarum polycephalum. Cell Str. Funct.* **4**, 181–190.

Tasaki, I., and Kamiya, N. (1964) A study on electrophysiological properties of carnivorous amoebae. *J. Cell Comp. Physiol.* **63**, 365–380.

Taub, E., Bacon, R. C., and Berman, A. J. (1965) Acquisition of a trace-conditioned avoidance response after deafferentation of the responding limb. *J. Comp. Physiol. Psych.* **59**, 275–279.

Taylor, D. L. (1976) Quantitative studies on the polarization optical properties of striated muscle. *J. Cell Biol.* **68**, 497–511.

Taylor, D. L., Rhodes, J. A., and Hammond, S. A. (1976) The contractile basis of amoeboid movement. II. Structure and contractility of motile extracts and plasmalemma-ectoplasm ghosts. *J. Cell Biol.* **70**, 123–143.

Taylor, F. L., Condeelis, J. S., Moore, P. L., and Allen R. D. (1973) The contractile basis of amoeboid movement. I. The chemical control of motility in isolated cytoplasm. *J. Cell Biol.* **59**, 378–394.

Taylor, D. L., Wang, Y. L., Heiple, J., and Hellewell, S. (1979) The contractile cytoskeleton of amoeboid cells *in vitro* and *in vivo.* In: *Motility in Cell Function,* F. A. Pepe, J. W. Sanger, and V. T. Nachmias, Eds., pp. 263–283. Academic, New York.

Taylor, S. R., Preiser, H., and Sandow, A. (1969) Mechanical threshold as a factor in excitation-contraction coupling. *J. Gen. Physiol.* **54**, 352–368.

Teräväinen, H., and Rovainen, C. M. (1971) Fast and slow motoneurons to body muscle of the sea lamprey. *J. Neurophysiol.* **34**, 990–998.

Tevaarwerk, G. J. M., Strickland, K. P., Lin, C. H., and Hudson, A. J. (1979) Studies on insulin resistance and insulin receptor binding in myotonia dystrophica. *J. Clin. Endocrin. Metab.* **49**, 216–222.

Thomas, J. S., Schmidt, E. M., and Hambrecht, F. T. (1978) Facility of motor unit control during tasks defined directly in terms of unit behaviors. *Exp. Neurol.* **59**, 384–395.

Thompson, C. M., and Wolpert, L. (1963) Isolation of motile cytoplasm from *Amoeba proteus. Exp. Cell Res.* **32**, 156–160.

Thurm, U. (1968) Steps in transducer process of mechanoreceptors. *Symp. Zool. Soc. Lond.* **23**, 199–216.

Tiegs, O. W. (1955) The flight muscles of insects – their anatomy and histology; with some observations on the structure of striated muscle in general. *Phil. Trans. R. Soc. Lond. Ser. B* **238**, 221–348.

Tilney, L. G. (1976) Nonfilamentous aggregates of actin and their association with membranes. In: *Cell Motility,* R. Goldman, T. Pollard, and J. Rosenbaum, Eds., pp. 513–528. Cold Spring Harbor Labs.

Tinbergen, N. (1951) *The Study of Instinct.* Oxford University Press, Oxford.

Titmus, M. (1977) Identified neuromuscular junctions of locust jumping muscle. Ph.D. thesis, University of Oregon.

Titmus, M. J. (1981) Ultrastructure of identified fast excitatory, slow excitatory and inhibitory neuromuscular junctions in the locust. *J. Neurocyt.* **10**, 363–385.

Titmus, M., and Hoyle, G. (1977) Morphology of identified neuromuscular junctions of locust jumping muscles. *7th Ann. Mtg. Soc. Neurosci.,* Abstr. 595.

Toda, N. (1968) Influence of sodium ions on the membrane potential of the sino-atrial node in response to sympathetic nerve stimulation. *J. Physiol.* **196**, 677–692.

Tomita, T., and Sakamoto, Y. (1977) Electrical and mechanical activities in the guinea-pig stomach muscle. In: *Excitation-Contraction Coupling in Smooth Muscle,* R. Casteels, T. Godfraind, and J. C. Rüegg, Eds., pp. 37–46. Elsevier/North Holland, Amsterdam, New York, and Oxford.

Toselli, P. A., and Pepe, F. A. (1968) The find structure of the ventral intersegmental abdominal muscles of the insect *Rhodnius prolixus* during the motling cycle. I. Muscle structure at molting. *J. Cell Biol.* **37**, 445–461.

Tregear, R. T., and Marston, S. B. (1979) The crossbridge theory. *Ann. Rev. Physiol.* **41**, 723–736.

Trendelenburg, P. (1917) Physiologische und pharmakologische Versuche über die Dünnarmperistaltik. *Arch. Exp. Path. Pharmakol.* **81**, 55–129.

Trombitás, K., and Tigyi-Sebes, A. (1974) Direct evidence for connecting C filaments in flight muscle of honey bee. *Acta Biochim. Biophys.* **9**, 243–253.

Trombitás, K., and Tigyi-Sebes, A. (1979) The continuity of thick filaments between sarcomeres in honey bee flight muscle. *Nature* **281**, 319–320.

Tsien, R. W., and Weingart, R. (1976) Inotropic effect of cyclic AMP in calf ventricular muscle studied by a cut end method. *J. Physiol.* **260**, 117–141.

Tsong, T. Y., Karr, T., and Harrington, W. F. (1979) Rapid helix-coil transitions in the S-2 region of myosin. *Proc. Nat. Acad. Sci.* **76**, 1109–1113.

Twarog, B. M. (1954) Responses of a molluscan smooth muscle to acetylcholine and 5-hydroxytryptamine. *J. Cell Comp. Physiol.* **44**, 141–164.

Twarog, B. M. (1960) Innervation and activity of a molluscan smooth muscle. *J. Physiol.* **152**, 220–235.

Twarog, B. M. (1967) Excitation of *Mytilus* smooth muscle. *J. Physiol.* **192**, 857–868.

Twarog, B. M. (1979) The nature of catch and its control. In: *Motility in Cell Function.* F. A. Pepe, J. W. Sanger, and V. T. Nachmias, Eds., pp. 231-241. Academic, New York.

Twarog, B. M., and Muneoka, Y. (1972) Calcium and the control of contraction and relaxation in a molluscan catch muscle. *C. Spr. Harb. Symp. Quant. Biol.* **37**, 489–503.

Twarog, B. M., Dewey, M. M., and Hidaka, T. (1973) The structure of *Mytilus* smooth muscle and the electrical constants of the resting muscle. *J. Gen. Physiol.* **61**, 207–221.

Tyrer, N. M. (1973) Functional development of the reticular system in an insect muscle with synchronously differentiating cells. *J. Cell Sci.* **12**, 197–215.

Uchizono, K. (1967) Characteristics of excitatory and inhibitory synapses in the central nervous system of the cat. *Nature* **207**, 642–643.

Uehara, Y., Campbell, G. R., and Burnstock, G. (1971) Cytoplasmic filaments in developing and adult vertebrate smooth muscle. *J. Cell Biol.* **50**, 484–497.

Uexküll, J. von. (1912) Studien über den Tonus. VI. Die Pilgermuschel. *Biol. Z.* **58**, 305–332.

Ullrick, W. C. (1967) A theory of contraction for striated muscle. *J. Theor. Biol.* **15**, 53–69.

Usherwood, P. N. R. (1962) The action of the alkaloid ryanodine on insect skeletal muscle. *Comp. Biochem. Physiol.* **6**, 181–199.

Usherwood, P. N. R. (1977) Neuromuscular transmission in insects. In: *Identified Neurons and Behavior of Arthropods,* G. Hoyle, Ed., pp. 31–48. Plenum, New York and London.

Usherwood, P. N. R. (1978) Amino acids as neurotransmitters. *Adv. Comp. Physiol. Biochem.* 7, 227–309.

Usherwood, P. N. R., and Cull-Candy, S. G. (1975) Pharmacology of somatic nerve-muscle synapses. In: *Insect Muscle,* P. N. R. Usherwood, Ed., pp. 207–280. Academic, London and New York.

Usherwood, P. N. R., and Grundfest, H. (1965) Peripheral inhibition in skeletal muscle in insects. *J. Neurophysiol.* **28**, 497–518.

Usherwood, P. N. R., and Runion, H. I. (1970) Analysis of the mechanical responses of metathoracic extensor tibiae muscles of free-walking locusts. *J. Exp. Biol.* **52**, 39–58.

Valvassori, R., de Equileor, M., and Lanzavecchia, G. (1978) Flight muscle differentiation in nymphs of a dragonfly *Anax imperator. Tissue & Cell* **10**, 167–178.

Vandenburgh, H., and Kaufman, S. (1979) In vitro model for stretch-induced hypertrophy of skeletal muscle. *Science* **203**, 265–268.

van der Kloot, W. G. (1966) Inhibitors of active Ca^{2+} uptake by fragments of the sarcoplasmic reticulum of lobster muscle. *Comp. Biochem. Physiol.* **17**, 75–86.

van Houten, J. (1979) Membrane potential changes during chemokinesis in *Paramecium. Science* **204**, 1101–1103.

Van Leeuwenhoek, A. (1674) More observations. *Phil. Trans. R. Soc. Lond. Ser. A* **9**, 186.

Van Winkle, W. B. (1976) Calcium release from skeletal muscle sarcoplasmic reticulum: site of action of dantrolene sodium. *Science* **193**, 1130–1131.

Vaughan, H. S., and Goldspink, G. (1979) Fibre number and fibre size in a surgically overloaded muscle. *J. Anat.* **129**, 293-303.

Vaughan-Jones, R. D. (1979) Non-passive chloride distribution in mammalian heart muscle: microelectrode measurement of the intracellular chloride activity. *J. Physiol.* **295**, 83–109.

Veratti, E. (1902) Richerche sulle fine struttura della fibra muscolare striata. *Mem. Ist. Lomb. Sci. Lett. Mat. Nat.* **19**, 87–133. (An English translation has been published in *J. Biophys. Biochem. Cytol.,* now *J. Cell Biol.,* **1961**, **10**, (Suppl.) 3–59.

Villafranca, G. de, and Haines, V. (1974) Paramyosin from arthropod cross-striated muscle. *Comp. Biochem. Physiol.* **47B**, 9–26.

Villafranca, G. W. de, and Maschhaus, C. E. (1962) Contraction of the A band. *J. Ultrastruct. Res.* **9**, 156–165.

Villafranca, G. W. de, Scheinblum, T. S., and Philpott, D. E. (1959) A study on the localization of contractile proteins in the muscle of the horseshoe crab (*Limulus polyphemus*). *Biochim. Biophys. Acta* **34**, 147–157.

Volkenstein, M. V. (1969) Muscular contraction. *Biochim. Biophys. Acta* **180**, 562–572.

von Ebner, S. B. (1895) Über den Bander Chordadorsalis des *Amphioxus lanceolatum. Akad. Wiss. Wien* **104**, 199–288.

Vrbová, G., Gordon, T., and Jones, R. (1978) *Nerve-Muslce Interaction.* Chapman and Hall, London.

Walcott, B., and Dewey, M. M. (1980) Length-tension relation in *Limulus* striated muscle. *J. Cell Biol.* 87, 204–208.

Walcott, B., and Ridgway, E. B. (1967) The ultrastructure of myosin-extracted striated muscle fibers. *Am. Zool.* 7, 499–504.

Wales, W., Clarac, F., Dando, M. R., and Laverack, M. S. (1970) Innervation of the receptors present at the various joints of the pereiopods and third maxilliped of *Homarus gammarus* (L) and other macruran decapods (Crustacea). *Z. Vergl. Physiol.* **68**, 345–384.

Walker, T. T. (1969) Systematics and acoustic behavior of United States crickets of the genus *Orocharis* (Orthoptera: Gryllidae). *Ann. Ent. Soc. Am.* **62**, 752–762.

Wallimann, T., Pelloni, G., Turner, D. C., and Eppenberger, H. M. (1978) Monovalent antibodies against MM-creatine kinase remove the M line from myofibrils. *Proc. Nat. Acad. Sci.* **75**, 4296–4300.

Wallimann, T., Pelloni, G., Turner, D. C., and Eppenberger, H. M. (1979) Removal of the M-line by treatment with FAB[1] fragments of antibodies against MM-creatine kinase. In: *Motility in Cell Function.* F. A. Pepe, J. W. Sanger, and V. T. Nachmias, Eds., pp. 415–417. Academic, New York.

Walmsley, B., Hodgson, J. A., and Burke, R. E. (1978) Forces produced by medial gastrocnemius and soleus muscles during locomotion in freely moving cats. *J. Neurophys.* **41**, 1203–1216.

Walsh, D. A., Brostrom, C. O., Brostrom, M. A., Chen, L., Corbin, J. D., Reimann, E., Soderling, T. R., and Krebs, E. G. (1972) Cyclic AMP-dependent protein kinases from skeletal muscle and liver. *Adv. Cyc. Nuc. Res.* **1**, 33–45.

Walther, C. (1980) Small motor axons in orthopteran insects. *J. Exp. Biol.* 87, 99–119.

Walther, C., and Rathmayer, W. (1974) The effect of *Habrobracon* venom on excitatory neuromuscular transmission in insects. *J. Comp. Physiol.* **89**, 33–39.

Wang, K. (1977) Filamin, a new high-molecular weight protein found in smooth muscle and non-muscle cells. Purification and properties of chicken gizzard filamin. *Biochemistry* **16**, 1857–1865.

Wang, K., and Ramirez-Mitchell, R. (1979) Titin: possible candidate as components of putative longitudinal filaments in striated muscle. *J. Cell Biol.* **83**, 389A.

Wang, K., Ash, F., and Singer, S. J. (1975) Filamin, a new high-molecular-weight protein found in smooth muscle and non-muscle cells. *Proc. Nat. Acad. Sci.* **72**, 4483–4486.

Wang, E., Cross, R. K., and Choppin, P. W. (1979a) Involvement of microtubules and 10 nM filaments in the movement and positioning of nuclei in syncytia. *J. Cell Biol.* **83**, 320–337.

Wang, K., McClure, J., and Tu. A. (1979b) Titin: major myofibrillar component of striated muscle. *Proc. Nat. Acad. Sci.* **76**, 3698–3702.

Ward, P. C. J., Edwards, C., and Benson, E. S. (1965) Relation between adenosinetriphosphate activity and sarcomere length in stretched glycerol-extracted frog skeletal muscle. *Proc. Nat. Acad. Sci.* **53**, 1377–1384.

Wareham, A. C. (1978) Effect of denervation and ouabain on the response of the resting membrane potential of rat skeletal muscle to potassium. **373**, 225–228. *Pflüg. Arch.*

Wareham, A. C., Duncan, C. J., and Bowler, K. (1974) Electrogenesis in cockroach muscle. *Comp. Biochem. Physiol.* **48**, 799–813.

Warmolts, J. R., and Engel, W. K. (1972) Open-biopsy electromyography. I. Correlation of motor unit behavior with histochemical muscle fiber type in human limb muscle. *Arch. Neurol.* **27**, 512–517.

Warren, R. H., and Porter, K. R. (1969) An electron miscoscope study of differentiation of the molting muscles of *Rhodnius prolixus. Am. J. Anat.* **124**, 1–30.

Washio, H. (1972a) The ionic requirements for the initiation of action potentials in insect muscle fibers. *J. Gen. Physiol.* **59**, 121–134.

Washio, H. (1972b) Calcium inward currents in insect muscle fibers. *Canad. J. Physiol. Pharmacol.*

50, 1114–1116.

Washio, H., Kakiuchi, Y., and Tamasige, M. (1961) Polarized light microscopy on the single frog muscle fibre. *J. Fac. Sci. Hokkaido U.* **14**, 499–510.

Washizu, Y. (1967) Electrical properties of leech dorsal muscle. *Comp. Biochem. Physiol.* **20**, 641–646.

Watanabe, K., Sasaki, F., Takahama, H., and Iseki, H. (1979) Histogenesis and distribution of red and white muscle fibres of urodelan larvae. *J. Anat.* **130**, 83–96,

Waterston, J. C., Epstein, H. F., and Brenner, S. (1974) Paramyosin of *Caenorhabditis elegans. J. Mol. Biol.* **90**, 285–290.

Weber, A., Herz, R., and Reiss, I. (1963) On the mechanism of the relaxing effect of fragmented sarcoplasmic reticulum. *J. Gen. Physiol.* **46**, 679–702.

Weber, A., Herz, R., and Reiss, I. (1964) Role of calcium in contraction and relaxation. *Fed. Proc.* **23**, 896.

Weeds, A. G. (1976) Light chains from slow-twitch muscle myosin. *Eur. J. Biochem.* **66**, 157–173.

Weeds, A. G., Trentham, D. R., Kean, C. J. C., and Buller, A. J. (1974) Myosin from cross-reinnervated cat muscles. *Nature* **247**, 135–139.

Weingart, R. (1974) The permeability to tetraethylammonium ions of the surface membrane and the intercalated disks of sheep and calf myocardium. *J. Physiol.* **240**, 741–762.

Weingart, R. (1977) The actions of ouabain on intercellular coupling and conduction velocity in mammalian ventricular muscle. *J. Physiol.* **264**, 341–365.

Weisblat, D. A., Byerly, L., and Russell, R. L. (1976) Ionic mechanisms of electrical activity in somatic muscle of the nematode *Ascaris lumbricoides. J. Comp. Physiol.* **111**, 93–112.

Weis-Fogh, T. (1960) A rubber-like protein in insect cuticle. *J. Exp. Biol.* **37**, 889–907.

Weis-Fogh, T., and Amos, W. B. (1972) Evidence for a new mechanism of cell motility. *Nature* **236**, 301–304.

Weiss, G. B. (1966) The effect of potassium on nicotine-induced contracture and Ca^{45} movements in frog sartorius muscle. *J. Pharmacol. Exp. Ther.* **154**, 595–604.

Weiss, K. R., Cohen, J. L., and Kupferman, I. (1978) Modulatory control of buccal musculature by a serotonergic neuron (metacerebral cell) in *Aplysia. J. Neurophysiol.* **41**, 181–203.

Wells, G. P. (1937) Studies on the physiology of *Arenicola marina* L. I. The pace-maker role of the oesophagus, and the action of adrenaline and acetylcholine. *J. Exp. Biol.* **14**, 117–157.

Wells, G. P. (1939) Studies on the physiology of *Arenicola marina* L. *J. Exp. Biol.* **14**, 117–157.

Wells, G. P. (1949) The behaviour of *Arenicola marina* L. In sand, and the role of spontaneous activity cycles. *J. Mar. Biol. Ass. U. K.* **28**, 465–478.

Welsh, J. H., and Moorhead, M. (1960) The quantitative distribution of 5-hydroxytryptamine in the invertebrates, especially in their nervous systems. *J. Neurochem.* **6**, 146–169.

West, D. L. (1978) The epitheliomuscular cell of *Hydra*: its fine structure, three-dimensional architecture and relation to morphogenesis. *Tissue & Cell* **10**, 629–646.

Westfall, J. (1973) Ultrastructural evidence for neuromuscular system in coelenterates. *Am. Zool.* **13**, 237–246.

Westfall, J., Yamataka, S., and Enos, P. D. (1971) Ultrastructural evidence of polarized synapses in the nerve net of *Hydra. J. Cell Biol.* **51**, 318–323.

Westort, C., and Hultin, H. O. (1966) A procedure for the preparation of empty cell segments from skeletal muscle using solutions of low ionic strength. *Analyt. Biochem.* **16**, 314–319.

White, D. C. S., and Thorson, J. (1972) Phosphate starvation and the nonlinear dynamics of insect fibrillar flight muscle. *J. Gen. Physiol.* **60**, 307–336.

White, D. C. S., and Thorson, J. (1973) The kinetics of muscle contraction. *Prog. Biophys. Mol. Biol.* **27**, 175–255.

White, J. G. (1976) The structure of the ventral nerve cord of *Caenorhabditis elegans. Phil. Trans. R. Soc. Lond. Ser. B* **275**, 327–348.

Wickstrom, C. E., and Castenholz, R. W. (1973) Thermophilic ostracod: Aquatic metazoan with the highest known temperature tolerance. *Science* **181**, 1063–1064.

Wiener, N. (1948) *Cybernetics.* MIT Press, Cambridge, Mass.

Wiersma, C. A. G. (1974) Introduction: invertebrate neurons and behavior. In: *The Neurosciences,* 3rd Study Program pp. 341–342. MIT Press, Cambridge, Mass.

Wiersma, C. A. G., and Ikeda, K. (1964) Interneurons commanding swimmeret movements in the crayfish, *Procambarns clarkii* (Girard). *Comp. Biochem. Physiol.* **12**, 509–525.

Wiersma, C. A. G., and Novitski, E. (1942) The mechanism of the nervous regulation of the crayfish heart. *J. Exp. Biol.* **19**, 255–265.

Wiersma, C. A. G., and Ripley, S. H. (1952) Innervation patterns of crustacean limbs. *Physiol. Comp. Ecol.* **2**, 391–405.

Wigglesworth, V. B. (1956) Formation and involution of striated muscle fibres during the growth and moulting cycles of *Rhodnius prolixus* (Hemiptera). *Q. J. Microsc. Sci.* **97**, 465–480.

Wilkie, D. R. (1956) The mechanical properties of muscle. *Brit. Med. Bull.* **12**, 177–182.

Wilkie, D. R. (1968) Heat, work and phosphorylcreatine breakdown in muscle. *J. Physiol.* **195**, 157–183.

Williams, T. L. (1975) The location of inward rectification in skeletal muscle. *J. Physiol.* **245**, 69–70P.

Willows, A. O. D., Dorsett, D. A., and Hoyle, G. (1973) The neuronal basis of behavior in *Tritonia.* III. Neuronal mechanisms of a fixed action pattern. *J. Neurobiol.* **4**, 255–285.

Wilson, D. F. (1969) The basis for double contractions and slow relaxations of non-striated muscle in a pelecypod mantle. *Comp. Biochem. Physiol.* **29**, 703–715.

Wilson, D. M. (1960) Nervous control of movement in cephalopods. *J. Exp. Biol.* **37**, 57–72.

Wilson, D. M. (1961) The central control of flight in a locust. *J. Exp. Biol.* **38**, 471–490.

Wilson, D. M. (1962) Bifunctional muscles in the thorax of grasshoppers. *J. Exp. Biol.* **39**, 669–677.

Wilson, D. M. (1968) The nervous control of insect flight and related behavior. *Adv. Ins. Physiol.* **8**, 289–338.

Wilson, F. J., Hirabayashi, T., and Irish, M. J. (1979) Ultrastructural location of antibodies to troponin components in dystrophic muscle. In: *Motility in Cell Function,* F. A. Pepe, J. W. Sanger, and V. T. Nachmias, Eds., pp. 377–380. Academic, New York.

Wilson, J. A., and Hoyle, G. (1978) Serially homologous neurones as concomitants of functional specialisation. *Nature* **274**, 377–379.

Wine, J. J., and Hagiwara, G. (1978) Durations of unitary synaptic potentials help time a behavioral sequence. *Science* **199**, 557–559.

Winegrad, S. (1965a) Autoradiographic studies of intracellular calcium in frog skeletal muscle. *J. Gen. Physiol.* **48**, 455–479.

Winegrad, S. (1965b) The location of muscle calcium with respect to the myofibrils. *J. Gen. Physiol.* **48**, 997–1002.

Winegrad, S. (1970) The intracellular site of calcium activation in frog skeletal muscle. *J. Gen. Physiol.* **55**, 77–88.

Winegrad, S. (1974) Resting sarcomere length-tension relation in living frog heart. *J. Gen. Physiol.* **64**, 343–355.

Winkelman, L. (1976) Comparative studies of paramyosins. *Comp. Biochem. Physiol.* **55B**, 391–397.

Winton, F. R. (1937) The changes in viscosity of an unstriated muscle (*Mytilus edulis*) during and after stimulation with alternating currents. *J. Physiol.* **88**, 492–511.

Wise, R. M., Rondinone, J. E., and Briggs, F. N. (1973) Effect of pCa on series elastic component of glycerinated skeletal muscle. *Amer. J. Physiol.* **224**, 576–579.

Wissocq, J. C. (1970) Évolution de la musculature longitudinale dorsale et ventrale au cours de la stolonisation de *Syllis amica* quatrefages (Annélide Polychete) I. Muscles du ver asexué et muscles due stolon. *J. Microscopic* **9**, 355–388.

Witaliński, W. (1974) Structure of muscle fibres and motor end-plates in the intercostal muscles of lizard, *Lacerta agilis* L. *Z. Mik. Anat.* **88**, 796–808.

Witaliński, W., and Loesch, A. (1975) Structure of muscle fibres and motor end plates in extraocular muscles of the grass snake, *Natrix natrix* L. *Z. Mikrosk. Anat.* **89**, 1133–1146.

Wohlfarth-Bottermann, K. E. (1965) Weitreichende fibrilläre Protoplasmadifferenzierungen und ihre

Bedeutung für die Protoplasmaströmung. III. Entstehung und experimentell induzierbare Musterbildungen. *Roux's Arch.* **156**, 371–403.

Woledge, R. C. (1961) The thermoelastic effect of change in tension in active muscle. *J. Physiol.* **155**, 187–208.

Woledge, R. C. (1963) Heat production and energy liberation in the early part of a muscular contraction. *J. Physiol.* **166**, 211–224.

Woledge, R. C. (1971) Heat production and chemical change in muscle. *Prog. Biophys. Mol. Biol.,* **22**, 37–74.

Wong, B. S., Miller, D. M., and Dunagan, T. T. (1979) Electrophysiology of acanthocephalan body wall muscles. *J. Exp. Biol.* **82**, 273–280.

Wood, D. W. (1957) The effect of ions upon neuromuscular transmission in a herbivorous insect. *J. Physiol.* **138**, 119–139.

Wood, D. W. (1963) The sodium and potassium composition of some insect skeletal muscle fibres in relation to their membrane potential. *Comp. Biochem. Physiol.* **15**, 303–312.

Wood, D. W. (1965) The relationship between chloride ions and resting potential in skeletal muscle fibres of the locust and cockroach. *Comp. Biochem. Physiol.* **15**, 302–312.

Woollacott, M., and Hoyle, G. (1977) Neural events underlying learning in insects: changes in pacemaker. *Proc. R. Soc. Lond. Ser. B* **195**, 395–415.

Worthington, C. R. (1961) X-ray diffraction studies on the large-scale molecular structure of insect muscle. *J. Mol. Biol.* **3**, 618–633.

Wray, J. S. (1979) X-ray diffraction studies of myosin filament structures in crustacean muscles. In: *Motility in Cell Function.* F. A. Pepe, J. W. Sanger, and V. T. Nachmias, Eds., pp. 347–350. Academic, New York.

Wray, J. S., Vibert, P. J., and Cohen, C. (1975) Diversity of cross-bridge configurations in invertebrate muscles. *Nature* **257**, 561–564.

Wróblewski, R., Anniko, M., and Edstrom, L. (1981) A new type of striated muscle in mammalian body: morphological, histochemical, and X-ray microanalytical observations of stapedius muscle in guinea pig. *J. Ultrastruct. Res.* **76**, 46–56.

Wu, K. S. (1939a) On the physiology and pharmacology of the earthworm gut. *J. Exp. Biol.* **16**, 184–250.

Wu, K. S. (1939b) The action of drugs, especially acetylcholines on the annelid body wall (*Lumbricus, Arenicola*). *J. Exp. Biol.* **16**, 251–257.

Wyman, R. J., Waldron, I., and Wachtel, G. M. (1974) Lack of fixed order of recruitment in cat motoneuron pools. *Exp. Brain Res.* **20**, 101–114.

Yagi, N., and Matsubara, I. (1980) Myosin heads do not move on activation in highly stretched vertebrate striated muscle. *Science* **207**, 307–308.

Yamada, K., Mashima, H., and Ebashi, S. (1976) The enthalpy change accompanying the binding of calcium to troponin relating to the activation heat production of muscle. *Proc. Jap. Acad.* **52**, 252–255.

Yamagishi, S., and Sano, J. (1966) Effect of tetrodotoxin on the pacemaker action potential of the sinus node. *Proc. Jap. Acad.* **42**, 1194–1196.

Yamamoto, I. (1961) An electron microscopic study on development of uterine smooth muscle. *J. Electronmic.* **10**, 145–160.

Yamamoto, D., and Fukami, J. (1977) Ionic requirements for non-synaptic electrogenesis in the muscle fibres of a lepidopterons insect. *J. Exp. Biol.* **70**, 41–47.

Yamamoto, D., and Washio, H. (1979) Permeation of sodium through calcium channels of an insect muscle membrane. *Can. J. Physiol. Pharmacol.* **57**, 220–222.

Yamamoto, D., and Washio, H. (1981) Voltage clamp studies on insect skeletal muscle. II. The outward currents. *J. Exp. Biol.* **92**, 13–22.

Yamamoto, D., Fukami, J., and Washio, H. (1978) Ca-electrogenesis in mealworm muscle: a voltage clamp study. *Experientia* **34**, 1603–1605.

Yamamoto, D., Fukami, J., and Washio, H. (1981) Voltage clamp studies on insect skeletal muscle. I. The inward current. *J. Exp. Biol.* **92**, 1–12.

Yanagida, T., Taniguchi, M., and Oosanjo, F. (1974) Conformational changes of F-actin in the thin filaments of muscle induced *in vivo* and *in vitro* by calcium ions. *J. Mol. Biol.* **90**, 509–522.

Yamauchi, A., and Burnstock, G. (1968) An electron microscopic study on the innervation of the trout heart. *J. Comp. Neurol.* **132**, 567–588.

Yano, Y., and Miki-Noumura, T. (1980) Sliding velocity between outer doublet microtubules of sea-urchin sperm axonemes. *J. Cell. Sci.* **44**, 169–186.

Yarom, R., Sherman, G., and Robin, G. C. (1980) Thick filament size changes in contraction of human muscles. *Experientia* **36**, 101–103.

Yentsch, C. S., and Pierce, C. S. (1955) 'Swimming' anemone from Puget Sound. *Science* **122**, 1231–1232.

Young, D. (1972) Neuromuscular mechanisms of sound production in Australian cicadas. *J. Comp. Physiol. Psychol.* 79, 343–362.

Young, J. H., McLick, J., and Korman, E. F. (1974) Pseudorotation mechanism of ATP hydrolysis in muscle contraction. *Nature* **249**, 474–476.

Young, J. Z. (1936) The innervation and reaction to drugs of the viscera of teleostean fish. *Proc. R. Soc. Lond. Ser. B* **120**, 303–318.

Younkin, S. G., Brett, R. S., Davey, B., and Younkin, L. H. (1978) Substances moved by axonal transport and released by nerve stimulation have an innervation-like effect on muscle. *Science* **200**, 1292–1295.

Zebe, E., and Rathmeyer, W. (1968) Elektronenmikroskopische Untersuchungen au Spinnenmuskeln. *Z. Zellforsch.* **92**, 377–387.

Zelená, J., and Jirmanova, I. (1977) Ultrastructure of slow chicken muscle after nerve cross-union. *Physiol. Bohem.* **22**, 107–116.

Zelená, J., and Jirmanova, I. (1974) Slow and fast motor end-plates and their effect upon differentiation of muscle fibre types in young chickens. In: *Structure and Function of Normal and Diseased Muscle and Peripheral Nerve,* Eds., I. Hansanowa – Petinsewicz and H. Jedrzejowska, pp. 113–117. Polish Medical Publishers, Warsaw.

Zenker, W., and Anzenbacher, H. (1964) On the two different forms of myo-neural junction in two types of muscle fiber from the external ocular muscles of the rhesus monkey. *J. Cell. Comp. Physiol.* **63**, 273–285.

Zill, S. N., and Moran, D. T. (1982) Suppression of reflex postural tonus: a role of peripheral inhibition in insects. *Science* **216**, 751–753.

Zmoray, I., and Guttekova, A. (1970) The ultrastructure of *Dictyocaulus* filaria and *Endorylaimus obtusicaudatus* (Nematoda). *Biologica* **25**, 65–73.

Zoond, A. (1932) The mechanism of projection of the chameleon's tongue. *J. Exp. Biol.* **13**, 174–185.

Zou, Y. and Gong, Z. (1979) Paramyosin filaments of Amphioxus. *Sci. Sinica.* **22**, 1329–1332.

Zucker, R. S. (1977) Synaptic plasticity at crayfish neuromuscular junctions. In: *Identified Neurons and Behavior of Arthropods.* G. Hoyle, Ed., pp. 49–65. Plenum, New York.

Zs.-Nagy, I., Salanki, J., and Garamvölgyi, N. (1971) The contractile apparatus of the adductor muscles in *Anodonta cygnea* L. (Mollusca, Pelecypoda). *J. Ultrastruct. Res.* **37**, 1–16.

Species Index

Subject Index